Element	Symbol	Atomic Number	Mass (amu)	Element	Symbol	Atomic Number	Mass (amu)
Neodymium	Nd	60	144.24	Silicon	Si	14	28.0855
Neon	Ne	10	20.179	Silver	Ag	47	107.8682
Neptunium	Np	93	237.0482	Sodium	Na	11	22.98977
Nickel	Ni	28	58.69	Strontium	Sr	38	87.62
Niobium	Nb	41	92.9064	Sulfur	S	16	32.066
Nitrogen	N	7	14.0067	Tantalum	Ta	73	180.9479
Nobelium	No	102	259.1	Technetium	Tc	43	98.9062
Osmium	Os	76	190.2	Tellurium	Te	52	127.60
Oxygen	O	8	15.9994	Terbium	Tb	65	158.9254
Palladium	Pd	46	106.42	Thallium	Tl	81	204.383
Phosphorus	P	15	30.97376	Thorium	Th	90	232.0381
Platinum	Pt	78	195.08	Thulium	Tm	69	168.9342
Plutonium	Pu	94	239.1	Tin	Sn	50	118.710
Polonium	Po	84	210.0	Titanium	Ti	22	47.88
Potassium	K	19	39.0983	Tungsten	W	74	183.85
Praseodymium	Pr	59	140.9077	Unnilseptium	Uns*	107	(262)
Promethium	Pm	61	144.9	Unnilhexium	Unh*	106	(263)
Protactinium	Pa	91	231.0359	Unnilennium	Une*	109	
Radium	Ra	88	226.0254	Unnilpentium	Unp*	105	(262)
Radon	Rn	86	222.0	Unnilquadium	Unq*	104	(261)
Rhenium	Re	75	186.207	Uranium	U	92	238.0289
Rhodium	Rh	45	102.9055	Vanadium	V	23	50.9415
Rubidium	Rb	37	85.4678	Xenon	Xe	54	131.29
Ruthenium	Ru	44	101.07	Ytterbium	Yb	70	173.04
Samarium	Sm	62	150.36	Yttrium	Y	39	88.9059
Scandium	Sc	21	44.95591	Zinc	Zn	30	65.39
Selenium	Se	34	78.96	Zirconium	Zr	40	91.224

CHEMISTRY
AN EXPERIMENTAL SCIENCE

CHEMISTRY
AN EXPERIMENTAL SCIENCE

GEORGE M. BODNER
Purdue University

HARRY L. PARDUE
Purdue University

WILEY

JOHN WILEY & SONS
New York   Chichester   Brisbane
Toronto   Singapore

Book production supervised by Elizabeth A. Austin.
Cover and text designed by Madelyn Lesure.
Photo editor: Stella Kupferberg.
Illustrations by John Balbalis, with the assistance
of the Wiley Illustration Department.
Manuscript editor was Beverly Peavler, under the
supervision of Richard Koreto.

Cover/Title Page Photograph by KEN KARP

Library of Congress Cataloging in Publication Data:

Bodner, George M.
    Chemistry, an experimental science.

    Bibliography
    1. Chemistry.   I. Pardue, Harry L.   II. Title.

QD33.B684 1989   540   88-27910
ISBN 0-471-87053-6

Printed in the United States of America

10 9 8 7 6 5 4 3 2

This book is dedicated to the teachers who inspired us to become chemists
and to the more than 20,000 students who taught us chemistry.

HOW TO LOOK AT THE ILLUSTRATIONS

• Color is used to identify, associate, and relate items dealt with in various contexts. In drawings dealing with orbital theory, for instance, each category of orbital (s, 2p, sp, d) has its own identifying color. These colors appear not only in three-dimensional pictorial depictions of the orbitals but also in the diagrams and charts that deal with them on an abstract level.

• Elements by themselves or as components of ball-and-stick and space-filling molecular models are identified by conventional colors throughout the book. Thus hydrogen is white, oxygen red, chlorine green, sulfur yellow, and so on.

• Various rays such as infrared rays, alpha, beta and gamma rays, ultraviolet rays and x-rays are shown in their characteristic colors. This is the case in illustrations of laboratory equipment generating these rays as well as in charts and diagrams expressing data about them.

• The relative sizes and configurations of the various orbitals are drawn to scale in accordance with their known mathematical data to ensure an authentic and consistent appearance.

• The space-filling molecular models are drawn to scale. Thus the relative diameters of the atomic elements comprising the molecules reflect their known mathematical data. In figures of the water molecule, for instance, the diameters of the component hydrogen and oxygen atoms are drawn in the ratio of 0.37 to 0.66.

• Figures of ball-and-stick and space-filling models show the bonding angles drawn to accurately reflect their known divergences. Thus in the water molecule the angle between the hydrogen atoms is shown at a true $104°5'$. The result is an authentic depiction and not a casual impression.

• Three-dimensional formula structures are drawn to emphasize their spatial geometry with clarity.

• Laboratory equipment (flasks, beakers, bunsen burners, etc.) has been drawn with reference to lab equipment catalogues to ensure accuracy in appearance.

• Classic experiments are depicted with careful attention to key aspects concerning the physical equipment used and its set-up.

• Crystal molecular structures are drawn to ensure that the positions of the atoms in space relative to each other and thus the geometry of the structures is unequivocally clear.

• Various items that repeat throughout the book are identified by their own unique colors: energy level diagrams; various diagrams and periodic tables that deal with the concept of the states of matter wherein gas, solid and liquid each have their own color; orbit diagrams; diagrams which deal with the concepts of metals, nonmetals and metalloids; symbols for the concepts of attraction and repulsion; and so on.

TO THE INSTRUCTOR

Those of us who teach chemistry argue about what topics should be emphasized and about the best order for presenting these topics, but chemists at widely differing institutions are in remarkable agreement about the material that should be covered in a given course.

This does not mean that the consensus is stable—there is abundant evidence to show that it changes with time. At one time, introductory chemistry courses focused on the preparation and properties of elements and compounds. More than half of a typical textbook at that time was devoted to topics such as the chemistry of the halogens, the Frasch process for mining sulfur, the preparation of phosphine, the manufacture of nitric acid, the production of pig iron in blast furnaces, and so on.

Courses based on these texts were inevitably displaced by those that focused on the principles of chemistry. These courses introduced discussions of the photoelectric effect, the Schrödinger model of the hydrogen atom, wave–particle duality, molecular orbitals, entropy, and free energy. They spent less time talking about the chemistry of the transition metals, and more time using atomic orbitals to explain this chemistry.

The title of this text—*Chemistry: An Experimental Science*—reflects our beliefs about the direction in which general chemistry courses should evolve. We agree with the Committee on Professional Training of the American Chemical Society, which recommends putting the "chemistry" back into introductory chemistry courses. But to us, this means more than just adding additional inorganic chemistry. It means returning to an experimental perspective, in which observations are made before they are explained.

Many texts, for example, place the descriptive chemistry at the end. This assumes that we have to introduce the students to all of the principles of chemistry—from the structure of the atom, to equilibria and free energy—before we can talk about the chemistry of the elements. We think this is a mistake, and have integrated the descriptive chemistry throughout the book. As much as possible, we start by noting what happens when chemi-

cal systems are observed. Once the observations are made, we then try to develop explanations for this behavior.

As early as Chapter 2, we describe the differences between the chemical and physical properties of metals and nonmetals and between ionic and covalent compounds. When the periodic properties of the elements are discussed in Chapter 6, they are immediately used to explain the existence of semimetals and to discuss the reactivity of the active metals. The chemistry of the main-group metals is introduced in Chapter 7 as a basis for discussions of oxidation-reduction reactions and ionic compounds, rather than vice versa. Ionic bonds are then contrasted with covalent bonds, thermochemistry is introduced, and the chemistry of the nonmetals is described in Chapter 10. Chapter 11 then focuses on the qualitative chemistry of acids, bases, and salts.

To us, descriptive chemistry is more than inorganic reaction chemistry. It includes an understanding of the chemistry of aqueous solutions, particularly acid–base, solubility, and complex-ion equilibria. We also believe that one of the best ways of learning this chemistry is through the qualitative analysis scheme. We have therefore included a truncated "qual" scheme that can be covered in four or perhaps five three-hour laboratory periods.

Most general chemistry texts introduce kinetics before equilibria, because we can explain why reactions come to equilibrium from a kinetic perspective. But you don't need to understand first-order and second-order reactions, rate constants, instantaneous rates of reaction, reaction mechanisms, activation energies, and so on, to understand equilibrium. We therefore begin the equilibrium chapters with a discussion of the collision-theory model of chemical reactions. The various tricks that are needed to handle equilibrium calculations are then introduced in terms of relatively simple, gas-phase reactions. (We find that when students understand how these techniques can be applied to reactions in which the solvent plays no role, they are much better at reactions in aqueous solution.) In a later chapter, we then return to chemical kinetics — after the student has achieved some facility with simpler calculations.

Our belief that observations should precede explanations also influenced our placement of the chapters on electrochemistry and thermodynamics. Many texts develop the concept of free energy and then relate it to electrochemical cells. We start with the system that can be observed — the electrochemical cell — and then develop the theoretical explanation for this behavior.

Many instructors who adopt this text will disagree with some of our ideas about order. We have therefore tried to write each section so that it could stand alone. The chapters on gases and thermochemistry, for example, can be taught at almost any time after the basics of stoichiometry have been introduced. Chapter 6 (the periodic table) builds on the discussion of the structure of the atom in Chapter 5. However, Chapter 7, on main-group metals and their salts, can be grouped with the material in Chapter 10 on nonmetals at almost any time during the year.

The chapter on electrochemistry builds on the preceding discussion of oxidation-reduction reactions, but these two chapters can be covered at essentially any point in the second-half of the course. Chapters 20 through 25, on thermodynamics, kinetics, transition-metal chemistry, nuclear chemistry, organic chemistry, and polymer chemistry were written so that they could be used at any point, in essentially any order.

The following supplements to this textbook are available:

1. SOLUTIONS MANUAL: Contains solutions to all end-of-chapter problems.

2. STUDY GUIDE: Contains statements of objectives, a review of significant topics, additional worked examples, self-test questions and problems, and a list of new terms for each chapter.

3. TRANSPARENCIES: A package of full-color transparencies that reproduce key illustrations from the text.

4. INSTRUCTOR'S MANUAL: Discusses the objectives and rationale for each chapter, possible course outlines, and suggestions for exams.

5. TEST BANK: A printed test bank with answers that includes a variety of test formats.

6. MICROTEST: A computerized version of the test bank available for use with IBM PC or Apple Macintosh computers.

7. LECTURE DEMONSTRATION MANUAL: Contains short, simple lecture demonstrations keyed to the text material.

TO THE STUDENT

Students enroll in general chemistry courses for many reasons. For some, it is a direct result of the major they select, because the language and critical thinking skills chemists use has been judged to be a valuable tool for success in that field. Some students choose chemistry to fulfill a science elective; others take it because it is a required course for professional schools in medicine, dentistry, or veterinary medicine. Some of you might even be considering a career in chemistry.

No matter why you are taking general chemistry, this course is designed to meet some of the following objectives: to introduce you to some of the language that chemists use to describe the world around us, a language that has been adopted by professionals in such diverse fields as political science and astronomy; to introduce you to concepts and skills that are needed in later courses in your major; to foster problem solving skills that can be transferred to your profession or to life in general.

Developing problem solving skills is such an impor-

tant component of chemistry courses that it is useful to summarize some of the differences research has found between good and poor problem solvers.

Good Problem Solvers:

1. Believe they can solve almost any problem if they work long enough.

2. Are persistent; they don't give up easily.

3. Read carefully, and reread a problem, until they understand what information is given and what they are asked to solve for.

4. Break problems into small steps, which they solve one at a time.

5. Organize their work so that don't lose sight of what they've accomplished, and can follow the steps they've taken so far.

6. Check their work, not only at the end of the problem but at various points along the way.

7. Build models, or representations, of the problem, which can take the form of a list of relevant information, a picture of the system under consideration, or a concrete example.
8. Try to solve a simpler, related problem when faced with a problem they can't solve.
9. Guess and test; they try out several approaches to a problem until they are successful.

Poor Problem Solvers:

1. Don't believe they can solve problems; they believe that you either know the answer or you don't.
2. Give up easily if they don't seem to get the answer.
3. Are careless readers, who often misread what is written. They tend to jump into the problem before they understand what it asks for.
4. Seldom check their work to see if it makes sense.
5. Organize their work carelessly.
6. Have only one approach to a given problem. When they can't recall a formula, for example, they give up.

This text contains worked examples designed to help you: (1) determine what the problem asks for, (2) select relevant information, (3) keep track of this information, (4) check the results of calculations, (5) work problems that contain too much information, (6) work problems that don't seem to contain enough information, (7) work backwards, and (8) make assumptions or approximations that turn complex problems into simpler ones.

It might be useful to distinguish between two closely related concepts: *problems* and *exercises*. Hayes defined a problem as follows.

> Whenever there is a gap between where you are now and where you want to be, and you don't know how to find a way to cross that gap, you have a problem.[1]

If you know what to do when you read a question, it's an exercise not a problem. Status as a problem is not an innate characteristic of a question, it is a subtle interaction between the question and the individual trying to answer the question. It reflects experience with that type of question more than intellectual ability.

When you go to class, you may find that your instructor has developed an impressive repertoire of techniques that can be used to turn problems into exercises. We like to call these techniques "algorithms", which are defined as "rules for calculating something, especially by ma-

chine." Algorithms are useful for solving routine questions or exercises. In fact, the existence of an algorithm constructed from prior experience may be what turns a question from a problem into an exercise.

Students who have not built algorithms for at least some of the steps in a problem will have difficulty solving the problem. There is more to working problems, however, than applying algorithms in the correct order. Problem solving has been defined as "What you do, when you don't know what to do." By definition, there is no clear cut answer to what you should do when faced with a novel problem. We believe, however, that successful problem solvers, when faced with a problem,

1. Read the problem.
2. Read the problem again.
3. Write down what they hope is the relevant information.
4. Read the problem again.
5. Draw a picture or make a list to help build a model, or representation, of the problem.
6. Try something.
7. Try something else.
8. See where this gets them.
9. Read the problem again.
10. Try something else.
11. See where that gets them.
12. Test an intermediate result.
13. Repeat this process until they get an answer that might be correct.
14. Test the answer to see if it makes sense.
15. Start over if they have to, celebrate if they don't.

We have two suggestions for improving your problem-solving skills. First, recognize the importance of practice. The more problems you work, the better you will become at solving problems. Second, recognize the importance of working with other students. Take turns working problems out loud, explaining each step in the problem to the others in your group. While you do this, they should listen carefully to make sure they understand each step you take, to check each step to make sure that you aren't making any errors, to identify errors when they perceive them (without giving any hints about what they believe is the correct answer), and to insure that you vocalize each of the major steps in the problem. Research has shown that this approach can significantly improve the problem solving skills of each member of the group.

[1] J. Hayes, "The Complete Problem Solver", Franklin Institute Press, Philadelphia, PA 1980.

ACKNOWLEDGMENTS

The authors would like to express their appreciation to the people at Wiley who made this book possible. To Beverly Peavler, Gilda Stahl, and Richard Koreto, who struggled mightily to teach us the difference between "that" and "which" during copy editing. To John Balbalis, who took simplistic, hand-drawn figures and—without laughing at our limited artistic ability—turned them into beautiful artwork. To Anita Duncan, Stella Kupferberg, Gina Velasquez, and Anne Manning, who developed the photo program for this text. To Madelyn Lesure, who designed this text, and to Elizabeth Austin, who saw it through production. To Cliff Mills, who offered us the initial contract for this book, and to Dennis Sawicki, whose patience was sorely tested during its evolution.

We are also indebted to Professor Grohman at Hunter College, who provided laboratory space in which many of the photographs were shot, to Carlos Ava at Hunter College and Kurt Keyes at Purdue University, who set up the demonstrations photographed, and most importantly to Ken Karp, who shot most of the photographs in this text.

We wish to thank the following list of individuals who were involved in the reviewing process for this text. Their knowledge of chemistry and insight into the process of teaching chemistry helped prevent us from making many errors of both omission and commission. Errors that remain are solely our fault, and we apologize for them.

Finally, we would like to thank Connie, who knows too well the cost of writing a textbook, and our co-workers who carried much of the burden we neglected during the development of this book.

George M. Bodner
Harry L. Pardue

Prof. Raymond Butcher
Howard University

Prof. Lawrence Conroy
University of Minnesota

Prof. Raymond Crawford
San Jacinto College South

Prof. Toby Block
Georgia Institute of Technology

Prof. Robert Reeves
Rensselaer Polytechnic Institute

Dr. R. Carl Stoufer
University of Florida

Prof. John DeKorte
Northern Arizona University

Prof. Roy Garvey
North Dakota State University

Prof. Mark Wicholas
Western Washington University

Prof. Roger Millikan
University of California

Prof. Brad Mundy
Montana State University

Prof. James Spain
Clemson University

Prof. Ralph Steinhaus
Western Michigan University

Prof. Norman Kulersky
University of North Dakota

Prof. Roger Barry
Northern Michigan University

Prof. Mary Baily
Ohio State University

Prof. John Thayer
University of Cincinnati

Prof. T. P. Forrest
Dalhousie University

Prof. Neil Kestner
Louisiana State University

Prof. Robert Kowerski
College of San Mateo

Prof. Michael Pavelich
Colorado School of Mines

Prof. R. Kent Murmann
University of Missouri

Prof. Donna Alway
St. Louis Community College

Tamar "Uni" Susskind
Oakland Community College

Prof. Carl von Frankenberg
University of Delaware

Prof. Philip Jaffe
Oakton Community College

Prof. Jo Beran
Texas A&I University

Kim Cohn
California State College

Marshall Bishop

CONTENTS

CHAPTER 5

THE STRUCTURE OF THE ATOM   155

CHAPTER 6

THE PERIODIC TABLE   196

CHAPTER 7

THE MAIN-GROUP METALS AND THEIR SALTS   228

CHAPTER 8

THE COVALENT BOND   271

CHAPTER 9

THERMOCHEMISTRY   326

CHAPTER 10

THE CHEMISTRY OF THE NONMETALS   366

CHAPTER 11

ACIDS, BASES, AND SALTS   426

CHAPTER 12

THE STRUCTURE OF SOLIDS   467

CHAPTER 13

LIQUIDS AND SOLUTIONS   506

CHAPTER 14

GAS-PHASE REACTIONS: AN INTRODUCTION TO KINETICS AND EQUILIBRIUM   549

CHAPTER 15

ACID-BASE EQUILIBRIA   595

CHAPTER 16

SOLUBILITY PRODUCT EQUILIBRIA   647

CHAPTER 21

KINETICS 824

CHAPTER 22

TRANSITION METAL COMPLEXES 864

CHAPTER 23

NUCLEAR CHEMISTRY 902

CHAPTER 24

THE ORGANIC CHEMISTRY OF CARBON 941

CHAPTER 1

THE FUNDAMENTALS OF MEASUREMENT

CHAPTER CONTENTS

1.1 CHEMISTRY AS AN EXPERIMENTAL SCIENCE

Chemistry is easy to define:

> **Chemistry is the science that deals with the composition and properties of substances and the reactions by which one substance is converted into another.**

But it is harder to appreciate what this definition means. It can be argued that the purpose of this text is to provide the background necessary to fully understand this definition and its implications.

One of the best ways of teaching a new word or concept is to provide both examples and non-examples of the concept. Perhaps the best way to begin defining chemistry is to provide you with examples of what it is not.

In 1921, a group from the American Museum of Natural History began excavations at an archaeological site on Dragon-Bone Hill, near the town of Chou-k'ou-tien (Zhoukoudian), 34 miles southwest of Beijing, China. Fossils found at this site were assigned to a new species, *Homo erectus pekinensis,* commonly known as Peking man. These excavations suggest that for at least 500,000 years, people have known enough about the properties of stone to make tools, and they have been able to take advantage of the chemical reactions involved in combustion in order to cook food. But even the most liberal interpretation would not allow us to call this chemistry, because of the absence of any control over these reactions or processes.

The ability to control the transformation of one substance into another can be traced back to the origin of two different technologies: brewing and metallurgy. There is ample evidence that people have been brewing beer for at least 12,000 years, since the time when the first cereal grains were cultivated. Metallurgy, the process of extracting metals from their ores, has been practiced for at least 6000 years, since copper was first produced by heating the ore malachite.

A sample of the ore malachite from the Campbell Cole Shaft in Bisbee, Arizona.

But brewing beer by burying barley until it germinates and then allowing the barley to ferment in the open air was not chemistry. Nor was extracting copper metal from one of its ores. People carried out both processes in the same fashion time after time without any understanding of what was happening or why. Even the discovery around 3500 B.C. that copper mixed with 10% to 12% tin gave a new metal that was harder than copper, and yet easier to melt and cast, was not chemistry. The preparation of bronze was a major technical breakthrough in metallurgy, but it did not provide people with an understanding of how to make other metals.

Between the sixth century B.C. and the third century B.C., the Greek philosophers (literally, lovers of wisdom) tried to build a theoretical model for the behavior of the natural world. They argued that the world was made up of four primary, or *elementary,* substances: fire, air, earth, and water (see Figure 1.1). These substances differed in two properties: hot versus cold and dry versus wet.

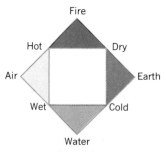

FIG. 1.1 The ancient Greeks assumed the world was composed of four elements—fire, air, earth, and water—that differed in two properties—hot versus cold and dry versus wet.

Fire:	**hot and dry**
Air:	**hot and wet**
Earth:	**cold and dry**
Water:	**cold and wet**

Their model was able to explain some common observations. Water (cold and wet) evaporates or turns into air (hot and wet) when it is heated. A piece of wood, which contains a great deal of earth (cold and dry), bursts into flame (hot and dry) when it is heated.

This model was the first step toward the goal of understanding the properties and compositions of different substances and the reactions that convert one substance to another. But some elements of modern chemistry were still missing. The model could explain certain observations of how the natural world behaved, but it could not be used to predict new observations or behaviors. The model was based on pure speculation; it could not be proved or disproved. In fact, its proponents were not interested in using the results of experiments to test the model.

Modern chemistry is based on certain principles.

1. ***One of the goals of chemistry is to recognize patterns in the behaviors of different substances.*** An example of this might be Antoine Lavoisier's discovery that every substance that burns in air gains weight.

2. ***Once a pattern is recognized, it should be possible to develop a model that explains these observations.*** Lavoisier concluded that any substance that burns in air combines with the oxygen in the air to form a product that weighs more than the starting material.

3. ***These models should allow us to predict the behavior of other substances.*** In 1869, Dmitri Mendeléeff used his model for the behavior of the known elements to predict the properties of elements that had not yet been discovered.

4. ***When possible, the models should be quantitative.*** They should not only predict what happens, but by how much.

5. ***Models should be able to make predictions that can be tested experimentally.*** Mendeléeff's periodic table was accepted by other chemists because of the degree of agreement between his predictions and the results of experiments based on these predictions.

In essence, chemistry is an experimental science. Experiment serves two important roles. It forms the basis of observations that define the problems that theories must explain, and it provides a way of checking the validity of new theories. This text emphasizes an experimental approach to chemistry. As often as possible it presents the experimental basis of chemistry before the theoretical explanations of these observations.

According to the Greek model of the elements, wood that is rich in earth (cold and dry) bursts into flame (hot and dry) when heated.

1.2 THE NORMAL PRACTICES OF SCIENCE

Sociologists and philosophers of science have identified certain normal practices that scientists follow. Alexander Kohn summarized these practices more or less as follows.

1. **Scientific facts and theories should be judged in terms of intellectual criteria valid in that branch of science. They should not be accepted or rejected because of the personal attributes of their author.**

2. **Scientists should direct their activities and efforts toward an extension of scientific knowledge, not the personal interests of an individual or group of scientists.**

3. **Science is a collaborative effort, which requires the open exchange of information.**

4. **Science should be approached rationally.**

5. **The validity of a scientific fact or theory should be judged from an emotionally neutral perspective.**

6. **Scientists must be honest, objective, tolerant, and unselfish.**

In the discussion of the ethics of science, Hans Mohr summarized these rules as follows.

> **Be honest; never manipulate data; be precise; be fair with regard to priority of ideas; be without bias with regard to data and ideas of your rival; do not make compromises in trying to solve a problem.**

1.3 CURIOSITY, OBSERVATION, AND MEASUREMENT

Young children are inherently curious. They want to know answers to questions ranging from why the sky is blue to why they have to go to bed at 9:00 p.m. This book is dedicated to those who retain some of this innate curiosity. It is written for people who at one time or another wondered why a hot-air balloon rises, or why water evaporates at room temperature but doesn't boil until it reaches 212°F. It is written for those who may not have noticed that the sand on a damp beach dries out for an instant when they step on it, but who would love to know why this happens when it is brought to their attention.

Explanations for these observations are all based on models scientists have constructed to explain the results of quantitative experiments. Our first goal, therefore, is to understand what quantities are important to measure in chemistry, how these quantities are measured, and what uncertainty or error accompanies measurement.

Eventually, we will use the results of these measurements to build models that help explain the world around us. These models exist at three different levels of sophistication. At the first level, we propose a *hypothesis.*

> **Hypothesis: A suggested or proposed explanation that has not be subjected to extensive testing.**

Once we have a hypothesis, we can design experiments that test it. A common cliché states that "the exception *proves* the rule." Nothing could be further from the truth. The following is a more accurate translation of the source of this cliché: "The exception *probes* the rule."

Once the hypothesis has been tested, and modified to fit new experimental results, it may reach the level of a *theory.*

> **Theory: A systematic statement of a principle that has been verified by repeated experimentation.**

Most of the time, development of the model stops at this stage. Every once in a while, a model reaches the level of sophistication at which it becomes a scientific *law.*

> **Law: A statement to which there are no known exceptions.**

Examples of this rare phenomenon are the laws of thermodynamics and the law of conservation of mass in a chemical reaction.

1.4 ENGLISH UNITS OF MEASUREMENT

All measurements contain at least three elements: a *number* that indicates the size or magnitude of the quantity being measured, *units* that provide a basis for comparing this quantity with a standard reference, and some *uncertainty* or *error.*

TABLE 1.1

The English System of Units

Length: inch (in.), foot (ft), yard (yd), mile (mi)
12 in. = 1 ft
3 ft = 1 yd
5280 ft = 1 mi
1760 yd = 1 mi

Volume: fluid ounce (oz), cup (c), pint (pt), quart (qt), gallon (gal)
2 c = 1 pt
2 pt = 1 qt
32 oz = 1 qt
4 qt = 1 gal

Weight: ounce (oz), pound (lb), ton
16 oz = 1 lb
2000 lb = 1 ton

Time: second (s), minute (min), hour (hr), day (d), year (yr)
60 s = 1 min
60 min = 1 hr
24 hr = 1 d
$365\frac{1}{4}$ d = 1 yr

There are several systems of units, and each contains units of measurement for properties such as length, volume, weight, and time. In the English system in use in the United States, the individual units are defined in a purely arbitrary way. There are 12 inches in a foot, 3 feet in a yard, and 1760 yards in a mile. There are 2 cups in a pint and 2 pints in a quart, but 4 quarts in a gallon. There are 16 ounces in a pound, but 32 ounces in a quart. The relationships between some of the common units in the English system are given in Table 1.1.

1.5 SIMPLE UNIT CONVERSIONS

A little over ten years ago, freshmen engineering students at a major university were asked to translate the following sentence into an equation.

There are six times as many students as professors at this university.

The following — the most common wrong answer — occurred 25 to 30% of the time.

$$6\,S = P$$

You can prove that this equation is wrong by using it in a concrete example. If you substitute $S = 2$ into the equation you will come to the conclusion that there are 12 professors for every 2 students.

The source of the error here is confusion between equations and equalities, which are often written in the same way but which mean different things. The following is an example of an *equation.*

$$S = 6\,P$$

It states that the number of students can be calculated by multiplying the number of professors by 6.

The relationship between students and professors can also be described in terms of the following **equality,** which states that one professor is equivalent to six students.

$$1\,P = 6\,S$$

It is tempting to write the relationships between different sets of units in the English system in terms of equalities, such as the following.

$$1\ \text{ft} = 12\ \text{in.}$$

It is important to recognize, however, that this is not an equation for converting from feet into inches. It is an equality that can be used to construct a **unit factor** that allows us to do this conversion.

Exercise 1.1

Convert 6.5 feet into inches.

Solution

We can start by writing an equality for the relationship between feet and inches.

$$\mathbf{1\ ft = 12\ in.}$$

We can then divide both sides of the equality by 1 foot.

$$\frac{1\ \cancel{ft}}{1\ \cancel{ft}} = \frac{12\ \text{in.}}{1\ \text{ft}}$$

The result is a ratio, or factor, that is equal to 1—a **unit factor.**

$$1 = \frac{\mathbf{12\ in.}}{\mathbf{1\ ft}}$$

This is not the only unit factor that can be constructed from this equality. We can divide both sides by 12 inches to create another unit factor.

$$\frac{1\ \text{ft}}{12\ \text{in.}} = 1$$

Both unit factors are valid, but only one is useful. If we choose the right unit factor, the units of feet will cancel when we multiply the original measurement by the unit factor.

$$6.5\ \cancel{ft} \times \frac{12\ \text{in.}}{1\ \cancel{ft}} = \mathbf{78\ in.}$$

A standard sample that has a mass of one kilogram.

1.6 THE METRIC SYSTEM

More than 300 years ago, the Royal Society of London discussed replacing the irregular English system of units with one based on decimals. It was not until the French Revolution, however, that a rational, decimal-based system of units was adopted. This so-called **Metric System** is based on fundamental units of measurement for length, volume, and mass, shown in Table 1.2 and Figure 1.2.

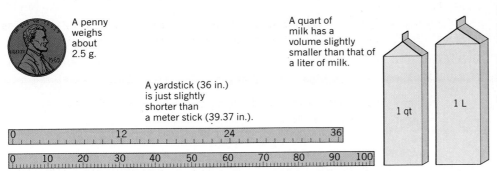

FIG. 1.2 The fundamental units in the Metric System are the gram for mass, the meter for length, and the liter for volume. A penny weighs about 2.5 grams, a meter is about 3 inches longer than a yard, and a liter is about 5% larger than a quart.

One of the principal advantages of the Metric System is the ease with which any of the base units can be converted into another unit that is more appropriate for the quantity being measured. This conversion is accomplished by addition of a prefix to the name of the base unit.

The prefix *kilo-* (k), for example, denotes multiplication by a factor of 1000. Thus, a kilometer is equal to 1000 meters.

$$1 \text{ km} = 1000 \text{ m}$$

The prefix *milli-* (m), on the other hand, means division by a factor of 1000. A milliliter (mL) is equal to 0.001 liters.

$$1 \text{ mL} = 0.001 \text{ L}$$

The common metric prefixes are given in Table 1.3. The prefixes you will encounter most often in chemistry are highlighted in color.

TABLE 1.2
The Fundamental Units of the Metric System

Length: meter (m)
1 m = 1.094 yd
1 yd = 0.9144 m

Volume: liter (L)
1 L = 1.057 qt
1 qt = 0.9464 L

Mass: gram (g)
1 g = 0.002205 lb
1 lb = 453.6 g

TABLE 1.3

Metric System Prefixes

Prefix	Symbol	Meaning
femto-	f	$\times$ 1/1,000,000,000,000,000 (10^{-15})
pico-	p	$\times$ 1/1,000,000,000,000 (10^{-12})
nano-	n	$\times$ 1/1,000,000,000 (10^{-9})
micro-	μ	$\times$ 1/1,000,000 (10^{-6})
milli-	m	$\times$ 1/1,000 (10^{-3})
centi-	c	$\times$ 1/100 (10^{-2})
deci-	d	$\times$ 1/10 (10^{-1})
kilo-	k	$\times$ 1,000 (10^{3})
mega-	M	$\times$ 1,000,000 (10^{6})
giga-	G	$\times$ 1,000,000,000 (10^{9})
tera-	T	$\times$ 1,000,000,000,000 (10^{12})

Exercise 1.2

Convert 0.135 kilometers into meters.

Solution

We can start with the definition of a kilometer.

$$1 \text{ km} = 1000 \text{ m}$$

We can then turn this equality into two unit factors.

$$\frac{1 \text{ km}}{1000 \text{ m}} = 1 \qquad \frac{1000 \text{ m}}{1 \text{ km}} = 1$$

By choosing the correct unit factor, we can get the units of kilometers in the original measurement to cancel.

$$0.135 \cancel{\text{ km}} \times \frac{1000 \text{ m}}{1 \cancel{\text{ km}}} = 135 \text{ m}$$

Exercise 1.3

Convert 0.00005 centimeters into nanometers.

Solution

We can start by noting that a meter contains 100 centimeters and 1,000,000,000 nanometers.

$$1 \text{ m} = 100 \text{ cm}$$
$$1 \text{ m} = 1,000,000,000 \text{ nm}$$

We can use this information to construct a unit factor to convert centimeters into meters.

$$0.00005 \cancel{\text{ cm}} \times \frac{1 \text{ m}}{100 \cancel{\text{ cm}}} = 0.0000005 \text{ m}$$

We can then construct another unit factor to convert from meters into nanometers.

$$0.0000005 \cancel{\text{ m}} \times \frac{1,000,000,000 \text{ nm}}{1 \cancel{\text{ m}}} = 500 \text{ nm}$$

Instead of doing the calculation in two separate steps, we can string the unit factors together, as follows.

$$0.00005 \cancel{\text{ cm}} \times \frac{1 \cancel{\text{ m}}}{100 \cancel{\text{ cm}}} \times \frac{1,000,000,000 \text{ nm}}{1 \cancel{\text{ m}}} = 500 \text{ nm}$$

There is nothing wrong with reporting the original measurement in centimeters — as 0.00005 centimeters — but 500 nanometers strikes most people as a more convenient way of reporting the data.

Samples of laboratory equipment used to measure the volume of a liquid.

A second advantage of the Metric System is the link between the base units of length and volume.

A liter is equal to the volume of a cube exactly 10 cm tall, 10 cm long, and 10 cm wide.

The volume of this cube is 10 centimeters × 10 centimeters × 10 centimeters, or 1000 cubic centimeters. Since 1 liter contains 1000 milliliters, 1 milliliter is equivalent to 1 cubic centimeter.

$$1 \text{ mL} = 1 \text{ cm}^3$$

The third advantage of the Metric System is the link between the base units of volume and weight.

The gram was originally defined as the mass of 1 mL of water at 4°C.

(It was important to specify the temperature in this definition, because water expands or contracts as the temperature changes.)

1.7 MASS VERSUS WEIGHT

There is a subtle but important difference between the concepts of weight and mass. *Mass* is a measure of the amount of matter in an object, so the mass of an object is constant. *Weight* is a measure of the force of attraction of the earth acting on an object. The weight of an object is not constant. There is a slight difference between the weight of an object resting on the earth's surface at sea level and the weight of the same object at the top of the Himalaya mountains.

Because the mass of an object is constant, it is a more fundamental quantity than weight. Most scientific balances are therefore designed to measure mass, by comparing the object with samples of known mass. Unfortunately, there is no English equivalent to the verb *weigh* that can be used to describe what happens when the mass of an object is measured. It doesn't sound right to talk about *massing* an object. This book therefore follows the common convention of using the terms *weigh* and *weight* when referring to operations and quantities that are more accurately associated with the term *mass*.

1.8 SI UNITS OF MEASUREMENT

A series of international conferences on weights and measures has been held periodically since 1875 to refine the Metric System. At the 11th conference, in 1960, a new system of units known as the *International System of Units* (abbreviated *SI* in all languages) was proposed as a replacement for the Metric System. The key to this system is the set of seven basic units given in Table 1.4.

DERIVED SI UNITS

The units of every measurement in the SI system, no matter how simple or complex, must be derived from one or more of the seven base units. The preferred unit

TABLE 1.4

SI Base Units

Physical Quantity	Name of Unit	Symbol
length	meter	m
mass	kilogram	kg
time	second	s
temperature	kelvin	K
electric current	ampere	A
amount of substance	mole	mol
luminous intensity	candela	cd

TABLE 1.5

Common Derived SI Units in Chemistry

Physical Quantity	Name of Unit	Symbol
density	kilograms per cubic meter	kg/m^3
electric charge	coulomb	$C (A \cdot s)$
electric potential	volt	$V (J/C)$
energy	joule	$J (kg\ m^2/s^2)$
force	newton	$N (kg\ m/s^2)$
frequency	hertz	$Hz (s^{-1})$
pressure	pascal	$Pa (N/m^2$ or $kg/m\ s^2)$
velocity (speed)	meters per second	m/s
volume	cubic meter	m^3

for volume is the cubic meter, for example, because volume has units of length cubed and the SI unit for length is the meter.

$$\text{SI unit of volume: } \mathbf{m^3}$$

The preferred unit for speed is meters per second, because speed is the distance traveled divided by the time it takes to cover this distance.

$$\text{SI unit of speed: } \mathbf{m/s}$$

Some of the common derived SI units are given in Table 1.5.

NON-SI UNITS

The size of common derived units in SI is not always convenient. Strict adherence to SI units, for example, would require changing the directions for an experiment from "add 250 milliliters of water to a 1-liter beaker" to "add 0.00025 cubic meters of water to an 0.001-cubic-meter container." Because of this, a number of common units of measurement that are not strictly acceptable under the SI convention are still in use. Some of these non-SI units are given in Table 1.6.

Legitimate arguments can be made both for and against SI units. The SI units are more internally consistent. But there is an enormous volume of literature written by previous generations of scientists and engineers that does not conform to the SI convention. This text uses both SI and non-SI units, because the authors believe that students must be familiar with both systems.

TABLE 1.6

Non-SI Units in Common Use

Physical Quantity	Name of Unit	Symbol
volume	liter	$L (10^{-3}\ m^3)$
length	angstrom	$Å (0.1\ nm)$
pressure	atmosphere	atm (101,325 Pa)
	torr	mmHg (133.32 Pa)
energy	calorie	cal (4.184 J)
	electron volt	$eV (1.601 \times 10^{-19}\ J)$
temperature	degree Celsius	$°C (K - 273.16)$
concentration	molarity	M (mol/L)

1.9 UNCERTAINTY IN MEASUREMENT

There is a fundamental difference between stating that there are 12 inches in a foot and stating that the circumference of the earth at the equator is 24,903.01 miles. The first relationship is based on a *definition*. By convention, there are exactly 12 inches in 1 foot. The second relationship is based on a *measurement*. It reports the circumference of the earth to within the limits of experimental error in an actual measurement.

Many unit factors are based on definition. There are exactly 5280 feet in a mile and 2.54 centimeters in an inch, for example. Unit factors based on definitions are known with complete certainty — there is no error or uncertainty associated with these numbers.

Measurements, however, are always accompanied by an element of uncertainty or error that reflects the care with which they were made. The first measurement of the circumference of the earth, in the third century B.C., for example, gave a value of 250,000 stadia — 29,000 miles — based on the assumption that the distance between Alexandria and Aswan, where the measurements were made, was roughly 5000 stadia. As the quality of the instruments used to make this measurement improved, the amount of error gradually decreased. But it never disappeared. Regardless of how carefully measurements are made, they always involve an element of uncertainty.

ACCURACY (VALIDITY) AND PRECISION (RELIABILITY)
1.10 SYSTEMATIC AND RANDOM ERRORS

There are two primary sources of error or uncertainty in a measurement: (1) limitations in the sensitivity of the instruments used and (2) imperfections in the techniques of the person making the measurement. These errors can be divided into two classes: systematic and random.

The idea of **systematic error** can be understood in terms of the bull's-eye analogy shown in Figure 1.3. Imagine what would happen if you aimed at a target with a rifle whose sights were not properly adjusted. Instead of hitting the bull's-eye, you would systematically hit the target at another point. Your results would be influenced by a systematic error caused by an imperfection in the equipment being used. Systematic error can also result from imperfections in the technique of the individual making the measurement. In the bull's-eye analogy, a systematic error of this kind might occur if you flinched each time the rifle was fired.

To understand **random error,** imagine what might happen if you closed your eyes for an instant just as you fired the rifle. The bullets would hit the target more

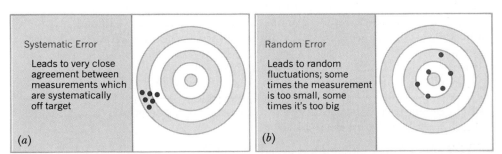

FIG. 1.3 (*a*) Systematic errors give results that are systematically too small or too large. (*b*) Random errors give results that fluctuate between being too small and being too large.

FIG. 1.4 The amount of uncertainty or error in a measurement depends in part on the equipment used. A 50-milliliter beaker, for example, can measure the volume of a liquid to the nearest 5 milliliters. A 50-milliliter graduated cylinder can measure the volume to the nearest 1 milliliter. The uncertainty can be reduced further by using a 50-milliliter buret, which is capable of delivering a volume to within ± 1 drop, or ± 0.05 milliliter. No matter how carefully the measurement is made, however, it will always contain some uncertainty.

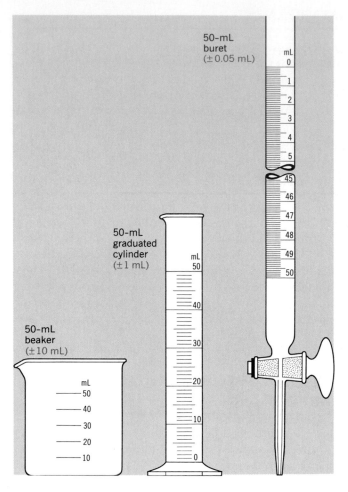

or less randomly — some too high, too low, too far to the right, too far to the left. Instead of an error that systematically gives a result too far in one direction, you now have a random error, which gives random fluctuations.

Random errors most often result from limitations in the equipment or techniques used to make a measurement. Suppose, for example, that you wanted to measure out 25 milliliters of a solution. You could use a beaker, a graduated cylinder, or a buret (see Figure 1.4). Volume measurements made with a 50-milliliter beaker are accurate to within ± 5 milliliters. That is, you would be as likely to pour out 20 milliliters of solution (5 milliliters too little) as 30 milliliters (5 milliliters too much). You could decrease the magnitude of this error by using a graduated cylinder, which is capable of measurements accurate to within ± 1 milliliter. The error could be decreased even further by use of a buret, which is capable of delivering a volume accurate to within ± 1 drop, or 0.05 milliliter.

Exercise 1.4

Which of the following procedures would lead to systematic errors, and which would produce random errors?

(a) Using a balance that always reads 5 grams too high.

(b) Using a balance that is sensitive to ± 0.1 g to weigh out 250 milligrams of vitamin C.

(c) Using a 100-milliliter graduated cylinder to measure 2.5 milliliters of solution.

(d) Using a 1-quart milk carton to measure 1-liter samples of milk.

Solution

Procedures (a) and (d) would result in systematic errors. In (a), the weight would always be too large. In (d), the volume would always be too small, because a quart is slightly smaller than a liter. Procedures (b) and (c) would lead to random errors, because the equipment used to make the measurements is not sensitive enough.

ACCURACY AND PRECISION

To most people, *accuracy* and *precision* are synonyms — words that have the same, or nearly the same, meaning. In the physical sciences, there is a subtle difference between these terms.

Accuracy: The agreement between a measurement and the true value of the quantity being measured.

Precision: The agreement between individual measurements of the same quantity.

The difference between these terms is similar to the difference between two terms from statistics: *validity* and *reliability*. Precision means the same thing as reliability. A measurement is precise, or reliable, if we get essentially the same result each time we make the measurement — if it is reproducible. Accuracy is the same as validity. A measurement is accurate, or valid, only if we get the correct answer.

To understand how systematic and random errors affect accuracy and precision, let's return to the bull's-eye analogy (see Figure 1.5). Systematic errors influence the accuracy, but not the precision, of a measurement. It is possible to get measurements that are consistently the same, but systematically wrong. Random errors influence both the accuracy and the precision of the measurement.

REDUCING SYSTEMATIC AND RANDOM ERRORS

Systematic errors can be reduced by increased care and patience on the part of the person making the measurement or improved sensitivity of the equipment used. Random errors are reduced when the results of many measurements of the same quantity are averaged, because the precision of the average of a series of measurements increases with the square root of the number of measurements.

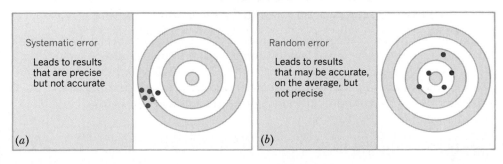

FIG. 1.5 (*a*) Systematic errors affect the accuracy of a measurement. The measurement may still be precise — there may be an excellent agreement between the results of different experiments — but the results are systematically different from the correct or true answer. (*b*) Random errors can affect both the accuracy and the precision of a measurement. Random errors can be reduced, however, by averaging of the results of many experiments.

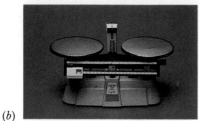

(a)

(b)

(c)

FIG. 1.6 Three balances of increasing precision: *(a)* a postage scale (±1 g), *(b)* a two-pan laboratory balance (±0.01 g), and *(c)* an analytical balance (±0.001 g).

1.11 SIGNIFICANT FIGURES

It is important to be honest when reporting a measurement so that it does not seem that the measurement is more accurate than the equipment used to make it would allow. We can achieve this goal by controlling the number of digits — or *significant figures* — used to describe the measurement.

Imagine what would happen if you weighed a copper penny on a postage scale, a two-pan laboratory balance, and an analytical balance (see Figure 1.6). The postage scale would give a weight of about 3 grams. This result means that the weight of the penny is closer to 3 grams than either 2 grams or 4 grams. In other words, the postage scale can measure the weight of the penny to within ±1 gram.

Postage scale: 3 ± 1 g

The two-pan laboratory balance is inherently more sensitive; it can weigh an object to the nearest hundreth of a gram — ±0.01 gram. This instrument would probably show that the penny weighs about 2.53 grams.

Two-pan balance: 2.53 ± 0.01 g

The analytical balance is even more sensitive. It can easily weigh an object to the nearest ±0.001 grams, in which case you might find that the penny weighs 2.531 grams.

Analytical balance: 2.531 ± 0.001 g

The postage scale gave a measurement that had only one reliable digit. This measurement is therefore said to be good to only one significant figure. The two-pan balance gave three significant figures (2.53), while the analytical balance gave four significant figures (2.531). The number of significant figures in a measurement — such as 2.531 — is equal to the number of digits that are known with some degree of confidence — 2, 5, and 3 — plus the last digit — 1 — which is generally an estimate or approximation. As we improve the sensitivity of the equipment used to make a measurement, the number of significant figures in the measurement becomes larger.

At first glance, it might seem that all we have to do to determine the number of significant figures is count the digits in the measurement. Unfortunately, zeros represent a problem, because they serve two different functions in mathematics. A zero can act as a counter, allowing us to distinguish between 304 and 324, for example. Or it can be used to set the decimal point, so that we can distinguish between 0.0045, 0.045, 0.45, 4.5, 45, 450, 4500, and so on. In the first case the zero is a significant figure; in the second, it isn't. The rules for deciding whether or not a zero is significant are summarized as follows.

Zeros used as counters (304) are significant.

Zeros used to set the decimal point (0.045) are not significant.

To understand why zeros used to set the decimal point are not significant, think about the relative accuracy of measurements of 0.045 inches and 45 inches. Both measurements are known to an accuracy of ±1 part in 45. If 0.045 — or 45/1000 — is no more accurate than 45, it can't contain more significant figures.

The rules for counting significant figures are summarized below.

RULE 1. Zeros *within* a number are always significant. Both 4308 and 40.05 contain four significant figures.

RULE 2. Zeros that appear *before* a number are never significant. Both 0.000098 and 0.98 contain two significant figures.

RULE 3. Zeros that *follow* a number may be significant.
  a. Zeros that do nothing but set the decimal point are not significant. Thus, 470,000 has two significant figures.
  b. Zeros that aren't needed to hold the decimal point are significant. For example, 4.00 has three significant figures.

If you are not sure whether a digit is significant, assume that it isn't. For example, if the directions for an experiment read: "Add the sample to 400 mL of water," assume the volume of water is known to one significant figure.

Exercise 1.5

Determine the number of significant figures in the following measurements.

  (a) 0.00098 g   (c) 0.003210 m   (e) 400
  (b) 23.07 in.   (d) 4005 L   (f) $400.00

Solution

  (a) Two significant figures — 9 and 8 — because the zeros are used to set the decimal point.
  (b) Four significant figures, because the zero is a counter.
  (c) Four significant figures — 3210 — because the last zero is not needed to set the decimal point.
  (d) Four significant figures, because the zeros are used as counters.
  (e) One significant figure — 4 — because the zeros are needed to set the decimal point.
  (f) Five significant figures, because this number has more zeros than needed to set the decimal point.

ADDITION AND SUBTRACTION WITH SIGNIFICANT FIGURES

What is the weight of a solution prepared from 0.507 grams of salt added to 150.0 grams of water? If we attacked this problem without considering significant figures, we would simply add the two measurements.

$$\begin{array}{r} 150.0 \quad \text{g } H_2O \\ +\underline{\quad 0.507 \text{ g salt}\quad} \\ 150.507 \text{ g solution} \end{array}$$ (without using significant figures)

But this answer doesn't make sense. Because we only know the weight of the water to the nearest tenth of a gram, we can only know the total weight of the solution to within ± 0.1 gram. Taking significant figures into account, we find that adding 0.507 grams of salt to 150.0 grams of water gives a solution that weighs 150.5 grams.

$$\begin{array}{r} 150.0 \quad \text{g } H_2O \\ +\underline{\quad 0.507 \text{ g salt}\quad} \\ 150.5 \quad \text{g solution} \end{array}$$ (using significant figures)

Many of the calculations in this book involve combining measurements with different degrees of accuracy and precision. The guiding principle in carrying out these calculations is easily stated.

The accuracy of the final answer can be no greater than the least accurate measurement.

This principle can be translated into a simple rule for addition and subtraction.

RULE FOR ADDING OR SUBTRACTING SIGNIFICANT FIGURES

When measurements are added or subtracted, the answer can contain no more *decimal places* than the least accurate measurement.

MULTIPLICATION AND DIVISION WITH SIGNIFICANT FIGURES

The same general principle governs the use of significant figures in multiplication and division—the final result can be no more accurate than the least accurate measurement. In this case, however, we count the significant figures in each measurement, not the number of decimal places.

RULE FOR MULTIPLYING OR DIVIDING SIGNIFICANT FIGURES

When measurements are multiplied or divided, the answer can contain no more *significant figures* than the least accurate measurement.

To illustrate this rule, let's calculate the cost of the copper in a typical penny. Let's assume that a typical penny weighs 2.531 grams, that it is essentially pure copper, and that the price of copper is 67 cents per pound.

We can start by converting the weight of the penny from grams to pounds.

$$2.531 \text{ g} \times \frac{1 \text{ lb}}{453.6 \text{ g}} = 0.005580 \text{ lb}$$

We can then use the price of a pound of copper to calculate the cost of this weight of copper metal.

$$0.005580 \text{ lb} \times \frac{67¢}{\text{lb}} = 0.37¢$$

There are four significant figures in both the weight of the penny (2.531) and the number of grams in a pound (453.6). But there are only two significant figures in the price of copper, so the final answer can only have two significant figures.

Exercise 1.6

Calculate the length in inches of a piece of wood 1.2450 feet long.

Solution

This problem is easy to set up.

$$1.2450 \text{ ft} \times \frac{12 \text{ in.}}{1 \text{ ft}} = 14.940 \text{ in.}$$

It is not as easy to decide how many significant figures the final answer should have. The original measurement (1.2540 feet) has five significant figures, but there seem to be only two significant figures in the number of inches in a foot. Thus, it might seem that the answer should contain only two significant figures.

We can clear up this confusion by remembering that only *measurements* involve error or uncertainty. Many unit factors are based on *definitions.* For example, 1 foot is defined as measuring exactly 12 inches. Unit factors based on definitions have an infinite number of significant figures. The answer to this problem therefore contains five significant figures.

1.12 ROUNDING OFF

When the answer to a calculation contains too many significant figures, it must be rounded off. Assume that the answer to a calculation is 1.247 and that the least accurate measurement has only three significant figures. The simplest way to round off this number would be to ignore the final digit and report the first three: 1.24. This approach has the disadvantage of introducing a systematic error into our calculations. Each time we rounded off, we would underestimate the value of the final answer.

There are ten digits that can occur in the last decimal place in a calculation. One way of rounding off involves underestimating the answer for five of these digits (0, 1, 2, 3, and 4) and overestimating the answer for the other five (5, 6, 7, 8, and 9).

RULES FOR ROUNDING OFF

RULE 1. If the digit is smaller than 5, drop this digit and leave the remaining number unchanged. Thus, 1.684 becomes 1.68.

RULE 2. If the digit is 5 or larger, drop this digit and add 1 to the preceding digit. Thus, 1.247 becomes 1.25.

1.13 THE DIMENSIONAL ANALYSIS, OR FACTOR-LABEL, APPROACH TO UNIT CONVERSIONS

Section 1.5 introduced unit factors as a tool for doing simple conversions from one set of units to another.

This section introduces a technique known as **dimensional analysis,** or the **factor-label approach,** which can be used to extend the use of unit factors to more difficult tasks. Dimensional analysis simply asks us to watch what happens to the units of the factors (or dimensions) in a calculation to make sure the calculation is done correctly. If the units cancel as expected, the calculation was set up properly.

Exercise 1.7

The record for the Kentucky Derby is held by Secretariat, who ran the 10 furlongs in 1 minute, 59.4 seconds. Calculate his average speed in miles per hour.

Solution

The key to this calculation is keeping track of the units with which the dimensions of the problem are expressed. The problem gives the distance for the Kentucky

Derby in furlongs and the length of the race in minutes and seconds. It then asks us to convert these data into units of miles and hours.

We can start by looking up the definition of a furlong: 1/8 mile. We can then turn this definition into an equality.

$$1 \text{ mi} = 8 \text{ furlongs}$$

The next step involves transforming this equality into a unit factor. Two unit factors are possible. One describes the number of furlongs per mile.

$$\frac{8 \text{ furlongs}}{1 \text{ mi}} = 1$$

The other tells us the number of miles in a furlong.

$$\frac{1 \text{ mi}}{8 \text{ furlongs}} = 1$$

The question is, which one should we use?

If we use the second unit factor, the units of furlongs cancel and the distance comes out in the desired units, miles.

$$10 \text{ furlongs} \times \frac{1 \text{ mi}}{8 \text{ furlongs}} = \frac{10 \text{ furlongs} \times \text{mi}}{8 \text{ furlongs}} = \textbf{1.25 miles}$$

(There are an infinite number of significant figures in this result because none of the quantities in this calculation are measurements. The distance of the Kentucky Derby is defined as 10 furlongs, and a furlong is defined as 1/8 mile.)

Let's see what happens if we use the wrong unit factor in this calculation. We'll multiply the distance of 10 furlongs by the unit factor that describes the number of furlongs per mile.

$$10 \text{ furlongs} \times \frac{8 \text{ furlongs}}{1 \text{ mi}} = \frac{80 \text{ furlongs}^2}{\text{mi}}$$

The answer comes out in units that don't make sense — furlongs squared per mile — which tells us we've made a mistake in setting up the calculation.

To complete this problem we have to convert 1 minute, 59.4 seconds into units of hours. We might start by calculating the total number of seconds.

$$\left[1 \text{ min} \times \frac{60 \text{ s}}{1 \text{ min}} \right] + 59.4 \text{ s} = \textbf{119.4 s}$$

We can then use the appropriate unit factors to convert seconds into minutes and minutes into hours. In each case, we choose the correct unit factor by keeping track of the units with which the dimensions of the problem are expressed. The first factor allows the units of seconds to cancel and introduces the units of minutes. The second factor cancels the units of minutes and introduces the desired units of hours.

$$119.4 \text{ s} \times \frac{1 \text{ min}}{60 \text{ s}} \times \frac{1 \text{ hr}}{60 \text{ min}} = \textbf{0.03317 hr}$$

To calculate the average speed in miles per hour we divide the distance traveled in miles by the time it took in hours.

$$\frac{1.25 \text{ mi}}{0.03317 \text{ hr}} = \textbf{37.68} \; \frac{\textbf{mi}}{\textbf{hr}}$$

Exercise 1.8

Calculate the volume in liters of a cubic container 8 feet tall.

Solution

We can start by calculating the volume of the container in units of cubic feet. The volume of a cube is the product of its height times its length times it width.

$$V = (8 \text{ ft} \times 8 \text{ ft} \times 8 \text{ ft}) = (8 \text{ ft})^3 = \textbf{512 ft}^3$$

It isn't always obvious what should be done next. So let's look at the goal of the calculation and work backwards. The goal is to determine the volume in liters. What do we know about this unit? Remember that one of the advantages of the Metric System is the link between units of length and volume. One liter contains 1000 milliliters. But a milliliter is equivalent to a cubic centimeter. Thus, 1 liter contains 1000 cubic centimeters.

$$1 \text{ L} = 1000 \text{ cm}^3$$

If we can convert the volume of the container in cubic feet into cubic centimeters, we can then convert cubic centimeters into liters.

Converting cubic feet into cubic centimeters requires two equalities.

$$1 \text{ ft} = 12 \text{ in.}$$
$$1 \text{ in.} = 2.54 \text{ cm}$$

Each unit factor must be cubed, because the goal is to convert cubic feet into cubic inches and then cubic inches into cubic centimeters.

$$512 \text{ ft}^3 \times \left[\frac{12 \text{ in}}{1 \text{ ft}} \right]^3 \times \left[\frac{2.54 \text{ cm}}{1 \text{ in.}} \right]^3 = 14{,}500{,}000 \text{ cm}^3$$

Once we know the volume in cubic centimeters, we can calculate the volume in liters.

$$14{,}500{,}000 \text{ cm}^3 \times \frac{1 \text{ L}}{1000 \text{ cm}^3} = \textbf{14{,}500 L}$$

Exercise 1.9

What is the value of a gold ingot 1.00 foot long by 6.00 inches wide by 2.75 inches tall, if a cubic centimeter of gold weighs 19.3 grams and the price of gold is $408 per ounce?

Solution

Complex problems like this must be worked one step at a time. Since the problem gives the length, height, and width of the gold ingot, we might start by calculating the volume of the ingot. Before we can do this, of course, we have to convert the length, height, and width of the ingot to a common set of units—such as feet.

$$1.00 \text{ ft} \times \left[6.00 \text{ in.} \times \frac{1 \text{ ft}}{12 \text{ in.}} \right] \times \left[2.75 \text{ in.} \times \frac{1 \text{ ft}}{12 \text{ in.}} \right] = \textbf{0.1146 ft}^3$$

Now what? We know how to convert from cubic feet into cubic centimeters, so let's see where that gets us.

$$0.1146 \, \cancel{ft^3} \times \left[\frac{12 \, \cancel{in.}}{1 \, \cancel{ft}} \right]^3 \times \left[\frac{2.54 \, cm}{1 \, \cancel{in.}} \right]^3 = 3245 \, cm^3$$

We did this because we know something about the weight of a cubic centimeter of gold.

$$1 \, cm^3 = 19.3 \, g$$

This equality allows us to calculate the weight of the ingot in grams.

$$3245 \, \cancel{cm^3} \times \frac{19.3 \, g}{1 \, \cancel{cm^3}} = 62{,}630 \, g$$

We can now convert this weight into units of pounds.

$$62{,}630 \, \cancel{g} \times \frac{1 \, lb}{453.6 \, \cancel{g}} = 138.1 \, lb$$

Once we have the weight of the ingot in pounds, we can calculate the weight in ounces.

$$138.1 \, \cancel{lb} \times \frac{16 \, oz}{1 \, \cancel{lb}} = 2210 \, oz$$

Finally, we can calculate the value of the ingot in dollars.

$$2210 \, \cancel{oz} \times \frac{\$408}{1 \, \cancel{oz}} = \mathbf{\$902{,}000}$$

Look at the number of significant figures in each step in this calculation. The final answer was rounded off to three significant figures, which reflects the accuracy of the least accurate measurement. However, four significant figures were carried in each intermediate step in the calculation. The technique of carrying an extra significant figure until the very end of a multistep calculation helps us avoid the error that can creep in if each step is rounded off to the number of significant figures allowed in the final answer.

1.14 SCIENTIFIC NOTATION

Chemists routinely work with numbers that are extremely large or extremely small. There are 24,000,000,000,000,000,000,000 copper atoms in a penny, and each weighs 0.000,000,000,000,000,000,000,1055 gram. In theory, the product of these numbers is the weight of the copper in a penny. It is impossible to multiply these numbers with most calculators, however, because they can't accept either number as it is written here. In order to do a calculation like this, it is necessary to express these numbers in *scientific notation*—as a number between 1 and 10 multiplied by 10 raised to some power, or exponent.

Before we discuss how to translate numbers into scientific notation, it might be useful to review some of the basics of exponential mathematics.

RULES FOR EXPONENTIAL MATHEMATICS

RULE 1. Any number raised to the zero power is equal to 1.

$$1^0 = 1 \qquad 2^0 = 1 \qquad 10^0 = 1 \qquad x^0 = 1$$

RULE 2. Any number raised to the first power is equal to itself.

$$1^1 = 1 \qquad 2^1 = 2 \qquad 10^1 = 10 \qquad x^1 = x$$

RULE 3. Any number raised to the second power is equal to the product of the number times itself.

$$2^2 = 2 \times 2 = 4 \qquad 10^2 = 10 \times 10 = 100$$

RULE 4. Any number raised to the nth power is equal to the product of that number times itself $n - 1$ times.

$$2^4 = 2 \times 2 \times 2 \times 2 = 16$$
$$10^5 = 10 \times 10 \times 10 \times 10 \times 10 = 100,000$$

RULE 5. Dividing by a number raised to some exponent is the same as multiplying by that number raised to an exponent of the opposite sign.

$$\frac{1}{10^2} = 1 \times 10^{-2} = 0.01 \qquad \frac{3}{4^{-3}} = 3 \times 4^3 = 192$$

These rules form the basis for writing numbers in scientific notation. Consider the following examples.

$$1645 = 1.645 \times 1000 = 1.645 \times 10 \times 10 \times 10 = \mathbf{1.645 \times 10^3}$$
$$34.2 = \mathbf{3.42 \times 10^1}$$
$$247,000 = 2.47 \times 100,000 = 2.47 \times 10 \times 10 \times 10 \times 10 \times 10 = \mathbf{2.47 \times 10^5}$$
$$0.015 = \frac{1.5}{100} = \frac{1.5}{10^2} = \mathbf{1.5 \times 10^{-2}}$$
$$0.00269 = \frac{2.69}{1000} = \frac{2.69}{10^3} = \mathbf{2.69 \times 10^{-3}}$$

Careful consideration of these examples leads to the following generalization.

> **The exponent in scientific notation is equal to the number of times the decimal point must be moved to produce a number between 1 and 10.**

According to the 1980 census, the population in metropolitan Chicago was 7,103,624. To convert this number to scientific notation we must move the decimal point to the left six times.

$$7 \, 1 \, 0 \, 3 \, 6 \, 2 \, 4 = \mathbf{7.103624 \times 10^6}$$

To translate 24,000,000,000,000,000,000,000 copper atoms into scientific notation, we have to move the decimal point to the left 22 times.

$$2 \, 4 \, 0 \, 0 \, 0 \, 0 \, 0 \, 0 \, 0 \, 0 \, 0 \, 0 \, 0 \, 0 \, 0 \, 0 \, 0 \, 0 \, 0 \, 0 \, 0 \, 0 \, 0$$
$$= \mathbf{2.4 \times 10^{22}}$$

A penny therefore contains 2.4×10^{22} copper atoms.

To convert numbers smaller than 1 into scientific notation, we have to move the decimal point to the right. The decimal point in 0.000985, for example, must be moved to the right four times. This number is therefore equal to 9.85×10^{-4}.

$$0 . 0 \, 0 \, 0 \, 9 \, 8 \, 5 = \mathbf{9.85 \times 10^{-4}}$$

Converting 0.000,000,000,000,000,000,000,1055 grams per copper atom into scientific notation involves moving the decimal point to the right 22 times.

$$0 . 0 \, 0 \, 0 \, 0 \, 0 \, 0 \, 0 \, 0 \, 0 \, 0 \, 0 \, 0 \, 0 \, 0 \, 0 \, 0 \, 0 \, 0 \, 0 \, 0 \, 0 \, 1 \, 0 \, 5 \, 5$$
$$= \mathbf{1.055 \times 10^{-22}}$$

The weight of a copper atom is therefore 1.055×10^{-22} grams.

The primary reason for converting numbers into scientific notation is to make calculations with unusually large or small numbers less cumbersome. But there is another important advantage to scientific notation. Since zeros are no longer used to set the decimal point, all of the digits in a number in scientific notation are significant, as shown by the following examples.

7.103624×10^6	7 significant figures
2.4×10^{22}	2 significant figures
9.85×10^{-4}	3 significant figures
1.055×10^{-22}	4 significant figures

Exercise 1.10

Convert each of the following numbers into scientific notation.

(a) 0.0046940	(c) 11.98	(e) 1.98	(g) 0.189
(b) 4,679,000	(d) 0.000007967	(f) 212.6	(h) 16,200

Solution

(a) 4.6940×10^{-3}	(c) 1.198×10^1	(e) 1.98×10^0	(g) 1.89×10^{-1}
(b) 4.679×10^6	(d) 7.967×10^{-6}	(f) 2.126×10^2	(h) 1.62×10^4

The mathematics behind the addition, subtraction, multiplication, and division of numbers in scientific notation is discussed in the appendix.

1.15 THE GRAPHICAL TREATMENT OF DATA

If the basis of science is a natural curiosity about the world that surrounds us, an important step in doing science is trying to find patterns in the observations and measurements that result from this curiosity.

Anyone who has ever played with a prism or seen a rainbow has watched what happens when white light is split into a spectrum of different colors. Physics tells us that the difference between blue and green or red light is the result of differences in the frequency and wavelength of the light. It also provides us with instruments that can measure the frequency and wavelength of light of different colors such as the data in Table 1.7.

There is an obvious pattern in these data—as the wavelength of the light becomes larger, the frequency becomes smaller. But recognizing this pattern is not

TABLE 1.7

Relationship Between the Characteristic Wavelengths and Frequencies of Light of Different Colors

Color	Wavelength (λ) (m)	Frequency (v) (1/s)
violet	4.100×10^{-7}	7.312×10^{14}
blue	4.700×10^{-7}	6.379×10^{14}
green	5.200×10^{-7}	5.765×10^{14}
yellow	5.800×10^{-7}	5.169×10^{14}
orange	6.000×10^{-7}	4.997×10^{14}
red	6.500×10^{-7}	4.612×10^{14}

enough. It would be even more useful to construct a mathematical equation that fits these data. This would allow us to calculate the frequency of light of any wavelength, such as blue-green light with a wavelength of 4.950×10^{-7} m, or to calculate the wavelength of light of a known frequency, such as blue-violet light with a frequency of 6.850×10^{14} cycles per second.

The first step toward constructing an equation that fits these data involves plotting the data in different ways until we get a straight line. We might decide, for example, to plot the wavelengths on the vertical axis and the frequencies on the horizontal axis, as shown in Figure 1.7. When we construct a graph, we should keep the following rules in mind.

RULES FOR CONSTRUCTING GRAPHS

RULE 1. The scales of the graph should be chosen so that the data fill as much of the available space as possible.

RULE 2. It isn't necessary to include the origin (0,0) on the graph. In fact, it may be more efficient to leave off the origin, so that the data fill the available space.

RULE 3. Once the scales have been chosen and labeled, the data are plotted one point at a time.

RULE 4. A straight line or a smooth curve is then drawn through as many points as possible. Because of experimental error, the line or curve may not pass through every data point.

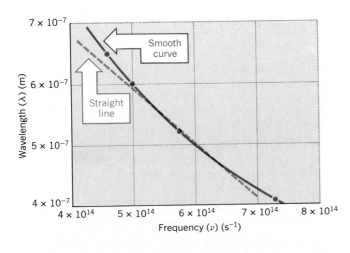

FIG. 1.7 A plot of wavelength (λ) in meters versus frequency (v) in cycles per second for light of different colors. Note that the curve is almost, but not quite, a straight line.

If this process gives a straight line, we can conclude that the quantity plotted on the vertical axis (y) is directly proportional to the quantity on the horizontal axis (x). We can then fit the graph to the following equation for a straight line.

$$y = mx + b$$

Here, y is the quantity on the vertical axis, x is the quantity on the horizontal axis, m is the slope of the line, and b is the intercept — the value of y when x is equal to zero.

Unfortunately, the graph in Figure 1.7 is not quite a straight line. Since the data are known to four significant figures, the deviation from a straight line is not the result of experimental error. We must therefore conclude that the wavelength and frequency of light are not directly proportional.

We might now look for an *inverse* proportionality between the two sets of measurements — x and y. In other words, we might try to fit the data to the following equation.

$$y = m\left[\frac{1}{x}\right] + b$$

We can test this relationship by calculating the reciprocal of the frequencies of the different colors of light in Table 1.7. (This is done by dividing each frequency into 1.) We can then plot the wavelengths in Table 1.7 versus the inverse of the frequencies, as shown in Figure 1.8.

This graph gives a beautiful straight-line relationship of the following form.

$$\lambda = m\left[\frac{1}{\nu}\right] + b$$

We can now calculate the slope of the line (m) and the intercept (b), as shown in Figure 1.9. In order to minimize the error in our calculations, it is important that we choose points for this calculation that are *not* original data points. When this is done for the data in Figure 1.9, the slope of the line is found to be equal to the speed of light — 2.998×10^8 meters per second (see Appendix A-2).

There are two ways of calculating the value of the intercept, b. In some cases, we can extrapolate (extend) the straight line and simply read the value on the vertical axis that corresponds with a value of zero on the horizontal axis, as shown in Figure 1.9. This isn't possible for the data in Figure 1.8, however, because the origin was left off this graph.

Another approach starts by selecting a point on the straight line. The values of y and x for this point are estimated and then combined with the slope of the line (m) to calculate the value of the intercept (b). In other words, if the data in the graph fit the equation

$$y = mx + b$$

then the value of b for this straight line can be calculated as follows.

$$b = y - mx$$

When this approach is applied to the data in Figure 1.8, we find that the intercept is equal to zero.

The data in Figure 1.8 therefore fit the following equation.

$$\nu = (2.998 \times 10^8 \text{ m/s})\frac{1}{\lambda}$$

Rearranging this equation, we find that the product of the frequency times the

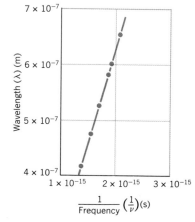

FIG. 1.8 A plot of wavelength (λ) in meters versus the inverse of the frequency ($1/\nu$) in seconds for light of different colors gives a straight line, within experimental error.

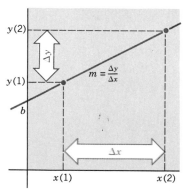

FIG. 1.9 To calculate the slope (m) of a straight line that fits the equation $y = mx + b$, divide the distance between two points on the vertical axis by the corresponding distance between two points along the horizontal axis.

wavelength of light is equal to the speed of light.

$$\nu\lambda = 2.998 \times 10^8 \text{ m/s}$$

Section 5.10 will explain why this is true.

Exercise 1.11

The following data were obtained from a study of the relationship between the volume (V) and temperature (T) of a gas at constant pressure.

Volume (mL):	273.0	277.4	282.7	287.8	293.1	298.1
Temperature (°C):	0.0	5.0	10.0	15.0	20.0	25.0

Extrapolate these data to determine the temperature at which the volume of the gas should become equal to zero.

Solution

A plot of these data gives a straight line, as shown in Figure 1.10. The mathematical equation for this straight line can be written as follows.

$$T = mV + b$$

By choosing any two points on this curve—such as $T = 7.5°C$ and $22.5°C$—and estimating the volumes at these temperatures from the straight line that passes through the points, we can calculate the slope of the line.

$$m = \frac{\Delta T}{\Delta V} = \frac{(22.5°C - 7.5°C)}{(295.5 \text{ ml} - 280.1 \text{ mL})} = 0.974 \frac{°C}{mL}$$

The value of b for this equation can be calculated from the data for any point on the straight line—such as the point at which the temperature is $22.5°C$ and the volume is 295.5 mL.

$$\begin{aligned} b &= T - mV \\ &= (22.5°C) - (0.974°C/mL)(295.5 \text{ mL}) \\ &= -265°C \end{aligned}$$

According to these data, the volume of the gas becomes zero when the gas is cooled until the temperature reaches about $-265°C$.

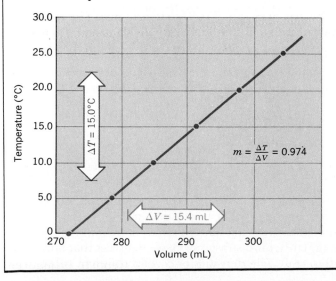

FIG. 1.10 A plot of the temperature-versus-volume data from Exercise 1.11.

1.16 EXTENSIVE AND INTENSIVE QUANTITIES

Many of the quantities whose measurements are discussed in this chapter are examples of *extensive properties,* which depend on the size of the sample. Mass and volume, for example, are extensive properties because they depend on the size of the sample. The larger the sample, the bigger the mass or volume. Other extensive properties include distance—length, height, width, and so on—and time.

At first glance, it might seem that anything you measure would be an extensive property. But that isn't true. There are also *intensive properties,* which do not depend on the size of the sample. Temperature and pressure are both examples of intensive properties. On the average, the temperature of the air in your immediate vicinity is the same as the average temperature of the air throughout the room. The same can be said about the pressure of the gas in one portion of the room compared with the pressure of the gas throughout the room.

Many intensive properties are the ratios of a pair of extensive properties that are proportional to each other. If you measure the number of times your heart beats and divide by the length of time during which you counted, the result is an intensive quantity known as the heart rate. You might count 35 beats in 30 seconds or 175 beats in 2.5 minutes. In either case, the heart rate would be 70 beats per minute. Speed is another example of an intensive property equal to the ratio of a pair of extensive properties that are proportional to each other. Regardless of whether you drive 3.90 miles in 216 seconds or 45.5 miles in 42 minutes, your speed is 65 miles per hour.

Ice floats because water is denser.

Because lead is denser than copper, a piece of lead seems "heavy" when compared with a piece of copper.

1.17 DENSITY AS AN EXAMPLE OF AN INTENSIVE QUANTITY

If you ask people, "Why does ice float on water?" the most common answer is, "Ice is lighter than water." If you give them a piece of lead and ask them to compare it with a piece of copper of about the same size, they often say that lead is heavier than copper. Regardless of the popularity of these responses, both are wrong. They result from confusion between extensive and intensive properties.

The key to creating an intensive property is to find two extensive properties that are proportional to each other. The mass and volume of a substance—such as sugar—offer a perfect example of this phenomenon. Figure 1.11 shows how the mass of a sample of sugar depends on its volume. Both mass and volume are extensive properties—the bigger the sample, the more it weighs. But the ratio of the mass of the sample to its volume is constant. This ratio is an intensive property known as *density,* which is a characteristic property of the substance.

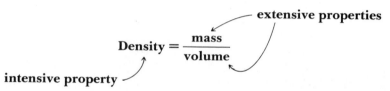

$$\underset{\text{intensive property}}{\text{Density}} = \frac{\overset{}{\text{mass}}}{\text{volume}} \quad \text{extensive properties}$$

The density of water, for example, is about 1 gram per milliliter, no matter whether we measure the water by the drop or by the gallon.

Densities of solids and liquids are given in units of grams per milliliter (g/mL) or grams per cubic centimeter (g/cm³); most substances have densities that fall in the range between 0.10 and 10 g/cm³. The densities of certain common substances

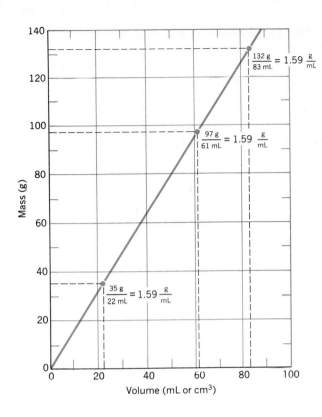

FIG. 1.11 Mass and volume are extensive properties. The larger the sample, the larger the mass or volume. Density is an intensive property. No matter how large or small the sample, the ratio of mass to volume is the same. This graph shows that we get the same value for the density of sugar regardless of the size of the sample.

are given in Table 1.8, and the densities of a number of common metals are given in Table 1.9.

Ice doesn't float on water because it is lighter than water, it floats because it is less dense. A given volume of ice weighs less than an equivalent volume of water, so the

TABLE 1.8			
Densities of Some Common Substances			
Substance	Density (g/cm³)	Substance	Density (g/cm³)
air at 25°C	0.0408	limestone	2.68–2.76
asbestos	2.0–2.8	milk	1.028–1.035
beeswax	0.96–0.97	olive oil	0.918
bone	1.7	quartz	2.65
cardboard	0.69	salt	2.18
cement	2.7–3.0	sawdust	0.19
chalk	1.9–2.8	sugar	1.59
charcoal	0.28–0.57	the sun	1.410
clay	1.8–2.6	talc	2.7–2.8
coal	1.2–1.8	turpentine	0.87
cork	0.22–0.26	water	1.00
diamond	3.5–3.52	wood	
the earth	5.519	balsa	0.11–0.14
gasoline	0.66–0.69	bamboo	0.31–0.40
graphite	2.30–2.72	ebony	1.11–1.33
ice	0.917	oak	0.60–0.90
ivory	1.83–1.92	walnut	0.64–0.70

TABLE 1.9			
Densities of Some Common Metals			
Substance	Density (g/cm³)	Substance	Density (g/cm³)
aluminum	2.70	mercury	13.60
brass	8.40	nickel	8.90
calcium	1.54	osmium	22.48
chromium	7.20	platinum	21.45
cobalt	8.9	potassium	0.856
copper	8.92	rhodium	12.4
gold	19.3	silver	10.5
iridium	22.42	sodium	0.97
iron	7.86	steel	7.8
lead	11.34	titanium	4.5
lithium	0.534	tungsten	19.35
magnesium	1.74	uranium	19.05
manganese	7.20	zinc	7.14

water settles to the bottom, and the ice floats on top. A piece of lead may feel heavy when we pick it up. But it isn't so much the weight that we notice as the difference between what we feel and what we expect for an object that size. Lead isn't heavier than copper, it is more dense.

Exercise 1.12

Calculate the density of concentrated sulfuric acid if a 175-milliliter sample weighs 322 grams.

Solution

Density is calculated by dividing the mass of a sample by its volume. The density of concentrated sulfuric acid is therefore 1.84 g/mL, or 1.84 g/cm³.

$$\text{Density} = \frac{322 \text{ g}}{165 \text{ mL}} = 1.84 \text{ g/mL}$$

The key to this calculation, of course, is recognizing that the measurements of mass and volume must be made on the same sample.

Exercise 1.13

Archaeologists use a technique they call *flotation* to isolate small fragments of charcoal or bone. They start by dissolving large quantities of a soluble salt — such as zinc chloride — in water to produce a solution with a density of about 1.8 g/cm³. The mixture of sand, soil, small rocks, and fragments of charcoal and bone they wish to separate is then added to this solution. Use the data in Table 1.8 and common sense to predict which components of the mixture float on top of the solution and which sink to the bottom.

Solution

Charcoal (0.28 – 0.57 g/cm³) will definitely float on top of this solution, because it is much less dense. Bone (1.7 g/cm³) should also float. Sand, soil, and rocks should all settle to the bottom, because they tend to have densities greater than 2 g/cm³.

1.18 TEMPERATURE AS AN INTENSIVE QUANTITY

There are certain words in any language that everyone understands but few can define. **Temperature** is an example of such a word. The best that most dictionaries can do is to say that temperature is a measure of the degree of hotness or coldness. A more popular, but less useful, definition suggests that temperature is the quantity measured with a thermometer.

The origin of the thermometer can be traced to the work of Galileo, who constructed the first ''thermoscope'' in 1592 by trapping air in a large glass bulb with a long narrow neck inverted over a container of water or wine. Any change in the temperature of the room would cause the air in the bulb to expand or contract, forcing the liquid to travel up or down the tube. It was not until almost 20 years later, however, that a colleague of Galileo suggested adding a scale to the thermoscope to make the first thermometer.

RELATIVE TEMPERATURE SCALES

By the early 1700s, at least 35 different temperature scales had been proposed. At that time, a Dutch instrument maker by the name of Daniel Gabriel Fahrenheit became famous for his mercury thermometers. The Fahrenheit scale he developed is still the most widely used temperature scale in the United States.

The problem that Fahrenheit faced in developing his scale is still a common one today. It is easy to make relative measurements—to determine that one object is hotter than another, for example. But it is much harder to make absolute measurements, to state the temperature of an object on an absolute scale.

The approach Fahrenheit took to this problem is still used for most instruments. He calibrated his thermometer against a pair of references. He defined the temperature of a mixture of salt and ice as 0°F, and he defined body temperature as 96°F. On this scale, the temperature of ice by itself is 32°F, and the temperature of boiling water is 212°F.

As early as the second century A.D., the Greek physician Galen suggested using ice and boiling water as the basis of a temperature scale. It was not until 1742, however, that a scale based on that principle was developed. In that year, the Swedish astronomer Anders Celsius introduced the temperature scale now in use in almost every country in the world. Celsius defined the temperature of ice and boiling water as 0°C and 100°C, respectively. His scale was known as the centigrade scale, because 1 degree on this scale is 1/100 of the difference between the temperatures of ice and boiling water. In 1948, the name was officially changed to the Celsius scale.

The Fahrenheit and Celsius scales are both relative temperature scales. They define two reference points (such as 0°C and 100°C), divide the range of temperatures between these points into degrees, and then compare all temperatures to these arbitrary references.

ABSOLUTE TEMPERATURE SCALES

At the beginning of the 1800s, the relationship between the volume and temperature of a gas that was the basis of Exercise 1.9 was discovered. The data given in this exercise were obtained with an apparatus that can be built in any general chemistry laboratory. When more accurate equipment is used, the data extrapolate to a volume of zero at a temperature of −273.15°C. In 1848, the British physicist William Thompson, who later became Lord Kelvin, suggested that this observation could be used as the basis for an absolute temperature scale. On the Kelvin scale, absolute zero—0 K—is the temperature at which the volume of a gas becomes zero. It is therefore the lowest possible temperature, or the absolute zero on any temperature scale.

Zero on the Kelvin scale is equal to −273.15°C.

$$0 \text{ K} = -273.15°\text{C}$$

Each unit on this scale, or each *kelvin,* is equal to 1 degree on the Celsius scale. There is a subtle difference between the units on these scales, however. Because the Celsius scale is based on two arbitrary references points, the difference between the temperatures of these two points is divided into *degrees.* The Kelvin scale, however, is an absolute scale. Zero is not arbitrarily defined; it is defined as the lowest possible temperature that can be achieved. Thus, temperatures on the Kelvin scale are not divided into degrees. Temperatures on this scale are reported in units of kelvin, not in degrees kelvin.

TEMPERATURE CONVERSIONS

Figure 1.12 shows the relationship between the Fahrenheit, Celsius, and Kelvin scales. The simplest temperature scales to interconvert are the Celsius and Kelvin scales. All you have to remember is that zero on the Kelvin scale is equivalent to $-273.15°C$

$$0 \text{ K} = -273.15°C$$

and 1 K is equal to 1°C. Thus, the temperature on the Kelvin scale is equal to the temperature in degrees Celsius plus 273.15. This relationship is expressed in the following equation.

$$T_K = °C + 273.15$$

Converting from degrees Celsius to degrees Fahrenheit, or vice versa, is a bit more difficult. We can start by noting that the difference between the temperatures of ice and boiling water is 180°F or 100°C. This means that a degree on the Fahrenheit scale is smaller than a degree on the Celsius scale. Their relationship can be expressed as a factor of 180°F/100°C, or 9/5.

There is another problem, however. The temperature at which the two scales reach zero is not the same. Zero on the Celsius scale is equivalent to 32°F.

$$0°C = 32°F$$

To convert from Celsius into Fahrenheit we therefore have to multiply by 9/5 and then add 32°.

$$°F = \frac{9}{5}(°C) + 32$$

This equation can be rearranged to convert from Fahrenheit to Celsius.

$$°C = \frac{5}{9}[°F - 32]$$

These equations are often difficult to remember. There are two ways of overcoming this problem. You can remember how they are derived, in which case you no longer have to remember the equations. Or you can check to make sure the equation you think you remember gives the right results. The equations are correct if they convert 212°F into 100°C and 32°F into 0°C.

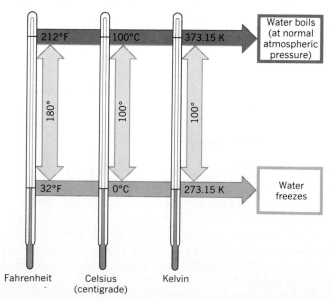

FIG. 1.12 The three common temperature scales. Note that water boils at 212°F, 100°C, or 373.15 K and freezes at 32°F, 0°C, or 273.15 K.

SUMMARY

There are no rules that we can follow blindly to ensure that we are using a scientific method. But there are certain principles that help us define what we mean by science, and there are normal practices that guide those who call themselves scientists. One of the principal goals of science is to recognize and then explain patterns in the behavior of the world around us. One route toward this goal involves making qualitative observations; another involves taking measurements.

All measurements contain at least three elements: a number that indicates the size of the quantity being measured, units that provide a basis for comparing this quantity with a standard reference, and some uncertainty or error. The existence of different systems of units—and different units within each system—makes it essential to be able to convert from one set of units to another. The error or uncertainty associated with a mea-surement requires that we: (1) pay attention to the accuracy and/or precision of the measurement, (2) try to minimize random and systematic errors involved in the measurement, and (3) estimate the magnitude of the error involved in a measurement through the use of significant figures

The quantities measured in this chapter can be divided into two categories: extensive and intensive properties. Extensive properties—such as mass and volume—depend on the size of the sample. Intensive properties—such as density and temperature—don't depend on the size of the sample. Intensive properties are therefore more useful for characterizing substances. An intensive property is often the ratio of a pair of extensive properties that are related to each other. The density of a substance, for example, is equal to the ratio of its mass to its volume.

PROBLEMS

Chemistry As an Experimental Science and the Nature of Science

1-1 Describe the difference between a hypothesis, a theory, and a law. Give an example of each taken from a field other than chemistry.

1-2 In what ways does the Greek model of four elements fit the five principles of modern chemistry outlined in Section 1.1? In what ways doesn't it fit these principles?

1-3 Use the six normal practices of science outlined in Section 1.2 to discuss the difference between science and pseudo-science, such as astrology, ESP, creation science, or the like.

English Units of Measurement

1-4 Describe the three elements that are present in all mea-surements.

1-5 Find a source that explains the origin of one or more units in the English system of units and describe how the system seemed rational to its inventors.

1-6 A gallon in the United States contains 128 ounces. Canada once used the imperial gallon, which contains 160 ounces. How many quarts are in an imperial gallon?

Unit Conversions in the English System

1-7 Calculate the number of seconds in a year.

1-8 Calculate the number of ounces in a ton.

1-9 Calculate the number of grains in a pound if there are 16 ounces in a pound, 16 drams in an ounce, and $437\frac{1}{2}$ grains in a dram.

1-10 Calculate the number of ounces (by weight) in a fluid ounce if a pint of water weighs 1.04 pounds.

1-11 A British gallon contains 0.027777 barrel, 0.125 bushel, 0.111111 firkin, 160 British fluid ounces, 0.5 peck, 0.019063 hogshead, or 0.05555 kilderkin. Calculate the number of:

(a) hogsheads in a kilderkin,

(b) barrels in a bushel,

(c) kilderkins in a firkin,

(d) British fluid ounces in a peck,

(e) bushels in a peck.

1-12 A bushel is a dry measure equal to 4 pecks. There are 8 dry quarts in a peck and 2 dry pints in a quart. If a dry pint contains 33.6 cubic inches, what is the volume of a bushel in cubic inches?

1-13 A light year is the distance light travels in a year. If a light year is 5.87851×10^{12} miles, what is the length of a distance of one foot in units of light years?

The Metric System

1-14 Describe three advantages of the Metric System.

1-15 Define the following prefixes from the Metric System.

(a) nano- (b) micro- (c) milli- (d) centi- (e) kilo-

Unit Conversions in the Metric System

1-16 Make the following conversions.

  (a) 0.043 kilogram into grams

  (b) 2.45 liters into milliliters

  (c) 0.00814 liter into milliliters

  (d) 346.8 millimeters into centimeters

1-17 The diameter of a helium atom is 1.9 angstroms (Å), where an angstrom is 10^{-8} centimeters. Calculate the diameter of a helium atom in units of centimeters, meters, nanometers, and picometers.

1-18 Light is the small portion of the electromagnetic spectrum that is visible to the naked eye. It has wavelengths between about 4×10^{-5} and 7×10^{-5} centimeters. Calculate the range of wavelengths of light in units of "microns" —micrometers—and in units of millimicrons, nanometers, and angstroms.

Unit Conversions Between the English and Metric Systems

1-19 Calculate the number of pounds in a kilogram, assuming a kilogram weighs 2.205 pounds.

1-20 Calculate the number of liters in a gallon, assuming a liter is 0.2542 gallons.

1-21 Make the following conversions.

  (a) 32.0 liters into gallons

  (b) 8.76 pounds into grams

  (c) 135 meters into miles

  (d) 53.67 meters into feet

  (e) 5.00 quarts into milliliters

  (f) 6500 grams of water into liters

1-22 Calculate the weight in milligrams of the aspirin in a tablet that contains 5.0 grains of aspirin assuming 1 pound contains 7000 grains.

1-23 Calculate the volume in milliliters of a bottle of typewriting correction fluid that contains 0.6 fluid ounces.

1-24 Which takes longer to run at a constant speed, a 100-meter dash or 100-yard dash?

1-25 A barrel of liquor contains 31 gallons, but a barrel of oil or gasoline contains 42 gallons. Calculate the number of liters per barrel for both liquor and gasoline.

1-26 Liquor, which used to be sold in "fifths," is now sold in 750-milliliter bottles. If a fifth is one-fifth of a gallon, which is the better buy: a fifth of scotch selling for $12.50 or a 750-milliliter bottle selling for the same price?

1-27 The radius of an atom is about 10^{-8} centimeters, and the radius of the nucleus of the atom is about 10^{-13} centimeters. If the atom were expanded until it filled a football field, how big would the nucleus be?

1-28 Determine the number of gallons in 1.00 cubic meter of water and the weight in pounds of 1.00 cubic foot of water.

1-29 Calculate the number of cubic feet of peat moss in a bag marked 113 cubic decimeters.

1-30 The largest Harley-Davidson motorcycle has an engine with a displacement of 80 cubic inches. The largest motorcycle Honda makes has a displacement of 1500 cubic centimeters. Which motorcycle has the larger engine?

1-31 Calculate the number of cubic inches in a liter and the number of cubic meters in a cubic foot.

Unit Conversions Involving Ratios of Two Quantities

1-32 Estimate the speed of sound in miles per hour, assuming the sound of thunder takes 5 seconds to travel 1 mile.

1-33 Calculate the speed limit on the nation's interstate highways in units of meters per second, kilometers per hour, and feet per second.

1-34 The speed of light is 2.9979×10^8 meters per second. Calculate the speed of light in centimeters per second, miles per hour, and miles per year.

1-35 What is the speed in miles per hour of an electron traveling across a cathode ray tube if the electron travels 1 yard in 12 nanoseconds?

1-36 If your heart beats an average of 70 times per minute, how many times will it beat in a 75-year lifetime? The Jarvik-7 artificial heart beats 40,000,000 times per year. How does this compare with a normal human heart?

1-37 Air flow is measured in units of cubic feet per minute (CFM). Convert 100 CFM into units of cubic meters per second.

1-38 Convert atmospheric pressure from 14.7 pounds per square inch (lb/in^2) into units of tons per square foot.

1-39 Lightning travels at a speed of 87,000 miles per hour. How does this compare with the average velocity of an oxygen molecule at room temperature, which is 440 meters per second, or the speed of light, which is 3×10^8 meters per second.

1-40 Isopropyl alcohol, or rubbing alcohol, has a density of $0.786 \ g/cm^3$. Calculate the weight of the rubbing alcohol in a quart bottle.

1-41 Calculate the density of corn oil in grams per cubic centimeter, assuming a liter of this oil weighs 0.91 kilograms.

1-42 Calculate the density of water in units of kilograms per cubic meter.

SI Units of Measurement

1-43 Calculate the derived SI base unit for the following quantities.

  (a) speed (b) area (c) volume (d) density

1-44 Explain why the liter, angstrom, atmosphere, and calorie are exceptions to the rules for derived SI units.

1-45 The same prefixes—kilo-, milli-, nano-, and so on—are used in both the SI and Metric System. There is only one difference. In SI, the prefixes are applied to the unit of grams —to give milligrams or micrograms—instead of the base unit of mass in this system. Explain why.

Uncertainty in Measurement

1-46 Describe the difference between systematic and random error and give examples of each.

1-47 Describe the difference between accuracy and precision. Give examples of measurements that are accurate but not precise, precise but not accurate, and both accurate and precise.

1-48 Describe how systematic and random errors affect the accuracy and precision of measurements.

1-49 Four nickels selected at random were found to weigh 5.0601 grams, 4.8881 grams, 4.9238 grams, and 5.0603 grams. Calculate the average weight of the four nickels and then calculate the difference between this weight and the weight of each nickel. Use the largest of these differences to estimate the error involved in counting nickels by weighing them. How many nickels can we weigh at a time before the error involved in the measurement is equal to the weight of the average nickel?

Significant Figures

1-50 Determine the number of significant figures in each of the following numbers.

(a) 0.00641   (b) 0.07850   (c) 0.00043891   (d) 0.56

1-51 Determine the number of significant figures in each of the following numbers.

(a) 36   (b) 4890   (c) 5801   (d) 500   (e) 50,003

1-52 Determine the number of significant figures in each of the following numbers.

(a) 3.4×10^{-2} (b) 5.98521×10^3 (c) 8.709×10^{-6}
(d) 1.93×10^{-20} (e) 7.00×10^8

1-53 Round off each of the following numbers to three significant figures.

(a) 474.53   (b) 0.067981
(c) 9.463×10^{10} (d) 30.0974

Dimensional Analysis

1-54 A 20-pound bag of 20:10:5 fertilizer is 20% nitrogen by weight. Calculate the number of grams of nitrogen per square foot if this bag of fertilizer covers 5000 square feet of lawn.

1-55 Calculate the fraction of the body weight that is water in an average human body, which weighs 70 kilograms and contains 10 gallons of water.

1-56 The Pacific Ocean has a volume of 6.96189×10^{23} liters and a surface area of 1.66241×10^{11} square kilometers. Calculate the average depth of the Pacific Ocean.

1-57 Calculate the weight of CO_2 in the atmosphere if the atmosphere is 338 parts per million (ppm) CO_2 by weight and the weight of the atmosphere is 5.2×10^{15} tons.

1-58 The LD_{50} for a drug is the dose that would be lethal for 50% of the population. LD_{50} for aspirin in rats is 1.75 grams per kilogram of body weight. Calculate how many aspirin tablets containing 325 milligrams of aspirin each a 70-kilogram human would have to consume to achieve this dose.

1-59 LD_{50} for sodium cyclamate in mice is 17.0 grams per kilogram of body weight. Calculate how many cans of diet soda a 70-kilogram human would have to drink to achieve this dose, assuming each can of diet soda contains 0.096 grams of sodium cyclamate.

1-60 Ipecac syrup is used to induce vomiting in people who have swallowed a poison. The syrup contains 7 grams of ipecac per 100 milliliters, and the recommended dosage is 1 tablespoon, or 15 milliliters. Calculate the amount of ipecac in the average dose.

1-61 The typical bar buys a 750-milliliter bottle of Jack Daniels bourbon for about $8.00 and pours drinks that use a jigger, or 1.5 ounces, of the whiskey. Estimate the profit per drink if the bar sells a jigger of bourbon for $2.50.

1-62 In 1773, Benjamin Franklin observed that one teaspoon of oil spilled on a pond near London spread out to form a film that covered an area of about 22,000 square feet. If a teaspoon of oil has a volume of about 5 cubic centimeters and the oil spread out to form a film roughly 1 molecule tall, what is the average height of an oil molecule?

1-63 A gardener wants to improve the soil in her vegetable garden. The Purdue extension agent recommended incorporating 6 inches of loamy topsoil into the existing soil. How many cubic yards of topsoil will she need for a garden 30 feet long and 20 feet wide? What is the density of topsoil if 1 cubic yard weighs between 1.1 and 1.5 tons, depending on how wet?

1-64 The height-and-weight charts that used to hang in doctors' offices are being replaced by formulas for estimating when someone has "medically significant obesity." For men, the weight in kilograms is divided by the square of the height in meters. For women, the height in meters is raised to the 1.5 power and then divided into the weight in kilograms. If this ratio is larger than 30, the person is obese. Assume that one of the authors of this text is 6 feet, 1 inch tall. What is the maximum weight he could carry before he would be considered obese? How tall would he have to be if he weighed 250 pounds to avoid being obese?

1-65 One thousand cubic feet of natural gas delivers 1×10^6 British thermal units (BTU) of energy at an average cost of $6.63. If a gallon of #2 fuel oil delivers 0.139×10^6 BTU of energy at an average cost of $1.01, which is the cheaper fuel for heating a home: gas or oil?

1-66 It takes 1000 ft^3 cubic feet of natural gas or 293 kilowatt hours of electricity to generate 1.0×10^6 BTU of energy. What is the ratio of the cost of electricity to the cost of natural gas if electricity sells for $0.058 per kilowatt hour and natural gas sells for $0.663 per 100 cubic feet?

1-67 Calculate the energy released per gram of solar fuel burned if the sun releases 3.6×10^{26} joules of energy per second and loses 1.25×10^{14} metric tons per year. If the earth absorbs 5.4×10^{24} joules per year of energy from the sun, what fraction of the sun's energy hits the earth?

1-68 Gold is the most malleable element. An ounce of gold can be beaten into a thin sheet that covers an area of 300 square feet. Calculate the thickness of this sheet in centimeters, assuming the density of gold is 19.3 g/cm^3. Calculate the thickness of the sheet in terms of the number of gold atoms if the radius of an individual gold atom is 1.4×10^{-8} centimeters.

Scientific Notation

1-69 Convert the following numbers to scientific notation.

(a) 11.98   (b) 0.0046940   (c) 4,679,000
(d) 0.000007967   (e) 1.98

1-70 Convert the following numbers to scientific notation.

(a) 212.6   (b) 0.189   (c) 16,221   (d) 0.5807   (e) 132.76

1-71 Convert the following numbers from scientific notation.

(a) 5.60×10^{-3} (b) 7.025×10^{5} (c) 8.216×10^{-2}
(d) 9.00×10^{2} (e) 2×10^{-6}

1-72 Do the following calculations. (Keep track of significant figures.)

(a) $132.76 + 21.16071$
(b) $32 + 0.9767$
(c) $3.02 \times 10^{4} + 1.69 \times 10^{3}$
(d) $4.18 \times 10^{-2} + 1.29 \times 10^{-3}$
(e) $8.17 \times 10^{5} - 1.20 \times 10^{4}$
(f) $6.49 \times 10^{-10} - 1.23 \times 10^{-12}$

1-73 Do the following calculations. (Keep track of significant figures.)

(a) 2.8×4.80
(b) $32.1/0.75$
(c) $8.16 \times 10^{-4} \times 4.78 \times 10^{15}$
(d) $5.00 \times 10^{4}/1.60 \times 10^{-2}$
(e) $3.62 \times 10^{-3}/5.23 \times 10^{6}$
(f) $1.39 \times 10^{7}/1.10 \times 10^{-3}$

The Graphical Treatment of Data

1-74 The following data show how the pressure (P) of a gas depends on its volume (V) when the amount of gas and the temperature of the gas are held constant. Plot these data in terms of P versus V, P versus $1/V$, and V versus $1/P$. Write an equation of the form $y = mx + b$ for each straight-line relationship you find.

Pressure (atm):	0.0586	0.0856	0.142	0.289
Volume (L):	1.28	0.876	0.528	0.259

1-75 ^{14}C nuclei are radioactive, decaying to ^{14}N with a half-life of 5730 years. The following data show the change in the activity of a sample containing ^{14}C in units of counts per minute (cpm) versus the age of the sample in years. Plot these data in terms of A versus t, $1/A$ versus t, and $\log A$ versus t. Write an equation of the form $y = mx + b$ for each straight-line relationship you find.

Activity (cpm):	1000	800	600	400	200	100
Age (year):	0	1845	4224	7576	13,307	19,039

1-76 Some of the light that shines on a colored solution is absorbed as it passes through the solution. The intensity of the light that passes through the sample (I) is therefore lower than the intensity of the light that shines on the sample (I_o). The following data show how the ratio of I_o/I depends on the concentration (measured as molarity, M) of the light-absorbing compound. Plot the I_o/I ratio versus concentration, the I_o/I ratio versus the log of the concentration, and log I_o/I versus concentration for the following data. Write an equation of the form $y = mx + b$ for each straight-line relationship you find.

I_o/I	2	3	4	5	6
Concentration (M):	0.250	0.396	0.500	0.581	0.646

Extensive and Intensive Quantities

1-77 Describe the difference between an extensive property —such as mass or volume—and an intensive property—such as density.

1-78 Describe why intensive properties are more useful than extensive properties for comparing two systems.

1-79 Describe why intensive properties are a characteristic of a system, whereas extensive properties are not.

1-80 List five quantities whose measurements were discussed in this chapter that are extensive properties. List as many intensive properties as you can.

Density

1-81 A 55-gallon drum holds 309.8 pounds of gasoline or 459.0 pounds of water. Which is denser: gasoline or water?

1-82 Calculate the weight of a cubic yard of peat moss, assuming the density of peat is 0.84 g/cm^3.

1-83 Calculate the weight of a gallon of milk, assuming the density of milk is 1.032 g/cm^3.

1-84 Liquid mercury is sold in units of *flasks*, which weigh 76 pounds each. What is the volume of a flask of mercury?

1-85 Calculate the weight of the mercury that would fill a barometer tube 760 millimeters tall and 1.00 centimeters in diameter.

1-86 What is the radius of the moon if it weighs 7.4×10^{22} kilograms and its density is 3.3 g/cm^3? Assume the moon is spherical and that the volume of a sphere is given by the following equation: $V = \frac{4}{3}\pi r^3$.

1-87 The density of Earth is 5.5 g/cm³, and its radius averages 6.4×10^6 meters. Calculate the weight of the planet.

1-88 So-called neutron stars have been estimated to have an average radius of 6 miles and a density of 1 billion tons per cubic inch. How does this compare with the density of water? How does it compare with the density of the sun (1.41 g/cm³)?

1-89 A 2-cubic-foot bag of mulch sold at lawn care centers weighs 30 pounds. Will the mulch float on water?

1-90 Which of the following liquids would float on top of water and which would sink?

(a) benzene (C_6H_6: density = 0.877 g/cm³)
(b) carbon disulfide (CS_2: density = 1.263 g/cm³)
(c) chloroform ($CHCl_3$: density = 1.483 g/cm³)
(d) ether [($C_2H_5)_2O$: density = 0.714 g/cm³]
(e) gasoline: density = 0.675 g/cm³

1-91 The iron in steel has a density of 7.86 g/cm³. When iron rusts, the density decreases to 5.12 g/cm³. Does iron expand or contract when it rusts?

1-92 If a beaker containing 250 milliliters of chloroform ($CHCl_3$: density = 1.483 g/cm³) weighs 524 grams, what is the weight of the beaker?

1-93 A student was given two metal cubes that looked similar. One was 1.05 centimeters on an edge and weighed 8.33 grams.

The other was 2.84 centimeters on an edge and weighed 164.9 grams. Were the cubes made of the same metal?

1-94 An irregularly shaped piece of metal weighs 54.6 grams and displaces 2.83 cubic centimeters of water when dropped into a graduated cylinder filled with water. What is the metal?

1-95 Archimedes is reputed to have run through the streets hollering "Eureka!" after realizing that he could determine the amount of gold in a crown by measuring the volume of water displaced. If a crown weighs 0.583 kilograms and displaces 36.9 milliliters of water, what is the percent by weight of gold in the crown?

Temperature

1-96 Explain why we need two reference points (such as 32°F and 212°F or 0°C and 100°C) to define a temperature scale.

1-97 Normal body temperature is 98.6°F. What is the equivalent temperature in degrees Celsius? In kelvins?

1-98 At what temperature are the readings on the Celsius and Fahrenheit scales the same? At what temperature are the magnitudes the same but the signs different?

1-99 Fahrenheit defined zero on his scale as the lowest temperature that could be achieved by adding salt to ice and found that ice by itself freezes at 32° on this scale. Unfortunately, Fahrenheit's measurements were wrong. The lowest temperature that can be achieved by adding salt to ice is −21.2°C. Calculate the equivalent temperature on the Fahrenheit scale.

CHAPTER 2

ELEMENTS AND COMPOUNDS

CHAPTER CONTENTS

Until recently, one of the central themes of developmental biology was summarized as follows: *ontogeny recapitulates phylogeny.* Phylogeny is the process by which a species evolves. Ontogeny is the process by which an individual member of the species develops. This statement therefore suggests that the development of an individual proceeds through a series of steps that summarizes, or restates, the sequence of events in the evolution of the species. Although the specifics of this idea are open to debate among biologists, there is general agreement that it is still a useful model for thinking about developmental biology.

This theme has also become popular among people who do research on learning. When applied to learning, it suggests that individuals have to go through many of the same stages in constructing their knowledge and understanding of a particular subject that the practitioners of the field went through as the field developed.

This chapter tries to reflect the idea that ontogeny recapitulates phylogeny in the mind of the individual trying to learn chemistry. Instead of starting with a statement of the modern model for the structure of atoms, it starts by trying to answer the question of why we believe in atoms. It then introduces some of the concepts that chemists have constructed to explain the pattern of observations they make about the world around them.

2.1 ELEMENTS AND COMPOUNDS

The history of many cultures contain myths that try to explain the creation of the earth, the origin of men and women, and the acquisition of food and fire. The main characters in these myths are gods or cultural heroes who overcame other gods, monsters, or beings to bring one of these essentials of life to humankind.

Historians trace the development of European science to the fifth century B.C., when the early Greek philosophers began to develop an explanation of the origin of the world that no longer relied on mythical ingredients such as gods or heroes. Their model, as you may recall from Chapter 1, was based on the assumption that there were four elements — fire, air, earth, and water.

This model provided two ways of explaining why different substances were found in the natural world. One substance could be changed into another by changes in the relative proportions of the four elements. Heating clay in an oven, for example, could be thought to drive off water and add more fire, thereby transforming the clay into a pot. Another way of changing one substance into another involved changing one of the properties that made the elements different from each other. A piece of wood, for example, which is rich in earth (cold and dry), bursts into flame (hot and dry) when heated.

The Greek concept of elements was popular for almost 2200 years and was the guiding force behind the alchemists' search for ways to turn cheap metals such as lead into gold. In 1661 the English scientist Robert Boyle argued that there was something wrong with the Greek model of the elements. He noted that it was impossible to combine the four Greek elements to form any substance and that it was equally impossible to extract these elements from a substance. In his book, *The Sceptical Chymist,* Boyle proposed a new definition of a chemical element that became the basis for the modern definition of this concept.

The modern definition of an element is based on the observation that many substances can be decomposed into simpler substances. Water, for example, decomposes into a mixture of hydrogen and oxygen when an electric current is passed through the liquid. Table salt decomposes into a shiny metal (sodium) and a toxic gas (chlorine) when the solid is melted and an electric current is passed through the molten liquid.

Some substances — such as hydrogen, oxygen, sodium, and chlorine — cannot be decomposed into simpler substances. These are therefore the elementary, or simplest, chemical substances. Thus, as Boyle pointed out, an **element** can be defined as follows.

> **An element is any substance (such as hydrogen or oxygen) that cannot be decomposed into a simpler substance by a chemical reaction.**

At the time this book was written, 108 elements had been discovered. They include a number of materials with which you are familiar, such as the oxygen in the atmosphere, the aluminum metal in aluminum foil, the iron in nails, the copper in electrical wires, the mercury in thermometers, and the silver and gold in jewelry.

Once we understand the definition of an element, it is easy to understand the next level of complexity — **compounds.**

> **A compound is a substance (such as water or table salt) that contains more than one element.**

Water is a compound formed by the reaction between the elements hydrogen and oxygen, and table salt is a compound formed by a reaction between the elements sodium and chlorine.

It is not obvious from their appearance which of these liquids is an element, and which is a compound.

2.2 THE LAW OF CONSTANT COMPOSITION

The preceding section defined a chemical compound as a substance that contains more than one element. This raises an interesting question: What is the difference between a mixture of two elements and a compound formed by a reaction between these elements?

Let's start by looking at one of the properties of chemical compounds. Between 1798 and 1808, Joseph Louis Proust analyzed the composition of a number of compounds from different sources and found that they always contained the same ratio by weight of their elements.

Proust found, for example, that table salt always contained the same ratio by weight of the elements we now know as sodium and chlorine. No matter where the table salt came from, it always contained 1.5 times as much chlorine by weight as sodium. Similarly, he found that water always contained 8 times as much oxygen by weight as hydrogen and that the mineral known as cinnabar always contained 6.25 times as much mercury by weight as sulfur. Repeated confirmation of these results eventually led to what is known as the **law of constant composition.**

The mineral cinnabar, a compound, (shown on the left) always contains 6.25 times as much mercury as sulfur by weight. But brass, a mixture, can be made with widely differing ratios of copper and zinc.

> **Law of constant composition: Chemical compounds always contain the same ratio by weight of their elements, regardless of the source of the compound.**

This conclusion provides us with one way of distinguishing between compounds and mixtures of elements. Compounds — such as table salt, water, and cinnabar — have a constant composition. Mixtures do not. Brass is an example of a mixture of two elements: copper and zinc. It can contain as little as 10% or as much as 45% zinc and still have about the same properties. Bronze is another example of a mixture of two elements, this time copper and tin. Once again, the composition of this mixture can vary over a wide range.

Another difference between chemical compounds and mixtures of elements is the ease with which the elements can be separated. A *mixture* can be defined as follows.

A mixture contains two or more elements or compounds that can be separated by some physical process.

The atmosphere is a good example of a mixture. At 20°C, it consists, by weight, of 76.2% nitrogen, 20.5% oxygen, 2.4% water, 0.9% argon, and traces of carbon dioxide and other contaminants. The individual components of this mixture can be physically separated from each other. For water to be removed, the gas is passed over a dehydrating agent; for oxygen to be separated from nitrogen and argon, the gas must be cooled until it becomes a liquid and then the more volatile nitrogen and argon can be boiled off; and so on. The elements in a chemical compound are bound together. They can only be separated by destroying the compound.

Some of the differences between chemical compounds and mixtures of elements are illustrated, in a general way, by the following exercise.

Exercise 2.1

At breakfast one morning, the author of this section was faced with a choice between two cereals. One was raisin bran, and the other was a cereal known as Crispix, which is made of flakes of rice fused with flakes of corn. Describe the characteristic properties of these cereals that make one an analogy for a mixture of elements and the other an analogy for a chemical compound.

Solution

Raisin bran is an analogy for a mixture because it has the following characteristic properties.

1. The ratio of raisins to bran flakes changes from sample to sample. The cereal does not have a constant composition.
2. It is possible to physically separate the two "elements"—to pick out the raisins and eat them separately.

Crispix has some of the characteristic properties of a compound.

1. The ratio of rice flakes to corn flakes is constant—it is always 1:1 in every sample.
2. There is no way to separate the "elements" without breaking the bond that holds them together.

Raisin bran can be thought of as a mixture of "elements"—raisins and bran flakes—while Crispix can be thought of as a "compound" in which each particle consists of a rice flake fused with a corn flake.

2.3 EVIDENCE FOR THE EXISTENCE OF ATOMS

Most students in beginning chemistry courses already believe in atoms. If asked to describe the evidence on which they base this belief, however, they hesitate. They often invoke authority—they justify their belief with what they have been told in other courses or what they have read in textbooks. If pressed further, they may use the existence of atomic bombs or nuclear reactors as evidence of atoms. A devil's advocate, however, could argue that neither "atomic" bombs nor "nuclear" reactors are evidence for the existence of atoms. All they prove is that certain substances, such as highly purified uranium metal, give off extraordinary amounts of energy when handled under a particular set of conditions.

Our senses argue against the existence of atoms. The atmosphere in which we live feels like a continuous fluid. We don't feel bombarded by collisions with individual particles in the air. The water we drink looks like a continuous fluid. We can take a glass of water, pour out half, divide the remaining water in half, and repeat this process again, and again, and again, without ever appearing to reach the point at which it is impossible to divide it one more time.

It isn't surprising that the debate about the existence of atoms goes back as far as we can trace and continues until well into this century. The first proponents of an atomic theory were the Greek philosophers Leucippus and Democritus, who proposed the following model in the fifth century B.C.

1. **Matter is composed of atoms separated by empty space through which the atoms move.**
2. **Atoms are solid, homogeneous, indivisible, and unchangeable.**
3. **All apparent changes in matter result from changes in the groupings of atoms.**
4. **Different kinds of atoms differ in size and shape.**
5. **The properties of matter reflect the properties of the atoms the matter contains.**

This model attracted few supporters among later generations of Greek philosophers. Aristotle, in particular, refused to accept the idea that the natural world could be reduced to a random assortment of atoms moving through a vacuum.

Until very recently, all of our evidence for atoms was indirect and circumstantial. The first indication that matter might be composed of indivisible particles came from Boyle's work on chemical elements. Boyle argued that the fact that chemical elements cannot be decomposed into simpler substances is consistent with the idea that there are elementary particles (or corpuscles, as he called them) that combine to form compounds, but cannot be subdivided.

Further evidence for the existence of atoms was the law of definite proportions, proposed by Jeremias Benjamin Richter in 1792. Richter found that the ratio of the weights of two chemical compounds consumed in a chemical reaction was always the same. Not long after that, Proust reported his work on the constant composition of chemical compounds, and the time was ripe for the reinvention of an atomic theory. The laws of definite proportions and constant composition do not prove that atoms exist, but these laws are difficult to explain without the assumption that chemical compounds are formed when atoms of different weights combine in constant proportions.

These results, combined with those of experiments with gases that had first become possible at the turn of the 19th century, eventually led John Dalton in 1803 to develop a modern theory of the atom based on the following assumptions.

1. **Matter is made up of atoms that are indivisible and indestructible.**
2. **All atoms of an element are identical.**
3. **Atoms of different elements have different weights and different chemical properties.**
4. **Atoms of different elements combine in ratios of simple whole numbers to form compounds.**
5. **Atoms cannot be created or destroyed; when a compound decomposes, the atoms are recovered unchanged.**

Although arguments about the existence of atoms continued well into the 20th century, active debate has now ended. The existence of atoms is universally accepted by scientists.

The concept of elementary particles or atoms provides another way of distinguishing between elements and compounds. *Elements* are substances that contain only one kind of atom. The oxygen we breathe, for example, contains only oxygen atoms. *Compounds* contain atoms of more than one element. Water, for example, contains both hydrogen and oxygen atoms.

2.4 ATOMS AND MOLECULES

Imagine cutting an infinitesimally small piece of gold metal in half and then repeating this process again, and again, and again. In theory, we should eventually end up with a single gold atom. If we tried to split this atom in half we would end up with something that no longer retains any of the properties of gold. An *atom,* in other words, can be defined as follows.

> **An atom is the smallest particle of an element that has any of the properties of the element.**

What would happen if we did the same thing to a compound, such as water? Eventually, we would end up with a single molecule of water, which contains two hydrogen atoms and an oxygen atom. In theory, we could break this molecule into its individual atoms. But at this point all of the properties of water would be lost (see Figure 2.1). A *molecule,* then, can be defined as follows.

> **A molecule is the smallest particle of a chemical compound that has any of the properties of the compound.**

Elements can also form molecules, but these molecules are composed of identical atoms. The oxygen we breathe, for example, consists of molecules that contain two oxygen atoms. Elemental phosphorus contains molecules of four phosphorus atoms, and elemental sulfur contains molecules of eight sulfur atoms (see Figure 2.2).

Chemists use a shorthand notation to save both time and space when describing molecules. Each element is represented by a unique symbol, such as those in Table 2.1. Many of these symbols come from the first letter of the name of the element. Examples of this convention include the following.

H = hydrogen   B = boron
C = carbon   N = nitrogen
O = oxygen   P = phosphorus

There are 108 different elements, however, and only 26 letters in the alphabet; so two-letter combinations are used as well. Most of these make sense, since they are derived from the name of the element.

Se = *se*lenium   Si = *si*licon
Po = *po*lonium   Br = *br*omine
Ti = *ti*tanium   Ca = *ca*lcium
Cr = *c*h*r*omium   Zn = *z*i*n*c

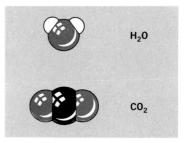

FIG. 2.1 A molecule is the smallest particle of a chemical compound that has the properties of the compound. Water molecules, for example, consist of two hydrogen atoms bound to an oxygen atom. Carbon dioxide molecules contain two oxygen atoms bound to a carbon atom.

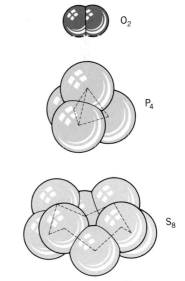

FIG. 2.2 Some nonmetallic elements also form molecules. At room temperature, oxygen exists as linear O_2 molecules, phosphorus forms tetrahedral P_4 molecules, and sulfur forms cyclic S_8 molecules.

TABLE 2.1

Some Common Elements and Their Symbols

Element	Symbol	Element	Symbol
aluminum	Al	lithium	Li
antimony	Sb	magnesium	Mg
argon	Ar	manganese	Mn
arsenic	As	mercury	Hg
barium	Ba	neon	Ne
beryllium	Be	nickel	Ni
bismuth	Bi	nitrogen	N
boron	B	oxygen	O
bromine	Br	phosphorus	P
cadmium	Cd	platinum	Pt
calcium	Ca	potassium	K
carbon	C	selenium	Se
cesium	Cs	silicon	Si
chlorine	Cl	silver	Ag
chromium	Cr	sodium	Na
copper	Cu	strontium	Sr
fluorine	F	sulfur	S
gold	Au	tin	Sn
helium	He	titanium	Ti
hydrogen	H	tungsten	W
iodine	I	vanadium	V
iron	Fe	xenon	Xe
krypton	Kr	zinc	Zn
lead	Pb		

Some of the symbols seem to make no sense, because they come from the Latin or German names of the elements. Fortunately, there are only a handful of elements in this category.

Ag = silver (argentum)	Na = sodium (natrium)
l2.Au = gold (aurum)	Pb = lead (plumbum)
Cu = copper (cuprum)	Sb = antimony (stibium)
Fe = iron (ferrum)	Sn = tin (stannum)
Hg = mercury (hydrargyrum)	W = tungsten (wolfram)
K = potassium (kalium)	

The shorthand notation for a compound identifies not only the elements that are present in the compound but the number of atoms of each element. This is indicated by a subscript written after the symbol for the element. The formulas for molecules of oxygen, phosphorus, and sulfur are therefore O_2, P_4, and S_8. By convention, no subscript is written when the molecule contains only one atom of an element. Thus, water is H_2O and carbon dioxide is CO_2.

The advantage of this shorthand notation becomes obvious when we look at more complex molecules, such as vitamin B-12. Each molecule of vitamin B-12 contains 63 carbon atoms, 88 hydrogen atoms, 14 nitrogen atoms, 14 oxygen

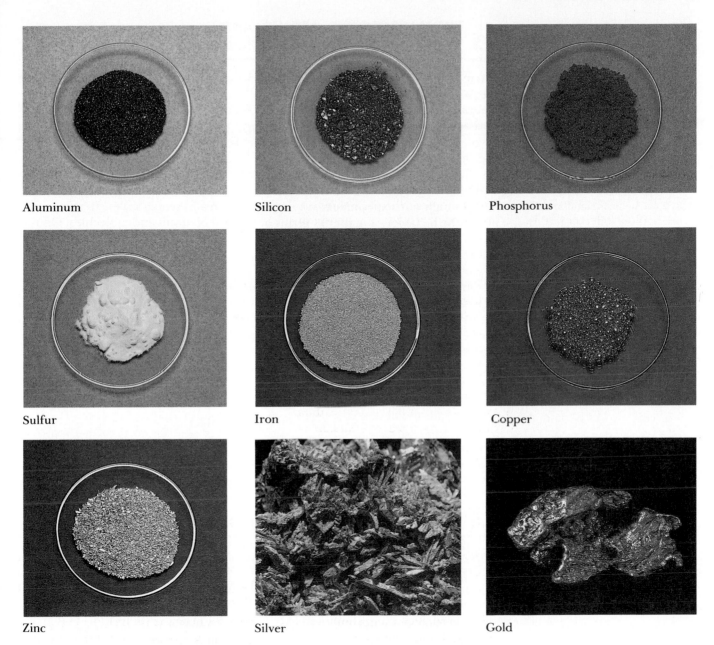

Aluminum

Silicon

Phosphorus

Sulfur

Iron

Copper

Zinc

Silver

Gold

atoms, 1 cobalt atom, and 1 phosphorus atom. The formula for vitamin B-12 is therefore $C_{63}H_{88}N_{14}O_{14}CoP$.

Although molecules such as vitamin B-12 seem very large, it is useful to keep the size of these molecules in perspective. A teaspoon of vitamin B-12 contains about 2.5×10^{21}, or 2,500,000,000,000,000,000,000, vitamin B-12 molecules. This number helps us understand why it was so difficult to prove the existence of atoms. Atoms and molecules are so extraordinarily small that they are virtually impossible to detect by direct measurements. The number also provides a hint of one of the most fascinating aspects of chemistry — the challenge of determining the structure of a world that exists on a scale so small it is almost impossible to detect.

THE MACROSCOPIC, ATOMIC, AND
2.5 SYMBOLIC WORLDS OF THE CHEMIST

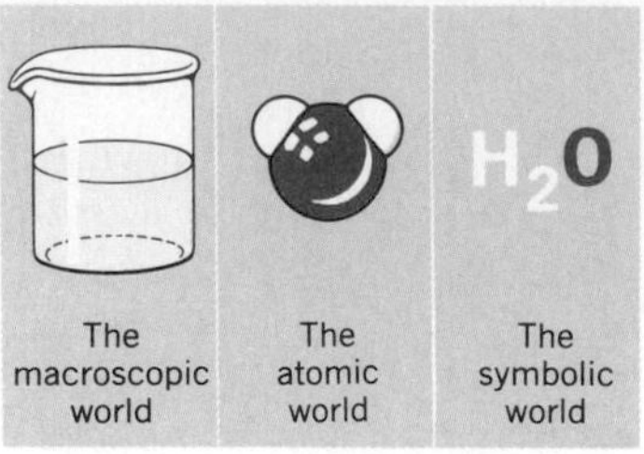

FIG. 2.3 Chemists simultaneously work in three different worlds: (1) the macroscopic world of objects visible to the naked eye, which can be represented by a beaker of water, (2) the atomic world, in which water is thought of as molecules that contain two hydrogen atoms bound to an oxygen atom, and (3) the symbolic world, in which water is represented as H_2O.

Chemists simultaneously work in three very different worlds (see Figure 2.3). Most measurements are done on objects that are visible to the naked eye—in the *macroscopic* world. When you walk into a chemical laboratory, you find a variety of bottles, tubes, flasks, and beakers designed to study samples of liquids and solids large enough to be seen. You may also find sophisticated instruments that can be used to detect microgram, nanogram, or even picogram quantities of materials, but the samples injected into these instruments are still large enough to be seen with the naked eye.

Although our experiments are done on the macroscopic scale, chemists think about the behavior of matter in terms of a world of atoms and molecules. On this *atomic* scale, water is not thought of as a liquid that freezes at 0°C and boils at 100°C but as individual molecules that contain two hydrogen atoms and an oxygen atom.

One of the challenges that students face when they encounter chemistry for the first time is understanding the process by which chemists perform experiments on the macroscopic scale that can be interpreted in terms of the structure of matter on the atomic scale. The task of bridging the gap between the atomic and macroscopic worlds is made more difficult by the fact that chemists work in a third world as well—a *symbolic* world, in which we represent water as H_2O and write equations such as the following to represent what happens when hydrogen and oxygen react to form water.

$$2\,H_2 + O_2 \longrightarrow 2\,H_2O$$

Chemists use the same symbols to describe what happens on both the macroscopic scale and the atomic scale. The symbol H_2O, for example, is used to represent both a single water molecule and a beaker full of water.

2.6 THE CHEMISTRY OF THE ELEMENTS

At least 10 elements have been known since ancient times: antimony (Sb), carbon (C), copper (Cu), gold (Au), iron (Fe), lead (Pb), mercury (Hg), silver (Ag), sulfur (S), and tin (Sn). By 1669, alchemists had isolated two more: arsenic (As) and phosphorus (P). Between 1735 and 1865, more than 50 additional elements were discovered, including: aluminum (Al), boron (B), bromine (Br), chlorine (Cl), hydrogen (H), magnesium (Mg), nitrogen (N), oxygen (O), platinum (Pt), potassium (K), silicon (Si), sodium (Na), and zinc (Zn). By 1940, a total of 91 elements had been isolated from natural sources. Another 17 elements have been synthesized since then.

As the number of elements increased, chemists began to look for ways to group these elements on the basis of their chemical and physical properties. The first major step in this direction was taken by the German chemist Johann Wolfgang Döbereiner. In 1829, Döbereiner noted that there were groups of three elements —or *triads*—with similar chemical properties. Chlorine (Cl), bromine (Br), and iodine (I) formed one triad. Another was composed of lithium (Li), sodium (Na), and potassium (K). A third contained calcium (Ca), strontium (Sr), and barium (Ba). Other chemists soon found other families, or groups, of elements with similar chemical properties.

Groups

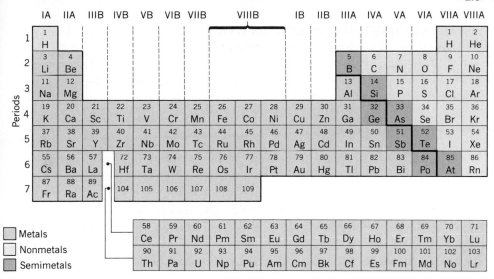

FIG. 2.4 The 108 known elements can be organized into a periodic table in which elements with similar chemical properties are placed in vertical columns, or groups. More than 75% of the known elements are metals. Another 15%, clustered primarily in the upper right-hand corner of the table, are nonmetals. Along the dividing line between these two categories are a handful of semimetals, which have properties that lie between the extremes of metals and nonmetals.

In 1869, the Russian chemist Dmitri Ivanovich Mendeléeff found that it was possible to generate a table in which elements with similar chemical properties were grouped together. A modern version of this table is shown in Figure 2.4. The majority of the elements in this table are divided into 18 vertical columns and seven horizontal rows.

The vertical columns are known as **groups,** or **families.** Traditionally, the groups, or families, have been differentiated by a shorthand notation consisting of a Roman numeral — I, II, III, and so on — followed by either an *A* or a *B*. This is the **group number.** In the United States, the elements in the first column on the left-hand side of the periodic table have been labeled Group IA; the next column, IIA; the next, IIIB; and so on.

Unfortunately, the same notation wasn't used in other countries. The elements known as Group VIA in the United States were Group VIB in Europe. A new convention for the periodic table has therefore been proposed, which numbers the columns from 1 to 18, reading from left to right. The new convention has obvious advantages. It is perfectly regular and therefore unambiguous. The advantages of the old format are less obvious, but they are equally real. This book therefore introduces the new convention but retains the old.

The key to understanding the periodic table is summarized in the following rule.

The elements in a column of the periodic table have similar chemical properties.

Elements in the first column, for example, combine with chlorine to form compounds with the same generic formula — HCl, LiCl, NaCl, KCl, RbCl, CsCl, and FrCl. Elements in the second column combine with chlorine to form compounds with the generic formula MCl_2 — $BeCl_2$, $MgCl_2$, $CaCl_2$, $SrCl_2$, $BaCl_2$, and $RaCl_2$.

The horizontal rows in this table are called **periods.** The first period contains only hydrogen (H) and helium (He). The second period contains eight elements, Li, Be, B, C, N, O, F, and Ne. Although there are nine horizontal rows in the periodic table shown in Figure 2.4, there are only seven periods. If you examine the atomic number of each element in the periodic table you will see that the two rows at the bottom of the table should be inserted into the sixth and seventh periods.

The 108 elements in the periodic table can be divided into three classes—metals, nonmetals, and semimetals, or metalloids. As you can see from Figure 2.4, more than 75% of the elements are metals. These elements are found toward the bottom and the left-hand side of the periodic table.

Only 17 elements are nonmetals. With only one exception—H—these elements are clustered in the upper-right-hand corner of the periodic table. The nonmetals include the following elements: H, He, C, N, O, F, Ne, P, S, Cl, Ar, Se, Br, Kr, I, Xe, and Rn.

The dividing line between the metals and the nonmetals in Figure 2.4 is marked with a heavy diagonal line. A cluster of elements that, strictly speaking, are neither metals nor nonmetals can be found on either side of this line. These elements are called the semimetals, or metalloids, and include the following elements: B, Si, Ge, As, Sb, Te, Po, and At.

Exercise 2.2

Classify each of the elements in Group IVA as a metal, a nonmetal, or a semimetal.

Solution

Group IVA consists of the following elements: carbon (C), silicon (Si), germanium (Ge), tin (Sn), and lead (Pb). According to Figure 2.4, these elements fall into the following categories.

Nonmetal: carbon

Semimetal: silicon and germanium

Metal: tin and lead

2.7 METALS, NONMETALS, AND SEMIMETALS

Every element or compound has a unique set of chemical and physical properties, which can be used to identify that substance. Physical properties are characteristics of the substance itself. Examples include the melting point, boiling point, color, and density of the substance. Chemical properties describe the way the substance interacts with other elements and compounds.

The tendency to divide elements into metals, nonmetals, and semimetals is based on differences between the chemical and physical properties of these three classes of elements.

PHYSICAL PROPERTIES

Metals have some or all of the following physical properties.

1. **They have a metallic shine or luster. They look like metals!**
2. **They are solids at room temperature. The only exception is mercury, which is a liquid at room temperature.**
3. **They are *malleable* (from the Latin for hammer). They can be hammered, pounded, or pressed into different shapes without breaking.**
4. **They are *ductile*. They can be drawn into thin sheets or wires without breaking.**

Metallic elements, like the aluminum shown here, have characteristic physical properties that are very different from nonmetallic elements, such as bromine.

5. **They conduct heat and electricity. Electrical wires, for example, are made from either copper or aluminum. Pots and pans are made from copper, cast iron, aluminum, stainless steel, or stainless steel coated with a thin layer of copper metal.**

Nonmetals have almost exactly the opposite physical properties.

1. **They seldom have a metallic luster. They tend to be colorless, like oxygen and nitrogen, or brilliantly colored, like bromine.**
2. **They are often gases at room temperature. Examples include: H_2, He, N_2, O_2, F_2, Ne, Cl_2, Ar, Kr, Xe, and Rn.**
3. **The handful of nonmetallic elements that are solids at room temperature are neither malleable nor ductile. Carbon, phosphorus, sulfur, selenium, and iodine are all solids at room temperature. None of these solids can be either shaped with a hammer or drawn into sheets or wires.**
4. **They are poor conductors of heat and electricity. Nonmetals tend to be insulators, rather than conductors.**
5. **They often form molecules in the elemental form. Examples include: H_2, N_2, O_2, F_2, P_4, S_8, Cl_2, Br_2, and I_2.**

The ***semimetals,*** or ***metalloids,*** have properties that lie between these extremes. They often look like metals but are brittle like nonmetals. They are neither conductors nor insulators but make excellent semiconductors.

Samples of bromine (a nonmetal), copper (a metal), and silicon (a semimetal).

CHEMICAL PROPERTIES

The chemical properties of metals and nonmetals are also very different. Chemical reactions that involve metals or nonmetals can be divided into three classes.

1. **A metal can combine with another metal.**
2. **A metal can combine with a nonmetal.**
3. **A nonmetal can combine with another nonmetal.**

Each of these reactions leads to a different kind of product.

Metals combine with other metals to form ***alloys.*** Brass, for example, is an alloy of copper and zinc, while bronze is an alloy of copper and tin. Alloys retain the physical properties of metal. They are almost always solids at room temperature, with a metallic shine or luster. They are usually malleable and ductile, and they conduct both heat and electricity. Because alloys can be made with a wide range of compositions, they are usually thought of as mixtures, not chemical compounds.

As a general rule, metals react with nonmetals to form ***ionic compounds,*** or ***salts.*** Sodium, for example, reacts with chlorine to form sodium chloride, or table salt. NaCl is a white, crystalline solid. It has a high melting point and readily dissolves in water. It therefore has none of the properties of either the metal or the nonmetal from which it is made.

As a general rule, nonmetals combine with each other to form ***covalent compounds*** — such as water (H_2O) and carbon dioxide (CO_2) — that have many of the properties of the nonmetallic elements. Like the nonmetallic elements, they exist as molecules. They are often gases at room temperature, or liquids with low boiling points. They are neither malleable nor ductile, and they are poor conductors of heat and electricity.

The reaction between aluminum (a metal) and liquid bromine (a nonmetal) to form aluminum bromide.

Exercise 2.3

Predict whether each of the following compounds should be ionic or covalent.

(a) Manganese(II) oxide, MnO   (c) Methanol, CH_3OH

(b) Carbon tetrachloride, CCl_4 (d) Calcium nitride, Ca_3N_2

Solution

The discussion of ionic and covalent compounds in this section can be summarized in the following general rules.

Metals generally combine with nonmetals to form ionic compounds or salts.

Nonmetals generally combine with other nonmetals to form covalent compounds.

(a) MnO should be an ionic compound, because it consists of a metal—manganese—and a nonmetal—oxygen.

(b) CCl_4 should be a covalent compound, because it contains two nonmetals—carbon and chlorine.

(c) CH_3OH should be a covalent compound, because it contains only nonmetallic elements—carbon, hydrogen, and oxygen.

(d) Ca_3N_2 should be an ionic compound, because it contains both a metal—calcium—and a nonmetal—nitrogen.

2.8 IONIC AND COVALENT COMPOUNDS

There are important differences between ionic and covalent compounds. As evidence of this, let's compare two of the best-known examples of these classes of chemical compounds—table salt and cane sugar.

Table salt is an ionic compound, known as sodium chloride, that has the formula NaCl. It is a white, crystalline solid with a very high melting point (801°C) and an even higher boiling point (1465°C). Cane sugar is a covalent compound, known as sucrose, that has the formula $C_{12}H_{22}O_{11}$. It is also a white, crystalline solid. But, as anyone who has made candy knows, it melts at a fairly low temperature (185°C) and decomposes if the temperature gets much higher.

Exercise 2.4

Use the following physical data to propose one way of distinguishing between ionic and covalent compounds.

Compound	Melting Point (°C)	Boiling Point (°C)	Density (g/cm³)
MnO	1785	?	5.44
CCl_4	−22.9	76.6	1.59
CH_3OH	−97.8	64.7	0.792
Ca_3N_2	1195	?	2.63

Solution

In general, ionic compounds—such as MnO and Ca_3N_2—tend to have much higher melting points and boiling points than covalent compounds—such as CCl_4 and CH_3OH.

One of the key differences between ionic and covalent compounds becomes apparent when these compounds are dissolved in water. Table salt is fairly soluble in water—up to 36 grams of NaCl will dissolve in 100 grams of water. Cane sugar is even more soluble—more than 180 grams of this compound will dissolve in 100 grams of water. These solutions differ significantly, however, in their ability to conduct an electric current.

A simple conductivity apparatus is shown in Figure 2.5. It consists of a light bulb connected to a pair of thin electrical wires that do not touch. Nothing happens when the light bulb is plugged into a wall socket, because there is a gap in the electrical circuit. When something that conducts an electric current is used to bridge this gap, however, the light bulb begins to glow.

The light bulb does not glow when the wires are immersed in a solution of cane sugar dissolved in water. But it glows brightly when the wires are immersed in a solution of table salt dissolved in water. Repeating this experiment with a variety of ionic and covalent compounds dissolved in water gives the following rule of thumb.

> **Ionic compounds, or salts, that dissolve in water give solutions that conduct an electric current.**
>
> **Solutions of covalent compounds in water usually do not conduct an electric current.**

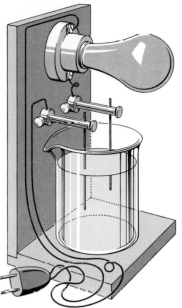

FIG. 2.5 A conductivity apparatus, such as the one shown here, can be used to determine whether a compound can conduct an electric current when it is dissolved in water.

An explanation for this behavior was proposed by Svante Arrhenius while he was a graduate student at the University of Uppsala in Sweden. In 1884, Arrhenius suggested that when NaCl dissolves in water it **dissociates** into units called ions—atoms or groups of atoms, that have an electrical charge. NaCl, according to Arrhenius, dissociates into positively charged Na^+ ions and negatively charged Cl^- ions. Although his professors questioned this theory, and only reluctantly granted him his degree, we now know that this theory is correct. Sodium chloride consists of positively charged Na^+ ions and negatively charged Cl^- ions, which are released into the solution when NaCl dissolves in water. It is the presence of these charged ions that allows the solution to conduct an electric current. Solutions of cane sugar dissolved in water do not conduct electricity, because this compound consists of neutral $C_{12}H_{22}O_{11}$ molecules that do not dissociate into positive and negative ions when they dissolve in water.

The simplest statement of the difference between ionic and covalent compounds is the following rule of thumb.

The light bulb glows brightly because NaCl dissociates in water to form Na^+ and Cl^- ions that carry the electric current through the solution.

> **Ionic compounds contain ions, which carry either a positive or a negative charge. Sodium chloride, for example, contains positively charged Na^+ ions and negatively charged Cl^- ions.**
>
> **Covalent compounds contain neutral molecules, which are not electrically charged. Cane sugar, for example, consists of neutral $C_{12}H_{22}O_{11}$ molecules.**

Exercise 2.5

Which of the following compounds should conduct an electric current when dissolved in water?

(a) Carbon dioxide, CO_2

(b) Strontium fluoride, SrF_2

(c) Cesium iodide, CsI

(d) Methanol, CH_3OH

Solution

The first step in answering this question involves predicting whether each compound is more likely to be ionic or covalent. Once we know this, we can predict whether the compound is likely to dissociate into ions when it dissolves in water.

(a) CO_2 is a covalent compound, because it contains two nonmetals—carbon and oxygen. It therefore consists of neutral CO_2 molecules and should not conduct an electric current when dissolved in water.

(b) SrF_2 is an ionic compound, because it contains a metal—strontium—and a nonmetal—fluorine. SrF_2 should therefore dissociate into positive and negative ions when it dissolves in water, and the resulting solution should conduct an electric current.

(c) CsI is also an ionic compound, because it contains both a metal—cesium—and a nonmetal—iodine. It should therefore dissociate into positive and negative ions when it dissolves in water, and we should expect this solution to conduct an electric current.

(d) CH_3OH is a covalent compound, because it contains three nonmetallic elements—carbon, hydrogen, and oxygen. This compound consists of neutral CH_3OH molecules that do not dissociate into ions when this compound dissolves in water. Therefore, the resulting solution should not conduct an electric current.

2.9 THE STRUCTURE OF ATOMS

To understand what it means to say that NaCl consists of positively charged Na^+ ions and negatively charged Cl^- ions, it is necessary to take a closer look at the assumption that atoms are indivisible.

At about the time that Dalton was reinstating the atomic theory, the British chemist Humphry Davy was studying the effect of electricity on chemical compounds. Davy found that passing an electric current through water decomposed this compound into its elements—hydrogen and oxygen. He also found that he could prepare sodium and potassium metal by passing an electric current through molten samples of sodium or potassium hydroxide.

The discovery that chemical compounds decomposed into their elements when an electric current passed through them led Davy to suggest that the force of attraction between the elements in a compound was electrical in nature. It took almost a hundred years, however, before the particles that carried this electrical charge were identified. Today, we recognize that atoms are not indivisible. They are composed of three fundamental subatomic particles—electrons, protons, and neutrons. (See Table 2.2).

Electrons are the negatively charged particles that carry an electric current when electricity flows through an electric wire or circuit. Atoms are electrically neutral. This means that the negatively charged electrons within the atom must be balanced by positively charged particles, which are called *protons*. The charge on the proton has the same size or magnitude as the charge on the electron, but it has the opposite sign. One proton exactly balances the charge on an electron, and vice versa.

As noted, atoms are electrically neutral, which means that they must contain the same number of electrons and protons. If you look at the periodic table in Figure 2.4, you will find that each element has been assigned an *atomic number* between 1

Sodium metal—when ignited by heating—reacts with chlorine gas to form sodium chloride, giving off energy in the form of both heat and light.

TABLE 2.2

Fundamental Subatomic Particles

Particle	Symbol	Charge
electron	e^-	-1
proton	p^+	$+1$
neutron	n^0	0

and 109. The atomic number of the element is equal to the number of protons and electrons in an atom of that element.

Exercise 2.6

Calculate the number of electrons and protons in a lead atom.

Solution

According to Figure 2.4, lead — Pb — is the 82nd element in the periodic table. It is therefore assigned an atomic number of 82. This means that a lead atom contains 82 protons and 82 electrons.

When this model for the structure of the atom was first proposed, it had a major flaw. The weights of the electrons and protons didn't add up to the weights of the atoms. For example, the weight of the average lead atom is about 2.5 times as large as the sum of the weights of the electrons and protons in the atom. This problem was resolved in 1930 by the discovery of a neutral subatomic particle, known as the **neutron.** There are just enough neutrons in an atom to make up for the difference between the weight of the atom and the sum of the weights of the electrons and protons.

2.10 THE DIFFERENCE BETWEEN ATOMS AND IONS

We are now ready to propose a model that explains the difference between a neutral Na atom and a positively charged Na^+ ion. This model is based on the following definition of an **ion.**

> **An ion is an electrically charged particle produced when one or more electrons are removed from an atom or molecule to give a positive ion or when electrons are added to an atom or molecule to give a negative ion.**

Neutral atoms can be turned into positively charged ions by removal of one or more electrons, as shown in Figure 2.6. By removing an electron from a sodium atom — which contains 11 electrons and 11 protons — we can produce an Na^+ ion that has 10 electrons and 11 protons. The net charge on this ion is therefore $+1$. Ions with larger positive charges can be produced by removal of more electrons. A neutral aluminum atom, for example, has 13 electrons and 13 protons. If 3 electrons are removed from this atom, we get a positively charged Al^{3+} ion that has 10 electrons and 13 protons, for a net charge of $+3$.

Atoms that gain extra electrons become negatively charged, as shown in Figure 2.7. A neutral chlorine atom, for example, has 17 protons and 17 electrons. By

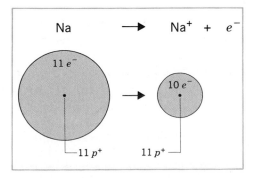

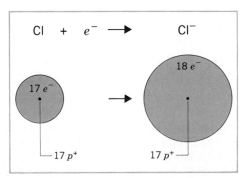

FIG. 2.6 A neutral sodium atom contains 11 electrons and 11 protons. Removing an electron from this atom forms an Na^+ ion with 10 electrons and 11 protons and a net charge of $+1$.

FIG. 2.7 A neutral chlorine atom contains 17 electrons and 17 protons. Adding an extra electron to this atom forms a Cl^- ion with 18 electrons and 17 protons and a net charge of -1.

adding one more electron to this atom, we can produce a Cl^- ion that has 18 electrons and 17 protons, for a net charge of -1.

Exercise 2.7

Calculate the number of electrons and protons in the Mn^{2+} and O^{2-} ions in manganese(II) oxide, MnO.

Solution

We can start by determining the number of electrons and protons in neutral atoms of these elements. The atomic number of oxygen is 8, while manganese has an atomic number of 25. A neutral oxygen atom therefore has 8 electrons and 8 protons, and a neutral manganese atom has 25 electrons and 25 protons.

Two electrons have to be removed from a manganese atom to form a Mn^{2+} ion, which means that this ion has only 23 electrons and 25 protons.

$$Mn^{2+}: \text{23 electrons, 25 protons}$$

An O^{2-} ion has two more electrons than an oxygen atom, which means this ion has 10 electrons and 8 protons.

$$O^{2-}: \text{10 electrons, 8 protons}$$

The gain or loss of electrons by an atom to form negative or positive ions has an enormous impact on the chemical and physical properties of the atom. Sodium metal, which consists of neutral sodium atoms, often bursts into flame when it comes in contact with water. But positively charged Na^+ ions are so unreactive they are essentially inert—lacking in chemical activity. Neutral chlorine atoms react instantly with each other to form Cl_2 molecules, which are still so reactive that entire communities are evacuated when trains carrying chlorine gas derail, because of the toxic effects of this gas's reaction with the tissues of the lungs. Negatively charged Cl^- ions, though, are essentially inert to chemical reactions.

The enormous difference between the chemistry of neutral atoms and that of their ions means that you will have to pay particular attention to the formulas you read, to make sure that you never confuse one with the other.

Exercise 2.8

Describe how the following symbols differ.

(a) H   (b) H_2 (c) H^+ (d) H^-

Solution

H represents a hydrogen atom, which contains one electron and one proton. H_2 stands for a hydrogen molecule formed when two hydrogen atoms combine. H^+ is the symbol for the positive ion formed when a hydrogen atom loses an electron; this ion contains one proton and no electrons. H^- is the symbol for the negative ion formed when a hydrogen atom picks up an extra electron; this ion contains one proton and two electrons.

2.11 FORMULAS OF COMMON IONIC COMPOUNDS, OR SALTS

The existence of positive and negative ions raises two important questions.

1. **How do we decide whether an element forms a positive ion or a negative ion?**
2. **How do we know how many electrons are gained or lost when an atom forms an ion?**

The first question is relatively easy to answer.

RULES CONCERNING ION FORMATION

RULE 1. Metals form positive ions, such as the Na^+, K^+, Ag^+, Mg^{2+}, Ba^{2+}, Zn^{2+}, Hg^{2+}, Sc^{3+}, Fe^{3+}, and Cr^{3+} ions.

RULE 2. Nonmetals tend to form negative ions, such as the F^-, Cl^-, Br^-, I^-, O^{2-}, S^{2-}, N^{3-}, and P^{3-} ions.

The second question is more challenging, and it will take a substantial portion of Chapters 5, 6, 7, 8, and 10 to develop a satisfactory answer. For now, we can use the group numbers in the periodic table — Group IA, IIA, IIIA, and so on — and the following general rules to predict the charge on common ions that contain a single atom.

RULE 3. Metals form positive ions by losing the number of electrons equal to the group number. Thus, sodium, in Group IA, forms Na^+ ions; magnesium, in Group IIA, forms Mg^{2+} ions; and aluminum, in Group IIIA, forms Al^{3+} ions.

RULE 4. Nonmetals form negative ions by gaining the number of electrons equal to the group number minus 8. Thus, chlorine, in Group VIIA, forms Cl^- ions, because $7 - 8 = -1$, and sulfur, in Group VIA, forms S^{2-} ions, because $6 - 8 = -2$.

Exercise 2.9

Predict what ions will form when magnesium metal reacts with nitrogen to form magnesium nitride.

Solution

The product of this reaction is an ionic compound, because it is formed by the reaction between a metal — magnesium — and a nonmetal — nitrogen. According to the rules just given, we expect the metal to form a positive ion and the nonmetal to form a negative ion. All we have to do now is predict the number of electrons gained or lost when these ions are formed.

According to the rules, the magnesium atom will lose the number of electrons equal to its group number. Magnesium is a member of Group IIA of the periodic table, so it forms Mg^{2+} ions. The nitrogen atom will gain the number of electrons equal to the group number minus 8. Since nitrogen is in Group VA, and $5 - 8$ is -3, this means that phosphorus will form N^{3-} ions.

The product of this reaction therefore contains Mg^{2+} and N^{3-} ions.

Exercise 2.9 helps explain why both conventions for labeling the groups in the periodic table are used in this text. As mentioned, the new convention, which numbers the columns from 1 to 18, is both logical and unambiguous. But the old convention is also useful, because the number of electrons gained or lost when an atom forms ions can often be predicted from the group number of the column in which the element is found.

The charges on positive and negative ions dictate the formulas of the ionic compounds they form. Ionic compounds have no net electric charge, which means they must contain just as many positive as negative charges. If table salt contains Na^+ and Cl^- ions, it must form a $1:1$ compound—$NaCl$.

Exercise 2.10

Predict the formula for magnesium nitride, if this compound consists of Mg^{2+} and N^{3-} ions.

Solution

If magnesium nitride contains Mg^{2+} and N^{3-} ions, there must be three Mg^{2+} ions for every two N^{3-} ions, to balance the charge.

$$\text{Total charge} = +6 \qquad \text{Total charge} = -6$$
$$3\ Mg^{2+} \qquad\qquad 2\ N^{3-}$$
$$Mg_3N_2$$

The formula for this compound is therefore Mg_3N_2.

Exercise 2.11

Aluminum chlorhydrate is added to antiperspirants to stop people from sweating. This compound contains Al^{3+}, OH^-, and Cl^- ions. What is the value of x, if the formula for the compound is $Al_x(OH)_5Cl$?

Solution

The compound contains five OH^- ions and one Cl^- ion, for a net charge of -6. The total charge on the Al^{3+} ions must therefore be $+6$, which means that there must be two Al^{3+} ions in the formula of the compound. The value of x in the above formula is therefore equal to 2.

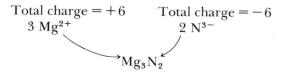

$$Al_2(OH)_5Cl$$
$$2(+3) + 5(-1) + 1(-1) = 0$$

2.12 POLYATOMIC IONS

Simple ions—such as the Mg^{2+} and N^{3-} ions—are formed when electrons are added to or subtracted from neutral atoms. It is also possible to add or subtract electrons from neutral molecules. The resulting ions contain more than one atom and are therefore called *polyatomic ions.*

TABLE 2.3

Common Polyatomic Negative Ions

		−1 *ions*		
HCO_3^-	bicarbonate		HSO_4^-	bisulfate
$CH_3CO_2^-$	acetate		ClO_4^-	perchlorate
NO_3^-	nitrate		ClO_3^-	chlorate
NO_2^-	nitrite		ClO_2^-	chlorite
MnO_4^-	permanganate		ClO^-	hypochlorite
CN^-	cyanide		OH^-	hydroxide

		−2 *ions*		
CO_3^{2-}	carbonate		O_2^{2-}	peroxide
SO_4^{2-}	sulfate		CrO_4^{2-}	chromate
SO_3^{2-}	sulfite		$Cr_2O_7^{2-}$	dichromate
$S_2O_3^{2-}$	thiosulfate		HPO_4^{2-}	hydrogen phosphate

		−3 *ions*		
PO_4^{3-}	phosphate		AsO_4^{3-}	arsenate
PO_3^{3-}	phosphite		BO_3^{3-}	borate

Polyatomic negative ions, such as the CO_3^{2-} and SO_4^{2-} ions, are far more common than polyatomic positive ions, such as the NO^+ ion. Some of the more important polyatomic negative ions are given in Table 2.3.

Exercise 2.12

The bone and tooth enamel in your body contains ionic compounds such as calcium phosphate and hydroxyapatite. Predict the formula of calcium phosphate, if this compound contains Ca^{2+} and PO_4^{3-} ions. Calculate the value of x, if the formula of hydroxyapatite is $Ca_x(PO_4)_5(OH)$.

Solution

It takes three Ca^{2+} ions to balance the charge on two PO_4^{3-} ions. The formula for calcium phosphate is therefore $Ca_3(PO_4)_2$.

The formula for hydroxyapatite contains five PO_4^{3-} ions and one OH^- ion. The total charge on the negative ions in this formula is therefore −10. It takes five Ca^{2+} ions to provide enough positive charge to balance the charge on the negative ions. The value of x is therefore 5, and the formula for hydroxyapatite is $Ca_3(PO_4)_5(OH)$.

2.13 OXIDATION NUMBERS

Some compounds are composed of positive and negative ions; others are not. The salt we use to season our food contains Na^+ and Cl^- ions, but the sugar we put in coffee or tea consists of neutral $C_{12}H_{22}O_{11}$ molecules. Manganese(II) oxide is an ionic compound that contains Mn^{2+} and O^{2-} ions, but carbon monoxide is a covalent compound that consists of neutral CO molecules.

Sometimes it is useful to ignore the differences between ionic and covalent compounds and emphasize the similarities. This can be done through the concept of *oxidation number,* or *oxidation state.* By convention, the manganese atom in MnO and the carbon atom in CO are both assigned oxidation numbers of +2 or described as being in the +2 oxidation state.

> **The oxidation number, or oxidation state, of an atom is equal to the charge that would be present on an atom if the compound was composed of ions.**

Oxidation numbers serve two functions. They provide a basis for naming chemical compounds, as we will see in the next section. They also provide a way of keeping track of how electrons are transferred from one atom to another in a chemical reaction.

Oxidation numbers, or oxidation states, are assigned according to a set of explicit rules.

RULES FOR ASSIGNING OXIDATION NUMBERS

1. The oxidation number of an atom is 0 in any neutral substance that contains atoms of only one element. The oxygen atoms in O_2 and O_3, the phosphorus atoms in P_4, the sulfur atoms in S_8, and the aluminum atoms in aluminum metal all have an oxidation number of 0.

2. The oxidation number of positive or negative ions that contain a single atom is equal to the charge on the ion. For example, the oxidation number of sodium in the Na^+ ion is +1, while the oxidation number of chlorine in the Cl^- ion is −1.

3. The oxidation number of hydrogen is +1 whenever it is combined with another nonmetal. Hydrogen is therefore in the +1 oxidation state in all of the following compounds: CH_4, NH_3, PH_3, H_2O, H_2S, HF, HCl, and HBr.

4. The oxidation number of hydrogen is −1 whenever it is combined with a metal. Hydrogen is in the −1 oxidation state in the following compounds: LiH, NaH, CaH_2, and $LiAlH_4$.

5. The metals in Group IA (Li, Na, K, Rb, Cs, and Fr) always form compounds in the +1 oxidation state. Lithium is +1 in Li_3N; sodium is +1 in Na_2S; potassium is +1 in KH_2PO_4; and so on.

6. The elements in Group IIA (Be, Mg, Ca, Sr, Ba, and Ra) always form compounds in the +2 oxidation state. Beryllium is +2 in BeF_2; magnesium is +2 in Mg_3N_2; calcium is +2 in $CaCO_3$; and so on.

7. Oxygen almost always has an oxidation number of −2. Exceptions include molecules that contain oxygen-to-oxygen bonds, such as O_2, O_3, H_2O_2, and the O_2^{2-} ion.

8. The nonmetals in Group VIIA (F, Cl, Br, I, and At) usually form compounds in the −1 oxidation state. Fluorine is in the −1 oxidation state in AlF_3; chlorine is in the −1 oxidation state in HCl; bromine is in the −1 oxidation state in $ZnBr_2$; and so on.

9. The sum of the oxidation numbers of the atoms in a molecule or ion is equal to the charge on the molecule or ion. The sum of the oxidation numbers in a neutral molecule is 0.

$$\overset{\displaystyle H_2O}{\overset{\displaystyle \frown}{2(+1)+(-2)=0}}$$

The sum of the oxidation numbers in a polyatomic ion is equal to the charge on the ion. For example, the oxidation number of the sulfur atom in the SO_4^{2-} ion must be $+6$, because the sum of the oxidation numbers of the atoms in this ion must equal the charge on the ion, -2.

$$\overset{\displaystyle SO_4^{2-}}{\overset{\displaystyle \frown}{(+6)+4(-2)=-2}}$$

10. Elements toward the bottom and the left-hand side of the periodic table are more likely to form positive oxidation states than elements toward the top and right-hand side of the table. Sulfur carries a positive oxidation state in SO_2, for example, because it is below oxygen in the periodic table.

$$\overset{\displaystyle SO_2}{\overset{\displaystyle \frown}{(+4)+2(-2)=0}}$$

Sulfur also carries a positive oxidation state in SF_6, because it is both below and to the left of fluorine in the periodic table.

$$\overset{\displaystyle SF_6}{\overset{\displaystyle \frown}{(+6)+6(-1)=0}}$$

Exercise 2.13

Assign the oxidation numbers of the atoms in the following compounds.

(a) Al_2O_3 (b) CH_3OH (c) XeF_4 (d) $K_2Cr_2O_7$

Solution

(a) The sum of the oxidation numbers in Al_2O_3 must be 0, because the compound is neutral. If we assume that oxygen is present in the -2 oxidation state, we can calculate the oxidation state of the aluminum atoms from the following equation.

$$\overset{\displaystyle Al_2O_3}{\overset{\displaystyle \frown}{2(x)+3(-2)=0}}$$

$$x=+3$$

(b) Once again, the sum of the oxidation numbers is 0. We can assume that hydrogen has an oxidation number of $+1$, because it is combined with nonmetals. We can also assume that oxygen has an oxidation number of -2.

This means that the carbon atom must have an oxidation number of -2.

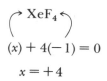

$$(x) + 3(+1) + (-2) + (+1) = 0$$
$$x = -2$$

(c) If we assume that the oxidation number of fluorine is -1, the xenon atom must be present in the $+4$ oxidation state.

$$XeF_4$$
$$(x) + 4(-1) = 0$$
$$x = +4$$

(d) Assigning oxidation numbers in this compound is simplified if you recognize that $K_2Cr_2O_7$ is an ionic compound, which contains K^+ and $Cr_2O_7^{2-}$ ions. The oxidation state of the potassium is $+1$ in the K^+ ion. The oxidation state of the chromium in the $Cr_2O_7^{2-}$ ion is $+6$.

$$Cr_2O_7^{2-}$$
$$2(x) + 7(-2) = -2$$
$$x = +6$$

Ammonium dichromate, $(NH_4)_2Cr_2O_7$, another salt of the $Cr_2O_7^{2-}$, or dichromate ion, which is an orange solid that decomposes violently when heated to form a dark-green solid (Cr_2O_3) by giving off N_2 and H_2O as a gas.

Exercise 2.14

Arrange the following compounds in order of increasing oxidation state for the carbon atom.

(a) CO (b) CO_2 (c) H_2CO (d) CH_3OH (e) CH_4

Solution

$$CH_4 < CH_3OH < H_2CO < CO < CO_2$$
$$\;-4\qquad -2\qquad\quad 0\qquad +2\qquad +4$$

Figure 2.8 shows the common oxidation numbers for many of the elements in the periodic table. There are several clear patterns in these data.

1. **Elements in the same group often have the same oxidation numbers. The elements in Group IA, for example, all form compounds with a $+1$ oxidation state, while the elements in Group IIA form compounds with an oxidation state of $+2$.**

2. **The largest, or most positive, oxidation state of an atom is equal to the group number of the element. Phosphorus is in Group VA, for example, and the largest oxidation number for phosphorus is $+5$.**

3. **The smallest, or most negative, oxidation number of an element can be found by subtracting 8 from the group number. The most negative oxidation state of phosphorus, for example, is $5 - 8$, or -3.**

	IA	IIA	IIIB	IVB	VB	VIB	VIIB	VIIIB			IB	IIB	IIIA	IVA	VA	VIA	VIIA	VIIIA
1	H +1 −1																H +1 −1	He 0
2	Li +1	Be +2											B +3	C −4 −2 +2 +4	N −3 +3 +5	O −2	F −1	Ne 0
3	Na +1	Mg +2											Al +3	Si +4	P −3 +3 +5	S −2 +2 +4 +6	Cl −1 +1 +3 +5 +7	Ar 0
4	K +1	Ca +2	Sc +3	Ti +2 +4	V +2 +4 +5	Cr +2 +3 +6	Mn +2 +4 +7	Fe +2 +3	Co +2 +3	Ni +2	Cu +1 +2	Zn +2	Ga +3	Ge −4 +2 +4	As −3 +3 +5	Se −2 +4 +6	Br −1 +1 +5	Kr 0
5	Rb +1	Sr +2	Y +3	Zr +4	Nb +3 +4 +5	Mo +2 +3 +4 +5 +6	Tc +4 +7	Ru +2 +3 +4	Rh +1 +3	Pd +2 +4	Ag +1	Cd +2	In +1 +3	Sn +2 +4	Sb −3 +3 +5	Te −2 +4 +6	I −1 +1 +5 +7	Xe 0
6	Cs +1	Ba +2	La +3	Hf +4	Ta +3 +4 +5	W +2 +3 +4 +5 +6	Re +2 +3 +4 +7	Os +3 +4	Ir +3	Pt +2 +4	Au +1 +3	Hg +1 +2	Tl +1 +3	Pb +2 +4	Bi +3 +5	Po +2 +4	At −1 +1 +3 +5 +7	Rn 0
7	Fr +1	Ra +2	Ac	104	105	106	107	108	109									

(Periods)

FIG. 2.8 This table summarizes some of the oxidation numbers of the common elements.

2.14 NOMENCLATURE

Long before chemists knew the formulas for chemical compounds, they had developed a system of **nomenclature** (from Latin *nomen*, name, and *calare*, to call) that gave each compound a unique name. Today we often use chemical formulas—such as $NaCl$, $C_{12}H_{22}O_{11}$, and $[Co(NH_3)_6](ClO_4)_3$—to describe chemical compounds. But we still need unique names that unambiguously identify each compound.

COMMON NAMES

Some compounds are so common or have been known for so long that no systematic nomenclature can compete with their well-established common names. Examples of compounds for which common names are still used include the following.

H_2O	water
NH_3	ammonia
CH_4	methane

NAMING IONIC COMPOUNDS, OR SALTS

The name of an ionic compound is written as the name of the positive ion followed by the name of the negative ion.

$NaCl$	sodium chloride
$(NH_4)_2SO_4$	ammonium sulfate
Fe_2O_3	iron(III) oxide

$NaHCO_3$ sodium bicarbonate
$Al(ClO_4)_3$ aluminum perchlorate

We therefore need a series of rules that allow us to unambiguously name positive and negative ions before we can name the salts these ions form.

POSITIVE IONS   Positive ions that consist of a single atom carry the name of the element from which they are formed.

Na^+ sodium	Zn^{2+} zinc
Ca^{2+} calcium	H^+ hydrogen
K^+ potassium	Sr^{2+} strontium

Some metals form two positive ions with different oxidation numbers. One of the earliest methods of distinguishing between these ions used the suffixes -*ic* and -*ous* added to the Latin name of the element to represent the higher and lower oxidation states, respectively.

Fe^{2+} ferrous	Fe^{3+} ferric
Sn^{2+} stannous	Sn^{4+} stannic
Cu^+ cuprous	Cu^{2+} cupric

Most chemists now use a simpler method, in which the charge on the ion is indicated by a Roman numeral in parentheses immediately after the name of the element.

Fe^{2+} iron(II)	Fe^{3+} iron(III)
Sn^{2+} tin(II)	Sn^{4+} tin(IV)
Cu^+ copper(I)	Cu^{2+} copper(II)

Polyatomic positive ions often have common names ending with the suffix -*onium*.

H_3O^+	hydronium
NH_4^+	ammonium

NEGATIVE IONS   Negative ions that consist of a single atom use the suffix -*ide* added to the stem of the name of the element.

F^- fluoride	O^{2-} oxide
Cl^- chloride	S^{2-} sulfide
Br^- bromide	N^{3-} nitride
I^- iodide	P^{3-} phosphide
H^- hydride	C^{4-} carbide

The nomenclature of some polyatomic negative ions was given in Table 2.3. At first glance, it may seem hopeless. There are several rules, however, that can bring some order out of this apparent chaos.

RULES FOR NAMING POLYATOMIC NEGATIVE IONS

RULE 1. The name of the ion usually ends in either -*ite* or -*ate*.

RULE 2. The -*ate* ending indicates a high oxidation state. For example, the NO_3^- ion is known as the nitr*ate* ion.

RULE 3. The *-ite* ending indicates a lower oxidation state. For example, the NO_2^- ion is known as the nitr*ite* ion.

RULE 4. The prefix *hypo-* is used to indicate the very lowest oxidation state. The ClO^-, or hypochlorite, ion is an example.

RULE 5. The prefix *per-* is used to indicate the very highest oxidation state. An example is the ClO_4^-, or perchlorate, ion.

There are only a handful of exceptions to these generalizations. The hydroxide (OH^-), cyanide (CN^-), and peroxide (O_2^{2-}) ions have the *-ide* ending because for many years they were thought to be monatomic (single-atom) ions.

Exercise 2.15

Name the following ionic compounds.

  (a) $Fe(NO_3)_3$ (b) $SrCO_3$ (c) Na_2SO_3 (d) $Ca(ClO)_2$

Solution

  (a) According to Figure 2.8, iron exists in two oxidation states — $+2$ or $+3$. It is therefore important to specify that this compound consists of the Fe^{3+} and NO_3^- ions. Thus, the compound is known as iron(III) nitrate.
  (b) There is no need to specify the oxidation state of strontium in this compound, because strontium is always in the $+2$ oxidation state in its compounds. $SrCO_3$ is therefore named strontium carbonate.
  (c) There is no need to specify the oxidation state of sodium in this compound, because sodium is always in the $+1$ oxidation state in its compounds. The negative ion in this compound is the SO_3^{2-}, or sulfite, ion, so the compound is named sodium sulfite.
  (d) This compound consists of Ca^{2+} and ClO^- ions. Once again, the metal is found in only one oxidation state, so all we have to do is recognize that the ClO^- ion is the hypochlorite ion to know that this compound is calcium hypochlorite.

NAMING SIMPLE COVALENT COMPOUNDS

Oxidation states also play an important role in the naming of simple covalent compounds. The name of the atom in the positive oxidation state is listed first. Then the suffix *-ide* is added to the stem of the name of the atom in the negative oxidation state.

  HCl   hydrogen chloride
  NO   nitrogen oxide
  BrCl   bromine chloride

The number of atoms of a given element in a simple covalent compound is indicated by one of the following Greek prefixes added to the name of the element.

✳️ | 1 mono- | 6 hexa- |
2 di-	7 hepta-
3 tri-	8 octa-
4 tetra-	9 nona-
5 penta-	10 deca-

The prefix *mono-* is seldom used, because it is redundant. For example, there is no need to name HCl *hydrogen monochloride*, because the name *hydrogen chloride* already indicates the presence of a single chlorine atom. The principal exception to this rule is carbon monoxide (CO).

Exercise 2.16

Name the following covalent compounds.

(a) N_2O (b) NO (c) NO_2 (d) N_2O_3 (e) N_2O_4 (f) N_2O_5

Solution

Because nitrogen is to the left of oxygen in the periodic table, we assume that nitrogen is present in a positive oxidation state in all of these compounds. We therefore add the *-ide* ending to the stem of the name of the element oxygen and specify the number of nitrogen and oxygen atoms in each of these compounds as follows.

(a) N_2O, dinitrogen oxide
(b) NO, nitrogen oxide
(c) NO_2, nitrogen dioxide
(d) N_2O_3, dinitrogen trioxide
(e) N_2O_4, dinitrogen tetroxide
(f) N_2O_5, dinitrogen pentoxide

NAMING ACIDS

Simple covalent compounds that contain hydrogen — such as HCl, HBr, and HCN — often dissolve in water to produce solutions that are known as *acids*. The names of these solutions add the prefix *hydro-* to the name of the compound and replace the suffix *-ide* with *-ic*. For example, hydrogen chloride (HCl) dissolves in water to form hydrochloric acid; hydrogen bromide (HBr) forms hydrobromic acid; hydrogen cyanide (HCN) forms hydrocyanic acid; and so on.

Many of the oxygen-rich polyatomic negative ions listed in Table 2.2 form acids; their names replace the suffix *-ate* with *-ic* and the suffix *-ite* with *-ous.*

$CH_3CO_2^-$	acetate	CH_3CO_2H	acetic acid
CO_3^{2-}	carbonate	H_2CO_3	carbonic acid
BO_3^{3-}	borate	H_3BO_3	boric acid
NO_3^-	nitrate	HNO_3	nitric acid
NO_2^-	nitrite	HNO_2	nitrous acid
SO_4^{2-}	sulfate	H_2SO_4	sulfuric acid
SO_3^{2-}	sulfite	H_2SO_3	sulfurous acid

ClO_4^-	perchlorate	$HClO_4$	perchloric acid
ClO_3^-	chlorate	$HClO_3$	chloric acid
ClO_2^-	chlorite	$HClO_2$	chlorous acid
ClO^-	hypochlorite	$HClO$	hypochlorous acid
PO_4^{3-}	phosphate	H_3PO_4	phosphoric acid
PO_3^{3-}	phosphite	H_3PO_3	phosphorus acid
MnO_4^-	permanganate	$HMnO_4$	permanganic acid
CrO_4^{2-}	chromate	H_2CrO_4	chromic acid

The names of complex acids indicate the presence of an acidic hydrogen as follows.

$NaHCO_3$	sodium *hydrogen* carbonate (also known as sodium bicarbonate)
$NaHSO_4$	sodium *hydrogen* sulfate (also known as sodium bisulfate)
KH_2PO_4	potassium *dihydrogen* phosphate
K_2HPO_4	potassium *hydrogen* phosphate

Exercise 2.17

Name the following compounds.

(a) $NaClO_3$ (b) $Al_2(SO_4)_3$ (c) Cr_2O_3 (d) P_4S_3 (e) ClF_3 (f) SCl_4

Solution

The key to naming these compounds is recognizing that the first three are ionic compounds, while the last three are covalent compounds.

(a) This salt contains the Na^+ and ClO_3^- ions; it is therefore sodium chlorate.

(b) This salt contains the Al^{3+} and SO_4^{2-} ions and is therefore aluminum sulfate.

(c) This salt contains the Cr^{3+} and O^{2-} ions. Because chromium forms several oxidation states, it is important to specify the oxidation state of the chromium atom in this compound as follows: chromium(III) oxide.

(d) The only trick to naming covalent compounds is recognizing which atom is present in a positive oxidation state and listing that atom first. P_4S_3 is tetraphosphorus trisulfide.

(e) ClF_3 is chlorine trifluoride.

(f) SCl_4 is sulfur tetrachloride.

SUMMARY

For at least 2500 years, from the 5th century B.C. until well into the 20th century, scientists and philosophers argued about whether matter was composed of elementary substances. Today, we define an element as any substance that cannot be decomposed into simpler substances by a chemical reaction. We explain their existence by assuming that an element is a substance that consists of only one kind of atom. Since atoms cannot be created or destroyed in a chemical reaction, elements cannot be broken down into simpler substances by these reactions.

The elements can be divided into three categories: metals, nonmetals, and semimetals. Most of the known elements are metals; they are found on the left and toward the bottom of the periodic table. A handful of nonmetals are clustered in the upper right-hand corner of the table. The semimetals, or metalloids, can be found along the dividing line between the metals and the nonmetals.

Elements combine to form chemical compounds, which are generally divided into two categories: ionic and covalent compounds. Ionic compounds — or salts, as they are also known — are formed when metals react with nonmetals. These compounds are composed of positive and negative ions formed when electrons are added to or subtracted from neutral atoms and molecules. Covalent compounds, which exist as neutral molecules, are formed when two or more nonmetals combine.

When we focus on the individual properties of a chemical compound, it is useful to distinguish between ionic and covalent compounds. When we consider how these compounds interact, however, it is often useful to follow these reactions in terms of the oxidation numbers, or oxidation states, of the atoms in each compound. The oxidation number of an atom is the charge that atom would possess if the compound was composed of ions. Oxidation numbers also play an important role in the systematic nomenclature of chemical compounds.

PROBLEMS

Elements and Compounds

2-1 Define the following terms: *element, compound,* and *mixture.*

2-2 Use examples to describe the difference between elements and compounds on the macroscopic scale.

2-3 Give examples of mixtures of elements encountered in daily life.

2-4 Give examples of mixtures of compounds encountered in daily life.

2-5 Classify each of the following substances into the categories of element, compound, mixture, metal, nonmetal, and semimetal. Use as many labels as necessary to classify each substance. (Use whatever outside reference you need.)

(a) diamond (b) brass (c) soil (d) glass (e) cotton (f) milk of magnesia (g) salt (h) iron (i) steel

2-6 Granite consists of four minerals: feldspar, magnetite, mica, and quartz. If one of these minerals can be physically separated from the others, is granite an element, a compound, or a mixture?

2-7 List the symbols for the following elements.

(a) antimony (b) gold (c) iron (d) mercury (e) potassium (f) silver (g) tin (h) tungsten

2-8 Name the elements with the following symbols.

(a) H (b) Li (c) Be (d) B (e) F (f) Ne

2-9 Name the elements with the following symbols.

(a) Na (b) Mg (c) Al (d) Si (e) P (f) Cl (g) Ar

2-10 Name the elements with the following symbols.

(a) Ti (b) V (c) Cr (d) Mn (e) Fe (f) Co (g) Ni (h) Cu (i) Zn

2-11 Name the elements with the following symbols.

(a) Mo (b) W (c) Rh (d) Ir (e) Pd (f) Pt (g) Ag (h) Au (i) Hg

The Law of Constant Composition

2-12 Describe the law of constant composition and show how it can be used to distinguish between compounds and mixtures.

2-13 Explain why the law of constant composition is consistent with Dalton's theory of the atom.

Evidence for the Existence of Atoms

2-14 Describe some of the evidence for the existence of atoms.

2-15 Describe some of the evidence from our senses that seems to deny the existence of atoms.

2-16 Compare Dalton's theory of the atom with the theory proposed by Leucippus and Democritus 2300 years earlier. In what ways are these theories the same? In what ways are they different?

2-17 Describe some of the evidence for the assumption that atoms are small.

Atoms and Molecules

2-18 Describe what the formula P_4S_3 tells us about this compound.

2-19 Describe the difference between the symbols in each of the following pairs.

(a) Co and CO (b) Cs and CS_2 (c) Ho and H_2O (d) 4 P and P_4

2-20 Describe the difference between the symbols in each of the following pairs.

(a) H and H^+ (b) H and H^- (c) 2 H and H_2 (d) H^+ and H^-

The Macroscopic, Atomic, and Symbolic Worlds of the Chemist

2-21 Explain the difference between H^+ ions, H atoms, and H_2 molecules on the atomic scale; on the macroscopic scale.

2-22 Describe some of the properties of water on the macroscopic scale and the atomic scale.

2-23 Draw a picture of what you think H_2, O_2, and H_2O molecules might look like.

2-24 Which of the following properties describe elements and compounds on the macroscopic scale? Which describe elements and compounds on the atomic scale? Which can be used to describe them on both scales?

(a) temperature   (b) pressure   (c) volume   (d) kinetic energy   (e) the shape of molecules   (f) elemental symbols, such as Au and Pt   (g) weight   (h) color   (i) melting point   (j) boiling point

The Chemistry of the Elements

2-25 Classify each of the following elements as a metal, a nonmetal, or a semimetal.

(a) Na   (b) Mg   (c) Al   (d) Si   (e) P   (f) S   (g) Cl   (h) Ar

2-26 Classify each of the following elements as a metal, a nonmetal, or a semimetal.

(a) S   (b) Sb   (c) Sc   (d) Se   (e) Si   (f) Sm   (g) Sn   (h) Sr

2-27 Describe the difference between a group and a period. Give examples of both.

2-28 List the elements in Group VIA of the periodic table.

2-29 List the elements in the third period.

2-30 Classify the elements in Group VA as either metals, nonmetals, or semimetals. Describe what happens to the properties of these elements as we go down this column of the periodic table.

2-31 Classify the elements in the third period as either metals, nonmetals, or semimetals. Describe what happens to the properties of these elements as we go across this period from left to right.

Metals, Nonmetals, and Semimetals

2-32 Describe what happens to the physical properties of a metal when it combines with another metal to form an alloy.

2-33 Describe what happens to the physical properties of a nonmetal when it combines with another nonmetal to form a covalent compound.

2-34 Describe what happens to the physical properties of a metal and a nonmetal when they combine to form an ionic compound, or salt.

2-35 Describe the differences between the chemical and physical properties of covalent compounds such as CO_2 and ionic compounds such as NaCl.

2-36 Which of the following compounds should be ionic?

(a) ZnS   (b) $AlCl_3$ (c) SnF_2 (d) BH_3 (e) H_2S

2-37 Which of the following compounds should be covalent?

(a) CH_4 (b) CO_2 (c) $SrCl_4$ (d) NaH   (e) SF_4

2-38 Which of the following pairs of elements should combine to give ionic compounds?

(a) Mg + O_2 (b) S_8 + F_2 (c) P_4 + Na   (d) Na + Hg   (e) K + I_2

2-39 Which of the following pairs of elements should combine to give covalent compounds?

(a) N_2 + O_2 (b) Cl_2 + F_2 (c) Cl_2 + Cr   (d) S_8 + Na

Ionic and Covalent Compounds

2-40 Which of the following compounds should conduct an electric current when dissolved in water?

(a) $MgCl_2$ (b) CO_2 (c) CH_3OH (d) KNO_3 (e) Ca_3P_2

2-41 Many covalent compounds have very distinctive odors. Examples include the H_2S given off by rotten eggs and the CH_3CO_2H in vinegar. Ionic compounds such as NaCl or Al_2O_3 have no odor. What difference between the physical properties of ionic and covalent compounds might be responsible for this observation?

2-42 One of the simplest ways of distinguishing between two covalent compounds, such as naphthalene and camphor, is to measure their melting points or boiling points. Naphthalene melts at 80.5°C and camphor melts at 178.8°C, for example. Explain why this procedure is not as useful in distinguishing between ionic compounds, such as NaCl and Al_2O_3.

2-43 Hydrogen chloride is a covalent compound that is a gas at room temperature. When cooled, it condenses to form a liquid. Neither HCl gas nor liquid HCl conducts electricity. But when HCl is dissolved in water, the result is hydrochloric acid, which conducts electricity very well. Explain why.

2-44 Which of the following substances would you expect to conduct an electric current?

(a) solid Na metal   (b) liquid Na metal   (c) solid NaCl   (d) liquid NaCl   (e) NaCl dissolved in water

The Structure of Atoms and Ions

2-45 Calculate the number of protons and electrons in atoms of the following elements.

(a) Ne   (b) Ti   (c) Cu   (d) Sn   (e) U

2-46 Calculate the number of protons and electrons in the following ions.

(a) Ca^{2+} (b) Cr^{3+} (c) Co^{2+} (d) Cu^{2+} (e) Cd^{2+}

2-47 Calculate the number of protons and electrons in the following ions.

(a) N^{3-} (b) S^{2-} (c) Br^- (d) Te^{2-}

2-48 Calculate the number of protons and electrons in the following ions. Describe any pattern you observe in these data.

(a) N^{3-} (b) O^{2-} (c) F^- (d) Na^+ (e) Mg^{2+}

2-49 Sodium metal reacts violently with water, often bursting into flame. Sodium ions dissolve peacefully in water. What is the difference between Na atoms and Na^+ ions?

2-50 Just as there is a difference between an orange and an orange peel, there is a difference between Fe metal and Fe^{3+} ions. If you are told that you have a shortage of iron, calcium, or zinc in your diet, which would you add to your diet, pieces of the metal or salts of their ions?

Formulas of Common Ionic Compounds, or Salts

2-51 Calculate the charge on the positive ions formed by the following elements.

(a) Mg (b) Al (c) Si (d) Cs (e) Ba

2-52 Calculate the charge on the negative ions formed by the following elements.

(a) C (b) P (c) S (d) I

2-53 Fluoride toothpastes convert the mineral apatite in tooth enamel into fluoroapatite, $Ca_5(PO_4)_3F$. If fluoroapatite contains Ca^{2+} and PO_4^{3-} ions, what is the charge on the fluoride ion in this compound?

2-54 Verdigris is a green pigment used in paint. The simplest formula for this compound is $Cu_3(OH)_2(CH_3CO_2)_4$. What is the charge on the copper ions in this compound, if the other ions are the OH^- and $CH_3CO_2^-$ ions?

2-55 Predict the formulas for neutral compounds containing the following pairs of ions.

(a) Mg^{2+} and NO_3^- (b) Fe^{3+} and SO_4^{2-} (c) Na^+ and CO_3^{2-}

2-56 Predict the formulas for neutral compounds containing the following pairs of ions.

(a) Na^+ and O_2^{2-} (b) Zn^{2+} and PO_4^{3-} (c) K^+ and $PtCl_6^{2-}$

2-57 Predict the formulas for potassium nitride and aluminum nitride, if the formula for magnesium nitride is Mg_3N_2.

2-58 Compounds that contain the O^{2-} ion are called oxides, while those that contain the O_2^{2-} ion are called peroxides. If the formula for potassium oxide is K_2O, what is the formula for potassium peroxide?

2-59 What is the value of x in the $[Co(NO_2)_x]^{3-}$ ion if this ion contains the Co^{3+} and NO_2^- ions?

2-60 Magnetic iron oxide has the formula Fe_3O_4. Explain this formula by assuming that Fe_3O_4 contains both Fe^{2+} and Fe^{3+} ions.

Oxidation Numbers

2-61 An area of active research interest in recent years has involved compounds, such as $Re_2Cl_8^{2-}$, $Cr_2Cl_9^{3-}$, and $Mo_2Cl_8^{4-}$, that contain bonds between metal atoms. Calculate the oxidation number of the metal atom in each of these compounds.

2-62 Determine the oxidation state of phosphorus in the following compounds.

(a) K_3P (b) $AlPO_4$ (c) PO_3^{3-} (d) P_2Cl_4

2-63 Calculate the oxidation number of the aluminum atom in the following compounds.

(a) AlH_3 (b) AlH_4^- (c) $LiAlH_4$ (d) $[Al(H_2O)_6]^{3+}$
(e) Al_2O_3 (f) $Al(OH)_4^-$ (g) Li_3AlH_6

2-64 The active ingredient in Rolaids has the formula $NaAl(OH)_2CO_3$. Calculate the oxidation state of the aluminum atom in this compound.

2-65 Which of the following compounds contain hydrogen in a negative oxidation state?

(a) H_2S (b) H_2O (c) NH_3 (d) PH_4^+ (e) $LiAlH_4$
(f) HF (g) CaH_2 (h) CH_4

2-66 Calculate the oxidation number of the chlorine atom in the following compounds.

(a) Cl_2 (b) Cl^- (c) ClO^- (d) ClO_2^- (e) ClO_3^- (f) ClO_4^-

2-67 Calculate the oxidation number of the iodine atom in the following compounds. Group compounds in which iodine has the same oxidation number.

(a) HI (b) KI (c) I_2 (d) HOI (e) KIO_3 (f) I_2O_5
(g) KIO_4 (h) H_5IO_6

2-68 Calculate the oxidation number of the boron atom in the following compounds and describe any trends in the oxidation number.

(a) BH_3 (b) BF_3 (c) BH_4^- (d) BO_2^- (e) H_3BO_3
(f) $B_4O_7^{2-}$ (g) HBO_2 (h) $H_2B_4O_7$ (i) B_2O_3 (j) HBF_4
(k) B_2H_6 (l) $NaBH_4$

2-69 Calculate the oxidation number of the xenon atom in the following compounds and describe any trends in these oxidation states.

(a) XeF_2 (b) XeF_4 (c) $XeOF_2$ (d) XeF_6 (e) $CsXeF_7$
(f) Cs_2XeF_8 (g) $XeOF_4$ (h) XeO_3 (i) XeO_3F^-
(j) XeO_4 (k) XeO_6^{4-}

2-70 Calculate the oxidation number of barium in BaO_2. Does your answer make sense? If it doesn't, determine the consequence of assuming that barium is present in its usual oxidation state in this compound.

2-71 Carbon can have any oxidation number between -4 and $+4$. Calculate the oxidation number of carbon in the following compounds.

(a) CF_4 (b) COF_2 (c) CO (d) CO_2 (e) CS_2 (f) CH_3Li
(g) CH_3^+ (h) CH_4 (i) CH_3^- (j) H_2CO (k) CO_3^{2-}
(l) H_2CO_3 (m) HCO_2H

2-72 Sulfur can have any oxidation number between $+6$ and -2. Calculate the oxidation number of sulfur in the following compounds. Group compounds with the same oxidation number and describe any trends.

(a) S_8 (b) H_2S (c) ZnS (d) Na_2S_2 (e) SF_4 (f) SF_5^- (g) SF_6 (h) SF_3^+ (i) SO_2 (j) SO_3 (k) SO_3^{2-} (l) SO_4^{2-} (m) $S_2O_3^{2-}$ (n) $S_2O_6^{2-}$ (o) $S_2O_8^{2-}$ (p) H_2SO_3 (q) H_2SO_4 (r) $H_2S_2O_3$

2-73 Calculate the oxidation number of manganese in the following compounds.

(a) MnO (b) Mn_2O_3 (c) MnO_2 (d) MnO_3 (e) Mn_2O_7 (f) $Mn(OH)_2$ (g) $Mn(OH)_3$ (h) H_2MnO_2 (i) H_2MnO_4 (j) $HMnO_4$ (k) $CaMnO_3$ (l) $MnSO_4$

2-74 Calculate the oxidation number of nitrogen in the following compounds. Group compounds with the same oxidation number and describe any patterns you observe.

(a) N_2 (b) N^{3-} (c) NH_3 (d) N_2O (e) NO (f) NO_2 (g) N_2O_3 (h) N_2O_4 (i) N_2O_5 (j) HNO_2 (k) HNO_3 (l) $NaNO_2$ (m) $NaNO_3$ (n) Mg_3N_2 (o) NH_4OH (p) NH_2OH

2-75 Prussian blue is a pigment with the formula $Fe_4[Fe(CN)_6]_3$. If this compound contains the $[Fe(CN)_6]^{4-}$ ion, what is the oxidation state of the other four iron atoms? Turnbull's blue is a pigment with the formula $Fe_3[Fe(CN)_6]_2$. If this compound contains the $[Fe(CN)_6]^{3-}$ ion, what is the oxidation state of the other three iron atoms?

Nomenclature

2-76 Describe what is wrong with the common names for the following compounds and write a better name for each compound.

(a) phosphorus pentoxide (P_2O_5) (b) iron oxide (Fe_2O_3) (c) sodium bicarbonate ($NaHCO_3$) (d) chlorine monoxide (Cl_2O) (e) copper bromide ($CuBr_2$)

2-77 Explain why calcium bromide is a satisfactory name for $CaBr_2$ but $FeBr_2$ has to be called iron(II) bromide.

2-78 Write the formulas for the following compounds.

(a) tetraphosphorus trisulfide (b) silicon dioxide (c) carbon disulfide (d) carbon tetrachloride (e) phosphorus pentafluoride

2-79 Write the formulas for the following compounds.

(a) silicon tetrafluoride (b) sulfur hexafluoride (c) oxygen difluoride (d) dichlorine heptoxide (e) chlorine trifluoride

2-80 Write the formulas for the following compounds.

(a) tin(II) chloride (b) mercury(II) nitrate (c) tin(IV) sulfide (d) chromium(III) oxide (e) iron(II) phosphide

2-81 Write the formulas for the following compounds.

(a) beryllium fluoride (b) magnesium nitride (c) barium peroxide (d) potassium carbonate

2-82 Write the formulas for the following compounds.

(a) cobalt(III) nitrate (b) iron(III) sulfate (c) gold(III) chloride (d) manganese(IV) oxide (e) tungsten(VI) chloride

2-83 Name the following compounds.

(a) KNO_3 (b) Li_2CO_3 (c) $BaSO_4$ (d) Na_2SO_3 (e) PbI_2

2-84 Name the following compounds.

(a) $AlCl_3$ (b) Na_3N (c) Ca_3P_2 (d) Li_2S (e) MgO

2-85 Name the following compounds.

(a) NH_4OH (b) H_2O_2 (c) $Mg(OH)_2$ (d) $Ca(OCl)_2$ (e) $NaCN$

2-86 Name the following compounds.

(a) Sb_2S_3 (b) $SnCl_2$ (c) SF_4 (d) $SrBr_2$ (e) $SiCl_4$

2-87 Write the formulas of the following common acids.

(a) acetic acid (b) hydrochloric acid (c) sulfuric acid (d) phosphorous acid (e) nitric acid

2-88 Write the formulas of the following less common acids.

(a) carbonic acid (b) hydrocyanic acid (c) boric acid (d) phosphorus acid (e) nitrous acid

2-89 If sodium carbonate is Na_2CO_3 and sodium bicarbonate is $NaHCO_3$, what are the formulas for sodium bisulfide and sodium bisulfite?

2-90 The prefix *thio-* describes compounds in which sulfur replaces oxygen—for example, cyanate (OCN^-) and thiocyanate (SCN^-). If SO_4^{2-} is the sulfate ion, what is the formula for the thiosulfate ion?

2-91 A mineral is a chemical compound found in the earth's crust. Name the compound in each of the following minerals.

(a) fluorite (CaF_2) (b) galena (PbS) (c) pyrite (FeS_2) (d) rutile (TiO_2) (e) hematite (Fe_2O_3)

2-92 Name the compound in each of the following minerals.

(a) magnetite (Fe_3O_4) (b) calcite ($CaCO_3$) (c) barite ($BaSO_4$) (d) quartz (SiO_2)

CHAPTER 3

STOICHIOMETRY: COUNTING ATOMS AND MOLECULES

CHAPTER CONTENTS

The idea that matter consists of elementary particles—atoms—that combine in ratios of simple whole numbers to form elements and compounds is the basis of modern chemistry. Chapter 2 looked at how this idea could be used to explain some of the properties of an element or compound. This chapter takes the first step toward applying the concept of atoms and molecules to predicting what will happen in a chemical reaction. As you will see, the concept of atoms and molecules provides a way of predicting exactly how much carbon dioxide (CO_2) and water (H_2O) will be given off when the body burns a known weight of sugar and how much oxygen (O_2) gas will have to be inhaled to carry out this reaction.

Predictions of this kind require that we find a way to count the number of atoms, ions, or molecules in a sample. This chapter focuses on how we can obtain this information from either the weight of the sample or the volume of a solution. It also introduces a shorthand notation for describing chemical reactions that is more efficient than saying "The oxygen we breathe reacts with sugar in the body to form water, carbon dioxide gas (which must be exhaled from the body), and enough energy to support the organism that carries out these reactions."

3.1 THE RELATIVE MASSES OF ATOMS

Atoms are so small it is difficult to appreciate their size. There are about 5×10^9 people on this planet. There are about 1×10^{11} stars in the Milky Way. The diameter of our galaxy is about 1.4×10^{17} miles. But there are 24,000,000,000,000,000,000,000, or 2.4×10^{22}, atoms in an object as small as a penny.

A sliver of copper metal just big enough to be detected on a good analytical balance contains about 1,000,000,000,000,000,000 (1×10^{18}) atoms. This means that it is impossible to measure the absolute weight of an atom. Chemists therefore have to be satisfied with measurements of their relative mass.

Figure 3.1 shows a block diagram of an instrument known as a ***mass spectrometer***, which can be used to determine the relative masses of atoms or molecules. The sample enters the instrument through a valve (A) into an evacuated chamber. The atoms or molecules in the sample flow past a filament (B), where they collide with

The instrument panel of a mass spectrometer.

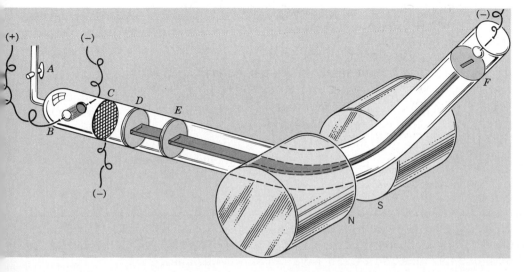

FIG. 3.1 A block diagram of a mass spectrometer. The sample enters the instrument through a valve, or port. Electrons are removed from the atoms or molecules to form positively charged ions. The ions are accelerated, focused, and then allowed to pass between the poles of a magnetic field, which bends the path of the ions. The lighter the ion, the larger the deflection. Eventually, the ions enter a detector, where they are counted.

high-energy electrons. As a result of these collisions, the neutral atoms or molecules lose electrons to form positively charged ions. These ions are then accelerated by a high-voltage accelerator (C), focused by a series of slits (D and E), and passed between the north and south poles of a magnet. The interaction between the magnetic field and the charge on the ion bends the path along which the ions travel. The larger the mass of the ion, the smaller the angle through which its path is bent before it enters the detector (F).

The mass spectrometer was invented by F. W. Aston shortly after World War I. By 1927, Aston had built an instrument that was accurate to more than 1 part in 10,000, and mass spectrometery became the method of choice for measuring the relative mass of an atom or molecule. It can tell us, for example, that the mass of a single fluorine atom is 2.010808 times the mass of a beryllium atom.

The most logical units for reporting the mass of an atom are **atomic mass units,** or **amu.** We can create a scale of atomic masses by finding the lightest atom, assigning that atom a mass of 1 amu, and then expressing the masses of all other atoms in terms of this standard. The first element in the periodic table — hydrogen — has the lightest atoms, so we can assign hydrogen a relative mass of 1 amu and report the mass of any other atom as a multiple of this unit. If oxygen atoms are 16 times heavier than hydrogen atoms, then oxygen should be assigned a relative mass of 16 amu.

3.2 ISOTOPES

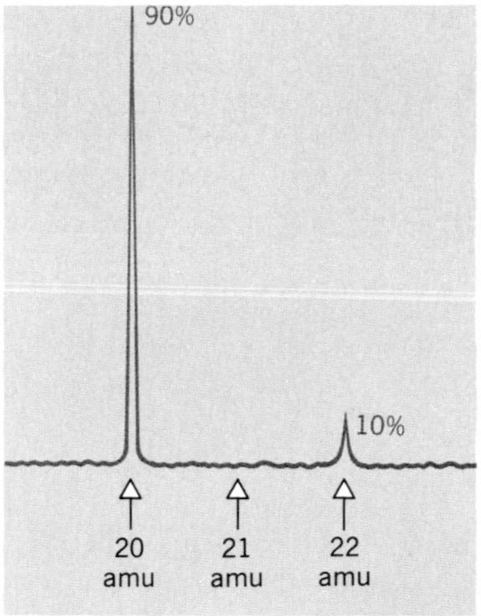

FIG. 3.2 The mass spectrum of neon shows two lines, corresponding to two different isotopes of neon. The relative intensities of these lines suggest that about 90% of naturally occurring neon atoms weigh 20 amu and the other 10% weigh 22 amu.

Long before Aston's work, chemists had used elaborate experiments to put together tables of relative atomic weights. Shortly after he began his work with the mass spectrometer, Aston made a surprising observation. According to the best estimates at the time, the atomic weight of neon was 20.2 amu. But when Aston injected a sample of neon into his instrument, he observed two peaks in the **mass spectrum** with relative masses of 20.0 and 22.0 amu, as shown in Figure 3.2.

Aston explained these results by assuming there are two different kinds of neon atoms, one with a mass of 20 amu and the other with a mass of 22 amu. He explained the difference between the height of the two peaks in the spectrum by suggesting that about 90% of the neon atoms have a mass of 20 amu and 10% have a mass of 22 amu. Thus, the atomic mass of neon obtained in other experiments would be a weighted average of 90% of the lighter atom and 10% of the heavier atom.

$$20.2 \text{ amu} = (0.90 \times 20 \text{ amu}) + (0.10 \times 22 \text{ amu})$$

Aston called these different neon atoms **isotopes,** using a term introduced by Frederick Soddy in 1911. *Isotope* literally means "same place" and is used to describe atoms that should be put into the same place in the periodic table. All of the chemical properties, and most of the physical properties, of the isotopes of neon are identical. The principal difference between them is their relative atomic mass.

Exercise 3.1

Aston also found that chlorine is a mixture of two isotopes with masses of 34.97 and 36.97 amu. Calculate the average mass of chlorine if 75.77% of the atoms have a mass of 34.97 amu and 24.23% have a mass of 36.97 amu.

Solution

Percent literally means "per hundred." The problem therefore describes a mixture of chlorine atoms for which 75.77 parts per hundred have a mass of 34.97 amu and 24.23 parts per hundred have a mass of 36.97 amu. The average mass of a chlorine atom can be calculated as follows.

$$(0.7577 \times 34.97 \text{ amu}) + (0.2423 \times 36.97 \text{ amu}) = \textbf{35.34 amu}$$

The average mass of a chlorine atom — to four significant figures — is 35.34 amu.

A useful convention has been proposed for distinguishing between the isotopes of an atom. According to this convention, the mass number of the isotope is written in the upper left-hand corner of the symbol for the element. Thus, we can talk about the ^{20}Ne and ^{22}Ne isotopes of neon or the ^{35}Cl and ^{37}Cl isotopes of chlorine. Examples of isotopes of some of the lighter elements in the periodic table are given in Table 3.1. The heading "Percent Natural Abundance" describes the relative abundance of this isotope in nature.

The discovery of isotopes eventually led to a small, but embarrassing, discrepancy between the tables of atomic masses used by chemists and those used by physicists. Physicists used atomic masses based on the assumption that the most common isotope of oxygen — ^{16}O — had a mass of exactly 16.00... amu. (The ellipsis points that follow 16.00 indicate a series of zeros.) The tables used by chemists, however, were based on a weighted average of ^{16}O, ^{17}O, and ^{18}O atoms. Differences between the two scales were small, corresponding to differences in the fifth or sixth significant figure, but it was obvious that a new standard for atomic mass was needed.

This discrepancy was resolved when the SI units were created. By convention, atomic masses are now based on the assumption that the mass of the ^{12}C isotope of carbon is exactly 12.000... amu. Note, however, that naturally occurring carbon is a mixture of two isotopes, ^{12}C (98.89%) and ^{13}C (1.11%). The average mass of a carbon atom is therefore slightly larger than the mass of the most common isotope. The average mass of a carbon atom is 12.011 amu.

TABLE 3.1

Common Isotopes of Some of the Lighter Elements

Isotope	Mass (amu)	Percent Natural Abundance
^{1}H	1.007825	99.985
^{2}H	2.0140	0.015
^{6}Li	6.01512	7.42
^{7}Li	7.01600	92.58
^{10}B	10.0129	19.7
^{11}B	11.00931	80.3
^{12}C	12.00000	98.89
^{13}C	13.00335	1.11
^{16}O	15.99491	99.76
^{17}O	16.99913	0.04
^{18}O	17.99916	0.20
^{20}Ne	19.99244	90.51
^{21}Ne	20.99395	0.27
^{22}Ne	21.99138	9.22

This book will differentiate between the mass of an individual atom and the average mass of the atoms of an element with the following convention. When discussing an individual atom — such as a ^{12}C atom — we will talk about the mass of the atom. When we want to focus attention on an average atom of the element, we will talk about the ***atomic weight*** of the element. Thus, an individual ^{12}C atom has a mass of 12.000... amu. But the atomic weight of carbon is 12.011 amu. A table of atomic weights can be found on the inside front cover of the book.

3.3 THE MOLE AS A COLLECTION OF ATOMS

Samples that contain one mole of atoms of the following elements: bromine, mercury, sulfur, silicon, and zinc.

Atoms are so small that it takes an enormous number to give a sample large enough to be seen with the naked eye. It is therefore useful to create a unit that represents a collection of atoms or molecules that can serve as a bridge between chemistry on the atomic and macroscopic scales.

The unit we will use is called a ***mole*** (from the Latin, meaning "a huge mass"). It can be defined as follows.

> **A mole is the amount of any substance that contains the same number of elementary particles as there are carbon atoms in exactly 12.000... grams of the ^{12}C isotope of carbon.**

Note that a single ^{12}C atom has a mass of 12.000... amu and a mole of ^{12}C atoms weighs 12.000... grams.

$$1 \ ^{12}C \text{ atom} = 12.000... \text{ amu}$$
$$1 \text{ mol } ^{12}C \text{ atoms} = 12.000... \text{ grams}$$

Exercise 3.2

Use the relationship between the atomic weight of magnesium and the mass of a ^{12}C atom to calculate the weight of a mole of magnesium atoms.

Solution

According to the table of atomic weights, the average weight of a magnesium atom is 24.305 amu. This is 2.0254 times the mass of a ^{12}C atom.

$$\frac{1 \text{ Mg atom}}{1 \ ^{12}C \text{ atom}} = \frac{24.305 \text{ amu}}{12.000... \text{ amu}} = 2.0254$$

If a mole of magnesium atoms contains the same number of elementary particles as a mole of ^{12}C atoms, the weight of a mole of magnesium atoms must also be 2.0254 times the weight of a mole of ^{12}C atoms. This means that a mole of magnesium atoms must weigh 24.305 grams.

$$\frac{1 \text{ mol Mg}}{1 \text{ mol } ^{12}C} = \frac{24.305 \text{ g}}{12.000... \text{ g}} = 2.0254$$

Exercise 3.2 makes an important point.

> **A mole of atoms of any element has a weight in grams equal to the atomic weight of the element in amu.**

If the atomic weight of magnesium is 24.305 amu, then a mole of magnesium atoms must weigh 24.305 grams.

3.4 THE MOLE AS A COLLECTION OF MOLECULES

The term *mole* can be applied to any particle. We can talk about a mole of Mg atoms, a mole of Na^+ ions, a mole of electrons, or a mole of CO_2 molecules. Each time we use the term, we refer to the number of elementary particles equal to the number of ^{12}C atoms in 12.000... grams of the ^{12}C isotope of carbon.

What would be the weight of a mole of a compound, such as carbon dioxide (CO_2) or cane sugar ($C_{12}H_{22}O_{11}$)? Before we can answer this question, we need to calculate the weight of an individual molecule of CO_2 or $C_{12}H_{22}O_{11}$ by adding the weights of the atoms in a molecule of each compound. The result of this calculation is most often described as the ***molecular weight*** of the compound.

Samples that contain one mole of each of the following compounds: $(NH_4)_2Cr_2O_7$, HgS, PbI_2, and $CuSO_4$ 5 H_2O.

Exercise 3.3

Calculate the average weight of molecules of CO_2 and $C_{12}H_{22}O_{11}$ and the weight of a mole of each compound.

Solution

The atomic weight of carbon is 12.011 amu, and the atomic weight of oxygen is 15.9994 amu. Carbon dioxide therefore has a molecular weight of 44.010 amu.

CO_2:

$$
\begin{aligned}
1 \text{ C atom} &= 1(12.011) \text{ amu} = 12.011 \text{ amu} \\
2 \text{ O atoms} &= 2(15.9994) \text{ amu} = \underline{31.9988 \text{ amu}} \\
& \phantom{= 2(15.9994) \text{ amu} =} 44.010 \text{ amu}
\end{aligned}
$$

This means that a mole of CO_2 molecules weighs 44.010 grams.

The weight of an average $C_{12}H_{22}O_{11}$ molecule is equal to the sum of the atomic weights of the 12 carbon atoms, 22 hydrogen atoms, and 11 oxygen atoms each molecule contains. The molecular weight of this compound is therefore 342.30 amu.

$C_{12}H_{22}O_{11}$:

$$
\begin{aligned}
12 \text{ C atoms} &= 12(12.011) \text{ amu} = 144.13 \text{ amu} \\
22 \text{ H atoms} &= 22(1.00794) \text{ amu} = 22.1747 \text{ amu} \\
11 \text{ O atoms} &= 11(15.9994) \text{ amu} = \underline{175.993 \text{ amu}} \\
& \phantom{= 11(15.9994) \text{ amu} =} 342.30 \text{ amu}
\end{aligned}
$$

A mole of $C_{12}H_{22}O_{11}$ molecules therefore weighs 342.30 grams.

Exercise 3.4

Calculate the molecular weight of vitamin B_{12}, which has the formula $C_{63}H_{88}N_{14}O_{14}CoP$.

Solution

The molecular weight of vitamin B_{12} is equal to the sum of the atomic weights of the 181 atoms in the molecule.

$C_{63}H_{88}N_{14}O_{14}CoP$:

$$63 \text{ C atoms} = 63(12.011) \text{ amu} = 756.69 \text{ amu}$$
$$88 \text{ H atoms} = 88(1.00794) \text{ amu} = 88.6987 \text{ amu}$$
$$14 \text{ N atoms} = 14(14.0067) \text{ amu} = 196.094 \text{ amu}$$
$$14 \text{ O atoms} = 14(15.9994) \text{ amu} = 223.992 \text{ amu}$$
$$1 \text{ Co atom} = 1(58.9332) \text{ amu} = 58.9332 \text{ amu}$$
$$1 \text{ P atom} = 1(30.97376) \text{ amu} = 30.97376 \text{ amu}$$
$$1355.38 \text{ amu}$$

The molecular weight of vitamin B_{12} is 1355.38 amu per molecule, or 1355.38 grams per mole.

A common source of confusion for students working with moles is the difference between a mole of atoms of an element and a mole of molecules of the element.

Exercise 3.5

Describe the difference between the weights of a mole of oxygen atoms and a mole of O_2 molecules.

Solution

The atomic weight of oxygen is 15.9994 amu. This means that the average weight of an oxygen atom is 15.9994 amu and a mole of oxygen atoms weighs 15.9994 grams.

$$1 \text{ mol O} = 15.9994 \text{ g}$$

Each oxygen molecule contains two oxygen atoms, so the molecular weight of an oxygen molecule is twice as large as the atomic weight of the element. The average weight of an oxygen molecule is 31.9988 amu, so a mole of O_2 molecules weighs 31.9988 grams.

$$1 \text{ mol O}_2 = 31.9988 \text{ g}$$

3.5 CONVERTING GRAMS INTO MOLES: USING MEASUREMENTS OF WEIGHT TO COUNT ATOMS

A one-carat diamond.

The mole is a powerful tool, which enables chemists armed with nothing more than a table of atomic weights and a balance to count the number of atoms, ions, or molecules in a sample. The first step toward this goal involves understanding the relationship between grams and moles of a substance. To show how grams can be converted into moles, we'll calculate the number of moles of carbon atoms in a 1.00-carat diamond.

Diamond is an element, not a compound. A diamond can be thought of as a giant crystal that contains only carbon atoms. The weight of a diamond is expressed in units of carats. In 1877, the weight of a carat was defined as 205.3 milligrams. A 1.00-carat diamond therefore weighs 0.2053 grams.

$$1.00 \text{ carat} \times \frac{205.3 \text{ mg}}{1 \text{ carat}} \times \frac{1 \text{ g}}{1000 \text{ mg}} = 0.2053 \text{ g}$$

Carbon has an atomic weight of 12.011 amu, so a mole of carbon atoms weighs 12.011 grams. We can represent this in terms of the following equality.

$$1 \text{ mol C} = 12.011 \text{ g}$$

We can also represent it in terms of a pair of unit factors, as discussed in Section 1.5.

$$\frac{1 \text{ mol C}}{12.011 \text{ g C}} = 1 \qquad \frac{12.011 \text{ g C}}{1 \text{ mol C}} = 1$$

One of these unit factors will be useful in converting grams of carbon into moles of carbon. If we pay attention to what happens to the units during the calculation, it shouldn't be difficult to choose the correct one.

Let's multiply the number of grams of carbon in a 1.00-carat diamond by the unit factor on the left.

$$0.2053 \text{ g C} \times \frac{1 \text{ mol C}}{12.011 \text{ g C}} = \textbf{0.01709 mol C}$$

Note that the units of grams cancel and the answer now has the units of moles. According to this calculation, there is 0.01709 mole of carbon atoms in a 1.00-carat diamond.

Let's look at this calculation, to see why it works. This time we will group the terms a little differently.

$$\frac{0.2053 \text{ g C}}{12.011 \text{ g C}} \times 1 \text{ mol C} = \textbf{0.01709 mol C}$$

The equality between grams of carbon and moles of carbon says that one mole weighs 12.011 grams. But this sample contains 0.01709 times as much carbon by weight and therefore must contain 0.01709 times as many moles of carbon.

Exercise 3.6 shows how the ability to convert grams of a substance into the equivalent number of moles can be used to determine the formula of a compound.

Exercise 3.6

What is the formula of magnesium chloride if 2.55 grams of magnesium combines with 7.45 grams of chlorine to form 10.00 grams of this compound?

Solution

The first step in any problem based on weights of elements or compounds is to convert grams into moles. To do this, we need to know the relationship between the number of grams and the number of moles of the substance. In this problem, then, we start by looking up magnesium and chlorine in a table of atomic weights.

$$1 \text{ Mg atom} = 24.305 \text{ amu}$$
$$1 \text{ Cl atom} = 35.453 \text{ amu}$$

There are three significant figures in each measurement, so the final answer can contain only three significant figures. But it is going to take several steps to reach the answer. It is therefore a good idea to carry an extra significant figure until the last step of the calculation, to minimize the error that can result from rounding off too often.

It doesn't matter which element we start with in this problem, since we eventually have to work with both elements. Let's arbitrarily start with magnesium. If the atomic weight of magnesium to four significant figures is 24.31 amu per atom, a mole of magnesium weighs 24.31 grams.

$$1 \text{ mol Mg} = 24.31 \text{ g Mg}$$

We can turn this equality into two unit factors.

$$\frac{1 \text{ mol Mg}}{24.31 \text{ g Mg}} = 1 \qquad \frac{24.31 \text{ g Mg}}{1 \text{ mol Mg}} = 1$$

Converting grams of magnesium into moles requires a factor that has units of moles in the numerator and grams in the denominator.

$$\cancel{\text{g Mg}} \times \frac{\text{mol Mg}}{\cancel{\text{g Mg}}} = \text{mol Mg}$$

The factor-label method suggests that the problem should be set up as follows.

$$2.55 \cancel{\text{ g Mg}} \times \frac{1 \text{ mol Mg}}{24.31 \cancel{\text{ g Mg}}} = 0.1049 \text{ mol Mg atoms}$$

The same format can be used to convert grams of chlorine atoms into moles of chlorine atoms. Since the atomic weight of chlorine is 35.45 amu, a mole of chlorine atoms weighs 35.45 grams.

$$7.45 \cancel{\text{ g Cl}} \times \frac{1 \text{ mol Cl}}{35.45 \cancel{\text{ g Cl}}} = 0.2102 \text{ mol Cl atoms}$$

Everything we have done so far is strictly mechanical, using a well-defined set of rules to convert from one set of units (grams) into another (moles). Once this is done, we write down the information we have obtained.

The product of this reaction contains:
0.1049 mol Mg atoms
0.2102 mol Cl atoms

We then reread the problem and ask: "Have we made any progress toward the solution?"

In this case, we are trying to find the formula for magnesium chloride. The formula of this compound gives us the ratio of magnesium atoms to chlorine atoms. If the formula is MgCl, there are just as many magnesium atoms as chlorine atoms in the compound. If the formula is $MgCl_2$, there are twice as many chlorine atoms in the compound.

What is the relationship between moles of magnesium and moles of chlorine in our sample?

$$\frac{0.2102 \text{ mol Cl}}{0.1049 \text{ mol Mg}} = 2.003$$

Within experimental error, there are twice as many moles of chlorine atoms as moles of magnesium atoms in this sample. Since a mole of atoms always contains the same number of atoms, the only possible conclusion is that there are twice as many chlorine atoms as magnesium atoms in the compound. In other words, the simplest formula for magnesium chloride is $MgCl_2$.

The same procedure has been used throughout this section to convert the weight of a sample of an element into the number of moles of atoms in the sample.

A similar technique can be used to convert the weight of a compound into the number of moles of molecules in the sample. In this situation, however, the unit factor is based on the molecular weight of the compound.

Exercise 3.7

Calculate the number of moles of H_2O molecules in 1.000 liter of water at 0°C if the density of water at this temperature is 0.9998 g/cm³.

Solution

Before we can calculate the number of moles of H_2O molecules we need to know the weight of this volume of water.

$$1.000 \; \cancel{L} \times \frac{1000 \; \cancel{cm^3}}{1 \; \cancel{L}} \times \frac{0.9998 \; g \; H_2O}{\cancel{cm^3}} = 999.8 \; g \; H_2O$$

The molecular weight of water is equal to the sum of the atomic weights of two hydrogen atoms and one oxygen atom. To four significant figures, the weight of a water molecule is 18.02 amu.

H_2O:

$$
\begin{aligned}
2 \; \text{H atoms} &= 2(1.008) \; \text{amu} = & 2.016 \; \text{amu} \\
1 \; \text{O atom} &= 1(16.00) \; \text{amu} = & \underline{16.00 \; \text{amu}} \\
& & 18.02 \; \text{amu}
\end{aligned}
$$

A mole of H_2O molecules therefore weighs 18.02 grams.

By paying attention to the units in this calculation, we can show that there are 55.48 moles of H_2O molecules in 1.000 liter of water at 0°C.

$$999.8 \; \cancel{g \; H_2O} \times \frac{1 \; \text{mol} \; H_2O}{18.02 \; \cancel{g \; H_2O}} = \textbf{55.48 mol } H_2O$$

3.6 DETERMINING EMPIRICAL FORMULAS

Exercise 3.6 showed one way to determine the formula of a compound. By carefully measuring the weights of the magnesium and chlorine that react to form magnesium chloride, it was possible to show that the formula for this compound is $MgCl_2$. Let's look at another way to approach this problem. This time we will examine a compound once known as "marsh gas," because it was first collected above certain swamps, or marshes, in Britain.

Marsh gas — or methane, as it is now known — is a member of a family of compounds known as ***hydrocarbons,*** which only contain the elements hydrogen and carbon. One of the ways chemists characterize a compound is by determining the percent by weight of the elements in the compound. Methane can be shown by experiment to be 74.9% carbon and 25.1% hydrogen by weight. In other words, a 100-gram sample of methane contains 74.9 grams of carbon and 25.1 grams of hydrogen.

$$100 \; g \; \text{methane} \times 74.9\% \; C = \textbf{74.9 g C}$$
$$100 \; g \; \text{methane} \times 25.1\% \; H = \textbf{25.1 g H}$$

The next step in this calculation involves converting grams of carbon and hydrogen into moles of these elements. In this case, the conversion is based on the fact that a mole of carbon atoms weighs 12.01 grams and a mole of hydrogen atoms weighs 1.008 grams (to four significant figures).

$$74.9 \text{ g C} \times \frac{1 \text{ mol C}}{12.01 \text{ g C}} = 6.236 \text{ mol C}$$

$$25.1 \text{ g H} \times \frac{1 \text{ mol H}}{1.008 \text{ g H}} = 24.90 \text{ mol H}$$

Let's stop at this point to review the information obtained so far.

100 grams of methane contains 6.24 moles of C.

100 grams of methane contains 24.9 moles of H.

Now let's play with this information to see what we have learned. One of the things we might do is look at the ratio of the moles of carbon and hydrogen in our sample.

$$\frac{24.9 \text{ mol H}}{6.24 \text{ mol C}} = 3.99$$

Within experimental error, our 100-gram sample of methane contains four times as many moles of hydrogen atoms as moles of carbon atoms. Since a mole of atoms always contains the same number of atoms, this means there are four times as many hydrogen atoms as carbon atoms in this sample. We can explain these results by assuming that the simplest formula for methane is CH_4, just as Exercise 3.6 led us to conclude that the simplest formula for magnesium chloride is $MgCl_2$. The simplest formula for a compound is called the *empirical formula.*

> **Empirical formula: the simplest formula for a compound; the lowest whole-number ratio of atoms of each element in the compound.**

This experiment doesn't tell us the formula for a molecule of methane. These results are consistent with molecules that contain one carbon atom and four hydrogen atoms — CH_4 — but they are also consistent with formulas of C_2H_8, C_3H_{12}, C_4H_{16}, and so on. All we know at this point is that the formula for the molecule is some multiple of an empirical formula that can be written as CH_4.

It is possible to write a sequence of steps that can be followed more or less automatically to determine the empirical formula for a compound from percent-by-weight data for the various elements.

STEPS FOR DETERMINING EMPIRICAL FORMULAS

1. Start with a 100-gram sample of the compound.

2. Use the percent-by-weight data to calculate the weight of each element in the 100-gram sample.

3. Convert each of these weights into the corresponding number of moles of atoms of the element.

4. Divide through by the element with the smallest number of moles of atoms.

5. Multiply the ratio of the number of atoms by small whole numbers until all of the coefficients in the formula are integers.

Exercise 3.8

Calculate the empirical formula of aspirin, if this compound is 60.0% C, 4.48% H, and 35.5% O by weight.

Solution

1. Start with a 100-gram sample of the compound.
2. Use the percent-by-weight data to calculate the weight of each element in the sample.

$$100 \text{ g aspirin} \times 60.0\% \text{ C} = 60.0 \text{ g C}$$
$$100 \text{ g aspirin} \times 4.48\% \text{ H} = 4.48 \text{ g H}$$
$$100 \text{ g aspirin} \times 35.5\% \text{ O} = 35.5 \text{ g O}$$

3. Convert each of these weights into the corresponding number of moles of atoms of the element.

$$60.0 \text{ g C} \times \frac{1 \text{ mol C}}{12.01 \text{ g C}} = 5.00 \text{ mol C}$$

$$4.48 \text{ g H} \times \frac{1 \text{ mol H}}{1.008 \text{ g H}} = 4.44 \text{ mol H}$$

$$35.5 \text{ g O} \times \frac{1 \text{ mol O}}{16.00 \text{ g O}} = 2.22 \text{ mol O}$$

4. Divide through by the element with the smallest number of moles of atoms.

$$\frac{5.00 \text{ mol C}}{2.22 \text{ mol O}} = 2.25 \qquad \frac{4.44 \text{ mol H}}{2.22 \text{ mol O}} = 2.00$$

5. Multiply the ratio of the number of atoms by small whole numbers until all of the coefficients in the formula are integers.

   It doesn't make sense to write the ratio of C to H to O atoms as $C_{2.25}H_2O$, because there is no such thing as a quarter of a carbon atom. We therefore multiply this ratio by small whole numbers until we get a formula in which all of the coefficients are integers.

$$2(C_{2.25}H_2O) = C_{4.5}H_4O_2$$
$$3(C_{2.25}H_2O) = C_{6.75}H_6O_3$$
$$4(C_{2.25}H_2O) = C_9H_8O_4$$

The empirical formula for aspirin is $C_9H_8O_4$.

3.7 ELEMENTAL ANALYSIS

Percent-by-weight data for a compound are often obtained by a process known as *elemental analysis.* If the compound contains carbon and hydrogen, a small sample is burned and the CO_2 and H_2O released in this reaction are captured. This experiment is done in the microanalysis apparatus diagramed in Figure 3.3. A sample of the compound weighing a few milligrams is placed in a tiny platinum "boat," which is placed in a furnace heated to about 850°C. A stream of oxygen (O_2) gas is passed over the sample, which is allowed to burn.

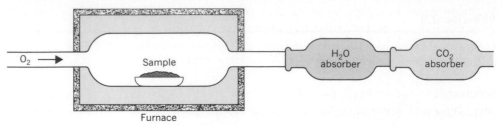

FIG. 3.3 A block diagram of a microanalysis apparatus. A very small sample (2 to 3 milligrams) of the compound of known weight is placed inside a furnace and burned in a stream of oxygen. The CO_2 and H_2O given off in this reaction are absorbed and weighed. Calculating the weight of the carbon and hydrogen atoms in the CO_2 and H_2O given off in this reaction and comparing these values with the weight of the original sample yield the percent by weight of carbon and hydrogen in the sample.

Any carbon in the sample reacts with oxygen to form carbon dioxide. Any hydrogen in the compound combines with the oxygen to form water. Compounds that only contain carbon and hydrogen burn to form a mixture of CO_2 and H_2O. If elements besides carbon and hydrogen are present, other gases may be formed as well.

The CO_2 and H_2O produced in this combustion reaction are swept out of the furnace by the stream of oxygen gas and trapped in a pair of absorbers. The water vapor is absorbed onto a sample of magnesium perchlorate, $Mg(ClO_4)_2$, of known weight. The carbon dioxide is absorbed onto a known weight of the mineral ascharite, $Mg_2B_2O_4 \cdot 2\,H_2O$.

Although this analysis is run on samples as small as 3 milligrams, it is easier to follow the calculations for larger samples. Let's assume that we analyzed a 1.00-gram sample of aspirin and found that 2.20 grams of CO_2 and 0.400 gram of H_2O were formed. Our goal is to convert this information into data that describe the percent by weight of carbon and hydrogen in aspirin.

We can start by converting the weights of the CO_2 and H_2O given off in this reaction into moles of these compounds, because that is the starting point for virtually every calculation based on measurements of the weights of chemical compounds.

$$2.20 \;\cancel{\text{g } CO_2} \times \frac{1 \text{ mol } CO_2}{44.01 \;\cancel{\text{g } CO_2}} = 0.0500 \text{ mol } CO_2$$

$$0.400 \;\cancel{\text{g } H_2O} \times \frac{1 \text{ mol } H_2O}{18.02 \;\cancel{\text{g } H_2O}} = 0.0222 \text{ mol } H_2O$$

We then note that there is one carbon atom in each CO_2 molecule, which means there is a mole of carbon atoms in a mole of CO_2 molecules.

$$0.0500 \;\cancel{\text{mol } CO_2} \times \frac{1 \text{ mol C}}{1 \;\cancel{\text{mol } CO_2}} = 0.0500 \text{ mol C}$$

Since all of the carbon came from the original sample, we must conclude that the sample contained 0.0500 mole of carbon atoms.

We then note that there are two hydrogen atoms in each H_2O molecule and therefore 2 moles of hydrogen atoms in a mole of H_2O molecules.

$$0.0222 \;\cancel{\text{mol } H_2O} \times \frac{2 \text{ mol H}}{1 \;\cancel{\text{mol } H_2O}} = 0.0444 \text{ mol H}$$

Thus, the aspirin sample contained 0.0444 mole of hydrogen atoms.

Let's summarize our results so far.

A 1.00-gram sample of aspirin contains 0.0500 mole of carbon and 0.0444 mole of hydrogen.

If we know the number of moles of carbon and hydrogen in this sample, we should be able to calculate the weights of these elements. Converting from moles into grams involves multiplying the number of moles by a unit factor that has units of grams in the numerator and moles in the denominator.

$$\cancel{\text{mol}} \times \frac{\text{grams}}{\cancel{\text{mol}}} = \text{grams}$$

We can therefore determine the weight of carbon in the sample by multiplying the number of moles of carbon by the number of grams per mole of this element.

$$0.0500 \, \cancel{\text{mol C}} \times \frac{12.01 \text{ g C}}{1 \, \cancel{\text{mol C}}} = 0.600 \text{ g C}$$

The weight of hydrogen in the sample can be calculated in a similar fashion.

$$0.0444 \, \cancel{\text{mol H}} \times \frac{1.008 \text{ g H}}{1 \, \cancel{\text{mol H}}} = 0.0448 \text{ g H}$$

According to this calculation, a 1.00-gram sample of aspirin contains 0.600 gram of carbon and 0.0448 gram of hydrogen. Aspirin is therefore 60.0% C and 4.48% H by weight.

$$\frac{0.600 \text{ g C}}{1.00 \text{ g aspirin}} \times 100\% = 60.0\% \text{ C by weight}$$

$$\frac{0.0448 \text{ g H}}{1.00 \text{ g aspirin}} \times 100\% = 4.48\% \text{ H by weight}$$

Carbon and hydrogen add up to only 64.5% of the total weight of the aspirin. The remaining weight (35.5%) must be due to the third element in aspirin — oxygen. Aspirin is therefore 60.0% C, 4.48% H, and 35.5% O by weight. As was shown in Exercise 3.8, these data can be used to determine that the empirical formula for aspirin is $C_9H_8O_4$.

USING ELEMENTAL ANALYSIS TO IDENTIFY CHEMICAL COMPOUNDS

Elemental analysis is a powerful technique for identifying compounds. To show this, we'll use it to distinguish between the white, crystalline powder known as aspirin, which is used to cure headaches, and the white, crystalline powder known as cocaine, which is more likely to cause headaches.

Cocaine is a naturally occurring substance that can be extracted from the leaves of the *coca* plant, which grows in South America. If the chemical formula for cocaine is $C_{17}H_{21}O_4N$, what is the percent by weight of carbon, hydrogen, oxygen, and nitrogen in this compound?

We can start by calculating the weight of a cocaine molecule. Since elemental analysis data are known to only three or perhaps four significant figures, we will use four significant figures in this calculation.

$C_{17}H_{21}O_4N$:

$$
\begin{array}{rll}
17 \text{ C atoms} = & 17(12.01) \text{ amu} = & 204.17 \text{ amu} \\
21 \text{ H atoms} = & 21(1.008) \text{ amu} = & 21.168 \text{ amu} \\
4 \text{ O atoms} = & 4(16.00) \text{ amu} = & 64.00 \text{ amu} \\
1 \text{ N atom} = & 1(14.01) \text{ amu} = & \underline{14.01 \text{ amu}} \\
& & 303.4 \text{ amu}
\end{array}
$$

The percent by weight of the elements in this molecule can be calculated as follows.

$$\% \text{ C} = \frac{204.17 \text{ amu}}{303.4 \text{ amu}} \times 100\% = 67.3\%$$

$$\% \text{ H} = \frac{21.168 \text{ amu}}{303.4 \text{ amu}} \times 100\% = 6.98\%$$

$$\% \text{ O} = \frac{64.00 \text{ amu}}{303.4 \text{ amu}} \times 100\% = 21.1\%$$

$$\% \text{ N} = \frac{14.01 \text{ amu}}{303.4 \text{ amu}} \times 100\% = 4.62\%$$

Let's now compare the percent-by-weight data for aspirin and cocaine.

Aspirin	Cocaine
60.0% C	67.3% C
4.48% H	6.98% H
35.5% O	21.1% O
	4.62% N

The difference between the percent by weight of carbon and hydrogen in these compounds would be immediately apparent from elemental analysis of the type described in this section. Elemental analysis, then, is useful for several purposes. It is used to determine the empirical formula of a compound that has been synthesized in the laboratory, or isolated from nature, for the first time. But it can also be used to confirm the identity of a compound whose formula is already suspected.

3.8 EMPIRICAL VERSUS MOLECULAR FORMULAS

The preceding section showed how elemental analysis data can be obtained, and the section before that showed how these data can be used to determine the empirical formula of a compound. Next, let's look at the results of an analysis of vitamin C.

Exercise 3.9

Calculate the empirical formula for vitamin C if this compound is 40.9% C, 54.5% O, and 4.58% H by weight.

Solution

A 100-gram sample of vitamin C contains 40.9 grams of carbon, 54.5 grams of oxygen, and 4.58 grams of hydrogen. This corresponds to 3.41 moles of carbon, 3.41 moles of oxygen, and 4.54 moles of hydrogen.

$$40.9 \text{ g C} \times \frac{1 \text{ mol C}}{12.01 \text{ g C}} = 3.41 \text{ mol C}$$

$$54.5 \text{ g O} \times \frac{1 \text{ mol O}}{16.00 \text{ g O}} = 3.41 \text{ mol O}$$

$$4.58 \text{ g H} \times \frac{1 \text{ mol H}}{1.008 \text{ g H}} = 4.54 \text{ mol H}$$

Dividing through by 3.41 moles of C atoms gives the following results.

$$\frac{4.54 \text{ mol H}}{3.41 \text{ mol C}} = 1.33$$

$$\frac{3.41 \text{ mol O}}{3.41 \text{ mol C}} = 1$$

The ratio of C to H to O atoms in vitamin C is therefore 1 : 1.33 : 1. Multiplying this ratio by 3 gives the simplest formula, $C_3H_4O_3$.

The *empirical weight* of a compound can be calculated from its empirical formula. The empirical formula for vitamin C is $C_3H_4O_3$, so the empirical weight of vitamin C is 88.06 amu, or 88.06 grams per mole, calculated as follows.

$C_3H_4O_3$:

$$3 \text{ C atoms} = 3(12.01) \text{ amu} = 36.03 \text{ amu}$$
$$4 \text{ H atoms} = 4(1.008) \text{ amu} = \underline{4.032 \text{ amu}}$$
$$3 \text{ O atoms} = 3(16.00) \text{ amu} = \underline{48.00 \text{ amu}}$$
$$88.06 \text{ amu}$$

This is as far as we can go with the data from an elemental analysis experiment. It is possible, however, to determine the molecular weight of a compound in a separate experiment. When this is done, the molecular weight of vitamin C is found to be about 176 amu per molecule, or 176 grams per mole.

How do we explain the difference between the molecular weight (176 grams per mole) and the empirical weight (88.1 grams per mole) of vitamin C? We can start by dividing the empirical weight of vitamin C into the molecular weight.

$$\frac{\text{Molecular weight}}{\text{Empirical weight}} = \frac{176 \text{ g/mol}}{88.1 \text{ g/mol}} = 2$$

Within experimental error, the molecular weight of this compound is twice as large as the empirical weight. The only possible conclusion is that a molecule of vitamin C is twice as large as the empirical formula. In other words, the *molecular formula* of vitamin C is $C_6H_8O_6$.

Molecular formula: The formula that gives the number of atoms of each element in a molecule of the compound.

Elemental analysis provides only the simple ratio of atoms in a compound — the empirical formula. We also need to measure the molecular weight of the compound to determine the molecular formula. Frequently, the molecular formula of a compound (such as aspirin, $C_9H_8O_4$) is the same as the empirical formula. Sometimes, it isn't. The polyethylene used to make plastic sandwich bags, for example, is composed of molecules that are 10,000 to 100,000 times larger than the empirical formula.

AVOGADRO'S CONSTANT: THE NUMBER
3.9 OF ELEMENTARY PARTICLES IN A MOLE

Section 3.3 defined the mole as a collection of elementary particles equal to the number of ^{12}C atoms in 12.000... grams of the ^{12}C isotope of carbon without specifying the number of elementary particles in a mole. There are several reasons for this. As the discussion that followed Section 3.3 showed, we don't really need to know the number of particles in a mole to use this concept productively. Furthermore, this approach helps us understand how chemists were able to use the idea behind the mole concept for many years before they were first able to measure the number of particles it contains.

The only way to determine the number of particles in a mole is to measure the same quantity, in the same unit of measurement, on both the atomic and macroscopic scales. In 1910, Robert Millikan measured the charge on an electron—1.59×10^{-19} C, as we would express it today. Since the charge on a mole of electrons (96,484.56 C) was already known, it was possible to estimate the number of electrons in a mole of the first time. Using more recent—and more accurate—data, this procedure yields the following results.

$$\frac{96,484.56 \ \cancel{C}}{1 \ \text{mol electrons}} \times \frac{1 \ \text{electron}}{1.6021892 \times 10^{-19} \ \cancel{C}} = 6.022045 \times 10^{23} \ \frac{\text{electrons}}{\text{mol}}$$

The number of particles in a mole is called Avogadro's number or, more accurately, *Avogadro's constant,* in recognition of the contribution of the Italian scientist Amadeo Avogadro to our belief in the existence of atoms. For most calculations, three (6.02×10^{23}) or at most four (6.022×10^{23}) significant figures for Avogadro's constant will be enough.

Now that we know the number of atoms in a mole, we can calculate the weight of a single atom for the first time. If a mole of carbon atoms weighs 12.011 grams, and there are 6.0220×10^{23} atoms per mole, a single carbon atom must weigh 1.9945×10^{-23} grams.

$$\frac{12.011 \ \text{g C}}{1 \ \cancel{\text{mol C}}} \times \frac{1 \ \cancel{\text{mol C}}}{6.0220 \times 10^{23} \ \text{C atoms}} = 1.9945 \times 10^{-23} \ \frac{\text{g C}}{\text{atom}}$$

We can use this information and the atomic weight of carbon, 12.011 amu, to calculate the number of grams per amu

$$\frac{1.9945 \times 10^{-23} \ \text{g}}{12.011 \ \text{amu}} = 1.6606 \times 10^{-24} \ \frac{\text{g}}{\text{amu}}$$

or the number of amu per gram.

$$\frac{12.011 \ \text{amu}}{1.9945 \times 10^{-23} \ \text{g}} = 6.0220 \times 10^{23} \ \frac{\text{amu}}{\text{g}}$$

Note that Avogadro's constant is equal to the number of amu in 1 gram.

Avogadro's constant provides us with another way of defining the mole.

A mole of any substance contains the number of elementary particles equal to Avogadro's constant.

It doesn't matter whether we talk about a mole of atoms, a mole of molecules, a mole of electrons, or a mole of ions. By definition, a mole always contains 6.022×10^{23} elementary particles.

There are 6.022×10^{23} oxygen atoms in a mole of oxygen.

There are 6.022×10^{23} H_2O molecules in a mole of water.

There are 6.022×10^{23} Na^+ ions in a mole of sodium ions.

There are 6.022×10^{23} Cl^- ions in a mole of chlorine ions.

There are 6.022×10^{23} electrons in a mole of electrons.

Avogadro's constant is so large it is difficult to comprehend. It would take 6 million million galaxies the size of the Milky Way to yield 6×10^{23} stars. At the speed of light, it would take 102 billion years to travel 6×10^{23} miles. There are only about 40 times this number of drops of water in all of the oceans on earth. When you consider the size of this constant, it isn't surprising that it took so long to prove that atoms exist.

Exercise 3.10

Calculate to three significant figures the number of carbon atoms in a pound of cane sugar, which consists of $C_{12}H_{22}O_{11}$ molecules.

Solution

To solve this problem we have to convert a pound of sugar into grams and then convert grams of sugar into an equivalent number of moles of sugar molecules. We can start by noting that a pound of sugar weighs 453.6 grams.

$$1 \text{ lb} = 453.6 \text{ g}$$

According to the calculations in Section 3.4, the molecular weight of sugar is 342.30 grams per mole. Thus, a pound of sugar contains 1.325 moles of sugar molecules.

$$453.6 \text{ g } C_{12}H_{22}O_{11} \times \frac{1 \text{ mol } C_{12}H_{22}O_{11}}{342.30 \text{ g } C_{12}H_{22}O_{11}} = 1.325 \text{ mol } C_{12}H_{22}O_{11}$$

There are 12 carbon atoms in each sugar molecule and therefore 12 moles of carbon atoms in a mole of sugar.

$$1.325 \text{ mol } C_{12}H_{22}O_{11} \times \frac{12 \text{ mol C}}{1 \text{ mol } C_{12}H_{22}O_{11}} = 15.90 \text{ mol C}$$

Since there are 6.022×10^{23} atoms in a mole, there are 9.57×10^{24} carbon atoms in a pound of sugar.

$$15.90 \text{ mol C} \times 6.022 \times 10^{23} \frac{\text{atoms}}{\text{mol}} = 9.57 \times 10^{24} \text{ C atoms}$$

3.10 CHEMICAL REACTIONS AND THE LAW OF CONSERVATION OF MATTER

Our attention so far has been focused on individual compounds, such as salt (NaCl), carbon dioxide (CO_2), aspirin ($C_9H_8O_4$), and vitamin C ($C_6H_8O_6$). Much of the fascination of chemistry, however, revolves around chemical reactions. The first breakthrough in the study of chemical reactions resulted from the work of the French chemist Antoine Laurent Lavoisier between 1772 and 1794. As a result of his measurements of what happens during a chemical reaction, Lavoisier discovered one of the fundamental laws of chemical behavior: *the law of conservation of matter.* Lavoisier was the first to note that the sum of the weights of the starting materials in a chemical reaction is always equal to the sum of the weights of the products. In other words, matter is conserved in a chemical reaction.

We now understand why this is so. Atoms are neither created nor destroyed in a chemical reaction. Hydrogen atoms in an H_2 molecule can combine with oxygen atoms in an O_2 molecule to form H_2O, for example. But the number of hydrogen and oxygen atoms before and after the reaction is the same. Since the number of atoms does not change during a chemical reaction, the total weight of the products of a reaction must be the same as the total weight of the reactants.

3.11 CHEMICAL EQUATIONS AS A REPRESENTATION OF CHEMICAL REACTIONS

It is possible to describe a chemical reaction in words, but it is much easier (and more useful) to describe it with a ***chemical equation.*** Chemical equations are always written from left to right, even in languages that do not read from left to right. The formulas of the starting materials, or ***reactants,*** are always written on the left-hand side of the equation, and the formulas of the ***products*** are written on the right. Instead of an equal sign, the reactants and products are separated by an arrow that shows the direction of the reaction. The reaction between hydrogen and oxygen to form water, for example, can be represented by the following equation.

$$2 \text{ H}_2 + \text{O}_2 \longrightarrow 2 \text{ H}_2\text{O}$$
$$\underset{\text{reactants}}{\phantom{2 \text{ H}_2 + \text{O}_2}} \qquad \underset{\text{product}}{\phantom{2 \text{ H}_2\text{O}}}$$

The reaction between aluminum and iron oxide can be written as follows.

$$2 \text{ Al} + \text{Fe}_2\text{O}_3 \longrightarrow \text{Al}_2\text{O}_3 + 2 \text{ Fe}$$
$$\underset{\text{reactants}}{\phantom{2 \text{ Al} + \text{Fe}_2\text{O}_3}} \qquad \underset{\text{products}}{\phantom{\text{Al}_2\text{O}_3 + 2 \text{ Fe}}}$$

It is often useful to indicate whether the reactants or products are solids, liquids, or gases. We do this by writing an s, l, or g in parentheses after the formula for each of the reactants or products, as shown in the following equations.

$$2 \text{ H}_2(g) + \text{O}_2(g) \longrightarrow 2 \text{ H}_2\text{O}(g)$$
$$2 \text{ Al}(s) + \text{Fe}_2\text{O}_3(s) \longrightarrow \text{Al}_2\text{O}_3(s) + 2 \text{ Fe}(l)$$

Although these symbols make the equations more complex, they also provide useful information.

Reactions between gases often happen so fast they are dangerous. Reactions between solids are usually so slow they seem to take forever. Reactions between liquids, however, often occur fast enough to give products in a reasonable amount of time, without occurring so fast they produce explosions. To conduct reactions between compounds that are not liquids at room temperature, we often first dissolve each compound in a liquid to give a mixture known as a ***solution.***

The vast majority of the reactions you will encounter in this course will occur when two substances dissolved in water are mixed. These ***aqueous*** solutions (from the Latin *aqua*, "water") are so important we will identify them with the special symbol *aq*. This way we can distinguish between sugar as a solid — $C_{12}H_{22}O_{11}(s)$ — and sugar dissolved in water — $C_{12}H_{22}O_{11}(aq)$ or between salt as a solid — $NaCl(s)$ — and salt dissolved in water — $NaCl(aq)$. The process in which a sample is dissolved in water will be indicated by equations such as the following.

$$C_{12}H_{22}O_{11}(s) \xrightarrow{\text{H}_2\text{O}} C_{12}H_{22}O_{11}(aq)$$

Chemical equations provide such a powerful shorthand device for describing chemical reactions that we tend to think about reactions in terms of these equations. This causes a slight problem, because some of the grammar of a reaction is lost in the chemical equation. It is important to remember that a chemical equation

The thermite reaction used to weld iron rails while building railroads involves the reaction between powdered aluminum and iron oxide to form aluminum oxide and molten iron metal.

is a statement of what *can* happen, not necessarily what *will* happen. For example, the following equation does not guarantee that hydrogen will react with oxygen to form water.

$$2\ H_2(g) + O_2(g) \longrightarrow 2\ H_2O(g)$$

It is possible to fill a balloon with a mixture of hydrogen and oxygen and find that no reaction occurs, no matter how long we wait, until we touch the balloon with a torch. This chemical equation should be translated as follows.

If, or *when*, hydrogen reacts with oxygen to form water, two molecules of hydrogen and one molecule of oxygen are consumed for every two molecules of water produced.

Similarly, the following equation does not say that aluminum metal must react with iron oxide to form aluminum oxide and iron metal.

$$2\ Al(s) + Fe_2O_3(s) \longrightarrow Al_2O_3(s) + 2\ Fe(l)$$

At Purdue University, we have 50 pounds of a mixture of powdered aluminum and iron oxide that is at least 10 years old, and there is no trace of any reaction having occurred in this sample. All the equation tells us is what would happen if, or when, the reaction got started.

3.12 TWO VIEWS OF CHEMICAL EQUATIONS: MOLECULES VERSUS MOLES

Section 2.5 suggested that chemists simultaneously work in three different worlds —atomic, macroscopic, and symbolic. Chemical reactions occur on the atomic scale. It is on this scale that hydrogen molecules combine with oxygen molecules to form water molecules, as illustrated in Figure 3.4. The only place we can see these reactions, however, is on the macroscopic scale.

The symbolic world in which we write chemical equations such as the following can represent what happens on either the atomic or the macroscopic scale.

$$2\ H_2(g) + O_2(g) \longrightarrow 2\ H_2O(g)$$

This equation can be read in either of the following ways.

If, or when, hydrogen reacts with oxygen, two molecules of hydrogen will combine with a molecule of oxygen to form two molecules of water.

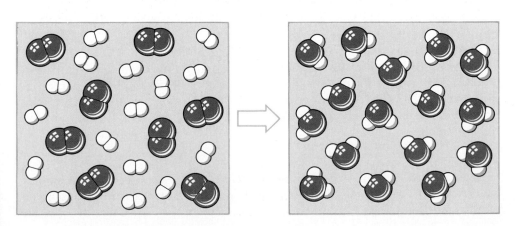

FIG. 3.4 Chemical reactions can be seen on the macroscopic scale, but they occur on the atomic scale. This figure represents what happens in a small corner of a container in which hydrogen and oxygen molecules react to form water molecules.

If, or when, hydrogen reacts with oxygen, two moles of hydrogen will be consumed for every mole of oxygen to form two moles of water.

Regardless of whether we think of the reaction in terms of molecules or moles, chemical equations must be ***balanced:*** they must have the same number of atoms of each element on both sides. There are four hydrogen atoms and two oxygen atoms on each side of the equation for the reaction between H_2 and O_2, for example.

$$2\ H_2(g) + O_2(g) \longrightarrow 2\ H_2O(g)$$

If the same number of atoms of each element are present on both sides of the equation, the total weight of the reactants must be equal to the total weight of the products of the reaction. On the atomic scale, this equation is balanced because the weight of the reactants is equal to the weight of the products.

$$2\ H_2(g) + O_2(g) \longrightarrow 2\ H_2O(g)$$
$$\underbrace{2 \times 2\ \text{amu} + 32\ \text{amu}}_{36\ \text{amu}} = \underbrace{2 \times 18\ \text{amu}}_{36\ \text{amu}}$$

On the macroscopic scale, the equation is balanced because the weight of 2 moles of hydrogen molecules and 1 mole of oxygen molecules is equal to the weight of 2 moles of water molecules.

$$2\ H_2(g) + O_2(g) \longrightarrow 2\ H_2O(g)$$
$$\underbrace{2 \times 2\ \text{g} + 32\ \text{g}}_{36\ \text{g}} = \underbrace{2 \times 18\ \text{g}}_{36\ \text{g}}$$

3.13 BALANCING CHEMICAL EQUATIONS

There is no sequence of rules that, when blindly followed, result in a balanced chemical equation. Thus, this section starts by specifying what can't be done when balancing an equation. For example, we can't balance the following equation

$$\underline{}\ H_2 + \underline{}\ O_2 \longrightarrow \underline{}\ H_2O$$

by changing the subscript on the oxygen atom in the product.

$$H_2 + O_2 \longrightarrow H_2O_2$$

That's cheating. There is a world of difference between H_2O and H_2O_2. H_2O is the formula for water; H_2O_2 is the formula for the hydrogen peroxide we use to disinfect cuts or bleach hair.

We can't balance an equation by changing the identity of the reactants or products of the reaction. We can, however, manipulate the coefficients written in front of the formulas for the reactants and products. For example, we can recognize that each H_2O molecule contains only one oxygen atom and therefore assume that two water molecules must be produced for every molecule of O_2 consumed in the reaction.

$$\underline{}\ H_2 + \mathbf{1}\ O_2 \longrightarrow \mathbf{2}\ H_2O$$

We can then work backwards, noting that two H_2O molecules contain a total of four hydrogen atoms and that the only way to get four hydrogen atoms among the products is to start with four hydrogen atoms, or two H_2 molecules, among the reactants.

$$2\ H_2 + 1\ O_2 \longrightarrow 2\ H_2O$$

The equation is now balanced, with four hydrogen atoms and two oxygen atoms on each side. By convention, we ignore all coefficients in a balanced equation that are equal to 1. The equation for the reaction between hydrogen and oxygen is therefore written as follows.

$$2\ H_2(g) + O_2(g) \longrightarrow 2\ H_2O(g)$$

The first thing to look for when balancing equations is relationships between the two sides. In the following equation, for example, we can see that at least three lithium atoms will be needed to make lithium nitride, Li_3N.

$$\underline{\quad}\ Li + \underline{\quad}\ N_2 \longrightarrow \underline{\quad}\ Li_3N$$

It therefore seems reasonable to start with three lithium atoms on the left-hand side of the equation and one Li_3N on the right.

$$3\ Li + \underline{\quad}\ N_2 \longrightarrow 1\ Li_3N$$

Now half the problem is solved — the lithium atoms are balanced.

We can now try to balance the nitrogen atoms. There are at least two nitrogen atoms on the left and only one on the right, so we have to assume that the reaction produces two Li_3N for each N_2 consumed.

$$3\ Li + 1\ N_2 \longrightarrow 2\ Li_3N$$

Now the nitrogen atoms are balanced, but the lithium atoms are not. There are three lithium atoms on the left and six (2×3) on the right. We can correct this, however, by doubling the number of lithium atoms on the left.

$$6\ Li + 1\ N_2 \longrightarrow 2\ Li_3N$$

Both sides of the equation now contain six lithium atoms and two nitrogen atoms. The balanced equation for this reaction is written as follows.

$$6\ Li(s) + N_2(g) \longrightarrow 2\ Li_3N(s)$$

The key to balancing a chemical equation is to keep exploring the equation until you have the same number of atoms of each element on both sides. As an example, let's balance the equation for the reaction that fuels the propane burners that campers use. Propane (C_3H_8) is a hydrocarbon that burns in air to form CO_2 and H_2O.

$$\underline{\quad}\ C_3H_8 + \underline{\quad}\ O_2 \longrightarrow \underline{\quad}\ CO_2 + \underline{\quad}\ H_2O$$

If you look at this equation carefully, you may conclude that it is going to be easier to balance the carbon and hydrogen atoms than the oxygen atoms in this reaction. All of the carbon atoms in propane end up in CO_2, and all of the hydrogen atoms end up in H_2O, but some of the oxygen atoms end up in each compound. This means that we can't predict how many O_2 molecules will be consumed in this reaction until we know how many CO_2 and H_2O molecules will be produced.

The most reasonable approach to balancing this equation is to start with the carbon and hydrogen atoms and leave the more difficult oxygen atoms for last. We can start by noting that there are three carbon atoms in each C_3H_8 molecule. Therefore, three CO_2 molecules are formed for every C_3H_8 molecule consumed.

$$1\ C_3H_8 + \underline{\quad}\ O_2 \longrightarrow 3\ CO_2 + \underline{\quad}\ H_2O$$

The flame of a propane torch.

Since there are eight hydrogen atoms in each C_3H_8 molecule, there must be eight hydrogen atoms, or four H_2O molecules, on the right-hand side of the equation.

$$1\ C_3H_8 + \underline{\quad}\ O_2 \longrightarrow 3\ CO_2 + 4\ H_2O$$

Now that the carbon and hydrogen atoms are balanced, we can try to balance the oxygen atoms. There are 6 oxygen atoms in 3 CO_2 molecules and 4 oxygen atoms in 4 H_2O molecules. To balance the 10 oxygen atoms among the products of this reaction we need 10 oxygen atoms, or 5 O_2 molecules, among the reactants.

$$1\ C_3H_8 + 5\ O_2 \longrightarrow 3\ CO_2 + 4\ H_2O$$

There are now 3 carbon atoms, 8 hydrogen atoms, and 10 oxygen atoms on each side of the equation. The balanced equation for this reaction is therefore is written as follows.

$$C_3H_8(g) + 5\ O_2(g) \longrightarrow 3\ CO_2(g) + 4\ H_2O(g)$$

Exercise 3.11

Write a balanced equation for the reaction that occurs when ammonia burns in air to form nitrogen oxide and water.

Solution

In this exercise we are expected to balance the following equation.

$$\underline{\quad}\ NH_3 + \underline{\quad}\ O_2 \longrightarrow \underline{\quad}\ NO + \underline{\quad}\ H_2O$$

A glance at the equation suggests that if we start with one molecule of ammonia and form one molecule of NO, the nitrogen atoms are balanced.

$$1\ NH_3 + \underline{\quad}\ O_2 \longrightarrow 1\ NO + \underline{\quad}\ H_2O$$

We can then turn to the hydrogen atoms. We have three hydrogen atoms on the left and two hydrogen atoms on the right in this equation. One way of balancing the hydrogen atoms is to look for the lowest common denominator: $2 \times 3 = 6$. Let's set up this equation so that there are six hydrogen atoms on both sides.

$$2\ NH_3 + \underline{\quad}\ O_2 \longrightarrow 1\ NO + 3\ H_2O$$

Now the hydrogen atoms are balanced, but the nitrogen atoms are not. We can correct this, however, by multiplying the nitrogen atoms on the right by 2.

$$2\ NH_3 + \underline{\quad}\ O_2 \longrightarrow 2\ NO + 3\ H_2O$$

The only task left is to balance the oxygen atoms. There are five oxygen atoms on the right side of this equation, so we need five oxygens on the left. Unfortunately, that's impossible, because the oxygen atoms come from O_2 molecules that contain two oxygen atoms each. If we insist that chemical equations work on both the atomic and macroscopic scale we must not leave the equation as follows, because there is no such thing as half an O_2 molecule.

$$2\ NH_3 + 2\tfrac{1}{2}\ O_2 \longrightarrow 2\ NO + 3\ H_2O$$

The only alternative is to multiply the entire equation by 2.

$$4\ NH_3 + 5\ O_2 \longrightarrow 4\ NO + 6\ H_2O$$

The balanced equation for this reaction can be written as follows.

$$4\ NH_3(g) + 5\ O_2(g) \longrightarrow 4\ NO(g) + 6\ H_2O(g)$$

3.14 MOLE RATIOS AND CHEMICAL EQUATIONS

Why worry about balancing chemical equations? What do we get in return that justifies the time and effort that goes into performing this task? The answer is simple. There are two fundamental goals of science: (1) explaining observations about the world around us and (2) predicting what will happen under a particular set of conditions. Any chemical equation explains something about the world, but a balanced chemical equation has the added advantage of allowing us to predict what will happen when the reaction takes place.

Let's see, for example, how a balanced chemical equation can predict the amount of O_2 consumed when one of the fireworks that brightens the sky each Fourth of July is ignited. These fireworks are based on the reaction between magnesium metal and oxygen to form magnesium oxide.

$$2\ Mg(s) + O_2(g) \longrightarrow 2\ MgO(s)$$

Let's assume that a typical flare contains 0.40 mole of magnesium metal. How much O_2 is consumed when this amount of magnesium burns?

There is a direct relationship between the number of moles of Mg and the number of moles of O_2 consumed in the reaction. The balanced equation states that 2 moles of magnesium are consumed for every mole of O_2. For this particular reaction, the following equality holds.

$$1\ mol\ O_2 = 2\ mol\ Mg$$

We ought to be able to turn this information into a unit factor similar to the unit factors introduced in Section 1.5.

Our goal is to convert 0.40 mole of Mg into an equivalent number of moles of O_2. Focusing on the units of this problem suggests that we need a factor that has units of moles of O_2 divided by moles of Mg.

$$\cancel{mol\ Mg} \times \frac{mol\ O_2}{\cancel{mol\ Mg}} = mol\ O_2$$

The following is therefore the correct unit factor for this calculation. This unit factor is known as a **mole ratio**.

$$\frac{1\ mol\ O_2}{2\ mol\ Mg} = 1$$

Multiplying 0.40 mole of magnesium by this conversion factor gives 0.20 mole of O_2.

$$0.40\ \cancel{mol\ Mg} \times \frac{1\ mol\ O_2}{2\ \cancel{mol\ Mg}} = 0.20\ mol\ O_2$$

According to this calculation, 0.20 mole of O_2 is needed to consume the 0.40 mole of magnesium in a typical flare.

Exercise 3.12

Calculate the number of moles of carbon dioxide given off when the 1.20 moles of butane (C_4H_{10}) in a can of butane lighter fluid burn.

Solution

This calculation involves two steps: writing a balanced equation for the reaction and then using the balanced equation to predict the number of moles of carbon dioxide produced when 1.20 moles of C_4H_{10} burn.

A gas burner fueled by butane.

Balancing the equation is fairly easy. Each molecule of butane contains 4 carbon atoms and 10 hydrogen atoms. For each molecule of butane consumed, 4 molecules of CO_2 and 5 molecules of H_2O are produced.

$$1 \text{ } C_4H_{10} + \underline{\hspace{1cm}} O_2 \longrightarrow 4 \text{ } CO_2 + 5 \text{ } H_2O$$

There are 13 oxygen atoms among the products of this reaction, so we need 13 oxygen atoms among the reactants. Since 13 oxygen atoms corresponds to $6\frac{1}{2}$ O_2 molecules, we multiply the equation by 2 and try again.

$$2 \text{ } C_4H_{10} + \underline{\hspace{1cm}} O_2 \longrightarrow 8 \text{ } CO_2 + 10 \text{ } H_2O$$

Now there are 26 oxygen atoms among the products, so we need 26 oxygen atoms, or 13 O_2 molecules, among the reactants.

$$2 \text{ } C_4H_{10} + 13 \text{ } O_2 \longrightarrow 8 \text{ } CO_2 + 10 \text{ } H_2O$$

According to this equation, we get 8 moles of CO_2 for every 2 moles of butane that burn. We can therefore convert 1.20 moles of butane into an equivalent number of moles of carbon dioxide as follows.

$$1.20 \text{ } \cancel{\text{mol } C_4H_{10}} \times \frac{8 \text{ mol } CO_2}{2 \text{ } \cancel{\text{mol } C_4H_{10}}} = 4.80 \text{ mol } CO_2$$

We conclude that 4.80 moles of carbon dioxide will be given off when the contents of the typical can of butane lighter fluid burn.

STOICHIOMETRY: PREDICTING THE WEIGHT OF REACTANTS
3.15 CONSUMED OR PRODUCTS GIVEN OFF IN A CHEMICAL REACTION

By now, you have encountered all the steps necessary to do calculations of the sort that are grouped under the heading **chemical stoichiometry**. The word *stoichiometry* comes from Greek stems meaning "quantities" and "to measure." The goal of these calculations is to use a balanced chemical equation to predict the relationships between the weights of the reactants and the products of a chemical reaction. All you have to do to master these calculations is learn how to put the techniques introduced in this chapter together.

To illustrate, we'll show how to predict how much oxygen we would have to inhale to digest 10.0 grams of sugar. We can assume, for the moment, that the sugar in our diet comes to us as cane sugar, $C_{12}H_{22}O_{11}$, and that our bodies burn this sugar according to the following equation.

$$C_{12}H_{22}O_{11}(s) + 12 \text{ } O_2(g) \longrightarrow 12 \text{ } CO_2(g) + 11 \text{ } H_2O(l)$$

There are many ways of doing this calculation. The following list introduces an approach based on a few relatively easy ground rules.

GROUND RULES FOR STOICHIOMETRY CALCULATIONS

1. Identify the goal of the problem.

2. Write down the key elements of the problem or draw a simple picture that summarizes the key information in the problem.

3. Try to do what can be done.

4. Don't try to do the impossible.

5. Never lose sight of your goal.

6. Don't try to work the problem in your head; write down all of the intermediate steps.

7. If you get lost, go back and read the question again.

8. Don't give up; explore the problem until you get an answer that makes sense.

Let's apply this set of rules to predicting the weight of O_2 needed to burn 10.0 grams of $C_{12}H_{22}O_{11}$. We can start by asking: "What are we trying to find?" Our goal is the following.

Goal: Find out how many grams of O_2 are consumed when 10.0 grams of cane sugar are digested.

We can then summarize the important pieces of information in the problem.

Fact: We start with 10.0 grams of cane sugar.

Fact: Cane sugar has the formula $C_{12}H_{22}O_{11}$.

Fact: The balanced equation for this reaction can be written as follows.

$$C_{12}H_{22}O_{11} + 12\ O_2 \longrightarrow 12\ CO_2 + 11\ H_2O$$

Having written down the important facts, we can turn to doing what can be done. One of the things that can be done is to convert the 10.0 grams of sugar into an equivalent number of moles.

To do this we need the molecular weight of sugar. How many significant figures should we use in this calculation? Eventually we will have to round off the answer to three significant figures, since the weight of sugar was given as 10.0 grams. If we carry an extra significant figure through the calculation, however, we can minimize the error that creeps into a calculation each time we round off. According to Exercise 3.3, the molecular weight of sugar — to four significant figures — is 342.3 amu per molecule, or 342.3 grams per mole. We can use this information to convert the 10.0 grams of sugar into moles of sugar.

$$10.0\ \text{g}\ \cancel{C_{12}H_{22}O_{11}} \times \frac{1\ \text{mol}\ C_{12}H_{22}O_{11}}{342.3\ \text{g}\ \cancel{C_{12}H_{22}O_{11}}} = 0.02921\ \text{mol}\ C_{12}H_{22}O_{11}$$

According to this calculation, 10.0 grams of sugar contains 0.02921, or 2.921×10^{-2}, mole of sugar.

Before we continue, let's take a look at the fourth of our general rules: Don't try to do the impossible. There is no way to get from moles of sugar directly to grams of oxygen in one step. So instead of trying to do this, we look for something that we can do. In other words, we continue exploring the problem.

We now have more information, so it might be a good idea to look again at what we know.

Goal: Find out how many grams of O_2 are consumed when 10.0 grams of cane sugar are digested.

Fact: We start with 10.0 grams of cane sugar.

Fact: Cane sugar has the formula $C_{12}H_{22}O_{11}$.

Fact: The balanced equation for this reaction can be written as follows.

$$C_{12}H_{22}O_{11} + 12\ O_2 \longrightarrow 12\ CO_2 + 11\ H_2O$$

Fact: We start with 0.02921 moles of sugar.

What else can we do? We have a balanced chemical equation, and we know the number of moles of sugar in the sample. Our goal is to calculate the number of grams of O_2 consumed. As a step toward that goal, we might calculate the number of moles of O_2 consumed in this reaction. According to the equation, 12 moles of O_2 are consumed for every mole of sugar consumed. We can therefore calculate the number of moles of oxygen needed to consume 0.02921 mole of sugar as follows.

$$0.02921 \text{ mol } C_{12}H_{22}O_{11} \times \frac{12 \text{ mol } O_2}{1 \text{ mol } C_{12}H_{22}O_{11}} = 0.3505 \text{ mol } O_2$$

Can we get to the goal from this stage? Yes! We know the amount of O_2 consumed in this reaction — in units of moles — and we can calculate the weight of 0.3505 mole of O_2 from the molecular weight of oxygen.

$$0.3505 \text{ mol } O_2 \times \frac{31.99 \text{ g } O_2}{1 \text{ mol } O_2} = 11.21 \text{ g } O_2$$

The goal has now been reached. According to this calculation, it takes 11.2 grams of O_2 (to three significant figures) to burn 10.0 grams of sugar.

Most students think the problem-solving process is over at this point. Chemists know it is not. It is essential to check the calculation at this stage to see whether we made any mistakes punching numbers into the calculator. There are two ways of checking a calculation. We can go back through the same mathematical operations and see whether we get the same answer. Or we can see if there is another way to solve the problem.

According to the balanced equation, it takes 12 moles, or 384 grams, of O_2 to burn 1 mole, or 342.3 grams, of sugar. Thus, it takes 1.12 times as much O_2 by weight as sugar to run this reaction. That means it takes 11.2 grams of O_2 to consume 10.0 grams of sugar.

Students often ask: "Since it is so much easier to get the answer to this problem by the shortcut used to check the answer, why didn't we use this approach to solve the problem in the first place?" If you can see a shortcut to the answer, and you are confident that it will work, try the shortcut. However, if you're not sure the shortcut can be trusted, don't use it. Solve the problem one step at a time.

Let's review the steps used to solve this problem. The first step used the molecular weight of sugar to convert grams of sugar into moles.

$$\text{Grams of sugar} \xrightarrow{\text{molecular weight of sugar}} \text{moles of sugar}$$

The next step used the balanced chemical equation to convert moles of sugar into moles of oxygen.

$$\text{Moles of sugar} \xrightarrow{\text{balanced chemical equation}} \text{moles of oxygen}$$

The last step used the molecular weight of oxygen to convert moles of oxygen into grams.

$$\text{Moles of oxygen} \xrightarrow{\text{molecular weight of oxygen}} \text{grams of oxygen}$$

The problem was therefore solved by the following sequence of steps, which can be used as a road map, or algorithm, for solving similar problems.

$$\text{Grams of } A \xrightarrow{\text{MW of } A} \text{moles of } A \xrightarrow{\text{equation}} \text{moles of } B \xrightarrow{\text{MW of } B} \text{grams of } B$$

If the road map doesn't make sense at first, or if you can't see how to apply it to a problem, you can always go back to the general rules outlined in this section and solve the problem one step at a time.

Exercise 3.13

Calculate the weight of ammonia and oxygen needed to prepare 3.00 grams of nitrogen oxide by the following reaction.

$$4\ NH_3(g) + 5\ O_2(g) \longrightarrow 4\ NO(g) + 6\ H_2O(g)$$

Solution

Our goal is to calculate the weight of ammonia and oxygen needed to prepare 3.00 grams of NO. By now, the first step in this calculation should be almost automatic. We start by converting 3.00 grams of NO into an equivalent number of moles of this compound. To do this, we need to calculate the molecular weight of NO, which is 30.0 amu per molecule, or 30.0 grams per mole.

$$1\ mol\ NO = 30.0\ g\ NO$$

The number of moles of NO formed in this reaction can be calculated as follows.

$$3.00\ \cancel{g\ NO} \times \frac{1\ mol\ NO}{30.0\ \cancel{g\ NO}} = 0.100\ mol\ NO$$

The balanced equation for this reaction states that we need 4 moles of NH_3 and 5 moles of O_2 to produce 4 moles of NO. These mole ratios can be used to calculate the number of moles of NH_3 and O_2 needed to produce 0.100 mole of NO.

$$0.100\ \cancel{mol\ NO} \times \frac{4\ mol\ NH_3}{4\ \cancel{mol\ NO}} = 0.100\ mol\ NH_3$$

$$0.100\ \cancel{mol\ NO} \times \frac{5\ mol\ O_2}{4\ \cancel{mol\ NO}} = 0.125\ mol\ O_2$$

We can now use the molecular weights of NH_3 (17.0 grams per mole) and O_2 (32.0 grams per mole) to calculate the weight of each of these reactants consumed in this reaction.

$$0.100\ \cancel{mol\ NH_3} \times \frac{17.0\ g\ NH_3}{1\ \cancel{mol\ NH_3}} = 1.70\ g\ NH_3$$

$$0.125\ \cancel{mol\ O_2} \times \frac{32.0\ g\ O_2}{1\ \cancel{mol\ O_2}} = 4.00\ g\ O_2$$

We can check this calculation by predicting how much H_2O will be formed in the reaction. The balanced equation can be used to calculate the amount of H_2O that will be produced when 0.100 mole of NH_3 reacts with 0.125 mole of O_2.

$$0.100\ \cancel{mol\ NH_3} \times \frac{6\ mol\ H_2O}{4\ \cancel{mol\ NH_3}} = 0.150\ mol\ H_2O$$

$$0.125\ \cancel{mol\ O_2} \times \frac{6\ mol\ H_2O}{5\ \cancel{mol\ O_2}} = 0.150\ mol\ H_2O$$

Since the molecular weight of water is 18.0 grams per mole, this amount of water should weigh 2.70 grams.

$$0.150\ \cancel{mol\ H_2O} \times \frac{18.0\ g\ H_2O}{1\ \cancel{mol\ H_2O}} = 2.70\ g\ H_2O$$

Our calculations seem to be correct, because the sum of the weights of the reactants (1.70 g NH_3 + 4.00 g O_2) is equal to the sum of the weights of the products (3.00 g NO + 2.70 g H_2O).

3.16 THE NUTS AND BOLTS OF LIMITING REAGENTS

Exercise 3.13 showed that it takes 1.70 grams of ammonia and 4.00 grams of oxygen to make 3.00 grams of nitrogen oxide by the following reaction.

$$4 NH_3(g) + 5 O_2(g) \longrightarrow 4 NO(g) + 6 H_2O(g)$$

What would happen if we started with 1.70 grams of NH_3 and only 2.00 grams of O_2 (half as much O_2 as is needed to consume all of the NH_3)? The reaction would run out of O_2 before all of the NH_3 was consumed. When that happened, the reaction would stop—long before it made 3.00 grams of NO.

When there isn't enough O_2 to consume all the NH_3, the amount of O_2 limits the amount of NO that can be produced. In this example, oxygen is the *limiting reagent.* Another way of looking at the reaction between 1.70 grams of NH_3 and 2.00 grams of O_2 suggests that there is an excess of NH_3 in the system. Thus, in this example, NH_3 can be considered the *excess reagent.*

The concept of limiting reagent is important, because chemists frequently run reactions in which only a limited amount of one of the reactants is present. An analogy should help clarify what's going on in limiting reagent problems.

Let's start with exactly 10 nuts and 10 bolts, as shown in Figure 3.5. Now, let's assume that we can assemble NB "molecules" by screwing a nut (N) onto each bolt (B). How many of these molecules can we make? The answer is obvious—we can make exactly 10. After that, we run out of both nuts and bolts. Since we run out of both nuts and bolts at the same time, neither is a limiting reagent. Furthermore, since no nuts or bolts are left over, neither is present in excess.

Let's assume that we try to assemble N_2B molecules by screwing 2 nuts onto each bolt. Starting with 10 nuts and 10 bolts, we can only make 5 N_2B molecules, as shown in Figure 3.6. Since we run out of nuts, they must be the limiting reagent. Since 5 bolts are left over, they are the excess reagent.

FIG. 3.5 Starting with 10 nuts (N) and 10 bolts (B) we can make exactly 10 NB molecules, with no nuts or bolts left over.

FIG. 3.6 Starting with 10 nuts and 10 bolts, we can make only 5 N_2B molecules, and we will have 5 bolts left over. The number of N_2B molecules we can make is limited by the number of nuts. The nuts can therefore be identified as the limiting reagent in this analogy.

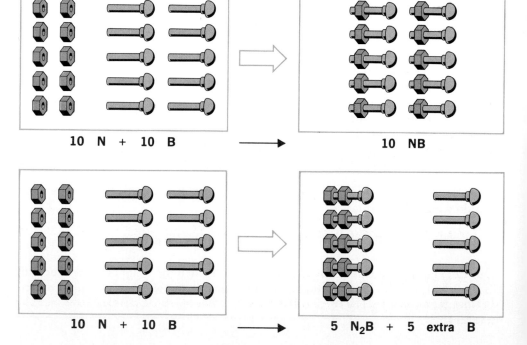

10 N + 10 B $\longrightarrow$ 10 NB

10 N + 10 B $\longrightarrow$ 5 N_2B + 5 extra B

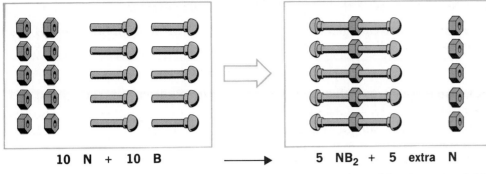

$$10 \ \text{N} \ + \ 10 \ \text{B} \ \longrightarrow \ 5 \ \text{NB}_2 \ + \ 5 \ \text{extra} \ \text{N}$$

FIG. 3.7 Starting with 10 nuts and 10 bolts, we can make only 5 NB_2 molecules, and we will have 5 nuts left over. In this case, the bolts are the limiting reagent.

It is also possible to put 2 bolts into a single nut. Starting with 10 nuts and 10 bolts, we can only make 5 NB_2 molecules, as shown in Figure 3.7. This time bolts are the limiting reagent, and nuts are present in excess.

Now let's extend this analogy to a slightly more difficult problem, in which we assemble as many N_3B molecules as possible from a collection of 1740 nuts and 747 bolts. There are three possibilities: (1) we have too many nuts and not enough bolts, (2) we have too many bolts and not enough nuts, or (3) we have just the right number of nuts and bolts.

One way to solve this problem is to pick one of these alternatives and test it. Let's assume, for the sake of argument, that we have too many nuts and not enough bolts. In other words, let's assume that bolts are the limiting reagent in this problem. Now let's test this assumption. According to the formula, N_3B, we need 3 nuts for every bolt. Thus, we need 2241 nuts to consume 747 bolts.

$$747 \ \text{bolts} \times \frac{3 \ \text{nuts}}{1 \ \text{bolt}} = \textbf{2241 nuts}$$

According to this calculation, we need 2241 nuts to use up all the bolts, and we only have 1740 nuts. Our original assumption must be wrong. We don't run out of bolts, we run out of nuts.

Since our initial assumption is wrong, let's turn it around and try again. Now let's assume that nuts are the limiting reagent and calculate the number of bolts we need.

$$1740 \ \text{nuts} \times \frac{1 \ \text{bolt}}{3 \ \text{nuts}} = \textbf{580 bolts}$$

Do we have enough bolts to use up all the nuts? Yes, we only need 580 bolts to use up the 1740 nuts, and we have 747 bolts to choose from.

Our second assumption is correct. The limiting reagent in this case is nuts, and the excess reagent is bolts. We can now calculate the number of N_3B molecules that can be assembled from 1740 nuts and 747 bolts. Since the limiting reagent is nuts, the number of nuts limits the number of N_3B molecules we can make. We get an N_3B molecule for every 3 nuts, so we can make a total of 580 of these molecules.

$$1740 \ \text{nuts} \times \frac{1 \ N_3B \ \text{molecule}}{3 \ \text{nuts}} = \textbf{580 } N_3B \textbf{ molecules}$$

The key to solving limiting reagent problems is the following sequence of steps.

STEPS TO SOLVING LIMITING REAGENT PROBLEMS

1. Recognize that you have a limiting reagent problem, or at least consider the possibility.

2. Assume that one of the reactants is the limiting reagent.

3. See if you have enough of the other reactant to consume all the material you have labeled the limiting reactant.

4. If you do, your original assumption was correct.

5. If you don't, assume that another reagent is the limiting reagent and test this assumption.

6. Once you have identified the limiting reagent, calculate the amount of product formed.

Exercise 3.14

Calculate the weight of magnesium oxide (MgO) formed when 10.0 grams of magnesium reacts with 10.0 grams of O_2.

Solution

We start, as always, by writing a balanced equation for the reaction.

$$2\ Mg(s) + O_2(g) \longrightarrow 2\ MgO(s)$$

We then pick one of the reactants and assume it is the limiting reagent. For the sake of argument, let's assume that magnesium is the limiting reagent and O_2 is present in excess. Our immediate goal is to test the validity of this assumption. If it is correct, we will have more O_2 than we need to burn 10.0 grams of magnesium. If it is wrong, O_2 is the limiting reagent.

We can start by converting grams of magnesium into moles of magnesium.

$$10.0\ \text{g Mg} \times \frac{1\ \text{mol Mg}}{24.31\ \text{g Mg}} = 0.4114\ \text{mol Mg}$$

We can then use the balanced equation to predict how many moles of O_2 will be needed to burn this much magnesium. According to the equation, it takes 1 mole of O_2 to burn 2 moles of magnesium. We therefore need 0.2057 mole of O_2 to consume all of the magnesium.

$$0.4114\ \text{mol Mg} \times \frac{1\ \text{mol }O_2}{2\ \text{mol Mg}} = 0.2057\ \text{mol }O_2$$

We can then calculate the weight of this amount of O_2.

$$0.2057\ \text{mol }O_2 \times \frac{31.99\ \text{g }O_2}{1\ \text{mol }O_2} = 6.580\ \text{g }O_2$$

According to this calculation, we need 6.580 grams of O_2 to burn all of the magnesium. Since we have 10.0 grams of O_2, our original assumption was correct. We have more than enough O_2 and only a limited amount of magnesium.

We can now calculate the amount of magnesium oxide formed when all of the magnesium is consumed. The balanced equation states that 2 moles of MgO are produced for every 2 moles of magnesium consumed.

$$0.4114\ \text{mol Mg} \times \frac{2\ \text{mol MgO}}{2\ \text{mol Mg}} = 0.4114\ \text{mol MgO}$$

We can use the weight of a mole of MgO to calculate what weight of MgO can be formed in this reaction.

$$0.4114 \cancel{\text{mol MgO}} \times \frac{40.31 \text{ g MgO}}{1 \cancel{\text{mol MgO}}} = 16.6 \text{ g MgO}$$

We can check this calculation by noting that 10.0 grams of magnesium combine with 6.6 grams of O_2 to form 16.6 grams of MgO. Weight is therefore conserved in the reaction, and we can feel confident that our calculations are correct.

3.17 SOLUTIONS AND CONCENTRATION

All the calculations done so far have been based on measurements of the weight of a sample of an element or compound. Weight is a convenient measurement for solids, which can be piled on top of a piece of weighing paper on the pan of a balance. When working with liquids or gases, however, it is often much easier to measure the volume. We therefore need a way to do the same thing with measurements of volume that we can do with weights — count the number of atoms, ions, or molecules in a sample.

For a pure liquid or gas, we can count the number of atoms, ions, or molecules in a sample by measuring the volume and density of the sample.

Exercise 3.15

Calculate to four significant figures the number of atoms in a liter of liquid mercury (density = 13.60 g/cm³).

Solution

The volume of mercury is given in liters and the density is given in grams per cubic centimeter. We can therefore start by calculating the volume of mercury in cubic centimeters.

$$1.000 \cancel{L} \times \frac{1000 \cancel{\text{mL}}}{\cancel{L}} \times \frac{1 \text{ cm}^3}{1000 \cancel{\text{mL}}} = 1000 \text{ cm}^3$$

We then combine the volume of the sample in cubic centimeters with the density in grams per cubic centimeters to calculate the weight of the sample.

$$1000 \cancel{\text{cm}^3} \times \frac{13.60 \text{ g Hg}}{1 \cancel{\text{cm}^3}} = 1.360 \times 10^4 \text{ g Hg}$$

We can now convert grams of mercury into moles of mercury.

$$1.360 \times 10^4 \cancel{\text{g Hg}} \times \frac{1 \text{ mol Hg}}{200.6 \cancel{\text{g Hg}}} = 67.80 \text{ mol Hg}$$

We then multiply by the number of atoms per mole to calculate the number of mercury atoms in a liter of liquid mercury.

$$67.80 \cancel{\text{mol}} \text{ Hg} \times 6.022 \times 10^{23} \frac{\text{atoms}}{\cancel{\text{mol}}} = 4.083 \times 10^{25} \text{ Hg atoms}$$

It is even possible to use measurements of density and volume to count atoms, ions, or molecules in a mixture, if the percent by weight of each of the components in the mixture is known.

Exercise 3.16

Concentrated sulfuric acid is 96.0% H_2SO_4 by weight. (The remaining 4.0% is water.) Calculate the number of moles of H_2SO_4 in a liter of concentrated sulfuric acid, assuming that the density of this liquid is 1.84 g/cm³.

Solution

We can best solve problems like this by keeping track of what we know and trying to relate it to the goal of the problem. What do we know about this solution?

Fact: It has a density of 1.84 g/cm³.

Fact: We have a liter of the solution.

Fact: 96.0% of the weight of the solution is sulfuric acid.

Fact: The rest of the weight is due to water.

What is the goal?

Goal: Calculate the number of moles of H_2SO_4 in a liter of this solution.

Since we know the density in units of grams per cubic centimeter, we might start by calculating the volume of the solution in cubic centimeters.

$$1.00 \cancel{L} \times \frac{1000 \cancel{mL}}{1 \cancel{L}} \times \frac{1 \text{ cm}^3}{1000 \cancel{mL}} = 1000 \text{ cm}^3$$

We can then calculate the weight of this solution.

$$1000 \cancel{cm^3} \times \frac{1.84 \text{ g}}{1 \cancel{cm^3}} = 1.84 \times 10^3 \text{ g}$$

We can use the fact that this solution is 96.0% H_2SO_4 by weight to calculate the weight of H_2SO_4 molecules in a liter of the solution.

$$1.84 \times 10^3 \text{ g} \times 96.0\% \text{ } H_2SO_4 = 1.77 \times 10^3 \text{ g } H_2SO_4$$

Once we know the weight of H_2SO_4 in the sample, we can use the molecular weight of this compound to calculate the number of moles of H_2SO_4.

$$1.77 \times 10^3 \cancel{\text{g } H_2SO_4} \times \frac{1 \text{ mol } H_2SO_4}{98.1 \cancel{\text{ g } H_2SO_4}} = 18.0 \text{ mol } H_2SO_4$$

According to this calculation, concentrated sulfuric acid contains 18.0 moles of H_2SO_4 per liter.

Another way of counting the number of atoms, ions, or molecules by measuring the volume of a mixture is based on the concept of concentration. To define concentration, it is necessary to first define three closely related terms: **solute, solvent,** and **solution.** Suppose we dissolve a solid, such as copper(II) sulfate—$CuSO_4$—in water, as shown in Figure 3.8.

The substance that dissolves ($CuSO_4$) is called the solute.

The substance in which the solute dissolves (water) is called the solvent.

The mixture of solute and solvent is called the solution.

How do we decide which substance is the solute and which is the solvent?

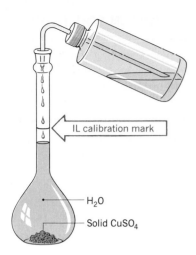

FIG. 3.8 To make a solution of $CuSO_4$ in water, we dissolve the solute ($CuSO_4$) in a solvent (H_2O).

RULES FOR IDENTIFYING SOLUTES AND SOLVENTS

1. There are three states of matter: solids, liquids, and gases. The reagent that undergoes a change in state when it forms a solution is the solute.

2. If neither reagent changes state, the reagent present in the smallest quantity is the solute.

Table 3.2 gives examples of solutions. Since solid $CuSO_4$ dissolves in water to give a liquid, $CuSO_4$ is the solute in Figure 3.8. When H_2 gas dissolves in platinum metal to form a solid solution, H_2 is the solute. When solid sodium metal reacts with liquid mercury to form a solid solution, mercury is the solute. Wine that is 12% alcohol by volume is a solution of a small quantity of alcohol (the solute) in a larger volume of water (the solvent).

The amount of solute or solvent in a solution is an extensive property of the solution (see Section 1.16). So is the amount of solution formed when the solute and solvent are mixed. However, the ratio of the amount of solute to the amount of either the solvent or the solution is an intensive property. This ratio, which is known as the **concentration** of the solution, does not depend on the size of the sample.

$$\text{Concentration} = \frac{\text{amount of solute}}{\text{amount of solvent or solution}}$$

The concept of concentration is a common one. We talk about *concentrated* orange juice, which must be *diluted* with water. We even describe certain laundry products as *concentrated,* which means that we don't have to use as much.

TABLE 3.2

Examples of Solutions

Solute	Solvent	Solution
$CuSO_4(s)$	$H_2O(l)$	$CuSO_4(aq)$
alcohol	water	wine
$Hg(l)$	$Na(s)$	$Na/Hg(s)$
$H_2(g)$	$Pt(s)$	$H_2/Pt(s)$

3.18 MOLARITY AS A WAY OF COUNTING ATOMS IN SOLUTIONS

All concentration units have one thing in common: they describe the ratio of the amount of solute to the amount of either solvent or solution. Chemists use one concentration unit more than any other — *molarity* (**M**). Molarity is defined as the number of moles of solute per liter of solution. We can calculate it by dividing the number of moles of solute in the solution by the volume of the solution in liters.

$$\text{Molarity (M)} = \frac{\text{moles of solute}}{\text{liters of solution}}$$

Exercise 3.17

Copper sulfate ($CuSO_4$) is available as blue crystals that contain water molecules trapped in holes within the crystal. Because these crystals contain five water molecules per Cu^{2+} or SO_4^{2-} ion, the compound is called a pentahydrate, and the formula is written as $CuSO_4 \cdot 5\ H_2O$. Calculate the molarity of a solution of 1.248 grams of this compound dissolved in enough water to give 50.00 milliliters of solution.

Solution

A useful problem-solving strategy involves trying to work the problem backwards. In this case, the strategy might involve looking at the goal of the problem — finding the molarity of the solution — and asking what information we need to reach this goal. The molarity of a solution is the number of moles of solute divided by the volume of the solution. Thus, we need two pieces of information to reach the goal of this exercise — the number of moles of solute and the volume of the solution in liters.

The volume of solution is easy to calculate.

$$50.00\ \cancel{mL} \times \frac{1\ L}{1000\ \cancel{mL}} = 0.05000\ L$$

The number of moles of solute can be calculated from the weight of solute used to prepare the solution and the weight of a mole of this compound.

$$1.248\ \cancel{g}\ CuSO_4 \cdot 5\ H_2O \times \frac{1\ mol}{249.68\ \cancel{g}} = 0.005000\ mol\ CuSO_4 \cdot 5\ H_2O$$

The solution therefore has a molarity of 0.1000 *M*.

$$\frac{0.005000\ mol\ CuSO_4 \cdot 5\ H_2O}{0.0500\ L} = \mathbf{0.1000\ \textit{M}\ CuSO_4}$$

The concept of molarity is important because it allows us to count the number of atoms, ions, or molecules of the solute in a solution from measurements of the volume of the solution.

Exercise 3.18

Calculate what volume of 1.50 M HCl would react with 25.0 grams of $CaCO_3$ according to the following balanced equation.

$$CaCO_3(s) + 2\ HCl(aq) \longrightarrow CaCl_2(aq) + CO_2(g) + H_2O(l)$$

Solution

We have only one piece of information about the HCl (its concentration is 1.50 M) and only one piece of information about the $CaCO_3$ (it weighs 25.0 grams). Which of these numbers do we start with? The answer is based on two points that were made in Section 3.15. Do what can be done, and don't try to do the impossible.

There is nothing we can do with the concentration of the HCl solution unless we know the number of moles of HCl consumed in the reaction or the volume of the solution. We can, however, work with the weight of $CaCO_3$ consumed in the reaction. We can start by converting grams of $CaCO_3$ into moles of $CaCO_3$.

$$25.0 \text{ g CaCO}_3 \times \frac{1 \text{ mol CaCO}_3}{100.1 \text{ g CaCO}_3} = 0.250 \text{ mol CaCO}_3$$

Now what? Let's see whether this information gets us any closer to our goal of calculating the volume of HCl consumed in this reaction. We know the number of moles of $CaCO_3$ consumed in the reaction, and we have a balanced equation that states that two moles of HCl are consumed for every mole of $CaCO_3$. We can therefore calculate the number of moles of HCl consumed in the reaction.

$$0.250 \text{ mol CaCO}_3 \times \frac{2 \text{ mol HCl}}{1 \text{ mol CaCO}_3} = 0.500 \text{ mol HCl}$$

We now have the number of moles of HCl (0.500) and the molarity of the solution (1.50). How do we put these two facts together? We can start by remembering that molarity has the units of moles per liter. The product of the molarity of a solution times its volume in liters is therefore equal to the number of moles of solute.

$$\frac{\text{mol}}{L} \times L = \text{mol}$$

We can write this as a generic equation as follows.

$$M \times V = n$$

In this equation, where M stands for the concentration of the solution in units of moles per liter, V is the volume of the solution in liters, and n is the number of moles of solute. Substituting the known values of the concentration of the HCl solution and the number of moles of HCl into this equation gives the following result.

$$1.50 \frac{\text{mol}}{L} \times V = 0.500 \text{ mol HCl}$$

This equation can be solved for the volume of the solution that would contain this amount of HCl.

$$V = 0.333 \text{ L}$$

According to this calculation, we need 333 milliliters of 1.50 M HCl to consume 25.0 grams of $CaCO_3$.

With a little imagination, we can use the concept of concentration to do far more interesting calculations.

Exercise 3.19

Calculate the atomic weight of the metal that reacts with hydrochloric acid according to the following balanced equation if 125 milliliters of 0.200-M HCl consumes 0.304 grams of the metal.

$$M(s) + 2\ HCl(aq) \longrightarrow MCl_2(aq) + H_2(g)$$

Solution

At first glance, it seems that we don't have enough information to solve this problem. The only way to proceed with a question like this is to follow the rough guidelines for problem solving outlined in Section 3.15. Identify what you know, do what can be done, and see where this leads you. In other words, start by exploring the problem.

What do we know?

Goal: Find the atomic weight of a metal.

Fact: The metal reacts with hydrochloric acid according to the following balanced equation.

$$M + 2\ HCl \longrightarrow MCl_2 + H_2$$

Fact: We start with 0.304 grams of the metal.

Fact: It takes 125 milliliters of 0.200-M HCl to consume the metal.

What can we do? We know the volume (125 milliliters) and the concentration (0.200 M) of a solution. We might therefore start by calculating the number of moles of solute in this solution.

$$\frac{0.200\ mol\ HCl}{\cancel{L}} \times 0.125\ \cancel{L} = 0.0250\ mol\ HCl$$

Now what? We know the number of moles of HCl, and we have a balanced equation. Furthermore, we are interested in one of the properties of the metal. It seems reasonable to convert moles of HCl consumed in this reaction into moles of metal consumed.

$$0.0250\ \cancel{mol\ HCl} \times \frac{1\ mol\ M}{2\ \cancel{mol\ HCl}} = 0.0125\ mol\ M$$

It is important never to lose sight of the goal of the problem. In this case, the problem asks for the atomic weight of the metal. It might be useful to go to the end of the problem and work backward. Atomic weight has units of amu per atom or grams per mole. If we knew both the number of grams and the number of moles of metal in a sample, the ratio of these numbers would be the atomic weight.

But we already have that information. We know the number of moles of metal (0.0125 mole) in a sample of known weight (0.304 gram). The ratio of these numbers is the atomic weight of the metal.

$$\frac{0.304\ g\ M}{0.0125\ mol\ M} = 24.3\ g/mol$$

By looking at a table of atomic weights we can deduce that the metal must be magnesium.

Once you understand the exercises in this section, you can use an equation or algorithm to solve similar problems. The algorithm states that molarity (M) is defined as the number of moles of solute (n) divided by the volume of the solution in liters (V).

$$M = \frac{n}{V}$$

Therefore, the product of the molarity of the solution (M) times the volume in liters (V) must be equal to the number of moles of solute (n).

$$M \times V = n$$

3.19 DILUTION AND TITRATION CALCULATIONS

Adding more solvent to a solution to make the concentration smaller is known as *dilution*. Starting with a known volume of a solution of known molarity, we should be able to prepare a more dilute solution of any desired concentration.

Exercise 3.20

Describe how you would prepare 2.50 liters of a 0.360-M solution of sulfuric acid starting with concentrated sulfuric acid that is 18.0 M.

Solution

We have one piece of information about the concentrated H_2SO_4 solution (the concentration is 18.0 moles per liter) and two pieces of information about the dilute solution (the volume is 2.50 liters and the concentration is 0.360 mole per liter). It seems reasonable to start with the solution about which we know the most.

Since we know the concentration (0.360 M) and the volume (2.50 liters) of the dilute H_2SO_4 solution we are trying to prepare, we can calculate the number of moles of H_2SO_4 it contains.

$$0.360 \; \frac{\text{mol } H_2SO_4}{\cancel{L}} \times 2.50 \; \cancel{L} = 0.900 \; \text{mol } H_2SO_4$$

All of the H_2SO_4 molecules in the dilute solution come from the concentrated solution. What volume of concentrated H_2SO_4 contains this many moles of H_2SO_4? We can start with the equation that describes the relationship between the molarity (M) of a solution, the volume of the solution (V), and the number of moles of solute in the solution (n).

$$M \times V = n$$

We can then substitute into this equation the molarity of the concentrated sulfuric acid solution and the number of moles of H_2SO_4 molecules needed to prepare the dilute solution.

$$18.0 \; \frac{\text{mol } H_2SO_4}{L} \times V = 0.900 \; \text{mol } H_2SO_4$$

We can then solve for the volume of this solution.

$$V = 0.0500 \; L$$

According to this calculation, we can prepare 2.50 liters of a 0.360-M H_2SO_4 solution by adding 50.0 milliliters of concentrated sulfuric acid to enough water to give a total volume of 2.50 liters.

We can measure the concentration of a solution by a technique known as *titration.* The solution being studied is slowly added to a known quantity of a reagent, with which it reacts until an *indicator* tells us that exactly equivalent numbers of moles of the reagents are present.

Phenolphthalein is an example of an indicator for acid-base reactions. When added to solutions that contain an acid, phenolphthalein is colorless. In the presence of a base, it has a light pink color. Phenolphthalein can therefore indicate when exactly enough base has been added to consume all of the acid in a solution, because it turns from colorless to pink just after the point when equivalent numbers of moles of acid and base have been added to the solution.

We can illustrate the calculations behind a titration experiment by looking at the reaction between oxalic acid ($H_2C_2O_4$) and sodium hydroxide (NaOH).

$$H_2C_2O_4(aq) + 2\,NaOH(aq) \longrightarrow Na_2C_2O_4(aq) + 2\,H_2O(l)$$

Exercise 3.21

Calculate the concentration of an oxalic acid solution if a titration experiment shows that it takes 34.0 milliliters of a 0.200 M NaOH solution to consume all of the oxalic acid in a 25.0-milliliter sample.

Solution

Which solution should we start with, NaOH or oxalic acid? We only know the volume of the oxalic acid solution, but we know both the volume and the concentration of the NaOH solution. Thus, it seems reasonable to start by calculating the number of moles of NaOH in this solution.

$$0.200\,\frac{\text{mol NaOH}}{\text{L}} \times 0.0340\,\text{L} = 6.80 \times 10^{-3}\,\text{mol NaOH}$$

Now what? We have a balanced chemical equation for the reaction that occurs when the two solutions are mixed. We also know the number of moles of NaOH consumed in the reaction. We can therefore calculate the number of moles of $H_2C_2O_4$ needed to consume this much NaOH.

$$6.80 \times 10^{-3}\,\text{mol NaOH} \times \frac{1\,\text{mol } H_2C_2O_4}{2\,\text{mol NaOH}} = 3.40 \times 10^{-3}\,\text{mol } H_2C_2O_4$$

We now know the number of moles of $H_2C_2O_4$ in the original oxalic acid solution, and we also know the volume of this solution. We can therefore calculate the number of moles of oxalic acid per liter, or the molarity of the solution.

$$\frac{3.40 \times 10^{-3}\,\text{mol } H_2C_2O_4}{0.0250\,\text{L}} = 0.136\,\text{M } H_2C_2O_4$$

The concentration of the oxalic acid solution is 0.136 mole per liter.

Titration involves adding a solution of one reagent to another until exactly equivalent numbers of moles of the two reagents are present in solution. One of the indicators commonly used in acid-base titrations is phenolphthalein, which turns from colorless to pink in the presence of excess base.

SUMMARY

Atoms are so infinitesimally small that we can't measure the absolute weight of a single atom. All we can determine is the relative mass of one atom compared with another. The mass of an atom is reported in atomic mass units, or amu, a scale on which the lightest atom is defined as having a mass of 1 amu. The atoms of an element do not all have the same mass. Because isotopes exist, the average mass of the atoms of an element — the atomic weight — is different from the mass of a single atom of the element.

Atoms are so small they can't be counted individually. We therefore need a unit that describes a collection of atoms large enough to be visible to the naked eye — the mole. By convention, the mole is defined as the number of atoms whose weight in grams is equal to the atomic weight in amu. If the atomic weight of carbon is 12.011 amu, a mole of carbon atoms weighs 12.011 grams. The concept of a mole can also be applied to other elementary particles, such as molecules. If the molecular weight of a CO_2 molecule is 44.01 amu, then a mole of CO_2 weighs 44.01 grams.

The concept of the mole allows us to predict the relationship between the weights of the reactants and products of a chemical reaction. We can convert grams of one of the reactants into moles of that reactant. We can then use a balanced equation to convert from moles of that reactant to moles of another reactant or moles of a product. We can also use the mole concept to convert from moles of the second reactant or product into grams of this material.

The mole concept allows us to count atoms, ions, or molecules in a sample by measuring the weight of the sample. Weight is a convenient property for measuring the size of a solid. But volume is a more convenient measurement for gases and liquids. The concept of concentration — especially the particular unit of concentration known as the molarity of a solution — enables us to count atoms, ions, or molecules in a solution from measurements of the volume of the solution.

PROBLEMS

The Relative Masses of Atoms

3-1 Calculate what the atomic weight of neon would be if the atomic mass scale were redefined so that ^{12}C atoms had a mass of exactly 1.000... amu.

3-2 Identify the element that has an atomic weight 4.33 times as large as that of carbon.

3-3 Identify the element that contains atoms that have an average weight of 4.48×10^{-23} grams.

3-4 By weight, naturally occurring bromine is 50.69% ^{79}Br atoms, which have a mass of 78.9183 amu, and 49.31% ^{81}Br atoms, which have a mass of 70.9163 amu. Calculate the atomic weight of bromine.

3-5 Naturally occurring zinc is 48.6% ^{64}Zn atoms (63.9291 amu), 27.9% ^{66}Zn atoms (65.9260 amu), 4.1% ^{67}Zn atoms (66.9721 amu), 18.8% ^{68}Zn atoms (67.9249 amu), and 0.6% ^{70}Zn atoms (69.9253 amu). Calculate the atomic weight of zinc.

The Mole as a Collection of Atoms

3-6 If the average chromium atom has a mass of 51.996 amu, what is the weight of a mole of chromium atoms?

3-7 If the average sulfur atom is approximately twice as heavy as the average oxygen atom, what is the ratio of the weight of a mole of sulfur atoms to the weight of a mole of oxygen atoms?

3-8 Calculate the weight in grams of a mole of atoms of each of the following elements.

(a) C (b) P (c) Ni (d) Hg (e) U

3-9 If oranges sell for 50¢ each, what is the cost of a bag of 24 oranges? If eggs sell for 90¢ a dozen, what does it cost to buy 4 dozen eggs? If a mole of carbon atoms weighs 12.011 grams, what is the weight of 2.5 moles of carbon atoms?

The Mole as a Collection of Molecules

3-10 Which pair of samples contains the same number of hydrogen atoms?

(a) 1 mole of NH_3 and 1 mole of N_2H_4 (b) 2 moles of NH_3 and 1 mole of N_2H_4 (c) 2 moles of NH_3 and 3 moles of N_2H_4 (d) 4 moles of NH_3 and 3 moles of N_2H_4

3-11 Calculate the molecular weights of formic acid, HCO_2H, and formaldehyde, H_2CO.

3-12 Calculate the molecular weights of the following compounds.

(a) methane, CH_4 (b) glucose, $C_6H_{12}O_6$ (c) ether, $(CH_3CH_2)_2O$ (d) thioacetamide, CH_3CSNH_2

3-13 Calculate the molecular weights of the following compounds.

(a) tetraphosphorus decasulfide, P_4S_{10} (b) nitrogen dioxide, NO_2 (c) zinc sulfide, ZnS (d) potassium permanganate, $KMnO_4$

3-14 Calculate the molecular weights of the following compounds.

(a) chromium hexacarbonyl, $Cr(CO)_6$ (b) iron(III) nitrate, $Fe(NO_3)_3$ (c) potassium dichromate, $K_2Cr_2O_7$ (d) calcium phosphate, $Ca_3(PO_4)_2$

3-15 Root beer hasn't tasted the same since the FDA outlawed the use of sassafras oil as a food additive because sassafras oil is 80% safrole, and safrole has been shown to cause cancer in rats and mice. Calculate the molecular weight of safrole, $C_{10}H_{10}O_2$.

3-16 Monosodium glutamate, or MSG, is a spice commonly used in Chinese cooking that causes some people to feel lightheaded (a disorder commonly known as Chinese restaurant syndrome). Calculate the molecular weight of MSG, $C_5H_8NNaO_4$.

3-17 Calculate the molecular weights of the active ingredients in the following prescription drugs.

(a) Darvon, $C_{22}H_{30}ClNO_2$ (b) Valium, $C_{16}H_{13}ClN_2O$ (c) Antabuse, $C_{10}H_{20}N_2S_4$ (d) Tetracycline, $C_{22}H_{24}N_2O_8$

Using Measurements of Mass to Count Atoms and Molecules

3-18 Calculate the number of moles of atoms in each of the following samples.

(a) 12.011 grams of carbon (b) 32.06 grams of sulfur
(c) 63.546 grams of copper (d) 207.2 grams of lead

3-19 Calculate the number of moles of atoms in each of the following samples.

(a) 48.1 grams of chromium (b) 15.8 grams of boron
(c) 1.803 grams of iron (d) 526 grams of arsenic

3-20 Calculate the number of moles of carbon in 0.244 gram of calcium carbide, CaC_2.

3-21 Calculate the number of moles of phosphorus in 15.95 grams of tetraphosphorus decaoxide, P_4O_{10}.

3-22 Calculate the atomic weight of platinum, assuming 0.8170 mole of this metal weighs 159.4 grams.

3-23 Calculate the weight of 2.587 moles of sulfur hexafluoride, SF_6.

3-24 Calculate the weight of 0.0582 mole of carbon tetrachloride, CCl_4.

3-25 Calculate the density of carbon tetrachloride, assuming 1 mole of CCl_4 occupies a volume of 96.94 cubic centimeters.

3-26 Calculate the volume of a mole of aluminum, assuming the density of this metal s 2.70 g/cm^3.

Empirical Formulas

3-27 Stannous fluoride, or "Fluoristan," is added to toothpaste to help prevent tooth decay. What is the empirical formula for stannous fluoride if this compound is 24.25% F and 75.75% Sn by weight?

3-28 Iron reacts with oxygen to form three compounds: FeO, Fe_2O_3, and Fe_3O_4. One of these compounds, known as magnetite, is 72.36% Fe and 27.64% O by weight. What is the formula of magnetite?

3-29 The chief ore of manganese is an oxide known as pyrolusite, which is 36.8% O and 63.2% Mn by weight. Which of the following oxides of manganese is pyrolusite?

(a) MnO (b) MnO_2 (c) Mn_2O_3 (d) MnO_3 (e) Mn_2O_7

3-30 Nitrogen combines with oxygen to form a variety of compounds, including N_2O, NO, NO_2, N_2O_3, N_2O_4, and N_2O_5. One of these compounds is called nitrous oxide, or "laughing gas." What is the formula of nitrous oxide if this compound is 63.65% N and 36.35% O by weight?

3-31 Chalcopyrite is a bronze-colored mineral that is 34.67% Cu, 30.43% Fe, and 34.94% S by weight. Calculate the empirical formula for this mineral.

3-32 Calculate the atomic weight of the metal that forms a compound with the formula MCl_2 that is 74.5% Cl by weight.

3-33 A compound of xenon and fluorine is found to be 53.5% xenon by weight. What is the oxidation number of the xenon atom in this compound?

3-34 In 1914, E. Merck and Company synthesized and patented a compound known as MDMA as an appetite suppressant. Although it was never marketed, it has reappeared in recent years as a street drug known as Ecstasy. What is the empirical formula of this compound if it contains 68.4% C, 7.8% H, 7.2% N, and 16.6% O by weight?

3-35 A compound that is 31.9% K and 28.9% Cl by weight decomposes when heated to give O_2 and a compound that is 52.4% K and 47.6% Cl by weight. Write a balanced chemical equation for this reaction.

Elemental Analysis: Percent-by-Weight Calculations

3-36 Calculate the percent by weight of chromium in each of the following oxides.

(a) CrO (b) Cr_2O_3 (c) CrO_3

3-37 Calculate the percent by weight of nitrogen in the following fertilizers.

(a) $(NH_4)_2SO_4$ (b) KNO_3 (c) $NaNO_3$ (d) $(H_2N)_2CO$

3-38 Calculate the percent by weight of carbon, hydrogen, and chlorine in DDT, $C_{14}H_9Cl_5$.

3-39 Emeralds are gem-quality forms of the mineral beryl, $Be_3Al_2(SiO_3)_6$. Calculate the percent by weight of silicon in beryl.

3-40 Osteoporosis is a disease common in older women who have not had enough calcium in their diets. Calcium can be added to the diet by tablets that contain either calcium carbonate, $CaCO_3$; calcium sulfate, $CaSO_4$; or calcium phosphate, $Ca_3(PO_4)_2$. Which is the most efficient way of getting Ca^{2+} ions into the body?

Elemental Analysis: Experimental Results

3-41 The methane in natural gas, the propane used in camping stoves, and the butane used in butane lighters are all members of a family of compounds known as the alkanes, which have the generic formula C_nH_{2n+2}. Calculate the value of n for butane, assuming 0.3149 gram of butane burns in air to form 0.9537 gram of CO_2 and 0.4881 gram of H_2O.

3-42 In small quantities, the nicotine in tobacco is addictive. In large quantities, it is a deadly poison. Calculate the molecular formula of nicotine, $C_xH_yN_z$, assuming that its molecular weight is 162.2 grams per mole and that 0.438 gram of this compound burns to form 1.188 grams of CO_2 and 0.341 gram of water.

3-43 How accurately would you have to measure the percent by weight of carbon and hydrogen to tell the difference between diazepam (Valium), with the formula $C_{16}H_{13}ClN_2O$, and chlordiazepoxide (Librium), with the formula $C_{16}H_{14}ClN_3O$?

Empirical and Molecular Formulas

3-44 β-Carotene is the protovitamin from which nature builds vitamin A. It is widely distributed in the plant and animal kingdoms, always occurring in plants together with chlorophyll. Calculate the molecular formula for β-carotene, assuming that this compound is 89.49% C and 10.51% H by weight and that its molecular weight is 536.85 grams per mole.

3-45 The phenolphthalein used as an indicator in acid-base titrations is also the active ingredient in laxatives such as Ex-Lax. Calculate the molecular formula for phenolphthalein, assuming that this compound is 75.46% C, 4.43% H, and 20.10% O by weight and that its molecular weight is 222.4 grams per mole.

3-46 Caffeine is a central nervous system stimulant found in coffee, tea, and cola nuts. Calculate the molecular formula of caffeine, assuming that this compound is 49.48% C, 5.19% H, 28.85% N, and 16.48% O by weight and that its molecular weight is 194.2 grams per mole.

3-47 Aspartame, also known as Nutra-Sweet, is 160 times sweeter than cane sugar when dissolved in water. The true name for this artificial sweetener is N-L-α-aspartyl-L-phenylalanine methyl ester. Calculate the molecular formula of aspartame, assuming that this compound is 57.14% C, 6.16% H, 9.52% N, and 27.18% O by weight and that its molecular weight is 294.30 grams per mole.

3-48 Halothane is an anaesthetic that is 12.17% C, 0.51% H, 40.48% Br, 17.96% Cl, and 28.87% F by weight. What is the molecular formula of this compound if each molecule contains one hydrogen atom?

Avogadro's Constant

3-49 Calculate the number of atoms in 16 grams of O_2, 31 grams of P_4, and 32 grams of S_8.

3-50 Calculate the number of iron atoms in a paper clip that weighs 0.08 gram.

3-51 Calculate the number of chlorine atoms in 0.756 gram of K_2PtCl_6.

3-52 Calculate the number of H_2O_2 molecules in 10.0 grams of hydrogen peroxide.

3-53 Calculate the number of oxygen atoms in each of the following.

(a) 0.100 mole of potassium permanganate, $KMnO_4$
(b) 0.25 mole of dinitrogen pentoxide, N_2O_5 (c) 0.45 mole of penicillin, $C_{16}H_{17}N_2O_5SK$

3-54 Benzaldehyde has the pleasant, distinctive odor of almonds. What is the molecular weight of benzaldehyde if a single molecule weighs 1.762×10^{-22} gram.

Balancing Chemical Equations

3-55 Balance the following chemical equations.

(a) $Sn(s) + O_2(g) \rightarrow SnO_2(s)$ (b) $Cr(s) + O_2(g) \rightarrow Cr_2O_3(s)$ (c) $P_4(s) + O_2(g) \rightarrow P_4O_{10}(s)$ (d) $S_8(s) + O_2(g) \rightarrow SO_2(g)$

3-56 Balance the following chemical equations.

(a) $UH_3(s) \rightarrow U(s) + H_2(g)$ (b) $SiH_4(g) \rightarrow Si(s) + H_2(g)$ (c) $SO_3(g) \rightarrow SO_2(g) + O_2(g)$

3-57 Balance the following chemical equations.

(a) $ZnCO_3(s) \rightarrow ZnO(s) + CO_2(g)$ (b) $Pb(NO_3)_2(s) \rightarrow PbO(s) + NO_2(g)$ (c) $NH_4NO_2(s) \rightarrow N_2(g) + H_2O(g)$ (d) $(NH_4)_2Cr_2O_7(s) \rightarrow N_2(g) + Cr_2O_3(s)$

3-58 Balance the following chemical equations.

(a) $CH_4(g) + O_2(g) \rightarrow CO_2(g) + H_2O(g)$ (b) $H_2S(g) + O_2(g) \rightarrow H_2O(g) + SO_2(g)$ (c) $CS_2(l) + O_2(g) \rightarrow CO_2(g) + SO_2(g)$ (d) $B_5H_9(g) + O_2(g) \rightarrow B_2O_3(s) + H_2O(g)$

3-59 Balance the following chemical equations.

(a) $Na(s) + H_2O(l) \rightarrow Na^+(aq) + OH^-(aq) + H_2(g)$
(b) $PF_3(g) + H_2O(l) \rightarrow H_3PO_3 + HF(aq)$ (c) $Na_2O(s) + H_2O(l) \rightarrow Na^+(aq) + OH^-(aq)$ (d) $P_4O_{10}(s) + H_2O(l) \rightarrow H_3PO_4(aq)$

3-60 Balance the following chemical equations.

(a) $C_3H_8(g) + O_2(g) \rightarrow CO_2(g) + H_2O(g)$ (b) $C_4H_{10}(g) + O_2(g) \rightarrow CO_2(g) + H_2O(g)$ (c) $C_2H_5OH(l) + O_2(g) \rightarrow CO_2(g) + H_2O(g)$ (d) $C_6H_{12}O_6(s) + O_2(g) \rightarrow CO_2(g) + H_2O(l)$

Mole Ratios and Chemical Equations

3-61 Carbon disulfide burns in oxygen to form carbon dioxide and sulfur dioxide.

$$CS_2(l) + 3 O_2(g) \longrightarrow CO_2(g) + 2 SO_2(g)$$

Calculate the number of O_2 molecules it would take to consume 500 molecules of CS_2. Calculate the number of moles of O_2 molecules it would take to consume 5.00 moles of CS_2.

3-62 Calculate the number of moles of oxygen produced when 6.75 moles of manganese dioxide decomposes to Mn_3O_4 and O_2.

$$3 MnO_2(s) \longrightarrow Mn_3O_4(s) + O_2(g)$$

3-63 Calculate the number of moles of oxygen consumed when 2.50 moles of magnesium burns in air.

$$2 Mg(s) + O_2(g) \longrightarrow 2 MgO(s)$$

3-64 Calculate the number of moles of carbon monoxide needed to reduce 3.00 moles of iron(III) oxide to iron metal.

$$Fe_2O_3(s) + 3 CO(g) \longrightarrow 2 Fe(s) + 3 CO_2(g)$$

Stoichiometry

3-65 Calculate the weight of oxygen released when enough mercury(II) oxide decomposes to give 25 grams of liquid mercury.

$$2 HgO(s) \longrightarrow 2 Hg(l) + O_2(g)$$

3-66 Calculate the weight of CO_2 produced and the weight of oxygen consumed when 10.0 grams of methane, CH_4, is burned.

3-67 Calculate the weight of sulfur that will react with 10.0 pounds of zinc to form zinc sulfide, ZnS.

3-68 Calculate the weight of oxygen that can be prepared by the decomposition of 25.0 grams of potassium chlorate.

$$2 KClO_3(s) \longrightarrow 2 KCl(s) + 3 O_2(g)$$

3-69 Calculate the number of water molecules in the formula for the hydrate of copper sulfate, assuming that 2.47 grams of $CuSO_4$ picks up water to form 3.86 grams of a compound with the formula $CuSO_4 \cdot x\ H_2O$.

3-70 Predict the formula of the compound produced when 1.00 gram of chromium metal reacts with 0.923 gram of oxygen.

3-71 Calculate the weight of oxygen that would be consumed when 0.985 gram of hydrogen sulfide, H_2S, burns to form SO_2 and H_2O.

3-72 Ethanol, or ethyl alcohol, is produced by the fermentation of sugars such as glucose.

$$C_6H_{12}O_6(aq) \longrightarrow 2 C_2H_5OH(aq) + 2 CO_2(g)$$

Calculate the number of pounds of alcohol that can be produced from a pound of glucose.

3-73 Calculate the number of pounds of aluminum metal that can be obtained from a ton of bauxite, $Al_2O_3 \cdot 2\ H_2O$.

3-74 A 3.500-gram sample of an oxide of manganese contains 1.288 grams of oxygen. What is the empirical formula of this compound?

3-75 Calculate what weight of phosphine, PH_3, can be prepared when 10.0 grams of calcium phosphide, Ca_3P_2, reacts with water.

$$Ca_3P_2(s) + 6 H_2O(l) \longrightarrow 3 Ca(OH)_2(aq) + 2 PH_3(g)$$

3-76 Hydrogen chloride can be made by reacting phosphorus trichloride with water and then boiling the HCl gas out of the solution.

$$PCl_3(g) + 3 H_2O(l) \longrightarrow 3 HCl(aq) + H_3PO_3(aq)$$

Calculate the weight of HCl gas that can be prepared from 15.0 grams of PCl_3.

3-77 Nitrogen reacts with red-hot magnesium to form magnesium nitride

$$3 Mg(s) + N_2(g) \longrightarrow Mg_3N_2(s)$$

which reacts with water to form magnesium hydroxide and ammonia.

$$Mg_3N_2(s) + 6 H_2O(l) \longrightarrow 3 Mg(OH)_2(aq) + 2 NH_3(aq)$$

Calculate the number of grams of magnesium that would be needed to prepare 15.0 grams of ammonia.

3-78 Nitrogen reacts with hydrogen to form ammonia

$$N_2(g) + 3 H_2(g) \longrightarrow 2 NH_3(g)$$

which burns in the presence of oxygen to form nitrogen oxide

$$4 NH_3(g) + 5 O_2(g) \longrightarrow 4 NO(g) + 6 H_2O(g)$$

which reacts with excess oxygen to form nitrogen dioxide

$$2 NO(g) + O_2(g) \longrightarrow 2 NO_2(g)$$

which dissolves in water to give nitric acid.

$$3 NO_2(g) + H_2O(l) \longrightarrow 2 HNO_3(aq) + NO(g)$$

Calculate the weight of nitrogen needed to make 15 moles of nitric acid.

Limiting Reagents

3-79 Calculate the number of water molecules that can be prepared from 500 H_2 molecules and 500 O_2 molecules.

$$2 H_2(g) + O_2(g) \longrightarrow 2 H_2O(l)$$

What would happen to the potential yield of water molecules if the amount of O_2 was doubled? What if the amount of H_2 was doubled?

3-80 Calculate the number of moles of P_4S_{10} that can be produced from 0.500 mole of P_4 and 0.500 mole of S_8.

$$4\ P_4(s) + 5\ S_8(s) \longrightarrow 4\ P_4S_{10}(s)$$

What would happen to the potential yield of P_4S_{10} if the amount of P_4 was doubled? What if the amount of S_8 was doubled?

3-81 Calculate the number of moles of nitrogen dioxide that could be prepared from 0.35 mole of nitrogen oxide and 0.25 mole of oxygen.

$$2\ NO(g) + O_2(g) \longrightarrow 2\ NO_2(g)$$

Identify the limiting reagent and the excess reagent in this reaction. What would happen to the potential yield of NO_2 if the amount of NO was doubled? What if the amount of O_2 was doubled?

3-82 Calculate the weight of hydrogen chloride that can be produced from 10.0 grams of hydrogen and 10.0 grams of chlorine.

$$H_2(g) + Cl_2(g) \longrightarrow 2\ HCl(g)$$

What would have to be done to increase the amount of hydrogen chloride produced in this reaction?

3-83 Calculate the weight of calcium nitride, Ca_3N_2, that can be prepared from 54.9 grams of calcium and 43.2 grams of nitrogen.

$$3\ Ca(s) + N_2(g) \longrightarrow Ca_3N_2(s)$$

3-84 Trimethyl aluminum, $Al(CH_3)_3$, must be handled in an apparatus from which oxygen has been rigorously excluded, because it bursts into flame in the presence of oxygen. Calculate the weight of trimethyl aluminum that can be prepared from 5.00 grams of aluminum metal and 25.0 grams of dimethyl mercury.

$$2\ Al + 3\ Hg(CH_3)_2 \longrightarrow 2\ Al(CH_3)_3 + 3\ Hg$$

3-85 The thermite reaction, used to weld rails together in the building of railroads, is described by the following equation.

$$Fe_2O_3(s) + 2\ Al(s) \longrightarrow Al_2O_3(s) + 2\ Fe(l)$$

Calculate the weight of molten iron that can be prepared from 150 grams of aluminum and 250 grams of iron(III) oxide.

Solutions and Concentration

3-86 Silver chloride is only marginally soluble in water. Only 0.00019 gram of AgCl dissolves in 100 mL of water. Calculate the molarity of this solution.

3-87 Ammonia (NH_3) is much more soluble in water. Calculate the molarity of a solution that contains 252 grams of NH_3 per liter.

3-88 Ethyl alcohol, or ethanol, C_2H_5OH, is infinitely soluble in water. The concentration of a solution of this alcohol in water is often expressed in units of "proof." Pure alcohol is 200 proof, and a 50:50 mixture is 100 proof. Calculate the molar-ity of a 90-proof solution of ethyl alcohol in water. Assume that the density of this solution is 0.894 g/cm^3.

3-89 During a recent physical exam, one of the authors was found to have a cholesterol level of 160 milligrams per deciliter. If the molecular weight of cholesterol is 386.67 grams per mole and a deciliter is 1/10 of a liter, what is the cholesterol level in his blood in units of moles per liter?

3-90 Concentrated hydrochloric acid is 38.0% HCl by weight. Calculate the molarity of this solution if it has a density of 1.1977 g/cm^3.

3-91 At 25°C, 5.77 grams of chlorine gas dissolves in a liter of water. Calculate the molarity of Cl_2 in this solution.

3-92 Calculate the weight of Na_2SO_4 needed to prepare 0.500 liter of an 0.150 M solution.

3-93 Calculate the weight of the $Cu(NH_3)_4{}^{2+}$ ion in 54.6 milliliters of a 0.238 M solution.

3-94 Calculate the concentration of an aqueous KCl solution if 25.00 milliliters of this solution gives 0.430 gram of AgCl when treated with excess $AgNO_3$.

$$KCl(aq) + AgNO_3(aq) \longrightarrow AgCl(s) + KNO_3(aq)$$

3-95 Calculate the volume of 0.25 M NaI that would be needed to precipitate all of the Hg^{2+} ion from 45 milliliters of a 0.10 M $Hg(NO_3)_2$ solution.

$$2\ NaI(aq) + Hg(NO_3)_2(aq) \longrightarrow HgI_2(s) + 2\ NaNO_3(aq)$$

Dilution Calculations

3-96 Calculate the volume of 17.4 M acetic acid needed to prepare 1.00 liter of 3.00 M acetic acid.

3-97 Calculate the concentration of the solution formed when 15.0 milliliters of 6.00 M HCl is diluted with 25.0 milliliters of water.

3-98 Describe in detail the steps you would take to prepare 125 milliliters of 0.745 M oxalic acid, $H_2C_2O_4 \cdot 2\ H_2O$. Describe the glassware you would need, the chemicals, the amounts of each chemical, and the sequence of steps you would take.

3-99 Describe how you would prepare 0.200 liter of 1.25 M nitric acid from a solution that is 5.94 M HNO_3.

Titration Calculations

3-100 Calculate the molarity of an acetic acid solution, assuming 34.57 milliliters of this solution is needed to neutralize 25.19 milliliters of 0.1025 M sodium hydroxide.

$$CH_3CO_2H(aq) + NaOH(aq) \longrightarrow$$
$$Na^+(aq) + CH_3CO_2{}^-(aq) + H_2O(l)$$

3-101 Calculate the molarity of a sodium hydroxide solution, assuming 10.42 milliliters of this solution is needed to neutralize 25.00 milliliters of 0.2043 M oxalic acid.

$$H_2C_2O_4(aq) + 2\ NaOH(aq) \longrightarrow Na_2C_2O_4(aq) + 2\ H_2O$$

3-102 Calculate the volume of 0.0985 M sulfuric acid that would be needed to neutralize 10.89 milliliters of a 0.01043 M aqueous ammonia solution.

$$H_2SO_4(aq) + 2\ NH_3(aq) \longrightarrow (NH_4)_2SO_4(aq)$$

3-103 α-D-Glucopyranose reacts with the periodate ion as follows.

$$C_6H_{12}O_6(aq) + 5\ IO_4^-(aq) \longrightarrow$$
$$5\ IO_3^-(aq) + 5\ HCO_2H(aq) + H_2CO(aq)$$

Calculate the molarity of the glucopyranose solution, assuming 25.0 milliliters of 0.750 M IO_4^- is required to consume 10.0 milliliters of this solution.

3-104 Oxalic acid reacts with the chromate ion in acidic solution as follows.

$$3\ H_2C_2O_4(aq) + 2\ CrO_4^{2-}(aq) + 10\ H^+(aq) \longrightarrow$$
$$6\ CO_2(g) + 2\ Cr^{3+}(aq) + 8\ H_2O(l)$$

Calculate the molarity of the oxalic acid solution, assuming 10.0 milliliters of this solution consumes 40.0 milliliters of 0.0250 M CrO_4^{2-}.

3-105 A 2.50-gram sample of bronze was dissolved in sulfuric acid. The copper in this alloy reacted with sulfuric acid as follows.

$$Cu(s) + 2\ H_2SO_4(aq) \longrightarrow CuSO_4(aq) + SO_2(g) + 2\ H_2O(l)$$

The $CuSO_4$ formed in this reaction was mixed with KI to form CuI.

$$2\ CuSO_4(aq) + 5\ I^-(aq) \longrightarrow 2\ CuI(s) + I_3^-(aq) + 2\ SO_4^{2-}(aq)$$

The I_3^- formed in this reaction was then titrated with $S_2O_3^{2-}$.

$$I_3^-(aq) + 2\ S_2O_3^{2-}(aq) \longrightarrow 3\ I^-(aq) + S_4O_6^{2-}(aq)$$

Calculate the percent by weight of copper in the original sample if 31.5 milliliters of 1.00 M $S_2O_3^{2-}$ was consumed in this titration.

GASES

FIG. 4.1 Each of the three states of matter has characteristic properties. Solids have a distinct shape. When they melt, the resulting liquid conforms to the shape of its container. Gases expand to fill their container.

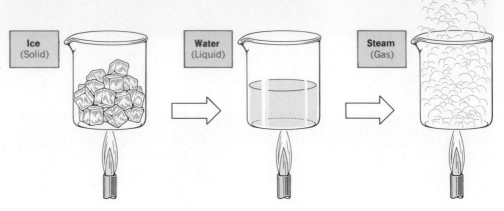

One way of defining a *state* is as a set of circumstances or conditions that characterize a person or thing at a given time. Scientists use this sense of the word when they divide matter into three states — *gases, liquids,* and *solids* — that have characteristic sets of properties, as illustrated in Figure 4.1. We will examine the behavior of gases before that of liquids or solids for several reasons.

1. In many ways, gases are simpler than liquids or solids. Most of the properties of gases discussed in this chapter are independent of the identity of the gas. We can therefore develop a model for a gas without worrying about whether the gas is O_2, N_2, H_2, He or a mixture of these gases.
2. The behavior of gases was understood long before that of liquids or solids.
3. A relatively simple, yet powerful, model — known as the kinetic molecular theory — is available to explain most of the experimental behavior observed for gases.

4.1 ELEMENTS AND COMPOUNDS THAT ARE GASES AT ROOM TEMPERATURE

Before examining the chemical and physical properties of gases, it might be useful to ask: What kinds of elements or compounds are gases at room temperature? To help answer this question, we list some common gases at room temperature in Table 4.1.

There are several recognizable patterns in these data.

1. Common gases at room temperature include both elements (such as H_2 and O_2) and compounds (such as CO_2 and NH_3).
2. All elements that are gases at room temperature are *nonmetals* (such as He, Ne, Ar, Kr, Xe, H_2, N_2, O_2, F_2, and Cl_2).
3. All compounds that are gases at room temperature are *covalent compounds* (such as CO_2, SO_2, and NH_3) that contain two or more nonmetals.
4. With only rare exception, these gases all have relatively small molecular weights.

Patterns can also be seen in the kinds of elements and compounds that are not on this list.

1. None of the *metals* (such as Al, Mg, Fe, and so on) are gases at room temperature.

TABLE 4.1	
Common Gases at Room Temperature	

Element or Compound	Atomic or Molecular Weight (g/mol)
He (helium)	4.00
Ne (neon)	20.18
Ar (argon)	39.95
Kr (krypton)	83.80
Xe (xenon)	131.30
H_2 (hydrogen)	2.02
CH_4 (methane)	16.04
NH_3 (ammonia)	17.03
HCN (hydrogen cyanide)	27.03
CO (carbon monoxide)	28.01
N_2 (nitrogen)	28.01
NO (nitrogen oxide)	30.01
C_2H_6 (ethane)	30.07
O_2 (oxygen)	32.00
PH_3 (phosphine)	34.00
H_2S (hydrogen sulfide)	34.08
HCl (hydrogen chloride)	36.46
F_2 (fluorine)	38.00
CO_2 (carbon dioxide)	44.01
N_2O (dinitrogen oxide)	44.01
C_3H_8 (propane)	44.10
NO_2 (nitrogen dioxide)	46.01
O_3 (ozone)	48.00
C_4H_{10} (butane)	58.12
SO_2 (sulfur dioxide)	64.06
BF_3 (boron trifluoride)	67.80
Cl_2 (chlorine)	70.91
CF_2Cl_2 (dichlorodifluoromethane)	120.91
SF_6 (sulfur hexafluoride)	146.05

2. No *ionic compounds* (such as NaCl or $CaCO_3$) are gases at room temperature.

3. Covalent compounds with relatively large molecular weights (such as cane sugar, $C_{12}H_{22}O_{11}$, and the octane in gasoline, C_8H_{18}) are not gases at room temperature.

As a rule, compounds that consist of relatively light, covalent molecules are most likely to be gases at room temperature.

Exercise 4.1

Which of the following would you expect to be gases at room temperature?

NO_2	SF_4	LiI
ClF_3	$CaCO_3$	MgO
$Cu(NO_3)_2$	HBr	Ni
$PbCl_2$	K_2SO_4	NF_3

> **Solution**
>
> NO_2, ClF_3, SF_4, HBr, and NF_3 are all covalent compounds with reasonably small molecular weights. They are therefore likely to be gases at room temperature. The others are either metals (Ni) or ionic compounds [$Cu(NO_3)_2$, $PbCl_2$, $CaCO_3$, K_2SO_4, LiI, and MgO]. Neither metals nor ionic compounds are gases at room temperature.

4.2 THE PROPERTIES OF GASES

Gases have three characteristic properties.

1. Gases are easy to *compress* — they can be forced into smaller containers.
2. Gases also *expand* easily — they expand to fill their containers.
3. Gases occupy much more space than the liquids or solids from which they form.

Let's examine these properties one at a time.

COMPRESSIBILITY

The internal combustion engine found in the vast majority of the cars on the road today offers a good example of the ease with which gases can be compressed. In a typical four-stroke engine, the piston is first pulled out of the cylinder to create a partial vacuum that draws a mixture of gasoline vapors and air into the cylinder (see Figure 4.2). The piston is then pushed into the cylinder, compressing the gasoline-and-air mixture to a fraction of its original volume.

The ratio of the volume of gas in the cylinder after the first stroke to its volume after the second stroke is the compression ratio of the engine. Modern car engines run at compression ratios of 9 : 1 or more. That means the gasoline-and-air mixture in the cylinder is compressed by a factor of at least 9 in the second stroke.

After the gasoline-and-air mixture is compressed, the spark plug at the top of the cylinder fires, and the resulting explosion pushes the piston out of the cylinder in the third stroke. Finally, the piston is pushed back into the cylinder in the fourth stroke, clearing out the exhaust gases.

FIG. 4.2 The operation of a typical four-stroke combustion engine can be divided into four cycles: intake, compression, power, and exhaust. During the compression stage, the gas in the cylinder is compressed by a factor of 9, or more.

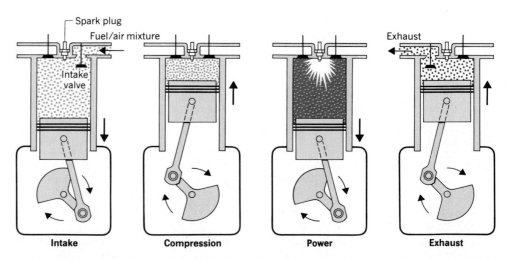

Intake   Compression   Power   Exhaust

Gas can be compressed by either applying an external pressure or by cooling. Note what happens to the gas in a balloon when the balloon is immersed in liquid nitrogen at a temperature of $-196°C$.

Liquids are much harder to compress than gases. They are so hard to compress that the hydraulic brake systems used in most cars operate on the principle that there is essentially no change in the volume of the brake fluid when pressure is applied to this liquid.

Most solids are even harder to compress than liquids. The only exceptions belong to a rare class of compounds that includes natural and synthetic rubber. Most rubber balls that seem easy to compress — such as a racquetball — are filled with air, which is compressed when the ball is squeezed. Solid rubber balls — such as a squash ball — can be compressed to some extent when they bounce, for reasons that will be explained in Chapter 25.

EXPANDIBILITY

Anyone who has walked into a kitchen where bread was baking has experienced the fact that gases expand to fill their containers, as the air in the kitchen becomes filled with wonderful odors. Unfortunately, the same thing happens whenever someone breaks open a rotten egg and the characteristic odor of hydrogen sulfide (H_2S) rapidly diffuses through the room. As a result of the tendency of gases to expand to fill their containers, the volume of a gas is always assumed to be equal to the volume of the container.

VOLUME OF GASES VERSUS VOLUME OF LIQUIDS OR SOLIDS

The difference between the volume of a gas and the volume of the liquid or solid from which it forms can be illustrated by the following examples. One gram of liquid O_2 at its boiling point has a volume of 0.894 milliliters. The same weight of O_2 gas at 0°C and atmospheric pressure has a volume of 700 milliliters — almost 800 times as large. One gram of solid CO_2 has a volume of 0.641 milliliter. At 0°C and atmospheric pressure, the same weight of CO_2 gas has a volume of 556 milliliters — 867 times as large.

The consequences of this enormous change in volume are frequently used to do work. The steam engine, which helped bring about the industrial revolution, is based on the fact that water boils to form a gas (steam) that has a much larger volume. The gas therefore escapes from the container in which it was generated, and the escaping steam can be made to do work.

The same principle is at work when dynamite blasts rock. In 1867, the Swedish chemist Alfred Nobel discovered that the highly dangerous liquid explosive known as nitroglycerin could be absorbed onto clay or sawdust to produce a solid that was much more stable and therefore safer to use. Nobel named this material *dynamite,* from the Greek *dynamis,* which means "power." When dynamite is detonated, the nitroglycerin decomposes to produce a mixture of H_2O, CO_2, O_2, and N_2 gases.

$$4\ C_3H_5N_3O_9(l) \longrightarrow 12\ CO_2(g) + 10\ H_2O(g) + 6\ N_2(g) + O_2(g)$$

Since 29 moles of gas molecules are produced for every 4 moles of liquid that decomposes, and each mole of gas occupies a volume roughly 800 times larger than a mole of liquid, it is not surprising that this reaction produces a shock wave that destroys anything in its vicinity.

The same phenomenon occurs on a much smaller scale when we pop popcorn. As the kernels of popcorn are heated, the liquids inside the kernel turn into gases. The pressure that builds up inside the kernel is enormous, and the kernel eventually explodes, or pops.

Exercise 4.2

Ammonia (NH_3), a gas at room temperature, condenses to become a liquid at temperatures below $-33°C$. The density of the gas is 0.719 grams per liter at $20°C$ and atmospheric pressure. The density of the liquid at its boiling point is 0.648 grams per milliliter. Calculate the ratio of the volumes of liquid and gaseous ammonia.

Solution

In order to compare the volumes of NH_3 gas and liquid NH_3 we have to start with samples of the same size. We can start by using the densities of the gas and the liquid to calculate the volumes of 1-gram samples of each.

$$1.00\ g\ NH_3(g) \times \frac{1\ L}{0.719\ g} = \textbf{1.391 L}$$

$$1.00\ g\ NH_3(l) \times \frac{1\ mL}{0.648\ g} = \textbf{1.543 mL}$$

Before these volumes can be compared, they must be converted to a consistent set of units, such as milliliters.

$$1.391\ L \times \frac{1000\ mL}{1\ L} = \textbf{1391 mL}$$

We can now calculate the ratio of the volume of NH_3 gas to the volume of NH_3 liquid.

$$\frac{V_{NH_3(g)}}{V_{NH_3(l)}} = \frac{1391\ mL}{1.543\ mL} = \textbf{901}$$

According to this calculation, NH_3 as a gas has a volume 901 times as large as the volume of an equivalent amount of liquid NH_3.

4.3 PRESSURE VERSUS FORCE

The volume of a gas is one of its characteristic properties. Another is the pressure the gas exerts on its surroundings. Many of us got our first exposure to this property when we rode to the neighborhood gas station to check the pressure of our bicycle tires. Depending on the kind of bicycle we had, we added air to the tires until the pressure gauge read between 30 and 70 pounds per square inch (lb/in.^{2} or psi).

Two important properties of pressure can be inferred from this example.

1. The pressure of a gas becomes larger as more gas is added to the container.
2. Pressure is measured in units (such as pounds per square inch) that describe the weight exerted by the gas divided by the area over which this weight is distributed.

The first conclusion can be summarized in the following relationship, where P is the pressure of the gas and n is the amount of gas in the container.

$$P \propto n$$

Since the pressure becomes larger as gas is added to the container, P is directly proportional to n.

The second conclusion describes the relationship between **pressure** and **force**.

Pressure is defined as the force exerted on an object divided by the area over which the force is distributed.

$$\text{Pressure} = \frac{\text{force}}{\text{area}}$$

The difference between pressure and force can be illustrated by an analogy based on a 10-penny nail, a hammer, and a piece of wood, as shown in Figure 4.3. By resting the nail on its point and hitting the head with the hammer, we can drive the nail into the wood. But what happens if we turn the nail over and rest the head of the nail against the wood? If we hit the nail with the same force, we can't get it to stick.

When we hit the nail on the head, the force of the blow is applied to the very small area of the wood in contact with the sharp point of the nail, and the nail slips easily into the wood. But when we turn the nail over and hit it on the point, the force is distributed over a much larger area — the surface of the wood that touches any part of the nail head. As a result, the pressure applied to the wood is much smaller, and the nail just bounces off the surface.

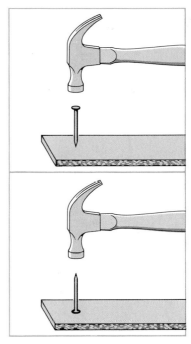

FIG. 4.3 The force exerted by a hammer hitting a nail is the same no matter whether the hammer hits the nail on the head or on the point. But the pressure exerted on the wood is very different. When the nail is hit on the head, all of the force acts on the very small portion of the wood that touches the point of the nail. When the nail is hit on the point, the force is distributed over a much larger area.

Exercise 4.3

(a) Calculate the pressure exerted by a 200-pound man wearing size-10 shoes, assuming the area of each shoe in contact with the floor is 20 square inches.
(b) Calculate the pressure exerted by the heels of a 100-pound woman in high heels, assuming the area beneath the heel of each shoe is 0.25 square inches.

Solution

(a) To calculate the pressure in units of pounds per square inch we divide the force by the surface area over which this force is distributed. Since half of the weight of the man is applied to each shoe, the pressure in this case is about 5 lb/in.²

$$\text{Pressure} = \frac{\text{force}}{\text{area}} = \frac{100 \text{ lb}}{20 \text{ in.}^2} = 5 \text{ lb/in.}^2$$

(b) We can assume that about one-quarter of the weight of the woman is applied to each heel, because half of her weight is divided between the heel and sole of each shoe.

$$\text{Pressure} = \frac{\text{force}}{\text{area}} = \frac{25 \text{ lb}}{0.25 \text{ in.}^2} = 100 \text{ lb/in.}^2$$

The pressure exerted by the heels of this 100-pound woman is 20 times the pressure exerted by the man, who weighs twice as much.

4.4 ATMOSPHERIC PRESSURE

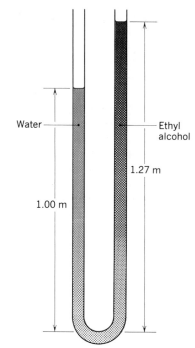

FIG. 4.4 Because of the difference between the densities of water and ethyl alcohol, the weight of a column of water 1.00 meter long in one arm of a U-tube balances the weight of a column of ethyl alcohol 1.27 meters long in the other arm of the tube.

What would happen if we bent a long piece of glass tubing into the shape of the letter U and then carefully filled one arm of this U-tube with water and, at the same time, filled the other arm with ethyl alcohol? Most people expect that the height of the columns of liquid in the two arms of the tube would be the same. Experimentally, though, we find the results shown in Figure 4.4. A 100-centimeter column of water counterbalances a 127-centimeter column of ethyl alcohol, regardless of the diameter of the glass tubing.

At first glance, this might not make sense. If we think about the densities of water (1.00 g/cm³) and ethyl alcohol (0.789 g/cm³), however, we can explain this behavior. A column of water 100 centimeters tall exerts a pressure of 100 grams per square centimeter (g/cm²).

$$100 \text{ cm} \times 1.00 \frac{g}{cm^3} = 100 \frac{g}{cm^2}$$

A column of ethyl alcohol 127 centimeters tall exerts exactly the same pressure.

$$127 \text{ cm} \times 0.789 \frac{g}{cm^3} = 100 \frac{g}{cm^2}$$

Because the pressure of the water pushing down on one arm of the U-tube is equal to the pressure of the alcohol pushing down on the other arm of the tube, the system is in balance.

THE DISCOVERY OF THE BAROMETER

In the early 1600s, Galileo argued that suction pumps were able to draw water from a well because of the "force of vacuum" inside the pumps. After Galileo's death, the Italian mathematician and physicist Evangelista Torricelli (1608–1647) proposed another explanation. He suggested that the air in our atmosphere has weight and that the force of the atmosphere pushing down on the surface of the water drives the water into the suction pump when it is evacuated.

In 1646, Torricelli described an experiment in which a glass tube about a meter long was sealed at one end, filled with mercury, and then inverted into a dish filled with mercury, as shown in Figure 4.5. Some, but not all, of the mercury drained out of the glass tube into the dish. Torricelli explained this as follows. He argued that mercury drained from the glass tube until the force of the column of mercury pushing down on the inside of the tube exactly balanced the force of the atmosphere pushing down on the surface of the liquid outside the tube.

Torricelli predicted that the height of the mercury column would change from day to day as the pressure of the atmosphere changed. Today, his apparatus is known as a **barometer,** from the Greek *baros,* meaning "weight," because it literally measures the weight of the atmosphere. Repeated experiments showed that the average pressure of the atmosphere at sea level was equal to the pressure of a column of mercury 760 millimeters tall. Thus, a standard unit of pressure known as the **atmosphere** was defined as follows.

1 atm = 760 mmHg

To recognize Torricelli's contributions, some scientists describe pressure in units of **torr,** which are defined as follows.

1 torr = 1 mmHg

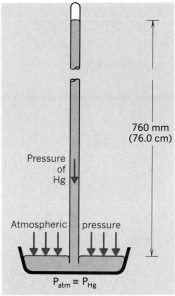

FIG. 4.5 The line of reasoning applied to Figure 4.4 can be used to explain how a barometer works. In this case, the weight of a column of mercury 760 millimeters tall inside the glass tube balances the weight of the air in the atmosphere pushing down on the pool of mercury that surrounds the tube. A unit of 1 atmosphere can therefore be said to be equivalent to 760 millimeters of mercury (mmHg).

Exercise 4.4

On the day this section was written, atmospheric pressure in West Lafayette, Indiana, was 745.8 mmHg. Calculate the pressure in units of atmosphere.

Solution

The conversion between mmHg and atmospheres is based on the following definition.

$$1 \text{ atm} = 760 \text{ mmHg}$$

Using this equality to generate a unit factor gives the following result.

$$745.8 \text{ }\cancel{\text{mmHg}} \times \frac{1 \text{ atm}}{760 \text{ }\cancel{\text{mmHg}}} = 0.9813 \text{ atm}$$

There are four significant figures in the answer because the equality is based on a definition known to an infinite number of significant figures.

Although chemists still work with pressures in units of either atmospheres or mmHg, neither unit is accepted in the SI system. The SI unit of pressure is the pascal (Pa). The relationship between atmospheres and pascals is given by the following equalities.

$$1 \text{ atm} = 101,325 \text{ Pa}$$
$$1 \text{ atm} = 1.01325 \times 10^5 \text{ Pa}$$
$$1 \text{ atm} = 101.325 \text{ kPa}$$

Exercise 4.5

Calculate the weight of the atmosphere if 1 atm is 14.7 lb/in² and the surface area of the planet is 5.1×10^8 km².

Solution

Since atmospheric pressure is given in units of pounds per square inch, we might start by converting the surface area of the planet from square kilometers into square inches.

$$5.1 \times 10^8 \text{ }\cancel{\text{km}^2} \times \left[\frac{1000 \text{ }\cancel{\text{m}}}{\cancel{\text{km}}} \right]^2 \times \left[\frac{39.4 \text{ in.}^2}{\cancel{\text{m}}} \right]^2 = 7.9 \times 10^{17} \text{ in.}^2$$

The pressure of the atmosphere can be demonstrated by evacuating an empty 1-gallon paint can. Within seconds of turning on the vacuum pump, the can collapses.

Multiplying the pressure in pounds per square inch by the surface area in square inches gives us the weight of the atmosphere in pounds.

$$\frac{14.7 \text{ lb}}{\text{in.}^2} \times 7.9 \times 10^{17} \text{ in.}^2 = 1.2 \times 10^{19} \text{ lb}$$

Converting to units of tons gives a value of 5.8×10^{15} tons.

Although pounds per square inch is not a commonly used unit for pressure among chemists, it has two advantages. First, it reminds us that we calculate pressure by dividing a force by the area on which it acts. Second, most people have a feeling for what is meant by a weight of 14.7 pounds and an area of 1 square inch.

Because they can't feel the pressure of the atmosphere, most people assume it is negligible. The results of Exercises 4.3 and 4.5, however, suggest that atmospheric pressure is equivalent to the pressure of a 600-pound man wearing size-10 shoes standing on every portion of the outer surface of our bodies!

The pressure of the atmosphere can be demonstrated with a 1-gallon paint can attached to a vacuum pump. Normally, the pressure of the gas inside the can balances the pressure of the atmosphere pushing on the outside of the can. When the vacuum pump is turned on, however, the can rapidly collapses as it is evacuated.

The surface area of a 1-gallon paint can is about 250 square inches. At 14.7 lb/in.^{2}, this corresponds to a total force over the surface of the can of about 3700 pounds. For the sake of comparison, it might be noted that each of the 18 wheels of a 70,000-lb truck carries about 3900 pounds.

We don't feel the pressure of the atmosphere because the pressure within our bodies counterbalances the pressure of the gas in the atmosphere. The consequences of this inner pressure have been shown quite graphically in several movies in recent years. The puncture of a space suit in the vacuum of outer space immediately leads to the rupture of the body, because there is nothing outside to balance the body's inner pressure.

THE DIFFERENCE BETWEEN THE PRESSURE OF A GAS AND PRESSURE DUE TO WEIGHT

There is an important difference between the pressure of a gas and the other examples of pressure discussed in this section. The pressure exerted by a woman in high heels or a 70,000-pound truck is directional. A truck, for example, exerts all of its pressure on the surface beneath its wheels. In contrast, gas pressure is the same in all directions.

The pressure of the gas in the atmosphere is sufficient to support the water in a graduated cylinder covered with a glass plate when the cylinder is inverted.

To demonstrate this, we can fill a glass cylinder with water and rest a glass plate on top of the cylinder. When we turn the cylinder over, the plate doesn't fall to the floor, because the pressure of the air outside the cylinder pushing up on the bottom of the plate is larger than the pressure exerted by the water in the cylinder pushing down on the plate. It would take a column of water 33.9 feet tall to produce a pressure equivalent to the pressure of the gas in the atmosphere.

4.5 BOYLE'S LAW ($PV =$ CONSTANT)

Torricelli's experiment did more than show that air has weight; it also provided a way of creating a vacuum. The space above the column of mercury at the top of Figure 4.5 is almost completely empty; it is free of air and other gases, except an infinitesimally small amount of mercury vapor. Torricelli's work with vacuums soon caught the eye of the British scientist Robert Boyle.

Boyle's most famous experiments with gases dealt with what he called the "spring of air." His experiments were based on the observation that gases are elastic—they return to their original size and shape after being stretched or squeezed. Boyle studied the elasticity of gases in a J-tube similar to the apparatus shown in Figure 4.6. By adding mercury to the open end of the tube he was able to trap a small volume of air in the sealed end.

Boyle studied what happened to the volume of the gas in the sealed end of the tube as he added mercury to the open end. Table 4.2 contains some of the experimental data he reported in his book *New Experiments Physico-Mechanicall, Touching the Spring of Air, and its Effects . . .* , published in 1662. The first column is the volume of the gas in the sealed end of the J-tube, in arbitrary units. The second column is the difference between the height of the mercury in the sealed and open arms of the J-tube, to the nearest ±1/16 inch, added to the height of mercury needed to balance the pressure of the atmosphere. The third column is the product of the volume of the gas times the pressure in units of inches of mercury.

Boyle noticed that the product of the pressure times the volume for any measurement in this table was equal to the product of the pressure times the volume for any other measurement.

$$P_1V_1 = P_2V_2$$

This expression, or its equivalent,

$$PV = \text{constant}$$

is now known as *Boyle's law.*

Exercise 4.6

Calculate the pressure in atmospheres in a motorcycle engine at the end of the compression stroke. Assume that at the start of this stroke, the pressure of the mixture of gasoline and air in the cylinder is 0.9813 atm and that the volume of the cylinder is 246.8 milliliters. Assume that the volume of the cylinder is 24.2 milliliters at the end of the compression stroke.

Solution

The most important step in this problem is recognizing what we know and what we don't know. We know the pressure and the volume at the start of the compression stroke and the volume at the end of the stroke.

Before Compression:	After Compression:
$P_1 = 0.9813$ atm	$P_2 = ?$
$V_1 = 246.8$ mL	$V_2 = 24.2$ mL

According to Boyle's law, the product of the pressure times the volume at the start of the compression stroke is equal to the product of the pressure times the volume at the end of this stroke.

$$P_1V_1 = P_2V_2$$

The next step in solving this problem involves substituting the known information into this equation.

$$(0.9813 \text{ atm})(246.8 \text{ mL}) = (P_2)(24.2 \text{ mL})$$

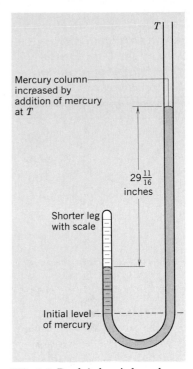

FIG. 4.6 Boyle's law is based on data obtained with a J-tube apparatus such as this. The pressure of the gas in the sealed arm of the tube balances the weight of the excess mercury in the open arm of the tube. If more mercury is added to the open arm, mercury flows into the sealed arm of the J-tube until the pressure of the gas becomes large enough to once again balance the weight of the column of mercury. The volume of the gas in the sealed arm of the tube therefore decreases as the pressure of the gas increases.

Labels in figure: *T* · Mercury column increased by addition of mercury at *T* · $29\frac{11}{16}$ inches · Shorter leg with scale · Initial level of mercury

TABLE 4.2

Boyle's Original (1662)
Data on the Dependence
of the Volume of a Gas on
the Pressure of the Gas

Volume	Pressure	$P \times V$
48	$29\frac{2}{16}$	1398
46	$30\frac{9}{16}$	1406
44	$31\frac{15}{16}$	1405
42	$33\frac{8}{16}$	1407
40	$35\frac{5}{16}$	1413
38	37	1406
36	$39\frac{4}{16}$	1413
34	$41\frac{10}{16}$	1415
32	$44\frac{3}{16}$	1414
30	$47\frac{1}{16}$	1412
28	$50\frac{5}{16}$	1409
26	$54\frac{5}{16}$	1412
24	$58\frac{13}{16}$	1412
22	$64\frac{1}{16}$	1409
20	$70\frac{11}{16}$	1414
18	$77\frac{14}{16}$	1401
16	$87\frac{14}{16}$	1406
14	$100\frac{7}{16}$	1406
12	$117\frac{9}{16}$	1410

We then rearrange this equation and solve for the unknown pressure.

$$P_2 = \frac{(0.9813 \text{ atm})(246.8 \text{ mL})}{(24.2 \text{ mL})} = 10.0 \text{ atm}$$

The pressure in each cylinder at the end of the compression stroke is 10.0 atm.

Exercise 4.7

Because of the elasticity of gases, the volume of a gas depends on the pressure at which it is collected. Assume that a sample of O_2 gas has a volume of 425.5 milliliters when collected at a pressure of 742.3 mmHg. Calculate the volume this quantity of gas would have at a pressure of 1 atm.

Solution

Once again, we can start by listing what we know and what we don't know. We know the pressure and volume of the gas when it was collected and the pressure at which we want to have the gas.

Initial Conditions:	Final Conditions:
$P_1 = 742.3$ mmHg	$P_2 = 1.000$ atm
$V_1 = 425.5$ mL	$V_2 = ?$

This leads us to the conclusion that the problem involves Boyle's law.

$$P_1V_1 = P_2V_2$$

Before we can substitute the known information into the equation for Boyle's law we have to make sure that the pressures are expressed in a consistent set of units. We could do this by converting P_1 into atmospheres, but it is easier to convert P_2 into mmHg.

Initial Conditions:	Final Conditions:
$P_1 = 742.3$ mmHg	$P_2 = 760.00$ mmHg
$V_1 = 425.5$ mL	$V_2 = ?$

We can substitute this information into the Boyle's law equation.

$$(742.3 \text{ mmHg})(425.5 \text{ mL}) = (760.0 \text{ mmHg})(V_2)$$

We can then rearrange this equation and solve for the unknown volume.

$$V_2 = \frac{(742.3 \text{ mmHg})(425.5 \text{ mL})}{(760.0 \text{ mmHg})} = 415.6 \text{ mL}$$

According to this calculation, the volume of the gas will become slightly smaller when the pressure of the gas is increased to 1 atm.

4.6 AMONTONS'S LAW ($P \propto T$)

Toward the end of the 1600s, the French physicist Guillaume Amontons built a thermometer based on the fact that the pressure of a gas is directly proportional to its temperature.

$$P \propto T$$

This relationship is known as *Amontons's law.*

The link between the pressure and temperature of a gas explains why car manufacturers recommend that you adjust the pressure of your tires before you start on a trip. The flexing of the tires as you drive inevitably raises the temperature of the air inside. When this happens, the pressure of the gas inside the tires becomes larger.

Amontons's law can be demonstrated with the apparatus shown in Figure 4.7. Data obtained with this apparatus at room temperature (24°C) and when the sphere was immersed in boiling water (100°C), hot water (74°C), ice water (0°C), and a low-temperature bath of dry ice and isopropyl alcohol (−47°C) are given in Table 4.3.

Figure 4.8 shows the straight-line relationship between the temperature and pressure data in this table. In 1779, Joseph Lambert proposed a definition for *absolute zero* on the temperature scale based on this relationship.

Absolute zero is the temperature at which the pressure of a gas seems to become zero when the data in a plot of pressure versus temperature for a gas are extrapolated.

According to the data in Table 4.3, the pressure of a gas approaches zero when the temperature is about −270°C. More accurate experiments indicate that the pressure approaches zero when the temperature is −273.15°C, as noted in Section 1.18.

The relationship between the temperature and pressure data in Table 4.3 can be greatly simplified by use of the following equation, which converts the temperatures from the Celsius to the Kelvin scale.

$$T_K = °C + 273.15$$

When this is done, a plot of the temperature versus the pressure of a gas gives a straight line that passes through the origin. Because the line passes through the origin, any two points along the line will fit the following equation.

$$\frac{P_1}{P_2} = \frac{T_1}{T_2}$$

It is important to remember that this equation is only valid if the temperatures are converted from the Celsius to the Kelvin scale before any calculations are done.

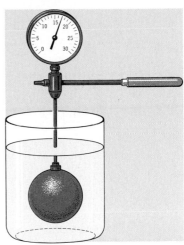

FIG. 4.7 Amontons's law can be demonstrated with an apparatus that consists of a pressure gauge connected to a metal sphere of constant volume, which can be immersed in solutions that have different temperatures.

TABLE 4.3

The Dependence of the Pressure of a Gas on Its Temperature

Temperature (°C)	Pressure (lb/in.²)
100	18.1
74	16.7
24	14.5
0	13.2
−47	10.8

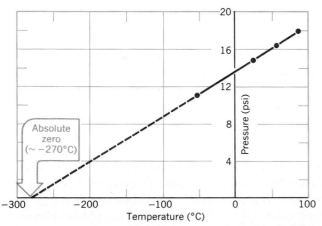

FIG. 4.8 A graph of the data on pressure versus temperature obtained with the apparatus in Figure 4.7. When these data are extrapolated, the pressure of the gas seems to approach zero when the temperature of the gas is about −270°C.

Exercise 4.8

If the pressure in the tires of your car is 30 lb/in.² at 40°C, what is the pressure when the tires cool to 20°C?

Solution

We can start, once again, by listing what we know and what we don't know about this problem.

Initial Conditions:	Final Conditions:
$P_1 = 30$ lb/in.²	$P_2 = ?$
$T_1 = 40°C$	$T_2 = 20°C$

This leads us to conclude that the problem involves Amontons's law. We can therefore start with one of the equations for this law.

$$\frac{P_1}{P_2} = \frac{T_1}{T_2}$$

Before we can use this equation, however, we have to convert the temperatures from °C to K.

Initial Conditions:	Final Conditions:
$P_1 = 30$ lb/in.²	$P_2 = ?$
$T_1 = 313$ K	$T_2 = 293$ K

We can now substitute the known information into the equation.

$$\frac{(30 \text{ lb/in.}^2)}{P_2} = \frac{(313 \text{ K})}{(293 \text{ K})}$$

Rearranging this equation and solving for the unknown pressure gives the following result.

$$P_2 = \frac{(30 \text{ lb/in.}^2)(293 \text{ K})}{(313 \text{ K})} = 28 \text{ lb/in.}^2$$

This answer is consistent with Amontons's law, which suggests that the pressure should become smaller as the tires cool down.

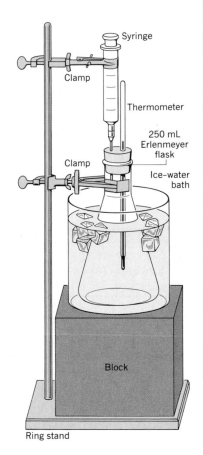

FIG. 4.9 Charles's law can be demonstrated with the apparatus shown here. When the flask is removed from the ice bath and placed in a warm-water bath, the gas in the flask expands, slowly pushing up on the piston of the syringe. Adding the volume of gas in the syringe to the volume of gas in the flask makes it possible to study the effect of changes in the temperature of a gas on its volume.

4.7 CHARLES'S LAW ($V \propto T$)

The relationship between the volume of a gas and its temperature was discovered by Jacques-Alexandre-César Charles. Charles became interested in the behavior of gases when he heard about the invention of the hot-air balloon. On June 5, 1783, Joseph and Étienne Montgolfier used a fire to inflate a spherical balloon about 30 feet in diameter, which traveled about a mile and a half before it came back to earth. News of this remarkable achievement rapidly spread through France, and Charles immediately tried to duplicate the performance. As a result of his work with balloons, Charles noticed that the volume of a gas is directly proportional to its temperature.

$$V \propto T$$

Charles's law provides us with an explanation of how hot-air balloons work. At least since the third century B.C., it has been known that an object floats when it weighs less than the fluid it displaces. If a gas expands when heated, then a given weight of hot air occupies a larger volume than the same weight of cold air. Hot air

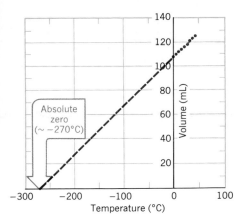

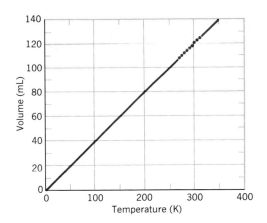

FIG. 4.10 A graph of the data on volume versus temperature obtained with the apparatus in Figure 4.9. When these data are extrapolated, the volume of the gas seems to approach zero when the temperature of the gas is about $-270°C$.

FIG. 4.11 When the data in Figure 4.10 are converted from the Celsius to the Kelvin scale, a plot of volume versus temperature gives a straight line that passes through the origin.

is therefore less dense than cold air. Once the air in a balloon gets hot enough, the net weight of the balloon plus the hot air is less than the weight of an equivalent volume of cold air, and the balloon starts to rise. When the gas in the balloon is allowed to cool, the balloon returns to the ground.

Charles's law can be demonstrated with the apparatus shown in Figure 4.9. A 30-milliliter syringe and a thermometer are inserted through a rubber stopper into a flask that has been cooled to $0°C$ in an ice bath. The ice bath is then removed, and the flask is immersed in a warm-water bath. The gas in the flask expands as it warms, slowly pushing the piston out of the syringe. The total volume of the gas in the system is equal to the volume of the flask plus the volume of the syringe. Table 4.4 contains typical data obtained with this apparatus.

Figure 4.10 shows a plot of the data in Table 4.4. This graph provides us with another way of defining ***absolute zero*** on the temperature scale.

> **Absolute zero is the temperature at which the volume of a gas seems to become zero when the data in a plot of volume versus temperature for a gas are extrapolated.**

The value of absolute zero obtained by extrapolation of the data in Table 4.4 is essentially the same as the value obtained from the graph of pressure versus temperature in the preceding section. Absolute zero can therefore be more accurately defined as the temperature at which the pressure and the volume of a gas approach zero when the data in graphs of pressure versus temperature and volume versus temperature are extrapolated.

When the temperatures in Table 4.4 are converted to kelvins, a plot of the volume versus the temperature of a gas becomes a straight line that passes through the origin, as shown in Figure 4.11. Any two points along this line can be used to construct the following equation.

$$\frac{V_1}{V_2} = \frac{T_1}{T_2}$$

Before using this equation, however, it is important to remember that temperatures must be converted to the Kelvin scale.

Exercise 4.9

Assume that the volume of a balloon filled with H_2 is 1.00 liter at $25°C$. Calculate the volume of the balloon when it is cooled to $-78°C$ in a low-temperature bath of dry ice added to acetone.

TABLE 4.1

The Dependence of the Volume of a Gas on Its Temperature

Temperature ($°C$)	Volume (mL)
0	107.9
5	109.7
10	111.7
15	113.6
20	115.5
25	117.5
30	119.4
35	121.3
40	123.2

Solution

We can start, once again, by listing what we know and what we don't know.

Initial Conditions: *Final Conditions:*

$V_1 = 1.00$ L $V_2 = ?$

$T_1 = 25°C$ $T_2 = -78°C$

This leads us to conclude that the problem involves Charles's law.

$$\frac{V_1}{V_2} = \frac{T_1}{T_2}$$

But we can't use this equation until we convert the temperatures to kelvins.

Initial Conditions: *Final Conditions:*

$V_1 = 1.00$ L $V_2 = ?$

$T_1 = \mathbf{298}$ **K** $T_2 = \mathbf{195}$ **K**

We can now substitute this information into the Charles's law equation.

$$\frac{(1.00 \text{ L})}{V_2} = \frac{(298 \text{ K})}{(195 \text{ K})}$$

We can then solve for the unknown volume.

$$V_2 = \frac{(1.00 \text{ L})(195 \text{ K})}{(298 \text{ K})} = 0.654 \text{ L}$$

According to this calculation, the volume of the balloon will shrink by about 35%, which is consistent with what Charles's law leads us to expect: the volume of the gas should become smaller as the gas is cooled.

4.8 GAY-LUSSAC'S LAW

Joseph Louis Gay-Lussac (1778–1850) began his career in 1801 by very carefully showing the validity for a number of different gases of the principle known as Charles's law. Gay-Lussac's major contributions to the study of gases, however, were the experiments he performed on the ratio of the volumes of gases involved in a chemical reaction.

Gay-Lussac studied the volume of gases consumed or produced in a chemical reaction because he was interested in the reaction between hydrogen and oxygen to form water. He argued that measurements of the *weights* of hydrogen and oxygen consumed in this reaction could be influenced by the moisture present in the reaction flask but that this moisture would not affect the *volumes* of hydrogen and oxygen gases consumed in the reaction.

Gay-Lussac's goal was to measure the ratio of the volumes of hydrogen and oxygen consumed in this reaction. Much to his surprise, he found that 199.89 parts by volume of hydrogen were consumed for every 100 parts by volume of oxygen. Thus, hydrogen and oxygen combined in a simple 2:1 ratio by volume.

Hydrogen + oxygen $\longrightarrow$ water

2 volumes   1 volume

Gay-Lussac found similar whole number ratios for the reaction between other pairs of gases. The compound we now know as hydrogen chloride (HCl) combined with ammonia (NH_3) in a simple 1 : 1 ratio.

$$\text{Hydrogen chloride} + \text{ammonia} \longrightarrow \text{ammonium chloride}$$
$$\underset{\text{1 volume}}{} \qquad \underset{\text{1 volume}}{}$$

Carbon monoxide combined with oxygen in a 2 : 1 ratio.

$$\text{Carbon monoxide} + \text{oxygen} \longrightarrow \text{carbon dioxide}$$
$$\underset{\text{2 volumes}}{} \qquad \underset{\text{1 volume}}{}$$

Gay-Lussac obtained similar results when he analyzed the ratio of the volumes of gases given off when compounds decomposed. Ammonia, for example, decomposed to give three times as much hydrogen by volume as nitrogen.

$$\text{Ammonia} \longrightarrow \text{nitrogen} + \text{hydrogen}$$
$$\underset{\text{1 volume}}{} \qquad \underset{\text{3 volumes}}{}$$

On December 31, 1808, Gay-Lussac announced his law of combining volumes to a meeting of the Société Philomatique in Paris. At that time, he summarized the law as follows: gases combine among themselves in very simple proportions. Today, *Gay-Lussac's law* is stated as follows.

The ratio of the volumes of gases consumed or produced in a chemical reaction is a ratio of simple whole numbers.

Exercise 4.10

Use the following balanced chemical equations to explain the results of Gay-Lussac's experiments.

$$2\,H_2(g) + O_2(g) \longrightarrow 2\,H_2O(g)$$
$$HCl(g) + NH_3(g) \longrightarrow NH_4Cl(s)$$
$$2\,CO(g) + O_2(g) \longrightarrow 2\,CO_2(g)$$
$$2\,NH_3(g) \longrightarrow N_2(g) + 3\,H_2(g)$$

Solution

In each case, the ratio of the number of moles of gases consumed or produced in the reaction is the same as the ratio of the volumes of gases involved in the reaction. For example, Gay-Lussac found that a 2 : 1 ratio by volume of hydrogen and oxygen react to form water, which is consistent with the 2 : 1 ratio of moles of H_2 and O_2 in the balanced equation for this reaction

4.9 AVOGADRO'S HYPOTHESIS ($V \propto n$)

Gay-Lussac's law of combining volumes was announced only a few years after John Dalton proposed his atomic theory. The link between these two ideas was first recognized by the Italian physicist Amadeo Avogadro three years later, in 1811. Avogadro argued that the law of combining volumes could be explained with an assumption that has since become known as *Avogadro's hypothesis.*

Equal volumes of different gases collected under the same conditions of temperature and pressure contain the same number of particles.

Thus, HCl and NH_3 combine in a 1 : 1 ratio by volume, because one molecule of HCl is consumed for every molecule of NH_3 in this reaction and equal volumes of these gases contain the same number of molecules.

$$NH_3(g) + HCl(g) \longrightarrow NH_4Cl(s)$$

So far we have used the mole concept and measurements of weight, or molarity and measurements of volume, to count the number of atoms, ions, or molecules in a sample. Avogadro's hypothesis provides a way of counting atoms or molecules in a gas by measuring the volume of the gas.

It is relatively easy to show that the volume of a gas is proportional to the number of particles in the gas.

$$V \propto n$$

Anyone who has ever blown up a balloon knows that the volume of a gas is directly proportional to the amount of gas in the balloon. The more air you add to the balloon, the bigger it gets.

Unfortunately, this demonstration does not test Avogadro's hypothesis that equal volumes of *different* gases contain the same number of gas molecules. The best way to prove the validity of this hypothesis is to measure the number of gas molecules in a given volume of different gases. This can be done with the apparatus shown in Figure 4.12.

A small hole is drilled through the plunger of a 50-milliliter plastic syringe. The plunger is then pushed into the syringe until there is no gas in the syringe, and the syringe is sealed with a syringe cap. Next, the plunger is pulled out of the syringe until the volume reads 50 milliliters, and a nail is inserted through the hole in the plunger so that the plunger is not sucked back into the barrel of the syringe. The "empty" syringe is then weighed, filled with 50 milliliters of a gas, and reweighed. The difference between these measurements is the weight of 50 milliliters of the gas.

The results of experiments with six different gases are given in Table 4.5. The number of molecules in a 50-milliliter sample of any one of these gases can be calculated from the weight of the sample, the molecular weight of the gas, and the number of molecules in a mole.

$$0.005 \text{ g } H_2 \times \frac{1 \text{ mol}}{2.02 \text{ g}} \times \frac{6.02 \times 10^{23} \text{ molecules}}{1 \text{ mol}} = 1 \times 10^{21} \text{ } H_2 \text{ molecules}$$

FIG. 4.12 Avogadro's hypothesis can be demonstrated with a plastic syringe that has a hole drilled through the plunger. The syringe is first sealed with a syringe cap. The plunger is then pulled out of the syringe to a volume of 50 milliliters, a nail is inserted through the hole in the plunger, and the "empty" syringe is weighed. The weight of 50 milliliters of a gas can be obtained by subtracting the weight of the empty syringe from the weight of the syringe filled with a 50-milliliter sample of the gas.

The last column in Table 4.5 summarizes the results obtained when this calculation is repeated for each gas. The number of significant figures in the answer changes from one calculation to the next. But within experimental error, the number of gas molecules in each sample is the same. We can therefore conclude that equal volumes of different gases collected under the same conditions of temperature and pressure do in fact contain the same number of gas particles.

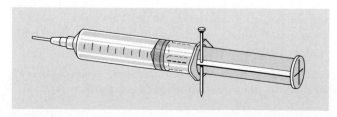

TABLE 4.5

Experimental Data for the Weight of 50-mL
Samples of Different Gases

Compound	Weight of 50 mL of Gas (g)	Molecular Weight of Gas (g/mol)	Number of Gas Molecules
H_2	0.005	2.02	1×10^{21}
N_2	0.055	28.01	1.2×10^{21}
O_2	0.061	32.00	1.1×10^{21}
CO_2	0.088	44.01	1.2×10^{21}
C_4H_{10}	0.111	58.12	1.15×10^{21}
CCl_2F_2	0.228	120.91	1.14×10^{21}

Exercise 4.11

The pleasant odor that sometimes accompanies summer thunderstorms is due to trace quantities of ozone (O_3) formed when lightning passes through the atmosphere. Calculate the volume of oxygen formed when 1.00 liter of ozone decomposes back to oxygen according to the following equation.

$$2 \, O_3(g) \longrightarrow 3 \, O_2(g)$$

Solution

The key to solving this problem is remembering that equal volumes of different gases measured at the same temperature and pressure contain the same number of molecules. This means that we can translate the balanced equation into a relationship between the volumes of gases involved in the reaction. The balanced equation states that we get 3 moles of O_2 for every 2 moles of O_3 that decompose — or 1.5 times as many moles of O_2. This means that 1.50 liters of O_2 will be produced for every 1.00 liter of O_3 consumed.

4.10 THE IDEAL GAS EQUATION ($PV = nRT$)

So far, gases have been described in terms of four variables: pressure (P), volume (V), temperature (T), and amount (n). In the course of this chapter, five relationships between pairs of these variables have been discussed. In each case, two of the variables were allowed to change while the other two were held constant. The discussion of the bicycle tire, for example, showed that the pressure of a gas is directly proportional to the amount of gas when the temperature and volume of the gas are held constant.

$$P \propto n \quad (T \text{ and } V \text{ constant})$$

Other relationships between pairs of these variables include the following.

Boyle's law:	$PV = \text{constant}$	(T and n constant)
Amontons's law:	$P \propto T$	(V and n constant)
Charles's law:	$V \propto T$	(P and n constant)
Avogadro's hypothesis:	$V \propto n$	(P and T constant)

Each of these relationships is a special case of a more general relationship known as the *ideal gas equation.*

$$PV = nRT$$

In this equation, R is a proportionality constant known as the *ideal gas constant,* and T is the absolute temperature. The value of R depends on the units used to express the four variables P, V, n, and T. By convention, most chemists use the following set of units.

P: **atmospheres**	*T*: **kelvin**
V: **liters**	*n*: **moles**

Exercise 4.12

Calculate the value of the ideal gas constant, R, if 1.000 mole of an ideal gas occupies a volume of 22.414 liters at 0°C and 1.000 atm.

Solution

The ideal gas law summarizes the relationship among the pressure, volume, temperature, and amount of an ideal gas. This law states that the product of the pressure times the volume of an ideal gas divided by the product of the amount of gas times the absolute temperature is a constant.

$$\frac{PV}{nT} = R$$

We can calculate the value of R for any set of units of P, V, n, and T by simply substituting the known values of these quantities into this equation.

$$\frac{(1.00 \text{ atm})(22.414 \text{ L})}{(1.000 \text{ mol})(273.15 \text{ K})} = 0.08206 \ \frac{\text{L-atm}}{\text{mol-K}}$$

4.11 IDEAL GAS CALCULATIONS: PART I

The ideal gas equation can be used to predict the value of one of the variables that describe a gas from known values of the other three.

Exercise 4.13

Many gases are available for use in the laboratory in compressed gas cylinders, in which they are stored at high pressures. Calculate the weight of O_2 that could be stored at 21°C and 170 atm in a cylinder with a volume of 60.0 liters.

Compressed gas cylinders used to store gases at high pressures.

Solution

The key to solving this problem is recognizing that we know three of the four variables in the ideal gas equation.

$$P = 170 \text{ atm}$$
$$V = 60.0 \text{ L}$$
$$T = 21°C$$
$$n = ?$$

This suggests that the ideal gas equation will play an important role in solving this problem.

$$PV = nRT$$

Before we can use this equation, however, we have to convert the temperature to kelvins.

$$T_K = °C + 273 = 294 \text{ K}$$

Substituting the known information into the ideal gas equation gives the following.

$$(170 \text{ atm})(60.0 \text{ L}) = (n)(0.08206 \text{ L-atm/mol-K})(294 \text{ K})$$

Rearranging this equation to solve for the number of moles of gas in the container gives the following result.

$$n = \frac{(170 \cancel{\text{ atm}})(60.0 \cancel{\text{ L}})}{(0.08206 \cancel{\text{ L-atm}}/\text{mol-}\cancel{\text{K}})(294 \cancel{\text{ K}})} = 422.8 \text{ mol}$$

We now note that the question asked for the *weight* of the O_2 that could be stored in this cylinder, which can be calculated from the number of moles of O_2 in the cylinder and the molecular weight of the gas.

$$422.8 \cancel{\text{ mol}} \text{ } O_2 \times \frac{32.00 \text{ g } O_2}{1 \cancel{\text{ mol}}} = 1.35 \times 10^4 \text{ g } O_2$$

According to this calculation, 13.5 kilograms—or 29.7 pounds—of O_2 can be stored in a 60-liter compressed gas cylinder at 21°C and 170 atm.

The key to solving ideal gas problems often involves recognizing what is known and deciding how to use this information.

Exercise 4.14

Calculate the weight in pounds of the air in a spherical hot-air balloon that has a volume of 14,100 cubic feet when the temperature of the gas is 86°F and the pressure is 748 mmHg. Assume that the average molecular weight of air is 29.0 grams per mole.

Solution

In this problem, we know the pressure, volume, and temperature of a gas.

$$P = 748 \text{ mmHg}$$
$$V = 1.41 \times 10^4 \text{ ft}^3$$
$$T = 86°F$$

We might therefore think about using the ideal gas equation to calculate the number of moles of gas in this balloon.

$$PV = nRT$$

Before we can do this, we have to convert the pressure from mmHg into atmospheres.

$$748 \cancel{\text{ mmHg}} \times \frac{1 \text{ atm}}{760 \cancel{\text{ mmHg}}} = 0.9842 \text{ atm}$$

We also have to convert the temperature to kelvins. We can start by converting from °F to °C.

$$\frac{5}{9}[86°F - 32] = 30°C$$

An etching of the first Montgolfier ascent at Versailles, on September 19, 1783.

We can then convert from °C to K.

$$T_K = \text{°C} + 273 = 303 \text{ K}$$

Finally, we have to convert the volume from cubic feet to liters.

$$1.41 \times 10^4 \text{ ft}^3 \times \left[\frac{12 \text{ in.}}{1 \text{ ft}} \right]^3 \times \left[\frac{2.54 \text{ cm}}{1 \text{ in.}} \right]^3 \times \frac{1 \text{ mL}}{1 \text{ cm}^3} \times \frac{1 \text{ L}}{1000 \text{ mL}} = 3.993 \times 10^5 \text{ L}$$

Once we know the pressure, temperature, and volume of the gas in the appropriate units, we can solve the ideal gas equation for the variable we don't know and substitute this information into the equation.

$$n = \frac{PV}{RT} = \frac{(0.9842 \text{ atm})(3.993 \times 10^5 \text{ L})}{(0.08206 \text{ L-atm}/\text{mol-K})(303 \text{ K})} = 1.581 \times 10^4 \text{ mol}$$

The problem asks for the weight in pounds of the air in this balloon. By combining the number of moles of gas from the preceding calculation with the average molecular weight of air given in the problem, we can calculate the weight of the air in units of grams.

$$1.581 \times 10^4 \text{ mol} \times \frac{29.0 \text{ g}}{1 \text{ mol}} = 4.585 \times 10^5 \text{ g}$$

We can then convert this weight into units of pounds.

$$4.585 \times 10^5 \text{ g} \times \frac{1 \text{ lb}}{453.6 \text{ g}} = 1.01 \times 10^3 \text{ lb}$$

The air in this balloon weighs a little more than 1000 pounds.

The ideal gas equation can be applied to problems that don't seem to ask for one of the variables in the equation.

Exercise 4.15

Calculate the molecular weight of butane, assuming that 0.5836 gram of this gas fills 250.0-milliliter flask at a temperature of 23.5°C and a pressure of 742.6 mmHg.

Solution

Once again, we know something about the pressure, volume, and temperature of a gas.

$$P = 742.6 \text{ mmHg}$$
$$V = 250.0 \text{ mL}$$
$$T = 23.5\text{°C}$$

We might therefore start by calculating the number of moles of gas in the sample. In order to do this, we need to convert the pressure to atmospheres.

$$742.6 \text{ mmHg} \times \frac{1 \text{ atm}}{760 \text{ mmHg}} = 0.9771 \text{ atm}$$

We also have to convert the temperature to kelvins

$$T_K = \text{°C} + 273.15 = 296.7 \text{ K}$$

and the volume to liters.

$$250.0 \text{ mL} \times \frac{1 \text{ L}}{1000 \text{ mL}} = 0.2500 \text{ L}$$

Rearranging the ideal gas equation to solve for the number of moles of gas and substituting the known values of the pressure, volume, and temperature into this equation give the following result.

$$n = \frac{PV}{RT} = \frac{(0.9771 \text{ atm})(0.2500 \text{ L})}{(0.08206 \text{ L-atm}/\text{mol-K})(296.7 \text{ K})} = 0.01003 \text{ mol}$$

The problem asks for the molecular weight of the gas. We know that a sample that weighs 0.5836 gram contains 0.01003 mole of gas. We can therefore calculate the molecular weight of the gas by dividing the weight of the sample in grams by the number of moles of butane in the sample.

$$\frac{0.5836 \text{ g}}{0.01004 \text{ mol}} = 58.13 \frac{\text{g}}{\text{mol}}$$

The molecular weight of butane is 58.13 grams per mole.

The ideal gas equation can even be used to solve problems that don't seem to contain enough information.

Exercise 4.16

Calculate the density in grams per liter of O_2 gas at $0°C$ and 1.00 atm.

Solution

This time we have information about only two of the four variables in the ideal gas equation.

$$P = 1.00 \text{ atm}$$
$$T = 0°C$$
$$V = ?$$
$$n = ?$$

What can we calculate from this information? The only way to answer this question is to rearrange the ideal gas equation, putting the knowns on one side and the unknowns on the other.

$$\frac{n}{V} = \frac{P}{RT}$$

According to this equation, we can calculate the number of moles of gas per liter. So let's do that.

$$\frac{n}{V} = \frac{P}{RT} = \frac{(1.00 \text{ atm})}{(0.08206 \text{ L-atm}/\text{mol-K})(273 \text{ K})} = 0.04464 \frac{\text{mol } O_2}{\text{L}}$$

The problem asks for the density of the gas in grams per liter. Since we know the number of moles of O_2 per liter, we can use the molecular weight of O_2 to calculate the number of grams per liter.

$$\frac{0.04464 \text{ mol } O_2}{1 \text{ L}} \times \frac{32.00 \text{ g } O_2}{1 \text{ mol}} = 1.43 \frac{\text{g } O_2}{\text{L}}$$

The density of O_2 gas at $0°C$ and 1.00 atm is 1.43 grams per liter.

4.12 IDEAL GAS CALCULATIONS: PART II

Gas law problems often ask you to predict what will happen when one or more changes are made in the variables that describe the gas. There are two ways of working such problems. One approach was used in Exercises 4.6 through 4.9. The answer to Exercise 4.6, for example, was based on Boyle's law, which states that the product of the pressure times the volume of the gas is constant if the temperature and the amount of gas are held constant. Thus, the product of the pressure times the volume before any change is made must be equal to the product of the pressure times the volume after the change is made.

$$P_1 V_1 = P_2 V_2$$

Another way of solving these problems takes advantage of the fact that the ideal gas constant, R, is truly constant. We start by solving the ideal gas equation for R.

$$R = \frac{PV}{nT}$$

We then note that PV/nT at any one time must be equal to this ratio at any other time.

$$\frac{P_1 V_1}{n_1 T_1} = \frac{P_2 V_2}{n_2 T_2}$$

We then substitute the known values of pressure, temperature, volume, and amount of gas into this equation and solve for whatever unknown we want. This approach has two advantages. First, only one equation has to be remembered. Second, it can be used to handle problems in which more than one variable changes at a time.

Exercise 4.17

A sample of NH_3 gas fills a 27.0-liter container at $-15°C$ and 2.58 atm. Calculate the volume of this gas at $21°C$ and 751 mmHg.

Solution

Let's start by listing what we know and what we don't know about this system.

Initial Conditions:	*Final Conditions:*
$P_1 = 2.58$ atm	$P_2 = 751$ mmHg
$V_1 = 27.0$ L	$V_2 = ?$
$T_1 = -15°C$	$T_2 = 21°C$
$n_1 = ?$	$n_2 = ?$

Now let's convert these data to a consistent set of units for use in the ideal gas equation.

Initial Conditions:	*Final Conditions:*
$P_1 = 2.58$ atm	$P_2 = 0.988$ atm
$V_1 = 27.0$ L	$V_2 = ?$
$T_1 = 258$ K	$T_2 = 294$ K
$n_1 = ?$	$n_2 = ?$

We can then substitute this information into the following equation.

$$\frac{P_1 V_1}{n_1 T_1} = \frac{P_2 V_2}{n_2 T_2}$$

When this is done we get one equation with three unknowns—n_1, n_2, and V_2.

$$\frac{(2.58 \text{ atm})(27.0 \text{ L})}{(n_1)(258 \text{ K})} = \frac{(0.988 \text{ atm})(V_2)}{(n_2)(298 \text{ K})}$$

It is impossible to solve one equation in three unknowns. But a careful reading of the problem leads to the conclusion that the number of moles of NH_3 is the same before and after the temperature and pressure change.

$$n_1 = n_2$$

Substituting this equality into the unknown equation gives the following.

$$\frac{(2.58 \text{ atm})(27.0 \text{ L})}{(n_1)(258 \text{ K})} = \frac{(0.988 \text{ atm})(V_2)}{(n_1)(298 \text{ K})}$$

Multiplying both sides of this equation by n_1 gives one equation in one unknown, which can be solved for that unknown.

$$\frac{(2.58 \text{ atm})(27.0 \text{ L})}{(258 \text{ K})} = \frac{(0.988 \text{ atm})(V_2)}{(298 \text{ K})}$$

$$V_2 = \frac{(2.58 \text{ atm})(27.0 \text{ L})(298 \text{ K})}{(258 \text{ K})(0.988 \text{ atm})} = 81.4 \text{ L}$$

According to this calculation, the volume of the gas increases from 27.0 liters at $-15°C$ and 2.58 atm to 81.4 liters at 21°C and 751 mmHg.

4.13 DALTON'S LAW OF PARTIAL PRESSURES ($P_t = P_1 + P_2 + \ldots$)

The *CRC Handbook of Chemistry and Physics* describes the atmosphere as 78.084% N_2, 20.946% O_2, 0.934% Ar, and 0.033% CO_2 by volume when the water vapor has been removed. What image does this description evoke in your mind? Do you believe that only 20.946% of the room you are sitting in contains O_2? Or do you believe that the atmosphere in your room is a homogeneous mixture of N_2, O_2, H_2O, Ar, and CO_2?

Section 4.2 stated that gases expand to fill their containers. Thus, the volume of O_2 in your room is the same as the volume of N_2. Both gases expand to fill the room. When we describe the atmosphere as 20.946% O_2 by volume, what we mean is that the volume of the atmosphere would shrink by 20.946% if the O_2 were removed.

Are the pressures of the O_2 and N_2 in the atmosphere also the same? We can answer this question by rearranging the ideal gas equation as follows.

$$P = n \times \frac{RT}{V}$$

According to this equation, the pressure of a gas is proportional to the number of moles of gas, if the temperature and volume are held constant. Since the temperature and volume of the O_2 and N_2 in the atmosphere are the same, the pressure of each gas must be proportional to the number of moles of the gas. The ideal gas law therefore provides another way of counting atoms or molecules in a gas. The pressure of a gas is a measure of the amount of the gas, at constant temperature and volume.

The total pressure of a mixture of gases is the sum of the contributions from the *partial pressures* of the individual components.

> **Partial pressure: The part of the total pressure of a mixture that results from one component of the mixture.**

The total pressure of the atmosphere, for example, is the sum of the partial pressures of all of the components.

$$P_t = P_{N_2} + P_{O_2} + P_{Ar} + P_{CO_2} + P_{H_2O} + \ldots$$

The idea that the total pressure of a mixture of gases is the sum of the partial pressures of its components was first recognized by John Dalton and is often called *Dalton's law of partial pressures.*

Exercise 4.18

Calculate the total pressure of a mixture that contains 1.00 gram of H_2 and 1.00 gram of He in a 5.00-liter container at 21°C.

Solution

We can start by listing what we know and what we don't know about the problem. In this case, we know the volume and temperature of the gas, but not the number of moles of gas or the pressure of the gas.

$$P = ?$$
$$V = 5.00 \text{ L}$$
$$T = 21°C$$
$$n = ?$$

We also know the weights of H_2 and He in the flask, so we can calculate the number of moles of each gas.

$$1.00 \text{ g } H_2 \times \frac{1 \text{ mol } H_2}{2.016 \text{ g}} = 0.496 \text{ mol } H_2$$

$$1.00 \text{ g He} \times \frac{1 \text{ mol He}}{4.003 \text{ g}} = 0.250 \text{ mol He}$$

We can now rearrange the ideal gas equation as follows.

$$P = \frac{nRT}{V}$$

We can then use this equation to calculate the partial pressure of each component of the mixture.

$$P_{H_2} = \frac{(0.496 \text{ mol } H_2)(0.08206 \text{ L-atm/mol-K})(294 \text{ K})}{(5.00 \text{ L})} = 2.39 \text{ atm}$$

$$P_{He} = \frac{(0.250 \text{ mol } H_2)(0.08206 \text{ L-atm/mol-K})(294 \text{ K})}{(5.00 \text{ L})} = 1.21 \text{ atm}$$

The total pressure in this mixture is the sum of the partial pressures of the two components.

$$P_t = P_{H_2} + P_{He} = 3.60 \text{ atm}$$

Dalton derived the law of partial pressures from his work in meteorology, particularly his experiments with the amount of water vapor that could be absorbed by air at different temperatures. It is therefore fitting that this law is used most often to correct for the amount of water vapor picked up when a gas is collected by displacement of water.

Suppose, for example, that we want to collect a sample of O_2. We can prepare this gas by heating potassium chlorate until it decomposes.

$$2 \text{ KClO}_3(s) \longrightarrow 2 \text{ KCl}(s) + 3 \text{ O}_2(g)$$

To collect the gas given off in this reaction, we can fill a flask with water, invert the flask in a trough, and then let the gas bubble into the flask, as shown in Figure 4.13. Try to imagine what the gas in this flask would look like on the atomic scale. Would it be pure, or even relatively pure, O_2? Or would it be a mixture of gases?

Some of the water in the flask will evaporate during the experiment. The gas that collects in this flask is therefore a mixture of O_2 and water vapor. The total pressure of this gas will be the sum of the partial pressures of these two components.

$$P_t = P_{O_2} + P_{H_2O}$$

We can use the concept of the *vapor pressure* of water to estimate the partial pressure of O_2 in this mixture.

Vapor pressure: The pressure of a gas that collects in a closed container above a liquid.

Table A-4 in the appendix shows the relationship between the temperature of water and its vapor pressure. As water becomes warmer, more of it evaporates, and the vapor pressure becomes larger. If we know the temperature at which a gas is collected by displacement of water, and we assume that the gas is saturated with water vapor, we can calculate the partial pressure of the gas by subtracting the vapor pressure of water from the total pressure of the mixture of gases.

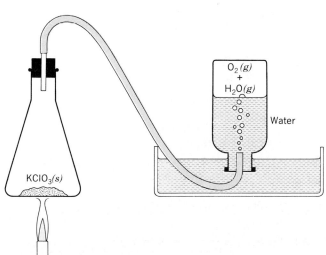

FIG. 4.13 Gases can be collected by displacement of water from a container. We can collect O_2, for example, by bubbling the gas given off when KClO$_3$ decomposes into a bottle or flask that was originally filled with water and inverted into a trough or pail of water.

Exercise 4.19

Calculate the weight of O_2 collected by displacement of water from a 250-milliliter flask at 21°C and 746.2 mmHg.

Solution

Let's work this problem backwards. Our goal is the weight of O_2 in the flask. To reach this goal we need to calculate the number of moles of O_2 in the sample. To do this, we need to know the pressure, volume, and temperature of the O_2. We already know the volume of the O_2, because we can always assume that a gas expands to fill its container.

$$V_{O_2} = 250 \text{ mL}$$

We also know the temperature of the O_2.

$$T_{O_2} = 21°C$$

But we don't know the pressure of the O_2. All we know is the total pressure of the oxygen plus the water vapor that collects in the flask.

$$P_t = P_{O_2} + P_{H_2O} = 746.2 \text{ mmHg}$$

Table A-4 gives a vapor pressure of water at 21°C of 18.7 mmHg. If we assume that the gas collected in this experiment is saturated with water vapor, the partial pressure of O_2 must be smaller than the total pressure in the flask by 18.7 mmHg. Substituting what we know into the equation for the total pressure gives the following result.

$$P_t = P_{O_2} + P_{H_2O}$$
$$746.2 \text{ mmHg} = P_{O_2} + 18.7 \text{ mmHg}$$

The partial pressure of O_2 in this flask is therefore 727.5 mmHg.

$$P_{O_2} = P_t - P_{H_2O}$$
$$= 746.2 \text{ mmHg} - 18.7 \text{ mmHg}$$
$$= 727.5 \text{ mmHg}$$

We now know the pressure (727.5 mmHg), volume (250 mL), and temperature (21°C) of the O_2 collected in this experiment. Before we can calculate the number of moles of O_2, however, we have to convert these measurements to appropriate units.

$$727.5 \text{ mmHg} \times \frac{1 \text{ atm}}{760 \text{ mmHg}} = 0.9572 \text{ atm}$$

$$21°C + 273 = 294 \text{ K}$$

$$250 \text{ mL} \times \frac{1 \text{ L}}{1000 \text{ mL}} = 0.250 \text{ L}$$

We can now substitute this information into the ideal gas equation and calculate the number of moles of O_2 given off in this experiment.

$$n = \frac{PV}{RT} = \frac{(0.9572 \text{ atm})(0.250 \text{ L})}{(0.08206 \text{ L-atm/mol-K})(294 \text{ K})} = 0.009919 \text{ mol } O_2$$

From this information, we can calculate the weight of the O_2.

$$0.009919 \text{ mol } O_2 \times \frac{32.00 \text{ g } O_2}{1 \text{ mol}} = 0.317 \text{ g } O_2$$

4.14 GRAHAM'S LAWS OF DIFFUSION AND EFFUSION (RATE $\propto 1/\sqrt{MW}$)

Although most of the physical properties of gases are independent of the identity of the gas, a few depend on whether the gas is H_2, O_2, CO_2, SF_6, and so on. One of the properties that depends on the identity of the gas can be seen when the movement of gases is studied.

In 1829, Thomas Graham used an apparatus similar to the one shown in Figure 4.14 to study the *diffusion* of gases — the rate at which two gases mix. This apparatus consists of a long glass tube sealed at one end with plaster that has holes large enough to allow gas molecules to enter or leave the tube. When the tube is filled with H_2 gas, the level of water in the tube slowly rises, because the H_2 molecules inside the tube escape through the holes in the plaster faster than the molecules in air can enter the tube. By studying the rate at which the water level in this apparatus changed, Graham was able to obtain data on the rate at which different gases diffused, or mixed with air.

Graham found that the rates at which gases diffuse is inversely proportional to the square root of their densities.

$$Rate \propto \frac{1}{\sqrt{density}}$$

This relationship eventually became known as *Graham's law of diffusion.*

To understand the importance of this discovery we have to remember that equal volumes of different gases contain the same number of particles. The number of moles of gas per liter at a given temperature and pressure is therefore constant. This means that the density of a gas is directly proportional to its molecular weight (*MW*). Graham's law of diffusion therefore can also be written as follows.

$$Rate \propto \frac{1}{\sqrt{MW}}$$

Graham also studied the rate of *effusion* of a gas — the rate at which the gas escapes through a pinhole into a vacuum. Basically the same results were obtained from these experiments. The rate of effusion of a gas is inversely proportional to the square root of either the density or the molecular weight of the gas. *Graham's*

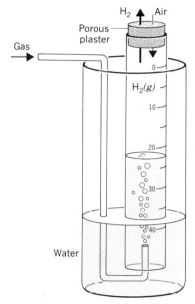

FIG. 4.14 The rate of diffusion, or mixing of a gas with air, can be studied with the apparatus shown here. One end of a glass tube is sealed with plaster that has pores large enough for the gas molecules to pass through. A given volume of gas is collected inside the glass tube by the displacement of water. If the gas escapes from the tube faster than air enters it, the amount of water in the tube will become larger. If air enters the tube faster than the gas escapes, the gas will be displaced from the tube. The rate at which the gas in the tube mixes with air can be studied by watching water either enter or leave the tube.

TABLE 4.6

Experimental Data Giving the Time Required for 25-mL Samples of Gases with Different Molecular Weights to Escape through a 0.006-in. Hole into a Vacuum

Compound	Time (s)	Molecular Weight (g/mol)
H_2	5.1	2.02
He	7.2	4.00
NH_3	14.2	17.0
air	18.2	29.0
O_2	19.2	32.0
CO_2	22.5	44.0
SO_2	27.4	64.1

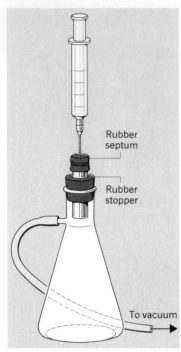

FIG. 4.15 The rate of effusion of a gas can be demonstrated with the apparatus shown here. A vacuum pump is used to evacuate a thick-walled filter flask. A syringe is filled with a 25-milliliter sample of a gas and the time required for the gas to escape into the evacuated flask is measured. The faster the gas molecules effuse, the less time it takes for gas to escape into the flask.

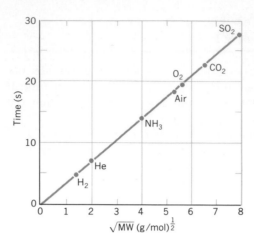

FIG. 4.16 A graph of the time required for 25-milliliter samples of different gases to escape into an evacuated flask versus the square root of the molecular weight of the gas. As gas molecules become heavier, they move more slowly and it takes more time for the gas to escape.

law of effusion can be demonstrated with the apparatus in Figure 4.15. A thick-walled filter flask is evacuated with a vacuum pump. A syringe is filled with 25 milliliters of a gas, and the time required for the gas to escape through the syringe needle into the evacuated filter flask is measured with a stopwatch. The experimental data in Table 4.6 were obtained by use of a special needle with a very small (0.006-inch) hole through which the gas could escape.

As we can see in Figure 4.16, which graphs these data, the *time* required for a 25-milliliter sample of a gas to escape into a vacuum is proportional to the square root of the molecular weight of the gas. The *rate* at which the gas effuses is therefore inversely proportional to the square root of the molecular weight. Graham's observations about the rate at which gases diffuse (mix) and effuse (escape through a pinhole into a vacuum) can be explained by assuming that relatively light gas particles such as H_2 molecules and He atoms move faster than relatively heavy gas particles such as CO_2 and SO_2 molecules.

4.15 THE KINETIC MOLECULAR THEORY

One of the most fascinating things about gases is the existence of a simple theoretical model for an ideal gas that can explain every one of the experimental observations about the behavior of gases discussed so far. This model is called the *kinetic molecular theory,* and it is based on the following postulates, or assumptions.

1. **Gases are composed of a large number of particles that behave like hard, spherical objects.**
2. **These particles are in constant, random motion. They move in a straight line until they collide with either another particle or the walls of the container.**
3. **The gas particles are very much smaller than the distance between particles. Most of the volume of a gas is therefore empty space.**
4. **There is no force of attraction between gas particles or between the particles and the walls of the container.**
5. **Collisions between gas particles or collisions with the walls of the container are perfectly *elastic*. None of the energy of a gas particle is lost when it collides with another particle or with the walls of the container.**

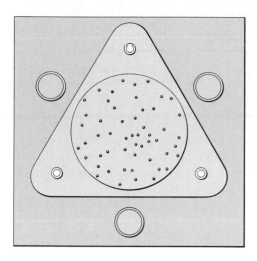

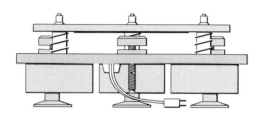

FIG. 4.17 The six postulates of the kinetic molecular theory can be demonstrated with the molecular dynamics simulator shown here. Steel ball bearings are placed on top of a glass plate supported on three vibrating motors. When the motors are turned on, the plate vibrates, and the steel ball bearings move across the glass plate until they collide either with other ball bearings or with the walls of the container.

6. The average kinetic energy of a collection of gas particles depends on the temperature of the gas, and nothing else.

The assumptions behind the kinetic molecular theory can be illustrated with the apparatus shown in Figure 4.17, which consists of a glass plate surrounded by walls mounted on top of three vibrating motors. A handful of steel ball bearings is placed on top of the glass plate to represent the gas particles (Postulate 1).

When the motors are turned on, the glass plate vibrates, which makes the ball bearings move in a constant, random fashion. Each ball moves in a straight line until it collides with another ball or with the walls of the container (Postulate 2). Although collisions are frequent, the average distance between the ball bearings is much larger than the diameter of the balls (Postulate 3). There is no force of attraction between the individual ball bearings or between the ball bearings and the walls of the container (Postulate 4).

The collisions that occur in this apparatus are very different from those that occur when a rubber ball is dropped on the floor. The rubber ball loses a portion of its energy each time it hits the floor, until it eventually rolls to a stop. In this apparatus, the collisions are perfectly elastic. The ball bearings seem to have just as much energy after a collision as before (Postulate 5).

Any object in motion has a kinetic energy (KE) that is defined as one-half of the product of its mass (m) times its velocity (v) squared.

$$KE = \tfrac{1}{2}mv^2$$

At any time, some of the ball bearings on this apparatus are moving faster than others. However, when we increase the "temperature" of the system by increasing the voltage to the motors, we find that the *average* kinetic energy of the ball bearings becomes larger (Postulate 6).

HOW THE KINETIC MOLECULAR
4.16 THEORY EXPLAINS THE GAS LAWS

The kinetic molecular theory can be used to explain all of the experimentally determined gas laws.

THE LINK BETWEEN P AND n

The pressure of a gas results from collisions between the gas particles and the walls of the container. Each time a gas particle hits the wall, it exerts a force on the wall. Any increase in the number of gas particles in the container increases the number of collisions with the walls and therefore the pressure of the gas.

AMONTONS'S LAW ($P \propto T$)

The last postulate of the kinetic molecular theory states that the average kinetic energy of a gas particle depends on the temperature of the gas, and nothing else. The average kinetic energy of the gas particles becomes larger as the gas becomes warmer. Since the mass of the gas particles is constant, this means that the average velocity of the particles must increase. The faster these particles are moving when they hit the wall, the greater the force they exert on the wall. Since the force per collision becomes larger as the temperature increases, the pressure of the gas must increase as well.

BOYLE'S LAW ($PV =$ CONSTANT)

Gases can be compressed because most of the volume of a gas is empty space. If we compress a gas without changing its temperature, the average kinetic energy of the gas particles stays the same. There is no change in the speed with which the particles move, but the container is smaller. Thus, the particles travel from one end of the container to the other in a shorter period of time. This means that they hit the walls more often. Any increase in the number of collisions with the walls must lead to an increase in the pressure of the gas. Thus, the pressure of a gas becomes larger as the volume of the gas becomes smaller.

CHARLES'S LAW ($V \propto T$)

The average kinetic energy of the particles in a gas is proportional to the temperature of the gas. Since the mass of these particles is constant, this means that the particles must move faster as the gas becomes warmer. Because they move faster, the particles exert a greater force on the container each time they hit the walls, and the pressure of the gas therefore increases. If the walls of the container are flexible, it will expand until the pressure of the gas once more balances the pressure of the atmosphere. The volume of the gas therefore becomes larger as the temperature of the gas increases.

AVOGADRO'S HYPOTHESIS ($V \propto n$)

As the number of gas particles becomes larger, the number of collisions with the walls of the container must increase. This, in turn, leads to an increase in the pressure of the gas. Flexible containers, such as balloons, will expand until the pressure of the gas inside balances the pressure of the gas outside. Thus, the volume of the gas is proportional to the number of gas particles.

DALTON'S LAW OF PARTIAL PRESSURES ($P_t = P_1 + P_2 + \ldots$)

There are 40 ball bearings on the apparatus shown in Figure 4.17. Imagine what would happen if 6 ball bearings of a different size were added to the apparatus. The total pressure would become larger, because there would be more collisions with the walls of the container. But the pressure due to the collisions between the 40 ball bearings and the walls of the container would remain the same.

There is so much empty space in the container that each type of ball bearing hits the walls of the container pretty much as often in the mixture as it did when there was only one kind of ball bearing on the glass plate. The total number of collisions with the wall is therefore equal to the sum of the collisions that would occur if each size of ball bearing were present by itself. In other words, the total pressure of a mixture of gases is equal to the sum of the partial pressures of the individual gases.

GRAHAM'S LAWS OF DIFFUSION AND EFFUSION

The average kinetic energy of a gas is proportional to the temperature of the gas, and nothing else. Two gases at the same temperature therefore have the same average kinetic energy. Samples of H_2 and O_2 gas at 0°C, for example, have the same average kinetic energy.

$$\tfrac{1}{2}m_{H_2}v_{H_2}^2 = \tfrac{1}{2}m_{O_2}v_{O_2}^2$$

We can simplify this equation by multiplying both sides by 2.

$$m_{H_2}v_{H_2}^2 = m_{O_2}v_{O_2}^2$$

We can then rearrange it to give the following.

$$\frac{v_{H_2}^2}{v_{O_2}^2} = \frac{m_{O_2}}{m_{H_2}}$$

Taking the square root of both sides of this equation gives a relationship between the ratio of the velocities at which the two gases move and the square root of the ratio of their molecular weights.

$$\frac{v_{H_2}}{v_{O_2}} = \sqrt{\frac{m_{O_2}}{m_{H_2}}}$$

This equation is a modified form of Graham's law. It suggests that the velocity (or rate) at which gas molecules move is inversely proportional to the square root of their molecular weights.

Exercise 4.20

Calculate the average velocity of an H_2 molecule at 0°C if the average velocity of an O_2 molecule at this temperature is 500 meters per second.

Solution

The relative velocities of the H_2 and O_2 molecules at a given temperature are described by the following equation.

$$\frac{v_{H_2}}{v_{O_2}} = \sqrt{\frac{m_{O_2}}{m_{H_2}}}$$

Substituting the weights of H_2 and O_2 molecules and the average velocity of an O_2 molecule into this equation gives the following result.

$$\frac{v_{H_2}}{500 \text{ m/s}} = \sqrt{\frac{32.0 \text{ amu}}{2.0 \text{ amu}}}$$

Solving this equation for the average velocity of an H_2 molecule gives a value of 2000 meters per second—or about 4500 miles per hour.

$$v_{H_2} = 2000 \text{ m/s}$$

It is very easy to make mistakes when setting up a ratio problem such as this. It is therefore important to check the answer to see whether it makes sense. Graham's law states that light molecules move faster on the average than heavy molecules. In this case, the answer does make sense, because H_2 molecules are much lighter than O_2 molecules and should therefore travel much faster.

DEVIATIONS FROM IDEAL GAS LAW BEHAVIOR:
4.17 THE VAN DER WAALS EQUATION

The behavior of real gases usually agrees with the predictions of the ideal gas equation to within $\pm 5\%$ at normal temperatures and pressures. At low temperatures or high pressures, however, real gases deviate from ideal gas behavior. In 1873, while searching for a way to link the behavior of liquids and gases, the Dutch physicist Johannes van der Waals developed an explanation for these deviations and an equation that was able to fit the behavior of real gases over a much wider range of pressures.

Van der Waals realized that two of the assumptions of the kinetic molecular theory were questionable. The kinetic theory assumes that gas particles are so small compared with the distance between them that they occupy a negligible fraction of the total volume of the gas. This theory also assumes that the force of attraction between gas molecules is zero.

The first assumption works well at pressures close to 1 atm. But something happens to the legitimacy of this assumption as the gas is compressed. Imagine for the moment that the atoms or molecules in a gas were all clustered in one corner of a cylinder, as shown in Figure 4.18. At normal pressures, the volume occupied by these particles is a negligibly small fraction of the total volume of the gas. But at high pressures, this is no longer true. At high pressures, the gas is not as compressible as an ideal gas. Thus, at high pressures, the volume of a real gas is often larger than what is expected from the ideal gas equation.

How can we correct for the fact that the volume of a real gas at high pressures is too large to fit the ideal gas equation? Van der Waals proposed that we subtract a term from the volume of the real gas before substituting it into the ideal gas equation. He therefore defined a new constant—b—that was equal to the volume actually occupied by a mole of gas particles. Since the volume of the gas particles depends on the number of moles of gas in the container, the term that is subtracted from the real volume of the gas is equal to n times b.

$$P(V - nb) = nRT$$

When the pressure is relatively small and the volume is reasonably large, the nb term is too small to make any difference in the calculation. But at high pressures,

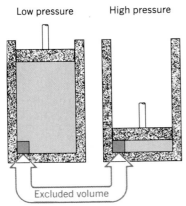

Low pressure   High pressure

Excluded volume

FIG. 4.18 The volume actually occupied by particles in a gas is relatively small at low pressures. But it can be a significant fraction of the total volume at high pressure. In O_2, for example, the gas molecules occupy 0.13% percent of the total volume at 0°C and 1.00 atm but 17% of the volume at 0°C and 100 atm.

TABLE 4.7		
van der Waals Constants for Several Gases		
Compound	*a (L^2-atm / mol^2)*	*b (L / mol)*
He	0.03412	0.02370
Ne	0.2107	0.01709
H_2	0.2444	0.02661
Ar	1.345	0.03219
O_2	1.360	0.03803
N_2	1.390	0.03913
CO	1.485	0.03985
CH_4	2.253	0.04278
CO_2	3.592	0.04267
NH_3	4.170	0.03707

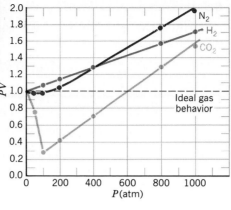

FIG. 4.19 A plot of the product of the pressure times the volume of H_2, N_2, and CO_2 gases versus the pressure of these gases in atmospheres. The product of pressure times volume for an ideal gas should be constant. At low pressures, most gases behave as expected from the ideal gas equation. As the pressure becomes larger, significant deviations from ideal gas behavior are observed. For some gases, PV becomes smaller than expected when the pressure is increased, because of the force of attraction between the gas molecules. Once the pressure gets large enough, PV becomes larger than expected, because gas particles no longer occupy a negligibly small fraction of the total volume of the gas.

when the volume of the gas is small, the *nb* term corrects for the fact that the volume of a real gas is larger than expected from the ideal gas equation.

If the assumption that there is no force of attraction between gas particles was true, gases would never condense to form liquids. In reality, there is a small force of attraction between gas particles that tends to hold the particles together. This force of attraction has two consequences: (1) gases condense to form liquids at low temperatures, and (2) the pressure of a real gas is sometimes smaller than expected for an ideal gas.

To correct for the fact that the pressure of a real gas is smaller than expected from the ideal gas equation, van der Waals added a term to the pressure in this equation. This term contains a second constant, *a*, and has the form an^2/V^2. The complete **van der Waals equation** is written as follows.

$$\left(P + \frac{an^2}{V^2}\right)(V - nb) = nRT$$

This equation is something of a mixed blessing. It provides a much better fit with the behavior of a real gas than the ideal gas equation. But it does this at the price of a loss in generality. The ideal gas equation is equally valid for any gas, while the van der Waals equation contains a pair of constants — *a* and *b* — that change from gas to gas.

The values of *a* and *b* for a given gas are calculated by fitting the experimental pressure and volume data for the gas to the van der Waals equation. Experimental data for the product of *P* times *V* versus the pressure in atmospheres for H_2 and N_2 gas at 0°C and CO_2 at 40°C are given in Figure 4.19. Values of the van der Waals constants for these and other gases are given in Table 4.7.

Exercise 4.21

(a) Calculate the pressure of 1.00 mole of CO_2 in a 22.4-liter container at 0°C using the ideal gas equation. Repeat this calculation using the van der Waals equation and compare the results.

(b) Repeat these calculations assuming the volume of the gas is 0.200 liter.

(c) Repeat these calculations assuming the volume of the gas is 0.0500 liter.

Solution

(a) According to the ideal gas equation, the pressure of this gas should be 1.00 atm.

$$P = \frac{nRT}{V} = \frac{(1.00 \text{ mol})(0.08206 \text{ L-atm/mol-K})(273 \text{ K})}{(22.4 \text{ L})} = 1.00 \text{ atm}$$

Substituting what we know about the gas into the van der Waals equation gives a much more complex equation.

$$\left[P + \frac{(3.592 \text{ L}^2\text{-atm/mol})(1.00 \text{ mol})^2}{(22.4 \text{ L})^2} \right]$$
$$[22.4 \text{ L} - (1.00 \text{ mol})(0.04267 \text{ L/mol})]$$
$$= (1.00 \text{ mol})(0.08206 \text{ L-atm/mol-K})(273 \text{ K})$$

But this equation can be solved for the pressure of the gas.

$$P = 0.995 \text{ atm}$$

At normal temperatures and pressures, the ideal gas equation and the van der Waals equation give essentially the same results.

(b) According to the ideal gas equation, the pressure would have to increase to 112 atm to compress 1.00 mole of CO_2 at 0°C to a volume of 0.200 liters.

$$P = \frac{nRT}{V} = \frac{(1.00 \text{ mol})(0.08206 \text{ L-atm/mol-K})(273 \text{ K})}{(0.200 \text{ L})} = 112 \text{ atm}$$

The van der Waals equation, however, predicts that the pressure will only have to increase to 52.6 atm to achieve the same results.

$$\left[P + \frac{(3.592 \text{ L}^2\text{-atm/mol})(1.00 \text{ mol})^2}{(0.200 \text{ L})^2} \right]$$
$$[0.200 \text{ L} - (1.00 \text{ mol})(0.04267 \text{ L/mol})]$$
$$= (1.00 \text{ mol})(0.08206 \text{ L-atm/mol-K})(273 \text{ K})$$
$$P = 52.6 \text{ atm}$$

As shown in Figure 4.19, as the pressure on CO_2 increases, the van der Waals equation initially gives pressures that are *smaller* than the ideal gas equation because of the strong force of attraction between CO_2 molecules.

(c) The ideal gas equation predicts the pressure would have to increase (to 448 atm to condense the gas to a volume of 0.0500 liter.

$$P = \frac{nRT}{V} = \frac{(1.00 \text{ mol})(0.08206 \text{ L-atm/mol-K})(273 \text{ K})}{(0.0500 \text{ L})} = 448 \text{ atm}$$

The van der Waals equation, however, predicts the pressure will have to reach 1620 atm to achieve the same results.

$$\left[P + \frac{(3.592 \text{ L}^2\text{-atm/mol})(1.00 \text{ mol})^2}{(0.0500 \text{ L})^2} \right]$$
$$[0.0500 \text{ L} - (1.00 \text{ mol})(0.04267 \text{ L/mol})]$$
$$= (1.00 \text{ mol})(0.08206 \text{ L-atm/mol-K})(273 \text{ K})$$
$$P = 1620 \text{ atm}$$

As shown in Figure 4.19, the van der Waals equation gives results that are *larger* than the ideal gas equation at very high pressures because of the volume occupied by the CO_2 molecules.

4.18 ANALYSIS OF THE VAN DER WAALS CONSTANTS

The van der Waals equation contains two constants, a and b, that are characteristic of a particular gas. The first of these constants corrects for the force of attraction between gas particles. Compounds for which the force of attraction between particles is strong have large values for a. If you think about what happens when a liquid boils, you might expect that compounds with large values of a would have large boiling points. As the force of attraction between gas particles becomes stronger, higher and higher temperatures are required to break the bonds between the molecules in the liquid to form a gas. It therefore isn't surprising to find a correlation between the value of the a constant in the van der Waals equation and the boiling points of a number of simple compounds, as shown in Figure 4.20. Gases with very small values of a, such as H_2 and He, must be cooled almost to absolute zero before they condense to form liquid.

The other van der Waals constant, b, is a rough measure of the size of a gas particle. According to the data in Table 4.7, the volume of a mole of argon atoms is 0.03219 liters. This number can be used to estimate the volume of an individual argon atom.

$$0.03219 \, \frac{L}{mol} \times \frac{1 \, mol}{6.022 \times 10^{23} \, atoms} = 5.345 \times 10^{-26} \, \frac{L}{atom}$$

The volume of an argon atom can then be converted into cubic centimeters by use of the appropriate unit factors.

$$5.345 \times 10^{-26} \, \frac{L}{atom} \times \frac{1000 \, mL}{L} \times \frac{1 \, cm^3}{mL} = 5.345 \times 10^{-23} \, \frac{cm^3}{atom}$$

If we assume that argon atoms are spherical, we can use this estimate of the volume of an individual argon atom to estimate the radius of one of these atoms. We start by noting that the volume of a sphere is related to its radius by the following formula.

$$V = \frac{4}{3} \pi r^3$$

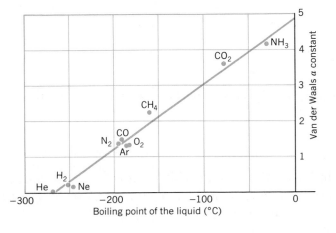

Boiling point of the liquid (°C)

FIG. 4.20 The boiling point of a liquid is an indirect measure of the force of attraction between its molecules. The first van der Waals constant —a— is a measure of the force of attraction between molecules in a gas. It isn't surprising, therefore, to find a correlation between these quantities for a number of compounds that are gases at room temperature.

We then assume that the volume of an argon atom is 5.345×10^{-23} cubic centimeters and calculate the radius of the atom.

$$\frac{4}{3}\pi r^3 = 5.345 \times 10^{-23} \text{ cm}^3$$

$$r = 2.3 \times 10^{-8} \text{ cm}$$

According to this calculation, an argon atom has a radius of about 2×10^{-8} centimeters.

SUMMARY

The word *gas* comes from the Greek stem *chaos*, because a gas is literally the most chaotic of all chemical systems —it consists of elementary particles in constant, random motion. The volume of a typical gas at room temperature and atmospheric pressure is about 800 times the volume of the liquid or solid from which it forms. Under these conditions, the particles in the gas occupy a negligibly small fraction of the total volume of the gas. This model explains many of the properties of gases. It helps us understand why gases are easy to compress—most of their volume is empty space. It explains why they expand to fill their containers—the particles move constantly and randomly until they collide with the walls of their container.

Four characteristic properties are used to describe a gas: pressure, volume, temperature, and the amount of the gas. By holding any two of these properties constant, we can determine the relationship between the other two. The discoveries known as Boyle's law, Charles's law, and so on are nothing more than special cases of a more general relationship among these four properties known as the ideal gas equation: $PV = nRT$.

The ideal gas equation works very well with almost any gas at reasonable temperatures and pressures. This equation summarizes an important characteristic of gases: many, if not most, of their properties are independent of the identity of the particles that form the gases.

One of the properties of a gas that does depend on its identity is the average speed or velocity as which the gas particles move. Because the average kinetic energy of a gas is proportional to the temperature of the gas, and nothing else, two gases at the same temperature have the same average kinetic energy. The fact that the kinetic energy of a particle is proportional to the product of its mass times its velocity squared means that relatively heavy gas particles must move slower than lighter gas particles at a given temperature and pressure.

Deviations from ideal gas behavior occur at low temperatures or high pressures. At low temperatures, gases condense to form liquids because the force of attraction between gas particles isn't zero. As the pressure increases, the molecules are compressed into a smaller volume, where the force of attraction between the gas molecules becomes even more important. As the pressure continues to increase, however, the product of the pressure times the volume of the gas eventually becomes larger than what is expected from the ideal gas equation. We can understand this by noting that the volume of the gas particles is no longer a negligibly small fraction of the total volume of the gas at very high pressures. Real gases are therefore not quite as easy to compress as an ideal gas at high pressures.

PROBLEMS

Gases at Room Temperature

4-1 Which of the following elements and compounds are most likely to be gases at room temperature?

(a) Ar (b) CO (c) CH_4 (d) $C_{10}H_{22}$ (e) Cl_2 (f) Fe_2O_3
(g) Na (h) NaCl (i) Pt (j) S_8

The Properties of Gases

4-2 Describe in detail how you would measure the following properties of a gas. In each case, describe the equipment you would use and the steps you would have to take to make these measurements.

(a) volume   (b) weight   (c) pressure   (d) temperature
(e) density

4-3 Describe the contribution of the following individuals to our understanding of the properties of gases.

(a) Boyle   (b) Amontons   (c) Charles   (d) Gay-Lussac
(e) Avogadro   (f) Graham   (g) van der Waals

4-4 Describe two ways in which the behavior of gases can be used to show that fluorine consists of F_2 molecules, not F atoms.

4-5 Imagine two identical flasks at the same temperature. One contains 2 grams of H_2 and the other contains 28 grams of N_2. Which of the following properties are the same for the two flasks?

(a) Pressure   (b) Average kinetic energy   (c) Density
(d) The number of molecules per container   (e) The weight of the container

4-6 Explain why the difference between the radius of a helium atom and that of a xenon atom has no effect on the volume of a mole of these gases.

Pressure versus Force

4-7 Describe the relationship between pressure and force.

4-8 Explain why women in high-heeled shoes can exert more force on the surface of a floor than men who weigh two to three times as much.

4-9 Calculate the force in pounds on the bottom of the column of mercury in a barometer when atmospheric pressure is 1.00 atm, assuming the diameter of the tube is 1.00 centimeter and the density of mercury is 13.6 g/cm^3. Calculate the pressure in pounds per square inch.

Atmospheric Pressure

4-10 What is the pressure in units of atmospheres when a barometer reads 745.8 mmHg? What is the pressure in units of pascals?

4-11 Convert a pressure of 20.0 lb/in.² into units of atmospheres, mmHg, and pascals.

4-12 One atm pressure will support a column of mercury 760 millimeters tall in a barometer with a tube 1.00 centimeter in diameter. What would be the height of the column of mercury if the diameter of the tube was twice as large?

4-13 Atmospheric pressure is announced in weather reports in the United States in units of inches of mercury. How many inches of mercury would exert a pressure of 1.00 atm?

4-14 If a pressure of 1 atm can support a column of mercury 760 millimeters tall, how tall a column of water will 1 atm support? (*Hint:* look at the difference between the densities of the two liquids.)

4-15 If the directions that come with your car tell you to inflate the tires to 200 kPa pressure and you have a tire pressure gauge calibrated in pounds per square inch (psi), what pressure in psi will you use?

4-16 The vapor pressure of the mercury gas that collects at the top of a barometer is 2×10^{-3} mmHg. Calculate the vapor pressure of this gas in atmospheres.

Boyle's Law

4-17 A 425-milliliter sample of O_2 gas was collected at 742.3 mmHg. What would be the pressure in mmHg if this gas was allowed to expand to 975 milliliters at constant temperature?

4-18 What happens to the volume of a balloon filled with 0.357 liter of H_2 gas collected at 741.3 mmHg when the atmospheric pressure increases to 758.1 mmHg?

4-19 What is the volume of a scuba tank if it takes 2000 liters of air collected at 1 atm to fill the tank to a pressure of 150 atm?

4-20 Calculate the volume of a balloon that could be filled with the helium in a 2.50-liter compressed gas cylinder that has a pressure of 200 atm at 25°C.

Amontons's Law

4-21 A butane lighter fluid can is filled with 5.00 atm of C_4H_{10} gas at 21°C. Calculate the pressure in the can when it is stored in a warehouse on a hot summer's day when the temperature reaches 38°C.

4-22 An automobile tire was inflated to a pressure of 32 lb/in.² at 21°C. At what temperature would the pressure reach 50 psi?

4-23 At 25°C, four-fifths of the pressure of the atmosphere is due to N_2 and one-fifth is due to O_2. What fraction of the pressure is due to N_2 at 100°C?

Charles's Law

4-24 Use Charles's law to explain how a hot-air balloon works.

4-25 Calculate the percent change in the volume of a toy balloon as the gas inside is heated from 22°C to 75°C in a hot-water bath.

4-26 A sample of O_2 gas with a volume of 0.357 liter was collected at 21°C. Calculate the volume of this gas when it is cooled to 0°C.

Gay-Lussac's Law

4-27 Calculate the ratio of the volumes of sulfur dioxide and oxygen produced when sulfuric acid decomposes.

$$2 \, H_2SO_4(aq) \longrightarrow 2 \, SO_2(g) + O_2(g) + 2 \, H_2O(l)$$

4-28 Calculate the volume of H_2 and N_2 gas formed when 1.38 liters of NH_3 decomposes at a constant temperature and pressure.

$$2 \, NH_3(g) \longrightarrow N_2(g) + 3 \, H_2(g)$$

4-29 Ammonia burns in the presence of oxygen to form nitrogen oxide and water.

$$4 \, NH_3(g) + 5 \, O_2(g) \longrightarrow 4 \, NO(g) + 6 \, H_2O(g)$$

What volume of NO can prepared when 15.0 liters of ammonia reacts with excess oxygen, assuming that all measurements are made at the same temperature and pressure?

4-30 Acetylene burns in oxygen to form CO_2 and H_2O.

$$2 C_2H_2(g) + 3 O_2(g) \longrightarrow 4 CO_2(g) + 2 H_2O(g)$$

Calculate the total volume of the products formed when 15.0 liters of C_2H_4 burns in the presence of 15.0 liters of O_2, if all measurements are made at the same temperature and pressure.

4-31 Methane reacts with steam to form hydrogen and carbon monoxide.

$$CH_4(g) + H_2O(g) \longrightarrow CO(g) + 3 H_2(g)$$

It can also react with steam to form carbon dioxide.

$$CH_4(g) + 2 H_2O(g) \longrightarrow CO_2(g) + 4 H_2(g)$$

What is the product of this reaction if 1.50 liters of methane is found by experiment to react with 1.50 liters of water vapor?

Avogadro's Hypothesis

4-32 Which of the following samples would have the largest volume at 25°C and 75) mmHg?

(a) 100 g CO_2 (b) 100 g CH_4 (c) 100 g NO (d) 100 g SO_2

4-33 Nitrous oxide decomposes to form nitrogen and oxygen. Use Avogadro's hypothesis to determine the formula for nitrous oxide if 2.36 liters of this compound decomposes to form 2.36 liters of N_2 and 1.18 liters of O_2.

The Ideal Gas Equation

4-34 Predict the shape of the following graphs for an ideal gas.

(a) Pressure versus volume (b) Pressure versus temperature (c) Volume versus temperature (d) Kinetic energy versus temperature (e) Pressure versus the number of moles of gas

4-35 Which of the following graphs does not give a straight line for an ideal gas?

(a) V versus T (b) T versus P (c) P versus $1/V$ (d) n versus $1/T$ (e) n versus $1/P$

4-36 Which of the following statements is always true for an ideal gas?

(a) If the temperature and volume of a gas both increase at constant pressure, the amount of gas must also increase. (b) If the pressure increases and the temperature decreases for a constant amount of gas, the volume must decrease. (c) If the volume and the amount of gas both decrease at constant temperature, the pressure must decrease.

4-37 Calculate the value of the ideal gas constant in units of mL-psi/mol-K if 1.00 mole of an ideal gas at 0°C occupies a volume of 22,400 milliliters at 14.7 lb/in.^{2} pressure.

Ideal Gas Calculations: Part I

4-38 Nitrogen gas sells for roughly 50 cents per 100 cubic feet at 0°C and 1 atm. What is the price per gram of nitrogen?

4-39 Calculate the pressure in atmospheres of 80 grams of CO_2 in a 30-liter container at 23°C.

4-40 Calculate the volume of 5.0 grams of CH_4 gas at 23°C and 758 mmHg.

4-41 Calculate the temperature at which 1.5 grams of O_2 has a pressure of 740 mmHg in a 1-liter container.

4-42 Calculate the number of kilograms of O_2 gas that can be stored in a compressed gas cylinder with a volume of 40 liters when the cylinder is filled at 150 atm and 21°C.

4-43 Calculate the pressure in a 250-milliliter container at 0°C when the O_2 in 1 cubic centimeter of liquid oxygen (density = 1.118 g/cm^3) evaporates.

4-44 Assume that a 1.00-liter flask was evacuated, 5.0 grams of liquid ammonia were added to the flask, and the flask was sealed with a cork. When the NH_3 in the flask warms to 21°C, will the cork be blown out of the mouth of the flask?

4-45 What is the volume of the Goodyear blimp if it contains 1.26×10^{29} He atoms at 25°C and 1 atm? Calculate the density of the gas in the blimp. If the density of air is 1.18 g/cm^3 under the same conditions, what weight can this volume of helium lift?

Ideal Gas Calculations: Part II

4-46 What is the volume of the gas in a balloon at −195°C if the balloon was filled to a volume of 5.0 liters at 25°C.

4-47 CO_2 gas with a volume of 25.0 liters was collected at 25°C and 0.982 atm. Calculate the pressure of this gas if it is compressed to a volume of 0.15 liters and heated to 350°C.

4-48 Calculate the pressure of 4.80 grams of ozone, O_3, in a 2.46-liter flask at 25°C and 0.985 atm. When the temperature was raised to 125°C, the ozone decomposed to molecular oxygen. Calculate the pressure inside the flask once this reaction was complete.

$$2 O_3(g) \longrightarrow 3 O_2(g)$$

4-49 One mole of an ideal gas occupies a volume of 22.4 liters at 0°C and 1.00 atm. What would be the volume of a mole of an ideal gas at 22°C and 748.8 mmHg?

4-50 Assume that 10.0 liters of O_2 gas were collected at 120°C and 749.3 mmHg. Calculate the volume of this gas when it is cooled to 0°C and stored in a container at 1.00 atm.

4-51 Assume that 5.0 liters of CO_2 gas were collected at 25°C and 2.5 atm. At what temperature would this gas have to be stored to fill a 10.0-liter flask at 0.978 atm?

4-52 Assume that two 10-liter samples of O_2 collected at 120°C and 749.3 mmHg were combined and stored in a 1.25-liter flask at 27°C. Calculate the pressure of this gas.

Ideal Gas Calculations Involving the Density of a Gas

4-53 Calculate the density of CH_2Cl_2 in the gas phase at 40°C and 1.00 atm. Compare this with the density of liquid CH_2Cl_2 (1.336 g/cm³).

4-54 Calculate the density of methane gas, CH_4, in kilograms per cubic meter at 25°C and 956 mmHg.

4-55 Calculate the density of helium at 0°C and 1.00 atm and compare this with the density of air (1.29 g/cm³) at 0°C and 1.00 atm. Explain why 1.00 cubic feet of helium can lift a weight of 0.076 pounds under these conditions.

4-56 Calculate the ratio of the densities of H_2 and O_2 at 0°C and 100°C.

4-57 Which of the rare gases in Group VIIIA of the periodic table has a density of 3.7493 grams per liter at 0°C?

4-58 Calculate the average molecular weight of air assuming a sample of air weighs 1.700 times as much as an equivalent volume of ammonia, NH_3.

4-59 Calculate the molecular formula of diazomethane, assuming this compound is 28.6% C, 4.8% H, and 66.6% N by weight and the density of this gas is 1.72 grams per liter at 25°C and 1 atm.

4-60 What are the molecular formulas for phosphine, PH_x, and diphosphine, P_2H_y, if the densities of these gases are 1.517 and 2.944 grams per liter, respectively, at 0°C and 1 atm?

Dalton's Law of Partial Pressures ($P_t = P_1 + P_2 + \ldots$)

4-61 Calculate the partial pressure of propane in a mixture that contains equal weights of propane (C_3H_8) and butane (C_4H_{10}) at 20°C and 746 mmHg.

4-62 Calculate the partial pressure of helium in a 1.00-liter flask that contains equal numbers of moles of N_2, O_2, and He at a total pressure of 7.5 atm.

4-63 Calculate the total pressure in a 10.0-liter flask at 27°C of a sample of gas that contains 6.0 grams of H_2, 15.2 grams of N_2, and 16.8 grams of He.

4-64 A 1-liter flask is filled with carbon monoxide at 27°C until the pressure is 0.200 atm. Calculate the total pressure after 0.450 grams of carbon dioxide has been added to this flask.

4-65 A few milliliters of water were added to a 1-liter flask at 25°C and the vapor pressure of the water that evaporated was found to be 23.8 mmHg at this temperature. What would be the vapor pressure of the water if this experiment was repeated in a 2-liter flask?

4-66 Calculate the volume of NO obtained when the water vapor is removed from 125 milliliters of NO gas collected by displacement of water from a flask at 25°C and 758 mmHg.

4-67 Calculate the volume of the hydrogen obtained when the water vapor is removed from 289 milliliters of H_2 gas collected by displacement of water from a flask at 15°C and 0.988 atm.

4-68 What would happen to the volume of 4.5 liters of air saturated with water vapor at 22°C and 752 mmHg if the water vapor was removed?

Graham's Law of Diffusion and Effusion (Rate $\propto 1/\sqrt{MW}$)

4-69 List the following gases in order of increasing rate of diffusion.

(a) Ar   (b) Cl_2 (c) CF_2Cl_2 (d) SO_2 (e) SF_6

4-70 Bromine vapor is roughly five times as dense as oxygen gas. Calculate the relative rates at which $Br_2(g)$ and $O_2(g)$ diffuse.

4-71 Two flasks with the same volume are connected by a valve. One gram of hydrogen is added to one flask, and 1 gram of oxygen is added to the other. What happens to the weight of the gas in the flask filled with hydrogen when the valve is opened?

4-72 What happens to the relative amounts of N_2, O_2, Ar, CO_2, and He in air as it diffuses from one flask to another through a pinhole?

4-73 N_2O and NO are often known by the common names nitrous oxide and nitric oxide. Associate the correct formula with the appropriate common name if nitric oxide diffuses through a pinhole 1.21 times as fast as nitrous oxide.

4-74 If it takes 6.5 seconds for 25.0 cubic centimeters of helium gas to effuse through a pinhole into a vacuum, how long would it take for 25.0 cubic centimeters of CH_4 to escape under the same conditions?

4-75 Calculate the molecular weight of an unknown gas if it takes 60.0 seconds for 250 cubic centimeters of this gas to escape through a pinhole in a flask into a vacuum and it takes 84.9 seconds for the same volume of oxygen to escape under identical conditions.

4-76 A lecture hall has 50 rows of seats. If laughing gas (N_2O) is released from the front of the room at the same time hydrogen cyanide (HCN) is released from the back of the room, in which row (counting from the front) will students first begin to die laughing?

4-77 The atomic weight of radon was first estimated by comparing its rate of diffusion with that of mercury vapor. What is the atomic weight of radon if mercury vapor diffuses 1.082 times as fast?

The Kinetic Molecular Theory

4-78 What would happen to a balloon if the collisions between gas molecules were not perfectly elastic?

4-79 What would happen to a balloon if the gas molecules were in a state of constant motion, but the motion was not random?

4-80 Explain how the pressure of a gas is evidence for the assumption that gas particles are in a state of constant random motion.

4-81 Use the kinetic molecular theory to explain why the pressure of a gas is proportional to the number of gas particles and the temperature of the gas but inversely proportional to the volume of the gas.

4-82 Which of the following statements explains why a hot-air balloon rises when the air in the balloon is heated?

(a) The average kinetic energy of the molecules increases, and the collisions between these molecules and the walls of the balloon make it rise. (b) The pressure of the gas inside the balloon increases, pushing up on the balloon. (c) The gas expands, forcing some of it to escape from the bottom of the balloon, and the decrease in the density of this gas lifts the balloon. (d) The balloon expands, causing it to rise. (e) The hot air rising inside the balloon produces enough force to lift the balloon.

Deviations from Ideal Gas Behavior

4-83 Does the force of attraction between particles cause the volume of a real gas to be larger or smaller than an ideal gas?

4-84 Does the fact that the volume of gas particles is not zero cause the volume of a real gas to be larger or smaller than an ideal gas?

4-85 Which term in the van der Waals equation would be used to explain why gases become cooler when they are allowed to expand rapidly?

4-86 Significant deviations from ideal gas behavior are observed under what conditions?

4-87 Calculate the fraction of empty space in CO_2 gas, assuming 1 liter of this gas at 0°C and 1.00 atm can be compressed until it changes to a liquid with a volume of 1.26 cubic centimeters.

4-88 The following data were obtained in a study of the pressure and volume of a sample of acetylene (*Ann. Chm. Phys., 19,* 345).

P (atm):	1	45.8	84.2	110.5	176.0	282.2	398.7
V (L):	1	0.01705	0.00474	0.00411	0.00365	0.00333	0.00313

Calculate the product of the pressure times the volume for each measurement. Plot *PV* versus *P* and explain the shape of this curve.

Combined Stoichiometry and Gas Law Problems

4-89 Equal volumes of oxygen and an unknown gas weigh 3.00 grams and 7.50 grams, respectively. Which of the following is the unknown gas?

(a) CO (b) CO_2 (c) NO (d) NO_2 (e) SO_2 (f) SO_3

4-90 What is the molecular weight of acetone if 0.520 grams of acetone occupies a volume of 275.5 milliliters at 100°C and 756 mmHg.

4-91 Boron forms a number of compounds with hydrogen, including B_2H_6, B_4H_{10}, B_5H_9, B_5H_{11}, and B_6H_{10}. For which compound would a 1.00-gram sample occupy a volume of 390 cubic centimeters at 25°C and 0.993 atm?

4-92 Cyclopropane is an anaesthetic that is 85.63% carbon and 14.37% hydrogen by weight. What is the molecular formula of this compound if 0.45 liter of cyclopropane reacts with excess oxygen at 120°C and 0.72 atm to form 1.35 liters of carbon dioxide and 1.35 liters of water vapor?

4-93 Calculate the weight of magnesium that would be needed to generate 500 milliliters of hydrogen gas at 0°C and 1.00 atm.

$$Mg(s) + 2\ HCl(aq) \longrightarrow Mg^{2+}(aq) + 2\ Cl^-(aq) + H_2(g)$$

4-94 Calculate the formula of the oxide formed when 10.0 grams of chromium metal reacts with 6.98 liters of O_2 at 20°C and 0.994 atm.

4-95 Calculate the volume of CO_2 gas measured at 756 mmHg and 23°C given off when 150 pounds of limestone is heated until it decomposes.

$$CaCO_3(s) \longrightarrow CaO(s) + CO_2(g)$$

4-96 Calculate the volume of O_2 that would have to be inhaled at 20°C and 1.00 atm to consume 1.00 pound of fat— $C_{57}H_{110}O_6$.

$$2\ C_{57}H_{110}O_6(s) + 163\ O_2(g) \longrightarrow 114\ CO_2(g) + 110\ H_2O(aq)$$

4-97 Calculate the volume of CO_2 gas collected at 23°C and 0.991 atm that can be prepared by reaction of 10.0 grams of calcium carbonate with excess acid.

$$CaCO_3(aq) + 2\ H^+(aq) \longrightarrow Ca^{2+}(aq) + CO_2(g) + H_2O$$

4-98 Calculate the volume of phosphine at 19°C and 750 mmHg that could be prepared by addition of 100 grams of calcium phosphide to water.

$$Ca_3P_2(s) + 6\ H_2O(aq) \longrightarrow 3\ Ca(OH)_2(aq) + 2\ PH_3(g)$$

4-99 Determine the identity of an unknown metal if 1.00 gram of this metal reacts with excess acid according to the following equation to produce 374 milliliters of H_2 gas at 25°C and 1.00 atm.

$$M(s) + 2\ H^+(aq) \longrightarrow M^{2+}(aq) + H_2(g)$$

THE STRUCTURE OF THE ATOM

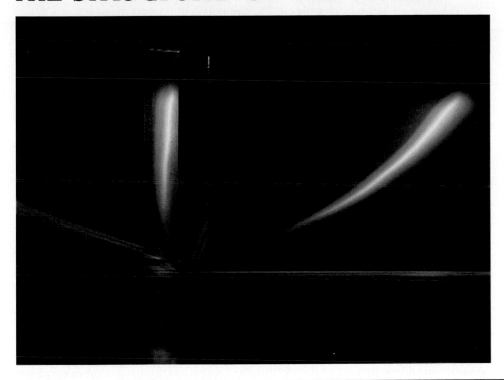

One of the most exciting aspects of science to those who practice it is the way each advance in our understanding raises new questions to be answered. Questions, in fact, invariably increase faster than answers. The topics discussed in the first four chapters raise a number of questions.

1. Why are elements such as N_2 or O_2 and covalent compounds such as CO_2 or HCl gases at room temperature?
2. Why are metals such as silver and gold and salts such as NaCl and $CaCO_3$ solids at room temperature?
3. Why does sugar dissolve in water when other covalent compounds such as gasoline do not?
4. Why does NaCl dissolve in water when other salts such as $CaCO_3$ do not?
5. Why is the O_2 in the atmosphere so reactive that almost everything we see in the world around us either will react or has reacted with O_2, whereas the N_2 in the atmosphere is essentially inert?
6. Why does sodium react with chlorine to form NaCl, not $NaCl_2$ or Na_2Cl?

It is going to take at least nine chapters to build a model that can explain these observations. The first step toward this model involves building a basic understanding of the structure of atoms.

5.1 ELECTRICITY AND MATTER

If you believe TV commercials, one of the most devastating problems of modern society is the static cling that makes socks stick to other pieces of clothing in a clothes dryer. If pressed to explain static cling, people often suggest that the constant, tumbling motion of the clothing in a warm, moist environment produces a static electric charge that makes the clothing stick together, without necessarily understanding any of the key words: *static, electric,* and *charge.* Where do these words come from?

As early as the fourth century B.C., Plato noted that a yellow substance then known as *elektron* attracted lightweight objects when rubbed against a piece of fur. This material is now known as amber, and it has been shown to be the solid, fossilized resin from pine trees. By 1600, William Gilbert was able to show that a number of different "electrics"—including diamonds, sapphires, opals, glass, sulfur, and sealing wax—produced a similar effect.

In 1733, Charles François de Cisternay du Fay noticed that objects that had been rubbed sometimes attracted each other and sometimes repelled each other. He explained this by proposing that two different kinds of electricity existed. ***Vitreous electricity*** (from the Latin for glass) was produced when glass or gems were rubbed. ***Resinous electricity*** (from the Latin for resin or amber) was obtained when amber, silk, or paper were rubbed. Du Fay argued that objects with different kinds of electricity attract each other, whereas objects with the same kind of electricity repel.

Du Fay's discovery soon led to a theory of electricity that assumed the existence of two fluids. Objects that were not electrified had equal amounts of these fluids, which neutralized each other. Rubbing an object removed one of these fluids, leaving an excess of the other.

An opposing school of thought was developed by Benjamin Franklin, who argued that there was only one kind of electric fluid, which was present in all

Pieces of paper are attracted to amber that has been rubbed with fur.

bodies. According to this theory, an object picked up an electric charge when some of this fluid was transferred from one body to another. When an object was rubbed, it either gained or lost electricity and thereby picked up either a positive or negative charge. Objects that had opposite charges attracted each other; objects with the same charge repelled.

Both Du Fay's two-fluid theory of electricity and Franklins one-fluid model were based on the assumption that it is possible to **charge** an object with these fluids, much as one can load, or charge, a cannon with gunpowder. To this day, we talk about an object that has been electrified as if it carries an electric "charge." Furthermore, we describe the charge as **static,** because it does not move.

5.2 THE DISCOVERY OF THE ELECTRON

Many of the early experiments with electricity focused on liquids and solids. By about 1830, however, Michael Faraday and others had begun to study the effect of an electric current on a gas. Faraday's apparatus consisted of a pair of metal plates sealed in a glass tube as shown in Figure 5.1a. The glass tube was filled with a gas, the metal plates were connected to a series of batteries, and the gas was slowly pumped out of the tube.

Faraday noted that when the pressure of the gas in the tube became small enough, the gas began to glow. The neon lights that illuminate the streets in many commercial areas and the mercury vapor lamps that light our highways are both examples of what happens when an electric current is passed through a gas at low pressure.

In 1858, Julius Plücker noted that when the pressure of the gas inside the tube is very small, the *glass* at one end of the tube emits light. He also found that he could change the position of the patch of glass that glowed by bringing a magnet close to

FIG. 5.1 (*a*) Making cathode-ray tubes involves sealing a pair of metal plates into a glass tube, filling the tube with a gas, and then pumping out the tube until only a small residual pressure is left. If the pressure is small enough, when the metal plates are connected to a series of batteries, the glass at the end of the tube across from the cathode will glow. (*b*) The rays given off by the cathode in this tube can be deflected by a magnetic field. The direction in which the beam is deflected suggests that these cathode rays are negatively charged. (*c*) A solid object placed in the path of the cathode rays will cast a shadow on the wall of the tube across from the cathode. (*d*) The cathode-ray beam can be deflected by an electric field. The fact that the beam is attracted toward the positive end of the electric field suggests that the cathode rays are negatively charged.

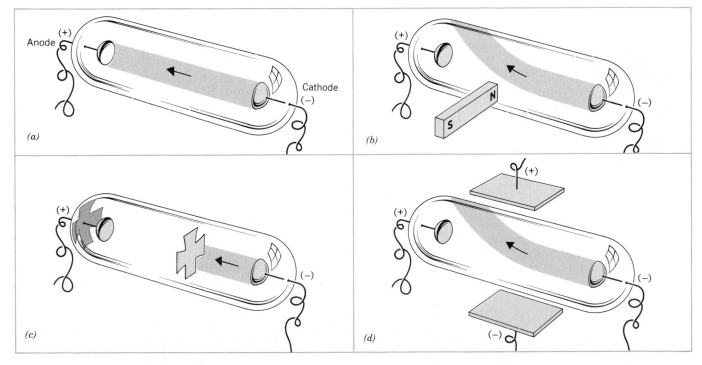

Neon lights are modern examples of what happens when an electric current is passed through a gas in a cathode-ray tube at low pressure.

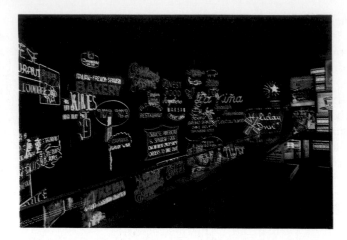

the tube, as shown in Figure 5.1*b*. Plücker interpreted the effect of a magnetic field as evidence that whatever produced this glow was electrically charged.

Much of the language used to describe electricity was proposed by either Benjamin Franklin or an 18th-century British philosopher named William Whewell. Franklin introduced the words *plus, minus, positive, negative, charge,* and *battery.* Whewell coined a number of other terms, including *cathode* and *anode.* By convention, the **cathode** was the metal plate connected to the negative terminal of the series of batteries used to study the effect of an electric current on a gas at low pressure. The other plate was the **anode.**

In 1869, Johannes Hittorf showed that when a solid object was placed between the cathode and anode of this apparatus, a shadow was cast on the end of the tube across from the cathode, as shown in Figure 5.1*c*. This suggested that some beam, or ray, was given off by the cathode, and these tubes soon became known as **cathode-ray tubes.**

The definitive experiments with cathode-ray tubes were reported by William Crookes in 1879. Crookes' major contribution was the development of a better vacuum pump, which allowed him to produce cathode ray tubes with a smaller residual gas pressure. Crookes not only confirmed the previous work by Plücker, Hittorf, and others, but he was able to show that cathode rays were *negatively* charged by studying the direction in which these rays were deflected by a magnetic field.

The key question at this point was whether cathode rays were really "rays" (that is, waves) or whether they were composed of negatively charged particles. The answer to this question was obtained by J. J. Thomson in 1897. Thomson found that the cathode rays could be deflected by an electric field, as shown in Figure 5.1*d*. By balancing the effect of a magnetic field on a cathode-ray beam with an electric field, Thomson was able to show that cathode rays were actually composed of particles, and he was able to estimate the ratio of the charge to the mass of these particles.

Thomson found that the same charge-to-mass ratio was observed no matter what metal was used to make the cathode and the anode. He also found the same charge-to-mass ratio regardless of what gas was used to fill the tube. He therefore concluded that the particles given off by the cathode in this experiment were a universal component of matter. Although Thomson called these particles "corpuscles," the name **electron,** which had been proposed by George Stoney several years earlier for the fundamental unit of negative electricity, was soon accepted.

5.3 THE DISCOVERY OF X-RAYS

In 1895, William Conrad Roentgen became interested in the ultraviolet (UV) radiation emitted by cathode-ray tubes. Since the eye cannot detect ultraviolet radiation, experiments of this type require a UV detector. Roentgen used a screen coated with barium tetracyanoplatinate, because $BaPt(CN)_4$ is one of a family of compounds that emit light, or ***fluoresce,*** when they are exposed to UV radiation.

One evening in November 1895, Roentgen was working with a cathode-ray tube that had been carefully wrapped with black cardboard. Much to his surprise, the $BaPt(CN)_4$ screen next to the cathode-ray tube gave off light when the tube was switched on. Obviously, something had hit the screen to make it emit light. It was equally obvious, however, that it couldn't have been either UV radiation or cathode rays, because neither of these substances could pass through the opaque cardboard.

In an intense series of experiments over a seven-week period, Roentgen found that this new kind of radiation—which he called ***x-rays***—passed through solid objects placed between the cathode-ray tube and the detector. He even found that he could see the image of the bones in his hand when he held it between the tube and the screen. Roentgen eventually discovered that he could capture such images on film; one of his first x-ray images is shown in Figure 5.2.

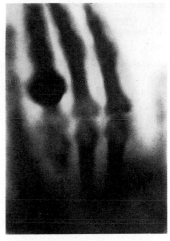

FIG. 5.2 One of the earliest x-ray images. There is some reason to believe this is an x-ray of Roentgen's wife's hand.

5.4 RADIOACTIVITY

The discovery of a form of radiation that could pass through solid matter fired the imagination of a generation of scientists, who rushed to study this phenomenon. Roentgen had noted that x-rays were emitted from cathode-ray tubes at the spot that emitted light, or fluoresced, when the cathode rays hit the glass walls of the tube. This observation caught the attention of the French physicist Henri Becquerel.

Becquerel decided to investigate the connection between x-rays and the fluorescence of the glass walls of the cathode-ray tube. He knew that salts of uranium, such as potassium uranyl sulfate, $K_2UO_2(SO_4)_2 \cdot 2\,H_2O$, emitted light when exposed to the UV radiation in sunlight. He therefore wrapped a photographic plate in black paper, placed crystals of this salt on top of the plate, and exposed the crystals to sunlight. When the plates were developed, black spots were found beneath the crystals, suggesting that some form of radiation had been emitted by the uranium salts that passed through the paper and fogged the photographic plate.

Becquerel found the same results, however, when the crystals and photographic plate were prepared and kept in the dark. He also noted that much better images were obtained with pure uranium metal, which did not fluoresce when exposed to sunlight. Apparently, the radiation given off by uranium metal and its compounds had nothing to do with whether they were exposed to sunlight. It soon became evident that a new form of radiation had been discovered, for which Marie and Pierre Curie suggested the name ***radioactivity.***

In 1899, Ernest Rutherford studied the absorption of radioactivity by thin sheets of metal foil and found that there were two components: ***alpha*** (α) radiation, which was absorbed by a few thousandths of a centimeter of metal foil, and ***beta*** (β) radiation, which could pass through 100 times as much foil before it was absorbed. Shortly thereafter, a third form of radiation, named ***gamma*** (γ) rays, was discovered that could penetrate as much as several centimeters of lead. The three kinds of

FIG. 5.3 (*a*) Soon after the discovery of radioactivity, three different forms of decay (α, β, and γ) were discovered. In her thesis, Marie Curie reported the drawing on the left, which showed the effect of a magnetic field on these three forms of radioactivity. α-particles were deflected more slowly than β-particles by the magnetic field, which suggested that α-particles were heavier than β-particles. γ-rays were not affected by a magnetic field. (*b*) The effect of an electric field on the different forms of radioactivity showed that α-particles and β-particles were both electrically charged but carried charges with opposite signs. γ-rays were not affected by an electric field and therefore had no electrical charge.

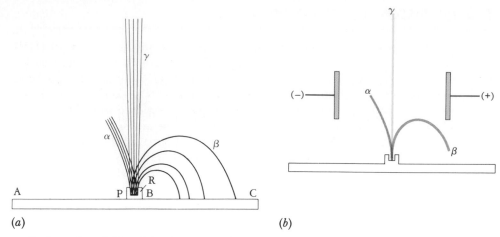

(*a*)   (*b*)

radiation also differed in the ways they were affected by electric and magnetic fields (see Figure 5.3).

Exercise 5.1

Use Figure 5.3 and the rule that opposite charges attract to determine the charge on α-particles, β-particles, and γ-rays.

Solution

α-particles are attracted to the negative pole of an electric field and must therefore carry a positive charge. β-particles are attracted to the positive pole of the field and therefore carry a negative charge. γ-rays are not affected by an electric field and are therefore electrically neutral.

5.5 MILLIKAN'S OIL DROP EXPERIMENT

Thomson's experiments with cathode-ray tubes suggested that the charge-to-mass ratio for the electron was about 10^8 coulombs per gram (C/g). This result was disconcerting, because it was at least 1000 times as large as any previous measurement of the charge-to-mass ratio of a particle. Until this experiment was done, the largest charge-to-mass ratio had been associated with the hydrogen — H^+ — ion.

There are two ways of explaining the unusually large charge-to-mass ratio of an electron. We can assume that the charge on the electron is 1000 times the charge on a hydrogen ion, which doesn't make sense. Or we can assume that in magnitude the charge on the electron is the same as the charge on a hydrogen ion, but the mass of the electron is only one-thousandth that of the hydrogen ion. Thomson preferred the second alternative, believing that the electron was at least 1000 times lighter than the smallest atom.

This hypothesis has important consequences. If electrons are a universal component of matter very much smaller than atoms, Dalton's model of the atom (described in Section 2.3) needed to be revised. Atoms are not indivisible. It must be possible to strip one or more negatively charged electrons away from an atom if enough energy is applied to the system.

The way to test this hypothesis is to measure either the charge on an electron or its mass. The charge on an electron was first measured by Robert Millikan between

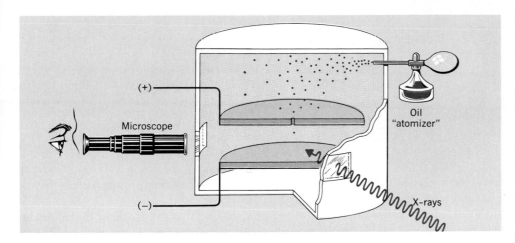

FIG. 5.4 The apparatus Millikan used to measure the charge on an electron used an atomizer from a perfume bottle to spray water or oil droplets through the hole in an electrical plate into a viewing chamber. The droplets picked up one or more electrons from molecules in the air that were ionized by a source of x-rays. Droplets that picked up electrons were attracted to the positively charged plate at the top of the viewing chamber. Measurements of the rate at which these drops fell through the chamber could be used to estimate the charge on each droplet. Since this charge must be an integral multiple of the charge on a single electron, the charge on an electron was deduced from this experiment to be about 1.6×10^{-19} coulomb.

1908 and 1917 using the apparatus shown in Figure 5.4. In these experiments, the atomizer from a perfume bottle was used to spray water or oil droplets into the sample chamber. Some of these droplets fell through a pinhole between two plates of an electric field, where they could be observed through a short-focus telescope.

A source of x-rays was then used to ionize the air in the chamber by removing electrons from the molecules in the air. Droplets that didn't capture one of these electrons fell to the bottom of the chamber because of the force of gravity. Droplets that captured one or more electrons were attracted to the positive plate at the top of the viewing chamber and either fell more slowly or rose toward the top. By carefully studying individual droplets, Millikan was able to show that the charge on a drop was always an integral multiple of 4.77×10^{-10} electrostatic units (esu) — or, as we would now express it, 1.59×10^{-19} coulombs, where a coulomb is the charge carried by a current of 1 ampere flowing for 1 second.

5.6 THOMSON'S RAISIN PUDDING MODEL OF THE ATOM

J. J. Thomson clearly recognized one of the consequences of the discovery of the electron. Since matter is electrically neutral, there must be a positively charged particle that balances the negative charge on the electrons in an atom. Furthermore, since electrons are very much lighter than atoms, these positively charged particles must carry the mass of the atom. Thomson therefore proposed a model that assumed atoms were spheres of positive charge in which small, light, negatively charged electrons were embedded, much as raisins might be embedded in the surface of a pudding (Figure 5.5).

At the time Thomson proposed this model, evidence for the existence of positively charged particles was available from cathode-ray tube experiments. In 1886, Eugen Goldstein noted that cathode-ray tubes with a perforated cathode (Figure 5.6) emitted a glow from the end of the tube near the cathode. Goldstein concluded that in addition to the electrons, or cathode rays, that travel from the negatively charged cathode toward the positively charged anode, there is another ray that travels in the opposite direction, from the anode toward the cathode. Because these rays pass through the holes, or channels, in the cathode, Goldstein called them *canal rays.*

By 1898, Wilhelm Wien was able to show that these canal rays could be deflected by both magnetic and electric fields, as would be expected for particles that carried

FIG. 5.5 J. J. Thomson proposed a model of the atom in which negatively charged electrons are embedded in a postively charged spherical atom, much as raisins are embedded in the surface of dish known in Britain as raisin pudding.

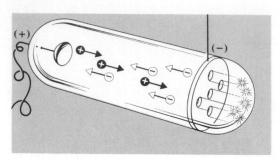

FIG. 5.6 Thomson's model suggests that positively charged particles are formed when electrons are removed from a neutral atom. Evidence for these positive particles was obtained by Goldstein. When the cathode of a cathode-ray tube was perforated, Goldstein observed rays he called "canal rays." The rays passed through the holes, or channels, in the cathode to strike the glass walls of the tube at the end near the cathode. Since canal rays travel in the opposite direction from cathode rays, they must carry the opposite charge.

an electric charge. They were deflected in the direction expected for positively charged particles, however.

These observations can be explained by assuming that the atoms or molecules in the gas in a cathode-ray tube lose one or more electrons when excited by the voltage applied to the electrodes of the tube. The negatively charged electrons flow toward the anode. The positive ions formed when the atoms or molecules lose electrons flow in the opposite direction.

The charge-to-mass ratio for cathode rays is always the same, because they always consist of negatively charged electrons. The charge-to-mass ratio for the positively charged canal rays, however, depends on the gas used to fill the tube, because the nature of the ions produced when the atoms or molecules lose electrons depends on the gas.

5.7 THE RUTHERFORD MODEL OF THE ATOM

J. J. Thomson (left) and Ernst Rutherford (right).

Thomson's model of the atom was relatively short-lived. It was replaced within 10 years by a model developed by one of his students, Ernest Rutherford. Rutherford began his graduate work by studying the effect of x-rays on various materials. Shortly after the discovery of radioactivity, he turned to the study of the α-particles given off by compounds of uranium.

Before he could study the effect of α-particles on matter, Rutherford had to develop a way of counting individual α-particles. He found that a screen coated with zinc sulfide emitted a flash of light each time it was hit by an α-particle. Rutherford and his technician, Hans Geiger, would sit in the dark until their eyes became sensitive enough and then try to count the flashes of light given off by the ZnS screen. (It is not surprising that Geiger was motivated to develop the electronic radioactivity counter that still carries his name.)

When a thin piece of gold foil about 0.00004 centimeter thick was bombarded with α-particles, Geiger and one of Rutherford's students, Ernest Marsden, found that virtually all of the α-particles were able to penetrate the foil, much as a rifle bullet would penetrate a bag of sand. Marsden noted, however, that a small fraction of the α-particles were deflected as they passed through the gold foil. When the experiment was repeated, Marsden and Geiger found that an even smaller fraction of the α-particles (perhaps 1 in 20,000) were scattered through angles larger than 90° (see Figure 5.7a).

By 1911, Rutherford had proposed a model for the structure of the atom consistent with these observations. He assumed that all of the positive charge and essentially all of the mass of an atom was concentrated in an infinitesimally small fraction of the total volume. He called this region the *nucleus* (from the Latin for little nut). Most of the α-particles were able to pass through the gold foil without hitting anything large enough to deflect them (see Figure 5.7b).

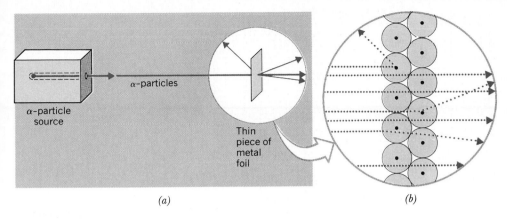

FIG. 5.7 (*a*) A block-diagram representation of the Ruther-ford-Marsden-Geiger experiment. A thin (0.00004-centimeter) piece of gold foil was bombarded with α-particles. Most of these particles passed through the gold foil unde-flected. A few were deflected by small angles, and a very small fraction of the particles were de-flected through angles of 90° or more. (*b*) These results can be explained by assuming that the mass of the atom is concen-trated in a very small fraction of the volume. Most of the α-particles pass through the empty space. A few come fairly close to the positively charged nucleus — close enough to be repelled by it — and these particles are deflected through small angles. Occasionally, an α-particle travels along a path that would lead to a direct hit with the nucleus. These particles are deflected through large angles by the force of repulsion between the α-particle and the nucleus of the atom.

A small fraction of the α-particles, however, came close to the nucleus of a gold atom as they passed through the foil. When this happened, the force of repulsion between the positively charged α-particle and the nucleus deflected the α-particle by a small angle. Occasionally, an α-particle travelled along a path that led to a direct collision with the nucleus of one of the 2000 or so atoms it had to pass through. When this happened, repulsion between the nucleus and the α-particle deflected the α-particle through an angle of 90° or more.

By carefully measuring the fraction of the α-particles deflected through large angles, Rutherford was able to estimate the size of the nucleus. According to his calculations, the radius of the nucleus was at least 10,000 times smaller than the radius of the atom. The vast majority of the volume of an atom was therefore empty space.

5.8 THE DISCOVERY OF THE PROTON

Rutherford's model assumed that all of the positive charge and most of the mass of an atom was concentrated in the nucleus. The remaining volume of the atom was occupied by lightweight, negatively charged electrons. There was still a problem with this model, however, the number of electrons or positively charged particles in an atom was not known.

This problem was solved shortly before the first world war through the work of H. G. J. Moseley. Moseley found that the frequencies of the x-rays given off by cathode-ray tubes depended on the metal used as the anode. He organized his data by assigning *atomic numbers* to each element, which specified the position of the element in the periodic table. Aluminum was the lightest metal for which Moseley was able to obtain results. Since it was the 13th element in the periodic table, it was assigned an atomic number of 13. By convention, the atomic number of an element is symbolized by the letter Z. Thus, we write: Al ($Z = 13$).

Moseley found a linear relationship between the square root of the frequencies of the x-rays given off by a cathode-ray tube and the atomic number of the element (see Figure 5.8).

$$\nu^{1/2} \propto (Z - 1)$$

Moseley argued that the frequencies of the x-rays given off by the anode when it was hit by electrons in the cathode-ray tube depends on the charge on the nucleus of the atom emitting these x-rays. He therefore concluded that the atomic number of an element is equal to the positive charge on the nucleus of its atoms. Aluminum, for example, must have a net charge of $+13$ on the nucleus of each atom.

FIG. 5.8 The plot of atomic number versus the square root of the frequencies of the x-rays given off when a beam of electrons in a cathode-ray tube strikes the metal surface of the anode. This plot accompanied the paper Moseley published in the *Philosophical Magazine* in 1914.

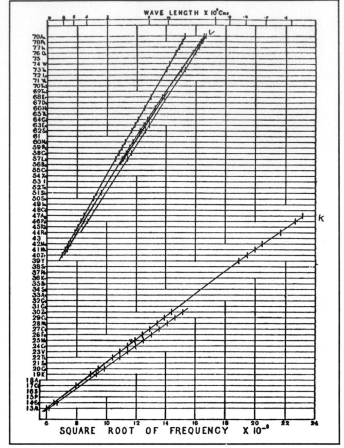

Less than six months after completing this work, Moseley volunteered to serve in the British army during World War I and became one of the more than 8.5 million casualties of this war. Shortly after the war, in 1920, Rutherford proposed the name ***proton*** for the positively charged particles in the nucleus of an atom.

5.9 THE DISCOVERY OF THE NEUTRON

According to the model of the atom that resulted from Moseley's work, the nucleus of a hydrogen atom ($Z = 1$) has a charge of $+1$ and therefore contains one proton. The remaining volume of the atom is occupied by an electron that balances the charge on the nucleus. Since the mass of a hydrogen atom is about 1 amu and the mass of an electron is negligibly small, the mass of a proton must be about 1 amu.

Unfortunately, this creates a problem, because the mass of an atom is usually at least twice as large as the mass predicted from its atomic number. Aluminum, for example, has an atomic number of 13, but the mass of the average aluminum atom is about 27 amu. There are two possibilities.

1. We can assume that the nucleus of an Al atom contains 27 protons and 14 electrons, for a net positive charge of $+13$. The nucleus would then be surrounded by another 13 electrons that would balance the net positive charge on the nucleus.

2. We can assume the nucleus of an Al atom contains 13 protons and 14 other particles that have the same weight as a proton but are electrically neutral.

TABLE 5.1

Fundamental Subatomic Particles

Particle	Symbol	Charge	Mass (amu)
electron	e^-	-1	0.0005486
proton	p^+	$+1$	1.007276
neutron	n^0	0	1.008665

At the same time that Rutherford proposed the name *proton* for the positively charged particle in the nucleus of an atom, he proposed that the nucleus also contained a neutral particle, eventually named the **neutron.** It was not until 1932, however, that James Chadwick was able to prove that these neutral particles exist.

Today, we assume that atoms are composed of the three fundamental subatomic particles in Table 5.1 — electrons, protons, and neutrons. The protons and neutrons are concentrated in the nucleus of the atom, and this positively charged nucleus is surrounded by a sea of negatively charged electrons.

The number of protons, neutrons, and electrons in an atom can be determined from a set of simple rules.

RULES FOR DETERMINING THE NUMBER OF PROTONS, NEUTRONS, AND ELECTRONS IN AN ATOM

1. The number of protons is equal to the atomic number (Z).

2. The number of electrons in a neutral atom is equal to the number of protons.

3. The mass number (M) of the atom is equal to the sum of the number of protons and neutrons in the nucleus.

4. The number of neutrons is equal to the difference between the mass number (M) and the atomic number (Z).

Exercise 5.2

Calculate the number of electrons, protons, and neutrons in neutral atoms of the following isotopes.
(a) ^{12}C (b) ^{13}C (c) ^{14}C (d) ^{14}N

Solution

Section 3.2 stated that the different isotopes of an element could be distinguished from one another by the mass number, which is written in the upper left-hand corner of the symbol for the element. ^{12}C, ^{13}C, and ^{14}C are all isotopes of carbon ($Z = 6$) and therefore all contain six protons. If the atoms are neutral, they must contain the same number of electrons. The only difference between these isotopes is the number of neutrons in the nucleus.

^{12}C: 6 electrons, 6 protons, and 6 neutrons
^{13}C: 6 electrons, 6 protons, and 7 neutrons
^{14}C: 6 electrons, 6 protons, and 8 neutrons

Neutral nitrogen atoms ($Z = 7$) contain seven protons and seven electrons. The remaining mass of the atom ($M = 14$) comes from neutrons.

^{14}N: 7 electrons, 7 protons, and 7 neutrons

Exercise 5.3

Calculate the number of electrons in the Cl^- and Fe^{3+} ions.

Solution

The atomic number of chlorine is 17. The number of protons in both a neutral Cl atom and a negatively charged Cl^- ion is exactly the same — 17. The difference between these particles is the number of electrons that surround the nucleus. The Cl^- ion has one more electron than a neutral Cl atom, for a total of 18 electrons.

Atoms can gain or lose electrons, but the number of protons in the nucleus of the atom cannot change unless the identity of the atom changes. The only way an iron atom ($Z = 26$) with 26 protons and 26 electrons can form a $+3$ ion is to lose three electrons. The Fe^{3+} ion therefore has 23 electrons.

This model explains the observations made in the early experiments on electricity. Matter can pick up an electrical charge because electrons can be transferred from one neutral atom or molecule to another. Substances, such as glass, that exhibit "vitreous electricity" lose electrons when they are rubbed and are therefore left with a net positive charge. Substances, such as amber, that exhibit "resinous electricity" gain electrons when rubbed and therefore have a net negative charge.

5.10 PARTICLES AND WAVES

According to Rutherford's model, essentially all of the mass of an atom is concentrated in an infinitesimally small nucleus surrounded by a sea of lightweight, negatively charged electrons that occupy the vast majority of the volume of the atom. Our next goal is to develop a picture of how electrons are distributed around the nucleus. Before we can do this, we have to understand the nature of the probes used to obtain this information.

Most of what we know about the distribution of electrons in an atom came from studies of the interaction between **electromagnetic radiation** and matter. Our first goal is therefore to understand what we mean when we say that electromagnetic radiation has some of the properties of both particles and waves.

Scientists since the time of Galileo have divided matter into two categories: particles and waves. **Particles** are easy to understand — they have a definite mass (or weight) and they occupy space. **Waves** are more challenging. They have no mass, and yet they carry energy as they travel through space. The best way to demonstrate that waves carry energy is to watch what happens when a pebble is tossed into a lake. As the waves produced by this action travel across the lake, they set in motion any leaves that lie on the surface.

In addition to the ability to carry energy, waves have four characteristic properties: velocity, frequency, wavelength, and amplitude. As we watch waves travel across the surface of a lake, we can see that they move through space at a certain *speed*, or **velocity**. By watching waves strike a pier at the edge of the lake we can see that they are also characterized by a **frequency** (ν), which is the number of waves or cycles that hit the pier per unit time. The frequency of a wave is reported in units of cycles per second (cps) or hertz (Hz).

$$1 \text{ Hz} = 1 \text{ cps}$$

Waves carry energy as they travel through space as illustrated by the ripples formed when a pebble is dropped into a lake move past a leaf.

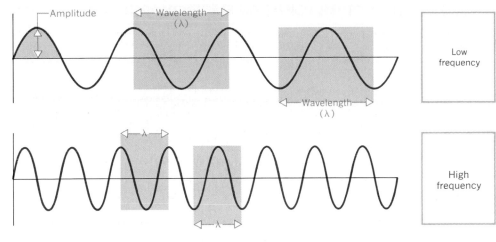

FIG. 5.9 The wavelength (λ) is the distance between two repeating points on a wave. The amplitude is the difference between the height of the highest point on the wave and its center of gravity. The frequency (ν) is the number of waves that pass a fixed point per unit of time. The product of the frequency (ν) times the wavelength (λ) is the speed, or velocity, at which the wave moves through space.

The idealized drawing of a wave in Figure 5.9 illustrates the definitions of amplitude and wavelength. The **wavelength** (λ) is the distance between any two repeating points on the wave. The **amplitude** of the wave is the distance between the highest (or lowest) point on the wave and the center of the wave.

If we measure the frequency (ν) of a wave in cycles per second and the wavelength (λ) in meters per wave, the product of these two numbers has the units of meters per second.

$$\nu\lambda = \frac{\text{waves}}{\text{s}} \times \frac{\text{m}}{\text{wave}} = \frac{\text{m}}{\text{s}}$$

The product of the frequency times the wavelength of a wave is therefore the speed, or velocity, at which the wave travels through space.

Frequency $\times$ wavelength = speed

It might be useful to imagine a concrete example of this relationship. What is the speed of a wave that has a wavelength of 1 meter and a frequency of 60 cycles per second? If 60 of these waves pass each second, and each wave is 1 meter long, the speed of the wave must be the product of these numbers, of 60 meters per second.

$$\frac{60 \text{ waves}}{1 \text{ s}} \times \frac{1 \text{ m}}{\text{wave}} = 60 \text{ m/s}$$

Exercise 5.4

Orchestras in the United States tune their instruments to an A that has a frequency of 440 cycles per second, or 440 hertz. If the speed of sound is 1116 feet per second, what is the wavelength of this A in feet?

Solution

The product of the frequency times the wavelength of any wave is equal to the speed, or velocity, with which the wave travels through space.

$$\text{Frequency} \times \text{wavelength} = \text{speed}$$

$$440 \frac{\text{cycles}}{\text{s}} \times \quad \lambda \quad = 1116 \frac{\text{ft}}{\text{s}}$$

Solving this equation for the wavelength of this note gives a value of 2.54 feet per cycle.

5.11 LIGHT AND OTHER FORMS OF ELECTROMAGNETIC RADIATION

White light can be separated into a spectrum of colors by passing the light through a glass prism.

In 1865, James Clerk Maxwell proposed a theory that assumed that light is a wave with both electric and magnetic components. Light is therefore a form of *electromagnetic radiation.*

Because it is a wave, light is bent when it enters a glass prism. When white light is focused on a prism, the light rays of different wavelengths are bent by differing amounts and the light is transformed into a spectrum of colors. Starting from the side of the spectrum where the light is bent by the smallest angle, the colors are red, orange, yellow, green, blue, and violet.

Light contains the narrow band of frequencies and wavelengths in the electromagnetic spectrum that our eyes can detect. It includes radiation with wavelengths between 400 nanometers (violet) and 700 nanometers (red). Since the wavelength of electromagnetic radiation can be as long as 40 meters or as short as 10^{-5} nanometers, the visible spectrum is only a small portion of the total range of electromagnetic radiation.

The total electromagnetic spectrum includes radio waves, television waves, microwaves, infrared, visible light, ultraviolet, x-rays, γ-rays, and cosmic rays, as shown in Figure 5.10. These different forms of radiation all travel at the same speed, the speed of light, c. They differ, however, in their frequencies and wavelengths. The product of the frequency times the wavelength of electromagnetic radiation is always equal to the speed of light.

$$c = \nu\lambda$$

Electromagnetic radiation that has a long wavelength therefore has a low frequency, and radiation with a high frequency has a short wavelength.

Exercise 5.5

What is the frequency of red light with a wavelength of 700.0 nanometers if the speed of light is 2.998×10^8 meters per second?

FIG. 5.10 The visible spectrum is the small portion of the total electromagnetic spectrum that our eyes can detect. Other forms of electromagnetic radiation include radio waves, microwaves, and infrared and ultraviolet radiation, as well as x-rays, γ-rays, and cosmic rays.

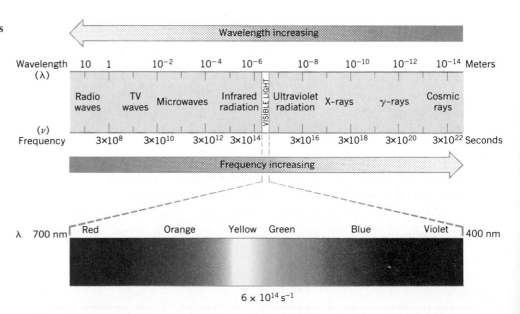

Solution

The product of the frequency times the wavelength of any wave is equal to the speed with which the wave travels through space. In this case, the wave travels at the speed of light.

$$v\lambda = 2.998 \times 10^8 \text{ m/s}$$

Before we can calculate the frequency of this radiation we have to convert the wavelength into units of meters.

$$700.0 \text{ nm} \times \frac{1 \text{ m}}{10^9 \text{ nm}} = 7.000 \times 10^{-7} \text{ m}$$

We can then substitute the wavelength into the following equation.

$$c = v\lambda$$
$$2.998 \times 10^8 \text{ m/s} = (v)(7.000 \times 10^{-7} \text{ m})$$

We can then solve this equation for the frequency of this light in units of cycles per second (s^{-1}).

$$v = 4.283 \times 10^{14} \text{ s}^{-1}$$

5.12 ATOMIC SPECTRA

In 1814, Joseph von Fraunhofer noted that the light from the sun does not give a continuous spectrum. By using an unusually good prism, Fraunhofer was able to observe a series of dark lines in the sun's spectrum, which he labeled A through H. Fraunhofer's experiment is an example of an **absorption spectrum,** because some of the light given off by the sun is apparently absorbed before the light reaches the earth.

For more than 200 years, chemists have known that sodium salts produce a yellow color when added to a flame. Robert Bunsen, however, was the first to systematically study this phenomenon. Bunsen went so far as to design a new burner that would produce a colorless flame for this work. Between 1855 and 1860, Bunsen and his colleague Gustav Kirchhoff developed a spectroscope, which focused the light from the burner flame onto a prism that separated this light into its spectrum. Using this device, Bunsen and Kirchhoff were able to show that the **emission spectrum** of sodium salts consists of two narrow bands of radiation emitted in the yellow portion of the spectrum.

Kirchhoff noticed that the wavelength of the light given off when sodium salts were added to a flame was the same as the wavelength of the D line in Fraunhofer's spectrum of sunlight. He therefore concluded that absorption and emission spectra were related. Certain substances, when heated, give off light that has the same frequencies and wavelengths as the light they absorb under other conditions.

Chemists and physicists soon began using the spectroscope to catalog the wavelengths of the light either emitted or absorbed by a variety of compounds. These data were used to identify elements in everything from mineral water to sunlight, but no obvious patterns were discovered until 1885, when Johann Jacob Balmer analyzed the spectrum of hydrogen.

When an electric current is passed through a glass tube that contains hydrogen gas at low pressure, some of the H_2 molecules dissociate to give individual hydro-

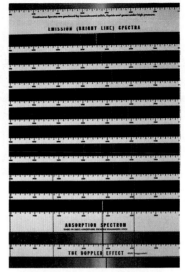

The light given off by the sun is not continuous because certain narrow bands of wavelengths are absorbed by elements present in the sun. These dark bands occur at the same wavelengths as the bright bands in the emission spectrum of these elements when they are heated in a flame, or excited with an electric current, on earth.

The characteristic color of the light given off when sodium atoms are excited can be seen by placing crystals of a sodium salt into the flame of a bunsen burner.

FIG. 5.11 The blue light given off when a tube filled with H_2 gas is excited with an electric discharge can be separated into four narrow bands of light when it is passed through a prism.

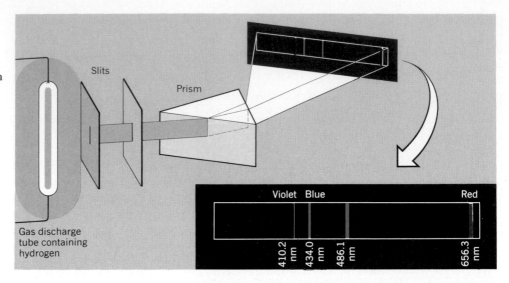

TABLE 5.2

The Characteristic Lines in the Visible Spectrum of Hydrogen

Wavelength (nm)	Color
656.3	red
486.1	blue
434.0	blue-violet
410.1	violet

gen atoms, and the tube gives off blue light. When this light is passed through a prism (as shown in Figure 5.11), four distinct bright lines are observed against a black background. These lines have the characteristic wavelengths and colors shown in Table 5.2. Balmer noticed that these data fit an equation that could be written as follows.

$$\frac{1}{\lambda} = R_H \left[\frac{1}{2^2} - \frac{1}{n^2} \right]$$

In this equation, R_H is a constant equal to 1.09678×10^{-2} nm^{-1}, and n is an integer between 3 and 6.

Between 1908 and 1924, searches of the infrared spectrum at longer wavelengths and the ultraviolet spectrum at shorter wavelengths revealed four more series of lines in the emission spectrum of H_2. Each of these lines fits the following general equation, where n and m are integers, m is larger than n, and R_H is 1.09678×10^{-2} nm^{-1}.

$$\frac{1}{\lambda} = R_H \left[\frac{1}{n^2} - \frac{1}{m^2} \right]$$

All five series of lines are shown in Figure 5.12.

FIG. 5.12 The four lines in the visible spectrum of hydrogen analyzed by Balmer are only a portion of the emission spectrum of this gas. Lyman discovered another series of lines in the ultraviolet portion of the spectrum, and three more series of lines were discovered in the infrared spectrum by Paschen, Brackett, and Pfund.

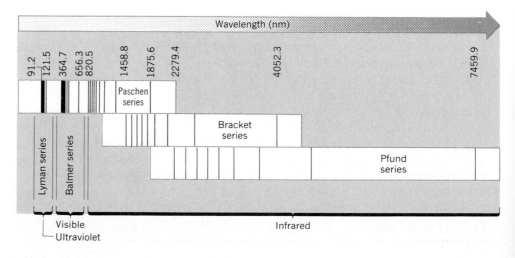

5.13 QUANTIZATION OF ENERGY

The hydrogen atom spectrum raises several important questions.

1. Why do hydrogen atoms give off only a handful of narrow bands of radiation when they emit light?
2. Why do hydrogen atoms, when excited, emit light that has the same wavelengths as the light they absorb at room temperature?

Answers to these questions came from work on a related topic.

It is common knowledge that objects give off light when heated. Examples range from the gentle red glow of an electric burner on a stove to the bright white light given off when the tungsten wire in a light bulb is heated by passing an electric current through the wire. What is less well known is a phenomenon discovered by Thomas Wedgwood in 1792. Wedgwood, whose father started the famous porcelain factory, noted that all objects give off a red glow when heated to the same temperature. The bottom of a crucible and the iron triangle on which the crucible rests, for example, both glow red when heated with a bunsen burner. Wedgwood found that the color of the light given off changes as the object is heated to even higher temperatures, but the spectrum of light given off at a particular temperature is approximately the same for any object.

At the turn of the 20th century, a model developed by Max Planck and Albert Einstein explained the spectrum of light given off when an object was heated. This model was based on two assumptions.

1. Light is composed of *photons, which* are small, discrete bundles of energy.
2. The energy of a photon is proportional to its frequency.

$$E = h\nu$$

In this equation, h is a constant known as Planck's constant, which is equal to 6.626×10^{-34} joule-seconds (J-s).

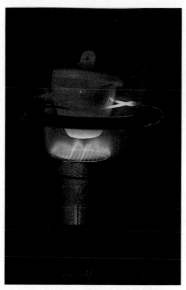

Wedgwood's notion that objects heated to the same temperature give off light of the same color can be illustrated by comparing the light given off by a crucible and the iron triangle on which it rests when the crucible is heated.

Exercise 5.6

Calculate the energy of a single photon of red light with a wavelength of 700.0 nanometers and the energy of a mole of these photons.

Solution

In Exercise 5.5 we found that red light with a wavelength of 700.0 nm has a frequency of 4.283×10^{14} cycles per second. Substituting this frequency into Planck's equation gives the following result.

$$E = h\nu$$
$$E = (6.626 \times 10^{-34} \text{ J-s})(4.283 \times 10^{14} \text{ s}^{-1})$$
$$= 2.838 \times 10^{-19} \text{ J}$$

A single photon of red light carries an insignificant amount of energy. A mole of these photons, however, carries about 171,000 joules of energy.

$$2.838 \times 10^{-19} \frac{\text{J}}{\text{photon}} \times 6.022 \times 10^{23} \frac{\text{photons}}{\text{mol}} = 1.709 \times 10^5 \text{ J/mol}$$

This is enough energy to raise the temperature of a gallon of water by almost 11°C.

This hypothesis was revolutionary. At a time when everyone agreed that light was a wave — and therefore continuous — the work of Planck and Einstein suggested that it behaved as if it was a stream of small bundles, or packets, of energy.

In its most naive form, this hypothesis explains why an object gives off red light when heated until it just starts to glow. Red light has the longest wavelength and the smallest frequency of any form of visible radiation. Since the energy of electromagnetic radiation is proportional to its frequency, red light carries the smallest amount of energy of any form of visible radiation. The light given off when an object just starts to glow therefore has the color characteristic of the lowest energy form of electromagnetic radiation our eyes can detect. As the object becomes hotter it starts emitting light at higher frequencies as well, until it eventually appears to glow white hot.

Planck's hypothesis that the energy of a photon was proportional to its frequency served as the basis for a model of the structure of the atom that explained why hydrogen atoms emit exactly four narrow bands of radiation in the visible spectrum when excited. In retrospect, we now recognize that this hypothesis was the first step in what is known as the **quantum revolution.**

5.14 THE BOHR MODEL OF THE ATOM

Although he was never able to incorporate electrons into his model of the atom, Rutherford favored a model in which the electrons circled the nucleus in much the same way planets revolve around the sun. Unfortunately, this model violates one of the classic principles of physics. Classical physics suggests that a charged particle — such as an electron — revolving in a circular orbit around the nucleus of an atom would enter into a spiral that would eventually lead to its collapse into the nucleus.

In 1913, one of Rutherford's students, Niels Bohr, proposed a model for the hydrogen atom that was consistent with Rutherford's model of the atom and yet also explained the spectrum of the hydrogen atom. The **Bohr model** was based on the following assumptions.

1. **The electron in a hydrogen atom travels around the nucleus in a circular orbit.**
2. **The energy of the electron in an orbit is proportional to its distance from the nucleus. The farther the electron is from the nucleus, the more energy it has.**
3. **Only a limited number of orbits with certain energies are allowed. In other words, the orbits are countable,** *quantized.*
4. **The only orbits that are allowed are those for which the angular momentum of the electron is an integral multiple of Planck's constant (h) divided by 2π.**
5. **Light is absorbed when an electron jumps to a higher-energy orbit.**
6. **Light is emitted when an electron falls into a lower-energy orbit.**
7. **The energy of the light emitted or absorbed is equal to the difference between the energies of the two orbits.**

Some of the key elements of this hypothesis are illustrated in Figure 5.13. Three points deserve particular attention.

First, Bohr recognized that his first assumption violated the principles of classical mechanics. But he also knew that it was impossible to solve this problem within the limits of classical physics. For that reason, he was willing to assume that one or more of the principles from classical physics that operate on matter at the macroscopic scale might not be valid on the atomic scale.

Second, he assumed there were only a limited number of orbits in which the electron can reside. He based this assumption on the fact that there are only a limited number of lines in the spectrum of the hydrogen atom and on his belief that these lines were the result of light emitted or absorbed as an electron moved from one orbit to another in the atom.

Third, Bohr restricted the number of orbits on the hydrogen atom by limiting the values of the angular momentum of the electron. Any object moving along a straight line has a **momentum** equal to the product of its mass (m) times its velocity (v). An object moving in a circular orbit has an **angular momentum** equal to the mass (m) times the velocity (v) times the radius of the orbit (r). Bohr assumed that the angular momentum of the electron could only take on certain values, equal to an integer times Planck's constant divided by 2π.

$$mvr = n\left[\frac{h}{2\pi}\right] \qquad \text{(where } n = 1, 2, 3, 4, 5, \ldots)$$

These assumptions enabled Bohr to calculate the radius of the allowed orbits for the electron in a hydrogen atom.

$$r = \left[\frac{h^2}{4\pi^2 me^2}\right] n^2$$

In this equation, h is Planck's constant, m is the mass of the electron, e is the charge on the electron, and n is the integer that distinguishes between the first, second, third, and succeeding orbits. This equation can be simplified by replacing the ratio of the physical constants—$h^2/4\pi^2 me^2$—with a new constant—a_0—which has a value of 0.0529 nanometer.

$$r = a_0 n^2 \qquad \text{(where } n = 1, 2, 3, 4, 5, \ldots)$$

Bohr started with an equation that described the relationship between the energy of each orbit and the radius of the orbit.

$$E_n = \frac{-e^2}{2r}$$

He then substituted the relationship between the radius of the orbit and a_0 times n^2 into this equation.

$$E_n = \frac{-e^2}{2a_0}\left[\frac{1}{n^2}\right]$$

If this equation gives the energy of any individual orbit around the hydrogen atom, then the difference between the energies of any two orbits (ΔE) should be given by the following equation.

$$\Delta E = R_H\left[\frac{1}{n^2} - \frac{1}{m^2}\right]$$

In this equation, n and m are both integers, m is larger than n, and R_H is a proportionality constant equal to $e^2/2a_0$.

Planck's equation states that the energy of a photon is proportional to its frequency.

$$E = h\nu$$

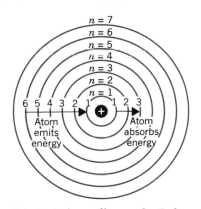

FIG. 5.13 According to the Bohr model, hydrogen atoms absorb light when an electron is excited from a low-energy orbit (such as $n = 2$) into a higher-energy orbit (such as $n = 4$). The energy of the photon absorbed is equal to the difference between the energies of these two orbits. Atoms that have been excited by an electric discharge or by being heated can give off light when an electron drops from a high-energy orbit (such as $n = 6$) into a lower-energy orbit (such as $n = 2$). The energy of the photon emitted when the electron falls from one orbit to another is equal to the difference between the energies of the two orbits.

The energy of a photon is therefore inversely proportional to its wavelength.

$$\mathbf{E} = \frac{hc}{\lambda}$$

By properly defining the units of the constant, R_{H}, Bohr was able to show that the wavelengths of the light given off or absorbed by a hydrogen atom should be given by the following equation.

$$\frac{1}{\lambda} = R_{\mathrm{H}} \left[\frac{1}{n^2} - \frac{1}{m^2} \right]$$

This is the equation obtained from the analysis of the hydrogen atom spectrum discussed in Section 5.12. Bohr was therefore able to show that the wavelengths in the UV spectrum of hydrogen discovered by Lyman correspond to transitions from one of the higher-energy orbits in his model into the $n = 1$ orbit, as shown in Figure 5.14. Similarly, the wavelengths in the visible spectrum of hydrogen ana-

FIG. 5.14 According to the Bohr model, the wavelengths in the ultraviolet spectrum of the hydrogen atom discovered by Lyman were the result of electrons dropping from higher-energy orbits into the $n = 1$ orbit. The Balmer, Paschen, Brackett, and Pfund series resulted from electrons falling from high-energy orbits into the $n = 2$, $n = 3$, $n = 4$, and $n = 5$ orbits, respectively.

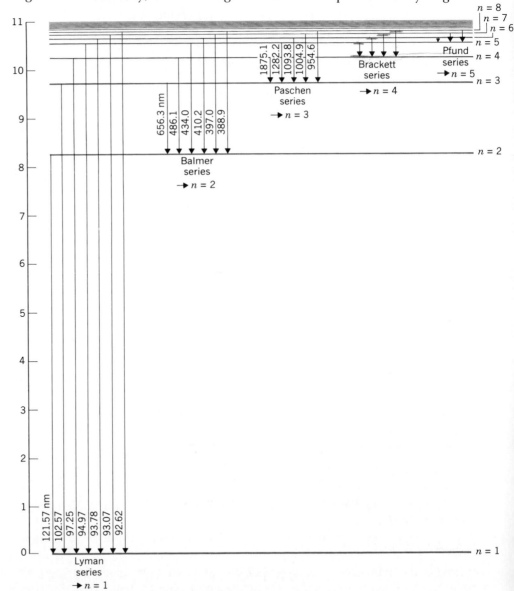

lyzed by Balmer were shown to be the result of transitions from one of the higher-energy orbits into the $n = 2$ orbit. The Paschen, Brackett, and Pfund series of lines in the infrared spectrum of hydrogen resulted from electrons dropping into the $n = 3$, $n = 4$, and $n = 5$ orbits, respectively.

Exercise 5.7

Calculate the wavelength of the light given off by a hydrogen atom when an electron falls from the $n = 4$ to the $n = 2$ orbit in the Bohr model.

Solution

According to the Bohr model, the wavelength of the light emitted by a hydrogen atom when the electron falls from a high-energy ($n = 4$) orbit into a lower-energy ($n = 2$) orbit is given by the following equation.

$$\frac{1}{\lambda} = R_H \left[\frac{1}{n^2} - \frac{1}{m^2} \right]$$

In this equation, m and n are integers, m is larger than n, and R_H is a constant equal to 1.09678×19^{-2} nm^{-1}. Substituting the appropriate values of n and m into this equation gives the following result.

$$\frac{1}{\lambda} = (1.09678 \times 10^{-2} \text{ nm}^{-1}) \left(\frac{1}{2^2} - \frac{1}{4^2} \right) = 2.0565 \times 10^{-3} \text{ nm}^{-1}$$

Solving for the wavelength of this light gives a value of 486.3 nanometers.

$$\lambda = \textbf{486.3 nm}$$

This result agrees with the experimental value of 486.1 nanometers for the blue line in the visible spectrum of the hydrogen atom given in Table 5.2.

5.15 REASONS FOR THE SUCCESS AND FAILURE OF THE BOHR MODEL

The Bohr model did an excellent job of explaining the spectrum of a hydrogen atom. By incorporating a Z^2 term into the equation to adjust for the increase in the attraction between an electron and the nucleus of the atom as the atomic number (Z) increased, it could even explain the spectra of ions that contain one electron, such as the He$^+$, Li^{2+}, and Be^{3+} ions. Nothing could be done, however, to make this model fit the spectra of atoms with more than one electron.

The Bohr model left two important questions unanswered.

1. Why are there only a limited number of orbits in which an electron can reside in a hydrogen atom?
2. Why can't this model be extended to many-electron atoms?

The theory of **wave-particle duality** developed by Louis-Victor de Broglie eventually explained the success of the Bohr model with atoms or ions that contained one electron and provided a basis for understanding why this model failed for more complex systems.

WAVE-PARTICLE DUALITY

De Broglie started with the conflicting observation that light acts as both a particle and a wave. In many ways, light acts as a wave, with a characteristic frequency, wavelength, and amplitude. Planck and Einstein, however, argued that light carries energy as if it contains discrete photons, or packets of energy. In his doctoral thesis at the Sorbonne in 1924, de Broglie looked at the consequences of simultaneously assuming that matter has the properties of both a particle and a wave.

When it behaves as a wave, matter has an energy that is proportional to its frequency.

$$E = h\nu = \frac{hc}{\lambda}$$

When it behaves as a particle, it has an energy that is related to its mass.

$$E = mc^2$$

By assuming that matter could be simultaneously both a particle and a wave, de Broglie could set up the following equation.

$$mc^2 = \frac{hc}{\lambda}$$

By rearranging this equation, he derived a relationship between one of the wavelike properties of matter—its wavelength—and one of the properties of a particle—its momentum.

$$\lambda = \frac{h}{mc}$$

De Broglie concluded that most particles are too heavy for their wave properties to be observed. When the mass of an object is very small, however, the wave properties can be detected experimentally. De Broglie predicted that electrons were light enough to exhibit the properties of both particles and waves. In 1927, this prediction was confirmed when the diffraction of electrons was observed experimentally.

De Broglie applied his theory of wave-particle duality to the Bohr model to explain why only certain orbits are allowed for the electron. He argued that only certain orbits allow the electron to display its particle and wave properties at the same time, because only certain orbits have a circumference that is an integral multiple of the wavelength of the electron, as shown in Figure 5.15.

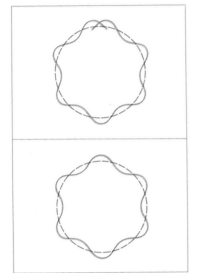

FIG. 5.15 The fact that only certain orbits are allowed in the Bohr model of the atom can be explained by assuming that the circumference of the orbit must be an integral ($n = 1, 2, 3, 4 \ldots$) multiple of the wavelength of the electron. When it is not, the electron cannot simultaneously satisfy its wave and particle behavior.

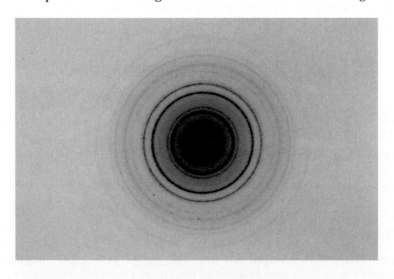

The mass of an electron is small enough that its wave behavior can be detected. This photograph shows the diffraction of electrons by a crystal of cubic zirconia.

CONSEQUENCES OF THE WAVE PROPERTIES OF ELECTRONS

At first glance, the Bohr model looks like a two-dimensional model of the atom, because it restricts the motion of the electron to a circular orbit in a two-dimensional plane. In reality, the Bohr model is a one-dimensional model, because a circle can be uniquely defined by specifying only one dimension — its radius, r. Only one coordinate (n) is therefore needed to describe the orbits in this model.

Unfortunately, electrons are not particles that can be restricted to a one-dimensional circular orbit. They act to some extent as waves, and they exist in three-dimensional space. The Bohr model works for one-electron atoms or ions only because certain factors present in more complex atoms are not present in these atoms or ions. In order to construct a model that describes the distribution of electrons in atoms that contain more than one electron, we have to allow the electrons to occupy three-dimensional space. We therefore need a model that uses three coordinates to describe the distribution of electrons in these atoms.

5.16 QUANTUM NUMBERS

Why do we still talk about the Bohr model of the atom 75 years after its discovery, if the only thing this model can do is explain the spectrum of the hydrogen atom? The answer is simple: It was the last model of the atom for which a simple physical picture can be constructed. It is easy to imagine an atom that consists of solid electrons revolving around the nucleus in circular orbits.

A more powerful model of the atom was developed by Erwin Schrödinger in 1926. Schrödinger combined the mathematical equations for the behavior of waves with the de Broglie equation to generate a mathematical model for the distribution of electrons in an atom. The advantage of this model is that it consists of mathematical equations known as **wave functions** that satisfy the requirements placed on the behavior of electrons. The disadvantage is that it is difficult to imagine a physical model of electrons as waves.

How does Schrödinger's model differ from Bohr's? The Bohr model assumed the electron was a particle that occupied space and tried to describe the location of this particle by telling us the distance between the electron and the nucleus at any moment in time. Since this was a one-dimensional model, it took only one coordinate, or one **quantum number,** to describe each orbit in the Bohr model.

> **Quantum number: A number used to indicate the magnitude of a property of an atom that is *quantized*, or countable.**

The number of orbits in the Bohr model was quantized, and these orbits were distinguished from each other by the value of the n quantum number. The $n = 1$ orbit was different from the $n = 2$ or $n = 3$ orbit, and so on.

The Schrödinger model assumed the electron was a wave and tried to describe the regions in space, or **orbitals,** where electrons are most likely to be found. Instead of trying to tell us where the electron is at any time, the Schrödinger model describes the probability that an electron can be found in a given region of space at a given time. This model no longer tells us where the electron is, it only tells us where it might be.

The Bohr model was a one-dimensional model, which used one quantum number to describe the distribution of electrons in the atom. The only information that was important was the *size* of the orbit, which was described by the n quantum number. Schrödinger's model allowed the electron to occupy three-dimensional space. It therefore required three quantum numbers, to describe the orbitals, or regions in space where electrons can be found.

The three coordinates that come from Schrödinger's wave equations are the *principal* (n), *angular* (l), and *magnetic* (m) *quantum numbers.* These quantum numbers describe the size, shape, and orientation in space of the orbitals in an atom.

The principal quantum number, n, describes the size and therefore the energy of the orbital.

The larger the value of n, the larger the orbital. Orbitals for which $n = 1$ are smaller than those for which $n = 2$, and so on. Because they have opposite electrical charges, electrons are attracted to the nucleus of the atom. Energy must therefore be absorbed to excite an electron from an orbital in which the electron is close to the nucleus ($n = 1$) into an orbital in which it is farther from the nucleus ($n = 2$).

In addition to the size of an orbital, we also have to describe its shape.

The angular quantum number, l, describes the shape of the orbital.

Orbitals have shapes that are best described as spherical ($l = 0$), polar ($l = 1$), or cloverleaf ($l = 2$), as shown in Figure 5.16. They can even take on more complex shapes as the value of the angular quantum number becomes larger.

There is only one way in which a sphere ($l = 0$) can be oriented in space. However, orbitals that have polar ($l = 1$) or cloverleaf ($l = 2$) shapes can point in several different directions, which are specified by the magnetic quantum number.

The magnetic quantum number, m, describes the orientation of the orbital in space.

(This is called the *magnetic* quantum number because the effect of different orientations of orbitals was first observed in the presence of an external magnetic field.)

Atomic spectra suggest that the orbitals on an atom are quantized; they are countable. We therefore need a set of rules that limit the number of possible combinations of the three quantum numbers: n, l, and m.

RULES GOVERNING THE ALLOWED COMBINATIONS OF QUANTUM NUMBERS

1. The three quantum numbers (n, l, and m) that describe an orbital are all integers —0, 1, 2, 3, 4, and so on.

2. The principal quantum number, n, cannot be zero. The allowed values of n are therefore 1, 2, 3, 4, and so on.

3. The angular quantum number (l) can be any integer between 0 and $n - 1$. If $n = 3$, for example, l can be either 0, 1, or 2.

4. The magnetic quantum number (m) can be any integer between $-l$ and $+l$. When $l = 2$, for example, m can be either -2, -1, 0, $+1$, or $+2$.

FIG. 5.16 The angular quantum number specifies the shape of the orbital in which electrons can be found. When $\ell = 0$, the orbital is spherical. When $\ell = 1$, the orbital is polar. When $\ell = 2$, the orbital typically has the shape of a cloverleaf.

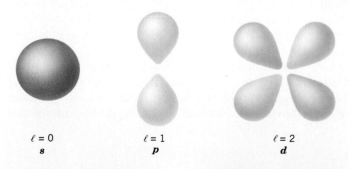

$\ell = 0$
s

$\ell = 1$
p

$\ell = 2$
d

Exercise 5.8

Describe the allowed combinations of the n, l, and m quantum numbers when $n = 3$.

Solution

The angular quantum number (l) can take on any integral value between 0 and $n - 1$. If $n = 3$, l can be either 0, 1, or 2. The magnetic quantum number (m) can take on any integral value between $-l$ and $+l$. Thus, when $l = 0$, m must be 0. When $l = 1$, m can be either -1, 0, or $+1$. When $l = 2$, m can be -2, -1, 0, $+1$, or $+2$. The allowed combinations of n, l, and m for which $n = 3$ are therefore restricted to the following.

n	l	m
3	0	0
3	1	-1
3	1	0
3	1	$+1$
3	2	-2
3	2	-1
3	2	0
3	2	$+1$
3	2	$+2$

5.17 SHELLS AND SUBSHELLS OF ORBITALS

Orbitals with the same principal quantum number are often said to form a **_shell_** of orbitals.

A shell consists of the orbitals on an atom that have the same principal quantum number (n).

Orbitals within a shell are divided into **_subshells._**

A subshell consists of the orbitals on an atom that have the same values for both the principal quantum number (n) and the angular quantum number (l).

The subshell in which an orbital belongs is often described by a two-character code such as $2p$ or $4f$. The first character indicates the shell ($n = 2$ or $n = 4$). The second character identifies the subshell. By convention, the following lowercase letters are used to indicate different subshells.

$$
\begin{array}{ll}
s\colon & l = 0 \\
p\colon & l = 1 \\
d\colon & l = 2 \\
f\colon & l = 3 \\
g\colon & l = 4
\end{array}
$$

Although there is no pattern in the first four letters (s, p, d, f), from $l = 3$ on, the letters increase alphabetically (f, g, h, and so on). Some of the allowed combinations of the n and l quantum numbers are shown in Figure 5.17.

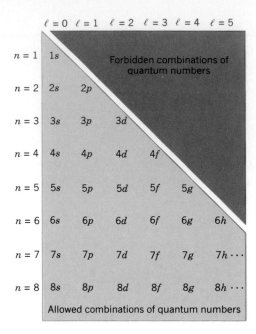

	$\ell = 0$	$\ell = 1$	$\ell = 2$	$\ell = 3$	$\ell = 4$	$\ell = 5$
$n = 1$	1s					
$n = 2$	2s	2p				
$n = 3$	3s	3p	3d			
$n = 4$	4s	4p	4d	4f		
$n = 5$	5s	5p	5d	5f	5g	
$n = 6$	6s	6p	6d	6f	6g	6h
$n = 7$	7s	7p	7d	7f	7g	7h ⋯
$n = 8$	8s	8p	8d	8f	8g	8h ⋯

Forbidden combinations of quantum numbers

Allowed combinations of quantum numbers

FIG. 5.17 According to the selection rules for quantum numbers, only certain combinations of the principal (n) and angular (ℓ) quantum numbers are allowed.

The third rule governing allowed combinations of the n, l, and m quantum numbers has an important consequence. It forces the number of subshells in a shell to be equal to the principal quantum number for the shell. The $n = 3$ shell, for example, contains three subshells—the $3s$, $3p$, and $3d$ orbitals.

Let's look at some of the possible combinations of the three quantum numbers: n, l, and m. There is only one orbital in the $n = 1$ shell, because there is only one way in which a sphere can be oriented in space. The only allowed combination of quantum numbers for which $n = 1$ is the following.

n	l	m	
1	0	0	**1s**

There are four orbitals in the $n = 2$ shell.

n	l	m	
2	0	0	**2s**
2	1	-1, 0, or $+1$	**2p**

FIG. 5.18 When the angular quantum number (ℓ) is 1, there are three possible values of the magnetic quantum number (m): -1, 0, and $+1$. These three values of m correspond to the three different orientations of these orbitals in space.

There is only one orbital in the $2s$ subshell. There are three orbitals in the $2p$ subshell, however, because there are three unique directions in which a polar ($l = 1$) orbital can point. One of these orbitals can be oriented along the x axis, another along the y axis, and the third along the z axis of a coordinate system, as shown in Figure 5.18. These orbitals are known as the $2p_x$, $2p_y$, and $2p_z$ orbitals.

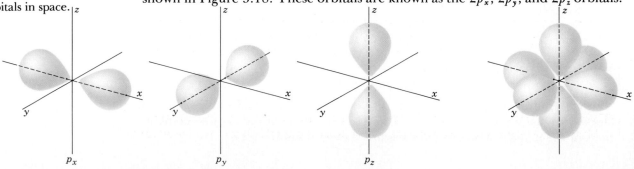

p_x p_y p_z

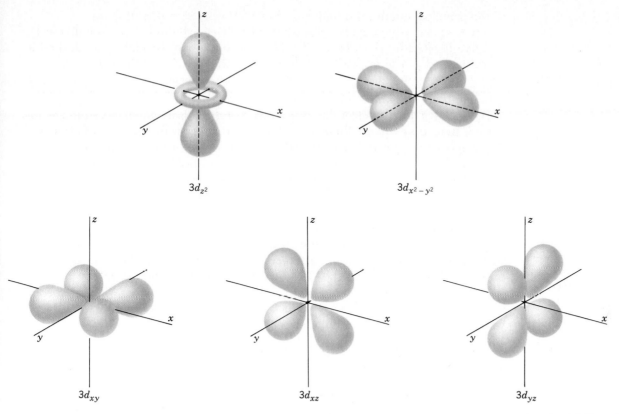

FIG. 5.19 When the angular quantum number (ℓ) is 2, there are five allowed values of the magnetic quantum number (m), corresponding to the five different orientations of these orbitals in space.

There are nine orbitals in the $n = 3$ shell, as already seen in Exercise 5.8.

n	l	m	
3	0	0	**3s**
3	1	-1, 0 or 1	**3p**
3	2	-2, -1, 0, 1 or 2	**3d**

There is one orbital in the 3s subshell, and there are three orbitals in the 3p subshell. The $n = 3$ shell, however, also includes 3d orbitals. The five different orientations of orbitals in the 3d subshell are shown in Figure 5.19. One of these orbitals lies in the xy plane between the x and y axes of a coordinate system and is called the $3d_{xy}$ orbital. The $3d_{xz}$ and $3d_{yz}$ orbitals have the same shape, but they lie between the axes of the coordinate system in the xz and yz planes. The fourth orbital in this subshell lies along the x and y axes and is called the $3d_{x^2-y^2}$ orbital. Most of the space occupied by the fifth orbital lies along the z axis, and this orbital is called the $3d_{z^2}$ orbital.

Exercise 5.9

Determine the relationships between: (1) the number of orbitals in a shell and the principal quantum number of the shell, and (2) the number of orbitals in a subshell and the angular quantum number for the subshell.

Solution

There is one orbital in the $n = 1$ shell, four orbitals in the $n = 2$ shell, and nine orbitals in the $n = 3$ shell. The number of orbitals in a shell is therefore the square

FIG. 5.20 It takes three quantum numbers to describe an orbital, but a fourth quantum number must be added to differentiate between the two electrons that can occupy the orbital. Because the two electrons that can occupy an orbital behave as if they were spinning in different directions, this fourth quantum number is called the spin (s) quantum number. By convention, the allowed values of the spin quantum number are $+\frac{1}{2}$ and $-\frac{1}{2}$.

of the principal quantum number — $1^2 = 1$, $2^2 = 4$, $3^2 = 9$, and so on.

There is one orbital in an s subshell ($l = 0$), three orbitals in a p subshell ($l = 1$), and five orbitals in a d subshell ($l = 2$). The number of orbitals in a subshell is therefore $2(l) + 1$. For example, $2(0) + 1 = 1$; $2(1) + 1 = 3$; $2(2) + 1 = 5$; and so on.

All we have done so far is describe a series of orbitals in which electrons can be placed. Before we can use these orbitals we need to know the number of electrons that can occupy an orbital and how they can be distinguished from one another. Experimental evidence suggests that an orbital can hold no more than two electrons. In order to distinguish between the two electrons in an orbital we need a fourth quantum number. This is called the **spin** (s) quantum number, because electrons behave as if they were spinning in either a clockwise or counterclockwise fashion (Figure 5.20). One of the electrons in an orbital is arbitrarily assigned an s quantum number of $+\frac{1}{2}$, the other is assigned an s quantum number of $-\frac{1}{2}$. It therefore takes three quantum numbers to define an orbital, but four quantum numbers to identify one of the electrons that can occupy the orbital.

The allowed combinations of n, l, and m quantum numbers for the first four shells are given in Table 5.3. For each of these orbitals, there are two allowed values of the spin quantum number, s.

5.18 THE RELATIVE ENERGIES OF ATOMIC ORBITALS

Because of the force of attraction between objects of opposite charge, the most important factor influencing the energy of either an orbit or an orbital is its size and therefore the value of the principal quantum number, n. For an atom that contains only one electron, there is no difference between the energies of the different subshells within a shell. The $3s$, $3p$, and $3d$ orbitals, for example, have the same energy in a hydrogen atom. Since the only factor that influences the energies of the orbitals in the hydrogen atom is the size of the orbital, the Bohr model, which specifies the energies of orbits in terms of nothing more than the distance between the electron and the nucleus, works for this atom.

TABLE 5.3

Summary of Allowed Combinations of Quantum Numbers

n	l	m	Subshell Notation	Number of Orbitals in the Subshell	Number of Electrons Needed to Fill Subshell	
1	0	0	$1s$	1	2	
2	0	0	$2s$	1	2	
2	1	1, 0, −1	$2p$	3	6	total: 8
3	0	0	$3s$	1	2	
3	1	1, 0, −1	$3p$	3	6	
3	2	2, 1, 0, −1, −2	$3d$	5	10	total: 18
4	0	0	$4s$	1	2	
4	1	1, 0, −1	$4p$	3	6	
4	2	2, 1, 0, −1, −2	$4d$	5	10	total: 8
4	3	3, 2, 1, 0, −1, −2, −3	$4f$	7	14	total: 32

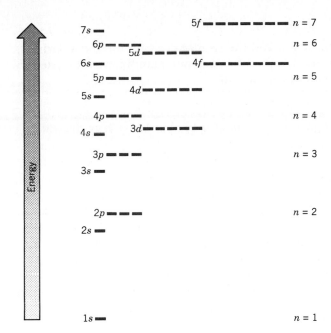

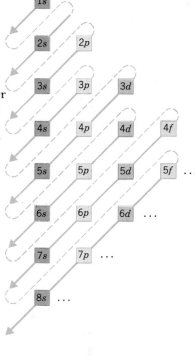

FIG. 5.21 The energy of an orbital increases as either the principal quantum number (n) or the angular quantum number (ℓ) becomes larger. The energy of an orbital in a given shell is therefore often higher than the energy of orbitals that are formally in a larger shell.

The hydrogen atom is unusual, however. For atoms and ions that contain more than one electron, the subshells no longer have the same energy. Within a given shell, the s orbitals always have the lowest energy. The energy of the subshells gradually increases as the value of the angular quantum number increases.

$$\text{Relative energies: } s < p < d < f$$

For most atoms, two factors control the energy of an orbital: size and shape, as shown in Figure 5.21.

A very simple device can be constructed to estimate the relative energies of atomic orbitals. The allowed combinations of the n and l quantum numbers are organized in a table, as shown in Figure 5.22, and arrows are drawn at 45° angles pointing toward the bottom left corner of the table. We can read off the order of increasing energy of the orbitals by following these arrows, starting at the top of the first line and then proceeding on to the second, third, and fourth lines, and so on. This diagram predicts the following order of increasing energy for atomic orbitals.

FIG. 5.22 The order of increasing energy for atomic orbitals can be predicted from the direction of the arrows in this diagram — $1s < 2s < 2p < 3s < 3p < 4s < 3d < 4p$ . . .

$$1s < 2s < 2p < 3s < 3p < 4s < 3d < 4p < 5s < 4d < 5p$$
$$< 6s < 4f < 5d < 6p < 7s < 5f < 6d < 7p < 8s \cdots$$

5.19 ELECTRON CONFIGURATIONS

We are now ready to predict which orbitals are occupied by electrons in an atom — the *electron configuration*. The basis of this prediction is the *aufbau* (literally, build-up) *principle.*

> **Aufbau principle: A guide for filling orbitals on an atom, which assumes that electrons are added to the atom starting with the lowest-energy orbital until all of the electrons have been placed in an appropriate orbital.**

A hydrogen atom ($Z = 1$) has only one electron, and this electron goes into the lowest-energy orbital, the $1s$ orbital. This is indicated by writing a superscript 1 after the symbol for the orbital.

$$\text{H } (Z = 1): \qquad 1s^1$$

The next element has two electrons, and the second electron fills the $1s$ orbital, because there are only two possible values for the spin quantum number used to distinguish between the electrons in an orbital.

$$\text{He } (Z = 2): \qquad 1s^2$$

The third electron therefore goes into the next orbital in the energy diagram — the $2s$ orbital.

$$\text{Li } (Z = 3): \qquad 1s^2 \, 2s^1$$

The fourth electron fills this orbital.

$$\text{Be } (Z = 4): \qquad 1s^2 \, 2s^2$$

After the $1s$ and $2s$ orbitals have been filled, the next-lowest-energy orbitals are the three $2p$ orbitals. The fifth electron goes into one of these orbitals.

$$\text{B } (Z = 5): \qquad 1s^2 \, 2s^2 \, 2p^1$$

When the time comes to add a sixth electron, the electron configuration is obvious.

$$\text{C } (Z = 6): \qquad 1s^2 \, 2s^2 \, 2p^2$$

But this configuration raises an interesting question. There are three orbitals in the $2p$ subshell. Does the second electron go into the same orbital as the first electron, or does it go into one of the other orbitals in this subshell?

Before we can answer this question we need to understand the concept of **degenerate orbitals.**

Degenerate orbitals: Two or more orbitals that have the same energy.

The energy of an orbital depends on both its size and its shape. As the orbital becomes larger or the shape becomes more complex, the electron spends more of its time further from the nucleus of the atom. In an isolated atom, however, the energy of an orbital doesn't depend on the direction in which it points in space. Orbitals that differ only in their orientation in space are therefore degenerate.

The three $2p$ orbitals in a carbon atom all have the same size and shape. The only difference between these orbitals is the direction in which they point. These orbitals therefore have the same energy — they are degenerate.

Electrons fill degenerate orbitals according to rules first stated by Friedrich Hund.

HUND'S RULES

1. One electron is added to each of the degenerate orbitals in a subshell before a second electron is added to any orbital in the subshell.

2. Electrons added to a subshell have the same value for the spin quantum number until each orbital in the subshell has at least one electron.

When the time comes to place two electrons into the $2p$ subshell we put one electron into each of two of these orbitals.

$$\text{C } (Z = 6): \qquad 1s^2 \, 2s^2 \, 2p_x^{\,1} \, 2p_y^{\,1}$$

We can show that both of the electrons in the $2p$ subshell have the same spin quantum number by representing an electron for which $s = +\frac{1}{2}$ with an arrow pointing up and an electron for which $s = -\frac{1}{2}$ with an arrow pointing down. The outermost electrons on the carbon atom can be represented as follows.

$$\uparrow \quad \uparrow \quad \underline{}$$

When we get to N ($Z = 7$), we have to put one electron into each of the three degenerate $2p$ orbitals.

$$\text{N } (Z = 7): \qquad 1s^2\, 2s^2\, 2p^3 \qquad \uparrow \quad \uparrow \quad \uparrow$$

Since each orbital in this subshell now contains one electron, the next electron added to the subshell must have the opposite spin quantum number, thereby filling one of the $2p$ orbitals.

$$\text{O } (Z = 8): \qquad 1s^2\, 2s^2\, 2p^4 \qquad \uparrow\downarrow \quad \uparrow \quad \uparrow$$

The ninth electron fills a second orbital in this subshell.

$$\text{F } (Z = 9): \qquad 1s^2\, 2s^2\, 2p^5 \qquad \uparrow\downarrow \quad \uparrow\downarrow \quad \uparrow$$

The tenth electron completes the $2p$ subshell.

$$\text{Ne } (Z = 10): \qquad 1s^2\, 2s^2\, 2p^6 \qquad \uparrow\downarrow \quad \uparrow\downarrow \quad \uparrow\downarrow$$

As we will soon see, there is something unusually stable about atoms, such as He and Ne, that have electron configurations with filled shells of orbitals. By convention, we therefore write abbreviated electron configurations in terms of the number of electrons beyond the previous element with a filled-shell electron configuration. Electron configurations of the next two elements in the periodic table, for example, can be written as follows.

$$\textbf{Na } (Z = 11): \qquad \text{[Ne] } 3s^1$$
$$\textbf{Mg } (Z = 12): \qquad \text{[Ne] } 3s^2$$

Exercise 5.10

Predict the electron configuration for a neutral tin atom (Sn, $Z = 50$).

Solution

We can predict the order of increasing energy of the atomic orbitals on tin from the diagram in Figure 5.22. We can then add electrons to these orbitals, starting with the lowest-energy orbital, until all 50 electrons have been used.

The key to success with the aufbau process is remembering that each orbital can hold 2 electrons. Since an s subshell contains only one orbital, only 2 electrons can be added to this subshell. There are three orbitals in a p subshell, however, so a set of p orbitals can hold up to 6 electrons. There are five orbitals in a d subshell, so a set of these orbitals can hold up to 10 electrons. The following is the complete electron configuration for tin.

$$\text{Sn } (Z = 50): \qquad 1s^2\, 2s^2\, 2p^6\, 3s^2\, 3p^6\, 4s^2\, 3d^{10}\, 4p^6\, 5s^2\, 4d^{10}\, 5p^2$$

The aufbau process can be used to predict the electron configuration for an element, but the actual configuration used by the element has to be determined experimentally. The experimentally determined electron configurations for the 108 known elements are given in Table 5.4.

TABLE 5.4

The Electron Configurations of the Elements

Atomic Number	Symbol	Electron Configuration	Atomic Number	Symbol	Electron Configuration
1	H	$1s^1$	49	In	$[Kr]\,5s^2\,4d^{10}\,5p^1$
2	He	$1s^2 = [He]$	50	Sn	$[Kr]\,5s^2\,4d^{10}\,5p^2$
3	Li	$[He]\,2s^1$	51	Sb	$[Kr]\,5s^2\,4d^{10}\,5p^3$
4	Be	$[He]\,2s^2$	52	Te	$[Kr]\,5s^2\,4d^{10}\,5p^4$
5	B	$[He]\,2s^2\,2p^1$	53	I	$[Kr]\,5s^2\,4d^{10}\,5p^5$
6	C	$[He]\,2s^2\,2p^2$	54	Xe	$[Kr]\,5s^2\,4d^{10}\,5p^6 = [Xe]$
7	N	$[He]\,2s^2\,2p^3$	55	Cs	$[Xe]\,6s^1$
8	O	$[He]\,2s^2\,2p^4$	56	Ba	$[Xe]\,6s^2$
9	F	$[He]\,2s^2\,2p^5$	57	La	$[Xe]\,6s^2\,5d^1$
10	Ne	$[He]\,2s^2\,2p^6 = [Ne]$	58	Ce	$[Xe]\,6s^2\,4f^2$
11	Na	$[Ne]\,3s^1$	59	Pr	$[Xe]\,6s^2\,4f^3$
12	Mg	$[Ne]\,3s^2$	60	Nd	$[Xe]\,6s^2\,4f^4$
13	Al	$[Ne]\,3s^2\,3p^1$	61	Pm	$[Xe]\,6s^2\,4f^5$
14	Si	$[Ne]\,3s^2\,3p^2$	62	Sm	$[Xe]\,6s^2\,4f^6$
15	P	$[Ne]\,3s^2\,3p^3$	63	Eu	$[Xe]\,6s^2\,4f^7$
16	S	$[Ne]\,3s^2\,3p^4$	64	Gd	$[Xe]\,6s^2\,4f^7\,5d^1$
17	Cl	$[Ne]\,3s^2\,3p^5$	65	Tb	$[Xe]\,6s^2\,4f^9$
18	Ar	$[Ne]\,3s^2\,3p^6 = [Ar]$	66	Dy	$[Xe]\,6s^2\,4f^{10}$
19	K	$[Ar]\,4s^1$	67	Ho	$[Xe]\,6s^2\,4f^{11}$
20	Ca	$[Ar]\,4s^2$	68	Er	$[Xe]\,6s^2\,4f^{12}$
21	Sc	$[Ar]\,4s^2\,3d^1$	69	Tm	$[Xe]\,6s^2\,4f^{13}$
22	Ti	$[Ar]\,4s^2\,3d^2$	70	Yb	$[Xe]\,6s^2\,4f^{14}$
23	V	$[Ar]\,4s^2\,3d^3$	71	Lu	$[Xe]\,6s^2\,4f^{14}\,5d^1$
24	Cr	$[Ar]\,4s^1\,3d^5$	72	Hf	$[Xe]\,6s^2\,4f^{14}\,5d^2$
25	Mn	$[Ar]\,4s^2\,3d^5$	73	Ta	$[Xe]\,6s^2\,4f^{14}\,5d^3$
26	Fe	$[Ar]\,4s^2\,3d^6$	74	W	$[Xe]\,6s^2\,4f^{14}\,5d^4$
27	Co	$[Ar]\,4s^2\,3d^7$	75	Re	$[Xe]\,6s^2\,4f^{14}\,5d^5$
28	Ni	$[Ar]\,4s^2\,3d^8$	76	Os	$[Xe]\,6s^2\,4f^{14}\,5d^6$
29	Cu	$[Ar]\,4s^1\,3d^{10}$	77	Ir	$[Xe]\,6s^2\,4f^{14}\,5d^7$
30	Zn	$[Ar]\,4s^2\,3d^{10}$	78	Pt	$[Xe]\,6s^1\,4f^{14}\,5d^9$
31	Ga	$[Ar]\,4s^2\,3d^{10}\,4p^1$	79	Au	$[Xe]\,6s^1\,4f^{14}\,5d^{10}$
32	Ge	$[Ar]\,4s^2\,3d^{10}\,4p^2$	80	Hg	$[Xe]\,6s^2\,4f^{14}\,5d^{10}$
33	As	$[Ar]\,4s^2\,3d^{10}\,4p^3$	81	Tl	$[Xe]\,6s^2\,4f^{14}\,5d^{10}\,6p^1$
34	Se	$[Ar]\,4s^2\,3d^{10}\,4p^4$	82	Pb	$[Xe]\,6s^2\,4f^{14}\,5d^{10}\,6p^2$
35	Br	$[Ar]\,4s^2\,3d^{10}\,4p^5$	83	Bi	$[Xe]\,6s^2\,4f^{14}\,5d^{10}\,6p^3$
36	Kr	$[Ar]\,4s^2\,3d^{10}\,4p^6 = [Kr]$	84	Po	$[Xe]\,6s^2\,4f^{14}\,5d^{10}\,6p^4$
37	Rb	$[Kr]\,5s^1$	85	At	$[Xe]\,6s^2\,4f^{14}\,5d^{10}\,6p^5$
38	Sr	$[Kr]\,5s^2$	86	Rn	$[Xe]\,6s^2\,4f^{14}\,5d^{10}\,6p^6 = [Rn]$
39	Y	$[Kr]\,5s^2\,4d^1$	87	Fr	$[Rn]\,7s^1$
40	Zr	$[Kr]\,5s^2\,4d^2$	88	Ra	$[Rn]\,7s^2$
41	Nb	$[Kr]\,5s^1\,4d^4$	89	Ac	$[Rn]\,7s^2\,6d^1$
42	Mo	$[Kr]\,5s^1\,4d^5$	90	Th	$[Rn]\,7s^2\,6d^2$
43	Tc	$[Kr]\,5s^2\,4d^5$	91	Pa	$[Rn]\,7s^2\,5f^2\,6d^1$
44	Ru	$[Kr]\,5s^1\,4d^7$	92	U	$[Rn]\,7s^2\,5f^3\,6d^1$
45	Rh	$[Kr]\,5s^1\,4d^8$	93	Np	$[Rn]\,7s^2\,5f^4\,6d^1$
46	Pd	$[Kr]\,4d^{10}$	94	Pu	$[Rn]\,7s^2\,5f^6$
47	Ag	$[Kr]\,5s^1\,4d^{10}$	95	Am	$[Rn]\,7s^2\,5f^7$
48	Cd	$[Kr]\,5s^2\,4d^{10}$	96	Cm	$[Rn]\,7s^2\,5f^7\,6d^1$

TABLE 5.4 continued

The Electron Configurations of the Elements

Atomic Number	Symbol	Electron Configuration
97	Bk	$[Rn]\ 7s^2\ 5f^9$
98	Cf	$[Rn]\ 7s^2\ 5f^{10}$
99	Es	$[Rn]\ 7s^2\ 5f^{11}$
100	Fm	$[Rn]\ 7s^2\ 5f^{12}$
101	Md	$[Rn]\ 7s^2\ 5f^{13}$
102	No	$[Rn]\ 7s^2\ 5f^{14}$
103	Lr	$[Rn]\ 7s^2\ 5f^{14}\ 6d^1$
104		$[Rn]\ 7s^2\ 5f^{14}\ 6d^2$
105		$[Rn]\ 7s^2\ 5f^{14}\ 6d^3$
106		$[Rn]\ 7s^2\ 5f^{14}\ 6d^4$
107		$[Rn]\ 7s^2\ 5f^{14}\ 6d^5$
109		$[Rn]\ 7s^2\ 5f^{14}\ 6d^7$

5.20 EXCEPTIONS TO PREDICTED ELECTRON CONFIGURATIONS

There are several patterns in the electron configurations listed in Table 5.4. One of the most striking is the remarkable level of agreement between these experimentally determined electron configurations and the configurations we would predict from the diagram in Figure 5.22. There are only two exceptions among the first 40 elements — chromium and copper. These exceptions are interesting, however, because they provide the first glimpse at a factor that will play an important role in the next chapter.

By strictly adhering to the rules of the aufbau process we would predict the following electron configurations for chromium and copper.

$$Cr\ (Z = 24):\qquad [Ar]\ 4s^2\ 3d^4$$
$$Cu\ (Z = 29):\qquad [Ar]\ 4s^2\ 3d^9 \qquad \text{predicted electron configurations}$$

The experimentally determined electron configurations for these elements are slightly different.

$$Cr\ (Z = 24):\qquad [Ar]\ 4s^1\ 3d^5$$
$$Cu\ (Z = 29):\qquad [Ar]\ 4s^1\ 3d^{10} \qquad \text{actual electron configurations}$$

In each case, one electron has been transferred from the $4s$ orbital to a $3d$ orbital, even though the $3d$ orbitals are supposed to be at a higher level than the $4s$ orbital.

These exceptions can be explained by an important principle that will guide our discussion of electron configurations in the next few chapters.

Electron configurations in which subshells are either full or half-full are unusually stable.

By transferring an electron from the $4s$ orbital into a $3d$ orbital, it is possible to produce an electron configuration for chromium in which the $3d$ subshell is half-filled and a configuration for copper in which the $3d$ subshell is filled.

$$Cr\ (Z = 24):\qquad [Ar]\ 4s^1\ 3d^5 \qquad \uparrow\ \ \uparrow\ \ \uparrow\ \ \uparrow\ \ \uparrow$$
$$Cu\ (Z = 29):\qquad [Ar]\ 4s^1\ 3d^{10} \qquad \uparrow\downarrow\ \ \uparrow\downarrow\ \ \uparrow\downarrow\ \ \uparrow\downarrow\ \ \uparrow\downarrow$$

Once we get beyond atomic number 40, the difference between the energies of adjacent orbitals in the following sequence is small enough that it becomes much easier to transfer an electron from one orbital to another.

$$1s < 2s < 2p < 3s < 3p < 4s < 3d < 4p \cdots$$

Most of the exceptions to the electron configuration predicted from the diagram in Figure 5.22 therefore occur among the elements with atomic numbers larger than 40. Although it is tempting to focus attention on the handful of elements in Table 5.4 that have electron configurations that differ from those predicted with Figure 5.22, the amazing thing is that this simple diagram works for so many elements.

5.21 ELECTRON CONFIGURATIONS AND THE PERIODIC TABLE

Table 5.4 lists the electron configurations of the elements in order of increasing atomic number. When these data are arranged so that we can compare elements in one of the horizontal rows of the periodic table, we find that these rows typically correspond to the filling of a shell of orbitals. The second row, for example, contains elements in which the orbitals in the $n = 2$ shell are filled.

Li (Z = 3)	$[He]\, 2s^1$
Be (Z = 4)	$[He]\, 2s^2$
B (Z = 5)	$[He]\, 2s^2\, 2p^1$
C (Z = 6)	$[He]\, 2s^2\, 2p^2$
N (Z = 7)	$[He]\, 2s^2\, 2p^3$
O (Z = 8)	$[He]\, 2s^2\, 2p^4$
F (Z = 9)	$[He]\, 2s^2\, 2p^5$
Ne (Z = 10)	$[He]\, 2s^2\, 2p^6$

There is an obvious pattern within the vertical columns, or groups, of the periodic table as well. The elements in a group have similar configurations for their outermost electrons. This relationship can be seen in the electron configurations of elements in columns on the left-hand side of the periodic table.

Group IA	*Group IIA*
H $1s^1$	
Li $[He]\, 2s^1$	**Be** $[He]\, 2s^2$
Na $[Ne]\, 3s^1$	**Mg** $[Ne]\, 3s^2$
K $[Ar]\, 4s^1$	**Ca** $[Ar]\, 4s^2$
Rb $[Kr]\, 5s^1$	**Sr** $[Kr]\, 5s^2$
Cs $[Xe]\, 6s^1$	**Ba** $[Xe]\, 6s^2$
Fr $[Rn]\, 7s^1$	**Ra** $[Rn]\, 7s^2$

It can also be seen in the configurations of elements in columns on the right-hand side of the table.

Group VIA	*Group VIIA*
O $[He]\, 2s^2\, 2p^4$	**F** $[He]\, 2s^2\, 2p^5$
S $[Ne]\, 3s^2\, 3p^4$	**Cl** $[Ne]\, 3s^2\, 3p^5$
Se $[Ar]\, 4s^2\, 3d^{10}\, 4p^4$	**Br** $[Ar]\, 4s^2\, 3d^{10}\, 4p^5$
Te $[Kr]\, 5s^2\, 4d^{10}\, 5p^4$	**I** $[Kr]\, 5s^2\, 4d^{10}\, 5p^5$
Po $[Xe]\, 6s^2\, 4f^{14}\, 5d^{10}\, 6p^4$	**At** $[Xe]\, 6s^2\, 4f^{14}\, 5d^{10}\, 6p^5$

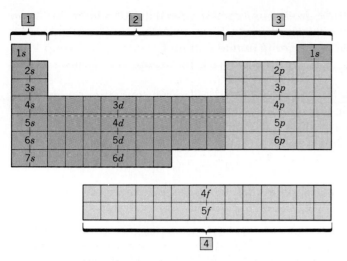

FIG. 5.23 The shape of the periodic table reflects the order in which atomic orbitals are filled. The s orbitals are filled in the two columns on the far left, and the p orbitals are filled in the six columns on the right. The d orbitals are filled along the transition between the s and p orbitals, and the f orbitals are filled in the two long rows of elements at the bottom of the table.

Figure 5.23 shows the relationship between the periodic table and the orbitals being filled during the aufbau process. The 2 columns on the left side of the periodic table correspond to the filling of an s orbital. The next 10 columns include elements in which the five orbitals in a d subshell are filled. The 6 columns on the right represent the filling of the three orbitals in a p subshell. Finally, the 14 columns at the bottom of the table correspond to the filling of the seven orbitals in an f subshell.

Exercise 5.11

Predict the electron configurations for calcium ($Z = 20$) and zinc ($Z = 30$) from their positions in the periodic table.

Solution

Calcium is in the second column and the fourth row of the table. The second column corresponds to the filling of an s orbital. The 1s orbital is filled in the first row, the 2s orbital in the second row, and so on. By the time we get to calcium, we are filling the 4s orbital. Calcium therefore has all of the electrons of argon, plus a filled 4s orbital.

$$\text{Ca } (Z = 20): \quad [\text{Ar}]\, 4s^2$$

Zinc is the tenth element in the region of the periodic table where d orbitals are filled. Zinc therefore has a filled subshell of d orbitals. The only question is which set of d orbitals are filled. Although zinc is in the fourth row of the periodic table, it must be remembered that the first time d orbitals occur is in the $n = 3$ shell. Furthermore, the diagram in Figure 5.21 predicts that the 3d orbitals will fill after the 4s orbitals. The following is therefore the abbreviated electron configuration for zinc.

$$\text{Zn } (Z = 30): \quad [\text{Ar}]\, 4s^2\, 3d^{10}$$

5.22 ELECTRON CONFIGURATIONS OF IONS

Atoms form ions by either gaining or losing electrons. Section 2.11 suggested that we can predict the charge on many of these ions from the number of the column, or group, in which the element is found in the periodic table. The maximum charge on a positive ion is equal to the group number of the element. Aluminum, in Group IIIA, for example, loses three electrons to form an Al^{3+} ion. The charge on a negative ion is often equal to the group number minus 8. Oxygen, in Group VIA, forms O^{2-} ions $(6 - 8 = -2)$, and phosphorus, in Group VA, forms P^{3-} ions $(5 - 8 = -3)$.

These rules can be understood in terms of the electron configurations in Table 5.4 and the generalization that there is something unusually stable about configurations in which a subshell is either full or half-full. A neutral aluminum atom has the following electron configuration.

$$\text{Al } (Z = 13): \qquad 1s^2\, 2s^2\, 2p^6\, 3s^2\, 3p^1$$

When an aluminum atom loses three of these electrons to form an Al^{3+} ion, it takes on the electron configuration of the rare gas neon.

$$\mathbf{Al^{3+} (Z = 13): \qquad 1s^2\, 2s^2\, 2p^6 = [Ne]}$$

The same line of reasoning can be used to explain why oxygen gains two electrons to form an O^{2-} ion. A neutral oxygen atom has the following electron configuration.

$$\text{O } (Z = 8): \qquad 1s^2\, 2s^2\, 2p^4$$

By picking up two more electrons, an oxygen atom can achieve the stable electron configuration of the rare gas neon.

$$\mathbf{O^{2-} (Z = 8): \qquad 1s^2\, 2s^2\, 2p^6 = [Ne]}$$

SUMMARY

At the beginning of the 20th century, scientists could still argue in good faith about whether atoms existed. As we approach the end of this century, the issue has been resolved. As the result of theoretical and experimental accomplishments of some of the greatest minds of our time, definitive evidence for the existence of atoms has been gathered and a detailed model of their internal structure has been developed.

Atoms contain a small, massive, positively charged nucleus, which occupies a negligibly small fraction of the total volume of the atom. The nucleus contains the number of positively charged protons equal to the atomic number of the element and enough neutral neutrons to make up the remainder of the mass of the atom.

The nucleus is embedded in a sea of lightweight, negatively charged electrons, which occupy the remaining space in the atom. These electrons have the properties of both particles and waves, which means that we can't specify their exact location in three-dimensional space. We therefore turn to a quantum mechanical model, which tries to describe the regions in space — or orbitals — in which electrons reside and the probability of finding an electron at a particular point in space at a given time.

The three coordinates chosen to describe the three-dimensional model of the structure of the electrons in an atom specify the size, shape, and orientation in space of the orbitals in which electrons can be found. These orbitals are organized into shells, which all have the same value of the n, or principal, quantum number. They are also divided into subshells, which have the same values of the n and l quantum numbers.

Each orbital can hold a maximum of two electrons. A fourth quantum number — known as the spin quantum number — is therefore used to identify the individual electrons in an orbital.

The electron configuration for an element can be predicted from the aufbau process, which involves adding electrons to orbitals starting with the lowest-energy orbitals on the atom. The results of these predictions agree to a remarkable extent with the experimentally determined electron configurations of the elements.

The electron configuration of the ions these elements form can often be predicted from the number of electrons the element has to either gain or lose to reach an electron configuration in which the subshells of orbitals are either full or empty.

PROBLEMS

Electricity and Matter

5-1 Use the present model for the structure of the atom to explain static electricity.

5-2 Describe the ways in which our present model for the structure of the atom is consistent with du Fay's two-fluid model of electricity. Describe how it is consistent with Franklin's one-fluid model.

The Discovery of the Electron

5-3 Describe the evidence that led to the belief that cathode rays consist of a beam of negatively charged particles that flow from the cathode to the anode of the tube.

5-4 Use the present model for the structure of the atom to propose an explanation for what happens in a cathode-ray tube.

The Discovery of X-Rays

5-5 The discovery of x-rays in 1895 set off what has been described as a scientific gold rush to investigate the properties of new forms of radiation. What was it about x-rays that captured the imagination of so many scientists?

5-6 X-rays and light are both forms of electromagnetic radiation. What difference between x-rays and light allows one to pass through solid bodies while the other is absorbed?

Radioactivity

5-7 Describe the difference between α-particles, β-particles, and γ-rays.

5-8 Describe the role that x-rays played in the discovery of radioactivity.

Millikan's Oil Drop Experiment

5-9 Use the present model for the structure of the atom to explain why the H^+ ion had the largest charge-to-mass ratio of any particle until the electron was discovered.

5-10 Which of the following particles has the largest charge-to-mass ratio?

(a) protons (b) neutrons (c) α-particles (d) electrons
(e) x-rays

The Structure of the Atom

5-11 Describe the three fundamental particles needed to explain the chemical behavior of atoms.

5-12 Describe the difference between a hydrogen atom and a proton, between a hydrogen atom and a neutron, and between a proton and a neutron.

5-13 Describe the present model for the structure of the atom.

5-14 Choose an example of an element that forms positive ions. Describe the relationship between the atomic number, mass number, number of protons, number of neutrons, and number of electrons in the atom and its positively charged ion. Do the same for an element that forms negative ions.

5-15 Write the symbol for the atom or ion that contains 24 protons, 21 electrons, and 28 neutrons.

5-16 Calculate the number of protons and neutrons in the nucleus and the number of electrons surrounding the nucleus of a $^{39}K^+$ ion. What is the atomic number and the mass number of this ion?

5-17 Calculate the number of protons and neutrons in the nucleus and the number of electrons surrounding the nucleus of a $^{127}I^-$ ion. What is the atomic number and the mass number of this ion?

5-18 Identify the element that forms atoms with a mass number of 20 that contain 11 neutrons.

5-19 Give the symbol for the atom or ion that has 34 protons, 45 neutrons, and 36 electrons.

5-20 Calculate the number of electrons in a $^{134}Ba^{2+}$ ion.

5-21 A chemistry text published in 1922 proposed a model of the atom in which the nucleus was composed of protons and electrons. The number of protons in the nucleus was equal to the atomic weight. The charge on the nucleus, or the atomic number, was equal to the number of protons minus the number of electrons in the nucleus. Use this model to calculate the number of protons and electrons in a neutral fluorine atom, F, and a fluoride ion, F^-. Compare this calculation with the results obtained by assuming that the nucleus is composed of protons and neutrons.

5-22 The charge on a single electron is 1.602×10^{-19} coulomb. Calculate the charge on a mole of electrons and compare the results of this calculation with Faraday's constant, given in Table A-2 in the Appendix.

Particles and Waves

5-23 Describe the difference between particles and waves.

5-24 Describe the relationship among the frequency of a wave, its wavelength, and the speed or velocity at which the wave moves through space.

5-25 An octave in a musical scale corresponds to a change in the frequency of a note by a factor of 2. If a note with a frequency of 440 hertz is an A, then the A one octave above this note has a frequency of 880 hertz. What happens to the wavelength of the sound as the frequency increases by a factor of 2? What happens to the speed at which the sound travels to your ear?

5-26 The human ear is capable of hearing sound waves with frequencies between about 20 and 20,000 hertz. If the speed of sound is 340.3 meters per second at sea level, what is the wavelength in meters of the shortest wave the human ear can hear?

Light and Other Forms of Electromagnetic Radiation

5-27 Calculate the wavelength in meters of green light that has a frequency of 5.0×10^{14} cycles per second.

5-28 Calculate the frequency of red light that has a wavelength of 700 nanometers.

5-29 Which has the longer wavelength: red light or blue light?

5-30 Which has the larger frequency: radio waves or x-rays?

5-31 In a magnet with a magnetic field of 23.490 kilogauss, ^{13}C nuclei absorb electromagnetic radiation that has a frequency of 25.147 megahertz. Calculate the wavelength of this radiation. In which region of the electromagnetic spectrum does this radiation fall?

5-32 The meter has been defined as the length of 1,650,763.73 wavelengths in a vacuum of the orange-red line of the emission spectrum of ^{86}Kr. Calculate the frequency of this radiation. In what portion of the electromagnetic spectrum does this radiation fall?

Atomic Spectra

5-33 Soap bubbles pick up colors because they reflect light with wavelengths equal to the thickness of the walls of the bubble. What frequency of light is reflected by a soap bubble 6 nanometers thick?

5-34 Methylene blue, $C_{16}H_{18}ClN_3S$, is a dye that absorbs light most intensely at wavelengths of 668 and 609 nanometers. What color is the light absorbed by this dye?

5-35 Sodium salts give off a characteristic yellow-orange light when added to the flame of a bunsen burner. This yellow-orange color is due to two narrow bands of radiation with wavelengths of 588.9953 nanometers and 589.5923 nanometers. Calculate the frequencies of these emission lines.

Quantization of Energy

5-36 Which carries more energy: ultraviolet or infrared radiation?

5-37 Which carries more energy: yellow light with a wavelength of 580 nanometers or green light with a wavelength of 560 nanometers?

5-38 List the four lines in the emission spectrum of hydrogen in order of increasing energy.

5-39 Calculate the energy in joules of the line in the emission spectrum of the hydrogen atom that has a wavelength of 656.3 nanometers.

5-40 Calculate the energy in joules of a single particle of radiation broadcast by an amateur radio operator who transmits at a wavelength of 10 meters.

5-41 O_2 molecules can dissociate to form oxygen atoms by absorbing energy in the form of light. If it takes 498 kilojoules of energy to break the bonds in a mole of O_2 molecules, what is the wavelength of the radiation that has just enough energy to decompose O_2 molecules to oxygen atoms? In what portion of the electromagnetic spectrum is this wavelength found?

The Bohr Model of the Atom

5-42 Describe the conditions under which atoms or molecules emit light and compare them with the conditions under which they absorb light.

5-43 Explain why a hydrogen atom emits only four narrow bands of light in the visible spectrum instead of a broad band of energies.

5-44 Explain why energy is emitted when an electron falls from the $n = 3$ into the $n = 2$ orbit in the Bohr model.

5-45 Calculate the wavelength of the radiation emitted by a hydrogen atom when an electron falls from the $n = 3$ to the $n = 2$ orbit in the Bohr model.

5-46 Identify the transition between orbits in the Bohr model that gives rise to the line in the emission spectrum of the hydrogen atom that has a wavelength of 410.1 nanometers.

5-47 Which of the following transitions in the spectrum of the hydrogen atom results in the emission of light with the longest wavelength?

(a) $n = 3$ to $n = 2$ (b) $n = 3$ to $n = 1$ (c) $n = 5$ to $n = 4$
(d) $n = 2$ to $n = 3$ (e) $n = 1$ to $n = 3$ (f) $n = 4$ to $n = 5$

5-48 Which of the following transitions in a hydrogen atom results in the absorption of a photon with the largest energy?

(a) $n = 2$ to $n = 3$ (b) $n = 2$ to $n = 4$ (c) $n = 1$ to $n = 4$
(d) $n = 3$ to $n = 1$ (e) $n = 7$ to $n = 1$

5-49 Use Figure 5.14 to explain why the transitions in the Lyman series are in the ultraviolet portion of the electromagnetic spectrum and those in the Paschen, Brackett, and Pfund series are in the infrared portion of this spectrum.

5-50 According to the Bohr model, the energy absorbed when an electron on a hydrogen atom jumps from the $n = 1$ to the $n = 2$ quantum level is 1.63×10^{-18} joules. Calculate the energy of the photon emitted when an electron falls from the $n = 3$ to the $n = 2$ quantum level.

Quantum Numbers

5-51 Describe the function of each of the four quantum numbers: n, l, m, and s.

5-52 Describe the selection rules for the n, l, m, and s quantum numbers.

5-53 Identify the quantum number that specifies each of the following.

(a) The size of the orbital (b) The shape of the orbital
(c) The way the orbital is oriented in space (d) The spin of the electrons that occupy an orbital

5-54 Determine the allowed values of the angular quantum number, l, when the principal quantum number is 4. Describe the difference between orbitals that have the same principal quantum number and different angular quantum numbers.

5-55 Determine the allowed values of the magnetic quantum number, m, when the angular quantum number is 2. Describe the difference between orbitals that have the same angular quantum number and different magnetic quantum numbers.

5-56 Determine the allowed values of the spin quantum number, s, when $n = 5$, $l = 2$, and $m = -1$.

5-57 Determine the number of allowed values of the magnetic quantum number when $n = 3$ and $l = 2$.

5-58 Determine the maximum value for the angular quantum number, l, when the principal quantum number is 3.

5-59 Which of the following is a legitimate set of n, l, m, and s quantum numbers?

(a) $4, -2, -1, \frac{1}{2}$ (b) $4, 2, 3, \frac{1}{2}$ (c) $4, 3, 0, 1$ (d) $4, 0, 0, -\frac{1}{2}$

5-60 Which of the following is a legitimate set of n, l, m, and s quantum numbers?

(a) $0, 0, 0, \frac{1}{2}$ (b) $8, 4, -3, -\frac{1}{2}$ (c) $3, 3, 2, +\frac{1}{2}$
(d) $2, 1, -2, -\frac{1}{2}$ (e) $5, 3, 3, -1$

5-61 Calculate the maximum number of electrons that can have the quantum numbers $n = 4$ and $l = 3$.

5-62 Calculate how many electrons in an atom can simultaneously possess the quantum numbers $n = 4$ and $s = +\frac{1}{2}$.

5-63 Write the combination of quantum numbers for every electron in the $n = 1$ and $n = 2$ shells using the selection rules outlined in this chapter.

5-64 Write the combination of quantum numbers for every electron in the $n = 1$ and $n = 2$ shell assuming that the selection rules for assigning quantum numbers are changed to the following.

1. The principal quantum number can be any integer greater than or equal to 1.
2. The angular quantum number can have any value between 0 and n.
3. The magnetic quantum number can have any value between 0 and l.
4. The spin quantum number can have a value of either $+1$ or -1.

Shells and Subshells of Orbitals

5-65 Describe what happens to the difference between the energies of subshells of orbitals as the value of the principal quantum number, n, becomes larger.

5-66 Identify the symbol used to describe orbitals for which $l = 1$.

5-67 Determine the number of f orbitals in the $n = 3$, $n = 4$, and $n = 5$ shells.

5-68 Which of the following sets of n, l, m, and s quantum numbers can be used to describe an electron in a $2p$ orbital?

(a) $2, 1, 0, -\frac{1}{2}$ (b) $2, 0, 0, \frac{1}{2}$ (c) $2, 2, 1, \frac{1}{2}$
(d) $3, 2, 1, -\frac{1}{2}$ (e) $3, 1, 0, \frac{1}{2}$

5-69 Which of the following orbitals cannot exist?

(a) $6s$ (b) $3p$ (c) $2d$ (d) $4f$ (e) $17f$

5-70 Calculate the maximum number of electrons in the $n = 1$, $n = 2$, $n = 3$, $n = 4$, and $n = 5$ shells of orbitals.

5-71 Calculate the maximum number of electrons that can fit into a $4d$ subshell.

5-72 Calculate the maximum number of unpaired electrons that can be placed in a $5d$ subshell.

5-73 Explain why the difference between the atomic numbers of any pair of elements in a vertical column, or group, of the periodic table is either 8, 18, or 32.

5-74 Show that the number of orbitals in a shell is always equal to n^2, where n is the principal quantum number.

5-75 Show that the maximum number of electrons in a shell of orbitals is equal to $2n^2$, where n is the principal quantum number.

5-76 Show that the number of orbitals in a subshell is always equal to $2(l) + 1$, where l is the angular quantum number.

The Relative Energies of Atomic Orbitals

5-77 Which pair of quantum numbers determines the energy of an electron in an orbital?

(a) n and l (b) n and m (c) n and s (d) l and m
(e) l and s (f) m and s

5-78 Which of the following sets of orbitals is arranged in order of increasing energy?

(a) $3d < 4s < 4p < 5s < 1d$ (b) $3d < 4s < 4p < 4d < 5s$
(c) $4s < 3d < 4p < 5s < 4d$ (d) $4s < 3d < 4p < 4d < 5s$
(e) $3d < 4s < 4p < 4d < 5s$

5-79 Which of the following orders of increasing energies is incorrect?

(a) $3s < 4s < 5s$ (b) $5s < 5p < 5d$ (c) $5s < 4d < 5p$
(d) $6s < 4f < 5d$ (e) $6s < 5f < 6p$

5-80 As atomic orbitals are filled according to the aufbau principle, the $6p$ orbitals are filled immediately after which of the following orbitals?

(a) $4f$ (b) $5d$ (c) $6s$ (d) $7s$

Electron Configurations

5-81 Which of the following characteristics describes orbitals that are degenerate?

(a) They contain the same number of electrons. (b) They have the same value of the angular quantum number, l, and different values of the principal quantum number, n. (c) They have the same set of three quantum numbers, n, l, and m, but have different values of the s quantum number. (d) They have the same energy.
(e) They contain the same number of unpaired electrons.

5-82 Describe the regions of the periodic table in which the s, p, d, and f subshells are filled.

5-83 Write the electron configurations for the eight elements in the third row of the periodic table.

5-84 Write the electron configurations for the 18 elements in the fourth row of the periodic table.

5-85 Use straight lines and arrows to draw the orbital diagram for the electron configuration of the nitrogen atom.

5-86 Use straight lines and arrows to draw the orbital diagram for the electron configuration for nickel.

5-87 The electron configuration of Si is $1s^2\ 2s^2\ 2p^6\ 3s^2\ 3p^x$, where x is which of the following?

(a) 1 (b) 2 (c) 3 (d) 4 (e) 6

5-88 Which of the following is the electron configuration for carbon that satisfies Hund's rules?

(a) $1s^2\ 2s^2\ 2p_x^2\ 2p_y^0\ 2p_z^0$ (b) $1s^2\ 2s^2\ 2p_x^0\ 2p_y^2\ 2p_z^2$
(c) $1s^2\ 2s^2\ 2p_x^1\ 2p_y^1\ 2p_z^0$ (d) $1s^2\ 2s^1\ 2p_x^1\ 2p_y^1\ 2p_z^1$

5-89 Which of the following is the correct electron configuration for the P^{3+} ion?

(a) [Ne] (b) [Ne] $3s^2$ (c) [Ne] $3s^2\ 3p^3$ (d) [Ne] $3s^2\ 3p^6$

5-90 Which of the following is the correct electron configuration for the Te^{2+} ion?

(a) [Kr] $5s^2\ 4d^{10}\ 5p^4$ (b) [Kr] $5s^2\ 4d^{10}\ 5p^2$
(c) [Kr] $5s^2\ 4d^{10}\ 4p^4$ (d) [Kr] $5s^2\ 4d^{10}\ 4p^6$
(e) [Kr] $5s^2\ 4d^{10}\ 5p^6$

5-91 Which of the following is the correct electron configuration for the bromide ion, Br^-?

(a) [Ar] $4s^2\ 4p^5$ (b) [Ar] $4s^2\ 3d^{10}\ 4p^7$
(c) [Ar] $4s^2\ 3d^{10}\ 4p^5$ (d) [Ar] $4s^2\ 3d^{10}\ 4p^6$
(e) [Ar] $4s^2\ 3d^{10}\ 3p^6$

5-92 Determine the number of electrons in a vanadium atom for which the principal quantum number, n, is 3.

5-93 Determine the number of electrons in s orbitals in the V^{2+} ion.

5-94 Which of the following elements has the largest number of electrons for which the angular quantum number is 1?

(a) He (b) F (c) S (d) As (e) Zn

5-95 Which of the following elements has the largest number of electrons for which the principal quantum number is 3?

(a) Na (b) Al (c) Si (d) Cl (e) Ar (f) Zn

5-96 Which of the following is a possible set of n, l, m, and s quantum numbers for the last electron added to form a gallium atom ($Z = 31$)?

(a) $3, 1, 0, -\frac{1}{2}$ (b) $3, 2, 1, \frac{1}{2}$ (c) $4, 0, 0, \frac{1}{2}$ (d) $4, 1, 1, \frac{1}{2}$
(e) $4, 2, 2, \frac{1}{2}$

5-97 Which of the following is a possible set of n, l, m, and s quantum numbers for the last electron added to form an As^{3+} ion?

(a) $3, 1, -1, \frac{1}{2}$ (b) $4, 0, 0, -\frac{1}{2}$ (c) $3, 2, 0, \frac{1}{2}$
(d) $4, 1, -1, \frac{1}{2}$ (d) $5, 0, 0, \frac{1}{2}$

5-98 Theoreticians predict that the element with atomic number 114 will be more stable than the elements with atomic numbers between 103 and 114. On the basis of its electron configuration, in which group of the periodic table should Element 114 be placed?

5-99 Theoreticians predict that Element 120 will be more stable than the elements recently discovered with atomic numbers between 103 and 109. On the basis of the electron configuration of this element, in which group of the periodic table should it be found?

5-100 Which is the first element to have $4d$ electrons in its electron configuration?

   (a) Ca (b) Sc (c) Rb (d) Y (e) La

5-101 The elements starting with lanthanum (La) in the first of the two long rows of elements at the bottom of the periodic table are called "lanthanides." The lanthanides correspond to the filling of orbitals with what value of the angular quantum number?

5-102 Which of the following neutral atoms has the largest number of unpaired electrons?

   (a) Na (b) Al (c) Si (d) P (e) S

5-103 Which of the following ions has five unpaired electrons?

   (a) Ti^{4+} (b) Co^{2+} (c) V^{3+} (d) Fe^{3+} (e) Zn^{2+}

CHAPTER 6

THE PERIODIC TABLE

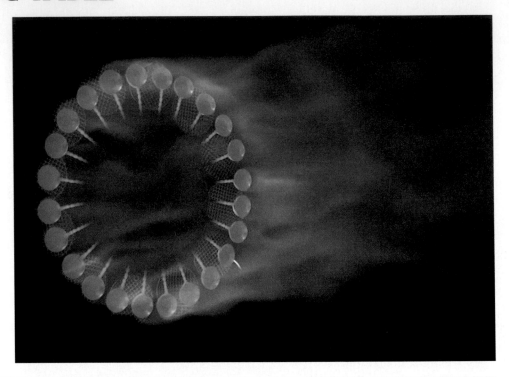

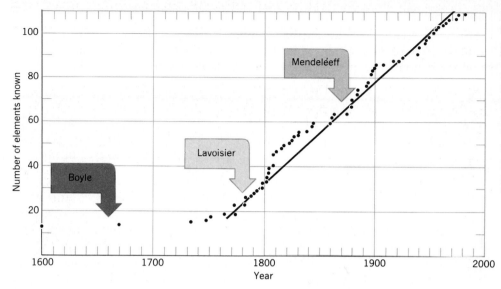

FIG. 6.1 A plot of the number of elements versus time from about 1600 to the present. When Boyle introduced the first modern definition of an element in 1661, only 13 elements were known. In 1789, when Lavoisier published a list of elements, a total of 24 were known. By the time Mendeléeff proposed his periodic table, the number had grown to 62. In 1987, there were 108 known elements.

6.1 THE SEARCH FOR PATTERNS IN THE CHEMISTRY OF THE ELEMENTS

In 1661, Robert Boyle proposed the following definition of an element.

Element: Any substance that cannot be decomposed into a simpler substance by a chemical reaction.

At that time, only 13 elements were known — antimony (Sb), arsenic (As), bismuth (Bi), carbon (C), copper (Cu), gold (Au), iron (Fe), lead (Pb), mercury (Hg), silver (Ag), sulfur (S), tin (Sn), and zinc (Zn). By the time of Lavoisier — toward the end of the 18th century — another 11 elements had been discovered: chlorine (Cl), cobalt (Co), hydrogen (H), manganese (Mn), molybdenum (Mo), nickel (Ni), nitrogen (N), oxygen (O), phosphorus (P), platinum (Pt), and tungsten (W). Since that time, a new element has been discovered on the average of every two and a half years, as shown in Figure 6.1.

As the number of elements increased, chemists inevitably began searching for patterns or trends in their properties. As noted in Section 2.6, one of the first attempts was made by Johann Wolfgang Döbereiner in 1829. Döbereiner discovered the existence of families of elements with similar chemical properties. Since there always seemed to be three elements in these families, he called them *triads.* Each vertical column in the following list represents one of these triads.

Li	**Ca**	**Cl**	**S**	**Mn**
Na	**Sr**	**Br**	**Se**	**Cr**
K	**Ba**	**I**	**Te**	**Fe**

Exercise 6.1

Which triads contain elements that are still grouped in the same column of the modern periodic table?

Solution

All of these triads *except* the one that contains Mn, Cr, and Fe are still grouped in the same column of the periodic table. The relationship between the remaining triads and the columns, or groups, of the periodic table is summarized below.

Li, Na, K:	Group IA
Ca, Sr, Ba:	Group IIA
S, Se, Te:	Group VIA
Cl, Br, I:	Group VIIA

The triads grouped elements that had similar chemical properties. For example, lithium, sodium, and potassium all react with water at room temperature. They react with chlorine to form compounds with the same empirical formula — LiCl, NaCl, KCl. They combine with hydrogen to form compounds with the same empirical formula — LiH, NaH, KH. They form oxides — Li_2O, Na_2O, K_2O — and hydroxides — LiOH, NaOH, KOH — with the same formulas, and so on.

Döbereiner also found patterns in the physical properties of the elements in a triad. He noticed that the atomic weight of the middle element in each triad was about equal to the average of the atomic weights of the first and third elements. Sodium, for example, has an atomic weight of 22.99 grams per mole, which is remarkably close to the average of the atomic weights of lithium (6.94 grams per mole) and potassium (39.10 grams per mole).

He also found that the density of the middle element in most triads is roughly equal to the average of the densities of the other elements. The density of strontium, for example, is 2.60 g/cm^3, which is close to the average of the densities of calcium (1.55 g/cm^3) and barium (3.51 g/cm^3).

Döbereiner's work focused on the relationships among the elements in a triad but gave no hint of any relationships between triads. The next major step toward an understanding of the patterns in the chemistry of the elements was taken by the British chemist John Newlands in 1865. Newlands found that when he listed the elements in order of increasing atomic weight, they seemed to fall into seven families that contained elements with similar chemical properties. His table listed those families in horizontal rows, as shown in Table 6.1.

Newlands was the first to recognize that the elements fall into a pattern in which their properties repeat at regular intervals, or *periods,* when they are listed in order of increasing atomic weight. He was also the first to assign atomic numbers to the elements. Unfortunately, his work was not accepted by his peers, and the paper in which he described his table was rejected by the *Journal of the Chemical Society.*

In fact, there are problems with Newlands's table of the elements. The first row,

TABLE 6.1

Newlands's Table of the Elements

H 1	F 8	Cl 15	Co & Ni 22	Br 29	Pd 36	I 42	Pt & Ir 50
Li 2	Na 9	K 16	Cu 23	Rb 30	Ag 37	Cs 44	Os 51
Be 3	Mg 10	Ca 17	Zn 24	Sr 31	Cd 38	Ba & V 45	Hg 52
B 4	Al 11	Cr 19	Y 25	Ce & La 33	U 40	Ta 46	Tl 53
C 5	Si 12	Ti 18	In 26	Zr 32	Sn 39	W 47	Pb 54
N 6	P 13	Mn 20	As 27	Di & Mo 34	Sb 41	Nb 48	Bi 55
O 7	S 14	Fe 21	Se 28	Rh & Ru 35	Te 43	Au 49	Th 56

for example, groups elements with similar chemical properties (such as F, Cl, Br, and I), but it also includes elements that have totally different chemical properties (such as Co, Ni, Pd, Pt, and Ir). Furthermore, at a time when elements were being discovered with some regularity, Newlands failed to leave room in his table for new elements.

These mistakes might have resulted from Newlands's attempt to link the periodicity of the chemical elements with the periodicity found in music. Newlands saw a pattern in which intervals of seven elements often separated two elements with similar chemical properties. There were seven elements between fluorine and chlorine, for example, and between sodium and potassium. This reminded him of a musical scale, in which any note in a key is separated from its octave by an interval of seven notes. Newlands apparently was so enthralled with this "law of octaves" that he made the mistake of trying to force all of the elements into this pattern.

6.2 THE DEVELOPMENT OF THE PERIODIC TABLE

In 1869, Dmitri Ivanovitch Mendeléeff[1] published the first of a series of papers outlining a *periodic table* of the elements. While trying to organize a discussion of the properties of the elements for a chemistry course at the Technological Institute in Petrograd, Mendeléeff listed the properties of each element on a different card. He then played with the cards, arranging them in different orders, until he noticed that the properties of the elements repeated in a periodic fashion when the elements are listed roughly in order of increasing atomic weight.

A copy of the table Mendeléeff published in 1871 is shown in Figure 6.2. Elements with similar chemical properties were listed in columns. All of the elements in the first column, for example, form compounds with the generic formula R_2O—such as H_2O, Li_2O, and Na_2O—while elements in the second group form compounds with the generic formula RO—such as BeO, MgO, and CaO.

Mendeléeff was not the first to propose a periodic table in which the elements were arranged in order of increasing atomic weight. Newlands had done the same thing four years earlier. It is therefore interesting to ask why Mendeléeff's contribution to the development of the periodic table is recognized as being so much

Mendeléeff had a striking appearance that was the result of his tendency to have his hair cut only once each year.

Tabelle II.

Reihen	Gruppe I. — R²O	Gruppe II. — RO	Gruppe III. — R²O³	Gruppe IV. RH⁴ RO²	Gruppe V. RH³ R²O⁵	Gruppe VI. RH² RO³	Gruppe VII. RH R²O⁷	Gruppe VIII. — RO⁴
1	H = 1							
2	Li = 7	Be = 9,4	B = 11	C = 12	N = 14	O = 16	F = 19	
3	Na = 23	Mg = 24	Al = 27,3	Si = 28	P = 31	S = 32	Cl = 35,5	
4	K = 39	Ca = 40	— = 44	Ti = 48	V = 51	Cr = 52	Mn = 55	Fe = 56, Co = 59, Ni = 59, Cu = 63.
5	(Cu = 63)	Zn = 65	— = 68	— = 72	As = 75	Se = 78	Br = 80	
6	Rb = 85	Sr = 87	?Yt = 88	Zr = 90	Nb = 94	Mo = 96	— = 100	Ru = 104, Rh = 104, Pd = 106, Ag = 108.
7	(Ag = 108)	Cd = 112	In = 113	Sn = 118	Sb = 122	Te = 125	J = 127	
8	Cs = 133	Ba = 137	?Di = 138	?Ce = 140	—	—	—	— — —
9	(—)							
10	—	—	?Er = 178	?La = 180	Ta = 182	W = 184	—	Os = 195, Ir = 197, Pt = 198, Au = 199.
11	(Au = 199)	Hg = 200	Tl = 204	Pb = 207	Bi = 208	—	—	— — —
12	—	—		Th = 231	—	U = 240	—	— — —

FIG. 6.2 This 1871 version of Mendeléeff's periodic table was published in the journal *Annalen der Chemie*.

[1] There are at least a half-dozen different ways of spelling Mendeléeff's name, because of disagreements about how to translate from the Cyrillic alphabet used in Russia. The version used here is the spelling that Mendeléeff himself used when he visited England in 1887.

greater than the contribution of any of the other chemists who made similar observations.

1. Mendeléeff grouped the elements on the basis of similarities in their chemical properties, not an arbitrary mathematical rule such as the law of octaves. As a result, he didn't have to force the elements to fit a preconceived structure.

2. Mendeléeff realized that additional elements would inevitably be discovered and left blank spaces in this table for these elements.

3. Mendeléeff was perceptive enough to predict where elements were missing and therefore left blank spaces at the appropriate places in his table.

4. Mendeléeff was willing to question accepted values of atomic weights when they disagreed with the pattern of elements in his table. He was able to show, for example, that the accepted values for the atomic weights of beryllium, indium, and uranium had to be wrong, because they would result in these elements' being placed in the wrong groups in the table.

5. Mendeléeff was willing to accept minor inversions in the order of increasing atomic weight when these inversions resulted in elements' being placed in the correct groups or families. Thus, tellurium was placed before iodine in the periodic table, even though Te had a slightly larger atomic weight than I.

6. Mendeléeff was both willing and able to predict the properties of elements that had not yet been discovered. This was important, because it allowed his theory to be tested.

Based on empty spaces in his periodic table, Mendeléeff predicted the discovery of 10 elements, which he tentatively named by adding the prefix *eka-* to the name of the element immediately above each empty space. He then tried to predict in some detail the properties of four of these elements: *eka*-aluminum, *eka*-boron, *eka*-silicon, and *eka*-tellurium.

The remarkable agreement between Mendeléeff's predictions for *eka*-aluminum and the observed properties of the element gallium, which was discovered in 1875, is shown in Table 6.2. A similar comparison between the predicted properties of *eka*-silicon and the observed properties of germanium, discovered in 1886, is

TABLE 6.2

*The Predicted Properties of **Eka-aluminum** Compared with the Observed Properties of Gallium*

Mendeléeff's Predictions	Observed Properties
Atomic weight: about 68	Atomic weight: 69.72
Density: 5.9	Density: 5.94
Melting point: low	Melting point: 30.15°C
Formula of oxide: Ea_2O_3	Formula of oxide: Ga_2O_3
Formula of chloride: $EaCl_3$	Formula of chloride: $GaCl_3$
Chemistry: Hydroxide should dissolve in acids and bases; metal should form basic salts; the sulfide should precipitate with H_2S or $(NH_4)_2$; the chloride should be more volatile than $ZnCl_2$.	Chemistry: The hydroxide dissolves in acids and bases; the metal forms basic salts; Ga_2S_3 is precipitated by either H_2S or $(NH_4)_2S$; $GaCl_3$ is more volatile than $ZnCl_2$.
The element will probably be discovered by spectroscopy.	The element was discovered with the aid of the spectroscope.

TABLE 6.3

The Predicted Properties of Eka-silicon Compared
with the Observed Properties of Germanium

Mendeléeff's Predictions	Observed Properties
Atomic weight: 72	Atomic weight: 72.59
Density: 5.5	Density: 5.47
Formula of oxide: EsO_2	Formula of oxide: GeO_2
Density of oxide: 4.7	Density of oxide: 4.703
Formula of chloride: $EsCl_4$	Formula of chloride: $GeCl_4$
Boiling point of $EsCl_4$: under 100°C	Boiling point of $GeCl_4$: 86°C

given in Table 6.3. It was the extraordinary success of Mendeléeff's predictions that led chemists not only to accept the periodic table but to recognize Mendeléeff more than anyone else as the originator of the concept on which it was based.

6.3 MODERN VERSIONS OF THE PERIOD TABLE

More than 700 versions of the periodic table were published in the first 100 years after the publication of the table in Figure 6.2. Some of these tables grouped elements on the basis of similar chemical properties. Copper and silver, for exam-

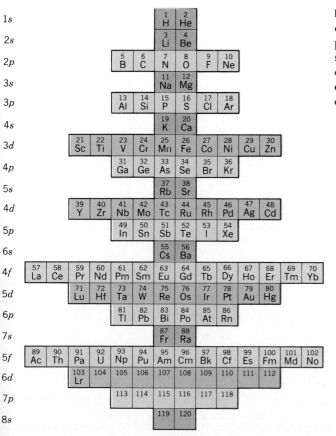

FIG. 6.3 More than 700 versions of the periodic table have been published. This chart emphasizes the pattern of the filling of orbital subshells that goes into determining the electron configuration of an element.

FIG. 6.4 The most popular form of the periodic table separates the elements into three classes: (1) the main-group elements, (2) the transition metals, and (3) the lanthanides and actinides.

Groups

	IA	IIA	IIIB	IVB	VB	VIB	VIIB		VIIIB		IB	IIB	IIIA	IVA	VA	VIA	VIIA	VIIIA
1	1 H																1 H	2 He
2	3 Li	4 Be											5 B	6 C	7 N	8 O	9 F	10 Ne
3	11 Na	12 Mg											13 Al	14 Si	15 P	16 S	17 Cl	18 Ar
4	19 K	20 Ca	21 Sc	22 Ti	23 V	24 Cr	25 Mn	26 Fe	27 Co	28 Ni	29 Cu	30 Zn	31 Ga	32 Ge	33 As	34 Se	35 Br	36 Kr
5	37 Rb	38 Sr	39 Y	40 Zr	41 Nb	42 Mo	43 Tc	44 Ru	45 Rh	46 Pd	47 Ag	48 Cd	49 In	50 Sn	51 Sb	52 Te	53 I	54 Xe
6	55 Cs	56 Ba	57 La	72 Hf	73 Ta	74 W	75 Re	76 Os	77 Ir	78 Pt	79 Au	80 Hg	81 Tl	82 Pb	83 Bi	84 Po	85 At	86 Rn
7	87 Fr	88 Ra	89 Ac	104	105	106	107	108	109									

(Periods)

58 Ce	59 Pr	60 Nd	61 Pm	62 Sm	63 Eu	64 Gd	65 Tb	66 Dy	67 Ho	68 Er	69 Tm	70 Yb	71 Lu
90 Th	91 Pa	92 U	93 Np	94 Pu	95 Am	96 Cm	97 Bk	98 Cf	99 Es	100 Fm	101 Md	102 No	103 Lr

ple, can be found in Group I of Mendeléeff's table because they form compounds with oxygen (Ag_2O or Cu_2O) and chlorine (AgCl or CuCl) that have the same formula as the compounds formed by other elements in this group, such as lithium (Li_2O and LiCl) and sodium (Na_2O and NaCl). Other versions were based on the observation that elements with similar chemical properties often have similar electron configurations. The table shown in Figure 6.3, for example, emphasizes the order in which subshells of orbitals are filled.

The most successful tables try to achieve both goals, listing elements in the same group, or column, when they have similar chemical properties and similar electron configurations. The most popular version of the periodic table is shown in Figure 6.4. This version of the periodic table separates the elements into three classes: (1) the main-group elements, which correspond to the filling of s and p orbitals; (2) the transition metals, which bridge the main-group elements on either side of the table; and (3) the lanthanides and actinides at the bottom of the table, which correspond to the filling of f orbitals.

Each group, or column, in the periodic table in Figure 6.4 contains compounds with similar chemical properties. All of the elements in Group IIA, for example, form oxides with the generic formula RO and chlorides with the formula RCl_2. In most cases, all of the elements in a column also have similar electron configurations. The two most important exceptions are hydrogen and helium.

On the basis of electron configuration, hydrogen should be placed in Group IA, along with lithium, sodium, potassium, and so on.

Group IA

H:	$1s^1$	Rb:	$[Kr]\, 5s^1$
Li:	$[He]\, 2s^1$	Cs:	$[Xe]\, 6s^1$
Na:	$[Ne]\, 3s^1$	Fe:	$[Rn]\, 7s^1$
K:	$[Ar]\, 4s^1$		

But the other elements in Group IA are metals, and hydrogen is not. Thus, it might be appropriate to include hydrogen in Group VIIA, with the other nonmetals that have one less electron than a filled-shell configuration. Many periodic tables therefore list hydrogen in both Group IA and VIIA.

On the basis of electron configurations, helium ($1s^2$) should be placed in Group IIA of the table, along with beryllium, magnesium, calcium, and so on.

Group IIA

Be:	[He] $2s^2$		Sr:	[Kr] $5s^2$
Mg:	[Ne] $3s^2$		Ba:	[Xe] $6s^2$
Ca:	[Ar] $4s^2$		Ra:	[Rn] $7s^2$

But the elements in Group IIA are metals, and helium is not. Helium behaves more like the elements with a filled-shell electron configuration in the last column of the periodic table. Virtually every periodic table therefore includes helium among the elements in Group VIIIA.

6.4 THE SIZE OF ATOMS: METALLIC RADII

Two properties of the elements that repeat in a periodic fashion have already been discussed. One of these is a chemical property — the formation of compounds with the same generic formula, such as H_2O, Li_2O, Na_2O, and so on. The other, electron configuration, might be described as a physical property. A number of other chemical and physical properties repeat periodically when the elements are arranged more or less in order of increasing atomic weight. The remainder of this chapter will be devoted to a discussion of the physical properties of the elements that play an important role in determining the chemical properties of the elements, which will be discussed in the next few chapters.

One of the physical properties that influence the chemical behavior of the elements is the relative size of their atoms and ions. It is impossible to measure the size of an isolated atom or ion, because it is impossible to determine the location of the electrons that surround the nucleus. What we can measure is the distance between the nuclei of adjacent atoms in a solid. We can then estimate the size of an atom by assigning half of this distance to the radius of each atom.

This technique is best suited to elements — such as the metals — that form solids composed of extended planes of atoms of that element. The result of these measurements are therefore described as *metallic radii.*

Metallic radius: An estimate of the size of an atom equal to half the distance between adjacent nuclei in a metal.

Fortunately, more than 75% of the elements are metals. Metallic radii are therefore available for most of the elements in the periodic table.

Metallic radii, in units of nanometers (nm), for the elements in Groups IA and IIA are given in Table 6.4 and graphed in Figure 6.5.

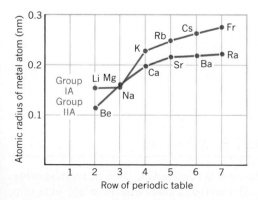

FIG. 6.5 Metallic radii become larger as we go down a column of the periodic table because the outermost electrons are placed in larger orbitals. Metallic radii become smaller from left to right across a row of the table because the number of protons increases, increasing the force of attraction between the nucleus and the electrons.

TABLE 6.4

Metallic Radii of Group IA and Group IIA Elements

	Element	Metallic Radius (nm)	Electron Configuration
Group IA:	Li	0.152	[He] $2s^1$
	Na	0.1537	[Ne] $3s^1$
	K	0.2272	[Ar] $4s^1$
	Rb	0.2475	[Kr] $5s^1$
	Cs	0.2654	[Xe] $6s^1$
	Fr	0.27	[Rn] $7s^1$
Group IIA:	Be	0.1113	[He] $2s^2$
	Mg	0.160	[Ne] $3s^2$
	Ca	0.197	[Ar] $4s^2$
	Sr	0.215	[Kr] $5s^2$
	Ba	0.217	[Xe] $6s^2$
	Ra	0.220	[Rn] $7s^2$

Exercise 6.2

Describe the trends in the metallic radii data in Table 6.4 and Figure 6.5.

Solution

There are two general trends in these data.

1. The metallic radius becomes larger as we go down a column of the periodic table. Beryllium atoms are smaller than magnesium atoms, which are smaller than calcium atoms, and so on.

2. The metallic radius becomes smaller as we go from left to right across a row of the periodic table. With only one exception, atoms of elements in Group IA are larger than atoms of the adjacent elements in Group IIA. Lithium atoms are larger than beryllium atoms, potassium atoms are larger than calcium atoms, and so on.

The only exception to these trends is the metallic radius of sodium, which is anomalously small.

The first trend in metallic radii is easy to understand. As we go down a column of the periodic table, electrons are placed in larger orbitals. The size of the atoms should therefore increase.

The second trend is a bit more surprising. We might expect that atoms would become larger as we go across a row of the periodic table, because each element has one more electron than the preceding element. But the additional electrons are added to the same shell of orbitals, not to larger orbitals. At the same time, the force of attraction between the nucleus and the electrons gradually becomes larger across the row, because each succeeding nucleus contains more protons. Each succeeding nucleus therefore tends to hold electrons more tightly, and the atoms become smaller.

6.5 THE SIZE OF ATOMS: COVALENT RADII

Another way of estimating the size of an atom is to measure the distance between adjacent atoms in covalent compounds. The size of a chlorine atom, for example,

H								H	He
0.037		Covalent radius						0.037	
Li	Be			B	C	N	O	F	Ne
0.123	0.089			0.088	0.077	0.070	0.066	0.064	
Na	Mg			Al	Si	P	S	Cl	Ar
0.157	0.136			0.125	0.117	0.110	0.104	0.099	
K	Ca			Ga	Ge	As	Se	Br	Kr
0.203	0.174			0.125	0.122	0.121	0.117	0.114	
Rb	Sr			Im	Sn	Sb	Te	I	Xe
0.216	0.192			0.150	0.140	0.141	0.137	0.133	
Cs	Ba			Tl	Pd	Bi	Po	At	Rn
0.235	0.198			0155	0.154	0.152	0.153	–	
Fr	Ra								
–	–								

FIG. 6.6 The covalent radii for the main-group elements. Once again, atoms become larger as we go down a column of the periodic table and smaller as we go from left to right across a row of the table. Note that the covalent radii of the elements in Group IA and IIA are slightly smaller than the metallic radii of these elements.

can be estimated by measuring the length of the bond between the two chlorine atoms in a Cl_2 molecule. Half of this distance can then be assigned to the ***covalent radius*** of each atom.

> **Covalent radius: An estimate of the size of an atom equal to half the distance between identical atoms linked by a covalent bond.**

We can check the validity of these data by comparing measurements of the bond lengths in covalent compounds with the sum of the covalent radii of the atoms that form the bond. Within experimental error, for example, the C—Cl bond length in CCl_4 (0.1766 nm) is equal to the average of the C—C bond length in diamond (0.1542 nm) and the Cl—Cl bond length in a Cl_2 molecule (0.1988 nm).

Covalent radii are available for most elements in the periodic table, except the elements in Group VIIIA, which form very few (if any) covalent compounds. The covalent radii of the main-group elements are given in Figure 6.6. These data confirm the trends observed for metallic radii.

1. **Atoms become larger down a column of the periodic table.**
2. **Atoms become smaller across a row of the table.**

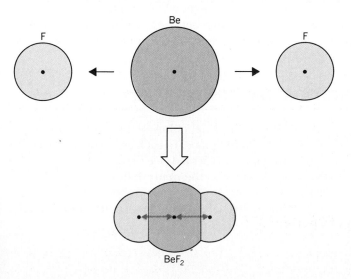

FIG. 6.7 Covalent radii are smaller than metallic radii because the force of attraction between the atoms in a covalent molecule tends to squeeze these atoms together.

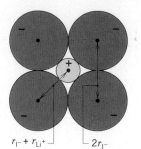

$r_{I^-} + r_{Li^+}$ $2r_{I^-}$

FIG. 6.8 The sizes of the Li^+ and I^- ions in lithium iodide can be calculated from measurements of the distance between the nuclei of adjacent ions if we assume that the Li^+ ions are so small that they fit into the holes left when the I^- ions pack together so that adjacent I^- ions touch. The distance between the nuclei of neighboring I^- ions is therefore twice the ionic radius of an I^- ion. Once this distance is known, the radius of the Li^+ ion can be estimated by subtracting the radius of the I^- ion from the distance between the nuclei of adjacent Li^+ and I^- ions.

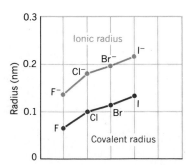

FIG. 6.9 This figure compares the radii of the F^-, Cl^-, Br^-, and I^- ions with the covalent radii of the corresponding neutral atoms. In each case, the negatively charged ion is larger than the neutral atom.

Table A-8 in the appendix contains both covalent and metallic radii for a number of elements. The covalent radius for an element is usually a little smaller than the metallic radius. That is because covalent bonds tend to squeeze the atoms together, as shown in Figure 6.7.

6.6 THE SIZE OF ATOMS: IONIC RADII

The relative size of atoms can also be studied by measuring their *ionic radii.*

> **Ionic radius: The radius of the positive or negative ions in ionic compounds, such as NaCl.**

The first ionic radii were obtained from examination of a crystal that contained a relatively small positive ion — the Li^+ ion — and a relatively large negative ion — the I^- ion. The structure of this crystal was analyzed on the basis of the following assumptions.

1. The lithium ions would pack in the holes between the very much larger iodide ions, as shown in Figure 6.8.
2. The I^- ions would touch each other, which means that the radius of one of these ions could be estimated from the distance between the nuclei of adjacent I^- ions.
3. The Li^+ ions would just touch the I^- ions, so the radius of the Li^+ ion could be estimated by subtracting the radius of the I^- ion from the distance between the nuclei of a pair of adjacent Li^+ and I^- ions.

In theory, the radii of the Li^+ and I^- ions obtained in this experiment could be combined with measurements of the distance between the ions in LiCl and NaI to estimate the size of the Cl^- and Na^+ ions. The validity of the listed assumptions could then be tested by comparing the sum of the ionic radii of the Na^+ and Cl^- ions with the distance between these ions in NaCl.

Unfortunately, only two of the three assumptions that were made for LiI are correct. The Li^+ ions are very much smaller than the I^- ions, and the I^- ions are in contact in this crystal. But the Li^+ ions don't quite touch the I^- ions. This experiment therefore overestimated the size of the Li^+ ion.

Repeating this analysis with a large number of ionic compounds has made it possible to obtain a set of more accurate ionic radii, which can be found in Table A-8 in the appendix. The experimental distance between the Na^+ and Cl^- ions in NaCl (0.281 nm) is reasonably close to the sum of the ionic radii of the Na^+ (0.095 nm) and Cl^- (0.181 nm) ions in this table.

6.7 THE RELATIVE SIZE OF ATOMS AND THEIR IONS

Table 6.5 compares the covalent radii of fluorine, chlorine, bromine, and iodine atoms with the radii of their F^-, Cl^-, Br^-, and I^- ions. In each case, the negative ion is much larger than the atom from which it was formed (see Figure 6.9). In fact, the negative ion can be more than twice as large as the neutral atom.

The only difference between an atom and its ions is the number of electrons that surround the nucleus. A neutral chlorine atom, for example, contains 17 electrons, while a Cl^- ion contains 18 electrons.

Cl: [Ne] $3s^2\,3p^5$
Cl^-:   [Ne] $3s^2\,3p^6$

TABLE 6.5		
A Comparison of the Covalent Radii of Neutral Group VIIA Atoms Versus the Ionic Radii of Their Negative Ions		
Element	Covalent Radii (nm)	Ionic Radii (nm)
F	0.064	0.136
Cl	0.099	0.181
Br	0.1142	0.196
I	0.1333	0.216

TABLE 6.6		
A Comparison of the Covalent Radii of Neutral Group IA Atoms Versus the Ionic Radii of Their Positive Ions		
	Covalent Radii (nm)	Ionic Radii (nm)
Li	0.123	0.068
Na	0.157	0.095
K	0.2025	0.1333
Rb	0.216	0.148
Cs	0.235	0.169

The nucleus, with its constant charge, can't hold the 18 electrons in the negatively charged Cl^- ion as tightly as it can hold the 17 electrons in the neutral atom. As a result, the negative ion is larger than the neutral atom.

Extending this line of reasoning would result in the prediction that positive ions should be smaller than the neutral atoms from which they are formed. The 11 protons in the nucleus of an Na^+ ion should be able to hold the 10 electrons on this ion more tightly than it can hold the 11 electrons that surround the nucleus of a neutral sodium atom. The Na^+ ion should therefore be smaller than the Na atom. Table 6.6 provides data to test this hypothesis. It lists the covalent radii for lithium, sodium, potassium, rubidium, and cesium atoms and the ionic radii for the corresponding Li^+, Na^+, K^+, Rb^+, and Cs^+ ions. In each case, the positive ion is indeed much smaller than the atom from which it forms (see Figure 6.10).

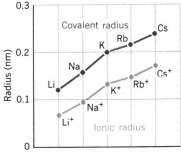

FIG. 6.10 This figure compares the radii of the Li^+, Na^+, K^+, Rb^+, and Cs^+ ions with the covalent radii of the corresponding neutral atoms. In each case, the positively charged ion is smaller than the neutral atom.

Exercise 6.3

Compare the sizes of neutral sodium and chlorine atoms and their Na^+ and Cl^- ions.

Solution

A neutral sodium atom is significantly larger than a neutral chlorine atom.

$$Na = 0.157 \text{ nm} \qquad Cl = 0.099 \text{ nm} \qquad \text{(covalent radii)}$$

But a positively charged Na^+ ion is only about half as big as a negatively charged Cl^- ion.

$$Na^+ = 0.095 \text{ nm} \qquad Cl^- = 0.181 \text{ nm} \qquad \text{(ionic radii)}$$

The fact that positive ions are much smaller than neutral atoms, while negative ions are much larger, has important implications for the structure of ionic compounds. When the time comes to build models of these compounds in Chapter 12, it will be important to remember that positive ions are usually much smaller than negative ions. The positive ions are often so small they pack in the holes between planes of adjacent negative ions. In NaCl, for example, the Na^+ ions are so small that the Cl^- ions almost touch, as shown in Figure 6.11.

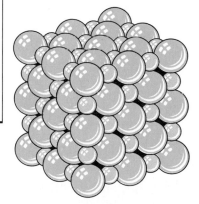

FIG. 6.11 The Na^+ ions in NaCl are so much smaller than the Cl^- ions that they fit in the holes between planes of Cl^- ions.

6.8 PATTERNS IN IONIC RADII

The ionic radii in Tables 6.5 and 6.6 confirm one of the patterns observed for both metallic and covalent radii: atoms become larger as we go down a column of the

TABLE 6.7

*Ionic and Covalent Radii for Isoelectronic
Second- and Third-Row Atoms and Ions*

Atom or Ion	Radius (nm)	Electron Configuration
C^{4-}	0.260	$1s^2\, 2s^2\, 2p^6$
N^{3-}	0.171	$1s^2\, 2s^2\, 2p^6$
O^{2-}	0.140	$1s^2\, 2s^2\, 2p^6$
F^-	0.136	$1s^2\, 2s^2\, 2p^6$
Ne	0.112	$1s^2\, 2s^2\, 2p^6$
Na^+	0.095	$1s^2\, 2s^2\, 2p^6$
Mg^{2+}	0.065	$1s^2\, 2s^2\, 2p^6$
Al^{3+}	0.050	$1s^2\, 2s^2\, 2p^6$

periodic table. In this case, we can see that Li^+ ions are much smaller than Na^+ ions which are smaller than K^+ ions , and so on. These data also show that F^- ions are much smaller than Cl^- ions which are smaller than Br^- ions, and so on.

We can examine trends in ionic radii across a row of the periodic table by comparing data for atoms and ions that are *isoelectronic.*

Atoms or ions are isoelectronic when they have the same number of electrons and therefore the same electron configuration.

Table 6.7 summarizes data on the radii of a series of isoelectronic ions and atoms of second- and third-row elements.

The data in Table 6.7 are easy to explain if we note that these atoms or ions all have exactly the same number of electrons but that the number of protons depends on the atomic number of the element. These atoms or ions all have 10 electrons, but the number of protons increases from 6 in the C^{4-} ion to 13 in the Al^{3+} ion. The larger the charge on the nucleus, the more tightly the electrons are held. As a result, the atoms or ions become smaller.

Exercise 6.4

For each of the following pairs of atoms or ions, predict which is larger.

   (a) S^{2-} or O^{2-} (b) Na or Al (c) C^{4-} or F^- (d) As^{3-} or As

Solution

   (a) The S^{2-} ion should be larger than the O^{2-} ion, because ions become larger as we go down a column of the periodic table.

   (b) A sodium atom should be larger than an aluminum atom, because size generally decreases across a row of the periodic table.

   (c) The C^{4-} ion should be larger than the F^- ion. Both have 10 electrons, but the C^{4-} ion has only 6 protons and the F^- ion has 9 protons. The 9 protons in the nucleus of the F^- ion should hold the electrons tighter than the 6 protons in the nucleus of the C^{4-} ion.

   (d) The As^{3-} ion should be larger than the As atom. The nucleus of both particles holds 33 protons; this number of protons should be able to hold the 33 electrons in the As atom tighter than the 36 electrons in the As^{3-} ion.

6.9 THE FIRST IONIZATION ENERGY

Another physical property of atoms that influences their chemical behavior is the energy needed to remove one or more electrons from a neutral atom to form a positively charged ion. Ionization energies are classified according to levels. The *first ionization energy* of an element can be defined as follows.

First ionization energy (1st IE): The energy needed to remove the outermost, or highest-energy, electron from a neutral atom in the gas phase.

This process is illustrated in Figure 6.12, which shows a single electron being removed from a hydrogen atom in the gas phase. A balanced chemical equation for this process could be written as follows.

$$H(g) + \text{energy} \longrightarrow H^+(g) + e^-$$

Experimentally we find that it takes 1312.0 kilojoules of energy to remove the electrons from a mole of hydrogen atoms.

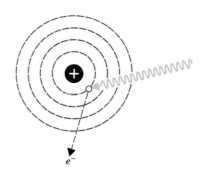

FIG. 6.12 The first ionization energy of an element is the energy needed to remove the outermost, or highest-energy, electron from a neutral atom in the gas phase. This figure illustrates the ionization of a hydrogen atom that occurs when this atom absorbs energy in the form of a photon with a wavelength of 91.127 nanometers.

Exercise 6.5

Use the Bohr model of the hydrogen atom to explain why the first ionization energy of hydrogen is 1312.0 kilojoules per mole (kJ/mol).

Solution

The Bohr model suggests that it takes energy to excite an electron from an orbit close to the nucleus into an orbit farther away. It also suggests that the wavelength of radiation required to excite the electron from one orbit to another is given by the following equation.

$$\frac{1}{\lambda} = R_H \left[\frac{1}{n^2} - \frac{1}{m^2} \right]$$

Ionizing the hydrogen atom is equivalent to exciting the electron from the $n = 1$ into the $m = \infty$ orbit. That is equivalent to taking the electron from the orbit in which it is as close as possible to the nucleus into the orbit in which it is as far as possible from the nucleus.

The value of R_H (the Rydberg constant) can be obtained from the table of fundamental constants in the appendix. Substituting this constant and the values of n and m into the equation gives the following result.

$$\frac{1}{\lambda} = (1.09737 \times 10^{-2} \text{ nm}^{-1}) \left[\frac{1}{1^2} - \frac{1}{\infty^2} \right]$$

This equation can now be solved for the wavelength of the photon that would have to be absorbed to ionize the hydrogen atom.

$$\lambda = \textbf{91.127 nm}$$

Once we convert this wavelength to meters, we can calculate the frequency of this photon from the relationship among the frequency, wavelength, and speed of light.

$$\nu = \frac{c}{\lambda} = \frac{2.99792 \times 10^8 \text{ m/s}}{9.1127 \times 10^{-8} \text{ m}} = 3.2898 \times 10^{15} \text{ s}^{-1}$$

We can then use Planck's equation to calculate the energy of the photon.

$$E = h\nu = (6.6262 \times 10^{-34} \text{ J-s})(3.2898 \times 10^{15} \text{ s}^{-1})$$
$$= 2.1799 \times 10^{-18} \text{ J}$$

The number of particles in a mole can then be used to calculate the energy of a mole of these photons.

$$\frac{2.1799 \times 10^{-18} \text{ J}}{1 \text{ photon}} \times \frac{6.0220 \times 10^{23} \text{ photons}}{1 \text{ mol}} = 1.3127 \times 10^6 \text{ J/mol}$$

A mole of these photons carries 1312.7 kilojoules of energy. Within experimental error, this calculation agrees with the first ionization energy of hydrogen.

Let's try to put the magnitude of the first ionization energy of hydrogen into perspective. How does it compare with the energy given off in a chemical reaction? When we burn natural gas, 802.4 kilojoules of energy is released per mole of methane consumed.

$$CH_4(g) + 2 \ O_2(g) \longrightarrow CO_2(g) + 2 \ H_2O(g)$$

The thermite reaction encountered in Section 3.10, which is used to weld iron rails, gives off about 850 kilojoules of energy per mole of Fe_2O_3 consumed.

$$Fe_2O_3(s) + 2 \ Al(s) \longrightarrow Al_2O_3(s) + 2 \ Fe(l)$$

The first ionization energy of hydrogen is more than half again as large as the energy given off in either of these reactions.

Exercise 6.6

Determine what electron is removed from a neutral lithium atom when the first ionization energy of lithium is measured.

Solution

The first ionization energy is defined as the energy needed to remove the outermost, or highest-energy, electron from a neutral atom in the gas phase. A neutral lithium atom has three electrons.

Li: $1s^2 \ 2s^1$

The outermost, or highest-energy, electron in a lithium atom is in the $2s$ orbital. The process defined as the first ionization energy of lithium can therefore be represented by the following equation.

$$\underset{1s^2 \ 2s^1}{Li(g)} + \text{energy} \longrightarrow \underset{1s^2}{Li^+(g)} + e^-$$

The first ionization energy of hydrogen can be brought into perspective by comparing its magnitude (1312.0 kJ/mol) with the energy given off when natural gas burns (802.4 kJ/mol), or when Fe_2O_3 reacts with powdered aluminum to give off aluminum oxide and molten iron (850 kJ/mol).

6.10 PATTERNS IN THE FIRST IONIZATION ENERGIES

It should be harder to remove an electron from a helium atom than a hydrogen atom, because each electron in helium feels the attractive force of two protons instead of one. The first ionization energy for helium is in fact just slightly less than twice the ionization energy for hydrogen. It takes 2372.3 kJ/mol to remove an electron from a neutral helium atom in the gas phase.

$$\underset{1s^2}{He(g)} + 2372.3 \text{ kJ/mol} \longrightarrow \underset{1s^1}{He^+(g)} + e^-$$

How much energy will it take to remove an electron from a lithium atom, which has three protons in its nucleus? Interestingly enough, it takes far less energy to ionize a lithium atom. The first ionization energy for Li is only 572.3 kJ/mol.

$$\underset{1s^2 2s^1}{Li(g)} + 572.3 \text{ kJ/mol} \longrightarrow \underset{1s^2}{Li^+(g)} + e^-$$

This is explained by the fact that the outermost, or highest-energy, electron on a lithium atom is in the $2s$ orbital. Electrons would like to be as close as possible to the nucleus of the atom. Because it is closer to the nucleus, an electron in a $1s$ orbital therefore has a lower energy than one in a $2s$ orbital. Because the electron in a $2s$ orbital is already at a higher energy, it takes less energy to remove this electron from the atom.

The first ionization energies for the main-group elements are given in Figure 6.13 and plotted in Figure 6.14.

Exercise 6.7

Describe the trends in the first ionization energy data in Figures 6.13 and 6.14.

Solution

Two trends are apparent from these data.

1. In general, the first ionization energy becomes larger as we go from left to right across a row of the periodic table.
2. The first ionization energy becomes smaller as we go down a column of the periodic table.

The first trend isn't surprising. We might expect the first ionization energy to become larger across a row of the periodic table, because the force of attraction

FIG. 6.13 The first ionization energies of the main-group elements. In general, the elements with the largest first ionization energies are found in the upper-right-hand corner of the periodic table.

FIG. 6.14 A plot of the first ionization energies of the first 18 elements in the periodic table shows that there is a gradual increase in the first ionization energy across a row of the periodic table from H to He, from Li to Ne, and from Na to Ar. There is a gradual decrease in the first ionization energy as we go down a column of the periodic table, as can be seen from the values for He, Ne, and Ar.

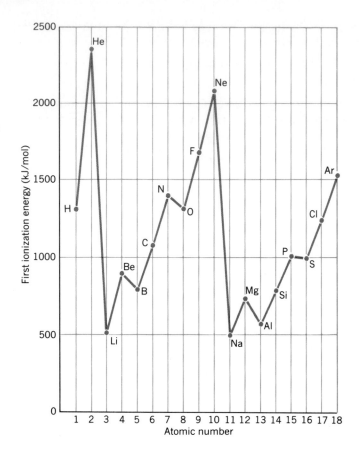

between the nucleus and an electron becomes larger as the number of protons in the nucleus becomes larger.

The second trend results from the fact that the principal quantum number of the orbital holding the outermost electron becomes larger as we go down a column of the periodic table. Although the number of protons in the nucleus also becomes larger as we go down a column, the electrons in smaller shells and subshells tend to screen the outermost electron from some of the force of attraction of the nucleus. Furthermore, the electron being removed when the first ionization energy is measured spends less of its time near the nucleus of the atom, and it therefore takes less energy to remove this electron from the atom.

EXCEPTIONS TO THE GENERAL PATTERN
6.11 OF FIRST IONIZATION ENERGIES

Figure 6.15 shows the first ionization energies for elements in the second row of the periodic table, from Li through Ne. Although there is a general trend toward an increase in the first ionization energy as we go from left to right across this row of the table, there are two minor inversions in this pattern. Boron has a first ionization energy that is smaller than that of beryllium, and the first ionization energy of oxygen is smaller than that of nitrogen.

To explain these observations, we must look at the electron configuration of these elements. The electron that is removed when a Be atom is ionized comes from the $2s$ orbital, but a $2p$ electron is removed when B is ionized.

Be: [He] $2s^2$

B: [He] $2s^2\, \mathbf{2p^1}$

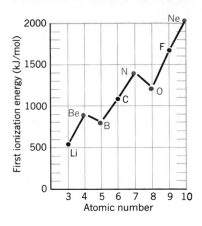

FIG. 6.15 There are minor exceptions to the general trend of increasing first ionization energy from left to right across a row of the periodic table.

An electron in a $2p$ orbital has more energy than an electron in a $2s$ orbital (see Figure 5.21). It therefore takes slightly less energy to remove an electron from the $2p$ orbital.

The electron removed when nitrogen and oxygen are ionized also comes from a $2p$ atomic orbital.

N: [He] $2s^2\, 2p^3$

O: [He] $2s^2\, 2p^4$

But there is an important difference between the way electrons are distributed in these atoms. Hund's rules predict that the three electrons in the $2p$ orbitals of a nitrogen atom should all have the same spin. But electrons are paired in one of the $2p$ orbitals on an oxygen atom.

$$\text{N} \qquad \uparrow \ \uparrow \ \uparrow$$
$$\text{O} \qquad \uparrow\downarrow \ \uparrow \ \uparrow$$

We can understand this phenomenon by assuming that electrons, which all have the same electric charge, try to stay as far apart as possible to minimize the force of repulsion between them. The three electrons in the $2p$ orbitals on nitrogen therefore enter different orbitals, with their spins aligned in the same direction. In oxygen, two electrons must occupy one of the $2p$ orbitals. The force of repulsion between these electrons is minimized to some extent by pairing of the electrons. But there is still some residual repulsion, which makes it slightly easier to remove an electron from a neutral oxygen atom than we would expect from the number of protons in the nucleus of the atom.

Exercise 6.8

Predict which element in each of the following pairs should have the larger first ionization energy.

(a) Na or Mg   (b) Mg or Al   (c) F or Cl

Solution

(a) We would expect magnesium to have the larger first ionization energy, because this quantity tends to increases across a row of the periodic table from left to right.

(b) Even though aluminum is to the right of magnesium in the periodic table, we might expect the first ionization energy for Al to be slightly smaller than that of Mg for the same reason that the first ionization energy of B is slightly smaller than that of Be. When an electron is removed from Mg it comes from a $3s$ orbital. The outermost, or highest-energy, electron on an Al atom is in a $3p$ orbital, so less energy is needed to remove this electron.

(c) Fluorine should have a larger first ionization energy than chlorine, because it takes less energy to remove an electron from one of the $3p$ orbitals on Cl than it does to remove one from the $2p$ orbitals on F.

6.12 SECOND, THIRD, FOURTH, AND HIGHER IONIZATION ENERGIES

By now, you know that sodium forms Na^+ ions, magnesium forms Mg^{2+} ions, and aluminum forms Al^{3+} ions. But have you ever asked: "Why can't sodium form Na^{2+} ions, or even Na^{3+} ions?" The answer to questions such as this can be obtained from data for the second, third, and higher ionization energies of the elements. The first ionization energy was defined as the energy needed to remove the outermost, or highest-energy, electron from a neutral atom in the gas phase. The *second ionization energy* can be defined in a similar fashion.

> **Second ionization energy: The energy needed to remove the outermost, or highest-energy, electron from a +1 ion in the gas phase to form a +2 ion.**

Exercise 6.9

Write a balanced equation that describes what happens when the second ionization energy of lithium is measured.

Solution

The first ionization energy of lithium is the energy it takes to accomplish the following reaction.

$$\underset{1s^2\,2s^1}{Li(g)} + energy \longrightarrow \underset{1s^2}{Li^+(g)} + e^-$$

The second ionization energy of lithium corresponds to the energy it takes to remove another electron to form an Li^{2+} ion in the gas phase.

$$\underset{1s^2}{Li^+(g)} + energy \longrightarrow \underset{1s^1}{Li^{2+}(g)} + e^-$$

By logical extension of this argument, we can see that the *third ionization energy* of lithium can be represented by the following equation.

$$\underset{1s^1}{Li^{2+}(g)} + energy \longrightarrow \underset{1s^0}{Li^{3+}(g)} + e^-$$

The overall energy required to form an Li^{3+} ion in the gas phase is the sum of the first, second, and third ionization energies of the element.

A complete set of experimental data for the ionization energies of the elements is given in Table A-5 in the appendix. Let's look for the moment at the first, second, third, and fourth ionization energies of sodium, magnesium, and aluminum, listed in Table 6.8.

It takes almost ten times as much energy to remove the second electron from a

TABLE 6.8

First, Second, Third, and Fourth Ionization Energies of Sodium, Magnesium, and Aluminum

	1st IE	2nd IE	3rd IE	4th IE	
Na	495.8	4562.4	6912	9543	kJ/mol
Mg	737.7	1450.6	7732.6	10,540	
Al	577.6	1816.6	2744.7	11,577	

sodium atom as it does to remove the first. To keep these numbers in perspective, note that the second ionization energy of sodium is almost six times as large as the energy given off when a mole of natural gas burns.

We can explain this observation by looking at the electron configuration of a neutral sodium atom.

Na: $1s^2\ 2s^2\ 2p^6\ 3s^1$

It doesn't take much energy to remove one electron from this atom to form an Na$^+$ ion.

Na$^+$: $1s^2\ 2s^2\ 2p^6 = $ [Ne]

But there is something inherently stable about electron configurations such as that of the Na$^+$ ion, which contain filled shells of electrons. It is not impossible to remove a second electron from a sodium atom. But it takes more energy than is likely to be given off in a chemical reaction. There is no reagent in the chemist's repertoire strong enough to remove the second electron from a sodium atom. Thus, sodium can react with other elements to form compounds that contain Na$^+$ ions, but not Na^{2+} or Na^{3+} ions.

A similar pattern is observed when the ionization energies of magnesium are analyzed. The first ionization energy of Mg is larger than that of Na, because Mg has one more proton in its nucleus to hold onto the electrons in the $3s$ orbital.

Mg: [Ne] $3s^2$

The second ionization energy of Mg is larger than the first, because it always takes more energy to remove an electron from a positively charged ion than from a neutral atom. The third ionization energy of magnesium is enormous, however, because the Mg^{2+} ion has a filled-shell electron configuration.

Mg^{2+}:   [Ne]

The same pattern can be seen in the ionization energies of aluminum. The first ionization energy of Al is smaller than that of Mg, as predicted in Exercise 6.8. The second ionization energy of Al is larger than the first, and the third ionization energy is even larger. It therefore takes a considerable amount of energy to remove three electrons from an aluminum atom to form an Al^{3+} ion.

Al^{3+}:   [Ne]

The energy required to break into the filled-shell configuration of the Al^{3+} ion is so large, however, that it would be foolish to look for an Al^{4+} ion as the product of a chemical reaction.

Exercise 6.10

Determine the group in the periodic table in which an element with the following ionization energies would most likely be found.

1st IE = 786 kJ/mol

2nd IE = 1577

3rd IE = 3232

4th IE = 4355

5th IE = 16,091

6th IE = 19,784

Solution

There is a gradual increase in the amount of energy needed to remove the first, second, third, and fourth electrons from this element. Then there is an abrupt increase in the amount of energy required to remove one more electron. This is consistent with an electron configuration in which the element has four more electrons than a filled-shell configuration, and we might expect that the element is in Group IVA of the periodic table.

These are in fact the ionization energies of silicon.

Si: [Ne] $3s^2\ 3p^2$

Exercise 6.11

Use the trends in the ionization energies of the elements to explain the following observations.

(a) Elements on the left-hand side of the periodic table are more likely than those on the right to form positive ions.

(b) The maximum positive charge on an ion is often equal to the group number of the element.

Solution

(a) The ionization energies of elements on the left-hand side of the table tend to be much smaller than the ionization energies of those on the right. Look at the first ionization energies for sodium and chlorine, for example.

Na: 1st IE = 495.8 kJ/mol
Cl: 1st IE = 1251.1 kJ/mol

Because elements on the left side of the periodic table have smaller ionization energies, they are more likely to form positive ions.

(b) The only electrons that can be removed from an atom in a chemical reaction are the electrons that aren't present in the filled-shell electron configuration of the previous rare gas, such as the $3s^2$ electrons in magnesium. The number of these electrons is equal to the group number, so the maximum positive charge on an ion is also equal to the group number. If aluminum is in Group VIIIA, for example, it can only lose three electrons before it reaches a stable filled-shell electron configuration. Thus the maximum possible charge on an aluminum ion is Al^{3+}.

6.13 ELECTRON AFFINITY

Ionization energies measure the tendency of a neutral atom to resist the loss of electrons. *Electron affinities* measure the affinity of a neutral atom for an extra electron.

Electron affinity (EA): The energy given off when a neutral atom in the gas phase gains an extra electron to form a negatively charged ion.

A fluorine atom in the gas phase, for example, gives off energy when it gains an electron to form a negatively charged fluoride ion.

$$F(g) + e^- \longrightarrow F^-(g) + 328.0 \text{ kJ/mol}$$

Electron affinity
H
73.5
Li
60.4
Be
*
Na
53.2
Mg
*
K
48.9
Ca
*
Rb
47.4
Sr
*
Cs
46.0
Ba
*
Fr
44.5
Ra
*
H
73.5
He
*
B
27
C
123.4
N
−7
O
142.5
F
331.4
Ne
*
Al
45
Si
135.0
P
72.4
S
202.5
Cl
352.4
Ar
*
Ga
30
Ge
120
As
78
Se
197.0
Br
327.9
Kr
*
In
29
Sn
122
Sb
102
Te
192.1
I
298.4
Xe
*
Tl
30
Pb
110
Bi
110
Po
190
At
270
Rn
*

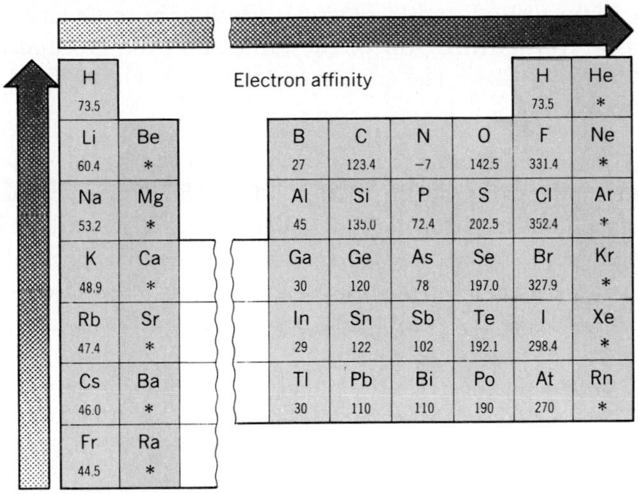

FIG. 6.16 In general, elements toward the top of the periodic table have larger electron affinities than those toward the bottom. Elements toward the right-hand side of the periodic table tend to have higher electron affinities than those toward the left-hand side. A number of these elements have no affinity for extra electrons. These elements are marked with an asterisk.

We can explain this by examining the electron configuration of a neutral fluorine atom and an F^- ion.

F: $1s^2\ 2s^2\ 2p^5$

F^-: $1s^2\ 2s^2\ 2p^6 = [Ne]$

In light of our generalization that there is something unusually stable about filled-shell configurations, it isn't surprising that a considerable amount of energy is released when a fluorine atom gains an electron to form an ion that has this configuration.

Electron affinities are more difficult to measure than ionization energies and are usually known to fewer significant figures. The electron affinities of the main-group elements are shown in Figure 6.16. (A more complete set of data can be found in Table A-6 in the appendix.) There are several patterns in these data.

1. Electron affinities gradually become smaller as we go down a column on the left-hand side of the periodic table, as shown in Figure 6.17.

100
80
60
40
20
Electron affinity (kJ/mol)
1st
2nd
3rd
4th
5th
6th
7th
Row of periodic table

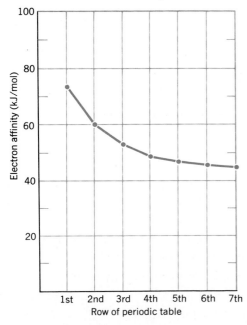

FIG. 6.17 The electron affinities of metals—such as the elements in Group IA—gradually decrease as we go down a column of the periodic table.

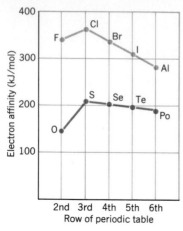

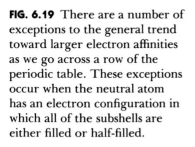

FIG. 6.18 This figure illustrates what happens to the electron affinity of nonmetals—such as the elements in Groups VIA and VIIA—as we go down a column of the periodic table. There is a slight increase in the electron affinity from the second to the third row and then a gradual decrease as we continue down the column.

2. On the right-hand side of the periodic table, electron affinities become larger as we go from the second to the third row of the table and then gradually become smaller as we continue down the column, as shown in Figure 6.18.

3. Electron affinities seem to oscillate as we go across a row of the periodic table, as shown in Figure 6.19.

Electron affinities become smaller as we go down a column on the left side of the periodic table for two reasons. First, the electron being added to the atom is placed in a larger orbital, where it spends less time near the nucleus of the atom. Second, the number of electrons on an atom becomes larger as we go down a column, so the force of repulsion between the electron being added and the electrons already present on a neutral atom becomes larger.

The same factors are responsible for the trend in electron affinities among the nonmetals on the right side of the periodic table. But these data are complicated by the fact that there are two ways of increasing the repulsion between the electron being added to the atom and the electrons already present on the atom. This repulsion depends on both the number of electrons and the volume of the atom. The repulsion increases when more electrons are added to approximately the same volume. But it also increases when the electrons are concentrated in a smaller volume.

Among the nonmetals in Group VIA and VIIA, this force of repulsion is largest for the very smallest atoms—oxygen and fluorine. As a result, these elements have a smaller electron affinity than the elements below them in these columns. From that point on, however, the electron affinities decrease as we continue down these columns.

At first glance, there appears to be no pattern in electron affinity across a row of the periodic table. When these data are listed in Table 6.9 along with the electron configurations of the elements, however, they begin to make sense, for the reasons described in the next paragraph.

We can start by noting that electron affinities are much smaller than ionization energies. Even chlorine, which has the largest affinity for electrons, gives off less

FIG. 6.19 There are a number of exceptions to the general trend toward larger electron affinities as we go across a row of the periodic table. These exceptions occur when the neutral atom has an electron configuration in which all of the subshells are either filled or half-filled.

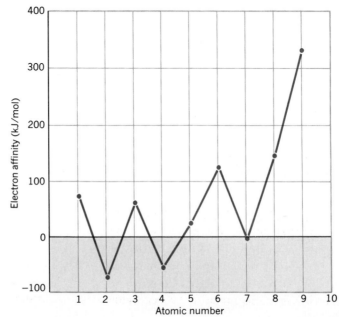

TABLE 6.9

Electron Affinities and Electron Configurations for the First 10 Elements in the Periodic Table

Element	EA (kJ/mol)	Electron Configuration
H	72.8	$1s^1$
He	less than 0	$1s^2$
Li	59.8	$1s^2\,2s^1$
Be	less than 0	$1s^2\,2s^2$
B	27	$1s^2\,2s^2\,2p^1$
C	122.3	$1s^2\,2s^2\,2p^2$
N	less than 0	$1s^2\,2s^2\,2p^3$
O	141.1	$1s^2\,2s^2\,2p^4$
F	328.0	$1s^2\,2s^2\,2p^5$
Ne	less than 0	$1s^2\,2s^2\,2p^6$

energy when it gains an electron than is required to remove an electron from the elements with the very smallest ionization energies. As a result, elements that have unusually stable electron configurations, such as He, Be, N, and Ne, have such small affinities for extra electrons that no energy is given off when a neutral atom of these elements picks up an electron. These configurations are so stable that it actually takes energy to force one of these elements to pick up an extra electron to form a negative ion.

Exercise 6.12

Compare the first ionization energy and the electron affinity for sodium and chlorine.

Solution

According to data from Tables A-5 and A-6 in the appendix, the first ionization energy for sodium is almost 10 times as large as the electron affinity for this element.

$$\text{Na:} \quad \text{1st IE} = 495.8 \text{ kJ/mol}$$
$$\text{EA} = 52.7 \text{ kJ/mol}$$

The first ionization energy and the electron affinity for chlorine are larger than the values for sodium.

$$\text{Cl:} \quad \text{1st IE} = 1251.1 \text{ kJ/mol}$$
$$\text{EA} = 348.8 \text{ kJ/mol}$$

But the first ionization energy is still much larger than the electron affinity.

6.14 CONSEQUENCES OF THE DIFFERENCE BETWEEN THE SIZE OF IONIZATION ENERGIES AND ELECTRON AFFINITIES

Beginning chemistry students are often told that sodium reacts with chlorine because chlorine atoms "like" electrons more than sodium atoms do and an electron is therefore transferred from a sodium atom to a chlorine atom to form Na^+ and Cl^- ions.

Sodium metal—when ignited by heating—reacts with chlorine gas to form sodium chloride, giving off energy in the form of both heat and light.

There is no doubt that sodium reacts vigorously with chlorine to form NaCl.

$$2 \text{ Na}(s) + \text{Cl}_2(g) \longrightarrow 2 \text{ NaCl}(s)$$

Furthermore, the ease with which solutions of NaCl in water conduct electricity is evidence for the fact that the product of this reaction is a salt, which contains Na^+ and Cl^- ions.

$$\text{NaCl}(s) \xrightarrow{\text{H}_2\text{O}} \text{Na}^+(aq) + \text{Cl}^-(aq)$$

The only question is whether it is legitimate to assume this reaction occurs because chlorine atoms "like" electrons more than sodium atoms do.

The data in Exercise 6.12 clearly show that this is not true. It takes more energy to remove an electron from a neutral sodium atom (1st IE = 495.8 kJ/mol) than is given off when the electron is picked up by a neutral chlorine atom (EA = 348.8 kJ/mol). The transfer of an electron from sodium to chlorine actually consumes energy and cannot possibly be used to explain why this reaction gives off energy.

Exercise 6.13

Compare the energy required to remove two electrons from a neutral magnesium atom to form an Mg^{2+} ion with the energy given off when a pair of chlorine atoms picks up electrons to form Cl^- ions. On the basis of this calculation, would you expect Mg to react with Cl_2 to form MgCl_2?

Solution

It takes 737.7 kJ/mol to remove the first electron from neutral magnesium atoms and 1450.6 kJ/mol to remove the second electron. Thus, it takes a total of 2188.3 kJ/mol to form Mg^{2+} ions. The electron affinity of chlorine is 348.8 kJ/mol. Twice this energy, or 697.6 kJ, is released when 2 moles of Cl^- ions are formed.

Therefore, the energy needed to remove electrons from magnesium atoms to form Mg^{2+} ions is three times the energy released when these electrons are transferred to chlorine atoms to form Cl^- ions. On the basis of this simple calculation, we would not expect magnesium to react with chlorine to form MgCl_2.

We are obviously going to have find another explanation for why sodium reacts with chlorine to form NaCl and why magnesium reacts with chlorine to form MgCl_2. Before we can do this, however, we need to know more about the chemistry of ionic compounds. The explanation for these reactions will therefore be delayed until the end of the next chapter.

6.15 METALS VERSUS NONMETALS IN THE PERIODIC TABLE: WHY DO SEMIMETALS EXIST?

The tendency to divide the elements into three groups—metals, nonmetals, and semimetals—on the basis of large differences between the chemical and physical properties of these elements was first discussed in Section 2.7.

Metals, as we have seen, have some or all of the following properties.

1. They have a metallic shine or luster.
2. With only one exception, they are solids at room temperature.
3. They are malleable and ductile.

4. They conduct heat and electricity.
5. They exist as extended arrays of atoms. There are no molecules in a metal.
6. They combine with other metals to form alloys, which still behave like metals.
7. They combine with nonmetals to form ionic compounds, or salts, which have the properties of neither metals nor nonmetals.
8. They form positive ions, such as the Na^+, Mg^{2+}, Fe^{3+}, and Cu^{2+} ions.

Nonmetals have almost exactly the opposite properties.

1. They seldom have a metallic luster.
2. They are often gases at room temperature.
3. They are neither malleable nor ductile.
4. They are poor conductors of both heat and electricity.
5. They often form molecules in their elemental form.
6. They combine with other nonmetals to form covalent compounds.
7. They combine with metals to form ionic compounds, or salts, which have the properties of neither metals nor nonmetals.
8. They tend to form negative ions, such as the F^-, Cl^-, O^{2-}, S^{2-}, P^{3-}, SO_4^{2-}, PO_4^{3-}, and NO_3^- ions.

The differences between the chemical and physical properties of metals and nonmetals can be traced to differences in their electron configurations, radii, ionization energies, and electron affinities.

Exercise 6.14

Use the general trends discussed in this chapter to describe the difference between the electron configurations, the size of the atoms, the ionization energies, and the electron affinities of a typical metal and a typical nonmetal.

Solution

As we go across a row of the periodic table from left to right, the number of electrons in the outermost, or highest-energy, orbitals of the atoms increase, the atoms become smaller, their ionization energies tend to increase, and in general their electron affinities also increase. As we go down a column, atoms become larger, their ionization energies become smaller, and their electron affinities tend to become smaller.

By looking at the relative positions of the metals and nonmetals in the periodic table, we can conclude the following.

In general, metals tend to have relatively few electrons in the outermost shell of orbitals, larger atoms, lower ionization energies, and smaller electron affinities than nonmetals.

There are no abrupt changes in the properties summarized in Exercise 6.14 as we go across a row of the periodic table or down a column. Since electron configurations, the size of atoms, ionization energy, and electron affinity change gradually as we go down a column or across a row of the periodic table, the change from metal to nonmetal must also be gradual.

Instead of arbitrarily dividing elements into metals and nonmetals, it might be

The elements in Group IVA become more metallic as we go down this column.

better to describe certain elements as being more metallic and other elements as more nonmetallic. Elements toward the left side of the periodic table have the most *metallic character.* Elements on the far right side of the table have the most *nonmetallic character.* Metallic character therefore becomes smaller as we go across a row of the periodic table from left to right, while nonmetallic character becomes larger.

Metallic character decreases ⟶

Na Mg Al Si P S Cl Ar

Nonmetallic character increases ⟶

For the same reason, elements toward the top of a column of the periodic table are the most nonmetallic, and elements toward the bottom of a column are the most metallic. This is best illustrated by Group IVA, where a gradual transition is seen from a distinctly nonmetallic element—carbon—toward the well-known metals tin and lead.

$$
\begin{array}{c}
\text{C} \\
\text{Si} \\
\text{Ge} \\
\text{Sn} \\
\text{Pb}
\end{array}
$$

Metallic character increases ↓   ↑ Nonmetallic character increases

Metallic character therefore increases as we go toward the bottom left-hand corner of the periodic table, and nonmetallic character reaches a maximum in the upper right-hand corner of the table.

If metallic character decreases across a row of the periodic table from left to right and nonmetallic character increases, we should encounter elements somewhere in the middle of the table that have properties that lie between the extremes of metals and nonmetals. The eight elements in this class (B, Si, Ge, As, Sb, Te, Po, and At) often look metallic, but they are brittle—like nonmetals. They are neither good conductors nor good insulators but serve as the basis for the semiconductor industry. These eight elements are often known as *metalloids,* but the authors prefer the name *semimetals.*

6.16 THE ACTIVE (REACTIVE) METALS

What do we mean when we say that sodium is more metallic than aluminum, or that barium is more metallic than magnesium? The principal difference between metals is the ease with which they enter into chemical reactions—their *activity,* or *reactivity.*

Metals differ widely in this property. Sodium reacts so vigorously with water that one of the products of this reaction often bursts into flame. Yet there is no apparent reaction between aluminum and water when we boil water in an aluminum pot. Magnesium burns in air when finely divided to give the flares and fireworks that make Fourth of July celebrations so beautiful. But barium, which is in the same column of the periodic table, is much more reactive.

The elements toward the bottom left-hand corner of the periodic table, which have the most metallic character, are the metals that are the most "active" in the sense of being the most reactive. Lithium, sodium, and potassium all react with

TABLE 6.10

Common Metals Divided into Classes on the Basis of Their Activity, or Reactivity

Class I Metals: The Active Metals
Li, Na, K, Rb, Cs (Group IA)
Ca, Sr, Ba (Group IIA)

Class II Metals: The Less Active Metals
Mg, Al, Zn, and Mn

Class III Metals: The Structural Metals
Cr, Fe, Sn, Pb, Cu

Class IV Metals: The Coinage Metals
Ag, Au, Pt, Hg

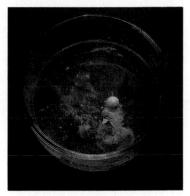

Lithium reacts slowly, sodium (shown here) reacts more rapidly, while potassium reacts violently with water. These metals therefore become more active in the sense of being more reactive as we go down this column of the periodic table.

water. However, Li reacts relatively slowly, Na reacts much more rapidly, and K reacts explosively.

The metals are often divided into classes on the basis of their activity, or reactivity, although it is important to recognize that the dividing lines between these classes are far from clear. The most active metals are so reactive that they readily combine with the O_2 and H_2O vapor in the atmosphere and must therefore be stored under an inert liquid, such as mineral oil. These metals are found exclusively in Groups IA and IIA of the periodic table, as shown in Table 6.10.

Metals in the second class are slightly less active. Although they don't react with water at room temperature, they react rapidly with acids. These less active metals can be used for a variety of purposes. Aluminum, for example, is used for everything from car bumpers to aluminum foil and beverage cans. It is important to protect these metals from exposure to acids, however, so aluminum cans are lined with a plastic coating to prevent contact with the acidic soft drinks or fruit juices they contain.

The third class contains metals such as chromium, iron, tin, and lead, which only react with strong acids. It also contains even less active metals such as copper, which only dissolves when treated with acids that can oxidize the metal, as we will see in the next chapter.

Metals in the fourth class are so unreactive they are essentially inert. These metals are ideal for making jewelry or coins because they do not react with the vast majority of the substances with which they come into daily contact.

The less active metals, such as aluminum and zinc, don't react with water, but they dissolve in strong acids. This photograph shows the reaction between aluminum foil and concentrated hydrochloric acid.

SUMMARY

An enormous amount of information about the chemical and physical properties of the elements and their compounds is summarized in the periodic table. The formulas for many chemical compounds, for example, can be predicted from the locations in the periodic table of the elements that form these compounds. The periodic table also summarizes information about a number of physical properties, including the relative size of atoms and the ions they form and the relative ease with which atoms gain or lose electrons.

There is a general trend toward an increase in the size of atoms or ions as we go down a column of the periodic table. Atoms and their ions become smaller, however, as we go across a row of the periodic table from left to right.

The number of protons in the nucleus of an atom remains constant as the atom gains or loses electrons to form an ion. As a result, positive ions are typically much smaller than neutral atoms. The radius of an Na^+ ion for example, is only about 60% of the radius of an Na atom. Negative ions, on the other hand, tend to be much larger than neutral atoms. A Cl^- ion, for example, has a radius that is almost twice as large as that of a neutral Cl atom.

The amount of energy needed to remove electrons from a neutral atom becomes smaller as we go down a column of the periodic table. In general, however, it becomes larger as we go from left to right across a row of the table. As a result, elements toward the bottom left-hand corner of the table are the most likely to form positive ions.

There is a pattern in the magnitude of the successive ionization energies of an element that is related to its electron configuration. As electrons are removed from an atom, the energy needed to remove the next electron always becomes larger. But there is a point at which the energy becomes so large it is impossible to remove another electron in a chemical reaction. This point usually corresponds to an electron configuration in which all the subshells are filled.

Because of the repulsion between electrons, the electron affinity of an element is much smaller than its ionization energy. In other words, it takes more energy to remove an electron from an atom than we get back when the electron is transferred to another atom.

The physical properties of the individual atoms play an important role in controlling the macroscopic properties of an element. For example, elements with relatively small ionization energies, low electron affinities, and large radii tend to be metals. Nonmetals, on the other hand, tend to have relatively large ionization energies, higher electron affinities, and smaller radii. There are no abrupt changes in the physical properties of atoms from one end of the periodic table to the other, which means that there is no abrupt change from metal to nonmetal as we go across a row or down a column of the table. As a result, elements that do not strictly belong in the classification of either metals or nonmetals can be found along the boundaries between these two categories.

As an element becomes more metallic, it becomes more active, or reactive. Metals are often divided into different categories on the basis of their activity.

PROBLEMS

The Search for Patterns in the Chemistry of the Elements

6-1 Eleven elements were listed in Section 2.4 as having symbols that were taken from the Latin or German name of the element. Compare the elements on this list with the thirteen elements known in 1661 and comment on any pattern you observe.

6-2 Describe the difference between families, periods, and groups of elements in the periodic table.

6-3 Use the graph in Figure 6.1 as the basis for explaining why the first attempts to recognize regular trends or patterns in the properties of the elements did not occur until about 1830.

6-4 Describe some of the similarities between the chemistry of chlorine, bromine, and iodine that might have led Döbereiner to suggest that these elements form a triad.

The Development of the Periodic Table

6-5 Describe the advantages of Mendeléeff's periodic table. What were the keys to the rapid acceptance of this model by his peers?

6-6 Find the three places in the periodic table where the elements are not listed in order of increasing atomic weights.

6-7 Mendeléeff placed both silver and copper in the same column, or group, as lithium and sodium. Describe some of the similarities among these four elements that allow them to be classified in the same group on the basis of their chemical properties.

6-8 Silver and copper are no longer in the same column, or group, as lithium and sodium in modern periodic tables. But the relationship between the chemical properties of these elements is still retained in periodic tables that use the symbols IA, IIA, and so on to identify groups. Describe how this is done.

6-9 From their positions in Mendeléeff's periodic table, predict the formulas for the oxides of the following elements.

(a) K (b) Zn (c) In (d) Si (e) V

Modern Versions of the Periodic Table

6-10 Describe some of the evidence that could be cited to justify the argument that the modern periodic table is based on similarities in the chemical properties of the elements.

6-11 Describe some of the evidence that could be used to justify the argument that the modern periodic table is also based on similarities in the electron configurations of the elements.

6-12 Describe some of the evidence that could be used to argue against the position that the modern periodic table groups elements with similar electron configurations.

6-13 Determine the row and column of the periodic table in which you would expect to find the first element to have $4d$ electrons in its electron configuration.

6-14 Determine the row and column of the periodic table in which you would expect to find the element that has five more electrons than the rare gas krypton.

6-15 Determine the group of the periodic table in which an element with the following electron configuration belongs.

$$1s^2\ 2s^2\ 2p^6\ 3s^2\ 3p^6\ 3d^{10}\ 4s^2\ 4p^6\ 5s^2\ 4d^{10}\ 5p^3$$

6-16 In which group of the periodic table should Element 119 belong if and when it is discovered?

6-17 Which of the following ions do not have the electron configuration of the rare gas Ar?

(a) Ga^{3+} (b) Cl^- (c) P^{3-} (d) Sc^{3+} (e) K^+

6-18 Which of the following ions has the electron configuration $[Ar]3d^4$?

(a) Ca^{2+} (b) Ti^{2+} (c) Cr^{2+} (d) Mn^{2+} (e) Fe^{2+}

The Size of Atoms: Metallic Radii

6-19 Describe what happens to the size of an atom as we go down a column of the periodic table. Explain why.

6-20 Describe what happens to the size of an atom as we go across a row of the periodic table from left to right. Explain why.

6-21 At one time, the size of an atom was given in units of angstroms, because the radius of a typical atom was about 1 angstrom. Now they are given in a variety of units, including centimeters, nanometers, and picometers. If the radius of a gold atom is 1.442 angstroms and 1 angstrom is equal to 10^{-8} centimeters, what is the radius of this atom in units of centimeters, nanometers, and picometers?

6-22 Which of the following atoms has the smallest radius?

(a) Na (b) Mg (c) Al (d) K (e) Ca

The Size of Atoms: Covalent Radii

6-23 Explain why the covalent radius of an atom is almost always smaller than the metallic radius of the atom.

6-24 Which of the following atoms has the largest covalent radius?

(a) N (b) O (c) F (d) P (e) S

The Size of Atoms: Ionic Radii

6-25 Describe the assumptions that one has to make to determine the size of the Li^+ and I^- ions from measurements of the distance between the nuclei of adjacent ions in a lithium iodide crystal.

6-26 Describe what happens to the radius of an atom when electrons are removed to form a positive ion. Describe what happens to the radius of the atom when electrons are added to form a negative ion.

6-27 Predict the order of increasing ionic radius for the following ions: H^-, F^-, Cl^-, Br^-, and I^-. Compare your predictions with the data for these ions in Table A-8 of the appendix. Explain any differences between your predictions and experiment.

The Relative Size of Atoms and Their Ions

6-28 Arrange the following atoms or ions in order of increasing radius.

(a) Na (b) Na^+ (c) Cl (d) Cl^-

6-29 Look up the covalent radii for magnesium and sulfur atoms and the ionic radii of Mg^{2+} and S^{2-} ions in the appendix. Explain why Mg^{2+} ions are smaller than S^{2-} ions even though magnesium atoms are larger than sulfur atoms.

6-30 Explain why Pb^{2+} ions have a radius (0.120 nanometer) that is very much larger than that of Pb^{4+} ions (0.084 nanometer).

6-31 Predict the relative size of the Sn^{2+} and Sn^{4+} ions.

Patterns in Ionic Radii

6-32 Sort the following atoms or ions into isoelectronic groups.

(a) N^{3-} (b) O^{2-} (c) F^- (d) Ne (e) Na^+ (f) Mg^{2+}
(g) Al^{3+} (h) Si^{4+} (i) P^{3-} (j) S^{2-} (k) Cl^- (l) Ar (m) K^+
(n) Ca^{2+}

6-33 Which of the following contains sets of atoms or ions that are all isoelectronic?

(a) B^{3+}, C^{4+}, H^+, He (b) Na^+, Ne, N^{3+}, O^{2-} (c) Mg^{2+}, F^-, Na^+, O^{2-} (d) Ne, Ar, Xe, Kr (e) O^{2-}, S^{2-}, Se^{2-}, Te^{2-}

6-34 Predict which of the following ions is smaller: Al^{3+} or Mg^{2+}. Explain why.

6-35 Which of the following ions has the largest radius?

(a) Na^+ (b) Mg^{2+} (c) S^{2-} (d) Cl^- (e) Se^{2-}

6-36 Which of the following atoms or ions is the smallest?

(a) Na (b) Mg (c) Na^+ (d) Mg^{2+} (e) O^{2-}

6-37 Which of the following ions has the smallest radius?

(a) K^+ (b) Li^+ (c) Be^{2+} (d) O^{2-} (e) F^-

6-38 Which of the following isoelectronic ions is the largest?

(a) Mn^{+7} (b) P^{3-} (c) S^{2-} (d) Sc^{3+} (e) Ti^{4+}

The First Ionization Energy

6-39 Write balanced chemical equations for the reactions that occur when the first, second, third, and fourth ionization energies of aluminum are measured.

6-40 Explain why it takes energy to remove an electron from an isolated atom in the gas phase.

6-41 Describe the general trend in first ionization energies from left to right across the second row of the periodic table.

6-42 Describe the general trend in first ionization energies from top to bottom of a column of the periodic table.

6-43 Explain why the first ionization energy of B is smaller than that of Be and why the first ionization energy of O is smaller than that of N.

6-44 Explain why the first ionization energy of hydrogen is so much larger than the first ionization energy of sodium.

6-45 Describe a possible set of n, l, m, and s quantum numbers for the electron removed from an isolated sodium atom in the gas phase when the first ionization energy of this element is measured.

6-46 List the following elements in order of increasing first ionization energy.

(a) Li   (b) Be   (c) F   (d) Na   (e) Si

6-47 Which of the following elements should have the largest first ionization energy?

(a) B   (b) C   (c) N   (d) Mg   (e) Al

6-48 Which of the following elements should have the smallest first ionization energy?

(a) Mg   (b) Ca   (c) Si   (d) S   (e) Se

Second, Third, Fourth, and Higher Ionization Energies

6-49 Which of the following atoms or ions has the largest ionization energy?

(a) P   (b) P^+ (c) P^{2+} (d) P^{3+} (e) P^{4+}

6-50 Explain why the second ionization energy of sodium is so much larger than the first ionization energy of this element.

6-51 Assume that an element has the following ionization energies.

$$1\text{st IE} = 578 \text{ kJ/mol}$$
$$2\text{nd IE} = 1817$$
$$3\text{rd IE} = 2745$$
$$4\text{th IE} = 11{,}577$$
$$5\text{th IE} = 14{,}831$$

Which of the following is the most probable electron configuration for this element?

(a) [Ar]   (b) [Ar] $3s^1$ (c) [Ar] $3s^2$ (d) [Ar] $3s^2\,3p^1$
(e) [Ar] $3s^2\,3p^2$ (f) [Ar] $3s^2\,3p^3$

6-52 Which of the following ionization energies is the largest?

(a) 1st IE of Ba   (b) 1st IE of Mg   (c) 2nd IE of Ba
(d) 2nd IE of Mg   (e) 3rd IE of Al   (f) 3rd IE of Mg

6-53 Which of the following elements should have the largest second ionization energy?

(a) Na   (b) Mg   (c) Al   (d) Si   (e) P

6-54 Which of the following elements should have the largest third ionization energy?

(a) B   (b) C   (c) N   (d) Mg   (e) Al

6-55 List the following elements in order of increasing second ionization energy.

(a) Li   (b) Be   (c) Na   (d) Mg   (e) Ne

6-56 Use the ionization energies in Table A-5 in the appendix to predict the largest — most positive — oxidation state of each of the following elements.

(a) C   (b) N   (c) O   (d) Si   (e) S   (f) Ca

6-57 Use the ionization energies in Table A-5 in the appendix to predict the largest — most positive — oxidation states of the following elements.

(a) Sc   (b) Ti   (c) V   (d) Cr   (e) Mn   (f) Zr   (g) Mo

6-58 Some elements, such as tin and lead, have more than one common oxidation state. Use the ionization energies in Table A-5 of the appendix to predict the most likely oxidation states of these metals.

Electron Affinity

6-59 Compare the chemical equations for the reactions that occur when the first ionization energy and the electron affinity of an element are measured. Show that the process in which the electron affinity of an element is measured is *not* the reverse of the process used to measure the first ionization energy of the element.

6-60 Explain why energy is usually released when the electron affinity of an element is measured.

6-61 Describe in general terms the trends in electron affinity across a row (from left to right) or down a column of the periodic table.

6-62 Explain why the electron affinities of He, Be, N, Ne, Mg, and Ar are all less than zero. What do these elements have in common?

6-63 Compare the magnitudes of the first ionization energies and the electron affinities of a variety of elements in the periodic table. In general, which is larger?

6-64 The first ionization energy increases more or less regularly across a row of the periodic table, but the electron affinity reaches its peak in the second to the last column. Explain why.

6-65 Which of the following elements has the largest electron affinity?

(a) Na   (b) K   (c) N   (d) O   (e) Ne

6-66 Which of the following elements has the largest electron affinity?

(a) Li: $1s^2\,2s^1$ (b) Be: $1s^2\,2s^2$ (c) B: $1s^2\,2s^2\,2p^1$
(d) C: $1s^2\,2s^2\,2p^2$ (e) N: $1s^2\,2s^2\,2p^3$

6-67 Which of the following elements has the largest electron affinity?

(a) Li   (b) Be   (c) N   (d) F   (e) Ne

Consequences of the Difference Between the Sizes of Ionization Energies and Electron Affinities

6-68 What's wrong with the popular misconception that sodium reacts with chlorine because "chlorine likes electrons more than sodium does"?

6-69 Compare the energy required to remove three electrons from a neutral aluminum atom to form the Al^{3+} ion with the energy given off when electrons are added to neutral bromine atoms to form Br^- ions. Is it legitimate to say that aluminum metal reacts with bromine to form $AlBr_3$ because bromine atoms like electrons more than aluminum atoms do?

Metals Versus Nonmetals

6-70 List the elements in the third row of the periodic table in order of decreasing metallic character. List these elements in order of increasing nonmetallic character. Explain the similarity between these trends. Identify each element as either a metal, a nonmetal, or a semimetal.

6-71 List the elements in Group VA of the periodic table in terms of increasing metallic character. Identify each element in this group as either a metal, a nonmetal, or a semimetal.

6-72 Which of the following sets of elements is arranged in order of increasing nonmetallic character?

(a) Sr, Al, Ga, N (b) K, Mg, Rb, Si (c) Ge, P, As, N
(d) Al, B, N, F

6-73 In general, which of the following statements about metals are true?

(a) They have large atomic radii (b) They have large ionization energies (c) They have large electron affinities (d) They are found on the left and toward the bottom of the periodic table.

The Active (Reactive) Metals

6-74 Describe the general trends in the activity, or reactivity, of the metals.

6-75 What do we mean when we say that sodium is more metallic than lithium?

6-76 In each of the following pairs of metals, which is more active?

(a) Mg or Ca (b) Na or Mg (c) Na or Ca (d) K or Mg
(e) Mg or Al

6-77 Which metal in each of the following pairs would you expect to react more rapidly with water?

(a) Na or K (b) Na or Mg (c) Mg or Ca (d) Ca or Al
(e) Al or Zn

CHAPTER 7

THE MAIN-GROUP METALS AND THEIR SALTS

CHAPTER CONTENTS

Chapter 6 concluded with the observation that most of the metals in Groups IA and IIA on the left side of the periodic table are highly reactive. These metals are so susceptible to chemical reactions they are called the **active metals.**

The reactivity of these metals can be attributed to two factors. First, these elements have one or two electrons beyond a filled-shell electron configuration. Second, they have relatively small ionization energies. Most of the chemistry of these metals can be explained by the fact that they lose one or two electrons relatively easily.

This chapter begins with a discussion of the chemistry of the active metals in Groups IA and IIA and then goes on to discuss the less active main-group metals, such as magnesium, aluminum, tin, and lead. Next, a model is developed that explains the relationship between metals and the salts they form when they react with nonmetal elements and compounds. This model is used as the basis for explaining how metals are prepared from their ores. The chapter concludes by returning to the question of why sodium and magnesium react with chlorine to form NaCl and $MgCl_2$, and not $NaCl_2$ or $MgCl$.

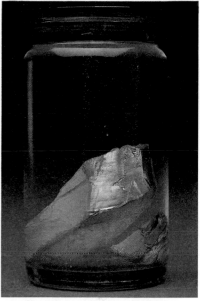

Sodium has a shiny metallic luster when freshly cut. Alkali metals are so active they are stored under an inert solvent, such as mineral oil, so that they don't react with the oxygen and water in the atmosphere.

7.1 GROUP IA: THE ALKALI METALS

The metals in Group IA include lithium (Li), sodium (Na), potassium (K), rubidium (Rb), cesium (Cs), and francium (Fr). These elements are called the **alkali metals,** because they all form hydroxides (such as LiOH, NaOH, and KOH) that were once known as *alkalies.*

Na and K are relatively common elements. In fact, they are among the eight most abundant elements in the earth's crust, as shown in Table 7.1. Li, Cs, Rb, and Fr are much less abundant. Discussions of the chemistry of alkali metals therefore generally focus on Na and K, and sometimes Li.

Solutions of solvated electrons prepared by dissolving sodium metal in liquid ammonia have a characteristic blue color. More concentrated solutions are so intensely colored they almost look like metals.

TABLE 7.1	
The Percent by Weight of the 10 Most Common Elements in the Earth's Crust	
Element	Percent by Weight
O	46.60
Si	27.72
Al	8.13
Fe	5.00
Ca	3.63
Na	2.83
K	2.59
Mg	2.09
Ti	0.44

The alkali metals all have similar electron configurations.

Li: [He] $2s^1$
Na: [Ne] $3s^1$
K: [Ar] $4s^1$
Rb: [Kr] $5s^1$
Cs: [Xe] $6s^1$
Fr: [Rn] $7s^1$

Their chemistry is therefore dominated by the fact that they tend to give up one electron to form a positively charged ion (Li^+, Na^+, K^+) in their chemical reactions. Every time we encounter an alkali metal in a chemical compound, we can assume it has an oxidation number of $+1$. This assumption is so safe it was incorporated into the rules for assigning oxidation numbers in Section 2.13.

The alkali metals lose electrons so easily that sodium actually dissolves in liquid ammonia to give Na^+ ions and electrons.

$$Na \xrightarrow{NH_3(l)} Na^+ + e^-$$

The electrons given off in this reaction are surrounded by and bound to a number of adjacent solvent (NH_3) molecules. They are therefore called *solvated* electrons. Dilute solutions of solvated electrons in ammonia have a characteristic blue color.

7.2 GROUP IA: HALIDES, HYDRIDES, SULFIDES, NITRIDES, AND PHOSPHIDES

The alkali metal halides are typically colorless or white solids, much like the sodium chloride shown here.

The alkali metals react with the nonmetals in Group VIIA (F_2, Cl_2, Br_2, I_2, and At_2) to form ionic compounds, or salts. Chlorine, for example, reacts with sodium metal to produce sodium chloride—table salt.

$$2\ Na(s) + Cl_2(g) \longrightarrow 2\ NaCl(s)$$

Because they form salts with so many metals, the elements in Group VIIA are known as the *halogens*. This name comes from the Greek word for salt—*hals*—and the Greek stem meaning "to produce"—*-genes*. The salts formed by the halogens are called *halides*. These salts include *fluorides* (LiF), *chlorides* (NaCl), *bromides* (KBr), and *iodides* (NaI).

The alkali metals react with hydrogen to form compounds known as *hydrides.* Potassium, for example, reacts with hydrogen to form potassium hydride.

$$2\ K(s) + H_2(g) \longrightarrow 2\ KH(s)$$

The alkali metals react with sulfur to form *sulfides,* such as sodium sulfide, Na_2S.

$$16\ Na(s) + S_8(s) \longrightarrow 8\ Na_2S(s)$$

N_2 is virtually inert to chemical reactions, but the most active metals react with nitrogen to form *nitrides,* such as lithium nitride.

$$6\ Li(s) + N_2(g) \longrightarrow 2\ Li_3N(s)$$

Elemental phosphorus exists as P_4 molecules, which are much more reactive than N_2 molecules. Phosphorus reacts with the alkali metals to form *phosphides,* such as sodium phosphide.

$$12\ Na(s) + P_4(s) \longrightarrow 4\ Na_3P(s)$$

Exercise 7.1

Describe the rule that gives us the names *fluoride, chloride, bromide, iodide, hydride, sulfide, nitride,* and *phosphide* for compounds that contain the F^-, Cl^-, Br^-, I^-, H^-, S^{2-}, N^{3-}, and P^{3-} ions, respectively.

Solution

We form the name of a monatomic negative ion (an ion containing a single atom) (such as the H^- or P^{3-} ion) by adding the suffix *-ide* to the stem of the name of the element. Thus, fluorine (F_2) becomes fluoride (F^-), sulfur (S_8) becomes sulfide (S^{2-}), and so on.

Exercise 7.2

Name the following compounds.

   (a) RbI (b) Li_2S (c) K_3N

Solution

   (a) rubidium iodide   (b) lithium sulfide   (c) potassium nitride

7.3 GROUP IA: PREDICTING THE PRODUCT OF ALKALI METAL REACTIONS

We can build a very simple model for predicting the product of the reaction between an alkali metal and a nonmetal by looking at what happens to the electron configurations of the elements in the reactions described in the preceding section.

Let's start with the reaction between sodium and chlorine to form sodium chloride. The electron configurations of sodium and chlorine can be written as follows.

   Na: $[Ne] \, 3s^1$
   Cl: $[Ne] \, 3s^2 \, 3p^5$

The product of this reaction is an ionic compound, which contains Na^+ and Cl^- ions that have the following configurations.

   Na^+: $[Ne]$
   Cl^-: $[Ne] \, 3s^2 \, 3p^6 = [Ar]$

The net effect of this reaction is to transfer one electron from a neutral sodium atom to a neutral chlorine atom to form Na^+ and Cl^- ions that have filled-shell electron configurations.

The following are the electron configurations for potassium and hydrogen.

   K: $[Ar] \, 4s^1$
   H: $1s^1$

These elements also react to form an ionic compound, so an electron has to be transferred from one of these elements to the other. We can decide which element should lose an electron by comparing the first ionization energy for potassium (418.8 kJ/mol) with that for hydrogen (1312.0 kJ/mol). Potassium is more likely to lose an electron in this reaction, while hydrogen gains an electron, to form K^+ and H^- ions.

K^+: [Ar]
H^-: $1s^2 = $ [He]

Once again, the product of this reaction contains ions that have filled-shell electron configurations.

When sodium reacts with sulfur to form sodium sulfide, we start with elements with the following electron configurations.

Na: [Ne] $3s^1$
S: [Ne] $3s^2\ 3p^4$

The product of this reaction contains Na^+ and S^{2-} ions, which have filled-shell configurations

Na^+: [Ne)
S^{2-}: [Ne] $3s^2\ 3p^6 = $ [Ar]

Exercise 7.3

Use electron configurations to explain why lithium reacts with nitrogen to form lithium nitride.

$$6\ Li(s) + N_2(g) \longrightarrow 2\ Li_3N(s)$$

Solution

Lithium and nitrogen atoms have the following electron configurations.

Li: [He] $2s^1$
N: [He] $2s^2\ 2p^3$

Active metals tend to lose electrons in their chemical reactions. Lithium can be expected to lose one electron per atom to form Li^+ ions.

Li^+: [He]

Each nitrogen atom has to gain three electrons to reach a filled-shell electron configuration.

N^{3-}: [He] $2s^2\ 2p^6 = $ [Ne]

The product of this reaction is therefore lithium nitride — Li_3N.

When applied with care, this model can be used to predict the products of simple chemical reactions.

Exercise 7.4

Write balanced equations for each of the following reactions.

(a) $K(s) + P_4(s) \longrightarrow$ (b) $Cs(s) + H_2(g) \longrightarrow$ (c) $Li(s) + O_2(s) \longrightarrow$

Solution

(a) We start by writing the electron configurations of the elements.

K: [Ar] $4s^1$
P: [Ne] $3s^2\ 3p^3$

Next, we predict what ions are likely to form when these elements react to produce an ionic compound.

K^+: [Ar]
P^{3-}: $[Ne]\, 3s^2\, 3p^6 = [Ar]$

We then combine these ions to predict the product of the reaction — K_3P — and write a balanced equation for this reaction.

$$12\ K(s) + P_4(s) \longrightarrow 4\ K_3P(s)$$

(b) Once again, we start with the electron configurations of the elements.

Cs: $[Xe]\, 6s^1$
H: $1s^1$

We then predict the ions they are most likely to form when they react.

Cs^+: [Xe]
H^-: $1s^2 = [He]$

We then write a balanced equation for the reaction that would produce these ions.

$$2\ Cs(s) + H_2(g) \longrightarrow 2\ CsH(s)$$

(c) Again, we start with the electron configurations of the elements.

Li: $[He]\, 2s^1$
O: $[He]\, 2s^2\, 2p^4$

We predict what ions they form when they react.

Li^+: [He]
O^{2-}: $[He]\, 2s^2\, 2p^6 = [Ne]$

We then combine the ions and write a balanced equation.

$$4\ Li(s) + O_2(g) \longrightarrow 2\ Li_2O(s)$$

This model for predicting — or at least explaining — the products of chemical reactions between metals and nonmetals might be the source of the misconception discussed in Section 6.14. It is easy to fall into the trap of assuming that this model explains *why* sodium reacts with chlorine. However, this would be a mistake. The model does not explain *why* these elements react, it only explains *what* happens when they react.

7.4 GROUP IA: OXIDES, PEROXIDES, AND SUPEROXIDES

The method just used to explain the products of the reactions of the alkali metals is simple, yet remarkably powerful. Many, if not most, of the chemical reactions between main-group metals and nonmetal elements can be understood in terms of this model. Exceptions to its predictions arise, however, when very active metals react with oxygen, which is one of the most reactive nonmetals.

Lithium is well behaved. It reacts with O_2 to form Li_2O, an *oxide*, as predicted in Exercise 7.4.

$$4\ Li(s) + O_2(g) \longrightarrow 2\ Li_2O(s)$$

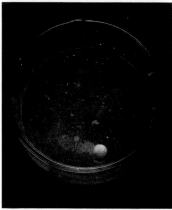

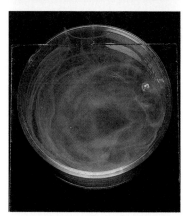

The alkali metals all react with water, but they do so at very different rates. Lithium (shown at the top) reacts relatively slowly. Sodium (in the middle) reacts much more rapidly. Potassium (at the bottom) is so reactive the metal often explodes off the surface of the water, and the crystallizing dish in which this reaction is run is therefore covered with a piece of plate glass.

Sodium, however, reacts with O_2 under normal conditions to form a compound that contains twice as much oxygen.

$$2\,Na(s) + O_2(g) \longrightarrow Na_2O_2(s)$$

Compounds, such as Na_2O_2, that are unusually rich in oxygen are called **peroxides.** The prefix *per-* means "above normal" or "excessive." Na_2O_2 is a peroxide because it contains more oxygen than is usual for an oxide.

To understand the difference between oxides and peroxides, remember the basic assumption behind alkali metal chemistry — these elements tend to form $+1$ ions. Li_2O therefore consists of Li^+ and O^{2-} ions, but Na_2O_2 contains Na^+ and O_2^{2-} ions.

The formation of sodium peroxide can be explained by assuming that sodium is so reactive that the metal is consumed before each O_2 molecule can combine with enough sodium to form Na_2O. The reaction therefore stops prematurely at Na_2O_2. This explanation is supported by the fact that sodium reacts with O_2 in the presence of a large excess of the metal — or a limited amount of O_2 — to form the oxide expected when this reaction goes to completion.

$$4\,Na(s) + O_2(g) \longrightarrow 2\,Na_2O(s)$$

It is also consistent with the fact that potassium, rubidium, and cesium — which are even more active than sodium — react with O_2 to form **superoxides,** which are even richer in oxygen. Potassium reacts with oxygen, for example, to give a salt that contains the K^+ and O_2^-, or superoxide, ions.

$$K(s) + O_2(g) \longrightarrow KO_2(s)$$

This explanation assumes that the very active alkali metals react so rapidly with oxygen they form superoxides, in which the alkali metal reacts with O_2 in a $1:1$ mole ratio.

$$K(s) + O_2(g) \longrightarrow KO_2(s)$$

Alkali metals that are less active react with oxygen to form peroxides, which have a $2:1$ ratio of moles of metal consumed per mole of O_2.

$$2\,Na(s) + O_2(g) \longrightarrow Na_2O_2(s)$$

When the least active metal in this column reacts with oxygen, the reaction forms the oxide, in which the maximum number of moles of metal are consumed per mole of O_2.

$$4\,Li(s) + O_2(g) \longrightarrow 2\,Li_2O(s)$$

Exercise 7.5

Describe the relationship between oxides, peroxides, and superoxides.

Solution

Oxides are compounds, such as Li_2O, that contain the O^{2-} ion. *Peroxides* are compounds, such as Na_2O_2, that contain the O_2^{2-} ion. *Superoxides* are compounds, such as KO_2, that contain the O_2^- ion.

7.5 GROUP IA: REACTIONS WITH H_2O AND NH_3

THE REACTION OF ALKALI METALS WITH WATER

One of the most common demonstrations of the reactivity of the active metals involves dropping pieces of lithium, sodium, and potassium into beakers of water.

Students who watch this demonstration often come away with the following observations: lithium reacts slowly with water, sodium reacts much more rapidly, and potassium often reacts violently.

A number of other observations can be gleaned from this demonstration.

1. Each metal floats on the surface of the water, which suggests that these metals have densities less than 1 g/cm³.
2. Bubbles form when lithium reacts with water, indicating that a gas is given off in this reaction.
3. The bubbles form on the metal surface, which suggests that the reaction occurs between the metal and water and is not the result of the reaction of water with an atom or ion released when the metal dissolves.
4. Sodium reacts so rapidly that no bubbles of gas are observed. Instead, a stream of smoke can often be seen rising off the surface of the liquid.
5. Enough heat is given off when sodium reacts with water to melt the metal.
6. The gas given off when sodium reacts with water often catches fire. When this happens, the flame has the distinct yellow color of excited Na atoms discussed in Section 5.12.
7. Potassium reacts so violently with water that the metal often explodes off the surface of the liquid. When this happens, the flame has the distinct violet color of excited K atoms.

WRITING A BALANCED EQUATION FOR THE REACTION OF ALKALI METALS WITH WATER

The model used to predict the products of reactions between alkali metals and nonmetals can be extended to predict what will happen when these alkali metals react with covalent compounds. Let's start by applying this model to the reaction between sodium metal and water.

$$Na(s) + H_2O(l) \longrightarrow ?$$

One thing is certain, the products of this reaction are going to contain sodium in the $+1$ oxidation state. In the course of this reaction, each sodium atom will lose an electron to form an Na^+ ion.

$$Na \longrightarrow Na^+ + e^-$$

Since electrons are neither created nor destroyed in chemical reactions, the electrons given off by the sodium atoms must be picked up by either the hydrogen or oxygen atoms in water. We can predict which of these elements will gain electrons in this reaction by looking at the oxidation numbers of the hydrogen and oxygen atoms in water.

Section 2.13 defined *oxidation number* as follows.

The oxidation number of an atom is the charge the atom would have if the compound that contained this atom was composed of ions.

The rules for assigning oxidation number are summarized, once again, in Table 7.2. According to these rules, the oxidation number of hydrogen is always $+1$ when it is bound to a nonmetal, and that of oxygen is almost always -2. Hydrogen therefore has an oxidation number of $+1$ in water, while oxygen has an oxidation number of -2.

$$\overset{+1}{H_2}\overset{-2}{O}$$

TABLE 7.2

Rules for Assigning Oxidation Numbers

1. The oxidation number of an atom is zero in any neutral substance that contains atoms of only one element. The oxygen atoms in O_2 and O_3, the phosphorus atoms in P_4, the sulfur atoms in S_8, and so on have oxidation numbers of zero.

2. The oxidation number of positive or negative ions that contain only a single atom is equal to the charge on the ion. For example, $Na^+ = +1$, $Mg^{2+} = +2$, $Cl^- = -1$, $O^{2-} = -2$, and so on.

3. The oxidation number of hydrogen is $+1$ whenever it is combined with another nonmetal. Hydrogen is therefore $+1$ in the following compounds: CH_4, NH_3, PH_3, H_2O, H_2S, HF, HCl, and HBr.

4. The oxidation number of hydrogen is -1 whenever it is combined with a metal. Hydrogen is therefore -1 in the following compounds: LiH, NaH, CaH_2, and $LiAlH_4$.

5. The metals in Group IA (Li, Na, K, Rb, Cs, and Fr) always form compounds in the $+1$ oxidation state.

6. The elements in Group IIA (Be, Mg, Ca, Sr, Ba, and Ra) always form compounds in the $+2$ oxidation state.

7. Oxygen almost always has an oxidation number of -2. Exceptions include molecules that contain oxygen-to-oxygen bonds, such as O_2, O_3, H_2O_2, and the O_2^{2-} ion.

8. The nonmetals in Group VIIA (F, Cl, Br, I, and At) usually form compounds in the -1 oxidation state.

9. The sum of the oxidation numbers of the atoms in a molecule or polyatomic ion is equal to the charge on the molecule or ion.

10. Elements toward the bottom and the left-hand side of the periodic table are more likely to form positive oxidation states than elements toward the top and right-hand side of the table.

The electron configurations of the hypothetical H^+ and O^{2-} ions in water provide a way of predicting what will happen when water gains electrons.

$$O^{2-}: \quad [Ne]$$
$$H^+: \quad 1s^0$$

An O^{2-} ion already has a filled-shell configuration, and so it has no affinity for additional electrons. But H^+ ions can gain electrons to form neutral hydrogen atoms, which combine to form H_2 molecules.

$$2\,H^+ + 2\,e^- \longrightarrow H_2$$

Subtracting a positively charged H^+ ion from a neutral H_2O molecule leaves an OH^- ion. The reaction that occurs when water molecules gain electrons can therefore be written as follows.

$$2\,H_2O + 2\,e^- \longrightarrow H_2 + 2\,OH^-$$

In retrospect, we can see that this discussion of the reaction between sodium and water divides the reaction into two halves. One half-reaction describes what happens when sodium atoms lose electrons.

$$Na \longrightarrow Na^+ + e^-$$

The other half-reaction describes what happens when water molecules gain these electrons.

$$2\ H_2O + 2\ e^- \longrightarrow H_2 + OH^-$$

By combining these half-reactions so that electrons are neither created nor destroyed we can obtain an overall equation for the reaction.

$$2[Na \longrightarrow Na^+ + e^-]$$
$$\underline{2\ H_2O + 2\ e^- \longrightarrow H_2 + 2\ OH^-}$$
$$2\ Na + 2\ H_2O \longrightarrow 2\ Na^+ + 2\ OH^- + H_2$$

The balanced equation for this reaction can therefore be written as follows.

$$2\ Na(s) + 2\ H_2O(l) \longrightarrow 2\ Na^+(aq) + 2\ OH^-(aq) + H_2(g)$$

WRITING A BALANCED EQUATION FOR THE REACTION OF ALKALI METALS WITH AMMONIA

The same line of reasoning can be used to predict what will happen when an alkali metal, such as potassium, reacts with ammonia (NH_3). We can start by noting that the potassium atoms will lose one electron each.

$$K \xrightarrow{NH_3(l)} K^+ + e^-$$

When these electrons react with the ammonia in the solution, they must be gained by either the nitrogen or the hydrogen atoms in NH_3. Applying the rules for assigning oxidation numbers to NH_3 gives the following results.

$$\overset{\displaystyle NH_3}{\underset{-3 \qquad +1}{\curvearrowright \qquad \curvearrowleft}}$$

The electron configurations of the hypothetical N^{3-} and H^+ ions in the NH_3 molecule provide a way of predicting what will happen to the electrons when they react with ammonia.

N^{3-}:   [Ne]
H^+: $1s^0$

N^{3-} ions have a filled-shell configuration and therefore no affinity for additional electrons. H^+ ions, however, can gain electrons to form neutral hydrogen atoms, which combine to form H_2 molecules.

$$2\ H^+ + 2\ e^- \longrightarrow H_2$$

Removing an H^+ ion from a neutral NH_3 molecule leaves a negatively charged NH_2^- ion. The following equation therefore describes what happens when ammonia gains electrons.

$$2\ NH_3 + 2\ e^- \longrightarrow H_2 + 2\ NH_2^-$$

Combining the two halves of this reaction so that electrons are conserved gives the balanced equation for the reaction.

$$2[K \longrightarrow K^+ + e^-]$$
$$\underline{2\ NH_3 + 2\ e^- \longrightarrow H_2 + 2\ NH_2^-}$$
$$2\ K + 2\ NH_3 \longrightarrow 2\ K^+ + 2\ NH_2^- + H_2$$

7.6 GROUP IIA: THE ALKALINE EARTH METALS

The elements in Group IIA (Be, Mg, Ca, Sr, Ba, and Ra) are all metals, and all but Be and Mg are active metals. These elements are often called the **alkaline earth metals.** The term *alkaline* reflects the fact that many of the compounds of these metals are basic or alkaline. The term *earth* was historically used to describe the fact that many compounds of these elements are insoluble in water.

Most of the chemistry of the alkaline earth metals (Group IIA) can be predicted from the behavior of the alkali metals (Group IA). Three points, however, should be kept in mind.

1. The alkaline earth metals tend to lose two electrons to form M^{2+} ions (Be^{2+}, Mg^{2+}, Ca^{2+}, and so on).
2. Each alkaline earth metal is less reactive than the neighboring alkali metal. Magnesium is less active than sodium; calcium is less active than potassium; and so on.
3. These metals become more active as we go down the column. Magnesium is more active than beryllium; calcium is more active than magnesium; and so on.

Reactions between an alkaline earth metal, such as magnesium, and a halogen, such as chlorine, give the products expected from the electron configurations of the elements.

$$\text{Mg:} \quad [Ne]\, 3s^2 \qquad \longrightarrow \qquad \text{Mg}^{2+}: \quad [Ne]$$
$$\text{Cl:} \quad [Ne]\, 3s^2\, 3p^5 \qquad \qquad \text{Cl}^-: \quad [Ne]3s^23p^6 = [Ar]$$

The reaction between magnesium and chlorine can therefore be described by the following equation.

$$Mg(s) + Cl_2(g) \longrightarrow MgCl_2(s)$$

The more active members of this group react with H_2 at high temperature. Calcium, for example, reacts with H_2 at temperatures above 600°C to form a compound that contains Ca^{2+} and H^- ions.

$$Ca(s) + H_2(g) \longrightarrow CaH_2(s)$$

Reactions between the alkaline earth metals and sulfur also give the products expected from the electron configurations of the elements. Strontium, for example, loses two electrons to form Sr^{2+} ions, while sulfur gains two electrons to form S^{2-} ions.

$$\text{Sr:} \quad [Kr]\, 5s^2 \qquad \longrightarrow \qquad \text{Sr}^{2+}: \quad [Kr]$$
$$\text{S:} \quad [Ne]\, 3s^2\, 3p^4 \qquad \qquad \text{S}^{2-}: \quad [Ar]$$

A balanced equation for this reaction can be written as follows.

$$8\, Sr(s) + S_8(s) \longrightarrow 8\, SrS(s)$$

The alkaline earth metals are not as active as the alkali metals. Most of the elements in this group therefore form oxides.

$$2\, Mg(s) + O_2(g) \longrightarrow 2\, MgO(s)$$

But calcium, strontium, and barium can also form peroxides.

$$Ba(s) + O_2(g) \longrightarrow BaO_2(s)$$

By now it should be easy to predict what will happen when a Group IIA metal reacts with nitrogen or phosphorus. Magnesium, for example, reacts with N_2 to form Mg^{2+} and N^{3-} ions.

$$\begin{array}{llll} \text{Mg:} & [\text{Ne}]\,3s^2 & \longrightarrow & \text{Mg}^{2+}: \quad [\text{Ne}] \\ \text{N:} & [\text{He}]\,2s^2\,2p^3 & & \text{N}^{3-}: \quad [\text{Ne}] \end{array}$$

The product of this reaction is therefore $[Mg^{2+}]_3[N^{3-}]_2$, or Mg_3N_2, and the balanced equation for the reaction is written as follows.

$$3\,\text{Mg}(s) + \text{N}_2(g) \longrightarrow \text{Mg}_3\text{N}_2(s)$$

Alkaline earth metals react with phosphorus to form compounds with formulas analogous to that of magnesium nitride, Mg_3N_2. Calcium, for example, reacts with phosphorus to form calcium phosphide, Ca_3P_2, and the balanced equation for this reaction is written as follows.

$$6\,\text{Ca}(s) + \text{P}_4(s) \longrightarrow 2\,\text{Ca}_3\text{P}_2(s)$$

REACTION OF ALKALINE EARTH METALS WITH WATER

Only the more active members of Group IIA (Ca, Sr, and Ba) react with water at room temperature. The products of these reactions, for the most part, are what we might expect. Calcium, for example, loses two electrons to form Ca^{2+} ions when it reacts with water.

$$\text{Ca} \longrightarrow \text{Ca}^{2+} + 2\,e^-$$

These electrons are picked up by the water molecules to form H_2 gas and OH^- ions.

$$2\,\text{H}_2\text{O} + 2\,e^- \longrightarrow \text{H}_2 + 2\,\text{OH}^-$$

Combining the two halves of the reaction so that electrons are conserved gives the following result.

$$\begin{array}{l} \text{Ca} \longrightarrow \text{Ca}^{2+} + 2\,e^- \\ \underline{2\,\text{H}_2\text{O} + 2\,e^- \longrightarrow \text{H}_2 + 2\,\text{OH}^-} \\ \text{Ca} + 2\,\text{H}_2\text{O} \longrightarrow \text{Ca}^{2+} + 2\,\text{OH}^- + \text{H}_2 \end{array}$$

A balanced equation for this reaction can be written as follows.

$$\text{Ca}(s) + 2\,\text{H}_2\text{O}(l) \longrightarrow \text{Ca}^{2+}(aq) + 2\,\text{OH}^-(aq) + \text{H}_2(g)$$

Although Mg does not react with water at room temperature, it will react with steam. The products of this reaction, however, are not going to be aqueous Mg^{2+} and OH^- ions, because there is no liquid water around to stabilize these ions. The products of this reaction are H_2 gas and magnesium oxide, MgO.

$$\text{Mg}(s) + \text{H}_2\text{O}(g) \longrightarrow \text{MgO}(s) + \text{H}_2(g)$$

When heated, magnesium metal bursts into flame as it reacts with the oxygen and the nitrogen in the atmosphere to form magnesium oxide and magnesium nitride. This reaction gives off both heat and a brilliant white light that is the basis for a number of products ranging from flash bulbs to military flares.

Exercise 7.6

Magnesium reacts with hydrogen gas to form compound A, which is a white solid at room temperature. It also reacts with hydrochloric acid to form gas B and an aqueous solution of compound C. Identify the products of these reactions—A, B, and C—and write balanced equations for each reaction.

Solution

Magnesium and hydrogen have the following electron configurations.

Mg: [Ne] $3s^2$
H: $1s^1$

When they react, magnesium should lose electrons to form Mg^{2+} ions, and hydrogen should gain electrons to form H^- ions. The product of the first reaction is therefore magnesium hydride, MgH_2.

$$Mg(s) + H_2(g) \longrightarrow MgH_2(s)$$
$$A$$

Magnesium also loses two electrons to form Mg^{2+} ions when it reacts with hydrochloric acid — HCl(aq).

$$Mg \longrightarrow Mg^{2+} + 2\ e^-$$

To understand what happens when the hydrochloric acid gains electrons, let's look at the oxidation numbers of the hydrogen and chlorine atoms in this compound.

$$\overset{\frown}{HCl}$$
$$+1 \qquad -1$$

The H^+ and Cl^- ions in this compound would have the following electron configurations.

H^+: $1s^0$
Cl^-:   [Ne] $2s^2\ 2p^6$ = [Ar]

The Cl^- ion has a filled-shell configuration and therefore no affinity for additional electrons. The H^+ ion can gain an electron, however, to form hydrogen atoms, which combine to give H_2 molecules.

$$2\ H^+ + 2\ e^- \longrightarrow H_2$$

The products of the reaction between magnesium metal and hydrochloric acid are therefore H_2 gas and an aqueous solution of Mg^{2+} and Cl^- ions.

$$Mg(s) + 2\ HCl(aq) \longrightarrow Mg^{2+}(aq) + 2\ Cl^-(aq) + H_2(g)$$
$$C \qquad\qquad\qquad B$$

7.7 GROUP IIIA: THE CHEMISTRY OF ALUMINUM

The elements in Group IIIA (B, Al, Ga, In, and Tl) can be divided into three classes.

1. Boron is the only element in this group that is not a metal. It behaves like a semimetal or even a nonmetal.

2. Aluminum is the third most abundant element in the earth's crust. It is just slightly less reactive than the active metals.

3. The other three elements in this group are active metals, but they are so scarce they are of limited interest. Gallium, indium, and thallium combined total less than $10^{-10}\%$ of the earth's crust.

(a)   (b)   (c)

(d)   (e)   (f)

Aluminum is the most abundant of the metallic elements, totalling about 8.13% by weight of the earth's crust. But aluminum is far too active to be found in nature as the metal. It is found in a variety of minerals — such as bauxite, clay, corundum, cryolite, feldspar, granite, and mica — as well as in gemstones — such as emerald, garnet, jade, lapis lazuli, ruby, sapphire, and turquoise, where it has an oxidation number of $+3$.

The chemistry of aluminum can be explained by assuming that the metal reacts to form compounds in which it has an oxidation number of $+3$. Aluminum reacts with the halogens, for example, to form compounds with the empirical formula AlX_3 and the molecular formula Al_2X_6.

$$2\ Al(s) + 3\ Br_2(l) \longrightarrow Al_2Br_6(s)$$

It reacts with oxygen and sulfur to form compounds with the formula Al_2X_3.

$$4\ Al(s) + 3\ O_2(g) \longrightarrow 2\ Al_2O_3(s)$$
$$16\ Al(s) + 3\ S_8(s) \longrightarrow 8\ Al_2S_3(s)$$

It also forms a variety of other compounds, such as AlH_3, $LiAlH_4$, Al_4C_3, AlP, and $Al_2(CH_3)_6$, in which the oxidation number of aluminum is $+3$.

Aluminum is less metallic and therefore less active than magnesium, and it does not react with water under normal conditions. Neither magnesium nor aluminum is active enough to remove an H^+ ion from a neutral water molecule. But both metals will react with solutions that are rich in H^+ ions.

Magnesium reacts with most acids to give Mg^{2+} ions and H_2 gas.

$$Mg(s) + 2\ H^+(aq) \longrightarrow Mg^{2+}(aq) + H_2(g)$$

(a) Rubies consist of aluminum oxide (Al_2O_3) with trace amounts of Cr_2O_3, which gives rise to their red color.

(b) Emeralds are the mineral beryl ($Be_3Al_2(SiO_3)_6$) with a trace of Cr_2O_3.

(c) Lapis lazuli is a deep-blue mineral with the formula $Na_4Al_3Si_3O_{12}Cl$.

(d) Turquoise is an opaque mineral with the formula $CuAl_6(PO_4)_4(OH)_8 4H_2O$.

(e) Garnets are a group of minerals with the general formula $M_3Al_2(SiO_4)_3$, where M is the Fe^{2+}, Ca^{2+}, Mg^{2+}, or Mn^{2+} ion.

(f) Jade is a term used to describe two minerals, jadeite ($NaAlSi_2O_6$) and nephrite ($Ca_2M_5Si_8O_{22}X_2$, where M is either Mg^{2+} or Fe^{2+} and X is either F^- or OH^-).

An aluminum can.

Aluminum reacts with concentrated acids to give Al^{3+} ions and H_2 gas.

$$2\ Al(s) + 6\ H^+(aq) \longrightarrow 2\ Al^{3+}(aq) + 3\ H_2(g)$$

It also reacts with concentrated bases to give H_2 gas and the aluminate ion, $Al(OH)_4^-$, in which aluminum is in the $+3$ oxidation state.

$$2\ Al(s) + 2\ OH^-(aq) + 6\ H_2O(l) \longrightarrow 2\ Al(OH)_4^-(aq) + 3\ H_2(g)$$

You can experience the reaction between aluminum metal and acid by using an aluminum pot to cook with a recipe that calls for large quantities of vinegar. The inside of the pot always looks like new when the pot is washed, because a small amount of the aluminum metal on the inner surface of the pot dissolves when it reacts with the acid in vinegar.

The reaction between aluminum metal and concentrated base is even easier to experience. Solid drain cleaners, such as Drano, are mixtures of aluminum metal and sodium hydroxide. When they are mixed with water, reaction between the aluminum metal and the OH^- ion from NaOH generates H_2 gas and gives off a significant amount of heat.

MAGNESIUM AND ALUMINUM AS STRUCTURAL METALS

Aluminum is becoming increasingly important as a structural metal. It is used for soft drink cans, electrical wire, pots and pans, and siding on houses. It is slowly replacing steel in cars and trucks and is an important metal in commercial aircraft. The property that makes aluminum so attractive as a structural metal is its density ($2.70\ g/cm^3$), which is only a third of the density of copper ($8.92\ g/cm^3$) or steel ($7.8\ g/cm^3$).

Aluminum is an excellent conductor of heat and electricity, which is why it is used so often in cooking. Aluminum is also unusually malleable and ductile. It can therefore be drawn into electrical wire or rolled into thin sheets of aluminum foil for cooking.

Magnesium has a density ($1.74\ g/cm^3$) that is even smaller than that of aluminum ($2.70\ g/cm^3$). It is therefore used when reducing the weight of an object is particularly important. Magnesium alloys can be found in the frames of luggage, in hand trucks, ladders, camera bodies, airplane parts, and the wheels of race cars.

HOW CAN THE LESS ACTIVE METALS BE USED AS STRUCTURAL METALS?

Magnesium is so reactive that a variety of products ranging from flash bulbs to fireworks and military flares are based on the following reaction.

$$2\ Mg(s) + O_2(g) \longrightarrow 2\ MgO(s)$$

Aluminum has such a high affinity for oxygen that it can strip oxygen atoms from Fe_2O_3 and still give off large quantities of energy.

$$Fe_2O_3(s) + 2\ Al(s) \longrightarrow Al_2O_3(s) + 2\ Fe(l)$$

Everyone should be familiar with the rust that forms when iron reacts with oxygen dissolved in water.

$$4\ Fe(s) + 3\ O_2(g) + 6\ H_2O(l) \longrightarrow 2\ Fe_2O_3 \cdot 3\ H_2O(s)$$

They should also appreciate the effect this has on iron's use as a structural metal. If aluminum has a higher affinity for oxygen than iron — and magnesium is even more reactive than aluminum — how can these metals be used to make products, such as ladders, pots and pans, aluminum foil, and "mag" wheels, that seem to be chemically inert?

The answer is subtle. Magnesium and aluminum are so active they react with O_2 in the air to form a thin coating of MgO or Al_2O_3. MgO then slowly reacts with carbon dioxide in the air to form magnesium carbonate.

$$MgO(s) + CO_2(g) \longrightarrow MgCO_3(s)$$

Although the coating may be only 5 nanometers thick, it covers the metal surface so evenly that the metal beneath this layer is protected from further reaction.

If we immerse these metals in an acid, however, the acid washes off the $MgCO_3$ or Al_2O_3 coating. This exposes a fresh layer of metal that can react with either oxygen or acid. If the acid is strong enough, we can watch these metals dissolve within a few minutes.

7.8 GROUP IVA: TIN AND LEAD

The elements in Group IVA can be divided into three classes: (1) carbon, which is a nonmetal; (2) silicon and germanium, which are semimetals; and (3) tin and lead, which are metals. Tin and lead are among the oldest metals known. Lead was first used to glaze pottery in Egypt more than 700 years ago, and tin was combined with copper to form bronze as early as 3500 B.C.

Tin and lead are much less reactive than any of the main-group metals discussed so far. According to the argument that elements become more metallic — and therefore more active — as we go down a column of the periodic table, lead should be more reactive than tin.

Tin does not react with either air or water at room temperature. Lead reacts with air to form a thin coating of PbO and/or $PbCO_3$, which protects the metal from further reaction.

$$2\ Pb(s) + O_2(g) \longrightarrow 2\ PbO(s)$$
$$PbO(s) + CO_2(g) \longrightarrow PbCO_3(s)$$

When it is finely divided, lead is **pyrophoric** — it bursts into flame in the presence of oxygen.

When heated until white hot, tin reacts with air to form SnO_2.

$$Sn(s) + O_2(g) \longrightarrow SnO_2(s)$$

At high temperatures it also reacts with steam to give SnO_2.

$$Sn(s) + 2\ H_2O(g) \longrightarrow SnO_2(s) + 2\ H_2(g)$$

Tin and lead are both less active than aluminum. Neither metal reacts with either dilute hydrochloric acid or dilute sulfuric acid at room temperature. Tin, when heated, reacts with either concentrated hydrochloric acid or concentrated sulfuric acid.

$$Sn(s) + 2\ HCl(aq) \longrightarrow SnCl_2(aq) + H_2(g)$$
$$Sn(s) + H_2SO_4(aq) \longrightarrow SnSO_4(aq) + H_2(g)$$

Lead reacts slowly with hydrochloric acid at room temperature and with concentrated sulfuric acid at temperatures above 200°C.

$$Pb(s) + 2\ HCl(aq) \longrightarrow PbCl_2(aq) + H_2(g)$$
$$Pb(s) + H_2SO_4(aq) \longrightarrow PbSO_4(aq) + H_2(g)$$

Tin and lead are less active than aluminum. Lead is a bit more active than tin, however, and finely divided lead metal, which has a dark black color, will glow, or even burst into flame, when it comes in contact with air.

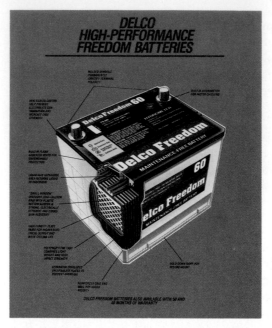

The fact that lead storage batteries include plates of lead metal immersed in sulfuric acid is evidence for the fact that lead is less reactive than other main-group metals, such as magnesium and aluminum.

Tin and lead are the first main-group metals we have encountered that form compounds in more than one oxidation state. $SnCl_2$, $SnSO_4$, PbO, $PbCO_3$, $PbCl_2$, and $PbSO_4$ are all examples of the $+2$ oxidation state. SnO_2, SnH_4, Sn_3N_4, PbO_2, and $PbCl_4$ are examples of the $+4$ oxidation state. The $+4$ oxidation state is easy to understand on the basis of the electron configurations of these metals.

Sn: $[Kr] 5s^2 4d^{10} 5p^2$

Pb: $[Xe] 6s^2 4f^{14} 5d^{10} 6p^2$

The hypothetical Sn^{4+} and Pb^{4+} ions in the $+4$ oxidation state of these elements would have electron configurations in which the orbital subshells were all filled (or empty).

Sn^{4+}: $[Kr] 4d^{10}$

Pb^{4+}: $[Xe] 4f^{14} 5d^{10}$

The $+2$ oxidation state of these elements, however, also corresponds to hypothetical ions with electron configurations in which orbital subshells are either filled or empty.

Sn^{2+}: $[Kr] 5s^2 4d^{10}$

Pb^{2+}: $[Xe] 6s^2 4f^{14} 5d^{10}$

The smaller of a pair of oxidation states becomes more stable as we go down a column of the periodic table. Lead, for example, is more likely to be found in the $+2$ oxidation state than tin. Thus, while tin reacts with oxygen at high temperatures to form SnO_2, lead forms PbO. (PbO_2 is never formed by the direct reaction between lead and oxygen.)

7.9 SALTS OF THE MAIN-GROUP METALS

Section 2.7 noted that metals react with nonmetals to form ionic compounds, or salts. With the exception of aluminum, tin, and lead, the average citizen never comes in direct contact with the main-group metals. The salts of these elements, however, have seemingly endless uses in our daily lives.

GROUP IA

The "lithium" used to treat manic-depressives is not lithium metal, but a lithium salt such as lithium carbonate (Li_2CO_3).

Sodium bicarbonate ($NaHCO_3$) is sold as baking soda for use in cooking and as bicarbonate of soda for use as an antacid. It is even mixed with kitty litter, where it serves as a deodorizer.

Ordinary table salt (NaCl) is so important it has entered our language in the form of clichés such as "salt of the earth" and "not worth his salt." It is the source for the word *salary*, which comes from the Latin *salarium*, a sum of money given to Roman soldiers to buy salt. NaCl is used to preserve meat and fish, to melt snow or ice, and to regenerate water softeners.

Monosodium glutamate, or MSG ($NaC_5H_8NO_4$), is another sodium salt used in cooking. Sodium benzoate ($NaC_7H_5O_2$) is used as a preservative to slow food spoilage. Its use has been partially responsible for the 10-fold decrease in the frequency of stomach cancer during the last 50 years.

Soap is the sodium salt of fatty acids, such as stearic acid ($C_{18}H_{36}O_4$), which is obtained by reaction of animal fat with sodium hydroxide, NaOH. Detergents are sodium salts of synthetic analogs of these fatty acids. Other sodium salts, such as sodium phosphate (Na_3PO_4) and sodium triphosphate ($Na_5P_3O_{10}$), can be found in a variety of cleaning agents.

Sodium nitrate ($NaNO_3$) has been used historically as the starting material for the synthesis of products ranging from explosives to fertilizers. Sodium thiosulfate ($Na_2S_2O_3$) is the "fixer" that plays such an important role in capturing photographic images. Sodium pentothal ($NaC_{11}H_{17}N_2O_2S$), the "truth serum" in so many spy thrillers, is an example of a wide spectrum of pharmaceutical agents that are sodium salts.

Potassium salts are distributed throughout nature. The only potassium salt commonly used in cooking is cream of tartar, or potassium hydrogen tartrate ($KHC_4H_4O_6$), which is used to transform baking soda into baking powder. Potassium salts are essential to plant nutrition, however, and potassium chloride (KCl), potassium nitrate (KNO_3), and potassium sulfate (K_2SO_4) are used extensively in agriculture.

GROUP IIA

Beryllium was once known as glucinium because so many of its compounds taste sweet. Exposure to even small quantities of beryllium salts can be fatal, however, so beryllium plays no significant role in our daily lives.

The best known magnesium salts are magnesium hydroxide, $Mg(OH)_2$, and magnesium sulfate, $MgSO_4 \cdot 7\ H_2O$. $Mg(OH)_2$ is used as milk of magnesia to treat acid indigestion, and $MgSO_4 \cdot 7\ H_2O$ is Epsom salts.

There are a number of important calcium salts. Calcium carbonate ($CaCO_3$) is found in chalk, coral, egg shells, limestone, marble, pearl, and clam shells. $CaCO_3$ decomposes on heating to form calcium oxide, or lime (CaO).

$$CaCO_3(s) \longrightarrow CaO(s) + CO_2(g)$$

Lime has a variety of industrial and agricultural uses. It is commonly used to neutralize soil that is too acidic, and it is the principal ingredient of cement.

Calcium sulfate occurs naturally as gypsum, $CaSO_4 \cdot 2 H_2O$, a salt that has two molecules of water trapped within the crystal structure for each Ca^{2+} ion. When heated, gypsum loses some of its water to form plaster of Paris.

$$2 \underset{\text{gypsum}}{CaSO_4 \cdot 2 H_2O(s)} \longrightarrow \underset{\text{plaster of Paris}}{(CaSO_4)_2 \cdot H_2O(s)} + 3 H_2O(l)$$

When plaster of Paris is mixed with water, the reaction is reversed. This makes plaster of Paris useful for everything from plastering the interior of buildings to making casts to support broken bones.

The principal component of both bone and tooth enamel is a mineral known as hydroxyapatite, $Ca_5(PO_4)_3OH$. The idea behind fluoridating water or adding fluorides to toothpaste is to convert a portion of the hydroxyapatite in tooth enamel into fluoroapatite, $Ca_5(PO_4)_3F$, which is a much harder mineral and therefore more resistant to decay.

GROUP IIIA

Aluminum oxide (Al_2O_3) is found in nature as corundum, one of the hardest known minerals. Corundum is commonly used as an abrasive in sandpaper and jeweler's rouge. Rubies are crystals of corundum that contain small quantities of chromium impurities, while sapphires are crystals of corundum with trace quantities of iron or titanium.

An aluminum salt with which many people have daily contact is "aluminum chlorhydrate," or more accurately, aluminum hydroxychloride, $Al_2(OH)_5Cl \cdot 2 H_2O$. This compound is an astringent, which contracts body tissues. When applied to underarms, its tendency to contract the pores of the skin makes it an excellent antiperspirant.

GROUP IVA

Tin(II) fluoride (SnF_2), or stannous fluoride, was the first source of the fluoride ion used in toothpastes. Tin(IV) oxide, SnO_2, is mixed with a variety of transition metal oxides, such as V_2O_5 and Cr_2O_3, to form glazes of different colors for use with ceramics. Trialkyltin compounds such as $(C_4H_9)_3SnOH$ are important biocides used to control fungi, bacteria, insects, and weeds. These compounds have also been used to inhibit the formation of barnacles on ships' hulls.

Lead(II) oxide (PbO), or litharge, is often added to glass to increase its density, brilliance, and strength. PbO is also an important component of the lead storage batteries used in cars and trucks. At one time, lead chromate ($PbCrO_4$) and lead dichromate ($PbCr_2O_7$) were important yellow and orange pigments for use in paint. As our awareness of the toxicity of lead compounds has increased, these pigments have gradually been replaced by others. Until recently, tetraethyl lead, $Pb(C_2H_5)_4$, was routinely added to gasoline to increase its octane number. Concern over high levels of lead emissions from automobile exhaust has resulted in a switch to "unleaded" gasolines.

7.10 OXIDATION-REDUCTION REACTIONS

Balanced equations for more than 30 different reactions between a main-group metal and a nonmetal element (such as O_2 or Cl_2) or compound (such as H_2O or

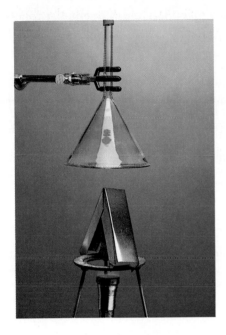

The oxidation of copper metal when the metal is heated with a bunsen burner is shown. Copper metal on the surface reacts with oxygen to form a black film of copper(II) oxide, CuO. The black CuO film that forms on the surface of the metal can be reduced to copper metal by turning off the burner, lowering the funnel in this photograph, and then blowing H_2 gas over the metal surface while it is still hot.

NH_3) have been presented in this chapter. These reactions all have at least one thing in common—they are all examples of *oxidation-reduction reactions.*

The term *oxidation* was initially used to describe reactions in which an element combined with oxygen. By definition, the reaction between magnesium metal and oxygen to form magnesium oxide therefore involves the oxidation of magnesium.

$$2 \ Mg(s) + O_2(g) \longrightarrow 2 \ MgO(s)$$

$$\underset{\text{oxidation}}{\underline{\hspace{3cm}}}\uparrow$$

The term *reduction* comes from the Latin stem meaning "to lead back." Anything that reverses this process—that leads back to magnesium metal—therefore involves reduction. The reaction between magnesium oxide and carbon at 2000°C to form magnesium metal and carbon monoxide is therefore an example of the reduction of magnesium oxide to magnesium metal.

$$MgO(s) + C(s) \longrightarrow Mg(s) + CO(g)$$

$$\underset{\text{reduction}}{\underline{\hspace{3cm}}}\uparrow$$

After electrons were discovered, chemists became convinced that oxidation-reduction reactions involved the transfer of electrons from one atom to another. From this perspective, the reaction between magnesium and oxygen is written as follows.

$$2 \ Mg + O_2 \longrightarrow 2 \ [Mg^{2+}][O^{2-}]$$

In the course of this reaction, the following happens.

1. Each magnesium atom loses two electrons to form an Mg^{2+} ion.

$$Mg \longrightarrow Mg^{2+} + 2 \ e^2$$

2. Each O_2 molecule gains four electrons to form two O^{2-} ions.

$$O_2 + 4 \ e^- \longrightarrow 2 \ O^{2-}$$

We can therefore generalize the concepts of oxidation and reduction as follows.

Oxidation **occurs when an atom** *loses* **electrons.**

Reduction **occurs when an atom** *gains* **electrons.**

When magnesium reacts with oxygen, the magnesium atoms lose two electrons and are therefore oxidized. The O_2 molecules gain electrons and are therefore reduced. Since electrons are neither created nor destroyed in a chemical reaction, oxidation and reduction are linked. It is impossible to have one without the other.

$$2\ Mg + O_2 \longrightarrow 2[Mg^{2+}][O^{2-}]$$

oxidation

reduction

Exercise 7.7

Determine which element is oxidized and which is reduced when lithium reacts with nitrogen to form lithium nitride.

$$6\ Li(s) + N_2(g) \longrightarrow 2\ Li_3N(s)$$

Solution

In the course of this reaction, the lithium atoms each lose an electron to form Li^+ ions, and the nitrogen atoms gain three electrons to form N^{3-} ions. Lithium is therefore oxidized in this reaction, and nitrogen is reduced.

$$Li + N_2 \longrightarrow [Li^+]_3[N^{3-}]$$

oxidation

reduction

THE ROLE OF OXIDATION NUMBERS
7.11 IN OXIDATION-REDUCTION REACTIONS

Chemists eventually realized that oxidation-reduction reactions don't always involve the transfer of electrons. When carbon reacts with oxygen to form CO_2, for example, the product of this reaction is a neutral molecule — it doesn't contain C^{4+} and O^{2-} ions.

$$C(s) + O_2(g) \longrightarrow CO_2(g)$$

They therefore introduced the concept of oxidation number to extend the idea of oxidation and reduction to reactions in which electrons are not really gained or lost.

The oxidation number of an atom was defined as the charge the atom would have *if* the element or compound contained ions. The oxidation number of carbon atoms in elemental carbon and the oxygen atoms in O_2 molecules are both zero. But the oxidation number of the carbon atom in a CO_2 molecule is $+4$, and the oxidation number of the oxygen atoms is -2.

$$\underset{0}{C} + \underset{0}{O_2} \longrightarrow \underset{+4\ -2}{CO_2}$$

In the course of this reaction, the following happens.

1. The oxidation number of carbon becomes larger; it increases from 0 to $+4$.
2. The oxidation number of oxygen becomes smaller; it decreases from 0 to -2.

By definition, carbon is oxidized in this reaction, which means that O_2 must be reduced. The concepts of oxidation and reduction can therefore be generalized as follows.

Oxidation occurs when the oxidation number of an atom becomes larger.

Reduction occurs when the oxidation number of an atom becomes smaller.

Exercise 7.8

Copper metal turns black when it is held over the flame of a bunsen burner as the metal reacts with oxygen in the air to form copper(II) oxide.

$$2\ Cu(s) + O_2(g) \longrightarrow 2\ CuO(s)$$

If hydrogen gas is passed over the surface of the metal while it is still hot, the copper(II) oxide can be converted back to copper metal.

$$CuO(s) + H_2(g) \longrightarrow Cu(s) + H_2O(g)$$

Determine which atom is oxidized and which is reduced in each of these reactions.

Solution

We can start by assigning oxidation numbers to the first reaction and looking for elements that undergo a change in oxidation number.

$$Cu + O_2 \longrightarrow CuO$$

The oxidation number of copper increases from 0 to $+2$, which means that copper metal is oxidized in this reaction. The oxidation number of oxygen decreases from 0 to -2, which means that oxygen is reduced.

$$Cu + O_2 \longrightarrow CuO$$

We can then assign oxidation numbers to the second reaction.

$$CuO + H_2 \longrightarrow Cu + H_2O$$

The oxidation number of oxygen is -2 on both sides of this equation. That means oxygen is neither oxidized nor reduced in this reaction. The oxidation number of copper, however, decreases from $+2$ to 0, which means that copper is reduced in this reaction. The oxidation number of hydrogen increases from 0 to $+1$, which means it is oxidized.

$$CuO + H_2 \longrightarrow Cu + H_2O$$

Exercise 7.9

Determine which atom is oxidized and which is reduced in the following reaction.

$$Sr(s) + 2\ H_2O(l) \longrightarrow Sr^{2+}(aq) + 2\ OH^-(aq) + H_2(g)$$

Solution

Once again, we can begin by looking for elements that undergo a change in oxidation number.

$$\underset{0}{Sr} + 2\ \underset{+1\ -2}{H_2O} \longrightarrow \underset{+2}{Sr^{2+}} + 2\ \underset{-2\ +1}{OH^-} + \underset{0}{H_2}$$

Nothing happens to the oxidation number of the oxygen in this reaction; it is neither oxidized nor reduced. The oxidation number of strontium becomes larger (from 0 to +2), which means Sr is oxidized. The oxidation number of some (but not all) of the hydrogen atoms becomes smaller (from +1 to 0). Hydrogen is therefore reduced in this reaction.

$$\underset{0}{Sr} + 2\ \underset{+1}{H_2O} \longrightarrow \underset{+2}{Sr^{2+}} + 2\ OH^- + \underset{0}{H_2}$$

oxidation

reduction

Oxidation numbers can be used in reactions that involve the transfer of electrons, even though they aren't needed for these reactions.

$$2\ \underset{0}{Na} + \underset{0}{Cl_2} \longrightarrow [\underset{+1}{Na^+}][\underset{-1}{Cl^-}]$$

oxidation

reduction

Oxidation numbers play an essential role, however, in oxidation-reduction reactions that occur by the transfer of an atom, such as an oxygen atom

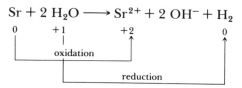

$$\underset{+2}{MgO} + \underset{0}{C} \longrightarrow \underset{0}{Mg} + \underset{+2}{CO}$$

reduction

oxidation

or a hydrogen atom.

$$\underset{0}{N_2} + 3\ \underset{0}{H_2} \longrightarrow \underset{-3\ +1}{NH_3}$$

reduction

oxidation

7.12 OXIDIZING AGENTS AND REDUCING AGENTS

There are two ways of looking at oxidation-reduction reactions. We can focus on what happens when a particular element gains or loses electrons.

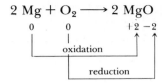

Or we can focus on the role that each element plays in the reaction.

Ignore for the moment what happens to the magnesium in this reaction and concentrate on the role that it plays. Magnesium atoms donate electrons to O_2 molecules in this reaction and thereby reduce the oxygen. Magnesium therefore acts as a ***reducing agent*** in this reaction.

$$2\,\underset{\text{reducing agent}}{\mathbf{Mg}} + O_2 \longrightarrow 2\,MgO$$

The O_2 molecules, on the other hand, gain electrons from magnesium atoms and thereby oxidize the magnesium. Oxygen is therefore an ***oxidizing agent*** in this reaction.

$$2\,Mg + \underset{\textbf{oxidizing agent}}{\mathbf{O_2}} \longrightarrow 2\,MgO$$

In general, reducing agents and oxidizing agents can be defined as follows.

Reducing agents lose electrons.

Oxidizing agents gain electrons.

Exercise 7.10

Identify the oxidizing agent and the reducing agent in the following reaction.

$$Ca(s) + H_2(g) \longrightarrow CaH_2(g)$$

Solution

We can start by assigning oxidation numbers to this reaction.

$$\underset{0}{Ca} + \underset{0}{H_2} \longrightarrow \underset{+2\,-1}{CaH_2}$$

We can then decide which element is oxidized and which is reduced. In the course of this reaction, calcium atoms are formally transformed into Ca^{2+} ions by the loss of a pair of electrons, which means that calcium is oxidized. The hydrogen atoms, on the other hand, formally gain an electron to form H^- ions. Hydrogen is therefore reduced.

$$\underset{0}{Ca} + \underset{0}{H_2} \longrightarrow \underset{+2\,-1}{CaH_2}$$

In the course of this reaction, calcium reduces H_2 molecules to form H^- ions. Calcium is therefore the reducing agent. The H_2 molecules, on the other hand, oxidize calcium atoms to form Ca^{2+} ions. H_2 is therefore the oxidizing agent in this reaction.

Reducing agent: Ca   Oxidizing agent: H_2

Table 7.3 identifies the reducing agent and the oxidizing agent for many of the reactions discussed in this chapter. One trend is immediately obvious.

The main-group metals act as reducing agent in their chemical reactions.

TABLE 7.3

Typical Reactions of Main-Group Metals

Reaction	Reducing Agent	Oxidizing Agent
$2\ Na + Cl_2 \longrightarrow 2\ NaCl$	Na	Cl_2
$2\ K + H_2 \longrightarrow 2\ KH$	K	H_2
$16\ Na + S_8 \longrightarrow 8\ Na_2S$	Na	S_8
$6\ Li + N_2 \longrightarrow 2\ Li_3N$	Li	N_2
$4\ Li + O_2 \longrightarrow 2\ Li_2O$	Li	O_2
$2\ Na + O_2 \longrightarrow Na_2O_2$	Na	O_2
$K + O_2 \longrightarrow KO_2$	K	O_2
$2\ Na + 2\ H_2O \longrightarrow 2\ Na^+ + 2\ OH^- + H_2$	Na	H_2O
$2\ K + 2\ NH_3 \longrightarrow 2\ KNH_2 + H_2$	K	NH_3
$Mg + Cl_2 \longrightarrow MgCl_2$	Mg	Cl_2
$Ca + H_2 \longrightarrow CaH_2$	Ca	H_2
$8\ Sr + S_8 \longrightarrow 8\ SrS$	Sr	S_8
$2\ Mg + O_2 \longrightarrow 2\ MgO$	Mg	O_2
$Ba + O_2 \longrightarrow BaO_2$	Ba	O_2
$3\ Mg + N_2 \longrightarrow Mg_3N_2$	Mg	N_2
$6\ Ca + P_4 \longrightarrow 2\ Ca_3P_2$	Ca	P_4
$Ca + 2\ H_2O \longrightarrow Ca^{2+} + 2\ OH^- + H_2$	Ca	H_2O
$Mg + H_2O \longrightarrow MgO + H_2$	Mg	H_2O
$2\ Al + 3\ Br_2 \longrightarrow Al_2Br_6$	Al	Br_2
$Mg + 2\ H^+ \longrightarrow Mg^{2+} + H_2$	Mg	H^+
$Fe_2O_3 + 2\ Al \longrightarrow Al_2O_3 + 2\ Fe$	Al	Fe_2O_3
$2\ Pb + O_2 \longrightarrow 2\ PbO$	Pb	O_2
$Sn + 2\ HCl \longrightarrow SnCl_2 + H_2$	Sn	HCl

7.13 CONJUGATE OXIDIZING AGENT/REDUCING AGENT PAIRS

In the preceding section, we concluded that the main-group metals are reducing agents in all of their chemical reactions. In fact, it is safe to conclude that all metals act as reducing agents in their chemical reactions. When copper is heated over a flame, for example, the surface slowly turns black as the copper metal reduces oxygen in the atmosphere to form copper(II) oxide.

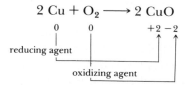

If we turn off the flame, however, and blow H_2 gas over the hot metal surface, the black CuO that formed on the surface of the metal is slowly converted back to copper metal. In the course of this reaction, the oxidation number of the copper decreases from $+2$ to 0 as the CuO is reduced to copper metal. Thus, H_2 is the reducing agent in this reaction, and CuO acts as an oxidizing agent.

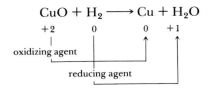

The first reaction transforms a reducing agent — Cu — into an oxidizing agent — CuO. The second reaction turns an oxidizing agent — CuO — into a reducing agent — Cu. This leads to the conclusion that oxidizing agents and reducing agents are not independent of each other. Every reducing agent is linked or coupled to a *conjugate* oxidizing agent.

Every time a reducing agent loses electrons, it forms an oxidizing agent that could gain electrons if the reaction was reversed.

$$Cu \longrightarrow Cu^{2+} + 2\,e^-$$

$$\underset{\substack{\text{reducing} \\ \text{agent}}}{} \qquad \underset{\substack{\text{oxidizing} \\ \text{agent}}}{\phantom{Cu^{2+}}}$$

Similarly, every time an oxidizing agent gains electrons, it forms a reducing agent that could lose electrons if the reaction went in the opposite direction.

$$O_2 + 4\,e^- \longrightarrow 2\,O^{2-}$$

$$\underset{\substack{\text{oxidizing} \\ \text{agent}}}{} \qquad \qquad \underset{\substack{\text{reducing} \\ \text{agent}}}{\phantom{2\,O^{2-}}}$$

In general, we can summarize the link between a reducing agent and its conjugate oxidizing agent as follows.

Reducing agent $\longrightarrow$ oxidizing agent + electrons

Oxidizing agent + electrons $\longrightarrow$ reducing agent

The idea that oxidizing agents and reducing agents are linked, or coupled, is not difficult to understand. By why call them *conjugate* oxidizing agents and reducing agents? Most people who recognize the term *conjugate* associate this word with the way it is used in grammar to mean systematically listing the different forms of a verb. *Conjugate* actually comes from the Latin stem meaning "to join together." It is therefore used to describe things that are linked or coupled, such as oxidizing agents and reducing agents.

The main-group metals are all reducing agents. More importantly, they tend to be "good," or strong, reducing agents. The active metals in Group IA, for example, give up electrons better than any other elements in the periodic table. They are therefore among the strongest chemical reducing agents.

The fact that an active metal such as Na is a strong reducing agent should tell us something about the relative strength of the Na^+ ion as an oxidizing agent. If Na is unusually good at giving up electrons, then we can conclude that Na^+ ions must be unusually bad at picking up electrons. If Na is a strong reducing agent, then the Na^+ ion must be a weak oxidizing agent.

$$Na \longrightarrow Na^+ + e^-$$

$$\underset{\substack{\text{strong} \\ \text{reducing} \\ \text{agent}}}{} \qquad \underset{\substack{\text{weak} \\ \text{oxidizing} \\ \text{agent}}}{}$$

Conversely, if O_2 has such a high affinity for electrons that it is unusually good at accepting them from other elements, it should be able to hang onto these electrons once it picks them up. In other words, if O_2 is a strong oxidizing agent, then the O^{2-} ion must be a weak reducing agent.

$$O_2 + 4\,e^- \longrightarrow 2\,O^{2-}$$

strong   weak
oxidizing   reducing
agent   agent

In general, the relationship between conjugate oxidizing and reducing agents can be described as follows.

> **Every strong reducing agent (such as Na or Mg) has a weak conjugate oxidizing agent (such as the Na^+ or Mg^{2+} ion).**

> **Every strong oxidizing agent (such as O_2 or F_2) has a weak conjugate reducing agent (such as the O^{2-} or F^- ion).**

7.14 THE RELATIVE STRENGTHS OF METALS AS REDUCING AGENTS

We can determine the relative strengths of a pair of metals—such as iron and aluminum—as reducing agents by determining whether a reaction occurs when one of the metals is mixed with a salt of the other metal. Although nothing seems to happen when we mix powdered aluminum metal with iron(III) oxide, if we place this mixture in a crucible and get the reaction started by applying a little heat, an extremely vigorous reaction takes place to give aluminum oxide and iron metal.

$$2\,Al(s) + Fe_2O_3(s) \longrightarrow Al_2O_3(s) + 2\,Fe(l)$$

By assigning oxidation numbers we can pick out the oxidation and reduction halves of the reaction.

According to this diagram, aluminum is oxidized to Al_2O_3 in this reaction, which means that Fe_2O_3 must be the oxidizing agent. Conversely, Fe_2O_3 is reduced to iron metal, which means that aluminum must be the reducing agent. Since a reducing agent is always transformed into its conjugate oxidizing agent—and vice versa—in an oxidation-reduction reaction, the products of this reaction include a new oxidizing agent (Al_2O_3) and a new reducing agent (Fe).

$$2\,Al + Fe_2O_3 \longrightarrow Al_2O_3 + 2\,Fe$$

reducing oxidizing oxidizing reducing
agent   agent   agent   agent

What can we conclude about the relative strengths of these oxidizing and reducing agents from the fact that the reaction proceeds in this direction? It seems reasonable to assume that the stronger of a pair of reducing agents will react with the stronger of a pair of oxidizing agents to form the weaker reducing agent and the weaker oxidizing agent.

$$2\,Al + Fe_2O_3 \longrightarrow Al_2O_3 + 2\,Fe$$

stronger stronger weaker weaker
reducing oxidizing oxidizing reducing
agent   agent   agent   agent

The simple observation that aluminum reduces Fe_2O_3 to form Al_2O_3 and iron metal suggests that aluminum is a stronger reducing agent than iron.

What can we conclude from the fact that aluminum cannot reduce sodium chloride to form sodium metal?

$$Al(s) + NaCl(s) \not\longrightarrow$$

It seems reasonable to assume that the starting materials in this reaction are the *weaker* oxidizing agent and the *weaker* reducing agent.

$$\underset{\substack{\text{weaker} \\ \text{reducing} \\ \text{agent}}}{Al} + \underset{\substack{\text{weaker} \\ \text{oxidizing} \\ \text{agent}}}{3\ NaCl} \not\longrightarrow \underset{\substack{\text{stronger} \\ \text{oxidizing} \\ \text{agent}}}{AlCl_3} + \underset{\substack{\text{stronger} \\ \text{reducing} \\ \text{agent}}}{3\ Na}$$

In other words, aluminum is a weaker reducing agent than sodium. We can test this hypothesis by determining whether sodium metal is strong enough to reduce an aluminum salt. Sodium metal does in fact react with aluminum chloride to form aluminum metal and sodium chloride when the reaction is run at temperatures high enough to melt the reactants.

$$3\ Na(l) + AlCl_3(l) \longrightarrow 3\ NaCl(l) + Al(l)$$

If sodium is strong enough to reduce Al^{3+} salts to aluminum metal, and aluminum is strong enough to reduce Fe^{3+} salts to iron metal, then the relative strengths of these reducing agents can be summarized as follows.

$$Na > Al > Fe$$

Exercise 7.11

Use the results of experiments summarized in the following equations to determine the relative strengths of sodium, magnesium, aluminum, and calcium metal as reducing agents.

$$2\ Na + MgCl_2 \longrightarrow 2\ NaCl + Mg$$
$$Al + MgBr_2 \not\longrightarrow$$
$$Ca + MgI_2 \longrightarrow CaI_2 + Mg$$
$$Ca + 2\ NaCl \not\longrightarrow$$

Solution

The first reaction suggests that sodium is a stronger reducing agent than magnesium. The second reaction suggests that aluminum is not as strong a reducing agent as magnesium, or, conversely, that magnesium is a stronger reducing agent than aluminum. The first two reactions therefore give the following sequence.

$$Na > Mg > Al$$

The third and fourth reactions suggest that calcium is a stronger reducing agent than magnesium but a weaker reducing agent than sodium. Thus, the overall sequence of metals in order of decreasing strength as reducing agents is the following.

$$Na > Ca > Mg > Al$$

The results of a large number of experiments of this nature are summarized in Table 7.4. By convention, each reaction in this table is written in the direction of

TABLE 7.4

The Relative Strengths of the Common Metals as Reducing Agents

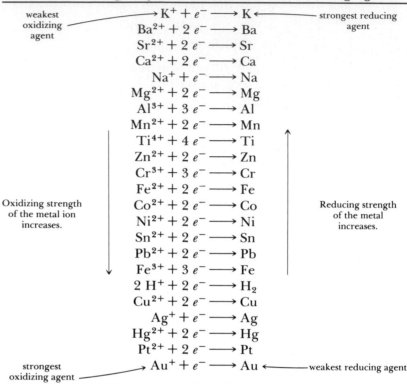

reduction. Furthermore, the table is organized so that the strongest reducing agents are found in the upper right-hand corner and the strongest oxidizing agents, in the bottom left-hand corner. The active metals are therefore found at the top of the table.

We can use this table to predict whether a reaction should occur. Any metal ion in this table can be reduced by metals listed *above* it. According to this table, for example, aluminum metal should be able to reduce Cr_2O_3 to chromium metal.

$$Cr_2O_3(s) + 2\ Al(s) \longrightarrow Al_2O_3(s) + 2\ Cr(s)$$

Exercise 7.12

Use Table 7.4 to predict whether copper metal should react with Ag^+ ions to form silver metal and Cu^{2+} ions.

$$Cu(s) + 2\ Ag^+(aq) \longrightarrow Cu^{2+}(aq) + 2\ Ag(s)$$

Solution

Copper metal is listed above silver in this table, which means that copper is the stronger of the two reducing agents. The Ag^+ ion is listed below the Cu^{2+} ion in the table, which means that the Ag^+ ion is the stronger of the two oxidizing agents. We would therefore expect copper metal to react with Ag^+ ions to form silver metal and Cu^{2+} ions.

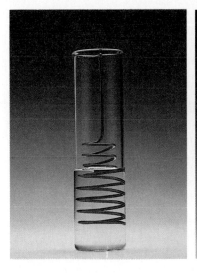 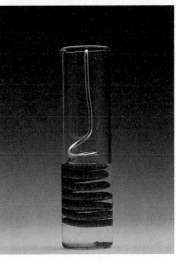

Copper metal reacts with a solution that contains the Ag^+ ion to form silver metal and a solution of the Cu^{2+} ions as predicted by Exercise 7.12.

Exercise 7.13

A strong acid can be defined as anything that acts as a good source of the H^+ ion. Use Table 7.4 to identify the metals that are not likely to dissolve in strong acids.

Solution

The H^+ ion can be reduced by any metal listed above it in Table 7.4. Zinc, for example, can reduce H^+ ions to elemental hydrogen.

$$Zn(s) + 2\ H^+(aq) \longrightarrow Zn^{2+}(aq) + H_2(g)$$

In the course of this reaction, the zinc metal dissolves. The only metals in Table 7.4 that cannot reduce H^+ ions are Cu, Ag, Hg, Pt, and Au. These are therefore the only metals in this table that should not dissolve in a strong acid.

7.15 THE PREPARATION OF METALS: CHEMICAL MEANS

In Section 6.16, the metals were divided into four categories on the basis of their activity, or reactivity. These four categories are reflected in the positions of these metals in Table 7.4. The active metals (such as Li, Na, K, and Ca) are at the top of the table. The less active metals (such as Mg, Al, and Zn) can be found in the top half. The structural metals (such as Cr, Fe, Sn, Pb, and Cu) are usually found in the lower half of the table, while the metals used to make coins and jewelry (such as Ag, Au, and Pt) are at the bottom.

It is also possible to classify metals on the basis of the relative ease with which they can be prepared from their salts or ores. As we proceed up the column of relative reducing strength in Table 7.4, the difficulty of preparing the metal from one of its ores increases.

A number of metals, including copper, iron, lead, tin, and zinc, are prepared from naturally occurring ores by a five-step process that includes: (1) concentrating, (2) sintering and/or calcining, (3) roasting, (4) smelting, and (5) refining.

The first step in preparing any metal involves ***concentrating*** the ore by separating it from the rock in which it is embedded. The rock is crushed and then ground

The pouring of a mixture of ore and silica during the smelting of iron.

into fine particles if necessary. The finely divided ore is then poured into a large tank of water, which contains a wetting agent that adheres to the ore particles. Air blown into the bottom of the tank causes the wetting agent to form bubbles that carry the ore to the surface of the tank, while the unwanted rock settles to the bottom. The ore is then skinned off the top of the tank.

The next step often involves **sintering,** a procedure in which the ore particles are heated until they come together to form larger particles. Some ores also undergo **calcining** when they are heated, a process in which a gas is given off. Sintering does not change the chemical composition of the ore, but calcining does. An example of an ore that undergoes calcining is bauxite, a mixture of $Al(OH)_3$ and $AlO(OH)$. The $Al(OH)_3$ in bauxite gives off water to form aluminum oxide when the ore is heated.

$$2\ Al(OH)_3(s) \longrightarrow Al_2O_3(s) + 3\ H_2O(g)$$

Many commercially important ores are sulfides, including chalcopyrite ($CuFeS_2$), cinnabar (HgS), galena (PbS), pyrite (FeS_2), and zincblende (ZnS). Before these ores can be reduced to the metal, they must first be **roasted.** This process involves heating the ore in the presence of O_2 to convert the sulfide into the corresponding oxide.

$$2\ ZnS(s) + 3\ O_2(g) \longrightarrow 2\ ZnO(s) + 2\ SO_2(g)$$
$$2\ PbS(s) + 3\ O_2(g) \longrightarrow 2\ PbO(s) + 2\ SO_2(g)$$

Cinnabar (HgS) is slightly different. When it is roasted, the S^{2-} ion is oxidized to SO_2, and the Hg^{2+} ion is reduced to mercury metal, which boils off.

$$HgS(s) + O_2(g) \longrightarrow Hg(g) + SO_2(g)$$

The steps so far are all designed to prepare the ore for **smelting,** the process in which the ore is reduced to the metal. Copper, iron, lead, tin, and zinc are just a few of the metals produced by the smelting of metal oxides in a blast furnace. The reducing agent in this process is almost always "coke," an amorphous form of carbon produced when coal is heated in the absence of air. Some metals react directly with the carbon in coke.

$$SnO_2(s) + C(s) \longrightarrow Sn(l) + CO_2(g)$$

Other metals are reduced by the carbon monoxide formed when coke reacts with oxygen.

$$2\ C(s) + O_2(g) \longrightarrow 2\ CO(g)$$
$$ZnO(s) + CO(g) \longrightarrow Zn(g) + CO_2(g)$$

These reactions are run at such high temperatures that the metal comes off as a liquid, or even a gas.

Figure 7.1 shows a drawing of a typical blast furnace used to make iron metal. The blast furnace is essentially a large chemical reactor as much as 100 feet tall and 22 feet in diameter in which the temperature varies from about 200°C at the top to 2000°C at the bottom. The iron ore is usually one of the oxides of iron, such as magnetite (Fe_3O_4) or hematite (Fe_2O_3). This ore is mixed with coke and loaded at the top of the furnace. Preheated compressed air or O_2 is blown into the furnace at the bottom, where it reacts with the coke to produce a mixture of CO and CO_2.

$$2\ C(s) + O_2(g) \longrightarrow 2\ CO(g)$$
$$C(s) + O_2(g) \longrightarrow CO_2(g)$$
$$C(s) + CO_2(g) \longrightarrow 2\ CO(g)$$

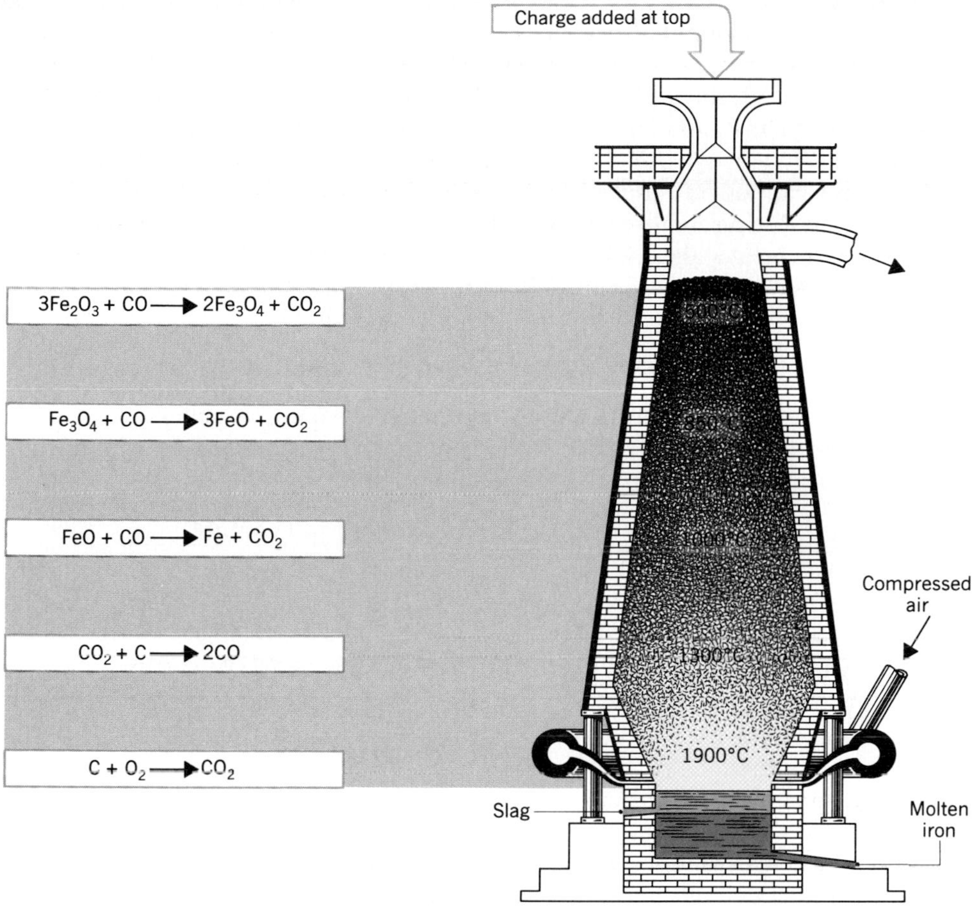

FIG. 7.1 A typical blast furnace for smelting iron.

$3Fe_2O_3 + CO \longrightarrow 2Fe_3O_4 + CO_2$

$Fe_3O_4 + CO \longrightarrow 3FeO + CO_2$

$FeO + CO \longrightarrow Fe + CO_2$

$CO_2 + C \longrightarrow 2CO$

$C + O_2 \longrightarrow CO_2$

These reactions generate the heat that fires the furnace.

At the top of the furnace, where the temperatures are below 700°C, the iron oxides are reduced to iron metal by carbon monoxide gas.

$$Fe_2O_3(s) + 3\ CO(g) \longrightarrow 2\ Fe(l) + 3\ CO_2(g)$$
$$Fe_3O_4(s) + 4\ CO(g) \longrightarrow 3\ Fe(l) + 4\ CO_2(g)$$

Any iron oxide left as the ore drops through the reactor will eventually reach temperatures above 700°C and react directly with the carbon in coke.

$$Fe_3O_4(s) + 3\ C(s) \longrightarrow 2\ Fe(l) + 3\ CO(g)$$
$$Fe_3O_4(s) + 4\ C(s) \longrightarrow 3\ Fe(l) + 4\ CO_2(g)$$

The mixture of iron oxide and coke is added more or less continuously at the top of the furnace, and molten iron, or pig iron, is drained off at the bottom every few hours.

The product of this reaction often contains impurities that must be removed during *refining*. Pig iron, for example, contains 3 to 5% carbon, 0.5 to 4% silicon, 0.15 to 2.5% manganese, 0.025 to 2.5% phosphorus, and 0.2% sulfur by weight. Historically, pig iron was heated until white-hot and then worked (or "wrought") with a hammer to drive out the impurities. The wrought iron produced this way was harder than bronze, and yet it could be hammered into different shapes to make anything from a gate or a railing to a horseshoe. Wrought iron usually has less

A wrought-iron fence.

than 0.1% carbon by weight. Steel, which is harder than wrought iron, can be produced by refinement of pig iron so that 1 to 2% by weight of carbon remains.

The primary reason for using coke as the reducing agent for the smelting of copper, iron, lead, tin, and zinc is its low cost. It is so cheap that part of the coke is used as a reducing agent and the remainder as a fuel to heat the blast furnace.

Eventually, as we proceed up the column in Table 7.4, we encounter metals whose ores cannot be reduced by coke. Such ores require stronger (and more expensive) reducing agents. Titanium, for example, has many attractive properties as a metal. Its principal drawback is the fact that it is prepared commercially by the reduction of titanium tetrachloride with magnesium metal, which is relatively expensive.

$$TiCl_4(l) + 2\ Mg(s) \longrightarrow Ti(s) + 2\ MgCl_2(s)$$

Other industrial processes that rely on relatively expensive reducing agents are the reduction of cobalt(III) oxide with H_2 gas and the use of powdered aluminum to reduce chromium(III) oxide to chromium.

$$Co_2O_3(s) + 3\ H_2(g) \longrightarrow 2\ Co(s) + 3\ H_2O(g)$$
$$Cr_2O_3(s) + 2\ Al(s) \longrightarrow Al_2O_3(s) + 2\ Cr(l)$$

7.16 THE PREPARATION OF ACTIVE METALS: ELECTROLYSIS

Magnesium can be used to reduce $TiCl_4$ to prepare titanium metal, and aluminum can be used to reduce Cr_2O_3 to chromium metal. But this leaves us with the task of preparing magnesium and aluminum metal. Obviously, we need a stronger reducing agent.

Magnesium was first prepared in bulk quantities from a mixture of magnesium chloride and potassium metal heated in a glass tube.

$$MgCl_2(s) + 2\ K(s) \longrightarrow Mg(s) + 2\ KCl(s)$$

Aluminum was first prepared from a mixture of aluminum chloride and potassium metal heated in a platinum crucible.

$$AlCl_3(s) + 3\ K(s) \longrightarrow Al(s) + 3\ KCl(s)$$

Eventually, it was possible to prepare magnesium and aluminum by use of sodium metal, which is cheaper than potassium. But this still leaves us with the challenge of preparing sodium metal.

There aren't any chemical reducing agents strong enough to prepare sodium metal. We therefore have to resort to physical means, such as passing an electric current through the salt. Sir Humphrey Davy first prepared potassium metal (in October 1807) by connecting a battery to a small piece of "potash," or potassium carbonate (K_2CO_3). A few days later he repeated this procedure with crystals of "caustic soda" (NaOH) and was able to prepare small quantities of sodium metal.

Only small quantities of metal can be prepared this way, because the Na^+ or K^+ ions are locked in place within the salt crystals. This problem can be overcome, however, if the salt is heated until it melts. The ions are then free to move through the molten salt, and significant quantities of the metal eventually collect at the electrode connected to the negative end of the battery.

Figure 7.2 shows a diagram of an electrical cell in which a pair of metal electrodes connected to a high-voltage battery are immersed in a molten sample of sodium chloride. One of these electrodes gains electrons from the battery and

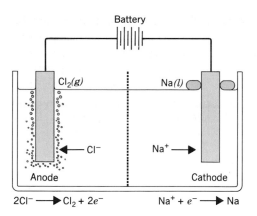

FIG. 7.2 A cell that can be used for the electrolysis of molten NaCl comprises a pair of metal electrodes connected to a battery immersed in molten NaCl. Na^+ ions migrate toward the negative electrode, or cathode, where they are reduced to sodium metal, which collects as a liquid at the top of the cell. Cl^- ions migrate toward the positive electrode, or anode, where they are oxidized to form Cl_2 gas, which bubbles out of the cell.

becomes negatively charged. The other electrode loses electrons to the battery and becomes positively charged. Since particles of opposite charge attract, the Na^+ ions (or **cations**) migrate toward the negatively charged electrode, or cathode. Conversely, the Cl^- ions (or **anions**) move toward the positively charged anode.

The Na^+ ions gain electrons at the cathode to form sodium metal, while the Cl^- ions lose electrons at the anode to form Cl_2 gas.

Cathode:
$$Na^+ + e^- \longrightarrow Na \qquad \text{(reduction)}$$
Anode:
$$2\,Cl^- \longrightarrow Cl_2 + 2\,e^- \qquad \text{(oxidation)}$$

Thus, reduction occurs at the cathode, and oxidation occurs at the anode. This process is called **electrolysis** (literally, "splitting with electrons"), because it uses electrons to split a compound into its elements.

$$2\,NaCl(l) \longrightarrow 2\,Na(l) + Cl_2(g)$$

(The sodium produced in this reaction is a liquid because the cell operates at temperatures much higher than the melting point of sodium metal.)

The same process can be used to make magnesium metal. In this case, an electric current is passed through a molten sample of magnesium chloride. The positive Mg^{2+} ions, or cations, are reduced to magnesium metal at the cathode, and the negative Cl^- ions, or anions, are oxidized at the anode.

Cathode:
$$Mg^{2+} + 2\,e^- \longrightarrow Mg \qquad \text{(reduction)}$$
Anode:
$$2\,Cl^- \longrightarrow Cl_2 + 2\,e^- \qquad \text{(oxidation)}$$

The overall electrolysis reaction can therefore be written as follows.

$$MgCl_2(l) \longrightarrow Mg(l) + Cl_2(g)$$

In theory, aluminum metal could be made the same way. A problem is presented, however, by the extremely high melting points of aluminum ores. Al_2O_3, for example, melts at temperatures above 2000°C. In 1886, Charles Martin Hall, a 21-year old graduate of Oberlin College, discovered a way to make aluminum by electrolyzing a mixture of Al_2O_3 dissolved in cryolite, Na_3AlF_6. This mixture melts at a lower temperature than either of the individual ores and allows the electrolysis to be done at more reasonable temperatures of 960 to 980°C.

Electrolysis of mixtures of Al_2O_3 and Na_3AlF_6 is still the principal source of aluminum metal. The Al_2O_3 is obtained from the mineral bauxite. The formula

The first few samples of aluminum metal prepared by Charles Hall, these globules of aluminum metal are referred to as the "crown jewels" of the Aluminum Corporation of America—ALCOA.

for bauxite is often written as $Al_2O_3 \cdot 2 H_2O$. This is misleading, however, because bauxite does not consist of Al_2O_3 with water molecules trapped in holes in the crystal. Bauxite is actually a mixture of two salts: $Al(OH)_3$ and $AlO(OH)$. The formula of bauxite should therefore be written as either $Al(OH)_3/AlO(OH)$ or $Al_2H_4O_5$.

The first step in making aluminum is to dissolve bauxite in NaOH to form sodium aluminate, $NaAlO_2$, which is then filtered to remove impurities.

$$Al(OH)_3/AlO(OH)(s) + 2 NaOH(aq) \longrightarrow 2 NaAlO_2(aq) + 3 H_2O(l)$$

The $NaAlO_2$ is allowed to react with water to form aluminum hydroxide, which precipitates from solution.

$$2 NaAlO_2(aq) + 4 H_2O(l) \longrightarrow 2 Al(OH)_3(s) + 2 NaOH(aq)$$

Part of the $Al(OH)_3$ formed in this reaction is used to make cryolite.

$$6 HF(aq) + Al(OH)_3(s) + 3 NaOH(aq) \longrightarrow Na_3AlF_6(aq) + 6 H_2O(l)$$

The rest is filtered out, washed, and calcined at 1100 to 1200°C.

$$2 Al(OH)_3(s) \longrightarrow Al_2O_3(s) + 3 H_2O(g)$$

A small quantity of Al_2O_3 (2 to 8%) is then added to Na_3AlF_6, heated to 960 to 980°C, and electrolyzed to form aluminum metal.

7.17 LATTICE ENERGIES AND THE STRENGTH OF THE IONIC BOND

Ionic compounds are held together by the force of attraction between the positive and negative ions they contain. The first measurements of the force of attraction between oppositely charged particles were reported by Charles Augustine Coulomb between 1785 and 1791. Coulomb found that this force of attraction was directly proportional to the product of the charges on the two objects (q_1 and q_2) and inversely proportional to the square of the distance between the objects (r).

$$F = \frac{q_1 \times q_2}{r^2}$$

The strength of the bond between the ions of opposite charge in an ionic compound therefore depends on the charges on the ions and the distance between the centers of the ions when they pack to form a crystal.

An estimate of the strength of the bond between ions in an ionic compound can be obtained by measuring the *lattice energy* of the compound.

Lattice energy: The energy given off when oppositely charged ions in the gas phase come together to form a solid.

The lattice energy of NaCl, for example, is the energy given off when Na^+ and Cl^- ions in the gas phase come together to form the lattice of alternating Na^+ and Cl^- ions in a NaCl crystal. This process is illustrated by Figure 7.3.

The lattice energies of ionic compounds are relatively large. The lattice energy of NaCl, for example, is 787 kilojoules per mole.

$$Na^+(g) + Cl^-(g) \longrightarrow NaCl(s) + 787 \text{ kJ/mol}$$

This is only slightly smaller than the energy given off when natural gas burns or the energy released in the thermite reaction.

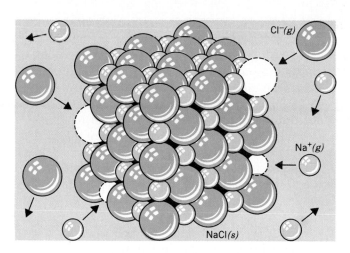

FIG. 7.3 The lattice energy of a salt such as NaCl is the energy released when Na$^+$ and Cl$^-$ ions in the gas phase come together to form a solid crystalline lattice of alternating Na$^+$ and Cl$^-$ ions.

$$CH_4(g) + 2\ O_2(g) \longrightarrow CO_2(g) + 2\ H_2O(g) + 802\ kJ/mol$$
$$Fe_2O_3(s) + 2\ Al(s) \longrightarrow Al_2O_3(s) + 2\ Fe(l) + 850\ kJ/mol$$

The bond between ions of opposite charge should be strongest when the ions are small, because small ions come closer together. The lattice energies for alkali metal halides is therefore largest for LiF and smallest for CsI, as shown by the data in Table 7.5. The ionic bond should also become stronger as the charge on the ions becomes larger. The data in Table 7.6 show that the lattice energies for salts of the F$^-$, OH$^-$, and O^{2-} ions increase rapidly as the charge on either ion becomes larger.

LATTICE ENERGIES AND SOLUBILITY

What happens when a salt, such as NaCl, dissolves in water? On the macroscopic scale, the crystal disappears. On the atomic scale, the Na$^+$ and Cl$^-$ ions in the crystal are released into solution.

$$NaCl(s) \xrightarrow{\text{H}_2\text{O}} Na^+(aq) + Cl^-(aq)$$

The lattice energy of a salt should give some indication of the solubility of the salt in water, because the lattice energy reflects the amount of energy needed to separate the positive and negative ions in a salt.

Sodium and potassium salts are very soluble in water, because they have relatively small lattice energies. Magnesium and aluminum salts are often much less soluble, because it takes more energy to separate the positive and negative ions in these salts. NaOH, for example, is very soluble in water (420 grams per liter). But Mg(OH)$_2$ only dissolves in water to the extent of 0.009 gram per liter, and Al(OH)$_3$ is essentially insoluble in water.

TABLE 7.5				
Lattice Energies of Alkali Metal Halides (kJ/mol)				
	F$^-$	Cl$^-$	Br$^-$	I$^-$
Li$^+$	1036	853	807	757
Na$^+$	923	787	747	704
K$^+$	821	715	682	649
Rb$^+$	785	689	660	630
Cs$^+$	740	659	631	604

TABLE 7.6			
Lattice Energies of Salts of the F$^-$, OH$^-$, and O^{2-} Ions (kJ/mol)			
	F$^-$	OH$^-$	O^{2-}
Na$^+$	923	900	2481
Mg^{2+}	2957	3006	3791
Al^{3+}	5492	5627	15,916

7.18 WHY DOES Na REACT WITH Cl_2 TO FORM NaCl AND NOT $NaCl_2$?

Section 6.14 noted that many beginning chemistry students believe that sodium reacts with chlorine because chlorine atoms like electrons more than sodium atoms do. This is obviously wrong, because it takes more energy to remove an electron from a sodium atom (1st IE = 495.8 kJ/mol) than is released when the electron is added to a chlorine atom (EA = 348.8 kJ/mol).

This chapter introduced a model that might explain the source of this misconcept. The product of the reaction between sodium and chlorine can be predicted — or at least explained — by the assumption that an electron is transferred from a sodium atom to a chlorine atom to form Na^+ and Cl^- ions, which have filled-shell electron configurations.

$$Na^+: \ [Ne] \qquad Cl^-: \ [Ne]\ 3s^2\ 3p^6$$

We are now ready to build a model that can explain *why* an electron is transferred from sodium to chlorine in this reaction, even though the transfer takes energy. The driving force behind the reaction can be understood when the reaction is divided into a number of steps and the amount of energy given off or absorbed in each step is determined.

The starting materials for this reaction are solid sodium metal and chlorine molecules in the gas phase, and the product of the reaction is solid sodium chloride.

$$2\ Na(s) + Cl_2(g) \longrightarrow 2\ NaCl(s)$$

Let's imagine that the reaction takes place by the following sequence of steps.

1. A mole of sodium is converted from a solid to a gas.

$$Na(s) + energy \longrightarrow Na(g)$$

2. An electron is then removed from each sodium atom to form a mole of Na^+ ions.

$$Na(g) + energy \longrightarrow Na^+(g) + e^-$$

3. A mole of chlorine atoms is formed by the breaking of the Cl—Cl bonds in $\frac{1}{2}$ mole of Cl_2.

$$\tfrac{1}{2}\ Cl_2(g) + energy \longrightarrow Cl(g)$$

4. An electron is then added to each chlorine atom to form a Cl^- ion.

$$Cl(g) + e^- \longrightarrow Cl^-(g) + energy$$

5. The Na^+ and Cl^- ions in the gas phase then come together to form solid NaCl.

$$Na^+(g) + Cl^-(g) \longrightarrow NaCl(s) + energy$$

The first three steps all consume energy. Only the last two steps give off energy.

The energy associated with each of the steps has been determined. Let's look at the total energy associated with the first four steps.

$$
\begin{aligned}
Na(s) + 107.3\ kJ &\longrightarrow Na(g) \\
Na(g) + 495.8\ kJ &\longrightarrow Na^+(g) + e^- \\
\tfrac{1}{2}\ Cl_2(g) + 121.7\ kJ &\longrightarrow Cl(g) \\
\underline{Cl(g) + e^- \longrightarrow Cl^-(g) + 348.8\ kJ} & \\
\hline
Na(s) + \tfrac{1}{2}\ Cl_2(g) + \mathbf{376.0\ kJ} &\longrightarrow Na^+(g) + Cl^-(g)
\end{aligned}
$$

According to this calculation, it takes 376.1 kJ/mol to form Na^+ and Cl^- ions from sodium metal and chlorine gas.

When we include the last step in the calculation, however, the lattice energy of NaCl is large enough to compensate for all of the steps in this reaction that consume energy, as shown in Figure 7.4.

$$Na(s) + 107.3 \text{ kJ} \longrightarrow Na(g)$$
$$Na(g) + 495.8 \text{ kJ} \longrightarrow Na^+(g) + e^-$$
$$\tfrac{1}{2} Cl_2(g) + 121.7 \text{ kJ} \longrightarrow Cl(g)$$
$$Cl(g) + e^- \longrightarrow Cl^-(g) + 348.8 \text{ kJ}$$
$$Na^+(g) + Cl^-(g) \longrightarrow NaCl(s) + 787.3 \text{ kJ}$$

$$\overline{Na(s) + \tfrac{1}{2} Cl_2(g) \longrightarrow NaCl(s) + \textbf{411.3 kJ/mol}}$$

The primary driving force behind this reaction is the force of attraction between the Na^+ and Cl^- ions formed in the reaction, not the affinity of a chlorine atom for electrons.

Why does the reaction stop at NaCl? Why doesn't it keep going to form $NaCl_2$ or $NaCl_3$, since the lattice energy would increase as the charge on the sodium atom increased from Na^+ to Na^{2+} or Na^{3+}?

The lattice energy of $MgCl_2$ is 2526 kJ/mol. Let's assume that the lattice energy

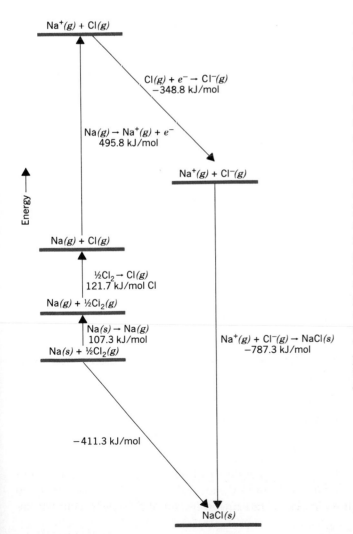

FIG. 7.4 This figure illustrates what happens to the energy of the system as solid sodium metal and chlorine gas are converted step by step into solid sodium chloride. The first step involves investing enough energy in the system to convert the solid sodium into a gas. Energy is then added to the system to break the bonds that hold the Cl_2 molecules together to form Cl atoms in the gas phase. Finally, energy is added to the system to remove an electron from the sodium atoms in the gas phase. When the electron is transferred to the Cl atom to form a Cl^- ion, some energy is given off. But the primary driving force behind this reaction is the energy released when the Na^+ and Cl^- ions produced in the reaction come together to form a solid NaCl crystal.

of $NaCl_2$ would be about the same. This is more than three times as large as the lattice energy of NaCl. But in order to form an Na^{2+} ion, we have to remove a second electron from the sodium atom, and the second ionization energy of sodium (4562.4 kJ/mol) is almost 10 times as large as the first ionization energy.

The increase in the lattice energy that would result from forming an Na^{2+} ion doesn't begin to compensate for the energy needed to break into the filled-shell configuration of the Na^+ ion to remove a second electron. The reaction between sodium and chlorine therefore stops at NaCl—it doesn't go on to form $NaCl_2$ or $NaCl_3$.

7.19 WHY DOES Mg REACT WITH Cl_2 TO FORM $MgCl_2$ AND NOT MgCl?

If the reaction between sodium and chlorine stops at NaCl, why does the reaction between magnesium and chlorine go on to form $MgCl_2$? To answer this question, let's break the reaction into the following steps.

1. A mole of magnesium is converted from a solid to a gas.

$$Mg(s) + energy \longrightarrow Mg(g)$$

2. An electron is removed from each magnesium atom to form a mole of Mg^+ ions.

$$Mg(g) + energy \longrightarrow Mg^+(g)$$

3. A second electron is then removed to form a mole of Mg^{2+} ions.

$$Mg^+(g) + energy \longrightarrow Mg^{2+}(g)$$

4. Two moles of chlorine atoms are formed by the breaking of the Cl—Cl bonds in a mole of Cl_2.

$$Cl_2(g) + energy \longrightarrow 2\ Cl(g)$$

5. An electron is then added to each chlorine atom to form 2 moles of Cl^- ions.

$$2\ Cl(g) + 2\ e^- \longrightarrow 2\ Cl^-(g) + energy$$

6. The Mg^{2+} and Cl^- ions in the gas phase then come together to form solid $MgCl_2$.

$$Mg^{2+}(g) + 2\ Cl^-(g) \longrightarrow MgCl_2(s) + energy$$

Let's now look at the energy associated with each of these steps.

$$Mg(s) + 147.7\ kJ \longrightarrow Mg(g)$$
$$Mg(g) + 737.7\ kJ \longrightarrow Mg^+(g) + e^-$$
$$Mg^+(g) + 1450.6\ kJ \longrightarrow Mg^{2+}(g) + e^-$$
$$Cl_2(g) + 243.4\ kJ \longrightarrow 2\ Cl(g)$$
$$2\ Cl(g) + 2\ e^- \longrightarrow 2\ Cl^-(g) + 697.4\ kJ$$
$$\underline{Mg^{2+}(g) + 2\ Cl^-(g) \longrightarrow MgCl_2(s) + 2526\ kJ}$$
$$Mg(s) + Cl_2(g) \longrightarrow MgCl_2(s) + \mathbf{644\ kJ}$$

The lattice energy for $MgCl_2$ is large enough to compensate for the energy it takes to remove the second electron from a magnesium atom, because we don't have to break into a filled-shell electron configuration to form Mg^{2+} ions. The reaction stops at $MgCl_2$, however, because it would take an enormous amount of energy to break into the filled-shell electron configuration of the Mg^{2+} ion to remove another electron.

SUMMARY

The products of many reactions of main-group metals can be predicted from a simple model in which the metal is assumed to lose electrons until it reaches a filled-shell electron configuration, or at least an electron configuration in which the orbital subshells are filled.

Nonmetal elements with which the main-group metals react tend to gain electrons until they reach a filled-shell configuration. The products of these reactions therefore tend to contain hydride (H^-), fluoride (F^-), chloride (Cl^-), bromide (Br^-), iodide (I^-), oxide (O^{2-}), sulfide (S^{2-}), nitride (N^{3-}), and phosphide (P^{3-}) ions.

Exceptions to the predictions of this model are encountered when the most active main-group metals react with oxygen. These exceptions can be explained by assuming that the most active metals react so rapidly with oxygen that the reaction stops prematurely at the stage of either the peroxide (such as Na_2O_2) or the superoxide (KO_2).

This model can be extended to reactions between main-group metals and nonmetal compounds by using oxidation numbers to determine which atom or ion in the compound is most likely to gain the electrons the metal loses.

The chemistry of the main-group metals revolves around the concept of oxidation-reduction reactions. Oxidation occurs when an atom or ion either loses electrons or undergoes an increase in its oxidation number.

Reduction involves either the gain of electrons or a decrease in oxidation number.

Metals undergo oxidation when they react with non-metal elements or compounds. They therefore act as reducing agents in these reactions. The relative strengths of a pair of metals as reducing agents can be determined by noting whether one of these metals is strong enough to reduce the salt of the other metal. By definition, the stronger of a pair of reducing agents should always react with the stronger of a pair of oxidizing agents to produce a weaker oxidizing agent and a weaker reducing agent.

Metals become harder to prepare from their salts as we move up the table of relative reducing strengths. Metals at the bottom of this table can sometimes be obtained by nothing more than the heating of one of their ores. Metals toward the middle of the table are often prepared by reduction of one of their ores with carbon. As we move toward the top of the table, it takes stronger (and usually more expensive) reducing agents to prepare the metal. The active metals at the top of the table can be prepared in useful quantities only when an electric current is passed through a molten salt.

The ionic compounds the metals form when they react with nonmetals are held together by the strong force of attraction between particles that have opposite charges. The strength of this bond is reflected by the lattice energy of the compound. The lattice energy becomes larger as the charge on the ions becomes larger.

PROBLEMS

Group IA: The Alkali Metals

7-1 Define the terms *alkali metal, halide, hydride, sulfide, nitride, phosphide, oxide, peroxide,* and *superoxide*. Give at least one example of each.

7-2 Write the formula for the halide, hydride, sulfide, nitride, phosphide, oxide, and peroxide of rubidium.

7-3 Describe the difference between the hydrogen atoms in metal hydrides such as LiH and the hydrogen atoms in nonmetal hydrides such as CH_4 and H_2O.

7-4 Phosphorus bursts spontaneously into flame in the presence of air and so is stored under water. Cesium metal also bursts spontaneously into flame in the presence of air, but it can't be stored under water. Why not?

7-5 Cesium metal is so active it reacts explosively with cold water and even with ice at temperatures as low as $-116°C$. Explain why cesium is more active than sodium.

Group IIA: The Alkaline Earth Metals

7-6 Magnesium metal burns rapidly in air to form magnesium oxide. This reaction gives off an enormous amount of energy in the form of light and is used in both flares and fireworks. What would happen if you tried to extinguish a magnesium flare by pouring water on the reaction?

Group IIIA: The Chemistry of Aluminum

7-7 High-school students often believe that the advantage of replacing steel with aluminum is that aluminum is "lighter."

Explain why it is wrong to say that aluminum is "lighter" than steel.

7-8 Chemists argue that aluminum is much more reactive than iron. Most people believe the opposite. Explain the common properties of aluminum metal that make most people believe aluminum is less reactive than iron.

7-9 Explain why an even coating of Al_2O_3 about 5 nanometers thick provides resistance to corrosion of aluminum and makes aluminum seem less active than it really is.

Group IVA: Tin and Lead

7-10 The first practical storage battery was built by Gaston Planté in 1859. His battery consisted of two sheets of lead separated by a strip of rubber and immersed in a 10% solution of sulfuric acid in water. Lead storage batteries can still be found in every car and truck on the road. Use the existence of these batteries to describe the difference between the activity of lead and that of the other main-group metals.

7-11 The "tin" cans found on the shelves of groceries stores are actually made out of steel that has been thinly coated with tin. Explain why these cans are coated with tin. What would happen if Del Monte tried to package fruit cocktail in a steel can?

7-12 When aluminum is used to make cans, the metal has to be coated with plastic to keep it from dissolving on contact with the acids in food. Why don't tin cans have to be coated with plastic?

7-13 Children often believe that metals such as iron lose weight when they rust. Explain why they might believe this and why they are wrong.

Identifying Main-Group Metals from the Products of Their Reactions

7-14 Which of the following elements is most likely to form an oxide with the formula XO and a hydride with the formula XH_2?

    (a) Na (b) Mg (c) Al (d) Si (e) P

7-15 A newly discovered main-group metal reacts with hydrogen and oxygen to form compounds with the formulas XH_4 and XO_2. In which group of the periodic table does this element belong?

7-16 In which column of the periodic table do we find main-group metals that form sulfides with the formula M_2S_3, form fluorides with the formula MF_3, and react with acid to form M^{3+} ions and H_2 gas?

Predicting the Products of Reactions Between Main-Group Metals and Nonmetal Elements

7-17 The electron configuration of aluminum is [Ne] $3s^2\, 3p^1$, and the configuration of sulfur is [Ne] $3s^2\, 3p^4$. Use this information to predict the formula for the product of the reaction between these elements.

7-18 Predict the product of the reaction of strontium metal with phosphorus.

7-19 The light meters in automatic cameras are based on the sensitivity of gallium arsenide to light. Predict the formula of gallium arsenide.

7-20 Write balanced equations for the reaction of sodium metal with each of the following elements or compounds.

    (a) F_2 (b) O_2 (c) H_2 (d) S_8 (e) P_4

7-21 Write balanced equations for the reaction of calcium metal with each of the following elements.

    (a) H_2 (b) O_2 (c) S_8 (d) F_2 (e) N_2 (f) P_4.

7-22 Predict the product of the reaction of F_2 with each of the following main-group metals.

    (a) Zn (b) Al (c) Sn (d) Mg (e) Bi

Predicting the Products of Reactions Between Main-Group Metals and Nonmetal Compounds

7-23 Which of the following describes the products of the reaction of calcium metal and water?

    (a) Ca^{2+}, OH^-, and H_2 (b) Ca^{2+}, H^-, and O_2 (c) Ca^{2+}, H^-, and OH^- (d) Ca^{2+} and OH^- (e) Ca^{2+}, OH^-, and H^+

7-24 Which of the following is *not* formed when potassium metal reacts with phosphine?

$$K(s) + PH_3(g) \longrightarrow$$

    (a) K^+ ions (b) PH_2^- ions (c) H_2 gas (d) H^+ ions (e) OH^- ions

7-25 Predict the products of the following reactions.

    (a) $Mg(s) + HCl(aq) \rightarrow$ (b) $Cu(s) + HCl(aq) \rightarrow$ (c) $Mg(s) + H_2O(g) \rightarrow$ (d) $Cs(s) + H_2O(l) \rightarrow$

7-26 Which of the following reactions does not make sense?

    (a) $Mg(s) + 2\,NH_3(l) \rightarrow Mg(NH_2)_2(s) + H_2(g)$ (b) $Mg(s) + N_2H_4(g) \rightarrow Mg(NH_2)_2(s)$ (c) $Mg(s) + 2\,HNO_3(aq) \rightarrow Mg(NO_3)_2 + H_2(g)$ (d) $Mg(s) + 2\,NO_2(g) + 2\,OH^-(aq) \rightarrow Mg(NO_3)_2(aq) + 2\,H^+(aq)$

7-27 Predict the products of the following reactions.

    (a) $Mg(s) + I_2(s) \rightarrow$ (b) $Al(s) + H_2SO_4(aq) \rightarrow$ (c) $Fe_2O_3(s) + H_2(g) \rightarrow$ (d) $Mg(s) + TiCl_4(l) \rightarrow$

7-28 Predict the products of the following reactions.

    (a) $K(s) + H_2O(l) \rightarrow$ (b) $NaH(s) + H_2O(l) \rightarrow$ (c) $Ag^+(aq) + Cu(s) \rightarrow$ (d) $Zn(s) + CuSO_4(aq) \rightarrow$

Oxidation-Reduction Reactions

7-29 Define *oxidation* and *reduction* in terms of the transfer of electrons and in terms of changes in oxidation number.

7-30 Explain why carbon is said to be oxidized when it reacts with sulfur at high temperatures to form carbon disulfide even though O_2 is not involved in the reaction.

$$4\,C(s) + S_8(l) \longrightarrow 4\,CS_2(l)$$

7-31 We can remove the Ag_2S that forms when silver tarnishes by polishing the silver object with a source of the cyanide ion or by wrapping it in aluminum foil and immersing it in salt water.

$$Ag_2S(s) + 4\ CN^-(aq) \longrightarrow 2\ Ag(CN)_2^-(aq) + S^{2-}(aq)$$

$$3\ Ag_2S(s) + 2\ Al \longrightarrow 6\ Ag(s) + Al_2S_3(s)$$

Which of these reactions involves oxidation-reduction?

7-32 Decide whether each of the following reactions involves oxidation-reduction. If it does, identify what is oxidized and what is reduced.

(a) $CO_2(g) + H_2O(l) \rightarrow H_2CO_3(aq)$ (b) $Fe_2O_3(s) + 3\ CO(g) \rightarrow 2\ Fe(s) + 3\ CO_2(g)$ (c) $SiO_2(s) + 3\ C(s) \rightarrow SiC(s) + 2\ CO(g)$ (d) $CO_2(g) + H_2(g) \rightarrow CO(g) + H_2O(g)$ (e) $CO(g) + 2\ H_2(g) \rightarrow CH_2OH(l)$

7-33 Decide whether each of the following reactions involves oxidation-reduction. If it does, identify what is oxidized and what is reduced.

(a) $Mg(s) + 2\ HCl(aq) \rightarrow MgCl_2(aq) + H_2(g)$ (b) $I_2(s) + 3\ Cl_2(g) \rightarrow 2\ ICl_3(l)$ (c) $HCl(aq) + NaOH(aq) \rightarrow NaCl(aq) + H_2O$ (d) $2\ Na(s) + 2\ H_2O(l) \rightarrow 2\ NaOH(aq) + H_2(g)$

7-34 Decide whether each of the following reactions involves oxidation-reduction. If it does, identify what is oxidized and what is reduced.

(a) $Ca_3P_2(s) + 6\ H_2O(l) \rightarrow 3\ Ca(OH)_2(aq) + 2\ PH_3(g)$
(b) $2\ PH_3(g) + 4\ O_2(g) \rightarrow 2\ H_3PO_4(s)$ (c) $PH_3(g) + HCl(g) \rightarrow PH_4Cl(s)$ (d) $P_4(s) + 5\ O_2(g) \rightarrow P_4O_{10}(s)$

7-35 Decide whether each of the following reactions involves oxidation-reduction. If it does, identify what is oxidized and what is reduced.

(a) $Mg(s) + H_2SO_4(aq) \rightarrow MgSO_4(aq) + H_2(g)$
(b) $2\ KBr(aq) + Cl_2(aq) \rightarrow 2\ KCl(aq) + Br_2(l)$
(c) $CaO(s) + SO_3(g) \rightarrow CaSO_4(g)$ (d) $2\ H_2(g) + Cl_2(g) \rightarrow 2\ HCl(g)$ (e) $Sb_2S_3(s) + 3\ S^{2-}(aq) \rightarrow 2\ SbS_3^{3-}(aq)$

7-36 Examine the results of Problems 34 through 37 and see if you can find a single example of a reaction in which one of the main-group metals undergoes any reaction other than oxidation. If you can't, summarize the obvious conclusion in a concise statement.

Reducing Agents and Oxidizing Agents

7-37 Define the terms *oxidizing agent* and *reducing agent*.

7-38 Identify the reducing agent and the oxidizing agent in the following reaction.

$$3\ S_8(s) + 16\ KClO_3(s) \longrightarrow 24\ SO_2(g) + 16\ KCl(s)$$

7-39 Identify the reducing agent and the oxidizing agent in the following reaction.

$$H_2O_2(aq) + 2\ HI(aq) \longrightarrow 2\ H_2O(l) + I_2(s)$$

7-40 Identify the reducing agent and the oxidizing agent in the following reaction.

$$Cu(s) + 2\ Ag^+(aq) \longrightarrow Cu^{2+}(aq) + 2\ Ag(s)$$

7-41 Identify the reducing agent and the oxidizing agent in each of the following reactions.

(a) $2\ Al(s) + 3\ CO(g) \rightarrow Al_2O_3(s) + 3\ C(s)$ (b) $2\ Al(s) + Cr_2O_3(s) \rightarrow Al_2O_3(s) + 2\ Cr(s)$ (c) $2\ Al(s) + 6\ H^+(aq) \rightarrow 2\ Al^{3+}(aq) + 3\ H_2(g)$ (d) $2\ Al(s) + 3\ I_2(s) \rightarrow 2\ AlI_3(s)$

7-42 Identify the reducing agent and the oxidizing agent in each of the following reactions.

(a) $2\ Mg(s) + CO_2(g) \rightarrow 2\ MgO(s) + C(s)$ (b) $Mg(s) + 2\ HCl(aq) \rightarrow Mg^{2+}(aq) + 2\ Cl^-(aq) + H_2(g)$ (c) $Mg(s) + H_2O(g) \rightarrow MgO(s) + H_2(g)$ (d) $3\ Mg(s) + N_2(g) \rightarrow Mg_3N_2(s)$ (e) $3\ Mg(s) + 2\ NH_3(g) \rightarrow Mg_3N_2(s) + 3\ H_2(g)$

7-43 Examine the results of Problems 43 and 44 and see if you can find an example of a reaction in which one of the main-group metals acts as anything other than a reducing agent. If you can't, summarize the obvious conclusion in a concise statement.

7-44 Identify the conjugate oxidizing agents for each of the following reducing agents.

(a) Na   (b) Zn   (c) H_2 (d) Sn^{2+} (e) H^-

7-45 Identify the conjugate reducing agents for each of the following oxidizing agents.

(a) Al^{3+} (b) Hg^{2+} (c) H^+ (d) H_2 (e) Sn^{2+}

7-46 Explain why every reducing agent has a conjugate oxidizing agent and why every oxidizing agent has a conjugate reducing agent.

7-47 Explain the difference between a strong reducing agent such as sodium or potassium metal and a weak reducing agent such as silver or gold.

7-48 Explain why every strong reducing agent has a weak conjugate oxidizing agent and why every strong oxidizing agent has a weak conjugate reducing agent.

7-49 Explain why the fact that sodium metal is a stronger reducing agent than aluminum metal implies that the Al^{3+} ion must be a stronger oxidizing agent than the Na^+ ion.

The Relative Strengths of Metals as Reducing Agents

7-50 Intact samples of copper metal about 5000 years old have been recovered by archaeologists. What does this say about the activity of copper compared with that of other metals, such as iron, aluminum, sodium, and magnesium?

7-51 Use Table 7.4 to predict which of the following elements are strong enough to reduce Fe_2O_3 to iron metal.

(a) Na   (b) Mg   (c) Al   (d) Ag   (e) H_2

7-52 Use Table 7.4 to predict which of the following oxides can be reduced to the metal with H_2.

(a) Na_2O (b) MgO   (c) Al_2O_3 (d) PbO   (e) Fe_2O_3
(f) HgO

7-53 Which of the following elements should be able to reduce Sn^{2+} ions to tin metal?

(a) Na   (b) Mg   (c) Al   (d) Fe   (e) Hg

7-54 Sodium metal reacts with ammonia to form sodium amide and hydrogen gas.

$$2 Na(s) + 2 NH_3(l) \longrightarrow 2 NaNH_2(s) + H_2(g)$$

Use this observation to determine the relative strengths of sodium metal and H_2 gas as reducing agents.

7-55 HgO decomposes to mercury metal when heated. CuO does not decompose on heating, but it can be reduced to copper metal with elemental carbon or hydrogen. Al_2O_3 cannot be reduced to aluminum metal with either carbon or hydrogen. Arrange these three metals in order of increasing activity.

7-56 Powdered aluminum reacts with iron oxide to give aluminum oxide and enough heat to melt the iron metal produced in this reaction.

$$Fe_2O_3(s) + 2 Al(s) \longrightarrow Al_2O_3(s) + 2 Fe(l)$$

Use this fact to predict the relative strengths of iron and aluminum as reducing agents.

7-57 Use Table 7.4 to predict whether powdered aluminum should be able to reduce chromium(III) oxide to chromium metal.

$$Cr_2O_3(s) + 2 Al(s) \longrightarrow Al_2O_3(s) + 2 Cr(l)$$

7-58 Titanium metal was first produced commercially by the following reaction.

$$TiCl_4(l) + Mg(s) \longrightarrow Ti(s) + MgCl_2(s).$$

What does this reaction tell us about the relative strengths of titanium and magnesium metal as reducing agents?

7-59 Use Table 7.4 to determine which of the following reactions should occur.

(a) $Zn^{2+}(aq) + Cu(s) \rightarrow Cu^{2+}(aq) + Zn(s)$ (b) $2 Ag^+(aq) + Cu(s) \rightarrow 2 Ag(s) + Cu^{2+}(aq)$ (c) $Cu^{2+}(aq) + Fe(s) \rightarrow Fe^{2+}(aq) + Cu(s)$ (d) $2 Cu^{2+}(aq) + Ti(s) \rightarrow 2 Cu(s) + Ti^{4+}(aq)$

7-60 Use the following results to decide where the reduction of Cd^{2+} to Cd metal should be placed in Table 7.4.

$$Cd(s) + Sn^{2+}(aq) \longrightarrow Cd^{2+}(aq) + Sn(s)$$
$$Cd^{2+}(aq) + Fe(s) \longrightarrow Fe^{2+}(aq) + Cd(s)$$
$$Cd(s) + Zn^{2+} \not\longrightarrow$$
$$Cd^{2+}(aq) + Co(s) \not\longrightarrow$$

7-61 Explain why the strongest reducing agents in Table 7.4 are elements toward the bottom left-hand corner of the periodic table.

The Preparation of Metals

7-62 On January 21, 1941, Dow Chemical Company produced the first ingot of magnesium metal from sea water, which was the first time a commercial ingot of metal had been taken from the sea. Describe the process Dow used to transform the Mg^{2+} ions in sea water into magnesium metal.

7-63 Which of the following metals is not manufactured by electrolysis?

(a) Na (b) Mg (c) Al (d) Zn (e) Fe

7-64 Metal ores are roasted to convert sulfides into the corresponding oxides.

$$2 ZnS(s) + 3 O_2(g) \longrightarrow 2 ZnO(s) + 2 SO_2(g)$$

Is this an oxidation-reduction reaction? What is oxidized or reduced?

7-65 List the metals in Table 7.4 that are prepared by electrolysis. Why are these metals prepared by electrolysis instead of by reaction of the corresponding salts with metals that are stronger reducing agents?

7-66 Describe the reactions that take place at the cathode and the anode of an electrolytic cell when $MgCl_2$ is electrolyzed. Which reaction occurs at the cathode of an electrolysis cell: oxidation or reduction?

Lattice Energies and the Strength of the Ionic Bond

7-67 Define the term *lattice energy*.

7-68 The lattice energy of NaCl is a measure of the energy given off in which of the following reactions?

(a) $Na(s) + Cl_2(s) \rightarrow 2 NaCl(s)$ (b) $Na^+(s) + Cl^-(s) \rightarrow NaCl(s)$ (c) $Na(g) + Cl(g) \rightarrow NaCl(g)$ (d) $Na^+(g) + Cl^-(g) \rightarrow NaCl(g)$ (e) $Na^+(g) + Cl^-(g) \rightarrow NaCl(s)$

7-69 Which of the following salts has the largest lattice energy?

(a) LiF (b) LiCl (c) LiBr (d) LiI

7-70 Which of the following salts has the largest lattice energy?

(a) NaCl (b) NaI (c) KI (d) MgO (e) MgS

7-71 Explain the following trends in lattice energies.

$$MgF_2 > MgCl_2 > MgBr_2 > MgI_2$$
$$BeF_2 > MgF_2 > CaF_2 > SrF_2 > BaF_2$$

Why are NaCl and MgCl₂ the Products of Reactions between Na or Mg Metal and Cl₂?

7-72 Most beginning chemistry students believe that sodium reacts with chlorine because chlorine atoms like electrons more than sodium atoms do. Explain why they are wrong.

7-73 Explain why sodium reacts with chlorine.

7-74 Explain why sodium reacts with chlorine to form NaCl and not $NaCl_2$ or $NaCl_3$.

7-75 Explain why magnesium reacts with chlorine to form $MgCl_2$ and not MgCl or $MgCl_3$.

7-76 Describe why lithium reacts with nitrogen to give Li_3N and not a compound with another formula.

7-77 What are the most important factors in determining the formula of an ionic compound such as NaCl or $MgCl_2$?

THE COVALENT BOND

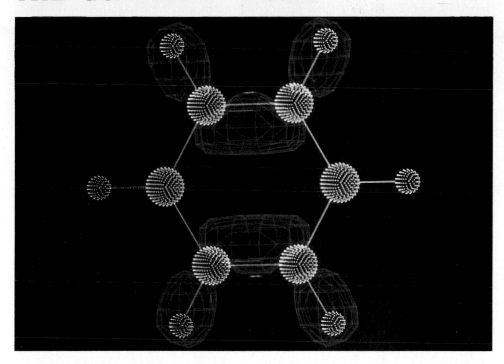

Ever since 1803, when Dalton proposed the atomic theory described in Section 2.3, chemists have tried to understand what force or forces hold atoms together in chemical compounds. Davy's experiments in 1807 on the effect of an electric current on crystals of potash (K_2CO_3) and caustic soda (NaOH), discussed in Section 7.16, were based on his belief that compounds were held together by the force of attraction between objects that carried opposite electric charges.

Thomson's discovery of the electron at the turn of the century inevitably led chemists to regard all compounds as the result of the transfer of electrons from one atom to another. In Chapter 7, we took advantage of this idea to develop a model for ionic compounds, which successfully predicts the products of reactions between main-group metals such as magnesium and nonmetals such as oxygen.

$$2\ Mg(s) + O_2(g) \longrightarrow 2\ MgO(s)$$

According to this model, the magnesium atoms lose two electrons to form Mg^{2+} ions, while the oxygen atoms pick up two electrons to form O^{2-} ions.

Mg: [Ne] $3s^2$ Mg^{2+}:   [Ne]

$\longrightarrow$

O: [He] $2s^2\ 2p^5$ O^{2-}:   [Ne]

This model assumes that the product of this reaction is an ionic compound, which contains Mg^{2+} and O^{2-} ions held together by the force of attraction between objects of opposite charge.

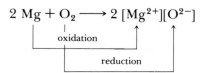

$$2\ Mg + O_2 \longrightarrow 2\ [Mg^{2+}][O^{2-}]$$

This model was so powerful that chemists applied it to reactions that form covalent compounds by inventing the concept of oxidation numbers. Even though there is no evidence to suggest that C^{4+} and O^{2-} ions are produced when carbon reacts with oxygen to form carbon dioxide, this reaction is classified as an oxidation-reduction reaction.

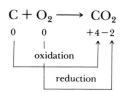

$$C + O_2 \longrightarrow CO_2$$

The carbon is assumed to be oxidized from the 0 to the $+4$ oxidation state, while oxygen is reduced from the 0 to the -2 oxidation state. In other words, we treat this reaction as if the product were an ionic compound, $[C^{4+}][O^{2-}]_2$, even though we know that these ions do not exist, because it provides the basis for one of the most important unifying themes of chemistry, the idea that electrons are transferred, to at least some extent, from one atom to another in oxidation-reduction reactions.

The goal of this chapter is to build a model that explains the difference between ionic compounds, such as NaCl, which contain positive and negative ions, and covalent compounds, such as CO_2, which do not contain ions. The basis for this model will be an explanation for the bond between atoms in covalent compounds.

8.1 VALENCE ELECTRONS

In 1902, while trying to find a way to explain some of the ideas behind the periodic table to a beginning chemistry class, G. N. Lewis discovered that the chemistry of the main-group elements shown in Figure 8.1 could be explained by a model in which electrons are arranged in atoms in the form of concentric cubes.

Lewis assumed that the number of electrons in the outermost cube on an atom was equal to the number of electrons lost when the atom formed positive ions. (Magnesium atoms, for example, have two electrons in the outermost cube because they lose two electrons to form Mg^{2+} ions.) Furthermore, he assumed that each neutral atom had one more electron in the outermost cube than the atom immediately preceding it in the periodic table. Finally, he assumed it took eight electrons —an octet— to complete a cube. Once an atom had an octet of electrons in its outermost cube, this cube became part of the core, or kernel, of electrons about which the next cube was built. Figure 8.2 shows a reproduction of a memo Lewis used in 1902 to summarize the key points in this model.

Lewis used this model to explain the formulas of simple ionic compounds by assuming that atoms gain electrons if the outermost cube is more than half full and lose electrons if the cube is less than half full, until the cube is either full or empty. Sodium, for example, loses the only electron in its outermost cube at the same time that chlorine gains the electron it needs to fill its outermost cube.

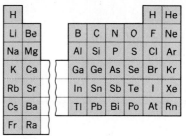

FIG. 8.1 The model developed by G. N. Lewis was first applied to the main-group elements, which are found on either side of the periodic table. Although the model can be extended for use in transition-metal compounds, this chapter will focus on compounds formed by combinations of main-group elements.

Na + Cl $\longrightarrow$ Na^+ + Cl^-

Lewis was the first to describe the pattern of compounds formed by the main-group elements in terms of the number of electrons in the outermost shell on an atom. The magnitude of this achievement is underlined by the fact that this model was generated only five years after Thomson's discovery of the electron and nine years before Rutherford proposed the model of the atom in which the mass was concentrated in a nucleus surrounded by a sea of electrons.

As our understanding of the structure of the atom developed, it soon became apparent why the magic number of electrons for main-group elements was eight. The outermost atomic orbitals for these elements are the *s* and *p* orbitals in a given shell, and it takes eight electrons to fill a set of these orbitals. A neutral nitrogen atom, for example, has five electrons in its outermost shell.

N: [He] $2s^2\ 3p^3$

It can therefore lose up to five electrons to form an N^{5+} ion or gain three electrons to fill this shell.

N^{5+}:   [He]

N^{3-}:   [He] $2s^2\ 2p^6$ = [Ar]

The electrons in the outermost shell eventually became known as the ***valence electrons.*** The origin of this name can be traced to the fact that the number of bonds an element can form is called its *valence* (from the Latin *valens,* "to be strong"). Because the number of electrons in the outermost cube in the Lewis

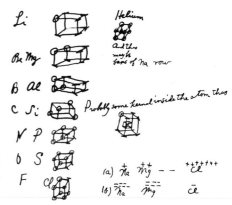

FIG. 8.2 This copy of notes G. N. Lewis made while working on his octet theory was first reproduced in his book, *Valence and the Structure of Atoms and Molecules,* published by the American Chemical Society in 1923.

model controls the number of bonds the atom can form, these outermost electrons are obviously the valence electrons.

The valence electrons are the electrons that can be gained or lost in a chemical reaction. Since filled d or f subshells are seldom disturbed in a chemical reaction, we can define valence electrons as follows.

> **Valence electrons: The electrons on an atom that were not present in the previous Group VIIIA element, ignoring filled d or f subshells.**

Gallium, for example, has the following electron configuration.

Ga: [Ar] $4s^2 \, 3d^{10} \, 4p^1$

The $4s$ and $4p$ electrons can be lost in a chemical reaction, but not the electrons in the filled $3d$ subshell. Gallium therefore has three valence electrons: $4s^2$ and $4p^1$.

So far, we have written electron configurations that emphasize the order of increasing energy of the atomic orbitals. It is also possible to write these configurations so that the valence electrons are emphasized. When this is done, the configuration of gallium is written as follows.

Ga: [Ar] $3d^{10} \, 4s^2 \, 4p^1$

Exercise 8.1

Determine the number of valence electrons in neutral atoms of each of the following elements.

(a) Si   (b) Mn   (c) Sb   (d) Pb

Solution

We start by writing the electron configuration for each element.

Si: [Ne] $3s^2 \, 3p^2$
Mn: [Ar] $4s^2 \, 3d^5$
Sb: [Kr] $4d^{10} \, 5s^2 \, 5p^3$
Pb: [Xe] $4f^{14} \, 5d^{10} \, 6s^2 \, 6p^2$

Ignoring filled d and f subshells, we conclude that neutral atoms of these elements contain the following numbers of valence electrons.

(a) Si = 4   (b) Mn = 7   (c) Sb = 5   (d) Pb = 4

8.2 THE COVALENT BOND

By 1916, Lewis had realized that there was another way atoms could combine to achieve an octet of valence electrons—they could share electrons. Two fluorine atoms, for example, could share a pair of electrons and thereby form a stable F_2 molecule in which each atom had an octet of valence electrons.

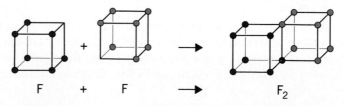

F   +   F   ⟶ F_2

A pair of oxygen atoms can form an O_2 molecule in which each atom had a total of eight valence electrons by sharing two pairs of electrons.

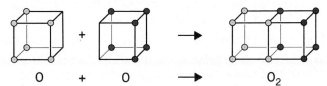

Lewis noted that when he applied this model to covalent compounds, the atoms always seemed to share *pairs* of electrons. He also found that all but a handful of compounds contained an even number of electrons, which suggested that electrons existed in pairs. He therefore replaced his cubic model of the atom, in which the eight electrons were oriented toward the corners of a cube, with a model based on pairs of electrons.

Lewis replaced the rigid structures of F_2 and O_2 given above by a simpler notation in which dots were used to indicate individual valence electrons. Each atom in this notation was surrounded by up to four pairs of dots, corresponding to the eight possible valence electrons. The **Lewis structures** of F_2 and O_2 were therefore written as follows.

$$:\!\overset{..}{\text{F}}\!:\!\overset{..}{\text{F}}\!: \qquad \overset{.}{\underset{.}{\text{O}}}::\overset{.}{\underset{.}{\text{O}}}\overset{.}{.}$$

This symbolism is still in use today. The only significant change is the use of lines to indicate covalent bonds formed by the sharing of a pair of electrons.

$$:\!\overset{..}{\text{F}}\!-\!\overset{..}{\text{F}}\!: \qquad \overset{..}{\underset{.}{\text{O}}}\!=\!\overset{..}{\underset{.}{\text{O}}}\overset{..}{.}$$

Lewis provided us with the first model that explained why covalent compounds form. The prefix *co-* is commonly used to indicate when things are joined or equal (for example, coalesce, coexist, cohere, cooperate, coordinate, and so on). It is therefore appropriate that the term **covalent bond** is used to describe the bonds in covalent compounds that result from the sharing of one or more pairs of electrons.

Covalent bond: A bond between two atoms formed by the sharing of a pair of electrons.

Exercise 8.2

Write the Lewis structure of the N_2 molecule.

Solution

Each nitrogen atom has five valence electrons, the $2s^2$ and $2p^3$ electrons.

N: [He] $2s^2\ 2p^3$

We can represent the valence electrons on a pair of isolated nitrogen atoms as follows.

$$\cdot\overset{..}{\underset{.}{\text{N}}}\cdot \qquad \cdot\overset{..}{\underset{.}{\text{N}}}\cdot$$

If each nitrogen atom contributes one electron to form a pair of electrons shared by the two atoms, each atom now has a total of six valence electrons.

$$\cdot\overset{..}{\underset{.}{\text{N}}}\!-\!\overset{..}{\underset{.}{\text{N}}}\cdot$$

If each nitrogen atom contributes one more electron, so that two pairs are shared, each atom has a total of seven valence electrons.

$$\cdot \ddot{N} = \ddot{N} \cdot$$

By contributing one more electron, so that three pairs are shared, each nitrogen atom can achieve an octet of valence electrons. The Lewis structure of the N_2 molecule is therefore written as follows.

$$:N \equiv N:$$

8.3 HOW DOES SHARING ELECTRONS BOND ATOMS?

To understand how sharing a pair of electrons can hold atoms together, let's look at the simplest of all covalent bonds—the bond that forms when two isolated hydrogen atoms come together to form an H_2 molecule.

$$H \cdot + \cdot H \longrightarrow H - H$$

Each isolated hydrogen atom consists of one proton and one electron held together by the electrostatic, or coulombic, force of attraction between oppositely charged particles. This force of attraction (F) is equal to the product of the charge on the electron (q_e) times the charge on the proton (q_p) divided by the square of the distance between these particles (r^2).

$$F = \frac{q_e \times q_p}{r^2}$$

There are two forces of attraction in a pair of isolated hydrogen atoms—the force of attraction between the proton in the nucleus of each atom and the electron that surrounds the atom.

When these isolated atoms are brought together, two new forces of attraction appear (Figure 8.3a), because of the attraction between the electron on one atom and the proton on the other. But two repulsive forces are also created (Figure 8.3b)—the two negatively charged electrons repel each other, as do the two positively charged protons.

At first glance, it might seem that the new repulsive forces would balance the new attractive forces. In that case, the H_2 molecule would be no more stable than the isolated hydrogen atoms. But there are ways in which the forces of repulsion can be minimized.

As noted in Section 5.17, electrons behave as if they were tops spinning on an axis. Just as there are two ways in which a top can spin—clockwise and counterclockwise—there are two possible states for the spin of an electron, $s = +\frac{1}{2}$ and $s = -\frac{1}{2}$. When electrons are paired so that they have opposite spins, they don't sense each other's presence as much as when they have the same spin. The force of repulsion between these electrons is therefore minimized.

The force of repulsion between the two protons is minimized if the pair of electrons occupies the space directly between the two nuclei. The distance between the electron on one atom and the nucleus of the other is now smaller than the distance between the two nuclei, as shown in Figure 8.4. The force of attraction between each electron and the nucleus of the other atom is therefore much larger than the force of repulsion between the two nuclei.

The net result is a system in which the attractive forces between the two atoms are larger than the repulsive forces. Thus, the H_2 molecule is more stable than a

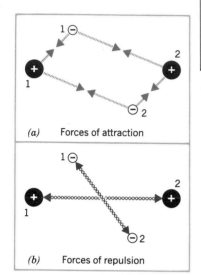

(a) Forces of attraction

(b) Forces of repulsion

FIG. 8.3 (a) There are two forces of attraction that act to bring a pair of hydrogen atoms together —the force of attraction between the electron on atom 1 and the proton on atom 2 and the force of attraction between the electron on atom 2 and the proton on atom 1. (b) There are two forces that tend to drive a pair of hydrogen atoms apart— the force of repulsion between the two protons and the force of repulsion between the two electrons.

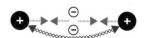

FIG. 8.4 By pairing the electrons, and restricting them to the region directly between the two nuclei, it is possible to construct a geometry in which the distance between the electrons and the protons is smaller than the distance between the two protons. As a result, the force of attraction between the electrons and protons is much larger than the force of repulsion between the two protons.

TABLE 8.1

Some Physical Properties of NaCl and Cl₂

	NaCl	*Cl₂*
phase at room temperature	solid	gas
density	2.165 g/cm³	0.003214 g/cm³
melting point	801°C	−100.98°C
boiling point	1413°C	−34.6°C
solubility in H₂O	35.8 g/100 mL	1.46 g/100 mL
ability of aqueous solution to conduct electricity	conducts	does not conduct

pair of isolated hydrogen atoms, and the hydrogen atoms are held together, or bonded, by the sharing of a pair of electrons.

8.4 SIMILARITIES AND DIFFERENCES BETWEEN IONIC AND COVALENT BONDS

There is a marked difference between the physical properties of NaCl and those of Cl₂, as shown in Table 8.1. These differences result from the difference between the *ionic bonds* in NaCl and the *covalent bonds* in Cl₂.

NaCl consists of positive (Na^+) and negative (Cl^-) ions held together by the force of attraction between these ions. To maximize this force of attraction, each positive ion is surrounded by as many negative ions as possible. The structure of NaCl in Figure 8.5a shows that each Na^+ ion is surrounded by six Cl^- ions, and vice versa. Removing an ion from this compound therefore involves breaking at least six bonds, which is not an easy task. Some of these bonds would have to be broken to melt NaCl, and they would all have to be broken to boil this compound, so ionic compounds tend to have high melting points and boiling points. They are therefore invariably solids at room temperature.

Cl₂ consists of molecules in which one atom is tightly bound to another, as shown in Figure 8.5b. The covalent bonds that hold these molecules together are at least as strong as an ionic bond. But we don't have to break these bonds to separate one Cl₂ molecule from another. It is therefore much easier to melt Cl₂ to form a liquid or boil it to form a gas, and Cl₂ is a gas at room temperature.

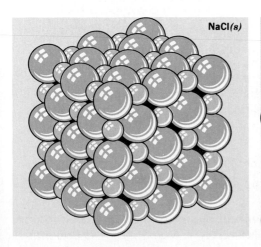

NaCl(*s*)

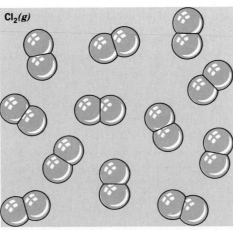

Cl₂(*g*)

FIG. 8.5 (*a*) Each Na^+ ion in NaCl is surrounded by six Cl^- ions, and vice versa. It therefore takes a great deal of energy to remove either Na^+ or Cl^- ions from this crystal. As a result, the melting point and boiling point of NaCl are relatively large. (*b*) Although there is a strong force of attraction between the chlorine atoms within a Cl₂ molecule, the force of attraction between the Cl₂ molecules is much smaller. As a result, it is relatively easy to separate Cl₂ molecules, and this element is a gas at room temperature.

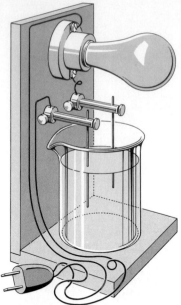

FIG. 8.6 Aqueous solutions of ionic compounds such as NaCl readily conduct electricity because these compounds dissociate when they dissolve in water to give positive and negative ions that can carry an electric current through the solution. Aqueous solutions of covalent substances, such as Cl_2 or $C_{12}H_{22}O_{11}$, don't conduct electricity, because these substances dissociate to give neutral molecules when they dissolve in water.

The difference between ionic and covalent bonds also explains why aqueous solutions of ionic compounds conduct electricity, while aqueous solutions of covalent compounds usually do not. When a salt dissolves in water, the ions are released into solution.

$$NaCl(s) \xrightarrow{H_2O} Na^+(aq) + Cl^-(aq)$$

These ions can flow through the solution, producing an electric current that completes the circuit, as shown in Figure 8.6. When a covalent compound dissolves in water, neutral molecules are released into the solution, and these molecules cannot carry an electric current.

$$C_{12}H_{22}O_{11}(s) \xrightarrow{H_2O} C_{12}H_{22}O_{11}(aq)$$

The difference between the physical properties of ionic and covalent compounds is so large that it is easy to fall into the trap of believing that ionic and covalent bonds are completely different phenomena. Lewis was the first to realize that this is not true. In the paper in which he first described bonds based on the sharing of electrons, Lewis argued that words such as *ionic* and *covalent* referred to the extremes at either end of a continuous spectrum of bonding.

To see how he came to this conclusion, let's compare what happens when covalent and ionic bonds form. When two chlorine atoms come together to form a covalent bond, each atom contributes one electron to form a pair of electrons shared equally by the two atoms.

$$:\ddot{C}l\cdot\ \cdot\ddot{C}l: \longrightarrow :\ddot{C}l{-}\ddot{C}l:$$

When a sodium atom combines with a chlorine atom to form an ionic bond, each atom still contributes one electron to form a pair of electrons. But this pair of electrons is not shared by the two atoms. The pair of electrons spends most, if not all, of its time on the chlorine atom.

$$Na\cdot\ \cdot\ddot{C}l: \longrightarrow Na^+ :\ddot{C}l:^-$$

The difference between ionic and covalent bonds is the extent to which the electrons are shared by the atoms that form the bond. When one of the atoms is much better at drawing electrons toward itself than the other, the bond is ionic. When the atoms are about equal in their ability to draw electrons in a bond toward themselves, the atoms share the pair of electrons equally, and the bond is covalent.

The following rule of thumb for predicting whether a compound is ionic or covalent was introduced in Section 2.7.

Metals combine with nonmetals to form ionic compounds, and nonmetals combine with other nonmetals to form covalent compounds.

This rule of thumb is useful, but it is also naive, for two reasons.

1. The only way to tell whether a compound should be classified as ionic or covalent is to measure the relative ability of the atoms to draw electrons in a bond toward themselves.
2. Any attempt to divide compounds into just two classes (ionic and covalent) is doomed to fail, because the bonding in many compounds falls between these two extremes.

The first limitation is the basis of the concept of electronegativity. The second serves as the basis for the concept of polarity.

8.5 ELECTRONEGATIVITY AND POLARITY

The relative ability of an atom to draw electrons in a bond toward itself is called the *electronegativity* of the atom. Atoms that have large electronegativities (such as F, Cl, and O) attract the electrons in a bond better than those (such as Na and Mg) that have small electronegativities.

The electronegativities of the main-group elements are given in Figure 8.7. (A more complete set of data can be found in Table A-7 in the Appendix.) There are several clear patterns in the electronegativity data for main-group elements.

1. Electronegativity becomes larger in a regular fashion from left to right across a row of the periodic table.
2. Electronegativity becomes smaller down a column of the periodic table.
3. There are no electronegativity values for elements in Group VIIIA, because these elements form very few compounds.

When the difference between the electronegativities of the elements in a compound is relatively large, the compound is best classified as ionic. NaCl, LiF, and $SrBr_2$ are good examples of ionic compounds. In each, the electronegativity of the nonmetal is at least 2 units larger than that of the metal.

NaCl:		*LiF:*		*SrBr₂:*	
Cl	EN = 3.16	F	EN = 3.98	Br	EN = 2.96
Na	EN = 0.93	Li	EN = 0.98	Sr	EN = 0.95
	ΔEN = 2.23		ΔEN = 3.00		ΔEN = 2.01

We can therefore assume a net transfer of electrons from the metal to the nonmetal to form positive and negative ions and write the Lewis structures of these compounds as follows.

$$[\text{Na}^+][:\ddot{\text{Cl}}:^-] \qquad [\text{Li}^+][:\ddot{\text{F}}:^-] \qquad [\text{Sr}^{2+}][:\ddot{\text{Br}}:^-]_2$$

These compounds all have high melting points and boiling points, as might be expected for ionic compounds,

	NaCl	*LiF*	*SrBr₂*
MP	801°	846°C	657°C
BP	1413°	1717°C	2146°C

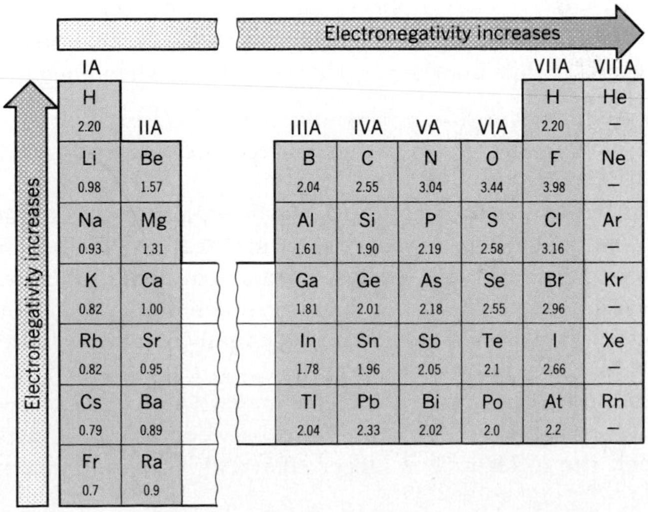

FIG. 8.7 The electronegativities of the main-group elements. Note that with only rare exception, electronegativity increases regularly from left to right across the periodic table and decreases regularly from top to bottom of most columns.

They also dissolve in water to give aqueous solutions that conduct electricity, as would be expected.

$$LiF(s) \xrightarrow{H_2O} Li^+(aq) + F^-(aq)$$

When the electronegativities of the elements in a compound are about the same, the atoms share electrons, and the substance is best classified as covalent. Examples of covalent compounds include the methane (CH_4) in natural gas, the nitrogen oxide (NO) and nitrogen dioxide (NO_2) pollutants formed when gasoline is burned in an internal combustion engine, and the sulfur dioxide (SO_2) formed when coal rich in sulfur is burned.

CH₄:	*NO/NO₂:*	*SO₂:*
C EN = 2.55	O EN = 3.44	O EN = 3.44
H EN = 2.20	N EN = 3.04	S EN = 2.58
ΔEN = 0.35	ΔEN = 0.40	ΔEN = 0.86

These compounds have relatively low melting points and boiling points, as might be expected for covalent compounds, and they are all gases at room temperature.

	CH₄	*NO*	*SO₂*
MP	−182.5°	−163.6°C	−75.5°C
BP	−161.5°	−151.8°C	−10°C

Inevitably, there must be compounds that fall between these extremes—compounds in which the difference between the electronegativities of the elements is large enough to be significant but not large enough to justify classifying the compound as ionic. Examples of such compounds include the following.

AlBr₃:	*H₂O:*	*MgI₂:*
Br EN = 2.96	O EN = 3.44	I EN = 2.66
Al EN = 1.61	H EN = 2.20	Mg EN = 1.31
ΔEN = 1.35	ΔEN = 1.24	ΔEN = 1.35

These compounds are neither purely ionic nor purely covalent. They do not contain positive and negative ions, as indicated by the Lewis structure on the left below. But the electrons are not shared equally, as indicated by the Lewis structure on the right.

$$[H^+]_2[\,:\!\ddot{O}\!:^{2-}] \qquad H—\ddot{O}—H$$

These compounds are best described as polar. One end, or pole, of the molecule has a partial positive charge (δ^+), while another end, or pole, has a partial negative charge (δ^-).

$$\overset{\delta+}{H}—\overset{\delta-}{\underset{..}{\overset{..}{O}}}—\overset{\delta+}{H}$$

The *polarity* of a compound describes the magnitude of the separation of charge. When the difference between the electronegativities of the two atoms in a bond is small, the polarity is so small the bond can be described as covalent. When the electronegativity difference is large, the electrons spend so much time on the more electronegative atom we can assume that an electron has actually been transferred from one atom to another to form an ionic compound.

As a rule, when the difference between the electronegativities (EN) of two elements is less than 1.2, we can assume that the bond between atoms of these elements is covalent. When the difference is larger than 1.8, the bond can be assumed to be ionic. Compounds, such as water, for which the electronegativity difference is between about 1.2 and 1.8 are best described as polar, or polar covalent.

Covalent: $\Delta EN < 1.2$

Polar: $1.2 < \Delta EN < 1.8$

Ionic: $\Delta EN > 1.8$

Exercise 8.3

Use electronegativities to classify the bonds in each of the following compounds as covalent, ionic, or polar.

(a) Sodium cyanide (NaCN)

(b) Tetraphosphorus decasulfide (P_4S_{10})

(c) Carbon monoxide (CO)

(d) Silicon tetrachloride ($SiCl_4$)

Solution

(a) When there are more than two elements in a compound, it is useful to look at the difference between the least and most electronegative atoms. The difference between the electronegativities of sodium and nitrogen is 2.11 electronegativity units. We can therefore safely assume that this compound is ionic. In this case, it contains the Na^+ and CN^- ions.

(b) The difference between the electronegativities of phosphorus and sulfur is only 0.39, and the bonds in P_4S_{10} are best described as covalent.

(c) The electronegativity difference in this compound is slightly larger ($\Delta EN = 0.89$), but the bonds are still covalent.

(d) The electronegativity difference in this compound is significant ($\Delta EN = 1.26$) but too small to allow us to assume that the bonds are ionic. $SiCl_4$ is best described as polar, or polar covalent.

8.6 LIMITATIONS OF THE ELECTRONEGATIVITY CONCEPT

Electronegativity is a powerful concept that summarizes the tendency of an element to gain, lose, or share electrons when it combines with another element. But there are limits to the success with which it can be applied. BF_3 ($\Delta EN = 1.94$) and SiF_4 ($\Delta EN = 2.08$), for example, have electronegativity differences that lead us to expect these compounds to behave as if they were ionic. Both compounds are covalent, however. Both compounds are gases at room temperature, and their boiling points are $-99.9°C$ and $-86°C$, respectively.

The source of this problem is that, by convention, each element is assigned only one electronegativity value, which is used for all of its compounds. But fluorine seems to be less electronegative when it bonds to semimetals (such as B or Si) or nonmetals (such as C) than when it bonds to metals (such as Na, Mg, Hg, or Pb).

This problem surfaces again when we look at compounds such as $TiCl_2$ and $TiCl_4$ or MnO and Mn_2O_7, in which one of the elements is present in more than one oxidation state. $TiCl_2$ and MnO have many of the properties of ionic com-

TiCl$_2$ is an ionic solid that melts at very high temperatures (MP = 1035°C). TiCl$_4$ is a covalent compound with a very low melting point (MP = −24.1°C) that is a liquid at room temperature. The difference between these compounds can be explained by assuming that the titanium atom becomes more electronegative as the oxidation state of this atom increases.

pounds. Both compounds are solids at room temperature, for example, and they have very high melting points, as expected for ionic compounds.

TiCl$_2$:   *MnO:*
MP = 1035°C   MP = 1785°C

TiCl$_4$ and Mn$_2$O$_7$, on the other hand, are both liquids at room temperature, with melting points below 0°C and relatively low boiling points, as might be expected for covalent compounds.

TiCl$_4$:   *Mn$_2$O$_7$:*
MP = −24.1°C   MP = −20°C
BP = 136.4°C   BP = 25°C

The principal difference between TiCl$_2$ and TiCl$_4$, or between MnO and Mn$_2$O$_7$, is the oxidation number of the metal. The oxidation number of titanium is +2 in TiCl$_2$ and +4 in TiCl$_4$, while the oxidation number of manganese is +2 in MnO and +7 in Mn$_2$O$_7$. As the oxidation number of an atom becomes larger, its ability to draw electrons in a bond toward itself also becomes larger. In other words, titanium atoms with a +4 oxidation number and manganese atoms with a +7 oxidation number are more electronegative than titanium and manganese atoms with an oxidation number of +2.

As the metal becomes more electronegative, the difference between the electronegativities of the metal and the nonmetal with which it combines becomes smaller. The bonds in the compounds these elements form therefore become less ionic (or more covalent), which gives rise to the difference between the physical properties of the compounds discussed in this section.

8.7 WRITING LEWIS STRUCTURES BY TRIAL AND ERROR

There are two approaches to generating the Lewis structure of a compound. One approach is based on trial and error. We start by writing symbols that contain the correct number of valence electrons for the atoms in the molecule. We then manipulate these symbols, combining electrons to form covalent bonds, until we come up with a Lewis structure in which all of the elements (with the exception of the hydrogen atoms) have an octet of valence electrons.

Exercise 8.4

Explain why hydrogen atoms are an exception to the general rule that main-group elements gain, lose, or share electrons until they have eight electrons in their valence shell.

Solution

Most of the main-group elements gain, lose, or share electrons until they have an octet of valence electrons because it takes eight electrons to fill the valence shell of orbitals. A neutral oxygen atom, for example, has six valence electrons.

O: [He] $2s^2 2p^4$

One way this atom can attain a filled-shell electron configuration is to pick up two electrons to form an O^{2-} ion, which has an octet of valence electrons.

O^{2-}:   [He] $2s^2 2p^6$

Alternatively, it can share pairs of electrons until it has an octet of valence electrons.

$$\overset{..}{:}O=O\overset{..}{\underset{..}{:}}$$

Hydrogen is an exception to this rule because the valence shell of a hydrogen atom only contains the $1s$ orbital.

H: $1s^1$

It therefore takes only two electrons to fill the valence shell of a hydrogen atom.

H^-: $1s^2$

Let's apply this approach to generating the Lewis structure of carbon dioxide, CO_2. We start by determining the number of valence electrons on each atom from the electron configurations of the elements. Carbon has four valence electrons and oxygen has six.

C: [He] $2s^2\ 2p^2$
O: [He] $2s^2\ 2p^4$

A faster route to this conclusion is the following general rule.

The number of valence electrons on a neutral atom of a main-group element is equal to the group number.

Thus, carbon, in Group IVA, has four valence electrons; and oxygen, in Group VIIA, has six valence electrons. Regardless of how we obtain this information, we can symbolize it as follows.

$$:\overset{..}{O}\cdot \qquad \cdot\overset{.}{C}\cdot \qquad \cdot\overset{..}{O}:$$

We now combine one electron from each atom to form covalent bonds between the atoms.

$$:\overset{..}{O}\cdot\,\cdot\overset{.}{C}\cdot\,\cdot\overset{..}{O}: \longrightarrow :\overset{..}{O}-\overset{.}{C}-\overset{..}{O}:$$

Each of the oxygen atoms now has a total of seven valence electrons, two that it shares with carbon and five more that are not involved in bonding. The carbon atom has a total of six valence electrons, two pairs of electrons that it shares with the neighboring oxygen atoms and two more that are not involved in bonding.

Since none of the atoms has an octet of valence electrons, we might combine another electron on each atom to form two more bonds.

$$:\overset{..}{O}-\overset{.}{C}-\overset{..}{O}: \longrightarrow :\overset{..}{O}=C=O\overset{..}{:}$$

We now have a Lewis structure in which each atom has an octet of valence-shell electrons. The oxygen atoms share four electrons with the carbon atom in the center of the molecule and have another four electrons that are not involved in bonding.

$$\overset{..}{:}O=C=O\overset{..}{\underset{..}{:}}$$

The carbon atoms share four electrons with each oxygen atom, for a total of eight valence electrons.

$$\ddot{\text{O}}=\text{C}=\ddot{\text{O}}$$

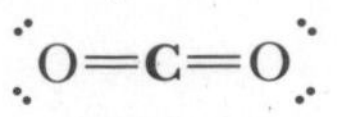

8.8 A STEP-BY-STEP APPROACH TO WRITING LEWIS STRUCTURES

Although the trial-and-error method for writing Lewis structures works, it is usually very time consuming. (Try, for example, to work out the Lewis structure of oxalic acid, $H_2C_2O_4$.) Thus, for all but the simplest molecules, the following step-by-step process is preferable.

STEP 1: Determine the total number of valence electrons.

The first step involves counting the total number of valence electrons in the molecule or ion. For a neutral molecule this is nothing more than the sum of the valence electrons on each atom.

Exercise 8.5

Determine the number of valence electrons in each of the following neutral compounds.

(a) Ammonia, NH_3 (b) Sulfur dioxide, SO_2 (c) Xenon tetrafluoride, XeF_4

Solution

(a) The number of valence electrons on a neutral atom is equal to the group number of the element. Ammonia therefore has a total of eight valence electrons, because it contains one nitrogen atom (Group VA) and three hydrogen atoms (Group IA).

NH_3: $5 + 3(1) = 8$

(b) Sulfur dioxide has 18 valence electrons, because sulfur and oxygen are both members of Group VIA.

SO_2: $6 + 2(6) = 18$

(c) Xenon tetrafluoride has 36 valence electrons, because xenon is in Group VIIIA and fluorine is in Group VIIA.

XeF_4: $8 + 4(7) = 36$

If the molecule carries an electric charge, we add one electron for each negative charge or subtract an electron for each positive charge.

Exercise 8.6

Determine the number of valence electrons in each of the following ions.

(a) Chlorate, ClO_3^- (b) Phosphate, PO_4^{3-} (c) Hydronium, H_3O^+

Solution

(a) A neutral ClO_3 molecule should have 25 electrons.

ClO_3: $7 + 3(6) = 25$

A negatively charged ClO_3^- ion must have one more electron, for a total of 26.

ClO_3^-: $7 + 3(6) + 1 = 26$

(b) A neutral PO_4 molecule should have 29 valence electrons.

PO_4: $5 + 4(6) = 29$

A negatively charged PO_4^{3-} ion has three more electrons, for a total of 32.

PO_4^{3-}: $5 + 4(6) + 3 = 32$

(c) The hydronium ion, H_3O^+, carries a charge of $+1$. That means this ion must have one less electron than a neutral H_3O molecule. The H_3O^+ ion therefore has a total of eight valence electrons.

H_3O^+: $3(1) + 6 - 1 = 8$

STEP 2: Write the skeleton structure of the molecule.

This step involves deciding which atoms in the molecule are connected by covalent bonds. That is not an easy task. Sometimes the formula of the compound provides a hint as to the skeleton structure. The formula for ethyl alcohol, for example, is often written as CH_3CH_2OH to indicate the following skeleton structure.

$$
\begin{array}{ccc}
\text{H} & \text{H} & \\
| & | & \\
\text{H}-\text{C}-\text{C}-\text{O}-\text{H} \\
| & | & \\
\text{H} & \text{H} &
\end{array}
$$

At one time, the formula of the thiocyanate ion was written as CNS^-. It is now written as SCN^- to indicate the following skeleton structure.

$$S-C-N^-$$

As you learn more chemistry, writing the skeleton structure for a molecule or ion will become easier. Once you learn, for example, that the prefix *thio-* is used to describe compounds in which an oxygen atom has been replaced by a sulfur atom, the relationship between the skeleton structures of the sulfate, SO_4^{2-}, and thiosulfate, $S_2O_3^{2-}$, ions will become clearer.

$$
\begin{array}{cc}
\text{O}\quad^{2-} & \text{O}\quad^{2-} \\
| & | \\
\text{O}-\text{S}-\text{O} & \text{S}-\text{S}-\text{O} \\
| & | \\
\text{O} & \text{O}
\end{array}
$$

The following generalizations may help you construct the skeleton structure of a compound.

1. Hydrogen usually forms bonds to only one atom in a molecule.
2. The atom in the center of the molecule is often the one that has the smallest electronegativity.
3. When a compound contains more of one element than the other, the element whose atoms are outnumbered is often in the center of the molecule.

STEP 3: Subtract two electrons from the total number of valence electrons for each bond in the skeleton structure.

Valence electrons that are used to form covalent bonds in the skeleton structure of a molecule are called *bonding electrons.* Any other electrons in the valence shell must be *nonbonding electrons.* The number of bonding electrons in most molecules can be calculated from the number of bonds that are needed to form the skeleton structure. Since it takes two electrons to form a covalent bond, we can usually calculate the number of nonbonding electrons in the molecule by subtracting two electrons from the total number of valence electrons for each bond in the skeleton structure.

Exercise 8.7

Calculate the number of bonding and nonbonding electrons in the phosphate (PO_4^{3-}) ion.

Solution

There are 32 valence electrons in this ion.

$$PO_4^{3-}: \qquad 5 + 4(6) + 3 = 32$$

The skeleton structure of the ion contains four covalent bonds.

Since each of these bonds is formed by the sharing of a pair of electrons, a total of 8 valence electrons are used as bonding electrons. The remainder ($32 - 8 = 24$) must be nonbonding electrons. This ion contains 8 bonding and 24 nonbonding electrons.

STEP 4: Try to satisfy the octets of all atoms in the molecule by distributing the remaining valence electrons as nonbonding electrons.

Once we have determined how many valence electrons are needed to form the covalent bonds that hold the molecule together, we can assume that the rest of the valence electrons are used as nonbonding electrons to satisfy the octets of the various atoms in the molecule.

The phosphate ion, for example, contains 24 nonbonding electrons. The phosphorus atom already has an octet of valence electrons—the 8 electrons in the four covalent bonds. The nonbonding electrons are therefore distributed among the four oxygen atoms until each has a total of 8 valence electrons.

Following these four steps should result in a satisfactory Lewis structure for most covalent molecules.

Exercise 8.8

Write the Lewis structure for the chlorate (ClO_3^-) ion.

Solution

STEP 1: There are 26 valence electrons in this ion.

ClO_3^-: $7 + 3(6) + 1 = 26$

STEP 2: The following is the most reasonable skeleton structure for the ion.

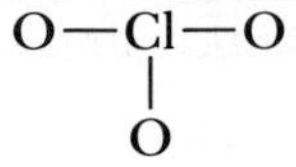

$$O—Cl—O$$
$$|$$
$$O$$

STEP 3: There are three covalent bonds in the skeleton structure. Thus, a total of 6 electrons must be used as bonding electrons. This leaves 20 nonbonding electrons in the valence shell.

26 valence electrons

— 6 bonding electrons
_____

20 nonbonding electrons

STEP 4: Each oxygen atom already has 2 electrons—the electrons in the Cl—O covalent bond. Each oxygen atom therefore needs 6 nonbonding electrons to satisfy its octet. Thus, it takes a total of 18 nonbonding electrons to satisfy the three oxygen atoms.

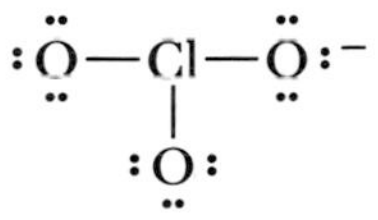

$$:\ddot{O}—Cl—\ddot{O}:^-$$
$$|$$
$$:\ddot{O}:$$

This leaves one pair of nonbonding electrons, which can be used to fill the octet of the central atom.

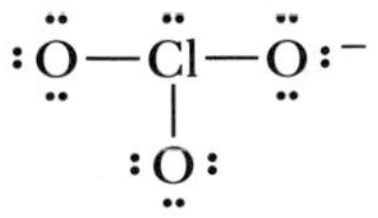

$$:\ddot{O}—\ddot{Cl}—\ddot{O}:^-$$
$$|$$
$$:\ddot{O}:$$

The valence electrons have all been used, and all of the atoms have an octet of valence electrons. This is a satisfactory Lewis structure for the ClO_3^- ion.

Occasionally, we encounter a molecule that seems to have too few electrons. When this happens, we proceed to Step 5.

STEP 5: When it seems there aren't enough valence electrons, assume that two of the atoms are linked by a double bond, or even a triple bond, if these atoms are C, N, O, P, or S atoms.

Exercise 8.9

Write the Lewis structure for the formaldehyde (H_2CO) molecule.

Solution

STEP 1: There are 12 valence electrons in this molecule.

H_2CO: $2(1) + 4 + 6 = 12$

STEP 2: The formula hints at the following skeleton structure.

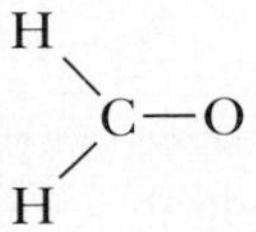

$$H$$
$$\backslash$$
$$C—O$$
$$/$$
$$H$$

STEP 3: There are three covalent bonds in this skeleton structure, and thus 6 of the 12 valence electrons must be used as bonding electrons. This leaves 6 nonbonding electrons.

12 valence electrons

− 6 bonding electrons

6 nonbonding electrons

STEP 4: It is impossible to satisfy the octets of all of the atoms in this molecule with only six nonbonding electrons.

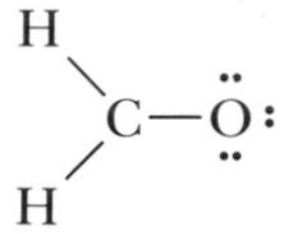

$$\text{H}\diagdown \atop \text{H}\diagup \text{C}-\overset{..}{\underset{..}{\text{O}}}:$$

When the bonding electrons are used to satisfy the octet of the oxygen atom, the carbon atom has a total of only six valence electrons. We are therefore forced to apply Step 5.

STEP 5: Let's assume that the C and O atoms are linked by two covalent bonds—a double bond, indicated by a double line.

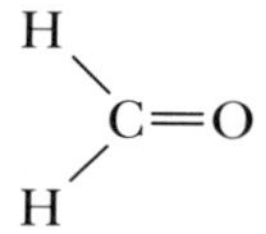

$$\text{H}\diagdown \atop \text{H}\diagup \text{C}=\text{O}$$

There are now four bonds in the skeleton structure and therefore eight bonding electrons. That leaves only four nonbonding electrons. But four nonbonding electrons are enough to satisfy the octets of the carbon and oxygen atoms.

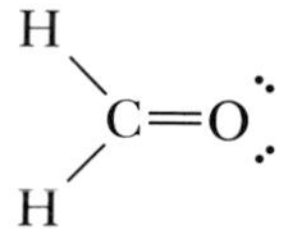

$$\text{H}\diagdown \atop \text{H}\diagup \text{C}=\overset{..}{\underset{..}{\text{O}}}$$

Every once in a while, we encounter a molecule for which it is impossible to write a satisfactory Lewis structure.

Exercise 8.10

Write the Lewis structure for boron trifluoride (BF_3).

Solution

STEP 1: There are 24 valence electrons in this molecule.

BF_3: $3 + 3(7) = 24$

STEP 2: The following is a reasonable skeleton structure for the molecule.

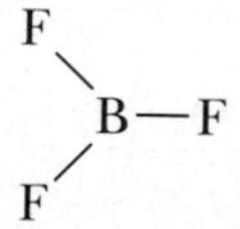

$$\text{F}\diagdown \atop \text{F}\diagup \text{B}-\text{F}$$

STEP 3: There are three covalent bonds and therefore 6 bonding electrons in the skeleton structure. This leaves 18 nonbonding electrons.

$$
\begin{array}{r}
24 \text{ valence electrons} \\
- \underline{6 \text{ bonding electrons}} \\
18 \text{ nonbonding electrons}
\end{array}
$$

STEP 4: Each fluorine atom needs six nonbonding electrons to satisfy its octet. Thus, all of the nonbonding electrons are consumed by the three fluorine atoms. As a result, we run out of electrons while the boron atom has only six valence electrons.

For reasons that will be discussed in the next chapter, the elements that form strong double or triple bonds are C, N, O, P, and S. Since neither B nor F falls in this category, we can't invoke Step 5. We have to stop with what appears to be an unsatisfactory Lewis structure.

It is also possible to encounter a molecule that seems to have too many valence electrons. When that happens, we go on to Step 6.

STEP 6: When it seems that there are too many valence electrons, expand the valence shell of the central atom.

Exercise 8.11

Write the Lewis structure for sulfur tetrafluoride (SF_4).

Solution

STEP 1: There are 34 valence electrons in this molecule.

SF_4: $6 + 4(7) = 34$

STEP 2: The best guess for the skeleton structure is the following.

STEP 3: There are four covalent bonds and therefore 8 nonbonding electrons in the skeleton structure of this molecule. This leaves 26 nonbonding electrons.

$$
\begin{array}{r}
34 \text{ valence electrons} \\
- \underline{8 \text{ bonding electrons}} \\
26 \text{ nonbonding electrons}
\end{array}
$$

STEP 4: Each fluorine atom needs 6 more electrons to satisfy its octet. Since there are four of these atoms, we need 24 nonbonding electrons for this purpose.

But there are 26 nonbonding electrons in this molecule. We have already satisfied the octets for all five atoms, and we still have one more pair of valence electrons. Whenever we have too many electrons, we invoke Step 6.

STEP 6: We place the last pair of electrons on the central atom, thus assuming that the valence shell of this atom can be expanded to hold a total of 10 electrons.

How does the sulfur atom in SF_4 hold 10 electrons in its valence shell? The electron configuration for a neutral sulfur atom seems to suggest that only 8 electrons will fit in the valence shell of this atom, because it takes 8 electrons to fill the $3s$ and $3p$ orbitals.

But let's look at the selection rules for atomic orbitals outlined in Section 5.16. According to these rules, when $n = 3$, there are three possible values of the angular quantum number — $l = 0$, $l = 1$, and $l = 2$. The $n = 3$ shell of orbitals therefore contains $3s$, $3p$, and $3d$ orbitals. The $3d$ orbitals on a neutral sulfur atom are all empty.

S: [Ne] $3s^2 \, 3p^4 \, \mathbf{3d^0}$

One of these orbitals can therefore be used to hold the extra pair of electrons on the sulfur atom in SF_4.

It is worth noting that elements in the first and second rows of the periodic table do not have d orbitals in their valence shells. Nitrogen and oxygen, for example, can't expand their valence shells, since there is no equivalent set of $2d$ orbitals on these elements.

Exercise 8.12

Write the Lewis structure for xenon tetrafluoride (XeF_4).

Solution

STEP 1: There are 36 valence electrons in this molecule.

XeF_4: $8 + 4(7) = 36$

STEP 2: The following is the most reasonable skeleton structure for the molecule.

F   F
\   /
Xe
/   \
F   F

STEP 3: There are four covalent bonds, or 8 bonding electrons, in this skeleton structure. This leaves 28 nonbonding electrons.

$$36 \text{ valence electrons}$$
$$- \; 8 \text{ bonding electrons}$$
$$\overline{28 \text{ nonbonding electrons}}$$

STEP 4: Each fluorine atom needs 6 nonbonding electrons, for a total of 24.

This leaves 4 nonbonding electrons unaccounted for. Since the octet of each atom appears to be satisfied, and we have electrons left over, we invoke Step 6.

STEP 6: The extra pairs of nonbonding electrons are placed on the central xenon atom to produce a Lewis structure in which the valence shell of the central atom contains 12 electrons.

8.9 RESONANCE HYBRIDS

Some molecules, such as sulfur dioxide, can be represented by more than one Lewis structure. Exercise 8.13 shows how to write one of the possible Lewis structures for this molecule.

Exercise 8.13

Write the Lewis structure for sulfur dioxide (SO_2).

Solution

STEP 1: There are 18 valence electrons in the SO_2 molecule.

$$SO_2: \qquad 6 + 2(6) = 18$$

STEP 2: The skeleton structure for this molecule is written as follows.

$$O{-}S{-}O$$

STEP 3: There are two covalent bonds in this skeleton structure, which leaves 14 nonbonding electrons.

$$18 \text{ valence electrons}$$
$$- \; 4 \text{ bonding electrons}$$
$$\overline{14 \text{ nonbonding electrons}}$$

STEP 4: Each oxygen atom needs six nonbonding electrons to fill its octet.

$$:\overset{..}{\underset{..}{O}} - S - \overset{..}{\underset{..}{O}}:$$

This leaves only one pair of nonbonding electrons for the sulfur atom.

$$:\overset{..}{\underset{..}{O}} - \overset{..}{S} - \overset{..}{\underset{..}{O}}:$$

We seem to run out of electrons before all of the atoms have an octet of valence electrons. But O and S are both members of the group of elements that form strong double bonds, so we can invoke Step 5.

STEP 5: Let's assume that the central atom shares two pairs of electrons with the O atom on the right.

$$O - S = O$$

There are now three covalent bonds in the skeleton structure, which leaves 12 nonbonding electrons. But that is enough to write a satisfactory Lewis structure for this molecule.

$$:\overset{..}{\underset{..}{O}} - \overset{..}{S} = O\overset{.}{\underset{.}{}}$$

In generating the Lewis structure in Exercise 8.13, we assumed that the sulfur atom formed a double bond with the oxygen atom on the right. What would happen if we assumed that the sulfur formed a double bond with the oxygen atom on the left? Obviously, we would get a slightly different, but equally satisfactory, Lewis structure for the molecule.

$$:\overset{..}{\underset{..}{O}} - \overset{..}{S} = O\overset{.}{\underset{.}{}} \qquad \overset{.}{\underset{.}{}}O = \overset{..}{S} - \overset{..}{\underset{..}{O}}:$$

Which of these structures is correct?

Interestingly enough, neither of these structures is correct. When we can write more than one Lewis structure for a molecule, such as SO_2, the molecule is said to be an average, or **resonance hybrid,** of these Lewis structures.

In music, the notes in a chord are often said to *resonate*—they mix to give something that is more than the sum of its parts. In a similar sense, the two Lewis structures for the SO_2 molecule are in resonance. They mix to give a hybrid that is more than the sum of its components.

The alternative Lewis structures for SO_2 suggest that one of the sulfur-oxygen bonds is stronger than the other. Experimentally, however, both bonds in SO_2 are found to be identical. Neither bond has the properties of either an S—O single bond or an S=O double bond. Both bonds have the properties that would be expected for a **bond order** of 1.5.

Bond order: The number of bonds between a pair of atoms.

Resonance hybridization is indicated by a double-headed arrow connecting the alternative Lewis structures for the molecule.

$$:\overset{..}{\underset{..}{O}} - \overset{..}{S} = O\overset{.}{\underset{.}{}} \longleftrightarrow \overset{.}{\underset{.}{}}O = \overset{..}{S} - \overset{..}{\underset{..}{O}}:$$

The relationship of the SO_2 molecule to its Lewis structures can be illustrated by an analogy. Suppose a knight of the Round Table returned to Camelot to describe a wondrous beast he had encountered during his search for the Holy Grail. He might have suggested that the animal looked something like a unicorn, because it had a large horn in the center of its face. But it also looked something like a dragon, because it was huge, heavy, ugly, and thick-skinned.

The beast was, in fact, a rhinoceros. The animal was real, but it was described as a hybrid of two mythical animals. The SO_2 molecule is also real, but we have to use two somewhat mythical Lewis structures to describe its properties.

The fact that we have to invoke resonance hybridization to explain the structure of molecules such as SO_2 suggests that there is a flaw in this model of covalent molecules. The molecule has properties slightly different from those predicted from its Lewis structures.

Fortunately, a more sophisticated model exists. This model, called the molecular orbital theory, can explain the experimental observations for molecules such as SO_2 that don't fit the Lewis theory. Unfortunately, the molecular orbital theory is more difficult to understand and use. We therefore often rely on the flawed but useful Lewis theory in place of the more accurate but less intuitively appealing molecular orbital approach.

Exercise 8.14

Write three alternative Lewis structures for the thiocyanate (SCN^-) ion.

Solution

STEP 1: There are 16 valence electrons in the SCN^- ion.

SCN^-: $6 + 4 + 5 + 1 = 16$

STEP 2: The most reasonable skeleton structure for this ion is written as follows.

$$S—C—N^-$$

STEP 3: There are two covalent bonds in this skeleton structure, which leaves 12 nonbonding electrons.

$$\begin{array}{r} 16 \text{ valence electrons} \\ -\ \ 4 \text{ bonding electrons} \\ \hline 12 \text{ nonbonding electrons} \end{array}$$

STEP 4: It takes all 12 nonbonding electrons to satisfy the octets of the sulfur and nitrogen atoms on either end of the molecule, which leaves no nonbonding electrons for the carbon atom.

$$:\!\overset{\cdot\cdot}{\underset{\cdot\cdot}{S}}\!—C—\overset{\cdot\cdot}{\underset{\cdot\cdot}{N}}\!:^-$$

We seem to run out of electrons before all the atoms have an octet of valence electrons. Since S, C, and N are members of the group of elements that form double bonds, we can invoke Step 5.

STEP 5: Let's assume, for the sake of argument, that the S and C atoms form a double bond.

$$S\!=\!C—N^-$$

There are now three covalent bonds, which leaves 10 nonbonding electrons. Unfortunately, that is not enough.

$$\overset{\cdot\cdot}{\underset{\cdot\cdot}{S}}\!=\!C—\overset{\cdot\cdot}{\underset{\cdot\cdot}{N}}\!:^-$$

So let's add another bond between the S and C atoms to form a triple bond.

$$S{\equiv}C{-}N^-$$

With four covalent bonds, we now have 8 nonbonding electrons. But these are enough to fill the octets of the three atoms.

$$:S{\equiv}C{-}\ddot{N}:^-$$

This process gives us one satisfactory Lewis structure for the SCN⁻ ion. We could get another by forming a triple bond between the C and N atoms and a third by assuming one S=C double bond and one C=N double bond. The following Lewis structures can therefore be written for the SCN⁻ ion.

$$:S{\equiv}C{-}\ddot{N}:^- \longleftrightarrow \ddot{S}{=}C{=}\ddot{N}^- \longleftrightarrow :\ddot{S}{-}C{\equiv}N:^-$$

8.10 FORMAL CHARGE

It is often useful to calculate the *formal charge* on each atom in a Lewis structure. The first step in calculating formal charge involves dividing the electrons in each covalent bond between the atoms that form the bond. The number of valence electrons formally assigned to each atom is then compared with the number of valence electrons on a neutral atom of the element. If the atom has more valence electrons than a neutral atom, it is assumed to carry a formal negative charge. If it has less valence electrons, it is assigned a formal positive charge. This procedure is demonstrated in Exercise 8.15.

Exercise 8.15

Calculate the formal charge on each atom in the following Lewis structure.

$$:S{\equiv}C{-}\ddot{N}:^-$$

Solution

We start by arbitrarily dividing pairs of bonding electrons so that each atom in a bond is formally assigned one of these electrons.

$$:S{\equiv}C{-}\ddot{N}:^- \longrightarrow :S: \ :C{\cdot} \ {\cdot}\ddot{N}:^-$$

This procedure gives the sulfur five valence electrons, which is one less than a neutral sulfur atom. Sulfur therefore carries a formal charge of $+1$ in this Lewis structure. The carbon has four valence electrons, which equals the number on a neutral carbon atom, so carbon carries no formal charge. The nitrogen atom is assigned seven valence electrons, which is two more than a neutral nitrogen atom; that means it carries a formal charge of -2 in this Lewis structure.

$$\overset{+1}{:S}{\equiv}C{-}\overset{-2}{\ddot{N}}:^-$$

The sum of the formal charges on the atoms in an ion is always equal to the net charge on the ion ($1 + 0 + -2 = -1$). We can demonstrate this by applying the same technique to the other Lewis structures for the thiocyanate ion.

Exercise 8.16

Calculate the formal charge on each atom in the following Lewis structures.

(a) $:\ddot{S}-C\equiv N:^-$

(b) $:\ddot{S}=C=\ddot{N}:^-$

Solution

(a) We start by formally dividing the electrons in each bond between the two atoms in the bond.

$$:\ddot{S}-C\equiv N:^- \longrightarrow :\ddot{S}\cdot \;\; \cdot\ddot{C}:\; :N:^-$$

This gives the sulfur atom seven valence electrons, one more than a neutral sulfur atom, so sulfur carries a formal charge of -1. Once again, the carbon atom has four valence electrons. Since this is the number on a neutral carbon atom, carbon has no formal charge. This time, the nitrogen atom has five valence electrons, which is the same as a neutral nitrogen atom, and it also carries no formal charge. The sum of the formal charges $(-1 + 0 + 0 = -1)$ is therefore equal to the net charge on the ion.

$$\overset{-1}{\underset{:\ddot{S}}{}}\;\;\overset{0}{\underset{-C}{}}\;\;\overset{0}{\underset{\equiv N:^-}{}}$$

(b) Once again, we start by formally dividing the electrons in each bond between the atoms in the bond.

$$\ddot{S}=C=\ddot{N}:^- \longrightarrow \;\;\ddot{S}:\; :C:\; :\ddot{N}:^-$$

The sulfur atom formally has six valence electrons and carries no formal charge. The carbon atom, with four valence electrons, also carries no formal charge. The nitrogen atom has six valence electrons, one more than a neutral atom, so nitrogen carries a formal charge of -1 in this Lewis structure. The sum of the formal charges $(0 + 0 + -1 = -1)$ is still equal to the net charge on the ion.

$$\overset{0}{\underset{\ddot{S}}{}}\;\;\overset{0}{\underset{=C}{}}\;\;\overset{-1}{\underset{=\ddot{N}}{}}$$

FORMAL CHARGE AND ALTERNATIVE LEWIS STRUCTURES

Alternative Lewis structures that are essentially the same contribute equally to the resonance hybrid for the molecule. The resonance hybrid for the nitrate (NO_3^-) ion, for example, consists of an equal contribution from three Lewis structures that differ only in which oxygen atom carries the double bond.

When the Lewis structures are significantly different, however, they do not contribute equally to the resonance hybrid.

Formal charge is one way of determining which Lewis structures are more important. The *electroneutrality rule* states that the best Lewis structures are those with the smallest separation of charge. We can illustrate this rule by looking at the formal charges on the three Lewis structures for the SCN⁻ ion.

$$:S\equiv C-\overset{..}{\underset{..}{N}}:^{-} \longleftrightarrow \overset{..}{S}=C=\overset{..}{N}:^{-} \longleftrightarrow :\overset{..}{\underset{..}{S}}-C\equiv N:^{-}$$

The first Lewis structures is less important than the others, because it contains the largest separation of positive and negative charge.

Whenever calculations suggest a large separation of charge, it is a good idea to see if it is possible to write a Lewis structure in which the charge separation is smaller. As an example, let's look at the Lewis structure of the sulfate (SO_4^{2-}) ion.

STEP 1: There are 32 valence electrons in this ion.

$$SO_4^{2-}: \qquad 6 + 4(6) + 2 = 32$$

STEP 2: The skeleton structure for this ion should be obvious by now.

$$\begin{array}{c} O \\ | \\ O-S-O \\ | \\ O \end{array} \quad ^{2-}$$

STEP 3: This skeleton structure requires 8 bonding electrons, leaving 24 nonbonding electrons.

STEP 4: Fortunately, 24 nonbonding electrons are all we need to write a satisfactory Lewis structure.

$$\begin{array}{c} :\overset{..}{O}: \\ | \\ :\overset{..}{O}-S-\overset{..}{O}: \\ | \\ :\overset{..}{O}: \end{array} \quad ^{2-}$$

To calculate the formal charges in this structure we have to divide the electrons in each bond between the atoms in that bond.

$$\begin{array}{c} :\overset{..}{O}: \\ | \\ :\overset{..}{O}-S-\overset{..}{O}: \\ | \\ :\overset{..}{O}: \end{array} \quad ^{2-} \longrightarrow \begin{array}{c} :\overset{..}{O}: \\ \cdot \\ :\overset{..}{O}\cdot \quad \cdot S\cdot \quad \cdot\overset{..}{O}: \\ \cdot \\ :\overset{..}{O}: \end{array} \quad ^{2-}$$

Each oxygen atom formally has seven valence electrons, one more than a neutral oxygen atom, and therefore carries a formal charge of -1. The sulfur atom has only four valence electrons, two less than a neutral sulfur atom, and it carries a formal charge of $+2$.

The sum of the formal charges is equal to the net charge on the ion $[4(-1) + 2 = -2]$. But there is a significant amount of charge separation in this structure, so we might look for a Lewis structure in which the charge separation is smaller.

Since there are empty $3d$ orbitals on a sulfur atom, it is possible to expand the valence shell on the central atom in the sulfate ion to include 10 or even 12 electrons by forming S=O double bonds. Each time we incorporate an S=O double bond into the structure, the charge separation becomes smaller. The electroneutrality rule suggests that the best Lewis structure for the SO_4^{2-} ion is the one with two S=O double bonds.

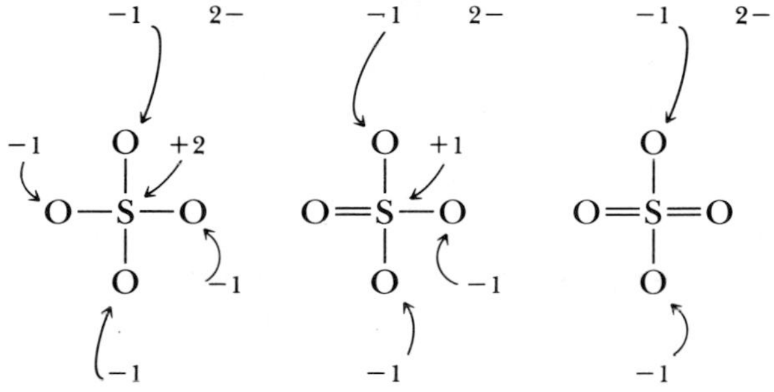

Experimental evidence supports the idea that S=O double bonds contribute to the resonance hybrid that describes the SO_4^{2-} ion. As a rule, double bonds are shorter than single bonds. The sulfur-oxygen bonds in this ion are experimentally found to be 0.149 nanometer long, or about 0.027 nanometer shorter than a standard S—O single bond.

8.11 THE SHAPES OF MOLECULES

The shape, or *geometry,* of a molecule can be an important factor controlling the chemistry of the compound. Two compounds with the formula $PtCl_2(NH_3)_2$, for example, can be prepared that differ only in the arrangement of Cl and NH_3 groups around the central atom. In one molecule, the Cl atoms are adjacent to each other, while in the other, they are on opposite sides of the platinum atom.

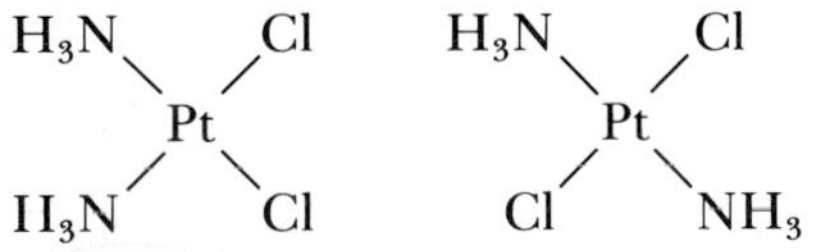

This subtle difference in shape has important consequences. The molecule on the left is a potent antitumor drug known as cisplatin. The molecule on the right is so toxic researchers can't tell whether the tumors have also been attacked.

Biological systems offer an endless number of examples that illustrate the importance of the shapes of molecules. Some of these examples will be discussed at greater detail in Chapter 25. For now, let's focus our attention on proteins, which comprise as much as half the dry weight of an organism.

Proteins are large molecules formed from amino acids. The Lewis structure of the simplest amino acid, glycine, is shown below.

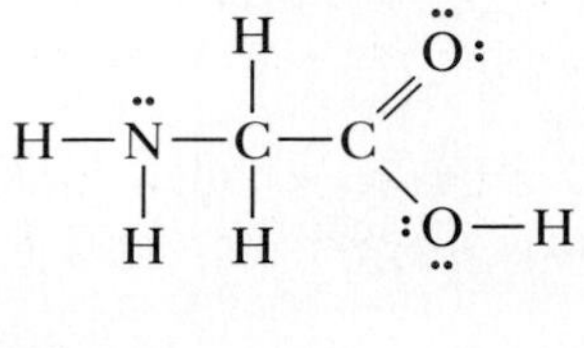

The amino acids in a protein are held together by peptide bonds formed when the
—NH_2 end of one amino acid reacts with the —CO_2H end of another. A portion of
the Lewis structure of a protein that focuses on one of the peptide bonds is shown
below.

$$-\overset{|}{\underset{|}{C}}-\overset{\cdot\cdot}{\underset{|}{N}}-\overset{\overset{\cdot\cdot}{\overset{O}{\parallel}}\cdot\cdot}{\underset{}{C}}-\overset{|}{\underset{|}{C}}-$$
$$\qquad\;\; H$$

One of the most important factors influencing the structure of proteins is the fact
that the six atoms that form these peptide bonds must all lie in the same two-dimensional plane.

Once each peptide bond is frozen into a planar geometry, the remainder of the
protein can fold to form a three-dimensional structure. The changes that occur in
this three-dimensional structure when an egg is heated are the primary reason for
the difference between a raw egg and a cooked one.

The importance of the shape of a molecule in biological systems is also illustrated
by hemoglobin, a protein with a molecular weight of 65,000 grams per mole that
carries oxygen through the body. Hemoglobin consists of four chains of amino
acids, two α chains and two β chains. The structure of one of the β chains in
hemoglobin is shown in Figure 8.8. The disease known as sickle-cell anemia is
caused by the replacement of a single amino acid among the 146 amino acids on
this chain. This substitution causes a subtle change in the structure of hemoglobin
that interferes with the ability of this molecule to pick up oxygen at low pressures.
The net result is so severe that children who inherit this disorder from both parents
seldom live past the age of two.

The goal of Sections 8.12 through 8.14 is to build a model that will allow us to
predict the shapes of small molecules and the geometries around individual portions of larger molecules, even those as large as hemoglobin. We will start by noting
that there is no obvious relationship between the formula of a compound and the
shape of its molecules, as shown by the four compounds in Figure 8.9. BeF_2 is a
linear molecule, while SnF_2 is best described as bent, or angular. GaF_3 is a planar

FIG. 8.8 Hemoglobin consists of
two long chains of amino acids
known as the α chains and two
long chains of amino acids
known as the β chains. A subtle
change in the identity of one of
the 146 amino acids along this
chain causes a large enough
change in the structure of the
molecule to interfere with its
ability to carry oxygen through
the blood.

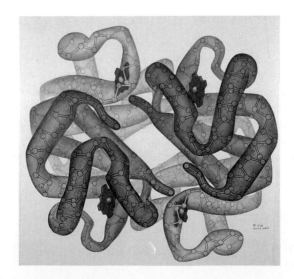

LINEAR	BENT OR ANGULAR	TRIGONAL PLANAR	TRIGONAL PYRAMIDAL
F —— Be —— F	F ⟍Sn⟋ F	F⟍ Ga —— F ⟋F	F⋯As⟋ F ⟍ F

FIG. 8.9 There is no obvious relationship between a chemical formula and the shape of a molecule. BeF_2 is linear, while SnF_2 is bent, or angular. GaF_3 is a planar molecule, while AsF_3 is pyramidal.

molecule with three fluorine atoms oriented toward the corners of a triangle, while AsF_3 has the shape of a pyramid, with the arsenic atom at the apex of the pyramid.

The shapes of these molecules can, however, be predicted, from their Lewis structures by a model developed about 30 years ago. This model is known as the *valence-shell electron-pair repulsion (VSEPR) theory.* Four points concerning its name are worth noting.

1. The theory is only concerned with the outermost, or *valence shell,* electrons on an atom.
2. It is based on the assumption that most electrons in a molecule are paired and is thus concerned with the distribution of *electron pairs* in the molecule.
3. It assumes that pairs of electrons repel each other and therefore predicts the shape of a molecule on the basis of the geometry that gives the smallest amount of *repulsion* between pairs of electrons.
4. It is a *theory.* The only evidence in its favor is the agreement between its predictions and the experimentally determined geometries of molecules.

8.12 PREDICTING THE SHAPES OF MOLECULES FROM THE VSEPR THEORY

The VSEPR theory can be described in terms of seven rules. The first rule states the goal of the model.

RULE 1. Electrons in the valence shell of an atom are distributed so as to minimize the force of repulsion between pairs of electrons.

The second rule suggests how this goal can be achieved.

RULE 2. The distribution of valence-shell electrons can be predicted from the number of places in the valence shell of the atom where electrons can be found.

Beryllium fluoride (BeF_2) can be used to demonstrate how these rules are applied. BeF_2 is an exception to the Lewis octet theory because there are only four electrons in the valence shell of the beryllium atom in the Lewis structure of this molecule.

$$: \overset{..}{\underset{..}{F}} — Be — \overset{..}{\underset{..}{F}} :$$

There are two places on the beryllium atom where valence electrons can be found, corresponding to the two pairs of bonding electrons. The repulsion between these pairs of electrons should be minimized if they are arranged so that they point in

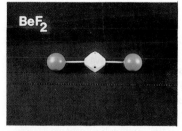

BeF_2 molecules in the gas phase have a linear structure in which the bond angle is 180°.

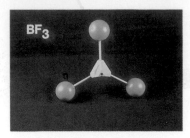

BF₃ has a trigonal planar structure in which the bond angle is 120°.

CH₄ has a tetrahedral structure in which the bond angle is 109°28′, or 109.5°.

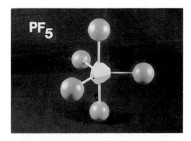

PF₅ has a trigonal bipyramidal structure. The five fluorine atoms are divided into two groups: those in the equatorial plane of the molecule (E) and those along the axis perpendicular to this plane (A). The bond angles in this molecule are 90°, 120°, and 180°.

opposite directions from the central beryllium atom. This results in a prediction that BeF_2 should be a ***linear*** molecule, with a 180° angle between the two Be—F bonds.

Boron trifluoride (BF_3) is an example of a molecule in which there are three places on the central atom where valence electrons can be found.

$$:\!\ddot{F}\!:$$
$$|$$
$$B$$
$$\overset{\cdot\cdot}{F}\quad\overset{\cdot\cdot}{F}$$

We can keep these pairs of electrons as far apart as possible by making the F—B—F bond angle as large as possible. The largest possible bond angle (120°) is obtained when the four atoms in BF_3 are kept in the same plane and the pairs of electrons are arranged toward the corners of an equilateral triangle. This results in a ***trigonal planar*** geometry for the BF_3 molecule.

Methane (CH_4) is an example of a compound in which valence electrons can be found in four places on the central atom.

$$\begin{array}{c} H \\ | \\ H\!-\!C\!-\!H \\ | \\ H \end{array}$$

All of the examples so far have been two-dimensional molecules, in which the atoms lie in the same plane. If we place the same restriction on methane, we end up with a ***square planar*** geometry in which the H—C—H bond angle is 90°. If we let this system expand into three dimensions, however, we end up with a ***tetrahedral*** molecule in which the H—C—H bond angle is 109°28′.

Electrons can be found in five places on the central atom in the Lewis structure of phosphorus pentafluoride (PF_5).

$$:\!\ddot{F}\!:$$
$$F\quad|\quad F$$
$$P$$
$$F\quad\quad F$$

Repulsion between the pairs of electrons in this molecule is minimized when these electrons are distributed toward the corners of a ***trigonal bipyramidal*** shape. To build a trigonal bipyramid, we start with a trigonal planar geometry, add one atom above this plane along an axis perpendicular to the plane to form a trigonal pyramid, and then add another atom beneath this plane to form the bipyramid.

The trigonal bipyramid differs from the other shapes of molecules described so far, because the points that define its geometry are not the same. Three of the atoms bound to the central atom in this geometry are described as ***equatorial***, because they lie along the "equator" of the molecule. The other two are called ***axial***, because they lie along an axis perpendicular to the equatorial plane. The angle between the three equatorial positions is 120°, while the angle between an axial and an equatorial position is 90°.

The Lewis structure of sulfur hexafluoride (SF_6) contains six places where valence electrons can be found on the central atom.

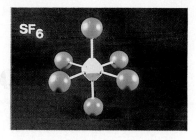

SF_6 has an octahedral structure in which the bond angle is 90°.

The repulsion between the six pairs of electrons in this system is minimized when they are distributed toward the corners of an **octahedral** shape. The term *octahedron* literally means "eight sides," but it is the six corners, or vertices, that interest us now. To imagine the geometry of an SF_6 molecule, locate fluorine atoms on opposite sides of the sulfur atom along the x, y, and z axes of an xyz coordinate system.

INCORPORATING DOUBLE AND TRIPLE
8.13 BONDS INTO THE VSEPR THEORY

We can understand why it is important to count the number of places where valence electrons can be found on an atom, instead of the number of pairs of valence electrons, by looking at the geometries of carbon dioxide (CO_2) and the carbonate (CO_3^{2-}) ion.

The Lewis structure of CO_2 suggests that there are four pairs of bonding electrons on the central atom.

$$\overset{..}{\underset{..}{O}} = C = \overset{..}{\underset{..}{O}}$$

But there are only two places where these electrons can be found. There are electrons in the C=O double bond on the left and electrons in the double bond on the right. The force of repulsion between these electrons is minimized when they are placed on opposite sides of the carbon atom. This results in a prediction that CO_2 is a linear molecule, just like BeF_2, with a bond angle of 180°.

The carbonate ion has the following Lewis structure.

Once again, there are four pairs of valence electrons on the carbon atom. But these electrons are concentrated in three places. Electrons can be found in the two C—O single bonds and the C=O double bond. Repulsions between these electrons is minimized when they are arranged toward the corners of an equilateral triangle. Thus, we predict a trigonal planar shape for the CO_3^{2-} ion, just like BF_3, with a 120° bond angle.

Compounds that contain double and triple bonds are incorporated into the VSEPR model by Rule 3.

TABLE 8.2

*Summary of the Relationship Between the Number
of Places Where Valence Electrons Can Be Found
and the Distribution of Electrons on the Atom*

Number of Places Where Electrons Can Be Found	Distribution of Electrons Around the Central Atom	Examples
2	linear	BeF_2, CO_2
3	trigonal planar	BF_3, CO_3^{2-}
4	tetrahedral	CH_4
5	trigonal bipyramidal	PF_5
6	octahedral	SF_6

RULE 3. Double and triple bonds are equivalent to a single pair of bonding electrons when counting the number of places in the valence shell of an atom where electrons can be found.

The results of Sections 8.12 and 8.13 are summarized in Table 8.2.

8.14 THE ROLE OF NONBONDING ELECTRONS IN THE VSEPR THEORY

The molecules listed in Table 8.2 contain only bonding electrons in the valence shell of the central atom. Rule 4 describes what to do when we encounter molecules, such as ammonia and water that have nonbonding electrons as well.

RULE 4. Both bonding and nonbonding electrons are counted in determining the number of places where electrons can be found in the valence shell of an atom.

According to this rule, both NH_3 and H_2O belong to the class of compounds with four places on the central atom where valence electrons can be found.

$$H-\overset{\cdot\cdot}{N}-H \qquad H-\overset{\cdot\cdot}{\underset{\cdot\cdot}{O}}-H$$
$$\overset{|}{\underset{H}{}}$$

In ammonia, these consist of three pairs of bonding electrons and one pair of nonbonding electrons. In water, they include two pairs of bonding and two pairs of nonbonding electrons. The VSEPR theory predicts that the valence electrons in these molecules will be distributed toward the corners of a tetrahedron, as shown in Figure 8.10.

Our goal is not predicting the distribution of electrons, however. Our goal is to use the distribution of electrons to predict the shape of the molecule. Until now, these two have been the same. The shapes of the molecules for the compounds in Table 8.2 were always the same as the distribution of electrons. But when we include nonbonding electrons in our model, that is no longer true.

In Section 5.5, we noted that electrons are almost 2000 times lighter than the nuclei of the smallest atom — the hydrogen atom. In Section 5.15, we encountered one of the consequences of this fact. The mass of an electron is so small that it has

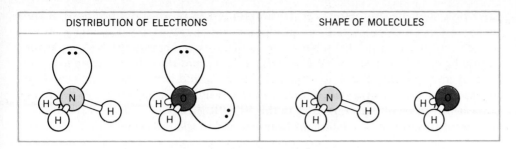

DISTRIBUTION OF ELECTRONS	SHAPE OF MOLECULES

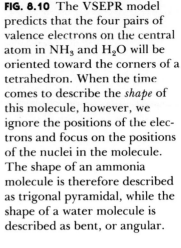

FIG. 8.10 The VSEPR model predicts that the four pairs of valence electrons on the central atom in NH_3 and H_2O will be oriented toward the corners of a tetrahedron. When the time comes to describe the *shape* of this molecule, however, we ignore the positions of the electrons and focus on the positions of the nuclei in the molecule. The shape of an ammonia molecule is therefore described as trigonal pyramidal, while the shape of a water molecule is described as bent, or angular.

some of the characteristics of a wave. This means that it isn't possible to determine the location of the electrons in an atom the way we can determine the positions of the more massive nuclei.

A number of techniques can locate the nuclei of atoms with some precision. Figure 8.11 shows the results of one of these techniques, which clearly locates the positions of the nuclei in the 10 carbon atoms in naphthalene ($C_{10}H_8$) and provides a reasonable idea of the location of the 8 hydrogen atoms. But the technique cannot locate individual pairs of electrons.

The VSEPR theory predicts that the valence electrons on the central atoms in ammonia and water point toward the corners of a tetrahedron. This prediction

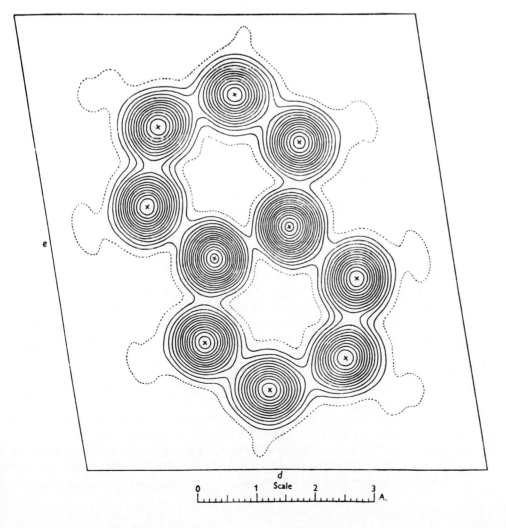

e

d

0		1	Scale	2		3

A.

FIG. 8.11 Molecular geometries are determined experimentally by techniques such as x-ray crystallography. This figure shows the results of an x-ray study of the structure of naphthalene, $C_{10}H_8$. The locations of the 10 carbon nuclei are clearly indicated by this contour diagram, and the 8 hydrogen nuclei are reasonably easy to locate. This technique cannot locate individual pairs of electrons, however. When the time comes to describe the shape of a molecule, we therefore tend to describe the locations of the nuclei in the molecule and not the way electrons in the valence shell of the central atom are distributed.

can't be tested directly. However, it can be used to predict the positions of the nuclei in these molecules, and this prediction can be tested experimentally. If we focus on the positions of the nuclei in ammonia, we predict that the NH_3 molecule should have a shape best described as *trigonal pyramidal*, with the nitrogen at the apex of the pyramid. Water, on the other hand, should have a shape that can best be described as *bent*, or *angular*. Both of these predictions have been shown to be correct, which reinforces our faith in the VSEPR theory.

The results of this discussion are summarized in the following rule.

RULE 5. The VSEPR theory predicts the distribution of electrons in the valence shell of an atom. This in turn allows us to predict the shape of the molecule, which describes the positions of the nuclei and does not include the location of either bonding or nonbonding electrons.

Exercise 8.17

Predict the shape of the following molecules.

(a) Tin(II) fluoride, SnF_2 (c) Iodine(III) chloride cation, ICl_2^+

(b) Hydronium ion, H_3O^+ (d) Phosphate ion, PO_4^{3-}

Solution

(a) The first step in predicting the shape of a molecule is always the same — we start by writing the Lewis structure of the molecule. There are 18 valence electrons in an SnF_2 molecule — $4 + 2(7) = 18$. There are two bonds in the skeleton structure of this molecule, which means there are 14 nonbonding electrons. SnF_2 is therefore another example of a compound whose Lewis structure doesn't include an octet of valence electrons for every atom.

$$\overset{..}{Sn}$$
$$..\diagup\diagdown..$$
$$.\overset{..}{F}..\overset{..}{F}.$$

There are three places where electrons can be found in the valence shell of the central atom, which means these electrons should be distributed toward the corners of an equilateral triangle. When the time comes to describe the shape of the molecule, however, we focus on the positions of the three nuclei. The molecule is therefore best described as bent, or angular.

(b) There are eight valence electrons in this ion — $3(1) + 6 - 1 = 8$. The Lewis structure of the ion is therefore written as follows.

$$H-\overset{..}{O}-H\ ^+$$
$$|$$
$$H$$

Electrons can be found in four places in the valence shell of the oxygen atom, and these electrons are therefore distributed toward the corners of a tetrahedron. When the time comes to describe the geometry around the oxygen, we focus on the positions of the three nuclei. The shape of this ion is therefore best described as trigonal pyramidal.

(c) There are 20 valence electrons in this ion — $7 + 2(7) - 1 = 20$. The Lewis structure of the ion therefore suggests that there are four places in the valence shell of the iodine atom where electrons can be found.

$$:\overset{\cdot\cdot}{\underset{\cdot\cdot}{Cl}}-\overset{\cdot\cdot}{I}-\overset{\cdot\cdot}{\underset{\cdot\cdot}{Cl}}:^+$$

The VSEPR theory predicts that these electrons should be distributed toward the corners of a tetrahedron. When we focus on the positions of the nuclei, however, the shape of the molecule is best described as bent, or angular.

(d) The Lewis structure of the phosphate ion can be predicted on the basis of the information in Exercise 8.7.

$$
\begin{array}{c}
:\overset{\cdot\cdot}{O}: \quad {}^{3-}\\
|\\
:\overset{\cdot\cdot}{O}-P-\overset{\cdot\cdot}{O}:\\
|\\
:\overset{\cdot\cdot}{O}:
\end{array}
$$

Since there are no nonbonding valence electrons on the phosphorus atom, the distribution of the electrons and the shape of the molecule can both be described as tetrahedral.

The VSEPR theory can also be applied to molecules, such as NO_2, that contain an odd number of valence electrons.

RULE 6. A single, unpaired electron is counted as one place where electrons can be found in the valence shell of an atom.

The odd number of valence electrons in NO_2 makes it impossible to write a Lewis structure in which the electrons are all paired.

$$
\cdot N \overset{\displaystyle \overset{\cdot\cdot}{O}:}{\underset{\displaystyle \overset{\cdot\cdot}{O}\overset{\cdot\cdot}{\cdot}}{\diagdown}}
$$

According to Rule 6, however, there are three places in the valence shell of the nitrogen atom where electrons can be found. There are electrons in the N—O single bond and the N=O double bond, and this Lewis structure also contains a single, unpaired, nonbonding electron. The geometry for this molecule is based on arranging these electrons toward the corners of a triangle, and the shape of the molecule is bent, or angular.

Table 8.3 summarizes the distribution of electrons and the shape of the molecule for all possible combinations of bonding and nonbonding electrons for molecules that have two, three, or four places where valence electrons can be found. When we try to extend this to molecules in which the electrons are distributed toward the corners of a trigonal bipyramid, we run into the question of where nonbonding electrons should be placed in this structure.

The Lewis structure of SF_4, for example, suggests that the five pairs of valence

TABLE 8.3

Geometries of Molecules with up to Four Places Where Valence Electrons Can be Found

Places Where Electrons Are Found	Places Due to Bonding Electrons	Places Due to Nonbonding Electrons	Distribution of Electrons	Molecular Geometry	Examples
1	1	0	linear	linear	H_2
2	2	0	linear	linear	BeF_2, CO_2
2	1	1	linear	linear	CO, N_2
3	3	0	trigonal planar	trigonal planar	BF_3, $CO_3{}^{2-}$
3	2	1	trigonal planar	bent	SnF_2, O_3, SO_2
3	1	2	trigonal planar	linear	O_2
4	4	0	tetrahedral	tetrahedral	CH_4, $NH_4{}^+$, $SO_4{}^{2-}$
4	3	1	tetrahedral	trigonal pyramidal	NH_3, H_3O^+, PF_3
4	2	2	tetrahedral	bent	H_2O, H_2S, $ICl_2{}^+$
4	1	3	tetrahedral	linear	HF, OH^-

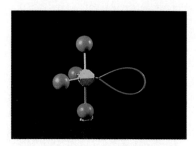

The electrons in the valence shell of the sulfur atom in SF_4 are oriented toward the corners of a trigonal bipyramid. The geometry of the molecule, however, might be best described as having the shape of a see-saw.

electrons on the sulfur atom in this molecule should be oriented toward the corners of a trigonal bipyramid.

$$\begin{array}{c} \ddot{F} \quad \ddot{F} \\ \diagdown\diagup \\ S \\ \diagup\diagdown \\ \ddot{F} \quad \ddot{F} \end{array}$$

But there are two different ways of placing the nonbonding electrons in a trigonal bipyramid. They can be placed in either the equatorial or the axial positions in this geometry, as shown in Figure 8.12.

Experimentally, we find that nonbonding electrons almost always end up in the equatorial positions of a trigonal bipyramid. To understand why, we have to recognize that nonbonding electrons take up more space than bonding electrons. Nonbonding electrons need to be close to only one nucleus, and there is a considerable amount of space in which nonbonding electrons can reside and still be near the nucleus of the atom. Bonding electrons, however, must be simultaneously as close as possible to two nuclei, and only a small region of space between the nuclei satisfies this restriction. The result of the difference in size of nonbonding and bonding pairs of electrons is summarized in Rule 7.

RULE 7. Repulsion between two pairs of nonbonding electrons is stronger than repulsion between a nonbonding pair and a bonding pair of electrons, which in turn is stronger than repulsion between two pairs of bonding electrons.

If the nonbonding electrons in SF_4 are placed in an axial position, they will be relatively close (90°) to three pairs of bonding electrons. But if the nonbonding electrons are placed in an equatorial position, they will be 90° away from only two pairs of bonding electrons. This means the repulsion between nonbonding and bonding electrons will be minimized if the nonbonding electrons are placed in an equatorial position in SF_4. The VSEPR theory therefore predicts the distribution of electrons shown on the left in Figure 8.12, and the molecule has a ***see-saw*** or ***teeter-totter*** shape.

FIG. 8.12 The nonbonding electrons on the sulfur atom in SF_4 can be placed in either an equatorial or an axial position. If they were placed in one of the equatorial positions, they would be relatively close (90°) to the two pairs of bonding electrons that bond the fluorines in the axial positions. If they were placed in one of the axial positions, they would be relatively close (90°) to three pairs of bonding electrons that hold the fluorines in the equatorial positions. Unfavorable repulsions between nonbonding and bonding electrons are therefore minimized when the nonbonding electrons are placed in the equatorial position.

The Lewis structure of chlorine trifluoride (ClF_3) suggests that the valence electrons on the chlorine should point toward the corners of a trigonal bipyramid.

$$F—Cl$$

Once again, it can be shown that repulsion between pairs of electrons can be minimized if the nonbonding electrons are placed in equatorial positions. ClF_3 therefore has a geometry that can be described as **_T-shaped._**

The triiodide (I_3^-) ion also has a Lewis structure that suggests a trigonal bipyramidal distribution of valence electrons.

The three pairs of nonbonding electrons are all placed in equatorial positions, and the molecule is therefore linear.

Molecular geometries based on an octahedral distribution of valence electrons are easier to predict, because the corners of an octahedron are all identical. No matter where we place a single pair of nonbonding electrons, the shape of the molecule is the same.

Exercise 8.18

Predict the geometry of bromine pentafluoride (BrF_5).

Solution

We start, as always, by predicting the Lewis structure of the molecule. There are 42 valence electrons in this molecule — $7 + 5(7) = 42$. The following is therefore the Lewis structure for the molecule.

The valence electrons on the bromine atom are distributed toward the corners of an octahedron. The molecule, however, has a geometry that is best described as **_square pyramidal._**

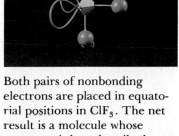

Both pairs of nonbonding electrons are placed in equatorial positions in ClF_3. The net result is a molecule whose geometry is best described as T-shaped.

When all three pairs of nonbonding electrons are placed in equatorial positions in the I_3^- ion, the result is a linear molecule.

The valence electrons on the bromine atom in BrF_5 are oriented toward the corners of an octahedron. When the nonbonding electrons that occupy one of the corners of the octahedron are ignored, the shape of the molecule is best described as square pyramidal.

TABLE 8.4

Geometries of Molecules with Five or Six Places Where Valence Electrons Can be Found

Places Where Electrons Are Found	Places Due to Bonding Electrons	Places Due to Nonbonding Electrons	Distribution of Electrons	Molecular Geometry	Examples
5	5	0	trigonal bipyramidal	trigonal bipyramidal	PF_5, $SbCl_5$, $SnCl_5^-$
5	4	1	trigonal bipyramidal	see-saw	SF_4, $TeCl_4$, IF_4^+
5	3	2	trigonal bipyramidal	T-shaped	ClF_3
5	2	3	trigonal bipyramidal	linear	I_3^-, XeF_2
6	6	0	octahedral	octahedral	SF_6, PF_6^-, SiF_6^{2-}
6	5	1	octahedral	square pyramidal	BrF_5, $SbCl_5^{2-}$
6	4	2	octahedral	square planar	XeF_4, ICl_4^-

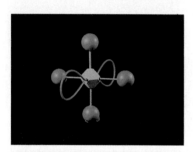

The valence electrons on the xenon atom in XeF_4 are also distributed toward the corners of an octahedron. The two nonbonding pairs of electrons on the xenon are kept as far apart as possible to minimize the force of repulsion between these electrons. The net result is a molecule whose shape is square planar.

When two of the six places in the valence shell of the central atom contain non-bonding electrons, these electrons are kept as far apart as possible to minimize the repulsion between the relatively large nonbonding pairs. The Lewis structure of XeF_4 written in Exercise 8.12 would therefore result in a square planar geometry for this molecule.

The distribution of electrons and the shape of molecules with five or six places where valence electrons can be found are summarized in Table 8.4.

8.15 HYBRID ATOMIC ORBITALS

A model has been developed which involves the construction of hybrid atomic orbitals that explains how molecules achieve the geometries predicted by the VSEPR theory. We can most easily explain this model by using an example, such as an isolated BeF_2 molecule in the gas phase.

Let's start by writing the Lewis structure of this molecule. There are two valence electrons on a neutral beryllium atom and seven valence electrons on each fluorine atom. The bonds that hold this molecule together can therefore be formed by combining an unpaired electron on each of the fluorine atoms with an unpaired electron on the beryllium atom.

$$:\ddot{F}\cdot \quad \cdot Be\cdot \quad \cdot \ddot{F}: \longrightarrow :\ddot{F}—Be—\ddot{F}:$$

The electron configuration of fluorine suggests that there is an unpaired valence electron in one of the $2p$ orbitals on this atom.

F: [He] $2s^2 \, 2p^5$

Let's arbitrarily assume that this electron lies in the $2p$ orbital oriented along the z axis of an *xyz* coordinate system.

F: [He] $2s^2 \, 2p_x^2 \, 2p_y^2 \, \mathbf{2p_z^1}$

The lowest-energy state for a neutral beryllium is described by the following electron configuration.

Be: [He] $2s^2$

There are no unpaired electrons in this configuration that can be used to form Be—F bonds. However, we can excite one of the $2s$ electrons into a $2p$ orbital by

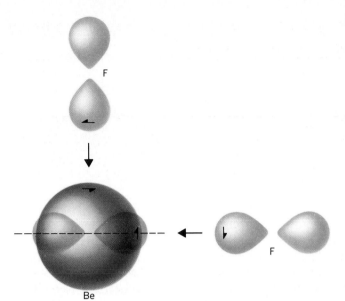

FIG. 8.13 Atomic orbitals can be used to explain the formation of a BeF_2 molecule. An electron is first excited from the $2s$ into the $2p$ orbital on the beryllium. The electron that remains in the $2s$ orbital overlaps with an electron in a $2p$ orbital on the fluorine. The electron in the $2p$ orbital on beryllium overlaps with an electron in a $2p$ orbital of a second fluorine atom. Unfortunately, this model predicts that the shape of the BeF_2 molecule should be bent, or angular, whereas the results of experiments suggest that this molecule is linear.

investing a little energy in the system to form a beryllium atom in an excited state — Be*.

Be*:   [He] $2s^1 2p_z^1$

We can then combine the $2p_z^1$ electron on one of the fluorine atoms with the electron in the $2s$ orbital on the beryllium atom to form a Be—F bond. The $2p_z^1$ electron on the other fluorine atom can then be combined with the $2p_z^1$ electron on the beryllium to form the second bond, as shown in Figure 8.13.

The energy it takes to excite an electron on the beryllium atom from the $2s$ to the $2p$ orbital is more than repaid by the energy given off when this excited beryllium atom combines with two fluorine atoms to form a pair of Be—F covalent bonds.

There are several problems with this model for BeF_2. It suggests that the two Be—F bonds will be different, because one is made from the $2s$ orbital and the other from the $2p_z$ orbital on the beryllium. But experimental results have shown that the two Be—F bonds in beryllium fluoride are identical. This model also leads us to expect that the BeF_2 molecule should be bent, or angular, because of the way the atomic orbitals overlap, as shown in Figure 8.13. But experiment has shown that the molecule is linear.

These problems were solved by Linus Pauling, who argued that the mathematical equations that describe the $2s$ and $2p_z$ orbitals in the Schrödinger model of the atom (discussed in Section 5.16) could be combined to form a new set of mathematical functions, known as **hybrid atomic orbitals,** which point in opposite directions from the nucleus of an atom. Because these orbitals are mixtures of a $2s$ orbital and a $2p$ orbital, they are called sp hybrids (see Figure 8.14).

The structure of a BeF_2 molecule can be explained by assuming that the valence orbitals on the beryllium atom are mixed, or hybridized, to form a set of two equivalent sp hybrid orbitals that point in opposite directions. One of the valence electrons on the beryllium atom is placed in each of these orbitals, and the orbitals are allowed to combine, or overlap, with half-filled $2p_z$ orbitals on a pair of fluorine atoms to form two equivalent Be—F covalent bonds.

Pauling also showed that the geometry of molecules such as BF_3 and the CO_3^{2-} ion could be explained by mixing a $2s$ orbital with both a $2p_x$ and a $2p_y$ orbital on the

FIG. 8.14 We can explain the observed geometry of BeF_2 by combining the $2s$ and $2p_z$ atomic orbitals on the beryllium atom to form a set of sp hybrid orbitals that point in opposite directions. The $2s$ orbital is described by a mathematical function that is positive at all points on the sphere. The equation for a $2p$ orbital, however, is positive in one direction and negative in the other. When we combine the $2s$ and $2p$ orbitals by adding these functions, we get an sp orbital that points in one direction. When we combine the $2s$ and $2p$ orbitals by subtracting these functions, we get an sp orbital that points in the opposite direction.

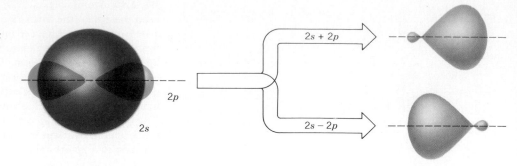

central atom to form three equivalent sp^2 hybrid orbitals that point toward the corners of an equilateral triangle. When he mixed a $2s$ orbital with all three $2p$ orbitals ($2p_x$, $2p_y$, and $2p_z$), Pauling obtained a set of four equivalent sp^3 orbitals that are oriented toward the corners of a tetrahedron. These sp^3 orbitals are therefore ideal for explaining the geometries of molecules such as CH_4 and the PO_4^{3-} ion.

This model can even be extended to account for molecules whose shapes are based on a trigonal bipyramidal or octahedral distribution of electrons. These molecules use empty valence-shell d orbitals on the central atom to hold the extra electrons in the valence shell of this atom. Pauling showed that when the $3d_{z^2}$ orbital is mixed with the $3s$, $3p_x$, $3p_y$, and $3p_z$ orbitals on an atom, the resulting sp^3d hybrid orbitals point toward the corners of a trigonal bipyramid. When both the $3d_{x^2-y^2}$ and $3d_{z^2}$ orbitals are mixed with the $3s$, $3p_x$, $3p_y$, and $3p_z$ orbitals, the result is a set of six sp^3d^2 hybrid orbitals that point toward the corners of an octahedron.

The geometries of the five different sets of hybrid atomic orbitals (sp, sp^2, sp^3, sp^3d, and sp^3d and sp^3d^2) are shown in Figure 8.15. The relationship between hybridization and the distribution of electrons in the valence shell of an atom is summarized in Table 8.5.

FIG. 8.15 The shapes of the sp, sp^2, sp^3, sp^3d, and sp^3d^2 sets of hybrid orbitals.

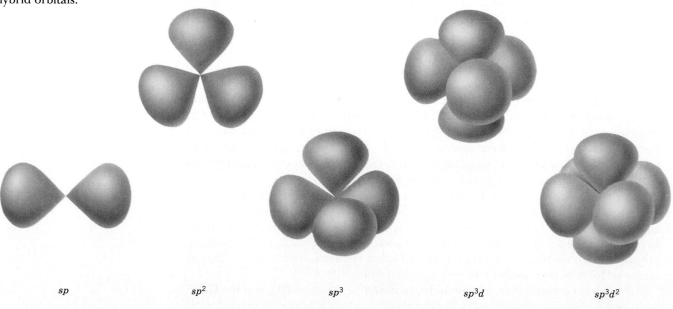

sp sp^2 sp^3 sp^3d sp^3d^2

TABLE 8.5

The Relationship Between the Distribution of Electrons
on an Atom and the Hybridization of That Atom

Places Where Electrons Are Found	Molecular Geometry	Hybridization	Examples
2	linear	sp	BeF_2, CO_2
3	trigonal planar	sp^2	BF_3, CO_3^{2-}
4	tetrahedral	sp^3	CH_4, SO_4^{2-}
5	trigonal bipyramidal	sp^3d	PF_5
6	octahedral	sp^3d^2	SF_6

Exercise 8.19

Use Table 8.5 to determine the hybridization of the central atom in each of the following compounds.

(a) O_2 (b) H_3O^+ (c) $TeCl_4$ (d) ICl_4^-

Solution

(a) The Lewis structure for the O_2 molecule suggests that the distribution of electrons around each oxygen atom should be trigonal planar.

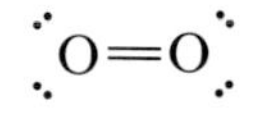

$$\overset{\cdot\cdot}{\underset{\cdot\cdot}{O}}{=}\overset{\cdot\cdot}{\underset{\cdot\cdot}{O}}$$

Orbitals that point toward the corners of an equilateral triangle can be created when the $2s$, $2p_x$, and $2p_y$ orbitals on each oxygen atom are mixed to form a set of sp^2 hybrid orbitals. The hybridization of the oxygen atoms in the O_2 molecule is therefore assumed to be sp^2.

(b) The Lewis structure suggests a tetrahedral distribution of electrons around the oxygen atom in the hydronium ion.

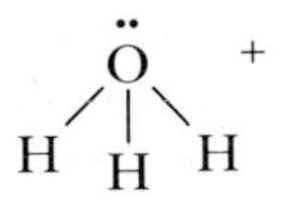

$$\overset{\cdot\cdot}{O}^+\underset{H\ \ H\ \ H}{\diagup\,|\,\diagdown}$$

A set of orbitals that point toward the corners of a tetrahedron can be constructed when the $2s$, $2p_x$, $2p_y$, and $2p_z$ atomic orbitals on the oxygen atom are mixed. The oxygen in this molecule is therefore described as sp^3 hybridized.

(c) The Lewis structure suggests a trigonal bipyramidal distribution of electrons.

$$\underset{\underset{Cl\ \ Cl}{|\diagdown}}{\overset{\overset{Cl\ \ Cl}{|\diagup}}{:Te}}$$

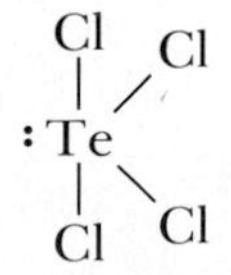

Orbitals pointing toward the corners of a trigonal bipyramid can be obtained when the $5s$, $5p_x$, $5p_y$, $5p_z$, and $5d_{z^2}$ orbitals on the tellurium atom are mixed. This atom is therefore sp^3d hybridized.

(d) The Lewis structure suggests an octahedral distribution of electrons.

$$\begin{array}{ccc} Cl & & Cl^{\,-} \\ \diagdown & \ddot{} & \diagup \\ & I & \\ \diagup & \ddot{} & \diagdown \\ Cl & & Cl \end{array}$$

Orbitals pointing toward the corners of an octahedron can be obtained when the $5s$, $5p_x$, $5p_y$, $5p_z$, $5d_{z^2}$, and $5d_{x^2-y^2}$ orbitals on the iodine atom are mixed. This atom is therefore sp^3d^2 hybridized.

8.16 A HYBRID ATOMIC ORBITAL DESCRIPTION OF MULTIPLE BONDS

The hybrid atomic orbital model can be used to explain the formation of double and triple bonds. As an example of this, let's consider the bonding in formaldehyde — H_2CO. The following Lewis structure for this molecule was derived in Exercise 8.9.

$$\begin{array}{c} H \\ \diagdown \\ C{=}\ddot{O}\!: \\ \diagup \\ H \end{array}$$

There are three places where electrons can be found in the valence shell of the carbon atom in this molecule — the two C—H single bonds and the C=O double bond. The VSEPR theory therefore predicts that the valence electrons on the carbon atom will be oriented toward the corners of an equilateral triangle. We can achieve this geometry by mixing the $2s$, $2p_x$, and $2p_y$ orbitals on the carbon atom to obtain a set of sp^2 hybrid orbitals.

There are also three places where electrons can be found in the valence shell of the oxygen atom — the two pairs of nonbonding electrons and the C=O double bond. The VSEPR theory predicts that these electrons should also be distributed toward the corners of an equilateral triangle. We can therefore assume that the oxygen atom is also sp^2 hybridized.

When the $2s$, $2p_x$, and $2p_y$ orbitals on an atom are mixed to form a set of sp^2 hybrid orbitals, the $2p_z$ orbital is left unchanged. Figure 8.16 shows how these orbitals can be used to form the formaldehyde molecule. Each of the hydrogen atoms contains a single electron in a $1s$ orbital. One of the electrons on the carbon atom is placed in each of the three sp^2 hybrid orbitals. The fourth is placed in the unhybridized $2p_z$ orbital.

There are six valence electrons on a neutral oxygen atom. A pair of these electrons is placed in each of two of the sp^2 hybrid orbitals. One electron is then

FIG. 8.16 The hybrid atomic orbital model can be used to explain the bonding in molecules, such as formaldehyde, that contain double or triple bonds.

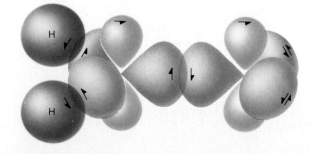

placed in the sp^2 hybrid orbital that points toward the carbon atom, and another is placed in the unhybridized $2p_z$ orbital.

The C—H bonds are formed when the electrons in two of the sp^2 hybrid orbitals on carbon interact with the $1s^1$ electrons on the hydrogen atoms. A C—O bond is formed when the electron in the other sp^2 hybrid orbital on carbon interacts with the unpaired electron in the sp^2 hybrid orbital on the oxygen atom. These bonds are all called *sigma* (σ) *bonds,* because they are formed by the overlap of atomic orbitals that contain at least some of the character of an *s* orbital.

Once the sigma framework is formed, the electron in the $2p_z$ orbital on the carbon atom can interact with the electron in the $2p_z$ orbital on the oxygen atom to form a second covalent bond between these atoms. This is called a *pi* (π) *bond,* because the orbitals that overlap to form this bond are *p* orbitals.

When double bonds were introduced in Section 8.8, we noted that they are most common in compounds that contain C, N, O, P, or S. There are two reasons for this. First, double bonds by their very nature are covalent bonds. They are therefore most likely to be found among the elements that form covalent compounds. Second, the interaction between $2p_z$ orbitals to form a π bond requires that the atoms come relatively close together, so these bonds tend to be the strongest for atoms that are relatively small.

8.17 MOLECULAR ORBITAL THEORY FOR A DIATOMIC MOLECULE

Ever since orbitals were first introduced in Chapter 5, our discussion has been based on atomic orbitals.

Atomic orbital: A region in space on an atom where electrons can be found.

An enormous amount of chemistry can be explained with nothing more sophisticated than this set of orbitals, which were originally defined for use with the hydrogen atom. But there are problems with this model. One was identified in Section 8.9. Our Lewis structure (or valence-bond) model can't explain the observation that SO_2 contains two equivalent bonds with a bond order halfway between that of an S—O single bond and that of an S=O double bond. The best we can do is write two Lewis structures for this molecule and argue that the molecule is some mixture, or hybrid, of these structures.

$$[:\overset{..}{\underset{..}{O}}-\overset{..}{S}=\overset{..}{\underset{..}{O}}: \longleftrightarrow :\overset{..}{\underset{..}{O}}=\overset{..}{S}-\overset{..}{\underset{..}{O}}:]$$

This problem, and many others, can be solved by a more sophisticated model of bonding based on *molecular orbitals.*

Molecular orbital: A region in space within a molecule in which electrons can be found.

Molecular orbital theory is inherently more powerful than valence-bond theory because the orbitals are tailored to reflect the geometries of the molecules to which they are applied. But this power carries a significant cost in terms of the ease with which the model can be visualized. To understand some of the features of molecular orbital theory, we will construct the molecular orbitals for the simplest kind of molecule—a diatomic molecule that contains two atoms of the same element.

The first rule of the molecular orbital theory describes how molecular orbitals are generated.

RULE 1. To construct molecular orbitals, we combine the mathematical functions that describe the atomic orbitals on the atoms in the molecule.

Molecular orbitals, like atomic orbitals, can hold up to two electrons. Since the number of electrons in the molecule is not affected by the creation of molecular orbitals, the number of orbitals available to hold these electrons must remain constant.

RULE 2. The number of molecular orbitals formed is always equal to the number of atomic orbitals combined.

These two rules are all we need to generate the molecular orbitals for the simplest diatomic molecule — H_2. According to the second rule, we should get two molecular orbitals when the two $1s$ atomic orbitals on a pair of hydrogen atoms are combined. One of these we construct by adding the mathematical functions for the two $1s$ atomic orbitals. The other results when we subtract one of these functions from the other, as shown in Figure 8.17.

One of the molecular orbitals is called a **bonding molecular orbital,** because electrons placed in it spend most of their time in the region directly between the two nuclei. It is called a sigma (σ) molecular orbital, because it is formed by the combination of a pair of s orbitals. Electrons placed in the other molecular orbital spend most of their time away from the region between the two nuclei. This orbital is therefore an **antibonding,** or sigma star ($\sigma*$), **molecular orbital.**

An electron in a $1s$ atomic orbital on a hydrogen atom spends some of its time in the region between a pair of hydrogen atoms. But an electron in a σ bonding molecular orbital spends most of its time in this area. In Section 8.3, we concluded that an H_2 molecule is more stable than a pair of isolated hydrogen atoms. Placing an electron in the σ bonding molecular orbital therefore stabilizes the molecule.

Since the $\sigma*$ antibonding molecular orbital forces the electron to spend most of its time away from the area between the nuclei, placing an electron in this orbital destabilizes the molecule. This principle is usually illustrated by the following diagram, which shows the relationship between the energies of the two $1s$ atomic orbitals and the energies of the σ and $\sigma*$ molecular orbitals formed when these atomic orbitals are combined.

FIG. 8.17 The interaction between a pair of $1s$ atomic orbitals leads to the formation of both a bonding (σ_{1s}) and an antibonding (σ_{1s}^*) molecular orbital. The bonding orbital is formed by adding the two $1s$ orbitals and the antibonding orbital is formed when the mathematical function for one of these orbitals is subtracted from the other.

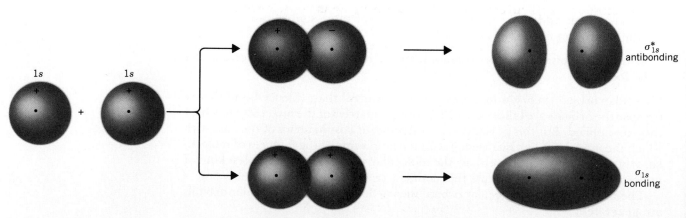

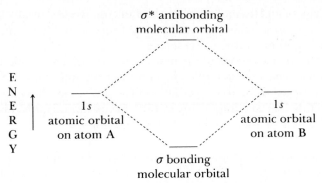

Rules 3 and 4 govern the way electrons are added to molecular orbitals.

RULE 3. Electrons are added to molecular orbitals, one at a time, starting with the lowest-energy molecular orbital.

RULE 4. Each molecular orbital can hold a maximum of two electrons.

The two electrons associated with a pair of hydrogen atoms are placed in the lowest-energy, or σ bonding, molecular orbital. Applying the convention of using arrows to represent the $+\frac{1}{2}$ and $-\frac{1}{2}$ spins of electrons to the molecular orbital diagram for H_2 gives the following result.

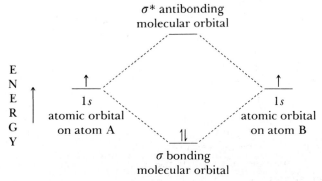

This diagram suggests that an H_2 molecule is more stable than a pair of isolated hydrogen atoms.

This model can be used to explain why He_2 molecules do not exist. Combining a pair of He atoms with $1s^2$ electron configurations would produce a molecule with two electrons in both the bonding and the antibonding molecular orbitals.

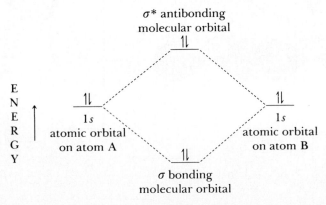

The total energy of an He$_2$ molecule would therefore be essentially the same as the energy of a pair of isolated helium atoms, and there would be nothing to hold the helium atoms together to form a molecule.

The fact that an He$_2$ molecule is neither more nor less stable than a pair of isolated helium atoms is an example of a general principle: the core orbitals on an atom make no contribution to the stability of the molecules that contain this atom. The only orbitals that are important in our discussion of molecular orbitals are those formed when valence-shell orbitals are combined.

RULE 5. Only valence-shell orbitals should be included in a molecular orbital model of a molecule.

A molecular orbital diagram for a diatomic molecule, such as O$_2$, formed from elements in the second row of the periodic table would ignore the $1s$ core electrons on both oxygen atoms and concentrate on the interactions between the $2s$ and $2p$ valence orbitals.

The $2s$ orbitals combine to form a σ_{2s} bonding and a σ_{2s}^* antibonding molecular orbital, just like the σ_{1s} and σ_{1s}^* molecular orbitals formed from the $1s$ atomic orbitals on hydrogen. If we define the z axis of the coordinate system for the two oxygen atoms as the axis along which the bond forms, the $2p_z$ orbitals on the adjacent atoms will meet head-on to form a σ_{2p} bonding and a σ_{2p}^* antibonding molecular orbital, as shown in Figure 8.18. These are called sigma orbitals, because they have the same basic cylindrical shape as those formed when $1s$ or $2s$ atomic orbitals combine.

The $2p_x$ orbitals on one atom interact with the $2p_x$ orbitals on the other to form molecular orbitals that have a different shape, as shown in Figure 8.19. These molecular orbitals are called pi orbitals. While σ and σ^* orbitals concentrate the electrons along the axis on which the nuclei of the atoms lie, π and π^* orbitals concentrate the electrons either above or below this axis. The $2p_x$ atomic orbitals combine to form a π_x bonding molecular orbital and a π_x^* antibonding orbital. The same happens when the $2p_y$ orbitals interact, only in this case we get a π_y bonding molecular orbital and a π_y^* antibonding molecular orbital. Since there is no difference between the energies of the $2p_x$ and $2p_y$ atomic orbitals, there is no difference between the energies of the π_x and π_y or π_x^* and π_y^* orbitals.

The interaction of four valence atomic orbitals on one atom ($2s$, $2p_x$, $2p_y$, and

FIG. 8.18 When a pair of $2p$ orbitals are combined in such a way that they meet head-on, bonding (σ_p) and antibonding (σ_p^*) molecular orbitals are formed. These are called σ molecular orbitals because they concentrate the electrons along the axis between the two nuclei, just like the σ_{1s} and σ_{2s} molecular orbitals.

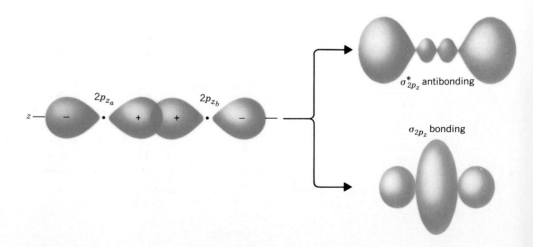

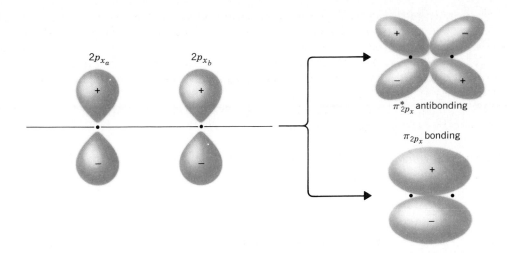

FIG. 8.19 When a pair of $2p_x$ orbitals are combined so that they meet edge-on, bonding (π_x) and antibonding (π_x^*) molecular orbitals are formed. The shape of these π molecular orbitals is different from the shape of the σ molecular orbitals. A similar set of bonding (π_y) and antibonding (π_y^*) molecular orbitals is formed when a pair of $2p_y$ atomic orbitals are combined.

$2p_z$) with a similar set of four atomic orbitals on another atom leads to the formation of a total of eight molecular orbitals — σ_{2s}, σ_{2s}^*, σ_p, σ_p^*, π_x, π_y, π_x^*, and π_y^*. The next question to be answered concerns the relative energies of these orbitals.

Both the bonding and antibonding molecular orbitals formed when the $2s$ orbitals are combined have less energy than the molecular orbitals formed when the $2p$ orbitals are combined, because the $2s$ atomic orbitals have a lower energy than the $2p$ orbitals in the isolated atoms. In order to sort out the relative energies of the six molecular orbitals formed when the $2p$ atomic orbitals on a pair of atoms are combined, we need to understand the relationship between the strength of the interaction between a pair of orbitals and the relative energies of the molecular orbitals they form.

RULE 6. The stronger the interaction between a pair of atomic orbitals, the larger the difference between the energies of the bonding and antibonding molecular orbitals they form.

Because they meet head-on instead of edge-on, the interaction between the $2p_z$ orbitals is stronger than the interaction between the $2p_x$ or $2p_y$ orbitals. The result is the relative energies shown in Figure 8.20.

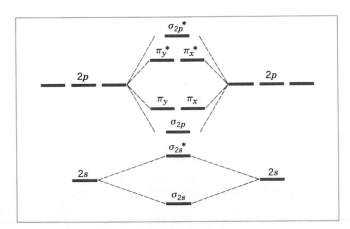

FIG. 8.20 The interaction between the $2p_z$ atomic orbitals that meet head-on is stronger than the interaction between the $2p_x$ or $2p_y$ orbitals that meet edge-on. As a result, the difference between the σ_p and σ_p^* orbitals is larger than the difference between the π_x and π_x^* or π_y and π_y^* orbitals.

FIG. 8.21 When hybridization is added to the molecular orbital model, the energies of the orbitals change slightly.

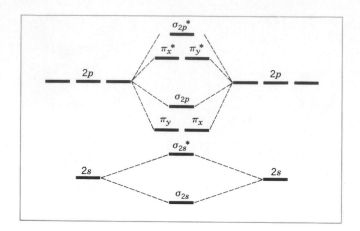

Unfortunately, one interaction is missing from this model. It is possible to envision an interaction between the $2s$ orbital on one atom and the $2p_z$ orbital on the other. This interaction introduces an element of s-p mixing, or hybridization, into the molecular orbital theory. The result is a slight change in the relative energies of the various molecular orbitals, to give the diagram shown in Figure 8.21.

Experiments have shown B_2, C_2, and N_2 are best described by a model that includes hybridization, such as shown in Figure 8.21. But O_2 and F_2 are best described by Figure 8.20.

Exercise 8.20

Construct a molecular orbital diagram for the O_2 molecule.

Solution

There are 6 valence electrons on a neutral oxygen atom and therefore 12 valence electrons in an O_2 molecule. These electrons are added to the diagram in Figure 8.20, one at a time, starting with the lowest-energy molecular orbital.

Because Hund's rules apply to the filling of molecular orbitals, just as to the filling of atomic orbitals, the molecular orbital model for oxygen predicts that there should be two unpaired electrons on this molecule—one electron each in the π_x^* and π_y^* orbitals.

8.18 BOND-ORDER CALCULATIONS FROM MOLECULAR ORBITAL DIAGRAMS

You may recall that the number of bonds between a pair of atoms is called the bond order. Bond orders can be calculated from Lewis structures, which are the heart of the valence-bond model of a molecule. For many molecules, the bond order is obvious. Oxygen, for example, has a bond order of 2.

$$\ddot{O} = \ddot{O}$$

When there is more than one Lewis structure for a molecule, the bond order is an average of these structures. The bond order in sulfur dioxide, for example, is $1\frac{1}{2}$—the average of an S—O single bond in one Lewis structure and an S=O double bond in the other.

$$[:\ddot{O}-\ddot{S}=\ddot{O} \longleftrightarrow \ddot{O}=\ddot{S}-\ddot{O}:]$$

The N—O bond order in the nitrate (NO_3^-) ion is $1\frac{1}{3}$, or the average of one N=O double bond and two N—O single bonds.

In molecular orbital theory, we calculate bond orders by assuming that two electrons in a bonding molecular orbital contribute one net bond and two electrons in an antibonding molecular orbital cancel the effect of one bond.

RULE 7. To calculate the bond order in a diatomic molecule, subtract the number of electrons in antibonding molecular orbitals from the number of electrons in bonding molecular orbitals and then divide by 2.

When writing the electron configuration of an atom, we usually list the orbitals in the order in which they fill.

Pb: $[Xe]\ 6s^2\ 4f^{14}\ 5d^{10}\ 6p^2$

We can write the electron configuration of a molecule, such as O_2, by doing the same thing. Concentrating only on the valence orbitals, we write the electron configuration of O_2 as follows.

$$O_2 = (\sigma_{2s})^2(\sigma_{2s}^*)^2(\sigma_p)^2(\pi_x)^2(\pi_y)^2(\pi_x^*)^1(\pi_y^*)^1$$

We can calculate the bond order in this molecule by noting that this configuration suggests there are eight valence electrons in bonding molecular orbitals and four valence electrons in antibonding molecular orbitals. Thus, the bond order is 2.

$$\text{Bond order} = \frac{\text{bonding electrons} - \text{antibonding electrons}}{2} = \frac{8-4}{2} = 2$$

Although the Lewis structure and molecular orbital models of oxygen yield the same bond order, there is an important difference between these models. The

One of the simplest tests of the molecular orbital theory involves confirmation of the fact that O_2 is paramagnetic. This photograph shows liquid O_2 bridging the gap between the poles of a horseshoe magnet.

electrons in the Lewis structure are all paired, but there are two unpaired electrons in the molecular orbital description of the molecule. This difference provides a basic test of the relative power of the valence-bond and molecular orbital theories for simple molecules such as O_2. The test is based on a characteristic property of molecules with unpaired electrons.

Atoms or molecules in which electrons are all paired are **_diamagnetic_**—they are repelled by both poles of a magnetic. Those that have one or more unpaired electrons are **_paramagnetic_**—they are attracted to a magnetic field. Liquid oxygen is paramagnetic. It is actively attracted to a magnetic field and can actually bridge the gap between the poles of a horseshoe magnet. Since O_2 is paramagnetic, it must contain unpaired electrons. The molecular orbital model of O_2 is therefore superior to the valence-bond model, which cannot explain this property of O_2.

Exercise 8.21

Use the molecular orbital diagram in Figure 8.21 to calculate the bond order in nitrogen oxide (NO). Compare the results of this calculation with the bond order obtained from the Lewis structure.

Solution

There are 11 valence electrons in this molecule, and there are two possible Lewis structures, depending on whether we locate the unpaired electron on the N atom or on the O atom.

$$\ddot{\text{N}}{=}\text{O}\!\cdot\!\!\cdot \longleftrightarrow \cdot\ddot{\text{N}}{=}\text{O}\!\cdot\!\!\cdot$$

Both Lewis structures contain an $\text{N}{=}\text{O}$ double bond, however, so the bond order in the valence bond model for NO is 2.

To calculate the bond order in the molecular orbital model of NO we have to predict the electron configuration of this molecule by adding electrons, one at a time, to the molecular orbitals in Figure 8.21.

$$\text{NO: } (\sigma_{2s})^2(\sigma_{2s}^*)^2(\pi_x)^2(\pi_y)^2(\sigma_p)^2(\pi_x^*)^1$$

There are eight electrons in bonding molecular orbitals and three electrons in antibonding molecular orbitals in this compound. The bond order in the molecular orbital model of NO is therefore $2\frac{1}{2}$.

$$\text{Bond order} = \frac{\text{bonding electrons} - \text{antibonding electrons}}{2} = \frac{8-3}{2} = 2\tfrac{1}{2}$$

The strength of the covalent bond in NO has been shown by experiment to be significantly stronger than a typical $\text{N}{=}\text{O}$ double bond, in agreement with the predictions of the molecular orbital theory for this molecule.

SUMMARY

Chapter 7 provided a model for the bonding in ionic compounds, which are held together by the force of attraction between particles that have opposite electrical charges. Chapter 8 has introduced a model for covalent compounds, which assumes that the atoms in these com-

pounds are held together by the sharing of one or more pairs of valence electrons.

The difference between ionic and covalent bonds is the extent to which electrons are shared by the atoms that form the bonds. When one atom in a bond is much more electronegative than the other, it draws the electrons in the bond closer to itself. The net result is a separation of charge that gives rise to positive and negative ions, which are held together by an ionic bond. When the two atoms in a bond have approximately the same electronegativities, neither atom can draw the electrons away from the other. They therefore share pairs of electrons more or less equally in a covalent bond. When the atoms in the bond have electronegativities that are between these extremes, the bonds are best described as polar, or polar covalent.

At the turn of the century, G. N. Lewis introduced a model for covalent compounds that assumed that the atoms of main-group elements gain, lose, or share electrons until they have an octet of valence-shell electrons (with the exception of hydrogen, which is satisfied with two valence electrons). The octet theory is the basis for writing so-called Lewis structures of molecules. The Lewis structure of a molecule can be used to predict the distribution of electrons in the valence shell of an atom in the molecule. This distribution of electrons can in turn be used to predict the shape, or geometry, of the molecule.

The atomic orbitals introduced in Chapter 5 were defined for use with an isolated atom in the gas phase.

When the time comes to explain the properties of the molecules that atoms form, it is necessary to modify these orbitals. One way of doing this is to mix the atomic orbitals on an atom to form a set of hybrid atomic orbitals. The hybrid orbitals on one atom interact with the orbitals on another atom to form the covalent bonds in a molecule.

Hybrid atomic orbitals and Lewis structures are part of the valence-bond model for covalent compounds. They focus attention on the bonds that can be formed when electrons in valence orbitals on different atoms are combined.

The bonding in covalent compounds can also be explained in terms of a molecular orbital theory. This model starts by mixing orbitals on the atoms to form orbitals that are characteristic of the molecule. Electrons are then added to these molecular orbitals. This theory has the advantage that the electrons can be delocalized over the entire molecule. Molecular orbital theory is a more powerful model than the valence-bond model, because it comes closer to predicting the properties of covalent compounds. It isn't as easy to use as the valence-bond model, however, because different sets of molecular orbitals have to be created for molecules with different shapes, or geometries.

PROBLEMS

Valence Electrons

8-1 Define the term *valence electrons*.

8-2 Describe the relationship between the number of valence electrons on an atom and the most positive oxidation state of the element.

8-3 Determine the number of valence electrons in neutral atoms of the following elements.

(a) Li (b) C (c) Ne (d) Mg (e) Si (f) Ar

8-4 Determine the number of valence electrons in neutral atoms of the following elements.

(a) Fe (b) Cu (c) Pb (d) Hg (e) I

8-5 Determine the number of valence electrons in the following negative ions and describe any general trends.

(a) C^{4-} (b) N^{3-} (c) S^{2-} (d) F^- (e) Cl^- (f) Br^- (g) I^-

8-6 Determine the number of valence electrons in the following positive ions and describe any general trends.

(a) Na^+ (b) Mg^{2+} (c) Al^{3+} (d) Sc^{3+} (e) Ga^{3+} (f) Hg^{2+}

8-7 Explain why filled d and f subshells are ignored when the number of valence electrons on an atom is counted.

8-8 Compare the number of valence electrons on a mercury atom in the $+1$ oxidation state with the number of valence electrons on a neutral hydrogen atom. Use this information to explain why mercury in the $+1$ oxidation state exists as Hg_2^{2+} ions.

8-9 In which group of the periodic table does an element belong if its compounds exhibit oxidation states between $+6$ and -2?

8-10 Calculate the largest positive oxidation state of a tantalum atom. The electron configuration for a neutral Ta atom is $[Xe] 6s^2 4f^{14} 5d^3$.

8-11 Determine the total number of valence electrons in the following molecules or ions.

(a) BF_3 (b) CH_4 (c) NH_3 (d) NH_4^+ (e) SO_4^{2-} (f) H_2SO_4

8-12 Determine the total number of valence electrons in the following molecules or ions.

(a) XeF^+ (b) KrF_2 (c) NF_3 (d) SF_4 (e) PF_5 (f) SiF_6^{2-}
(g) ZrF_7^{3-} (h) UF_7^{3-}

8-13 Which of the following molecules or ions contain the same number of valence electrons?

(a) CO_2 (b) N_2O (c) CNO^- (d) NO_2^+ (e) SO_2 (f) O_3
(g) NO_2^-

The Covalent Bond

8-14 Write electron dot symbols for the elements in the third row of the periodic table.

8-15 Describe how the sharing of a pair of electrons by two chlorine atoms makes the Cl_2 molecule more stable than a pair of isolated Cl atoms.

8-16 Describe the difference between the ionic bond in NaCl and the covalent bond in Cl_2.

8-17 Students sometimes assume that the distinction between ionic and covalent is as clear as black and white. Explain why chemists believe that ionic and covalent are the two extremes of a continuum of differences in the bonding between two atoms.

Electronegativity and Polarity

8-18 Predict whether the following compounds are ionic or covalent by using the general rule that metals combine with nonmetals to form ionic compounds, or salts, and on the basis of differences between the electronegativities of the elements.

(a) OF_2 (b) CS_2 (c) MgO (d) Al_2Cl_6 (e) ZnS

8-19 Predict whether the following compounds are ionic or covalent by using the general rule that metals combine with nonmetals to form ionic compounds, or salts, and on the basis of differences between the electronegativities of the elements.

(a) IF_3 (b) $SiCl_4$ (c) BF_3 (d) Na_2S (e) CaI_2

8-20 Describe the trends in the electronegativities of the elements across a row or down a column of the periodic table.

8-21 Which element in the periodic table has the largest electronegativity? Which has the smallest? Which compound is therefore the most ionic?

8-22 Which of the following elements is the most electronegative?

(a) S (b) As (c) P (d) Se (e) Cl (f) Br

8-23 Which of the following series of elements is arranged in order of decreasing electronegativity?

(a) C, Si, P, As, Se (b) O, P, Al, Mg, K (c) Na, Li, B, N, F (d) K, Mg, Be, O, N (e) Li, Be, B, C, N

8-24 Which of the following elements forms the least ionic bonds with fluorine?

(a) P (b) Ca (c) Al (d) O (e) Se

8-25 Which of the following elements forms the least ionic bonds with oxygen?

(a) Sr (b) In (c) Sb (d) Te (e) Se

8-26 Which of the following compounds is most likely to be on the borderline between ionic and covalent?

(a) MgO (b) $MgBr_2$ (c) Al_2O_3 (d) AlF_3 (e) $AlBr_3$

8-27 Which of the following compounds are best described as polar covalent?

(a) CO (b) H_2O (c) BeF_2 (d) $MgBr_2$ (e) AlI_3 (f) ZnS

8-28 Which of the following compounds are best described as polar covalent?

(a) CaH_2 (b) BrF_5 (c) NF_3 (d) $FeCl_4^-$ (e) UO_2^{2+} (f) $SiCl_4$

Limitations of the Electronegativity Concept

8-29 The electronegativity of an element depends on the oxidation number of the element. Which of the following compounds should be the most covalent?

(a) MnO (b) Mn_2O_3 (c) MnO_2 (d) Mn_2O_7

8-30 Which of the following compounds should be the most ionic?

(a) TiO (b) Ti_2O_3 (c) Ti_4O_7 (d) TiO_2

Writing Lewis Structures

8-31 Write Lewis structures for the following ionic compounds.

(a) NaCl (b) CaF_2 (c) MgO (d) Na_2S (e) AlF_3

8-32 Write Lewis structures for the following ionic compounds.

(a) $MgSO_4$ (b) NH_4Cl (c) Na_3PO_4 (d) $Ca(OCl)_2$

8-33 Write Lewis structures for the following ions or molecules.

(a) NH_3 (b) CH_3^+ (c) H_3O^+ (d) BH_4^- (e) H_2S

8-34 Write Lewis structures for the following ions.

(a) Cl^- (b) O^{2-} (c) O_2^{2-} (d) S_2^{2-} (e) ClO_3^-

8-35 Write Lewis structures for the following ions.

(a) NO_3^- (b) SO_3^{2-} (c) CO_3^{2-} (d) CN^- (e) NO_2^+

8-36 Write Lewis structures for the following compounds.

(a) HCl (b) $MgCl_2$ (c) ICl_3 (d) PCl_3 (e) XeF_4

8-37 Write Lewis structures for the following compounds and then describe how these compounds are unusual.

(a) NO (b) NO_2 (c) ClO_2 (d) ClO_3

8-38 Write Lewis structures for the following ions or molecules.

(a) C_2H_6 (b) C_2H_4 (c) C_2H_2 (d) C_2^{2-} (e) C_2F_4

8-39 Write Lewis structures for the following nitrogen compounds.

(a) N_2O (b) NO (c) NO_2 (d) N_2O_3 $(ON-NO_2)$ (e) N_2O_5 $(O_2N-O-NO_2)$

8-40 Write Lewis structures for the following nitrogen compounds.

(a) ClNO (b) $ClNO_2$ (c) NO^+ (d) NO_2^- (e) ONF_3

8-41 Write Lewis structures for the following ions or molecules.

(a) O_2 (b) O_2^{2-} (c) O^{2-} (d) O_2^- (e) O_3

8-42 Write Lewis structures for the following ions, molecules, or ionic compounds.

(a) S^{2-} (b) H_2S (c) S_2^{2-} (d) H_2S_2 (e) Na_2S_2

8-43 Write Lewis structures for the following ions or molecules.

(a) SO_2 (b) SO_3 (c) SO_3^{2-} (d) SO_4^{2-}
(e) $S_2O_3^{2-}$ ($S-SO_3$)

8-44 Write Lewis structures for the following ions.

(a) Carbonate, CO_3^{2-} (b) Thiocarbonate, SCO_2^{2-}
(c) Dithiocarbonate, OCS_2^{2-}

8-45 Write Lewis structures for the following ions or molecules.

(a) XeF_2 (b) XeF_4 (c) XeF_3^+ (d) XeO_3 (e) $OXeF_4$

8-46 Which of the following elements forms a compound with the following Lewis structure?

$$\overset{..}{\underset{..}{O}}=\overset{..}{X}-\overset{..}{\underset{..}{O}}$$

(a) Al (b) Si (c) P (d) S (e) Cl

8-47 Which of the following elements forms a compound with the following Lewis structure?

$$:\overset{..}{\underset{..}{Cl}}-\overset{..}{X}=\overset{..}{\underset{..}{O}}$$

(a) Be (b) B (c) C (d) N (e) O (f) F

8-48 A pair of NO_2 molecules can combine, or *dimerize*, to form N_2O_4.

$$2\ NO_2(g) \longrightarrow N_2O_4(g)$$

Use the Lewis structures of these compounds to explain why.

8-49 Chemical formulas usually give a good idea of the skeleton structure of a compound, but some formulas are misleading. The formulas of common acids are often written as follows: HNO_3, H_2SO_4, H_3PO_4, $HClO_4$, and so on. Write Lewis structures for these four acids, assuming their skeleton structures are more appropriately described by the following formulas: $HONO_2$, $(HO)_2SO_2$, $(HO)_3PO$, and $HOClO_3$.

8-50 Which of the following ions or molecules have the same electron configuration as the N_2 molecule?

(a) CO (b) NO (c) CN^- (d) NO^+ (e) NO^-

Exceptions to the Lewis Octet Rule

8-51 Which of the following are exceptions to the Lewis octet rule?

(a) CO_2 (b) BeF_2 (c) SF_4 (d) SO_3 (e) PCl_3

8-52 Which of the following are exceptions to the Lewis octet rule?

(a) BF_3 (b) H_2CO (c) XeF_4 (d) H_3O^+ (e) IF_3

Lewis Structures of Molecules with Double or Triple Bonds

8-53 Which of the following compounds does not contain a double bond?

(a) N_2 (b) CO_2 (c) C_2H_4 (d) NO_2 (e) SO_3^{2-}

8-54 Determine the number of S—O single bonds in the Lewis structures of thionyl chloride, $SOCl_2$, and sulfuryl chloride, SO_2Cl_2.

8-55 Which of the following contains only single bonds?

(a) CN^- (b) NO^+ (c) CO (d) O_2^{2-} (e) Cl_2CO

Lewis Structures and Nonbonding Electrons

8-56 Which of the following elements forms an ion with the formula XF_6^{2-} that has no nonbonding electrons in the valence shell of the central atom?

(a) Ca (b) C (c) Si (d) S (e) P

8-57 Determine the number of nonbonding pairs of electrons on the sulfur atom in the following molecules or ions.

(a) SO_2 (b) SO_3 (c) SO_3^{2-} (d) SO_4^{2-}

8-58 Determine the number of nonbonding pairs of electrons on the iodine atom in the following molecules or ions.

(a) I_2 (b) I_3^- (c) IF_3 (d) ICl_4^-

8-59 In which group of the periodic table does the element X belong if there are two pairs of nonbonding electrons in the valence shell of the central atom in the XF_4^- ion?

8-60 One reason for writing Lewis structures is to determine which ions or molecules can act as bases. Bases invariably have one or more pairs of nonbonding electrons. Which of the following ions or molecules can act as a base?

(a) OH^- (b) O_2 (c) CO_3^{2-} (d) Br^- (e) NH_3

8-61 Which of the following cannot act as a base?

(a) H_2S (b) NH_4^+ (c) AlH_3 (d) CH_3^- (e) NH_2^-

Resonance Hybrids

8-62 Draw all of the possible Lewis structures for the CO_3^{2-} ion.

8-63 Which of the following Lewis structures can contribute to the resonance hybrid that describes the electron structure of N_2O?

(a) $:N\equiv N=\overset{..}{\underset{..}{O}}$ (b) $\overset{..}{\underset{..}{N}}=N=\overset{..}{\underset{..}{O}}$ (c) $:\overset{..}{\underset{..}{N}}-N\equiv O:$
(d) $\overset{..}{\underset{..}{N}}=\overset{..}{N}-\overset{..}{\underset{..}{O}}:$ (e) $:N\equiv N-\overset{..}{\underset{..}{O}}:$

8-64 Which of the following molecules or ions are resonance hybrids?

(a) HCO_2^- (b) PH_3 (c) HCN (d) SO_4^{2-} (e) NO_2

8-65 Which of the following molecules or ions is not a resonance hybrid?

(a) CO_2 (b) NO_2^- (c) N_2O (d) SO_3 (e) SCN^-

8-66 Calculate the average S—O bond order in the SO_3 molecule.

8-67 Calculate the average N—O bond order in NO_2, NO_2^-, and NO_3^-.

Formal Charge

8-68 Calculate the formal charge on the sulfur atom in the following molecules or ions.

(a) SO_2 (b) SO_3 (c) SO_3^{2-} (d) SO_4^{2-}

8-69 Calculate the formal charge on the bromine atom in the following molecules or ions.

(a) HBr (b) Br_2 (c) NaBr (d) HOBr (e) BrF_5

8-70 Calculate the formal charge on the nitrogen atom in the following ions, molecules, or compounds.

(a) NH_3 (b) NH_4^+ (c) N_2H_4 (d) NH_2^- (e) Mg_3N_2

8-71 Calculate the formal charge on the nitrogen atom in the following molecules or ions.

(a) N_2O (b) NO (c) NO_2 (d) NO_2^- (e) N_2O_3
(f) NO_3^- (g) N_2O_5

8-72 Calculate the formal charge on the boron and nitrogen atoms in BF_3, NH_3, and the $F_3B—NH_3$ molecule formed when these two gases combine.

8-73 Calculate the formal charge on the two different sulfur atoms in the thiosulfate, $S_2O_3^{2-}$, ion.

Predicting the Shapes of Molecules

8-74 Predict the shape, or geometry, around the central atom in the following molecules or ions.

(a) PH_3 (b) GaH_3 (c) ICl_3 (d) XeF_3^+

8-75 Predict the shape, or geometry, around the central atom in the following molecules or ions.

(a) PO_4^{3-} (b) SO_4^{2-} (c) XeO_4 (d) MnO_4^- (e) ClO_4^-

8-76 Predict the shape, or geometry, around the central atom in the following molecules or ions.

(a) SnF_2 (b) SnF_3^- (c) SnF_4 (d) SnF_6^{2-}

8-77 Predict the shape, or geometry, around the central atom in the following molecules or ions.

(a) SF_3^+ (b) SF_4 (c) SF_5^- (d) SF_6

8-78 Predict the shape, or geometry, around the central atom in the following molecules or ions.

(a) IF_2^- (b) IF_3 (c) IF_4^+ (d) IF_4^- (d) IF_6^-

8-79 Predict the shape, or geometry, around the central atom in the following molecules.

(a) SO_2 (b) SO_3 (c) $SOCl_2$ (d) SO_2Cl_2

8-80 Predict the shape, or geometry, around the central atom in the following molecules or ions.

(a) N_2O (b) NO (c) NO_2 (d) NO_2^- (e) NO_3^-

8-81 Predict the shape, or geometry, around the central atom in the following molecules.

(a) $Hg(CH_3)_2$ (b) $Al_2(CH_3)_6$ (c) $Pb(CH_2CH_3)_4$

8-82 Which of the following compounds is best described as T-shaped?

(a) XeF_3^+ (b) NO_3^- (c) NH_3 (d) ClO_3^- (e) SF_4

8-83 Which of the following molecules or ions have the same shape, or geometry?

(a) NH_2^- and H_2O (b) NH_2^- and BeH_2 (c) H_2O and BeH_2 (d) NH_2^-, H_2O and BeH_2

8-84 Which of the following molecules or ions have the same shape or geometry?

(a) SF_4 and CH_4 (b) CO_2 and H_2O (c) CO_2 and BeH_2 (d) N_2O and NO_2 (e) PCl_4^+ and PCl_4^-

8-85 Which of the following molecules are best described as bent, or angular?

(a) H_2S (b) CO_2 (c) ClNO (d) NH_2^- (e) O_3

8-86 Which of the following molecules are planar?

(a) SO_3 (b) SO_3^{2-} (c) NO_3^- (d) PF_3 (e) BH_3

8-87 Which of the following molecules are tetrahedral?

(a) SiF_4 (b) CH_4 (c) NF_4^+ (d) BF_4^- (e) TeF_4

8-88 Which of the following molecules is linear?

(a) C_2H_2 (b) CO_2 (c) NO_2^- (d) NO_2^+ (e) H_2O

8-89 In which group of the periodic table does Element X belong if the distribution of electrons in the valence shell of the central atom in the XCl_4^- ion is octahedral?

8-90 In which group does Element X belong if the shape of the XF_2^- ion is linear?

8-91 Which of the following elements would be most likely to form a linear compound with the formula XO_2?

(a) Li (b) Be (c) C (d) O (e) F

Hybrid Atomic Orbitals

8-92 Determine the hybridization of the central atom in the following molecules.

(a) BeH_2 (b) BH_3 (c) CH_4 (d) PF_5 (e) SF_6

8-93 Determine the hybridization of the central atom in the following molecules or ions.

(a) NH_3 (b) NH_4^+ (c) NO (d) NO_2 (e) NO_2^+

8-94 Determine the hybridization of the central atom in the following molecules or ions.

(a) CH_4 (b) H_2CO (c) H_2CO_3 or $OC(OH)_2$ (d) HCO_2^-

8-95 Determine the hybridization of the central atom in the following molecules or ions.

(a) $TeCl_4$ (b) BrO_3^- (c) XeF_3^+ (d) Cl_2CO

A Hybrid Atomic Orbital Description of Multiple Bonds

8-96 Write the Lewis structures for molecules of ethylene, C_2H_4, and acetylene, C_2H_2. Use the hybrid atomic orbital model to describe the bonding in these compounds.

8-97 Write the Lewis structures for carbon monoxide and carbon dioxide. Use the hybrid atomic orbital model to describe the bonding in these compounds.

Molecular Orbital Theory for a Diatomic Molecule

8-98 Describe the difference between σ and π molecular orbitals.

8-99 Describe the difference between bonding and antibonding molecular orbitals.

8-100 Explain why the difference between the energies of the σ_{2p} bonding and σ_{2p}^* antibonding orbitals is larger than the difference between the π_{2p} bonding and π_{2p}^* antibonding orbitals.

8-101 Describe the molecular orbitals formed by the overlap of the following atomic orbitals. (Assume that the bond lies along the z axis of the coordinate system.)

(a) $2s + 2s$ (b) $2p_x + 2p_x$ (c) $2p_y + 2p_y$ (d) $2p_z + 2p_z$
(e) $2s + 2p_z$

8-102 Write the electron configuration for the following diatomic molecules and calculate the bond order in each molecule.

(a) H_2 (b) C_2 (c) N_2 (d) O_2 (e) F_2

8-103 Use the molecular orbital theory to predict whether the H_2^+, H_2^- and H_2^{2-} ions should be more or less stable than a neutral H_2 molecule.

8-104 Use the molecular orbital theory to explain why the oxygen-to-oxygen bond is stronger in the O_2 molecule than in the O_2^{2-} ion.

8-105 Use the molecular orbital theory to predict whether the bond order in the superoxide ion, O_2^-, should be stronger or weaker than the bond order in a neutral O_2 molecule.

8-106 Use the molecular orbital theory to predict whether the peroxide ion, O_2^{2-}, should be paramagnetic.

8-107 Write the electron configuration for the following diatomic molecules. Calculate the bond order in each molecule.

(a) HF (b) CO (c) CN^- (d) NO (e) NO^+

8-108 Which of the following molecules is paramagnetic?

(a) HF (b) CO (c) CN^- (d) NO (e) NO^+

CHAPTER 9

THERMOCHEMISTRY

CHAPTER CONTENTS

Every language contains words that are easy to understand but difficult to define. In English, a good example is the word *time.* Everyone understands what it means, but the only way most people can define it is in terms of words that are themselves based on the definition of time. They might say, "Time is something measured with a clock." But when asked to define a clock, they might then say, "A clock is something that measures time." This chapter will focus on four words that everyone uses but few can adequately define — *temperature, heat, energy,* and *work.*

These four concepts serve as the basis for **thermochemistry** — one portion of a more general subject known as **thermodynamics.** The prefix *thermo-* comes from the Greek word for heat. Thermochemistry therefore describes the interaction between heat (or other forms of energy) and chemistry. The suffix *-dynamics* comes from the Greek word for force. Thermodynamics therefore literally means an analysis of the force of heat (or other forms of energy).

We often use the terms *temperature, heat, energy,* and *work* without worrying about exact definitions. Neither the discussion of scales for measuring temperature in Section 1.18 nor the discussion of the effect of temperature on the behavior of gases in Chapter 4 was based on a specific definition of *temperature.* Similarly, Chapter 6 contained discussions of processes that absorb energy

$$Na(g) + energy \longrightarrow Na^+(g) + e^- \qquad \text{(1st IE)}$$

and processes that give off energy

$$Cl(g) + e^- \longrightarrow Cl^-(g) + energy \qquad \text{(EA)}$$

without precisely defining what is meant by the term *energy.*

Discussions of thermodynamics, however, require exact definitions of these four words. Since the concept of temperature lies at the heart of thermodynamics, we will begin our discussion of thermochemistry by defining what is meant by the temperature of an object.

9.1 TEMPERATURE AS AN INTENSIVE PROPERTY OF MATTER

We don't need to be told that the temperature is 95°F (35°C) on a summer's afternoon to know it is "hot." Nor do we need to be told that it is −15°F (−26°C) on a clear winter's night to realize it is "cold." For most purposes, we can rely on our senses to distinguish between hot and cold. But there are times when our senses can be misled.

You have probably had the experience of getting out of a "warm" bed in the middle of the night and stepping onto a "cold" floor. Your senses tell you that the floor is colder than the blanket on your bed. If you then step onto a small throw rug, your senses will tell you that the rug is warmer than the floor. Unfortunately, your senses are wrong. The floor is not colder than the rug or the blanket. All three objects are just as warm (or just as cold) as the air in the room.

Then why do they feel different? The blanket and rug are both examples of **thermal insulators,** substances that tend to prevent an object from becoming hotter or colder. But the floor is not a thermal insulator. Because the bottoms of your feet are usually warmer than the floor on a cold winter's night, your feet become colder for the same reason that a cup of coffee left at room temperature gradually becomes colder.

If you use both metal and plastic ice-cube trays you may encounter a similar phenomenon. When you reach into the freezer, you may notice that metal ice-cube trays feel colder than plastic trays. But this can't be true—all the trays in the freezer are equally cold. Metals are better **thermal conductors,** however. They are better at conducting heat away from our hands and therefore feel colder.

Because our senses can be misled, it is useful to have a reliable, quantitative measure of the degree to which an object is either hot or cold. This quantity is called the **temperature** of the object. Temperature can be measured on either relative or absolute scales (see Figure 1.12).

The Celsius (°C) and Fahrenheit (°F) scales measure *relative* temperatures—the temperature compared with arbitrary standards such as boiling water and an ice-water bath. The difference between the two reference points on these scales is divided into an arbitrary number of *degrees,* much as a circle is divided into 360 degrees.

The Kelvin (K) scale measures *absolute* temperatures—the temperature of an object compared with the temperature at which the volume and pressure of an ideal gas extrapolate to zero. Because absolute scales have only one reference point—absolute zero—these scales can't be divided into degrees. It is therefore a mistake to talk about "degrees kelvin." Absolute temperatures are reported in units of kelvin (K).

Common sense tells us that when two objects at different temperatures are brought together, they eventually reach a point at which their temperatures are the same. This point is called **thermal equilibrium.** The most fundamental law of thermodynamics—the **zeroth law of thermodynamics**—states this idea in more formal language.

> **Zeroth law of thermodynamics: Two objects that are in thermal equilibrium with a third object are in thermal equilibrium with each other.**

Temperature differs from the other quantities measured by the base units of the SI system. The other quantities (length, mass, time, electric current, luminous intensity, and amount of substance) are all *extensive* properties, which depend on the size of the sample (see Section 1.16). Temperature is an *intensive* property, which does not depend on the size of the sample. When we mix two 50-milliliter samples of water at 20°C, the mass doubles, and the volume doubles, but the temperature remains the same.

9.2 HEAT AND HEAT CAPACITY

Temperature is a quantitative measure of whether an object can be labeled "hot" or "cold." **Heat** is the phenomenon that causes a change in temperature. When we "heat" an object, it gets hotter. If we leave the object alone, it cools to room temperature by losing heat to its surroundings.

To explain the relationship between heat and temperature, we'll look at the units with which heat was first measured—the **British thermal unit,** or **Btu.**

> **One Btu is the amount of heat needed to raise the temperature of 1 pound of water by 1°F.**

Exercise 9.1

Calculate the cost of using electricity to heat the water in a 30-gallon water heater from 70°F (room temperature) to 170°F. Assume that 30 gallons of water weighs 250 pounds and that 1 kilowatt-hour of electricity, which sells for 5.8 cents, generates 3400 Btu of heat.

Solution

The definition of the Btu can be used to construct the following unit factor.

$$\frac{1 \text{ Btu}}{1 \text{ lb } H_2O \times 1°F} = 1$$

It therefore takes 25,000 Btu of heat to raise the temperature of 250 pounds of water by 100°F.

$$250 \text{ lb } H_2O \times 100°F \times \frac{1 \text{ Btu}}{1 \text{ lb } H_2O \times 1°F} = \textbf{25,000 Btu}$$

The relationship between a kilowatt-hour of electricity and the amount of heat it can generate can be used to calculate the amount of electricity consumed in this process.

$$25,000 \text{ Btu} \times \frac{1 \text{ kw-hr}}{3400 \text{ Btu}} = \textbf{7.35 kw-hr}$$

The cost of a kilowatt-hour of electricity can in turn be used to calculate the cost of heating the water in a 30-gallon water heater from 70°F to 170°F.

$$7.35 \text{ kw-hr} \times \frac{5.8¢}{\text{kw-hr}} = \textbf{43¢}$$

The fact that an intensive quantity is often related to the ratio of a pair of extensive quantities was introduced in Section 1.16.

intensive quantity

$$\textbf{Density} = \frac{\textbf{mass}}{\textbf{volume}}$$

extensive quantities

This idea can help explain why the temperature of the water in Exercise 9.1 — an intensive property — is related to both the heat that enters the water and the weight of the water, which are both extensive properties.

To explain why the definition of a Btu specifically refers to water requires an understanding of a series of experiments done by Joseph Black between 1759 and 1762. Black studied what happened when liquids at different temperatures were mixed. He noted that when water at 100°F was mixed with an equal volume of water at 150°F, the temperature after mixing was the average of the temperatures of the two samples (125°F). When water at 100°F was mixed with an equal volume of mercury at 150°F, however, the temperature of the liquids after mixing was only 115°F (see Figure 9.1). The temperature of the mercury fell by 35°F, but the temperature of the water increased by only 15°F.

Black assumed that the heat lost by the mercury as it cooled was equal to the heat gained by the water as it became warmer. He therefore concluded that the temperature of the water changed by a smaller amount than that of the mercury because water has a larger "capacity for heat." In other words, it takes more heat to produce a given change in the temperature of water than it does to produce the same change in the temperature of an equivalent volume of mercury.

More than 200 years after Black's experiments, we still talk about the **_heat capacity_** of a substance. Because water has a larger heat capacity than mercury, it takes more heat to raise the temperature of a pound of water by 1°F than it does to

FIG. 9.1 (*a*) When equal volumes of water at different temperatures are combined, the temperature of the mixture is the average of the temperatures of the two samples before mixing. (*b*) When equal volumes of water and mercury at different temperatures are mixed, the temperature of the mixture is much closer to the temperature of the water. We can explain this by assuming that water has a larger heat capacity than mercury—it takes more heat to raise the temperature of a given amount of water by 1°F than an equivalent amount of mercury.

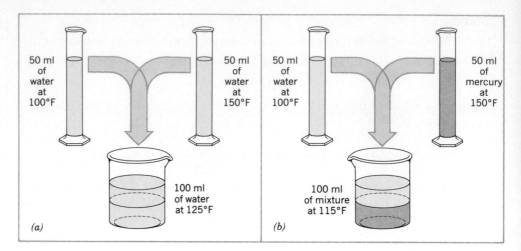

raise the temperature of a pound of mercury by 1°F. Any unit used to measure heat must therefore specify not only the mass of the sample and the change in the temperature of the sample being heated but also the identity of the substance being heated. The Btu was defined so that the heat capacity of water is exactly 1 Btu per pound per degree Fahrenheit (Btu/lb-°F).

In the metric system, the heat capacity of a substance is described in terms of either the specific heat or the molar heat capacity. The ***specific heat*** of a substance is the heat in joules needed to raise the temperature of 1 gram of the substance by 1°C or 1 K. The units of specific heat are therefore joules per gram per degree Celsius (J/g-°C) or per kelvin (J/g-K). The ***molar heat capacity*** is the heat in joules needed to raise the temperature of a mole of the substance by 1°C or 1 K. Molar heat capacity is measured in units of joules per mole per degree Celsius (J/mol-°C) or per kelvin (J/mol-K).

Exercise 9.2

Until the introduction of the SI system, heat was commonly measured in units of calories. The ***calorie*** was often defined as the amount of heat needed to raise the temperature of 1 gram of water from 14.5°C to 15.5°C. The SI unit for heat, the joule, is slightly smaller than a calorie. By definition, 1 calorie is equal to exactly 4.184 joules.

$$\textbf{4.184 J = 1 cal}$$

Use this information to calculate both the specific heat and the molar heat capacity of water.

Solution

It takes 1 calorie to raise the temperature of 1 gram of water by 1°C. The specific heat of water is therefore 1 calorie per gram per degree Celsius (cal/g-°C). Since there are 4.184 joules in a calorie, the specific heat of water is 4.184 J/g-°C.

$$1 \; \frac{\cancel{\text{cal}}}{\text{g-}°\text{C}} \times 4.184 \; \frac{\text{J}}{\cancel{\text{cal}}} = \textbf{4.184 J/g-°C}$$

Because a degree on the Celsius scale is equal to a unit on the Kelvin scale, the specific heat of water is also 4.184 J/g-K.

To find the molar heat capacity of water, we multiply the specific heat of water by its molecular weight.

$$4.184 \ \frac{J}{g\text{-}K} \times 18.015 \ \frac{g \ H_2O}{mol} = 75.376 \ J/mol\text{-}K$$

The molar heat capacity of water is 75.376 J/mol-K.

The specific heat and the molar heat capacities of a number of substances are given in Table 9.1. These data illustrate the fact that the heat capacity of water is unusually large, regardless of whether we compare specific heats or molar heat capacities. This has very important consequences for our weather: the water that covers so much of the surface of our planet acts as a heat sink, which moderates the changes in temperature that occur as the amount of energy absorbed from the sun varies with the sun's distance from the earth.

9.3 LATENT HEAT

The calculations in Exercises 9.1 and 9.2 are based on a hidden assumption—that water always becomes hotter when we heat it. If you believe this is a legitimate assumption, think about what happens when you heat a pot of water. At first, the water gets hotter, but eventually, it starts to boil. From that moment on, the temperature of the water remains the same (100°C), regardless of how fast you heat the water, until the liquid evaporates.

Try to imagine another experiment in which a thermometer is immersed in a snowbank on a day when the temperature finally gets above 32°F. The snow gradually melts as it gains heat from the air above it. But the temperature remains the same—0°C—until the last snow melts.

Joseph Black was the first to recognize the importance of these observations. He used them to differentiate between two kinds of heat—latent and sensible heat. Black defined **sensible heat** as heat that you can sense, or detect. When you heat water at any temperature between its melting point and boiling point, you can sense, or detect, the heat, because the temperature of the water increases.

TABLE 9.1

Specific Heats and Molar Heat Capacities of Common Substances

Substance	Specific Heat (J/g-K)	Molar Heat Capacity (J/mol-K)
$Al(s)$	0.901	24.3
$C(s)$	0.7197	8.644
$Cu(s)$	0.3844	24.43
$H_2O(l)$	4.1840	75.376
$Fe(s)$	0.449	25.1
$Hg(l)$	0.1395	27.98
$O_2(g)$	0.9172	29.35
$N_2(g)$	1.040	29.12
$NaCl(s)$	0.8641	50.50

FIG. 9.2 The data shown here were obtained by the authors when they heated a sample that contained equal amounts of ice and water in a 250-milliliter beaker on a hot plate. The temperature initially remained constant as the ice melted. It then gradually increased until the sample started to boil, at which point it leveled off again.

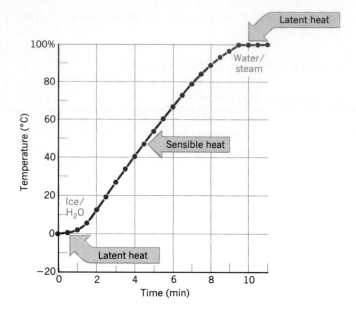

Heat that you can't sense or detect is **latent heat.** The word *latent* is sometimes used to describe things that exist but can't be seen. When a photograph is taken, for example, a latent image is captured. The image exists, but it isn't visible until the film is developed. Mystery novels often talk about latent fingerprints, which can't be seen until the surface of the object is sprayed with powder.

Latent heat is encountered whenever there is a change in the state of matter. Heat can enter or leave a sample without any detectable change in its temperature when a solid melts, when a liquid either freezes or boils, or when a gas condenses to form a liquid.

Figure 9.2 shows results obtained when the authors heated a sample made up of equal weights of ice and water on a hot plate and monitored its temperature. Initially, the heat that entered the system was used to melt the ice — there was no change in the temperature until essentially all of the ice was gone. The amount of heat required to melt the ice is called the **latent heat of fusion.** The word *fusion* is used because it literally refers to melting something. We "blow a fuse" in an electric circuit, for example, by melting the thin piece of metal in the center of the fuse.

Once the ice melted, the heat that entered the system became sensible, and the temperature of the water slowly increased from 0°C to 100°C. As soon as the sample started to boil, however, the heat that entered the sample was used to convert the liquid into a gas, and the temperature of the sample remained constant until the liquid evaporated. The amount of heat required to boil, or vaporize, the liquid is called the **latent heat of vaporization.**

Exercise 9.3

Describe each of the following as an example of either sensible or latent heat.

(a) Cooking a 3-minute egg in boiling water on the burner of a stove that has been turned to its highest setting.

(b) Using a microwave oven to reheat a cup of coffee that has cooled to room temperature.

Solution

(a) This is an example of latent heat. Heat is constantly being added to the boiling water. But once the egg reaches the temperature of boiling water, there is no change in the temperature of either the water or the egg.

(b) The energy in the microwave radiation emitted by the oven is absorbed by the water molecules in the coffee. The net effect is to raise the temperature of the water, which means that this is an example of sensible heat—at least until the water starts to boil.

9.4 THE CALORIC THEORY

Black's experiments served as the basis for the ***caloric theory of heat*** (from the Latin *calor,* "heat"). The caloric theory was based on a series of postulates, or assumptions.

1. Heat is a fluid that flows from a hot object to a cold object.
2. Heat is strongly attracted to matter.
   (Matter can therefore hold a great deal of heat.)
3. Particles of heat repel each other strongly.
   (This postulate was used to explain why gases expand when they are heated. The more heat, or caloric, in the gas particles, the larger the repulsion between particles, and therefore the larger the volume of the gas.)
4. Heat is indestructible.
   (It is therefore conserved in all processes.)
5. Heat can be either sensible or latent.
6. Sensible heat that flows into an object raises the temperature of the object.
7. Latent heat combines with the particles in matter.
   (This postulate was used to explain how a solid could be transformed into a liquid, or a liquid into a gas.)
8. Heat has no appreciable weight.

The caloric theory was so powerful that most of the basic structure of thermodynamics was developed during its lifetime. It also served as the basis for much of the language used to discuss heat. Heat is often described as "flowing" from a hot object to a cold object, as if it were a liquid. We talk about the "heat capacity" of a substance, as if heat were a fluid that could fill its container. Some scientists and engineers even talk about the "heat content" of a system, as if a given amount of heat were stored in the system like a fluid in its container.

Unfortunately, there is a problem with the caloric theory. It's wrong. Heat is *not* a fluid. More importantly, heat is *not* conserved.

9.5 HEAT AND THE KINETIC MOLECULAR THEORY

People who don't understand science often search for a single experiment that proves or disproves a major theory, such as the caloric theory. Some even claim to have found one. There are those, for example, who cite the cannon-boring experiment of Sir Benjamin Thompson (Count Rumford) in 1798 as having sounded the death knell for the caloric theory.

Thompson was an active opponent of the caloric theory. It was one of his ingenious experiments, for example, that showed that heat had no appreciable weight. Thompson set up a balance in the cellar of his laboratories when the temperature was just below freezing. He placed a container of water on one pan of the balance and a container of a salt-water solution on the other. He immersed thermometers in both containers and used small pieces of metal to adjust the weight until the pans were in perfect balance. He then locked the room so that the apparatus could not be disturbed.

Several weeks later, he carefully opened the door and made the following observations.

1. The water was frozen.
2. The salt water was not frozen.
3. Both thermometers showed the same temperature.
4. The system was still in perfect balance.

He concluded that the heat that had escaped from the water on one pan of the balance as it froze must have had no appreciable weight.

Thompson's most famous experiment was done while he was in charge of the military arsenal in Munich, Germany. He noticed that the brass barrel of a cannon invariably became hot while it was being bored. Furthermore, the metal chips that were scraped from the barrel as it was hollowed out were often hotter than the temperature of boiling water. Proponents of the caloric theory explained this by assuming that the friction between the boring tool and the brass cannon barrel "squeezed" heat out of the brass.

Thompson's opposition to the caloric theory led him to measure the amount of heat given off in an experiment in which the amount of metal removed from the cannon barrel was kept to a minimum. He immersed a cannon in water and used a dull tool to scrape the barrel of the cannon. In a period of $2\frac{1}{2}$ hours, only 270 grams of metal shavings were scraped from the cannon barrel. But enough heat was generated to boil $26\frac{1}{2}$ pounds of water.

Thompson believed that this experiment showed that friction was an inexhaustible source of heat. If this is true, then heat is not conserved. According to Thompson, the heat given off in this experiment was not squeezed from the metal; it was generated during the experiment. In hindsight, it is tempting to agree with Thompson. But his contemporaries didn't. They were willing to accept the fact that considerable amounts of heat were given off when metal was rubbed or hammered. But they explained Thompson's results by assuming that metals contained so much heat that this experiment gave off only a small fraction of the total. For as much as 50 years after this experiment was done, the caloric theory remained intact.

Although Thompson's experiments by no means disproved the caloric theory, they marked the turning point in the development of an alternative — the **kinetic theory of heat.** The kinetic theory divides the universe into a system and its surroundings. The **system** is that small portion of the universe in which we are interested. It might consist of a gas trapped in a piston and cylinder or the water in a beaker, as shown in Figure 9.3. The **surroundings** are everything else — in other words, the rest of the universe.

The system and its surroundings are separated by a **boundary.** The boundary can be real, such as the glass in a beaker or the walls of a balloon. It can also be imaginary, such as a line 200 nanometers from the surface of a metal that arbitrar-

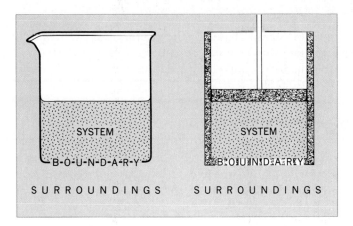

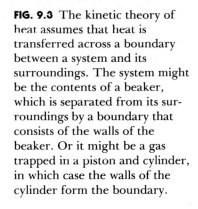

FIG. 9.3 The kinetic theory of heat assumes that heat is transferred across a boundary between a system and its surroundings. The system might be the contents of a beaker, which is separated from its surroundings by a boundary that consists of the walls of the beaker. Or it might be a gas trapped in a piston and cylinder, in which case the walls of the cylinder form the boundary.

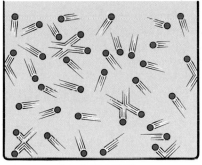

Low temperature

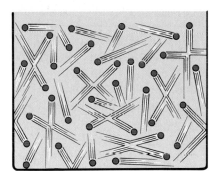

High temperature

FIG. 9.4 According to the kinetic theory of heat, a system becomes hotter because of an increase in the average kinetic energy of the atoms or molecules in the sample.

ily divides the air close to the metal surface from the rest of the atmosphere. The boundary can be rigid, or it can be elastic. It can be assumed to be an ideal conductor of heat or a perfect insulator.

In the kinetic theory, heat is no longer a liquid that can be stored in an object. Heat is something that is transferred across the boundary between a system and its surroundings. In the kinetic theory, heat is no longer conserved — it can be either created or destroyed. The kinetic theory agrees with Thompson's belief that a virtually inexhaustible amount of heat can be generated by the friction of a dull tool rubbing against the barrel of a brass cannon.

The last postulate of the kinetic molecular theory as described in Section 4.15 offers an excellent basis for understanding the kinetic theory of heat.

The average kinetic energy of a collection of gas particles depends on the temperature of the gas, and nothing else.

According to this postulate, a gas becomes warmer because the average kinetic energy of the gas particles becomes larger. When heat enters a system from its surroundings, the net effect is an increase in the kinetic energy of the atoms and molecules in the system. Heat, when it enters a system, is translated into an increase in the speed with which the particles of the system move, as shown in Figure 9.4.

Exercise 9.4

Summarize the principal difference between the caloric theory of heat and the kinetic theory of heat.

Solution

The caloric theory of heat was based on the assumption that heat is conserved. The kinetic theory assumes that heat can be either created or destroyed. The friction between two objects, for example, is a seemingly inexhaustible source of heat.

9.6 WORK

In physics, **work** is done when an object is moved. Two factors determine the amount of work done: (1) the force that must be applied to move the object (F), and (2) the distance the object is moved (d).

$$w = F \times d$$

Exercise 9.5

Calculate the amount of work that has to be done to lift a 10-pound bag of groceries a distance of $2\frac{1}{2}$ feet from the floor to the top of the kitchen cabinet.

Solution

It takes 25 foot-pounds of work to lift a 10-pound bag of groceries a distance of $2\frac{1}{2}$ feet.

$$w = (10 \text{ lb})(2.5 \text{ ft}) = \textbf{25 ft-lb}$$

In the course of this task, we (the system) do 25 foot pounds of work on our surroundings.

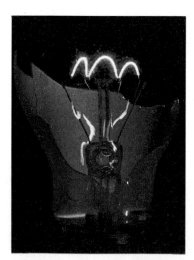

A flashlight is an example of a chemical reaction doing work: an electric current is forced to flow through the filament of the light bulb.

Heat and work both result from an interaction between a system and its surroundings. Both are transferred across the boundary between the system and its surroundings. The simplest way of distinguishing between heat and work is to ask, "What would happen if a thermal insulator, such as a blanket, was placed between the system and its surroundings?" If the thermal insulator would have no effect, the interaction between the system and its surroundings involves work.

Chemical reactions can do two kinds of work: electrical work and work of expansion. **Electrical work** occurs when an electric current is passed through a wire. A flashlight is an example of a device that uses a chemical reaction in the battery to do electrical work in the form of passing an electric current through the filament of the light bulb. **Work of expansion** occurs when the volume of a system changes during a chemical reaction.

To understand how a chemical reaction does work of expansion, imagine a system that consists of ammonia as a gas trapped in a piston and cylinder, as shown in Figure 9.5. Assume, for the moment, that the pressure of the gas pushing up on the piston just balances the weight of the piston, so that the volume of the gas is constant. Now assume that some of the ammonia decomposes to nitrogen and hydrogen.

$$2 \text{ NH}_3(g) \longrightarrow \text{N}_2(g) + 3 \text{ H}_2(g)$$

The net effect of this reaction is to increase the number of gas particles in the container. The volume of the gas must therefore increase, if the temperature and pressure of the gas are held constant.

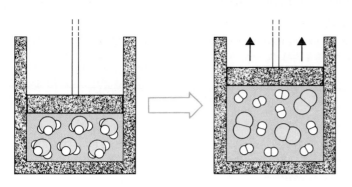

FIG. 9.5 One of the simplest examples of work of expansion involves a gas, such as ammonia, trapped in a piston and cylinder. When the gas decomposes, the number of gas particles increases.

$$2\ NH_3\ (g) \rightarrow N_2(g) + 3\ H_2(g)$$

This causes the gas to expand, pushing the piston partway out of the cylinder. The amount of work done by the system is equal to the product of the force applied to the piston times the distance the piston moves.

The volume of the gas can increase by pushing the piston partway out of the cylinder. This requires moving the piston against the force of gravity, so it involves work. Since the work is generated by the expansion of a gas, it is work of expansion. The amount of work done is equal to the product of the force exerted on the piston times the distance the piston is moved.

$$w = F \times d$$

The relationship between pressure and force was described in Section 4.3. The pressure (P) the gas exerts on the piston is equal to the force (F) with which it pushes up on the piston divided by the surface area (A) of the piston.

$$P = \frac{F}{A}$$

This equation can be rearranged to show that the force exerted by the gas is equal to the product of its pressure times the surface area of the piston.

$$F = P \times A$$

Substituting this expression into the equation defining work gives the following result.

$$w = (P \times A) \times d$$

The product of the area of the piston times the distance the piston moves is equal to the change (Δ) that occurs in the volume of the system when the gas expands.

$$A \times d = \Delta V$$

The magnitude of the work done when a gas expands ($|w|$) is therefore equal to the product of the pressure of the gas times the change in the volume of the gas.

$$|w| = P\ \Delta V$$

9.7 HEAT FROM WORK, AND VICE VERSA

Thompson's cannon-boring experiment suggested that heat and work were related. In a series of experiments begun 40 years later, James Prescott Joule measured the amount of heat that could be produced from a given amount of work.

Joule's most famous experiment used falling weights to do work by turning a paddle immersed in water, mercury, or oil, as shown in Figure 9.6. By measuring the resulting change in the temperature of the liquid, Joule was able to show that

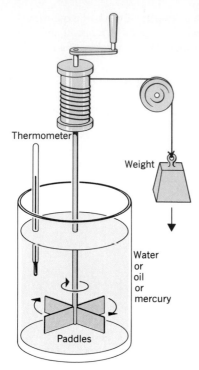

FIG. 9.6 Joule measured the heat that could be generated from a given amount of work by measuring the change in the temperature of a liquid that occurred when the work done by a set of falling weights was used to turn a paddle wheel immersed in the liquid. By measuring the change in the temperature of the liquid to the nearest $\pm 0.005\,^\circ$F, Joule was able to show that 772.5 foot pounds of work could generate 1 Btu of heat.

772.5 foot pounds of work could generate 1 Btu of heat. In recognition of his contributions to this field, the unit of work was eventually named the **joule** (J).

> **One joule is the amount of work done when a force of 1 newton is used to move an object 1 meter (1 J = 1 N-m).**

Since work can be converted into heat, and vice versa, the SI system uses the joule to measure energy in the form of both heat and work.

THE FIRST LAW OF THERMODYNAMICS:
9.8 CONSERVATION OF ENERGY

The relationship between heat and work was first established between 1850 and 1851 by the British physicist William Thomson (Lord Kelvin) and his German contemporary Rudolf Clausius. The result of their efforts was the first law of thermodynamics. Working independently, Thomson and Clausius realized that neither heat nor work is conserved in nature. Only energy is conserved. Their conclusions are summarized in the **first law of thermodynamics.**

> **First law of thermodynamics: Energy is conserved; it can be neither created nor destroyed.**

A system can gain or lose energy. But any change in the energy in the system must be accompanied by an equivalent change in the energy of its surroundings, because the total energy of the universe is constant.

By convention, the magnitude of the *change* in one of the properties of a system is represented by an uppercase Greek delta (Δ) added to the symbol for that property. The magnitude of the change in the energies of a system and its surroundings are therefore described by the following equation.

$$\Delta E_{sys} + \Delta E_{surr} = 0$$

(The subscripts *sys* and *surr* stand for the system and its surroundings.)

The energy of a system is often called its **internal energy,** because it is proportional to the sum of the kinetic and potential energies of the particles that form the system. For an ideal gas, the internal energy is directly proportional to the temperature of the gas, because the potential energy of an ideal gas is zero and the kinetic energy of the gas particles is proportional to the temperature of the gas, as noted in Section 4.15.

$$E = \tfrac{3}{2}RT$$

(In this equation, R is the ideal gas constant and T is the temperature of the gas in units of kelvin. The subscript *sys* has been dropped from this and all subsequent equations for the internal energy of a system to simplify these equations.)

The internal energy for more complex systems can't be measured directly. But changes in the internal energy of the system can be detected if the temperature of the system is monitored. The *change* in the internal energy of the system is defined as the difference between the *initial* and *final* values of this quantity.

$$\Delta E = E_f - E_i$$

Because the internal energy of a system is proportional to its temperature, ΔE is positive when the temperature of the system increases.

9.9 THE FIRST LAW OF THERMODYNAMICS: INTERCONVERSION OF HEAT AND WORK

According to the first law of thermodynamics, energy can be transferred between a system and its surroundings in the form of either heat or work. The change in the internal energy of the system is therefore a balance between the heat (q) and the work (w) that cross the boundary between the system and its surroundings.

$$\Delta E = q + w$$

This equation limits the amount of heat or work that can be extracted from a system. It suggests that an inexhaustible amount of heat can be obtained from the system only if we are willing to do enough work on the system to compensate for the drain on the system's internal energy. Alternatively, the system can do an inexhaustible amount of work only if we are willing to pump enough heat into it to make up for the load on its internal energy.

This equation tells us that if we try to do nothing else but extract either heat or work from a system, the internal energy of the system will decrease to a point at which the system can no longer provide either heat or work. The first law of thermodynamics has therefore been described as suggesting that you can't get something for nothing.

Exercise 9.6

The caloric theory introduced the concept of latent heat so that heat could be conserved. Use the equation for the first law of thermodynamics to explain how the kinetic theory accounts for the phenomenon of latent heat.

Solution

The first law of thermodynamics suggests that heat can do one of two things when it enters a system. It can increase the temperature of the system, or it can do work.

$$\Delta E = q + w$$

Heat that increases the temperature of the system (ΔE = positive) can be detected; it is therefore "sensible" heat. Heat that only does work on the system ($\Delta E = 0$) can't be detected; it is therefore "latent" heat.

The sign convention for the relationship between the internal energy of a system and the heat that crosses the boundary between the system and its surroundings is given on the right side of Figure 9.7.

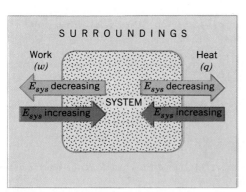

FIG. 9.7 The internal energy of a system becomes smaller (ΔE is negative) when the system loses heat to its surroundings or when the system does work on its surroundings. When heat enters the system from its surroundings or when the surroundings do work on the system, the internal energy of the system becomes larger (ΔE is positive).

1. When the heat that enters a system leads to an increase in the temperature of the system, the internal energy of the system becomes larger, and therefore ΔE is positive.
2. When heat leaves the system so that the temperature becomes smaller, ΔE is negative.

The relationship between work and the internal energy of a system can be explained in terms of the model of a gas trapped in a piston and cylinder shown in Figure 9.5. Work is done by the system when it expands. When this happens, the system becomes cooler — energy is lost to the surroundings. When work is done on the system by its surroundings, the system becomes warmer. This information is summarized in the sign convention for the relationship between work and internal energy on the left side of Figure 9.7.

1. When the system does work on its surroundings, energy is lost, and ΔE is negative.
2. When the surroundings do work on the system, the internal energy of the system becomes larger, so ΔE is positive.

In Section 9.6, the relationship between the magnitude of the work done by a system when it expands and the change in the volume of the system was described by the following equation.

$$|w| = P\,\Delta V$$

Figure 9.7 shows that the sign convention for work of expansion can be included by writing this equation as follows.

$$w = -P\,\Delta V$$

9.10 STATE FUNCTIONS

Every system can be described by certain measurable properties. A gas, for example, can be described in terms of the number of moles of particles it contains, its temperature, pressure, volume, mass, density, internal energy, and so on. These properties describe the *state* of the system, and equations that connect two or more of these properties are called *equations of state.* The ideal gas law is an example of an equation of state.

$$PV = nRT$$

The quantities that describe the state of a system can be sorted into two categories — extensive and intensive. They can also be sorted into categories depending on whether or not they are *state functions.* State functions have the following property.

X is a state function if it depends only on the state of the system, not on the path used to get the system to that state.

In other words, X is a state function if the difference between the initial and final values of this quantity

$$\Delta X = X_f - X_i$$

is the same regardless of the path used to go from the initial to the final state of the system.

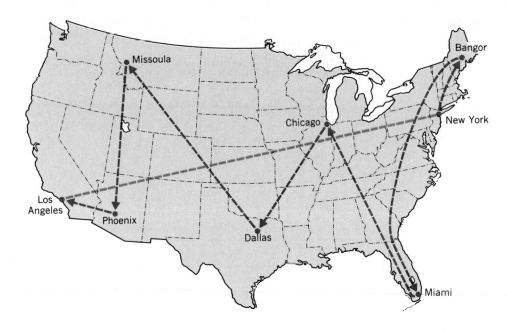

FIG. 9.8 Any property of a system whose value depends only on the state of the system is a state function. The value of a state function does not depend on the path used to take the system from one state to another. Altitude, latitude, and longitude are examples of state functions. The difference between the values of these quantities in New York City and Los Angeles does not depend on the path used to drive from one city to the other. The cost of the trip, the gasoline consumed, the tire wear, the length of the trip in either miles or hours, and the amount of work expended in driving between these cities are *not* state functions. Each of these quantities depends on the path used to go from the initial to the final state.

Imagine a trip from New York to Los Angeles, as shown in Figure 9.8. There are an infinite number of routes that can be taken from one of these cities to the other. For the moment, let's compare a more or less direct route between the two cities with a route that passes through Bangor, Maine; Miami Beach, Florida; Chicago, Illinois; Dallas, Texas; Missoula, Montana; and Phoenix, Arizona.

Any quantity that depends on the route used to go from one city to the other is *not* a state function. A number of extensive quantities fit this classification. The cost of the trip obviously depends on the route. So does the amount of gasoline used, the amount of tire wear, the distance traveled, and both the length of time and the amount of work it takes to drive this distance.

Any quantity that has the same value regardless of the route is a state function. At least three state functions describe this trip—the net change in latitude, longitude, and altitude. The path we take from New York to Los Angeles does not influence the value of any of these quantities.

Another test of a state function involves asking. "What happens if we return from the final state back to the initial state?" If the quantity is a state function, the overall change must be zero, because the path takes us back to the same initial state.

$$\Delta X = X_i - X_i = 0$$

If tire wear were a state function, rubber would wear off the tires as we drove from New York to Los Angeles and then reappear as we drove back to New York!

Exercise 9.7

Determine which of the following properties of an ideal gas are state functions.

(a) Temperature, T
(b) Volume, V
(c) Pressure, P
(d) Number of moles of gas, n
(e) Internal energy, E

Solution

Temperature is a state function. No matter how many times we heat, cool, reheat, recool, expand, compress, or otherwise change the system, the net change in the temperature only depends on the initial and final states of the system.

$$\Delta T = T_f - T_i$$

The same can be said for the volume, pressure, and the number of moles of gas in the sample. These quantities are all state functions.

The internal energy of a gas is directly proportional to the temperature of the gas, which is a state function.

$$E = \tfrac{3}{2}RT$$

This means that the internal energy of the gas is also a state function.

All of these properties of an ideal gas are therefore state functions.

Heat and work are not state functions. Work can't be a state function, because it is proportional to the distance an object is moved, which depends on the path used to go from the initial to the final state. If work isn't a state function, then heat can't be a state function either. The first law of thermodynamics states that the change in the internal energy of a system is equal to the sum of the heat and the work transferred between the system and its surroundings.

$$\Delta E = q + w$$

If ΔE is constant, regardless of the path used to go from the initial to the final state, but w can take different values depending on the path used, then q must also depend on the path.

By convention, the thermodynamic properties of a system that are state functions are usually symbolized by capital letters (T, V, P, E, and so on). Thermodynamic properties that are not state functions are often described by lowercase letters (q and w).

9.11 MEASURING HEAT WITH A CALORIMETER

Thermodynamics can be applied to any system. But the only systems of interest in this chapter are chemical reactions. Furthermore, the question that concerns us right now is, "How much heat is either absorbed or given off by a particular chemical reaction?" Before we can answer this question, we need to address one that is more practical: "How can we measure the heat given off or absorbed by a chemical reaction?"

The first law of thermodynamics suggests that both heat and work can affect the internal energy of a system.

$$\Delta E = q + w$$

But if we can find a way to prevent work from being transferred between the system and its surroundings, the amount of heat (q) given off or absorbed by a chemical reaction will be equal to the change in the system's internal energy (ΔE).

$$\mathbf{\Delta E = q} \qquad \text{(if and only if } w = 0)$$

We can therefore measure the heat given off or absorbed in the reaction by determining the change in temperature that occurs when the reaction is run under conditions in which no work is done by the system on its surroundings, or vice versa.

This raises a related question: "How can we prevent the system from doing work on its surroundings (or vice versa) while we measure the heat given off (or absorbed) by a reaction?" For most chemical reactions, the only kind of work we have to concern ourselves with is work of expansion.

According to Section 9.9, the work done by a system on its surroundings when it expands can be described by the following equation.

$$w = -P \, \Delta V$$

Substituting this relationship into the equation for the first law

$$\Delta E = q + w$$

gives the following result.

$$\Delta E = q - P \, \Delta V$$

We can therefore prevent the system from doing work on its surroundings, or vice versa, by holding the volume of the reaction constant. If there is no change in the volume of the system, there can be no work of expansion.

Heat measured under conditions of constant volume is given the symbol q_V. (The subscript V indicates that the measurement was made at constant volume.) When a chemical reaction is run in a container at constant volume, the change in the internal energy that accompanies the reaction is equal to the heat given off or absorbed by the reaction.

$$\Delta E = q_V \qquad \text{(when } V \text{ is constant and therefore } w = 0\text{)}$$

These measurements are made in an apparatus known as a **bomb calorimeter**. The term *bomb* refers to the fact that the reaction is assumed to take place rapidly, as if the reactants exploded in a bomb. The apparatus is a *calorimeter* because it is literally a meter for measuring calories.

After the reactants have been added to the bomb, it is sealed and then immersed in a container that holds a known amount of water, as shown in Figure 9.9. The

FIG. 9.9 A bomb calorimeter.

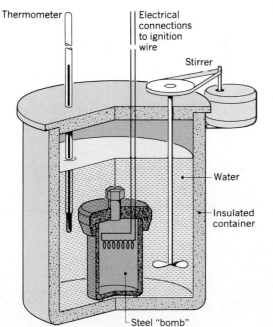

reaction is then initiated by an electric current passed through an ignition wire, and the heat given off or absorbed in the reaction is measured by determining what happens to the temperature of the water bath that surrounds the "bomb."

The heat given off or absorbed by a chemical reaction run in a bomb calorimeter can be calculated from the heat capacity of water (C), given in Table 9.1. As noted in Section 9.2, it takes 75.376 joules of heat to raise the temperature of 1 mole of water by 1°C or 1 K.

Heat capacity of water: **$C = 75.376$ J/mol-K**

The units of this constant suggest that we can calculate the heat absorbed by the water bath from the number of moles of water in the bath and the magnitude of the change in the temperature of the water.

$$\frac{J}{\text{mol-}K} \times \text{mol} \times K = J$$

The heat absorbed by the water is equal to the product of the number of moles of water in the bath times the heat capacity of water times the change in the temperature of the water in kelvin.

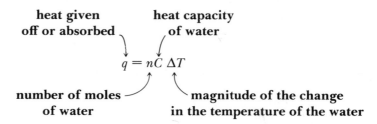

If we assume that the water bath captures all of the heat given off or absorbed by the chemical reaction, the value of q calculated from this equation is equal to the heat of reaction.

Exercise 9.8

The natural gas in methane reacts with oxygen to give carbon dioxide and water.

$$CH_4(g) + 2\ O_2(g) \longrightarrow CO_2(g) + 2\ H_2O(g)$$

Calculate the heat given off when 0.01000 mole of methane reacts with excess oxygen in a bomb calorimeter if the temperature of 1.00 kilogram of water in the bath surrounding the bomb increases by 1.918°C.

Solution

The amount of heat absorbed by the water can be calculated from the number of moles of water that captured this heat (n), the heat capacity of water (C), and the change in the temperature of the water (ΔT). We can therefore start by calculating the number of moles of water in 1.000 kilogram.

$$1.000 \times 10^3\ g \times \frac{1\ \text{mol H}_2\text{O}}{18.015\ g} = 55.51\ \text{mol H}_2\text{O}$$

We then substitute the known values for the amount of water, the heat capacity of water, and the change in the temperature of water into the following equation.

$$q = nC\,\Delta T$$

$$= (55.51\ \text{mol})\left(75.376\ \frac{\text{J}}{\text{mol-K}}\right)(1.918\ \text{K}) = 8025\ \text{J}$$

The reaction therefore gives off 8.025 kilojoules of energy in the form of heat.

Exercise 9.8 illustrates an important point—heat is an extensive quantity. The amount of heat given off when methane is burned depends on the size of the sample. (Doubling the amount of CH_4 in the calorimeter doubles the amount of heat given off.) In order to compare this reaction with other sources of heat, we have to find a way to convert this measurement into an intensive quantity.

We can do this by dividing the heat given off by another extensive quantity that describes the size of the sample. We could calculate the amount of heat given off *per gram* of CH_4 consumed, for example. Since the molecular weight of methane is 16.04 grams per mole, 0.01000 mole of this gas weighs 0.1604 gram. Burning methane therefore gives off 50.03 kilojoules of heat per gram of CH_4 consumed.

$$\frac{8.025\ \text{kJ}}{0.1604\ \text{g}} = 50.03\ \text{kJ/g}$$

The most common approach among chemists for converting measurements of heat into an intensive quantity is to calculate the heat of reaction in units of joules (or kilojoules) *per mole*. The heat of reaction for the combustion of methane, for example, is 802.5 kJ/mol.

$$\frac{8.025\ \text{kJ}}{0.01000\ \text{mol}} = 802.5\ \text{kJ/mol}$$

The concept of **molar heat of reaction** can therefore be defined as follows.

Molar heat of reaction: The amount of heat given off or absorbed by a chemical reaction expressed in units of joules per mole (or kilojoules per mole) of one of the reagents in the reaction.

9.12 ENTHALPY VERSUS INTERNAL ENERGY

Section 9.11 introduced one way of measuring the heat of reaction. We can run the reaction in a steel container immersed in a water bath and watch what happens to the temperature of the water. The heat transferred from the system (the chemical reaction) to its surroundings (the water bath) depends on the amount of water in the bath, the heat capacity of water, and the magnitude of the change in the temperature of the water.

$$q = nC\,\Delta T$$

Because the reaction is run under conditions of constant volume, the system cannot do any work of expansion on it surroundings (or vice versa), so the heat given off or absorbed by the reaction is equal to the change in the internal energy of the system.

$$\Delta E = q_v$$

Chemists, however, seldom run reactions in sealed containers at constant volume. They are much more likely to run reactions in an open flask. The system isn't at constant volume, because gas can enter or leave the flask during the reaction.

Bomb calorimeters measure the heat given off when a chemical reaction is run at constant volume. However, most chemical reactions are run in beakers and flasks in which the volume of the system is allowed to change, but the *pressure* is constant.

The system is at constant *pressure*, however, because the total pressure inside the flask is always equal to atmospheric pressure.

If a gas is driven out of the flask during the reaction, the system will do work of expansion on its surroundings. Alternatively, if the net effect of the reaction is to pull a gas into the flask, the surroundings will do work on the system. If the system does work on its surroundings (or vice versa) during a chemical reaction, the change in the internal energy of the system will be the sum of the heat and the work transferred between the system and its surroundings.

$$\Delta E = q + w$$

The heat of reaction (q) is no longer equal to the change in the internal energy of the system (ΔE). It is now equal to ΔE minus the amount of work done by the system.

$$q = \Delta E - w$$

We can still measure the heat given off or absorbed by the reaction (q). But we can no longer set it equal to the change in the system's internal energy.

Fortunately, there is a way out of this problem. We start by introducing a new concept known as the **enthalpy** of the system—H. Enthalpy is defined as the internal energy (E) of the system plus the product of the pressure (P) of the gas in the system times its volume (V).

$$\boldsymbol{H = E + PV}$$

The *change* in the enthalpy of the system that occurs during a chemical reaction is equal to the difference between the final and initial values for this quantity.

$$\boldsymbol{\Delta H = H_f - H_i}$$

By definition, the change in the enthalpy of a system is equal to the change in the value of the internal energy plus the pressure times the volume of the system.

$$\Delta H = \Delta(E + PV)$$

This in turn is equal to the change in the internal energy of the system plus the change in the product of the pressure times the volume of the system.

$$\Delta H = \Delta E + \Delta(PV)$$

If the pressure of the system is constant, the change in pressure times the volume is equal to the product of the constant pressure times the change in volume.

$$\Delta(PV) = P\,\Delta V \qquad \text{(at constant pressure)}$$

At constant pressure, the change in the enthalpy of the system (ΔH) is therefore the sum of the change in the internal energy (ΔE) plus the pressure times the change in volume ($P\,\Delta V$).

$$\boldsymbol{\Delta H = \Delta E + P\,\Delta V} \qquad \text{(at constant pressure)}$$

Substituting the first law of thermodynamics into this equation gives the following result.

$$\boldsymbol{\Delta H = (q + w) + P\,\Delta V} \qquad \text{(at constant pressure)}$$

Assuming that the only work done by the reaction is work of expansion gives the following equation.

$$\Delta H = (q - P\,\Delta V) + P\,\Delta V) \qquad \text{(at constant pressure)}$$

Canceling terms in this equation tells us that the heat given off or absorbed by a chemical reaction run under conditions of constant pressure is equal to the change in the enthalpy of the system.

$$\Delta H = q_P \qquad \text{(at constant pressure)}$$

The heat given off or absorbed by a chemical reaction run under conditions of constant pressure is therefore equal to the change in the enthalpy associated with this chemical reaction.

$$q_P = \Delta H$$

The change in enthalpy is defined as the final minus the initial value of this quantity.

$$q_P = H_f - H_i$$

When this equation is applied to a chemical reaction, the final state corresponds to the products of the reaction and the initial state, to the reactants. Thus, the heat given off or absorbed by a chemical reaction at constant pressure is equal to the difference between the sum of the enthalpies of the products of the reaction and the sum of the enthalpies of the reactants.

$$q_P = \Sigma H_{products} - \Sigma H_{reactants}$$

This change in the enthalpy of the system as the reactants are converted into the products of the reaction is known as the **enthalpy of reaction** — ΔH.

$$\Delta H = \Sigma H_{products} - \Sigma H_{reactants}$$

The following points summarize our discussion of the relationship between heat, internal energy, and the change in the enthalpy for a chemical reaction.

1. The heat given off or absorbed when a reaction is run at constant volume is equal to the change in the internal energy of the system.

$$\Delta E = q_V$$

2. The heat given off or absorbed when a reaction is run at constant pressure is equal to the change in the enthalpy of the system.

$$\Delta H = q_P$$

3. The change in the enthalpy of the system during a chemical reaction is equal to the change in the internal energy plus the change in the product of the pressure of the gas in the system times its volume.

$$\Delta H = \Delta E + \Delta(PV)$$

4. The difference between ΔH and ΔE is relatively small for reactions that involve only liquids and solids, because there is little if any change in the volume of the system during the reaction. The difference can be relatively large, however, for reactions that involve gases, if there is a change in the number of moles of gas in the course of the reaction.

Exercise 9.9

For which of the following reactions is ΔH about the same as ΔE?

(a) $CaCO_3(s) \longrightarrow CaO(s) + CO_2(g)$

(b) $2\ NH_3(g) \longrightarrow N_2(g) + 3\ H_2(g)$

(c) $CH_4(g) + 2\ O_2(g) \longrightarrow CO_2(g) + 2\ H_2O(g)$

(d) $Fe_2O_3(s) + 2\ Al(s) \longrightarrow Al_2O_3(s) + 2\ Fe(l)$

(e) $NaCl(s) \xrightarrow{H_2O} Na^+(aq) + Cl^-(aq)$

Solution

ΔH is roughly the same as ΔE for reactions (c), (d), and (e). Reactions (d) and (e) involve liquids and solids, so there is no significant change in the volume of the system during these reactions. Reaction (c) involves gases, but the number of moles of gas remains constant, so the volume of the system doesn't change during the reaction. The volume of the system does increase in reactions (a) and (b), however, so ΔH for these reactions is significantly larger than ΔE.

9.13 ENTHALPIES OF REACTION

Chemical reactions are divided into two classes on the basis of whether they give off or absorb heat from their surroundings.

> *Exothermic reactions* **give off heat.**
> *Endothermic reactions* **absorb heat.**

The prefixes *ex-* and *exo-* are often used to mean "out" or "outside," in words such as *exodus* (to go out), *exorcise* (to drive out), *expand* (to spread out), *extinguish* (to put out), *exude* (to give out), and so on. Reactions that give off heat are therefore *exo*thermic.

The prefix *endo-* comes from the Greek word meaning within. Thus, we have words such as *endogenous* (to grow or originate from within), *endoskeleton* (the internal bony structure of vertebrates), *endogamy* (the practice of marrying only within one's tribe), and *endoplasm* (the inner part of the cytoplasm of a cell). *Endo*thermic reactions therefore absorb, or take in, heat from their surroundings.

The heat given off or absorbed in a chemical reaction run at constant pressure is the enthalpy of reaction, ΔH. By definition, the enthalpy of reaction is the difference between the sum of the enthalpies of the products of the reaction and the sum of the enthalpies of the starting materials.

$$\Delta H = \Sigma H_{products} - \Sigma H_{reactants}$$

When a reaction gives off heat to its surroundings, the energy of the system decreases. Exothermic reactions are therefore always characterized by *negative* values of ΔH.

Exothermic reactions: ΔH **is negative**

Endothermic reactions, on the other hand, take in heat from their surroundings, and the energy of the system increases. Endothermic reactions are therefore characterized by *positive* values of ΔH.

Endothermic reactions: ΔH **is positive**

An enormous number of chemical reactions could be cited as examples of exothermic reactions. The reaction between sodium and chlorine to form sodium chloride, for example, gives off 411.15 kilojoules of energy per mole of NaCl formed. There are several ways this information can be added to the balanced equation for the reaction. One approach assumes that the balanced equation is written in terms of moles. When 2 moles of sodium react with a mole of chlorine, for example, 2 moles of sodium chloride are formed. Thus, a total of 822.30 kilojoules of energy is released.

$$2\ Na(s) + Cl_2(g) \longrightarrow 2\ NaCl(s) \qquad \Delta H = -822.30\ kJ$$

Another approach is based on reporting values of the *molar enthalpy of reaction* — the enthalpy of reaction per mole of one of the reactants or products. This approach is indicated as follows.

$$2\ Na(s) + Cl_2(g) \longrightarrow 2\ NaCl(s) \qquad \Delta H = -411.15\ kJ/mol\ NaCl$$

Another exothermic reaction occurs when a balloon filled with H_2 is ignited. In this case, 241.83 kilojoules of energy is released per mole of H_2O formed.

$$2\ H_2(g) + O_2(g) \longrightarrow 2\ H_2O(g) \qquad \Delta H = -241.83\ kJ/mol\ H_2O$$

The thermite reaction is also exothermic; it gives off 851.7 kilojoules per mole of Fe_2O_3 consumed.

$$Fe_2O_3(s) + 2\ Al(s) \longrightarrow Al_2O_3(s) + 2\ Fe(s) \qquad \Delta H = -851.7\ kJ/mol\ Fe_2O_3$$

These reactions all have one thing in common — they give off energy in the form of both heat and light. A few exothermic reactions give off most of their energy in the form of light. More than 300 years ago, Robert Boyle studied the glow given off by decaying wood and glowworms. This phenomenon is now known as *bioluminescence,* and the best known example is the light given off by fireflies. The compounds in fireflies and glowworms responsible for bioluminescence are known as luciferins (from the Latin *lucifer,* "light-bringing"). A similar phenomenon, known as *chemiluminescence,* involves a synthetic analog of luciferin known as luminol.

The reactions described so far are all examples of ***spontaneous reactions.***

Spontaneous reactions behave like a boulder balanced on the side of a mountain. Once a slight push gets them started, they continue spontaneously.

Reactions don't always give off excess energy in the form of heat; they sometimes give it off as light.

Most spontaneous reactions are exothermic. Most endothermic reactions have to be driven by some external force, much as work has to be done to roll a boulder uphill. An example of this phenomenon is the electrolysis of sodium chloride.

$$2\ NaCl(l) \xrightarrow{\text{electric current}} 2\ Na(s) + Cl_2(g)$$

By doing work on the system in the form of passing an electric current through molten sodium chloride, it is possible to drive this reaction "uphill" to form sodium metal and chlorine gas.

A handful of endothermic reactions are spontaneous. One example of a spontaneous endothermic reaction has been used as the basis of a commercial product — an ice pack that doesn't have to be kept in the freezer. These ice packs contain a small quantity of a salt such as ammonium nitrate (NH_4NO_3) or ammonium chloride (NH_4Cl) that is separated from a small sample of water by a thin membrane. When the pack is struck with the palm of the hand, the membrane is broken, and the salt dissolves spontaneously in the water.

$$NH_4NO_3(s) \xrightarrow{H_2O} NH_4^+(aq) + NO_3^-(aq) \qquad \Delta H = 25.7\ kJ/mol$$

Since the reaction is endothermic, it absorbs heat from its surroundings, and the ice pack can get cold enough to take the place of the ice commonly used to treat athletic injuries.

Not all spontaneous chemical reactions are exothermic. In commercial cold packs there is spontaneous reaction in which an ammonium salt such as NH_4NO_3 or NH_4Cl dissolves in water. In these photographs, the reaction in which NH_4NO_3 dissolves in water absorbed enough heat to lower the temperature of the water from 24°C to 1°C.

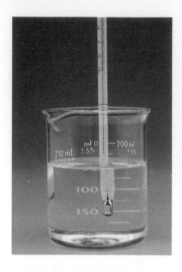 

Another spontaneous endothermic reaction can be demonstrated when ammonium thiocyanate, NH_4SCN, is mixed with barium hydroxide octahydrate, $Ba(OH)_2 \cdot 8\ H_2O$. The starting materials for this reaction are both solids. When they react, however, water is released.

$$2\ NH_4SCN(s) + Ba(OH)_2 \cdot 8\ H_2O(s) \longrightarrow$$
$$Ba^{2+}(aq) + 2\ SCN^-(aq) + 2\ NH_3(aq) + 10\ H_2O(l)$$
$$\Delta H = 130\ kJ/mol\ Ba(OH)_2$$

If a little water is poured on the outside of the beaker in which the reaction is run, stirring the contents of the beaker will produce a reaction that absorbs enough heat from its surroundings to freeze the beaker to a plywood board.

The reaction between NH_4SCN and $Ba(OH)$ is another example of a spontaneous endothermic reaction: $\Delta H = 130\ kJ/mol$ $Ba(OH)_2$. This reaction absorbs so much heat from its surroundings that it can freeze its container to a plywood board if the outside of the container is moistened.

Exercise 9.10

Predict which of the following reactions should be exothermic and which should be endothermic.

(a) $Na(s) \longrightarrow Na(g)$

(b) $Na(g) \longrightarrow Na^+(g) + e^-$

(c) $Cl(g) \longrightarrow 2\ Cl(g)$

(d) $Cl(g) + e^- \longrightarrow Cl^-(g)$

(e) $Na^+(g) + Cl^-(g) \longrightarrow NaCl(s)$

Solution

(a) This reaction should be endothermic, because it takes energy in the form of heat to convert any solid into a gas.

(b) This reaction should be endothermic, because it takes energy to overcome the force of attraction between the electron and the nucleus of the atom and remove an electron from the atom.

(c) This reaction should be endothermic, because it takes energy to break the bonds between atoms in a covalent compound.

(d) This reaction should be exothermic, because energy is released when an element with a high affinity for electrons picks up an extra electron.

(e) This reaction should be exothermic, because oppositely charged objects attract, and energy should therefore be given off when positive and negative ions in the gas phase come together to form a solid.

It takes heat to convert a liquid into a gas. At 100°C, for example, it takes 40.88 kilojoules of energy in the form of heat to convert a mole of liquid water into a mole of water vapor.

$$H_2O(l) \longrightarrow H_2O(g) \qquad \Delta H = 40.88 \text{ kJ/mol}$$

To obtain evidence for the fact that this reaction is endothermic, think about what happens when a moist cloth is placed on your forehead on a hot summer's day. The water gradually evaporates, and your forehead feels cooler. The system (the chemical reaction) absorbs heat from its surroundings (your forehead). The reaction is therefore endothermic.

What does the fact that the evaporation of water is an endothermic reaction tell us about the reverse reaction, the condensation of water vapor to form a liquid?

$$H_2O(g) \longrightarrow H_2O(l)$$

Reversing the direction in which a reaction is written can't change the *magnitude* of the enthalpy of reaction, but only the *sign* of ΔH. The sign of ΔH changes because the initial and final states of the system have been reversed. Heat is therefore given off when water vapor condenses, and ΔH for this reaction at 100°C must be -40.88 kJ/mol.

$$H_2O(g) \longrightarrow H_2O(l) \qquad \Delta H = -40.88 \text{ kJ/mol}$$

9.14 STANDARD-STATE ENTHALPIES OF REACTION

The amount of heat given off or absorbed by a chemical reaction depends on the conditions of the reaction, particularly the temperature and pressure. The combustion of methane can be used to illustrate the magnitude of the problem.

$$CH_4(g) + 2\ O_2(g) \longrightarrow CO_2(g) + 2\ H_2O(g)$$

Let's assume that we start with a mixture of CH_4 and O_2 in which the partial pressure of both gases is 1 atmosphere and the temperature of the system is 25°C. If we run this reaction and let the products cool to 25°C, this reaction will give off a total of 802.4 kilojoules of energy in the form of heat per mole of CH_4 consumed. But if we start with the reactants at 1000°C and 1 atmosphere pressure, and return the products to the same conditions, the reaction will give off only 792.4 kJ/mol. The difference between these numbers is small (10.0 kJ/mol), but it is at least 100 times larger than the experimental error (± 0.1 kJ/mol) with which the measurements were made.

One way out of this problem is to define a set of standard conditions for thermodynamic experiments, so that the results of different experiments can be compared. By definition, the *standard state* for thermodynamic measurements is 1 atmosphere pressure and 25°C. Measurements done under standard-state conditions are indicated by the superscript ° added to the symbol of the quantity being reported. The *standard enthalpy of reaction* for the combustion of natural gas is -802.4 kJ/mol, and the symbol for this quantity is $\Delta H°$.

$$CH_4(g) + 2\ O_2(g) \longrightarrow CO_2(g) + 2\ H_2O(g) \qquad \Delta H° = -802.4 \text{ kJ/mol } CH_4$$

9.15 HESS'S LAW

According to Exercise 9.7, the internal energy, pressure, and volume of a gas are all state functions. The enthalpy of a system is defined in terms of these quantities.

$$H = E + PV$$

Thus, it is also a state function. The difference between the initial and final values of the enthalpy of a system does not depend on the path used to go from one of these states to the other.

In 1840, Germain Henri Hess, professor of chemistry at the University and Artillery School in St. Petersburg, Russia, came to the same conclusion on the basis of experiment and proposed a general rule known as *Hess's law.*

Hess's law: The enthalpy of reaction ($\Delta H°$) is the same regardless of whether a reaction occurs in one step or in several steps.

The following statement is a consequence of Hess's law.

To calculate the enthalpy of reaction ($\Delta H°$), we can add together the enthalpies associated with a series of hypothetical steps into which the reaction can be broken.

Exercise 9.11

The standard enthalpies of reaction for the combustion of carbon to form both carbon monoxide and carbon dioxide have been measured.

$$C(s) + \tfrac{1}{2}O_2(g) \longrightarrow CO(g) \qquad \Delta H° = -110.52 \text{ kJ/mol } CO$$
$$C(s) + O_2(g) \longrightarrow CO_2(g) \qquad \Delta H° = -393.51 \text{ kJ/mol } CO_2$$

Use these data and Hess's law to calculate $\Delta H°$ for the following reaction.

$$CO(g) + \tfrac{1}{2}O_2(g) \longrightarrow CO_2(g)$$

Solution

The key to solving this problem is finding a way to combine the two reactions for which experimental data are known so that they give the reaction for which $\Delta H°$ is unknown. We can achieve this goal by reversing the direction in which the first of the known reactions is written and then adding this to the second known reaction.

$$
\begin{array}{ll}
CO(g) \longrightarrow C(s) + \tfrac{1}{2}O_2(g) & \Delta H° = 1 \text{ mol} \times 110.52 \text{ kJ/mol } CO \\
\underline{C(s) + O_2(g) \longrightarrow CO_2(g)} & \underline{\Delta H° = 1 \text{ mol} \times -393.51 \text{ kJ/mol } CO_2} \\
CO(g) + \tfrac{1}{2}O_2(g) \longrightarrow CO_2(g) & \Delta H° = -282.99 \text{ kJ}
\end{array}
$$

(Note that when we reverse the direction in which the first reaction is written we have to reverse the sign of $\Delta H°$.)

The standard molar enthalpy of reaction for the following reaction is therefore -282.99 kJ per mole of CO_2 produced.

$$CO(g) + \tfrac{1}{2}O_2(g) \longrightarrow CO_2(g) \qquad \Delta H° = -282.99 \text{ kJ/mol } CO_2$$

There was just enough information in Exercise 9.11 to solve the problem. Exercise 9.12 is more realistic because it forces us to choose the reactions that will be combined from a wealth of information.

Exercise 9.12

Before pipelines were built to deliver natural gas, individual towns and cities contained plants that produced a fuel known as "town gas" or "water gas" by passing steam over red-hot charcoal, or coke.

$$C(s) + H_2O(g) \longrightarrow CO(g) + H_2(g)$$

Calculate $\Delta H°$ for this reaction from the following information.

$$
\begin{aligned}
C(s) + \tfrac{1}{2}O_2(g) &\longrightarrow CO(g) & \Delta H° &= -110.52 \text{ kJ/mol } CO \\
C(s) + O_2(g) &\longrightarrow CO_2(g) & \Delta H° &= -393.51 \text{ kJ/mol } CO_2 \\
CO(g) + \tfrac{1}{2}O_2(g) &\longrightarrow CO_2(g) & \Delta H° &= -282.99 \text{ kJ/mol } CO_2 \\
H_2(g) + \tfrac{1}{2}O_2(g) &\longrightarrow H_2O(g) & \Delta H° &= -241.83 \text{ kJ/mol } H_2O
\end{aligned}
$$

Solution

The simplest way of doing this calculation uses the first reaction to generate CO and the reverse of the fourth reaction to generate H_2.

$$
\begin{aligned}
C(s) + \tfrac{1}{2}O_2(g) &\longrightarrow CO(g) & \Delta H° &= 1 \text{ mol} \times -110.52 \text{ kJ/mol } CO \\
\underline{H_2O(g) \longrightarrow H_2(g) + \tfrac{1}{2}O_2(g)} & & \underline{\Delta H° = 1 \text{ mol} \times +241.83 \text{ kJ/mol } H_2O} \\
C(s) + H_2O(g) &\longrightarrow CO(g) + H_2(g) & \Delta H° &= -131.31 \text{ kJ}
\end{aligned}
$$

Combining these two equations gives an overall equation that describes the reaction between charcoal and steam to form town gas. The standard molar enthalpy of the overall reaction is therefore the sum of the standard enthalpies of reaction of these two steps — $\Delta H° = -131.31$ kJ/mol CO.

9.16 ENTHALPIES OF FORMATION

Hess's law suggests that we can save a great deal of work measuring enthalpies of reaction by using a little imagination in choosing the reactions for which measurements are made. The question is, "What is the best set of reactions to study so that we get the greatest benefit from the smallest number of experiments?"

Exercise 9.11 predicted $\Delta H°$ for the reaction

$$CO(g) + \tfrac{1}{2}O_2(g) \longrightarrow CO_2(g)$$

by combining enthalpy of reaction measurements for two reactions.

$$
\begin{aligned}
C(s) + \tfrac{1}{2}O_2(g) &\longrightarrow CO(g) \\
C(s) + O_2(g) &\longrightarrow CO_2(g)
\end{aligned}
$$

The two reactions used to solve this exercise have one thing in common. Each leads to the formation of a compound from its elements in their most stable form at 25°C and 1 atmosphere pressure. The enthalpy of reaction for each of these reactions could therefore be called the *standard enthalpy of formation* of the compound, ΔH_f°.

> **Standard enthalpy of formation: The enthalpy associated with any reaction that forms a substance from its elements in their most stable states at 25°C and 1 atmosphere pressure.**

Exercise 9.13

Which of the following equations describes a reaction for which ΔH° is equal to the standard enthalpy of formation of a compound, ΔH_f°?

(a) $C(s) + O_2(g) \longrightarrow CO_2(g)$ (c) $MgO(s) + CO_2(g) \longrightarrow MgCO_3(s)$

(b) $Mg(s) + \frac{1}{2}O_2(g) \longrightarrow MgO(s)$ (d) $Mg(s) + C(s) + \frac{3}{2}O_2(g) \longrightarrow MgCO_3(s)$

Solution

Equations (a), (b), and (d) describe enthalpy of formation reactions. Each of these reactions results in the formation of a compound from the most stable form of its elements at 25°C and 1 atmosphere. Equation (c) is not an enthalpy of formation reaction, because the product of this reaction is not formed from its elements.

The enthalpy of reaction for any chemical reaction can be calculated from the enthalpies of formation of the products and reactants.

Exercise 9.14

Use Hess's law to calculate ΔH° for the reaction

$$MgO(s) + CO_2(g) \longrightarrow MgCO_3(s)$$

from the following enthalpy of formation data.

$Mg(s) + \frac{1}{2}O_2(g) \longrightarrow MgO(s)$ $\Delta H_f^\circ = -601.70 \text{ kJ/mol MgO}$

$C(s) + O_2(g) \longrightarrow CO_2(g)$ $\Delta H_f^\circ = -393.51 \text{ kJ/mol CO}_2$

$Mg(s) + C(s) + \frac{3}{2}O_2(g) \longrightarrow MgCO_3(s)$ $\Delta H_f^\circ = -1095.8 \text{ kJ/mol MgCO}_3$

Solution

MgO and CO_2 are both reactants in the unknown reaction, so we might start by reversing the first two enthalpy of formation reactions. $MgCO_3$ is a product of the unknown reaction, so we can write this reaction as given.

$MgO(s) \longrightarrow Mg(s) + \frac{1}{2}O_2(g)$ $\Delta H_f^\circ = 1 \text{ mol} \times 601.70 \text{ kJ/mol}$

$CO_2(g) \longrightarrow C(s) + O_2(g)$ $\Delta H_f^\circ = 1 \text{ mol} \times 393.51 \text{ kJ/mol}$

$\underline{Mg(s) + C(s) + \frac{3}{2}O_2(g) \longrightarrow MgCO_3(s)}$ $\underline{\Delta H_f^\circ = 1 \text{ mol} \times -1095.8 \text{ kJ/mol}}$

$MgO(s) + CO_2(g) \longrightarrow MgCO_3(s)$ $\Delta H^\circ = -100.6 \text{ kJ}$

Adding these three equations gives the desired unknown reaction, and ΔH° for this reaction is therefore the sum of the enthalpies of the three hypothetical steps — $\Delta H^\circ = -100.6 \text{ kJ/mol MgCO}_3$.

The procedure used in Exercise 9.14 works no matter how simple or complex the reaction. All we have to do as the reaction becomes more complex is add more hypothetical intermediate steps. Analysis of the results of Exercise 9.14, however, can be used to construct an algorithm that greatly simplifies problems of this nature.

We obtained the answer to this exercise by *adding* the enthalpy of formation of each of the products and *subtracting* the enthalpy of formation of each of the reactants. In general, the enthalpy of reaction for any chemical reaction is equal to the difference between the sum of the enthalpies of formation of the products and the sum of the enthalpies of formation of the reactants.

$$\Delta H° = [\Delta H_f° (MgCO_3)] - [\Delta H_f°(MgO) + \Delta H_f° (CO_2)]$$
$$= [1 \text{ mol } MgCO_3 \times -1095.8 \text{ kJ/mol}]$$
$$- [1 \text{ mol } MgO \times -601.70 \text{ kJ/mol} + 1 \text{ mol } CO_2 \times -393.51 \text{ kJ/mol}]$$
$$= [-1095.8 \text{ kJ}] - [-995.21 \text{ kJ}]$$
$$= -100.6 \text{ kJ}$$

The molar enthalpy of reaction is therefore -100.6 kJ/mol $MgCO_3$.

This formula works because enthalpy is a state function. Thus, $\Delta H°$ is the same regardless of the path used to get from the starting materials to the products of the reaction. Instead of running the reaction in a single step

$$MgO(s) + CO_2(g) \longrightarrow MgCO_3(s)$$

we can split it into two steps. In the first step, the starting materials are converted to the elements from which they form in their most stable states at 25°C and 1 atmosphere.

$$MgO(s) + CO_2(g) \longrightarrow Mg(s) + C(s) + \tfrac{3}{2}O_2(g)$$

In the second step, these elements combine to form the products of the reaction.

$$Mg(s) + C(s) + \tfrac{3}{2}O_2(g) \longrightarrow MgCO_3(s)$$

The first step is the opposite of the enthalpy of formation reactions for the two reactants. The second step corresponds to the enthalpy of formation reaction for the product of the original reaction. Combining these two steps is therefore equivalent to subtracting the sum of the enthalpies of formation of the reactants from the sum of the enthalpies of formation of the products, as shown in Figure 9.10.

Standard state enthalpy of formation data for a variety of elements and compounds can be found in Table A-15 in the appendix. One more point needs to be understood before this table can be used effectively.

FIG. 9.10 $\Delta H°$ for a chemical reaction is equal to the difference between the sums of the enthalpies of formation of the products and the enthalpies of formation of the reactants.

$$\Delta H° = \Sigma \Delta H_f°(\text{products}) - \Sigma \Delta H_f° \text{reactants}$$

This formula works because enthalpy is a state function— $\Delta H°$ is therefore the same regardless of whether the reaction goes directly from the starting materials to the products or through a hypothetical intermediate, which consists of the elements in their most stable states at 25°C and 1 atmosphere. Converting the reactants into their elements is the same as subtracting the enthalpies of formation of these substances, and converting the elements into their products is the same as adding the enthalpies of formation of the products.

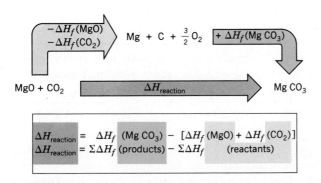

> **The standard-state enthalpy of formation (ΔH_f°) of any element in its most stable form at 25°C and 1 atmosphere pressure is zero.**

The most stable form of oxygen at 25°C and 1 atmosphere, for example, is the diatomic molecule in the gas phase, $O_2(g)$. The enthalpy of formation of this substance is equal to the enthalpy associated with the reaction in which it is formed from its elements in their most stable form at 25°C and 1 atmosphere. For O_2 molecules in the gas phase, ΔH_f° is therefore equal to the heat given off or absorbed in the following reaction.

$$O_2(g) \longrightarrow O_2(g)$$

Because the initial and final states of this reaction are identical, no heat can be given off, so ΔH_f° for $O_2(g)$ is zero.

Exercise 9.15

Predict which of the following substances should have an enthalpy of formation equal to zero.

(a) $Hg(l)$

(b) $Br_2(g)$

(c) $H(g)$

Solution

(a) The enthalpy of formation of liquid mercury is zero, because this element exists as a liquid at room temperature (25°C) and atmospheric pressure (1 atmosphere). This prediction can be confirmed by Table A-15.

(b) Bromine is a liquid at room temperature and atmospheric pressure, not a gas. Since it takes heat to boil a liquid, ΔH_f° for this compound should be positive. Table A-15 lists a value of 30.91 kJ/mol for the standard enthalpy of formation of Br_2 as a gas.

(c) Hydrogen exists as H_2 molecules in the gas phase at room temperature and atmospheric pressure. Because it takes energy to break the bonds in a covalent molecule, ΔH_f° for hydrogen atoms in the gas phase should be positive. Table A-15 lists a value of 217.97 kJ/mol for the standard enthalpy of formation of H atoms in the gas phase.

We are now ready to use standard enthalpy of reaction data to predict enthalpies of reaction.

Exercise 9.16

Pentaborane(9), B_5H_9, was once studied as a potential rocket fuel. Calculate the heat given off when 1 mole of B_5H_9 reacts with excess oxygen according to the following equation.

$$2\,B_5H_9(g) + 12\,O_2(g) \longrightarrow 5\,B_2O_3(s) + 9\,H_2O(g)$$

Solution

Table A-15 contains the following data for the reactants and products of this reaction.

Compound	$\Delta H_f^\circ (kJ/mol)$
$B_5H_9(g)$	73.2
$B_2O_3(s)$	-1272.8
$O_2(g)$	0
$H_2O(g)$	-241.83

The standard enthalpy of reaction (ΔH°) for this reaction is equal to the sum of the standard enthalpies of formation of the products minus the sum of the standard enthalpies of formation of the reactants.

$$\Delta H^\circ = \Sigma \Delta H_f^\circ \text{ (products)} - \Sigma \Delta H_f^\circ \text{ (reactants)}$$

We can therefore calculate ΔH° by multiplying the enthalpies of formation of the reactants and products by the number of moles of each reactant or product involved in the reaction.

$$\begin{aligned} \Delta H^\circ &= \Sigma \Delta H_f^\circ \text{ (products)} - \Sigma \Delta H_f^\circ \text{ (reactants)} \\ &= [5 \text{ mol } B_2O_3 \times -1272.8 \text{ kJ/mol} + 9 \text{ mol } H_2O \times -241.83 \text{ kJ/mol}] \\ &\quad - [2 \text{ mol } B_5H_9 \times 73.2 \text{ kJ/mol} + 12 \text{ mol } O_2 \times 0 \text{ kJ/mol}] \\ &= -8686.9 \text{ kJ} \end{aligned}$$

According to the balanced equation, this is the energy given off when 2 moles of B_5H_9 are consumed. The molar enthalpy of reaction is therefore -4343.5 kJ/mol B_5H_9.

$$\Delta H^\circ = -4343.5 \text{ kJ/mol } B_5H_9$$

9.17 BOND DISSOCIATION ENTHALPIES

When standard enthalpies of formation (ΔH_f°) are known for all the reactants and products of a chemical reaction, ΔH° for the reaction can be calculated with the following formula.

$$\Delta H^\circ = \Sigma \Delta H_f^\circ \text{ (products)} - \Sigma \Delta H_f^\circ \text{ (reactants)}$$

Only a limited number of enthalpies of formation have been measured, however, and it is easy to imagine a reaction for which ΔH_f° data are not available for one or more reagent.

Although we can no longer predict the exact value of ΔH° for the reaction, we can estimate the standard enthalpy of reaction using **bond dissociation enthalpies.**

Bond dissociation enthalpy: The enthalpy of the reaction in which an $X - Y$ bond is broken to give X and Y atoms in the gas phase.

Before we apply bond dissociation enthalpies to calculating ΔH° for a reaction, let's see how these numbers are obtained.

The bond dissociation enthalpy for a C—H single bond can be calculated from the following enthalpy of formation data.

$$\begin{aligned} C(s) + H_2(g) &\longrightarrow CH_4(g) & \Delta H_f^\circ &= -74.81 \text{ kJ/mol } CH_4 \\ C(s) &\longrightarrow C(g) & \Delta H_f^\circ &= 716.68 \text{ kJ/mol } C \\ H_2(g) &\longrightarrow 2 H(g) & \Delta H_f^\circ &= 217.97 \text{ kJ/mol } H \end{aligned}$$

These reactions can be combined to give an equation that corresponds to a reaction in which the four bonds in a CH_4 molecule are broken to give isolated carbon and hydrogen atoms in the gas phase.

$$CH_4(g) \longrightarrow C(s) + 2\,H_2(g) \qquad \Delta H_f^\circ = 1\ mol \times 74.81\ kJ/mol\ CH_4$$
$$C(s) \longrightarrow C(g) \qquad\qquad\qquad \Delta H_f^\circ = 1\ mol \times 716.68\ kJ/mol\ C$$
$$\underline{2\,H_2(g) \longrightarrow 4\,H(g)} \qquad\qquad \underline{\Delta H_f^\circ = 4\ mol \times 217.97\ kJ/mol\ H}$$
$$CH_4(g) \longrightarrow C(g) + 4\,H\,(g) \qquad \Delta H^\circ = \mathbf{1663.37\ kJ}$$

According to this calculation, it takes 1663.37 kilojoules of energy in the form of heat to break the 4 moles of C—H bonds in a mole of CH_4. We can estimate the average C—H bond strength in CH_4 by dividing this number by 4. The bond dissociation enthalpy for a C—H bond is therefore 415.8 kJ/mol.

The results of many calculations such as this are summarized in Table A-14 in the appendix. In order to use this table correctly, it is important to remember the sign convention for ΔH.

Exothermic reactions: ΔH is negative
Endothermic reactions: ΔH is positive

Bond dissociation enthalpies are always positive numbers, because it takes energy to break covalent bonds. When this table is used to estimate the enthalpy associated with the *formation* of a covalent bond, the sign becomes negative, because energy is released when covalent bonds form.

Exercise 9.17

Oxy-acetylene torches, which operate at temperatures as high as 3300°C, are fueled by the combustion of acetylene, C_2H_2.

$$2\,C_2H_2(g) + 5\,O_2(g) \longrightarrow 4\,CO_2(g) + 2\,H_2O(g)$$

(a) Estimate ΔH° for this reaction from the bond dissociation enthalpies in Table A-14.

(b) Compare the results of this calculation with the value of ΔH° obtained from the enthalpies of formation of the reactants and products.

Solution

(a) We can start by writing the Lewis structures of the reactants and products in order to determine how many bonds are broken and how many bonds are formed in the reaction.

$$2\ H-C\equiv C-H + 5\ \ddot{O}=\ddot{O} \longrightarrow 4\ \ddot{O}=C=\ddot{O} + 2\ H-\ddot{O}-H$$

We can transform the starting materials into isolated carbon, hydrogen, and oxygen atoms in the gas phase by breaking 4 moles of C—H single bonds, 1 mole of C≡C triple bonds, and 5 moles of O=O double bonds. We can estimate the energy it takes to break these bonds by multiplying the number of bonds of each kind by the bond dissociation enthalpy for that bond given in Table A-14.

$$4\ C-H\ bonds = (4\ mol)(415\ kJ/mol) = 1660\ kJ$$
$$1\ C\equiv C\ bonds = (1\ mol)(837\ kJ/mol) = 837\ kJ$$
$$5\ O=O\ bonds = (5\ mol)(498\ kJ/mol) = \underline{2490\ kJ}$$
$$total = \mathbf{4987\ kJ}$$

When these isolated atoms in the gas phase recombine to form the products of the reaction, 8 moles of C=O double bonds and 4 moles of H—O single bonds are formed. We can estimate the energy released when these bonds form by reversing the signs of the bond dissociation enthalpies for these bonds given in Table A-14.

$$8 \text{ C} = \text{O bonds} = (8 \text{ mol})(-745 \text{ kJ/mol}) = -5960 \text{ kJ}$$
$$4 \text{ H} - \text{O bonds} = (4 \text{ mol})(-464 \text{ kJ/mol}) = \underline{-1856 \text{ kJ}}$$
$$\text{total} = -7816 \text{ kJ}$$

Adding the energy consumed to break the bonds in the starting materials to the energy given off when the products are formed produces an estimate for the overall enthalpy of reaction of -2989 kJ.

$$4987 \text{ kJ}$$
$$\underline{-7816 \text{ kJ}}$$
$$-2829 \text{ kJ}$$

Since this is the energy given off when 2 moles of acetylene are burned, the estimated value of $\Delta H°$ for this reaction is -1414 kJ/mol C_2H_2.

(b) Bond dissociation enthalpies can only give estimated values for $\Delta H°$. In this case, we can compare the estimate with the actual value of $\Delta H°$, because the standard enthalpies of formation of the reactants and products are all listed in Table A-15.

Compound	$\Delta H_f°(kJ/mol)$
$C_2H_2(g)$	226.7
$O_2(g)$	0
$CO_2(g)$	-393.51
$H_2O(g)$	-241.83

The balanced equation for this reaction is the following.

$$2 \text{ } C_2H_2(g) + 5 \text{ } O_2(g) \longrightarrow 4 \text{ } CO_2(g) + 2 \text{ } H_2O(g)$$

The standard enthalpy of reaction for the combustion of acetylene can therefore be calculated as follows.

$$\Delta H° = \Sigma\Delta H_f°(\text{products}) - \Sigma\Delta H_f°(\text{reactants})$$
$$= [4 \text{ mol } CO_2 \times -393.51 \text{ kJ/mol} + 2 \text{ mol } H_2O \times -241.83 \text{ kJ/mol}]$$
$$- [2 \text{ mol } C_2H_2 \times 226.7 \text{ kJ/mol} + 5 \text{ mol } O_2 \times 0 \text{ kJ/mol}]$$
$$= -2511.10 \text{ kJ}$$

Since 2 moles of acetylene are consumed in the balanced equation for this reaction, the standard molar enthalpy of reaction is -1255.6 kJ/mol C_2H_2. The estimated value of $\Delta H°$ for this reaction, which we obtained from bond dissociation enthalpies in the first part of this problem, is a little more than 10% larger than the actual value of $\Delta H°$ obtained from the enthalpy of formation data.

Exercise 9.18

Dimethyl sulfoxide, or DMSO, is a liquid that has the remarkable ability to penetrate the skin. It therefore has potential as a drug for treating arthritis, because it can be applied directly to the area that causes discomfort. DMSO has not achieved its potential for many reasons; not the least is the fact that, because it is an excellent solvent, any impurities dissolved in it are also carried through the skin, with potentially dangerous side effects.

Assume that DMSO has the following skeleton structure.

$$
\begin{array}{ccc}
\text{H} & \text{O} & \text{H} \\
| & | & | \\
\text{H}-\text{C}-\text{S}-\text{C}-\text{H} \\
| & & | \\
\text{H} & & \text{H}
\end{array}
$$

Use bond dissociation enthalpies to estimate $\Delta H°$ for the combustion of DMSO in the gas phase.

$$2\,(CH_3)_2SO(g) + 9\,O_2(g) \longrightarrow 4\,CO_2(g) + 6\,H_2O(g) + 2\,SO_2(g)$$

Solution

We start by writing Lewis structures for this reaction in order to count the number of bonds that have to be broken among the starting materials and formed among the products of the reaction.

$$
2\,\text{H}-\overset{\overset{\displaystyle H}{|}}{\underset{\underset{\displaystyle H}{|}}{C}}-\overset{\cdot\cdot}{\underset{\cdot\cdot}{S}}-\overset{\overset{\displaystyle H}{|}}{\underset{\underset{\displaystyle H}{|}}{C}}-\text{H} + 9\,\overset{\cdot\cdot}{\underset{\cdot\cdot}{O}}{=}\overset{\cdot\cdot}{\underset{\cdot\cdot}{O}} \longrightarrow
$$

$$
4\,\overset{\cdot\cdot}{\underset{\cdot\cdot}{O}}{=}C{=}\overset{\cdot\cdot}{\underset{\cdot\cdot}{O}} + 6\,\text{H}-\overset{\cdot\cdot}{\underset{\cdot\cdot}{O}}-\text{H} + 2\,\overset{\cdot\cdot}{\underset{\cdot\cdot}{O}}{=}S{-}\overset{\cdot\cdot}{\underset{\cdot\cdot}{O}}{:}
$$

There are 12 moles of C—H single bonds, 4 moles of C—S single bonds, 2 moles of S—O single bonds, and 9 moles of O=O double bonds in the starting materials. The energy it takes to break these bonds can be estimated as follows.

$$
\begin{array}{lll}
12\ \text{C—H bonds} = (12\ \text{mol})(415\ \text{kJ/mol}) = & 4980\ \text{kJ} \\
4\ \text{C—S bonds} = (4\ \text{mol})(270\ \text{kJ/mol}) = & 1080\ \text{kJ} \\
2\ \text{S—O bonds} = (2\ \text{mol})(423\ \text{kJ/mol}) = & 846\ \text{kJ} \\
9\ \text{O=O bonds} = (9\ \text{mol})(498\ \text{kJ/mol}) = & \underline{4482\ \text{kJ}} \\
& \text{total} = \textbf{11,388 kJ}
\end{array}
$$

To estimate the energy released when the products of this reaction are formed, we note that the products contain 8 moles of C=O double bonds, 12 moles of H—O single bonds, 2 moles of S—O single bonds, and 2 moles of S=O double bonds.

$$
\begin{array}{lll}
8\ \text{C=O bonds} = (8\ \text{mol})(-745\ \text{kJ/mol}) = & -5960\ \text{kJ} \\
12\ \text{H—O bonds} = (12\ \text{mol})(-464\ \text{kJ/mol}) = & -5568\ \text{kJ} \\
2\ \text{S—O bonds} = (2\ \text{mol})(-423\ \text{kJ/mol}) = & -846\ \text{kJ} \\
2\ \text{S=O bonds} = (2\ \text{mol})(-523\ \text{kJ/mol}) = & \underline{-1046\ \text{kJ}} \\
& \text{total} = \textbf{-13,420 kJ}
\end{array}
$$

Adding the energy consumed in ripping apart the starting materials to the energy released when the products are formed gives an estimate of the enthalpy of reaction when 2 moles of DMSO in the gas phase burn.

$$
\begin{array}{r}
11,388\ \text{kJ} \\
\underline{-13,420\ \text{kJ}} \\
\textbf{-2032 kJ}
\end{array}
$$

According to this calculation, the enthalpy of combustion for DMSO is -1016 kJ/mol.

SUMMARY

Thermochemistry revolves around four concepts—temperature, heat, work, and energy. Temperature is a quantitative measure of the degree to which an object can be described as "cold" or "hot." Heat is a way of transferring energy between a system and its surroundings that changes the temperature of the system. For an ideal gas, the internal energy can be calculated from the temperature of the system—$E = \frac{3}{2}RT$. The internal energy of more complex systems can't be measured, but *changes* in internal energy can be detected as changes in the temperature of the system. *Work* can be defined as the product of the force used to move an object times the distance the object is moved.

The language used to describe heat was derived from the caloric theory of heat, which assumes that heat is a fluid that is conserved in natural processes. The heat needed to raise the temperature of a given amount of a substance by 1°C or 1 K is therefore described as the heat capacity of the substance.

The caloric theory has been replaced by a kinetic theory, which assumes that heat can be created or destroyed. An inexhaustible amount of heat can be generated by doing work, for example. Heat can also be converted into work. What is conserved in this model is energy. Heat and work are ways of transferring energy across the boundary between a system and its surroundings. The sum of the energies of the system and its surroundings, however, remains constant.

The heat given off or absorbed in a chemical reaction can be measured with a bomb calorimeter. Because the reaction occurs in a sealed container at constant volume, no work of expansion can be done during the reaction. As a result, the heat given off or absorbed by the reaction (q_V) is equal to the change in the internal energy of the system during the course of the reaction (ΔE).

Most chemical reactions, however, are run in open flasks or beakers, in which the volume of the system changes but the *pressure* is constant. By definition, the heat of reaction under conditions of constant pressure (q_P) is equal to the change in the enthalpy of the system during the course of the reaction.

Chemical reactions that give off heat are said to be exothermic. Those that absorb heat are described as endothermic. Most, but not all, spontaneous chemical reactions are exothermic. Usually, an external source of heat or work has to be used to drive endothermic chemical reactions.

The enthalpy of reaction depends on the conditions under which the reaction is run. A standard state for thermochemical measurements has therefore been defined, in which the temperature is 25°C and the pressure of any gas is 1 atmosphere. Any measurement made under the standard-state conditions is indicated a superscript ° added to the symbol for the quantity being measured. Thus, ΔH refers to the enthalpy of reaction run under any set of conditions, but $\Delta H°$ refers to measurements in which the reactants and products are handled in their standard state.

Because enthalpy is a state function, the magnitude of $\Delta H°$ for a reaction does not change when a reaction is broken up into a series of small steps. Enthalpy of reaction data can therefore be combined to predict the exact value of $\Delta H°$ for reactions that have not been studied experimentally. By convention, the data most often tabulated for use in predicting $\Delta H°$ for a reaction are standard-state enthalpies of formation, $\Delta H_f°$. The enthalpy of formation of a compound is the enthalpy associated with the reaction in which this compound is made from its elements in their most stable states at 25°C and 1 atmosphere pressure.

When the value of $\Delta H_f°$ hasn't been measured for one or more of the reactants or products of a chemical reaction, bond dissociation enthalpies can be used to estimate the value of $\Delta H°$ for the reaction. This process involves estimating the enthalpy associated with breaking all of the bonds in the starting materials (which is invariably an endothermic process) and adding the result to the enthalpy associated with the formation of the bonds in the products (which is always an exothermic process).

PROBLEMS

Temperature

9-1 When you touch a motorcycle on a hot August day, the metal feels even hotter than the seat. Explain why.

9-2 It can be a painful experience to sit on the vinyl seats of a car that has been out in the hot summer sun. Explain why it is much less painful if someone has left a towel on the seat or if the seats are covered with lambskin.

9-3 Describe the difference between relative temperature scales such as the Celsius (°C) and Fahrenheit (°F) scales and absolute temperature scales such as the Kelvin (K) scale.

9-4 Explain why temperature is an intensive rather than an extensive property.

9-5 Use the zeroth law of thermodynamics to explain why a cup of coffee cools down but a glass of iced tea warms up.

Heat and Heat Capacity

9-6 Children often confuse heat and temperature. Describe how you would explain the difference to them.

9-7 Describe the units for heat capacity when heat is measured in Btu, in calories, and in joules.

9-8 Calculate the number of joules in 1 Btu of heat.

9-9 Describe the difference between the specific heat of a substance and its molar heat capacity.

9-10 Use the heat capacities of mercury and water in Table 9.1 to explain why the temperature of mercury changes more than the temperature of water when equal volumes of the two liquids at different temperatures are mixed.

9-11 A piece of copper metal weighing 145.0 grams was heated to 100.0°C and then dropped into 250.0 grams of water at 25.0°C. The copper cooled down and the water become warmer until each had a temperature of 28.8°C. Calculate the amount of heat absorbed by the water. Assuming that the heat lost by the copper was absorbed by the water, what is the molar heat capacity of copper metal?

Latent Heat

9-12 Children who are beginning to understand the concepts of heat and temperature often believe that temperature always increases when you heat something. Explain why they are wrong, and describe the example you would use to convince them they are wrong.

9-13 Define the terms *latent heat of fusion* and *latent heat of vaporization,* and classify these quantities as either extensive or intensive.

The Caloric Theory

9-14 Describe the difference between the model of heat based on the caloric theory and the model based on the kinetic molecular theory.

9-15 The caloric theory is based on the assumption that heat is neither created nor destroyed. Give examples from daily experience that suggest that heat is conserved. Give examples that violate this hypothesis.

9-16 Explain how Thompson's cannon-boring experiment was inconsistent with the caloric theory and consistent with the kinetic theory of heat.

The Kinetic Theory of Heat

9-17 Use the kinetic theory of heat to explain what happens when each of the following becomes warmer.

(a) A beaker of water   (b) A balloon filled with helium
(c) A gold coin

9-18 Define the terms *system, surroundings,* and *boundary.* Give three examples of a system separated from its surroundings by a boundary, either real or imaginary.

9-19 Children often believe that things that are hot contain a lot of heat. Use the thermodynamic concepts of system, surroundings, and boundaries to explain why they are wrong.

Work

9-20 Describe the relationship among work, force, and the distance an object is moved.

9-21 Which of the following reactions could do work of expansion on their surroundings?

(a) $CH_4(g) + 2\ O_2(g) \longrightarrow CO_2(g) + 2\ H_2O(g)$
(b) $CaCO_3(s) \longrightarrow CaO(s) + CO_2(g)$
(c) $2\ CO(g) + O_2(g) \longrightarrow 2\ CO_2(g)$
(d) $2\ N_2O(g) \longrightarrow 2\ N_2(g) + O_2(g)$

Heat from Work, and Vice Versa

9-22 Describe a simple test that can be used to decide whether the interaction between a system and its surroundings involves the transfer of heat or of work.

9-23 Describe how each of the following interactions between a system and its surroundings involve the transfer of heat and/or work.

(a) Cooling a glass of lemonade by adding ice cubes
(b) A balloon expanding as more gas is forced into the system   (c) A gas in a cylinder being rapidly compressed by a piston   (d) The combustion of a mixture of gasoline and air forcing the piston out of the cylinder in an internal combustion engine   (e) An electric current being driven through a thin coil of copper wire

The First Law of Thermodynamics

9-24 What physical property of an ideal gas is directly proportional to the internal energy of the gas?

9-25 Describe how the internal energy of an ideal gas can be monitored. Describe how changes in the internal energy of more complex systems can be detected.

9-26 The first law of thermodynamics is often described as saying that energy is conserved. Describe why it is incorrect to assume that the first law suggests that the energy of a system is conserved.

9-27 Describe what happens to the internal energy of a system when the system does work on its surroundings. What happens to the internal energy of the system when it loses heat to its surroundings?

9-28 Give examples of a system doing work on its surroundings and a system losing heat to its surroundings. What happens to the temperature of the system?

9-29 Give examples of a system having work done on it by its surroundings and a system gaining heat from its surroundings. What happens to the temperature of the system?

9-30 Explain why the first law of thermodynamics is often described as suggesting that there is no such thing as a free lunch.

State Functions

9-31 Give examples of at least five physical properties that are state functions.

9-32 Which of the following descriptions of a trip are state functions?

(a) work done   (b) energy expanded   (c) cost   (d) distance traveled   (e) tire wear   (f) gasoline consumed   (g) location of the car   (h) elevation   (i) latitude   (j) longitude

9-33 Which of the following physical properties of a system are not state functions?

(a) temperature   (b) internal energy   (c) enthalpy
(d) pressure   (e) volume   (f) heat   (g) work

9-34 When X is a state function, ΔX can be defined as follows.

$$\Delta X = X_f - X_i$$

Explain why there is a unique value for ΔX for a given set of initial and final states when X is a state function.

Measuring Heat with a Calorimeter

9-35 Chemists often determine the heat given off in a chemical reaction by running the reaction in a bomb calorimeter and measuring the change in the temperature of a sample of water surrounding the calorimeter. How much heat is given off by a reaction that raises the temperature of 740.3 grams of water by 1.85°C?

9-36 Calculate the heat given off when 0.01248 mole of Fe_2O_3 reacts with an excess of powdered aluminum in a bomb calorimeter if the temperature of the 985 grams of water surrounding the calorimeter increases by 2.58°C.

$$Fe_2O_3(s) + 2\ Al(s) \longrightarrow Al_2O_3(s) + 2\ Fe(l)$$

9-37 Calculate the molar heat of reaction for the thermite reaction in Problem 9.36 in units of kilojoules per mole of Fe_2O_3 consumed and in units of kilojoules per mole of iron metal produced.

9-38 Describe the difference between q_V and q_P. When chemists measure the heat given off or absorbed in a reaction they often run the reaction in a steel container, or "bomb." Does this experiment measure q_V or q_P? When students measure the heat given off or absorbed in a reaction, they often run the reaction in a styrofoam cup. Does their experiment measure q_V or q_P?

9-39 The heat given off in a chemical reaction is an extensive quantity that depends on the amount of reactants used. It is converted to an intensive quantity when it is expressed in units of kilojoules per mole of one of the reactants or products of the reaction. Calculate the molar heat of reaction for the combustion of butane if 45.71 kilojoules of heat are released when 1.000 gram of butane is burned.

$$2\ C_4H_{10}(g) + 13\ O_2(g) \longrightarrow 8\ CO_2(g) + 10\ H_2O(g)$$

9-40 In theory, B_5H_9 should be an excellent rocket fuel because of the enormous amount of energy released when this compound burns.

$$2\ B_5H_9(g) + 12\ O_2(g) \longrightarrow 5\ B_2O_3(s) + 9\ H_2O(g)$$

Calculate the molar heat of reaction for the combustion of B_5H_9 if the reaction between 0.100 gram of B_5H_9 and excess oxygen in a bomb calorimeter raises the temperature of the 852 grams of water surrounding the calorimeter by 1.93°C.

Enthalpy versus Internal Energy

9-41 When will the change in the enthalpy associated with a chemical reaction (ΔH) be equal to the change in internal energy (ΔE)?

9-42 For which of the following reactions is ΔH roughly equal to ΔE?

(a) $2\ H_2(g) + O_2(g) \longrightarrow 2\ H_2O(g)$
(b) $Pb(NO_3)_2(s) + 2\ KI(s) \longrightarrow PbI_2(s) + 2\ KNO_3(s)$
(c) $HCl(aq) + NaOH(aq) \longrightarrow NaCl(aq) + H_2O(l)$
(d) $NaOH(s) + CO_2(g) \longrightarrow NaHCO_3(s)$

9-43 If the internal energy of a system is directly proportional to the temperature of the system, what is the advantage of introducing the concept of enthalpy?

Enthalpies of Reaction

9-44 Calculate the energy needed to convert 18 grams of ice at 0°C to steam at 100°C using some or all of the following data.

$2\ H_2(g) + O_2(g) \longrightarrow 2\ H_2O(g)$ $\Delta H° = -241.8$ kJ/mol H_2O
$H_2O(s) \longrightarrow H_2O(l)$ $\Delta H° = 6.03$ kJ/mol H_2O
$H_2O(l) \longrightarrow H_2O(g)$ $\Delta H° = 40.67$ kJ/mol H_2O
$H_2O(l, 0°C) \longrightarrow H_2O(l, 100°C)$ $\Delta H° = 7.53$ kJ/mol H_2O

9-45 Which of the following reactions would you expect to be endothermic?

(a) $H_2(g) \longrightarrow 2\ H(g)$ (b) $2\ H_2(g) + O_2(g) \longrightarrow 2\ H_2O(g)$
(c) $H_2O(g) \longrightarrow H_2O(l)$
(d) $HCl(aq) + NaOH(aq) \longrightarrow NaCl(aq) + H_2O(l)$

9-46 Which of the following reactions would you expect to be endothermic?

(a) $2\ Na(s) + 2\ H_2O(l) \longrightarrow 2\ Na^+(aq) + 2\ OH^-(aq) + H_2(g)$
(b) $2\ Mg(s) + O_2(g) \longrightarrow 2\ MgO(s)$
(c) $2\ NaCl(s) \longrightarrow 2\ Na(s) + Cl_2(g)$
(d) $Na^+(g) + e^- \longrightarrow Na(g)$

9-47 Humans sweat and dogs pant to keep cool. Explain how these processes help keep them cool.

Standard-State Enthalpies of Reaction

9-48 Describe the difference between the enthalpy of reaction — ΔH — and the standard enthalpy of reaction — $\Delta H°$ — for a chemical reaction.

9-49 Explain why there is only one value for the standard enthalpy of reaction for a chemical reaction, whereas there can be many different values for the enthalpy of reaction.

9-50 At which of the following temperatures are standard enthalpy of reaction measurements made?

(a) 0 K   (b) 273.15 K   (c) 0°C   (d) 25°C   (e) 100°C

9-51 How much heat is given off when 1 mole of nitrogen reacts with 2 moles of oxygen to give 2 moles of NO_2 gas, if $\Delta H°$ for the following reaction is 33.2 kilojoules per mole of NO_2?

$$N_2(g) + 2\ O_2(g) \qquad 2\ NO_2(g)$$

9-52 Calculate the standard molar enthalpy of reaction for the following reaction, assuming that 1.00 gram of magnesium gives off 46.22 kilojoules of heat when it reacts with excess fluorine.

$$Mg(s) + F_2(g) \longrightarrow MgF_2(s)$$

9-53 Calculate $\Delta H°$ for the following reaction, assuming that 1.00 gram of hydrogen gives off 4.65 kilojoules of heat when it reacts with 1.00 gram of calcium.

$$Ca(s) + H_2(g) \longrightarrow CaH_2(s)$$

Hess's Law

9-54 Explain how Hess's law is a direct consequence of the fact that the enthalpy of a system is a state function.

9-55 Use the following standard enthalpy of reaction data

$$C(graphite) + O_2(g) \longrightarrow CO_2(g) \quad \Delta H° = -393.51\ kJ/mol\ CO_2$$
$$C(diamond) + O_2(g) \longrightarrow CO_2(g) \quad \Delta H° = -395.94\ kJ/mol\ CO_2$$

to calculate $\Delta H°$ for the conversion of graphite into diamond.

9-56 Use the following standard enthalpy of reaction data

$$2\ H_2(g) + O_2(g) \longrightarrow 2\ H_2O(aq) \quad \Delta H° = -285.83\ kJ/mol\ H_2O$$
$$H_2(g) + O_2(g) \longrightarrow H_2O_2(aq) \qquad \Delta H° = -187.8\ kJ/mol\ H_2O_2$$

to calculate $\Delta H°$ for the decomposition of hydrogen peroxide.

$$2\ H_2O_2(aq) \longrightarrow 2\ H_2O(aq) + O_2(g)$$

9-57 Use the following standard enthalpy of reaction data

$$N_2(g) + 2\ O_2(g) \longrightarrow 2\ NO_2(g) \quad \Delta H° = 33.2\ kJ/mol\ NO_2$$
$$2\ NO(g) + O_2(g) \longrightarrow 2\ NO_2(g) \quad \Delta H° = -57.1\ kJ/mol\ NO_2$$

to calculate $\Delta H°$ for the following reaction.

$$N_2(g) + O_2(g) \longrightarrow 2\ NO(g)$$

9-58 Use the following standard enthalpy of reaction data

$$2\ C(s) + H_2(g) \longrightarrow C_2H_2(g) \quad \Delta H° = 226.7\ kJ/mol\ C_2H_2$$
$$6\ C(s) + 3\ H_2(g) \longrightarrow C_6H_6(g) \quad \Delta H° = 82.93\ kJ/mol\ C_6H_6$$

to calculate $\Delta H°$ for the following reaction.

$$3\ C_2H_2(g) \longrightarrow C_6H_6(g)$$

9-59 Use the following standard enthalpy of reaction data

$$N_2(g) + \tfrac{3}{2}\ O_2(g) + H_2(g) \longrightarrow 2\ HNO_3(aq)$$
$$\Delta H° = -207.4\ kJ/mol$$
$$N_2O_5(g) + H_2O(l) \longrightarrow 2\ HNO_3(aq) \quad \Delta H° = 218.4\ kJ/mol$$
$$2\ H_2(g) + O_2(g) \longrightarrow 2\ H_2O(g) \qquad \Delta H° = -285.8\ kJ/mol$$

to calculate $\Delta H°$ for the following reaction.

$$2\ N_2(g) + 5\ O_2(g) \longrightarrow 2\ N_2O_5(g)$$

9-60 Use the following standard enthalpy of reaction data to calculate the first ionization energy of hydrogen in kilojoules per mole.

$$H_2(g) + Cl_2(g) \longrightarrow 2\ HCl(g) \quad \Delta H° = -92.31\ kJ/mol\ HCl$$
$$H_2(g) \longrightarrow 2\ H(g) \qquad \Delta H° = 435.94\ kJ/mol\ H_2$$
$$Cl_2(g) \longrightarrow 2\ Cl(g) \qquad \Delta H° = 243.36\ kJ/mol\ Cl_2$$
$$Cl(g) + e^- \longrightarrow Cl^-(g) \qquad \Delta H° = -348.79\ kJ/mol\ Cl^-$$
$$H^+(g) + Cl^-(g) \longrightarrow HCl(g) \quad \Delta H° = -1395.38\ kJ/mol\ HCl$$

9-61 Use the following standard enthalpy of reaction data

$$3\ C(s) + 4\ H_2(g) \longrightarrow C_3H_8(g) \quad \Delta H° = -103.85\ kJ/mol\ C_3H_8$$
$$C(s) + O_2(g) \longrightarrow CO_2(g) \qquad \Delta H° = -393.51\ kJ/mol\ CO_2$$
$$H_2(g) + \tfrac{1}{2}\ O_2(g) \longrightarrow H_2O(g) \quad \Delta H° = -241.83\ kJ/mol\ H_2O$$

to calculate the heat of combustion of propane, C_3H_8.

$$C_3H_8(g) + O_2(g) \longrightarrow 3\ CO_2(g) + 4\ H_2O(g)$$

9-62 Use the following standard enthalpy of reaction data

$$C_4H_9OH(l) + 6\ O_2(g) \longrightarrow 4\ CO_2(g) + 5\ H_2O(g)$$
$$\Delta H° = -2456.1\ kJ/mol\ C_4H_9OH$$
$$(C_2H_5)_2O(l) + 6\ O_2(g) \longrightarrow 4\ CO_2(g) + 5\ H_2O(g)$$
$$\Delta H° = -2510.0\ kJ/mol\ (C_2H_5)_2O$$

to calculate $\Delta H°$ for the following reaction.

$$(C_2H_5)_2O(l) \longrightarrow C_4H_9OH(l)$$

Enthalpies of Formation

9-63 For which of the following substances is $\Delta H_f°$ equal to zero?

(a) $H_2(g)$ (b) $H_2O(g)$ (c) $H_2O(l)$ (d) $O_3(g)$ (e) $Cl(g)$
(f) $F_2(g)$ (g) $Na(g)$ (h) $P_4(s)$

9-64 Use the standard enthalpies of formation in the appendix to determine whether heat is given off or absorbed when limestone is converted to lime and carbon dioxide.

$$CaCO_3(s) \longrightarrow CaO(s) + CO_2(g)$$

9-65 Calculate $\Delta H°$ for the following reaction from the standard enthalpies of formation in the appendix.

$$2\ Na(s) + Cl_2(g) \longrightarrow 2\ NaCl(s)$$

9-66 Calculate $\Delta H°$ for the following reaction from the standard enthalpies of formation in the appendix.

$$CO(g) + NH_3(g) \longrightarrow HCN(g) + H_2O(g)$$

9-67 Calculate $\Delta H°$ for the following reaction from the standard enthalpies of formation in the appendix.

$$PH_3(g) + 2\ O_2(g) \longrightarrow H_3PO_4(s)$$

9-68 Calculate $\Delta H°$ for the following reaction from the standard enthalpies of formation in the appendix.

$$CS_2(l) + 3\ O_2(g) \longrightarrow CO_2(g) + 2\ SO_2(g)$$

9-69 Calculate $\Delta H°$ for the combustion of butane from the standard enthalpies of formation in the appendix.

$$2\ C_4H_{10}(g) + 13\ O_2(g) \longrightarrow 8\ CO_2(g) + 10\ H_2O(g)$$

9-70 Calculate $\Delta H°$ for the following reaction from the standard enthalpies of formation in the appendix.

$$4\ NH_3(g) + 5\ O_2(g) \longrightarrow 4\ NO(g) + 6\ H_2O(g)$$

9-71 Calculate $\Delta H°$ for the following reaction from the standard enthalpies of formation in the appendix.

$$P_4O_{10}(s) + 6\ H_2O(aq) \longrightarrow 4\ H_3PO_4(aq)$$

9-72 Calculate $\Delta H°$ for the following reaction from the standard enthalpies of formation in the appendix.

$$2\ KClO_3(s) \longrightarrow 2\ KCl(s) + 3\ O_2(g)$$

9-73 Which of the following reactions gives off the most heat per mole of aluminum consumed?

$$Fe_2O_3(s) + 2\ Al(s) \longrightarrow 2\ Fe(s) + Al_2O_3(s)$$
$$Cr_2O_3(s) + 2\ Al(s) \longrightarrow 2\ Cr(s) + Al_2O_3(s)$$

9-74 Use the standard enthalpies of formation of O_2, CO_2, and H_2O to calculate the standard enthalpy of formation of ethanol, CH_3CH_2OH, assuming that ethanol gives off 1,277.41 kJ/mol when burned.

$$CH_3CH_2OH(l) + 3\ O_2(g) \longrightarrow 2\ CO_2(g) + 3\ H_2O(g)$$

9-75 Calculate the enthalpy of formation of benzoic acid, $C_6H_5CO_2H$, assuming that this compound gives off 3,095.0 kJ/mol when burned.

$$2\ C_6H_5CO_2H(s) + 16\ O_2(g) \longrightarrow 14\ CO_2(g) + 6\ H_2O(g)$$

9-76 Use the standard enthalpy of combustion for methane, given below, to estimate the energy released when 100 cubic feet of natural gas is burned.

$$CH_4(g) + 2\ O_2(g) \longrightarrow CO_2(g) + 2\ H_2O(g)\ \Delta H°$$
$$= -802.36\ kJ/mol\ CH_4$$

Bond Dissociation Enthalpies

9-77 Estimate the enthalpy of formation of Cl atoms in the gas phase from the bond dissociation enthalpy of the Cl—Cl bond.

9-78 Use bond dissociation enthalpies to estimate $\Delta H°$ for the following reaction.

$$H_2C{=}CH_2(g) + H_2(g) \longrightarrow H_3C{-}CH_3(g)$$

Compare the results of this calculation with the value of $\Delta H°$ calculated from enthalpies of formation.

9-79 Calculate the bond dissociation enthalpy for the N≡N triple bond in the N_2 molecule from the N—H and H—H bond dissociation enthalpies and standard enthalpy of reaction for the following reaction.

$$N_2(g) + 3\ H_2(g) \longrightarrow 2\ NH_3(g)$$
$$\Delta H° = -46.1\ kJ/mol\ NH_3$$

9-80 Use bond dissociation enthalpies to estimate the energy given off when the iso-octane used in octane ratings is burned.

$$CH_3{-}\underset{\underset{CH_3}{|}}{\overset{\overset{CH_3}{|}}{CH}}{-}\underset{\underset{CH_3}{|}}{C}{-}CH_2{-}CH_3 + \tfrac{25}{2}\ O_2 \longrightarrow 8\ CO_2 + 9\ H_2O$$

9-81 Use bond dissociation enthalpies to estimate the standard enthalpy of formation of HCl.

$$H_2(g) + Cl_2(g) \longrightarrow 2\ HCl(g)$$

Compare your estimate with the value of $\Delta H_f°$ for HCl.

9-82 Calculate the H—O bond dissociation enthalpy from the H—H and O=O bond dissociation enthalpies and the standard enthalpy of formation of H_2O in the gas phase.

$$2\ H_2(g) + O_2(g) \longrightarrow 2\ H_2O(g)$$
$$\Delta H° = -241.83\ kJ/mol\ H_2O$$

9-83 Use bond dissociation enthalpies to estimate the enthalpy of reaction for the combustion of CS_2 in the gas phase.

$$CS_2(g) + 3\ O_2(g) \longrightarrow CO_2(g) + 2\ SO_2(g)$$

9-84 Methanol, CH_3OH or H_3COH, has been proposed as an alternative to gasoline for use as a fuel. Use bond dissociation enthalpies to estimate the energy given off when methanol burns in the gas phase.

$$2\ H_3C\ddot{O}H + 3\ \overset{\cdot\cdot}{\underset{\cdot\cdot}{O}}{=}\overset{\cdot\cdot}{\underset{\cdot\cdot}{O}} \longrightarrow 2\ \overset{\cdot\cdot}{\underset{\cdot\cdot}{O}}{=}C{=}\overset{\cdot\cdot}{\underset{\cdot\cdot}{O}} + 4\ H{-}\overset{\cdot\cdot}{\underset{\cdot\cdot}{O}}{-}H$$

9-85 Estimate the first ionization energy of hydrogen from the enthalpy of formation of $HCl(g)$ and $H^+(g)$ and the H—Cl, H—H, and Cl—Cl bond dissociation enthalpies.

CHAPTER 10

THE CHEMISTRY OF THE NONMETALS

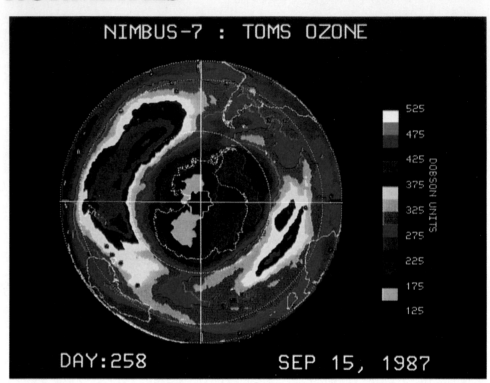

NIMBUS-7 : TOMS OZONE

DAY:258   SEP 15, 1987

H																H	He
Li	Be											B	C	N	O	F	Ne
Na	Mg											Al	Si	P	S	Cl	Ar
K	Ca	Sc	Ti	V	Cr	Mn	Fe	Co	Ni	Cu	Zn	Ga	Ge	As	Se	Br	Kr
Rb	Sr	Y	Zr	Nb	Mo	Tc	Ru	Rh	Pd	Ag	Cd	In	Sn	Sb	Te	I	Xe
Cs	Ba	La	Hf	Ta	W	Re	Os	Ir	Pt	Au	Hg	Tl	Pb	Bi	Po	At	Rn
Fr	Ra	Ac	104	105	106	107		109									

Metals
Nonmetals
Semimetals

Ce	Pr	Nd	Pm	Sm	Eu	Gd	Tb	Dy	Ho	Er	Tm	Yb	Lu
Th	Pa	U	Np	Pu	Am	Cm	Bk	Cf	Es	Fm	Md	No	Lr

FIG. 10.1 The elements can be divided into three classes — metals, semimetals, and nonmetals.

Roughly three-quarters of the elements in the periodic table are metals (see Figure 10.1). The characteristic properties of these elements were described in Section 2.7. They have a metallic luster, they are malleable and ductile, and they conduct heat and electricity. Metals also tend to combine with elements that aren't metals to form compounds in which the metal is present as a positive ion (such as NaCl), or at least has a positive oxidation number (such as MgI_2).

Eight other elements (B, Si, Ge, As, Sb, Te, Po, and At) are best described as semimetals, or metalloids. They often look like metals, but they tend to be brittle. They are more likely to be semiconductors than conductors of electricity. They form compounds in which the semimetal has a positive oxidation number, such as B_2O_3 and SiO_2. But they also form compounds in which the semimetal has a negative oxidation number, such as BH_3 and SiH_4.

Once the metals and semimetals are removed from the list of known elements, only 17 are left to be classified as nonmetals. Six of these elements belong to the family of rare gases in Group VIIIA of the periodic table, which are highly unreactive. Discussions of the chemistry of the nonmetals therefore tend to focus on the behavior of 12 elements: H, C, N, O, F, P, S, Cl, Se, Br, I, and Xe.

10.1 THE ROLE OF NONMETAL ELEMENTS IN CHEMICAL REACTIONS

There is a clear pattern in the chemistry of the main-group metals discussed in Chapter 7.

The main-group metals are oxidized in all of their chemical reactions.

Main-group metals (such as aluminum metal) are oxidized when they react with nonmetal elements (such as bromine).

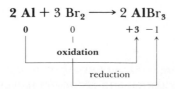

The reaction between aluminum and bromine gives off energy in the form of both heat and light, enough to boil the liquid bromine.

Nonmetals can also be reducing agents. In the manufacture of steel, elemental carbon is used to reduce Fe_2O_3 to iron metal.

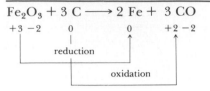

TABLE 10.1

Reactions in Which a Nonmetal Acts As an Oxidizing Agent and Is Therefore Reduced

$$2\,Mg + O_2 \longrightarrow 2\,MgO$$

$$CH_4 + 4\,Cl_2 \longrightarrow CCl_4 + 4\,HCl$$

$$4\,NH_3 + 5\,O_2 \longrightarrow 4\,NO + 6\,H_2O$$

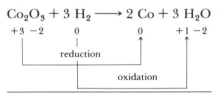

TABLE 10.2

Reactions in Which a Nonmetal Acts As a Reducing Agent and Is Therefore Oxidized

$$Fe_2O_3 + 3\,C \longrightarrow 2\,Fe + 3\,CO$$

$$Co_2O_3 + 3\,H_2 \longrightarrow 2\,Co + 3\,H_2O$$

They are also oxidized when they react with nonmetal compounds, such as water.

$$2\,Na + 2\,H_2O \longrightarrow 2\,Na^+ + 2\,OH^- + H_2$$

The chemistry of the nonmetals is much more interesting, because these elements can be either oxidized or reduced in their chemical reactions. Table 10.1 shows a series of reactions in which O_2 and Cl_2 act as oxidizing agents—in the course of these reactions, these nonmetals are therefore reduced. Table 10.2, on the other hand, contains examples of reactions in which carbon and H_2 act as reducing agents—in these reactions, the nonmetal elements are oxidized.

Exercise 10.1

Nonmetals can either oxidize or reduce other elements and compounds. Identify the role of the nonmetal in each of the following reactions and then describe what happens to the nonmetal.

(a) $CO(g) + Cl_2(g) \longrightarrow COCl_2(g)$

(b) $CuO(s) + H_2(g) \longrightarrow Cu(s) + H_2O(g)$

Solution

(a) We start by assigning oxidation numbers to each atom in the reaction and then identify the oxidation and reduction halves of the reaction.

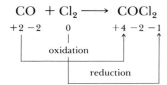

The nonmetal (Cl_2) oxidizes the carbon in carbon monoxide from an oxidation number of $+2$ to one of $+4$. Cl_2 is therefore an oxidizing agent in this reaction. In the course of the reaction, the oxidation number of chlorine is reduced from 0 to -1.

(b) Once again, we start by identifying the oxidation and reduction halves of the reaction.

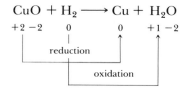

In this case, the nonmetal (H_2) reduces CuO to copper metal. H_2 is therefore a reducing agent in this reaction. In the course of the reaction, the hydrogen atoms are oxidized from an oxidation number of 0 to one of $+1$.

10.2 DECIDING WHETHER A NONMETAL IS OXIDIZED OR REDUCED

The key to deciding whether a nonmetal is oxidized or reduced in a chemical reaction involves comparing the electronegativities of the elements involved in the reaction. We can demonstrate this by considering the reaction between calcium and phosphorus to form calcium phosphide.

$$6 \ Ca(s) + P_4(s) \longrightarrow 2 \ Ca_3P_2(s)$$

The simplest model of chemical reactions was used in Exercise 2.9 to predict the product of a similar reaction. This model assumed that metals react with nonmetals to form ionic compounds, or salts. Chapter 7 tied this model to the electron configurations of the elements. Calcium is assumed to lose electrons to form Ca^{2+} ions, while phosphorus is assumed to gain electrons to form P^{3-} ions.

$$Ca \ = [Ar] \ 4s^2 \qquad P \ = [Ne] \ 3s^2 \ 3p^1$$
$$Ca^{2+} = [Ar] \qquad P^{3-} = [Ar]$$

When electronegativity was introduced in Section 8.5, however, it became obvious that this model is too simple. Phosphorus ($EN = 2.19$) is more electronegative than calcium ($EN = 1.00$). When these elements combine, they won't share electrons evenly, the electrons will be drawn toward the more electronegative phosphorus atoms. But the difference between the electronegativities of these elements ($\Delta EN = 1.19$) is too small to allow us to assume that the product of this reaction contains Ca^{2+} and P^{3-} ions.

By using the concept of oxidation number to describe this reaction, we can indicate that there is some separation of charge in the product without naively assuming that this compound contains full-blown Ca^{2+} and P^{3-} ions. Calcium is said to be oxidized in this reaction because the oxidation number becomes larger.

When two elements react, the more electronegative element can be assumed to oxidize the less electronegative element. This photograph shows the reaction in which phosphorus is oxidized by oxygen to form P_4O_{10}.

Phosphorus, on the other hand, is reduced.

$$6 \text{ Ca} + P_4 \longrightarrow 2 \text{ Ca}_3P_2$$

Let's now turn to the reaction between phosphorus and oxygen.

$$P_4(s) + 5 \text{ O}_2(g) \longrightarrow P_4O_{10}(s)$$

Phosphorus (EN = 2.19) is less electronegative than oxygen (EN = 3.44). When these elements react, the electrons will be drawn toward the more electronegative oxygen atoms. We can therefore envision this reaction in terms of an increase in the oxidation number of phosphorus and a decrease in the oxidation number of oxygen. Phosphorus is therefore oxidized in this reaction, while oxygen is reduced.

$$P_4 + 5 \text{ O}_2 \longrightarrow P_4O_{10}$$

The following list summarizes what has been said so far.

1. Nonmetals tend to oxidize metals.

$$2 \text{ Mg}(s) + \text{O}_2(g) \longrightarrow 2 \text{ MgO}(s)$$
$$8 \text{ Ba}(s) + \text{S}_8(s) \longrightarrow 8 \text{ BaS}(s)$$

2. Nonmetals with relatively large electronegativities (such as O_2 and Cl_2) tend to oxidize other substances.

$$2 \text{ H}_2\text{S}(g) + 3 \text{ O}_2(g) \longrightarrow 2 \text{ SO}_2(g) + 2 \text{ H}_2\text{O}(g)$$
$$\text{PH}_3(g) + 3 \text{ Cl}_2(g) \longrightarrow \text{PCl}_3(l) + 3 \text{ HCl}(s)$$

3. Nonmetals with relatively small electronegativities (such as C and H_2) can reduce other substances.

$$\text{SnO}_2(s) + \text{C}(s) \longrightarrow \text{Sn}(s) + \text{CO}_2(g)$$
$$\text{CuO}(s) + \text{H}_2(g) \longrightarrow \text{Cu}(s) + \text{H}_2\text{O}(g)$$

4. When two nonmetals react, the more electronegative element oxidizes the less electronegative element.

$$\text{S}_8(s) + 8 \text{ O}_2(g) \longrightarrow 8 \text{ SO}_2(g)$$
$$P_4(s) + 6 \text{ Cl}_2(g) \longrightarrow 4 \text{ PCl}_3(l)$$

These observations can be brought together in the form of a general rule.

When two elements react, the more electronegative element can be assumed to oxidize the less electronegative element.

Exercise 10.2

Determine which nonmetal is oxidized and which is reduced when sulfur vapor reacts with red-hot charcoal to form carbon disulfide.

$$4 \text{ C}(s) + \text{S}_8(g) \longrightarrow 4 \text{ CS}_2(g)$$

Solution

Sulfur (EN = 2.58) is just slightly more electronegative than carbon (EN = 2.55). We therefore assume that sulfur is reduced in this reaction, while carbon is oxidized.

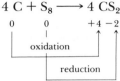

$$4 \, C + S_8 \longrightarrow 4 \, CS_2$$

10.3 THE CHEMISTRY OF HYDROGEN

The remainder of this chapter will be devoted to a discussion of the chemistry of the individual nonmetals. A reasonable place to start this discussion is with hydrogen, the lightest element in the periodic table. Hydrogen is the most abundant element in the universe; 90% of the atoms and 75% of the mass of the universe are hydrogen. Hydrogen is much less abundant on earth, however. Only 0.15% of the mass of the earth's crust is hydrogen. Even when the enormous number of hydrogen atoms in the earth's oceans is included, hydrogen makes up less than 1% of the mass of the planet.

Models for the evolution of the earth assume that the atmosphere once contained very little oxygen but significant amounts of hydrogen in the form of H_2 molecules and compounds such as methane (CH_4) and ammonia (NH_3). Today, the atmosphere is 21% O_2 by volume, with only traces of CH_4 (2 ppm) and H_2 (0.5 ppm).

What happened to the H_2 in the atmosphere? As the concentration of O_2 in the atmosphere became larger, so did the probability of reaction between H_2 and O_2 to form water.

$$2 \, H_2(g) + O(g) \longrightarrow 2 \, H_2O(g) \qquad \Delta H^\circ = -241.83 \text{ kJ/mol } H_2O$$

The very name *hydrogen* comes from the Greek stems *hydro-*, "water," and *gennan*, "to form or generate." Thus, hydrogen is literally the "water former."

Hydrogen that did not react with oxygen to form the water in the earth's oceans escaped from the planet into space. As noted in Section 4.14, the speed with which gas particles move is inversely proportional to the square root of their molecular weights.

$$\text{Rate} \propto \frac{1}{\sqrt{\text{MW}}}$$

Because they are so light, H_2 molecules move faster than any other molecules in the atmosphere. In Exercise 4.20, the average velocity of an H_2 molecule at 0°C was found to be 2000 meters per second. Most of the H_2 molecules in the atmosphere are therefore traveling too slowly to escape from the earth's gravity. (The speed required is 11.18 kilometers per second.) But a very small fraction of these molecules are moving fast enough to escape. Thus, over geological time periods, a significant amount of the hydrogen that was once in the atmosphere has escaped from the planet's surface.

Hydrogen combines with every element in the periodic table except the nonmetals in Group VIIIA (He, Ne, Ar, Kr, Xe, and Rn). Although it is often stated that more compounds contain carbon than any other element, this is not true.

The three main engines of the space shuttle develop 1.1 million pounds of thrust by burning a mixture of liquid hydrogen and liquid oxygen.

FIG. 10.2 Hydrogen combines with every element in the periodic table except the elements in Group VIIIA (He, Ne, Ar, and so on). The formulas of the hydrides of the main-group elements are shown here.

H_2							H_2	
LiH	BeH_2		B_2H_6	CH_4	NH_3	H_2O	HF	
NaH	MgH_2		AlH_3	SiH_4	PH_3	H_2S	HCl	
KH	CaH_2		GaH_3	GeH_4	AsH_3	H_2Se	HBr	
RbH	SrH_2			SnH_4	SbH_3	H_2Te	HI	
CsH	BaH_2			PbH_4	BiH_3	H_2Po	HAt	

Almost all of the compounds of carbon also contain hydrogen, and hydrogen forms compounds with virtually all of the other elements as well.

Compounds of hydrogen are frequently called *hydrides*, even though the name *hydride* should be restricted to compounds that contain an H^- ion. There is a regular trend in the formula of the hydrogen compounds across a row of the periodic table, as shown in Figure 10.2. (This trend is so regular that the combining power, or *valence*, of an element was once defined as the number of hydrogen atoms bound to the element in its hydride.)

The chemistry of hydrogen is unlike that of any other element. Hydrogen is the only element that forms compounds in which the valence electrons are in the $n = 1$ shell. As a result, hydrogen can have only three possible oxidation numbers, corresponding to the H^+ ion, a neutral H atom, and the H^- ion.

$$H^+ = 1s^0$$

$$H = 1s^1$$

$$H^- = 1s^2$$

The fact that hydrogen forms compounds with oxidation numbers of both $+1$ and -1 is one reason why many periodic tables include this element in both Group IA (Li, Na, K, Rb, Cs, Fr) and Group VIIA (F, Cl, Br, I, and At).

Hydrogen can be included among the elements in Group IA because it forms compounds (such as HCl, HNO_3, and H_2S) that are analogous to compounds of the alkali metals (such as NaCl, KNO_3, and Li_2S). It can also be included among the elements in Group VIIA, because it forms compounds (such as NaH and CaH_2) that are analogous to compounds of the halogens (such as NaF and $CaCl_2$).

Another reason that hydrogen appears to belong among the nonmetals on the right side of the periodic table is that it combines with other nonmetals to form covalent compounds, such as H_2O, CH_4, NH_3, and H_2S, the way a nonmetal should. Furthermore, the element is a gas at room temperature and atmospheric pressure, like other nonmetals, such as O_2 and N_2.

On the other hand, hydrogen has an oxidation number of $+1$ in most of its compounds, like the elements in Group IA. It has also been suggested that hydrogen has the properties of a metal under conditions of very high pressure. It has been argued, for example, that any hydrogen present at the center of the planet Jupiter is likely to be a metallic solid. Additional evidence for placing hydrogen among the metals in Group IA comes from the fact that this element combines with a handful of metals, such as scandium, titanium, chromium, nickel, palladium, and uranium, to form materials that behave as if they were alloys.

It is difficult to decide where hydrogen belongs in the periodic table because of the neutral properties of the element. The first ionization energy of hydrogen (1312 kJ/mol) is roughly halfway between the extremes of the elements with the largest (2372 kJ/mol) and smallest (376 kJ/mol) ionization energies. Hydrogen also has an electronegativity (EN = 2.20) halfway between the extremes of the most electronegative (EN = 3.98) and least electronegative (EN = 0.7) elements.

As expected from the general rule introduced in the preceding section, hydrogen is oxidized by elements that are more electronegative (such as Cl_2) to form compounds in which it has an oxidation number of +1.

$$H_2 + Cl_2 \longrightarrow 2\ HCl$$

$$\begin{array}{cccc} 0 & 0 & +1 & -1 \end{array}$$

oxidation

reduction

Hydrogen is reduced by elements that are less electronegative (such as sodium metal) to form compounds in which its oxidation number is −1.

$$2\ Na + H_2 \longrightarrow 2\ NaH$$

$$\begin{array}{cccc} 0 & 0 & +1 & -1 \end{array}$$

oxidation

reduction

At room temperature, hydrogen is a colorless, odorless gas with a density only $\frac{1}{14}$ the density of air. Small quantities of H_2 gas can be prepared in several different ways.

1. By reacting an active metal such as sodium or calcium with water.

$$2\ Na(s) + 2\ H_2O(l) \longrightarrow 2\ Na^+(aq) + 2\ OH^-(aq) + H_2(g)$$
$$Ca(s) + 2\ H_2O(l) \longrightarrow Ca^{2+}(aq) + 2\ OH^-(aq) + H_2(g)$$

2. By reacting a less active metal such as zinc or iron with a strong acid such as hydrochloric acid.

$$Zn(s) + 2\ HCl(aq) \longrightarrow Zn^{2+}(aq) + 2\ Cl^-(aq) + H_2(g)$$

3. By reacting an ionic metal hydride such as LiH or NaH with water.

$$NaH(s) + H_2O(l) \longrightarrow Na^+(aq) + OH^-(aq) + H_2(g)$$

4. By passing an electric current through water to decompose this compound into its elements.

$$2\ H_2O(l) \xrightarrow{\text{electrolysis}} H_2(g) + O_2(g)$$

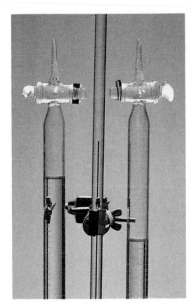

One way of preparing H_2 gas in the laboratory is by the electrolysis of water. Because water does not readily conduct an electric current, an inert salt such as Na_2SO_4 is dissolved in the water to increase its conductivity. Twice as much hydrogen gas collects at one electrode as oxygen gas at the other.

The covalent radius of a neutral hydrogen atom is 0.0371 nanometers, smaller than that of any other element. Because small atoms can come very close to each other, they tend to form strong covalent bonds. The bond dissociation enthalpy for the H—H bond is therefore relatively large (435 kJ/mol). As a result, H_2 tends to be unreactive at room temperature.

In the presence of a spark, however, a fraction of the H_2 molecules dissociate to form hydrogen atoms that are highly reactive.

$$H_2(g) \xrightarrow{\text{spark}} 2\ H(g)$$

The heat given off when these H atoms react with O_2 is enough to catalyze the dissociation of additional H_2 molecules. Mixtures of H_2 and O_2 that are infinitely stable at room temperature therefore explode in the presence of a spark or flame.

Exercise 10.3

Use oxidation numbers to explain what happens in the following reactions, which are used to prepare H_2 gas.

(a) $Ca(s) + 2\ H_2O(l) \longrightarrow Ca^{2+}(aq) + 2\ OH^-(aq) + H_2(g)$
(b) $Zn(s) + 2\ HCl(aq) \longrightarrow Zn^{2+}(aq) + 2\ Cl^-(aq) + H_2(g)$

Solution

(a) Calcium metal is oxidized in this reaction, and water is reduced.

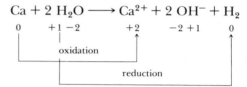

When water is reduced, the H^+ ion removed from a neutral H_2O molecule gains an electron to form a hydrogen atom. A pair of hydrogen atoms formed in this reaction combine to give a molecule of hydrogen, and the OH^- ions formed when H^+ ions are removed from water molecules are released into solution.

(b) Zinc metal is oxidized in this reaction, and the H^+ ions from hydrochloric acid are reduced.

$$Zn + 2\ HCl \longrightarrow Zn^{2+} + 2\ Cl^- + H_2$$

10.4 THE CHEMISTRY OF OXYGEN

ELEMENTAL FORMS OF OXYGEN: O_2 AND O_3

Oxygen is the most abundant element on this planet—the earth's crust is 46.6% oxygen by weight, the oceans are 86% oxygen by weight, and the atmosphere is 21% oxygen by volume. The name *oxygen* comes from the Greek stems *oxys,* "acid," and *gennan,* "to form or generate." Thus, *oxygen* literally means "acid former." This name was introduced by Lavoisier, who noticed that compounds rich in oxygen, such as SO_2 and P_4O_{10}, dissolve in water to give acids.

The electron configuration of an oxygen atom

$$O = [He]\ 2s^2\ 2p^4$$

suggests that neutral oxygen atoms can achieve an octet of valence electrons by sharing two pairs of electrons to form an O=O double bond.

$$\ddot{O}=\ddot{O}$$

This Lewis structure is somewhat misleading, however. According to this structure, all of the electrons in the O_2 molecule are paired, so the compound should be **diamagnetic**—it should be repelled by a magnetic field. Experimentally, however, O_2 is found to be **paramagnetic**—it is attracted to a magnetic field. As noted in Section 8.17, this can be explained by assuming that there are two unpaired electrons in the π^* antibonding molecular orbitals of the O_2 molecule.

At temperatures below $-183\,^\circ C$, O_2 condenses to form a liquid with a characteristic light blue color, which results from the absorption of light with a wavelength of 630 nanometers. This absorption is not seen in the gas phase and is relatively weak even in the liquid, because it requires that three bodies—two O_2 molecules and a photon—collide simultaneously. This is a very rare phenomenon, even in the liquid phase.

Liquid oxygen has a pale-blue color due to the absorption of photons at a wavelength of approximately 630 nanometers.

THE CHEMISTRY OF OZONE

The O_2 molecule, or *dioxygen*, as it is formally called, is not the only elemental form of oxygen. In the presence of lightning or another source of a spark, O_2 molecules dissociate to form oxygen atoms.

$$O_2(g) \xrightarrow{\text{spark}} 2\ O(g)$$

These O atoms can react with O_2 molecules to form **ozone**, O_3, whose Lewis structure is shown in Figure 10.3.

$$O_2(g) + O(g) \longrightarrow O_3(g)$$

Oxygen (O_2) and ozone (O_3) are examples of **allotropes** (from the Greek *allotropos*, "in another manner").

Allotropes: Two or more different forms of an element that have different structures and therefore different chemical and physical properties.

O_2 and O_3 are both elemental forms of oxygen because they contain only oxygen atoms. They have different chemical and physical properties, as shown in Table 10.3, because of their different structures.

FIG. 10.3 Ozone is a resonance hybrid of two Lewis structures, each of which contains one O=O double bond and one O—O single bond. The electrons in the valence shell of the central atom are distributed toward the corners of an equilateral triangle. The O_3 molecule is therefore bent, with an O—O—O bond angle of 116.5°.

TABLE 10.3		
Properties of Allotropes of Oxygen		
	Dioxygen (O_2)	*Ozone (O_3)*
melting point	$-218.75\,^\circ C$	$-192.5\,^\circ C$
boiling point	$-182.96\,^\circ C$	$-110.5\,^\circ C$
density (at 20°C)	1.331 g/L	1.998 g/L
O—O bond order	2	1.5
O—O bond length	0.1207 nm	0.1278 nm

Ozone is an unstable compound with a sharp, pungent odor, which slowly decomposes to oxygen.

$$3\ O_3(g) \longrightarrow 3\ O_2(g)$$

At low concentrations, exposure to ozone can be relatively pleasant. (The characteristic clean odor associated with summer thunderstorms is due to the formation of small amounts of O_3.) Exposure to O_3 at higher concentrations leads to coughing, rapid beating of the heart, chest pain, and general body pain. Ozone becomes toxic at concentrations above 1 ppm.

One of the characteristic properties of ozone is its ability to absorb radiation in the ultraviolet portion of the spectrum. The ozone in the atmosphere therefore acts as a filter that protects us from exposure to high-energy ultraviolet radiation emitted by the sun. We can understand the importance of this filter if we think about what happens when radiation from the sun is absorbed by our skin.

Electromagnetic radiation in the infrared, visible, and low-energy portions of the ultraviolet spectrum, with wavelengths longer than about 300 nanometers, carries enough energy to excite an electron on a molecule into a higher-energy orbital. This electron eventually falls back into the orbital from which it was excited, and energy is given off to the surrounding tissue in the form of heat. Anyone who has suffered from a sunburn can appreciate the fact that excessive amounts of this radiation can have painful consequences.

Radiation in the high-energy portion of the ultraviolet spectrum, with wavelengths below about 300 nm, has a slightly different effect when it is absorbed. This radiation can carry enough energy to ionize atoms or molecules. The ions formed in these reactions have an odd number of electrons and are extremely reactive. They can cause permanent damage to the cell tissue and even induce processes that eventually result in skin cancer.

In 1974, Molina and Rowland pointed out that chlorofluorocarbons, such as $CFCl_3$ and CF_2Cl_2, which had been used as propellants in aerosol cans and as refrigerants—were beginning to accumulate in the atmosphere. In the stratosphere, at altitudes of 10 to 50 km above the earth's surface, chlorofluorocarbons decompose to form Cl atoms and chlorine oxides such as ClO when they absorb sunlight.

Cl atoms and ClO molecules have an odd number of electrons.

$$\cdot \ddot{\underset{\cdot\cdot}{Cl}}\colon \qquad \cdot \ddot{\underset{\cdot\cdot}{Cl}}-\ddot{\underset{\cdot\cdot}{O}}\colon$$

They are therefore examples of a class of atoms and molecules known as **_free radicals,_** which have unpaired electrons and are unusually reactive. In the atmosphere, they react with ozone or with the oxygen atoms that are needed to form ozone.

$$Cl + O_3 \longrightarrow ClO + O_2$$
$$ClO + O \longrightarrow Cl + O_2$$

The net result is a depletion of the ozone shield, with potentially dangerous implications.

OXYGEN AS AN OXIDIZING AGENT

The chemistry of oxygen is dominated by the fact that this element has an unusually large electronegativity. (Fluorine is the only element that is more electronegative than oxygen.) As a result, oxygen gains electrons in virtually all its chemical reactions.

Each O_2 molecule must gain a total of four electrons to satisfy the octets of the two oxygen atoms without sharing electrons.

$$\ddot{O}\!\!=\!\!\ddot{O} + 4\,e^- \longrightarrow \, :\!\ddot{O}\!:^{2-} + \,:\!\ddot{O}\!:^{2-}$$

Oxygen therefore oxidizes metals, such as iron, to form salts in which the oxygen atoms are formally present as O^{2-} ions.

$$4\ Fe(s) + 3\ O_2(g) \longrightarrow \quad 2\ Fe_2O_3(s)$$
$$[Fe^{3+}]_2[O^{2-}]_3$$

It also oxidizes nonmetals, such as carbon, to form covalent compounds in which the oxygen formally has an oxidation number of -2.

$$C(s) + O_2(g) \longrightarrow \underset{+4\ -2}{CO_2(g)}$$

Oxygen is therefore the perfect example of an *oxidizing agent.*

> **Oxidizing agent: An atom, ion, or molecule that either gains electrons in a chemical reaction or produces an increase in the oxidation number of the substance with which it reacts.**

In the course of its reactions, oxygen is reduced. The substances it reacts with are therefore *reducing agents.*

> **Reducing agent: An atom, ion, or molecule that either loses electrons in a chemical reaction or produces a decrease in the oxidation number of the substance with which it reacts.**

Exercise 10.4

Identify the oxidizing agents and reducing agents in the following reactions.

(a) $CH_4(g) + 2\ O_2(g) \longrightarrow CO_2(g) + 2\ H_2O(g)$

(b) $Fe_2O_3(s) + 3\ C(s) \longrightarrow 2\ Fe(s) + 3\ CO(g)$

Solution

(a) We can start by identifying the oxidation and reduction halves of the reaction.

$$CH_4 + 2\ O_2 \longrightarrow CO_2 + 2\ H_2O$$

(with oxidation states: $-4\ +1$ for CH_4; 0 for O_2; $+4\ -2$ for CO_2; $+1\ -2$ for H_2O; the carbon undergoes *oxidation* and the oxygen undergoes *reduction*)

CH_4 is oxidized by O_2 in this reaction, so O_2 is the oxidizing agent. O_2 is reduced by CH_4, which means that CH_4 is the reducing agent.

It might be noted in passing that early models of the atmosphere that assume the presence of CH_4, NH_3, and H_2 are called reducing atmospheres, because they contain large quantities of compounds that are reducing agents.

(b) Once again, we start by identifying the oxidation and reduction halves of the reaction.

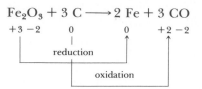

$$Fe_2O_3 + 3\,C \longrightarrow 2\,Fe + 3\,CO$$

In the course of this reaction, carbon reduces Fe_2O_3 to iron metal, so carbon is the reducing agent. Fe_2O_3 oxidizes carbon to CO and is therefore the oxidizing agent.

Each year, between 75 and 80 quads, or quadrillion (10^{15}) Btu, of energy is consumed in the United States. Less than 10% of this energy is provided by nuclear, solar, geothermal, or hydro power. More than 90% can be traced to a combustion reaction in which a fuel is oxidized by O_2.

The cars, trucks, and buses that fill our highways are powered by gasoline or diesel engines that burn hydrocarbons such as octane, C_8H_{18}, or by diesel engines that burn larger hydrocarbons such as cetane, $C_{16}H_{34}$.

$$2\,C_8H_{18}(l) + 25\,O_2(g) \longrightarrow 16\,CO_2(g) + 18\,H_2O(g)$$
$$2\,C_{16}H_{34}(l) + 49\,O_2(g) \longrightarrow 32\,CO_2(g) + 34\,H_2O(g)$$

We heat our homes by burning the methane (CH_4) in natural gas, the high-molecular-weight hydrocarbons in fuel oil, or the hydrocarbons in wood or by using electricity generated in a power plant that burns either oil or coal.

The energy we use to fuel our bodies also comes from combustion reactions. Energy enters our systems in the form of fats (lipids), proteins, and carbohydrates. These are converted into carbohydrates, such as glucose ($C_6H_{12}O_6$), which react with oxygen to produce the energy we need to survive.

$$C_6H_{12}O_6(aq) + 6\,O_2(g) \longrightarrow 6\,CO_2(g) + 6\,H_2O(l) \qquad \Delta H° = -2870 \text{ kJ/mol}$$

About 65% of the energy given off in this reaction is captured in the synthesis of the ATP (adenosine triphosphate) that fuels biological processes. The remaining 35% is released as the heat that keeps our body temperatures higher than the temperature of the surroundings.

There is an ever-growing appreciation that the earth contains a finite amount of fossil fuels, such as oil and coal, which will eventually run out. Nuclear, solar, and geothermal power will be increasingly important sources of energy. But they won't replace fossil fuels by themselves, because they are used to produce electrical energy, which is difficult to store. One possible solution to this problem is the so-called hydrogen economy.

The first step in the hydrogen economy is to use energy from nuclear, solar, or geothermal power to split water into its elements.

$$2\,H_2O(l) \longrightarrow 2\,H_2(g) + O_2(g)$$

The oxygen is then released to the atmosphere, and the hydrogen is either burned as a fuel when needed

$$2\,H_2(g) + O_2(g) \longrightarrow 2\,H_2O(g) \qquad \Delta H° = -241.83 \text{ kJ/mol}$$

or used to reduce carbon monoxide to methanol, or even gasoline, which can be stored and later burned as a fuel.

$$CO(g) + 2\ H_2(g) \longrightarrow CH_3OH(l)$$
$$8\ CO(g) + 17\ H_2(g) \longrightarrow C_8H_{18}(l) + 8\ H_2O(l)$$

PEROXIDES

Reactions in which O_2 captures electrons to form compounds in which the oxygen atoms have an oxidation number of -2 are common—so common that we used the following statement as one of the basic rules for assigning oxidation numbers in Section 2.13.

> **Oxygen almost always has an oxidation number of -2. The only exceptions are compounds that contain O—O bonds, such as O_2, O_3, the O_2^{2-} ion, the O_2^- ion, and H_2O_2.**

We will now examine the more important exceptions to this rule, the peroxide ion (O_2^{2-}) and hydrogen peroxide (H_2O_2).

It takes four electrons to reduce an O_2 molecule to a pair of O^{2-} ions.

$$:\!\overset{..}{O}\!\!=\!\!\overset{..}{O}\!: + 4\ e^- \longrightarrow 2\ :\!\overset{..}{\underset{..}{O}}\!:^{2-}$$

If the reaction stops after the O_2 molecule has gained only two electrons, the O_2^{2-} ion is produced.

$$:\!\overset{..}{O}\!\!=\!\!\overset{..}{O}\!: + 2\ e^- \longrightarrow :\!\overset{..}{\underset{..}{O}}\!\!-\!\!\overset{..}{\underset{..}{O}}\!:^{2-}$$

This ion has two more electrons than a neutral O_2 molecule, which means that the oxygen atoms must share only a single pair of bonding electrons to achieve an octet of valence electrons. The O_2^{2-} ion is called the *peroxide* ion because compounds that contain this ion are very rich in oxygen. They are not just oxides, they are (hy-)*peroxides*.

The easiest way to prepare a peroxide is to react sodium or barium metal with oxygen.

$$2\ Na(s) + O_2(g) \longrightarrow Na_2O_2(s)$$
$$Ba(s) + O_2(g) \longrightarrow BaO_2(s)$$

When the product of one of these reactions is treated with an acid, *hydrogen peroxide* (H_2O_2) is produced.

$$BaO_2(s) + 2\ H^+(aq) \longrightarrow Ba^{2+}(aq) + H_2O_2(aq)$$

The Lewis structure of hydrogen peroxide is shown in Figure 10.4. The VSEPR theory predicts that the geometry around each oxygen atom in H_2O_2 should be bent, or angular. But this theory cannot predict whether the four atoms should lie in the same plane or whether the molecule should be visualized as lying in two intersecting planes. The experimentally determined structure of H_2O_2 is also shown in Figure 10.4. The H—O—O bond angle in this molecule is only slightly larger than the angle between a pair of adjacent $2p$ atomic orbitals on the oxygen atom, and the angle between the planes that form the molecule is slightly larger than the tetrahedral angle.

Hydrogen peroxide is interesting because the oxygen atoms have an oxidation number of -1. H_2O_2 can therefore act as an oxidizing agent and capture two more

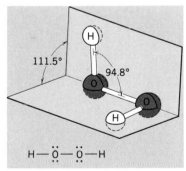

FIG. 10.4 The VSEPR theory predicts that the valence electrons on each oxygen atom in hydrogen peroxide should be distributed toward the corners of a tetrahedron. The geometry around each oxygen atom is therefore best described as bent, or angular, with an O—O—H bond angle of 94.8°. In order to keep the nonbonding pairs of electrons on the oxygen atoms as far apart as possible, the four atoms in this molecule lie in two planes that intersect at an angle of 111.5°.

electrons to form hydroxide ions, in which the oxygen has an oxidation number of -2.

$$H_2O_2 + 2e^- \longrightarrow 2\ OH^-$$

Or, it can act as a reducing agent and lose a pair of electrons to form an O_2 molecule.

$$H_2O_2 \longrightarrow O_2 + 2\ H^+ + 2\ e^-$$

Exercise 10.5

A reaction in which a compound simultaneously undergoes both oxidation and reduction is called **disproportionation.** Explain why the products of the disproportionation of hydrogen peroxide are oxygen and water.

$$2\ H_2O_2(aq) \longrightarrow 2\ H_2O(l) + O_2(g)$$

Solution

Adding the half-reaction for the oxidation of H_2O_2 to the half-reaction for the reduction of this compound gives the following results.

$$H_2O_2 + 2\ e^- \longrightarrow 2\ OH^-$$
$$\underline{H_2O_2 \longrightarrow O_2 + 2\ H^+ + 2\ e^-}$$
$$2\ H_2O_2 \longrightarrow O_2 + 2\ H^+ + 2\ OH^-$$

The H^+ and OH^- ions produced in the two halves of the reaction can combine to form water. The net result is the following reaction.

$$2\ H_2O_2 \longrightarrow 2\ H_2O + O_2$$

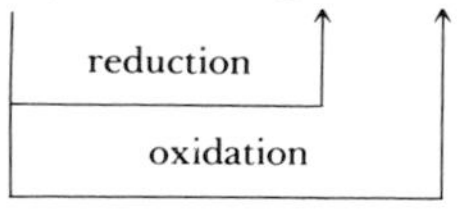

reduction

oxidation

The disproportionation of H_2O_2 is an exothermic reaction.

$$2\ H_2O_2(aq) \longrightarrow 2\ H_2O(l) + O_2(g) \qquad \Delta H^\circ = -94.6\ \text{kJ/mol}\ H_2O_2$$

The decomposition of hydrogen peroxide, however, is relatively slow in the absence of a **catalyst** such as dust, a metal ion, or one of the peroxidase enzymes found in living tissue.

> **Catalyst: Any substance that can speed up the rate of a chemical reaction when it is added to the reaction mixture.**

An interesting demonstration of the effect of a catalyst on the rate of a chemical reaction involves the addition of a drop of blood or a slice of horseradish or turnip to a sample of H_2O_2. The peroxidase enzymes in living tissue catalyze the decomposition of hydrogen peroxide to produce a rapid stream of oxygen gas bubbling out of the solution.

The principal uses of H_2O_2 revolve around its oxidizing ability. It is used in dilute (3%) solutions as a disinfectant and in more concentrated solutions as a bleaching agent for hair, fur, leather, or the wood pulp used to make paper. H_2O_2 has also been used as rocket fuel because of the ease with which it decomposes to give O_2.

METHODS OF PREPARING O_2

Small quantities of O_2 gas can be prepared in a number of ways.

1. By decomposing a dilute solution of hydrogen peroxide, with dust or a metal surface used as the catalyst.

$$2\ H_2O_2(aq) \longrightarrow 2\ H_2O(l) + O_2(g)$$

2. By reacting hydrogen peroxide with an even better oxidizing agent, such as the permanganate ion, MnO_4^-.

$$5\ H_2O_2(aq) + 2\ MnO_4^-(aq) + 6\ H^+(aq) \longrightarrow$$
$$2\ Mn^{2+}(aq) + 5\ O_2(g) + 8\ H_2O(l)$$

3. By passing an electric current through water.

$$2\ H_2O(l) \xrightarrow{\text{electrolysis}} 2\ H_2(g) + O_2(g)$$

4. By heating potassium chlorate ($KClO_3$) in the presence of a catalyst until it decomposes.

$$2\ KClO_3(s) \xrightarrow{MnO_2} 2\ KCl(s) + 3\ O_2(g)$$

Because H_2O_2 can undergo both oxidation and reduction it can react with itself in an oxidation-reduction reaction to form water and oxygen gas. The rate of this reaction can be increased by adding a catalyst, such as the peroxidase enzyme in blood. When this is done, the reaction is so rapid that the solution foams at the surface.

10.5 THE CHEMISTRY OF SULFUR

Sulfur is directly below oxygen in the periodic table. The two elements therefore have similar electron configurations.

$$O = [He]\ 2s^2\ 2p^4$$
$$S = [Ne]\ 3s^2\ 3p^4$$

As a result, sulfur forms many compounds that are analogs of oxygen compounds, as shown in Table 10.4. Examples cited in this table show how the prefix *thio-* is sometimes used to indicate compounds in which sulfur replaces an oxygen atom; the thiocyanate (SCN^-) ion, for instance, is analogous to the cyanate (OCN^-) ion.

TABLE 10.4	
Oxygen Compounds and their Sulfur Analogs	
Oxygen Compounds	*Sulfur Compounds*
Al_2O_3 (aluminum oxide)	Al_2S_3 (aluminum sulfide)
Na_2O (sodium oxide)	Na_2S (sodium sulfide)
MgO (magnesium oxide)	MgS (magnesium sulfide)
H_2O (water)	H_2S (hydrogen sulfide)
H_2O_2 (hydrogen peroxide)	H_2S_2 (hydrogen disulfide)
O_3 (ozone)	SO_2 (sulfur dioxide)
P_4O_{10} (tetraphosphorus decaoxide)	P_4S_{10} (tetraphosphorus decasulfide)
N_2O_5 (dinitrogen pentoxide)	N_2S_5 (dinitrogen pentasulfide)
CO_2 (carbon dioxide)	CS_2 (carbon disulfide)
OCN^- (cyanate)	SCN^- (thiocyanate)
CO_3^{2-} (carbonate)	CS_3^{2-} (trithiocarbonate)
$CO(NH_2)_2$ (urea)	$CS(NII_2)_2$ (thiourea)

There are four principal differences between sulfur and oxygen, which lead to different behaviors for the two elements.

1. **O=O double bonds are much stronger than S=S double bonds.**
2. **S—S single bonds are almost twice as strong as O—O single bonds.**
3. **Sulfur (EN = 2.58) is much less electronegative than oxygen (EN = 3.44).**
4. **Sulfur can expand its valence shell to hold more than eight electrons, but oxygen cannot.**

These seemingly minor differences have important consequences.

THE EFFECT OF DIFFERENCES IN THE X—X AND X=X BOND STRENGTHS

The covalent radius of a sulfur atom is 0.104 nm, which is about one and a half times the covalent radius of an oxygen atom (0.066 nm). Because it is harder for sulfur atoms to come close enough together to form π bonds, S=S double bonds are much weaker than O=O double bonds.

Elemental oxygen consists of O_2 molecules in which each atom completes its octet of valence electrons by sharing two pairs of electrons with a single neighboring atom. Because sulfur does not form strong S=S double bonds, elemental sulfur usually consists of cyclic S_8 molecules in which each atom completes its octet by forming single bonds to two neighboring atoms, as shown in Figure 10.5.

Double bonds between sulfur and oxygen or carbon atoms can be found in compounds such as SO_2 and CS_2.

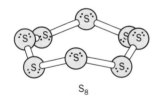
S_8

FIG. 10.5 Because they are too large to form stable S=S double bonds, sulfur atoms form single bonds with two different neighboring sulfur atoms to give cyclic S_8 molecules.

$$\ddot{O}=\ddot{S}-\ddot{O}\!: \qquad \!:\!\ddot{S}=C=\ddot{S}\!:$$

But these double bonds are much weaker than the equivalent double bonds to oxygen atoms in O_3 or CO_2. The bond dissociation enthalpy for a C=S double bond is 477 kJ/mol, for example, whereas the bond dissociation enthalpy for a C=O double bond is 745 kJ/mol.

The tendency of an element to form bonds to itself is called **catenation** (from the Latin *catena*, "chain"). Because sulfur forms unusually strong S—S single bonds, it is better at catenation than any other element except carbon. By varying the temperature, it is possible to obtain samples of sulfur that contain up to 10^6 sulfur atoms linked by S—S single bonds. These different forms of elemental sulfur are examples of allotropes. Oxygen has only two important allotropes, the O_2 and O_3 molecules, but sulfur has an almost endless number of allotropes.

One way in which allotropes of sulfur differ from each other involves the way S_8 molecules pack to form the crystal. The most stable form of sulfur consists of yellow orthorhombic crystals of S_8 molecules, which are often found near volcanos. If these crystals are heated until they melt and the molten sulfur is then cooled, an allotrope consisting of monoclinic crystals of S_8 molecules is obtained. The monoclinic crystals slowly transform themselves into the more stable orthorhombic structure over a period of time.

Another way in which allotropes of sulfur can differ is in the size of the sulfur molecules that form the crystal. Cyclic molecules that contain 6, 7, 8, 10, and 12 sulfur atoms are known.

Sulfur melts at 119.25°C to form a yellow liquid that is less viscous (that is, less resistant to flow) than water. If this liquid is heated to 159°C, however, it abruptly

Orthorhombic sulfur crystals.

turns into a dark red liquid that cannot be poured from its container. The viscosity of this dark red liquid is 2000 times greater than that of molten sulfur, because the cyclic S_8 molecules found in molten sulfur have opened up and linked together to form chains of as many as 100,000 sulfur atoms.

$$\text{(S-ring structure)} \longrightarrow [-S-S-S-S-S-S-S-S]_n$$

When sulfur reacts with a reducing agent, such as an active metal, it can gain a pair of electrons to form a variety of polysulfide ions with a charge of -2. These ions differ only in the number of sulfur atoms in the chain.

$$2\ K(s) + S_8(s) \longrightarrow K_2S_8 = [K^+]_2[^-S-S-S-S-S-S-S-S^-]$$
$$K_2S_6 = [K^+]_2[^-S-S-S-S-S-S^-]$$
$$K_2S_5 = [K^+]_2[^-S-S-S-S-S^-]$$
$$K_2S_4 = [K^+]_2[^-S-S-S-S^-]$$
$$K_2S_3 = [K^+]_2[^-S-S-S^-]$$
$$K_2S_2 = [K^+]_2[^-S-S^-]$$

Exercise 10.6

The oxidation number of iron is usually either $+2$ or $+3$. Use the tendency of sulfur to form polysulfide ions to explain why iron has an oxidation number of $+2$ in iron pyrite, FeS_2, one of the most abundant sulfur ores.

Solution

If the oxidation number of iron in FeS_2 is $+2$, the sulfur must be present as the S_2^{2-} ion.

$$FeS_2 = [Fe^{2+}][S_2^{2-}]$$

This disulfide ion is the sulfur analog of the peroxide ion, O_2^{2-}, and has an analogous Lewis structure.

$$:\overset{..}{S}-\overset{..}{S}:^{2-}$$

When heated gently, sulfur melts to form a yellow liquid that flows more easily than water. When heated further, the liquid turns a dark red color, and it becomes much more viscous as the individual S_8 molecules link together to form long polymeric chains.

Iron pyrite crystals.

THE EFFECT OF DIFFERENCES IN THE ELECTRONEGATIVITIES OF SULFUR AND OXYGEN

The fact that sulfur (EN = 2.58) is much less electronegative than oxygen (EN = 3.44) gives rise to one of the largest differences between these elements. As noted, oxygen usually has an oxidation number of either 0, -1, or -2. (The only important exception is OF_2, in which oxygen has an oxidation number of $+2$.) Because sulfur is much less electronegative, it is more likely to form compounds in which it has a positive oxidation number. Examples of the common oxidation numbers of sulfur are given in Table 10.5. Although there are examples of compounds with virtually every oxidation number between -2 and $+6$, the most important positive oxidation numbers for this element are $+4$ and $+6$.

TABLE 10.5

Common Oxidation Numbers for Sulfur

Oxidation Number	Examples
-2	Na_2S, H_2S
-1	Na_2S_2, H_2S_2
0	S_8
$+1$	S_2Cl_2
$+2$	SCl_2, $S_2O_3^{2-}$ (thiosulfate)
$+2\frac{1}{2}$	$S_4O_6^{2-}$ (tetrathionate)
$+3$	$S_2O_4^{2-}$ (dithionite)
$+4$	SF_4, SO_2, H_2SO_3, HSO_3^-, SO_3^{2-}
$+5$	$S_2O_6^{2-}$ (dithionate)
$+6$	SF_6, SO_3, H_2SO_4, HSO_4^-, SO_4^{2-}

The chemistry of sulfur provides us with the first example of the difference between the predictions of thermodynamics and the realities of what happens in a chemical reaction. On the basis of the electronegativities of the two elements and their electron configurations, we would predict that sulfur should react with oxygen to give either SO_2, in which sulfur has an oxidation number of $+4$, or SO_3, in which sulfur has an oxidation number of $+6$. The Lewis structures of these compounds are given in Figure 10.6.

Enthalpy of reaction measurements suggest that SO_3 is more stable thermodynamically than SO_2 because more energy is released when the reaction goes all the way to SO_3.

$$S_8(s) + 8\ O_2(g) \longrightarrow 8\ SO_2(g) \qquad \Delta H° = -296.83 \text{ kJ/mol } SO_2$$
$$S_8(s) + 12\ O_2(g) \longrightarrow 8\ SO_3(g) \qquad \Delta H° = -395.7 \text{ kJ/mol } SO_3$$

Thermodynamics therefore predicts that the reaction should continue until the reactants are converted into SO_3. Experiment, however, shows that the product of this reaction is SO_2, regardless of whether sulfur, hydrogen sulfide, carbon disulfide, or iron pyrite is burned.

$$S_8(s) + 8\ O_2(g) \longrightarrow \textbf{8}\ \textbf{SO}_\textbf{2}\textbf{(g)}$$
$$2\ H_2S(g) + 3\ O_2(g) \longrightarrow 2\ H_2O(g) + \textbf{2}\ \textbf{SO}_\textbf{2}\textbf{(g)}$$
$$CS_2(l) + 3\ O_2(g) \longrightarrow CO_2(g) + \textbf{2}\ \textbf{SO}_\textbf{2}\textbf{(g)}$$
$$3\ FeS_2(s) + 8\ O_2(g) \longrightarrow Fe_3O_4(s) + \textbf{6}\ \textbf{SO}_\textbf{2}\textbf{(g)}$$

This observation is relatively easy to explain. Although the SO_2 formed when these compounds burn should react with O_2 to form SO_3, the rate of this reaction is very slow.

$$2\ SO_2(g) + O_2(g) \longrightarrow 2\ SO_3$$

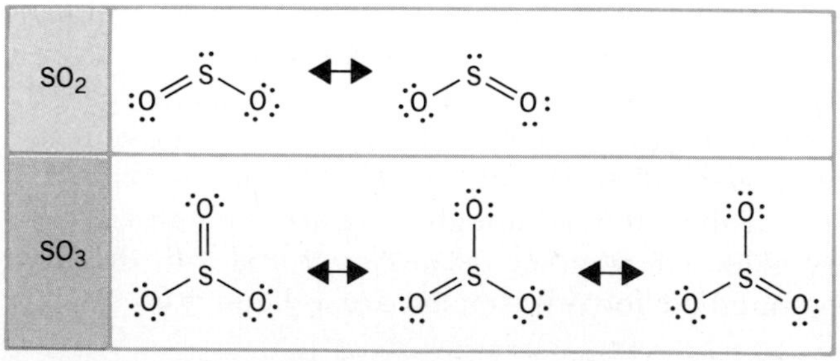

FIG. 10.6 SO_2 is a resonance hybrid of two Lewis structures analogous to the structure for ozone shown in Figure 10.3. SO_3 is a resonance hybrid of three Lewis structures, each of which consists of one S=O double bond and two S—O single bonds. SO_3 can be imagined to result from the donation of a pair of nonbonding electrons on the sulfur atom in SO_2 to a neutral oxygen atom to form a covalent bond.

The product isolated when sulfur compounds are burned is therefore SO_2, not SO_3.

The rate of the conversion of SO_2 into SO_3 can be greatly increased by adding a catalyst, such as platinum metal or a mixture of V_2O_5 and K_2O.

$$2\,SO_2(g) + O_2(g) \xrightarrow{\;V_2O_5/K_2O\;} 2\,SO_3(g)$$

Enormous quantities of SO_2 are produced by industry each year and then converted to SO_3, which can be used to produce sulfuric acid, H_2SO_4. In theory, sulfuric acid can be made by dissolving SO_3 gas in water.

$$SO_3(g) + H_2O(l) \longrightarrow H_2SO_4(aq)$$

In practice, this is not convenient. Instead, SO_3 is absorbed in 98% H_2SO_4, where it reacts with the water to form additional H_2SO_4 molecules. Water is then added, as needed, to keep the concentration of this solution between 96% and 98% H_2SO_4 by weight.

Sulfuric acid is by far the most important industrial chemical. It has even been argued that there is a direct relationship between the amount of sulfuric acid consumed and a country's standard of living. Sulfuric acid plays an important role in the manufacture of almost the entire spectrum of products of the chemical industry. More than 50% of the sulfuric acid produced each year is used to make fertilizers. The rest is used to make paper, synthetic fibers and textiles, insecticides, detergents, feed additives, dyes, drugs, antifreeze, paints and enamels, linoleum, synthetic rubber, printing inks, cellophane, photographic film, explosives, automobile batteries, and metals such as magnesium, aluminum, iron, and steel.

The formula for sulfuric acid, H_2SO_4, is misleading. The hydrogen atoms are not bound to sulfur in this compound. They are bound to oxygen atoms. The reaction between SO_3 and H_2O to form H_2SO_4 can be thought of as involving the addition of a molecule of water across an $S{=}O$ double bond.

It would be more accurate to write the formula for sulfuric acid as $O_2S(OH)_2$, because this formula indicates the correct skeleton structure for the molecule. But the formula H_2SO_4 has been in use for so long, there is no reasonable chance of replacing it with a more appropriate formula.

Sulfur dioxide dissolves in water to form sulfurous acid.

$$SO_2(g) + H_2O(l) \longrightarrow H_2SO_3(aq)$$

This reaction also involves the addition of a molecule of water across an $S{=}O$ double bond.

Sulfuric acid dissociates in water to give the H^+ ion and the HSO_4^- ion, which is known as the hydrogen sulfate, or bisulfate, ion.

$$H_2SO_4(aq) \longrightarrow H^+(aq) + HSO_4^-(aq)$$

Roughly 10% of the hydrogen sulfate ions formed in this reaction dissociate further to give additional H^+ ions and the SO_4^{2-}, or sulfate, ion.

$$HSO_4^-(aq) \longrightarrow H^+(aq) + SO_4^{2-}(aq)$$

A variety of salts can be formed by replacing the H^+ ions in sulfuric acid with positively charged ions, such as the Na^+ or K^+ ions.

$NaHSO_4$ = sodium hydrogen sulfate, or sodium bisulfate

Na_2SO_4 = sodium sulfate

Sulfurous acid is less likely to dissociate in water, but it is still possible to replace the H^+ ions in H_2SO_3 with positive ions to form salts.

$NaHSO_3$ = sodium hydrogen sulfite, or sodium bisulfite

Na_2SO_3 = sodium sulfite

Exercise 10.7

Which of the following reactions involves a change in the oxidation number of sulfur?

    (a) $2 SO_2(g) + O_2(g) \longrightarrow SO_3(g)$
    (b) $SO_3(g) + H_2O(l) \longrightarrow H_2SO_4(aq)$
    (c) $H_2SO_4(aq) \longrightarrow H^+(aq) + HSO_4^-(aq)$
    (d) $HSO_4^-(aq) \longrightarrow H^+(aq) + SO_4^{2-}(aq)$

Solution

Only the first reaction involves a change in the oxidation number of sulfur, from $+4$ (SO_2) to $+6$ (SO_3). The others involve compounds, such as SO_3, H_2SO_4, HSO_4^-, and SO_4^{2-}, in which the oxidation number of sulfur is $+6$.

Sulfuric acid and sulfurous acid are both examples of a class of compounds known as *oxyacids,* because they are literally acids that contain oxygen. The SO_4^{2-} and SO_3^{2-} ions can be described as *oxyanions.* The Lewis structures of some of the oxides of sulfur that form oxyacids or oxyanions are given in Figure 10.7. Only one of these oxides deserves special mention. Reaction between sulfur and the sulfite (SO_3^{2-}) ion can lead to the formation of the thiosulfate ion.

$$8 SO_3^{2-}(aq) + S_8(s) \longrightarrow 8 S_2O_3^{2-}(aq)$$

We can derive the Lewis structure of this ion from the structure of the sulfate ion by replacing one of the terminal oxygen atoms with sulfur.

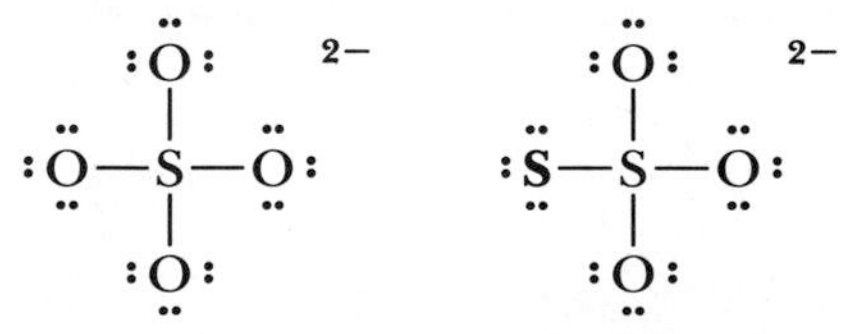

sulfate   thiosulfate

The thiosulfate ion is commonly used in the photographic industry as the "fixer" that permanently fixes the image onto the photographic film.

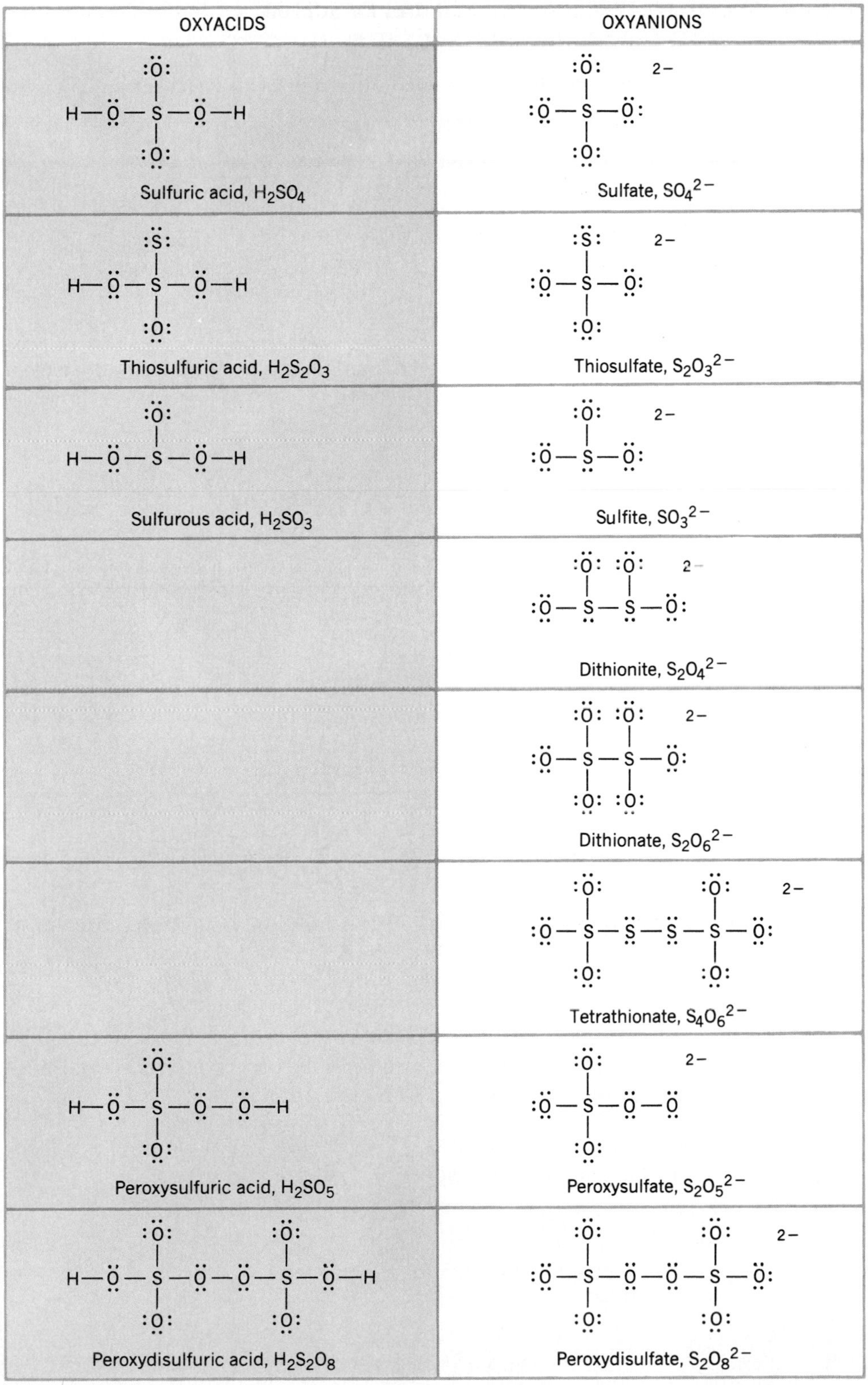

OXYACIDS	OXYANIONS
Sulfuric acid, H_2SO_4	Sulfate, SO_4^{2-}
Thiosulfuric acid, $H_2S_2O_3$	Thiosulfate, $S_2O_3^{2-}$
Sulfurous acid, H_2SO_3	Sulfite, SO_3^{2-}
	Dithionite, $S_2O_4^{2-}$
	Dithionate, $S_2O_6^{2-}$
	Tetrathionate, $S_4O_6^{2-}$
Peroxysulfuric acid, H_2SO_5	Peroxysulfate, $S_2O_5^{2-}$
Peroxydisulfuric acid, $H_2S_2O_8$	Peroxydisulfate, $S_2O_8^{2-}$

FIG. 10.7 Oxyacids of sulfur and their oxyanions.

THE EFFECT OF DIFFERENCES IN THE ABILITIES OF SULFUR AND OXYGEN TO EXPAND THE VALENCE SHELL

The electron configurations of oxygen and sulfur are usually written as follows.

$$O = [He] \, 2s^2 \, 2p^4$$
$$S = [Ne] \, 3s^2 \, 3p^4$$

Although this notation shows the similarity between the configurations of the two elements, it hides an important difference that allows sulfur to expand its valence shell to hold more than eight electrons.

Exercise 10.8

Explain why sulfur has empty valence-shell orbitals that allow it to hold more than eight valence electrons and why oxygen does not.

Solution

There are only four orbitals in the valence shell of an oxygen atom—the $2s$, $2p_x$, $2p_y$, and $2p_z$ orbitals.

$$O = [He] \, 2s^2 \, 2p^4$$

As a result, oxygen can hold no more than the eight valence electrons that would be present in an O^{2-} ion.

$$O^{2-} = [He] \, 2s^2 \, 2p^6$$

The valence orbitals of sulfur, however, are in the $n = 3$ shell, which includes $3s$, $3p$, and $3d$ orbitals. Sulfur therefore has empty $3d$ valence orbitals that can be used to expand its valence shell.

$$S = [Ne] \, 3s^2 \, 3p^4 \, 3d^0$$

Oxygen reacts with fluorine to form OF_2.

$$O_2(g) + 2 \, F_2(g) \longrightarrow 2 \, OF_2(g)$$

Reaction stops at this point because oxygen can hold only eight electrons in its valence shell.

$$: \overset{..}{\underset{..}{F}} - \overset{..}{\underset{..}{O}} - \overset{..}{\underset{..}{F}} :$$

Sulfur reacts with fluorine to form either SF_4 or SF_6, because sulfur can expand its valence shell to hold the 10 or even 12 electrons on the central atoms in the Lewis structures of these compounds, as shown in Figure 10.8.

$$S_8(s) + 16 \, F_2(g) \longrightarrow 8 \, SF_4(g)$$
$$S_8(s) + 24 \, F_2(g) \longrightarrow 8 \, SF_6(g)$$

FIG. 10.8 There are 10 valence electrons on the sulfur atom in SF_4, so the structure of this compound is based on a distorted trigonal bipyramid. There are 12 valence electrons on the central atom in SF_6, which are distributed toward the corners of an octahedron.

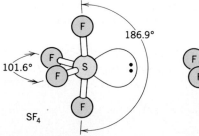
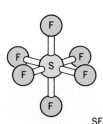

10.6 THE CHEMISTRY OF NITROGEN

One of the dominant factors influencing the chemistry of nitrogen is the ease with which nitrogen atoms form double and triple bonds. A neutral nitrogen atom contains five valence electrons.

$$N = [He]\ 2s^2\ 2p^3$$

Thus, a nitrogen atom can achieve an octet of valence electrons by sharing three pairs of electrons with another nitrogen atom.

$$:N \equiv N:$$

Because the covalent radius of a nitrogen atom is relatively small — only 0.070 nm — nitrogen atoms can come close enough together to form very strong π bonds. The bond dissociation enthalpy for the $N \equiv N$ triple bond is 946 kJ/mol, almost twice as large as that for $O = O$ double bond.

The strength of the $N \equiv N$ triple bond makes the N_2 molecule very unreactive. N_2 is so inert that lithium is one of the few elements it reacts with at room temperature.

$$6\ Li(s) + N_2(g) \longrightarrow 2\ Li_3N(s)$$

In the spite of the fact that the N_2 molecule is very unreactive, compounds containing nitrogen exist for virtually every element in the periodic table except the elements in Group VIIIA (He, Ne, Ar, and so on). This can be explained in two ways. First, N_2 becomes significantly more reactive as the temperature of the system increases. At high temperatures, nitrogen reacts with hydrogen to form ammonia and with oxygen to form nitrogen oxide.

$$N_2(g) + 3\ H_2(g) \longrightarrow 2\ NH_3(g)$$
$$N_2(g) + O_2(g) \longrightarrow 2\ NO(g)$$

Second, a number of catalysts found in nature overcome the inertness of N_2 at low temperature.

THE SYNTHESIS OF AMMONIA

It is difficult to imagine a living system that does not contain nitrogen, because this element is an essential component of the proteins, nucleic acids, vitamins, and hormones that make life possible. Animals pick up the nitrogen they need from the plants or other animals in their diet. Plants have to pick up their nitrogen from the soil or absorb it as N_2 from the atmosphere. The concentration of nitrogen in the soil is fairly small, so the process by which plants reduce N_2 to NH_3, or "fix" N_2, is extremely important. Estimates suggest that 200 million tons of NH_3 are produced by biological nitrogen fixation each year.

Plants, by themselves, cannot reduce N_2 to NH_3. This reaction is carried out by blue-green algae and bacteria. The best-understood example of nitrogen fixation involves the *rhizobium* bacteria found in the root nodules of legumes, such as clover, peas, and beans. These bacteria contain a nitrogenase enzyme, which is capable of the remarkable feat of reducing N_2 from the atmosphere to NH_3 at room temperature.

Ammonia is made on an industrial scale by a process first developed between 1909 and 1913 by Fritz Haber. In the ***Haber process,*** a mixture of N_2 and H_2 gas at 200–300 atm and 400–600°C is passed over a catalyst of finely divided iron metal.

$$N_2(g) + 3\ H_2(g) \xrightarrow{Fe} 2\ NH_3(g)$$

It has been estimated that more than 10 times as much N_2 is reduced to NH_3 by biological systems as the NH_3 produced by the Haber process. Furthermore, the nitrogenase enzyme biological systems use to catalyze this reaction gives good yields of nitrogen fixation at room temperature and atmospheric pressure.

Anhydrous ammonia being applied to open ground.

Almost 20 million tons of NH_3 is produced in the United States each year by this process. About 80% of it, worth more than $2 billion, is used to make fertilizers for plants that can't fix nitrogen from the atmosphere. On the basis of weight, ammonia is the second most important industrial chemical in the United States. Only sulfuric acid is produced in larger quantities.

Two-thirds of the ammonia used for fertilizers is converted into solids such as ammonium nitrate, NH_4NO_3; ammonium phosphate, $(NH_4)_3PO_4$; ammonium sulfate, $(NH_4)_2SO_4$; and urea, H_2NCONH_2.

$$NH_3(g) + HNO_3(aq) \longrightarrow NH_4NO_3(aq)$$
$$3\ NH_3(g) + H_3PO_4(aq) \longrightarrow (NH_4)_3PO_4(aq)$$
$$NH_3(g) + H_2SO_4(aq) \longrightarrow (NH_4)_2SO_4(aq)$$

The other third is applied directly to the soil as *anhydrous* (literally, "without water") ammonia. Ammonia is a gas at room temperature, but it can be handled as a liquid when dissolved in water to form an aqueous solution or cooled to temperatures below $-33°C$, in which case the gas condenses to form the anhydrous liquid, $NH_3(l)$.

THE SYNTHESIS OF NITRIC ACID

Most of the NH_3 produced by the Haber process that is not used as fertilizer is burned in oxygen to generate nitrogen oxide.

$$4\ NH_3(g) + 5\ O_2(g) \longrightarrow \mathbf{4\ NO}(g) + 6\ H_2O(g)$$

Nitrogen oxide — or nitric oxide, as it is also known — is a colorless gas that reacts rapidly with excess oxygen to produce nitrogen dioxide, a dark brown gas.

$$2\ NO(g) + O_2(g) \longrightarrow \mathbf{2\ NO_2}(g)$$

Nitrogen dioxide dissolves in water to give nitric acid and NO, which can be captured and recycled.

$$3\ NO_2(g) + H_2O(l) \longrightarrow \mathbf{2\ HNO_3}(aq) + NO(g)$$

The NO produced when ammonia burns is a colorless gas that is only marginally soluble in water. When it is exposed to air, it rapidly reacts with O_2 to form NO_2, a dark-brown gas. When the flask shown is shaken, the NO_2 dissolves in water to produce a mixture of nitric acid and NO gas.

Thus, by a process discovered by Friedrich Ostwald in 1908, ammonia can be converted into nitric acid.

$$4\ NH_3(g) + 5\ O_2(g) \longrightarrow 4\ NO(g) + 6\ H_2O(g)$$
$$2\ NO(g) + O_2(g) \longrightarrow 2\ NO_2(g)$$
$$3\ NO_2(g) + H_2O(l) \longrightarrow 2\ HNO_3(aq) + NO(g)$$

The Haber process for the synthesis of ammonia combined with the ***Ostwald process*** for the conversion of ammonia into nitric acid revolutionized the explosives industry. Nitrates have been important explosives ever since Friar Roger Bacon mixed sulfur, saltpeter, and powdered carbon to make gunpowder in 1245.

$$16\ KNO_3(s) + S_8(s) + 24\ C(s) \longrightarrow 8\ K_2S(s) + 24\ CO_2(g) + 8\ N_2(g)$$
$$\Delta H° = -571.9\ kJ/mol\ N_2$$

Before the Ostwald process was developed, the only source of nitrates for use in explosives was naturally occurring nitrate minerals, such as saltpeter, a mixture of $NaNO_3$ and KNO_3. Once a dependable supply of nitric acid became available from the Ostwald process, any number of nitrates could be made for use as explosives. Combining NH_3 from the Haber process with HNO_3 from the Ostwald process, for example, gives ammonium nitrate, which is both an excellent fertilizer and a cheap, dependable explosive commonly used in blasting powder.

$$2\ NH_4NO_3(s) \longrightarrow 2\ N_2(g) + O_2(g) + 4\ H_2O(g)$$

The destructive power of ammonium nitrate is apparent from a famous accident at Texas City, Texas, part of a large shipping complex on Galveston Bay. In 1947, a freighter loaded with NH_4NO_3 blew up in this harbor, killing almost 600 people and injuring 4000 others.

INTERMEDIATE OXIDATION NUMBERS

Nitric acid (HNO_3) and ammonia (NH_3) represent the maximum ($+5$) and minimum (-3) oxidation numbers for nitrogen. Nitrogen also forms compounds with every oxidation number between these extremes, as shown by the examples in Table 10.6. These intermediate oxidation numbers can be divided into negative and positive states.

TABLE 10.6	
Common Oxidation Numbers for Nitrogen	
Oxidation Number	Examples
-3	NH_3, NH_4^+, NH_2^-, Li_3N, Mg_3N_2
-2	N_2H_4, $N_2H_5^+$
-1	NH_2OH, H_2NF
$-\frac{1}{3}$	NaN_3, HN_3
0	N_2
$+1$	N_2O, $N_2O_2^{2-}$, HNF_2
$+2$	NO, N_2O_2, N_2F_4
$+3$	HNO_2, NO_2^-, N_2O_3, NOF, NF_3, NO^+
$+4$	NO_2, N_2O_4
$+5$	HNO_3, NO_3^-, N_2O_5, NOF_3, NO_2F

NEGATIVE OXIDATION NUMBERS OF NITROGEN BESIDES -3

At about the same time that Haber developed the process for making ammonia and Ostwald worked out the process for converting ammonia into nitric acid, a process was developed that oxidized ammonia with the hypochlorite (OCl^-) ion to produce hydrazine, N_2H_4.

$$2\ NH_3(aq) + OCl^-(aq) \longrightarrow N_2H_4(aq) + Cl^-(aq) + H_2O(l)$$

Pure hydrazine is a colorless liquid with a faint odor of ammonia that can be collected when the solution described above is heated until N_2H_4 vapor distills out of the reaction flask. Hydrazine has a structure similar to what might be obtained if a pair of ammonia molecules were fused together.

Many of the physical properties of hydrazine are similar to those of water.

	H_2O	N_2H_4
Density	1.000 g/cm^3	1.008 g/cm^3
Melting Point	$0.00°C$	$1.54°C$
Boiling Point	$100°C$	$113.8°C$

But there is a very large difference between the chemical properties of these compounds. Hydrazine burns when ignited in the presence of oxygen to give nitrogen gas, water vapor, and large amounts of energy.

$$N_2H_4(l) + O_2(g) \longrightarrow N_2(g) + 2\ H_2O(g) \qquad \Delta H° = -534.3\ \text{kJ/mol}\ N_2H_4$$

The principal use of hydrazine is as a rocket fuel. It is second only to liquid hydrogen in terms of the number of kilograms of thrust produced per kilogram of fuel burned. Hydrazine has several advantages over liquid H_2, however. It can be stored at room temperature, whereas liquid hydrogen must be stored at temperatures below $-253°C$. Hydrazine is also denser than liquid H_2 and therefore requires less storage space.

Pure hydrazine is seldom used as a rocket fuel, because it freezes at about 2°C and therefore becomes a solid at the temperatures encountered in the upper atmosphere. Hydrazine is therefore mixed with N,N-dimethylhydrazine, $(CH_3)_2NNH_2$, to form a solution that remains a liquid at low temperatures. Mixtures of hydrazine and N,N-dimethylhydrazine were used to fuel the Titan II rockets that carried the Project Gemini spacecraft, and the reaction between hydrazine derivatives and N_2O_4 is still used to fuel the small rocket engines that enable the space shuttle to maneuver in space.

The product of the combustion of hydrazine in oxygen is somewhat unusual. When carbon compounds burn, the carbon is oxidized to CO or CO_2. When sulfur compounds burn, SO_2 is produced. When hydrazine is burned, however, the product of the reaction has an oxidation number of 0, not $+3$ or $+5$, because of the unusually strong $N{\equiv}N$ triple bond in the N_2 molecule.

$$N_2H_4(l) + O_2(g) \longrightarrow N_2(g) + 2\ H_2O(g)$$

Hydrazine reacts with nitrous acid (HNO_2) to form a compound known as hydrogen azide, HN_3, in which the nitrogen atom formally has an oxidation number of $-\frac{1}{3}$.

$$N_2H_4(aq) + HNO_2(aq) \longrightarrow \mathbf{HN_3}(aq) + 2\ H_2O(l)$$

Pure hydrogen azide is an extremely dangerous substance — even dilute solutions should be handled with care because of the risk of explosions. Hydrogen azide is best described as a resonance hybrid of the following Lewis structures.

$$[H-\ddot{N}{=}N{=}\ddot{N} \longleftrightarrow H-\ddot{N}-N{\equiv}N\colon]$$

The corresponding azide ion, $N_3{}^-$, is a linear molecule that is a resonance hybrid of at least three Lewis structures.

$$\left[\ddot{N}{=}N{=}\ddot{N}^- \longleftrightarrow \colon\!\ddot{N}-N{\equiv}N\colon^- \longleftrightarrow \colon\!N{\equiv}N-\ddot{N}\colon^- \right]$$

POSITIVE OXIDATION NUMBERS FOR NITROGEN: THE NITROGEN HALIDES

Fluorine, oxygen, and chlorine are the only elements more electronegative than nitrogen. Positive oxidation numbers of nitrogen are therefore found in compounds that contain one or more of these elements.

In theory, N_2 could react with F_2 to form a compound with the formula NF_3. In practice, N_2 is far too inert to undergo this reaction at room temperature. Ammonia, however, can be oxidized by F_2 in the presence of a copper metal catalyst to form NF_3.

$$NH_3 + 3\ F_2 \longrightarrow \mathbf{NF_3} + 3\ HF$$

$$\underset{-3}{}\quad \underset{0}{}\qquad \underset{+3\ -1}{}\quad \underset{-1}{}$$

oxidation

reduction

The HF produced in this reaction combines with ammonia to form ammonium fluoride. The following is therefore the overall stoichiometry for the reaction.

$$4\ NH_3(g) + 3\ F_2(g) \xrightarrow{\ Cu\ } NF_3(g) + 3\ NH_4F(s)$$

The Lewis structure of NF_3 is analogous to the Lewis structure of NH_3, and the two molecules have similar shapes.

NF_3 is essentially inert at room temperature. At elevated temperatures, however, it can be oxidized by copper to form N_2F_4, in much the same way NH_3 can be oxidized to form N_2H_4.

$$2\ NF_3(g) + 2\ Cu(s) \longrightarrow N_2F_4(g) + 2\ CuF(s)$$

Ammonia reacts with chlorine to form NCl_3, which seems at first glance to be closely related to NF_3. But there is a significant difference between these compounds. NCl_3 is a shock-sensitive, highly explosive liquid that decomposes to form N_2 and Cl_2.

$$2\ NCl_3(l) \longrightarrow N_2(g) + 3\ Cl_2(g)$$

Exercise 10.9

Use the following enthalpy of formation data to explain why NF_3 is essentially inert at room temperature and NCl_3 decomposes violently to N_2 and Cl_2.

Compound	ΔH_f°
$NF_3(g)$	-124.7 kJ/mol
$NCl_3(l)$	$+230$ kJ/mol

Solution

The enthalpy of formation of NF_3 is negative, which means that the compound is more stable than its elements at 25°C and 1 atm. The enthalpy of formation of NCl_3 is positive, which means the compound is less stable than its elements. Considering both the sign and magnitude of ΔH_f° for NCl_3, it is not surprising that this compound decomposes to its elements so easily.

Although ammonia reacts with iodine, it is impossible to isolate the NI_3 produced. The product of this reaction is obtained as a solid that is a complex between NI_3 and NH_3. This material is the subject of a popular, but dangerous, demonstration in which freshly prepared samples of NI_3 in ammonia are poured onto filter paper, which is allowed to dry on a ring stand. After the ammonia evaporates, the NH_3/NI_3 crystals are touched with a feather attached to a meter stick, resulting in immediate detonation of this shock-sensitive solid, which decomposes to form a mixture of N_2 and I_2.

$$2\ NI_3(s) \longrightarrow N_2(g) + 3\ I_2(g)$$

TABLE 10.7

Enthalpy of Formation Data for Oxides of Nitrogen

Compound	$\Delta H_f^\circ(kJ/mol)$
$N_2O(g)$	62.05
$NO(g)$	90.25
$NO_2(g)$	35.98
$N_2O_3(g)$	50.29
$N_2O_4(g)$	9.16
$N_2O_5(g)$	11.3

POSITIVE OXIDATION NUMBERS FOR NITROGEN: THE NITROGEN OXIDES

Lewis structures for seven oxides of nitrogen with oxidation numbers ranging from $+1$ to $+5$ are given in Figure 10.9. These compounds all have two things in common—they contain $N=O$ double bonds, and they are less stable than their elements in the gas phase, as shown by the data on enthalpy of formation in Table 10.7.

Dinitrogen oxide, N_2O, also known as nitrous oxide, can be prepared by careful decomposition of ammonium nitrate.

$$NH_4NO_3(s) \xrightarrow{170-200°C} N_2O(g) + 2\ H_2O(g)$$

FIG. 10.9 Oxides of nitrogen.

$$\left[:\ddot{N}=N=\ddot{O}: \longleftrightarrow :N\equiv N-\ddot{O}: \right]$$

Dinitrogen oxide, N_2O
(nitrous oxide)

$$:\ddot{N}=\ddot{O}:$$

Nitrogen oxide, NO
(nitric oxide)

$$:\ddot{O}=\ddot{N}-\ddot{N}=\ddot{O}:$$

Dinitrogen dioxide, N_2O_2

$$:\ddot{O}=\ddot{N}-N\overset{\displaystyle :\ddot{O}:}{\underset{\displaystyle :\ddot{O}:}{}}$$

Dinitrogen trioxide, N_2O_3

$$\cdot\overset{\displaystyle :\ddot{O}:}{N}\underset{\displaystyle :\ddot{O}:}{}$$

Nitrogen dioxide, NO_2

$$\overset{\displaystyle :\ddot{O}:\quad\ddot{O}:}{N-N}\underset{\displaystyle :\ddot{O}:\quad:\ddot{O}:}{}$$

Dinitrogen tetroxide, N_2O_4

$$\overset{\displaystyle :\ddot{O}:\qquad\ddot{O}:}{N-\ddot{O}-N}\underset{\displaystyle :\ddot{O}:\qquad:\ddot{O}:}{}$$

Dinitrogen pentoxide, N_2O_5

The Royal Institute was founded in 1800 to provide a place where lectures on the latest developments in science could be given to "improve the lot of the working man." This etching by James Gillray is a caricature of Thomas Garrett and Humphrey Davy administering nitrous oxide or "laughing gas," with unfortunate consequences.

Nitrous oxide is a sweet-smelling, colorless gas best known to nonchemists as laughing gas. As early as 1800, Humphrey Davy noted that N_2O, inhaled in relatively small amounts, produced a state of apparent intoxication often accompanied by either convulsive laughter or crying. Inhaling laughing gas soon became a popular fad.

When taken in larger doses, nitrous oxide provides fast and efficient relief from pain. Thus, N_2O was used as the first anesthetic. Because relatively large doses are needed to produce anesthesia, however, and because continued exposure to the gas can be fatal, N_2O is used today only for relatively short operations. Furthermore, it is usually administered in combination with other drugs, such as barbiturates, which cause the patient to lose consciousness, and a muscle relaxant, such as curare.

Nitrous oxide has several other interesting properties. First, it is highly soluble in cream; for that reason, it is used as the propellant in most whipped-cream dispensers. Second, although it does not burn by itself, it is better than air at supporting the combustion of other objects. We can explain this by noting that N_2O can decompose to form an atmosphere that is one-third O_2 by volume (normal air is only 21% oxygen by volume).

$$2 N_2O(g) \longrightarrow 2 N_2(g) + O_2(g)$$

For many years, the endings *-ous* and *-ic* were used to distinguish between the lowest and highest of a pair of oxidation numbers. N_2O is *nitrous* oxide because the oxidation number of the nitrogen is $+1$. NO is *nitric* oxide because the oxidation number of the nitrogen is $+2$.

Enormous quantities of nitrogen oxide, or nitric oxide, are generated each year by a reaction between the N_2 and O_2 in the atmosphere, catalyzed by a stroke of lightning passing through the atmosphere or by the hot walls of an internal combustion engine.

$$N_2(g) + O_2(g) \longrightarrow \mathbf{2\ NO(g)}$$

One of the reasons for lowering the compression ratio of automobile engines in recent years, so that they run on lower-octane fuels, is to decrease the temperature of the combustion reaction, thereby decreasing the amount of NO emitted into the atmosphere as a pollutant.

As already mentioned, NO is made on an industrial scale by burning ammonia on a metal catalyst at 800–900°C—the first step in the Ostwald process.

$$4 NH_3(g) + 5 O_2(g) \longrightarrow \mathbf{4\ NO(g)} + 6 H_2O(g)$$

We can prepare it in the laboratory by reacting copper metal with *dilute* nitric acid.

$$3 Cu(s) + 8 HNO_3(aq) \longrightarrow Cu(NO_3)_2(aq) + \mathbf{NO(g)} + 4 H_2O(l)$$

NO is a fascinating molecule, because it contains an odd number of valence electrons. As a result, it is impossible to write a Lewis structure for this molecule in which all of the electrons are paired.

$$\left[\ddot{N}=\ddot{O} \longleftrightarrow \ddot{N}=\ddot{O} \right]$$

NO is therefore another example of a free radical (see Section 10.4). When a sample of NO gas is cooled, pairs of NO molecules combine in a reversible reaction to form a **dimer** (from the Greek, "two parts"), with the formula N_2O_2, in which all of the valence electrons are paired.

$$\left[\ddot{O}=\ddot{N} + \ddot{N}=\ddot{O} \longrightarrow \ddot{O}-\ddot{N}-\ddot{N}=\ddot{O} \right]$$

When NO_2 (left) is cooled in liquid air, as shown in the photograph on the right, the brown color disappears as the NO_2 dimerizes to form N_2O_4.

NO reacts rapidly with O_2 to form nitrogen dioxide (once known as nitrogen peroxide), which is a dark-brown gas at room temperature.

$$2\ NO(g) + O_2(g) \longrightarrow 2\ NO_2(g)$$

We can prepare NO_2 in the laboratory by heating certain metal nitrates until they decompose.

$$2\ Pb(NO_3)(s) \longrightarrow 2\ PbO(s) + 4\ NO_2(g) + O_2(g)$$

We can also make it by reacting copper metal with *concentrated* nitric acid,

$$Cu(s) + 4\ HNO_3(aq) \longrightarrow Cu(NO_3)_2(aq) + 2\ NO_2(g) + 2\ H_2O(l)$$

NO_2 also has an odd number of electrons and therefore contains at least one unpaired electron in all of its Lewis structures.

NO_2 dimerizes to form N_2O_4 molecules, in which all the electrons are paired.

Once again, low temperatures favor the formation of the dimer.

Mixtures of NO and NO_2 combine when cooled to form a pale blue liquid that consists of dinitrogen trioxide, N_2O_3.

The appearance of a blue liquid when either NO or NO_2 is cooled therefore implies the presence of at least a small portion of the other oxide, because N_2O_2 and N_2O_4 are both colorless.

By carefully removing water from concentrated nitric acid at low temperatures with a dehydrating agent such as P_4O_{10}, we can form dinitrogen pentoxide.

$$4\ HNO_3(aq) + P_4O_{10}(s) \longrightarrow 2\ N_2O_5(s) + 4\ HPO_3(s)$$

N_2O_5 is a colorless solid that decomposes in light or on warming to room temperature. It consists of molecules with the following Lewis structure.

As might be expected, N_2O_5 dissolves in water to form nitric acid.

$$N_2O_5(s) + H_2O(l) \longrightarrow 2\ HNO_3(aq)$$

10.7 THE CHEMISTRY OF PHOSPHORUS

White phosphorus stored under water.

Red phosphorus is less reactive than white phosphorus, and doesn't have to be stored under water.

Phosphorus is another element that is vital to life. It can be found in the nucleic acids that control the genetic code and the lipids that form cell membranes. As the phosphate ion, PO_4^{3-}, it plays an important role in the metabolism of sugars and other carbohydrates, and roughly 20% of the weight of bone is due to minerals such as calcium phosphate, $Ca_3(PO_4)_2$. Phosphorus compounds are important components of many fertilizers.

Phosphorus is the first element whose discovery can be traced to a single individual. In 1669, while searching for a way to convert silver into gold, Hennig Brand obtained a sample of a white, waxy solid that glowed in the dark and burst spontaneously into flame when exposed to air. He made this substance by evaporating the water from urine to form a black residue and allowing this substance to putrify for several months. He then mixed the black residue with sand, heated the mixture in the presence of a minimum of air, and collected under water the volatile products that distilled out of the reaction flask.

Phosphorus is directly below nitrogen in the periodic table, and we should therefore expect similarities in the chemistry of the two elements.

$$N = [He]\ 2s^2\ 2p^2$$
$$P = [Ne]\ 3s^2\ 3p^3$$

Phosphorus forms a number of compounds that are direct analogs of nitrogen-containing compounds, as shown by the examples cited in Table 10.8. The fact that elemental nitrogen is virtually inert at room temperature, whereas elemental phosphorus bursts spontaneously into flame when exposed to air, shows there are differences between the behaviors of these elements as well. Phosphorus often forms compounds with the same oxidation numbers as the analogous nitrogen compounds but with different formulas, as shown in Table 10.9.

TABLE 10.8

Nitrogen Compounds and Their Phosphorus Analogs

Nitrogen Compound	Phosphorus Compound
NH_3 (ammonia)	PH_3 (phosphine)
N_2H_4 (hydrazine)	P_2H_4 (diphosphine)
NF_3 (nitrogen trifluoride)	PF_3 (phosphorus trifluoride)
N_2F_4 (dinitrogen tetrafluoride)	P_2F_4 (diphosphorus tetrafluoride)
Ca_3N_2 (calcium nitride)	Ca_3P_2 (calcium phosphide)

TABLE 10.9

Nitrogen and Phosphorus Compounds with the Same Oxidation Numbers but Different Formulas

Nitrogen Compound	Phosphorus Compound	Oxidation Number
N_2	P_4	0
HNO_2 (nitrous acid)	H_3PO_3 (phosphorous acid)	+3
$NaNO_2$ (sodium nitrite)	Na_2HPO_3 (sodium phosphite)	+3
N_2O_3	P_4O_6	+3
HNO_3 (nitric acid)	H_3PO_4 (phosphoric acid)	+5
$NaNO_3$ (sodium nitrate)	Na_3PO_4 (sodium phosphate)	+5
N_2O_5	P_4O_{10}	+5

The same factors that explain the differences between sulfur and oxygen can explain the differences between phosphorus and nitrogen.

1. $N \equiv N$ **triple bonds are much stronger than** $P \equiv P$ **triple bonds.**
2. **P—P single bonds are stronger than N—N single bonds.**
3. **Phosphorus (EN = 2.19) is much less electronegative than nitrogen (EN = 3.04).**
4. **Phosphorus can expand its valence shell to hold more than eight electrons, but nitrogen cannot.**

THE EFFECT OF DIFFERENCES IN THE X—X AND X≡X BOND STRENGTHS

The covalent radius of a phosphorus atom is 1.57 times the covalent radius of a nitrogen atom.

$$\frac{\text{Covalent radius of phosphorus}}{\text{Covalent radius of nitrogen}} = \frac{0.110 \text{ nm}}{0.070 \text{ nm}} = 1.57$$

Within experimental error, this is the same as the ratio of the covalent radii of sulfur and oxygen atoms.

$$\frac{\text{Covalent radius of sulfur}}{\text{Covalent radius of oxygen}} = \frac{0.104 \text{ nm}}{0.066 \text{ nm}} = 1.58$$

In each case, the third-row element is slightly more than half again as large as the second-row element.

For the same reason that $S = S$ double bonds are much weaker than $O = O$ double bonds, $P \equiv P$ triple bonds are much weaker than $N \equiv N$ triple bonds. Phosphorus atoms are too big to come close enough together to form strong π bonds.

Elemental nitrogen consists of N_2 molecules in which each atom completes its octet of valence electrons by sharing three pairs of electrons with a single neighboring atom. Because phosphorus does not form strong multiple bonds with itself, elemental phosphorus consists of discrete tetrahedral P_4 molecules in which each atom completes its octet by forming single bonds with three neighboring atoms, as shown in Figure 10.10. Pure elemental phosphorus (often called white phosphorus) is a white or colorless solid with a waxy appearance, which melts at 44.1°C and boils at 287°C. Elemental phosphorus is made by reduction of calcium phosphate with carbon in the presence of silica (or sand) at very high temperatures.

$$2 \text{ Ca}_3(\text{PO}_4)_2(s) + 6 \text{ SiO}_2(s) + 10 \text{ C}(s) \longrightarrow 6 \text{ CaSiO}_3(s) + \textbf{P}_4(s) + 10 \text{ CO}(g)$$

White phosphorus is usually stored under water, because the element spontaneously bursts into flame in the presence of oxygen at temperatures only slightly above room temperature. Although phosphorus is insoluble in water, it is very soluble in carbon disulfide — almost 900 grams of phosphorus can dissolve in 100 grams of CS_2. Solutions of P_4 in CS_2 are reasonably stable. When a piece of filter paper is soaked in this solution, however, the paper bursts into flame as soon as the CS_2 evaporates.

If phosphorus atoms form strong P—P single bonds, why is white phosphorus so reactive? We can understand this reactivity by noting that the P—P—P bond angle is only 60° in a tetrahedral P_4 molecule. This very small bond angle produces a considerable amount of strain in the P_4 molecule, which can be relieved by the breaking of one of the P—P bonds.

Phosphorus forms other allotropes by opening up the P_4 tetrahedron. When

White phosphorus is so reactive that it will even burn under water when O_2 gas is bubbled into the solution at temperatures above 70°C.

P_4, white phosphorous

FIG. 10.10 Pure elemental phosphorus is a white or colorless waxy solid that consists of tetrahedral P_4 molecules in which the P—P—P bond angle is 60°.

White phosphorus dissolves in carbon disulfide to form solutions that are relatively stable, until the CS_2 evaporates. At this point, the phosphorus bursts into flame.

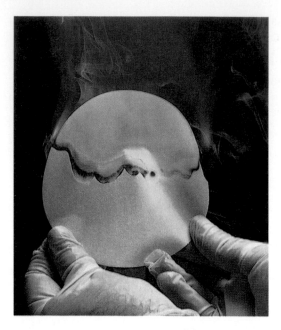

white phosphorus is heated to $270-300\,^{\circ}C$, one bond inside each P_4 tetrahedron is broken, and the P_4 molecules link together to form a ***polymer*** (from the Greek *poly,* "many," and *meros,* "parts") with the structure shown in Figure 10.11. This allotrope of phosphorus is dark red, and its presence in small traces often gives white phosphorus a light yellow color. Red phosphorus is denser ($2.16\,g/cm^3$) than white phosphorus ($1.82\,g/cm^3$) and much less reactive at normal temperatures.

By heating phosphorus at high temperature and high pressure, "black" phosphorus, which consists of an extended network of three-coordinate phosphorus atoms can be produced. This allotrope has a density of $2.69\,g/cm^3$ and is stable in air.

The ability of white phosphorus to burst into flame when it comes in contact with air made it the active ingredient of the first matches — slips of paper coated with white phosphorus and sealed in glass tubes, which were broken to ignite the matches. Later matches were mixtures of white phosphorus, potassium chlorate ($KClO_3$), chalk, and either sulfur or paraffin glued to the tips of wooden splints. $KClO_3$ was added so that the match lit when it was struck; phosphorus was present to start the match burning; and sulfur or paraffin was used to keep the fire going until the wooden splint ignited.

The toxicity of phosphorus eventually led to the use of safety matches, which consist of a mixture of $KClO_3$, sulfur, and ground glass glued to the tip of a piece of paper and coated with paraffin. These matches light when the tip is struck on a surface that contains a mixture of red phosphorus and glass powder.

FIG. 10.11 When white phosphorus is heated, one of the P—P bonds in each P_4 tetrahedron opens up, and bonds form between these tetrahedra to make a polymeric chain. The resulting red allotrope is much less reactive than white phosphorus.

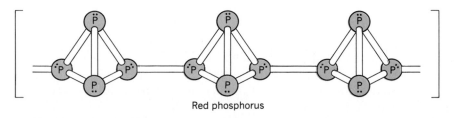

Red phosphorus

THE EFFECT OF DIFFERENCES IN THE STRENGTHS
OF P=X AND N=X DOUBLE BONDS

The size of a phosphorus atom interferes not only with its ability to form triple bonds to other phosphorus atoms but also with its ability to form multiple bonds to other elements, such as oxygen, nitrogen, and sulfur.

As a result, phosphorus tends to form compounds that contain two $P-O$ single bonds where nitrogen would form an $N=O$ double bond. For example, nitrogen forms the nitrate, NO_3^-, ion, in which it has an oxidation number of $+5$. When phosphorus forms an ion with an oxidation number of $+5$, it is the phosphate, PO_4^{3-}, ion.

Similarly, nitrogen forms nitric acid, HNO_3, which contains an $N=O$ double bond, while phosphorus forms phosphoric acid, H_3PO_4, which contains $P-O$ single bonds.

The difference between the formulas of nitric acid and phosphoric acid is equal to the formula for a molecule of water.

$$H_3XO_4 - HXO_3 = H_2O$$

Thus, phosphoric acid can be thought to result from the addition of a molecule of water across a hypothetical $P=O$ double bond.

We see a similar phenomenon whenever we compare second- and third-row elements. Carbon, for example, forms carbonic acid, H_2CO_3. The silicon analog is silicic acid, H_4SiO_4, which can be viewed as resulting from the addition of water across a hypothetical $Si=O$ double bond.

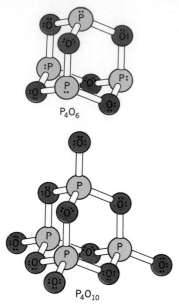

P_4O_6

P_4O_{10}

FIG. 10.12 P_4O_6 consists of a tetrahedron in which oxygen atoms have been inserted into each P—P bond in a P_4 molecule. P_4O_{10} consists of the same tetrahedral framework, with one more oxygen atom on each of the four phosphorus atoms.

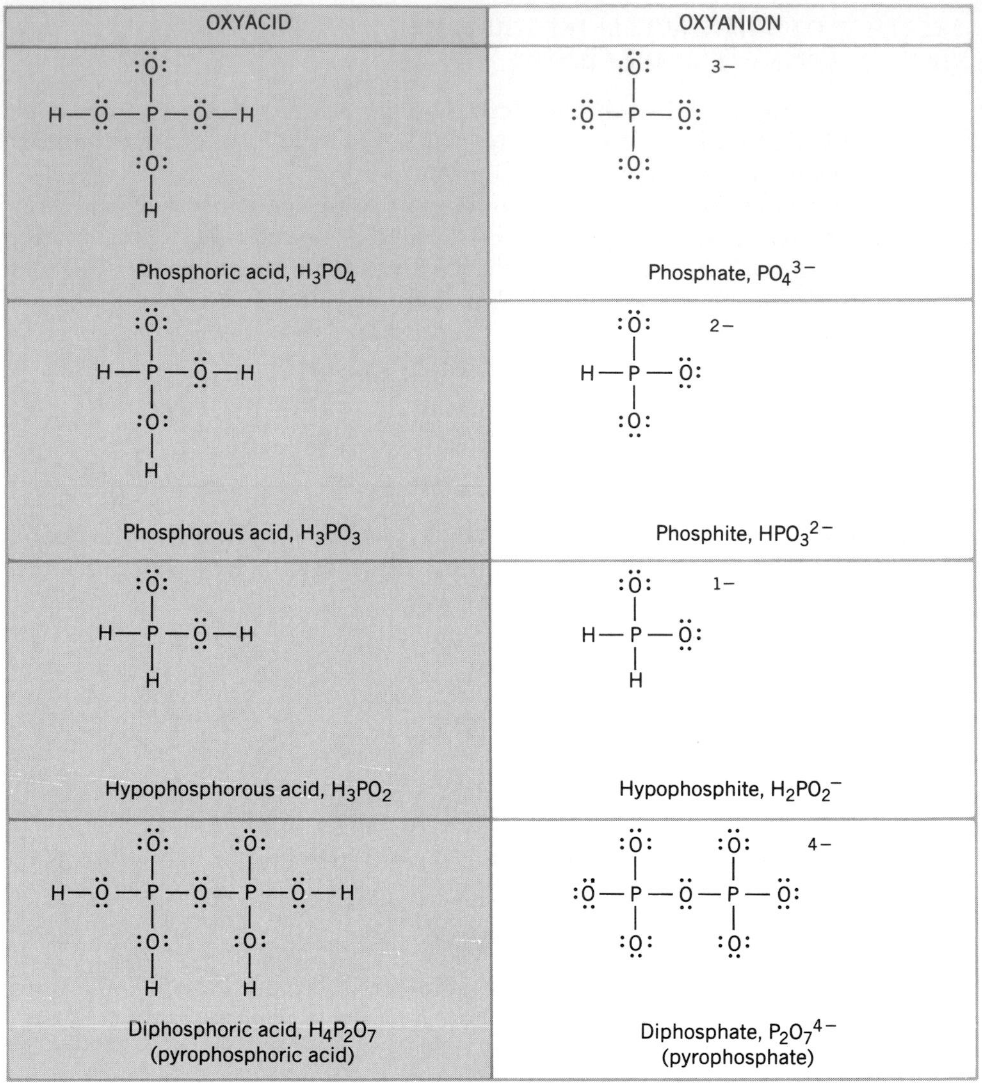

FIG. 10.13 Oxyacids of phosphorus and their oxyanions.

OXYACID	OXYANION
Phosphoric acid, H_3PO_4	Phosphate, PO_4^{3-}
Phosphorous acid, H_3PO_3	Phosphite, HPO_3^{2-}
Hypophosphorous acid, H_3PO_2	Hypophosphite, $H_2PO_2^{-}$
Diphosphoric acid, $H_4P_2O_7$ (pyrophosphoric acid)	Diphosphate, $P_2O_7^{4-}$ (pyrophosphate)

THE EFFECT OF DIFFERENCES IN THE ELECTRONEGATIVITIES OF PHOSPHORUS AND NITROGEN

Within experimental error, the difference between the electronegativity of phosphorus (2.19) and nitrogen (3.04) is equal to the difference between the electronegativities of sulfur (2.58) and oxygen (3.44). Just as sulfur is more likely than oxygen to form compounds in which it has a positive oxidation number, phosphorus is more likely than nitrogen to exhibit positive oxidation numbers in its compounds.

The most important oxidation numbers for phosphorus are −3, +3, and +5. Examples of compounds in which phosphorus has these oxidation numbers are given in Table 10.10. Because phosphorus is more electronegative than most metals, it reacts with metals at elevated temperatures to form phosphides, in which it has an oxidation number of −3.

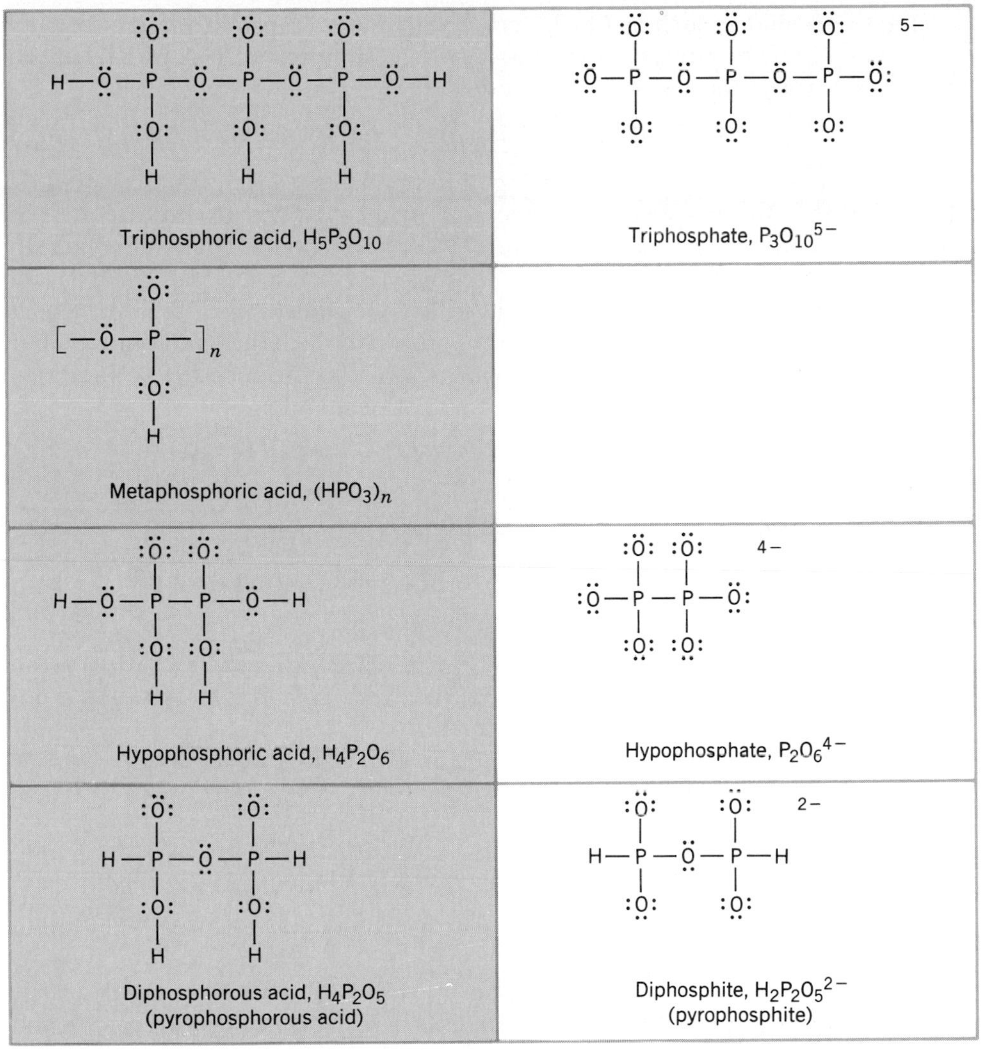

Triphosphoric acid, $H_5P_3O_{10}$

Triphosphate, $P_3O_{10}^{5-}$

Metaphosphoric acid, $(HPO_3)_n$

Hypophosphoric acid, $H_4P_2O_6$

Hypophosphate, $P_2O_6^{4-}$

Diphosphorous acid, $H_4P_2O_5$
(pyrophosphorous acid)

Diphosphite, $H_2P_2O_5^{2-}$
(pyrophosphite)

$$6 \, Ca(s) + P_4(s) \longrightarrow 2 \, Ca_3P_2(s)$$

Phosphides are also formed when the phosphorus in phosphates is reduced with carbon at high temperatures.

$$Ca_3(PO_4)_2(s) + 8 \, C(s) \longrightarrow Ca_3P_2(s) + 8 \, CO(g)$$

TABLE 10.10

Common Oxidation Numbers of Phosphorus

Oxidation Number	Examples
−3	Ca_3P_2, PH_3, PH_4^+, PH_2^-
+3	PF_3, H_3PO_3, NaH_2PO_3, Na_2HPO_3, P_4O_6
+5	PF_5, PF_6^-, OPF_3, H_3PO_4, NaH_2PO_4, Na_2HPO_4, Na_3PO_4, P_4O_{10}

Metal phosphides, such as Ca_3P_2, react with water to produce an extremely poisonous, highly reactive, colorless gas known as phosphine (PH_3), which has the foulest odor the authors have encountered.

$$Ca_3P_2(s) + 6 H_2O(l) \longrightarrow 2\ PH_3(g) + 3\ Ca^{2+}(aq) + 6\ OH^-(aq)$$

Samples of PH_3, the phosphorus analog of ammonia, are often contaminated by traces of P_2H_4, the phosphorus analog of hydrazine. As if the toxicity and odor of PH_3 were not enough, mixtures of PH_3 and P_2H_4 burst spontaneously into flame in the presence of oxygen.

Compounds such as Ca_3P_2 and PH_3, in which phosphorus has a negative oxidation number, are far outnumbered by compounds in which the oxidation number of phosphorus is positive. Phosphorus burns in O_2 to produce P_4O_{10} and truly extraordinary amounts of energy in the form of heat and light.

$$P_4(s) + 5\ O_2(g) \longrightarrow P_4O_{10}(s) \qquad \Delta H° = -2985 \text{ kJ/mol } P_4$$

When phosphorus burns in the presence of a limited amount of O_2, the product of the reaction is a white solid with the formula P_4O_6.

$$P_4(s) + 3\ O_2(g) \longrightarrow P_4O_6(s) \qquad \Delta H° = -1640 \text{ kJ/mol } P_4$$

The structures of P_4O_6 and P_4O_{10} are shown in Figure 10.12. P_4O_6 consists of a tetrahedron in which an oxygen atom has been inserted into each P—P bond in the P_4 molecule. P_4O_{10} has an analogous structure, with an additional oxygen atom bound to each of the four phosphorus atoms.

P_4O_6 reacts with water to form phosphorous acid, H_3PO_3, whereas P_4O_{10} reacts with water to form phosphoric acid, H_3PO_4.

$$P_4O_6(s) + 6 H_2O(l) \longrightarrow 4 H_3PO_3(aq)$$
$$P_4O_{10}(s) + 6 H_2O(l) \longrightarrow 4 H_3PO_4(aq)$$

P_4O_{10} has such a highly affinity for water that it is commonly used as a dehydrating agent.

Phosphorous acid, H_3PO_3, and phosphoric acid, H_3PO_4, are examples of a large class of oxyacids of phosphorus. Lewis structures for some of these oxyacids and their related oxyanions are given in Figure 10.13.

THE EFFECT OF DIFFERENCES IN THE ABILITIES OF PHOSPHORUS AND NITROGEN TO EXPAND THE VALENCE SHELL

The reaction between ammonia and fluorine in the presence of a copper metal catalyst has already been described.

$$4 NH_3(g) + 3 F_2(g) \xrightarrow{\text{Cu}} NF_3(g) + 3 NH_4F(s)$$

This reaction stops at NF_3 because nitrogen uses the $2s$, $2p_x$, $2p_y$, and $2p_z$ orbitals to hold valence electrons. A nitrogen atom can therefore hold a maximum of eight valence electrons.

$$:\ddot{F}—N—\ddot{F}:$$
$$|$$
$$:\ddot{F}:$$

Phosphorus, however, has empty $3d$ atomic orbitals that can be used to expand the valence shell to hold 10 electrons. Thus, phosphorus can form both PF_3 and PF_5. Phosphorus can even form the PF_6^- ion, in which there are 12 valence electrons on the central atom, as shown in Figure 10.14.

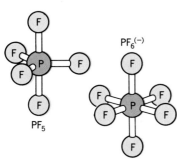

FIG. 10.14 Nitrogen can hold a maximum of 8 electrons in its valence shell. When it reacts with fluorine, reaction stops at NF_3, because forming NF_5 would require nitrogen to hold 10 valence electrons. Phosphorus has empty $3d$ orbitals that can be used to expand its valence shell. It therefore forms compounds such as PF_5, which has 10 valence electrons, and PF_6^-, which has 12 valence electrons.

10.8 THE CHEMISTRY OF THE HALOGENS

There are six elements in Group VIIA, the next-to-last column of the periodic table: hydrogen, fluorine, chlorine, bromine, iodine, and astatine. All six elements are listed in this column because they have certain properties in common. All form diatomic molecules, such as H_2, F_2, Cl_2, Br_2, I_2, and At_2, for example, and all form negatively charged ions, such as the H^-, F^-, Cl^-, Br^-, I^-, and At^- ions.

When the chemistry of these elements is discussed, however, hydrogen is separated from the others, and astatine is ignored because it is radioactive. The most stable isotopes of astatine have half-lives of less than a minute. As a result, the largest samples of astatine compounds studied to date have weighed less than 50 nanograms.

By convention, descriptions of the chemistry of the elements in Group VIIA tend to focus on four elements — fluorine, chlorine, bromine, and iodine. These elements are called the *halogens* (from the Greek *hals*, "salt," and *gennan*, "to form or generate") because they are literally salt formers. None of the halogens can be found in nature in their elemental form; they are invariably found as salts of the *halide* ions (F^-, Cl^-, Br^-, and I^-).

Fluorine is found in minerals such as fluorite (CaF_2) and cryolite (Na_3AlF_6). Chlorine is found in rock salt (NaCl). It is also found in the oceans, which are roughly 2% Cl^- ion by weight, and in lakes that have a high salt content, such as the Great Salt Lake in Utah, which is 9% Cl^- ion by weight. Both bromine and iodine are found at low concentrations as Br^- and I^- ions in the oceans, as well as in brine wells in Louisiana, California, and Michigan.

At room temperature, Cl_2 (chlorine) is a pale yellow gas, Br_2 (bromine) is a dark red-brown liquid, and I_2 is a solid that is so intensely colored it looks like a metal. F_2 (fluorine) is a colorless gas at room temperature, which bursts into flame when it comes in contact with glass. At_2 (astatine) is a solid at room temperature, but it is radioactive and so rare that no compounds of this element are listed in most handbooks of chemistry."

THE HALOGENS IN THEIR ELEMENTAL FORM

Fluorine (F_2), a highly toxic, colorless gas, is the most reactive element known — so reactive that asbestos, water, and silicon burst into flame in its presence. It is so reactive it even forms compounds with Kr, Xe, and Rn — elements in Group VIIIA that were once thought to be inert. Fluorine is such a powerful oxidizing agent it can coax other elements into unusually high oxidation numbers, as in AgF_2, BiF_5, XeF_6, PtF_6, IF_7, and ReF_7. It is so reactive it is difficult to find a container in which it can be stored. F_2 attacks both glass and quartz, for example, and causes most metals to burst into flame. Fluorine is handled in equipment built out of certain alloys of copper and nickel. It still reacts with these alloys, but it forms a layer of a fluoride on the surface that protects the metal from further reaction.

Fluorine has become an increasingly important industrial chemical in recent years. It is used in the manufacture of Teflon — or poly(tetrafluoroethylene), $(C_2F_4)_n$ — which is used for everything from linings for pots and pans to gaskets that are inert to chemical reactions. Large amounts of fluorine are also consumed each year to make the freons (such as CCl_2F_2) used in refrigerators. As noted in Section 7.16, cryolite (Na_3AlF_6) played an important role in the first industrial process for making aluminum, and it is still used for that purpose today. Fluorine has also played an important role in the development of atomic power. It reacts with uranium to form UF_6, which boils at 51°C. The relative rate of diffusion of $^{238}UF_6$ and $^{235}UF_6$ in the gas phase is used to separate the more abundant ^{238}U isotope from ^{235}U, which is used in nuclear reactors.

Chlorine (Cl_2) is a highly toxic gas with a pale green color. It is used commercially as a bleaching agent and as a disinfectant, because it is a very strong oxidizing agent — strong enough to oxidize the dyes that give wood pulp its yellow or brown color, for example, thereby bleaching out this color, and strong enough to destroy

TABLE 10.11

Some Properties of F$_2$, Cl$_2$, Br$_2$, and I$_2$

	Melting Point (°C)	Boiling Point (°C)	Color	Natural Abundance (ppm)	First Ionization Energy (kJ/mol)	Electron Affinity (kJ/mol)	Ionic Radius (nm)	Density (g/cm^3)
F$_2$	−218.6	−188.1	colorless	544	1680.6	322.6	0.133	1.513
Cl$_2$	−101.0	−34.0	pale green	126	1255.7	348.5	0.184	1.655
Br$_2$	−7.3	59.5	dark red-brown	2.5	1142.7	324.7	0.196	3.187
I$_2$	113.6	185.2	very dark violet, almost black	0.46	1008.7	295.5	0.220	3.960

Iodine does not melt when it is heated; it *sublimes*. It goes directly from the solid to the gas phase.

bacteria and thereby act as a germicide. Large quantities of chlorine are used each year to make solvents such as carbon tetrachloride (CCl$_4$), chloroform (CHCl$_3$), dichloroethylene (C$_2$H$_2$Cl$_2$), and trichloroethylene (C$_2$H$_3$Cl$_3$), which are important dry-cleaning fluids.

Bromine (Br$_2$) is a reddish orange liquid with an unpleasant, choking odor. In fact, the name of the element comes from the Greek stem *bromos,* "stench." Bromine is used to prepare a variety of compounds that act as flame retardants, fire-extinguishing agents, mild sedatives, anti-knock agents for gasoline, and insecticides.

Iodine is an intensely colored solid with an almost metallic luster. This solid is relatively volatile, however, and it readily sublimes when heated to form a violet-colored gas. Iodine has been used for many years as a disinfectant in a preparation known as tincture of iodine. Iodine compounds are used as catalysts, drugs, and dyes; and silver iodide (AgI) plays an important role in the photographic process and in attempts to make rain by seeding clouds. Finally, iodide is added to salt to protect against goiter, an iodine deficiency disease characterized by a swelling of the thyroid gland.

A number of the chemical and physical properties of the halogens are summarized in Table 10.11. We can see a regular increase in many of the properties of the halogens as we look down a column from fluorine to iodine, including the melting point, boiling point, intensity of the color of the halogen, element's natural abundance, first ionization energy of the element, radius of the corresponding halide ion, and density of the element. As a result, there is a regular decrease in the oxidizing strength of the halogens from fluorine to iodine.

$$\textbf{F}_2 > \textbf{Cl}_2 > \textbf{Br}_2 > \textbf{I}_2$$

oxidizing strength

This trend is mirrored by a regular increase in the reducing strength of the corresponding halides.

$$\textbf{I}^- > \textbf{Br}^- > \textbf{Cl}^- > \textbf{F}^-$$

reducing strength

Exercise 10.10

Use the fact that Cl$_2$ is a stronger oxidizing agent than Br$_2$ to devise a way of preparing elemental bromine from an aqueous solution of one of its salts.

Solution

In Section 7.15 we found that we could prepare a metal by reacting one of its salts with a metal that is a stronger reducing agent. Titanium metal, for example, is prepared by reacting $TiCl_4$ with magnesium metal.

$$TiCl_4(l) + 2\ Mg(s) \longrightarrow Ti(s) + 2\ MgCl_2(s)$$

Extending this argument to oxidizing agents suggests that we can produce Br_2 by reacting a solution that contains the Br^- ion with an even stronger oxidizing agent, such as Cl_2 dissolved in water.

$$2\ Br^-(aq) + Cl_2(aq) \longrightarrow \mathbf{Br_2}(aq) + 2\ Cl^-(aq)$$

METHODS OF PREPARING THE HALOGENS FROM THEIR HALIDES

We can prepare iodine by reacting a solution of the iodide ion with any substance that is a stronger oxidizing agent, such as bromine or chlorine.

$$2\ I^-(aq) + Br_2(aq) \longrightarrow \mathbf{I_2}(aq) + 2\ Br^-(aq)$$

As noted in Exercise 10.10, we can prepare bromine by adding Cl_2 dissolved in water to a solution that contains the bromine ion.

$$2\ Br^-(aq) + Cl_2(aq) \longrightarrow \mathbf{Br_2}(aq) + 2\ Cl^-(aq)$$

This is the reaction by which bromine was first prepared in 1826 by A. J. Balard. The reaction is still used today to isolate bromine from naturally occurring brine solutions that contain the Br^- ion.

The process by which chlorine was first prepared involves reacting an aqueous solution of the chloride ion with manganese dioxide, which is an even stronger oxidizing agent than Cl_2.

$$2\ Cl^-(aq) + MnO_2(aq) + 4\ H^+(aq) \longrightarrow \mathbf{Cl_2}(aq) + Mn^{2+}(aq) + 2\ H_2O(l)$$

This reaction is still used to prepare small quantities of chlorine in the lab. On the industrial scale, chlorine is made by the electrolysis of aqueous solutions of sodium chloride, as will be described in Chapter 19.

The synthesis of fluorine escaped the efforts of chemists for almost a hundred years. Part of the problem was finding an oxidizing agent strong enough to oxidize the F^- ion to F_2. The task of preparing fluorine was made even more difficult by the extraordinary toxicity of both F_2 and the hydrogen fluoride (HF) used to make it.

In Section 7.16, we noted that the best way of producing strong reducing agents, such as the active metals, is to pass an electric current through a salt of the metal. Sodium, for example, can be prepared by the electrolysis of molten sodium chloride.

$$2\ NaCl(l) \xrightarrow{\text{electrolysis}} \mathbf{2\ Na}(s) + Cl_2$$

In theory, the same process should be capable of generating substances, such as fluorine, that are very strong oxidizing agents.

Attempts to prepare fluorine by electrolysis, however, were initially unsuccessful. Humphry Davy, who prepared potassium, sodium, barium, strontium, calcium, and magnesium for the first time, repeatedly tried to prepare F_2 by the

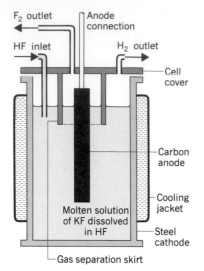

F₂ outlet

Anode connection

HF inlet

H₂ outlet

Cell cover

Carbon anode

Cooling jacket

Molten solution of KF dissolved in HF

Steel cathode

Gas separation skirt

FIG. 10.15 A schematic diagram for the electrolysis of KHF_2 to produce F_2 gas at the anode and H_2 gas at the cathode.

electrolysis of fluorite (CaF_2) and succeeded only in ruining his health. Joseph Louis Gay-Lussac and Louis Jacques Thenard, who prepared elemental boron for the first time, also tried to prepare fluorine and suffered from very painful exposures to hydrogen fluoride. George and Thomas Knox were badly poisoned during their attempts to make fluorine, and both Paulin Louyet and Jerome Nickles died from fluorine poisoning.

Finally, in 1886, Henri Moissan successfully isolated F_2 gas from the electrolysis of a mixed salt of KF and HF and noted that crystals of silicon burst into flame when mixed with this gas. Electrolysis of KHF_2 is still used to prepare fluorine today, as shown in Figure 10.15.

$$2\ KHF_2(s) \xrightarrow{\text{electrolysis}} H_2(g) + F_2(g) + 2\ KF(s)$$

COMMON OXIDATION NUMBERS FOR THE HALOGENS

Fluorine is the most electronegative element in the periodic table. As a result, it has an oxidation number of 0 as an element and -1 in all its compounds. Chlorine, bromine, and iodine are less electronegative; it is possible to prepare compounds in which these elements have oxidation numbers of $+1, +3, +5$, and $+7$, as shown in Table 10.12.

GENERAL TRENDS IN HALOGEN CHEMISTRY

There are several patterns in the chemistry of the halogens.

1. **Neither double nor triple bonds are needed to explain the chemistry of the halogens.**
2. **The chemistry of fluorine is simplified by the fact it is the most electronegative element in the periodic table and by the fact that it has no *d* orbitals in its valence shell, so it can't expand its valence shell.**
3. **Chlorine, bromine, and iodine have valence-shell *d* orbitals and can expand their valence shells to hold as many as 14 valence electrons.**
4. **The chemistry of the halogens is dominated by oxidation-reduction reactions.**

THE HYDROGEN HALIDES (H*X*)

The hydrogen halides are compounds that contain hydrogen attached to one of the halogens, such as HF, HCl, HBr, and HI. These compounds are all colorless gases

TABLE 10.12

Common Oxidation Numbers for the Halogens

Oxidation Number	Examples
-1	AgI, $NaBr$, HCl, $(C_2F_4)_n$, PtF_6
0	F_2, Cl_2, Br_2, I_2
$+1$	$HClO$, IF, ClF, ClF_2^-
$+3$	$HClO_2$, IF_3, ClF_2^+, ClF_3, ClF_4^-
$+5$	$HClO_3$, $KClO_3$, IF_4^+, IF_5, BrF_5, BrF_6^-
$+7$	$HClO_4$, $KClO_4$, BrF_6^+, IF_7

that are soluble in water. Up to 512 milliliters of HCl gas, for example, can dissolve in a single milliliter of water at 0°C and 1 atm. HCl ionizes when it dissolved in water.

$$HCl(g) \xrightarrow{H_2O} H^+(aq) + Cl^-(aq)$$

Exercise 10.11

Explain why students often think that HCl is an ionic compound and describe the evidence that demonstrates that it is not.

Solution

Students often believe HCl is an ionic compound because it forms ions when it dissolves in water.

$$HCl(g) \xrightarrow{H_2O} H^+(aq) + Cl^-(aq)$$

At first glance, this seems to be the same as the behavior displayed by an ionic compound, or salt, when it dissolves in water.

$$NaCl(s) \xrightarrow{H_2O} Na^+(aq) + Cl^-(aq)$$

The evidence that demonstrates that HCl is not an ionic compound is simple — HCl is a gas at room temperature, and ionic compounds are solids at room temperature. As noted in Section 8.4, the covalent bonds that hold molecules together are very strong. But the bonds between covalent molecules are much weaker. It is much easier to separate covalent molecules to form a gas than it is to separate the positive and negative ions in an ionic compound. As a result, covalent compounds are much more likely to be gases at room temperature.

Some chemists try to distinguish between the behavior of HCl and NaCl when they dissolve in water as follows. They argue that HCl *ionizes* when it dissolves in water, because ions are *created* by this reaction. NaCl, on the other hand, *dissociates* in water. NaCl already consists of Na^+ and Cl^- ions in the solid. When NaCl dissolves in water, these ions simply separate, or *dissociate.*

Several of the hydrogen halides can be prepared directly from the elements. Mixtures of H_2 and Cl_2, for example, react with explosive violence in the presence of light to form HCl.

$$H_2(g) + Cl_2(g) \longrightarrow 2\ HCl(g)$$

Since we are usually more interested in the aqueous solutions than the pure gases, however, these compounds are usually synthesized in water. Solutions of the hydrogen halides in water are often called **mineral acids,** because they are literally acids prepared from minerals. Hydrochloric acid is prepared by reacting table salt with sulfuric acid, for example, and hydrofluoric acid is prepared from fluorite and sulfuric acid.

$$2\ NaCl(s) + H_2SO_4(aq) \longrightarrow 2\ HCl(aq) + Na_2SO_4(aq)$$
$$CaF_2(s) + H_2SO_4(aq) \longrightarrow 2\ HF(aq) + CaSO_4(aq)$$

To purify these acids, chemists take advantage of the ease with which HF and HCl gas boil out of the solutions. The gas given off when one of these solutions is heated is collected and then redissolved in water to give relatively pure samples of the mineral acid.

These acids can also be prepared by reacting PCl_3, PBr_3, or PI_3 with water. Hydrobromic acid, for example, results from the reaction of PBr_3 and water.

$$PBr_3(l) + 3\ H_2O(l) \longrightarrow \textbf{3 HBr}\textbf{(}\textbf{\textit{aq}}\textbf{)} + H_3PO_3(aq)$$

It isn't necessary to isolate the PBr_3 intermediate in this reaction. It is possible to obtain the acid by mixing phosphorus, bromine, and water.

$$P_4(s) + 6\ Br_2(s) + 12\ H_2O(l) \longrightarrow \textbf{12 HBr}\textbf{(}\textbf{\textit{aq}}\textbf{)} + 4\ H_3PO_3(aq)$$

This reaction undoubtedly occurs in two steps. The first step involves the synthesis of PBr_3 from P_4 and Br_2; the second involves the reaction of PBr_3 with water to give HBr and H_3PO_3.

THE INTERHALOGEN COMPOUNDS

The compounds formed when different halogens combine are called the **interhalogen compounds**. All possible interhalogen compounds of the type *XY* are known. Bromine reacts with chlorine, for example, to give BrCl, which is a gas at room temperature.

$$Br_2(l) + Cl_2(g) \longrightarrow \textbf{2 BrCl}\textbf{(}\textbf{\textit{g}}\textbf{)}$$

A number of interhalogen compounds with the general formulas XY_3, XY_5, and even XY_7 are formed when pairs of halogens react. Chlorine reacts with fluorine, for example, to form chlorine trifluoride.

$$Cl_2(g) + 3\ F_2(g) \longrightarrow \textbf{2 ClF}_3\textbf{(}\textbf{\textit{g}}\textbf{)}$$

XY_3, XY_5, and XY_7 compounds, such as ClF_3, BrF_5, and IF_7, are easiest to form when *Y* is fluorine. Iodine is the only halogen that forms an XY_7 interhalogen compound, and it does so only with fluorine.

ClF_3 and BrF_5 are extremely reactive compounds. ClF_3 is so reactive that wood, asbestos, and even water spontaneously burn in its presence. These compounds are excellent fluorinating agents and tend to react with each other by the transfer of a fluorine atom to form positive ions such as ClF_2^+ and BrF_4^+ and negative ions such as IF_2^- and BrF_6^-.

$$2\ BrF_5(l) \longrightarrow \textbf{[BrF}_4^+\textbf{][BrF}_6^-\textbf{]}\textbf{(}\textbf{\textit{s}}\textbf{)}$$

NEUTRAL OXIDES OF THE HALOGENS

Under certain conditions, it is possible to isolate a handful of neutral oxides of the halogens, such as Cl_2O, Cl_2O_3, ClO_2, Cl_2O_4, Cl_2O_6, and Cl_2O_7. Cl_2O_7, for example, results from the dehydration of perchloric acid, $HClO_4$.

$$2\ HClO_4(aq) \longrightarrow \textbf{Cl}_2\textbf{O}_7\textbf{(}\textbf{\textit{l}}\textbf{)} + H_2O$$

These oxides are notoriously unstable compounds that explode when subjected to either thermal or physical shock. Some are so unstable they detonate when warmed to temperatures above $-40°C$.

OXYACIDS OF THE HALOGENS AND THEIR SALTS

Chlorine reacts with the OH^- ion in aqueous solution to form chloride ions and *hypochlorite* (OCl^-) ions.

$$Cl_2(aq) + 2\ OH^-(aq) \longrightarrow Cl^-(aq) + \textbf{OCl}^-\textbf{(}\textbf{\textit{aq}}\textbf{)} + H_2O(l)$$

TABLE 10.13

Oxyanions and Oxyacids of Chlorine

Oxyanions of Chlorine

		Chlorine Oxidation Number
ClO^-	hypochlorite	+1
ClO_2^-	chlorite	+3
ClO_3^-	chlorate	+5
ClO_4^-	perchlorate	+7

Oxyacids of Chlorine

HClO	hypochlorous acid	+1
$HClO_2$	chlorous acid	+3
$HClO_3$	chloric acid	+5
$HClO_4$	perchloric acid	+7

This is a disproportionation reaction in which half of the chlorine atoms are oxidized to hypochlorite ions and the other half are reduced to chloride ions.

$$Cl_2 + 2\ OH^- \longrightarrow Cl^- + OCl^- + H_2O$$

When the solution is hot, chlorine reacts with the OH^- ion to form a mixture of chloride and *chlorate* (ClO_3^-) ions.

$$3\ Cl_2(aq) + 6\ OH^-(aq) \longrightarrow 5\ Cl^-(aq) + \mathbf{ClO_3^-(aq)} + 3\ H_2O(l)$$

Under carefully controlled conditions, it is possible to convert a mixture of the chlorate and hypochlorite ions into a solution that contains the *chlorite* (ClO_2^-) ion.

$$ClO_3^-(aq) + ClO^-(aq) \longrightarrow \mathbf{2\ ClO_2^-(aq)}$$

The last member of this class of compounds, the *perchlorate* ion (ClO_4^-), is made by the electrolysis of aqueous solutions of the chlorate ion.

The names of four ions in this family use the endings *-ite* and *-ate* to indicate low and high oxidation numbers and the prefixes *hypo-* and *per-* to indicate the very lowest and very highest oxidation numbers, as shown in Table 10.13. Each of these ions can be converted into an oxyacid, which we name by replacing the *-ite* ending with *-ous* and the *-ate* ending with *-ic*.

10.9 THE CHEMISTRY OF THE RARE GASES

In 1892, Lord Rayleigh published the results of experiments that showed oxygen was always 15.882 times as dense as hydrogen no matter how it was prepared. When he tried to extend this work to nitrogen, however, Rayleigh found that nitrogen isolated from air was denser than nitrogen prepared from ammonia.

William Ramsey attacked this problem by purifying a sample of nitrogen gas to remove any moisture, carbon dioxide, and organic contaminants. He then passed the purified gas over hot magnesium metal, which reacts with nitrogen to form the nitride.

$$3\ Mg(s) + N_2(s) \longrightarrow Mg_3N_2(s)$$

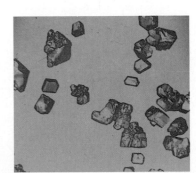

Crystals of the rare gas compound XeF_4.

When he was done, Ramsey was left with a small residue of gas that occupied roughly 1/80 of the original volume. He excited this gas in an electric discharge tube and found that the resulting emission spectrum contained lines that corresponded to no known gas. After repeated communications discussing the results of these experiments, Rayleigh and Ramsey jointly announced the discovery of a new element, which they named *argon,* from a Greek word meaning the "lazy one," because this gas refused to react with any element or compound they tested.

Argon did not fit into any of the known families of elements in the periodic table. Its atomic weight, however, suggested that it might belong to a new group that could be inserted between chlorine and potassium. Shortly after reporting the discovery of argon in 1894, Ramsey found another apparently unreactive inert gas when he heated a mineral of uranium. The lines in the spectrum of this gas also occurred in the spectrum of the sun, which led Ramsey to name the element *helium* (from the Greek *helios,* "sun").

Experiments with liquid air soon led Ramsey to a third gas, which he named *krypton* ("the hidden one"). Experiments with liquid argon led him to a fourth gas, *neon* ("the new one"), and finally a fifth gas, *xenon* ("the stranger"). Ramsey is the only individual to be involved in the discovery of essentially an entire family of elements.

The period during which these elements were discovered was 1894 through 1898. Since Moissan had only recently isolated fluorine for the first time, and fluorine was the most active of the known elements, Ramsey sent a sample of argon to Moissan to see whether it would react with fluorine. Moissan found that it did not.

The failure of Moissan's attempts to react argon with fluorine, coupled with repeated failures by other chemists to get the more abundant of these gases to undergo chemical reaction, eventually led to their being labeled *inert gases.* The development of the electronic theory of atoms did little to dispel this notion, because it was obvious that these gases had very symmetrical electron configurations.

As a result, these elements continued to be labeled *inert gases* in almost every textbook and periodic table until about 30 years ago. In 1962, Neil Bartlett found that PtF_6 was a strong enough oxidizing agent to remove an electron from an O_2 molecule.

$$PtF_6(g) + O_2(g) \longrightarrow [O_2^+][PtF_6^-]$$

Bartlett realized that the first ionization energy of Xe (1170 kJ/mol) was slightly smaller than the first ionization energy of the O_2 molecule (1177 kJ/mol). He therefore predicted that PtF_6 might also react with Xe. When he ran the reaction, he isolated the first compound of a Group VIIIA element.

$$Xe(g) + PtF_6(g) \longrightarrow [Xe^+][PtF_6^-]$$

A few months later, workers at the Argonne National Laboratory near Chicago found that Xe reacts with F_2 to form XeF_4. Since that time, more than 200 compounds of Kr, Xe, and Rn have been isolated. No compounds of the more abundant elements in this group—He, Ne, and Ar—have yet been isolated. However, the fact that elements in this family can undergo chemical reactions has led in recent years to the use of the terms **rare gases** and **noble gases,** rather than *inert gases,* to describe these elements.

Compounds of xenon are by far the most numerous of the rare-gas compounds. With the exception of $XePtF_6$, rare-gas compounds have oxidation numbers of +2, +4, +6, and +8, as shown by the examples cited in Table 10.14.

There is some controversy over whether the rare gases should be viewed as

TABLE 10.14

Compounds of Xenon and their Oxidation Numbers

Compound	Oxidation Number	Compound	Oxidation Number
Xe^+	$+1$	XeO_3	$+6$
XeF^+	$+2$	$HXeO_4^-$	$+6$
XeF_2	$+2$	$XeOF_4$	$+6$
XeF_3^+	$+4$	XeO_2F_2	$+6$
XeF_4	$+4$	XeO_3F^-	$+6$
$Xe_2F_3^+$	$+4$	XeO_4	$+8$
$XeOF_2$	$+4$	XeO_6^{4-}	$+8$
XeF_5^+	$+6$	XeO_3F_2	$+8$
XeF_6	$+6$	XeO_2F_4	$+8$
$Xe_2F_{11}^+$	$+6$	$XeOF_5^+$	$+8$

having the outermost shell of electrons filled (in which case they should be labeled Group VIIIA) or empty (in which case they should be labeled Group O). The authors believe these elements should be labeled Group VIIIA for one reason — when they form compounds, they behave as if they contribute eight valence electrons.

Exercise 10.12

Predict the shapes of the XeF_2 and XeF_4 molecules.

Solution

XeF_2 contains 22 valence electrons.

$$XeF_2: \qquad 8 + 2(7) = 22$$

The procedure outlined in Section 8.8 leads to the following Lewis structure.

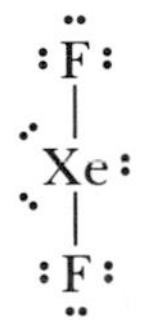

The VSEPR theory described in Chapter 8 predicts a trigonal bipyramidal distribution of valence electrons around the Xe atom, with nonbonding pairs of electrons in the equatorial positions. The molecule is therefore linear.

XeF_4 contains 36 valence electrons.

$$XeF_4: \qquad 8 + 4(7) = 36$$

Thus, it has the following Lewis structure.

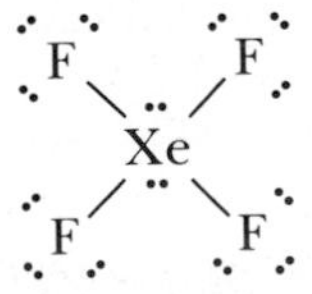

The VSEPR theory predicts an octahedral distribution of valence electrons on the Xe atom, with the two pairs of nonbonding electrons as far from each other as possible to minimize repulsion between these nonbonding electrons. The molecule is therefore best described as square planar.

FIG. 10.16 Lewis structures of a representative sample of xenon compounds.

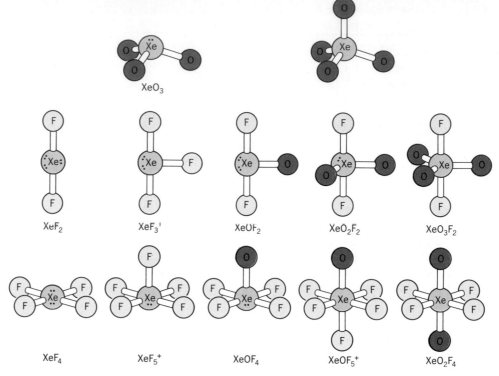

The distribution of electrons in the valence shells of a number of other xenon compounds is shown in Figure 10.16.

The synthesis of most xenon compounds starts with the reaction between Xe and F_2 at high temperatures ($250-400°C$) to form a mixture of XeF_2, XeF_4, and XeF_6.

$$Xe(g) + F_2(g) \longrightarrow \mathbf{XeF_2(s) + XeF_4(s) + XeF_6(s)}$$

The positively charged XeF_n^+ ion is then made by reaction of XeF_2, XeF_4, or XeF_6 with SbF_5, RuF_5, BiF_5, or AsF_5.

$$XeF_2(s) + SbF_5(l) \longrightarrow [XeF^+][SbF_6^-](s)$$
$$2\ XeF_2(s) + AsF_5(g) \longrightarrow [Xe_2F_3^+][AsF_6^-](s)$$
$$XeF_4(s) + BiF_5(s) \longrightarrow [XeF_3^+][BiF_6^-](s)$$
$$XeF_6(s) + AsF_5(g) \longrightarrow [XeF_5^+][AsF_6^-](s)$$
$$2\ XeF_6(s) + AsF_5(g) \longrightarrow [Xe_2F_{11}^+][AsF_6^-](s)$$

Oxides of xenon, such as $XeOF_2$, $XeOF_4$, XeO_2F_2, XeO_3F_2, XeO_2F_4, XeO_3, and XeO_4, are prepared by reacting XeF_4 or XeF_6 with water. $HXeO_4^-$ is produced in small quantities when XeO_3 reacts with the OH^- ion in aqueous solution.

$$XeO_3(s) + OH^-(aq) \longrightarrow \mathbf{HXeO_4^-(aq)}$$

The XeO_6^{4-} ion is produced when XeF_6 dissolves in strong base.

$$2\ XeF_6(s) + 16\ OH^-(aq) \longrightarrow$$
$$\mathbf{XeO_6^{4-}(aq) + Xe(g) + O_2(g) + 12\ F^-(aq) + 8\ H_2O(l)}$$

Some xenon compounds are relatively stable. XeF_2, XeF_4, and XeF_6, for example, are stable solids that can be purified by sublimation in a vacuum at $25°C$. $XeOF_4$ and Na_4XeO_6 are also reasonably stable. Others, such as XeO_3, XeO_4,

$XeOF_2$, XeO_2F_2, XeO_3F_2, and XeO_2F_4, are unstable compounds that can decompose violently.

The principal use of rare-gas compounds at present is as the light-emitting component in lasers. For example, mixtures of 10% Xe, 89% Ar, and 1% F_2 can be "pumped," or excited, with high-energy electrons to form excited XeF molecules, which emit a photon with a wavelength of 354 nm.

10.10 THE INORGANIC CHEMISTRY OF CARBON

For more than 200 years, chemists have divided naturally occurring compounds into two categories—inorganic and organic. Compounds isolated from plants or animals were called organic, while those extracted from ores and minerals were inorganic. Organic compounds, such as formic acid (HCO_2H), which was first isolated by distillation from ants, were once thought to be fundamentally different from inorganic compounds—organic compounds could only be synthesized by a "vital force" found in living systems.

Now that organic compounds are routinely made from inorganic starting materials, chemists no longer believe in a mystical vital force that is required for their preparation. But they still differentiate between organic and inorganic compounds. Organic chemistry has been defined as the chemistry of carbon. But this definition would include compounds such as calcium carbonate ($CaCO_3$) and graphite, which more closely resemble inorganic compounds. We will therefore suggest that *organic compounds* are compounds, such as formic acid (HCO_2H), methane (CH_4), and vitamin C ($C_6H_8O_6$), that contain both carbon and hydrogen. Chapter 24 will be devoted to a discussion of organic chemistry, and this section will focus on inorganic carbon compounds.

The chemistry of carbon is dominated by three factors.

1. **Carbon forms unusually strong C—C single bonds, C=C double bonds, and C≡C triple bonds.**

2. **The electronegativity of carbon (EN = 2.55) is too small to allow carbon to capture electrons from most elements to form C^{4-} ions and too large for carbon to lose electrons to form C^{4+} ions. Carbon therefore forms covalent bonds with many other elements.**

3. **Carbon forms strong double and triple bonds with a number of other nonmetals, including N, O, P, and S.**

ELEMENTAL FORMS OF CARBON: GRAPHITE, DIAMOND, COKE, AND CARBON BLACK

Carbon occurs in several different elemental forms. There are two crystalline forms—diamond and graphite—and a number of amorphous (noncrystalline) forms, such as charcoal, coke, and carbon black.

References to the characteristic hardness of diamond (from the Greek *adamas*, "invincible") date back at least 2600 years. It was not until 1797, however, that Smithson Tennant was able to show that diamonds consist solely of carbon. The properties of diamond are remarkable. It is among the least volatile substances known—it doesn't melt until the temperature exceeds 3550°C and doesn't boil until the temperature exceeds 4827°C. Diamond is also the hardest substance known, and it expands less on heating than any other material.

The two allotropic forms of elemental carbon: diamond and graphite.

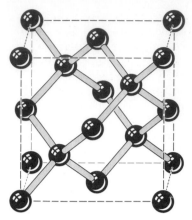

FIG. 10.17 Many of the remarkable properties of diamond can be traced to the fact that each sp^3-hybridized carbon atom is tightly bound to four neighboring carbon atoms arranged in a tetrahedral array, so that the crystal can be viewed as a single giant molecule.

The properties of diamond are a logical consequence of the structure of the solid. Carbon, with four valence electrons, forms covalent bonds to four neighboring carbon atoms arranged toward the corners of a tetrahedron, as shown in Figure 10.17. Each of these sp^3 hybridized atoms is then bound to four other carbon atoms, which form bonds to four other carbon atoms, and so on. As a result, a perfect diamond can be thought of as a single giant molecule. The strength of the individual C—C bonds and their arrangement in space give rise to the unusual properties of diamond.

In some ways, the properties of graphite are like those of diamond—both compounds boil at 4827°C, for example. But graphite is also very different from diamond. Diamond (3.514 g/cm³) is much denser than graphite (2.26 g/cm³). Whereas diamond is the hardest substance known, graphite is one of the softest. Diamond is an excellent insulator, with little or no tendency to carry an electric current. But graphite is such a good conductor of electricity that graphite electrodes are used in electrical cells.

The physical properties of graphite can also be understood from the structure of the solid, shown in Figure 10.18. Graphite consists of extended planes, or layers, of sp^2 hybridized carbon atoms in which each carbon is tightly bound to three other carbon atoms. (The strong bonds between carbon atoms within each plane explain the exceptionally high melting point and boiling point of graphite.) The distance between these planes of atoms, however, is very much larger than the distance between the atoms within the planes, which means that the bonds between planes are weak. This makes it easy to deform the solid by allowing one plane of atoms to move relative to another. As a result, graphite is soft enough to be used in pencils and as a lubricant in motor oil.

"Lead" pencils do not, incidentally, contain lead. This is fortunate, since many people chew pencils, and lead compounds are very toxic. Lead pencils contain graphite, or "black lead," as it was once known. The graphite is mixed with clay (20% to 60% by weight) and then baked to form a ceramic rod, which is encased in wood. The hardness of the pencil depends on the ratio of clay to graphite. Increasing the percentage of clay makes the pencil harder, so that less graphite is deposited on the paper.

FIG. 10.18 The high melting point and boiling point of graphite can be explained by the very strong C—C bonds within the extended planes of sp^2-hybridized carbon atoms. (The bond dissociation enthalpy for the C—C bonds within one of these planes is 477 kJ/mol.) The softness of graphite can be explained by the ease with which these planes slide past one another. (The bond between planes of atoms has been estimated to have a dissociation enthalpy of only 17 kJ/mol.)

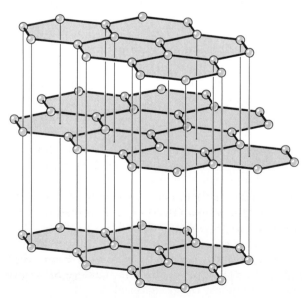
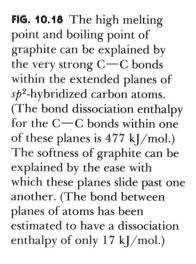

FIG. 10.19 Graphite has two properties that are unexpected for a nonmetal element—a metallic luster, or shine, and the ability to conduct an electric current. Both properties result from the structure of the individual planes of carbon atoms in the solid. Within these planes, the atoms are held together by alternating C—C single bonds and C=C double bonds. This figure shows one of the three Lewis structures necessary to describe the resonance hybridization that leads to delocalization of the electrons within a plane. Since the electrons are free to move through the plane, they can conduct an electric current. The metallic luster of graphite results from the ease with which the solid can absorb energy in the form of light and use this energy to move electrons through the solid.

Although carbon is one of the nonmetallic elements, graphite shows many of the properties usually associated with metals. It has a shiny, metallic luster, and it conducts electricity. Both the metallic luster of graphite and its ability to conduct an electric current result from the structure of the individual planes of carbon atoms, shown in Figure 10.19. This figure represents one of three resonance structures that can be drawn for a plane of atoms in graphite. The resonance hybridization in this structure tends to *delocalize* the electrons — free them to move from one carbon atom to another and thereby carry an electric current along the plane.

The characteristic properties of graphite and diamond might lead you to expect that diamond would be more stable than graphite. This is not what is observed experimentally. The enthalpy of formation of diamond (2.425 kJ/mol) is slightly larger than the enthalpy of formation of graphite, which is the most stable form of carbon at 25°C and 1 atm pressure.

At very high temperatures and pressures, however, diamond becomes more stable than graphite. In 1955, General Electric developed a process to make industrial-grade diamonds by treating graphite with a metal catalyst at temperatures of 2000 to 3000 K and pressures above 125,000 atm. Roughly 40% of industrial-quality diamonds are now synthetic. Although gem-quality diamonds can also be synthesized, the costs involved are prohibitive.

Both diamond and graphite occur as regularly packed crystals. Other forms of carbon are *amorphous* — they lack a regular structure. Charcoal, carbon black, and coke are all amorphous forms of carbon.

Charcoal results from the heating of wood in the absence of oxygen. To make carbon black, natural gas or other carbon compounds are burned in a limited amount of air to give a thick, black smoke that contains extremely small particles of carbon, which can be collected when the gas is cooled and passed through an electrostatic precipitator. Coke is a more regularly structured material, closer in structure to graphite than either charcoal or carbon black, which is made from coal.

Carbon black is one of the oldest industrial products. It was made for use as a pigment in ink as early as 3000 B.C. in China by a process that involved burning specially purified oils in a small lamp and then carefully scraping off the black deposit of carbon that collected on the lamp's ceramic cover. Carbon black is still used as a pigment in printing inks. It is also added to polymers to give them a dark color, which helps protect them from sunlight, and it is used as a strengthening or reinforcing agent in rubber tires.

Covalent carbides have many of the properties of diamond. They tend to be very hard and inert to chemical reactions. Silicon carbide, for example, is used as the abrasive in emery paper.

CARBIDES: COVALENT, IONIC, AND INTERSTITIAL

The suffix *-ide* is added to the stem of an element's name to describe the negative ion formed when an atom of the element gains electrons. Although carbon is essentially inert at room temperatures, it reacts with the less electronegative elements at high temperatures to form a series of compounds known as *carbides.*

There are three major classes of carbides: covalent, ionic, and interstitial. When carbon reacts with an element of similar size and electronegativity, a *covalent carbide,* such as silicon carbide, is produced. Silicon carbide is made by treating silicon dioxide (SiO_2) from quartz with an excess of carbon in an electric furnace at 2300 K.

$$SiO_2(s) + 3\ C(s) \longrightarrow SiC(s) + 2\ CO(g)$$

Covalent carbides such as SiC have properties very similar to those of diamond. Both SiC and diamond are inert to chemical reactions, except at very high temperatures. Both have very high melting points, and both are among the hardest substances known. SiC was first synthesized by Edward Acheson in 1891. Shortly thereafter, Acheson founded the Carborundum Company to market this material. Then, as now, materials in this class are most commonly used as abrasives.

Compounds that contain carbon and one of the more active metals are often called *ionic carbides,* or *salt-like carbides.*

$$2\ BeO(s) + 3\ C(s) \longrightarrow Be_2C(s) + 2\ CO(g)$$
$$CaO(s) + 3\ C(s) \longrightarrow CaC_2(s) + CO(g)$$

It is useful, but somewhat naive, to think about these compounds as if they contained C^{4-} and C_2^{2-} ions.

$$[Be^{2+}]_2[C^{4-}] \qquad [Ca^{2+}][C_2^{2-}]$$
$$[Al^{3+}]_4[C^{4-}]_3 \qquad [Na^+]_2[C_2^{2-}]$$

This model is useful, because it explains why these carbides are so reactive toward water that they often burst into flame when added to water.

$$Al_4C_3(s) + 12\ H_2O(l) \longrightarrow 3\ CH_4(g) + 4\ Al(OH)_3(aq)$$
$$CaC_2(s) + 2\ H_2O(l) \longrightarrow C_2H_2(g) + Ca^{2+}(aq) + 2\ OH^-(aq)$$

The ionic carbides that formally contain the C^{4-} ion can be thought to abstract H^+ ions from four water molecules to form methane. These compounds are therefore also known as *methanides.*

$$C^{4-} + 4\ H_2O \longrightarrow CH_4 + 4\ OH^-$$

The ionic carbides that formally contain the C_2^{2-} ion can be thought to abstract H^+ ions from a pair of water molecules to form acetylene.

$$C_2^{2-} + 2\ H_2O \longrightarrow C_2H_2 + 2\ OH^-$$

These compounds are therefore also known as *acetylides.* At one time, miners' lamps were fueled by the combustion of acetylene prepared from the reaction of calcium carbide with water.

As noted, however, the model in which ionic carbides are treated as if they contained C^{4-} and C_2^{2-} ions is naive. The difference between the electronegativities of carbon and the metals that form these compounds is not actually large enough to justify classifying the compounds as ionic. There is reason to believe that

The burning of acetylene produced from the reaction of calcium carbide with water.

the difference between covalent and ionic carbides is primarily a question of the extent to which the bond is best described as covalent or polar covalent.

Interstitial carbides, such as tungsten carbide (WC), form when carbon combines with a metal that has an intermediate electronegativity and a relatively large atomic radius. In these compounds, the carbon atoms pack in the holes *(interstices)* between planes of metal atoms. The interstitial carbides, which include TiC, ZrC, and MoC, retain the properties of metals. They act as alloys, rather than as either salts or covalent compounds.

THE OXIDES OF CARBON

Although graphite, diamond, charcoal, and other forms of carbon are essentially inert at room temperature, all combine with oxygen at high temperatures to produce a mixture of carbon monoxide and carbon dioxide and a considerable amount of energy in the form of heat.

$$2 \, C(s) + O_2(g) \longrightarrow 2 \, CO(g) \qquad \Delta H° = -110.52 \text{ kJ/mol CO}$$
$$C(s) + O_2(g) \longrightarrow CO_2(g) \qquad \Delta H° = -393.51 \text{ kJ/mol CO}_2$$

When carbon is burned in the presence of a limited amount of oxygen, CO is the major product. If an excess of oxygen is present, most of the carbon is oxidized to CO_2.

CO can also be obtained when red-hot carbon is treated with steam, in which case the carbon captures an oxygen atom from water to produce a mixture of hydrogen and carbon monoxide.

$$C(s) + H_2O(g) \longrightarrow CO(g) + H_2(g)$$

Since this mixture of gases is formed by the reaction of charcoal or coke with water, it is often referred to as water gas. It is also known as town gas, because it was once made by towns and cities for use as a fuel. Water gas, or town gas, was a common commercial fuel for both home and industrial use before natural gas became readily available. The H_2 in this fuel burns to form water, while the CO is oxidized to CO_2. For many industrial processes, water gas is still a more economical fuel than natural gas. Eventually, as our supply of natural gas is depleted, it will become economical to replace natural gas with other fuels, such as water gas, that can be produced from our abundant supply of coal.

CO and CO_2 are both colorless gases. CO boils at $-191.5°C$, and CO_2 sublimes at $-78.5°C$. Although CO has no odor or taste, CO_2 has a faint, pungent odor and a distinctly acidic taste. Both are dangerous substances, but at very different levels of exposure. Air contaminated with as little as 0.002 grams of CO per liter can be fatal, because CO binds tightly to the hemoglobin and myoglobin proteins that carry oxygen through the blood, thereby displacing the oxygen. CO_2 is not lethal until the concentration in the air approaches 15%. At that point, it has replaced so much oxygen that a person who attempts to breathe this atmosphere suffocates. The danger of CO_2 poisoning is magnified by the fact that CO_2 is roughly 1.5 times as dense as the air in our atmosphere. Thus, CO_2 can accumulate at the bottom of tanks or wells.

The chemistry of CO_2 has many important consequences. CO_2 helps regulate the temperature of the atmosphere through the greenhouse effect, which works as follows. CO_2 is transparent to the visible radiation from the sun, but it absorbs some of the sun's lower-energy, longer-wavelength infrared radiation that would otherwise by reflected back from the surface of the planet. Thus, CO_2 in the atmosphere helps trap heat. Although there are other factors at work as well, it is worth noting that Venus, whose atmosphere contains a great deal of CO_2, has a surface temperature of roughly $400°C$, whereas Mars, with little or no atmosphere, has a surface temperature of $-50°C$.

There are many sources of CO_2 in the atmosphere. Over geologic times scales, the single largest source has been volcanos. Within the last century, however, the combustion of petroleum, coal, and natural gas has made a significant contribution to atmospheric levels of CO_2. Between 1958 and 1978, the average annual level of CO_2 in the atmosphere increased by 6%, from 315.8 to 334.6 ppm.

At one time, the amount of CO_2 released to the atmosphere was not a matter for concern, because it was thought that natural processes that removed CO_2 from the atmosphere could compensate for the CO_2 produced by the combustion of carbon-based fuels. The vast majority of the CO_2 liberated by volcanic action, for example, was captured by calcium oxide (CaO) or magnesium oxide (MgO) to form calcium carbonate or magnesium carbonate.

$$CaO(s) + CO_2(g) \longrightarrow \textbf{CaCO}_3\textbf{(s)}$$
$$MgO(s) + CO_2(g) \longrightarrow \textbf{MgCO}_3\textbf{(s)}$$

$CaCO_3$ is found as limestone or marble or mixed with $MgCO_3$ as dolomite. The amount of CO_2 in deposits of carbonate minerals is at least several thousand times larger than the amount in the atmosphere.

CO_2 also dissolves, to some extent, in water.

$$CO_2(g) \xrightarrow{\text{H}_2\text{O}} \textbf{CO}_2\textbf{(aq)}$$

It then reacts with water to form carbonic acid, H_2CO_3.

$$CO_2(aq) + H_2O(l) \longrightarrow \textbf{H}_2\textbf{CO}_3\textbf{(aq)}$$

As a result of these reactions, the sea contains about 60 times as much CO_2 as the atmosphere.

Can the sea absorb more CO_2 from the atmosphere, or is it near its level of saturation? Is the rate at which the sea absorbs CO_2 greater than the rate at which we are adding it to the atmosphere? The observed increase in the concentration of CO_2 in recent years suggests pessimistic answers to these two questions. A gradual warming of the earth's atmosphere could result from continued increases in CO_2 levels, with the potential for significant adverse effects on the climate and therefore the agriculture of at least the Northern Hemisphere.

THE CHEMISTRY OF CARBONATES — CO_3^{2-} AND HCO_3^-

A number of interesting properties of carbonates revolve around their relative solubilities. Calcium carbonate ($CaCO_3$), for example, is insoluble in water. Evidence for this comes from noting that the shells of many marine organisms consist of calcium carbonate and that buildings constructed of limestone or marble do not dissolve when it rains. Calcium carbonate, however, will dissolve in water saturated with CO_2. Carbonated water (or carbonic acid) reacts with calcium carbonate to form calcium bicarbonate, $Ca(HCO_3)_2$, which is soluble in water.

$$CaCO_3(s) + H_2CO_3(aq) \longrightarrow Ca^{2+}(aq) + 2\ HCO_3^-(aq)$$

When water rich in carbon dioxide flows through limestone formations, part of the limestone dissolves. If the CO_2 escapes from this water, or if some of the water evaporates, solid $CaCO_3$ is redeposited. When this happens as water runs across the roof of a cavern, stalactites, which hang from the roof of the cave, are formed. If the water drops before the carbonate reprecipitates, stalagmites, which grow from the floor of the cave, are formed.

The chemistry of carbon dioxide dissolved in water is the basis of the soft drink industry. The first artificially carbonated beverages were introduced in Europe at the end of the 19th century. Carbonated soft drinks today consist of carbonated water, a sweetening agent (such as sugar, saccharin, or aspartame), an acid to impart a sour or tart taste, flavoring agents, coloring agents, and preservatives. As much as 3.5 liters of gaseous CO_2 is dissolved in each liter of soft drink. The CO_2 contributes the characteristic bite associated with carbonated beverages.

Carbonate chemistry plays an important role in other parts of the food industry as well. Baking soda, or bicarbonate of soda, is sodium bicarbonate, $NaHCO_3$, a weak base. It is added to recipes to neutralize the acidity of other ingredients. Baking powder is a mixture of baking soda and a weak acid, such as tartaric acid or calcium hydrogen phosphate $CaHPO_4$. When mixed with water, the acid reacts with the HCO_3^- ion to form CO_2 gas, which causes the dough or batter to rise.

$$HCO_3^-(aq) + H^+(aq) \longrightarrow H_2CO_3(aq) \longrightarrow H_2O(l) + CO_2(g)$$

Before commercial baking powders were available, cooks obtained the same effect by mixing roughly a teaspoon of baking soda with a cup of sour milk or buttermilk. The acids that give sour milk and buttermilk their characteristic tastes also react with the bicarbonate ion to give CO_2.

SUMMARY

The chemistry of the nonmetals is made more interesting by the fact that these elements can act as either oxidizing agents or reducing agents. In general, the more electronegative the element, the more likely it will act as an oxidizing agent to form compounds in which it has a negative oxidation number.

The chemistry of hydrogen is dominated by the fact that a neutral hydrogen atom has only one electron. Hydrogen therefore forms compounds in which it has an oxidation number of either $+1$, 0, or -1. In a formal sense, these oxidation numbers correspond to H^+ ions, neutral H atoms, and H^- ions.

The chemistry of oxygen revolves around the fact that this element is so electronegative it oxidizes virtually every substance it reacts with. Most of the time, oxygen forms compounds in which it has an oxidation number of -2. Under certain conditions, it forms peroxides, in which it has an oxidation number of -1. Peroxides can undergo disproportionation reactions, in which they are oxidized to O_2 at the same time they are reduced to form compounds in which the oxidation number of the oxygen is -2.

In many ways, the chemistry of sulfur resembles that of oxygen. There are four major differences. Sulfur is less likely to form strong double and triple bonds; it forms stronger single bonds; and it is less electronegative than oxygen and more likely to form compounds with positive oxidation numbers. Finally, sulfur (like the other elements in this column of the periodic table) contains valence-shell d orbitals that allow it to expand its valence shell to hold more than eight electrons.

The chemistry of nitrogen is dominated by the strength of the double and triple bonds this element forms with other nonmetals. The strength of these bonds makes N_2 remarkably inert at room temperature. It also explains why a number of nitrogen compounds decompose to form N_2 or react with other substances to form products that include N_2.

Phosphorus forms compounds that have oxidation numbers analogous to those of nitrogen compounds. The formulas of these analogous compounds are different, however, because phosphorus atoms are too large to form strong double and triple bonds. Where a nitrogen compound would contain an $N{=}O$ double bond, the analogous phosphorus compound contains two $P{-}O$ single bonds.

The chemistry of the halogens is dominated by the fact that these elements are all relatively strong oxidizing agents. None of the halogens can be found in nature —each must be prepared by oxidation of one of its salts. As we go up this column of the periodic table, each succeeding element requires a stronger oxidizing agent to be separated from one of its compounds. F_2 is such a strong oxidizing agent that it must be prepared by electrolysis.

For many years, chemists assumed that there was no chemistry of the rare gases—He, Ne, Ar, Kr, Xe, and Rn. Compounds of Xe were first prepared about 30 years ago. For the most part, the chemistry of this group of elements is still restricted to compounds formed by the group's heavier members and the most reactive nonmetals, such as fluorine and oxygen.

The chemistry of carbon is dominated by the strength of the covalent bonds it forms with itself and its ability to form strong double and triple bonds to other nonmetals. As a result, the chemistry of this element is more diverse than that of any other element in the periodic table.

PROBLEMS

Metals, Nonmetals, and Semimetals

10-1 List the elements that are nonmetals. Describe where these elements are found in the periodic table.

10-2 Explain why there are only 12 nonmetals but these elements form 99.98% of the known chemical compounds.

10-3 Explain why semimetals, such as B, Si, Ge, As, Sb, Te, Po, and At, exist, and describe some of their physical properties.

10-4 Which member of each of the following pairs of elements is more nonmetallic?

(a) As or Bi (b) As or Se (c) As or S (d) As or Ge
(e) As or P

The Chemistry of the Nonmetals

10-5 Which of the following elements can exist as a triatomic (three-atom) molecule?

(a) hydrogen (b) helium (c) sulfur (d) oxygen
(e) chlorine

10-6 Which of the following elements should form compounds with the formulas Na_2X, H_2X, XO_2, and XF_6?

(a) B (b) C (c) N (d) O (e) S

10-7 Which of the following elements should form compounds with the formulas XH_3, XF_3, and Na_3XO_4?

(a) Al (b) Ge (c) As (d) S (e) Cl

10-8 Which of the following can't be found in nature? Explain why.

(a) $MgCl_2$ (b) $CaCO_3$ (c) F_2 (d) Na_3AlF_6 (e) NaCl

The Role of Nonmetal Elements in Chemical Reactions

10-9 Explain why more electronegative elements tend to oxidize less electronegative elements.

10-10 Which of the following ions or molecules can be oxidized?

(a) H_2SO_3 (b) P_4 (c) Cl^- (d) SiO_2 (e) PO_4^{3-} (f) Mg^{2+}

10-11 Which of the following ions or molecules can be reduced?

(a) H_2O (b) H_2SO_3 (c) HCl (d) CO_2 (e) Mg^{2+} (f) Na

Deciding What Is Oxidized and What Is Reduced

10-12 For each of the following reactions, identify what is oxidized and what is reduced.

(a) $Fe_2O_3(s) + 3 CO(g) \rightarrow 2 Fe(s) + 3 CO_2(g)$ (b) $H_2(g) + CO_2(g) \rightarrow H_2O(g) + CO(g)$ (c) $CH_4(g) + 2 O_2(g) \rightarrow CO_2(g) + 2 H_2O(g)$ (d) $2 H_2S(g) + 3 O_2(g) \rightarrow 2 SO_2(g) + 2 H_2O(g)$

10-13 For each of the following reactions, identify what is oxidized and what is reduced.

(a) $PH_3(g) + 3 Cl_2(g) \rightarrow PCl_3(g) + 3 HCl(g)$
(b) $2 NO(g) + F_2(g) \rightarrow 2 NOF(g)$ (c) $2 Na(s) + 2 NH_3(l) \rightarrow 2 NaNH_2(s)$ (d) $3 NO_2(g) + H_2O(l) \rightarrow 2 HNO_3(aq) + NO(g)$

10-14 Hydrazine is made by a reaction known as the Raschig process.

$2 NH_3(aq) + NaOCl(aq) \longrightarrow N_2H_4(aq) + NaCl(aq) + H_2O(l)$

Decide whether this is an oxidation-reduction reaction. If it is, identify the compound oxidized and the compound reduced.

10-15 The thiosulfate ion, $S_2O_3^{2-}$, is prepared by a process that involves boiling solutions of sulfur dissolved in sodium sulfite.

$$8 SO_3^{2-}(aq) + S_8(s) \longrightarrow 8 S_2O_3^{2-}(aq)$$

Decide whether this is an oxidation-reduction reaction. If it is, identify the compound oxidized and the compound reduced.

10-16 Chlorine dioxide, ClO_2, is used commercially as a bleach or a disinfectant because of its excellent oxidizing ability. ClO_2 is prepared by decomposition of chlorous acid as follows.

$$8 HOClO(aq) \longrightarrow 6 ClO_2(g) + Cl_2(g) + 4 H_2O(l)$$

Decide whether this is an oxidation-reduction reaction. If it is, identify the compound oxidized and the compound reduced.

10-17 Nitric acid sometimes acts as an acid (as a source of the H^+ ion) and sometimes as an oxidizing agent. For each of the following reactions, decide whether HNO_3 is acting as an acid or an oxidizing agent.

(a) $Na_2CO_3(s) + 2\ HNO_3(aq) \longrightarrow$
$$2\ NaNO_3(aq) + CO_2(g) + H_2O$$

(b) $3\ P_4(s) + 20\ HNO_3(aq) + 8\ H_2O(l) \longrightarrow$
$$12\ H_3PO_4(aq) + 20\ NO(g)$$

(c) $Al_2O_3(s) + 6\ HNO_3(aq) \longrightarrow$
$$2\ Al(NO_3)_3(aq) + 3\ H_2O(l)$$

(d) $3\ Cu(s) + 8\ HNO_3(aq) \longrightarrow$
$$3\ Cu(NO_3)_2(aq) + 2\ NO(g) + 4\ H_2O(l)$$

10-18 Which of the following would you expect to be the best oxidizing agent?

(a) Na   (b) H_2 (c) N_2 (d) P_4 (e) O_2

10-19 Which of the following would you expect to be the best reducing agent?

(a) Na^+ (b) F^- (c) Na   (d) Br_2 (e) Fe^{3+}

10-20 For each of the following pairs of elements, determine which is the better reducing agent.

(a) P_4 or As   (b) As or S_8 (c) P_4 or S_8 (d) S_8 or Cl_2
(e) C or O_2

Predicting the Products of Chemical Reactions

10-21 Predict the products of the following reactions.

(a) $Mg(s) + N_2(g) \longrightarrow$ (c) $Br_2(l) + I^-(aq) \longrightarrow$
(b) $Li(s) + O_2(g) \longrightarrow$

10-22 Predict the products of the following reactions.

(a) $SO_2(g) + H_2O(l) \longrightarrow$ (c) $CO_2(g) + H_2O(l) \longrightarrow$
(b) $Cl_2(g) + H_2O(l) \longrightarrow$

10-23 Predict the products of the following reactions.

(a) $HCl(g) + H_2O(l) \longrightarrow$ (c) $NO_2(g) + H_2O(l) \longrightarrow$
(b) $P_4O_{10}(s) + H_2O(l) \longrightarrow$

10-24 Predict the products of the following reactions.

(a) $S_8(s) + O_2(g) \longrightarrow$ (c) $P_4(s) + F_2(g) \longrightarrow$
(b) $Al(s) + I_2(s) \longrightarrow$

The Chemistry of Hydrogen

10-25 Describe three ways of preparing small quantities of H_2 in the lab.

10-26 Give an example of a compound in which hydrogen has an oxidation number of $+1$; of 0; of -1.

10-27 Which of the following reactions result in a compound in which hydrogen has an oxidation number of -1?

(a) $Li + H_2 \longrightarrow$ (b) $O_2 + H_2 \longrightarrow$ (c) $S_8 + H_2 \longrightarrow$
(d) $Cl_2 + H_2 \longrightarrow$ (e) $Ca + H_2 \longrightarrow$

10-28 Use tables of first ionization energies and electronegativities to explain why it is so difficult to decide whether hydrogen belongs in Group IA or Group VIIA of the periodic table.

10-29 A recent catalog listed the following prices: $21.40 for 450 grams of sodium, $18.00 for 1 kilogram of zinc, and $52.80 for 250 grams of sodium hydride. Which reagent would be the least expensive source of H_2 gas?

10-30 The earth's atmosphere once contained significant amounts of H_2. Explain why only traces of H_2 are left in the earth's atmosphere, whereas the atmospheres of other planets, such as Jupiter, Saturn, and Neptune, contain large quantities of H_2.

The Chemistry of Oxygen and Sulfur

10-31 Lavoisier proposed the name *oxygen* from the Greek stems meaning "acid former" because he believed that all acids contained this element. List as many acids as you can that contain oxygen. List just a couple of exceptions that show Lavoisier was wrong.

10-32 Describe three ways of preparing small quantities of O_2 in the lab.

10-33 Describe the relationship among oxygen (O_2), the peroxide ion (O_2^{2-}), and the oxide ion (O^{2-}).

10-34 If a solution of hydrogen peroxide in water that is 30% hydrogen peroxide by weight sells for $15.95 per 500 milliliters and if potassium chlorate sells for $12.75 per 500 grams, is it cheaper to generate oxygen by decomposing H_2O_2 or $KClO_3$?

10-35 Which of the following elements or compounds could eventually produce O_2 when it reacts with water?

(a) Na   (b) Na_2O (c) Na_2O_2 (d) NaOH   (e) NaCl

10-36 Explain why the only compounds in which oxygen has a positive oxidation number are compounds, such as OF_2, which contain fluorine.

10-37 Explain why hydrogen peroxide can be either an oxidizing agent or a reducing agent. Describe at least one reaction in which H_2O_2 oxidizes another substance and one reaction in which it reduces another substance.

10-38 Write the Lewis structures for ozone, O_3, and sulfur dioxide, SO_2, and discuss the relationship between these compounds.

10-39 Explain why elemental oxygen exists as O_2 molecules, whereas elemental sulfur forms S_8 molecules.

10-40 Explain why sulfur forms compounds such as SF_4 and SF_6, but oxygen can only form OF_2.

10-41 Describe the relationship between the thiosulfate and sulfate ions, the thiocyanate and cyanate ions, and the trithiocarbonate and carbonate ions.

10-42 Write the Lewis structures of the following products of the reaction between sodium and sulfur.

(a) Na_2S (b) Na_2S_2 (c) Na_2S_3 (d) Na_2S_8

10-43 Explain why sulfur readily forms compounds in the $+2$, $+4$, and $+6$ oxidation states but only a handful of compounds exist in which oxygen is in a positive oxidation state.

10-44 Explain why sulfur-containing compounds such as FeS_2, CS_2, and H_2S form SO_2 instead of SO_3 when they burn.

10-45 Use Lewis structures to explain why solutions of the SO_3^{2-} ion react with sulfur to form thiosulfate, $S_2O_3^{2-}$.

10-46 Use Lewis structures to propose an explanation for the following reduction half-reactions.

$$S_2O_6^{2-} + 2\ e^- \longrightarrow 2\ SO_3^{2-}$$
$$S_4O_6^{2-} + 2\ e^- \longrightarrow 2\ S_2O_3^{2-}$$

10-47 Which of the following does not have a reasonable oxidation number for sulfur?

(a) Na_2S (b) H_2S (c) SO_3^2 (d) SO_4 (e) SF_4

10-48 Explain why SO_2 plays an important role in the phenomenon known as acid rain.

10-49 Explain why problems with acid rain would be much more severe if sulfur compounds burned to form SO_3 instead of SO_2.

The Chemistry of Nitrogen and Phosphorus

10-50 Nitrogen has a reasonable oxidation number in all of the following compounds, and yet one of them is still impossible. Which one?

(a) NF_3 (b) NF_5 (c) NO_3^- (d) NO_2^- (e) NO

10-51 The earth's atmosphere contains roughly 4×10^{16} tons of nitrogen, and yet the biggest problem facing agriculture in the world today is a lack of "nitrogen." Explain why.

10-52 Explain why elemental nitrogen is almost inert but nitrogen compounds such as NH_4NO_3, NaN_3, nitroglycerine, and trinitrotoluene (TNT) form some of the most dangerous explosives.

10-53 Which of the following oxides of nitrogen are paramagnetic?

(a) N_2O (b) NO (c) NO_2 (d) N_2O_3 (e) N_2O_4 (f) N_2O_5

10-54 Use Lewis structures to propose an explanation for the following reaction.

$$2\ NO + O_2 \longrightarrow 2\ NO_2$$

10-55 Use the fact that nitrous oxide decomposes to form nitrogen and oxygen to explain why a glowing splint bursts into flame when immersed in a container filled with N_2O.

$$2\ N_2O(g) \longrightarrow 2\ N_2(g) + O_2(g)$$

10-56 Describe ways of preparing small quantities of each of the following compounds in the laboratory.

(a) N_2O (b) NO (c) NO_2 (d) N_2O_4

10-57 Describe how to tell the difference between a flask filled with NO gas and a flask filled with N_2O.

10-58 Lightning catalyzes the reaction between nitrogen and oxygen in the atmosphere to form nitrogen oxide, NO.

$$N_2(g) + O_2(g) \longrightarrow 2\ NO(g)$$

Explain how lightning acts as one source of acid rain.

10-59 Write a sequence of reactions for the conversion of elemental nitrogen into nitric acid. Calculate the weight of nitric acid that can be produced from a ton of nitrogen gas.

10-60 Which of the following elements or compounds is not involved at some stage in the preparation of nitric acid?

(a) O_2 (b) N_2 (c) NO (d) NO_2 (e) H_2

10-61 Explain why phosphorus forms both PCl_3 and PCl_5 but nitrogen only forms NCl_3.

10-62 Explain why nitrogen is essentially inert at room temperature, whereas white phosphorus can burst spontaneously into flame when it comes into contact with air.

10-63 Explain why nitrogen forms extraordinarily stable N_2 molecules at room temperature but phosphorus only forms P_2 molecules at very high temperatures.

10-64 Explain why nitric acid has the formula HNO_3 and phosphoric acid has the formula H_3PO_4.

10-65 Write the Lewis structures for phosphoric acid, H_3PO_4, and phosphorous acid, H_3PO_3. Explain why phosphoric acid can lose three H^+ ions, to form a phosphate ion, PO_4^{3-}, whereas phosphorous acid can lose only two H^+ ions, to form the phosphite ion, HPO_3^{2-}.

10-66 Explain why only two of the four hydrogen atoms in $H_4P_2O_5$ are lost when this oxyacid forms an oxyanion.

10-67 Describe the role of carbon in the preparation of elemental phosphorus from calcium phosphate.

10-68 Predict the product of the reaction of phosphorus with excess oxygen and then predict what will happen when the product of this reaction is dissolved in water.

10-69 Explain why the most common oxidation states of antimony are $+3$ and $+5$.

10-70 At 1700°C, P_4 molecules decompose partially to form P_2.

$$P_4(g) \longrightarrow 2\ P_2(g)$$

If the average molecular weight of phosphorus at that temperature is 91 grams per mole, what fraction of the P_4 molecules decompose?

10-71 Which of the following compounds should not exist?

(a) Na_3P (b) $(NH_4)_3PO_4$ (c) PO_2 (d) PH_3 (e) $POCl_3$

The Chemistry of the Halogens

10-72 Which of the halogens is the most active, or reactive? Explain why.

10-73 Describe the difference between halogens and halides. Give examples of each.

10-74 Explain why chlorine reacts with sodium bromide, but bromine does not react with sodium chloride. Describe what you would observe if you slowly bubbled Cl_2 gas through a solution of NaBr dissolved in a mixture of water and CCl_4.

10-75 Explain why chlorine and fluorine react to form ClF_3 but not FCl_3.

10-76 Explain why reaction with fluorine is one of the best ways of coaxing an element into its most positive oxidation state (SF_6, PF_5, IF_7, PtF_6, and so on).

10-77 Describe how you would prepare small quantities of each of the following substances in the laboratory.

(a) $Cl_2(g)$ (b) $HCl(g)$ (c) $HCl(aq)$ (d) $BrCl(g)$
(e) $NaOCl(aq)$

10-78 Clorox is a 5.25% solution of sodium hypochlorite ($NaOCl$) in water. Muriatic acid is a 6 M solution of hydrochloric acid in water that is sold in many building supply stores as a cleaner for brick and cement. Use the following equation to describe why it is not a good idea to mix these chemicals.

$$NaOCl(aq) + 2\ HCl(aq) \longrightarrow$$
$$Cl_2(g) + Na^+(aq) + Cl^-(aq) + H_2O(l)$$

The Chemistry of the Rare Gases

10-79 Explain why none of the elements in Group VIIIA acts as an oxidizing agent.

10-80 Predict the geometry of the following rare-gas compounds.

(a) XeF_2 (b) XeF_4 (c) $OXeF_4$ (d) XeF_3^+ (e) XeF_5^+

10-81 The density of nitrogen is 1.257 grams per liter when it is isolated from air by removal of oxygen, carbon dioxide, and water, but 1.251 grams per liter when it is prepared by decomposition of ammonia. Explain why.

10-82 Use the ionization energy data in Table A-5 in the appendix to explain why xenon forms compounds more easily than the other elements in its family.

10-83 Explain why there are no electronegativity data for the rare gases in Table A-7 in the appendix.

The Inorganic Chemistry of Carbon

10-84 Describe the differences between the physical properties of graphite and those of diamond, and show how these properties are related to the crystal structures of these elemental forms of carbon.

10-85 As a rule, metals conduct electricity and nonmetals do not. Explain how and why graphite violates this rule.

10-86 When students are told that graphite and diamond are different forms of carbon, they often assume that diamonds are more stable than graphite. Use the ΔH_f° data in the appendix to decide whether they are right.

10-87 Describe the difference between covalent carbides, such as SiC; ionic carbides, such as CaC_2; and interstitial carbides, such as WC.

10-88 Explain why ionic carbides such as CaC_2 and Al_4C_3 burst into flame when they react with water.

10-89 Explain why a mixture of CO and H_2 can be used as a fuel.

10-90 Consider the chemical formulas Na_2CO_3 and $NaHCO_3$, and explain why $NaHCO_3$ was named sodium *bi*carbonate, which literally means "twice as much carbonate."

10-91 Each year the authors tell their students that limestone, marble, and eggshells are all made from $CaCO_3$. Explain why they feel frustrated when, immediately afterward, some students can't tell them whether $CaCO_3$ is soluble in water.

10-92 Write the Lewis structures for CO_2, the CO_3^{2-} ion, the HCO_3^- ion, and H_2CO_3. Use these Lewis structures to show the relationship between CO_2 and carbonic acid

$$CO_2(g) + H_2O(l) \longrightarrow H_2CO_3(aq)$$

between carbonic acid and the bicarbonate ion

$$H_2CO_3(aq) \longrightarrow H^+(aq) + HCO_3^-(aq)$$

and between the bicarbonate ion and the carbonate ion.

$$HCO_3^-(aq) \longrightarrow H^+(aq) + CO_3^{2}(aq)$$

CHAPTER 11

ACIDS, BASES, AND SALTS

CHAPTER CONTENTS

11.1 ACIDS, BASES, AND SALTS: A HISTORIC PERSPECTIVE

For more than 300 years, chemists have grouped substances into three classes: acids, alkalies (or bases), and salts. Compounds that behaved like vinegar were called *acids*, those with properties like wood ash were called *alkalies*, and those formed when acids and alkalies reacted were called *salts*.

As early as 1661, Robert Boyle summarized the properties of acids as follows.

1. **Acids have a sour taste.**
2. **Acids are corrosive.**
3. **Acids change the color of certain vegetable dyes, such as litmus and syrup of violets, from blue to red.**
4. **Acids lose their acidity when they combine with alkalies.**

The name **acid** comes from the Latin *acidus*, which means "sour," and refers to the sharp odor and sour taste of many acids. Vinegar, for example, tastes sour because it is a dilute solution of acetic acid (CH_3CO_2H) in water. Lemon juice tastes sour because it contains citric acid ($C_6H_8O_7$). Milk turns sour when it spoils because of the formation of lactic acid ($C_3H_6O_3$), and the unpleasant, sour odor of rotten meat or butter can be attributed to compounds, such as butyric acid ($C_4H_8O_2$), that form when fat spoils.

One of the characteristic properties of acids is their ability to corrode, or dissolve, a variety of substances, including most metals. Zinc metal, for example, rapidly dissolves in hydrochloric acid to form an aqueous solution of $ZnCl_2$ and hydrogen gas.

$$Zn(s) + 2\ HCl(aq) \longrightarrow Zn^{2+}(aq) + 2\ Cl^-(aq) + H_2(g)$$

This is such an important characteristic of acids that we still talk about subjecting something to the "acid test," a name that comes from a test that was once used to determine whether a piece of metal was gold.

Another characteristic property of acids is their ability to change the color of vegetable dyes such as litmus, which is a mixture of blue dyes extracted from several species of lichens native to the Netherlands. Litmus turns red in the presence of acid and has been used to test for acids for at least 300 years.

Blue litmus paper turns red in the presence of the citric acid in lemons.

Here, zinc metal can be seen to rapidly dissolve in concentrated hydrochloric acid to give an aqueous solution of Zn^{2+} ions and H_2 gas.

Vegetable dyes, such as litmus, have been used for more than 300 years to distinguish between acids and bases.

TABLE 11.1

*Common Substances Classified
on the Basis of the Litmus Test*

Acid (red)	Base (blue)
fruit juices	ammonia (NH_3)
pickles	lime (CaO)
salad dressings	lye (NaOH)
stomach fluid	magnesia (MgO)
vinegar	

Boyle attributed the following properties to alkalies.

1. Alkalies feel slippery when they come into contact with the skin.

2. Alkalies change the color of vegetable dyes such as litmus from red to blue.

3. Alkalies become less alkaline when they combine with acids.

In essence, Boyle defined *alkalies* as substances that consume, or neutralize, acids. Acids lose their characteristic sour taste and their ability to dissolve metals when mixed with alkalies. Alkalies even reverse the change in color that occurs when litmus comes in contact with an acid.

The products of reactions between acids and alkalies were called *salts* because they have many of the properties of table salt, NaCl. Sulfuric acid reacts with sodium hydroxide to form sodium sulfate, which is a salt.

$$H_2SO_4(aq) + 2\ NaOH(aq) \longrightarrow Na_2SO_4(aq) + 2\ H_2O(l)$$

Nitric acid reacts with ammonia to form ammonium nitrate, another salt.

$$HNO_3(aq) + NH_3(aq) \longrightarrow NH_4NO_3(aq)$$

Eventually, alkalies became known as *bases* because they served as the "base," or "basis," for making salts.

$$\text{Acid} + \text{base} \longrightarrow \text{salt}$$

Thus, in the strictest sense, bases are compounds that react with acids to form salts.

Table 11.1 classifies several common substances as acid or base as the result of their behavior toward litmus.

11.2 THE ARRHENIUS DEFINITION OF ACIDS AND BASES

Boyle's definitions provided a useful basis for classifying compounds, but they raised an important question: "What factor determines whether a compound is an acid, a base, or a salt?"

The first step toward answering this question was taken by Antoine Lavoisier, who found the element oxygen in every acid he analyzed. Oxygen occurs in a variety of common acids, including acetic acid (CH_3CO_2H), boric acid (H_3BO_3), carbonic acid (H_2CO_3), nitric acid (HNO_3), oxalic acid ($H_2C_2O_4$), phosphoric acid (H_3PO_4), and sulfuric acid (H_2SO_4), to name just a few. It is not surprising that

Lavoisier believed that all acids must contain oxygen and named this element from the Greek stems meaning "acid former."

In 1811, Humphry Davy was able to prove that hydrochloric acid does not contain oxygen. By 1830, at least 10 more acids that do not contain oxygen, including HF, HBr, HI, HCN, HSCN, and H_2S, had been discovered. In 1838, Justig Liebig proposed an alternative to Lavoisier's oxygen theory of acids.

Liebig suggested that acids contain one or more hydrogen atoms that can be replaced by metal atoms. According to this theory, HSCN is an acid because it contains a hydrogen atom that can be replaced by a metal atom to form a salt, such as NaSCN. Similarly, H_2S is an acid because it contains hydrogen atoms that can be replaced to form salts such as NaSH or Na_2S.

Liebig's theory provided a way to recognize acids, but it didn't explain what gave acids their characteristic properties or why acids had so many properties in common. The answers to these questions were the result of work by Svante Arrhenius. In 1884, Arrhenius suggested that salts such as NaCl dissociate when they dissolve in water to give particles he called *ions*.

$$\text{NaCl}(s) \xrightarrow{\text{H}_2\text{O}} \textbf{Na}^+(\textbf{\textit{aq}}) + \textbf{Cl}^-(\textbf{\textit{aq}})$$

Three years later, in an extension of this theory, Arrhenius proposed that acids are neutral molecules that **ionize** when they dissolve in water to give H^+ ions and a corresponding negative ion. According to this theory, hydrogen chloride is an acid because it ionizes when it dissolves in water to give hydrogen (H^+) and chloride (Cl^-) ions.

$$\text{HCl}(g) \xrightarrow{\text{H}_2\text{O}} \textbf{H}^+(\textbf{\textit{aq}}) + \text{Cl}^-(\textbf{\textit{aq}})$$

Arrhenius also said that bases are neutral compounds that either dissociate or ionize in water to give OH^- ions and a positive ion. NaOH is an Arrhenius base because it dissociates in water to give the hydroxide (OH^-) and sodium (Na^+) ions.

$$\text{NaOH}(s) \xrightarrow{\text{H}_2\text{O}} \text{Na}^+(\textbf{\textit{aq}}) + \textbf{OH}^-(\textbf{\textit{aq}})$$

According to the Arrhenius model, then, acids and bases are defined as follows.

An *Arrhenius acid* is any substance that ionizes when it dissolves in water to give the H^+, or hydrogen, ion.

An *Arrhenius base* is any substance that gives the OH^-, or hydroxide, ion when it dissolves in water.

Arrhenius acids include compounds such as HCl, HCN, and H_2SO_4, which ionize in water to give the H^+ ion. Arrhenius bases include ionic compounds that contain the OH^- ion, such as LiOH, NaOH, KOH, $Ca(OH)_2$, and so on.

This theory explained why Liebig's definition of an acid had been successful. Compounds that ionize in water to give the H^+ ion must contain hydrogen with an oxidation number of $+1$. These compounds therefore contain a hydrogen atom that can be replaced by a positively charged metal ion, such as the Na^+ or K^+ ion.

The theory also explained why acids have similar properties. The characteristic properties of acids all result from the presence of the H^+ ion generated when an acid dissolves in water. It also explained why acids neutralize bases, and vice versa. Acids are sources of the H^+ ion; bases are sources of the OH^- ion; and these ions combine to form water.

$$\text{H}^+(aq) + \text{OH}^-(aq) \longrightarrow \text{H}_2\text{O}(l)$$

Exercise 11.1

> Which of the following compounds can be classified as Arrhenius acids or bases?
>
> (a) HI   (b) $Ca(OH)_2$ (c) NaH   (d) Na_2CO_3
>
> **Solution**
>
> (a) HI is an Arrhenius acid, because it contains hydrogen with a $+1$ oxidation number and can therefore dissolve in water to give H^+ ions.
>
> $$HI(g) \xrightarrow{H_2O} H^+(aq) + I^-(aq)$$
>
> (b) $Ca(OH)_2$ is an Arrhenius base. It is an ionic compound that contains OH^- ions, which it releases into the solution when it dissolves in water.
>
> $$Ca(OH)_2(s) \xrightarrow{H_2O} Ca^{2+}(aq) + 2\ OH^-(aq)$$
>
> (c) NaH is neither an Arrhenius acid nor an Arrhenius base. It can't be an Arrhenius acid, because the first step in its reaction with water can be thought to involve dissociation to form Na^+ and H^- ions, not H^+ ions.
>
> $$NaH(s) \xrightarrow{H_2O} Na^+(aq) + H^-(aq)$$
>
> It can't be an Arrhenius base, because it doesn't contain OH^- ions.
>
> (d) Na_2CO_3 is neither an Arrhenius acid nor an Arrhenius base. It can't be an Arrhenius acid, because it doesn't contain hydrogen. It isn't an Arrhenius base, because it doesn't contain the OH^- ion.

The Arrhenius theory represents a major step toward our present understanding of acids and bases, but it has several disadvantages.

1. It can only be applied to reactions that occur in water, because it defines acids and bases in terms of what happens when compounds dissolve in water.

2. It doesn't explain why some compounds in which hydrogen has an oxidation number of $+1$ (such as HCl) dissolve in water to give solutions that are acidic, whereas others (such as CH_4) do not.

3. The only compounds that can be classified as Arrhenius bases are those that formally contain the OH^-, or hydroxide, ion. The Arrhenius theory can't explain why other compounds — Na_2CO_3, for example — have the characteristic properties of bases.

11.3 AN OPERATIONAL DEFINITION OF ACIDS AND BASES

The Lewis structure of water can help us understand why H^+ and OH^- ions play such an important role in the chemistry of aqueous solutions. At first glance, the Lewis structure suggests that the hydrogen and oxygen atoms are bound together by the sharing of a pair of electrons.

$$H-\overset{\cdot\cdot}{\underset{\cdot\cdot}{O}}-H$$

But we know that oxygen (EN = 3.44) is much more electronegative than hydrogen (EN = 2.20). Thus, the pairs of electrons in these covalent bonds are not shared equally by the hydrogen and oxygen atoms. The electrons are drawn toward the oxygen atom in the center of the molecule and away from the hydrogen atoms on either end. As a result, the water molecule is *polar,* as first noted in Section 8.5. The oxygen atom carries a partial negative charge (δ^-), and the hydrogen atoms carry a partial positive charge (δ^+).

$$\delta^+\text{H}-\ddot{\underset{..}{\text{O}}}{}^{\delta-}-\text{H}^{\delta+}$$

Water molecules can therefore dissociate to form ions. When a positively charged H^+ ion is removed from a neutral water molecule, a negatively charged OH^- ion is formed.

$$\text{H}-\ddot{\underset{..}{\text{O}}}-\text{H} \longrightarrow \text{H}^+ + {:}\ddot{\underset{..}{\text{O}}}-\text{H}^-$$

The opposite reaction, of course, can also occur. H^+ ions can combine with OH^- ions to form neutral water molecules.

$$\text{H}^+ + \text{OH}^- \longrightarrow \text{H}_2\text{O}$$

The fact that water molecules continuously dissociate to form H^+ and OH^- ions, which recombine to form water molecules, is indicated by the following equation.

$$\textbf{H}_2\textbf{O}(l) \rightleftharpoons \textbf{H}^+(aq) + \textbf{OH}^-(aq)$$

The pair of arrows pointing in opposite directions that separates the "starting materials" and the "products" of this reaction indicates that the reaction actually occurs in both directions.

At 25°C, the density of water is 0.9971 g/cm^3, or 0.9971 g/mL. The concentration of water is therefore 55.41 molar.

$$\frac{0.9971 \text{ g H}_2\text{O}}{1 \text{ mL}} \times \frac{1000 \text{ mL}}{1 \text{ L}} \times \frac{1 \text{ mol H}_2\text{O}}{18.015 \text{ g}} = \textbf{55.35 mol H}_2\textbf{O/L}$$

At the same temperature, the concentration of the H^+ and OH^- ions formed by the dissociation of neutral H_2O molecules is only 1.0×10^{-7} moles per liter. The ratio of the concentrations of the H^+ (or OH^-) ions and neutral H_2O molecules is therefore 1.8×10^{-9}.

$$\frac{1.0 \times 10^{-7} \, M \text{ H}^+}{55.41 \, M \text{ H}_2\text{O}} = \textbf{1.8} \times \textbf{10}^{-9}$$

Thus, only about 2 parts per billion (ppb) of the water molecules dissociate into ions at room temperature.

It is difficult to imagine what is meant by the statement that water dissociates to the extent of only 2 ppb. One way of visualizing these numbers is to note that there are roughly 2500 letters on a typical page of this book. If typographical errors occurred with a frequency of 2 ppb, this book would have to be 400,000 pages long to contain a pair of typographical errors.

Figure 11.1 shows a model of 20 water molecules, one of which has dissociated to form a pair of H^+ and OH^- ions. If this figure were the equivalent of a very-high-resolution photograph of the structure of water, we would have to take 28 million such photographs before we encountered another pair of H^+ and OH^- ions.

It is possible to combine the model for the dissociation of water with the theory first proposed by Arrhenius to come up with a slightly more comprehensive set of definitions of acid and base.

FIG. 11.1 This figure can be thought of as a high-resolution snapshot of the structure of water that focuses on 20 H_2O molecules, one of which has dissociated to form H^+ and OH^- ions. To realistically show the number of H^+ and OH^- ions in water, we would need an illustration with the same number of H^+ and OH^- ions but 28 million times as many H_2O molecules.

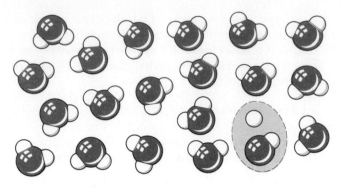

An operational definition of acids and bases:

An acid is any substance that dissolves in water to increase the concentration of the H^+ ion.

A base is any substance that dissolves in water to increase the concentration of the OH^- ion.

These are operational definitions because they tie the theory of acids and bases to a simple laboratory operation that can be used to test for acids and bases. To decide whether a compound is an acid or a base we dissolve the compound in water and test the solution to see whether the H^+ or OH^- ion concentration has in fact increased.

There are several ways of performing such tests. One of the oldest uses an acid–base indicator that consists of paper soaked in a dilute solution of litmus. Blue litmus paper turns red when immersed in solutions rich in the H^+ ion, and an excess of the OH^- ion causes red litmus paper to turn blue.

At first glance, the operational definitions of acids and bases look very much like the Arrhenius definitions given in Section 11.2. There is a subtle difference, however. CO_2 is not an Arrhenius acid, because it can't dissociate to give the H^+ ion in water. According to the operational definition introduced in this section, however, CO_2 is an acid, because it reacts with water to form a compound that dissociates to increase the H^+ ion concentration.

$$CO_2(g) + H_2O(l) \longrightarrow H_2CO_3(aq)$$
$$H_2CO_3(aq) \longrightarrow H^+(aq) + HCO_3^-(aq)$$

The only compounds that are bases according to the Arrhenius definition are salts (such as LiOH or NaOH) that contain the OH^- ion. The operational definition, however, expands the number of compounds that can be called bases to include salts, such as CaO or $CaCO_3$, that react with water to form the OH^- ion.

$$CaO(s) + H_2O(l) \longrightarrow Ca^{2+}(aq) + 2\ OH^-(aq)$$
$$CaCO_3(s) + H_2O(l) \longrightarrow Ca^{2+}(aq) + HCO_3^-(aq) + OH^-(aq)$$

Exercise 11.2

Which of the following compounds can be classified as acids or bases by the Arrhenius definition? Which can be classified as acids or bases by the operational definition?

(a) $HCN(g)$ (c) $P_4O_{10}(s)$ (e) $Na_2O_2(s)$

(b) $SO_3(g)$ (d) $Na_2O(s)$ (f) $NaOH(s)$

Solution

HCN is the only one of these compounds that can dissociate in water to give the H^+ ion, and so it is the only Arrhenius acid. NaOH is the only compound that contains the OH^- ion and is therefore the only Arrhenius base.

According to the operational definition, any compound that forms the H^+ ion when it dissolves in water is an acid. HCN, SO_3, and P_4O_{10} all satisfy this definition of an acid. HCN dissociates to give the H^+ ion when it dissolves in water.

$$HCN(g) \xrightarrow{H_2O} H^+(aq) + CN^-(aq)$$

SO_3 reacts with water to form sulfuric acid, which dissociates to give the H^+ ion.

$$SO_3(g) + H_2O(l) \longrightarrow H_2SO_4(aq)$$
$$H_2SO_4(aq) \longrightarrow H^+(aq) + HSO_4^-(aq)$$

P_4O_{10} reacts with water to form phosphoric acid, which then dissociates.

$$P_4O_{10}(s) + 6\ H_2O(l) \longrightarrow 4\ H_3PO_4(aq)$$
$$H_3PO_4(aq) \longrightarrow 4\ H^+(aq) + 4\ H_2PO_4^-(aq)$$

Na_2O, Na_2O_2, and NaOH all satisfy the operational definition of a base. NaOH dissolves in water to give the OH^- ion.

$$NaOH(s) \xrightarrow{H_2O} Na^+(aq) + OH^-(aq)$$

Na_2O and Na_2O_2 react with water to form the OH^- ion.

$$Na_2O(s) + H_2O(l) \longrightarrow 2\ Na^+(aq) + 2\ OH^-(aq)$$
$$Na_2O_2(s) + 2\ H_2O(l) \longrightarrow 2\ Na^+(aq) + 2\ OH^-(aq) + H_2O_2(aq)$$

11.4 TYPICAL ACIDS AND BASES

Ever since we introduced the chemistry of metals and nonmetals in Section 2.7, we have repeatedly noted that these two classes of elements have opposite properties. It is therefore tempting to assume that the opposite behavior of acids and bases might be traced to differences between the chemistry of metals and nonmetals. To test this hypothesis, let's look at the chemistry of three categories of compounds — hydrides, oxides, and hydroxides.

Nonmetal hydrides, such as HCl and H_2S, are often acids. These compounds all contain hydrogen with a $+1$ oxidation number and can therefore act as a source of the H^+ ion in water.

$$HCl(g) \xrightarrow{H_2O} H^+(aq) + Cl^-(aq)$$
$$H_2S(g) \xrightarrow{H_2O} H^+(aq) + HS^-(aq)$$

Conversely, metal hydrides, such as NaH and CaH_2, contain hydrogen with an oxidation number of -1. These compounds dissociate to give the H^-, or hydride, ion when they dissolve in water.

$$NaH(s) \xrightarrow{H_2O} Na^+(aq) + H^-(aq)$$

The H⁻ ion, with its pair of valence electrons, can combine with an H⁺ ion to form a hydrogen molecule.

$$H:^- + H^+ \longrightarrow H—H$$

Alternatively, the H⁻ ion can abstract an H⁺ ion from a water molecule.

$$H:^- + H—\overset{..}{\underset{..}{O}}—H \longrightarrow H—H + :\overset{..}{\underset{..}{O}}—H^-$$

Since removing H⁺ ions from water molecules is one way to increase the OH⁻ ion concentration in a solution, metal hydrides are bases.

$$NaH(s) + H_2O(l) \longrightarrow Na^+(aq) + OH^-(aq) + H_2(g)$$
$$CaH_2(s) + 2\ H_2O(l) \longrightarrow Ca^{2+}(aq) + 2\ OH^-(aq) + 2\ H_2(g)$$

A similar pattern can be found in the chemistry of nonmetal versus metal oxides. Nonmetal oxides dissolve in water to form acids. CO_2 dissolves in water to give carbonic acid, SO_3 gives sulfuric acid, and P_4O_{10} reacts with water to give phosphoric acid.

$$CO_2(g) + H_2O(l) \longrightarrow H_2CO_3(aq)$$
$$SO_3(g) + H_2O(l) \longrightarrow H_2SO_4(aq)$$
$$P_4O_{10}(s) + 6\ H_2O(l) \longrightarrow 4\ H_3PO_4(aq)$$

Acid rain that has corroded this bronze plaque, and the Civil War monument in the chapter opening photograph, is formed when nonmetal oxide pollutants in the atmosphere dissolve in rain to form solutions that can have a pH as low as 2.4.

Metal oxides such as CaO (lime), MgO (magnesia), and Na_2O are bases. "Milk of magnesia," for example, which is used to neutralize excess stomach acid, is MgO suspended in water. Metal oxides formally contain the O^{2-} ion, which reacts with water to give a pair of OH⁻ ions.

$$O^{2-}(aq) + H_2O(l) \longrightarrow 2\ OH^-(aq)$$

Metal oxides therefore fit the operational definition of a base, because they react with water to increase the concentration of the OH⁻ ion.

$$CaO(s) + H_2O(l) \longrightarrow Ca^{2+}(aq) + 2\ OH^-(aq)$$

We see the same pattern in the chemistry of compounds that contain the —OH, or hydroxide, group. Metal hydroxides, such as LiOH, NaOH, KOH, and $Ca(OH)_2$, are bases by any definition.

$$NaOH(s) \xrightarrow{H_2O} Na^+(aq) + OH^-(aq)$$

CO_2 dissolves in water to give a solution that tests acid. CaO and MgO aren't very soluble in water, but their solutions test basic.

TABLE 11.2

Typical Acids and Bases

Acids	Bases
Nonmetal Hydrides	Metal Hydrides
HF, HCl, HBr, HI,	LiH, NaH, KH,
HCN, HSCN, H_2S, H_2Se	MgH_2, CaH_2, SrH_2
Nonmetal Oxides	Metal Oxides
CO_2, SO_2, SO_3,	Li_2O, Na_2O, K_2O,
NO_2, P_4O_{10}	MgO, CaO
Nonmetal Hydroxides	Metal Hydroxides
HOCl, $HONO_2$,	LiOH, NaOH, KOH,
$O_2S(OH)_2$, $OP(OH)_3$	$Ca(OH)_2$, $Ba(OH)_2$

Nonmetal hydroxides, however, are acids. The simplest example of this class of compounds is hypochlorous acid, HOCl. Hypochlorous acid is a classical Arrhenius acid, because it ionizes in water to give the H^+ and OCl^-, or hypochlorite, ion.

$$HOCl(aq) \longrightarrow H^+(aq) + OCl^-(aq)$$

Table 11.2 summarizes the trends observed in these three categories of compounds. Metal hydrides, metal oxides, and metal hydroxides are bases. Nonmetal hydrides, nonmetal oxides, and nonmetal hydroxides tend to be acids.

The last category of compounds in the left-hand column of Table 11.2, the nonmetal hydroxides, are more important than you might expect. Students often look at the formulas for acids such as HNO_3, H_3PO_4, H_2SO_4, $HClO_4$ and assume that these acids belong to the category of nonmetal hydrides. Unfortunately, the chemical formulas used to represent these acids are misleading. Chemists tend to write formulas such as H_2SO_4, HNO_3, and H_3PO_4, which lead students to think that the acidic hydrogen atoms are bound to the sulfur, nitrogen, or phosphorus atoms. They aren't. Each of these compounds is a nonmetal hydroxide, in which the acidic hydrogen is attached to an oxygen atom.

Another reason students may be unfamiliar with the category of nonmetal hydroxides is that these oxygen-rich acids are often called ***oxyacids.*** Skeleton structures for a number of oxyacids are given in Figure 11.2. In theory, the formulas of these compounds should be written as $O_2S(OH)_2$, $OP(OH)_3$, $HONO_2$, $HOClO_3$, and $OC(OH)_2$, and so on, to indicate their skeleton structures. In practice, these formulas can't compete with the well-established formulas by which the acids have long been known — H_2SO_4, H_3PO_4, HNO_3, $HClO_4$, H_2CO_3, and so on.

You can sort acids into the categories of nonmetal hydrides (such as HCl and HCN) and nonmetal hydroxides (such as H_2SO_4 and H_3PO_4) by remembering that acids that contain oxygen almost always have skeleton structures in which the acidic hydrogens are attached to oxygen atoms.

Exercise 11.3

Write the Lewis structures of the following acids and classify these acids as either nonmetal hydrides (XH) or nonmetal hydroxides (XOH).

(a) HCN   (b) HNO_3 (c) $H_2C_2O_4$ (d) CH_3CO_2H

Solution

HCN is the only nonmetal hydride among these acids.

$$H\!-\!C\!\equiv\!N\!:$$

HNO_3, $H_2C_2O_4$, and CH_3CO_2H are all nonmetal hydroxides, or oxyacids.

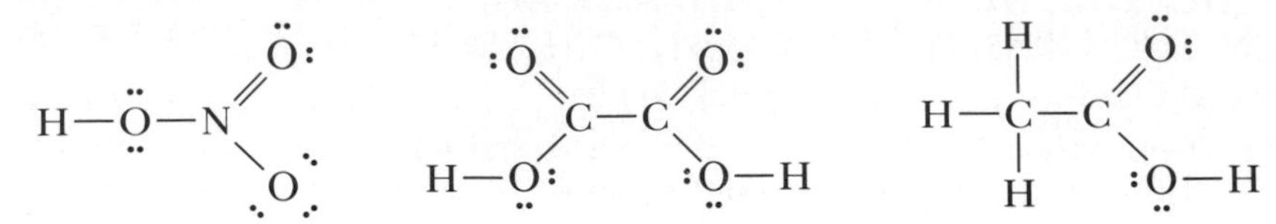

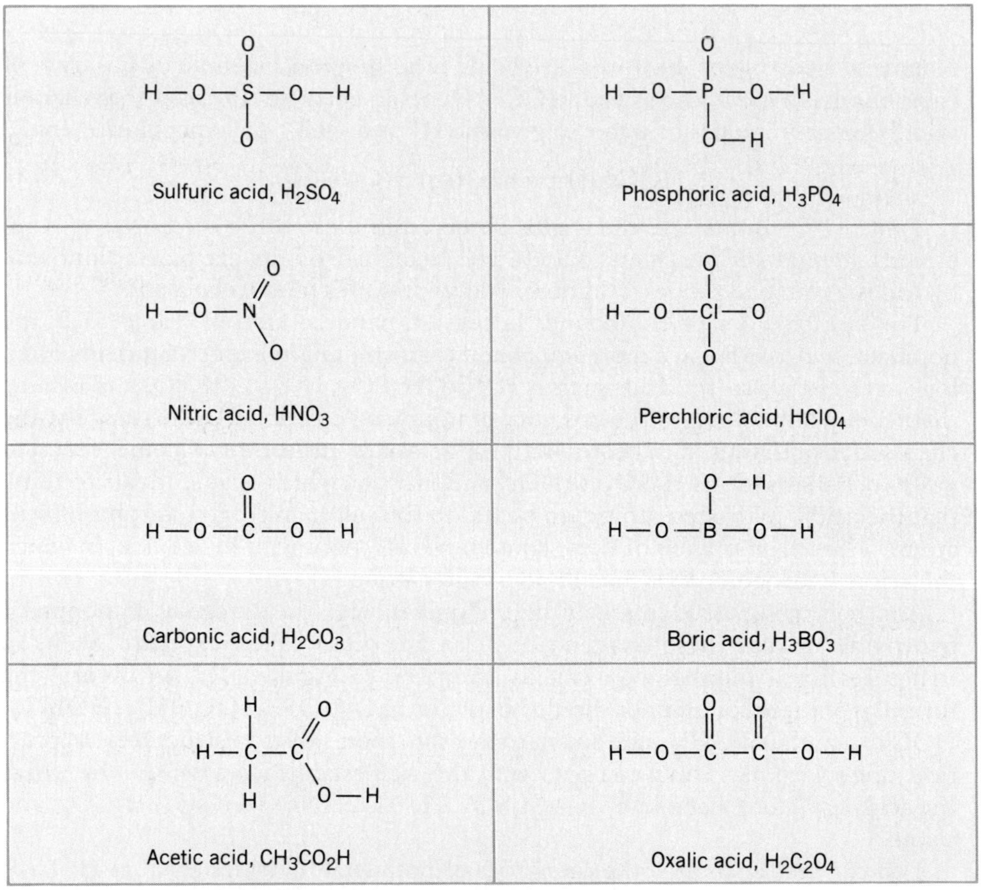

FIG. 11.2 The correct skeleton structures for some of the common nonmetal hydroxides, or oxyacids.

Sulfuric acid, H_2SO_4

Phosphoric acid, H_3PO_4

Nitric acid, HNO_3

Perchloric acid, $HClO_4$

Carbonic acid, H_2CO_3

Boric acid, H_3BO_3

Acetic acid, CH_3CO_2H

Oxalic acid, $H_2C_2O_4$

11.5 WHY ARE METAL HYDROXIDES BASES AND NONMETAL HYDROXIDES ACIDS?

It is easy to understand why nonmetal hydrides, such as HCl and H_2S, are acids and metal hydrides, such as NaH or CaH_2, are bases. Nonmetal hydrides contain hydrogen with an oxidation number of $+1$, so they can act as sources of the H^+ ion.

$$HCl(g) \xrightarrow{\text{H}_2\text{O}} H^+(aq) + Cl^-(aq)$$

Metal hydrides, on the other hand, contain hydrogen with an oxidation number of -1 and are therefore sources of the H^- ion, which reacts with water to give the OH^- ion.

Step 1: $$NaH(s) \xrightarrow{H_2O} Na^+(aq) + H^-(aq)$$

Step 2: $$H^-(aq) + H_2O(l) \longrightarrow H_2(g) + OH^-(aq)$$

Net reaction: $$NaH(s) + H_2O(l) \longrightarrow Na^+(aq) + \mathbf{OH^-(aq)} + H_2(g)$$

To understand why nonmetal hydroxides, such as HOCl, are acids and metal hydroxides, such as NaOH, are bases we have to look at the relative electronegativities of the atoms in these compounds. Let's start with a typical metal hydroxide — sodium hydroxide.

$$Na{-}O{-}H$$

Electronegativity: 0.93   3.44   2.20

The difference between the electronegativities of sodium and oxygen is so large ($\Delta EN = 2.51$) that it isn't reasonable to assume that the electrons in the Na—O bond are shared equally. The electrons in this bond are drawn toward the more electronegative oxygen atom. NaOH might be more accurately described by the following Lewis structure.

$$[Na^+][\!:\!\ddot{O}{-}H^-]$$

There is so much ionic character in this bond that we expect NaOH to dissociate to give Na^+ and OH^- ions when it dissolves in water.

$$NaOH(s) \xrightarrow{H_2O} Na^+(aq) + \mathbf{OH^-(aq)}$$

We get a very different pattern when we apply the same procedure to a typical nonmetal hydroxide — hypochlorous acid, HOCl.

$$Cl{-}O{-}H$$

Electronegativity: 3.16   3.44   2.20

Here, the difference between the electronegativities of the chlorine and oxygen atoms is small ($\Delta EN = 0.28$), so the electrons in the Cl—O bond are shared more or less equally by the two atoms. On the other hand, the O—H bond in this molecule is highly polar ($\Delta EN = 1.24$). The electrons in this bond are drawn toward the more electronegative oxygen atom. When this molecule ionizes, the electrons in the O—H bond remain with the oxygen atom, and OCl^- and H^+ ions are formed.

$$HOCl(aq) \longrightarrow \mathbf{H^+(aq)} + OCl^-(aq)$$

Since there is no abrupt change from metal to nonmetal across a row of the periodic table or down a column, we should expect to find compounds that are intermediate between the extremes of metal oxides and nonmetal oxides or metal hydroxides and nonmetal hydroxides. These compounds, such as Al_2O_3 and $Al(OH)_3$, are called **amphoteric** (literally, "either or both"), because they can act as either acids or bases. $Al(OH)_3$, for example, acts as an acid when it reacts with a base.

$$Al(OH)_3(s) + OH^-(aq) \longrightarrow Al(OH)_4^-(aq)$$
$$\quad\ \text{acid} \qquad\quad\ \text{base}$$

But it acts as a base when it reacts with an acid.

$$\underset{\text{base}}{\text{Al(OH)}_3(s)} + 3\ \underset{\text{acid}}{\text{H}^+(aq)} \longrightarrow \text{Al}^{3+}(aq) + 3\ \text{H}_2\text{O}(l)$$

11.6 THE BRØNSTED DEFINITION OF ACIDS AND BASES

The operational definitions of acids and bases, introduced in Section 11.3 as a logical extension of Arrhenius's original hypothesis, are useful for describing the chemistry of common acids and bases in aqueous solution. In 1923, Johannes Brønsted and Thomas Lowry independently proposed a more powerful set of definitions.

The Brønsted, or Brønsted–Lowry, model is based on the assumption that an acid donates H^+ ions to another ion or molecule, which acts as a base. According to this model, HCl doesn't dissociate in water to form H^+ and Cl^- ions, but instead the HCl molecules transfer H^+ ions to water molecules to form H_3O^+ and Cl^- ions.

$$\text{HCl}(g) + \text{H}_2\text{O}(l) \longrightarrow \textbf{H}_3\textbf{O}^+(\textbf{\textit{aq}}) + \textbf{Cl}^-(\textbf{\textit{aq}})$$

The Brønsted theory assumes that the dissociation of water can be explained by the following equation.

$$2\ \text{H}_2\text{O}(l) \rightleftharpoons \textbf{H}_3\textbf{O}^+(\textbf{\textit{aq}}) + \textbf{OH}^-(\textbf{\textit{aq}})$$

In this model, one water molecule donates an H^+, or hydrogen, ion to another water molecule.

Exercise 11.4

Use the model of the structure of the atom introduced in Chapter 5 to explain why chemists feel more comfortable with the Brønsted theory, in which H^+ ions are transferred from one molecule to another, than with the Arrhenius model, in which acids give off H^+ ions when they dissolve in water.

Solution

An isolated hydrogen atom has a single electron in the $1s$ orbital and a single proton in the nucleus of the atom. Hydrogen atoms are relatively small—the covalent radius of hydrogen (0.0371 nm) is less than half the covalent radius of a carbon atom (0.077 nm), for example. But an H^+ ion is a nothing more than a bare proton. It is many orders of magnitude smaller than the smallest atom or ion. The charge on an isolated H^+ ion would be distributed over such a small amount of space that this H^+ ion would be attracted toward any source of negative charge that existed in the solution.

In fact, there is no evidence for the existence of a bare H^+ ion in aqueous solution. The instant such an ion was created, it would immediately bond to a neighboring water molecule.

The Brønsted model, in which H^+ ions are transferred from one ion or molecule to another, therefore makes more sense than the Arrhenius theory, which assumes that H^+ ions exist in aqueous solution.

There is reason to believe that even the Brønsted model is a bit too naive. Each H^+ ion that an acid donates to water is actually bound to four neighboring water molecules, as shown in Figure 11.3. Although a more realistic formula for the substance produced when an H^+ ion is transferred from one water molecule to

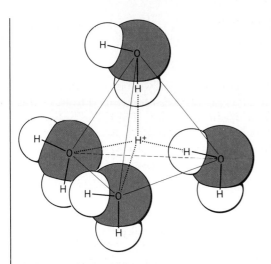

FIG. 11.3 There is no such thing as an H^+ ion in aqueous solution. The H^+ ion produced when an acid ionizes is attached to at least one water molecule to form an H_3O^+ ion. There is reason to believe that the H^+ produced in this reaction is actually shared by at least four H_2O molecules, to form an $H(H_2O)_4^+$, or $H_9O_4^+$, ion. For practical purposes, however, this ion can be thought of as an H_3O^+ ion.

another would be $H(H_2O)_4^+$, or perhaps $H_9O_4^+$, for all practical purposes, the product of this reaction can be represented as the H_3O^+ ion.

According to the Brønsted model, an acid is a substance that can donate H^+ ions and a base is a substance that can accept an H^+ ion. Acids are therefore called hydrogen-ion, or proton, donors and bases are hydrogen-ion, or proton, acceptors.

A *Brønsted acid* is any substance (such as HCl) that can donate an H^+ ion to a base. Acids are therefore *hydrogen-ion donors*, or *proton donors*.

A *Brønsted base* is any substance (such as H_2O) that can accept an H^+ ion from an acid. Bases are therefore *hydrogen-ion acceptors*, or *proton acceptors*.

From the perspective of the Brønsted model, reactions between acids and bases involve the transfer of an H^+ ion from an acid (proton donor) to a base (proton acceptor). Acids and bases can be neutral molecules.

$$\underset{\substack{\text{acid}\\ \text{(proton donor)}}}{HCl(aq)} + \underset{\substack{\text{base}\\ \text{(proton acceptor)}}}{NH_3(aq)} \longrightarrow Cl^-(aq) + NH_4^+(aq)$$

And they can be either positive or negative ions.

$$\underset{\substack{\text{acid}\\ \text{(proton donor)}}}{NH_4^+(aq)} + \underset{\substack{\text{base}\\ \text{(proton acceptor)}}}{OH^-(aq)} \longrightarrow NH_3(aq) + H_2O(l)$$

$$\underset{\substack{\text{acid}\\ \text{(proton donor)}}}{H_2PO_4^-(aq)} + \underset{\substack{\text{base}\\ \text{(proton acceptor)}}}{H_2O(l)} \longrightarrow HPO_4^{2-}(aq) + H_3O^+(aq)$$

The Brønsted theory expands the number of potential acids to include positive ions and negative ions, as well as neutral molecules. It also allows us to decide which compounds are acids from their chemical formulas. Any compound that contains hydrogen with an oxidation number of $+1$ has the potential to be an acid. Brønsted acids therefore include HCl, H_2S, H_2CO_3, H_2PtF_6, NH_4^+, HSO_4^-, $HMnO_4$, and so on.

Brønsted bases can be identified from their Lewis structures. According to the Brønsted model, a base is any ion or molecule that can accept a proton. To understand the implications of this definition, look at how the prototypical base, the OH^- ion, accepts a proton.

$$H^+ \;\curvearrowleft\; :\overset{..}{\underset{..}{O}}-H^- \longrightarrow H-\overset{..}{\underset{..}{O}}-H$$

The only way to accept an H^+ ion is to form a covalent bond with it. The only way to form a covalent bond with an H^+ ion that has no valence electrons is to have the base provide both the electrons needed to form the bond. Thus, the only compounds that can act as hydrogen-ion acceptors, or Brønsted bases, are those that have nonbonding pairs of electrons in the valence shell of one or more atoms.

All of the following compounds can act as Brønsted bases, because they all contain nonbonding pairs of electrons.

$$SO_4{}^{2-}: \quad :\overset{\displaystyle :\overset{..}{O}:}{\underset{\displaystyle :\underset{..}{O}:}{\overset{|}{\underset{|}{O}-S-\overset{..}{O}:}}} \qquad\qquad NH_3: \quad H-\overset{\displaystyle ..}{\underset{\displaystyle |}{N}}-H \atop H$$

$$H_2O: \quad H-\overset{..}{\underset{..}{O}}-H \qquad\qquad CO_3{}^{2-}: \quad :\overset{\displaystyle \overset{.\,.}{O}}{\overset{\displaystyle \|}{\underset{..}{O}-C-\underset{..}{O}:}} \; {}^{2-}$$

$$MnO_4{}^-: \quad :\overset{\displaystyle :\overset{..}{O}:}{\underset{\displaystyle :\underset{..}{O}:}{\overset{|}{\underset{|}{O}-Mn-\overset{..}{O}:}}} \; {}^- \qquad CH_3OCH_3: \quad \overset{\displaystyle H \quad\; H}{\underset{\displaystyle H \quad\; H}{H-\overset{|}{\underset{|}{C}}-\overset{..}{\underset{..}{O}}-\overset{|}{\underset{|}{C}}-H}}$$

The only bases in the Arrhenius model are compounds, such as NaOH or $Ca(OH)_2$, that contain the OH^- ion. The operational definition expands this list to include compounds, such as NaH or Na_2CO_3, that react with water to give off the OH^- ion. But the Brønsted model expands the list of potential bases to include any ion or molecule that contains one or more pairs of nonbonding valence electrons. The Brønsted definition of a base applies to so many ions and molecules it is almost easier to count substances, such as the following, that can't be Brønsted bases because they don't have pairs of nonbonding valence electrons.

$$BeH_2: \quad H-Be-H \qquad\qquad H_2: \quad H-H$$

$$CH_4: \quad \overset{\displaystyle H}{\underset{\displaystyle H}{H-\overset{|}{\underset{|}{C}}-H}} \qquad\qquad NH_4{}^+: \quad \overset{\displaystyle H}{\underset{\displaystyle H}{H-\overset{|}{\underset{|}{N}}-H}} \; {}^+$$

Exercise 11.5

Which of the following compounds can be Brønsted acids and which can be Brønsted bases?

(a) H_2O (b) NH_3 (c) $HSO_4{}^-$ (d) OH^-

Solution

All four of these compounds can be Brønsted acids, because they all contain hydrogen with an oxidation number of $+1$. Furthermore, all four compounds can be Brønsted bases, because they all contain at least one pair of nonbonding valence electrons.

$$H-\ddot{O}-H \qquad H-\overset{\displaystyle H}{\underset{\displaystyle |}{\overset{\displaystyle |}{\ddot{N}}}}-H \qquad H-\ddot{O}-\overset{\displaystyle :\ddot{O}:}{\underset{\displaystyle :\ddot{O}:}{\overset{\displaystyle |}{\underset{\displaystyle |}{S}}}}-\ddot{O}:^- \qquad :\ddot{O}-H^-$$

11.7 THE ROLE OF WATER IN THE BRØNSTED THEORY

The Brønsted theory explains the role water plays in acid-base reactions as follows.

1. Water dissociates to form ions by transferring an H^+ ion from one molecule acting as an acid to another molecule acting as a base.

$$H_2O(l) + H_2O(l) \rightleftharpoons H_3O^+(aq) + OH^-(aq)$$

2. Acids react with water by donating an H^+ ion to a neutral water molecule to form the H_3O^+ ion.

$$HCl(g) + H_2O(l) \longrightarrow H_3O^+(aq) + Cl^-(aq)$$

3. Bases react with water by accepting an H^+ ion from a water molecule to form the OH^- ion.

$$NH_3(aq) + H_2O(l) \longrightarrow NH_4^+(aq) + OH^-(aq)$$

4. Water molecules act as intermediates in acid-base reactions by gaining H^+ ions from the acid

$$HCl(g) + H_2O(l) \longrightarrow H_3O^+(aq) + Cl^-(aq)$$

and then losing these H^+ ions to the base.

$$NH_3(aq) + H_3O^+(aq) \longrightarrow NH_4^+(aq) + H_2O(l)$$

We can extend the Brønsted model to explain acid-base reactions in other solvents by analogy with the chemistry of water. For example, there is a very small tendency in liquid ammonia for an H^+ ion to be transferred from one NH_3 molecule to another to form the NH_4^+ and NH_2^- ions.

$$NH_3(l) + NH_3(l) \longrightarrow NH_4^+ + NH_2^-$$

Reasoning by analogy, we infer that acids in liquid ammonia include any source of the NH_4^+ ion and bases include any source of the NH_2^- ion. Thus, NH_4Cl is an acid and $NaNH_2$ is a base in liquid ammonia.

The Brønsted model can even be extended to reactions that do not occur in solution. A classic example of a gas-phase acid-base reaction is encountered when the stoppers from bottles of concentrated hydrochloric acid and aqueous ammonia are held next to each other. A white cloud of ammonium chloride soon forms as the HCl gas that escapes from one solution reacts with the NH_3 gas from the other. This reaction can be represented by the following equation.

$$\underset{\text{acid}}{HCl(g)} + \underset{\text{base}}{NH_3(g)} \longrightarrow NH_4Cl(s)$$

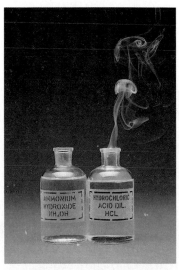

One advantage of the Brønsted theory is the ease with which this model can be extended to include gas-phase acid–base reactions, such as the formation of a white cloud of NH_4Cl when a stopper from a bottle of concentrated hydrochloric acid is held close to the stopper from a bottle of concentrated aqueous ammonia.

The product of this reaction is a salt that contains NH_4^+ and Cl^- ions.

$$H—\overset{..}{\underset{..}{Cl}}: \ + \ H—\overset{\textstyle H}{\underset{\textstyle H}{\overset{|}{\underset{|}{N}}}}—H \ \longrightarrow \ H—\overset{\textstyle H}{\underset{\textstyle H}{\overset{|\ +}{\underset{|}{N}}}}—H \ \ :\overset{..}{\underset{..}{Cl}}:^-$$

The reaction involves a net transfer of an H^+ ion from HCl to NH_3 and is therefore an example of a Brønsted acid-base reaction, even though it occurs in the gas phase.

11.8 CONJUGATE ACID-BASE PAIRS

Perhaps the most important consequence of the Brønsted theory was the recognition that acids and bases exist as *conjugate acid-base pairs.* *Conjugate* comes from the Latin stems meaning "joined together" and refers to things that are joined, particularly in pairs. It is therefore an ideal term for describing the relationship between acids and bases in the Brønsted theory.

Every time a Brønsted acid acts as an H^+-ion donor, it forms a conjugate base. Imagine a generic acid with the formula H*A* that can act as a hydrogen-ion donor. When the acid loses an H^+ ion, one of the products of the reaction is the A^- ion, which is a hydrogen-ion acceptor, or Brønsted base.

$$\underset{\text{acid}}{HA} \longrightarrow H^+ + \underset{\text{base}}{A^-}$$

Every time a base — which we can symbolize as A^- — gains an H^+ ion, the product is a Brønsted acid, H*A*.

$$\underset{\text{base}}{A^-} + H^+ \longrightarrow \underset{\text{acid}}{HA}$$

Acids and bases in the Brønsted model therefore exist as conjugate pairs, whose formulas are related by the gain or loss of a hydrogen ion.

> **Conjugate acid-base pair: Two substances related by the gain or loss of an H^+ ion. Every acid (HA) has a conjugate base (A^-) formed when the acid loses an H^+ ion. Every base (A^-) has a conjugate acid (HA) formed when the base gains an H^+ ion.**

Our use of the symbols H*A* and A^- for a conjugate acid–base pair does not mean that all acids are neutral molecules and all bases are negative ions. It only signifies that the acid contains an H^+ ion that isn't present in the conjugate base. As noted earlier, Brønsted acids or bases can be neutral molecules, positive ions, or negative ions. A number of Brønsted acids and their conjugate bases are given in Table 11.3.

TABLE 11.3

Typical Brønsted Acids and Their Conjugate Bases

Acid	Base
H_3O^+	H_2O
H_2O	OH^-
OH^-	O^{2-}
HCl	Cl^-
H_2SO_4	HSO_4^-
HSO_4^-	SO_4^{2-}
NH_4^+	NH_3
NH_3	NH_2^-

Exercise 11.6

Determine whether each of the following pairs of compounds is an example of a conjugate acid-base pair.

Acid	Base		Acid	Base
(a) H_3O^+	OH^-		(c) H_2O	OH^-
(b) H_3O^+	H_2O		(d) H_2O	H_2O

Solution

(a) No. The conjugate base of the H_3O^+ ion is the H_2O molecule, not the OH^- ion. The conjugate acid of the OH^- ion is the H_2O molecule, not the H_3O^+ ion.

(b) Yes. H_2O is the conjugate base of the H_3O^+ ion, and the H_3O^+ ion is the conjugate acid of H_2O.

(c) Yes. The OH^- ion is the conjugate base of H_2O, and H_2O is the conjugate acid of the OH^- ion.

(d) No. H_2O is both an acid and a base. However, the conjugate acid of H_2O is the H_3O^+ ion, and the conjugate base of H_2O is the OH^- ion.

A feature of the Brønsted theory that is sometimes difficult to understand is the fact that a compound can be both a Brønsted acid and a Brønsted base. H_2O, OH^-, HSO_4^-, and NH_3, for example, can be found in both the acid and base columns among the conjugate acid-base pairs listed in Table 11.3. Water is the perfect example of this behavior, because it simultaneously acts as an acid and a base when it forms the H_3O^+ and OH^- ions. The OH^- ion and NH_3 are usually bases, but they can act as acids if they come into contact with very strong bases. The HSO_4^- ion is usually an acid but can act as a base when it comes into contact with a very strong acid.

11.9 THE RELATIVE STRENGTHS OF ACIDS AND BASES

Many hardware stores and building supply companies sell "muriatic acid" to clean bricks and concrete, which is a moderately concentrated solution of hydrochloric acid in water. The solubility of HCl gas in water limits the concentration of hydrochloric acid to a maximum of about 12 moles per liter — 12 M. Muriatic acid is sold as a 6 M solution.

Grocery stores sell vinegar, which is a dilute solution of acetic acid (CH_3CO_2H) in water. Distilled white vinegar may have a concentration as low as 4% CH_3CO_2H by weight, but cider and malt vinegars have concentrations as high as 6% CH_3CO_2H. (A 6% solution of acetic acid has a concentration of about 1 M.)

You can demonstrate that vinegar is an acid by preparing sauerbraten in an aluminum pot. Sauerbraten is beef cooked in a marinade prepared by mixing vinegar with an equal volume of water. If you make it in an aluminum pot, you will find that the inside of the pot looks cleaner than before when you wash it after the meal. It should look cleaner. A very small layer of metal from the inner surface of the pot will have dissolved in the vinegar during cooking.

Although muriatic acid and vinegar are both acids, there is an obvious difference between them. You wouldn't use muriatic acid in salad dressing, and vinegar is ineffective in cleaning bricks or concrete. The difference between these acids is simple — muriatic acid is a strong acid and vinegar is a weak acid.

Muriatic acid is strong because it is very good at doing what an acid should do — transferring an H^+ ion to a water molecule. In a 6 M solution of hydrochloric acid, all but 1 in every 28,000 HCl molecules react with water to form H_3O^+ and Cl^- ions.

$$HCl(aq) + H_2O(l) \longrightarrow H_3O^+(aq) + Cl^-(aq)$$

Only 0.004% of the HCl molecules remain in solution. Muriatic acid is therefore a strong acid indeed.

Vinegar is a weak acid in the sense that it is not a very good hydrogen-ion donor—it is not very good at transferring H^+ ions to water. In a typical 1 M solution, less than 0.4% of the CH_3CO_2H molecules react with water to form H_3O^+ and $CH_3CO_2^-$ ions.

$$CH_3CO_2H(aq) + H_2O(l) \longrightarrow H_3O^+(aq) + CH_3CO_2^-(aq)$$

This means that more than 99.6% of the acetic acid molecules remain intact.

It is useful to have a quantitative measure of the relative strengths of acids to replace the labels *strong* and *weak*. For reasons that will be discussed in Chapter 15, the relative strengths of acids is described in terms of an **acid-dissociation equilibrium constant, K_a**.

The reaction between an acid and water can be represented by the following generic equation.

$$HA(aq) + H_2O(l) \longrightarrow H_3O^+(aq) + A^-(aq)$$

Let's assume that some fraction of the HA molecules react to form H_3O^+ and A^- ions. We can represent the final concentrations of these ions in units of moles per liter with the symbols $[H_3O^+]$ and $[A^-]$. The concentration of the HA molecules that remain in solution can be represented by the symbol $[HA]$. The value of the acid-dissociation equilibrium constant for this acid is calculated from the following equation.

$$K_a = \frac{[H_3O^+][A^-]}{[HA]}$$

Strong acids, which react extensively with water to form H_3O^+ and A^- ions, have values of K_a that are larger than 1. Weak acids, which react only slightly with water, have values of K_a that are much smaller than 1.

Strong acids, such as HCl:	$K_a > 1$
Weak acids, such as CH_3CO_2H:	$K_a < 1$

Hydrochloric acid, for example, has a K_a of roughly 1×10^6.

$$\frac{[H_3O^+][Cl^-]}{[HCl]} = 1 \times 10^6$$

Acetic acid, on the other hand, has a K_a of only 1.8×10^{-5}

$$\frac{[H_3O^+][CH_3CO_2^-]}{[CH_3CO_2H]} = 1.8 \times 10^{-5}$$

A list of common acids and their acid dissociation constants is given in Table 11.4.

11.10 THE RELATIVE STRENGTHS OF CONJUGATE ACID-BASE PAIRS

The acid-dissociation constant, K_a, is one indication of the strength of an acid. Acetic acid, for example, is a weak acid in an absolute sense because its value of K_a is quite small. However, acetic acid is a stronger acid than water, because the value of K_a for water is even smaller.

CH_3CO_2H:	$K_a = 1.8 \times 10^{-5}$
H_2O:	$K_a = 1.8 \times 10^{-16}$

TABLE 11.4

Common Acids and Their Acid-Dissociation Equilibrium Constants, K_a

Strong Acids	K_a
HI	3×10^9
HBr	1×10^9
HCl	1×10^6
H_2SO_4	1×10^3
$HClO_4$	1×10^3
HNO_3	28
H_2CrO_4	10

Intermediate Acids	
H_3PO_4	7.1×10^{-3}
HF	7.2×10^{-4}

Weak Acids	
citric acid	7.5×10^{-5}
CH_3CO_2H	1.8×10^{-5}
H_2S	1.0×10^{-7}
H_2CO_3	4.5×10^{-7}
H_3BO_3	5.8×10^{-10}
H_2O	1.8×10^{-16}

The Brønsted definitions of acids and bases provide a way to determine the relative strengths of conjugate bases from the values of K_a for the corresponding acids. To see how this is done, compare hydrochloric acid and acetic acid.

$$HCl: \qquad K_a = 1 \times 10^6$$
$$CH_3CO_2H: \qquad K_a = 1.8 \times 10^{-5}$$

The Brønsted theory suggests that every acid-base reaction converts an acid into its conjugate base and a base into its conjugate acid.

$$HCl(g) + H_2O(l) \longrightarrow H_3O^+(aq) + Cl^-(aq)$$

acid   base   acid   base

The fact that 99.996% of the HCl molecules in a 6 M solution react with water to give H_3O^+ and Cl^- ions tells us something about the relative strength of the acids and bases in this reaction. It suggests that HCl must be a stronger acid than the H_3O^+ ion and that H_2O must be a stronger base than the Cl^- ion.

$$HCl(g) + H_2O(l) \longrightarrow H_3O^+(aq) + Cl^-(aq)$$

stronger   stronger   weaker   weaker
acid   base   acid   base

If this weren't true — if the H_3O^+ ion were the stronger acid and the Cl^- ion the stronger base — we would expect the reaction to go in the opposite direction. We would expect that very few of the HCl molecules would ionize in water.

What about weak acids, such as acetic acid? Once again, according to the Brønsted theory, the reaction between the acid and water must convert the acid into its conjugate base and the base into its conjugate acid.

$$CH_3CO_2H(aq) + H_2O(l) \longrightarrow H_3O^+(aq) + CH_3CO_2^-(aq)$$

acid   base   acid   base

But acetic acid is a weak acid; only a very small fraction of the CH_3CO_2H molecules actually donate an H^+ ion to a water molecule. Why doesn't the reaction proceed to the right? The most reasonable answer is that the products of this reaction are a stronger acid and a stronger base than the starting materials.

$$CH_3CO_2H(aq) + H_2O(l) \longrightarrow H_3O^+(aq) + CH_3CO_2^-(aq)$$

weaker   weaker   stronger   stronger
acid   base   acid   base

We can turn the results of this discussion into a general rule.

When conjugate acid-base pairs are compared, the stronger acid always has the weaker conjugate base. Conversely, the stronger base always has the weaker conjugate acid.

The fact that HCl is a stronger acid than the H_3O^+ ion therefore implies that the Cl^- ion is a weaker base than water.

Acid strength: $HCl > H_3O^+$

Base strength: $Cl^- < H_2O$

If CH_3CO_2H is a weaker acid than the H_3O^+ ion, we must conclude that the $CH_3CO_2^-$ ion is a stronger base than water.

Acid strength:	$CH_3CO_2H < H_3O^+$
Base strength:	$CH_3CO_2^- > H_2O$

The relative strengths of a pair of bases can therefore be determined from the values of K_a for their conjugate acids. If HCl is a stronger acid ($K_a = 10^6$) than CH_3CO_2H ($K_a = 1.8 \times 10^{-5}$), the Cl^- ion must be a weaker base than the $CH_3CO_2^-$ ion.

Acid strength:	$HCl > CH_3CO_2H$
Base strength:	$Cl^- < CH_3CO_2^-$

Exercise 11.7

Which of the following ions is the stronger base, $CH_3CO_2^-$ or OH^-?

Solution

According to the data in Table 11.4, acetic acid is a stronger acid than water.

CH_3CO_2H:	$K_a = 1.8 \times 10^{-5}$
H_2O:	$K_a = 1.8 \times 10^{-16}$

The conjugate base of acetic acid must therefore be weaker than the conjugate base of water. This means that the $CH_3CO_2^-$, or acetate, ion is a weaker base than the OH^-, or hydroxide, ion.

The magnitude of K_a can also be used to explain why some compounds that are potential Brønsted acids or bases don't act like acids or bases when they dissolve in water. When its K_a is very large, an acid reacts with water until essentially all of the acid molecules have been consumed. Sulfuric acid ($K_a = 1 \times 10^3$), for example, reacts with water until 99.9% of the H_2SO_4 molecules in a 1 M solution have lost a proton to form HSO_4^- ions.

$$H_2SO_4(aq) + H_2O(l) \longrightarrow H_3O^+(aq) + HSO_4^-(aq)$$

As K_a becomes smaller, the extent to which the acid reacts with water becomes smaller. As long as the K_a for the acid is significantly larger than the K_a for water, the acid will ionize to some extent. Acetic acid, for example, reacts to some extent with water to form H_3O^+ and $CH_3CO_2^-$, or acetate, ions.

$$CH_3CO_2H(aq) + H_2O(l) \longrightarrow H_3O^+(aq) + CH_3CO_2^-(aq)$$

As the value of K_a for the acid approaches the K_a of water, the compound becomes more like water in its acidity. Although it is still a Brønsted acid, it is so weak that we might not be able to detect this acidity in aqueous solution.

Some potential Brønsted acids are so weak that they have values of K_a that are smaller than water's. Ammonia, for example, has a K_a of only 1×10^{-30}. Although NH_3 is a Brønsted acid, because it has the potential to act as a hydrogen-ion donor, there is no evidence of this acidity when NH_3 dissolves in water.

This discussion can be extended to explain an interesting property of strong acids and bases in aqueous solutions. All strong acids and bases seem to have the same strength when dissolved in water, regardless of the value of K_a. This phenomenon is known as the *leveling effect* of water—the tendency of water to limit the strength of strong acids and bases. We can explain this observation by noting that strong acids react strongly with water to form the H_3O^+ ion. As we have already seen, more than 99% of the HCl molecules in hydrochloric acid react with water to form H_3O^+ ions and Cl^- ions

$$HCl(g) + H_2O(l) \longrightarrow H_3O^+(aq) + Cl^-(aq)$$

and more than 99% of the H_2SO_4 molecules in a 1 M solution react with water to form H_3O^+ and HSO_4^- ions.

$$H_2SO_4(aq) + H_2O(l) \longrightarrow H_3O^+(aq) + HSO_4^-(aq)$$

Thus, the strength of strong acids is limited by the strength of the acid (H_3O^+) formed when water molecules pick up an H^+ ion.

Something similar happens in solutions of strong bases. Strong bases react quantitatively with water to form the OH^- ion. Once this has happened, the solution can't be made any more basic. The strength of strong bases is limited by the inherent strength of the base (OH^-) formed when water molecules lose an H^+ ion.

11.11 SIMILARITIES BETWEEN OXIDATION-REDUCTION AND ACID-BASE REACTIONS

In Section 7.14, we concluded that a spontaneous chemical reaction should convert a stronger reducing agent and a stronger oxidizing agent into a weaker oxidizing agent and a weaker reducing agent. A similar argument can be applied to acid-base reactions. Spontaneous acid-base reactions convert the stronger acid and the stronger base into a weaker acid and a weaker base.

$$HCl(aq) + H_2O(l) \longrightarrow H_3O^+(aq) + Cl^-(aq)$$
$$\text{stronger acid} \quad \text{stronger base} \quad\quad \text{weaker acid} \quad \text{weaker base}$$

The only time we don't expect to see an acid-base reaction is when we try to convert a weaker acid and a weaker base into a stronger acid and a stronger base. We don't expect salt to react with water, for example, to produce a mixture of hydrochloric acid and sodium hydroxide.

$$Cl^-(aq) + H_2O(l) \not\longrightarrow HCl(aq) + OH^-(aq)$$
$$\text{weaker base} \quad \text{weaker acid} \quad\quad \text{stronger acid} \quad \text{stronger base}$$

Nor do we expect acetic acid to be very good at ionizing to form H_3O^+ and $CH_3CO_2^-$ ions.

$$CH_3CO_2H(aq) + H_2O(l) \not\longrightarrow H_3O^+(aq) + CH_3CO_2^-(aq)$$
$$\text{weaker acid} \quad\quad \text{weaker base} \quad\quad \text{stronger acid} \quad\quad \text{stronger base}$$

We can therefore use the results of experiments that tell us whether acid-base reactions occur to determine the relative strengths of pairs of acids and bases.

Exercise 11.8

If $NaNH_2$ reacts with water to give an aqueous solution of NaOH and NH_3, which is the stronger base, the NH_2^- or OH^- ions?

$$NaNH_2(s) + H_2O(l) \longrightarrow Na^+(aq) + OH^-(aq) + NH_3(aq)$$

Solution

This reaction involves the net transfer of an H^+ ion from a water molecule acting as an acid to an NH_2^- ion acting as a base.

$$NH_2^-(aq) + H_2O(l) \longrightarrow NH_3(aq) + OH^-(aq)$$
$$\text{base} \qquad\quad \text{acid} \qquad\qquad \text{acid} \qquad\quad \text{base}$$

If the reaction occurs as written, the reactants must include the stronger acid and the stronger base. Thus, the NH_2^- ion is a stronger base than the OH^- ion.

$$NH_2^-(aq) + H_2O(l) \longrightarrow NH_3(aq) + OH^-(aq)$$
$$\text{stronger base} \quad \text{stronger acid} \qquad \text{weaker acid} \quad \text{weaker base}$$

11.12 THE ADVANTAGES OF THE BRØNSTED DEFINITION

The Brønsted definition of acids and bases offers many advantages over the Arrhenius and operational definitions.

1. It expands the list of potential acids to include both positive and negative ions as well as neutral molecules.
2. It expands the list of bases to include any molecule or ion with at least one pair of nonbonding valence electrons.
3. It explains the role of water in acid-base reactions; water accepts H^+ ions (or protons) from acids to form the H_3O^+ ion.
4. It can be expanded to include other solvents besides water.
5. It can be expanded to include reactions that occur in the gas or solid phase.
6. It links acids and bases.
7. It can be used to explain differences between the relative strengths of a pair of acids or a pair of bases.
8. It can be used to explain the relationship between the strength of an acid and the strength of its conjugate base.
9. It can be used to explain the leveling effect of water — the fact that strong acids and bases all have the same strength when dissolved in water.
10. It emphasizes the similarity between oxidation-reduction and acid-base reactions.

Because of these advantages, whenever chemists use the words *acid* or *base* without any further description, it is safe to assume they are referring to a *Brønsted* acid or a *Brønsted* base.

11.13 FACTORS THAT CONTROL THE RELATIVE STRENGTHS OF ACIDS AND BASES

A number of factors influence the probability of having a heart attack. Heart attacks have been said to hit those who smoke more often than those who don't, the elderly more often than the young, those who are overweight more than those who aren't, those who never exercise more often than those who exercise regularly, and so on. It is difficult to identify any one factor as the cause of a particular heart attack. However, it is possible to sort out the influence of these factors, one at a time, by trying to keep as many of the other factors constant as possible.

The same approach can be used to sort out the influence of factors that control the relative strengths of acids and bases. The key is considering each of these factors one at a time. We will look at four properties that influence the strength of acids and bases: (1) polarity, (2) size, (3) charge, and (4) oxidation number. As we look at each, we will assume that the others remain constant.

THE POLARITY OF THE X—H BOND

When all other factors are kept constant, acids become stronger as the X—H bond becomes more polar. The polarity of the second-row hydrides, for example, becomes larger in the following order.

$$CH_4 < NH_3 < H_2O < HF$$

| ΔEN | 0.4 | 0.8 | 1.2 | 1.8 |

The acidity of these hydrides mirrors this trend perfectly. HF is by far the strongest of these acids, and CH_4 is one of the weakest Brønsted acids known.

HF: $K_a = 7.2 \times 10^{-4}$

H_2O: $K_a = 1.8 \times 10^{-16}$

NH_3: $K_a = 1 \times 10^{-30}$

CH_4: $K_a = 1 \times 10^{-49}$

Polarity is an important factor in determining the relative strength of an acid because an X—H bond has to be broken in order for X^- and H^+ ions to be formed when the compound acts as an acid. The more polar this bond, the more easily it is broken to form X^- and H^+ ions. Thus, the more polar the bond, the stronger the acid.

THE SIZE OF THE X ATOM

At first glance, we might expect that acids such as HF, HCl, HBr, and HI would become weaker down a column of the periodic table, because the X—H bond becomes less polar. Experimentally, we find that these acids actually become stronger.

HF: $K_a = 7.2 \times 10^{-4}$

HCl: $K_a = 1 \times 10^6$

HBr: $K_a = 1 \times 10^9$

HI: $K_a = 3 \times 10^9$

Why does the size of the X atom influence the acidity of the X—H bond? Acids become stronger as the X—H bond becomes weaker, and bonds generally become weaker as the atoms get larger. The acidity data given above mirror the tendency of the X—H bond dissociation enthalpy to become smaller as the size of the X atom becomes larger.

Acid	Bond Dissociation Enthalpy (kJ/mol)
HF	569
HCl	431
HBr	370
HI	300

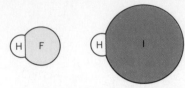

FIG. 11.4 HI is a much stronger acid than HF because of the difference in size between the iodine and fluorine atoms. The covalent radius of iodine is more than twice as large as that of fluorine, which means that the charge density on the F^- ion is very much larger than the charge density on the I^- ion. As a result, it is more difficult to separate the atoms in an HF molecule to form H^+ and F^- ions than it is to separate the atoms in an HI molecule to form H^+ and I^- ions.

The obvious question is: "Why does the X—H bond become weaker as the X atom becomes larger?" The simplest way of explaining this trend is to invoke the concept of charge density. The F^-, Cl^-, Br^-, and I^- ions all have the same charge. But the volume over which this charge is distributed becomes larger as we go down the column. Thus, the charge density becomes smaller as we go down this column, as shown in Figure 11.4. As a result, the force of attraction between the H^+ and X^- ions becomes smaller as well.

The trend in the bond dissociation enthalpies for HF through HI mirrors the trend in lattice energies for LiF through LiI. A relatively small F^- ion has a stronger force of attraction for either an H^+ ion or an Li^+ ion than for a relatively large I^- ion.

THE CHARGE ON THE ACID OR BASE

When all other factors are kept constant, acids become weaker as the negative charge on the ion or molecule becomes larger. Neutral sulfuric acid molecules lose one H^+ ion quite easily to form HSO_4^- ions. (More than 99% of the H_2SO_4 molecules in a 1 M solution lose one proton to form an HSO_4^- ion.)

$$H_2SO_4(aq) + H_2O(l) \longrightarrow H_3O^+(aq) + \mathbf{HSO_4^-(aq)}$$

But the HSO_4^- ion is a much weaker acid. Only about 10% of these ions go on to lose a second H^+ ion.

$$HSO_4^-(aq) + H_2O(l) \longrightarrow H_3O^+(aq) + \mathbf{SO_4^{2-}(aq)}$$

When all other factors are kept constant, bases become stronger as the negative charge becomes larger. The $H_2PO_4^-$ ion is a relatively weak base, the HPO_4^{2-} ion is a good base, and the PO_4^{3-} ion is a very strong base.

These trends can also be explained with the concept of charge density. There is a strong force of attraction between the negative charge on a PO_4^{3-} ion and the positive charge on an H^+ ion. As a result, this ion is a reasonably good base. The charge density is smaller on the HPO_4^{2-} ion, however, because this ion carries a smaller charge. It isn't surprising that the HPO_4^{2-} ion is therefore a weaker base. The charge density on the $H_2PO_4^-$ ion is even smaller, so this ion is an even weaker base.

$$\text{Basicity:} \qquad PO_4^{3-} > HPO_4^{2-} > H_2PO_4^-$$

The converse of this argument can be used to explain how the strengths of acids vary with the charge on the ion or molecule. It is much easier to remove a positive H^+ ion from a neutral H_3PO_4 molecule than it is to remove an H^+ ion from a negatively charged $H_2PO_4^-$ ion. It is even harder to remove an H^+ ion from a negatively charged HPO_4^{2-} ion. As a result, these acids become weaker in the following order.

$$\text{Acidity:} \qquad H_3PO_4 > H_2PO_4^- > HPO_4^{2-}$$

THE OXIDATION STATE OF THE CENTRAL ATOM

Once again, when all other factors are held constant, there is a gradual increase in the acidity of oxyacids as the oxidation number of the central atom becomes larger. H_2SO_4 is a much stronger acid than H_2SO_3, and HNO_3 is a much stronger acid than HNO_2. This trend is easiest to see in the four oxyacids of chlorine.

Oxyacid	K_a	Oxidation Number of Chlorine
HOCl	2.9×10^{-8}	$+1$
HOClO	1.1×10^{-2}	$+3$
HOClO$_2$	5.0×10^2	$+5$
HOClO$_3$	1×10^3	$+7$

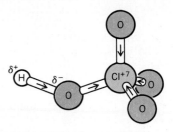

When you consider that all of the atoms are the same size and that the compounds all have the same charge, it might seem difficult to explain a factor of 10^{11} difference in the value of K_a for hypochlorous acid (HOCl) and perchloric acid (HClO$_4$ or HOClO$_3$). This difference can be traced to a point that was introduced in Section 8.6. There is only one value for the electronegativity of an element, and yet the tendency of an atom to draw electrons toward itself becomes larger as the oxidation number of the atom becomes larger.

As the oxidation number of the chlorine atom becomes larger, the atom becomes more electronegative. This tends to draw electrons away from the oxygen atoms that surround the chlorine, thereby making the oxygen atoms more electronegative as well, as shown in Figure 11.5. As the oxygen atoms become more electronegative, the O—H bond becomes more polar, and the compound becomes more acidic.

FIG. 11.5 The factor of 10^{11} difference between the K_a of hypochlorous acid (HClO or HOCl) and the K_a of perchloric acid (HClO$_4$ or HOClO$_3$) is the result of differences between the effective electronegativities of the chlorine atoms in these compounds. The increase in the oxidation number of the chlorine atom from $+1$ in HOCl to $+7$ in HOClO$_3$ results in an increase in the effective electronegativity of this atom. As the chlorine atom becomes more electronegative, it pulls electrons in the Cl—O bonds toward itself. The oxygen in the O—H bond can partially compensate for this by pulling the electrons in this bond toward itself. The net result is an increase in the polarity of the O—H bond, which leads to an increase in the acidity of the compound.

Exercise 11.9

For each of the following pairs of compounds, predict which compound should be the stronger acid.

(a) H$_2$O or NH$_3$ (b) NH$_3$ or NH$_2^-$ (c) NH$_3$ or PH$_3$ (d) HNO$_3$ or HNO$_2$

Solution

(a) H$_2$O is a stronger acid ($K_a = 1.8 \times 10^{-16}$) than NH$_3$ ($K_a = 1 \times 10^{-30}$), because the O—H bond is more polar than the N—H bond. Note that oxygen and nitrogen are roughly the same size (since both are second-row elements), that the two compounds have the same charge, and that the comparison does not involve compounds of the same element with different oxidation numbers.

(b) NH$_3$ is a much stronger acid than the NH$_2^-$ ion, because it is easier to remove an H$^+$ ion from a neutral NH$_3$ molecule than to remove a positively charged H$^+$ ion from a negatively charged NH$_2^-$ ion.

(c) PH$_3$ is a stronger acid than NH$_3$, because compounds become more acidic as the size of the atom holding the hydrogen atom becomes larger and the X—H bond becomes correspondingly weaker.

(d) HNO$_3$ is a stronger acid than HNO$_2$, because both are oxyacids (HONO$_2$ and HONO, respectively) and the nitrogen atom in HNO$_3$ is more electronegative than the nitrogen atom in HNO$_2$.

The relative strengths of Brønsted bases can often be predicted from the relative strengths of their conjugate acids combined with the general rule that the stronger of a pair of acids always has the weaker conjugate base.

Exercise 11.10

For each of the following pairs of compounds, predict which compound should be the stronger base.

(a) OH^- or NH_2^- (b) NH_3 or NH_2^- (c) NH_2^- or PH_2^- (d) NO_3^- or NO_2^-

Solution

(a) We can start by noting that the NH_2^- ion is the conjugate base of ammonia and the OH^- ion is the conjugate base of water. We can then note that H_2O is a stronger acid than NH_3, which means that the conjugate base of H_2O must be weaker than the conjugate base of NH_3. In other words, the NH_2^- ion is a stronger base than the OH^- ion.

(b) The NH_2^- ion is a stronger base than NH_3. There are two ways of reaching this conclusion. One is to note that the force of attraction between a negatively charged NH_2^- ion and an H^+ ion must be greater than the force of attraction between a neutral NH_3 molecule and a positively charged H^+ ion. The other is to note that NH_4^+ is a stronger acid than NH_3 and the conjugate base of NH_4^+ must therefore be weaker than the conjugate base of NH_3.

(c) The NH_2^- ion is a stronger base than the PH_2^- ion, because PH_3 is a stronger acid than NH_3.

(d) The NO_2^- ion is a stronger base than the NO_3^- ion, because HNO_3 is a stronger acid than HNO_2.

11.14 pH AS A MEASURE OF THE CONCENTRATION OF THE H_3O^+ ION

Pure water is both a weak acid and a weak base. By itself, water forms only a negligibly small number of the H_3O^+ and OH^- ions that characterize aqueous solutions of stronger acids and bases.

$$H_2O(l) + H_2O(l) \rightleftharpoons H_3O^+(aq) + OH^-(aq)$$
$$\text{base} \qquad\quad \text{acid} \qquad\qquad \text{acid} \qquad\quad \text{base}$$

Measurements of the ability of water to conduct electricity suggest that pure water at 25°C contains only 1.0×10^{-7} moles per liter of the H_3O^+ and OH^- ions.

$$[H_3O^+] = [OH^-] = 1.0 \times 10^{-7}\,M \qquad \text{(at 25°C)}$$

Although values of $1.0 \times 10^{-7}\,M$ for the concentrations of the H_3O^+ and OH^- ions in water have been etched into the memory of generations of chemistry students, it is useful to remember that the concentrations of these ions depend on the temperature of the solution. At 45°C the concentrations are about twice as large; at 80°C they are 5 times as large; and at 120°C they are 10 times as large. The concentrations of these ions in superheated steam at temperatures on the order of 600° to 700°C are large enough to eventually corrode metal pipes.

What happens when we add a strong acid such as HCl to water? Obviously, the H_3O^+ ion concentration becomes larger.

$$HCl(aq) + H_2O(aq) \longrightarrow H_3O^+(aq) + Cl^-(aq)$$

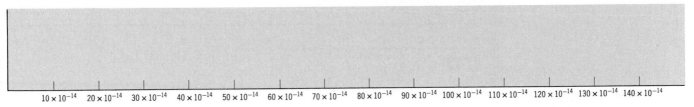

$$10 \times 10^{-14} \quad 20 \times 10^{-14} \quad 30 \times 10^{-14} \quad 40 \times 10^{-14} \quad 50 \times 10^{-14} \quad 60 \times 10^{-14} \quad 70 \times 10^{-14} \quad 80 \times 10^{-14} \quad 90 \times 10^{-14} \quad 100 \times 10^{-14} \quad 110 \times 10^{-14} \quad 120 \times 10^{-14} \quad 130 \times 10^{-14} \quad 140 \times 10^{-14}$$

FIG. 11.6 The range of H_3O^+ ion concentrations in aqueous solution is so large it would take a graph 10^{14} millimeters—or 62 million miles—long to reflect changes across the entire range. By calculating the pH of the solution, however, we can compress these data into a range from slightly less than 1 to slightly more than 14.

At the same time, the OH^- ion concentration becomes smaller, because the H_3O^+ ions produced in this reaction neutralize some of the OH^- ions in water.

$$H_3O^+(aq) + OH^-(aq) \longrightarrow 2\,H_2O(l)$$

Experimentally, we find that the product of the concentrations of the H_3O^+ and OH^- ions is constant, no matter how much acid or base is added to water. In pure water, at 25°C, the product of the concentrations of these ions is 1.0×10^{-14}.

$$[H_3O^+][OH^-] = [1.0 \times 10^{-7}][1.0 \times 10^{-7}] = 1.0 \times 10^{-14}$$

This constant is known as the **water-dissociation equilibrium constant, K_w.**

As the H_3O^+ ion concentration becomes larger, the concentration of the OH^- ion becomes smaller at the same rate, so that the product of the concentrations of these ions is always 1.0×10^{-14}. In 1.0 M hydrochloric acid, the H_3O^+ ion concentration is 1.0 M, and the concentration of the OH^- ion is only 1.0×10^{-14} M.

$$K_w = [H_3O^+][OH^-]$$
$$1.0 \times 10^{-14} = [1.0][1.0 \times 10^{-14}]$$

In 1.0 M sodium hydroxide, the OH^- ion concentration is 1.0 M, and the concentration of the H_3O^+ ion is only 1.0×10^{-14}.

$$K_w = [H_3O^+][OH^-]$$
$$1.0 \times 10^{-14} = [1.0 \times 10^{-14}][1.0]$$

The range of concentrations of the H_3O^+ and OH^- ions in aqueous solution is so large it is difficult to work with. Let's assume, for example, that we want to follow the concentration of the H_3O^+ ion as we add acids or bases to water. Furthermore, let's assume that we want to plot our data on a graph sensitive enough to register the smallest H_3O^+ ion concentrations (see Figure 11.6).

To do this we have to use a graph with intervals of 10^{-14} M. If we construct this graph so that each change of 10^{-14} M in the concentration of the H_3O^+ ion corresponds to 1 millimeter on the horizontal axis of the graph, we will need a piece of graph paper 10^{14} millimeters long to cover the entire range of H_3O^+ ion concentrations. Unfortunately, 10^{14} millimeters is about 62 million miles, or about two-thirds of the distance between the earth and the sun.

In 1909, the Danish biochemist S. P. L. Sørenson suggested reporting the concentration of the H_3O^+ ion on a logarithmic scale, which he named the **pH scale.** Because the H_3O^+ ion concentration in water is almost always smaller than 1, the log of these concentrations is a negative number. To avoid having to constantly work with negative numbers, Sørenson defined pH as the negative of the log of the H_3O^+ ion concentration.

$$\mathbf{pH = -log[H_3O^+]}$$

Exercise 11.11

Calculate the pH of a 0.10 M acetic acid solution if 1.34% of the CH_3CO_2H molecules in this solution have ionized to form H_3O^+ and $CH_3CO_2^-$ ions.

Solution

Acetic acid reacts with water to form the H_3O^+ and $CH_3CO_2^-$ ions.

$$CH_3CO_2H(aq) + H_2O(l) \longrightarrow H_3O^+(aq) + CH_3CO_2^-(aq)$$

The solution was initially 0.10 M in CH_3CO_2H, but 1.34% of the CH_3CO_2H molecules have dissociated. Since 1.34% of 0.10 M is 0.00134 M, the concentration of the H_3O^+ and $CH_3CO_2^-$ ions in this solution is 0.00134 M. The pH of the solution is therefore 2.87.

$$\begin{aligned} pH &= -\log[H_3O^+] \\ &= -\log(0.00134) \\ &= -(-2.87) = \mathbf{2.87} \end{aligned}$$

Exercise 11.12

Calculate the H_3O^+ ion concentration in Pepsi Cola if the pH is 2.46.

Solution

The pH of a solution is the negative of the log of the H_3O^+ ion concentration.

$$pH = -\log[H_3O^+]$$

Rearranging this equation gives the following.

$$\log[H_3O^+] = -pH$$

Since the log of a number is the exponent to which 10 must be raised to give that number, the H_3O^+ ion concentration is 10^{-pH}.

$$[H_3O^+] = 10^{-pH}$$

Thus, the concentration of the H_3O^+ ion is the antilog of the pH, after the sign has been changed. In this case, the H_3O^+ ion concentration is 10 raised to the power of −2.46.

$$[H_3O^+] = 10^{-2.46} = \mathbf{0.00347}$$

To two significant figures, the H_3O^+ ion concentration in Pepsi Cola is 0.0035 M.

TABLE 11.5

The Relationship Between the H_3O^+ Ion Concentration and the pH of an Aqueous Solution

H_3O^+ Ion Concentration (M)	pH	
1.0	0	
1.0×10^{-1}	1	
1.0×10^{-2}	2	
1.0×10^{-3}	3	ACID
1.0×10^{-4}	4	
1.0×10^{-5}	5	
1.0×10^{-6}	6	
1.0×10^{-7}	7	NEUTRAL
1.0×10^{-8}	8	
1.0×10^{-9}	9	
1.0×10^{-10}	10	
1.0×10^{-11}	11	BASE
1.0×10^{-12}	12	
1.0×10^{-13}	13	
1.0×10^{-14}	14	

The concept of pH compresses the range of H_3O^+ ion concentrations into a scale that is much easier to handle.

As the H_3O^+ ion concentration decreases from roughly 10^0 to 10^{-14}, the pH of the solution increases from 0 to 14.

The relationship between the H_3O^+ ion concentration and the pH of a solution is shown in Table 11.5.

If the concentration of the H_3O^+ ion in pure water at 25°C is 1.0×10^{-7} M, then the pH of pure water is 7.

$$\begin{aligned} pH &= -\log[H_3O^+] \\ &= -\log(1.0 \times 10^{-7}) \\ &= -(-7) = 7 \end{aligned}$$

When the pH of a solution is less than 7, the solution is acidic. When the pH is more than 7, the solution is basic.

Acidic: **pH < 7**

Basic: **pH > 7**

To decide whether a solution is acidic or basic, we need a way of determining its pH. Historically, this determination was made by use of a family of compounds known as acid-base indicators. To a large extent, these indicators have been replaced by pH meters, which are more accurate.

The actual measuring device in a pH meter is an electrode that consists of a resin-filled tube with a thin glass bulb at one end. The glass bulb is typically filled with a 0.1 M HCl solution in contact with a silver wire coated with a thin layer of silver chloride. The electrode is immersed in the solution to be measured, and the difference between the concentration of the H_3O^+ ion within the glass bulb and the H_3O^+ ion concentration in the surrounding solution gives rise to an electrical potential, measured in millivolts, whose magnitude is directly proportional to the difference between the two H_3O^+ ion concentrations.

Four points concerning pH meters should be kept in mind. First, the electrode is subject to errors of measurement in either highly acidic or highly basic solutions. Second, each electrode has a characteristic speed of response. It takes a certain amount of time for the electrode to respond to changes in pH. Third, the thin glass bulb at the tip of the electrode is easily broken if the electrode bumps against the bottom or sides of a beaker. Fourth, and perhaps most importantly, these instruments can only make relative measurements. They can't determine the absolute pH of a solution. All they can do is compare the pH of a solution with a known standard. Thus, pH meters must be calibrated against a solution of known pH before they are used.

Acid-base indicators are still routinely used in the laboratory for rough pH measurements. These indicators are weak acids or weak bases that change color

A digital pH meter testing the pH of a neutral solution.

The glass electrodes of a pH meter.

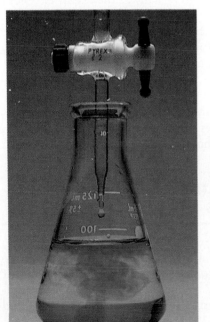

Phenolphthalein is an acid–base indicator that turns from colorless to pink as the pH increases from 8.0 to 10.0.

Red cabbage contains naturally occurring acid–base indicators that turn pink in the presence of acid and green in the presence of base.

TABLE 11.6

Acid-Base Indicators

Indicator	pH Range	Color Change
methyl violet	0.0–1.6	yellow to blue-violet
thymol blue	1.2–2.8	red to yellow
	8.0–9.6	yellow to blue
bromophenol blue	3.0–4.6	yellow to blue-violet
methyl orange	3.2–4.4	red to yellow-orange
bromocresol green	4.0–5.6	yellow to blue
methyl red	4.4–6.2	red to yellow
litmus	5–8	pink to blue
bromocresol purple	5.2–6.8	yellow to purple
bromophenol red	5.2–6.8	yellow to red
bromothymol blue	6.2–7.6	yellow to blue
cresol red	7.2–8.8	yellow to red
thymol blue	8.0–9.6	yellow to blue
phenolphthalein	8.0–10.0	colorless to pink
alizarin yellow	10.0–12.0	yellow to red-violet

Iced tea contains a variety of naturally occurring acid–base indicators that cause it to turn color when a source of citric acid, such as lemon juice, is added.

By mixing indicators it is possible to produce a universal indicator pH paper.

when they gain or lose an H^+ ion. The oldest examples of acid-base indicators are litmus and syrup of violets, which have been used for this purpose since Boyle described their behavior in 1661. Litmus turns pink in solutions whose pH is below 5 and blue in solutions whose pH is above 8. Another common indicator is phenolphthalein, which is colorless when the pH is less than 8.0 and pink when the pH is above 10.0.

Less well-recognized indicators are iced tea and red cabbage. There is a subtle change in the color of iced tea when the citric acid in lemon juice is added, and red cabbage ground with a small quantity of water in a blender makes an excellent acid-base indicator.

More than 200 acid-base indicators, which change color over a broad range of pH values, are known. A few common indicators, the pH range over which they change color, and their color changes are given in Table 11.6.

By mixing two or more of these compounds, we can develop indicators for use over almost the entire range of pH. A so-called universal indicator can be made from equal volumes of methyl red, methyl orange, phenolphthalein, and bromothymol blue. As the pH of the solution increases, the indicator changes color from red to orange to yellow to green to blue and finally to purple.

pH AS A MEASURE OF THE RELATIVE
11.15 STRENGTHS OF ACIDS AND BASES

Because the pH of a solution is proportional to the strength of an acid or base, measurements of the pH of dilute solutions are often good indicators of the relative strengths of acids and bases. Values of the pH of 0.10 *M* solutions of a number of common acids and bases are given in Table 11.7. These data demonstrate quantitatively many of the ideas about the relative strengths of acids and bases developed earlier in this chapter.

TABLE 11.7

pH of 0.10 M Solutions of Common Acids and Bases

Compound	pH
HCl (hydrochloric acid)	1.1
H_2SO_4 (sulfuric acid)	1.2
$NaHSO_4$ (sodium hydrogen sulfate)	1.4
H_2SO_3 (sulfurous acid)	1.5
H_3PO_4 (phosphoric acid)	1.5
$FeCl_3$ [iron(III) chloride]	2.0
$C_6H_8O_7$ (citric acid)	2.2
CH_3CO_2H (acetic acid)	2.9
$AlCl_3$ (aluminum chloride)	3.0
H_2CO_3 (carbonic acid)	3.8 (saturated solution)
$Cu(NO_3)_2$ [copper(II) nitrate]	4.0
H_2S (hydrogen sulfide)	4.1
NaH_2PO_4 (sodium dihydrogen phosphate)	4.4
NH_4Cl (ammonium chloride)	4.6
HCN (hydrocyanic acid)	5.1
H_3BO_3 (boric acid)	5.2
Na_2SO_4 (sodium sulfate)	6.1
NaCl (sodium chloride)	6.4
$Na_3C_6H_5O_7$ (sodium citrate)	8.2
$NaCH_3CO_2$ (sodium acetate)	8.4
$NaHCO_3$ (sodium bicarbonate)	8.4
K_2CrO_4 (potassium chromate)	9.2
Na_2HPO_4 (sodium hydrogen phosphate)	9.3
Na_2SO_3 (sodium sulfite)	9.8
MgO (magnesium oxide, magnesia)	10.5 (saturated solution)
NaCN (sodium cyanide)	11.0
NH_3 (aqueous ammonia)	11.1
Na_2CO_3 (sodium carbonate)	11.6
Na_3PO_4 (sodium phosphate)	12.0
CaO (calcium oxide, lime)	12.4 (saturated solution)
NaOH (sodium hydroxide, lye)	13.0

The data in Table 11.7 substantiate the claim that the stronger of a pair of acids has the weaker conjugate base. According to these data, hydrochloric acid is a stronger acid than acetic acid, but the acetate ion is a stronger base than the chloride ion.

Acid	Conjugate Base
0.1 M HCl: pH = 1.1 (stronger acid)	0.1 M NaCl: pH = 6.4 (weaker base)
0.1 M CH_3CO_2H: pH = 2.4 (weaker acid)	0.1 M $NaCH_3CO_2$: pH = 8.4 (stronger

Hydrochloric acid is a strong acid, and sodium hydroxide is a strong base. When these two reagents are mixed, they should form a solution that consists of a much weaker acid and a much weaker base.

$$HCl(aq) + NaOH(aq) \longrightarrow NaCl(aq) + H_2O(l)$$

| stronger | stronger | weaker | weaker |
| acid | base | base | acid |

Data from Table 11.7 comparing the pH of 0.1 M solutions of hydrochloric acid, sodium hydroxide, and sodium chloride confirm this prediction.

0.1 M HCl:	pH = 1.1	(HCl is a strong acid.)
0.1 M NaCl:	pH = 6.4	(H_2O is a weak acid; Cl^- is a weak base.)
0.1 M NaOH:	pH = 13.0	(OH^- is a strong base.)

Data in this table also show that the weaker of a pair of bases has the stronger conjugate acid. According to these data, ammonia is a weaker base than sodium hydroxide. But the ammonium ion is a stronger acid than water.

Base	*Conjugate Acid*
0.1 M NaOH: pH = 13.0 (stronger base)	H_2O: pH = 7 (weaker acid)
0.1 M NH_3: pH = 11.1 (weaker base)	0.1 M NH_4Cl: pH = 4.6 (stronger acid)

In Section 11.13, we concluded that acids become weaker and bases become stronger as the negative charge on the ion becomes larger. The pH data in Table 11.7 for H_3PO_4, NaH_2PO_4, Na_2HPO_4, and Na_3PO_4 support this hypothesis. The first member of this family is a strong acid, the second is a weak acid, the third is a weak base, and the last is a strong base.

0.1 M H_3PO_4:	pH = 1.5
0.1 M NaH_2PO_4:	pH = 4.4
0.1 M Na_2HPO_4:	pH = 9.3
0.1 M Na_3PO_4:	pH = 12.0

We see the same effect when we compare the pH of derivatives of H_2SO_4 and H_2CO_3. In each case, the compound becomes less acidic (or more basic) as the negative charge on the molecule increases. There is a significant difference between these two families, however. Sulfuric acid is such a strong acid that both H_2SO_4 and the HSO_4^- ion are acids in water. Carbonic acid is a much weaker acid—so weak that solutions of the HCO_3^- ion in water are actually basic.

0.1 M H_2SO_4:	pH = 1.2	saturated H_2CO_3:	pH = 3.8
0.1 M $NaHSO_4$:	pH = 1.4	0.1 M $NaHCO_3$:	pH = 8.4
0.1 M Na_2SO_4:	pH = 6.1	0.1 M Na_2CO_3:	pH = 11.6

According to the discussion in Section 11.13, oxides such as MgO and CaO should be stronger bases than hydroxides such as NaOH, because the O^{2-} ion should be a better base than the OH^- ion. But the data in Table 11.7 seem to suggest that NaOH is a stronger base than either MgO or CaO.

Saturated MgO:	pH = 10.5
Saturated CaO:	pH = 12.4
0.1 M NaOH:	pH = 13.0

These data are easy to explain. Section 7.14 noted that the lattice energy for compounds such as MgO and CaO is so large that they are almost insoluble in water. A saturated solution of MgO, for example, has a concentration of only 0.00015 M. As a result, the pH of saturated solutions of these bases is smaller than the pH of 0.1 M NaOH.

Three compounds in Table 11.7 seem to be far more acidic than we might expect.

$$0.1 \ M \ \text{FeCl}_3: \qquad \text{pH} = 2.0$$
$$0.1 \ M \ \text{AlCl}_3: \qquad \text{pH} = 3.0$$
$$0.1 \ M \ \text{Cu(NO}_3)_2: \qquad \text{pH} = 4.0$$

How can we explain the acidity of these compounds? The Cl^- and NO_3^- ions are both very weak bases. The acidity of these solutions must therefore result from the behavior of the Fe^{3+}, Al^{3+}, and Cu^{2+} ions.

The Fe^{3+}, Al^{3+}, and Cu^{2+} ions can't possibly be Brønsted acids, or proton donors, by themselves. They can only act as proton donors by influencing the ability of the neighboring water molecules to give up H^+ ions. They do this by first forming covalent bonds to six water molecules to form a **complex ion**, as shown in Figure 11.7.

$$\text{Al}^{3+}(aq) + 6 \ \text{H}_2\text{O}(l) \longrightarrow [\text{Al(H}_2\text{O})_6]^{3+}(aq)$$
$$\text{Fe}^{3+}(aq) + 6 \ \text{H}_2\text{O}(l) \longrightarrow [\text{Fe(H}_2\text{O})_6]^{3+}(aq)$$
$$\text{Cu}^{2+}(aq) + 6 \ \text{H}_2\text{O}(l) \longrightarrow [\text{Cu(H}_2\text{O})_6]^{2+}(aq)$$

Water molecules covalently bound to one of these metal ions are more acidic than normal. Thus, reactions such as the following occur.

$$[\text{Al(H}_2\text{O})_6]^{3+}(aq) + \text{H}_2\text{O}(l) \longrightarrow [\text{Al(H}_2\text{O})_5(\text{OH})]^{2+}(aq) + \textbf{H}_3\textbf{O}^+\textbf{(aq)}$$
$$[\text{Fe(H}_2\text{O})_6]^{3+}(aq) + \text{H}_2\text{O}(l) \longrightarrow [\text{Fe(H}_2\text{O})_5(\text{OH})]^{2+}(aq) + \textbf{H}_3\textbf{O}^+\textbf{(aq)}$$

These reactions give rise to a net increase in the H_3O^+ ion concentration in the solutions, thereby making the solutions acidic.

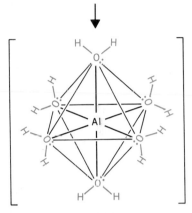

FIG. 11.7 The Al^{3+} ion is a Lewis acid, or electron-pair acceptor, that can combine with six H_2O molecules, which act as Lewis bases, or electron-pair donors, to form a complex ion with the formula $[\text{Al(H}_2\text{O})_6]^{3+}$.

Exercise 11.13

It has been estimated that as many as 25% of the books in the Library of Congress are in brittle condition, in part because of the acidity of the paper on which they were printed. Explain how adding aluminum sulfate to paper when it is manufactured makes paper acidic.

Solution

At first glance, $\text{Al}_2(\text{SO}_4)_3$ would appear to be neither an acid nor a base. H_2SO_4 is much too strong an acid for us to believe that the SO_4^{2-} ion acts as a base toward water. Al^{3+} ions, however, react with water to form $[\text{Al(H}_2\text{O})_6]^{3+}$ complex ions, in which the H_2O molecules bound to the Al^{3+} ion are much more acidic than normal.

As many as 25% of the books in the general collection of the Library of Congress are in brittle condition. Both inherent acidity and environmental pollution threaten the collection's stability.

11.16 THE LEWIS DEFINITIONS OF ACIDS AND BASES

The Brønsted definition of bases rests on the ability of a substance to accept an H^+ ion from an acid. The OH^- and CH_3CO_2^- ions, for example, are Brønsted bases because they contain a pair of nonbonding electrons that can be used to form a covalent bond to an H^+ ion.

$$H^+ \curvearrowleft \ddot{\underset{\cdot\cdot}{O}}-H^- \longrightarrow H-\ddot{\underset{\cdot\cdot}{O}}-H$$

base   conjugate acid

$$CH_3C \overset{\ddot{O}\!:}{\underset{\ddot{\underset{\cdot\cdot}{O}}\!:\curvearrowright H^+}{\diagup}} \longrightarrow CH_3C \overset{\ddot{O}\!:}{\underset{\ddot{\underset{\cdot\cdot}{O}}-H}{\diagup}}$$

base   conjugate acid

In essence, the Brønsted model expanded the number of potential bases to include any ion or molecule that can accept an H^+ ion the way an OH^- ion does.

In 1923, G. N. Lewis proposed a model that greatly expanded the number of potential acids by defining an acid as any ion or molecule that can act like an H^+ ion. Lewis suggested that there was another way of looking at the reaction between H^+ and OH^- ions.

$$H^+ + :\ddot{\underset{\cdot\cdot}{O}}-H^- \longrightarrow H_2O$$

In the Brønsted model, the OH^- ion is the active species in this reaction—it accepts an H^+ ion to form a covalent bond. In the Lewis model, the H^+ ion is the active species—it accepts a pair of electrons from the OH^- ion to form a covalent bond.

In the Lewis theory of acid-base reactions, bases donate pairs of electrons and acids accept pairs of electrons.

> A *Lewis acid* is any substance, such as the H^+ ion, that can accept a pair of nonbonding electrons. A Lewis acid is an *electron-pair acceptor*.

> A *Lewis base* is any substance, such as the OH^- ion, that can donate a pair of nonbonding electrons. A Lewis base is an *electron-pair donor*.

One of the advantages of the Lewis model is the way it complements the model of oxidation-reduction reactions introduced in Chapter 7. Oxidation-reduction reactions involve a transfer of electrons from one atom to another, with a net change in the oxidation number of one or more atoms.

$$Na\cdot + \cdot\ddot{\underset{\cdot\cdot}{C}l}\!: \longrightarrow [Na^+][:\ddot{\underset{\cdot\cdot}{C}l}\!:^-]$$

The Lewis theory suggests that acids react with bases to share a pair of electrons, with no change in the oxidation numbers of any atoms.

$$H^+ + :\ddot{\underset{\cdot\cdot}{O}}-H^- \longrightarrow H-\ddot{\underset{\cdot\cdot}{O}}-H$$

Any chemical reaction, no matter how simple or complex, can be sorted into one or the other of these classes: electrons are either transferred from one atom to another, or the atoms come together to share a pair of electrons.

The principal advantage of the Lewis theory is the way it expands the number of acids and therefore the number of acid-base reactions. The Lewis theory expands the category of acids to include any ion or molecule that can accept a pair of nonbonding valence electrons.

In the preceding section, we concluded that Al^{3+} ions form bonds to six water molecules to give a complex ion.

$$Al^{3+}(aq) + 6 H_2O(l) \longrightarrow [Al(H_2O)_6]^{3+}(aq)$$

This reaction is an example of a Lewis acid-base reaction. The Lewis structure of water suggests that the water molecule has nonbonding pairs of valence electrons and can therefore act as a Lewis base. The electron configuration of the Al^{3+} ion suggests that this ion has empty $3s$, $3p$, and $3d$ orbitals that can be used to hold pairs of nonbonding electrons donated by neighboring water molecules.

$$Al^{3+} = [Ne] \, 3s^0 \, 3p^0 \, 3d^0$$

Thus, the $[Al(H_2O)_6]^{3+}$ ion is formed when an Al^{3+} ion acting as a Lewis acid picks up six pairs of electrons from neighboring water molecules acting as Lewis bases to give an **acid-base complex,** or, in this case, a complex ion.

$$\underset{\text{Lewis acid}}{Al^{3+}(aq)} + \underset{\text{Lewis base}}{6 H_2O(l)} \longrightarrow [Al(H_2O)_6]^{3+}(aq)$$

Lewis acid-base complexes play a vital role in nature. Oxygen is carried from the lungs to the body in the form of a Lewis acid-base complex between O_2 molecules and an Fe^{3+} ion in either hemoglobin or myoglobin (see Figure 11.8). The CO_2 given off when O_2 is used to burn sugars or other carbohydrates is carried through the body to the lungs in the form of a Lewis acid-base complex with the same Fe^{3+} ion found in hemoglobin and myoglobin. The nitrogenase enzymes that catalyze the reduction of N_2 to NH_3 operate by first forming a Lewis acid-base complex between the N_2 molecule and a transition metal ion. Cytochromes, important components in the electron transport process by which the energy given off when carbohydrates are burned is captured, also operate by forming Lewis acid-base complexes. Indeed, any protein that carries a transition metal ion, such as the Cu^{2+}, Fe^{3+}, Mn^{2+}, or Zn^{2+} ions, operates by forming a Lewis acid-base complex.

Lewis acid-base complex chemistry plays an important role in industry as well. The processing of both black and white and color photographs is based on the fact that insoluble AgCl or AgBr salts dissolve in thiosulfate ($S_2O_3^{2-}$), or fixer, to form soluble complex ions.

$$AgCl(s) + 2 S_2O_3^{2-}(aq) \longrightarrow [Ag(S_2O_3)_2]^{3-}(aq) + Cl^-(aq)$$

In the course of this reaction, the Ag^+ ion acts as a Lewis acid, or electron-pair acceptor, and the $S_2O_3^{2-}$ ion acts as a Lewis base, or electron-pair donor.

$$\underset{\text{Lewis acid}}{Ag^+} + \underset{\text{Lewis base}}{2 S_2O_3^{2-}} \longrightarrow [Ag(S_2O_3)_2]^{3-}$$

Another industrial example of Lewis acid-base chemistry is provided by the use of lime to absorb, or scrub, SO_2 and SO_3 fumes emitted from smokestacks. This reaction can be described by the following equation.

$$CaO(s) + SO_2(g) \longrightarrow CaSO_3(s)$$

What actually happens is a reaction between an O^{2-} ion acting as an electron-pair donor, or Lewis base, and an SO_2 molecule acting as an electron-pair acceptor, or Lewis acid, to form the SO_3^{2-} ion, as shown in Figure 11.9.

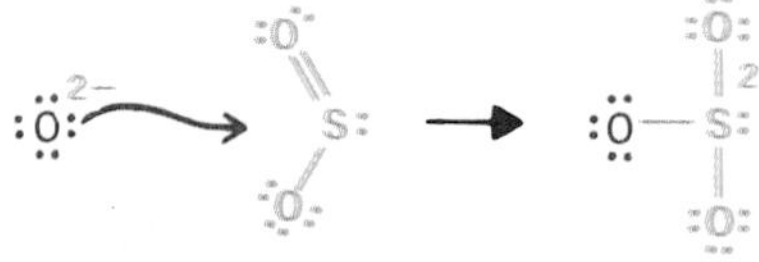

When the shape of molecules was first discussed in Chapter 8, boron trifluoride was given as an example of a compound with a trigonal planar geometry. That

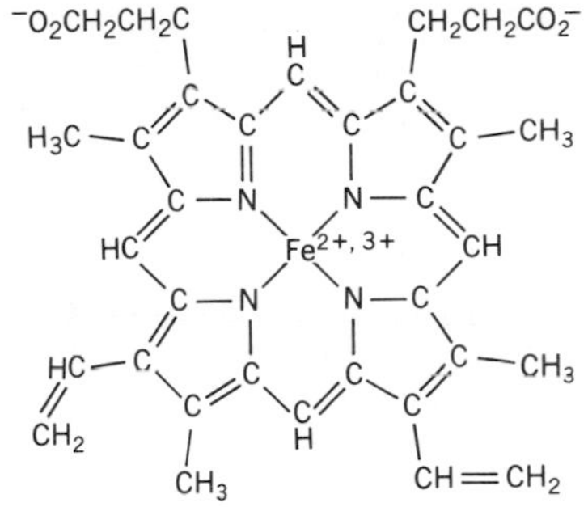

Heme

FIG. 11.8 Many biologically important compounds are Lewis acid-base complexes. This figure shows a small portion of the active site of the proteins hemoglobin and myoglobin. These proteins are often said to contain a "heme" group, which consists of either an Fe^{2+} ion or an Fe^{3+} ion bound to a planar molecule known as a porphyrin. The covalent bonds represented by lines between the iron and nitrogen atoms in this structure result from donating pairs of nonbonding electrons from the nitrogen atom into empty valence orbitals on the Fe^{2+} or Fe^{3+} ion.

FIG. 11.9 The reaction between CaO and SO_2 to form $CaSO_3$ is an example of a Lewis acid-base reaction. The O^{2-} ion in CaO acts as an electron-pair donor, or Lewis base, and SO_2 acts as an electron-pair acceptor, or Lewis acid, to form the SO_3^{2-} ion.

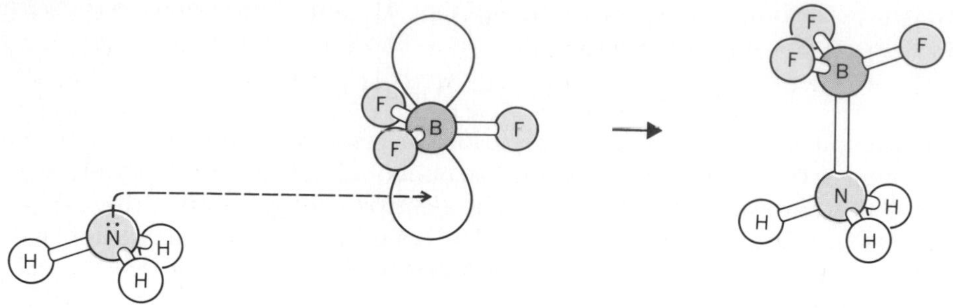

FIG. 11.10 BF_3 is a trigonal planar molecule that contains an sp^2-hybridized boron atom with only six valence electrons. The boron atom therefore has an empty $2p$ atomic orbital that can accept a pair of nonbonding electrons from an NH_3 molecule to form a covalent B—N bond. In the course of this reaction, the BF_3 acts as an electron-pair acceptor, or Lewis acid, while NH_3 acts as an electron-pair donor, or Lewis base.

discussion concluded that this compound is trigonal planar because electrons can be found in only three places in the valence shell of the boron atom.

According to the discussion of hybrid atomic orbitals in Section 8.15, the boron atom in BF_3 is sp^2 hybridized. This leaves an empty $2p$ atomic orbital on the boron atom, which was not used to form the Lewis structure of the BF_3 molecule. BF_3 can therefore act as an electron-pair acceptor, or Lewis acid. It can use the empty $2p$ orbital to pick up a pair of nonbonding electrons from a Lewis base to form a covalent bond. As a result, BF_3 reacts with Lewis bases such as NH_3 to form an acid-base complex in which all of the atoms have a filled shell of valence electrons, as shown in Figure 11.10. In the course of this reaction, the hybridization of the boron changes from sp^2 to sp^3.

The Lewis acid-base theory can also be used to explain why nonmetal oxides such as CO_2 dissolve in water to form acids, such as carbonic acid H_2CO_3.

$$CO_2(g) + H_2O(l) \longrightarrow H_2CO_3(aq)$$

In the course of this reaction, the water molecule acts as an electron-pair donor, or Lewis base. The electron-pair acceptor is the carbon atom in CO_2. When the carbon atom picks up a pair of electrons from the water molecule, it no longer needs to share two pairs of electrons with both of the other oxygen atoms (see Figure 11.11).

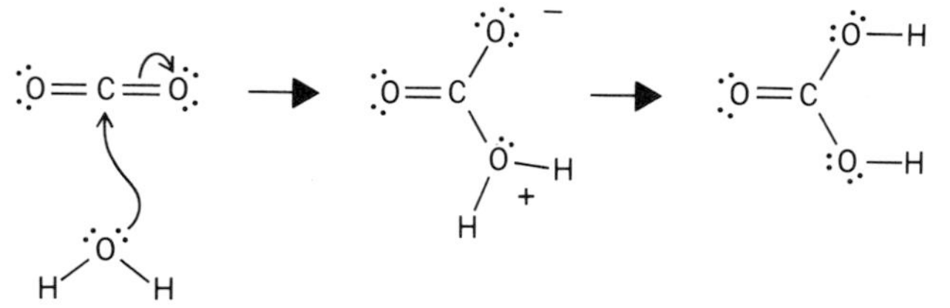

FIG. 11.11 The reaction between CO_2 and H_2O to form carbonic acid is another example of a Lewis acid-base reaction. In this case, the H_2O molecule is a Lewis base that donates a pair of electrons to the carbon atom of CO_2. The distribution of charge in the intermediate formed in this reaction is not symmetrical. An H^+ ion therefore migrates from one oxygen atom to another to form a symmetrical H_2CO_3 molecule.

According to the discussion of formal charge in Section 8.10, one of the oxygen atoms in the intermediate formed when water is added to CO_2 carries a positive charge, and another carries a negative charge. After an H^+ ion has been transferred from one of these oxygen atoms to the other, all of the oxygen atoms in the compound are electrically neutral. The net result of the reaction between CO_2 and water is carbonic acid, H_2CO_3.

Exercise 11.14

Predict whether each of the following ions or molecules can act as a Lewis acid or a Lewis base.

(a) Ag^+ (b) NH_3 (c) OH^- (d) Cu^{2+} (e) BeF_2

Solution

(a) Ag^+ has empty $5s$ and $5d$ valence orbitals that allow it to act as a Lewis acid, or electron-pair acceptor.

$$Ag^+ = [Kr]\ 4d^{10}\ 5s^0\ 5p^0$$

(b) There are no empty valence orbitals in the Lewis structure of NH_3, which suggests that it can't be a Lewis acid. There is a pair of nonbonding electrons, however, which means that ammonia can be an electron-pair donor, or Lewis base.

$$H-\overset{\displaystyle ..}{N}-H$$
$$|$$
$$H$$

(c) Once again, there are no empty valence orbitals in the Lewis structure, which means that the OH^- ion can't be a Lewis acid. There are nonbonding electrons, however, so the ion can be an electron-pair donor, or Lewis base.

$$H-\overset{\displaystyle ..}{\underset{\displaystyle ..}{O}}:^-$$

(d) Cu^{2+} has empty valence-shell orbitals that allow it to act as a Lewis acid, or electron-pair acceptor.

$$Cu^{2+} = [Ar]\ 3d^9\ 4s^0\ 4p^0$$

(e) There are only four electrons in the valence shell of the beryllium atom in the Lewis structure for BeF_2, which suggests that this compound can act as an electron-pair acceptor, or Lewis acid.

$$:\overset{\displaystyle ..}{\underset{\displaystyle ..}{F}}-Be-\overset{\displaystyle ..}{\underset{\displaystyle ..}{F}}:$$

The nonbonding electrons on the fluorine atoms allow this molecule to act as an electron-pair donor, or Lewis base, as well. In solid BeF_2, the fluorine atoms that would formally be assigned to one molecule act as a Lewis base toward the beryllium atom that would be assigned to another molecule, to give a structure that consists of an extended array of BeF_2 units held together by fluorine atoms that bridge adjacent beryllium atoms.

SUMMARY

Compounds have been classified as acids or bases (alkalies) for more than 300 years. Thus, it isn't surprising that a number of different models have been offered to explain the chemistry of these compounds. Arrhenius suggested that acids and bases could be viewed in terms of the ions that form when water dissociates.

$$H_2O(l) \longrightarrow H^+(aq) + OH^-(aq)$$

Brønsted suggested that water molecules don't dissociate into ions; instead, H^+ ions are transferred from one molecule to another to form H_3O^+ and OH^- ions.

$$2\ H_2O(l) \longrightarrow H_3O^+(aq) + OH^-(aq)$$

According to this model, any acid-base reaction can be thought of in terms of the transfer of an H^+ ion from an acid to a base to form the conjugate base and the conjugate acid.

The Brønsted definition of acids and bases significantly broadens the category of compounds that can be classified as bases. In the Brønsted model, a base is any compound that can accept an H^+ ion. It is therefore any compound with at least one pair of nonbonding valence electrons that can be used to form a bond to an H^+ ion.

Lewis suggested that there was another way of looking at acid-base reactions. Instead of thinking of the base as a hydrogen-ion acceptor, it is possible to look at it as an electron-pair donor.

This doesn't have much effect on the number of compounds that can be classified as bases. But it greatly expands the number of possible acids. A Lewis acid is any atom, ion, or molecule that has an empty valence orbital that can be used to accept a pair of electrons. Many reactions that couldn't be classified as acid-base reactions in the Brønsted model, become acid-base reactions in Lewis's model.

By convention, many chemists use the words *acid* and *base* to describe substances that are Brønsted acids and bases. They then use the terms *Lewis acid* and *Lewis base* to refer to acid-base reactions that can only be understood in terms of electron-pair donors and acceptors.

PROBLEMS

Acids, Bases, and Salts

11-1 Describe how acids, bases, and salts differ. Give examples of substances encountered in daily life that fit each of these categories.

11-2 Strong acids (such as hydrochloric acid and sulfuric acid) react with strong bases (such as sodium hydroxide and ammonia) to form salts (such as sodium chloride, sodium sulfate, ammonium chloride, and ammonium sulfate). Write balanced chemical equations for these reactions and describe how these salts differ from the acids and bases from which they are made.

The Arrhenius and Operational Definitions of Acids and Bases

11-3 Describe the Arrhenius definition of acids and bases, and give examples of acids and bases that fit this definition.

11-4 Which of the following compounds are not Arrhenius acids?

(a) HCl (b) H_2SO_4 (c) H_2O (d) $FeCl_3$ (e) H_2S

11-5 Which of the following compounds are not Arrhenius bases?

(a) NaOH (b) MgO (c) $Ca(OH)_2$ (d) H_2O (e) NH_3

11-6 The following compounds dissolve in water to give solutions that turn litmus from red to blue. Which of these compounds satisfy the operational definition of a base?

(a) Na_2O (b) $NaHCO_3$ (c) CaO (d) CaH_2 (e) NH_3

Typical Acids and Bases

11-7 Write balanced equations to show what happens when hydrogen bromide dissolves in water to form an acidic solution and when ammonia dissolves in water to form a basic solution.

11-8 Give an example of an acid whose formula carries a positive charge, an acid that is electrically neutral, and an acid that carries a negative charge.

11-9 Explain why compounds that contain hydrogen in the -1 oxidation state, such as NaH or CaH_2, are bases, whereas compounds that contain hydrogen in the $+1$ oxidation state, such as HCl or HNO_3, are more likely to be acids.

11-10 Would you expect metal peroxides, such as Na_2O_2 and BaO_2, to be acids or bases? What about nonmetal peroxides, such as H_2O_2?

11-11 Which of the following compounds are metal hydrides (bases) and which are nonmetal hydrides (acids)?

(a) H_2CO_3 (b) CH_3CO_2H (c) $LiAlH_4$ (d) H_2S (e) ZnH_2

11-12 Which of the following compounds are metal oxides (bases) and which are nonmetal oxides (acids)?

(a) CO_2 (b) SO_3 (c) P_4O_{10} (d) MgO

11-13 Which of the following compounds is not an acid when dissolved in water?

(a) SO_2 (b) HNO_3 (c) SrO (d) HI (e) K_2S

11-14 Which of the following compounds is not a base when dissolved in water?

(a) $Ca(OH)_2$ (b) Na_2O (c) NO_2 (d) Li_2O (e) CsOH

11-15 Write balanced equations for the reaction between lithium and nitrogen to form lithium nitride and the subsequent reaction of lithium nitride with water to form a basic solution.

11-16 Predict the products of the reaction between each of the following compounds and water.

(a) CO_2 (b) P_4O_{10} (c) CaO (d) SO_3 (e) Na_2O

11-17 Explain why lime (CaO) is used to neutralize soils that are too acidic. Use the fact that microorganisms in soil oxidize sulfur to SO_2 and SO_3 to explain why sulfur is added to soil that is too basic.

11-18 Which of the following elements is most likely to form an acidic oxide with the formula XO_2 and an acidic hydride with the formula XH_2?

(a) Na (b) Mg (c) Al (d) P (e) S (f) Cl

11-19 Which of the following substances is the most likely to be a white solid with a high melting point that dissolves in water to form a basic solution?

(a) O_2 (b) CO_2 (c) Na_2O (d) P_4O_{10} (e) Cl_2O_7

Why Are Metal Hydroxides Bases and Nonmetal Hydroxides Acids?

11-20 Which of the following compounds are metal hydroxides (bases) and which are nonmetal hydroxides (acids)?

(a) H_2CO_3 (b) HNO_3 (c) $Ca(OH)_2$ (d) $Zn(OH)_2$
(e) H_3PO_4

11-21 The following compounds are all oxyacids. In each case, the acidic hydrogen atoms are bound to oxygen atoms. Write the skeleton structures for these acids.

(a) H_3PO_4 (b) HNO_3 (c) $HClO_4$ (d) H_2SO_4
(e) H_3BO_3 (f) H_2CrO_4

11-22 Explain why metal hydroxides, such as LiOH, are bases and nonmetal hydroxides, such as HOBr, are acids.

11-23 Explain why metal oxides, such as CaO, are bases and nonmetal oxides, such as CO_2, are acids.

11-24 Many insects leave small quantities of formic acid, HCO_2H, behind when they bite. Explain why the itching sensation can be relieved by treating the bite with an aqueous solution of ammonia (NH_3), baking soda ($NaHCO_3$), or borax ($Na_2B_4O_7$).

11-25 Barium oxide (BaO) and "phosphorus pentoxide" (P_4O_{10}) are both white solids. What is the easiest way of distinguishing between these compounds?

11-26 Which of the following compounds is most likely to be amphoteric?

(a) Na_2O (b) CaO (c) Al_2O_3 (d) P_4O_{10} (e) Cl_2O_7

The Brønsted Definition of Acids and Bases

11-27 Describe how the Brønsted definition of an acid can be used to explain why compounds that contain hydrogen with an oxidation number of $+1$ are often acids.

11-28 Which of the following compounds can act as both a Brønsted acid and a Brønsted base?

(a) $NaHCO_3$ (b) Na_2CO_3 (c) H_2CO_3 (d) CO_2 (e) H_2O

11-29 Which of the following compounds cannot be Brønsted bases?

(a) H_3O^+ (b) MnO_4^- (c) BH_4^- (d) CN^- (e) S^{2-}

11-30 Which of the following compounds cannot be Brønsted bases?

(a) O_2^{2-} (b) CH_4 (c) PH_3 (d) SF_4 (e) CH_3^+

11-31 Label the Brønsted acids and bases in the following reactions.

(a) $HSO_4^-(aq) + H_2O(l) \rightarrow H_3O^+(aq) + SO_4^{2-}(aq)$
(b) $CH_3CO_2H(aq) + OH^-(aq) \rightarrow CH_3CO_2^-(aq) + H_2O(l)$
(c) $CaF_2(s) + H_2SO_4(aq) \rightarrow CaSO_4(aq) + 2\ HF(aq)$
(d) $HNO_3(aq) + NH_3(aq) \rightarrow NH_4NO_3(aq)$
(e) $LiCH_3(l) + NH_3(l) \rightarrow CH_4(g) + LiNH_2(s)$

Conjugate Acid-Base Pairs

11-32 Identify the conjugage acid-base pairs in the following reactions.

(a) $HCl(aq) + H_2O(l) \rightarrow H_3O^+(aq) + Cl^-(aq)$
(b) $HCO_3^-(aq) + H_2O(l) \rightarrow H_2CO_3(aq) + OH^-(aq)$
(c) $NH_3(aq) + H_2O(l) \rightarrow NH_4^+(aq) + OH^-(aq)$
(d) $CaCO_3(s) + 2\ HCl(aq) \rightarrow Ca^{2+}(aq) + 2\ Cl^-(aq) + H_2CO_3(aq)$ (e) $CaO(s) + H_2O(l) \rightarrow Ca^{2+}(aq) + 2\ OH^-(aq)$

11-33 Which of the following is not a Brønsted conjugate acid-base pair?

(a) NH_4^+/NH_3 (b) H_2O/OH^- (c) H_3O^+/OH^-
(d) CH_4/CH_3^-

11-34 Write the Lewis structure for formic acid, HCO_2H, and its conjugate base.

11-35 Write the Lewis structure for methanol, CH_3OH, and its conjugate base.

11-36 Write the Lewis structure for methylamine, CH_3NH_2, and its conjugate acid.

11-37 Identify the conjugate base of each of the following Brønsted acids.

(a) H_3O^+ (b) H_2O (c) OH^- (d) NH_4^+

11-38 Identify the conjugate base of each of the following Brønsted acids.

(a) HPO_4^{2-} (b) $Al(H_2O)_6^{3+}$ (c) HCO_3^- (d) HS^-

11-39 Identify the conjugate acid of each of the following Brønsted bases.

(a) O^{2-} (b) OH^- (c) H_2O (d) NH_2^-

11-40 Identify the conjugate acid of each of the following Brønsted bases.

(a) Na_2HPO_4 (b) $NaHCO_3$ (c) Na_2SO_4 (d) $NaNO_2$

11-41 Which of the following is the conjugate base of a strong acid?

(a) OH^- (b) HSO_4^- (c) NH_2^- (d) S^{2-} (e) H_3O^+

11-42 Predict the products of the following acid-base reaction. (*Hint:* Write the Lewis structures.)

$$NaOCH_3(aq) + NaHCO_3(aq) \longrightarrow$$

11-43 Predict the products of the following acid-base reaction. (*Hint:* Write the Lewis structures.)

$$NH_4Cl(aq) + NaSH(aq) \longrightarrow$$

11-44 Write balanced chemical equations for the following acid-base reactions.

(a) $Al_2O_3(s) + HCl(aq) \rightarrow$ (b) $CaO(s) + H_2SO_4(aq) \rightarrow$
(c) $CO_2(g) + NaOH(aq) \rightarrow$ (d) $MgCO_3(s) + HCl(aq) \rightarrow$
(e) $Na_2O_2(s) + H_3PO_4(aq) \rightarrow$

The Relative Strengths of Acids and Bases

11-45 In the Brønsted model of acid-base reactions, what does the statement that HCl is a stronger acid than H_2O mean in terms of the following reaction?

$$HCl(aq) + H_2O(l) \longrightarrow H_3O^+(aq) + Cl^-(aq)$$

11-46 In the Brønsted model of acid-base reactions, what does it mean to say that NH_3 is a weaker acid than H_2O?

11-47 What can you conclude from experiments that suggest that the following reaction proceeds to the right, as written?

$$HBr(aq) + H_2O(l) \longrightarrow H_3O^+(aq) + Br^-(aq)$$

11-48 What can you conclude from experiments that suggest that the following reaction proceeds to the right, as written?

$$HCl(aq) + NH_3(aq) \longrightarrow NH_4^+(aq) + Cl^-(aq)$$

11-49 What can you conclude from experiments that suggest that the following reaction does not proceed to the right, as written?

$$HCO_2^-(aq) + H_2O(l) \longrightarrow HCO_2H(aq) + OH^-(aq)$$

11-50 Describe the relationship between the acid-dissociation equilibrium constant for an acid, K_a, and the strength of the acid.

11-51 Use the table of acid-dissociation equilibrium constants in the appendix to classify the following acids as either strong or weak.

(a) acetic acid, CH_3CO_2H (b) boric acid, H_3BO_3
(c) chromic acid, H_2CrO_4 (d) formic acid, HCO_2H
(e) hydrobromic acid, HBr

11-52 Which of the following is the weakest Brønsted acid?

(a) $H_2S_2O_3$: $K_a = 0.3$ (b) H_2CrO_4: $K_a = 9.6$ (c) H_3BO_3: $K_a = 7.3 \times 10^{-10}$ (d) C_6H_5OH: $K_a = 1.0 \times 10^{-10}$

Factors That Control the Relative Strengths of Acids and Bases

11-53 Explain each of the following observations.

(a) H_2SO_4 is a stronger acid than HSO_4^-. (b) HNO_3 is a stronger acid than HNO_2. (c) H_2S is a stronger acid than H_2O. (d) H_2S is a stronger acid than PH_3.

11-54 Arrange the following acids in order of increasing strength.

(a) H_2O (b) HCl (c) CH_4 (d) NH_3

11-55 Arrange the following bases in order of increasing strength.

(a) NH_3 (b) PH_3 (c) H_2O (d) H_2S

11-56 Which of the following is the strongest Brønsted acid?

(a) H_3O^+ (b) HF (c) NH_3 (d) $NaHSO_4$ (e) NaOH

11-57 Which of the following is the weakest Brønsted acid?

(a) H_2Se (b) H_2S (c) H_2O (d) SH^- (e) OH^-

11-58 Which of the following is the weakest Brønsted acid?

(a) $HClO_4$ (b) H_3PO_4 (c) H_2CO_3 (d) H_2SiO_4
(e) $Al(OH)_3$

11-59 Which of the following is the strongest Brønsted base?

(a) NH_2^- (b) PH_2^- (c) CH_3^- (d) SiH_3^-

11-60 Which of the following is the strongest Brønsted base?

(a) H_2O (b) SH^- (c) OH^- (d) S^{2-} (e) O^{2-}

11-61 Which of the following has the strongest conjugate base?

(a) H_2O (b) H_2S (c) NH_3 (d) PH_3 (e) CH_4

pH as a Measure of the Relative Strengths of Acids and Bases

11-62 What happens to the concentration of the H_3O^+ ion when a strong acid is added to water? What happens to the pH of the solution?

11-63 What happens to the OH^- ion concentration when a strong base is added to water? What happens to the pH of the solution?

11-64 Which of the following solutions is most acidic?

(a) 0.10 M acetic acid: pH = 2.9 (b) 0.10 M hydrogen sulfide: pH = 4.1 (c) 0.10 M sodium acetate: pH = 8.4 (d) ammonia: pH = 11.1

11-65 Which of the following is the strongest acid?

(a) 0.10 M H_3PO_4: pH = 1.4 (b) 0.10 M $H_2PO_4^-$: pH = 4.4 (c) 0.10 M HPO_4^{2-}: pH = 9.3 (d) 0.10 M PO_4^{3-}: pH = 12.0

11-66 Explain why aqueous solutions of the Fe^{3+} ion are acidic.

11-67 Which of the following compounds could dissolve in water to give a solution with a pH of about 5?

(a) NH_3 (b) NaCl (c) HCl (d) KOH (e) NH_4Cl

11-68 Explain why pure (nonpolluted) rain water has a pH of 5.6. Explain why boiling water to drive off the CO_2 raises the pH. Explain why the presence of SO_2, SO_3, NO, NO_2, and similar compounds in the atmosphere gives rise to acid rain.

11-69 On April 10, 1974, at Pitlochny, Scotland, the rain was found to have a pH of 2.4. Calculate the H_3O^+ ion concentration in this rain and compare it with the H_3O^+ concentration in 0.10 M acetic acid, which has a pH of 2.9.

The Lewis Definition of Acids and Bases

11-70 Describe the difference between the Lewis and Brønsted definitions of acids and bases. Describe the advantages of the Lewis definition.

11-71 Which of the following compounds can be Lewis acids?

(a) CH_3^+ (b) CH_4 (c) CH_3^- (d) BF_3 (e) Ag^+ (f) Fe^{3+}

11-72 Which of the following compounds can be Lewis bases?

(a) H_2O (b) OH^- (c) O^{2-} (d) O_2^{2-} (e) MnO_4^- (f) CH_4

THE STRUCTURE OF SOLIDS

12.1 GASES, LIQUIDS AND SOLIDS

Three characteristic properties of gases were discussed in Section 4.2.

1. They expand to fill their containers.
2. They can be compressed to fit smaller containers.
3. Their volume at 25°C and 1 atm is typically 700–800 times the volume of the corresponding liquid or solid.

The kinetic molecular theory, described in Section 4.15, explains these properties by assuming that gas particles are in a state of constant random motion and that the diameter of a gas particle is very much smaller than the distance between particles. Most of the volume of a gas is therefore empty space; the simplest analogy might compare the particles of a gas to fruit flies in a jar (see Figure 12.1).

Solids have the opposite properties.

1. They tend to retain their shape regardless of the container in which they are stored.
2. It takes enormous pressures to compress a solid by even an infinitesimally small fraction of its volume.
3. Solids are denser than most liquids and all gases.

A simple analogy might compare particles in a solid to a brick wall in which the individual bricks form a regular structure and the amount of empty space is kept to a minimum.

Many of the properties of solids have been captured in the way the word is used in English. *Solid* describes something that holds its shape, such as a solidly constructed house. It implies continuity, as when a movie is described as running for three solid hours. It implies the absence of empty space, as in a solid chocolate Easter bunny. Finally, it describes things that occupy three dimensions, as in solid geometry.

Liquids have properties between the extremes of the properties of gases and solids.

1. Like gases, they flow to conform to the shape of their containers.
2. Like solids, they cannot expand to fill their containers.
3. Like solids, they are very difficult to compress.

An analogy for liquids might compare their structure to marbles in a bag.

FIG. 12.1 (*a*) Solids have structures in which atoms, ions, or molecules pack as tightly as possible, much as bricks stack to form a wall. (*b*) Individual particles in a liquid, like marbles in a bag, are free to move past each other, so liquids are able to conform to the shape of their containers. (*c*) The force of attraction between particles in a gas is so small that the individual particles are free to move in any direction, rather as fruit flies move about in a jar.

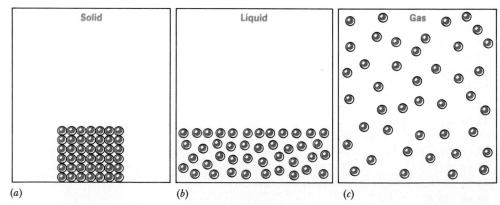

(a)   (b)   (c)

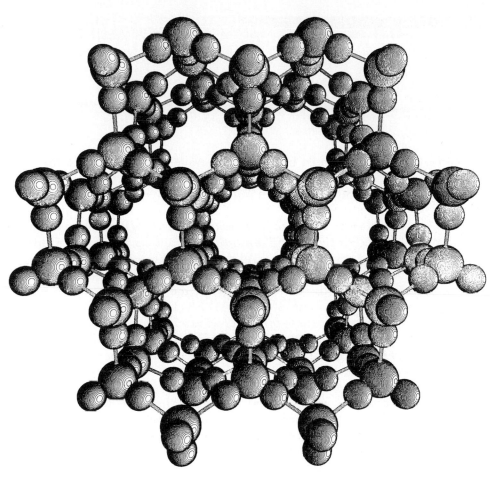

FIG. 12.2 At low temperatures, the force of attraction between water molecules is large enough to lock these molecules in place to form the characteristic structure of ice shown in this figure.

Water is the only substance that we encounter routinely as a solid, a liquid, and a gas. By focusing on the behavior of water at different temperatures we can develop a model for the differences in the properties of these three states of matter.

At low temperatures, water is a solid in which bonds between H_2O molecules lock these molecules into the rigid structure characteristic of ice, shown in Figure 12.2. The covalent bonds between hydrogen and oxygen atoms in an H_2O molecule are called **intramolecular bonds.** The prefix *intra-* comes from the Latin stem meaning "within or inside." Thus, intramural sports match teams from within the same college or university. The bonds between the neighboring water molecules in ice are called **intermolecular bonds,** from the Latin stem meaning "between." This far more common prefix is used in words such as *interact, interface, interdependent, intermediate,* and *international.*

The intramolecular bonds that hold together the atoms in H_2O molecules are almost 25 times as strong as the intermolecular bonds between water molecules. It takes 464 kJ/mol to break the H—O bonds within a water molecule but only 19 kJ/mol to separate water molecules.

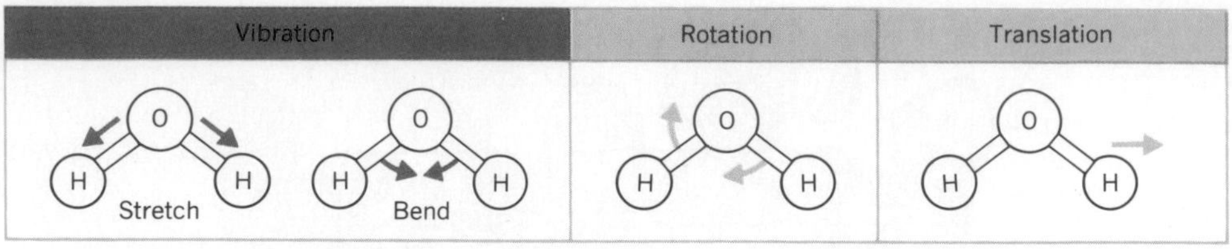

FIG. 12.3 A water molecule can move in three ways. (*a*) Vibration occurs when O—H bonds stretch or bend. (*b*) Rotation involves movement around the center of gravity of the molecule. (*c*) Translation occurs when a water molecule moves through space.

Ice melts when it is heated because the average kinetic energy of the water molecules increases with temperature; water molecules therefore move faster as the temperature increases. Molecules can move in three ways: (1) vibration, (2) rotation, and (3) translation, as shown in Figure 12.3. Water molecules *vibrate* when H—O bonds are stretched or bent. *Rotation* involves the motion of a molecule around its center of gravity. *Translation* literally means change from one place to another and describes motion of molecules through space.

All three modes of motion disrupt the bonds between water molecules. As the system becomes warmer, the thermal energy of the water molecules eventually becomes too large to allow these molecules to be locked into the rigid structure of ice. At this point, the solid melts to form a liquid in which intermolecular bonds are constantly broken and reformed as the molecules move through the liquid. Eventually, the thermal energy of the molecules becomes so large they move too rapidly to form intermolecular bonds, and the liquid boils to form a gas in which each particle moves more or less randomly through space.

The difference between solids, liquids, and gases therefore involves a competition between the strength of intermolecular bonds and the thermal energy of the system. At a given temperature, substances that contain strong intermolecular bonds are more likely to be solids. For a given intermolecular bond strength, the higher the temperature, the more likely that the substance will be a gas.

12.2 VAN DER WAALS FORCES

The simple fact that liquids and solids exist suggests that one of the postulates of the kinetic molecular theory described in Section 4.15 is wrong. This theory assumes that there is no force of attraction between gas particles. If this assumption were true, gases would never condense to form liquids and solids at low temperatures.

As noted in Section 4.17, the Dutch physicist Johannes van der Waals derived an equation that not only corrected for the force of attraction between gas particles but also corrected for the fact that the volume of these particles becomes a significant fraction of the total volume of the gas at high pressures. The *van der Waals equation* is often written as follows.

$$\left[P + \frac{an^2}{V^2}\right](V - nb) = nRT$$

(Here, as you may remember, a and b are parameters that adjust for the force of attraction between the gas particles and the size of the particles, respectively.)

The van der Waals equation is used today to give a better fit to the experimental P, V, n, and T data of real gases than can be obtained with the ideal gas equation. But that was not van der Waals's goal. He was trying to develop a model that would explain the behavior of liquids by including terms that reflected the size of the atoms or molecules in the liquid and the strength of the bonds between these atoms or molecules. As a result, the weak intermolecular bonds in liquids and solids are often called ***van der Waals forces.*** These forces can be divided into three categories: (1) dipole–dipole, (2) dipole–induced dipole, and (3) induced dipole–induced dipole.

DIPOLE–DIPOLE FORCES

Section 8.5 noted that many molecules contain bonds that fall between the extremes of ionic and covalent bonds. The difference between the electronegativities of the atoms in these molecules is large enough that the electrons are not shared equally and yet small enough that the electrons aren't drawn exclusively to one of the atoms to form positive and negative ions. The bonds in these molecules are said to be *polar,* because they have positive and negative ends, or poles. Because they have both positive and negative poles, these molecules are often said to have a ***dipole,*** or a ***dipole moment.*** HCl molecules, for example, have a dipole moment because the hydrogen atom has a small, or partial, positive charge and the chlorine atom has a partial negative charge. Because of the force of attraction between oppositely charged particles, there is a small dipole–dipole force of attraction between adjacent HCl molecules.

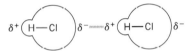

The dipole–dipole interaction in HCl is very weak, with a bond dissociation enthalpy of only 3.3 kJ/mol. (The covalent bonds between the hydrogen and chlorine atoms in HCl are 130 times as strong.) The force of attraction between HCl molecules is so small that this compound boils at $-85.0°C$.

DIPOLE–INDUCED DIPOLE FORCES

What would happen if we mixed a gas such as HCl, which contains molecules that have a dipole moment, with another gas, such as argon, that has no dipole moment?

Argon atoms are perfectly spherical. The electrons are therefore distributed homogeneously around the nucleus of the atom. But these electrons are in constant motion. When an argon atom comes close to a polar HCl molecule, it can shift its electrons to one side of the nucleus to produce a very small dipole moment that lasts for only an instant.

$$\delta^+ \left(H\!-\!Cl \right) \delta^- \cdots\cdots \delta^+ \left(Ar \right) \delta^-$$

By distorting the distribution of electrons around the argon atom, the polar HCl molecule induces a small dipole moment on this atom, which creates a weak dipole–induced dipole force of attraction between the HCl molecule and the Ar atom. This force is very weak, with a bond energy of about 1 kJ/mol.

INDUCED DIPOLE–INDUCED DIPOLE FORCES

What kind of van der Waals force can be used to explain the fact that helium becomes a liquid at temperatures below 4.2 K? By itself, a helium atom is perfectly symmetrical. But since the electrons that surround the nucleus of the atom are in constant motion, the movement of electrons around the nuclei of a pair of helium atoms can become synchronized so that each atom simultaneously obtains an induced dipole moment.

$$\delta^+ \left(\text{He}\right) \delta^- \text{------} \delta^+ \left(\text{He}\right) \delta^-$$

These fluctuations in electron density occur constantly, creating an induced dipole–induced dipole force of attraction between pairs of atoms.

This particular van der Waals force is relatively weak in helium, only 0.076 kJ/mol. But atoms or molecules become more polarizable as they become larger, because there are more electrons to be polarized. In argon, for example, the induced dipole–induced dipole force is 8.5 kJ/mol, which is larger than any other van der Waals force encountered so far. For atoms or molecules larger than argon, this force is even more important. It has been argued that the primary force of attraction between molecules in solid I_2 and in frozen CCl_4 is induced dipole–induced dipole attraction.

12.3 CRYSTALS, AMORPHOUS SOLIDS, AND POLYCRYSTALLINE SUBSTANCES

A polarized light micrograph of a thin section of brass showing the random distribution of microcrystals that gives rise to the grain structure of this alloy.

Solids can be divided into three categories on the basis of similarities in their structures at the microscopic level. *Crystalline solids* are three-dimensional analogs of brick walls, with the individual atoms, ions, or molecules oriented in a regular pattern from one edge of the solid to the other. *Amorphous solids,* (literally, solids without form), such as the polyethylene in plastic sandwich bags or the polystyrene in coffee cups, are noncrystalline; they have little or no long-range order. Many important solids, including aluminum and steel, have a *polycrystalline* structure, which is an aggregate of a large number of small crystals, or grains, arranged randomly.

The extent to which a solid is crystalline instead of amorphous has important effects on its physical properties. The polyethylene used to make sandwich bags and garbage bags is an amorphous solid that consists of more or less randomly oriented chains of (. . . —CH_2—CH_2— . . .) linkages. Milk bottles are made from a more crystalline form of polyethylene, and they have a much more rigid structure.

12.4 MOLECULAR, COVALENT, IONIC, AND METALLIC SOLIDS

Solids can also be classified on the basis of the bonds that hold the atoms or molecules together. This approach tends to categorize solids as either molecular, covalent, ionic, or metallic.

Molecular solids are held together by van der Waals forces. Iodine (I_2), cane sugar ($C_{12}H_{22}O_{11}$), and polyethylene are examples of compounds that form molecular solids at room temperature. Water and bromine are liquids that form molecular solids when cooled slightly—H_2O freezes at 0°C, and Br_2 freezes at −7°C. Molecular solids are characterized by very strong intramolecular bonds and much weaker intermolecular bonds. As a result, these solids tend to be soft substances with low melting points. The rare gases (He, Ne, Ar, Kr, Xe, and Rn) form solids at low temperatures that have many of the characteristics of molecular solids.

Dry ice, or solid carbon dioxide, is a perfect example of a molecular solid. The van der Waals forces holding the CO_2 molecules together are weak enough that dry ice passes from the solid to the gaseous phase at $-78°C$. As noted in Section 10.8, substances that go directly from the solid to the gas phase are said to sublime.

The CO_2 molecules in dry ice are held together by van der Waals's forces. This material is known as dry ice because it is cold (like ice) but dry (because it doesn't melt). Dry ice sublimes at $-78°C$—it goes directly from the solid to the gas phase.

Changes in the strength of the van der Waals forces that hold molecular solids together can have important consequences for the properties of the solid. Polyethylene (. . . CH_2—CH_2 . . .) is a soft plastic that melts at relatively low temperatures. Replacing one of the hydrogens on every other carbon atom with a chlorine atom yields a plastic known as poly(vinyl chloride), or PVC, which is hard enough to be used to make the plastic pipes that are slowly replacing metal pipes for plumbing.

Much of the strength of PVC can be attributed to an increase in the strength of the van der Waals force of attraction between the chains of molecules that form the solid.

When poly(vinyl chloride) (. . . CH_2—$CHCl$ . . .) and poly(vinylidene chloride) (. . . CH_2—CCl_2 . . . .) are combined, a substance is formed that is sold under the trade name Saran. The same increase in the force of attraction between chains that makes PVC harder than polyethylene gives thin films of Saran a tendency to cling. Saran wrap therefore clings to itself.

Covalent solids include substances, such as diamond, whose crystals can each be viewed as a single giant molecule made up of an almost endless number of covalent bonds. Each carbon atom in diamond is covalently bound to four other carbon atoms oriented toward the corners of a tetrahedron, as shown in Figure 12.4. Because all bonds in such structures are equally strong, covalent solids are often very hard, and they are notoriously difficult to melt. Diamond is the hardest natural substance, and it melts at temperatures above $3550°C$.

Ionic solids are salts, such as NaCl, that are held together by the strong force of attraction (F) between ions of opposite charge (q_1 and q_2).

$$F = \frac{q_1 \times q_2}{r^2}$$

Because this force of attraction depends on the distance between the positive and negative charges (r^2), the strength of an ionic bond depends on the radii of the ions that form the solid. As these ions become larger, the bond becomes weaker. But

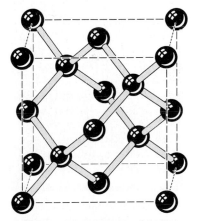

FIG. 12.4 The properties of diamond result from the fact that each carbon atom in the crystal is tightly bound to four neighboring carbon atoms arranged toward the corners of a tetrahedron. The crystal can be viewed as a single giant molecule held together by covalent bonds. It is therefore an example of a covalent solid.

FIG. 12.5 There are three different forms of chemical bonds: ionic, covalent, and metallic. Ionic bonds are formed between metals and nonmetals, covalent bonds are formed between nonmetals, and metals are held together by metallic bonds.

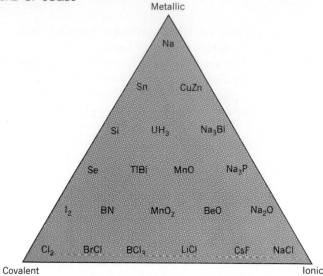

regardless of the size of the ions, the ionic bond is very strong, so salts have relatively high melting points and boiling points.

The fourth class we shall discuss is **_metallic solids._** To understand metallic solids we have to clear up a common misconception about chemical bonds. Ionic and covalent bonds are often imagined as if they were opposite ends of a two-dimensional model of bonding in which compounds that contain polar bonds fall somewhere between these two extremes.

ionic . . . . polar . . . . covalent

In reality, there are three kinds of bonds between adjacent atoms: ionic, covalent, and metallic, as shown in Figure 12.5. Nonmetals combine to form elements and compounds that contain covalent bonds, such as Cl_2, HCl, and CH_4. Metals combine with nonmetals to form salts, such as NaCl and Na_2SO_4, which are held together by ionic bonds. The force of attraction between atoms in metals, such as copper and aluminum, or alloys, such as brass and bronze, are metallic bonds.

Molecular, ionic, and covalent solids all have one thing in common. With only rare exceptions, the electrons in these solids are _localized_—they either reside on one of the atoms or ions or are shared by a pair of atoms or ions.

Metal atoms don't have enough electrons to fill their valence shells by sharing electrons with their immediate neighbors. Electrons in the valence shell are therefore shared by many atoms, instead of just two. In effect, the valence electrons are _delocalized_ over a number of metal atoms. Because these electrons aren't tightly bound to individual atoms, they are free to migrate through the metal. That's why metals are good conductors of electricity; electrons that enter the metal at one edge can displace other electrons to give rise to a net flow of electrons through the metal.

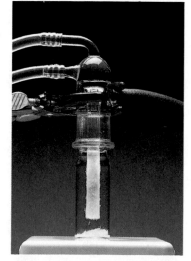

Many solids sublime at very low pressure. This process can be used to purify compounds. The product of a chemical reaction, for example, is placed in the bottom of a sublimation apparatus, which is evacuated and heated with an oil bath. The compound sublimes and is captured as it solidifies on a water-cooled glass finger that extends into the sublimation apparatus.

12.5 THE STRUCTURE OF METALS AND OTHER MONATOMIC SOLIDS

The crystal structures of pure samples of metals, such as aluminum and copper, are easy to describe, because the atoms are all identical perfect spheres. The same can be said about the structure of the rare gases (He, Ne, Ar, and so on) at low temperatures. These substances all crystallize in one of four basic structures, known as simple cubic (SC), body-centered cubic (BCC), hexagonal closest-packed (HCP), and cubic closest-packed (CCP).

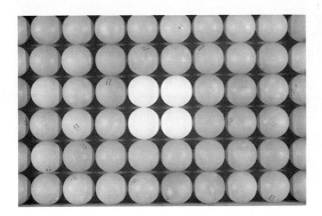

FIG. 12.6 A square-packed plane of spheres.

The principle that will guide our discussion of the structures of solids is based on the fact that solids are very difficult to compress, so that the space between atoms must be kept to a minimum.

The most probable structure for a solid is the structure that makes the most efficient use of space. When a solid crystallizes, the atoms, ions, or molecules pack as tightly as possible.

To illustrate this principle, let's try to imagine the best way of packing spheres, such as ping-pong balls, into an empty box.

One way to do this is to carefully pack the ping-pong balls to form a square-packed plane of spheres, as shown in Figure 12.6. By tilting the box to one side, we can stack a second plane of spheres directly on top of the first. The net result is a regular structure in which the simplest repeating unit is a cube of eight spheres, as shown in Figure 12.7. This structure is called a *simple cubic (SC) packing.*

Each sphere in this structure touches four identical spheres in the same plane. It also touches one sphere in the plane above and one in the plane below. Each sphere is therefore said to have a *coordination number* of 6. If these spheres represent atoms, each atom can form bonds to its six nearest neighbors.

Is a simple cubic structure an efficient way of packing spheres in three-dimensional space? One way of answering this question is to ask: What happens when we shake the box? Do the ping-pong balls stay in the same positions, or do they settle into a different structure? It is fairly easy to show that a simple cubic structure uses space inefficiently; only 52% of the available space is actually occupied by the spheres in a simple cubic structure. The rest is empty space. Of the 108 elements in the periodic table, the semimetal polonium is the only element that packs in a simple cubic structure.

How can we use space more efficiently? One approach starts by separating the spheres in a square-packed plane so that they do not quite touch each other, as shown in Figure 12.8. The spheres in the second plane pack above the holes in the first plane, as shown in Figure 12.9. Spheres in the third plane pack above holes in the second plane, spheres in the fourth plane pack above holes in the third plane, and so on. The net result is a structure in which all the odd-numbered planes of atoms are identical and all the even-numbered planes of atoms are identical. This *ABABABAB . . .* repeating structure is known as *body-centered cubic (BCC) packing.*

The name *body-centered cubic* results from the fact that each sphere touches four spheres in the plane above and four more in the plane below, arranged toward the corners of a cube. Thus, the simplest repeating unit in this structure is a cube of

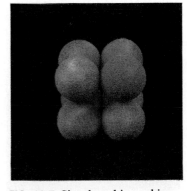

FIG. 12.7 Simple cubic packing.

FIG. 12.8 A square-packed plane in which the spheres do not quite touch each other.

FIG. 12.9 The spheres in a second plane pack above the holes of the plane shown in Figure 12.8.

FIG. 12.10 A body-centered cube.

eight spheres with a ninth identical sphere in the center of the body—in other words, a body-centered cube, as shown in Figure 12.10. The coordination number in this structure is 8.

With some use of solid geometry, it can be shown that 68% of the space in a body-centered cubic structure is filled. Body-centered cubic packing therefore uses space more efficiently than simple cubic packing. As a result, body-centered cubic packing is an important structure for metals. All of the metals in Group IA (Li, Na, K, and so on), the heavier metals in Group IIA (Ca, Sr, and Ba), and a number of the early transition metals (such as Ti, V, Cr, Mo, W, and Fe) pack in a body-centered cubic structure.

Two structures pack spheres so efficiently they are called closest-packed structures. Both start by packing the spheres in a given plane in a hexagonal structure in which each sphere touches six others oriented toward the corners of a hexagon, as shown in Figure 12.11. Spheres are then packed above the triangular holes in the bottom plane, as shown in Figure 12.12, and a second plane is formed.

What about the next plane of spheres? One possibility is to pack the spheres in the third plane so that they lie directly above the spheres in the first plane, to form an *ABABABAB* . . . repeating structure. Because this structure is composed of alternating planes of hexagonal closest-packed spheres, it is called a ***hexagonal closest-packed (HCP) structure.*** Each sphere touches three spheres in the plane above, three spheres in the plane below, and six spheres in the same plane, as shown in Figure 12.13. Thus, the coordination number in a hexagonal closest-packed structure is 12.

It is possible to show that 74% of the space in a hexagonal closest-packed structure is filled. No more efficient way of packing spheres is known, and the

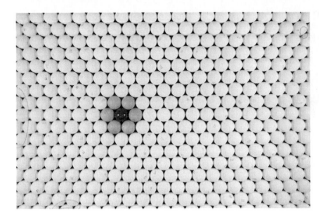

FIG. 12.11 A closest-packed plane in which each sphere touches six others oriented toward the corners of a hexagon.

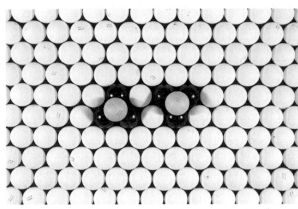

FIG. 12.12 A second plane is formed when spheres are packed above the triangular holes in the plane shown in Figure 12.11.

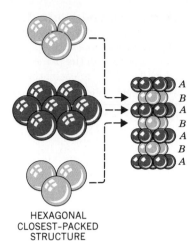

HEXAGONAL
CLOSEST-PACKED
STRUCTURE

FIG. 12.13 Each atom in a hexagonal closest-packed structure touches six atoms in the same plane, three in the plane above, and three in the plane below. These planes of atoms stack in an *ABABAB* . . . repeating pattern.

hexagonal closest-packed structure is important for metals such as Be, Co, Mg, and Zn, as well as the rare gas He at low temperatures.

There is another way of stacking hexagonal closest-packed planes of spheres. The atoms in the third plane can be packed above the holes in the first plane that were not used to form the second plane. The fourth hexagonal closest-packed plane of atoms then packs directly above the first. The net result is an ABCABCABC . . . repeating structure.

This structure is called **cubic closest-packed (CCP)** for reasons that will become apparent in Section 12.12. Each sphere in this structure still touches six others in the same plane, three in the plane above, and three in the plane below, as shown in Figure 12.14 and the coordination number is still 12. You can see the difference between hexagonal and cubic closest-packed structures by comparing Figures 12.13 and 12.14. In the hexagonal closest-packed structure, the atoms in the first and third planes lie directly above each other. In the cubic closest-packed structure, the atoms in these planes are oriented in opposite directions.

The cubic closest-packed structure is just as efficient as the hexagonal closest-packed structure. Both use 74% of the available space. Many metals, including Ag, Al, Au, Ca, Co, Cu, Ni, Pb, and Pt, crystallize in a cubic closest-packed structure, as do all the rare gases at low temperatures except He.

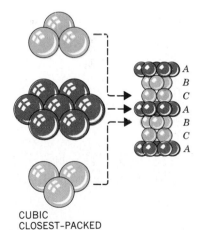

CUBIC
CLOSEST-PACKED
STRUCTURE

FIG. 12.14 Each atom in a cubic closest-packed structure touches six atoms in the same plane, three in the plane above, and three in the plane below. But the atoms in the top plane are rotated by 180° relative to the atoms in the bottom plane. These planes of atoms stack on top of each other in an ABCAB-CABC . . . repeating pattern.

12.6 COORDINATION NUMBERS AND THE STRUCTURES OF METALS

The coordination numbers of the four structures introduced in the preceding section are summarized in Table 12.1. It is easy to understand why metals pack in

FIG. 12.15 Each atom in a body-centered cubic structure touches four atoms in the plane above and four in the plane below. In addition, each atom almost touches six more atoms.

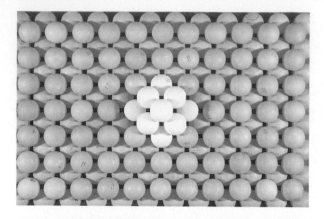

TABLE 12.1		
Coordination Numbers for Common Crystal Structures		
Structure	*Coordination Number*	*Stacking Pattern*
simple cubic	6	*AAAAAAA . . .*
body-centered cubic	8	*ABABABAB . . .*
hexagonal closest-packed	12	*ABABABAB . . .*
cubic closest-packed	12	*ABCABCABC . . .*

hexagonal or cubic closest-packed structures. Not only do these structures use space as efficiently as possible, they also have the largest coordination numbers, which allows each metal atom to form bonds to the largest possible number of neighboring metal atoms.

It is less obvious why one-third of the metals pack in a body-centered cubic structure, in which the coordination number is only 8. The popularity of this structure can be understood by referring to Figure 12.15. The coordination number for body-centered cubic structures given in Table 12.1 counts only the atoms that actually touch a given atom in this structure. Each atom touches four atoms in the plane above and four in the plane below. A careful look at Figure 12.15 shows that each atom also almost touches four neighbors in the same plane, a fifth neighbor two planes above, and a sixth two planes below.

The distance from each atom to the nuclei of these nearby atoms is only 15% larger than the distance to the nuclei of the atoms that it actually touches. Each atom in a body-centered cubic structure can therefore form a total of 14 bonds — 8 strong bonds to the atoms that it touches and 6 somewhat weaker bonds to atoms it almost touches.

This makes it easier to understand why a metal might prefer the body-centered cubic structure to the hexagonal or cubic closest-packed structure. Each metal atom in the closest-packed structures can form strong bonds to 12 neighboring atoms. In the body-centered cubic structure, each atom forms a total of 14 bonds to neighboring atoms, although 6 of these bonds are somewhat weaker than the other 8.

12.7 PHYSICAL PROPERTIES THAT RESULT FROM THE STRUCTURE OF METALS

According to Section 2.7, metals have the following characteristic physical properties.

1. They have a metallic shine or luster.
2. They are usually solid at room temperature.
3. They are *malleable* (from the Latin for "hammer"); they can be hammered, pounded, or pressed into different shapes.
4. They are *ductile;* they can be drawn into thin sheets or wires without breaking.
5. They conduct heat and electricity.

Many of these properties result from the structure of metals.

When asked why metals have a characteristic metallic shine, or luster, many people would say that metals reflect (literally, throw back) the light that shines on their surface. They might argue, in effect, that light bounces off a metal's surface the way a racquetball bounces off the walls of a racquetball court. But this is not what happens when light hits the surface of the metal. Metals actually absorb a significant fraction of the light that hits their surface.

A portion of the energy captured when the metal absorbs light is turned into thermal energy. (You can easily demonstrate this by placing your hand on the surface of a car that has spent several hours in the sun.) The rest of the energy is reradiated by the metal as "reflected" light. Silver is better than any other metal at reflecting light, and yet only 88% of the light that hits the surface of a silver mirror is reradiated. A better analogy for the interaction between light and a metal surface might compare the metal to a shortstop in a double play, who first catches the ball and then throws it off in another direction.

Why do metals absorb light when other solids, such as polyethylene, do not? Light is absorbed when the energy of the radiation is equal to the energy needed to excite an electron to a higher-energy excited state or when the energy can be used to move an electron through the solid. Because electrons are delocalized in metals and so are free to move through the solid, metals absorb light easily. Other solids, such as polyethylene and glass, don't have electrons that can move through the solid, so they can't absorb light the way metals do. These solids are colorless and can only be colored by addition of an impurity in which the energy associated with exciting an electron from one orbital to another falls in the visible portion of the spectrum. Glass is usually colored by addition of a small quantity of one of the transition metals. Cobalt produces a blue color, chromium makes the glass appear green, and traces of gold give a deep red color.

Why are metals solid? Nonmetals, such as hydrogen and oxygen, are gases at room temperature because these elements can achieve a filled shell of valence electrons by sharing pairs of electrons to form molecules, such as H_2 and O_2. Metals can't do this. There aren't enough electrons on a metal atom to allow it to fill its valence shell by sharing pairs of electrons with one or two nearest neighbors. The only way a metal can obtain the equivalent of a filled shell of valence electrons is by delocalizing the valence electrons—by allowing these electrons to be shared by a number of adjacent metal atoms. This is possible only if the metal atoms are kept close together, and metals are therefore solids at room temperature.

Why are metals malleable and ductile? Most metals pack in either body-centered cubic, hexagonal closest-packed, or cubic closest-packed structures. In theory, changing the shape of the metal is simply a matter of applying a force that makes the atoms in one of the planes slide past the atoms in an adjacent plane, as shown in Figure 12.16. In practice, it is easier to do this when the metal is hot.

Why are metals good conductors of heat and electricity? As you have already seen, the delocalization of valence electrons in a metal allows the solid to conduct an electric current. To understand why metals conduct heat, remember that temperature is a macroscopic property that reflects the kinetic energy of the individual

Solids, such as glass, that do not absorb light can be colored by adding a small quantity of an impurity that absorbs visible light.

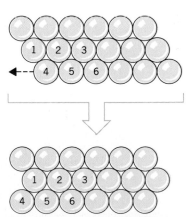

FIG. 12.16 Metals are malleable and ductile because planes of atoms can slip past each other to reach equivalent positions.

atoms or molecules. The tight packing of atoms in a metal means that kinetic energy can be transferred from one atom to another both rapidly and efficiently.

12.8 SOLID SOLUTIONS AND INTERMETALLIC COMPOUNDS

Most of the solutions chemists work with involve a solid such as NaCl dissolved in a liquid such as water. As noted in Section 3.18, however, it is also possible to prepare solutions in which a gas, a liquid, or a solid dissolves in a solid. Liquid mercury, for example, dissolves in sodium metal to form an alloy, or solid solution, that is an excellent reducing agent. Hydrogen gas dissolves in platinum metal to form another example of a solid solution. The most important class of solid solutions are those in which one solid is dissolved in another — for example, copper dissolved in aluminum or carbon dissolved in iron.

The solubility of one solid in another usually depends on temperature. At room temperature, for example, copper does not dissolve in aluminum. At 550°C, however, aluminum can form solutions that contain up to 5.6% copper by weight.

Aluminum metal that has been saturated with copper at 550°C will try to reject the copper atoms as it cools to room temperature. In theory, the solution could reject copper atoms by forming a polycrystalline structure composed of small crystals of more or less pure aluminum interspersed with small crystals of copper metal. In fact, however, the copper metal combines with aluminum as it cools to form an *intermetallic compound* with the formula $CuAl_2$. $CuAl_2$ is a perfect example of the difference between a mixture (such as a solution of copper dissolved in aluminum) and a compound. The solution, or alloy, can contain varying amounts of copper and aluminum. At 550°C, for example, the solution can be between 0 and 5.6% copper metal by weight. The intermetallic compound, however, has a fixed composition. $CuAl_2$ is always exactly 49.5% aluminum by weight.

Intermetallic compounds such as $CuAl_2$ are the key to a process known as *precipitation hardening.* Aluminum metal packs in a cubic closest-packed structure in which one plane of atoms can very easily slip past another. Pure aluminum metal is therefore too weak to be used as a structural metal in cars or airplanes. When carefully processed, however, alloys of copper in aluminum are five to six times as strong and make an excellent structural metal.

The first step in precipitation hardening of aluminum involves heating the metal to 550°C. Copper is then added to form a solid solution, which is quenched with cold water. The solution cools so fast that the copper atoms don't have time to come together to form microcrystals of copper metal. They are effectively trapped in solution.

Comparing a solid with a brick wall as we did in Figure 12.1 has one major disadvantage. It leads one to believe that atoms can't move through the metal. This is not quite true. Diffusion through the metal can in fact occur, although it occurs relatively slowly. Over a period of time, copper atoms can move through the quenched solution to form microcrystals of the $CuAl_2$ intermetallic compound that are so small they are hard to see with a microscope.

These $CuAl_2$ particles are both hard and strong — so hard that they inhibit the flow of the aluminum metal that surrounds them. The presence of these microcrystals of $CuAl_2$ strengthens the aluminum metal by interfering with the way planes of atoms slip past each other. The net result is a metal that is both harder and stronger than pure aluminum.

Copper dissolved in aluminum at high temperature is an example of a *substitu-*

tion solution, in which copper atoms pack in the positions normally occupied by aluminum atoms. There is another way in which a solid solution can be made. Very small atoms of one element can pack in the holes, or *interstices,* between atoms of the host metal, because even the most efficient crystal structures (HCP and CCP) use only 74% of the available space in the crystal. The result is an **interstitial solution.**

Steel at high temperatures is a good example of an interstitial solution. Steel is an alloy of carbon dissolved in iron. At very high temperatures, iron packs in a cubic closest-packed structure that leaves just enough space to allow carbon atoms to fit in the holes between the iron atoms. Below 910°C, iron metal packs in a body-centered cubic structure, in which the holes are too small to hold carbon atoms.

This has important consequences for the properties of steel. At temperatures above 910°C, carbon readily dissolves in iron to form a solid solution that contains as much as 1% carbon by weight. This material is both malleable and ductile, and it can be rolled into thin sheets or hammered into various shapes.

When this solution cools below 910°C, the iron changes to a body-centered cubic structure, and the carbon atoms are rejected from the metal. If the solution is allowed to cool gradually, the carbon atoms migrate through the metal to form a compound with the formula Fe_3C, which precipitates from the solution. These Fe_3C crystals serve the same role in steel that the $CuAl_2$ crystals play in aluminum. They inhibit the flow of the planes of metal atoms and thereby make the metal stronger.

12.9 HOLES IN CLOSEST-PACKED AND SIMPLE CUBIC STRUCTURES

Metals are not the only solids that pack in simple cubic, body-centered cubic, hexagonal closest-packed, and cubic closest-packed structures. A large number of ionic solids use these structures as well.

Sodium chloride (NaCl) and zinc sulfide (ZnS), for example, form crystals that can be thought of as cubic closest-packed arrays of negative ions (Cl^- or S^{2-}), with the positive ions (Na^+ or Zn^{2+}) packed in holes between the closest-packed planes of negative ions. There is a subtle difference between these structures, because the Na^+ ions in NaCl pack in holes that are different from those used by the Zn^{2+} ions in ZnS.

There are two kinds of holes in a closest-packed structure. So-called **tetrahedral holes** are shown in Figure 12.17. The solid lines in this figure represent one plane of closest-packed atoms, and the dashed lines correspond to a second plane of atoms, which pack above the holes in the first plane. Each of the holes marked with a *t* touches three atoms in the first plane and one atom in the second plane. They are called tetrahedral holes because positive ions that pack in these holes are surrounded by four negative ions arranged toward the corners of a tetrahedron.

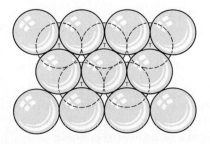

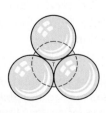

FIG. 12.17 Tetrahedral and octahedral holes are formed when one plane of closest-packed atoms (dashed lines) packs above the holes in another plane of closest-packed atoms (solid lines). Tetrahedral holes are surrounded by four atoms arranged toward the corners of a tetrahedron.

FIG. 12.18 Octahedral holes are surrounded by six atoms arranged toward the corners of an octahedron.

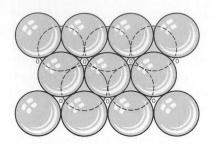 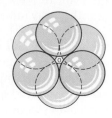

So-called *octahedral holes* in a closest-packed structure are shown in Figure 12.18. Once again, the solid lines represent one plane of closest-packed atoms, and the dashed lines correspond to a second plane, which packs above the holes in the first plane. Each of the holes marked with an *o* touches three atoms in the first plane and three atoms in the second plane. They are called octahedral holes because positive ions that pack in them are surrounded by six negative ions arranged toward the corners of an octahedron.

Tetrahedral holes are very small. The largest atom that can fit into a tetrahedral hole without distorting the tetrahedron has a radius only 0.225 times as large as the radius of the atoms that form the hole. Octahedral holes are almost twice as large as tetrahedral holes. The largest atom that can fit into an octahedral hole has a radius 0.414 times the size of the radius of the atoms that form the hole. Whether a particular crystal uses tetrahedral or octahedral holes therefore depends on the relative size of the atoms or ions that form the crystal.

Sometimes positive ions are too big to pack in either tetrahedral or octahedral holes in a closest-packed structure of negative ions. When this happens, the negative ions pack in a simple cubic structure, and the positive ions pack in *cubic holes* between the planes of negative ions. Cesium chloride (CsCl) is an example of a salt with this structure. According to the data in Table A-8 in the appendix, the radius of a Cl^- ion (0.181 nm) is only about 7% larger than that of a Cs^+ ion (0.169 nm). The Cs^+ ions are too big to fit in either tetrahedral or octahedral holes in a closest-packed structure. The Cl^- ions in this compound therefore pack in a simple cubic structure, and the Cs^+ ions pack in the cubic holes between planes of Cl^- ions.

12.10 THE STRUCTURE OF IONIC SOLIDS: GROUND RULES

The structure of many ionic solids can be explained on the basis of four general rules.

1. Ions of opposite charge pack as tightly as possible to maximize the force of attraction between these ions.
2. Positive ions tend to be much smaller than negative ions.
3. Positive ions seldom touch other positive ions in a crystal. (Negative ions may touch other negative ions.)
4. Each positive ion touches as many negative ions as possible, and vice versa.

The first rule explains how ionic solids satisfy the principle introduced in Section 12.5, which stated that the most probable structure for a solid is the one that makes the most efficient use of space. The most probable structure for an ionic solid is one in which positive and negative ions pack as tightly as possible.

The second rule summarizes a general trend first encountered in Section 6.13.

Positive ions are smaller than the neutral atoms from which they are made. Negative ions are larger than the corresponding neutral atoms.

The radius of an Na^+ ion is only about 60% as large as the radius of a neutral sodium atom.

$$Na = 0.157 \text{ nm} \qquad Na^+ = 0.095 \text{ nm}$$

A Cl^- ion, on the other hand, is almost twice as large as a chlorine atom.

$$Cl = 0.099 \text{ nm} \qquad Cl^- = 0.181 \text{ nm}$$

As a result, the positive ions in an ionic solid tend to be much smaller than the negative ions. In NaCl, the sodium ion is only half as big as the chloride ion.

$$Na^+ = 0.095 \text{ nm} \qquad Cl^- = 0.181 \text{ nm}$$

The third rule is a logical consequence of the second. Recall from Section 12.4 that the force of repulsion between ions with the same charge is inversely proportional to the square of the distance between the centers of the ions.

$$F = \frac{q_1 \times q_2}{r^2}$$

Small positive ions therefore repel each other better than larger, negative ions. The repulsion between ions with the same charge is minimized if the relatively small positive ions are kept as far apart as possible.

The fourth rule reminds us that it isn't the absolute size of the positive or negative ions that determines the structure of an ionic solid but the relative size of the ions. The relative size is given by the ***radius ratio,*** which is equal to the radius of the positive ion divided by the radius of the negative ion.

$$\textbf{Radius ratio} = \frac{\textbf{radius of the positive ion}}{\textbf{radius of the negative ion}} = \frac{r_+}{r_-}$$

As the radius ratio becomes larger, the number of negative ions that can pack around each positive ion increases.

A summary of the relationship between the coordination number of the positive ions in ionic crystals and the radius ratio of the ions is given in Table 12.2. When the radius ratio is between 0.225 and 0.414, positive ions tend to pack in tetrahedral holes between planes of negative ions in a cubic or hexagonal closest-packed structure. When the radius ratio is between 0.414 and 0.732, the positive ions tend to pack in octahedral holes between planes of negative ions in a closest-packed structure.

In Section 12.9, we noted that the largest positive ion that can fit into a tetrahe-

TABLE 12.2

Radius Ratio Rules

Radius Ratio	Coordination Number	Holes in Which Positive Ions Pack
0.225–0.414	4	tetrahedral holes
0.414–0.732	6	octahedral holes
0.732–1	8	cubic holes
1	12	closest-packed structure

dral hole without distorting the tetrahedron has a radius that is 0.225 times the radius of the negative ions that form that hole.

$$\text{Tetrahedral hole:} \qquad r_+ = 0.225 \, r_-$$

Tetrahedral holes are not used until the positive ion is large enough to touch all four of the negative ions that form this hole. As the radius ratio increases from 0.225 to 0.414, the positive ion distorts the structure of the negative ions away from a true closest-packed structure.

As soon as the positive ion is large enough to touch all six negative ions in an octahedral hole (radius ratio = 0.414), the positive ions start to pack in octahedral holes. These holes are used until the positive ion is so large that it can't fit into even a distorted octahedral hole.

Eventually a point is reached at which the positive ion can no longer fit into either the tetrahedral or octahedral holes in a closest-packed crystal. When the radius ratio is between about 0.732 and 1, ionic solids tend to crystallize in a simple cubic array of negative ions with positive ions occupying some or all of the cubic holes between these planes. When the radius ratio is about 1, the positive ions can be incorporated directly into the positions of the closest-packed structure.

Exercise 12.1

Predict whether tetrahedral, octahedral, or cubic holes should be used in each of the following crystals.

(a) NaCl   (b) ZnS   (c) TiO_2 (d) $TiCl_4$ (e) CsCl

Solution

(a) The ionic radii of the Na^+ (0.095 nm) and Cl^- (0.181 nm) ions give a radius ratio of 0.52. According to Table 12.2, the Na^+ ions therefore pack in octahedral holes between planes of closest-packed Cl^- ions.

(b) The ionic radii of the Zn^{2+} (0.074 nm) and S^{2-} (0.184 nm) ions give a radius ratio of 0.40, which suggests that the Zn^{2+} ions should pack in tetrahedral holes.

(c) The ionic radii of the Ti^{4+} (0.068 nm) and O^{2-} (0.140 nm) ions give a radius ratio of 0.49, so the Ti^{4+} ions should pack in octahedral holes.

(d) The ionic radii of the Ti^{4+} (0.068 nm) and Cl^- (0.181 nm) ions give a radius ratio of 0.38, which suggests that the Ti^{4+} ions should pack in tetrahedral holes.

(e) The ionic radii of the Cs^+ (0.169 nm) and Cl^- (0.181 nm) ions give a radius ratio of 0.933. The Cs^+ ions are too large to fit the tetrahedral or octahedral holes in a closest-packed structure but should pack in cubic holes in a simple cubic array of Cl^- ions.

12.11 THE FILLING OF TETRAHEDRAL, OCTAHEDRAL, AND CUBIC HOLES

Both the formula of an ionic solid and the radius ratio of the ions determine the structure of the compound. In the preceding section, we saw how the radius ratio determines whether the positive ions pack in tetrahedral, octahedral, or cubic holes. The formula of the compound determines what fraction of these holes are filled and what fraction are empty.

Figure 12.18 showed that six atoms or ions surround each octahedral hole in a closest-packed structure. Conversely, it can be shown that each atom or ion in a closest-packed structure is surrounded by six octahedral holes. If there are six atoms around each octahedral hole and six octahedral holes around each atom, there are just as many octahedral holes as atoms in a closest-packed structure. If all of the octahedral holes between planes of negative ions in a closest-packed structure contain positive ions, there must be just as many positive ions as negative ions in the crystal.

Exercise 12.2

Use the radius ratio for NaCl and the formula for this compound to predict the structure of the crystal.

Solution

According to the table of ionic radii in the appendix, the radius ratio of NaCl is 0.52.

$$\frac{r_+}{r_-} = \frac{0.095 \text{ nm}}{0.181 \text{ nm}} = 0.52$$

The Cl^- ions should therefore pack in a closest-packed structure, with Na^+ ions in octahedral holes.

There is one octahedral hole per Cl^- ion in a closest-packed array of these ions, and the formula for NaCl suggests that there is one Na^+ ion per Cl^- ion in this compound. Thus, in this crystal, all of the octahedral holes must be filled with Na^+ ions.

NaCl does in fact crystallize in a cubic closest-packed structure of Cl^- ions with Na^+ ions in all of the octahedral holes. A model of the NaCl crystal structure is shown in Figure 12.19. Note that not all of the holes are filled. Closest-packed structures have both octahedral and tetrahedral holes. The tetrahedral holes in NaCl are empty because the Na^+ ions are too large to fit into them. All of the octahedral holes are filled, however.

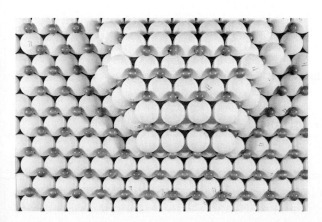

FIG. 12.19 A model of the crystal structure of NaCl.

Exercise 12.3

Titanium dioxide (TiO_2) is a pigment used to make white paint. Predict the structure of this compound.

Solution

If we assume that TiO_2 contains Ti^{4+} and O^{2-} ions, the radius ratio for this compound is 0.49.

$$\frac{r_+}{r_-} = \frac{0.068 \text{ nm}}{0.140 \text{ nm}} = 0.49$$

According to Table 12.2, the Ti^{4+} ions should pack in octahedral holes in a closest-packed aray of O^{2-} ions.

There is one octahedral hole per O^{2-} ion in this structure. If all of these holes were filled, there would be one titanium atom for every oxygen atom, and the formula of the compound would be TiO. Because the compound only contains one Ti^{4+} ion for every pair of O^{2-} ions, we must conclude that half of the octahedral holes are filled in this structure.

Experiment has shown that TiO_2 does in fact crystallize in a cubic closest-packed array of oxygen atoms with titanium atoms in half of the octahedral holes.

Figure 12.17 showed that each tetrahedral hole in a closest-packed structure is surrounded by four atoms or ions. Each of the atoms or ions that forms a closest-packed structure is surrounded by eight tetrahedral holes. If there are four atoms around each tetrahedral hole and eight tetrahedral holes around each atom, there must be twice as many tetrahedral holes as atoms in a closest-packed structure. If all of the tetrahedral holes in a closest-packed array of negative ions are filled with positive ions, there are twice as many positive ions in the crystal, and the compound has the formula M_2X.

Exercise 12.4

Predict the structure of the mineral zinc blende, ZnS.

Solution

If we assume that ZnS contains Zn^{2+} and S^{2-} ions, the radius ratio for this compound is 0.40.

$$\frac{r_+}{r_-} = \frac{(0.074 \text{ nm})}{(0.184 \text{ nm})} = 0.40$$

ZnS should therefore crystallize as a closest-packed array of S^{2-} ions with Zn^{2+} ions in the tetrahedral holes.

There are two tetrahedral holes per S^{2-} ion in this structure. But there is only one Zn^{2+} ion per S^{2-} ion in the formula of the compound. Thus, only half of the tetrahedral holes are filled.

Experiment has shown that ZnS does in fact crystallize in a cubic closest-packed array of sulfur atoms with zinc atoms in half of the tetrahedral holes.

Exercise 12.5

Titanium metal reacts with chlorine to form titanium tetrachloride ($TiCl_4$), which is a colorless liquid at room temperature. Predict the structure of the solid formed when this compound freezes at $-24°C$.

Solution

If we assume that $TiCl_4$ contains Ti^{4+} and Cl^- ions, the radius ratio for this compound is 0.38.

$$\frac{r_+}{r_-} = \frac{0.068 \text{ nm}}{0.181 \text{ nm}} = 0.38$$

The Ti^{4+} ions should therefore pack in tetrahedral holes in a closest-packed array of Cl^- ions.

There are two tetrahedral holes per Cl^- ion in this structure. Four Cl^- ions are therefore associated with a total of eight tetrahedral holes. There is only one Ti^{4+} ion for each four Cl^- ions, however. Therefore, only one-eighth of the tetrahedral holes should be used.

$TiCl_4$ has been found to crystallize in a cubic closest-packed array of chlorine atoms with titanium atoms in one-eighth of the tetrahedral holes.

Every cubic hole in a simple cubic structure is surrounded by eight atoms or ions, and every atom that forms this structure is surrounded by eight cubic holes. There are therefore just as many cubic holes as atoms in a simple cubic structure.

Exercise 12.6

Predict the crystal structure of cesium chloride, CsCl.

Solution

CsCl is an example of a compound whose radius ratio falls between 0.732 and 1.

$$\frac{r_+}{r_-} = \frac{(0.169 \text{ nm})}{(0.181 \text{ nm})} = 0.933$$

The Cl^- ions in CsCl should therefore pack in a simple cubic array, with Cs^+ ions in the holes between planes of Cl^- ions. Because there are just as many Cs^+ ions as Cl^- ions in this crystal and there is one cubic hole per atom in a simple cubic structure, all of the cubic holes in CsCl are filled.

Exercise 12.7

The mineral fluorite, CaF_2, is an important source of the element fluorine. Predict the structure of this mineral.

Solution

If we assume that fluorite contains Ca^{2+} and F^- ions, the radius ratio of the compound is 0.73.

$$\frac{r_+}{r_-} = \frac{0.099 \text{ nm}}{0.136 \text{ nm}} = 0.73$$

The Ca^{2+} ions are just a little too big to fit into octahedral holes. CaF_2 therefore packs as a simple cubic array of F^- ions with Ca^{2+} in the cubic holes between planes of F^- ions.

There is one cubic hole per atom in a simple cubic structure. Because there are only half as many Ca^{2+} ions as F^- ions in this compound, only half of these holes should be occupied, which is exactly what is found experimentally.

12.12 UNIT CELLS: THE SIMPLEST REPEATING UNIT IN A CRYSTAL

The model used so far to describe solids focuses on the way atoms, ions, or molecules pack to fill space. Another model used to describe solids treats crystals as if they were three-dimensional analogs of a piece of wallpaper. Wallpaper has a regular repeating design that extends from one edge to the other. Crystals have a similar repeating "design," but in this case the design extends in all three dimensions from one edge of the solid to the other.

We can unambiguously describe a piece of wallpaper by simply specifying the size, shape, and contents of the simplest repeating unit in the design. We can similarly describe a three-dimensional crystal by specifying the size, shape, and contents of the simplest repeating unit and the way these repeating units stack to form the crystal.

The simplest repeating unit in a crystal is called a *unit cell*. Each unit cell is defined in terms of *lattice points* — the points in space about which an atom, ion, or molecule is free to vibrate in a crystal. In 1850, Auguste Bravais showed that any crystal could be described in terms of one of 14 unit cells, which meet the following criteria.

1. The unit cell is the simplest repeating unit in the crystal.
2. Opposite faces of a unit cell are parallel.
3. The edge of the unit cell connects equivalent points.

The 14 Bravais unit cells are shown in Figure 12.20. These unit cells fall into seven categories, which differ in terms of three unit-cell edge lengths (a, b, and c) and three internal angles (α, β, and γ), as shown in Table 12.3.

This chapter will focus on the cubic category, which includes three types of unit cells: simple cubic, body-centered cubic, and face-centered cubic. These unit cells are important for two reasons. First, a number of metals, ionic solids, and intermetallic compounds crystallize in these unit-cell patterns. Second, it is relatively easy to do calculations related to these unit cells because the cell-edge lengths are all the same and the cell angles are all 90°.

The *simple cubic unit cell* is the simplest repeating unit in a simple cubic structure. Each corner of the unit cell is defined by a lattice point at which an atom, ion, or molecule can be found in the crystal. Since the unit cell's edge always connects equivalent points, each of the eight corners of the unit cell must contain an identical atom, ion, or molecule. Other atoms, ions, or molecules might be present on the edges or faces of the unit cell or within the body of the unit cell. But the minimum that must be present for the unit cell to be classified as simple cubic is eight equivalent lattice points on the eight corners.

The *body-centered cubic unit cell* is the simplest repeating unit in a body-centered cubic structure. Once again, there are eight identical atoms, ions, or molecules on the eight corners of the unit cell. However, this time there is a ninth identical atom, ion, or molecule in the center of the body of the unit cell.

Exercise 12.8

Iron metal and cesium chloride have similar structures. The simplest repeating unit in iron is a cube of eight Fe atoms with a ninth Fe atom in the center of the body of the cube. The simplest repeating unit in CsCl is a cube of Cl⁻ ions with a Cs⁺ ion in the center of the body. Explain why these substances are not examples of the same Bravais unit cell.

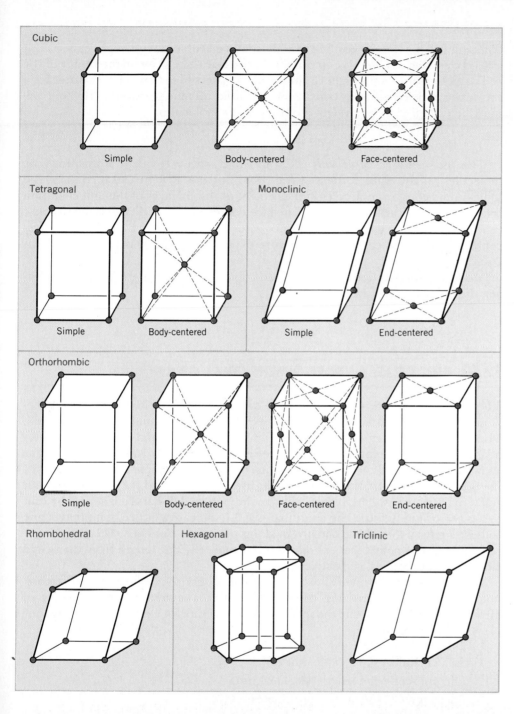

FIG. 12.20 The 14 Bravais unit cells: (1) simple cubic, (2) body-centered cubic, (3) face-centered cubic, (4) tetragonal, (5) body-centered tetragonal, (6) orthorhombic, (7) end-centered orthorhombic, (8) body-centered orthorhombic, (9) face-centered orthorhombic, (10) monoclinic, (11) end-centered monoclinic, (12) triclinic, (13) rhombohedral, and (14) hexagonal.

TABLE 12.3

The Seven Categories of Bravais Unit Cells

Cubic:	$(a = b = c)$	$(\alpha = \beta = \gamma = 90°)$	Triclinic:	$(a \neq b \neq c)$	$(\alpha \neq \beta \neq \gamma \neq 90°)$
Tetragonal:	$(a = b \neq c)$	$(\alpha = \beta = \gamma = 90°)$	Rhombohedral:	$(a = b = c)$	$(\alpha = \beta = \gamma \neq 90°)$
Orthorhombic:	$(a \neq b \neq c)$	$(\alpha = \beta = \gamma = 90°)$	Hexagonal:	$(a = b \neq c)$	$(\alpha = \beta = 90°, \gamma = 120°)$
Monoclinic:	$(a \neq b \neq c)$	$(\alpha = \beta = 90° \neq \gamma)$			

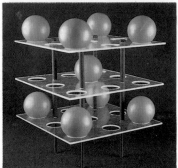

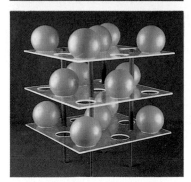

FIG. 12.21 Models of simple cubic, body-centered cubic, and face-centered cubic unit cells.

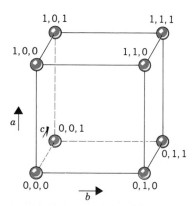

FIG. 12.22 The coordinates of the eight lattice points on the corners of a cubic unit cell.

Solution

The unit cell for iron is classified as body-centered cubic, as you might expect. But CsCl is classified as a simple cubic unit cell, because the Cs^+ ion in the center of the cell is not the same as the eight Cl^- ions at the corners of the unit cell. In order for a substance to be classified as body-centered cubic, all nine positions in the unit cell must be occupied by equivalent atoms, ions, or molecules.

The *face-centered cubic unit cell* also starts with equivalent atoms, ions, or molecules on the eight corners of the cube. But this structure also contains the same atoms, ions, or molecules in the centers of the six faces of the unit cell, for a total of 14 identical lattice points. The face-centered cubic unit cell is the simplest repeating unit in a cubic closest-packed structure. In fact, the presence of face-centered cubic unit cells in this structure explains why the structure is known as cubic closest-packed.

Models of simple cubic, body-centered cubic, and face-centered cubic unit cells are shown in Figure 12.21.

12.13 UNIT CELLS: A THREE-DIMENSIONAL GRAPH

The lattice points in a cubic unit cell can be described in terms of a three-dimensional graph. Because all three cell-edge lengths are equal in a cubic unit cell, it doesn't really matter what orientation is used for the *a*, *b*, and *c* axes. For the sake of argument, we'll define the *a* axis as the vertical axis of our coordinate system, as shown in Figure 12.22. The *b* axis will then describe movement across the front of the unit cell, and the *c* axis will represent movement toward the back of the unit cell. Furthermore, we'll arbitrarily define the bottom left-hand corner of the unit cell as the origin (0,0,0). The coordinates 1,0,0 indicate a lattice point that is one cell-edge length away from the origin along the *a* axis. Similarly, 0,1,0 and 0,0,1 represent lattice points that are displaced by one cell-edge length from the origin along the *b* and *c* axes, respectively.

Thinking about the unit cell as a three-dimensional graph allows us to unambiguously describe the structure of a crystal with a remarkably small amount of information. We can specify the shape of a CsCl crystal, for example, with only four pieces of information.

1. CsCl crystallizes in a cubic unit cell.
2. The unit-cell-edge length is 0.4123 nm.
3. There is a Cl^- ion at the coordinates 0,0,0.
4. There is a Cs^+ ion at the coordinates $\frac{1}{2},\frac{1}{2},\frac{1}{2}$.

Because the cell edge must connect equivalent lattice points, the presence of a Cl^- ion at one corner of the unit cell (0,0,0) implies that there must be a Cl^- ion at every corner of the cell. The coordinates $\frac{1}{2},\frac{1}{2},\frac{1}{2}$ correspond to a lattice point at the center of the cell. Because there is no other point in the unit cell that is one cell-edge length away from these coordinates, this is the only Cs^+ ion in the cell. CsCl is therefore a simple cubic unit cell of Cl^- ions with a Cs^+ in the center of the body of the cell, as shown in Figure 12.23.

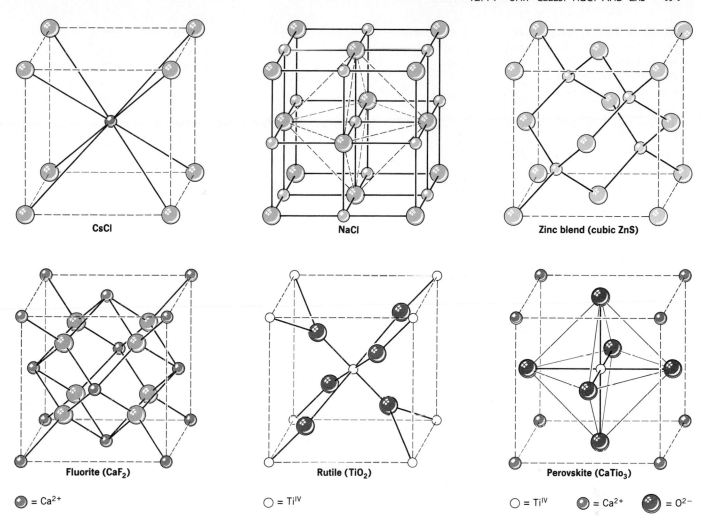

FIG. 12.23 The unit cells of NaCl (rock salt), CsCl, ZnS (zinc blende), CaF₂ (fluorite), TiO₂ (rutile), and CaTiO₃ (perovskite).

12.14 UNIT CELLS: NaCl AND ZnS

In Section 12.11, NaCl was described as a cubic closest-packed array of Cl⁻ ions with Na⁺ ions in all of the octahedral holes. We can translate this information into a unit-cell model for NaCl by remembering that the face-centered cubic unit cell is the simplest repeating unit in a cubic closest-packed structure.

There are four unique positions in a face-centered cubic unit cell. These positions are defined by the following sets of coordinates.

$$0,0,0; \quad 0,\tfrac{1}{2},\tfrac{1}{2}; \quad \tfrac{1}{2},0,\tfrac{1}{2}; \quad \tfrac{1}{2},\tfrac{1}{2},0$$

The presence of an atom at one corner of the unit cell (0,0,0) requires the presence of an equivalent atom on each of the eight corners of the unit cell. The coordinates $0,\tfrac{1}{2},\tfrac{1}{2}$ correspond to an atom in the center of the bottom face of the unit cell. Because a unit-cell edge must always connect equivalent points, the presence of an atom in the center of the bottom face implies the presence of an equivalent atom in the center of the top face $(1,\tfrac{1}{2},\tfrac{1}{2})$. Similarly, the presence of atoms in the center of the $\tfrac{1}{2},0,\tfrac{1}{2}$ and $\tfrac{1}{2},\tfrac{1}{2},0$ faces of the unit cell implies equivalent atoms in the centers of the $\tfrac{1}{2},1,\tfrac{1}{2}$ and $\tfrac{1}{2},\tfrac{1}{2},1$ faces.

FIG. 12.24 The octahedral holes in a face-centered cubic unit cell are in the center of the body (*a*) and on each of the edges of the unit cell (*b*).

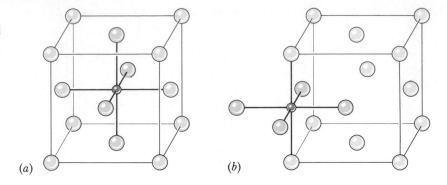

(*a*)   (*b*)

Figure 12.24*a* shows that there is an octahedral hole in the center of a face-centered cubic unit cell, at the coordinates $\frac{1}{2},\frac{1}{2},\frac{1}{2}$. Any atom at this point touches the atoms in the centers of the six faces of the unit cell. The other octahedral holes in a face-centered cubic unit cell are on the edges of the cell, as shown in Figure 12.24*b*. If Cl^- ions occupy the lattice points of a face-centered cubic unit cell and all of the octahedral holes are filled with Na^+ ions, we get the unit cell shown in Figure 12.23.

In Section 12.11, ZnS was described as a cubic closest-packed array of S^{2-} ions with Zn^{2+} ions in half of the tetrahedral holes. The S^{2-} ions in this crystal are at the same positions as the Cl^- ions in NaCl. The only difference between the crystals is the location of the positive ions. Figure 12.25 shows that the tetrahedral holes in a face-centered cubic unit cell are in the corners of the unit cell, at coordinates such as $\frac{1}{4},\frac{1}{4},\frac{1}{4}$. Any atom with these coordinates touches the atom at this corner as well as the atoms in the centers of the three faces that form this corner. Although it is difficult to see without a three-dimensional model, the four atoms that surround this hole are arranged toward the corners of a tetrahedron.

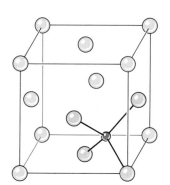

FIG. 12.25 The tetrahedral holes in a face-centered cubic unit cell are found in the eight corners of the unit cell. Each hole is surrounded by the atom on the corner of the unit cell and the three atoms in the centers of the faces that form this corner.

Because the corners of a cubic unit cell are identical, there must be a tetrahedral hole in each of the eight corners of the face-centered cubic unit cell. If S^{2-} ions occupy the lattice points of a face-centered cubic unit cell and Zn^{2+} ions are packed into every other tetrahedral hole, we get the unit cell of ZnS shown in Figure 12.23.

Section 12.11 noted that each atom in a closest-packed structure is surrounded by six octahedral holes. This can be seen in Figure 12.26*a*. Let's focus, for the moment, on the atom in the center of the top face of this unit cell. This atom is surrounded by six octahedral holes: four on the four edges of the face, one in the

FIG. 12.26 Every atom or ion that forms a face-centered cubic unit cell is surrounded by six octahedral holes and eight tetrahedral holes.

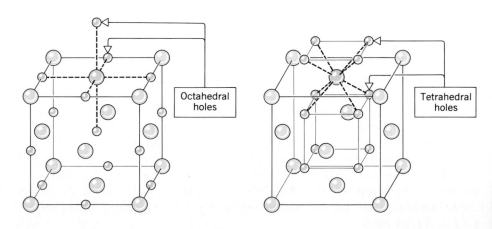

Octahedral holes

Tetrahedral holes

center of the unit cell beneath this face, and one in the center of the unit cell above this face.

Figure 12.26*b* can be used to demonstrate that each atom in a closest-packed structure is surrounded by eight tetrahedral holes, as noted in Section 12.11. Once again, focus on the atom in the center of the top face of the unit cell. There are four tetrahedral holes below this atom, at the four corners of the face. There should be an identical set of four tetrahedral holes above the atom, in the next unit cell. Thus, there are in fact eight tetrahedral holes around each atom in a face-centered cubic unit cell.

Now that we know where the octahedral and tetrahedral holes are located in a face-centered cubic unit cell, we can describe the crystal structure of NaCl in terms of the following information.

1. NaCl crystallizes in a cubic unit cell.
2. The cell-edge length is 0.5641 nm.
3. There are Cl^- ions at the positions $0,0,0$; $\frac{1}{2},\frac{1}{2},0$; $\frac{1}{2},0,\frac{1}{2}$; and $0,\frac{1}{2},\frac{1}{2}$.
4. There are Na^+ ions at the positions $\frac{1}{2},\frac{1}{2},\frac{1}{2}$; $\frac{1}{2},0,0$; $0,\frac{1}{2},0$; and $0,0,\frac{1}{2}$.

There is no need to report the positions of all 14 Cl^- ions in the unit cell. The 4 Cl^- ions specified above form a face-centered cubic unit cell. If there is a Cl^- ion at one corner of the unit cell $(0,0,0)$ there must be a Cl^- ion on each of the eight corners. Because the unit-cell edge connects equivalent positions, placing a Cl^- ion at the center of the three faces defined by the coordinates $\frac{1}{2},\frac{1}{2},0$; $\frac{1}{2},0,\frac{1}{2}$; and $0,\frac{1}{2},\frac{1}{2}$ requires the presence of a Cl^- in the center of all six faces of the cell.

The coordinates $\frac{1}{2},\frac{1}{2},\frac{1}{2}$ locate an Na^+ ion in the center of the unit cell. Because the cell edge must connect equivalent points, the presence of Na^+ ions on the cell edges with the coordinates $\frac{1}{2},0,0$; $0,\frac{1}{2},0$; and $0,0,\frac{1}{2}$ implies the presence of an equivalent Na^+ ion on all of the edges of the unit cell.

The structure of ZnS can be described as follows.

1. ZnS crystallizes in a cubic unit cell.
2. The cell-edge length is 0.5411 nm.
3. There are S^{2-} ions at the positions $0,0,0$; $\frac{1}{2},\frac{1}{2},0$; $\frac{1}{2},0,\frac{1}{2}$; and $0,\frac{1}{2},\frac{1}{2}$.
4. There are Zn^{2+} ions at the positions $\frac{1}{4},\frac{1}{4},\frac{1}{4}$; $\frac{1}{4},\frac{3}{4},\frac{3}{4}$; $\frac{3}{4},\frac{1}{4},\frac{3}{4}$; and $\frac{3}{4},\frac{3}{4},\frac{1}{4}$.

The positions of the four S^{2-} ions form a face-centered cubic unit cell. The coordinates of the four Zn^{2+} ions place positive ions in tetrahedral holes at every other corner of the cell.

12.15 UNIT CELLS: MEASURING THE DISTANCE BETWEEN PLANES OF ATOMS OR IONS

In Section 12.5, nickel was said to be one of the metals that crystallize in a cubic closest-packed structure. When you consider that a nickel atom weighs 9.75×10^{-23} gram and has an ionic radius of 1.24×10^{-10} meter, it is a remarkable achievement to be able to describe the structure of this metal. The obvious question is: How do we know that nickel packs in a cubic closest-packed structure?

The only way to determine the structure of matter on an atomic scale is to use a probe that is even smaller. One of the most useful probes for studying matter on this scale is electromagnetic radiation (see Section 5.11). In 1912, Max van Laue found that x-rays that struck the surface of a crystal were diffracted into patterns

FIG. 12.27 A model for the diffraction of x-rays by the first and second planes of atoms in a crystal.

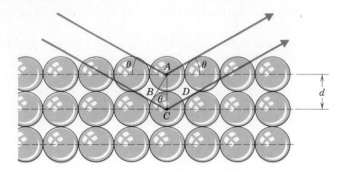

that resembled the patterns produced when light is diffracted as it passes through a very narrow slit. Shortly thereafter, William Lawrence Bragg, who was just completing his undergraduate degree in physics at Cambridge University, developed a model that explained van Laue's results.

Bragg argued that x-rays were reflected from planes of atoms near the surface of the crystal, as shown in Figure 12.27. X-rays that were reflected from different planes of atoms would no longer be in phase, and they would tend to cancel each other. They could stay in phase only if the extra distance ($BC + CD$) the x-ray moved when it struck the second plane of atoms was equal to an integer (n) times the wavelength (λ) of the radiation. Thus, Bragg argued that the sum of the distances labeled BC and CD in Figure 12.27 was equal to an integer times the wavelength of the radiation that remained in phase.

$$BC + CD = n\lambda$$

Bragg defined the angle between the beam of x-rays and the planes of the crystal as θ. The BAC angle is also equal to θ. The BC distance is therefore equal to the AC distance times the sine of θ. Because the AC distance is equal to the distance between planes of atoms, d, the relationship between the BC and AC distance can be written as follows.

$$BC = d \sin \theta$$

Because the BC distance is equal to the CD distance, the following must be true.

$$BC + CD = 2d \sin \theta$$

Substituting the relationship between $BC + CD$ and $n\lambda$ gives the following.

$$n\lambda = 2n \sin \theta$$

This is known as the **Bragg equation,** and it allows us to calculate the distance between planes of atoms in a crystal from the pattern of diffraction of x-rays of known wavelength.

12.16 UNIT CELLS: DETERMINING THE UNIT CELL OF A CRYSTAL

The pattern by which x-rays are diffracted by nickel metal suggests that this metal packs in a cubic unit cell with a distance between planes of atoms of 0.3524 nm. Thus, the cell-edge length in this crystal must also be 0.3524 nm.

Knowing that nickel crystallizes in a cubic unit cell is not enough. We still have to decide whether it is a simple cubic, body-centered cubic, or face-centered cubic unit cell. Fortunately, we can do this by measuring the density of the metal.

Atoms on the corners, edges, and faces of a unit cell are shared by more than one

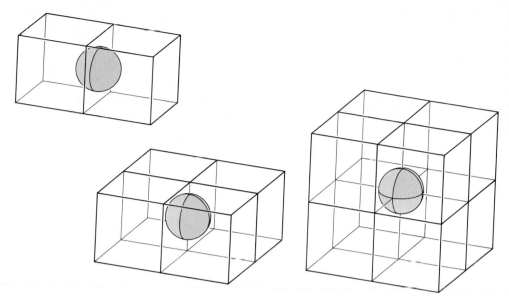

FIG. 12.28 Because an atom on the face of a unit cell is shared by two unit cells, only half of this atom belongs to each of these cells. For similar reasons, only one-quarter of an atom on an edge and one-eighth of an atom on a corner of a unit cell belong to this unit cell.

unit cell, as shown in Figure 12.28. An atom on a face is shared by two unit cells, so only half of the volume, or mass, of the atom belongs to each of these cells. An atom on an edge is shared by four unit cells, and an atom on a corner is shared by eight unit cells. Thus, only one-quarter of an atom on an edge and one-eighth of an atom on a corner can be attributed to each of the unit cells that share these atoms.

If nickel crystallized in a simple cubic unit cell, there would be eight nickel atoms on the eight corners of the cell. Because only one-eighth of the volume, or mass, of each of these atoms could be attributed to a given unit cell, each unit cell in a simple cubic structure would have one net nickel atom.

Simple cubic structure:

$$8 \text{ corners} \times \frac{1}{8} = 1 \text{ net atom}$$

A body-centered cubic structure for nickel would have two net atoms per unit cell, because of the nickel atom in the center of the body of the unit cell, which wouldn't be shared with any other unit cells.

Body-centered cubic structure:

$$\left(8 \text{ corners} \times \frac{1}{8}\right) + 1 \text{ body} = 2 \text{ net atoms}$$

If nickel crystallized in a face-centered cubic structure, the six atoms on the faces of the unit cell would contribute three net nickel atoms, for a total of four atoms per unit cell.

Face-centered cubic structure:

$$\left(8 \text{ corners} \times \frac{1}{8}\right) + \left(6 \text{ faces} \times \frac{1}{2}\right) = 4 \text{ net atoms}$$

We now have enough information to calculate the density of nickel under three different assumptions: that it crystallizes in a simple cubic unit cell, a body-centered cubic unit cell, and a face-centered cubic unit cell.

Recall that the cell-edge length is 0.3524nm and that density is the mass or weight of a sample divided by its volume. If nickel crystallized in a simple cubic unit cell, there would be one net nickel atom per unit cell. The density of nickel in this structure would be equal to the weight of a single nickel atom divided by the volume of the unit cell in cubic centimeters.

The volume (V) of the unit cell is equal to the cell-edge length (a) cubed.

$$V = a^3 = (0.3524 \text{ nm})^3 = 0.04376 \text{ nm}^3$$

Since there are 10^9 nanometers in a meter and 100 centimeters in a meter, there must be 10^7 nanometers in a centimeter.

$$10^9 \, \frac{\text{nm}}{\text{m}} \times \frac{1 \text{ m}}{100 \text{ cm}} = 10^7 \, \frac{\text{nm}}{\text{cm}}$$

Converting the volume of the unit cell to cubic centimeters gives the following result.

$$4.376 \times 10^{-2} \text{ nm}^3 \times \left(\frac{1 \text{ cm}}{10^7 \text{ nm}} \right)^3 = 4.376 \times 10^{-23} \text{ cm}^3$$

The weight of a nickel atom can be calculated from the atomic weight of this metal and Avogadro's constant.

$$\frac{58.69 \text{ g Ni}}{1 \text{ mol}} \times \frac{1 \text{ mol}}{6.022 \times 10^{23} \text{ atoms}} = 9.746 \times 10^{-23} \text{ g/atom}$$

Thus, the density of nickel if it crystallized in a simple cubic structure would be 2.227 g/cm³.

Simple cubic structure:
$$\frac{9.746 \times 10^{-23} \text{ g/unit cell}}{4.376 \times 10^{-23} \text{ cm}^3/\text{unit cell}} = 2.227 \text{ g/cm}^3$$

Because there would be twice as many nickel atoms per unit cell if nickel crystallized in a body-centered cubic structure, the density of nickel in this structure would be twice as large.

Body-centered cubic structure:
$$\frac{2(9.746 \times 10^{-23}) \text{ g/unit cell}}{4.376 \times 10^{-23} \text{ cm}^3/\text{unit cell}} = 4.454 \text{ g/cm}^3$$

There would be four nickel atoms per unit cell in a face-centered cubic structure, and so the density of nickel in this structure would be four times as large.

Face-centered cubic structure:
$$\frac{4(9.746 \times 10^{-23}) \text{ g/unit cell}}{4.376 \times 10^{-23} \text{ cm}^3/\text{unit cell}} = 8.909 \text{ g/cm}^3$$

The experimental value for the density of nickel is 8.90 g/cm³. The obvious conclusion is that nickel crystallizes in a face-centered cubic unit cell and therefore has a cubic closest-packed structure.

12.17 UNIT CELLS: CALCULATING METALLIC OR IONIC RADII OF ATOMS AND IONS

In Section 6.4, we discussed what happens to the size of metal atoms across a row or down a column of the periodic table, and estimates of the radii of most metal atoms can be found in Table A-8 in the appendix. Where do these data come from? How do we know, for example, that the radius of a nickel atom is 0.1246 nm?

Exercise 12.9

Use the fact that nickel crystallizes in a face-centered cubic unit cell with a cell-edge length of 0.3524 nm to calculate the radius of a nickel atom.

Solution

One of the faces of a face-centered cubic unit cell is shown in Figure 12.29. According to the figure, the diagonal across the face of this unit cell is equal to four times the radius of a nickel atom.

$$d_{face} = 4 \, r_{Ni}$$

The Pythagorean theorem states that the diagonal across a right triangle is equal to the sum of the squares of the other sides. The diagonal across the face of the unit cell is therefore related to the unit-cell edge length by the following equation.

$$d_{face}^{2} = a^2 + a^2 = 2a^2$$

Taking the square root of both sides gives the following result.

$$d_{face} = \sqrt{2} \, a$$

Since the diagonal across the face is four times the radius of a nickel atom, the following substitution can be made.

$$4 \, r_{Ni} = \sqrt{2} \, a$$

Thus, the radius of a nickel atom is 0.1246 nm.

$$r_{Ni} = \frac{\sqrt{2} \, a}{4} = \frac{\sqrt{2} \times (0.3524 \text{ nm})}{4} = 0.1246 \text{ nm}$$

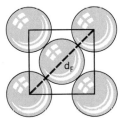

FIG. 12.29 The diagonal across the face of a face-centered cubic unit cell is equal to four times the radius of the atoms that form this cell.

Estimates of the size of a metal atom come from measurements of the size of the unit cell in which the metal crystallizes. A similar approach can be taken to estimating the size of an ion.

Exercise 12.10

Use the fact that the cell-edge length in cesium chloride is 0.4123 nm to calculate the Cs—Cl bond length in this crystal.

Solution

CsCl crystallizes in a simple cubic unit cell of Cl^- ions with a Cs^+ ion in the center of the body of the cell, as shown in Figure 12.23. According to the ground rules for predicting the structures of ionic solids, the Cs^+ ion in the center of the unit cell must touch the Cl^- ions at the corners.

The diagonal across the body of the CsCl unit cell contains two Cs—Cl bonds. This diagonal is therefore equal to the sum of the radii of two Cl^- ions and two Cs^+ ions.

$$d_{body} = 2 \, r_{Cs^+} + 2 \, r_{Cl^-}$$

The three-dimensional equivalent of the Pythagorean theorem suggests that the square of the diagonal across the body of a cube is the sum of the squares of the three sides.

$$d_{body}^2 = a^2 + a^2 + a^2 = 3\,a^2$$

Taking the square root of both sides of this equation gives the following result.

$$d_{body} = \sqrt{3}\,a$$

If the cell-edge length in CsCl is 0.4123 nm, the diagonal across the body in this unit cell is 0.7141 nm.

$$d_{body} = \sqrt{3}\,a = \sqrt{3} \times (0.4123 \text{ nm}) = \textbf{0.7141 nm}$$

The sum of the ionic radii of Cs^+ and Cl^- ions is half this distance, or 0.3571 nanometer.

$$r_{Cs^+} + r_{Cl^-} = \frac{d_{body}}{2} = \frac{0.7141 \text{ nm}}{2} = 0.3571 \text{ nm}$$

If we had an estimate of the size of either the Cs^+ or Cl^- ion, we could use the results of Exercise 12.10 to calculate the radius of the other ion. Ever since Chapter 6, we have used a value of 0.181 nm for the ionic radius of the Cl^- ion. Substituting this value into the last equation in Exercise 12.10 gives a value of 0.176 nm for the radius of the Cs^+ ion.

$$r_{Cs^+} = 0.3571 \text{ nm} - r_{Cl^-}$$
$$= 0.3571 \text{ nm} - 0.181 = \textbf{0.176 nm}$$

The results of this calculation are in reasonable agreement with the value of 0.169 nanometer given in Table A-8 for the radius of the Cs^+ ion. The discrepancy between two values reflects the fact that ionic radii seem to vary from one crystal to another. The tabulated values are averages of the results of a number of calculations of this type.

12.18 METALS, SEMICONDUCTORS, AND INSULATORS

A significant fraction of the gross national product of the United States, and all of the contribution to the GNP from the high-technology industries, can be traced to efforts to harness differences in the way metals, semiconductors, and insulators conduct electricity. This difference can be expressed in terms of *electrical conductivity* (σ), which measures the ease with which materials conduct electricity. It can also be expressed in terms of *electrical resistivity* (ρ), the inverse of conductivity, which measures the resistance of a material to carrying an electric charge.

Silver and copper metal are among the best conductors of electricity, with a resistance of only 10^{-6} ohms per centimeter (ohm-cm). (That is why copper is the metal most often used to carry the massive flow of current associated with electric wires.) The resistance of semiconductors, which include semimetals such as silicon and germanium, is 10^8 to 10^{10} times as large. When pure, these semimetals have a resistivity of 10^2 to 10^4 ohm-cm. Insulators include glass (10^{10} ohm-cm), diamond (10^{14} ohm-cm), and quartz (10^{18} ohm-cm); these materials have an extremely large resistance to carrying an electric current.

The 10^{24}-fold range of resistance is not the only difference among metals, semiconductors, and insulators. Metals become better conductors when they are cooled to lower temperatures. Some metals are such good conductors at very low temperatures that they no longer have a measurable resistance and therefore become superconductors.

Semiconductors show the opposite behavior—they become much better conductors as the temperature increases. The difference in the temperature dependence between metals and semiconductors is so large that it is often the best criterion for distinguishing between these two classes of materials.

Semiconductors are also very sensitive to impurities. The conductivity of silicon or germanium can be increased by a factor of as much as 10^6 by the addition of as little as 0.01% of an impurity. Metals, on the other hand, are fairly insensitive to impurities. It takes a lot of impurity to change the conductivity of a metal by as much as a factor of 10; and unlike semiconductors, metals become poorer conductors when impure.

The difference among metals, semiconductors, and insulators can be explained by building a model of the bonding in solids. The simplest element in the periodic table that is a solid at room temperature is lithium, whose electron configuration is $1s^2 2s^1$. In order to describe what happens when many lithium atoms interact to form a solid, we'll start by bringing together a pair of lithium atoms.

The molecular orbital diagram for an Li_2 molecule is shown in Figure 12.30a. The $1s$ orbitals interact to form a pair of σ_{1s} and σ_{1s}^* molecular orbitals. The same thing happens to the $2s$ orbitals. In each case, one of the molecular orbitals is lower in energy than the atomic orbitals, and the other molecular orbital is higher in energy than the atomic orbitals.

Imagine what happens when a great many—let's say 10^{21}—lithium atoms combine to form a crystal, which weighs about 0.01 gram. The $1s$ orbitals on the 10^{21} atoms overlap to form a band of 10^{21} orbitals with energies between the extremes of the σ_{1s} and σ_{1s}^* molecular orbitals in the Li_2 molecule, as shown in Figure 12.30b. Similarly, the $2s$ orbitals on the 10^{21} atoms overlap to form a band of 10^{21} orbitals with energies between the extremes of the σ_{2s} and σ_{2s}^* molecular orbitals.

The 10^{21} orbitals in the $1s$ band are filled with electrons. The $2s$ band, however, has only 1×10^{21} electrons—it is half-filled. It takes little, if any, energy to excite one of the electrons in the $2s$ band from one orbital to another in this band, so the electrons are free to move from one end of the crystal to the other. This band of orbitals is called the **conduction band,** and it explains how lithium metal conducts electricity.

How does a metal such as magnesium conduct electricity? The electron configuration of magnesium is $[Ne] 3s^2$. The $3s$ orbitals on 10^{21} magnesium atoms overlap to form a band of 10^{21} orbitals. But there are two electrons in each $3s$ orbital, and this band of orbitals is totally filled with 2×10^{21} electrons. The empty $3p$ orbitals on magnesium, however, also interact to form a band of 3×10^{21} orbitals. This empty $3p$ band overlaps the filled $3s$ band in magnesium, so that the combined band is only partially filled, allowing magnesium to conduct electricity.

The difference among metals, semiconductors, and insulators is shown in Figure 12.31. Metals have partially filled bands of orbitals that allow electrons to move from one end of the crystal to the other. All of the bands in an insulator are either filled or empty. Furthermore, the gap between the highest-energy filled band and the lowest-energy empty band is so large in an insulator that it is difficult to excite electrons from one of these bands to the other. Semiconductors also have a band structure that consists of filled and empty bands. The gap between the highest-energy filled band and the lowest-energy empty band is small enough, however,

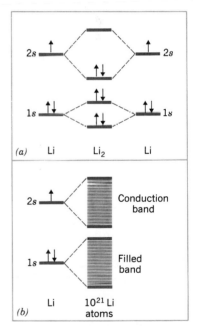

FIG. 12.30 (a) The overlap of atomic orbitals on one atom with those on another forms a limited number of molecular orbitals with distinct energies. (b) The overlap of atomic orbitals on a large number of atoms forms a continuous band of orbitals.

FIG. 12.31 Metals have bands or combinations of bands of orbitals that are only partially filled. Electrons can be excited from one orbital to another in these bands and can therefore move from one end of the crystal to the other. Insulators have totally filled and completely empty bands that are so far apart in energy that it is difficult to excite electrons from one band to the other. The gap between filled and empty bands in a semiconductor is small enough to allow some electrons in the filled band to cross this gap and enter the empty band. The partially filled band can now conduct an electric current by allowing electrons to flow through the crystal. The partially empty band can also conduct a current by allowing electrons to fill one hole, thereby creating another, so that the holes seem to flow through the crystal.

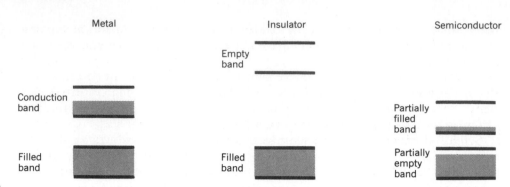

that electrons can be excited into the empty band by the thermal energy associated with electrons at room temperature.

To understand why metals become better conductors at low temperature, remember that temperature is a macroscopic reflection of the kinetic energy of the individual particles. Much of the resistance of a metal to an electric current at room temperature is the result of scattering of the electrons by the thermal motion of the metal atoms. As the metal is cooled and the thermal motion slows down, there is less scattering, and the metal becomes a better conductor.

Semiconductors become better conductors at high temperatures for exactly the opposite reason. As the temperature increases, the number of electrons with enough thermal energy to be excited from the filled band to the empty band becomes larger, and the semiconductor becomes a better conductor of electricity.

To understand why semiconductors, such as germanium and silicon, are sensitive to impurities, let's look at what happens when we add a small amount of a Group VA element, such as arsenic, to one of these Group IVA semiconductors. Arsenic atoms have one more valence electron than germanium and silicon atoms. Arsenic atoms can therefore lose an electron to form As^+ ions, which can occupy some of the lattice points in the crystal where silicon or germanium atoms are normally found.

If the amount of arsenic is kept very small, the distance between these atoms is so large that they do not interact. As a result, the extra electrons from the arsenic atoms occupy orbitals in a very narrow band of energies that lie between the filled and empty bands of the semiconductor, as shown in Figure 12.32. This structure decreases the amount of energy required to excite an electron into the lowest-energy empty band in the semiconductor and therefore increases the number of electrons that have enough energy to cross this gap. As a result, this "doped" semiconductor becomes a very much better conductor of electricity. Because the electric charge is carried by a flow of negative particles, these semiconductors are called *n-type*.

It is also possible to dope a Group IVA semiconductor with one of the elements in Group IIIA, such as indium. Indium atoms have one less valence electron than

FIG. 12.32 The energy required to excite an electron to the empty band of a semiconductor can be decreased if the semiconductor is "doped" with an impurity that contains electrons in orbitals that are closer in energy to the empty band. The net result is an increase in the number of electrons that are excited, which increases the conductivity of the semiconductor. Semiconductors can also be doped with an impurity that has empty orbitals that are closer in energy to the filled band. This increases the number of electrons that are excited out of the filled band, thereby increasing the conductivity of the semiconductor.

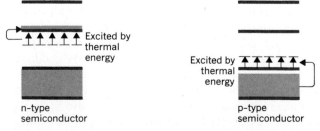

silicon and germanium atoms, and they can capture electrons from the highest-energy filled band of orbitals to form holes in this band. The presence of holes in a filled band has the same effect as the presence of electrons in an empty band — it allows the solid to carry an electric current. Because the electric charge is now carried by a flow of positive particles, or holes, these semiconductors are called *p-type.*

Semiconductors have many advantages over conductors such as metals. Because they tend to carry a much smaller electric current, it is easier to control the current's flow. Furthermore, bringing n-type and p-type semiconductors together produces a device that has a natural one-directional flow of electrons, which can be turned off by application of a small voltage in the opposite direction. This junction between n-type and p-type semiconductors was the basis of the revolution in industrial technology that followed the discovery of the transistor by William Shockley, John Bardeen, and Walter Brattain at Bell Laboratories in 1948.

A photograph of a small magnet floating above a liquid-nitrogen cooled specimen of the recently discovered high-temperature superconducting ceramics.

12.19 LIQUID CRYSTALS

Molecules that are particularly large, rigid, and linear often form solids that have two melting points. Instead of going directly from the solid to the liquid phase, these compounds pass through an intermediate phase best described as a **liquid crystal.** Liquid crystals have some of the structure of solids and some of the freedom of motion associated with liquids.

Liquid crystals were discovered in 1888, but they were primarily a laboratory curiosity until about 30 years ago. They are now used commercially in the displays of electrical devices such as digital watches and calculators and as radiation, pressure, or temperature sensors. Their commercial applications result from the fact that the weak bonds that hold the molecules together in a liquid crystal are easily affected by changes in pressure, temperature, or electromagnetic fields.

Liquid crystals are often divided into three categories: smectic, nematic, and cholesteric. **Smectic** liquid crystals have a structure that resembles a handful of cigars, as shown in Figure 12.33a. Not only do the molecules all point in the same direction, they are so well ordered that they form planes perpendicular to the axes of the molecules. **Nematic** liquid crystals are slightly less ordered. The molecules still point in the same direction, but they start and stop at different positions within the liquid, as shown in Figure 12.33b.

Cholesteric liquid crystals have a structure similar to nematic liquid crystals, but each plane of molecules is twisted slightly in relation to the plane above or below. These liquid crystals received their name from the fact that many derivatives of

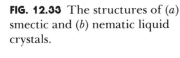

FIG. 12.33 The structures of (a) smectic and (b) nematic liquid crystals.

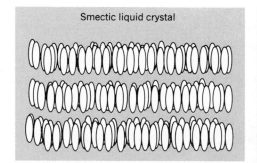

Smectic liquid crystal

(a)

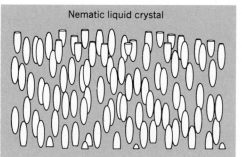

Nematic liquid crystal

(b)

cholesterol form this structure. The slight twist in their structures tends to make these liquid crystals colored, and the fact that changes in the amount of twisting lead to changes in color make these crystals sensitive indicators of changes in temperature or pressure. The most sensitive cholesteric liquid crystals show a detectable color change with temperature changes as small as $0.001°C$.

SUMMARY

Solids retain their shape, are very difficult to compress, and are denser than liquids and gases. These characteristic properties suggest that solids contain atoms, ions, or molecules packed as tightly as possible. Ionic compounds form solids to maximize the force of attraction between the ions of opposite charge by keeping these ions as close together as possible. Covalent solids, such as diamond, are held together by extended arrays of relatively strong covalent bonds. Metals form solids to maximize the number of metallic bonds that can form. Molecular solids form because of the intermolecular van der Waals forces between molecules.

Some solids are crystalline; they have structures that repeat in a regular pattern. Some are polycrystalline; they contain small regions where the structure fits a regular pattern. Others are amorphous; their structures have little, if any, regularity.

There are two ways of describing the regular patterns in a crystal. One looks at the crystal as an extended array of planes of atoms, ions, or molecules. The other focuses attention on the simplest repeating unit, or unit cell, in the crystal. From the first perspective, solids are described in terms of simple cubic, body-centered cubic, hexagonal closest-packed, and cubic closest-packed structures. The other perspective classifies solids in terms of unit cells, such as simple cubic, body-centered cubic, and face-centered cubic unit cells.

The structure of ionic compounds, or salts, can be described in terms of either model. Ionic compounds can be thought of as extended arrays of relatively large negative ions, with the smaller positive ions packed in the holes between the planes of atoms. When the negative ions form a closest-packed structure, the positive ions can occupy either tetrahedral or octahedral holes, depending on their size. When the negative ions pack in a simple cubic structure, the positive ions can occupy cubic holes.

The unit-cell model describes the structure of ionic compounds in terms of a three-dimensional graph, in which the position of an atom on the corner of a unit cell is indicated by the coordinates 0,0,0. It isn't necessary to list the coordinates of all of the atoms that form a unit cell. Since the cell edge must connect equivalent points in the unit cell, the presence of an atom at the coordinates 0,0,0 implies the presence of equivalent atoms at all eight corners of the cell.

PROBLEMS

Gases, Liquids, and Solids

12-1 Describe the differences in the properties of gases, liquids, and solids on the atomic scale. Explain how these differences gives rise to the observed differences in the macroscopic properties of these three states of matter.

12-2 Describe the difference between intermolecular and intramolecular bonds, giving an example of each. Which is stronger?

Van der Waals Forces

12-3 Which term in the van der Waals equation corrects for the force of attraction between molecules?

12-4 Describe the difference among the three forms of van der Waals forces.

12-5 Propose an explanation for the fact that induced dipole–induced dipole forces become larger as the number of electrons on an atom increases.

Crystals, Amorphous Solids, and Polycrystalline Substances

12-6 Describe the difference between crystalline and amorphous solids. Give common examples of both kinds of solids.

12-7 Plastic sandwich bags are made out of polyethylene, whereas the plastic chairs used in classrooms are made out of polypropylene. Which of these plastics is more crystalline?

Molecular, Covalent, Ionic, and Metallic Solids

12-8 Classify the following solids as molecular, covalent, ionic, or metallic.

(a) $BaSO_4$ (b) NaOH (c) Xe (d) I_2 (e) aluminum (f) brass (g) P_4 (h) P_4O_{10}

12-9 Which of the following solids are held together by an extended network of covalent bonds?

(a) sodium chloride (b) graphite (c) gold (d) calcium carbonate (e) diamond (f) dry ice (solid CO_2)

12-10 Which force makes the most important contribution to the lattice energy of solid CO_2?

(a) metallic bonding (b) ionic bonding (c) covalent bonding (d) van der Waals forces

12-11 Which force makes the most important contribution to the lattice energy of solid argon?

(a) metallic bonding (b) ionic bonding (c) covalent bonding (d) van der Waals forces

12-12 Which of the following categories is most likely to contain a compound that is a poor conductor of electricity when solid but a very good conductor when molten?

(a) molecular solid (b) covalent solid (c) ionic solid (d) metallic solid

The Structure of Metals and Other Monatomic Solids

12-13 Describe the difference in the way planes of atoms stack to form hexagonal closest-packed, cubic closest-packed, body-centered cubic, and simple cubic structures.

12-14 Explain why the structure of polonium is called *simple cubic;* why the structure of iron is called *body-centered cubic;* and why the structure of copper is called *hexagonal closest-packed.*

12-15 Determine the coordination numbers of the metal atoms in each of the following structures.

(a) cubic closest-packed aluminum (b) hexagonal closest-packed magnesium (c) body-centered cubic chromium (d) simple cubic polonium

12-16 In which of the following structures would a xenon atom form the largest number of induced dipole–induced dipole interactions?

(a) simple cubic (b) body-centered cubic (c) cubic closest-packed (d) hexagonal closest-packed

12-17 Explain why a metal such as iron would pack in a body-centered cubic structure, in which the coordination number is 8, instead of a cubic closest-packed or hexagonal closest-packed structure, in which the coordination number is 12.

12-18 Sodium crystallizes in a structure in which the coordination number is 8. Which structure best describes this crystal?

(a) simple cubic (b) body-centered cubic (c) cubic closest-packed (d) hexagonal closest-packed

Physical Properties That Result from the Structure of Metals

12-19 Use the structure of metals to explain why they have a metallic luster or shine.

12-20 Use the structure of metals to explain why they are solids at room temperature.

12-21 Use the structure of metals to explain why they are malleable and ductile.

12-22 Use the structure of metals to explain why they conduct heat and electricity.

Solid Solutions and Intermetallic Compounds

12-23 Describe the difference between an intermetallic compound, such as $CuAl_2$, and an alloy, such as brass or bronze.

12-24 Describe how the formation of the intermetallic compounds $CuAl_2$ and Fe_3C hardens aluminum and steel.

Holes in Closest-Packed and Simple Cubic Structures

12-25 Tetrahedral holes and octahedral holes can be found in which of the following structures?

(a) simple cubic (b) body-centered cubic (c) cubic closest-packed (d) hexagonal closest-packed

12-26 What is the coordination number of a cation packed in each of the following holes?

(a) tetrahedral holes (b) octahedral holes (c) cubic holes

12-27 Cubic holes can be found in which of the following structures?

(a) simple cubic (b) body-centered cubic (c) cubic closest-packed (d) hexagonal closest-packed

12-28 Which is the smallest hole: tetrahedral, octahedral, or cubic? Which is the largest?

12-29 Prove that the cation that just fits into an octahedral hole formed by six identical anions has a radius 0.414 times the radius of the anions that form the hole.

12-30 Prove that the cation that just fits into a tetrahedral hole formed by four identical anions has a radius 0.225 times the radius of the anions that form the hole.

The Structure of Ionic Solids

12-31 In KF, the K^+ and F^- ions are almost exactly the same size; both have ionic radii of approximately 0.134 nm. Which is larger, a neutral potassium atom or a neutral fluorine atom?

12-32 Rutile is a mineral that contains titanium and oxygen. The crystal structure can be described as a closest-packed array of oxygen atoms with titanium atoms in half of the octahedral holes. What is the empirical formula of this mineral?

12-33 Titanium carbide is a covalent carbide that is essentially inert to chemical reactions, has a very high melting point, and is almost as hard as diamond. The crystal structure can be described as a closest-packed array of carbon atoms with titanium atoms in all of the octahedral holes. What is the empirical formula of this compound?

12-34 NaCl, AgCl, KH, LiH, MgO, MnS, and CaO all have the same crystal structure—a closest-packed array of negative ions with positive ions in all of the octahedral holes. This structure would be predicted for all but one of these compounds on the basis of their chemical formula and the radius ratios of the ions. For which compound is this structure not expected?

12-35 Pyrite, FeS_2, crystallizes in a structure that can be described as a closest-packed array of positive ions with negative ions in all of the octahedral holes. Which formulation describes this mineral better, $[Fe^{4+}][S^{2-}]_2$ or $[Fe^{2+}][S_2^{2-}]$?

12-36 Cadmium iodide crystallizes in a structure that can be described as a closest-packed array of iodide atoms with cadmium atoms in half of the octahedral holes. What is the oxidation state of cadmium in this compound?

12-37 What is the formula of an oxide of titanium that crystallizes in a structure that can be described as a closest-packed array of oxygen atoms with titanium atoms in two-thirds of the octahedral holes?

12-38 Use the relative size of the Ge^{4+} and O^{2-} ions to predict the structure of GeO_2. What holes are used? What fraction of the holes are occupied? What is the coordination number of the Ge^{4+} ions?

12-39 Zinc telluride crystallizes in a cubic closest-packed structure of tellurium atoms with zinc atoms in half of the tetrahedral holes. What is the empirical formula of this compound?

12-40 Which of the following compounds is most likely to crystallize in a structure in which the positive ions pack in tetrahedral holes in a closest-packed array of negative ions?

(a) Li_2S (b) Cs_2S (c) Li_3P (d) Cs_3P (e) $CsCl$

12-41 Which compound can crystallize in a structure that contains a closest-packed array of negative ions with lithium ions in all of the tetrahedral holes?

(a) Li_4C (b) Li_3P (c) Li_2S (d) LiI

12-42 Which structure would BeO be expected to most closely resemble?

(a) NaCl (b) CsCl (c) ZnS (d) CaF_2

12-43 An oxide of cobalt crystallizes in a closest-packed array of oxygen atoms with cobalt atoms in one-eighth of the tetrahedral holes and one-half of the octahedral holes. What is the oxidation state of the cobalt in this compound?

12-44 Perovskite is a mineral with the formula $CaTiO_3$. Which of the positive ions in this crystal—Ti^{4+} or Ca^{2+}—is more likely to pack in the octahedral holes?

12-45 Thallium cyanide, $Tl(CN)_x$, crystallizes as a simple cubic array of CN^- ions with thallium ions in all of the cubic holes. What is the oxidation state of thallium in this compound?

Unit Cells: The Simplest Repeating Unit in a Crystal

12-46 Define the term *unit cell*. Describe the common properties of all unit cells.

12-47 Describe the difference among simple cubic, body-centered cubic, and face-centered cubic unit cells.

12-48 Simple cubic unit cells are found in simple cubic structures, and body-centered cubic unit cells are found in body-centered cubic structures. Where are face-centered cubic unit cells found, in cubic closest-packed structures or hexagonal closest-packed structures?

12-49 Sodium hydride crystallizes in a face-centered cubic unit cell of H^- ions with Na^+ ions at the center of the unit cell and in the center of each edge of the unit cell. How many Na^+ ions does each H^- ion touch? How many H^- ions does each Na^+ ion touch?

12-50 Picture the smallest repeating unit of a simple cubic crystal. Since neighboring atoms touch in this crystal, the length of an edge of this simplest repeating unit is twice the radius of the atoms that form the crystal. If the volume of a sphere is $\frac{4}{3}\pi r^3$, what fraction of this crystal is empty space?

12-51 Calculate the fraction of empty space in a body-centered cubic unit cell and a face-centered cubic unit cell.

Unit Cells: A Three-Dimensional Graph

12-52 The positions of atoms in a unit cell can be described in terms of a three-dimensional graph in which the location 0,0,0 corresponds to one of the corners of the cell. Describe the positions indicated by the following sets of coordinates.

(a) $\frac{1}{2},0,0$ (b) $\frac{1}{2},\frac{1}{2},0$ (c) $\frac{1}{2},\frac{1}{2},\frac{1}{2}$ (d) $\frac{1}{4},\frac{1}{4},\frac{1}{4}$

12-53 Explain why stating that there is a Cs^+ ion at the coordinates 0,0,0 in the unit cell of CsCl implies that there must be an equivalent Cs^+ ion on each of the other corners of the unit cell.

12-54 Explain why stating that there are Cl^- ions in the centers of three of the faces of the unit cell of NaCl at the coordinates $0,\frac{1}{2},\frac{1}{2}$; $\frac{1}{2},0,\frac{1}{2}$; and $\frac{1}{2},\frac{1}{2},0$ implies that there must be an equivalent Cl^- ion in the center of each of the other faces of the unit cell.

12-55 At very low temperatures, argon crystallizes in a structure in which Ar atoms are located at the following positions: $0,0,0$; $0,\frac{1}{2},\frac{1}{2}$; $\frac{1}{2},0,\frac{1}{2}$; $\frac{1}{2},\frac{1}{2},0$. Is this unit cell simple cubic, body-centered cubic, or face-centered cubic?

12-56 The mineral cuprite crystallizes in a structure in which oxygen atoms are located at the coordinates 0,0,0 and $\frac{1}{2},\frac{1}{2},\frac{1}{2}$ and copper atoms are located at $\frac{1}{4},\frac{1}{4},\frac{1}{4}$; $\frac{1}{4},\frac{3}{4},\frac{3}{4}$; $\frac{3}{4},\frac{1}{4},\frac{3}{4}$; and $\frac{3}{4},\frac{3}{4},\frac{1}{4}$. Is this unit cell simple cubic, body-centered cubic, or face-centered cubic?

Unit Cells: Determining the Unit Cell of a Crystal

12-57 Calculate the number of chloride and ammonium ions per unit cell for NH_4Cl, which crystallizes in a structure that can be described as a simple cubic unit cell of NH_4^+ ions with a Cl^- ion in the center of the cell.

12-58 Gallium arsenide is a semiconductor with several advantages over silicon. It crystallizes in a structure in which there are gallium atoms at 0,0,0; $\frac{1}{2},\frac{1}{2},0$; $\frac{1}{2},0,\frac{1}{2}$; and $0,\frac{1}{2},\frac{1}{2}$ and arsenic atoms at $\frac{1}{4},\frac{1}{4},\frac{1}{4}$; $\frac{1}{4},\frac{3}{4},\frac{3}{4}$; $\frac{3}{4},\frac{1}{4},\frac{3}{4}$; and $\frac{3}{4},\frac{3}{4},\frac{1}{4}$. Describe the unit cell of this compound, the kind of holes in which the arsenic atoms are found, the fraction of holes occupied, and the empirical formula of the compound.

12-59 The mineral perovskite crystallizes in a cubic unit cell in which there is a titanium atom at 0, 0, 0, a calcium atom at $\frac{1}{2},\frac{1}{2},\frac{1}{2}$, and oxygen atoms at $\frac{1}{2},0,0$; $0,\frac{1}{2},0$; and $0,0,\frac{1}{2}$. Describe the unit cell and calculate the empirical formula of this compound.

Unit Cells: Calculating Metallic or Ionic Radii of Atoms and Ions

12-60 Chromium metal (density = 7.20 g/cm³) crystallizes in a body-centered cubic unit cell. Calculate the volume of this unit cell and the radius of a chromium atom.

12-61 The atomic radius of a titanium atom is 0.1448 nm. What is the density of titanium if this metal crystallizes in a body-centered cubic unit cell?

12-62 Calculate the atomic radius of an Ar atom, assuming that argon crystallizes at low temperature in a face-centered cubic unit cell with a density of 1.623 g/cm³.

12-63 Silver crystallizes in a face-centered cubic unit cell with an edge length of 0.40862 nm. Calculate the density of this metal in grams per cubic centimeter.

12-64 Potassium crystallizes in a cubic unit cell with an edge length of 0.5247 nm. The density of potassium is 0.856 g/cm³. Determine whether this element crystallizes in a simple cubic, body-centered cubic, or face-centered cubic unit cell.

12-65 Determine whether calcium crystallizes in a simple cubic, body-centered cubic, or face-centered cubic unit cell, assuming that the cell-edge length is 0.5582 nm and the density of this metal is 1.55 g/cm³.

12-66 Determine whether molybdenum crystallizes in a simple cubic, body-centered cubic, or face-centered cubic unit cell, assuming that the cell-edge length is 0.3147 nm and the density of this metal is 10.2 g/cm³.

12-67 Diamond crystallizes in a cubic unit cell in which there are carbon atoms at the positions $0,0,0$; $\frac{1}{2},\frac{1}{2},0$; $\frac{1}{2},0,\frac{1}{2}$; $0,\frac{1}{2},\frac{1}{2}$; $\frac{1}{4},\frac{1}{4},\frac{1}{4}$; $\frac{1}{4},\frac{3}{4},\frac{3}{4}$; $\frac{3}{4},\frac{1}{4},\frac{3}{4}$; and $\frac{3}{4},\frac{3}{4},\frac{1}{4}$. What is the cell-edge length of this crystal if the density of diamond is 3.515 g/cm³?

12-68 Which of the following metals crystallizes in a face-centered cubic unit cell with an edge length of 0.3608 nm if the density of the metal is 8.95 g/cm³?

(a) Na (b) Ca (c) Tl (d) Cu (e) Au

12-69 Iron (density = 7.86 g/cm³) crystallizes in a body-centered cubic unit cell at room temperature. Calculate the radius of an iron atom in this crystal. At temperatures above 910°C, iron prefers a face-centered cubic structure. If we assume that the change in the size of the iron atom is negligible when the metal is heated to 910°C, what is the density of iron in the face-centered cubic structure? Does iron expand or contract when it changes from the BCC to the FCC structure?

12-70 Barium crystallizes in a body-centered cubic structure in which the cell-edge length is 0.5025 nm. Calculate the shortest distance between neighboring barium atoms in this crystal.

12-71 NaH crystallizes in a structure similar to that of NaCl. If the cell-edge length in this crystal is 0.4880 nm, what is the average length of the Na—H bond in the crystal?

12-72 TlI crystallizes in a structure similar to that of CsCl with a cell-edge length of 0.4198 nm. Calculate the average Tl—I bond length in this crystal. If the ionic radius of an I⁻ ion is 0.216 nanometer, what is the ionic radius of the Tl⁺ ion?

12-73 Calculate the ionic radius of the Cs⁺ ion, assuming that the cell-edge length for CsCl is 0.4123 nm and the ionic radius of a Cl⁻ ion is 0.181 nm.

12-74 CdO crystallizes in a cubic unit cell with a cell-edge length of 0.4695 nm. Calculate the number of Cd^{2+} and O^{2-} ions per unit cell, assuming that the density of this crystal is 8.15 g/cm³.

12-75 LiF crystallizes in a cubic unit cell with a cell-edge length of 0.4017 nm. Calculate the number of Li⁺ and F⁻ ions per unit cell, assuming that the density of this salt is 2.640 g/cm³.

Metals, Semiconductors, and Insulators

12-76 Describe how metals, semiconductors, and insulators differ.

12-77 Use the information presented in Section 12.18 to explain how metals conduct an electric current.

12-78 Explain why metals become better conductors of electricity as the temperature decreases but semiconductors become better conductors of electricity as the temperature increases.

12-79 Explain why adding small quantities of arsenic or gallium increases the conductivity of silicon.

12-80 Describe the difference between n-type and p-type semiconductors.

CHAPTER 13

LIQUIDS AND SOLUTIONS

CHAPTER CONTENTS

13.1 THE STRUCTURE OF LIQUIDS

The particles in a solid are packed as tightly as possible, and they diffuse through the solid slowly — if at all. As a result, solids retain their shapes. Gases, on the other hand, are primarily empty space through which particles are free to move randomly. Gases therefore expand or contract to conform to the shape of their containers. Liquids lie between the extremes of solids and gases. The particles in a liquid are free to move past each other, so liquids take on the shape of the portion of the container they occupy. But they never expand or contract to fill the container.

Many of the physical properties of gases are the same regardless of the identity of the gas. One equation, for example, does a remarkably good job of describing the relationship among the pressure, volume, amount, and temperature of almost any gas. Solids are very different. The physical properties of a solid depend on the way the atoms, ions, or molecules pack to form the solid. Thus, the structure of each solid must be described individually. Once again, liquids lie between these extremes. They don't all have the same physical properties; NaCl dissolves in water, for example, but not in carbon tetrachloride (CCl_4). But liquids don't have the extended structure associated with solids, so they tend to fall into a limited number of categories with similar properties.

The difference in the structures of gases, liquids, and solids might best be described by comparing the densities of substances, such as argon, nitrogen, and oxygen, that can be studied in all three phases. As shown by the data in Table 13.1, in each case the solid is about 20% denser than the corresponding liquid, and the liquid is as much as 800 times as dense as the gas.

Figure 13.1 shows a model for the structure of a liquid that is consistent with these data. The key points of this model are summarized below.

1. The particles that form a liquid are relatively close together.
2. The particles in a liquid have more kinetic energy than the particles in the corresponding solid.
3. As a result, these particles move faster in terms of vibration, rotation, and translation.
4. Because they are moving faster, the individual particles in the liquid occupy more space, and the liquid is less dense than the corresponding solid.
5. Differences in kinetic energy alone cannot explain the relative densities of liquids and solids. The model therefore assumes that there are small, particle-sized holes randomly distributed through the liquid.
6. Particles that are close to one of these holes behave in much the same way as particles in a gas, whereas particles that are far from a hole act more like the particles in a solid.

FIG. 13.1 A model for the structure of liquids. The particles do not pack as tightly in a liquid as they do in a solid because they have more thermal energy and are moving faster about their lattice positions. This model assumes the presence of small, molecule-sized holes that enable the liquid to flow so that it can conform to the shape of its container.

TABLE 13.1			
Densities of Solid, Liquid, and Gaseous Forms of the Elements Argon, Nitrogen, and Oxygen			
	Solid (g/cm³)	Liquid (g/cm³)	Gas (g/cm³)
Ar	1.65	1.40	0.001784
N_2	1.026	0.8081	0.001251
O_2	1.426	1.149	0.001429

WHAT KINDS OF MOLECULES FORM
13.2 LIQUIDS AT ROOM TEMPERATURE?

Three factors determine whether an element or compound is a gas, a liquid, or a solid at room temperature and atmospheric pressure: (1) the strength of the bonds between the particles that form the substance, (2) the atomic or molecular weight of these particles, and (3) the shape of the particles.

The covalent, ionic, and metallic solids described in Section 12.4 all have one thing in common: the bonds between the atoms or ions that form these substances are relatively strong. The molecular solids described in that section have a similar property. The van der Waals forces that hold these molecules together are relatively strong—so strong that these substances are solids at room temperature. When the van der Waals forces that hold atoms or molecules together are relatively weak, the substance is much more likely to be a gas at room temperature. Compounds that are liquids at room temperature tend to include substances, such as H_2O and CCl_4, in which the intermolecular bonds are neither too strong nor too weak.

The role of atomic or molecular weight in determining whether a substance is a gas, a liquid, or a solid at room temperature can be understood in terms of the kinetic molecular theory outlined in Section 4.15, which includes the following assumption.

The average kinetic energy of a collection of gas particles depends on the temperature of the gas, and nothing else.

As noted in Section 4.16, this means that the average velocity at which different molecules move at the same temperature is inversely proportional to the square root of their molecular weights.

$$\frac{v_A}{v_B} = \sqrt{\frac{MW_B}{MW_A}}$$

Relatively light molecules (such as CH_4) move so rapidly at room temperature that they can easily break the bonds that hold them together in a liquid or solid (CH_4 is therefore a gas at room temperature). Heavier molecules must be heated to a higher temperature before they can move fast enough to escape from the liquid. They therefore tend to have higher boiling points and are more likely to be liquids at room temperature.

The relationship between the molecular weight of a compound and its boiling point is shown in Table 13.2. The compounds in this table are members of a family of compounds known as the **hydrocarbons** (literally, compounds of hydrogen and carbon), and they all have the same generic formula—C_nH_{2n+2}. The only difference in these compounds is their size and therefore their molecular weights. As shown by Figure 13.2, the relationship between the molecular weights of these compounds and their boiling points is not a straight line, but it is a remarkably smooth curve.

The data in Figure 13.3 show how the shape of a molecule influences the melting point and boiling point of a compound and therefore the probability that the compound is a liquid at room temperature. The three compounds in this figure are **isomers** (literally, equal parts); they have the same chemical formulas but different structures. One of these isomers—neopentane—is a very symmetrical molecule with four identical CH_3 groups arranged in a tetrahedral pattern around

TABLE 13.2

Melting Points and Boiling Points of Compounds
with the Generic Formula C_nH_{2n+2}

Compound	Melting Point (°C)		Boiling Point (°C)
CH_4	−182	−164	
C_2H_6	−183.3	−88.6	gases at room temperature
C_3H_8	−189.7	−42.1	
C_4H_{10}	−138.4	−0.5	
C_5H_{12}	−130	36.1	
C_6H_{14}	−95	69	
C_7H_{16}	−90.6	98.4	
C_8H_{18}	−56.8	125.7	liquids at room temperature
C_9H_{20}	−51	150.8	
$C_{10}H_{22}$	−29.7	174.1	
$C_{12}H_{26}$	−9.6	216.3	

a central carbon atom. This molecule is so symmetrical that it easily packs to form a solid, so neopentane has to be cooled only to −16.5°C before it crystallizes.

Pentane and isopentane molecules have zig-zag, or pleated, structures, which differ only in terms of whether the chain of C—C bonds is linear or branched. These less symmetrical molecules are harder to pack to form a solid, so these compounds must be cooled to much lower temperatures before they become solids. Pentane freezes at −130°C, and isopentane must be cooled to almost −160°C before it forms a solid.

The shape of the molecule also influences the boiling point. The symmetrical neopentane molecules escape from the liquid the way marbles might pop out of a box when it is shaken. The pentane and isopentane molecules tend to get tangled, like coat hangers, and they must be heated to higher temperatures before they can

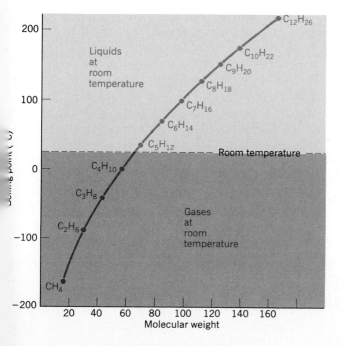

FIG. 13.2 There is a gradual increase in the boiling point of compounds with the generic formula C_nH_{2n+2} as the molecules become heavier.

FIG. 13.3 Melting points and boiling points for the three isomers with the formula C_5H_{12}.

Compound	Melting point (°C)	Boiling point (°C)
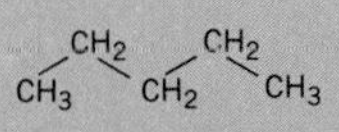 CH_3 CH_2 CH_2 CH_2 CH_3 (pentane)	−130	36.1
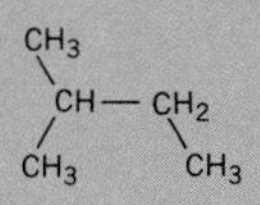 CH_3 CH CH_2 CH_3 CH_3 (isopentane)	−159.9	27.8
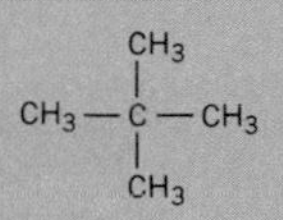 CH_3—C—CH_3 CH_3 CH_3 (neopentane)	−16.5	9.5

boil. Unsymmetrical molecules therefore tend to be liquids over a larger range of temperatures than symmetrical molecules.

13.3 VAPOR PRESSURE

A liquid doesn't have to be heated to its boiling point before it can become a gas, or vapor. Water, for example, evaporates from an open container at room temperature (20° to 25°C), even though the boiling point of water is 100°C.

We can explain this fact by noting that the kinetic theory assumes that the *average* kinetic energy of a collection of gas particles depends on the temperature of the gas; not all of the particles have the same kinetic energy. At a given temperature, some of the particles have more kinetic energy than others, as shown in Figure 13.4.

Liquids and solids are far more complex than gases. But the kinetic energy of their particles still depends on the temperature, and some particles still have more kinetic energy than others. Even at temperatures well below the boiling point of water, some water molecules are moving fast enough to escape from the liquid.

When this happens, the average kinetic energy of the liquid decreases. In essence, the liquid becomes cooler. It therefore absorbs energy from its surroundings until it returns to thermal equilibrium (see Section 9.1). But as soon as this happens, some of the water molecules once again have enough energy to escape from the liquid. In an open container, this process continues until all of the water evaporates.

What happens in a closed container? Once again, some of the molecules escape from the surface of the liquid to form a gas, or vapor, as shown in Figure 13.5. After a while, the rate at which the liquid evaporates to form a gas becomes equal to the rate at which the gas condenses to form the liquid. At this point, the system is said to be in **equilibrium** (from the Latin, "a state of balance"). The space above the liquid is saturated with water vapor, and no more water evaporates.

The pressure of the water vapor in a closed container at equilibrium is called the **vapor pressure.** The kinetic molecular theory suggests that the vapor pressure of a liquid depends on its temperature. As can be seen in Figure 13.4, the fraction of the molecules that have enough energy to escape from a liquid increases with the

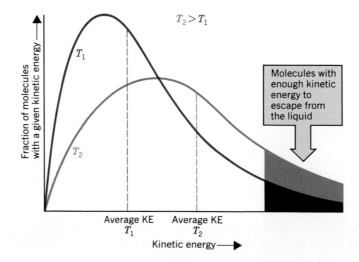

FIG. 13.4 At a given temperature, some of the molecules in a liquid have enough thermal energy to escape to form a gas. As the temperature increases, the fraction of the molecules moving fast enough to escape increases. Thus, the vapor pressure of a liquid increases with the temperature of the system.

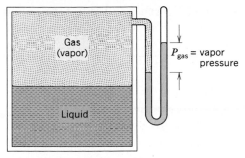

FIG. 13.5 The pressure of the gas or vapor that collects above a liquid in a closed container is called the *vapor pressure*.

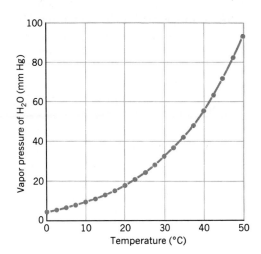

FIG. 13.6 A plot of the vapor pressure of water versus its temperature.

temperature of the liquid. As a result, the vapor pressure of a liquid should increase with temperature.

The vapor pressure of water at temperatures ranging from 0 to 50°C is given in Table A-4 of the appendix. Figure 13.6 shows that the vapor pressure of water increases more rapidly than the temperature of the system.

Exercise 13.1

The relative humidity given in weather reports is calculated by dividing the partial pressure of the water vapor in the atmosphere by the vapor pressure of water at that temperature and then expressing the result as a percent. Calculate the partial pressure of water in the atmosphere on a day when the relative humidity is 75% and the temperature is 88°F.

Solution

The temperature must first be converted from the Fahrenheit to the Celsius scale.

$$°C = 5/9 (°F - 32)$$
$$= 5/9 (88°F - 32°) = \mathbf{31°C}$$

According to Table A-4, the vapor pressure of water at 31°C is 33.7 mmHg. This means that if the atmosphere contained all of the water it could hold at this temperature, the pressure of this water vapor would be 33.7 mmHg. If the humidity is 75%, the actual pressure of the water in the atmosphere is 75% of this value, or 25.3 mmHg.

Exercise 13.2

The dew point is the temperature at which air is saturated with water vapor. If the temperature drops below this point, dew forms. What is the dew point on a day when the humidity is 46% at 21°C?

Solution

According to Table A-4, the vapor pressure of water at 21°C is 18.7 mmHg. If the humidity is 46%, the partial pressure of water in the atmosphere on that day is 46%

of this maximum value, or 8.6 mmHg. Once again referring to Table A-4, we find that air is saturated with water vapor at a pressure of 8.6 mmHg when the temperature is 9°C. The dew point is therefore 9°C.

13.4 MELTING POINT AND FREEZING POINT

Pure, crystalline solids have a characteristic **melting point,** which is the temperature at which the solid becomes a liquid. The transition between solid and liquid is so sharp for small samples of a pure solid that melting points can be measured to ± 0.1°C. The melting point of solid oxygen, for example, is -218.4°C.

Liquids have a characteristic temperature at which they turn into solids, the **freezing point.** In theory, the melting point of a solid should be equal to the freezing point of the liquid. In practice, small differences between these two temperatures can be observed.

It is difficult, if not impossible, to heat a solid above its melting point, because all of the heat that enters the solid at its melting point is used to convert the solid into a liquid. It is possible, however, to cool some liquids to temperatures below their freezing points without forming solids. A liquid cooled in this way is said to be **supercooled.**

We can make a supercooled liquid by heating solid sodium acetate trihydrate ($NaCH_3CO_2 \cdot 3\ H_2O$) until it melts. This compound is a salt that has water molecules trapped in the holes in the crystal lattice. When it melts at 58°C, the sodium acetate dissolves in the water that was trapped in the crystal to form a solution. In theory, when the solution cools to room temperature, it should solidify. But it doesn't. This supercooled liquid can be stored almost indefinitely at room temperature. If a small crystal of sodium acetate trihydrate is added to the liquid, however, the entire contents of the flask solidify within seconds, as shown in Figure 13.7.

Why can a liquid become supercooled? The particles in a solid are packed in a regular structure that is characteristic of that particular element or compound. Some of these solids form very easily; others do not. Some need a particle of dust, or a seed crystal, to act as a site on which the crystal can grow. In order to form crystals of sodium acetate trihydrate, Na^+ ions, $CH_3CO_2^-$ ions, and water molecules must come together in the proper orientation. It is difficult for these particles to organize themselves. A seed crystal, however, can provide the framework on which the proper arrangement of ions and water molecules can grow.

Because it is difficult to heat solids to temperatures above their melting points, and because pure solids tend to melt over a very small temperature range, melting points are often used to help identify compounds. We can distinguish between the three sugars glucose (MP = 150°C), fructose (MP = 103 – 105°C), and sucrose (MP = 185 – 186°C), for example, by determining the melting point of a small sample.

Measurements of the melting point of a solid can also provide information about the purity of the substance. Pure, crystalline solids melt over a very narrow range of temperatures, whereas mixtures melt over a broad temperature range. Mixtures also tend to melt at temperatures below the melting points of the pure solids.

13.5 BOILING POINT

When a liquid is heated, it eventually reaches a temperature at which the vapor pressure is large enough that bubbles form inside the body of the liquid. This

FIG. 13.7 It only takes a few seconds for a supersaturated solution of sodium acetate to crystallize after a seed crystal has been added.

temperature is called the ***boiling point.*** Once the liquid starts to boil, the temperature remains constant until all of the liquid has been converted to a gas.

The normal boiling point of water is 100°C. But if you try to cook an egg in boiling water while camping in the Rocky Mountains at an elevation of 10,000 feet above sea level, you will find that it takes longer for the egg to cook, because water boils at only 90°C at this elevation.

In theory, you shouldn't be able to heat a liquid to temperatures above its normal boiling point. Before microwave ovens became popular, however, pressure cookers were used to decrease the amount of time it took to cook food. In a typical pressure cooker, water can reach temperatures as high as 120°C, and food cooks in as little as one-third the normal time.

To explain why water boils at 90°C in the mountains and 120°C in a pressure cooker, even though the normal boiling point of water is 100°C, we have to understand why a liquid boils.

A liquid boils when the vapor pressure of the gas escaping from the liquid is equal to the pressure exerted on the liquid by its surroundings, as shown in Figure 13.8.

The normal boiling point of water is 100°C because this is the temperature at which the vapor pressure of water is equal to 760 mmHg, or 1 atm. Under normal conditions, when the pressure of the atmosphere is approximately 760 mmHg, water boils at 100°C.

At 10,000 feet above sea level, however, the pressure of the atmosphere is only 526 mmHg. At these elevations, water boils when its vapor pressure is 526 mmHg, which occurs at a temperature of 90°C.

Bubbles rising toward the surface of a liquid as it starts to boil.

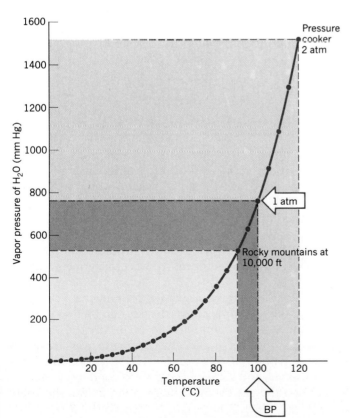

FIG. 13.8 A liquid boils when its vapor pressure is equal to the pressure exerted on it by its surroundings. The normal boiling point of water—at 1 atm—is 100°C. At a 10,000-foot elevation, atmospheric pressure is less than 1 atm, and water boils at temperatures below 100°C. In a pressure cooker at 2 atm, water doesn't boil until the temperature is 120°C.

Pressure cookers operate on the opposite principle. They are equipped with a valve that lets gas escape when the pressure inside the pot exceeds some fixed value. When the valve is set at 15 psi, for example, the water vapor inside the pot must reach a pressure of 2 atm before it can escape. Because water doesn't reach a vapor pressure of 2 atm until the temperature is 120°C, it boils in this container at 120°C.

Exercise 13.3

Use the definition of boiling point and the data in Table A-4 to describe how to make water boil at room temperature.

Solution

Water boils when the vapor pressure of the gas escaping from the liquid is equal to the pressure exerted on the liquid by its surroundings. The vapor pressure of water is roughly 20 mmHg at room temperature. We can therefore make water boil at room temperature by reducing the pressure in its container to less than 20 mmHg.

Figure 13.9 shows a sample of water boiling at room temperature. A 1-L flask was partially filled with water and sealed with a stopcock. A vacuum pump was then used to remove air from the flask until the pressure inside the flask was less than 20 mmHg, at which point the water began to boil.

FIG. 13.9 Water boils at room temperature when the pressure in its container is reduced to less than 20 mmHg.

Liquids often boil in an uneven fashion, or *bump.* They tend to bump when there aren't any air bubbles or scratches on the walls of the container, where bubbles can form. Bumping is easily prevented by adding a few boiling chips to the liquid. Air trapped in the pores of the boiling chips, as well as the rough surface of the chips, provides a place for bubbles to form. When boiling chips are used, essentially all of the bubbles that rise through the solution form on the surface of these chips.

13.6 CRITICAL TEMPERATURE AND CRITICAL PRESSURE

A liquid boils when it reaches a temperature equal to its boiling point. Conversely, we should be able to turn a gas into a liquid by cooling it to a temperature below its boiling point. This technique was used by Michael Faraday as early as 1823 to prepare the first sample of liquid chlorine (BP = −34°C).

Another way of condensing a gas to a liquid involves raising the pressure on the gas. This technique takes advantage of the fact that a liquid boils at the temperature at which its vapor pressure is equal to the pressure on the liquid from its surroundings. Raising the pressure on a gas therefore effectively increases the boiling point of the liquid.

Suppose that we have water vapor (or steam) in a closed container at 120°C and 1 atm. Since the temperature of the system is above the normal boiling point of water, there is no reason for the steam to condense to form a liquid. Nothing happens as we slowly compress the container — thereby raising the pressure on the gas — until the pressure reaches 2 atm. At this point — 120°C and 2 atm — the system is at the boiling point of water, and some of the gas will condense to form a liquid. As soon as the pressure on the gas becomes larger than 2 atm, the vapor pressure of water at 120°C is no longer large enough for the liquid to boil. As a result, the gas will condense to form a liquid.

In theory, we should be able to predict the pressure at which a gas condenses at a given temperature by consulting a plot of vapor pressure versus temperature, such as the one shown in Figure 13.8. This figure suggest that water vapor, for example, can be condensed to a liquid at 120°C if the pressure is raised to 2 atm. In practice, every compound has a ***critical temperature (T_c)***. If the temperature of the gas is above the critical temperature, the gas can't be condensed, regardless of the pressure applied.

The existence of a critical temperature was discovered by Thomas Andrews in 1869. While studying the effect of temperature and pressure on the behavior of carbon dioxide, Andrews found that he could condense CO_2 gas into a liquid by raising the pressure on the gas, as long as he kept the temperature below 31.0°C. At 31.0°C, for example, it takes a pressure of 72.85 atm to liquify CO_2 gas. Andrews found that it was impossible to turn CO_2 into a liquid above this temperature, no matter how much pressure was applied.

Gases can't be liquified at temperatures above the critical temperature because at this point the properties of gases and liquids become the same, and there is no basis on which to distinguish between gases and liquids. The vapor pressure of a liquid at the critical temperature is called the ***critical pressure (P_c)***. The vapor pressure of a liquid never gets larger than this critical pressure.

The critical temperatures, critical pressures, and boiling points of a number of compounds are given in Table 13.3. There is an obvious correlation between the critical temperatures and boiling points of these compounds. These properties are related because they are both indirect measures of the force of attraction between particles in the gas phase.

The experimental values of the critical temperature and pressure of a compound are used to calculate the a and b constants in the van der Waals equation (see Section 4.17). The value of the a constant is calculated from the following equation.

$$a = \frac{27 \, R^2 T_c^{\,2}}{64 \, P_c}$$

The b constant is calculated from the following equation.

$$b = \frac{RT_c}{8 \, P_c}$$

TABLE 13.3

Critical Temperatures, Critical Pressures, and Boiling Points of Common Liquids and Gases

Compound	T_c(°C)	P_c(atm)	BP(°C)	Compound	T_c(°C)	P_c(atm)	BP(°C)
He	−267.96	2.261	−268.935	O_2	−118.38	50.14	−182.96
H_2	−240.17	12.77	−252.76	CH_4	−82.60	45.44	−161.49
Ne	−228.71	26.86	−246.1	Kr	−63.75	54.20	−153.4
N_2	−146.89	33.54	−195.81	CO_2	31.04	72.85	−78.44
CO	−140.23	34.53	−191.49	NH_3	132.4	111.3	−33.42
F_2	−129.0	55	−188.20	Cl_2	144.0	78.1	−34.03
Ar	−122.44	48.00	−185.87	Br_2	311	102	58.75

FIG. 13.10 A steel sewing needle floats on a beaker of water, even though steel is almost eight times as dense as water.

13.7 SURFACE TENSION

Why do wet sheets of paper stick together, though dry sheets don't? Why do a pair of microscope slides with a drop of water between them stick together, though it is easy to separate a pair of dry slides? Why does a steel sewing needle float on top of a beaker of water, as shown in Figure 13.10, even though steel is almost eight times as dense as water? Why does water curve upward in a small-diameter glass tube while mercury curves downward, as shown in Figure 13.11? Why does liquid mercury break up into drops, or beads, when it spills on the floor? Why does rain form similar beads on the surface of a freshly waxed car?

The answers to these questions can all be traced to the force of attraction between molecules in liquids and the fact that liquids can flow until they take on the shape that maximizes this force of attraction. Below the surface of the liquid, the force of *cohesion* (literally, "sticking together") between molecules is the same in all directions, as shown in Figure 13.12. Molecules on the surface of the liquid feel a net force of attraction that tries to pull them back into the body of the liquid. As a result, the liquid tries to take on the shape that has the smallest possible surface area — the shape of a sphere. The magnitude of the force that controls the shape of the liquid is called the *surface tension.* The stronger the bonds between the molecules in the liquid, the larger the surface tension.

Figure 13.12 also shows the force of *adhesion* (literally, "sticking") between a liquid and the walls of the container. When the force of adhesion is more than half as large as the force of cohesion between the liquid molecules, the liquid is said to "wet" the solid. A good example of this phenomenon is the wetting of paper by water. The force of adhesion between paper and water and the force of cohesion between water molecules explains why sheets of wet paper stick together.

FIG. 13.12 Molecules that lie on the surface of a liquid feel a net force of *cohesion* that pulls them back into the body of the liquid. The net effect of this force is to minimize the surface area of the liquid. Molecules that lie along one of the walls of the container feel a net force of *adhesion* that binds them to the walls. When the force of adhesion is much smaller than the force of cohesion, the liquid will pull away from the walls of the container. But if the force of adhesion is more than half the force of cohesion, the liquid will "wet" the walls of the container.

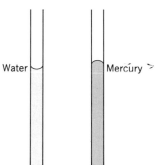

Water   Mercury

FIG. 13.11 The force of adhesion between water molecules is more than half the force of cohesion between water molecules. As a result, water "wets" glass, and it climbs the walls of a small-diameter glass tube to form a meniscus that curves upward. The force of attraction between mercury atoms and glass is much smaller than the force of attraction between the mercury atoms. As a result, the surface area in which mercury and glass are in contact is kept to a minimum, and mercury forms a meniscus that curves downward.

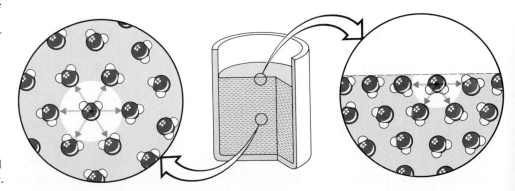

Water also wets glass because of the force of attraction between the positive ends of the polar water molecules and the negatively charged oxygen atoms in glass. As a result, water forms a *meniscus* that curves upward in a small-diameter glass tube, such as a buret. (The term *meniscus* comes from the Greek word for "moon" and is used to describe anything that has a crescent shape.) The meniscus that water forms in a buret results from a balance between the force of adhesion pulling up on the column of water to wet the walls of the glass tube and the force of gravity pulling down on the liquid.

The force of adhesion between water and wax is very small compared to the force of cohesion between water molecules. As a result, rain doesn't adhere to wax. It tends to form beads, or drops, with the smallest possible surface area, thereby maximizing the force of cohesion between the water molecules. The same thing happens when mercury is spilled on glass or poured into a narrow glass tube. The force of cohesion between mercury atoms is so much larger than the force of adhesion between mercury and glass that the area of contact between mercury and glass is kept to a minimum, with the net result being the meniscus shown in Figure 13.11.

13.8 VISCOSITY

One way of testing for a slow leak in a tire involves immersing the tire in a barrel of water and looking for the place on the tire surface where bubbles form. In theory, it should be possible to fill the tire with water and see where the water escapes. In practice, this doesn't work, because air escapes through holes that are much too small to pass water. It might be tempting to explain this observation by arguing that water molecules are bigger than the O_2 or N_2 molecules in air. But this can't be correct, because gasoline molecules, which are much larger than water molecules, can leak through holes that are too small to pass water.

These observations are the result of differences in the viscosities of air, water, and gasoline. *Viscosity* is a measure of the resistance to flow. For example, heavy motor oils are much more viscous than gasoline, and the maple syrup used on pancakes is much more viscous than the vegetable oils used in salad dressings.

We measure viscosity by determining the rate at which a liquid or gas flows through a small-diameter glass tube. In 1844, Jean Louis Marie Poiseuille showed that the volume of fluid (V) that flows down a small-diameter capillary tube per unit of time (t) is proportional to the radius of the tube (r), the pressure pushing the fluid down the tube (P), the length of the tube (l), and the viscosity of the fluid (η).

$$\frac{V}{t} = \frac{\pi r^4 P}{8 \eta l}$$

The unit in which viscosity is expressed is called the *poise* (pronounced "pwahz"). The viscosity of water at room temperature is roughly 1 centipoise, or 1 cP. Gasoline has a viscosity between 0.4 and 0.5 centipoise, and the viscosity of air is 0.018 centipoise.

Because the layer of molecules closest to the walls of a small-diameter tube adheres to the glass, viscosity measures the rate at which molecules in the middle of the stream of liquid or gas flow past this outer layer of more or less stationary molecules. Viscosity therefore depends on any factor that can influence the ease with which molecules slip past each other. Liquids tend to become more viscous as their molecules become larger or as the amount of intermolecular bonding increases. They become less viscous as the temperature increases. The viscosity of water, for example, decreases from 1.77 centipoise at 0°C to 0.28 centipoise at 100°C.

13.9 HYDROGEN BONDING AND THE ANOMALOUS PROPERTIES OF WATER

We are so familiar with the properties of water that it is difficult to appreciate the extent to which the behavior of water is unusual.

1. Most solids expand by up to 10% when they melt; water expands when it freezes.
2. Most solids are denser than the corresponding liquids; ice (0.917 g/cm³) is not as dense as water (0.998 g/cm³).
3. Water has a melting point at least 100°C higher than expected on the basis of the melting points of H_2S, H_2Se, and H_2Te.
4. Water has a boiling point almost 200°C higher than expected from the boiling points of H_2S, H_2Se, and H_2Te.
5. Water has the largest surface tension of any common liquid except liquid mercury.
6. Water has an unusually large viscosity.
7. Water is an excellent solvent; it can dissolve compounds, such as NaCl, that are insoluble in any other common liquid.
8. Water has an unusually high heat capacity. It takes more heat to raise the temperature of 1 gram of water by 1°C than any other liquid.

These anomalous properties all result from the strong intermolecular bonds in water. In Chapter 8, we concluded that water is best described as a polar molecule in which there is a partial separation of charge to give positive and negative poles.

$$^{\delta+}H\!-\!\overset{..}{\underset{..}{O}}\!^{\delta-}\!-\!H^{\delta+}$$

The force of attraction between a positively charged hydrogen atom on one water molecule and the negatively charged oxygen atom on another gives rise to an intermolecular bond, as shown in Figure 13.13. The dipole–dipole interaction between water molecules is an example of a special case of intermolecular bonds known as ***hydrogen bonds.***

Hydrogen bonds are separated from other examples of van der Waals forces because they are stronger than other intermolecular bonds. The strength of the hydrogen bond in water has been estimated to be between 10 and 12 kJ/mol. The hydrogen bonds in water are particularly important because of the dominant role that water plays in the chemistry of living systems. Hydrogen bonds are not limited to water, however.

Hydrogen-bond donors include substances that contain relatively polar H—X bonds, such as NH_3, H_2O, and HF. Hydrogen-bond acceptors include substances that have nonbonding pairs of valence electrons. The H—X bond must be polar to create the partial positive charge on the hydrogen atom that allows dipole–dipole interactions to exist. As the X atom in the H—X bond becomes less electronegative, hydrogen bonding between molecules becomes less important. Hydrogen bonding in HF, for example, is much stronger than in either H_2O or HCl.

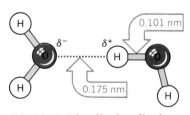

FIG. 13.13 The dipole–dipole force of attraction between water molecules is called hydrogen bonding. Because of the polarity of the water molecules, this is an unusually strong example of the van der Waals force of attraction between molecules.

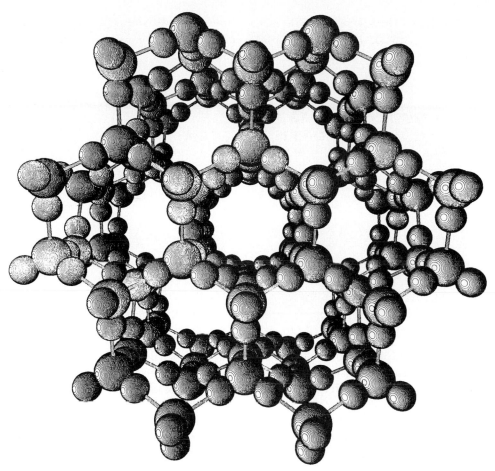

FIG. 13.14 The structure of ice. Note that the hydrogen atoms are closer to one of the oxygen atoms than the other in each of these hydrogen bonds.

The hydrogen bonds between water molecules in ice produce the open structure shown in Figure 13.14. When ice melts, some of these bonds are broken, and this structure collapses to form a liquid that is about 8% denser. This very unusual property of water has several important consequences. Water's expansion when it freezes is responsible for the cracking of concrete that forms potholes in streets and highways. But it also means that ice floats on top of rivers and streams. The ice that forms each winter therefore has a chance to melt during the summer.

Figure 13.15 shows another consequence of the strength of hydrogen bonds. There is a steady increase in boiling point in the series CH_4, GeH_4, SiH_4, and SnH_4, as expected from the discussion in Section 13.2. (There is a tendency for the boiling points of compounds to increase with molecular weight, because heavier molecules must be heated to higher temperatures before they can move fast enough to escape from the liquid.) The boiling points of H_2O and HF, however, are anomalously large because of the strong hydrogen bonds between molecules in these liquids. If this doesn't seem important, try to imagine what life would be like if water boiled at $-80°C$.

The surface tension and viscosity of water are also related to the strength of the hydrogen bonds between water molecules. The surface tension of water is respon-

FIG. 13.15 A plot of the melting points and boiling points of hydrides of elements in Groups IVA, VIA, and VIIA. The melting points and boiling points of both HF and H_2O are anomalously large because of the strength of the hydrogen bonds between molecules in these compounds.

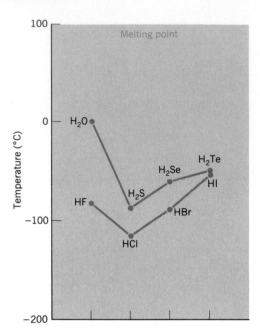

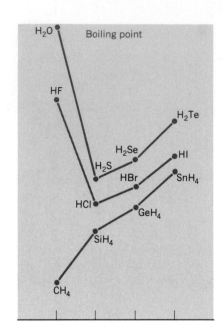

sible for the capillary action that brings water up through the root systems of plants. It is also responsible for the efficiency with which the wax that coats the surface of leaves can protect plants from excessive loss of water through evaporation.

The unusually large heat capacity of water is also related to the strength of the hydrogen bonds between water molecules. Anything that increases the motion of water molecules, and therefore the temperature of water, must interfere with the hydrogen bonds between these molecules. The fact that it takes so much energy to disrupt these bonds means that water can store enormous amounts of thermal energy. Although the water in lakes and rivers gets warmer in the summer and cooler in the winter, the large heat capacity of water limits the extreme temperatures that would threaten the life that flourishes in this environment. The heat capacity of water is also responsible for the way the oceans act as a thermal reservoir to moderate the swings in temperature that occur from winter to summer.

13.10 PHASE DIAGRAMS

Two factors determine whether a pure substance is a gas, a liquid, or a solid under a particular set of conditions—the temperature and the pressure. Figure 13.16 shows an example of a *phase diagram,* which summarizes the effect of temperature and pressure on a substance in a closed container. Every point in this diagram represents a possible combination of temperature and pressure for the system. The diagram is divided into three areas, which represent the solid, liquid, and gaseous states of the substance.

The best way to remember which area corresponds to each of these states is to remember the conditions of temperature and pressure that are most likely to be associated with a solid, a liquid, and a gas. Low temperatures and high pressures favor the formation of a solid, whereas gases are most likely to be found at high temperatures and low pressures. Liquids lie between these extremes. You can test this assignment by drawing a line from left to right across the top of the phase

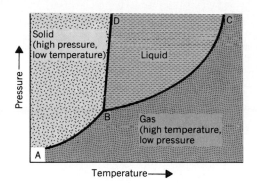

FIG. 13.16 A phase diagram describing the possible combinations of temperature and pressure that give rise to the solid, liquid, and gaseous phases of a typical substance.

diagram, which corresponds to an increase in the temperature of the system at constant pressure. When we heat a solid at constant pressure it melts to form a liquid, which eventually boils to form a gas.

Phase diagrams can be used in several ways. We can focus on the regions separated by the lines in the diagrams and get some idea of the conditions of temperature and pressure that are most likely to produce a gas, a liquid, or a solid. Or we can focus on an individual point and predict whether a substance will be a gas, a liquid, or a solid at this particular combination of temperature and pressure.

We can also focus on the lines that divide the diagram into states. The solid lines in this diagram represent the combinations of temperature and pressure at which two states are in equilibrium. For example, all of the points along the line connecting points A and B in this phase diagram represent temperatures and pressures at which the solid is in equilibrium with the gas. At these temperatures and pressures, the rate at which the solid sublimes to form a gas is exactly equal to the rate at which the gas condenses to form the solid.

Along AB line:

>   **rate at which solid sublimes to form a gas**
>>   **= rate at which gas condenses to form a solid**

The solid line between points B and C is identical to the plot of the temperature dependence of the vapor pressure of the liquid shown in Figure 13.8. It contains all of the combinations of temperature and pressure at which the liquid boils. At every point along this line, the liquid boils to form a gas and the gas condenses to form a liquid at the same rate.

Along BC line:

>   **rate at which liquid boils to form a gas**
>>   **= rate at which gas condenses to form a liquid**

The solid line between points B and D contains the combinations of temperature and pressure at which the solid and liquid are in equilibrium. At every point along this line, the solid melts at the same rate at which the liquid freezes.

Along BD line:

>   **rate at which solid melts to form a liquid**
>>   **= rate at which liquid freezes to form a solid**

This line is almost vertical because the melting point of a solid is not very sensitive to changes in pressure. The slope of this line changes from one substance to another, however. For many, if not most, compounds, it has a small positive slope, as shown in Figure 13.16. The slope of this line is slightly negative for water,

FIG. 13.17 A horizontal line drawn across a phase diagram at a pressure of 1 atm crosses the curve separating solids and liquids at the melting point of the solid, and it crosses the curve separating liquids and gases at the boiling point of the liquid.

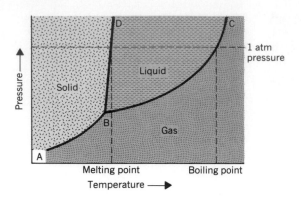

however. As a result, water can melt at temperatures near its freezing point when subjected to very high pressures. The ease with which ice skaters glide across a frozen pond can be explained by the fact that the pressure exerted by their skates melts a small portion of the ice that lies beneath the blades.

Point B in this phase diagram represents the only combination of temperature and pressure at which the compound can exist simultaneously as a solid, a liquid, and a gas. It is therefore called the *triple point* of the substance, and it represents the only point in the phase diagram in which all three states are in equilibrium. Point C is the critical temperature of the substance, which was defined in Section 13.6 as the highest temperature at which a gas can be condensed by an increase in the pressure on the gas.

Figure 13.17 shows what happens when we draw a horizontal line across a phase diagram at a pressure of exactly 1 atm. This line crosses the line between points B and D at the normal melting point of the substance, because solids normally melt at the temperature at which the solid and liquid are in equilibrium at 1 atm pressure. The line crosses the line between points B and C at the substance's normal boiling point, because the normal boiling point of a liquid is the temperature at which the liquid and gas are in equilibrium at 1 atm pressure and the vapor pressure of the liquid is equal to 1 atm.

13.11 SOLUTIONS: LIKE DISSOLVES LIKE

There are many different kinds of solutions, as first noted in Section 3.17. We can dissolve one gas in another to form solutions such as the atmosphere. We can dissolve one solid in another, such as copper in aluminum or carbon in iron. We can even dissolve gases in solids, such as H_2 in platinum. When chemists think about solutions, however, they tend to think about gases (such as HCl) or solids (such as NaCl) dissolved in liquids.

Chemists' interest in liquid solutions is easy to understand when you consider the rate at which reactions occur in the three states of matter. Solids will react with each other, but these reactions are slow, at least at room temperature. Gases also react with each other. But anyone who has watched what happens when a mixture of hydrogen and oxygen in a balloon is ignited should appreciate the fact that reactions in the gas phase are often too rapid to control.

Because atoms, ions, and molecules are free to move through liquids, reactions in liquids tend to occur much faster than reactions between solids. They are seldom as violent, however, as reactions between gases. Solutions of gases or solids dis-

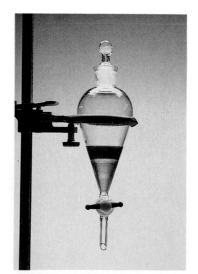

FIG. 13.18 Water and carbon tetrachloride form two separate liquid phases in a separatory funnel.

solved in liquids are therefore useful for controlling the rate at which chemical reactions occur.

In Section 3.17, the terms *solute, solvent,* and *solution* were defined as follows.

The substance that dissolves is called the *solute*—for example, HCl gas or NaCl crystals.

The substance in which the solute dissolves is called the *solvent* —for example, H₂O.

The mixture of solute and solvent is called the *solution*—for example, hydrochloric acid or salt water.

This section focuses on the question of which solutes dissolve in which solvents.

Let's start by trying to dissolve a pair of solutes—I_2 and $KMnO_4$—in a pair of solvents—H_2O and CCl_4. The solutes have two things in common. They are both solids, and they both have a deep violet or purple color. The solvents are colorless liquids that don't mix.

When water and carbon tetrachloride are poured into a separatory funnel, as shown in Figure 13.18, two separate liquid phases are clearly visible. We can use the relative densities of CCl_4 (density = 1.594 g/cm³) and H_2O (density = 1.0 g/cm³) to decide which phase is water and which is carbon tetrachloride. The denser CCl_4 should settle to the bottom of the funnel.

When a few crystals of iodine are added to the separatory funnel and the contents of the funnel are shaken, the I_2 dissolves in the CCl_4 layer to form an intensely violet-colored solution, as shown in Figure 13.19. The water layer stays essentially colorless, however, suggesting that little if any I_2 dissolves in water.

When this experiment is repeated with potassium permanganate, the water layer picks up the color of $KMnO_4$, and the CCl_4 layer remains colorless, as shown in Figure 13.20. This suggests that $KMnO_4$ dissolves in water but not in carbon tetrachloride. The results of this simple experiment are summarized in Table 13.4.

The difference between the solutes is easy to understand. Iodine consists of I_2 molecules held together by weak van der Waals forces. Potassium permanganate consists of K^+ and MnO_4^- ions held together by the strong force of attraction between ions of opposite charge. It is therefore much easier to separate the I_2 molecules in iodine than it is to separate the ions in $KMnO_4$.

There is also a significant difference between CCl_4 and H_2O. The difference between the electronegativities of the carbon and chlorine atoms in CCl_4 is so small ($\Delta EN = 0.56$) that there is very little ionic character in the C—Cl bonds. CCl_4 is therefore best described as nonpolar. Even if there were some separation of charge in these bonds, the CCl_4 molecule wouldn't be polar, because it has a highly

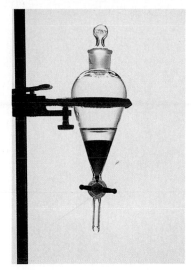

FIG. 13.19 I_2 dissolves in the CCl_4 layer to form an intensely colored solution.

FIG. 13.20 $KMnO_4$ dissolves in the water layer, leaving the CCl_4 colorless.

TABLE 13.4

Solubilities of Iodine and Potassium Permanganate in Carbon Tetrachloride and Water

Solutes	Solvents	
	H_2O	CCl_4
I_2	insoluble	very soluble
$KMnO_4$	very soluble	insoluble

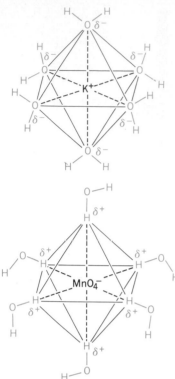

FIG. 13.21 KMnO₄ dissolves in water because the energy released when bonds form between the K⁺ ion and the negative end of the neighboring water molecule and between the MnO₄⁻ ion and the positive end of the H₂O dipole compensates for the energy needed to separate the K⁺ and MnO₄⁻ ions.

symmetrical shape in which the four chlorine atoms point toward the corners of a tetrahedron.

The difference between the electronegativities of the hydrogen and oxygen atoms in water is much larger ($\Delta EN = 1.24$), and the H—O bonds in this molecule are therefore polar. Furthermore, this molecule has a bent, or angular, shape that concentrates the hydrogen atoms that carry a partial positive charge on one side of the molecule. As a result, water molecules have a dipole moment—they have distinct positive and negative poles.

Why does $KMnO_4$ dissolve in water and not in carbon tetrachloride? It takes a lot of energy to separate the K^+ and MnO_4^- ions in potassium permanganate. But these ions can form bonds with neighboring water molecules, as shown in Figure 13.21. The energy released when these bonds form compensates for the energy that has to be invested to take apart the $KMnO_4$ crystal. No such bonds can form between the K^+ or MnO_4^- ions and the nonpolar CCl_4 molecules. As a result, $KMnO_4$ can't dissolve in CCl_4.

Why does I_2 dissolve in carbon tetrachloride but not in water? The I_2 molecules in iodine and the CCl_4 molecules in carbon tetrachloride are held together by weak van der Waals forces, which can also exist between I_2 and CCl_4 molecules in a solution. I_2 therefore readily dissolves in CCl_4. The molecules in water are held together by hydrogen bonds that are stronger than most van der Waals forces. No interaction between I_2 and H_2O molecules is strong enough to compensate for the hydrogen bonds that have to be broken to dissolve iodine in water, so little if any I_2 dissolves in H_2O.

We can summarize the results of this experiment by noting that *nonpolar* solutes (such as I_2) dissolve in *nonpolar* solvents (such as CCl_4), whereas *polar* solutes (such as $KMnO_4$) dissolve in *polar* solvents (such as H_2O). This can be summarized further in the following general rule.

Like dissolves like.

Exercise 13.4

Section 10.7 noted that elemental phosphorus is often stored under water because it isn't soluble in water. Elemental phosphorus is very soluble in carbon disulfide, however—up to 880 grams of phosphorus can dissolve in 100 grams of CS_2. Explain why P_4 is soluble in CS_2 but not in water.

Solution

The structure of the P_4 molecule was shown in Figure 10.10. This molecule is a perfect example of a nonpolar solute. It is therefore more likely to be soluble in nonpolar solvents than in polar solvents such as water.

The Lewis structure of CS_2 suggests that this molecule is linear.

$$:\!S\!=\!C\!=\!S\!:$$

Thus, even if there is some separation of charge in the $C\!=\!S$ double bond, the molecule has no net dipole moment, because of its symmetry. The electronegativities of carbon (EN = 2.55) and sulfur (EN = 2.58), however, suggest that the $C\!=\!S$ double bonds are almost perfectly covalent. CS_2 is therefore a nonpolar solvent, which should readily dissolve P_4.

Exercise 13.5

The iodide ion reacts with iodine in aqueous solution to form the I_3^-, or triiodide, ion.

$$I^-(aq) + I_2(aq) \longrightarrow I_3^-(aq)$$

What would happen if CCl_4 was added to an aqueous solution that contained a mixture of KI, I_2, and KI_3?

Solution

KI and KI_3 are both salts. One contains the K^+ and I^- ions; the other contains the K^+ and I_3^- ions. These salts are more soluble in polar solvents, such as water, than in nonpolar solvents, such as CCl_4. They would therefore remain in the aqueous solution.

I_2, however, is a nonpolar molecule, which is more soluble in a nonpolar solvent, such as carbon tetrachloride. The I_2 would therefore leave the aqueous layer and enter the CCl_4 layer, where it would exhibit the characteristic violet color of solutions of molecular iodine.

13.12 HYDROPHILIC AND HYDROPHOBIC MOLECULES

As noted in Section 13.2, there is a family of compounds, known as the *hydrocarbons,* whose members contain only the elements carbon and hydrogen. This family includes compounds such as the following, which is found in gasoline.

$$CH_3-\underset{\underset{CH_3}{|}}{\overset{\overset{CH_3}{|}}{C}}-CH_2-\underset{\overset{CH_3}{|}}{CH}-CH_3$$

isooctane

Because the difference between the electronegativities of carbon and hydrogen is very small ($\Delta EN = 0.40$), hydrocarbons are nonpolar. A number of commercial products are based on the fact that hydrocarbons are ***immiscible*** (literally, "not mixable") with water.

It is possible to replace one of the hydrogen atoms in a hydrocarbon with an $-OH$ group to form a substance that has properties between the extremes of gasoline and water. These compounds are called ***alcohols,*** and they include the following examples.

CH_3OH	methanol	$CH_3CH_2CH_2CH_2OH$	butanol
CH_3CH_2OH	ethanol	$CH_3CH_2CH_2CH_2CH_2CH_2OH$	hexanol

TABLE 13.5

Solubilities of Alcohols in Water

Formula	Name	Solubility in Water (g / 100 g)
CH_3OH	methanol	infinitely soluble
CH_3CH_2OH	ethanol	infinitely soluble
$CH_3(CH_2)_2OH$	propanol	infinitely soluble
$CH_3(CH_2)_3OH$	butanol	9
$CH_3(CH_2)_4OH$	pentanol	2.7
$CH_3(CH_2)_5OH$	hexanol	0.6
$CH_3(CH_2)_6OH$	heptanol	0.18
$CH_3(CH_2)_7OH$	octanol	0.054
$CH_3(CH_2)_9OH$	decanol	insoluble in water

When the hydrocarbon chain is short, the alcohol is soluble in water. Methanol (CH_3OH) and ethanol (CH_3CH_2OH), for example, are infinitely soluble in water —there is no limit on the amount of these alcohols that can dissolve in a given quantity of water. The alcohol in beer, wine, and hard liquors is actually ethanol, and mixtures of ethanol and water can vary in concentration between pure alcohol (200 proof) and pure water (0 proof).

As the hydrocarbon chain becomes longer, the alcohol becomes less soluble in water, as shown by the data in Table 13.5. One end of alcohol molecules has so much nonpolar character it is called **hydrophobic** (literally, "water-hating"). The other end contains an —OH group that can form hydrogen bonds to neighboring water molecules and is therefore said to be **hydrophilic** (literally, "water-loving"). One end of these molecules attracts water molecules, whereas the other end repels water. As the hydrocarbon chain becomes longer, the hydrophobic character of the molecule increases, and the solubility of the molecule gradually decreases until the alcohol becomes essentially insoluble in water.

People encountering the terms *hydrophilic* and *hydrophobic* for the first time sometimes have difficulty remembering which word stands for water-hating and which stands for water-loving. If you can remember that Hamlet's girlfriend was named Ophelia (not Ophobia), you might be able to remember that the prefix *philo-* is commonly used to describe love—for example, in *philanthropist, philharmonic, philosopher,* and so on.

The data in Table 13.5 shows one consequence of the general rule that like dissolves like. As molecules become more nonpolar, they become less soluble in water. Table 13.6 shows another example of this rule. NaCl is relatively soluble in water—35.92 grams of salt dissolves in 100 grams of water. As the solvent becomes more nonpolar, however, the solubility of this polar solute decreases.

TABLE 13.6

Solubility of Sodium Chloride in Water and in Alcohols

Formula of Solvent	Name	Solubility of NaCl (g / 100 g solvent)
H_2O	water	35.92
CH_3OH	methanol	1.40
CH_3CH_2OH	ethanol	0.065
$CH_3(CH_2)_2OH$	propanol	0.012
$CH_3(CH_2)_3OH$	butanol	0.005
$CH_3(CH_2)_4OH$	pentanol	0.0018

13.13 SOAPS, DETERGENTS, AND DRY-CLEANING AGENTS

The chemistry behind the manufacture of soap hasn't changed since soap was first made from animal fat and the ash from wood fires almost 5000 years ago. Solid animal fats, such as the tallow obtained during the butchering of sheep or cattle, and liquid plant oils, such as palm oil and coconut oil, are still heated in the presence of a strong base to form a soft, waxy material that enhances the ability of water to wash away the grease and oil that forms on our bodies and our clothes.

Animal fats and plant oils contain compounds known as fatty acids. Fatty acids —such as stearic acid, shown below—have small, polar, hydrophilic heads attached to long, nonpolar, hydrophobic tails.

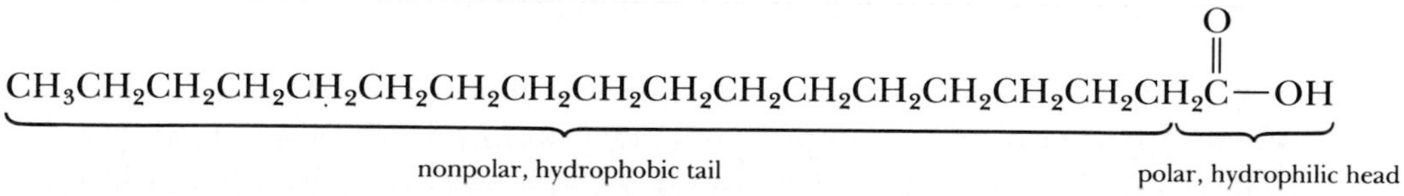

$$CH_3CH_2CH_2CH_2CH_2CH_2CH_2CH_2CH_2CH_2CH_2CH_2CH_2CH_2CH_2CH_2CH_2\overset{\overset{\displaystyle O}{\|}}{C}-OH$$

$\underbrace{\qquad\qquad}_{\text{nonpolar, hydrophobic tail}}$ $\underbrace{\qquad}_{\text{polar, hydrophilic head}}$

Fatty acids are seldom found by themselves in nature. They are usually bound to molecules of glycerol ($HOCH_2CHOHCH_2OH$) to form triglycerides.

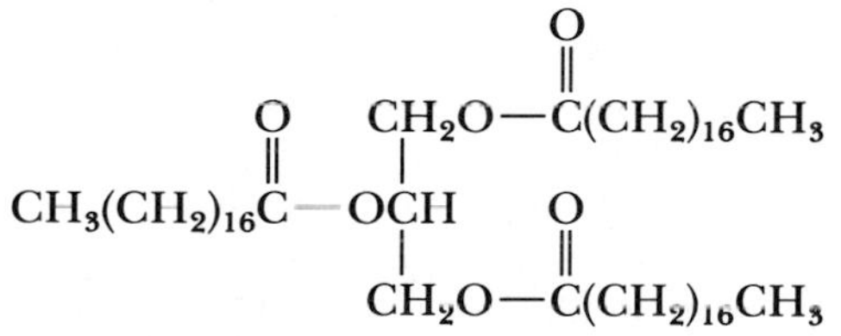

$$CH_3(CH_2)_{16}\overset{\overset{\displaystyle O}{\|}}{C}-OCH\begin{matrix}CH_2O-\overset{\overset{\displaystyle O}{\|}}{C}(CH_2)_{16}CH_3\\[1em]\\CH_2O-\overset{\overset{\displaystyle O}{\|}}{C}(CH_2)_{16}CH_3\end{matrix}$$

These triglycerides break down in the presence of a strong base to form the Na^+ or K^+ salt of the fatty acid, as shown in Figure 13.22. This reaction is called *saponification,* which literally means "the making of soap."

Part of the cleaning action of soap results from the fact that soap molecules are *surfactants*—they tend to concentrate on the surface of water. They cling to the surface because they try to orient their polar CO_2^- heads toward water molecules and their nonpolar $CH_3CH_2CH_2$ . . . tails away from neighboring water molecules.

Water can't wash the soil out of clothes by itself, because the soil particles that cling to textile fibers are covered by a layer of nonpolar grease or oil molecules, which repel water. The nonpolar tails of the soap molecules on the surface of a drop of water can dissolve in the grease or oil that surrounds a soil particle, as shown in Figure 13.23. The soap molecules therefore disperse, or *emulsify,* the soil particles, which makes it possible to wash these particles out of the clothes.

Most soaps are denser than water. They can be made to float, however, by the incorporation of air during their manufacture. Most soaps are also opaque—they absorb rather than transmit light. Translucent soaps can be made, however, by the addition of alcohol, sugar, and glycerol, which slow down the growth of soap

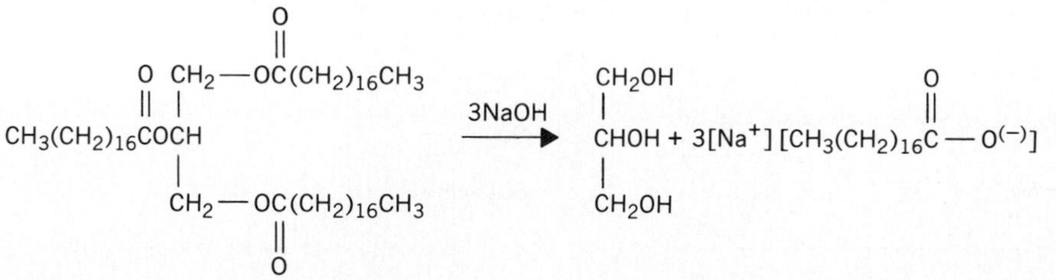

$$CH_3(CH_2)_{16}\overset{\overset{\displaystyle O}{\|}}{C}OCH\begin{matrix}CH_2-OC(CH_2)_{16}CH_3\\[1em]\\CH_2-OC(CH_2)_{16}CH_3\end{matrix}\ \xrightarrow{3NaOH}\ \begin{matrix}CH_2OH\\[1em]CHOH\\[1em]CH_2OH\end{matrix}+3[Na^+]\,[CH_3(CH_2)_{16}\overset{\overset{\displaystyle O}{\|}}{C}-O^{(-)}]$$

FIG. 13.22 The reaction in which a "fat molecule," or triglyceride, reacts with a strong base to form the sodium salt of a carboxylic acid is called *saponification* because it is the reaction used to make soap.

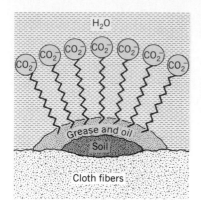

FIG. 13.23 Soap molecules disperse, or emulsify, soil particles coated with a layer of nonpolar grease or oil molecules.

crystals while the soap solidifies. Liquid soaps result when the sodium salts of the fatty acid are replaced with the more soluble K^+ or NH_4^+ salts.

Forty years ago, more than 90% of the cleaning agents sold in the United States were soaps. This percentage has steadily decreased, and today soap represents less than 20% of the market for cleaning agents. The primary reason for this decline in the popularity of soap is the reaction between soap and "hard" water.

The most abundant positive ions in tap water are Na^+, Ca^{2+}, and Mg^{2+} ions. Water that is particularly rich in Ca^{2+}, Mg^{2+}, or Fe^{3+} ions is said to be hard. Hard water interferes with the action of soap because these ions combine with soap molecules to form insoluble precipitates that have no cleaning power. These salts not only decrease the concentration of the soap molecules in solution, they actually bind soil particles to clothing, leaving a dull, gray film.

One way around this problem is to "soften" the water by replacing the Ca^{2+} and Mg^{2+} ions with Na^+ ions. Many water softeners, such as those sold by Culligan, are filled with a resin that contains $—SO_3^-$ ions attached to a polymer, as shown in Figure 13.24. The resin is treated with NaCl until each $—SO_3^-$ ion picks up an Na^+ ion. When hard water flows over this resin, Ca^{2+} and Mg^{2+} ions bind to the $—SO_3^-$ ions on the polymer chain, and Na^+ ions are released into solution. Periodically, the resin becomes saturated with Ca^{2+} and Mg^{2+} ions, and it has to be regenerated by being washed with a concentrated solution of NaCl.

There is another way to get around the problem of hard water. Instead of removing Ca^{2+} and Mg^{2+} ions from water, we can find a cleaning agent that doesn't form insoluble salts with these ions. Synthetic detergents are examples of such cleaning agents. Detergent molecules consist of long, hydrophobic hydrocarbon tails attached to polar, hydrophilic SO_3^- or OSO_3^- heads.

$$CH_3CH_2CH_2CH_2CH_2CH_2CH_2CH_2CH_2CH_2CH_2CH_2—O—\overset{\displaystyle O}{\underset{\displaystyle O}{\overset{|}{\underset{|}{S}}}}—O^-$$

By themselves, detergents don't have the cleaning power of soap. "Builders" are therefore added to synthetic detergents to increase their strength. These builders are often salts of highly charged ions, such as the triphosphate ($P_3O_{10}^{5-}$) ion.

Cloth fibers swell when they are washed in water. This leads to changes in the dimensions of the cloth that can cause wrinkles — which are local distortions in the structure of the fiber — or even more serious damage, such as shrinking. Such problems can be avoided by dry cleaning, which uses a nonpolar solvent that does not adhere to, or wet, the cloth fibers. The nonpolar solvents used in dry cleaning dissolve the nonpolar grease or oil layer that coats soil particles, freeing the soil particles to be removed by detergents added to the solvent or by the tumbling

FIG. 13.24 Water softeners contain $—SO_3^-$ ions bound to a polymer backbone. When the softener is "charged," it is washed with a concentrated solution of NaCl in water until all of the $—SO_3^-$ ions pick up an Na^+ ion. The softener then picks up Ca^{2+} and Mg^{2+} ions from hard water, replacing these ions with Na^+ ions. Recharging the softener involves washing the polymer once again with a concentrated solution of NaCl in water.

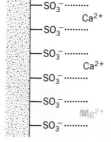

action inside a machine. Dry cleaning has the added advantage that it can remove oily soil at lower temperatures than soap or detergent dissolved in water, so it is safer for delicate fabrics.

When dry cleaning was first introduced in the United States between 1910 and 1920, the solvent was a mixture of hydrocarbons isolated from petroleum during the refining of gasoline. Over the years, these flammable hydrocarbon solvents have been replaced by halogenated hydrocarbons, such as carbon tetrachloride (CCl_4), trichloroethylene ($Cl_2C=CHCl$), trichloroethane (Cl_3C-CH_3), and perchloroethylene ($Cl_2C=CCl_2$).

13.14 UNITS OF CONCENTRATION: MOLARITY, MOLALITY, AND MOLE FRACTION

The concept of *concentration* was introduced in Section 3.18 to describe the amount of solute dissolved in a given amount of solvent or solution.

$$\text{Concentration} = \frac{\text{amount of solute}}{\text{amount of solvent or solution}}$$

There are at least seven units of concentration. So far we have focused on two—molarity and weight percent.

The *molarity* of a solution is the number of moles of solute in the solution divided by the volume of the solution in liters.

$$\text{Molarity } (M) = \frac{\text{moles of solute}}{\text{liters of solution}}$$

Exercise 13.6

At 25°C, 5.77 grams of chlorine gas dissolves in enough water to give a liter of solution. Calculate the molarity of this solution.

Solution

Before we can calculate the molarity of a solution we need to know the number of moles of solute and the volume of solution. The number of moles of solute in this solution is easy to calculate.

$$5.77 \text{ g } Cl_2 \times \frac{1 \text{ mol } Cl_2}{70.91 \text{ g}} = 0.0814 \text{ mol } Cl_2$$

We now divide this number by the volume of the solution.

$$\frac{0.08137 \text{ mol } Cl_2}{1 \text{ L}} = 0.0814 \text{ M}$$

The concentration of Cl_2 in this solution is 0.0814 moles per liter. The solution can also be described as 0.0814 *molar (M)* in Cl_2.

Weight percent is the percentage of the total weight of a solution that is due to the solute.

$$\text{Weight percent} = \frac{\text{weight of solute}}{\text{weight of solution}} \times 100$$

A 3.5% solution of hydrochloric acid, for example, has 3.5 grams of HCl in every 100 grams of solution.

Exercise 13.7

Concentrated sulfuric acid is 96.0% H_2SO_4 by weight, and it has a density of 1.84 g/cm³. Calculate the molarity of this solution.

Solution

In order to calculate the molarity of a solution we need to know the number of moles of solute in a given volume of solution. The volume of the solution in this case is not specified, so let's assume it is exactly 1 liter. The weight of a liter of this solution can be calculated from its density.

$$\frac{1.84 \text{ g solution}}{1 \text{ cm}^3} \times \frac{1 \text{ cm}^3}{1 \text{ mL}} \times \frac{1000 \text{ mL}}{1 \text{ L}} = 1840 \text{ g solution}$$

But the problem states that this solution is 96.0% H_2SO_4 by weight. Thus, the weight of the H_2SO_4 in this solution is 96.0% of the total weight of the solution.

$$1840 \text{ g solution} \times \frac{96.0 \text{ g } H_2SO_4}{100 \text{ g solution}} = 1766 \text{ g } H_2SO_4$$

We can now calculate the number of moles of H_2SO_4 in this sample.

$$1766 \text{ g } H_2SO_4 \times \frac{1 \text{ mol } H_2SO_4}{98.08 \text{ g}} = 18.0 \text{ mol } H_2SO_4$$

Dividing the number of moles of H_2SO_4 by the volume of the solution gives a concentration of 18.0 M.

$$\frac{18.0 \text{ mol } H_2SO_4}{1 \text{ L}} = 18.0 \text{ } M$$

A third concentration unit is ***volume percent.*** This unit is used to describe solutions of one liquid dissolved in another, or mixtures of gases. Wine labels, for example, describe the alcoholic content as 12% by volume, because 12% of the total volume is alcohol.

$$\textbf{Volume percent} = \frac{\textbf{volume of solute}}{\textbf{volume of solution}} \times \textbf{100}$$

Molarity is the concentration unit most commonly used by chemists. It has a major disadvantage, however. It tells us how much *solute* we need to make a solution, and it gives us the volume of the *solution* produced, but it gives no information about the amount of *solvent* needed to make the solution.

We can make a 0.100 M solution of $CuSO_4$, for example, by dissolving 0.100 mole of $CuSO_4$ in enough water to give one liter of solution. But how much water is enough? Because the $CuSO_4$ crystals occupy some volume, it takes less than a liter of water, but we have no idea how much less.

When it is important to know how much solute and solvent are present in a solution, chemists use two other concentration units: molality and mole fraction.

The *molality* of a solution is equal to the number of moles of solute in the solution divided by the weight in kilograms of solvent used to make the solution.

$$\text{Molality } (m) = \frac{\text{moles of solute}}{\text{kilograms of solvent}}$$

A 0.100 m solution of $CuSO_4$, for example, can be prepared by dissolving 0.100 mole of $CuSO_4$ in 1 kilogram of water. Because the density of water is roughly 1 gram per milliliter, the total volume of this solution is larger than 1 liter. A 0.100 m solution is therefore slightly more dilute than a 0.100 M solution of the same solute.

Exercise 13.8

Calculate the molality of a saturated solution of hydrogen sulfide in water at room temperature if 0.385 grams of H_2S gas dissolve in 100 grams of water at 20°C and 1 atm.

Solution

In order to calculate the molality of a solution we need to know the number of moles of solute and the number of kilograms of solvent. The number of moles of solute in this case is easy to calculate.

$$0.358 \text{ g } H_2S \times \frac{1 \text{ mol } H_2S}{34.08 \text{ g}} = 0.0113 \text{ mol } H_2S$$

Now all we have to do is divide the number of moles of solute by the number of kilograms of solvent to find that this solution is 0.113 molal (m).

$$\frac{0.0113 \text{ mol } H_2S}{0.100 \text{ kg } H_2O} = 0.113 \text{ } m$$

Molality has another important advantage over molarity. The molarity of an aqueous solution changes with temperature, because the density of water is sensitive to temperature. Because molality is defined in terms of the weight of the solvent, not its volume, the molality of a solution remains constant regardless of changes in temperature.

The ratio of solute to solvent in a solution can also be described by the mole fraction of the solute or the solvent in a solution.

> The *mole fraction* of any component of a solution is literally the fraction of the total number of moles of solute and solvent that come from that component of the solution.

For reasons that are not obvious, the symbol for mole fraction is a Greek capital letter chi, X. The mole fraction of the *solute* is the number of moles of solute divided by the sum of the number of moles of solute and the number of moles of solvent.

Mole fraction of solute:

$$X_{solute} = \frac{\text{moles of solute}}{\text{moles of solute} + \text{moles of solvent}}$$

Conversely, the mole fraction of the *solvent* is the number of moles of solvent divided by the total number of moles of solute and moles of solvent.

Mole fraction of solvent:

$$X_{solvent} = \frac{\text{moles of solvent}}{\text{moles of solute} + \text{moles of solvent}}$$

In a solution that consists of a single solute dissolved in a solvent, the sum of the mole fraction of the solute and that of the solvent must equal 1.

$$X_{solute} + X_{solvent} = 1$$

Exercise 13.9

Calculate the mole fractions of both the solute and the solvent in a solution of 0.385 grams of hydrogen sulfide dissolved in 100 grams of water at 20°C.

Solution

We can start by converting grams of solute and solvent into moles.

$$0.385 \text{ g } H_2S \times \frac{1 \text{ mol } H_2S}{34.08 \text{ g}} = 0.0113 \text{ mol } H_2S$$

$$100 \text{ g } H_2O \times \frac{1 \text{ mol } H_2O}{18.02 \text{ g}} = 5.55 \text{ mol } H_2O$$

The mole fraction of the solute is the number of moles of H_2S divided by the total number of moles of both H_2S and H_2O.

$$X_{solute} = \frac{(0.0113 \text{ mol } H_2S)}{(0.113 \text{ mol } H_2S + 5.55 \text{ mol } H_2O)} = 0.00203$$

The mole fraction of the solvent is the moles of H_2O divided by the moles of both H_2S and H_2O.

$$X_{solvent} = \frac{(5.55 \text{ mol } H_2O)}{(0.113 \text{ mol } H_2S + 5.55 \text{ mol } H_2O)} = 0.998$$

Note that the sum of the mole fractions is 1.

$$X_{solute} + X_{solvent} = 0.998 + 0.00203 = 1$$

13.15 COLLIGATIVE PROPERTIES: VAPOR PRESSURE DEPRESSION

In Section 1.16, we divided physical properties into two categories: extensive and intensive. Extensive properties — such as mass and volume — depend on the size of the sample. Intensive properties — such as density and concentration — depend on the identity of the atoms, ions, or molecules the sample contains but not on the size of the sample. This section will introduce a third category of physical properties — *colligative properties.*

> **Colligative property: Any physical property that depends only on the ratio of the number of particles of solute and solvent in a solution, not on the identity of the solute.**

As our first example of a colligative property, we'll consider what happens to the vapor pressure of a solvent when a solute is added to form a solution. We'll define $P°$ as the vapor pressure of the pure liquid (the solvent) and P as the vapor pressure of the solvent after a solute has been dissolved in it.

$$P° = \textbf{vapor pressure of the pure liquid, or solvent}$$
$$P = \textbf{vapor pressure of the solvent in a solution}$$

Exercise 13.10

When the temperature of a liquid is below its boiling point, we can assume that the only molecules that can escape from the liquid to form a gas are those that lie near the surface of the liquid. Use this assumption to predict whether the vapor pressure of a solvent should increase, decrease, or remain the same when a solute is added to the solvent.

Solution

When a solute is added to the solvent, some of the solute molecules occupy the space near the surface of the liquid, as shown in Figure 13.25. This has no effect on the rate at which the solvent molecules in the gas phase condense to form a liquid. But it decreases the rate at which the solvent molecules in the liquid can escape into the gas phase. As a result, the vapor pressure of the solvent becomes smaller when a solute is added. The vapor pressure of the solvent escaping from a solution is therefore smaller than the vapor pressure of the pure solvent.

$$P < P°$$

vapor pressure of the solvent above a solution   vapor pressure of the pure solvent

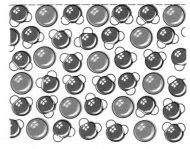

FIG. **13.25** The only molecules that can escape from a liquid to form a gas at temperatures below the boiling point of the liquid are those on or near the surface of the liquid. When a solute is dissolved in a solvent, the number of solvent molecules near the surfaces decreases, and the vapor pressure of the solvent therefore decreases.

Between 1887 and 1888, Francois-Marie Raoult showed that the vapor pressure of a solution is equal to the mole fraction of the solvent times the vapor pressure of the pure liquid.

$$P = X_{solvent}P°$$

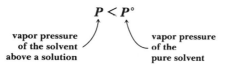

vapor pressure of the solvent above a solution   vapor pressure of the pure solvent

When the solvent is pure, and the mole fraction of the solvent is therefore equal to 1, P is equal to $P°$ — the vapor pressure of the pure solvent. As the fraction of the moles of solvent in the solution becomes smaller, the vapor pressure of the solvent escaping from the solution also becomes smaller.

The vapor pressure of a pure liquid at a particular temperature is one of its characteristic properties, because it depends on the identity of the liquid. But the *change* in the vapor pressure that occurs when a solute is added to the liquid is a colligative property. *Raoult's law* states that the difference between the vapor pressure of a pure solvent and the vapor pressure of the solvent after a solute has been added depends only on the mole fraction of the solvent. It therefore depends on the ratio of the number of particles of solute to solvent in the solution, but not on the identity of the solute.

COLLIGATIVE PROPERTIES: BOILING POINT
13.16 ELEVATION AND FREEZING POINT DEPRESSION

Figure 13.26 shows the consequences of the fact that solutes lower the vapor pressure of a solvent. The line connecting points B and C in this phase diagram contains all of the combinations of temperature and pressure at which the liquid and its corresponding gas are in equilibrium. Each point along this line describes a closed container in which the space above the liquid is saturated with vapor. Each point on this line therefore describes the vapor pressure of the liquid at that temperature.

In this figure, the behavior of the pure solvent is represented by a solid line. The dotted line describes the solution obtained when a solute is dissolved in this solvent. At any given temperature, the vapor pressure of the solvent escaping from the solution is smaller than the vapor pressure of the pure solvent. The dotted line therefore lies below the solid line.

According to this figure, the solution can't boil at the same temperature as the pure solvent. Because the vapor pressure of the solution is smaller than that of the pure solvent at any given temperature, the solution must be heated to a higher temperature before it boils. The lowering of the vapor pressure of the solvent that occurs when it is used to form a solution therefore inevitably results in an elevation of the boiling point of the liquid.

The triple point was defined in Section 13.10 as the only combination of temperature and pressure at which the gas, liquid, and solid can exist at the same time. Figure 13.26 shows that the triple point of the solution occurs at a lower temperature than the triple point of the pure solvent.

By itself, the change in the triple point is not important. But it results in a change in the temperature at which the solution freezes or melts. To understand why, we have to look carefully at the line that separates the solid and liquid regions in the phase diagram. This line is almost perfectly vertical, because the melting point of a substance is not very sensitive to pressure. (As you can see from the diagram, the melting point of this substance increases slightly as the pressure on the substance increases; but the change is fairly small.)

FIG. 13.26 The decrease in vapor pressure that occurs when a solute is added to a solvent causes an increase in the boiling point and a decrease in the melting point of the solution.

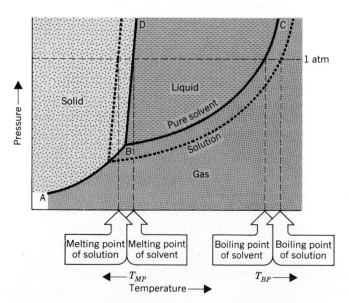

Adding a solute to a solvent doesn't change the way the melting point depends on pressure, and the line that separates the solid and liquid regions of the solution is parallel to the line that serves the same function for the pure solvent. This line must pass through the triple point, however. Therefore, the decrease in the triple point that occurs when a solute is dissolved in a solvent effectively lowers the melting point of the solution.

Figure 13.26 shows that the change in vapor pressure that occurs when a solute dissolves in a solvent causes changes in the melting point and the boiling point of the solvent as well. Because the change in vapor pressure is a colligative property that depends only on the relative number of solute and solvent particles, the changes in the boiling point and the melting point of the solvent are also colligative properties.

13.17 COLLIGATIVE PROPERTIES: CALCULATIONS

Perhaps the best way of demonstrating the importance of colligative properties is to examine the logical consequences of Raoult's law. Raoult found that the vapor pressure of the solvent escaping from a solution is proportional to the mole fraction of the solvent.

$$P = X_{solvent}P°$$

But the vapor pressure of a solvent is not a colligative property. Only the change in the vapor pressure that occurs when a solute is added to the solvent can be included among the colligative properties of a solution.

The change in the vapor pressure of the solvent can be defined as follows.

$$\Delta P = P° - P$$

This equation is based on the fact that pressure is a state function. The magnitude of the change in the vapor pressure of the solvent when a solute added is therefore equal to the vapor pressure of the pure solvent minus the vapor pressure of the solvent escaping from the solution. Substituting Raoult's law into this equation gives the following result.

$$\Delta P = P° - X_{solvent}P°$$

This equation can be rearranged as follows.

$$\Delta P = (1 - X_{solvent})P°$$

We can now remember the relationship between the mole fraction of the solute and the mole fraction of the solvent in a solution that was introduced in Section 13.14.

$$X_{solute} + X_{solvent} = 1$$

The mole fraction of the solute in the solution is therefore equal to 1 minus the mole fraction of the solvent.

$$X_{solute} = 1 - X_{solvent}$$

Substituting this into the equation that describes the relationship between ΔP and $P°$ gives the following result.

$$\Delta P = X_{solute}P°$$

The change in the vapor pressure of the solvent that occurs when a solute is added to the solvent is therefore proportional to the mole fraction of the solute. As more solute is dissolved in the solvent, the vapor pressure of the solvent becomes smaller, and the change in the vapor pressure of the solvent becomes larger.

The mole fraction of the solute in a solution is proportional to the molality of the solution. Thus, the change in the vapor pressure of the solvent that occurs when a solute is added is also proportional to the molality of the solution.

$$\Delta P \propto m$$

This proportionality can be turned into an equation by introducing a suitable proportionality constant, k_P.

$$\Delta P = k_P m$$

The value of k_P depends on the solvent—some solvents are more sensitive to changes in vapor pressure than others. For a given solvent, the change in vapor pressure that occurs when the solvent is used to prepare a solution is equal to k_P for that solvent times the molality of the solution.

Because changes in the boiling point of the solvent (ΔT_{BP}) that occur when a solute is added result from changes in the vapor pressure of the solvent, the magnitude of the change in the boiling point must also be proportional to the molality of the solution.

$$\Delta T_{BP} = k_B m$$

Here, ΔT_{BP} is the *boiling point elevation*—the change in the boiling point that occurs when a solute dissolves in the solvent—and k_B is a proportionality constant known as the *molal boiling point elevation constant* for the solvent.

A similar procedure can be used to derive the following equation, which describes what happens to the freezing point (or melting point) of a solvent when a solute is added.

$$\Delta T_{FP} = -k_F m$$

In this equation, ΔT_{FP} is the *freezing point depression*—the change in the freezing point that occurs when the solute dissolves in the solvent—and k_F is the *molal freezing point depression constant* for the solvent. A negative sign is used in this equation to indicate that the freezing point of the solvent becomes lower when a solute is added.

TABLE 13.7

Freezing Point Depression and Boiling Point Elevation Constants for Several Solvents

Compound	Freezing Point (°C)	k_F(°C/m)	Compound	Boiling Point (°C)	k_B(°C/m)
water	0°	1.853	water	100	0.515
acetic acid	16.66	3.90	ethyl ether	34.55	1.824
benzene	5.53	5.12	carbon disulfide	46.23	2.35
p-xylene	13.26	4.3	benzene	80.10	2.53
naphthalene	80.29	6.94	carbon tetrachloride	76.75	4.48
phenol	40.90	7.40	camphor	207.42	5.611
cyclohexane	6.54	20.0			
carbon tetrachloride	−22.95	29.8			
camphor	178.75	37.7			

Values of k_F and k_B as well as freezing points and boiling points for a number of pure solvents are given in Table 13.7. Because the change in the freezing point (or melting point) and the boiling point of a solvent is a colligative property that depends only on the ratio of the number of particles of solute and solvent in the solution, these constants are used most often to determine the molecular weight of an unknown solute.

Exercise 13.11

Calculate the molecular weight of elemental sulfur if 35.5 grams of sulfur dissolve in 100.0 grams of carbon disulfide to produce a solution that has a boiling point of 49.48°C.

Solution

The relationship between the boiling point of the solution and the molecular weight of sulfur is not immediately obvious. We might therefore start by asking: What do we know about this problem?

We know the boiling point of the solution, so we might start by calculating the difference between the boiling point of the solution and that of pure carbon disulfide.

$$\Delta T_{BP} = 49.48°C - 46.23°C = 3.25°C$$

We also know that the magnitude of the change in the boiling point is proportional to the molality of the solution.

$$\Delta T_{BP} = k_B m$$

We can therefore calculate the molality of the solution as follows.

$$m = \frac{\Delta T_{BP}}{k_B} = \frac{3.25°C}{2.35°C/m} = 1.38 \; m$$

Once we know the molality of the solution, we can calculate the number of moles of sulfur that must be present in 100.0 grams of carbon disulfide.

$$\frac{1.38 \text{ mol sulfur}}{1000 \text{ g } CS_2} \times 100.0 \text{ g } CS_2 = 0.138 \text{ mol sulfur}$$

We now know the number of moles of sulfur in this solution and the weight of the sulfur. We can therefore calculate the number of grams per mole, or the molecular weight, of sulfur.

$$\frac{35.5 \text{ g}}{0.138 \text{ mol}} = 257 \text{ g/mol}$$

The molecular weight of sulfur in this solution is eight times the atomic weight of sulfur.

$$\frac{257 \text{ g/mol}}{32 \text{ g/mol}} = 8$$

We can therefore conclude that elemental sulfur consists of S_8 molecules when dissolved in CS_2.

Exercise 13.12

Determine the molecular weight of acetic acid if a solution that contains 30.0 grams of acetic acid per kilogram of water freezes at $-0.93°C$. Do these results agree with the assumption that acetic acid has the formula CH_3CO_2H?

Solution

The freezing point depression for this solution is equal to the difference between the freezing point of the solution and the freezing point of pure water.

$$\Delta T_{FP} = -0.93°C - 0.0°C = -0.93°C$$

The molal freezing point depression constant for water given in Table 13.7 is $1.853°C/m$. We can therefore start with the following equation

$$\Delta T_{FP} = -k_F m$$

and solve for the molality of the solution.

$$m = -\frac{\Delta T_{FP}}{k_F} = \frac{0.93°C}{1.853°C/m} = 0.50 \ m$$

According to this calculation, there are 0.50 moles of acetic acid per kilogram of water in this solution. The problem stated that there were 30.0 grams of acetic acid per kilogram of water in the solution. Acetic acid therefore has a molecular weight of approximately 60 grams per mole.

$$\frac{30.0 \ g}{0.50 \ mol} = 60 \ g/mol$$

The results of this experiment are in good agreement with the molecular weight (60.05 g/mol) calculated from the formula CH_3CO_2H for acetic acid.

What would happen if the calculation in Exercise 13.12 were repeated with a stronger acid, such as hydrochloric acid? It is important to remember that colligative properties such as freezing point depression and boiling point elevation depend on the number of solute particles in a solution, not their identity. If the acid dissociates to an appreciable extent, the solution will contain more solute particles than we might expect from its molality.

Exercise 13.13

Calculate the freezing point that would be expected for a 0.56 m solution of hydrochloric acid, assuming that HCl is a strong acid that dissociates completely in water. Compare the results of this calculation with the experimental value for the freezing point of this solution, $-2.07°C$.

Solution

If HCl dissociates completely in water, the total concentration of solute particles (H_3O^+ and Cl^- ions) in the solution will be twice as large as the molality of the solution.

$$HCl(g) + H_2O(aq) \longrightarrow H_3O^+(aq) + Cl^-(aq)$$

The freezing point depression of a solution depends only on the number of solute particles, not their identity. The freezing point depression for HCl, since it is a strong acid, is therefore twice as large as the change that would be observed if HCl did not dissociate to an appreciable extent.

If we assume that 0.56 m HCl dissociates to give twice as many solute particles, the freezing point depression for this solution should be $-2.07°C$, which agrees with what is observed experimentally.

$$\Delta T_{FP} = -k_F m$$
$$= -(1.853°C/m)(2 \times 0.56\ m) = -2.07°C$$

Exercise 13.14

Explain why 0.60 grams of acetic acid dissolve in 200 grams of benzene to form a solution that lowers the freezing point of benzene to 5.40°C.

Solution

Because pure benzene freezes at 5.53°C, the freezing point depression in this experiment is $-0.13°C$.

$$\Delta T_{FP} = 5.40°C - 5.53°C = -0.13°C$$

Once again, we can start with the relationship between the freezing point depression and the molality of the solution

$$\Delta T_{FP} = -k_F m$$

and use the known molal freezing point depression constant for benzene to calculate the molality of the solution.

$$m = -\frac{\Delta T_{FP}}{k_F} = \frac{0.13°C}{5.12°C/m} = 0.025\ m$$

The concentration of this solution is therefore 0.025 m.

Multiplying the molality of the solution by the weight of solvent used to make the solution gives the number of moles of solute particles in the solution.

$$\frac{0.025\ \text{mol solute}}{1\ \text{kg benzene}} \times 0.200\ \text{kg benzene} = 0.0050\ \text{mol solute}$$

The problem, however, said that we started with 0.60 grams of acetic acid. The solution therefore simultaneously contains 0.60 grams of acetic acid and 0.0050 moles of solute. This would suggest that the molecular weight of acetic acid in this experiment is 120 grams per mole.

$$\frac{0.60\ \text{g solute}}{0.0050\ \text{mol solute}} = 120\ \text{g/mol}$$

The molecular weight of acetic acid in benzene is twice as large as what is expected from the molecular formula, CH_3CO_2H. The only way to explain this observation is to assume that the acetic acid molecules associate in benzene to form dimers. These dimers are held together by hydrogen bonds.

$$CH_3C \begin{array}{c} O \cdots H-O \\ \diagdown \quad \diagup \\ \diagup \quad \diagdown \\ O-H \cdots O \end{array} CCH_3$$

An egg can be used to demonstrate the osmotic pressure through a semipermeable membrane. The egg is first soaked overnight in acetic acid, which softens the $CaCO_3$ in the egg shell. The egg shell is then peeled from the egg, the egg is weighed on an analytical balance, immersed in distilled water for 20–30 minutes, and then reweighed. The difference between the concentration of the solution within the egg and the distilled water in the beaker generates an osmotic pressure that forces water to pass through the semipermeable membrane that lies beneath the shell of the egg, thereby increasing the weight of the egg by as much as 10%. Here, an egg that initially weighed 72.50 grams gained weight until it eventually weighed 78.14 grams.

FIG. 13.27 When a semipermeable membrane, such as a pig's bladder, is used to separate water in a beaker from a solution of sugar or alcohol in water in a tube, water flows through the membrane until the force of gravity pulling down on the column of water in the tube balances the osmotic pressure pushing the water through the membrane.

13.18 COLLIGATIVE PROPERTIES: OSMOTIC PRESSURE

In 1784, the French physicist and clergyman Jean Antoine Nollet discovered that a pig's bladder filled with a concentrated solution of alcohol in water expanded when it was immersed in water. The bladder acted as a *semipermeable membrane*—it allowed water molecules to enter the solution but kept alcohol molecules from moving in the other direction. Movement of one component of a solution through a membrane to dilute the solution is called *osmosis,* and the pressure this produces is called *osmotic pressure.*

Osmotic pressure can be demonstrated with the apparatus shown in Figure 13.27. A small piece of a semipermeable membrane is tied across the open end of a thistle tube. The tube is then partially filled with a solution of sugar or alcohol in water and immersed in a beaker of water. Water will flow into the tube until the pressure on the column of water due to the force of gravity balances the osmotic pressure driving water through the membrane.

The same year that Raoult discovered the relationship between the vapor pressure of a solution and the vapor pressure of a pure solvent, Jacobus Henricus van't Hoff found that the osmotic pressure of a dilute solution (π) obeyed an equation analogous to the ideal gas equation.

$$\pi = \frac{nRT}{V}$$

This equation has two important implications. First, it identifies osmotic pressure as another example of a colligative property, because this pressure depends only on the number of solute particles in a liter of solution and not on the identity of the solute particles. Second, it reminds us of the magnitude of osmotic pressure. According to this equation, a 1.00 M solution has an osmotic pressure of 22.4 atm at 0°C.

$$\pi = \frac{(1.00 \text{ mol})(0.08206 \text{ L-atm/mol-K})(273 \text{ K})}{(1.00 \text{ L})} = \textbf{22.4 atm}$$

That means that a 1.00 M solution should be able to support a column of water 9108 inches, or 759 feet, tall!

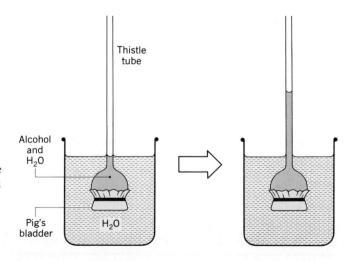

Thistle tube

Alcohol and H_2O

Pig's bladder

H_2O

Biologists and biochemists frequently take advantage of osmotic pressure when they try to isolate the components of a cell. When a cell is added to an aqueous solution that contains a much higher concentration of ions than the liquid within the cell, water leaves the cell by flowing through the cell membrane until the cell shrinks so much that the membrane is broken. Alternatively, when a cell is placed in a solution that has a much smaller ionic strength, water pours into the cell, and the cell expands to the point at which the cell membrane bursts.

13.19 CHEMICAL SEPARATIONS

One of the most important, and time-consuming, activities of chemists involves isolating, separating, or in one way or another purifying chemical compounds. No matter what the goals of the project, the techniques available to reach them are similar.

Extraction (literally, "taking out by force") is a useful technique for separating compounds, such as I_2 and $KMnO_4$, that have different polarities. The compounds to be separated are treated with a mixture of a polar solvent, such as water, and a nonpolar solvent, such as carbon tetrachloride. As noted in Section 13.11, the I_2 will dissolve in the CCl_4, while the $KMnO_4$ will dissolve in the H_2O. By simply separating these two phases and allowing the solvents to evaporate, we can cleanly separate I_2 and $KMnO_4$.

This technique can also be used to extract polar or nonpolar solutes from a solid. Coffee, for example, is decaffeinated by extraction. Most of the popularity of coffee can be attributed to the stimulating effect of caffeine, which makes up as much as 2.5% of the dry weight of coffee beans. But many people who have grown to like the flavor of coffee want to reduce the amount of caffeine they consume and therefore turn to decaffeinated coffee. There are many ways of removing caffeine from coffee beans. It can be extracted with hot water, for example. Unfortunately, this also tends to extract the oils that give coffee its flavor. An alternative approach involves treating the coffee beans with a nonpolar solvent such as trichloroethylene ($Cl_2C=CHCl$), which dissolves most of the caffeine without destroying the flavor of the coffee. (The caffeine is then sold to manufacturers of cola drinks, who add it to their products.)

Distillation is the most frequently used method for purifying liquids. At its simplest, this technique involves heating a mixture with a bunsen burner or heating mantle, as shown in Figure 13.28, until the liquid boils. The vapor that escapes is passed through a water-cooled condenser, where it condenses to form a liquid, which collects in a clean flask. Compounds that boil at very high temperatures at atmospheric pressure can often be distilled at lower temperatures when a vacuum pump is used to reduce the pressure in the flask and thereby reduce the boiling point.

Exercise 13.15

Assume that a mixture of hydrocarbons that contains equal amounts of pentane (C_5H_{12}) and octane (C_8H_{18}) is distilled. Describe the difference between the composition of the vapor that is given off when this mixture starts to boil and the composition of the liquid in the distillation flask.

Solution

According to the boiling points of these hydrocarbons, which were given in Table

13.2, pentane (BP = 36.1°C) is much more volatile than octane (BP = 125.7°C). The pentane will therefore begin to distill from this mixture long before the octane, and the partial pressure of the pentane that collects above the solution will be much larger than the partial pressure of the octane. The composition of the vapor that collects above the solution is therefore different from the composition of the liquid that forms the vapor. The liquid contains more or less equal amounts of the two hydrocarbons, but the vapor is much richer in pentane.

We can translate the results of Exercise 13.15 into a more general rule for what happens during distillation by adding another dimension to the phase diagram first introduced in Section 13.10. This third dimension shows what happens when we change the composition of a mixture of two or more substances. It is possible to construct three-dimensional phase diagrams that simultaneously show the effects of temperature, pressure, and percent composition. These diagrams are difficult to understand, however, and we will therefore restrict our attention to phase diagrams that show only two of these dimensions at a time.

A phase diagram that plots temperature versus percent composition for a mixture of pentane and octane is shown in Figure 13.29. The points T_P and T_O represent the boiling points of pure pentane and pure octane. These points are connected by two curves. At temperatures below the bottom curve, mixtures of pentane and octane are liquids. At temperatures above the top curve, these mixtures are gases. Points that lie between the curves describe systems that contain both a liquid and a gas phase with the same composition.

We have argued, however, that the liquid and the gas should not have the same

FIG. 13.28 A distillation apparatus.

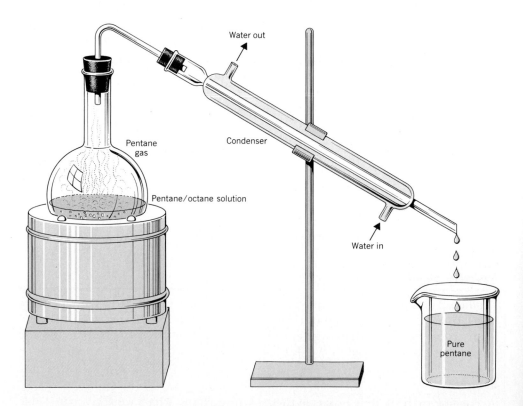

composition at a given temperature. Points between the two curves are therefore unstable. Any system at point B, for example, would come to equilibrium by forming a vapor with the composition described by point A and a liquid with the composition described by point C.

The vertical line in Figure 13.29 shows what happens to the system when a 50 : 50 mixture of pentane and octane is heated. Nothing changes until the temperature reaches point C — the temperature at which the liquid starts to boil. The vapor that collects above the boiling liquid will be richer in pentane, because pentane is more volatile than octane. The vapor will have the composition described by point A in this diagram; it will therefore be much richer in pentane than the liquid from which it boils.

If this vapor escapes, more pentane leaves the system than octane, and the percent by weight of octane in the system increases. This increases the boiling point of the mixture, as well as gradually increasing the amount of octane that collects in the vapor that escapes from the liquid.

Figure 13.29 provides a clue as to how we can separate compounds by a technique called *fractional distillation*. We start by heating the mixture and collecting the first few portions, or fractions, that distill over. These fractions will be rich in the component that has the lower boiling point. We might then discard the next few fractions, which contain a mixture of the lower- and higher-boiling compounds, and save the last few fractions, which should be rich in the compound that has the higher boiling point.

Distillation works best for purifying compounds that are liquids at room temperature. Solids are more likely to be purified by *fractional recrystallization.* The components of a mixture seldom have the same solubility in all solvents. It should therefore be possible to find a solvent that dissolves one of the components more readily than the others. The extent to which the components are separated can be enhanced by taking advantage of the fact that solids usually become more soluble as the temperature of the solvent increases.

We start by dissolving as much of the solid as possible in a sample of a hot solvent. This gives us some separation, because one component of the starting material is presumably more soluble in this solvent than the others. We then cool the solution back to room temperature, which gives us even more separation, because the less soluble component crystallizes out of the solution faster than the more soluble component. The result is a solution that is richer in the more soluble component and a solid that is richer in the less soluble component. Each time the sample is dissolved and recrystallized, it is purer than before. This was the technique used by Marie and Pierre Curie to isolate the first minuscule sample of radium from literally tons of the mineral pitchblende.

One of the fastest-growing methods of separating compounds was developed by a Russian botanist named Mikhail Tsvet in 1906. This technique was called *chromatography* (literally, "writing with color") because it was first used to separate the colored pigments in plants. The basic principle behind chromatography is simple. A liquid or a gas is allowed to flow over a solid support. Compounds that have a high affinity for the solvent are carried along with it as it moves along the stationary support. Compounds that have a higher affinity for the solid move more slowly.

In *thin-layer chromatography (TLC),* a glass plate or plastic support is coated with a thin layer of a solid such as alumina (Al_2O_3) or silica (SiO_2). Small samples of the compounds to be separated are then placed on the silica or alumina, and the TLC plate is immersed in a solvent until the solvent line is just below the point where the samples were applied (see Figure 13.30).

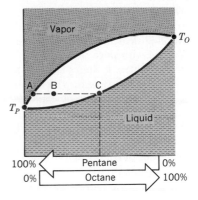

FIG. 13.29 A phase diagram for a mixture of pentane and octane that shows the effect of changes in the composition of the mixture on the boiling point of the solution.

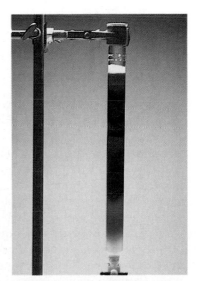

Chromatography literally means "writing with color." The name refers to the fact that this technique was first used to separate the colored pigments in plants.

FIG. 13.30 Thin-layer chromatography. (*a*) The solution is applied near one end of a plate coated with silica or alumina. (*b*) The plate is then placed in a container of solvent, which rises through the coating by capillary action. The mixture separates into its components on the basis of differences in the relative affinities of these compounds for the stationary SiO₂ or Al₂O₃ phase and the mobile solvent phase.

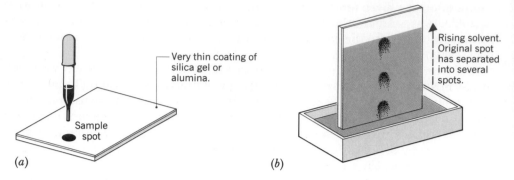

The solvent slowly moves up the plate, carrying the components of the mixture with it. Some of the components have a high affinity for the Al_2O_3 or SiO_2 support, and they move very slowly. Others have a high affinity for the solvent, and they move more rapidly. The net result is a separation of the mixture into its components.

Column chromatography extends the principles of TLC to the large-scale separation of mixtures. This technique involves filling a glass column with a solid support, applying up to several grams of the mixture to the top of the column, and then slowly washing the column with solvent. Mixtures can also be separated on the basis of differences between the rate at which the components of a mixture move down a column in equilibrium between the gas and liquid phases. This technique is known as *gas-phase chromatography.*

Zone refining is a technique used to obtain very pure (99.999%) samples of metals and semimetals. The metal or semimetal is drawn into a thin rod which is inserted into a heating coil that slowly moves along the length of the rod. As the coil moves, it constantly brings a new portion of the rod to the point at which the solid just melts. Because solutions melt at lower temperatures than pure solids, it is possible to concentrate the impurities in the small sample of liquid that moves just ahead of the heating coil, and crystallize almost perfectly pure metal or semimetal just behind the coil.

SUMMARY

Chapter 13 contains a model for the structure of liquids that explains many of their physical properties, including the boiling point, melting point, freezing point, vapor pressure, critical temperature, critical pressure, surface tension, and viscosity.

The vapor pressure of a liquid is the partial pressure of the gas that collects in a closed container at a particular temperature. The boiling point of the liquid is the temperature at which the vapor pressure is equal to atmospheric pressure. The melting point (or freezing point) of a substance is the temperature at which the solid and the liquid are in equilibrium at atmospheric pressure. The relationship between the gaseous, liquid, and solid states of a substance can be summarized in a phase diagram that describes the state of the substance in a closed container under all possible combinations of temperature and pressure.

Liquids are often used as solvents to control the rate of chemical reactions. Solvents can be divided into two categories: polar and nonpolar. As a rule, like dissolves like — nonpolar compounds are soluble in nonpolar solvents, and polar compounds are soluble in polar solvents.

There are many ways of describing the concentration of a solution, including molarity, molality, mole fraction, weight percent, and volume percent. The unit of concentration used most commonly by chemists is molarity — the number of moles of solute in the solution divided by the volume of the solution. There are times, however, when it is useful to know the ratio of the amount of solute to the amount of solvent in the solution. Under these circumstances, chemists turn to the units of mole fraction and molality. These units are particularly important in discussions of colligative properties—properties of a solution that depend on the ratio of the number of particles of solute and solvent but not on the identity of the solute.

Colligative properties include the changes in the vapor pressure, boiling point, and freezing point of a solvent that occur when a solute is dissolved in the solvent. Adding a solute to a solvent lowers the vapor pressure of the solvent. This results in an elevation in the boiling point of the solution and a depression in its freezing point. Another colligative property of a solution, which is not directly related to those just mentioned, is osmotic pressure, which causes water to flow across a semipermeable membrane when the concentrations of solute on either side of the membrane are unequal.

Because colligative properties depend only on the number of solute particles in a solution, not their identity, these properties can be used to measure the molecular weight of an unknown solute.

PROBLEMS

The Structure of Liquids

13-1 Describe how assuming that there are empty spaces, or holes, in the structure of a liquid explains the ease with which the liquid flows.

13-2 Predict the order in which the boiling points of the following compounds should increase if the intermolecular forces in these compounds are about the same.

(a) CH_4 (b) SiH_4 (c) GeH_4 (d) SnH_4

13-3 Explain why the boiling points of hydrocarbons, that have the generic formula C_nH_{2n+2} increase with molecular weight.

13-4 Explain why pentane is a liquid over a much larger range of temperatures than neopentane.

Vapor Pressure

13-5 Explain why the vapor pressure of water becomes larger as the temperature of the water increases.

13-6 Explain why a cloth soaked in water feels cool when it is placed on your forehead.

13-7 What would happen to the vapor pressure of liquid bromine, Br_2, at 20°C if the liquid was transferred from a narrow 10-milliliter graduated cylinder into a wide petri dish or crystallizing dish? Would it increase, decrease, or remain the same?

13-8 What would happen to the vapor pressure in a closed container if more liquid was added to the container? Would it increase, decrease, or remain the same?

13-9 Explain why each of the following increases the rate at which water evaporates from an open container.

(a) Increasing the temperature of the water (b) Increasing the surface area of the water (c) Blowing air over the surface of the water (d) Decreasing atmospheric pressure on the water

13-10 Calculate the relative humidity when the temperature is 31°C and the partial pressure of water vapor in the atmosphere is 29.8 mmHg.

13-11 Calculate the dew point on a day when the humidity is 85% and the temperature is 60°F.

Melting Point, Freezing Point, and Boiling Point

13-12 Explain why water boils when the pressure on the system is reduced.

13-13 At what temperature does water boil when the pressure is 50 mmHg?

13-14 Increasing the temperature of a liquid will do which of the following?

(a) Increase its boiling point (b) Increase its melting point (c) Increase its vapor pressure (d) Increase the amount of heat required to boil a mole of the liquid

13-15 Which compound would you expect to have the highest boiling point?

(a) Methane, CH_4 (b) Chloromethane, CH_3Cl (c) Dichloromethane, CH_2Cl_2 (d) Chloroform, $CHCl_3$ (e) Carbon tetrachloride, CCl_4

13-16 Liquid air is composed primarily of liquid oxygen (BP = −183°C) and liquid nitrogen (BP = −196°C). It can be purified by increasing the temperature until one of these gases boils off. Which gas boils off first?

13-17 Predict which liquid should have the lowest boiling point from the following vapor pressures at 0°C.

(a) Acetone: $P = 67$ mmHg (b) Benzene: $P = 24.5$ mmHg (c) Ether: $P = 183$ mmHg (d) Methyl alcohol: $P = 30$ mmHg (e) Water: $P = 4.6$ mmHg

Critical Temperature and Critical Pressure

13-18 Explain why water is a gas at 120°C and 1 atm but some of the water vapor in a closed container condenses to form a liquid at 120°C when the pressure on the vapor becomes larger than 2 atm.

13-19 What happens to the critical temperature of a series of compounds as the force of attraction between the particles becomes larger?

13-20 The properties of a liquid become more like those of a gas as the temperature of the liquid increases. The properties of a gas become more like those of a liquid as the pressure on the gas increases. Use these observations to explain why it is impossible to distinguish between a liquid and a gas when the pressure on the system is equal to the critical pressure and the temperature is above the critical temperature.

Surface Tension

13-21 The force of cohesion between mercury atoms is much larger than the force of cohesion between water molecules. Conversely, the force of adhesion between water molecules and glass is much larger than the force of adhesion between mercury atoms and glass. Use these observations to explain why the meniscus in a water-filled buret curves upward, whereas the meniscus in a barometer curves downward.

13-22 Describe how the fact that water has an unusually large surface tension can be used to explain the fact that a steel sewing needle floats on the surface of water, even though it is much denser than water.

13-23 If a steel sewing needle floats on the surface of water, why does a steel ball bearing sink to the bottom?

Viscosity

13-24 Gasoline molecules are very much larger than water molecules, and yet gasoline can escape through holes that are much too small to pass water. Explain why.

13-25 Use the viscosities of water, gasoline, and air given in Section 13.8 to predict the order in which the force of adhesion between these compounds and glass increases.

13-26 Explain why liquids become less viscous when they are heated and more viscous when they are cooled.

Hydrogen Bonding and the Anomalous Properties of Water

13-27 Explain why the boiling point and melting point of water are much higher than you would expect from the boiling points and melting points of H_2S, H_2Se, and H_2Te.

13-28 Explain why hydrogen bonding is very strong in HF and H_2O. Explain why hydrogen bonding is much weaker in HCl and H_2S.

13-29 What makes a compound a good hydrogen-bond donor? What makes a compound a good hydrogen-bond acceptor?

13-30 Explain why the strength of hydrogen bonds decreases in the following order.

$$HF > H_2O > NH_3$$

Phase Diagrams

13-31 Use a phase diagram to explain why a liquid boils when the pressure on the liquid is reduced.

13-32 Use a phase diagram to explain why a gas condenses to form a liquid when the pressure on the gas is increased.

13-33 Use a phase diagram to explain why most liquids become solids at temperatures near their freezing points when the pressure on the liquids is increased. Explain why ice melts at temperatures near its melting point when the pressure is increased.

13-34 Use a phase diagram to explain why the boiling point of a liquid is relatively sensitive to changes in pressure but the melting point is not.

Solutions: Like Dissolves Like

13-35 Define the terms *solution, solvent,* and *solute.*

13-36 Draw a picture that describes what happens when I_2 molecules dissolve in CCl_4 and when $KMnO_4$ dissolves in water.

13-37 One way of screening potential anaesthetics involves testing whether the compound dissolves in olive oil, because all common anaesthetics, including nitrous oxide (N_2O), cyclopropane (C_3H_6), and halothane (C_2HF_3ClBr), are soluble in olive oil. What property do these compounds have in common?

13-38 Carboxylic acids with the general formula $CH_3(CH_2)_nCO_2H$ have a nonpolar $CH_3{-}CH_2 \ldots$ tail and a polar $\ldots CO_2H$ head. What effect does increasing the value of n in the formula for carboxylic acids have on the solubility of these acids in polar solvents, such as water? What is the effect on their solubility in nonpolar solvents, such as CCl_4?

13-39 Which of the following compounds would be the most soluble in a nonpolar solvent, such as CCl_4?

(a) H_2O (b) CH_3OH (c) $CH_3CH_2CH_2OH$
(d) $CH_3CH_2CH_2CH_2CH_2OH$
(e) $CH_3CH_2CH_2CH_2CH_2CH_2CH_2OH$

13-40 Potassium iodide reacts with iodine in aqueous solution to form the triiodide, or $I_3{}^-$, ion.

$$KI(aq) + I_2(aq) \longrightarrow KI_3(aq)$$

What would happen if we added CCl_4 to this reaction mixture?

(a) The KI would dissolve in the CCl_4 layer. (b) The I_2 would dissolve in the CCl_4 layer. (c) Both KI and I_2 would dissolve in the CCl_4 layer. (d) Neither KI nor I_2 would dissolve in the CCl_4 layer. (e) No distinct CCl_4 layer would form because CCl_4 is soluble in water.

Units of Concentration: Molarity and Molality

13-41 Define *molarity* and *molality.* How are these units of concentration similar? How are they different?

13-42 Explain why the molality of a solution is proportional to the mole fraction of the solute for dilute solutions.

13-43 If liquids expand as the temperature increases, what effect should an increase in the temperature of a solution have on the molarity, molality, and mole fraction of the solute and solvent?

13-44 When asked to prepare a liter of 1.00 M K_2CrO_4, a student weighed out exactly 1.00 mole of K_2CrO_4 and added this solid to 1.00 liter of water in a volumetric flask. What did the student do wrong? How would *you* prepare the solution?

13-45 What is the molarity of an aqueous solution that is 18.0% $Pb(NO_3)_2$ by weight if the density of this solution is 1.18 g/cm³?

13-46 Benzene was once used as a common solvent in chemistry. Its use is now restricted because of fear that it might cause cancer. The recommended limit of exposure to benzene is 3.2 milligrams per cubic meter of air (3.2 mg/m³). Calculate the molarity of this solution. Calculate the number of parts per million by weight of benzene in this solution.

13-47 You can make the chromic acid bath commonly used to clean lab glassware by dissolving 92 grams of sodium dichromate ($Na_2Cr_2O_7 \cdot 2 H_2O$) in enough water to give 458 milliliters of solution and then adding 800 milliliters of concentrated sulfuric acid. Calculate the molarity of the $Cr_2O_7{}^{2-}$ ion in this solution.

Units of Concentration: Mole Fraction

13-48 Calculate the mole fraction of N_2 and O_2 in air, assuming that air is 78% nitrogen and 21% oxygen by volume.

13-49 Calculate the mole fraction of water in a solution that contains 10.0 grams of glucose ($C_6H_{12}O_6$) dissolved in 100 grams of water.

13-50 Chlorine gas can be detected in air at a concentration of 3.5 ppm by weight. Calculate the mole fraction of chlorine in air at this concentration.

13-51 Nichrome is an alloy that is 60% Ni, 24% Fe, 16% Cr, and 0.1% C by weight. Calculate the mole fraction of each element in this solution.

13-52 Babbitt metal is an alloy that is 69% Zn, 19% Sn, 4% Cu, 3% Sb, and 5% Pb by weight. Calculate the mole fraction of each metal in this alloy. Calculate the sum of the mole fractions of all five metals. Explain why the sum of these mole fractions is equal to 1, within experimental error.

13-53 Calculate the mole fraction of both the solute and the solvent in a 0.100 m $K_2Cr_2O_7$ solution. Explain why you don't need to know the volume of the solution to do this calculation. Explain why you can't perform the same calculation for a 0.100 M $K_2Cr_2O_7$ solution.

13-54 At high temperatures and pressures, ammonia decomposes to nitrogen and hydrogen according to the following equation.

$$2 NH_3(g) \longrightarrow N_2(g) + 3 H_2(g)$$

What is the mole fraction of N_2 when one-half of the NH_3 has decomposed?

13-55 What are the density, molarity, molality, and mole fraction of a solution prepared from 5.0 grams of NaCl dissolved in 65 grams of water if the total volume of the solution is 68 milliliters?

Units of Concentration: Weight Percent

13-56 Concentrated acetic acid is called glacial acetic acid because of the ease with which this solution freezes. (The term *glacial* comes from the Latin stem meaning "ice.") What is the weight percent of the CH_3CO_2H molecules in glacial acetic acid if this solution is 17.4 M and its density is 1.05 g/cm³.

13-57 Calculate the molarity of concentrated hydrochloric acid, assuming that this solution is 38% HCl by weight and that its density is 1.1977 g/cm³.

13-58 The solubility of NH_3 gas in water is 33.1% by weight. Calculate the molarity of a saturated solution of ammonia in water.

13-59 Calculate the molality of an aqueous solution that is 10.0% sodium chloride by weight.

13-60 Calculate the molality of carbon in austenite, one of the crystal forms of steel, if austenite is 0.8% C by weight dissolved in iron.

Colligative Properties: Vapor Pressure Depression

13-61 Explain how dissolving a solute in a solvent leads to a decrease in the vapor pressure of the solvent.

13-62 Explain why the pressure of the solvent escaping from a solution is equal to the mole fraction of the solvent times the vapor pressure of the pure solvent.

13-63 Predict what will happen to the rate at which water evaporates from an open flask when salt is dissolved in the water, and explain why the rate of evaporation changes.

13-64 If you place a beaker of pure water (I) and a beaker of a saturated solution of sugar in water (II) in a sealed bell jar, the level of water in beaker I will slowly decrease, and the level of the sugar solution in beaker II will slowly increase. Explain why.

13-65 Explain why the vapor pressure of a liquid at a particular temperature is an intensive property but the change in the vapor pressure of the liquid when a solute is added is a colligative property.

Colligative Properties: Boiling Point Elevation and Freezing Point Depression

13-66 Explain how the decrease in the vapor pressure of a solvent that occurs when a solute is added leads to an increase in the solvent's boiling point.

13-67 Explain how the decrease in the vapor pressure of the solvent that occurs when a solute is added leads to a decrease in the solvent's melting point.

13-68 Predict the shape of a plot of the freezing point of the solvent in a solution versus the molality of the solution.

Colligative Properties: Calculations

13-69 What is the freezing point of a saturated solution of caffeine ($C_8H_{10}O_2N_4 \cdot H_2O$) in water if it takes 45.6 grams of water to dissolve 1.00 gram of caffeine?

13-70 What is the molecular weight of nicotine if 3.62 grams of this deadly poison changes the freezing point of 73.4 grams of water by 0.563°C?

13-71 A 0.100 m solution of sulfuric acid in water freezes at -0.371°C. Which of the following statements agrees with this observation?

(a) H_2SO_4 does not dissociate in water (b) H_2SO_4 dissociates into H_3O^+ and HSO_4^- ions in water (c) H_2SO_4 dissociates in water to form two H_3O^+ ions and one SO_4^{2-} ion. (d) H_2SO_4 associates in water to form $(H_2SO_4)_2$ molecules.

13-72 The "Tip of the Week" in a recent newspaper suggested using a fertilizer such as ammonium nitrate or ammonium sulfate instead of salt to melt snow and ice on sidewalks, because salt can damage lawns. Which of the following compounds will give the largest freezing point depression when 100 grams of the compound are dissolved in 1 kilogram of water?

(a) NaCl (b) NH_4NO_3 (c) $(NH_4)_2SO_4$

13-73 A 5% solution of dextrose in water at 20°C exhibits an osmotic pressure of 6.68 atm. What is the molecular weight of dextrose? If the empirical formula is CH_2O, what is the molecular formula?

13-74 Ortho-dichlorobenzene (ODCB) is the active ingredient in some moth balls. Calculate the value of k_B for camphor if a 0.260 m solution of ODCB in camphor increases the boiling point of camphor by 9.8°C.

13-75 If an aqueous solution boils at 100.50°C, at what temperature does it freeze?

13-76 What is the boiling point of a solution of 10.0 grams of P_4 in 25.0 milliliters of carbon disulfide if the boiling point elevation constant for CS_2 is 2.35°C/m and the boiling point is 46.23°C?

13-77 We usually assume that salts such as KCl dissociate completely when they dissolve in water.

$$KCl(s) + H_2O \longrightarrow K^+(aq) + Cl^-(aq)$$

What percent of KCl actually dissociates in water if the freezing point of a 0.100 m solution of KCl in water is -0.345°C?

13-78 Calculate the freezing point of a 0.100 m solution of acetic acid in water if the CH_3CO_2H molecules are 1.33% ionized in this solution.

13-79 Calculate the boiling point of a solution of 4.39 grams of naphthalene, $C_{10}H_8$, in 99.5 grams of carbon tetrachloride, CCl_4, assuming that the boiling point of pure CCl_4 is 349.90 K and the boiling point elevation constant for CCl_4 is 4.48 K/m.

13-80 What fraction of chlorous acid, $HClO_2$, dissociates in water if a solution that contains 3.22 grams of $HClO_2$ in 47.0 grams of water has a freezing point of 271.10 K?

Colligative Properties: Osmotic Pressure

13-81 The osmotic pressure of a solution obeys an equation similar to the ideal gas law: $\pi = nRT/V$, where V is the volume of the solution in liters and n is the number of moles of solute. Use this equation to show why the osmotic pressure of a solution is proportional to temperature.

13-82 Osmotic pressure is so much more sensitive than freezing point depression or boiling point elevation it can be used to measure the molecular weights of molecules as large as the hemoglobin that carries oxygen through the blood. The concentration of hemoglobin in blood is roughly 15 grams per 100 milliliters. Assume that a solution contains 15 grams of hemoglobin dissolved in 100 grams of water and that the osmotic pressure of this solution is found to be 0.050 atm at 25°C. What is the molecular weight of hemoglobin? (The osmotic pressure of a 1 m solution at 25°C is 24.45 atm.)

Chemical Separations

13-83 Organic compounds are often purified by distillation. Describe the equipment you would use to distill a compound that boils at 156°C.

13-84 How would you distill a compound that has a boiling point of 156°C if the compound decomposes at temperatures above 100°C?

13-85 Describe the basic principle behind the separation of compounds by thin-layer, column, and gas-phase chromatography.

13-86 Describe the general principle that allows metals and semimetals to be purified by zone refining.

CHAPTER 14

GAS-PHASE REACTIONS: AN INTRODUCTION TO KINETICS AND EQUILIBRIUM

CHAPTER CONTENTS

14.1 HIDDEN ASSUMPTIONS ABOUT CHEMICAL REACTIONS

FIG. 14.1 This photograph shows what happens when a thin ribbon of magnesium metal is ignited in the flame of a bunsen burner.

Figure 14.1 shows a thin strip of magnesium metal being ignited in the flame of a bunsen burner. Suppose that when you went to class one day, your instructor asked you to calculate the weight of the finely divided white solid produced when a strip of magnesium metal weighing 2.000 grams is burned.

If asked to describe how you solved this problem, you might organize your work in the following steps.

1. **Start by assuming that the magnesium reacts with oxygen in the atmosphere when it burns.**
2. **Predict that the formula of the product is MgO.**
3. **Use this formula to generate the following balanced equation.**

$$2 \, Mg(s) + O_2(g) \longrightarrow 2 \, MgO(s)$$

4. **Convert grams of magnesium into moles of the metal.**

$$2.000 \, \text{g Mg} \times \frac{1 \, \text{mol Mg}}{24.305 \, \text{g}} = 0.08229 \, \text{mol Mg}$$

5. **Use the balanced equation to convert moles of magnesium into moles of magnesium oxide.**

$$0.08229 \, \text{mol Mg} \times \frac{1 \, \text{mol MgO}}{1 \, \text{mol Mg}} = 0.08229 \, \text{mol MgO}$$

6. **Convert moles of magnesium oxide into grams of MgO.**

$$0.08229 \, \text{mol MgO} \times \frac{40.304 \, \text{g MgO}}{1 \, \text{mol MgO}} = 3.317 \, \text{g MgO}$$

At this point, you might have checked your answer by deciding whether the sum of the weights of the magnesium and oxygen consumed in this reaction was equal to the weight of the product formed, as follows.

7. **Calculate the weight of oxygen needed to burn the magnesium.**

$$0.08229 \, \text{mol Mg} \times \frac{1 \, \text{mol O}_2}{2 \, \text{mol Mg}} = 0.04115 \, \text{mol O}_2$$

$$0.04115 \, \text{mol O}_2 \times \frac{31.999 \, \text{g O}_2}{1 \, \text{mol}} = 1.317 \, \text{g O}_2$$

8. **Compare the weight of magnesium and oxygen consumed in the reaction with the weight of magnesium oxide produced.**

$$2.000 \, \text{g Mg} + 1.317 \, \text{g O}_2 = 3.317 \, \text{g MgO}$$

If this is the answer you would get when faced with this task, you have every right to feel good about the chemistry you have learned so far. But, before reading any further, ask yourself the following questions. "How confident am I in this answer? Is there anything else that has to be done before I can safely conclude that I have the 'right' answer to the calculation?"

Before we can trust this answer we have to ask whether there are any hidden assumptions behind the calculation and then check the validity of these assumptions. Three hidden assumptions were made in this calculation.

1. **We assumed that the metal contained 2.000 grams of magnesium—in other words, we assumed that the strip was pure magnesium.**

2. **We assumed that the magnesium reacted only with the oxygen in the atmosphere, to form MgO, thereby ignoring the possibility that some of the magnesium might react with the nitrogen in the atmosphere to form Mg_3N_2.**

3. **We assumed the reaction doesn't stop until all of the magnesium metal has been consumed.**

We can correct for the fact that the starting material is actually 99% magnesium by weight and for the fact that perhaps as much as 5% of the product of this reaction is Mg_3N_2 instead of MgO. But it is the third assumption that is of particular importance for the discussion in this chapter.

14.2 CHEMICAL REACTIONS DON'T ALWAYS GO TO COMPLETION

At some point, most students believe that all chemical reactions go to completion. This belief can result from doing calculations such as predicting the amount of MgO that can be produced from a known amount of Mg—calculations done under the assumption that the reaction goes to completion. It is reinforced by demonstrations such as that in Figure 14.1, in which the reaction continues until all of the metal is gone, and that in Figure 14.2, which shows the reaction between copper metal and concentrated nitric acid to form NO_2. Once the reaction gets started, it doesn't stop until the copper penny disappears completely.

FIG. 14.2 The reaction between copper metal and concentrated nitric acid to form an aqueous solution of the Cu^{2+} ion and NO_2 gas.

Many students are therefore surprised to learn that chemical reactions don't always go to completion. The dimerization of the NO_2 formed by the reaction in Figure 14.2 is an example of a chemical reaction that seems to stop prematurely under normal conditions.

$$2\ NO_2(g) \rightleftharpoons N_2O_4(g)$$

At 25°C, for example, this reaction seems to stop when 92.2% of the NO_2 has been converted into N_2O_4. Once it has reached this point, the reaction doesn't go any further. As long as the reaction is left at 25°C, 7.8% of the NO_2 that was present initially will remain in the flask.

Reactions, such as the dimerization of NO_2, that seem to stop before completion are said to reach *equilibrium.*

Equilibrium: A state in which all reactants and products of a reaction exist simultaneously with no further change in the amounts or concentrations of any of the reactants or products.

It is important to recognize the difference between a reaction that comes to equilibrium and one that stops because it has run out of a limiting reagent (see Section 3.16). The reaction between magnesium and oxygen is an example of a reaction that stops when it has run out of the limiting reagent. Once the reaction gets started, it continues until it runs out of either magnesium or oxygen. We indicate this by writing the equation for the reaction as follows.

$$2\ Mg(s) + O_2(g) \longrightarrow 2\ MgO(s)$$

The dimerization of NO_2 is an example of a reaction that comes to equilibrium—the reaction seems to stop before it runs out of the limiting reagent. For reasons that will soon be apparent, we indicate this by writing a pair of arrows, one pointing left and one pointing right, between the reactants and the products of the reaction.

$$2 \; NO_2(g) \rightleftharpoons N_2O_4(g)$$

In order to work with reactions that reach equilibrium, we need a way to specify the amounts of reactant and product present at equilibrium. By convention, the equilibrium concentration of a given reactant or product in units of moles per liter is indicated by a symbol that consists of the formula for the reactant or product written in square brackets.

[NO$_2$] = concentration of NO$_2$ (in moles per liter)
in a reaction that is at equilibrium.

[N$_2$O$_4$] = concentration of N$_2$O$_4$ (in moles per liter)
in a reaction that is at equilibrium.

Experiments show that the equilibrium concentrations of NO$_2$ and N$_2$O$_4$ depend on how much NO$_2$ and N$_2$O$_4$ were present initially. But at a given temperature, the concentration of N$_2$O$_4$ at equilibrium divided by the square of the equilibrium concentration of NO$_2$ is equal to a constant.

equilibrium constant expression

$$K_c = \frac{[N_2O_4]}{[NO_2]^2}$$

equilibrium constant

This equation is known as the ***equilibrium constant expression,*** and K_c is the ***equilibrium constant*** for this reaction. The subscript c in the equilibrium constant indicates that the constant was calculated from the equilibrium concentrations of the reactants and products in units of moles per liter.

It doesn't matter whether we start with a great deal of NO$_2$ or a relatively small sample. When the reaction reaches equilibrium at 25°C, the concentration of N$_2$O$_4$ divided by the square of the NO$_2$ concentration is always the same. At 25°C, the ratio of the N$_2$O$_4$ concentration divided by the NO$_2$ concentration squared is always 20.5.

$$K_c = \frac{[N_2O_4]}{[NO_2]^2} = 20.5 \qquad (\text{at } 25°C)$$

Exercise 14.1

Calculate the equilibrium constant for the dimerization of NO$_2$ for each set of experimental data given below.

$$2 \; NO_2(g) \rightleftharpoons N_2O_4(g)$$

(a) A 5.00-mole sample of NO$_2$ at 25°C came to equilibrium when the NO$_2$ concentration was 0.337 M and the N$_2$O$_4$ concentration was 2.33 M.

(b) A 1.00-mole sample of NO$_2$ at 25°C came to equilibrium at an NO$_2$ concentration of 0.144 M and an N$_2$O$_4$ concentration of 0.428 M.

(c) A 0.100-mole sample of NO$_2$ at 25°C came to equilibrium at an NO$_2$ concentration of 0.0387 M and an N$_2$O$_4$ concentration of 0.0307 M.

Solution

(a) To calculate the equilibrium constant for this reaction, we substitute the observed equilibrium concentrations for NO_2 and N_2O_4 into the equilibrium constant expression for the reaction.

$$K_c = \frac{[N_2O_4]}{[NO_2]^2} = \frac{[2.33]}{[0.337]^2} = 20.5$$

(b) Although the concentrations of NO_2 and N_2O_4 when this reaction reaches equilibrium are different from those in Part a, the equilibrium constant is the same (within experimental error).

$$K_c = \frac{[N_2O_4]}{[NO_2]^2} = \frac{[0.428]}{[0.144]^2} = 20.6$$

(c) Once again, the concentrations of NO_2 and N_2O_4 at equilibrium are different from those in the preceding experiments, but the equilibrium constant for the reaction is the same.

$$K_c = \frac{[N_2O_4]}{[NO_2]^2} = \frac{[0.0307]}{[0.0387]^2} = 20.5$$

The fact that some reactions come to equilibrium raises a number of interesting questions.

1. **Why do these reactions seem to stop before they have reached completion?**
2. **What is the difference between reactions that go to completion and reactions that reach equilibrium?**
3. **Is there any way to predict whether a reaction will go to completion or reach equilibrium?**
4. **How does a change in the conditions of the reaction influence the amount of product formed?**
5. **At a given temperature, why does the equilibrium constant always have the same value when the reaction reaches equilibrium, regardless of the initial concentrations of the reactants and products?**

A significant part of the remainder of this text will be devoted to answering these questions. Before we can begin to understand how and why a chemical reaction comes to equilibrium, however, we will have to build a model of the factors that influence the rate of a chemical reaction.

14.3 THE RATES OF CHEMICAL REACTIONS

By now you should be familiar with acid-base titrations that use phenolphthalein as the endpoint indicator, as shown in Figure 14.3.

You might not have noticed, however, what happens when a solution that contains phenolphthalein in the presence of excess base is allowed to stand for a few minutes. The solution initially has a strong pink color. Over a period of several minutes, however, it gradually turns colorless, as shown in Figure 14.4. This figure demonstrates a reaction between phenolphthalein and base, in which the indicator is slowly destroyed by reacting with the OH^- ion in this solution.

The intensity of the pink color in Figure 14.4 is directly proportional to the

FIG. 14.3 The endpoint of an acid-base titration, at which the phenolphthalein indicator in the solution just turns pink.

FIG. 14.4 This photograph shows what happens to the characteristic pink color of phenolphthalein in the presence of excess base when the solution is allowed to stand. With time, this pink color gradually fades as the phenolphthalein is destroyed by a reaction with the OH⁻ ion in this solution.

concentration of phenolphthalein in the solution at any moment in time. As the concentration of phenolphthalein in the solution gradually decreases, so does the solution's ability to absorb light.

Figure 14.5 shows a spectrophotometer, which can be used to measure how much light at a given frequency is absorbed by a sample. The first step in using this instrument involves measuring the absorbance of solutions of known concentrations. The absorbance of the solution is then plotted against the concentration, and a straight line is drawn between these two points. The linear relationship between the absorbance of the solution and its concentration can be used to translate measurements of the intensity of the color of the solution into the concentration of the light-absorbing component at any moment in time.

Table 14.1 shows what happens to the concentration of phenolphthalein in a solution that was initially 0.005 M in phenolphthalein and 0.61 M in OH⁻ ion. Figure 14.6 is a graph of these data, in which the concentration of phenolphthalein is plotted on the vertical axis versus time on the horizontal axis. As you can see from these data, the phenolphthalein concentration decreases by a factor of 10 over a period of about 230 seconds, or just under 4 minutes.

Experiments such as the one that gave us the data in Figure 14.6 are classified as measurements of **chemical kinetics** (from a Greek stem meaning "to move"). One of the goals of these experiments is to describe the rate of a chemical reaction — the speed with which the reactants are transformed into the products of the reaction.

What do we mean by the term **rate of reaction?** The rate, or speed, at which an object travels through space has units that indicate the distance traveled per unit of time — such as miles per hour or kilometers per second. In chemical kinetics, the distance traveled can be thought of as the change in the concentration of one of the reactants or products of the reaction. The rate of reaction is therefore the change in the concentration per unit of time. We calculate this rate by dividing the change in the concentration by the time it takes for the change to occur.

Before we can write an equation that describes the rate of reaction, we need a way to describe the change in concentration that takes place during a given period of time. As noted earlier, square brackets are used to indicate the concentration of a substance in a reaction that is at equilibrium.

$$[X] = \text{concentration of } X \text{ in moles per liter in a reaction that is at equilibrium.}$$

We need another symbol to represent concentrations when a reaction isn't at equilibrium. In the remainder of this text, the concentration of a substance at any time before the reaction has reached equilibrium is represented by a symbol that consists of the formula of the substance written in parentheses.

$$(X) = \text{concentration of } X \text{ in moles per liter at any time before the reaction has reached equilibrium.}$$

A change in the concentration of X as the reaction comes to equilibrium can be represented by the symbol $\Delta(X)$. The period of time over which the measurement is made is indicated by the symbol Δt. The rate of a chemical reaction is therefore equal to the following ratio.

$$\text{Rate of reaction} = \frac{\Delta(X)}{\Delta t}$$

The rate of reaction is equal to the change in the concentration of one of the reactants — $\Delta(X)$ — that occurs during a given period of time — Δt.

FIG. 14.5 A spectrophotometer can be used to monitor changes in the concentration of a solution that absorbs light somewhere in the visible portion of the electromagnetic spectrum. The amount of light absorbed as it passes through the solution is directly proportional to the concentration of the light-absorbing substance.

Exercise 14.2

Use the data in Table 14.1 to calculate the rate of decomposition of phenol-phthalein during each of the following periods.

(a) During the first time interval, when the phenolphthalein concentration falls from 0.0050 to 0.0045 M.

(b) During the second time interval, when the concentration falls from 0.0045 to 0.0040 M.

(c) During the third time interval, when the concentration falls from 0.0040 to 0.0035 M.

Solution

The rate of reaction is equal to the change in the phenolphthalein concentration divided by the length of time over which this change occurs.

(a) During the first period for which data are available, the rate of reaction is 4.8×10^{-5} M/per second.

$$\text{Rate} = \frac{\Delta(X)}{\Delta t} = \frac{(0.0050\ M) - (0.0045\ M)}{(10.5\ \text{s} - 0\ \text{s})} = 4.8 \times 10^{-5}\ M/s$$

(b) During the second time period, the rate of reaction is slightly smaller.

$$\text{Rate} = \frac{\Delta(X)}{\Delta t} = \frac{(0.0045\ M) - (0.0040\ M)}{(22.3\ \text{s} - 10.5\ \text{s})} = 4.2 \times 10^{-5}\ M/s$$

(c) During the third time period, the rate of reaction is even smaller.

$$\text{Rate} = \frac{\Delta(X)}{\Delta t} = \frac{(0.0040\ M) - (0.0035\ M)}{(35.7\ \text{s} - 22.3\ \text{s})} = 3.7 \times 10^{-5}\ M/s$$

TABLE 14.1

Experimental Data for the Reaction Between Phenolphthalein and Excess Base

Concentration of Phenolphthalein (M)	Time (s)
0.0050	0.0
0.0045	10.5
0.0040	22.3
0.0035	35.7
0.0030	51.1
0.0025	69.3
0.0020	91.6
0.0015	120.4
0.0010	160.9
0.00050	230.3
0.00025	299.6
0.00015	350.7
0.00010	391.2

Exercise 14.2 raises an important point. The rate of this reaction isn't constant —it changes with time. The rate at which phenolphthalein is destroyed in the reaction becomes slower as more phenolphthalein is consumed. This means that the rate of reaction changes while it is being measured. In theory, the most accurate data for the rate of a chemical reaction would be obtained if we could measure the

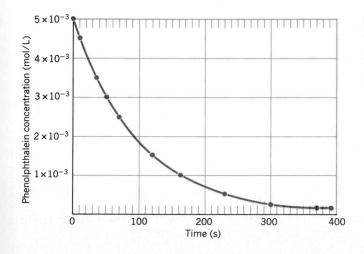

FIG. 14.6 A plot of the concentration of phenolphthalein versus time for the reaction between this indicator and excess OH$^-$ ion.

FIG. 14.7 The rate of a reaction, such as the decomposition of phenolphthalein, can be calculated at any moment in time from the slope of the line tangent to this curve at that moment in time.

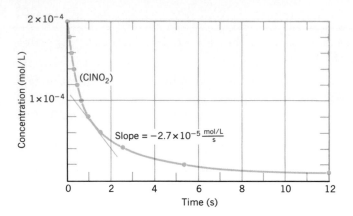

infinitesimally small change in concentration — $d(X)$ — that occurs over an infinitesimally small period of time — dt.

$$\text{Rate of reaction} = \frac{d(X)}{dt}$$

Figure 14.7 shows how the rate of reaction can be calculated from a graph of concentration versus time. The rate of reaction at any time is equal to the slope of a tangent drawn to this curve at that moment. When this technique is used to analyze the data in Figure 14.7, an interesting result is obtained. The rate of this reaction is directly proportional to the concentration of phenolphthalein in the solution at that moment in time.

$$\text{Rate} = k(\text{phenolphthalein})$$

This equation is known as the ***rate law,*** and the proportionality constant, k, is known as the ***rate constant*** for the reaction.

14.4 GAS-PHASE REACTIONS

The decomposition of phenolphthalein in the presence of excess base offers a good visual introduction to chemical kinetics, because it occurs on a time scale that allows us to observe it with the naked eye. But this reaction is too complex to be used as a model for explaining why the rate of a chemical reaction is proportional to the concentration of the reactants consumed.

Let's therefore turn to gas-phase reactions, which are much simpler. To simplify matters even further, let's focus our attention on a reaction that occurs in a single step — such as the transfer of a chlorine atom between $ClNO_2$ and NO to form ClNO and NO_2.

$$ClNO_2(g) + NO(g) \rightleftharpoons ClNO(g) + NO_2(g)$$

Exercise 14.3

Use Lewis structures to describe what happens when $ClNO_2$ reacts with NO to give ClNO and NO_2.

Solution

NO and NO_2 both contain an odd number of electrons. The Lewis structures of these molecules are usually written with the unpaired electrons on the nitrogen atoms (see Section 10.6). Both NO and NO_2 can therefore combine with a neutral chlorine atom to form a molecule in which all of the electrons are paired. This reaction can be thought to involve the transfer of a chlorine atom from one molecule to another.

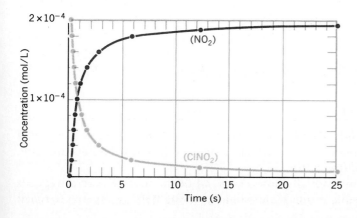

Figure 14.8 summarizes the results of experiments designed to study the rate of this reaction. This graph combines a plot of the disappearance of $ClNO_2$ with a plot of the appearance of NO_2. These data are consistent with the following rate law for this reaction.

$$\text{Rate} = k(\text{ClNO}_2)(\text{NO})$$

According to this rate law, the rate at which $ClNO_2$ and NO are converted into NO_2 and ClNO is proportional to the product of the concentrations of the two reactants. Initially, the rate of reaction is fast. As the reactants are converted into products, however, the $ClNO_2$ and NO concentrations become smaller, and the reaction slows down.

In theory, the reaction should stop when it runs out of either $ClNO_2$ or NO. In practice, however, the reaction seems to stop before this happens. This is a very fast reaction — the concentration of $ClNO_2$ drops by a factor of 2 in less than a second. And yet, no matter how long we wait, some residual $ClNO_2$ and NO always remain in the reaction flask. In other words, the reaction comes to equilibrium.

Plots such as Figure 14.8 are sometimes divided into a **kinetic region** and an **equilibrium region.**

Kinetic region: The period during which the concentrations of the reactants and products are constantly changing.

Equilibrium region: The period after which the reaction seems to stop — after which there is no further change in the concentrations of the reactants or products.

FIG. 14.8 A plot of the change in the concentration of $ClNO_2$ superimposed on a plot of the change in the concentration of NO_2 as $ClNO_2$ reacts with NO to produce NO_2 and ClNO according to the following equation.

$$\text{ClNO}_2(g) + \text{NO}(g) \rightleftharpoons \text{NO}_2(g) + \text{ClNO}(g)$$

14.5 A COLLISION THEORY MODEL FOR GAS-PHASE REACTIONS

A model based on the assumption that molecules must collide before they can react can be used to explain why the reaction just discussed seems to stop before all of the reactants have been consumed.

$$ClNO_2(g) + NO(g) \rightleftharpoons NO_2(g) + ClNO(g)$$

According to the collision theory of chemical reactions, $ClNO_2$ and NO molecules must collide before a chlorine atom can be transferred from one molecule to the other.

This assumption explains why the rate of the reaction is proportional to the concentration of both $ClNO_2$ and NO.

$$\text{Rate} = k(ClNO_2)(NO)$$

The number of collisions per second between $ClNO_2$ and NO molecules depends on their concentrations. As $ClNO_2$ and NO are consumed in the reaction, the number of collisions per second between these molecules becomes smaller, and the reaction therefore slows down.

Suppose that we start with a mixture of $ClNO_2$ and NO, but no NO_2 or ClNO. The only reaction that can occur at first is the transfer of a chlorine atom from $ClNO_2$ to NO.

$$ClNO_2(g) + NO(g) \longrightarrow NO_2(g) + ClNO(g)$$

Eventually, NO_2 and ClNO build up in the reaction flask, and these molecules begin to collide as well. Suppose that some of these collisions result in the transfer of a chlorine atom in the opposite direction.

$$ClNO_2(g) + NO(g) \longleftarrow NO_2(g) + ClNO(g)$$

In other words, suppose that the reaction is reversible.

The **collision theory model** of chemical reactions assumes that the rate of a simple, one-step reaction is proportional to the product of the concentrations of the ions or molecules consumed in that reaction. The rate of the forward reaction is therefore proportional to the product of the concentrations of the two reactants.

$$\text{Rate}_{forward} = k_f(ClNO_2)(NO)$$

The rate of the reverse reaction, on the other hand, is proportional to the concentrations of the products of the reaction.

$$\text{Rate}_{reverse} = k_r(NO_2)(ClNO)$$

Initially, the rate of the forward reaction is much larger than the rate of the reverse reaction, because the system contains $ClNO_2$ and NO but virtually no NO_2 and ClNO.

$$\textbf{Initially:} \qquad \textbf{rate}_{forward} \gg \text{rate}_{reverse}$$

As some of the $ClNO_2$ and NO initially present is consumed, however, the rate of the forward reaction slows down. At the same time, NO_2 and ClNO accumulate, and the reverse reaction speeds up.

If the forward reaction gradually slows down, while the reverse reaction speeds up, there is only one possible result. If we wait long enough, the system has to reach a point at which the rates of the forward and reverse reactions are the same.

Eventually: $rate_{forward} = rate_{reverse}$

At this point, the reaction will seem to stop. $ClNO_2$ and NO will be consumed in the forward reaction at the same rate at which they are produced in the reverse reaction. The same thing will happen to NO_2 and $ClNO$. When the rates of the forward and reverse reactions are the same, there is no longer any change in the concentrations of the reactants or products of the reaction. In other words, the reaction is at equilibrium.

We can now see that there are two definitions of equilibrium.

1. **The state that exists when there is no apparent change in the concentrations of the reactants and products of a reaction.**

2. **The state that exists when the rates of the forward and reverse reactions are equal.**

The first definition is based on the results of experiments that tell us that some reactions seem to stop—they reach a point at which no more reactants are converted into products. The other definition is based on a theoretical model of chemical reactions that tries to explain why reactions reach equilibrium.

We can now make a distinction between reactions that go to completion and those that reach equilibrium. Reactions that aren't reversible, or that strongly favor the products, are assumed to go to completion and are represented by equations that contain a single arrow.

$$2\,Mg(s) + O_2(g) \longrightarrow 2\,MgO(s)$$

Reversible reactions that reach equilibrium are indicated by a pair of arrows between the reactants and products of the reaction, as follows.

$$ClNO_2(g) + NO(g) \rightleftharpoons NO_2(g) + ClNO(g)$$

14.6 EQUILIBRIUM CONSTANT EXPRESSIONS

Reactions don't stop when they come to equilibrium. But the forward and reverse reactions are in balance at equilibrium, so there is no net change in the concentrations of the reactants or products, and the reaction appears to stop on the macroscopic scale. Chemical equilibrium is an example of a dynamic balance between opposing forces—the forward and reverse reactions—not a static balance.

Let's look at the logical consequences of the assumption that the reaction between $ClNO_2$ and NO eventually reaches equilibrium.

$$ClNO_2(g) + NO(g) \rightleftharpoons NO_2(g) + ClNO(g)$$

We have argued that the rates of the forward and reverse reactions are equal when this system comes to equilibrium.

At equilibrium: $rate_{forward} = rate_{reverse}$

Substituting the rate laws for the forward and reverse reactions into this equality gives the following result.

At equilibrium: $k_f(ClNO_2)(NO) = k_r(NO_2)(ClNO)$

But this equation is only valid when the system is at equilibrium, so we should replace the $(ClNO_2)$, (NO), (NO_2), and $(ClNO)$ terms with symbols that indicate that the reaction is at equilibrium. By convention, we use square brackets for this purpose. The equation describing the balance between the forward and reverse reactions when the system is at equilibrium should therefore be written as follows.

At equilibrium: $\qquad k_f[ClNO_2][NO] = k_r[NO_2][ClNO]$

Rearranging this equation gives the following result.

$$\frac{k_f}{k_r} = \frac{[NO_2][ClNO]}{[ClNO_2][NO]}$$

Since k_f and k_r are constants, the quotient of k_f divided by k_r must also be a constant. This quotient is the equilibrium constant for the reaction, K_c. The ratio of the concentrations of the reactants and products is therefore the equilibrium constant expression.

equilibrium constant expression

$$K_c = \frac{[NO_2][ClNO]}{[ClNO_2][NO]}$$

equilibrium constant

No matter what combination of concentrations of reactants and products we start with, the reaction will reach equilibrium when the quotient of the concentrations of these reagents as defined by the equilibrium constant expression is equal to the equilibrium constant for the reaction. We can start with a lot of $ClNO_2$ and very little NO, or a lot of NO and very little $ClNO_2$. It doesn't matter. When the reaction reaches equilibrium, the relationship among the concentrations of the reactants and products as described by the equilibrium constant expression will always be the same. At 25°C, this reaction always reaches equilibrium when the ratio of these concentrations is 1.3×10^4.

$$K_c = \frac{[NO_2][ClNO]}{[ClNO_2][NO]} = 1.3 \times 10^4$$

What happens if we approach equilibrium from the other direction? That is, what happens if we start with a system that contains the products of this reaction — NO_2 and ClNO — and then let the reaction come to equilibrium? Think about the balanced equation for the reaction.

$$ClNO_2(g) + NO(g) \rightleftharpoons NO_2(g) + ClNO(g)$$

The rate laws for the forward and reverse reactions will still be the same.

$$Rate_{forward} = k_f(ClNO_2)(NO)$$
$$Rate_{reverse} = k_f(NO_2)(ClNO)$$

Now, however, the rate of the forward reaction will initially be much smaller than the rate of the reverse reaction.

Initially: $\qquad rate_{forward} \ll rate_{reverse}$

But, as time passes, the rate of the reverse reaction will slow down and the rate of the forward reaction will speed up until they become equal. At that point, the reaction will have reached equilibrium.

At equilibrium: $\text{rate}_{forward} = \text{rate}_{reverse}$

$$k_f(\text{ClNO}_2)(\text{NO}) = k_r(\text{NO}_2)(\text{ClNO})$$

Rearranging this equation gives us the same equilibrium constant expression.

$$\frac{k_f}{k_r} = \frac{[\text{NO}_2][\text{ClNO}]}{[\text{ClNO}_2][\text{NO}]} = K_c$$

We get the same equilibrium constant expression and the same equilibrium constant no matter whether we start with only reactants, only products, or a mixture of reactants and products.

Exercise 14.4

The rate constants for the forward and reverse reactions in the following equilibrium have been measured.

$$\text{ClNO}_2(g) + \text{NO}(g) \rightleftharpoons \text{NO}_2(g) + \text{ClNO}(g)$$

At 25°C, k_f is about 7.3×10^3 L/mol-s (liters per mole/second) and k_r is about 0.55 L/mol-s. Calculate the equilibrium constant for this reaction.

Solution

We can start by assuming that the rates of the forward and reverse reactions are the same at equilibrium.

At equilibrium: $\text{rate}_{forward} = \text{rate}_{reverse}$

We can then substitute the rate laws for the forward and reverse reactions into this equality.

At equilibrium: $k_f[\text{ClNO}_2][\text{NO}] = k_r[\text{NO}_2][\text{ClNO}]$

The equation can be rearranged to give the equilibrium constant expression for this reaction.

$$\frac{k_f}{k_r} = \frac{[\text{NO}_2][\text{ClNO}]}{[\text{ClNO}_2][\text{NO}]} = K_c$$

The equilibrium constant for the reaction is equal to the rate constant for the forward reaction divided by the rate constant for the reverse reaction.

$$K_c = \frac{k_f}{k_r} = \frac{7300 \text{ L/mol-s}}{0.55 \text{ L/mol-s}} = 13{,}000$$

The procedure used in this section to derive the equilibrium constant only works with reactions that occur in a single step, such as the transfer of a chlorine atom from ClNO_2 to NO. As you will see in Chapter 21, some reactions take a number of steps to convert reactants into products. However, it can be shown that any reaction that reaches equilibrium, no matter how simple or complex, has an equilibrium constant expression that satisfies the following rules.

RULES FOR WRITING EQUILIBRIUM CONSTANT EXPRESSIONS

1. Even though chemical reactions that reach equilibrium occur in both directions, the reagents on the right-hand side of the equation are assumed to be the products of the reaction and the reagents on the left-hand side of the equation are assumed to be the reactants.

2. The products of the reaction are always written above the line — in the numerator.

3. The reactants are always written below the line — in the denominator.

4. The equilibrium constant expression contains a term for each reactant and each product of the reaction.

5. The numerator of the equilibrium constant expression is equal to the product of the concentrations of the products of the reaction.

6. The denominator of the equilibrium constant expression is equal to the product of the concentrations of the reactants.

7. The concentration of each reactant or product is therefore raised to a power equal to the coefficient for this reactant or product in the balanced equation for the reaction.

Exercise 14.5

Write equilibrium constant expressions for the following reactions.

(a) $2 NO_2(g) \rightleftharpoons N_2O_4(g)$
(b) $2 SO_3(g) \rightleftharpoons 2 SO_2(g) + O_2(g)$
(c) $N_2(g) + 3 H_2(g) \rightleftharpoons 2 NH_3(g)$

Solution

(a) The equilibrium constant expression is equal to the product of the concentrations of the species on the right side of the equation divided by the product of the concentrations on the left side of the equation. The equilibrium constant expression for this reaction is therefore written as follows.

$$K_c = \frac{[N_2O_4]}{[NO_2]^2}$$

(b) The equilibrium constant expression is equal to the product of the concentrations of the products of the reaction — in this case, two molecules of SO_2 and one of O_2 — divided by the product of the concentrations of the reactants — two molecules of SO_3.

$$K_c = \frac{[SO_2]^2[O_2]}{[SO_3]^2}$$

(c) Once again, the equilibrium constant expression is equal to the product of the concentrations of the products raised to the appropriate power divided by the product of the concentrations of the reactants raised to the appropriate power.

$$K_c = \frac{[NH_3]^2}{[N_2][H_2]^3}$$

14.7 ALTERING OR COMBINING EQUILIBRIUM REACTIONS

What happens to the magnitude of the equilibrium constant for a reaction when we turn the equation around? Consider the following reaction, for example.

$$ClNO_2(g) + NO(g) \rightleftharpoons NO_2(g) + ClNO(g)$$

The equilibrium constant expression for this equation is written as follows.

$$K_c = \frac{[NO_2][ClNO]}{[ClNO_2][NO]} = 1.3 \times 10^4 \qquad \text{(at 25°C)}$$

Because this is a reversible reaction, it can also be represented by the following equation.

$$NO_2(g) + ClNO(g) \rightleftharpoons ClNO_2(g) + NO(g)$$

The equilibrium constant expression is now written as follows.

$$K_c' = \frac{[ClNO_2][NO]}{[NO_2][ClNO]}$$

Each of these two equilibrium constant expressions is the inverse, or reciprocal, of the other. We can therefore calculate K_c' by dividing K_c into 1.

$$K_c' = \frac{1}{K_c} = \frac{1}{1.3 \times 10^4} = 7.7 \times 10^{-5}$$

We can also calculate equilibrium constants by combining two or more reactions for which the value of K_c is known. Assume, for example, that we know the equilibrium constants for the following gas-phase reactions at 200°C.

$$I_2(g) \rightleftharpoons 2 \, I(g) \qquad K_{c1} = \frac{[I]^2}{[I_2]} = 3 \times 10^{-16}$$

$$H_2(g) + 2 \, I(g) \rightleftharpoons 2 \, HI(g) \qquad K_{c2} = \frac{[HI]^2}{[H_2][I]^2} = 1.4 \times 10^{15}$$

We can combine these two reactions to find an overall equilibrium.

$$\begin{aligned} I_2(g) &\rightleftharpoons 2 \, I(g) \\ + H_2(g) + 2 \, I(g) &\rightleftharpoons 2 \, HI(g) \\ \hline H_2(g) + I_2(g) &\rightleftharpoons 2 \, HI(g) \end{aligned}$$

The equilibrium constant expression for the overall reaction is equal to the product of the equilibrium constant expressions for the two steps in this reaction.

$$\frac{[HI]^2}{[H_2][I_2]} = \frac{[I]^2}{[I_2]} \times \frac{[HI]^2}{[H_2][I]^2}$$

We can therefore conclude that the equilibrium constant for the overall reaction is equal to the product of the equilibrium constants for the two steps.

$$\begin{aligned} K_c &= K_{c1} \times K_{c2} \\ &= (3 \times 10^{-16}) \times (1.4 \times 10^{15}) \\ &= \mathbf{0.42} \end{aligned}$$

REACTION QUOTIENTS: A WAY TO DECIDE
14.8 WHETHER A REACTION IS AT EQUILIBRIUM

We now have a model that describes what happens when a reaction reaches equilibrium. At the molecular level, the rate of the forward reaction is equal to the rate of the reverse reaction. Since the reaction proceeds in both directions at the same rate, there is no apparent change in the concentrations of the reactants or the products on the macroscopic scale — the level of objects visible to the naked eye.

This model can also be used to predict the direction in which the reaction has to proceed to reach equilibrium. If the concentrations of the reactants are too large for the reaction to be at equilibrium, the rate of the forward reaction will be faster than that of the reverse reaction, and some of the reactants will be converted into products until equilibrium is achieved. Conversely, if the concentrations of the products are too large for the reaction to be at equilibrium, the rate of the reverse reaction will exceed that of the forward reaction, and the reaction will convert some of the excess products back into reactants until the system reaches equilibrium.

It would be useful to have a way to test a reaction at any moment in time to determine whether the system is at equilibrium. We can do this by introducing the concept of the *reaction quotient, Q_c*.

Reaction quotient: The product of the concentrations of the products of a reaction divided by the product of the concentrations of the reactants at any moment in time.

To illustrate how the reaction quotient can help us decide whether a reaction is at equilibrium, let's consider the following gas-phase reaction.

$$H_2(g) + I_2(g) \rightleftharpoons 2 HI(g)$$

The equilibrium constant expression for this reaction is written as follows.

$$K_c = \frac{[HI]^2}{[H_2][I_2]}$$

When this reaction is run at 200°C, the equilibrium constant for the reaction is 60.

$$K_c = \frac{[HI]^2}{[H_2][I_2]} = 60$$

By analogy, we can write an expression for the reaction quotient as follows.

$$Q_c = \frac{(HI)^2}{(H_2)(I_2)}$$

There are two differences between the equilibrium constant and the reaction quotient.

1. **We use brackets around the symbols for the reactants and products only when the reaction is at equilibrium. We use parentheses at all other times.**
2. **There is only one value of K_c for a given reaction at a given temperature, but there are an infinite number of values for the reaction quotient. Q_c can take on any value between zero and infinity.**

If the system contains a great deal of HI and very little H_2 and I_2, the reaction quotient is very large. If the system contains very little HI and a great deal of H_2 and I_2, the reaction quotient is very small.

At any moment in time, there are three possibilities.

1. **Q_c is smaller than K_c. This means that the system contains too much of the reactants and not enough products for the reaction to be at equilibrium. The value of Q_c must become larger in order for the reaction to reach equilibrium. Thus, the reaction has to convert some of the reactants into products to come to equilibrium.**
2. **Q_c is equal to K_c. If this is true, then the reaction is at equilibrium.**
3. **Q_c is larger than K_c. This means that the system contains too much of the products and not enough reactants for the reaction to be at equilibrium. The value of Q_c must become smaller before the reaction can come to equilibrium. Thus, the reaction must convert some of the products into reactants to reach equilibrium.**

Exercise 14.6

Assume that the concentrations of H_2, I_2, and HI can be measured for the following reaction at any moment in time.

$$H_2(g) + I_2(g) \rightleftharpoons 2\ HI(g) \qquad K_c = 60$$

For each of the following sets of concentrations, determine whether the reaction is at equilibrium. If it isn't at equilibrium, decide in which direction it must go to reach equilibrium.

(a) $(H_2) = (I_2) = (III) = 0.010\ M$
(b) $(HI) = 0.30\ M$; $(H_2) = 0.01\ M$; $(I_2) = 0.15\ M$
(c) $(H_2) = (HI) = 0.10\ M$; $(I_2) = 0.0010\ M$

Solution

(a) The only way to decide whether the reaction is at equilibrium is to compare the reaction quotient with the equilibrium constant for the reaction.

$$Q_c = \frac{(HI)^2}{(H_2)(I_2)} = \frac{(0.010)^2}{(0.010)(0.010)} = 1 < K_c$$

The reaction quotient in this case is smaller than the equilibrium constant. The only way to get this system to equilibrium is to increase the magnitude of the reaction quotient. The only way to do that is to convert some of the H_2 and I_2 into HI. The reaction therefore has to shift to the right to reach equilibrium.

(b) The reaction quotient for this set of concentrations is equal to the equilibrium constant for the reaction.

$$Q_c = \frac{(HI)^2}{(H_2)(I_2)} = \frac{(0.30)^2}{(0.010)(0.15)} = 60 = K_c$$

The reaction is therefore at equilibrium.

(c) The reaction quotient for this set of concentrations is larger than the equilibrium constant for the reaction.

$$Q_c = \frac{(HI)^2}{(H_2)(I_2)} = \frac{(0.10)^2}{(0.10)(0.001)} = 100 > K_c$$

In order to reach equilibrium, the concentrations of the reactants and products must be adjusted until the reaction quotient is equal to the equilibrium constant. This will involve converting some of the HI back into H_2 and I_2. In other words, the reaction has to shift to the left to reach equilibrium.

14.9 CHANGES IN CONCENTRATION THAT OCCUR AS A REACTION COMES TO EQUILIBRIUM

The reaction quotient, Q_c, allows us to decide whether a reaction is at equilibrium at any moment in time. If it isn't at equilibrium, the relative sizes of Q_c and K_c tell us in which direction the reaction must shift to reach equilibrium. Now we need a way of predicting how far the reaction has to go to reach equilibrium.

Suppose that you are faced with the following problem.

Phosphorus pentachloride decomposes to phosphorus trichloride and chlorine when heated.

$$PCl_5(g) \rightleftharpoons PCl_3(g) + Cl_2(g)$$

The equilibrium constant for this reaction at 250°C is 0.030. Calculate the concentrations of PCl_5, PCl_3, and Cl_2 at equilibrium if the initial concentration of PCl_5 is 0.100 moles per liter.

The first step toward solving this problem involves organizing the information so that it provides clues as to how to proceed. The problem contains four chunks of information: (1) a balanced equation, (2) an equilibrium constant for the reaction, (3) a description of the initial conditions, and (4) an indication of the goal of the calculation—the equilibrium concentrations of the three components of the reaction.

The following format offers a useful way to summarize this information.

$$PCl_5(g) \rightleftharpoons PCl_3(g) + Cl_2(g) \qquad K_c = 0.030$$

Initial:	0.100 M	0	0
Equilibrium:	?	?	?

We start with the balanced equation and the equilibrium constant for the reaction and then add what we know about the initial and equilibrium concentrations of the reactants and products of the reaction. Initially, the flask contains 0.100 moles per liter of PCl_5 and no PCl_3 or Cl_2. Our goal is to calculate the equilibrium concentrations of these three substances.

Before we can do anything else, we have to decide whether the reaction is at equilibrium. We can do this by comparing the reaction quotient for the initial conditions with the equilibrium constant for the reaction.

$$Q_c = \frac{(PCl_3)(Cl_2)}{(PCl_5)} = \frac{(0)(0)}{(0.100)} = 0 < K_c$$

Although the equilibrium constant is small ($K_c = 3.0 \times 10^{-2}$), the reaction quotient is even smaller ($Q_c = 0$). The only way for this reaction to get to equilibrium is for some of the PCl_5 to decompose into PCl_3 and Cl_2.

Since the reaction isn't at equilibrium, one thing is sure — the concentrations of PCl_5, PCl_3, and Cl_2 will all change as the reaction comes to equilibrium. Because the reaction has to shift to the right to reach equilibrium, the PCl_5 concentration will become smaller, while the PCl_3 and Cl_2 concentration will become larger.

At first glance, this problem appears to have three unknowns — the equilibrium concentrations of PCl_5, PCl_3, and Cl_2. Because it is difficult to solve a problem in three unknowns, we should look for relationships that can reduce the problem's complexity. One way of achieving this goal is to look at the relationship between the changes that occur in the concentrations of PCl_5, PCl_3, and Cl_2 as the reaction approaches equilibrium.

Exercise 14.7

Calculate the magnitude of the changes in the PCl_3 and Cl_2 concentrations that will occur as the following reaction comes to equilibrium if the concentration of PCl_5 decreases by 0.042 moles per liter.

$$PCl_5(g) \rightleftharpoons PCl_3(g) + Cl_2(g)$$

Solution

The decomposition of PCl_5 has a $1:1:1$ stoichiometry.

$$PCl_5(g) \rightleftharpoons PCl_3(g) + Cl_2(g)$$

For every mole of PCl_5 that decomposes, we get a mole of PCl_3 and a mole of Cl_2. Thus, the magnitude of the change in the concentration of PCl_5 that occurs as the reaction comes to equilibrium is equal to the magnitude of the changes in the PCl_3 and Cl_2 concentrations. We can therefore conclude that 0.042 moles per liter of both PCl_3 and Cl_2 are formed as this reaction comes to equilibrium.

Let's define $\Delta(PCl_5)$ as the change that occurs in the concentration of this reagent as the reaction goes from the initial conditions to equilibrium. The concentration of PCl_5 at equilibrium will be equal to the initial concentration of PCl_5 minus the amount of PCl_5 that is consumed as the reaction comes to equilibrium.

$$[PCl_5] = (PCl_5) - \Delta(PCl_5)$$

equilibrium concentration   initial concentration   amount consumed as reaction comes to equilibrium

We can then define $\Delta(PCl_3)$ and $\Delta(Cl_2)$ as the changes that occur in the PCl_3 and Cl_2 concentrations as the reaction comes to equilibrium. The concentrations of both these substances at equilibrium will be larger than their initial concentrations.

$$[PCl_3] = (PCl_3) + \Delta(PCl_3)$$
$$[Cl_2] = (Cl_2) + \Delta(Cl_2)$$

Generalizing the results of Exercise 14.7 gives the following equation.

$$\Delta(PCl_5) = \Delta(PCl_3) = \Delta(Cl_2)$$

Because of the $1:1:1$ stoichiometry of the reaction, the magnitude of the change in the concentration of PCl_5 as the reaction comes to equilibrium will be equal to the magnitudes of the changes in the concentrations of PCl_3 and Cl_2. We can therefore rewrite the equations that define the equilibrium concentrations of PCl_5, PCl_3, and Cl_2 in terms of a single unknown — Δ.

$$[PCl_5] = (PCl_3) - \Delta$$
$$[PCl_3] = (PCl_3) + \Delta$$
$$[Cl_2] = (Cl_2) + \Delta$$

Substituting what we know about the initial concentrations of PCl_5, PCl_3, and Cl_2 into these equations gives the following result.

$$[PCl_5] = 0.100 - \Delta$$
$$[PCl_3] = 0 + \Delta$$
$$[Cl_2] = 0 + \Delta$$

We can now summarize what we know about this reaction as follows.

$$PCl_5(g) \rightleftharpoons PCl_3(g) + Cl_2(g) \qquad K_c = 0.030$$

Initial:	0.100 M	0	0
Equilibrium:	0.100 − Δ	Δ	Δ

We have only one unknown — Δ — and we need only one equation to solve for one unknown. The obvious equation to turn to is the equilibrium constant expression for this reaction.

$$K_c = \frac{[PCl_3][Cl_2]}{[PCl_5]} = 0.030$$

Substituting what we now know about the equilibrium concentrations of PCl_5, PCl_3, and Cl_2 into this equation gives the following result.

$$\frac{[\Delta][\Delta]}{[0.100 - \Delta]} = 0.030$$

This equation can be expanded and then rearranged to give a quadratic equation

$$\Delta^2 + 0.030\Delta - 0.0030 = 0$$

which can be solved with the quadratic formula.

$$\Delta = \frac{-b \pm \sqrt{b^2 - 4\,ac}}{2a}$$

$$= \frac{-0.030 \pm \sqrt{(0.030)^2 - 4(1)(-0.0030)}}{2(1)}$$

$$= 0.042 \text{ or } -0.072$$

Although two answers come out of this calculation, only the positive root makes any physical sense. Thus, the magnitude of the change in the concentrations of PCl_5, PCl_3, and Cl_2 as this reaction comes to equilibrium is 0.042 moles per liter.

$$\Delta = 0.042 \, M$$

Plugging this value of Δ back into the equations that define the equilibrium concentrations of PCl_5, PCl_3, and Cl_2 gives the following results.

$$[PCl_5] = 0.100 - 0.042 = 0.058 \ M$$
$$[PCl_3] = 0 + 0.042 = 0.042 \ M$$
$$[Cl_2] = 0 + 0.042 = 0.042 \ M$$

Do these equilibrium concentrations work? Do they yield the equilibrium constant for this reaction when substituted into the equilibrium constant expression?

$$K_c = \frac{[PCl_3][Cl_2]}{[PCl_5]} = \frac{[0.042][0.042]}{[0.058]} = 0.030$$

Yes! The equilibrium constant calculated from these concentrations is equal to the value of K_c given in the problem, within experimental error.

Exercise 14.8

Calculate the equilibrium concentrations of PCl_5, PCl_3, and Cl_2 if the initial concentration of PCl_5 is 0.100 M, the initial concentration of Cl_2 is 0.020 M, and no PCl_3 is present initially. Assume that the equilibrium constant for the decomposition of PCl_5 is 0.030.

Solution

We can start by representing this problem as follows.

$$PCl_5(g) \rightleftharpoons PCl_3(g) + Cl_2(g) \qquad K_c = 0.030$$

Initial:	0.100 M	0	0.020
Equilibrium:	?	?	?

Before doing anything else, we need to compare the reaction quotient for the initial conditions with the equilibrium constant for the reaction.

$$Q_c = \frac{(PCl_3)(Cl_2)}{(PCl_5)} = \frac{(0)(0.20)}{(0.100)} = 0 < K_c$$

Once again, the reaction quotient ($Q_c = 0$) is smaller than the equilibrium constant ($K_c = 0.030$). Some of the PCl_5 present initially therefore has to decompose to PCl_3 and Cl_2 before the reaction can come to equilibrium.

Because of the stoichiometry of this reaction, the magnitude of the change in the PCl_5 concentration as the reaction comes to equilibrium will be equal to the magnitude of the changes in the concentrations of PCl_3 and Cl_2. The problem can therefore be represented as follows.

$$PCl_5(g) \rightleftharpoons PCl_3(g) + Cl_2(g) \qquad K_c = 0.030$$

Initial:	0.100 M	0	0.020
Change:	$-\Delta$	$+\Delta$	$+\Delta$
Equilibrium:	$0.100 - \Delta$	Δ	$0.020 + \Delta$

Substituting what we know about the concentrations of PCl_5, PCl_3, and Cl_2 into the equilibrium constant expression for the reaction gives the following equation.

$$K_c = \frac{[PCl_3][Cl_2]}{[PCl_5]} = \frac{[\Delta][0.020 + \Delta]}{[0.100 - \Delta]} = 0.030$$

Expanding this equation gives a quadratic equation.

$$\Delta^2 + 0.050\Delta - 0.0030 = 0$$

Solving this equation with the quadratic formula gives the following positive root.

$$\Delta = 0.035 \ M$$

This value of Δ can be used to calculate the equilibrium concentrations of the three components of the reaction.

$$[PCl_5] = 0.100 - \Delta = 0.065 \ M$$
$$[PCl_3] = 0 + \Delta = 0.035 \ M$$
$$[Cl_2] = 0.020 + \Delta = 0.055 \ M$$

We can check these results by substituting the values back into the equilibrium constant expression.

$$\frac{[PCl_3][Cl_2]}{[PCl_5]} = \frac{[0.035][0.055]}{[0.065]} = 0.030$$

Once again, the results of the calculation agree with the equilibrium constant for the reaction within experimental error.

14.10 HIDDEN ASSUMPTIONS THAT MAKE EQUILIBRIUM CALCULATIONS EASIER

Suppose that you are asked to solve a slightly more difficult problem.

Sulfur trioxide decomposes to give sulfur dioxide and oxygen with an equilibrium constant of 1.3×10^{-10} at 300°C.

$$2 \ SO_3(g) \rightleftharpoons 2 \ SO_2(g) + O_2(g)$$

Calculate the equilibrium concentrations of the three components of this system if the initial concentration of SO_3 is 0.130 M.

The first step still involves building a representation of the information in the problem.

$$2 \ SO_3(g) \rightleftharpoons 2 \ SO_2(g) + O_2(g) \qquad K_c = 1.3 \times 10^{-10}$$

Initial:	0.130 M	0	0
Equilibrium:	?	?	?

We then compare the reaction quotient for the initial conditions with the equilibrium constant for the reaction.

$$Q_c = \frac{(SO_2)^2(O_2)}{(SO_3)^2} = \frac{(0)^2(0)}{(0.130)} = 0 < K_c$$

Because the initial concentrations of SO_2 and O_2 are both zero, the reaction has to shift to the right to reach equilibrium — some of the SO_3 has to decompose to give SO_2 and O_2.

The stoichiometry of this reaction is a bit more complex than that in the example in Section 14.9. But the magnitudes of changes in the concentrations of the components — SO_3, SO_2, and O_2 — are still related. For every 2 moles of SO_3 that decompose we get 2 moles of SO_2 and 1 mole of O_2. We can incorporate this relationship into the format we used earlier by using the balanced equation for the reaction as a guide. Everything we know about the reaction can be summarized as follows.

$$2\ SO_3(g) \rightleftharpoons 2\ SO_2(g) + O_2(g) \qquad K_c = 1.3 \times 10^{-10}$$

Initial:	0.130 M	0	0
Change:	-2Δ	$+2\Delta$	$+\Delta$
Equilibrium:	$0.130 - 2\Delta$	2Δ	Δ

The signs of the Δ terms in this problem are determined by the fact that the reaction has to shift from left to right to reach equilibrium.

Note that the coefficients in the Δ terms mirror the coefficients in the balanced equation for the reaction. This is always the case and represents the simplest way to derive these expressions.

Because twice as many moles of SO_2 are produced as moles of O_2, the change in the concentration of SO_2 as the reaction comes to equilibrium has to be twice as large as the change in the concentration of O_2. Because 2 moles of SO_3 are consumed for every mole of O_2 produced, the magnitude of the change in the SO_3 concentration must be twice as large as the magnitude of the change in the concentration of O_2.

Substituting what we know about the problem into the equilibrium constant expression for the reaction gives the following equation.

$$K_c = \frac{[SO_2]^2[O_2]}{[SO_3]^2} = \frac{[2\Delta]^2[\Delta]}{[0.130 - 2\Delta]^2} = 1.3 \times 10^{-10}$$

This equation is a bit more of a challenge to expand, but it can be rearranged to give the following cubic equation.

$$4\Delta^3 - 5.2 \times 10^{-10}\Delta^2 + 6.76 \times 10^{-11}\Delta - 2.197 \times 10^{-12} = 0$$

This equation brings to mind an interesting question: How do we solve it for Δ?

This problem is the first example in this text of a family of problems that are difficult, if not impossible, to solve exactly. These problems are solved with a general strategy, or *heuristic*, that consists of making an assumption or an approximation that turns the problem into a simpler problem that *can* be solved. Let's start the discussion of methods of approximation with a pair of general rules.

RULES FOR USING APPROXIMATION METHODS

1. There is nothing wrong with making an assumption.

2. The cardinal sins are forgetting what assumptions were made and forgetting to check whether the assumptions are valid.

What assumption can be made to simplify this problem? Let's go back to the first thing we did after building a representation for the problem. We started our calculation by comparing the reaction quotient for the initial concentrations with the equilibrium constant for the reaction.

$$Q_c = \frac{(SO_2)^2(O_2)}{(SO_3)^2} = \frac{(0)^2(0)}{(0.130)} = 0 < K_c$$

We then concluded that the reaction quotient ($Q_c = 0$) was smaller than the equilibrium constant ($K_c = 1.3 \times 10^{-10}$) and decided that some of the SO_3 would have to decompose in order for this reaction to come to equilibrium.

But what about the relative sizes of the reaction quotient and the equilibrium constant for the reaction? The initial values of Q_c and K_c are both relatively small. This means that the initial conditions are reasonably close to equilibrium and so the reaction does not have very far to go to reach equilibrium. It is therefore reasonable to assume that Δ in this problem is relatively small.

It is essential to understand the nature of the assumption we are making. We are *not* assuming that Δ is zero. If we did that, all of the unknowns would disappear from the equation! We are only assuming that Δ is *small* — so small compared with the initial concentration of SO_3 that it doesn't make a significant difference when 2Δ is subtracted from this number. We can write this assumption as follows.

$$0.130 - 2\Delta \cong 0.130$$

Let's now go back to the equation we are trying to solve.

$$\frac{[2\Delta]^2[\Delta]}{[0.130 - 2\Delta]^2} = 1.3 \times 10^{-10}$$

By making the assumption that 2Δ is very much smaller than 0.130, we can replace this equation with the following approximate equation.

$$\frac{[2\Delta]^2[\Delta]}{[0.130]^2} \cong 1.3 \times 10^{-10}$$

Expanding this gives an equation that is much easier to solve for Δ.

$$4\Delta^3 \cong 2.20 \times 10^{-12}$$

$$\Delta \cong 0.000082 \ M$$

Before we can go any further, we have to check our assumption that 2Δ is so small compared with 0.130 that it doesn't make a significant difference when it is subtracted from this number.

$$0.130 - 2\Delta \cong 0.130$$

Is this assumption valid? Is 2Δ small enough compared with 0.130 to be ignored? Yes, 2Δ is smaller than the experimental error involved in the measurement of the initial concentration of SO_3.

$$0.130 - 2(0.000082) = 0.130$$

We can therefore use this approximate value of Δ to calculate the equilibrium concentrations of SO_3, SO_2, and O_2.

$$[SO_3] = 0.130 - 2\Delta = 0.130 \ M$$
$$[SO_2] = 2\Delta = 1.6 \times 10^{-4} \ M$$
$$[O_2] = \Delta = 8.2 \times 10^{-5} \ M$$

We can check these results by substituting these values into the equilibrium constant expression for the reaction.

$$K_c = \frac{[SO_2]^2[O_2]}{[SO_3]^2} = \frac{[1.6 \times 10^{-4}]^2[8.2 \times 10^{-5}]}{[0.130]^2} = 1.2 \times 10^{-10}$$

The value of the equilibrium constant that comes out of this calculation agrees with the value given in the problem, within experimental error. Our assumption that 2Δ is negligibly small compared with the initial concentration of SO_3 is therefore valid, and we can feel confident in the answers it provides.

14.11 A RULE OF THUMB FOR TESTING THE VALIDITY OF ASSUMPTIONS

There was no doubt about the validity of the assumption that Δ was small compared with the initial concentration of SO_3 in the calculation in the preceding section. The value of Δ was so small that 2Δ was smaller than the experimental error involved in measuring the initial concentration of SO_3.

We can get some idea of whether Δ is small enough to be ignored in a problem by comparing the initial reaction quotient with the equilibrium constant for the reaction. If Q_c and K_c are both much smaller than 1, or both much larger than 1, the reaction doesn't have very far to go to reach equilibrium, and the assumption that Δ is small enough to be ignored seems reasonable.

This raises an interesting question: How do we decide whether the assumption is valid? The answer to this question depends on how much error we are willing to let into our calculation before we no longer trust the results. As a rule of thumb, we will assume in this text that Δ is negligibly small as long as it is less than 5% of the initial concentrations of the reactants or products with which it is compared. The best way to decide whether the assumption meets this rule of thumb in a particular calculation is to try it and see if it works.

Exercise 14.9

Ammonia is made from nitrogen and hydrogen by the following reaction.

$$N_2(g) + 3\ H_2(g) \rightleftharpoons 2\ NH_3(g)$$

Calculate the equilibrium concentrations of the three components of this reaction if the initial concentration of N_2 was 0.050 moles per liter, the initial concentration of H_2 was 0.100 moles per liter, and no ammonia was present initially. Assume that the reaction is run at a temperature at which the equilibrium constant for the reaction is 0.040.

Solution

We can start by building a representation for the problem.

	$N_2(g)$	$+\ 3\ H_2(g)$	$\rightleftharpoons 2\ NH_3(g)$	$K_c = 0.040$
Initial:	0.050 M	0.100	0	
Equilibrium:	?	?	?	

We can then calculate the initial reaction quotient and compare it with the equilibrium constant for the reaction.

$$Q_c = \frac{(NH_3)^2}{(N_2)(H_2)^3} = \frac{(0)^2}{(0.050)(0.100)^3} = 0 < K_c$$

The reaction quotient ($Q_c = 0$) is smaller than the equilibrium constant ($K_c = 0.040$), so the reaction has to shift to the right to reach equilibrium. This will result in a decrease in the concentrations of N_2 and H_2 and an increase in the NH_3 concentration. The relationship among the magnitudes of the changes in the concentrations of the three components of this system is specified by the balanced equation for the reaction.

We can summarize what we know about the reaction as follows.

$$N_2(g) \quad + \quad 3\,H_2(g) \quad \rightleftharpoons \quad 2\,NH_3(g) \qquad K_c = 0.040$$

Initial:	0.050 M	0.100	0
Change:	$-\Delta$	-3Δ	$+2\Delta$
Equilibrium:	0.050 $-\Delta$	0.100 -3Δ	2Δ

Substituting this information into the equilibrium constant expression for the reaction gives the following equation.

$$K_c = \frac{[NH_3]^2}{[N_2][H_2]^3} = \frac{[2\Delta]^2}{[0.050 - \Delta][0.100 - 3\Delta]^3} = 0.040$$

Since Q_c for the initial concentrations and K_c for the reaction are both smaller than 1, let's try the assumption that Δ is small enough that subtracting it from 0.050 — or even subtracting 3Δ from 0.100 — doesn't make a significant change. This assumption gives the following approximate equation.

$$\frac{[2\Delta]^2}{[0.050][0.100]^3} \cong 0.040$$

Solving this equation for Δ gives the following result.

$$\Delta \cong 0.00071\ M$$

Now we have to check our assumption. Is Δ significantly smaller than 0.050? Is 3Δ significantly smaller than 0.100? Yes. Δ is about 1% of the initial concentration of N_2

$$\frac{0.00071}{0.050} \times 100\% = 1.4\%$$

and 3Δ is slightly more than 2% of the initial concentration of H_2.

$$\frac{3(0.00071)}{0.100} \times 100\% = 2.1\%$$

We can therefore use this approximate value of Δ to determine the equilibrium concentrations of N_2, H_2, and NH_3.

$$[NH_3] = 2\Delta = 0.0014\ M$$
$$[N_2] = 0.050 - \Delta = 0.049\ M$$
$$[H_2] = 0.100 - 3\Delta = 0.098\ M$$

We can check the validity of our calculation by substituting this information back into the equilibrium constant expression.

$$K_c = \frac{[NH_3]^2}{[N_2][H_2]^3} = \frac{[0.0014]^2}{[0.049][0.098]^3} = 0.042$$

Once again, the equilibrium constant calculated from these data agrees with the value of K_c given in the problem, within experimental error.

14.12 WHAT DO WE DO WHEN THE APPROXIMATION FAILS?

It is easy to envision a problem in which the assumption that Δ is small compared with the initial concentrations can't possibly be valid. Consider the following problem, for example.

Nitrogen oxide reacts with oxygen to form nitrogen dioxide.

$$2\,NO(g) + O_2(g) \rightleftharpoons 2\,NO_2(g)$$

The equilibrium constant for this reaction at 200°C is 3.4×10^6. Calculate the equilibrium concentrations of the three components of the reaction assuming initial concentrations of 0.050 M for NO and 0.100 M for O_2, with no NO_2 present initially.

We can start, once again, by representing the information in the problem as follows.

	$2\,NO(g) +$	$O_2(g)$	$\rightleftharpoons 2\,NO_2(g)$	$K_c = 3.4 \times 10^6$
Initial:	0.100 M	0.050 M	0	
Equilibrium:	?	?	?	

The first step is always the same: compare the initial value of the reaction quotient with the equilibrium constant.

$$Q_c = \frac{(NO_2)^2}{[NO]^2[O_2]} = \frac{(0)^2}{(0.100)^2(0.050)} = 0 < K_c$$

The relationship between the initial reaction quotient and the equilibrium constant tells us something we may already have suspected — the reaction must shift to the right to reach equilibrium. We can summarize what we know about the reaction as follows.

	$2\,NO(g) +$	$O_2(g)$	$\rightleftharpoons 2\,NO_2(g)$	$K_c = 3.4 \times 10^6$
Initial:	0.100 M	0.050 M	0	
Change:	-2Δ	$-\Delta$	$+2\Delta$	
Equilibrium:	$0.100 - 2\Delta$	$0.050 - \Delta$	2Δ	

Some might ask: "Why calculate the initial value of the reaction quotient for this reaction? Isn't it obvious that the reaction has to shift to the right to produce at least some NO_2?" Yes, it is. But calculating the value of Q_c for the reaction does more than tell us in which direction it has to shift to reach equilibrium. It also gives us an indication of how far the reaction has to go to reach equilibrium.

In this case, Q_c is so very much smaller than K_c for the reaction that we have to conclude that the initial conditions are very far from equilibrium. It would therefore be a mistake to assume that Δ is small. It might be instructive, however, to see what happens if we make this assumption — even though we expect it to be wrong. We can start by substituting what we know about the problem into the equilibrium constant expression.

$$K_c = \frac{[NO_2]^2}{[NO]^2[O_2]} = \frac{[2\Delta]^2}{[0.100 - 2\Delta]^2[0.050 - \Delta]} = 3.4 \times 10^6$$

Just to see what happens, let's assume that Δ is small compared with the initial concentrations of NO and O_2. That assumption gives us the following approximate equation.

$$\frac{[2\Delta]^2}{[0.100]^2[0.050]} \cong 3.4 \times 10^6$$

Rearranging and solving this equation for Δ gives the following result.

$$\Delta \cong 20.6 \, M$$

Since it is physically impossible for Δ to be this large, there must be something wrong with the approximation that was made in calculating the number. The value of Δ in this problem is obviously too large to ignore.

We can't assume that Δ is negligibly small in this problem, but we can redefine the problem so that this assumption becomes valid. The key to achieving this goal is to remember the conditions under which we can assume that Δ is small enough to be ignored. This assumption is only valid when Q_c is of the same order of magnitude as K_c—when Q_c and K_c are both much larger than 1 or much smaller than 1. We can solve problems for which Q_c isn't close to K_c by redefining the initial conditions so that Q_c becomes close to K_c. To show how this can be done, we'll look once again at the problem given in this section.

The equilibrium constant for the reaction between NO and O_2 to form NO_2 is much larger than 1, $K_c = 3.4 \times 10^6$. This means that the equilibrium favors the products of the reaction. The best way to handle this problem is to drive the reaction as far as possible to the right, and then let it come back to equilibrium. This problem was constructed so that there is just enough O_2 to consume all of the NO and vice versa. Let's therefore define an intermediate set of conditions that correspond to what would happen if all of the starting materials were converted into products.

	$2 \, NO(g) \, +$	$O_2(g)$	$\rightleftharpoons$	$2 \, NO_2(g)$	$K_c = 3.4 \times 10^6$
Initial:	0.100 M	0.050 M		0	
Intermediate:	0	0		0.100 M	

We can see where this gets us by calculating the reaction quotient for the intermediate set of conditions.

$$Q_c = \frac{(NO_2)^2}{[NO]^2[O_2]} = \frac{(0.100)^2}{(0)^2(0)} = \infty \gg K_c$$

The reaction quotient is now larger than the equilibrium constant, and the reaction has to shift back to the left to reach equilibrium. Some of the NO_2 must now decompose to form NO and O_2.

	$2 \, NO(g) +$	$O_2(g) \rightleftharpoons$	$2 \, NO_2(g)$	$K_c = 3.4 \times 10^6$
Intermediate:	0	0	0.100 M	
Change:	$+2\Delta$	$+\Delta$	-2Δ	
Equilibrium:	2Δ	Δ	$0.100 - 2\Delta$	

Because the reaction quotient for the intermediate conditions and the equilibrium constant are both relatively large, we assume that the reaction doesn't have very far to go to reach equilibrium. We can therefore substitute what we know about the reaction into the equilibrium constant expression.

$$K_c = \frac{[NO_2]^2}{[NO]^2[O_2]} = \frac{[0.100 - 2\Delta]^2}{[2\Delta]^2[\Delta]} = 3.4 \times 10^6$$

We can then assume that 2Δ is small compared with the intermediate concentration of NO_2 and derive the following approximate equation.

$$\frac{[0.100]^2}{[2\Delta]^2[\Delta]} \cong 3.4 \times 10^6$$

This equation can then be solved for an approximate value of Δ.

$$\Delta \cong 0.00090 \ M$$

Now we have to check our assumption that 2Δ is small enough compared with the intermediate concentration of NO_2 to be ignored.

$$\frac{2(0.0009)}{0.100} \times 100\% = 1.8\%$$

The value of 2Δ is less than 2% of the intermediate concentration of NO_2, which means that it can be legitimately ignored in this calculation.

Since the approximation is valid, we can use the resulting value of Δ to calculate the equilibrium concentrations of NO, NO_2, and O_2.

$$[NO_2] = 0.100 - 2\Delta = 0.098 \ M$$
$$[NO] = 2\Delta = 0.0018 \ M$$
$$[O_2] = \Delta = 0.0009 \ M$$

We can check our calculations by substituting these concentrations back into the equilibrium constant expression.

$$K_c = \frac{[NO_2]^2}{[NO]^2[O_2]} = \frac{[0.098]^2}{[0.0018]^2[0.0009]} = 3.3 \times 10^6$$

Once again, the value of the equilibrium constant that comes out of this calculation agrees with the value of K_c given in the problem, within experimental error.

14.13 SUCCESSIVE APPROXIMATIONS

Let's summarize what we have established so far.

1. If the difference between Q_c and K_c is small, the reaction doesn't have far to go to reach equilibrium. We can therefore assume that Δ is small compared with the initial concentrations and solve for an approximate value of Δ.

2. If the difference between Q_c and K_c is large, we can force the reaction to completion in the direction favored by the equilibrium constant and then let the reaction come back to equilibrium. In essence, this means redefining the problem so that Q_c is close to K_c.

What do we do when there is no way to redefine the problem so that Δ is small? Consider the following problem, for example.

Calculate the equilibrium concentrations of N_2, H_2, and NH_3 at 650°C if the initial concentration of N_2 is 0.250 moles per liter, the initial concentration of H_2 is 0.750 moles per liter, and the equilibrium constant for the following reaction at 650°C is 0.040.

$$N_2(g) + 3\ H_2(g) \rightleftharpoons 2\ NH_3(g)$$

We can start, as always, by setting up the problem as follows.

	$N_2(g)$	$+\ 3\ H_2(g)$	$\rightleftharpoons 2\ NH_3(g)$	$K_c = 0.040$
Initial:	0.250 M	0.750 M	0	
Equilibrium:	?	?	?	

We can then calculate the reaction quotient for the initial concentrations.

$$Q_c = \frac{(NH_3)^2}{(N_2)(H_2)^3} = \frac{(0)^2}{(0.250)(0.750)^3} = 0 < K_c$$

The reaction quotient is smaller than the equilibrium constant, so the reaction must shift to the right to produce NH_3, as we would expect.

	$N_2(g)$	$+\ \ 3\ H_2(g)$	$\rightleftharpoons 2\ NH_3(g)$	$K_c = 0.040$
Initial:	0.250 M	0.750 M	0	
Change:	$-\Delta$	-3Δ	$+2\Delta$	
Equilibrium:	$0.250 - \Delta$	$0.750 - 3\Delta$	2Δ	

We can substitute this information into the equilibrium constant expression for the reaction and obtain the following equation.

$$K_c = \frac{[NH_3]^2}{[N_2][H_2]^3} = \frac{(2\Delta)^2}{(0.250 - \Delta)(0.750 - 3\Delta)^3} = 0.040$$

Because Q_c and K_c are both smaller than 1, we can try the assumption that Δ is small compared with the initial concentration of N_2 and that 3Δ is small compared with the initial concentration of H_2.

$$\frac{[2\Delta]^2}{[0.250][0.750]^3} \cong 0.040$$

Solving this equation for Δ gives the following result.

$$\Delta \cong 0.032$$

Is our approximation valid? No, Δ is 12.8% of the initial concentration of N_2, and 3Δ is 12.8% of the initial concentration of H_2.

$$\frac{0.032}{0.250} \times 100\% = 12.8\% \qquad \frac{3(0.32)}{0.750} \times 100\% = 12.8\%$$

Can we redefine the problem to make Δ small? No — the initial conditions are already on the side of the equilibrium favored by the equilibrium constant. When the equilibrium constant for a reaction is small, we can set up the problem so that Q_c is small and therefore Δ is small. When the equilibrium constant is large, we can drive the reaction toward the products so that Q_c is also large and therefore Δ is small. But there is no way to redefine problems so that Δ is small when K_c is close to 1.

Problems like this can be solved by a technique known as *successive approximations.* We set up the problem as usual.

$$K_c = \frac{[NH_3]^2}{[N_2][H_2]^3} = \frac{(2\Delta)^2}{(0.250 - \Delta)(0.750 - 3\Delta)^3} = 0.040$$

We then assume that Δ is small

$$\frac{[2\Delta]^2}{[0.250][0.750]^3} \cong 0.040$$

and calculate a first approximation of the value of Δ.

$$\Delta \cong 0.0325$$

We then substitute this approximate value of Δ back into the equation

$$\frac{[2\Delta']^2}{[0.250 - 0.0325][0.750 - 3(0.0325)]^3} \cong 0.040$$

and solve this equation for a second approximation.

$$\Delta' \cong 0.0246$$

We then substitute this approximate value of Δ back into the equation and solve for a third approximation

$$\frac{[2\Delta'']^2}{[0.250 - 0.0246][0.750 - 3(0.0246)]^3} \cong 0.040$$

$$\Delta'' \cong 0.0264$$

and substitute this approximate value of Δ back into the equation and solve for a fourth approximation.

$$\frac{[2\Delta''']^2}{[0.250 - 0.0264][0.750 - 3(0.0264)]^3} \cong 0.040$$

$$\Delta''' \cong 0.0260$$

Note that the difference between the successive values of Δ obtained by this technique keeps getting smaller. The difference between the last two passes through this iteration is smaller than the number of significant figures to which we are entitled. We can therefore assume that this calculation converged to a value of Δ equal to 0.026 and use this value to calculate the equilibrium concentrations of N_2, H_2, and NH_3.

$$[N_2] = 0.250 - \Delta = 0.224 \ M$$
$$[H_2] = 0.750 - 3\Delta = 0.672 \ M$$
$$[NH_3] = 2\Delta = 0.052 \ M$$

We can check the validity of this calculation by substituting these values into the equilibrium constant expression for the reaction.

$$K_c = \frac{[NH_3]^2}{[N_2][H_2]^3} = \frac{(0.052)^2}{(0.224)(0.672)^3} = 0.0398$$

The value of K_c obtained from this calculation agrees with the experimental value of K_c within experimental error. We can therefore feel confident that successive approximations converged on the correct value of Δ.

14.14 EQUILIBRIA EXPRESSED IN PARTIAL PRESSURES

Gas-phase reactions were chosen for this introduction to kinetics and equilibrium because they are among the simplest chemical reactions. Some might question, however, why all of the problems in this chapter have been worked in terms of the *concentration* of the gases in units of moles per liter.

We used units of concentration to emphasize the relationship between chemical equilibria and the rates of chemical reactions, which are reported in terms of the concentrations of the reactants and products. This choice of units was indicated by the subscript c added to the symbols for the reaction quotients and equilibrium constants, to show that they were calculated from the concentrations of the reactants and products.

It is also possible to study gas-phase equilibria by following the *partial pressures* of the gases in the reaction. We can understand why this is possible by rearranging the ideal gas equation

$$PV = nRT$$

to give the following relationship between the pressure of a gas and its concentration in moles per liter.

$$P = \left(\frac{n}{V}\right) \times RT$$

We can therefore characterize a chemical reaction such as the following

$$N_2(g) + 3\ H_2(g) \rightleftharpoons 2\ NH_3(g)$$

with an equilibrium constant defined in terms of units of concentration

$$K_c = \frac{[NH_3]^2}{[N_2][H_2]^3}$$

or an equilibrium constant defined in terms of partial pressures.

$$K_p = \frac{P_{NH_3}^2}{P_{N_2}P_{H_2}^3}$$

What is the relationship between K_p and K_c for a gas-phase reaction? According to the rearranged version of the ideal gas equation, the pressure of a gas is equal to the concentration of the gas times the product of the ideal gas constant and the temperature in units of kelvin. We can therefore calculate the value of K_p for a reaction by multiplying each of the terms in the K_c expression by RT.

$$K_p = \frac{P_{NH_3}^2}{P_{N_2}P_{H_2}^3} = \frac{([NH_3] \times RT)^2}{([N_2] \times RT)([H_2] \times RT)^3}$$

Collecting terms in this example gives the following result.

$$K_p = K_c \times (RT)^{-2}$$

In general, the value of K_p for a reaction can be calculated from K_c with the following equation.

$$K_p = K_c \times (RT)^{\Delta n}$$

In this equation, Δn is the difference between the number of moles of products and the number of moles of reactants in the balanced equation. For the reaction in which ammonia is synthesized from nitrogen and hydrogen, the value of Δn is $2 - (1 + 3)$, or -2.

$$N_2(g) + 3\ H_2(g) \rightleftharpoons 2\ NH_3(g)$$

The balanced equation for this reaction contains 2 moles of products for every 4 moles of reactants, and therefore Δn is equal to -2.

The techniques for working problems using K_p expressions are the same as those described for K_c problems, except that partial pressures are used instead of concentrations to represent the amounts of starting materials and products that are present both initially and at equilibrium.

14.15 THE EFFECT OF TEMPERATURE ON A CHEMICAL REACTION

You may have noticed that when an equilibrium constant has been given in this chapter, the temperature at which the reaction was run has also been given. If the equilibrium constant is really constant, why do we have to worry about the temperature of the reaction?

The answer is simple. Both K_c and K_p for a reaction are constants at a given temperature, but they can change with temperature. Consider the equilibrium between NO_2 and its dimer, N_2O_4, for example.

$$2\ NO_2(g) \rightleftharpoons N_2O_4(g)$$

Figure 14.9 shows the effect of temperature on this equilibrium. When we cool a sealed tube containing NO_2 in a dry-ice/acetone bath at $-78\,°C$, the intensity of the brown color of NO_2 gas decreases significantly. If we warm the tube in a hot-water bath, the brown color becomes even more intense that it is at room temperature.

The equilibrium constant for this reaction changes with temperature, as shown in Table 14.2. At low temperatures, the equilibrium favors the dimer, N_2O_4. At high temperatures, the equilibrium favors NO_2. The fact that equilibrium con-

FIG. 14.9 The effect of temperature on the following equilibrium.

$$2\ NO_2(g) \rightleftharpoons N_2O_4(g)$$

When a tube filled with NO_2 is immersed in a dry-ice/acetone bath at $-78\,°C$, the intensity of the brown color decreases significantly, because cooling this system shifts the equilibrium toward the dimer. When the tube is immersed in a hot-water bath, the brown color becomes more intense, because an increase in the temperature of this system shifts the equilibrium toward NO_2.

TABLE 14.2

The Temperature Dependence of the Equilibrium Constant for the Dimerization of NO_2

Temperature (°C)	K_p	K_c
100	0.065	2.0
25	6.8	170
0	56	1300
−78	1,300,000	22,000,000

stants are temperature dependent explains why you may find different values for the equilibrium constant for the same chemical reaction.

14.16 LECHATELIER'S PRINCIPLE

In 1884, the French chemist and engineer Henry-Louis LeChatelier proposed one of the central concepts of chemical equilibria. What is now known as *LeChatelier's principle* can be stated in many different ways. The following is one way of expressing this idea.

A change in one of the variables that describe a system at equilibrium will cause a shift in the position of the equilibrium that counteracts the effect of this change.

LeChatelier's principle can also be stated as follows.

When a stress is applied to a chemical reaction at equilibrium, the reaction will try to minimize the effect of this stress.

All of our attention so far has been devoted to describing what happens when a system comes to equilibrium. LeChatelier's principle describes what happens to a system already at equilibrium when something momentarily takes it away from equilibrium. This section focuses on three ways in which we can change the conditions of a chemical reaction at equilibrium: changing the concentration of one or more of the reactants or products, changing the pressure on the system, and changing the temperature at which the reaction is run.

CHANGES IN CONCENTRATION

Let's start by calculating the equilibrium concentrations of N_2, H_2, and NH_3 that would result from heating a mixture that contains 0.100 moles per liter of both N_2 and H_2 to 650°C, where $K_c = 0.040$. We start, as always, by assembling the relevant information in this problem in the following format.

$$N_2(g) \;+\; 3\,H_2(g) \;\rightleftharpoons\; 2\,NH_3(g) \qquad K_c = 0.040$$

Initial:	0.100 M	0.100 M	0
Equilibrium:	?	?	?

We then compare the initial value of the reaction quotient with the equilibrium constant for the reaction

$$Q_c = \frac{(NH_3)^2}{(N_2)(H_2)^3} = \frac{(0)^2}{(0.100)(0.100)^3} = 0 < K_c$$

and decide that the reaction must shift to the right to reach equilibrium.

$$N_2(g) \quad + \quad 3\ H_2(g) \quad \rightleftharpoons 2\ NH_3(g) \qquad K_c = 0.040$$

Initial:	0.100 M	0.100 M	0
Equilibrium:	0.100 $- \Delta$	0.100 $- 3\Delta$	2Δ

Substituting this information into the equilibrium constant expression gives the following equation.

$$K_c = \frac{[NH_3]^2}{[N_2][H_2]^3} = \frac{[2\Delta]^2}{[0.100 - \Delta][0.100 - 3\Delta]^3} = 0.040$$

Assuming that Δ is negligibly small compared with the initial concentrations of N_2 and H_2 gives the following approximate equation.

$$\frac{[2\Delta]^2}{[0.100][0.100]^3} \cong 0.040$$

Solving this equation for Δ gives the following result.

$$\Delta \cong 0.0010$$

The assumption that Δ is negligibly small is valid for this calculation, because Δ is 1% of the initial concentration of N_2 and 3Δ is 3% of the initial concentration of H_2. The equilibrium concentrations of N_2, H_2, and NH_3 in this system can therefore be calculated as follows.

$$[NH_3] = 2\Delta = 0.0020\ M$$
$$[N_2] = 0.100 - \Delta = 0.099\ M$$
$$[H_2] = 0.100 - 3\Delta = 0.097\ M$$

The fact that Δ is small compared with the initial concentrations of N_2 and H_2 makes this calculation much easier to do. But it implies that very little ammonia is actually produced in the reaction. According to these calculations, between 1% and 3% of the starting materials are converted into ammonia at this temperature. Let's see what would happen if we added more N_2 to the reaction while it was at equilibrium. Suppose that enough N_2 was added to increase the initial concentration by a factor of 10.

The reaction can't be at equilibrium any more, because there is too much N_2 in the system. We can confirm this by calculating the new reaction quotient.

$$Q_c = \frac{(NH_3)^2}{(N_2)(H_2)^3} = \frac{(0.0020)^2}{(1.000)(0.097)^3} = 0.0044 < K_c$$

Because the reaction quotient ($Q_c = 0.0044$) is smaller than the equilibrium constant ($K_c = 0.040$), the reaction has to shift to the right to get back to equilibrium. Some of the excess N_2 must be used to make more ammonia.

$$N_2(g) \quad + \quad 3\ H_2(g) \quad \rightleftharpoons \quad 2\ NH_3(g) \qquad K_c = 0.040$$

Intermediate:	1.000 M	0.097 M	0.0010 M
Change:	$-\Delta$	-3Δ	$+2\Delta$
Equilibrium:	1.000 $- \Delta$	0.097 $- 3\Delta$	0.0010 $+ 2\Delta$

The fact that the reaction shifts to the right when excess N_2 is added to the system is exactly what LeChatelier's principle would predict. Adding an excess of one of the reactants places a stress on the system. The system responds by minimizing the effect of this stress — by shifting the equilibrium toward the products.

We can calculate the effect of adding this much excess N_2 on the equilibrium concentrations of N_2, H_2, and NH_3. It might be tempting to start by substituting what we know about the reaction at equilibrium into the equilibrium constant expression

$$K_c = \frac{[NH_3]^2}{[N_2][H_2]^3} = \frac{[0.0010 + 2\Delta]^2}{[1.000 - \Delta][0.097 - 3\Delta]^3} = 0.040$$

and then assuming that Δ is small compared with the intermediate concentrations of N_2, H_2, and NH_3.

$$\frac{[0.0010]^2}{[1.000][0.097]^3} = 0.0060$$

Unfortunately, if we do this, all of the Δ terms disappear from the equation, and it is difficult to solve an equation with no unknowns!

We can get around this by converting the NH_3 generated from the initial reaction back into N_2 and H_2 and then letting the reaction come to equilibrium once again.

$$N_2(g) \quad + \quad 3 H_2(g) \quad \rightleftharpoons \quad 2 NH_3(g) \quad K_c = 0.040$$

Original intermediate:	1.000 M	0.097 M	0.0010 M
New intermediate:	1.001 M	0.100 M	0
Equilibrium:	$1.001 - \Delta$	$0.100 - 3\Delta$	2Δ

We then substitute this information into the equilibrium constant expression.

$$K_c = \frac{[NH_3]^2}{[N_2][H_2]^3} = \frac{[2\Delta]^2}{[1.001 - \Delta][0.100 - 3\Delta]^3} = 0.040$$

Assuming that Δ is small compared with the new intermediate concentrations of N_2 and H_2 gives the following approximate equation.

$$\frac{[2\Delta]^2}{[1.001][0.100]^3} \cong 0.040$$

This can be solved for the following result.

$$\Delta \cong 0.0032$$

This value of Δ is too big to be ignored — 3Δ is almost 10% of the H_2 concentration. Fortunately, successive approximations converges after only two iterations to the following value of Δ, which is legitimate.

$$\Delta \cong 0.0028$$

We can use this value of Δ to calculate the equilibrium concentrations of N_2, H_2, and NH_3.

$$[NH_3] = 2\Delta = 0.0056 \ M$$
$$[N_2] = 1.001 - \Delta = 0.998 \ M$$
$$[H_2] = 0.100 - 3\Delta = 0.092 \ M$$

By comparing the new equilibrium concentrations with those obtained before excess N_2 was added to the system, we can see the magnitude of the effect of adding the excess N_2.

Before	After
[NH₃] = 0.0020 M	**[NH₃] = 0.0056 M**
[N₂] = 0.099 M	[N₂] = 0.998 M
[H₂] = 0.097 M	[H₂] = 0.092 M

The amount of NH_3 at equilibrium has increased by a factor of almost 3. Adding an excess of one of the reactants to a reaction at equilibrium shifts the position of the equilibrium toward the product, without changing the magnitude of the equilibrium constant for the reaction. Adding an excess of one of the products has the opposite effect; it shifts the equilibrium toward the reactants.

LeChatelier's principle provides us with two ways of driving a reaction toward the products. We can either add an excess of one of the reactants (usually the cheapest!) or remove one of the products of the reaction.

CHANGES IN PRESSURE

The effect of changing the pressure on a gas-phase reaction depends on the stoichiometry of the reaction. We can demonstrate this by looking at the result of compressing the following reaction at equilibrium.

$$N_2(g) + 3 H_2(g) \rightleftharpoons 2 NH_3(g)$$

Suppose that we start with a system that initially contains 2.5 atm of N_2 and 7.5 atm of H_2 at 600°C, allow the reaction to come to equilibrium, and then compress the system by a factor of 20. What happens to the equilibrium concentrations of the three components of the system?

We can start by calculating the equilibrium concentrations of N_2, H_2, and NH_3 before the system is compressed.

$$N_2(g) + 3 H_2(g) \rightleftharpoons 2 NH_3(g) \qquad K_p = 1.4 \times 10^{-5}$$

Initial:	2.5 atm	7.5 atm	0
Equilibrium:	?	?	?

The initial reaction quotient is zero, because there is no NH_3 initially. The reaction therefore has to shift to the right to reach equilibrium.

$$N_2(g) + 3 H_2(g) \rightleftharpoons 2 NH_3(g) \qquad K_p = 1.4 \times 10^{-5}$$

Initial:	2.5 atm	7.5 atm	0
Equilibrium:	$2.5 - \Delta$	$7.5 - 3\Delta$	2Δ

Substituting this information into the equilibrium expression gives the following equation.

$$K_p = \frac{[2\Delta]^2}{[2.5 - \Delta][7.5 - 3\Delta]^3} = 1.4 \times 10^{-5}$$

If we assume that Δ is small compared with the initial pressures of N_2 and H_2, we obtain the following approximate equation.

$$\frac{[2\Delta]^2}{[2.5][7.5]^3} \cong 1.4 \times 10^{-5}$$

Solving for Δ gives the following result.

$$\Delta \cong 0.061 \text{ atm}$$

The assumption that Δ is small enough to be ignored is valid, because Δ and 3Δ are only about 2% of the initial pressures of N_2 and H_2. We can therefore use this value of Δ to calculate the partial pressures of N_2, H_2, and NH_3 at equilibrium.

$$P_{NH_3} = 2\Delta = 0.122 \text{ atm}$$
$$P_{N_2} = 2.5 - \Delta = 2.4 \text{ atm}$$
$$P_{H_2} = 7.5 - 3\Delta = 7.4 \text{ atm}$$

Compressing the reaction by a factor of 20 will result in a 20-fold increase in the partial pressure of each component of this system.

$$N_2(g) + 3 H_2(g) \rightleftharpoons 2 NH_3(g) \qquad K_p = 1.4 \times 10^{-5}$$

Intermediate: 48 atm   148 atm   2.4 atm

Is the reaction still at equilibrium? There is only one way to tell—compare the reaction quotient under these conditions with the equilibrium constant for the reaction.

$$Q_p = \frac{(2.4)^2}{(48)(148)^3} = 3.7 \times 10^{-8} < K_p$$

The reaction quotient ($Q_p = 3.7 \times 10^{-8}$) after the system has been compressed is smaller than the equilibrium constant ($K_p = 1.4 \times 10^{-5}$). The reaction must therefore shift to the right to get back to equilibrium.

This is another example of LeChatelrium's principle. A reaction at equilibrium was subjected to a stress—an increase in the total pressure on the system. The reaction shifted in the direction that minimized the effect of the stress—in this case, toward the products. Shifting toward the products reduces the number of particles in the gas, thereby decreasing the total pressure on the system.

$$N_2(g) + 3 H_2(g) \rightleftharpoons 2 NH_3(g)$$

We can calculate the magnitude of the change caused by an increase in the pressure on the system. We start by forcing the reaction all the way back to the starting materials and then allowing it to come to equilibrium again.

$$N_2(g) + 3 H_2(g) \rightleftharpoons 2 NH_3(g) \qquad K_p = 1.4 \times 10^{-5}$$

Intermediate: 50 atm   150 atm   0 atm
Equilibrium: $50 - \Delta$ $150 - 3\Delta$ 2Δ

Substituting this information into the equilibrium constant expression,

$$K_p = \frac{[2\Delta]^2}{[50 - \Delta][150 - 3\Delta]^3} \cong 1.4 \times 10^{-5}$$

assuming that Δ is small compared with the intermediate pressures of N_2 and H_2 in the system,

$$K_p = \frac{(2\Delta)^2}{(50)(150)^3} \cong 1.4 \times 10^{-5}$$

and then solving for Δ gives the following result.

$$\Delta \cong 24 \text{ atm}$$

Unfortunately, our assumption that Δ is small isn't valid in this case. But we can substitute this approximate answer back into the equation, calculate a second approximate value of Δ, and repeat this process until successive approximations converges on the following value for Δ.

$$\Delta \cong 13.2 \text{ atm}$$

We can use this value of Δ to calculate the pressures of N_2, H_2, and NH_3 when the system comes back to equilibrium.

$$P_{NH_3} = 2\Delta = 26 \text{ atm}$$
$$P_{N_2} = 50 - \Delta = 37 \text{ atm}$$
$$P_{H_2} = 150 - 3\Delta = 110 \text{ atm}$$

We can now compare the pressures of N_2, H_2, and NH_3 before and after the reaction was compressed by a factor of 20.

Before	After
$P_{NH_3} = 0.12$ atm	$P_{NH_3} = 26$ atm
$P_{N_2} = 2.4$ atm	$P_{N_2} = 37$ atm
$P_{H_2} = 7.4$ atm	$P_{H_2} = 110$ atm

Before the system was compressed, the partial pressure of NH_3 was only about 1% of the total pressure. After the system was compressed, the partial pressure of NH_3 was 15% of the total.

CHANGES IN TEMPERATURE

Changes in the concentrations of the reactants or products of a reaction shift the position of the equilibrium but do not change the size of the equilibrium constant. Similarly, a change in the pressure on a gas-phase reaction shifts the position of the equilibrium but does not change the equilibrium constant. Changes in the temperature of the system, however, affect both the position of the equilibrium and the magnitude of the equilibrium constant.

The direction of change is easy to understand. Chemical reactions either give off heat to their surroundings or absorb heat from their surroundings. In fact, when we write chemical reactions such as

$$2 \text{ NO}_2(g) \rightleftharpoons \text{N}_2\text{O}_4(g)$$

we might write them as follows.

$$2 \text{ NO}_2(g) \rightleftharpoons \text{N}_2\text{O}_4(g) + 57.2 \text{ kJ}$$

If we consider heat to be one of the reactants or products of a chemical reaction, we can understand the effect of changes in temperature on the equilibrium. Increasing the temperature of a reaction that gives off heat is the same as adding more of one of the products of the reaction. It places a stress on the reaction, which must be alleviated by converting some of the products back to reactants. Since the dimerization of NO_2 is an exothermic reaction, increasing the temperature of this reaction will lead to a decrease in the equilibrium constant, as shown in Table 14.2.

Exercise 14.10

Predict the effect of the changes described below on the following reaction.

$$2\ SO_3(g) \rightleftharpoons 2\ SO_2(g) + O_2(g) \qquad \Delta H^\circ = 197.74\ kJ$$

(a) Increasing the temperature of the reaction.

(b) Increasing the pressure on the reaction.

(c) Adding more O_2 when the reaction is at equilibrium.

(d) Removing O_2 from the system when the reaction is at equilibrium.

Solution

(a) This is an endothermic reaction; it absorbs heat from its surroundings. An increase in the temperature of the reaction therefore leads to an increase in the equilibrium constant and a shift in the position of the equilibrium toward the products.

(b) There is a net increase in the number of molecules in the system as the reactants are converted into products, which leads to an increase in the pressure of the system. The system can minimize the effect of an increase in pressure by shifting the position of the equilibrium toward the reactants — by converting some of the SO_2 and O_2 into SO_3.

(c) Adding more O_2 to the system will result in a shift in the position of the equilibrium toward the reactants.

(d) Removing O_2 from the system has the opposite effect; it leads to a shift in the equilibrium toward the products of the reaction.

14.17 LECHATELIER'S PRINCIPLE AND THE HABER PROCESS

As noted in Section 10.6, ammonia has been produced commercially from N_2 and H_2 ever since 1913, when Badische Anilin und Soda Fabrik (BASF) built a plant that used the Haber process to make 30 metric tons of synthetic ammonia per day.

$$N_2(g) + 3\ H_2(g) \rightleftharpoons 2\ NH_3(g) \qquad \Delta H^\circ = -92.2\ kJ$$

Until that time, the principal source of nitrogen for use in farming had been animal and vegetable waste. Today, almost 20 million tons of ammonia worth 2.5 billion dollars is produced in the United States each year, and about 80% of it is used for fertilizers. Ammonia is usually applied directly to the fields as a liquid at or near its boiling point of $-33.35\,^\circ C$. By using this so-called "anhydrous ammonia," farmers can apply a fertilizer that contains 82% nitrogen by weight.

About 10% of the synthetic ammonia produced each year is used to make plastics and synthetic fibers. Another 5% is used to make explosives, such as the nitrocellulose (gun cotton) used in bullets and military shells, the nitroglycerin and ammonium nitrate used to make dynamite, and trinitrotoluene, or TNT. Because of its enormous impact on society, it isn't surprising that the Haber process was the first example of the use of LeChatelier's principle to optimize the yield of an industrial chemical.

In the preceding section we saw that an *increase* in the pressure at which this reaction is run favors the products of the reaction, because there is a net reduction in the number of molecules in the system as N_2 and H_2 combine to form NH_3. Because the reaction is exothermic, the equilibrium constant becomes larger as the temperature of the reaction *decreases*.

TABLE 14.3

Mole Percent of NH₃ at Equilibrium

		Pressure			
		200 atm	*300 atm*	*400 atm*	*500 atm*
Temperature	*400°C*	38.74	47.85	58.87	60.61
	450°C	27.44	35.93	42.91	48.84
	500°C	18.86	26.00	32.25	37.79
	550°C	12.82	18.40	23.55	28.31
	600°C	8.77	12.97	16.94	20.76

Table 14.3 shows the mole percent of NH_3 at equilibrium when the reaction is run at different combinations of temperature and pressure. The *mole percent* of NH_3 under a particular set of conditions is equal to the number of moles of NH_3 at equilibrium divided by the total number of moles of all three components of the reaction. As the data in Table 14.3 demonstrate, the best yields of ammonia are obtained at low temperatures and high pressures.

Unfortunately, low temperatures slow down the rate of this reaction, and the cost of building plants rapidly escalates as the pressure at which the reaction is run is increased. When commercial plants are designed, a temperature is chosen that allows the reaction to proceed at a reasonable rate without decreasing the equilibrium concentration of the product by too much. The pressure is also adjusted so that it favors the production of ammonia without excessively increasing the cost of building and operating the plant. The optimum conditions for running this reaction at present are a pressure between 140 and 340 atm and a temperature between 400 and 600°C.

SUMMARY

Chemical reactions often seem to stop before the limiting reagent has been consumed. This can be explained with the collision theory model of chemical reactions, which assumes that the rate of a simple, one-step reaction is proportional to the concentrations of the molecules that must collide in order for the reaction to occur. Initially, the rate of the forward reaction is much faster than the rate of the reverse reaction. As the starting materials are converted into the products of the reaction, however, the forward reaction slows down and the reverse reaction becomes faster. When the reaction proceeds at the same rate in both directions, it reaches equilibrium — there is no change in the concentration of the starting materials or the products of the reaction.

The collision theory model of chemical reactions can be used to explain why the quotient obtained by dividing the product of the concentrations of the products of the reaction by the product of the concentrations of the reactants is a constant, known as the equilibrium constant for the reaction. Regardless of the initial conditions, the ratio of the concentrations of the reactants and products as defined by the equilibrium constant expression is always equal to the equilibrium constant for the reaction at that temperature.

We can decide whether a reaction is at equilibrium by comparing the reaction quotient (Q_c) at that moment with the equilibrium constant (K_c). If Q_c is smaller than K_c, we can conclude that the concentrations of the reactants are too large and the concentrations of the products are too small for the reaction to be at equilibrium. The reaction has to shift to the right — converting some of the reactants into products — to reach equilibrium. If Q_c is larger than K_c, the system contains too much product and not enough reactant. It therefore has to shift to the left to reach equilibrium.

The stoichiometry of the reaction dictates the relationship between the magnitudes of the changes in the concentrations of the reactants and products as the reaction comes to equilibrium. This relationship allows us to write an algebraic expression that relates the equilibrium concentration to the initial concentration of reactants or products and a single variable — Δ.

Calculations of the concentrations of the reactants and products of a reaction at equilibrium can be simplified by assuming that Δ is small compared with the initial concentrations to which it is added or from which it is subtracted. By convention, we assume that this assumption is legitimate when Δ is less than 5% of the initial concentrations with which it is compared. This assumption is valid when Q_c is relatively close to K_c. When Q_c is very far from K_c, we can often redefine the problem to bring Q_c close to K_c. This usually involves creating a set of intermediate conditions in which the reaction is pushed as far as possible in the direction favored by the equilibrium constant.

At a given temperature, the partial pressure of a gas is directly proportional to its concentration in units of moles per liter. Gas-phase equilibrium constant expressions can therefore be written in terms of either the concentrations of the reactants and products at equilibrium (K_c) or their partial pressures (K_p).

Three kinds of stress that can be applied to a chemical reaction at equilibrium involve changes in the temperature, the pressure, and the concentration of one or more of the reactants or products. LeChatelier's principle states that a chemical reaction at equilibrium responds to these changes by shifting the position of the equilibrium in the direction that minimizes the effect of the changes.

PROBLEMS

Chemical Reactions Don't Always Go to Completion

14-1 Describe the difference between reactions that go to completion and reactions that come to equilibrium.

14-2 Define the terms *equilibrium, equilibrium constant, equilibrium constant expression,* and *reaction quotient.*

14-3 Describe the meaning of the following symbol: $[NO_2]$.

The Rates of Chemical Reactions

14-4 Describe why the intensity of the pink color of a solution of phenolphthalein in the presence of excess base is proportional to the concentration of phenolphthalein in this solution.

14-5 Define the term *rate of reaction.*

14-6 Describe how the rate of a chemical reaction is analogous to other rate processes, such as the rate at which a car travels or the rate of inflation.

14-7 Translate the following equation into a English sentence that carries the same meaning.

$$\text{Rate of reaction} = \Delta(X)/\Delta t$$

14-8 Sketch a graph of what happens to the concentrations of N_2, H_2, and NH_3 versus time as the following reaction comes to equilibrium.

$$N_2(g) + 3\ H_2(g) \rightleftharpoons 2\ NH_3(g)$$

Assume that the initial concentrations of N_2 and H_2 are both 1.00 moles per liter and that no NH_3 is present initially. Label the kinetic and the equilibrium regions of this graph.

A Collision Theory Model for Gas-Phase Reactions

14-9 Describe how the collision theory model can be used to explain the fact that the rate at which $ClNO_2$ reacts with NO to form ClNO and NO_2 is proportional to the product of the concentrations of $ClNO_2$ and NO_2.

14-10 Use the fact that the rate of a chemical reaction is proportional to the product of the concentrations of the reagents consumed in that reaction to explain why reversible reactions inevitably come to equilibrium.

Writing Equilibrium Constant Expressions

14-11 Which of the following is the correct equilibrium constant expression for the reaction $Cl_2(g) + 3\ F_2(g) \rightleftharpoons 2\ ClF_3(g)$?

(a) $K_c = \dfrac{2\ [ClF_3]}{[Cl_2] + 3\ [F_2]}$ (d) $K_c = \dfrac{[Cl_2][F_2]^3}{[ClF_3]^2}$

(b) $K_c = \dfrac{[Cl_2] + 3\ [F_2]}{2\ [ClF_3]}$ (e) $K_c = \dfrac{[ClF_3]}{[Cl_2][F_2]}$

(c) $K_c = \dfrac{[ClF_3]^2}{[Cl_2][F_2]^3}$

14-12 Which of the following is the correct equilibrium constant expression for the reaction $2\ NO_2(g) \rightleftharpoons 2\ NO(g) + O_2(g)$?

(a) $K_c = \dfrac{[NO_2]}{[NO][O_2]}$ (d) $K_c = \dfrac{[NO]^2[O_2]}{[NO_2]^2}$

(b) $K_c = \dfrac{[NO][O_2]}{[NO_2]}$ (e) $K_c = \dfrac{[2\ NO]^2[O_2]}{[2\ NO_2]^2}$

(c) $K_c = \dfrac{[NO_2]^2}{[NO]^2[O_2]}$

14-13 Write equilibrium constant expressions for the following reactions.

(a) $O_2(g) \rightleftharpoons 2\ O(g)$ (b) $O_2(g) + 2\ F_2(g) \rightleftharpoons 2\ OF_2(g)$
(c) $2\ SO_2(g) + O_2(g) \rightleftharpoons 2\ SO_3(g)$ (d) $2\ SO_3(g) + 2\ Cl_2(g) \rightleftharpoons 2\ SO_2Cl_2(g) + O_2(g)$

14-14 Write equilibrium constant expressions for the following reactions.

(a) $2 NO(g) + 2 H_2(g) \rightleftharpoons N_2(g) + 2 H_2O(g)$
(b) $2 NOCl(g) \rightleftharpoons 2 NO(g) + Cl_2(g)$ (c) $2 NO_2(g) \rightleftharpoons$
$N_2O_4(g)$ (d) $2 NO(g) + O_2(g) \rightleftharpoons 2 NO_2(g)$

14-15 Write equilibrium constant expressions for the following reactions.

(a) $2 CO(g) + O_2(g) \rightleftharpoons 2 CO_2(g)$ (b) $CO_2(g) +$
$H_2(g) \rightleftharpoons CO(g) + H_2O(g)$ (c) $CO(g) + 2 H_2(g) \rightleftharpoons$
$CH_3OH(g)$ (d) $CH_4(g) + 2 O_2(g) \rightleftharpoons CO_2(g) +$
$2 H_2O(g)$

Calculating Equilibrium Constants

14-16 Calculate K_c for the following reaction at 400 K assuming that 1.00 mole per liter of NOCl decomposes at this temperature to give equilibrium concentrations of 0.0222 M NO, 0.0111 M Cl_2, and 0.989 M NOCl.

$$2 NOCl(g) \rightleftharpoons 2 NO(g) + Cl_2(g)$$

14-17 Taylor and Crist [*J. Am. Chem. Soc.*, *63*, 1381 (1941)] studied the reaction between hydrogen and iodine to form hydrogen iodide

$$H_2(g) + I_2(g) \rightleftharpoons 2 HI(g)$$

and obtained the following data for the concentrations of H_2, I_2, and HI at equilibrium in units of moles per liter.

Trial	[H_2]	[I_2]	[HI]
I	0.0032583	0.0012949	0.015869
II	0.0046981	0.0007014	0.013997
III	0.0007106	0.0007106	0.005468

Calculate the value of K_c for each of these trials. Is K_c a constant for this reaction, within the limits of experimental error?

Combining Equilibrium Constant Expressions

14-18 Write equilibrium constant expressions for the following reactions.

(a) $2 NO_2(g) \rightleftharpoons 2 NO(g) + O_2(g)$
(b) $2 NO(g) + O_2(g) \rightleftharpoons 2 NO_2(g)$

Calculate the value of K_c at 500 K for reaction a assuming that the value of K_c for reaction b is 7.7×10^5.

14-19 Write equilibrium constant expressions for the following reactions.

(a) $NO(g) + \frac{1}{2} O_2(g) \rightleftharpoons NO_2(g)$

(b) $2 NO(g) + O_2(g) \rightleftharpoons 2 NO_2(g)$

Calculate the value of K_c at 500 K for reaction a assuming that the value of K_c for reaction b is 7.7×10^5.

14-20 Write equilibrium constant expressions for the following reactions.

(a) $N_2(g) + 3 H_2(g) \rightleftharpoons 2 NH_3(g)$
(b) $2 NH_3(g) \rightleftharpoons N_2(g) + 3 H_2(g)$

(c) $NH_3(g) \rightleftharpoons \frac{1}{2} N_2(g) + \frac{3}{2} H_2(g)$?

Calculate the value of K_p at 500°C for reaction b assuming that K_p at this temperature for reaction a is 7.1×10^{-5}. What is the value of K_p for reaction c at room temperature?

14-21 Use the equilibrium constants for reactions a and b at 200°C to calculate the equilibrium constant for reaction c at this temperature.

(a) $N_2(g) + O_2(g) \rightleftharpoons 2 NO(g)$ $K_c = 2.3 \times 10^{-19}$
(b) $2 NO(g) + O_2(g) \rightleftharpoons 2 NO_2(g)$ $K_c = 9.0 \times 10^4$
(c) $N_2(g) + 2 O_2(g) \rightleftharpoons 2 NO_2(g)$ $K_c = ?$

14-22 Use the equilibrium constants for reactions a and b at 1000 K to calculate the equilibrium constant for reaction c, the water-gas shift reaction, at this temperature.

(a) $CO(g) + \frac{1}{2} O_2(g) \rightleftharpoons CO_2(g)$ $K_c = 1.9 \times 10^{11}$

(b) $H_2O(g) \rightleftharpoons H_2(g) + \frac{1}{2} O_2(g)$ $K_c = 5.3 \times 10^{-12}$

(c) $CO(g) + H_2O(g) \rightleftharpoons CO_2(g) + H_2(g)$ $K_c = ?$

Reaction Quotients: A Way to Decide Whether a Reaction Is at Equilibrium

14-23 Explain why there is only one value for the equilibrium constant for a reaction at a given temperature but an infinite number of values for the reaction quotient.

14-24 Suppose that the reaction quotient (Q_c) for the following reaction at some moment in time is 1.0×10^{-8} and the equilibrium constant for this reaction (K_c) at the same temperature is 2.9×10^{-7}.

$$2 NO_2(g) \rightleftharpoons 2 NO(g) + O_2(g)$$

Which of the following is a valid conclusion?

(a) The reaction is at equilibrium   (b) The reaction must shift toward the products to reach equilibrium   (c) The reaction must shift toward the reactants to reach equilibrium

14-25 Which of the following statements correctly describes a system for which Q_c is smaller than K_c?

(a) The reaction is at equilibrium   (b) The reaction must shift to the right to reach equilibrium   (c) The reaction must shift to the left to reach equilibrium   (d) The reaction can never reach equilibrium

14-26 Under which set of conditions must the following reaction shift to the right to reach equilibrium?

$$2 SO_2(g) + O_2(g) \rightleftharpoons 2 SO_3(g)$$

(a) $K_c < 1$ (b) $K_c > 1$ (c) $Q_c < K_c$ (d) $Q_c = K_c$ (e) $Q_c > K_c$

14-27 Carbon monoxide reacts with chlorine to form phosgene.

$$CO(g) + Cl_2(g) \rightleftharpoons COCl_2(g)$$

The equilibrium constant, K_p, for this reaction is 520 at 350°C. Is the system at equilibrium at the following partial pressures: (a) 0.050 atm $COCl_2$, (b) 0.010 atm CO, and (c) 0.0050 atm Cl_2? If not, in which direction does the reaction have to shift to reach equilibrium?

Changes in Concentration That Occur as a Reaction Comes to Equilibrium

14-28 Describe the relationship among the initial concentration of a reactant (X), the concentration of this reactant at equilibrium $[X]$, and the change in the concentration of X that occurs as the reaction comes to equilibrium $\Delta(X)$.

14-29 Explain why the change in the N_2 concentration that occurs when the following reaction comes to equilibrium is related to the change in the H_2 concentration.

$$N_2(g) + 3 H_2(g) \rightleftharpoons 2 NH_3(g)$$

Derive an equation that describes the relationship between the changes in the concentrations of these two reagents.

14-30 When confronted with the task in Problem 14-29, students often write the following equation.

$$\Delta(N_2) = 3\Delta(H_2)$$

Explain why this equation is wrong.

14-31 Calculate the changes in the CO and Cl_2 concentrations as the following reaction comes to equilibrium assuming that the concentration of $COCl_2$ decreases by 0.250 moles per liter.

$$COCl_2(g) \rightleftharpoons CO(g) + Cl_2(g)$$

14-32 Calculate the changes in the N_2 and H_2 concentrations as the following reaction comes to equilibrium assuming that the concentration of NH_3 decreases by 0.234 moles per liter.

$$2 NH_3(g) \rightleftharpoons N_2(g) + 3 H_2(g)$$

14-33 Which of the following equations describes the relationship between the magnitudes of the changes in the NO_2 and O_2 concentrations as the following reaction comes to equilibrium?

$$2 NO(g) + O_2(g) \rightleftharpoons 2 NO_2(g)$$

(a) $\Delta(NO_2) = \Delta(O_2)$ (b) $\Delta(NO_2) = 2\Delta(O_2)$
(c) $\Delta(O_2) = 2\Delta(NO_2)$

14-34 Which of the following equations correctly describes the relationships between the magnitudes of the changes in the Cl_2 and F_2 concentrations as the following reaction comes to equilibrium?

$$Cl_2(g) + 3 F_2(g) \rightleftharpoons 2 ClF_3(g)$$

(a) $\Delta(Cl_2) = \Delta(F_2)$ (b) $\Delta(Cl_2) = 2\Delta(F_2)$ (c) $\Delta(Cl_2) = 3\Delta(F_2)$ (d) $\Delta(F_2) = 2\Delta(Cl_2)$ (e) $\Delta(F_2) = 3\Delta(Cl_2)$

14-35 How can we describe the change that occurs in the concentration of H_2O when ammonia reacts with oxygen to form nitrogen oxide and water according to the following equation if the change in the NH_3 concentration is Δ?

$$4 NH_3(g) + 5 O_2(g) \rightleftharpoons 4 NO(g) + 6 H_2O(g)$$

(a) Δ (b) 1.5Δ (c) 2Δ (d) 4Δ (e) 6Δ

14-36 Calculate the concentrations of H_2 and NH_3 at equilibrium assuming that a reaction that initially contained 1.00 M concentrations of both N_2 and H_2 is found to have an N_2 concentration of 0.922 M at equilibrium.

$$N_2(g) + 3 H_2(g) \rightleftharpoons 2 NH_3(g)$$

Initial:	1.00 M	1.00 M	0 M
Equilibrium:	0.922 M	?	?

14-37 Calculate the equilibrium constant for the reaction in Problem 14-36.

Hidden Assumptions That Make Equilibrium Calculations Easier

14-38 Some students have described the technique used in this chapter to simplify equilibrium problems as follows: "Assume that Δ is zero." Explain why they are wrong. What is the correct way of describing the assumption?

14-39 Describe the advantage of setting up equilibrium problems so that Δ is small compared with the initial concentrations.

14-40 Describe what happens if you make the assumption that Δ is zero in the following equation.

$$\frac{[0.125 - \Delta][2.40 - 2\Delta]^2}{[0.200 + \Delta]^2} = 1.3 \times 10^{-8}$$

Explain how to get around this problem.

A Rule of Thumb for Testing the Validity of Assumptions

14-41 Describe how to test whether Δ is small enough compared with the initial concentrations to be legitimately ignored.

14-42 Explain why Δ is relatively small when the reaction quotient (Q_c) is reasonably close to the equilibrium constant for the reaction (K_c).

14-43 Explain why the assumption that Δ is small compared with the initial concentrations of the reactants and products is doomed to fail when the reaction quotient (Q_c) is very different from the equilibrium constant for the reaction (K_c).

What Do We Do When the Approximation Fails?

14-44 Describe the technique used to solve problems for which the reaction quotient is very different from the equilibrium constant.

14-45 Before we can solve the following problem, we have to define a set of intermediate conditions under which the concentration of one of the reactants or products is zero.

$$2 NO_2(g) \rightleftharpoons 2 NO(g) + O_2(g) \quad K_c = 4.2 \times 10^{-6}$$

Initial:	0.10 M	0.10 M	0.005 M	(at 250°C)

Which of the following goals determines whether we push the reaction as far as possible to the right or as far as possible to the left?

(a) To make both Q_c and K_c large (b) To make both Q_c and K_c small (c) To bring Q_c as close as possible to K_c (d) To make the difference between Q_c and K_c as large as possible

Successive Approximations

14-46 Describe the kind of problem whose solution is most likely to require the use of successive approximations.

14-47 Describe the procedure you would use if you decided to solve an equilibrium problem by successive approximations.

Gas-Phase Equilibrium Problems

14-48 Calculate the concentrations of PCl_5, PCl_3, and Cl_2 that are present when the following gas-phase reaction comes to equilibrium.

$$PCl_5(g) \rightleftharpoons PCl_3(g) + Cl_2(g) \qquad K_c = 0.0013$$
Initial: 1.00 M   0   0   (at 450 K)

14-49 Calculate the percent of PCl_5 that has decomposed when the reaction in Problem 14-48 comes to equilibrium at 450 K.

14-50 Calculate the concentrations of PCl_5, PCl_3, and Cl_2 present when the following gas-phase reaction comes to equilibrium.

$$PCl_5(g) \rightleftharpoons PCl_3(g) + Cl_2(g) \qquad K_c = 0.0013$$
Initial: 1.00 M   0   0.20 M   (at 450 K)

14-51 Calculate the concentrations of NO, NO_2, and O_2 present when the following gas-phase reaction reaches equilibrium.

$$2\ NO_2(g) \rightleftharpoons 2\ NO(g) + O_2(g) \quad K_c = 1.1 \times 10^{-5}$$
Initial: 0.10 M   0   0   (at 200°C)

14-52 Calculate the concentrations of NO, NO_2, and O_2 present when the following gas-phase reaction reaches equilibrium.

$$2\ NO_2(g) \rightleftharpoons 2\ NO(g) + O_2(g) \quad K_c = 1.1 \times 10^{-5}$$
Initial: 0   0.10 M   0.10 M   (at 200°C)

14-53 Calculate the equilibrium concentrations of SO_3, SO_2 and O_2 present when 0.100 moles of SO_3 in a 250-milliliter flask at 500 K decomposes to form SO_2 and O_2.

$$2\ SO_3(g) \rightleftharpoons 2\ SO_2(g) + O_2(g) \qquad K_c = 3.5 \times 10^{-13}$$
$$\text{(at 500 K)}$$

14-54 Calculate the equilibrium concentrations of SO_3, SO_2, and O_2 present when a mixture of 0.100 moles of SO_2 and 0.050 moles of O_2 in a 250-milliliter flask at 500 K combine to form SO_3.

$$2\ SO_2(g) + O_2(g) \rightleftharpoons 2\ SO_3(g) \qquad K_c = 2.9 \times 10^{12}$$
$$\text{(at 500 K)}$$

14-55 Calculate the equilibrium concentration of NO_2 present when 0.100 M N_2O_4 decomposes to form NO_2 at $-40°C$.

$$N_2O_4(g) \rightleftharpoons 2\ NO_2(g) \qquad K_c = 1.2 \times 10^{-5}$$

14-56 Calculate the equilibrium concentration of NO_2 present when 1.00 M NO_2 dimerizes to form N_2O_4 at $-40°C$.

$$N_2O_4(g) \rightleftharpoons 2\ NO_2(g) \qquad K_c = 1.2 \times 10^{-5}$$

14-57 Calculate the equilibrium concentrations of N_2, H_2, and NH_3 present when a mixture that was initially 0.10 M N_2, 0.10 M H_2, and 0.10 M NH_3 comes to equilibrium at 650°C.

$$N_2(g) + 3\ H_2(g) \rightleftharpoons 2\ NH_3(g) \qquad K_c = 0.040$$
$$\text{(at 650°C)}$$

14-58 Calculate the equilibrium concentrations of CO, H_2O, CO_2, and H_2 present in the water-gas shift reaction at 800°C assuming that the initial concentrations of CO and H_2O were 1.00 M.

$$CO(g) + H_2O(g) \rightleftharpoons CO_2(g) + H_2(g) \qquad K_c = 0.64$$

14-59 What initial concentrations of CO and H_2O would be needed to reach an equilibrium concentration of 1.00 M CO_2 in the water-gas shift reaction described in Problem 14-57?

14-60 Calculate the equilibrium concentrations of N_2, O_2, and NO present when a mixture that was initially 0.100 M in N_2 and 0.090 M in NO comes to equilibrium at 600°C.

$$N_2(g) + O_2(g) \rightleftharpoons 2\ NO(g) \qquad K_c = 3.2 \times 10^{-10}$$

Equilibria Expressed in Partial Pressures

14-61 Explain why pressure can be used instead of concentration to describe equilibrium constant expressions for gas-phase reactions.

14-62 Which equation correctly describes the relationship between K_p and K_c for the following reaction?

$$Cl_2(g) + 3\ F_2(g) \rightleftharpoons 2\ ClF_3(g)?$$

(a) $K_c = K_p$ (b) $K_c = K_p \times (RT)^{-1}$ (c) $K_c = K_p \times (RT)^{-2}$ (d) $K_c = K_p \times (RT)$ (e) $K_c = K_p \times (RT)^2$

14-63 Which equation correctly describes the relationship between K_p and K_c for the following reaction?

$$2\ NO_2(g) \rightleftharpoons 2\ NO(g) + O_2(g)$$

(a) $K_p = K_c$ (b) $K_p = 1/K_c$ (c) $K_p = K_c(RT)$ (d) $K_p = K_c(RT)^{-1}$

14-64 Calculate K_p for the decomposition of NOCl at 500 K if 27.3% of a 1.00-atm sample of NOCl decomposes to NO and Cl_2 at equilibrium.

$$2\ NOCl(g) \rightleftharpoons 2\ NO(g) + Cl_2(g)$$

14-65 Sulfuryl chloride decomposes to sulfur dioxide and chlorine. Calculate what partial pressures of the three components of this system at equilibrium would result from the decomposition of 6.75 grams of SO_2Cl_2 in a 1.00-liter flask at 25°C.

$$SO_2Cl_2(g) \rightleftharpoons SO_2(g) + Cl_2(g) \qquad K_p = 3.4 \times 10^{-4}$$

14-66 Calculate the partial pressures of phosgene, carbon monoxide, and chlorine at equilibrium when 0.124 atm of $COCl_2$ decomposes at 300°C according to the following equation.

$$COCl_2(g) \rightleftharpoons CO(g) + Cl_2(g) \qquad K_p = 1.9 \times 10^{-3}$$

14-67 Calculate the partial pressures of SO_3, SO_2, and O_2 that would be present at equilibrium when a mixture that was initially 0.490 atm in SO_2 and 0.245 atm in O_2 comes to equilibrium at 700 K.

$$2 SO_2(g) + O_2(g) \rightleftharpoons 2 SO_3(g) \qquad K_p = 8.8 \times 10^4$$

14-68 Calculate the partial pressures of SO_3, SO_2, and O_2 that would be present at equilibrium when a mixture that was initially 0.490 atm in SO_2 and 0.345 atm in O_2 reacts at 700 K.

$$2 SO_3(g) \rightleftharpoons 2 SO_2(g) + O_2(g) \qquad K_p = 1.1 \times 10^{-5}$$

14-69 Calculate the partial pressures of SO_3, SO_2, and O_2 present at equilibrium when a mixture that was initially 0.30 atm in SO_2 and 0.12 atm in O_2 comes to equilibrium at 650°C, assuming that K_p is 24 for the reaction leading to the formation of SO_3 at this temperature.

14-70 Calculate the partial pressures of N_2, H_2, and NH_3 present at equilibrium when a mixture that was initially 0.50 atm in N_2, 0.60 atm in H_2, and 0.20 atm in NH_3 comes to equilibrium at a temperature at which K_p for the following reaction is 1.9×10^{-4}.

$$N_2(g) + 3 H_2(g) \rightleftharpoons 2 NH_3(g)$$

14-71 Calculate the partial pressures of NH_3, O_2, NO_2, and H_2O present at equilibrium when a mixture that was initially 0.50 atm in each of the four gases comes to equilibrium at 1000°C.

$$4 NO_2(g) + 6 H_2O(g) \rightleftharpoons$$
$$4 NH_3(g) + 7 O_2(g) \qquad K_p = 6.3 \times 10^{-28}$$

14-72 Calculate the partial pressures of NO, Cl_2, and NOCl at equilibrium assuming that 0.50 atm of NO combines with 0.25 atm of Cl_2 to form nitrosyl chloride with a K_p of 52 at 500 K.

$$2 NO(g) + Cl_2(g) \rightleftharpoons 2 NOCl(g)$$

14-73 Calculate the partial pressures of N_2, H_2, and NH_3 present at equilibrium when 4.00 atm of NH_3 decomposes at 1000 K, assuming that K_p for this reaction is 3.9×10^6.

$$2 NH_3(g) \rightleftharpoons N_2(g) + 3 H_2(g)$$

14-74 Calculate the concentrations of N_2, O_2, and NO present when a mixture of 0.40 M N_2 and 0.60 M O_2 reacts to form NO at 700°C, assuming that K_p at this temperature is 4.1×10^{-9}.

$$N_2(g) + O_2(g) \rightleftharpoons 2 NO(g)$$

(Read this problem carefully before answering.)

14-75 Industrial chemicals can be made from coal by a process that starts with the reaction between red-hot coal and steam to form a mixture of CO and H_2. This mixture, which is known as

synthesis gas, can be converted to methanol (CH_3OH) in the presence of a ruthenium/cobalt catalyst. The methanol produced in this reaction can be converted into a host of other products, ranging from acetic acid to gasoline. Calculate the percent yield of methanol when a mixture of 1.00 atm CO and 1.00 atm H_2 comes to equilibrium at 125°C.

$$CO(g) + 2 H_2(g) \rightleftharpoons CH_2OH(g) \qquad K_p = 2.3 \times 10^4$$
$$\text{(at 125°C)}$$

LeChatelier's Principle

14-76 LeChatelier's principle has been applied to many fields, ranging from economics to psychology to political science. Give an example of LeChatelier's principle in a field outside the physical sciences.

14-77 Predict the effect of increasing the pressure on the following reactions at equilibrium.

(a) $N_2(g) + 3 H_2(g) \rightleftharpoons 2 NH_3(g)$ (b) $2 SO_3(g) + 2 Cl_2(g) \rightleftharpoons 2 SO_2Cl_2(g) + O_2(g)$ (c) $O_2(g) + 2 F_2(g) \rightleftharpoons 2 OF_2(g)$ (d) $CH_4(g) + 2 O_2(g) \rightleftharpoons CO_2(g) + 2 H_2O(g)$

14-78 Predict the effect of decreasing the pressure on the following reactions at equilibrium.

(a) $N_2O_4(g) \rightleftharpoons 2 NO_2(g)$ (b) $2 NO_2(g) \rightleftharpoons 2 NO(g) + O_2(g)$ (c) $N_2(g) + O_2(g) \rightleftharpoons 2 NO(g)$ (d) $NO(g) + NO_2(g) \rightleftharpoons N_2O_3(g)$

14-79 Predict the effect of increasing the concentration of the underlined reagent on each of the following reactions at equilibrium.

(a) $CH_4(g) + 2 \underline{O_2}(g) \rightleftharpoons CO_2(g) + 2 H_2O(g)$
(b) $2 \underline{NO_2}(g) \rightleftharpoons N_2O_4(g)$ (c) $2 SO_3(g) \rightleftharpoons 2 SO_2(g) + \underline{O_2}(g)$ (d) $\underline{PF_5}(g) \rightleftharpoons PF_3(g) + F_2(g)$

14-80 Predict the effect of decreasing the concentration of the underlined reagent on each of the following reactions at equilibrium.

(a) $N_2(g) + O_2(g) \rightleftharpoons 2 \underline{NO}(g)$ (b) $3 O_2(g) \rightleftharpoons 2 \underline{O_3}(g)$
(c) $Cl_2(g) + 3 \underline{F_2}(g) \rightleftharpoons 2 ClF_3(g)$ (d) $2 H_2S(g) + 3 O_2(g) \rightleftharpoons 2 \underline{SO_2}(g) + 2 H_2O(g)$

14-81 Use LeChatelier's principle to predict the effect of an increase in pressure on the solubility of a gas in water.

14-82 List as many ways as possible of increasing the yield of ammonia in the Haber process.

$$N_2(g) + 3 H_2(g) \rightleftharpoons 2 NH_3(g)$$

14-83 Explain why an increase in pressure favors the formation of ammonia in the Haber process.

14-84 Predict how an increase in the volume of the container by a factor of 2 would affect the concentrations of ammonia and oxygen in the following reaction.

$$4 NH_3(g) + 5 O_2(g) \rightleftharpoons 4 NO(g) + 6 H_2O(g)$$

ACID-BASE EQUILIBRIA

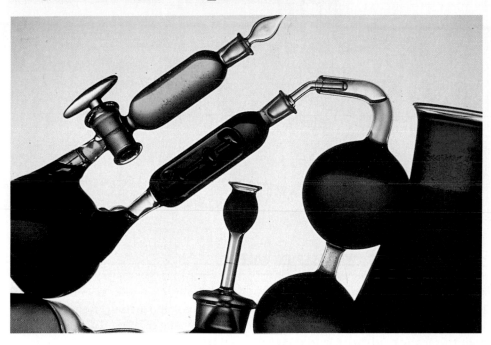

FIG. 15.1 This etching shows some of the apparatus used by Joseph Priestley for studying gases. It is reproduced from his book *Experiments and Observations on Different Kinds of Air*, Vol. 1, published in 1774.

15.1 SOLUTIONS OF GASES IN WATER

Gases, and the gas-phase reactions discussed in Chapter 14, are very important. It has even been argued that the most important factor in turning chemistry into a science was the development during the 18th century of apparatus designed to collect, measure, identify, and quantify gases (see Figure 15.1).

In a practical sense, however, gases are notoriously difficult to handle. Meaningful quantities at atmospheric pressure occupy an enormous amount of space. Gases are therefore stored at high pressures in compressed gas cylinders, or they are dissolved in an appropriate solvent, such as water.

If you've ever worked in a chemistry laboratory when concentrated solutions of both ammonia and hydrochloric acid were being used, you may have noticed that the air seemed to get cloudy. This is the result of a gas-phase reaction between HCl and NH_3. These compounds escape from the solutions to form finely divided NH_4Cl, which collects in the atmosphere.

$$HCl(g) + NH_3(g) \longrightarrow NH_4Cl(s)$$

Few chemists would intentionally run this reaction in the gas phase, however. They are more likely to run the reaction by mixing aqueous solutions of these compounds.

$$HCl(aq) + NH_3(aq) \longrightarrow NH_4Cl(aq)$$

As a result, the only exposure that most people get to ammonia and hydrogen chloride is as aqueous solutions. The best way of cleaning glass, for example, consists of spraying a dilute aqueous solution of ammonia—$NH_3(aq)$—on the glass and then rubbing with a crumpled piece of newspaper. Few chemists ever work with hydrogen chloride as a gas, but as an aqueous solution—$HCl(aq)$—it is one of the first chemicals people encounter in their exposure to chemistry laboratories.

Table 15.1 gives the solubility in water of some common gases. These data are given in terms of the number of liters of gas that will dissolve in a liter of water at $0°C$ under a pressure of 1 atm. For example, 1.713 liters of CO_2 gas dissolves in a

TABLE 15.1

The Solubilities of Common Gases in Water

Gas	Solubility (L of gas / L of H_2O at $0°C$ and 1 atm)
He	0.0094
H_2	0.02148
N_2	0.02354
CO	0.03537
O_2	0.04889
CH_4	0.05563
CO_2	1.713
H_2S	4.670
SO_2	79.8
HCl	512
NH_3	1130

liter of water under these conditions. This measurement is taken under conditions in which the following reaction is at equilibrium.

$$CO_2(g) \underset{}{\overset{H_2O}{\rightleftharpoons}} CO_2(aq)$$

The solubility of gases in water depends on both the temperature of the water and the pressure of the gas above the water. In 1801, William Henry first noticed that the amount of gas that dissolves in water is proportional to the pressure on the gas. Henry's law, as it is now known, is an example of LeChatelier's principle. An increase in the pressure of CO_2 above a sample of water, for example, can force more CO_2 into solution. This CO_2 will escape, of course, when the pressure is released, as shown in Figure 15.2. LeChatelier's principle can also be used to explain why a gas becomes less soluble as its solution is heated—as might be expected by anyone who can envision what will happen if the bottle in Figure 15.2 is heated in a hot-water bath before being opened.

Some of the gases in Table 15.1, such as He, H_2, N_2, and O_2, are only marginally soluble in water. Others, such as NH_3 and HCl, are much more soluble. The solubility of ammonia in water can be demonstrated with the apparatus shown in Figure 15.3. A large round-bottomed flask is filled with NH_3 gas and sealed with a rubber stopper. A second round-bottomed flask is filled with water, a few drops of phenolphthalein are added to the water, and the two flasks are connected with rubber tubing that is clamped shut. The demonstration involves opening the clamp and blowing just enough air into the open end of the hose to force a few drops of water into the upper flask, which contains NH_3. The NH_3 rapidly dissolves in this small quantity of water, which creates a vacuum that causes water to rush into the flask with a fountain-like effect.

The difference between the solubilities of O_2 and NH_3 in water can be explained, in part, by the fact that ammonia molecules can form hydrogen bonds with water, as shown in Figure 15.4, whereas O_2 molecules can't. The hydrogen bonds between ammonia and water molecules effectively pull NH_3 into solution, driving the following equilibrium to the right.

$$NH_3(g) \underset{}{\overset{H_2O}{\rightleftharpoons}} NH_3(aq)$$

FIG. 15.2 Solutions that are saturated with a gas are at equilibrium. The rate at which the gas escapes from the solution is equal to the rate at which it redissolves. By increasing the pressure on the gas, it is possible to increase the amount of gas that dissolves. However, when the pressure is released (in this case, when the bottle is opened), gas rapidly bubbles out of the solution.

FIG. 15.3 At atmospheric pressure and 0°C, 1130 liters of ammonia will dissolve in a liter of water. When a few drops of water from the bottom flask in the apparatus shown in the photo are forced into the upper flask, which is filled with NH_3 gas, the ammonia rapidly dissolves in the water. This creates a vacuum in the flask, forcing water to rush in with a fountain-like effect. The solution changes color because it contains the acid-base indicator phenolphthalein. Ammonia reacts with water to give a small quantity of the OH^- ion, which turns the phenolphthalein from colorless to pink.

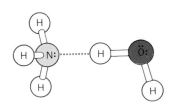

FIG. 15.4 Part of the solubility of ammonia in water can be explained by the fact that NH_3 molecules can form hydrogen bonds to neighboring H_2O molecules.

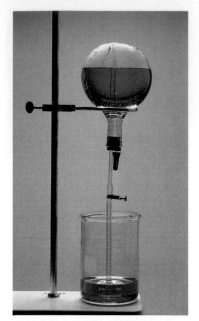

FIG. 15.5 Hydrogen chloride is very soluble in water. When a few drops of water are forced into a flask filled with this gas, the HCl dissolves, creating a vacuum, and water gushes into the flask in a fountain-like effect. The solution changes colors because it contains the acid-base indicator methyl violet. HCl reacts with water to give the H_3O^+ ion, which turns methyl violet from violet to blue.

But hydrogen bonds between NH_3 and water can't explain the observation that solutions of ammonia dissolved in water turn phenolphthalein pink. The color that forms in the upper flask in Figure 15.3 is the most important feature of this demonstration. It reminds us that water is more than just the solvent for this equilibrium. Ammonia not only dissolves in water, it also reacts with water. About 1% of the NH_3 molecules in this solution pick up H^+ ions from neighboring water molecules to form ammonium ions and hydroxide ions.

$$NH_3(aq) + H_2O(l) \rightleftharpoons NH_4^+(aq) + OH^-(aq)$$

It is the OH^- ions formed in the reaction that turn phenolphthalein pink.

Figure 15.5 shows a similar demonstration, in which the top flask is initially filled with HCl gas. The indicator used in this demonstration is methyl violet which is violet when dissolved in water. When a few drops of water are forced into the upper flask, the HCl rapidly dissolves in the water, once again creating a vacuum, which forces water to gush into the upper flask.

$$HCl(g) \underset{}{\overset{H_2O}{\rightleftharpoons}} HCl(aq)$$

Once again, the color of the indicator changes—this time because the HCl reacts with water to give hydronium and chloride ions.

$$HCl(aq) + H_2O(l) \longrightarrow H_3O^+(aq) + Cl^-(aq)$$

The remainder of this chapter is devoted to building a model that enables us to calculate the concentrations of the various ions formed when an acid such as HCl or a base such as NH_3 reacts with water.

15.2 THE ACID-BASE CHEMISTRY OF WATER

The chemistry of aqueous solutions is dominated by the following equilibrium between neutral water molecules and the ions they form.

$$2\,H_2O(l) \rightleftharpoons H_3O^+(aq) + OH^-(aq)$$

Strict application of the rules for writing equilibrium constant expressions produces the following result for this reaction.

$$K_c = \frac{[H_3O^+][OH^-]}{[H_2O]^2}$$

This is a perfectly legitimate equilibrium constant expression, but it fails to take into account the enormous difference between the concentrations of neutral H_2O molecules and H_3O^+ and OH^- ions at equilibrium. Measurements of the ability of water to conduct an electric current suggest that pure water at 25°C contains 1.0×10^{-7} moles per liter of each of these ions.

$$[H_3O^+] = [OH^-] = 1.0 \times 10^{-7}\,M \qquad \text{(at 25°C)}$$

Exercise 15.1

Calculate the equilibrium constant for the dissociation of water at 25°C, assuming that the density of water is 0.9971 g/cm³ at this temperature.

Solution

We already know the equilibrium concentrations of the H_3O^+ and OH^- ions—

$1.0 \times 10^{-7}M$—so the only other information we need to calculate K_c for this reaction is the H_2O concentration at equilibrium.

$$K_c = \frac{[H_3O^+][OH^-]}{[H_2O]^2}$$

We can start by calculating the weight of a liter of water.

$$0.9971 \frac{g\ H_2O}{cm^3} \times 1 \frac{cm^3}{mL} \times 1000 \frac{mL}{L} = 997.1 \frac{g\ H_2O}{L}$$

We then calculate the number of moles of water per liter.

$$997.1 \frac{g\ H_2O}{L} \times \frac{1\ mol\ H_2O}{18.015\ g} = 55.35 \frac{mol\ H_2O}{L}$$

Substituting what we know into the equilibrium constant expression gives the following result.

$$K_c = \frac{[H_3O^+][OH^-]}{[H_2O]^2} = \frac{[1.0 \times 10^{-7}][1.0 \times 10^{-7}]}{[55.35]^2} = 3.3 \times 10^{-18}$$

Exercise 15.2

Use the results of the preceding exercise to calculate the fraction of water molecules that dissociate at 25°C.

Solution

Two water molecules are consumed for every H_3O^+ and OH^- ion formed when water dissociates.

$$2\ H_2O(l) \rightleftharpoons H_3O^+(aq) + OH^-(aq)$$

Out of the 55.35 moles of H_2O molecules in a liter of water, only 2.0×10^{-7} moles dissociate into H_3O^+ and OH^- ions at equilibrium. This means that slightly less than four out of every billion H_2O molecules—or 4 parts per billion (ppb)—dissociate into H_3O^+ and OH^- ions at any moment in time.

$$\frac{2.0 \times 10^{-7}}{55.35} = 3.6 \times 10^{-9}$$

The equilibrium concentration of H_2O molecules is so much larger than the concentrations of the H_3O^+ and OH^- ions that it is effectively constant. (The change in the concentration of H_2O molecules as this reaction comes to equilibrium is smaller by a factor of 100,000 than the experimental error involved in the measurement of this concentration.)

If the $[H_2O]$ term is essentially constant in all acid-base reactions, we can build this term into the equilibrium constant for the reaction and thus greatly simplify equilibrium calculations. Rearranging the equilibrium constant expression for the dissociation of water

$$K_c = \frac{[H_3O^+][OH^-]}{[H_2O]^2}$$

to achieve this goal gives the following equation.

$$[H_3O^+][OH^-] = K_c \times [H_2O]^2$$

By convention, the product of K_c times the square of the equilibrium concentration of water is known as the **water-dissociation equilibrium constant, K_w.**

$$K_w = [H_3O^+][OH^-]$$

In pure water, at 25°C, the $[H_3O^+]$ and $[OH^-]$ ion concentrations are 1.0×10^{-7} M. The value of K_w at 25°C is therefore 1.0×10^{-14}.

$$\begin{aligned} K_w &= [H_3O^+][OH^-] \\ &= [1.0 \times 10^{-7}][1.0 \times 10^{-7}] = \mathbf{1.0 \times 10^{-14}} \end{aligned} \qquad \text{(at 25°C)}$$

Although K_w is defined in terms of the dissociation of water, the water-dissociation equilibrium constant expression is equally valid for solutions of acids and bases dissolved in water. Regardless of the source of the H_3O^+ and OH^- ions in water, the product of the concentrations of these ions when the reaction is at equilibrium at 25°C is always 1.0×10^{-14}.

What happens to the H_3O^+ and OH^- ion concentrations when we add a strong acid to water? Suppose, for example, that we add enough hydrochloric acid to a beaker of water to raise the H_3O^+ ion concentration to 0.010 M. As noted in Section 11.9, hydrochloric acid is a strong acid that ionizes almost completely in water.

$$HCl(g) + H_2O(l) \longrightarrow H_3O^+(aq) + Cl^-(aq)$$

If the concentration of the hydrochloric acid in this solution is 0.010 moles per liter, we can assume that the H_3O^+ ion concentration in the solution is approximately 0.010 M.

Before the acid was added, the concentrations of the H_3O^+ and OH^- ions were 1.0×10^{-7} M. After the acid is added, the H_3O^+ ion concentration is 100,000 times larger — $[H_3O^+] = 0.010$ M. According to LeChatelier's principle, this should drive the following equilibrium to the left, reducing the number of H_3O^+ and OH^- ions in the solution.

$$2\ H_2O(l) \rightleftharpoons H_3O^+(aq) + OH^-(aq)$$

The concentration of H_3O^+ ions in this solution is much larger than the OH^- ion concentration, however. As a result, the change in the H_3O^+ ion concentration as the equilibrium shifts to the left is too small to notice. When the reaction returns to equilibrium, the H_3O^+ ion concentration is still about 0.010 M.

When the reaction returns to equilibrium, the product of the H_3O^+ and OH^- ion concentrations is once again equal to K_w.

$$[H_3O^+][OH^-] = K_w$$

We can therefore calculate the OH^- ion concentration when the solution comes back to equilibrium from the following equation.

$$[OH^-] = \frac{K_w}{[H_3O^+]} = \frac{1.0 \times 10^{-14}}{[0.010]} = 1.0 \times 10^{-12}\ M$$

What happens to the H_3O^+ and OH^- ion concentrations when we add a base to water? Suppose, for example, that we add enough ammonia to a beaker of water to raise the OH^- ion concentration to 0.0010 M. The equilibrium between water and its ions once again shifts to the left, to reduce the number of ions in solution.

$$2\ H_2O(l) \rightleftharpoons H_3O^+(aq) + OH^-(aq)$$

This time, the OH^- ion concentration stays more or less constant as some of these ions combine with H_3O^+ ions to form H_2O molecules. When the reaction returns to equilibrium, the product of the H_3O^+ and OH^- ion concentrations is equal to K_w once again. We can therefore calculate the H_3O^+ ion concentration in this solution from the following equation.

$$[H_3O^+] = \frac{K_w}{[OH^-]} = \frac{1.0 \times 10^{-14}}{[0.0010]} = 1.0 \times 10^{-11} \, M$$

15.3 pH AND pOH

Adding an acid to water increases the H_3O^+ ion concentration and decreases the OH^- ion concentration. Adding a base does the opposite. Regardless of what is added to water, however, the product of the concentrations of these ions at equilibrium is always 1.0×10^{-14} at 25°C.

$$K_w = [H_3O^+][OH^-] = 1.0 \times 10^{-14}$$

Table 15.2 lists pairs of H_3O^+ and OH^- ion concentrations that can coexist at equilibrium in water at 25°C. The equilibrium concentrations of these ions vary over an enormous range (from roughly 1 mole per liter to 10^{-14} moles per liter). The factor of 10^{14} that separates one end of this range from the other is difficult, if not impossible, to conceptualize. The difference is comparable to the difference between the value of two pennies and the national debt of $2 trillion, or the difference between the radius of a gold atom and a distance of 8 miles.

Data from Table 15.2 are plotted in Figure 15.6 over a narrow range of concentrations between $1 \times 10^{-7} \, M$ and $1 \times 10^{-6} \, M$. The point at which the concentrations of the H_3O^+ and OH^- ions are equal is called the *neutral* point. Solutions in which the concentration of the H_3O^+ ion is larger than $1 \times 10^{-7} \, M$ are described as

FIG. 15.6 A graph of a small fraction of the data in Table 15.2. The vertical axis plots the H_3O^+ ion concentration at equilibrium over the narrow range of concentrations between 1×10^{-7} M and $1 \times 10^{-6} \, M$. The horizontal axis plots the equilibrium concentration of the OH^- ion over the same narrow range of concentrations. Any point that does not fall on the solid line represents a pair of H_3O^+ and OH^- ion concentrations for which the solution is not at equilibrium. When the concentrations of both ions are the same — 1.0×10^{-7} — the solution is said to be neutral. When the H_3O^+ ion concentration is larger than the OH^- ion concentration, the solution is said to be acidic. When there is more OH^- ion than H_3O^+ ion in this solution, it is said to be basic.

TABLE 15.2

Some Pairs of Equilibrium Concentrations of H_3O^+ and OH^- Ions That Can Coexist in Water

Concentration (mol/L)		
$[H_3O^+]$	$[OH^-]$	
1	1×10^{-14}	
1×10^{-1}	1×10^{-13}	
1×10^{-2}	1×10^{-12}	
1×10^{-3}	1×10^{-11}	Acidic solution
1×10^{-4}	1×10^{-10}	
1×10^{-5}	1×10^{-9}	
1×10^{-6}	1×10^{-8}	
1×10^{-7}	1×10^{-7}	Neutral solution
1×10^{-8}	1×10^{-6}	
1×10^{-9}	1×10^{-5}	
1×10^{-10}	1×10^{-4}	
1×10^{-11}	1×10^{-3}	Basic solution
1×10^{-12}	1×10^{-2}	
1×10^{-13}	1×10^{-1}	
1×10^{-14}	1	

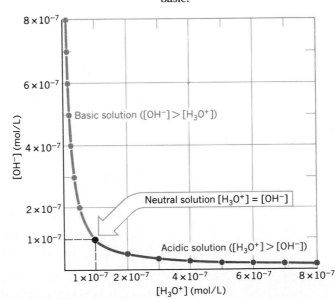

acidic. Those in which the concentration of the H_3O^+ ion is smaller than 1×10^{-7} *M* are *basic.*

There is no way to construct a plot similar to Figure 15.6 that includes all of the data from Table 15.2. If we tried to make the graph sensitive enough to reflect changes of 10^{-14} *M* in the H_3O^+ and OH^- ion concentrations, and if we used graph paper on which the scale divisions were 1 millimeter, each axis of the graph would have to be 62 million miles long — roughly two-thirds of the distance between the earth and the sun.

In 1909, the Danish biochemist S. P. L. Sørenson proposed a way around this problem. Sørenson worked at a laboratory set up by the Carlsberg brewery to apply scientific methods to the study of the fermentation reactions in brewing beer. Since the acidity of the fermentation malt influenced the behavior of the yeast responsible for this reaction, Sørenson spent a great deal of time measuring the effect of changes of acidity on the brewing process.

Faced with the task of constructing graphs of the activity of the malt versus the H_3O^+ ion concentration, Sørenson proposed condensing the range of H_3O^+ and OH^- ion concentrations to a more convenient scale by taking advantage of the mathematics of logarithms. By definition, the logarithm of a number is the power to which a base (such as 10) must be raised to obtain that number. The logarithm to the base 10 of 10^{-7}, for example, is -7.

$$\log(10^{-7}) = -7$$

Since the concentrations of the H_3O^+ and OH^- ions in aqueous solutions are usually smaller than 1 *M*, the logarithms of these concentrations are usually negative numbers. To convert the logarithms into positive numbers, which he considered more convenient, Sørenson suggested that the sign of the logarithm should be changed after it had been calculated. He introduced the symbol *p* to indicate the negative of the logarithm of a number. Thus, *pH* is the negative of the logarithm of the H_3O^+ ion concentration.

$$\mathbf{pH = -\log[H_3O^+]}$$

Similarly, *pOH* is the negative of the logarithm of the OH^- ion concentration.

$$\mathbf{pOH = -\log[OH^-]}$$

To show how the pH and pOH of an aqueous solution are related, we can start by taking the logarithm of both sides of the K_w expression given earlier.

$$[H_3O^+][OH^-] = 1.0 \times 10^{-14}$$
$$\log([H_3O^+][OH^-]) = \log(10^{-14})$$

The logarithm of the product of two numbers is equal to the sum of their logs. Thus, the sum of the logarithms of the H_3O^+ and OH^- ion concentrations is equal to the log of 10^{-14}.

$$\log[H_3O^+] + \log[OH^-] = -14$$

Both sides of this equation can now be multiplied by -1.

$$-\log[H_3O^+] - \log[OH^-] = 14$$

Substituting the definitions $pH = -\log(H_3O^+)$ and $pOH = -\log[OH^-]$ into this equation gives the following result.

$$\mathbf{pH + pOH = 14}$$

This equation can be used to convert from pH to pOH, or vice versa, for any

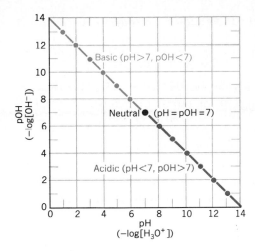

FIG. 15.7 We can fit the entire range of H_3O^+ and OH^- ion concentrations in Table 15.2 on a single graph by taking advantage of logarithmic mathematics. This figure shows a plot of pOH on the vertical axis versus pH on the horizontal axis. Any point that falls on the solid line corresponds to a solution at equilibrium. Any other point represents a system in which the H_3O^+ and OH^- ion concentrations are not in balance.

aqueous solution at 25°C, regardless of how much acid or base has been added to the solution.

By converting the H_3O^+ and OH^- ion concentrations in Table 15.2 into pH and pOH data, we can fit the entire range of concentrations onto a single graph, as shown in Figure 15.7.

15.4 ACID-DISSOCIATION EQUILIBRIUM CONSTANTS

One of the oldest systems for classifying chemical compounds divides them into acids and alkalies (or bases) on the basis of the properties of their aqueous solutions. We now know that acids such as hydrochloric acid (HCl), sulfuric acid (H_2SO_4), nitric acid (HNO_3), acetic acid (CH_3CO_2H), and citric acid ($C_6H_7O_8$) have one thing in common. They all react with water by donating an H^+ ion, or proton, to a neutral water molecule to form an H_3O^+ (hydronium) ion, as shown in the following examples.

$$HCl(aq) + H_2O(l) \longrightarrow H_3O^+(aq) + Cl^-(aq)$$
$$CH_3CO_2H(aq) + H_2O(l) \rightleftharpoons H_3O^+(aq) + CH_3CO_2^-(aq)$$

People who have had the misfortune of spilling hydrochloric acid or sulfuric acid on their clothes know that there is a big difference between these so-called strong acids and weak acids such as the acetic acid in vinegar or the citric acid in orange juice. But the difference between these acids is simply a difference of degree, as first noted in Section 11.9. Both hydrochloric acid and acetic acid satisfy the Brønsted definition of an acid—they are both H^+ ion, or proton, donors.

A strong acid is any substance that is good at donating an H^+ ion to water. In a 6 M solution of hydrochloric acid, for example, 99.996% of the HCl molecules dissociate when the following reaction comes to equilibrium. This equilibrium lies so far to the right that we write the equation for the reaction with a single arrow, suggesting that hydrochloric acid dissociates completely in aqueous solution.

$$HCl(aq) + H_2O(l) \longrightarrow H_3O^+(aq) + Cl^-(aq)$$

For all practical purposes, we can treat strong acids as if they were completely dissociated. Using the format introduced in Chapter 14, we can represent a 6.0 M solution of hydrochloric acid as follows.

$$HCl(aq) + H_2O(l) \longrightarrow H_3O^+(aq) + Cl^-(aq)$$

Initial:	6.0 M		
Equilibrium:	~ 0	$\sim 6.0\ M$	$\sim 6.0\ M$

Weak acids are compounds that are relatively poor H^+ ion donors. There is no doubt that acetic acid, for example, belongs in the family of Brønsted acids, because it can donate an H^+ ion to water.

$$CH_3CO_2H(aq) + H_2O(l) \rightleftharpoons H_3O^+(aq) + CH_3CO_2^-(aq)$$

But it isn't a very good H^+ ion donor. Only a small fraction of the acetic acid molecules in an aqueous solution actually lose a proton to water. In a 0.100 M solution, for example, only about 1.3% of the CH_3CO_2H molecules actually react with water.

$$CH_3CO_2H(aq) + H_2O(l) \rightleftharpoons H_3O^+(aq) + CH_3CO_2^-(aq)$$

Initial:	0.100 M		
Equilibrium:	0.099 M	0.0013 M	0.0013 M

It is easy to develop a qualitative feeling for what is meant by the terms *strong acid* and *weak acid*. Strong acids dissociate more or less completely in aqueous solution; weak acids dissociate only slightly. Only a few acids are strong. They include aqueous solutions of the following compound: HI, HBr, HCl, H_2SO_4, $HClO_4$, HNO_3, and H_2CrO_4. A few others are almost strong enough to fit this classification; they include $HClO_2$, and H_3PO_4. Most acids are categorized as weak.

A quantitative feeling for the difference between strong acids and weak acids can be obtained from the equilibrium constants for the reaction between acids and water. The equilibrium constant expression for the dissociation of acetic acid, for example, is written as follows.

$$K_c = \frac{[H_3O^+][CH_3CO_2^-]}{K[CH_3CO_2H][H_2O]}$$

Because it is somewhat time-consuming to write the formula CH_3CO_2H for acetic acid, chemists commonly abbreviate the formula as HOAc and describe the dissociation of the acid as follows.

$$HOAc(aq) + H_2O(l) \rightleftharpoons H_3O^+(aq) + OAc^-(aq)$$

Using this convention, we write the equilibrium constant expression for the dissociation of acetic acid as follows.

$$K_c = \frac{[H_3O^+][OAc^-]}{[HOAc][H_2O]}$$

Like the equilibrium constant expression for the dissociation of water, this is a perfectly legitimate equation. But most acids are weak, so the equilibrium concentration of H_2O is effectively the same after dissociation as before the acid was added. Since the [H_2O] term is essentially constant, we can build it into the equilibrium constant for the reaction as follows.

$$\frac{[H_3O^+]OAc^-]}{[HOAc]} = K_c \times [H_2O]$$

The equilibrium constant for this equation is known as the ***acid-dissociation equilibrium constant, K_a.***

$$K_a = \frac{[H_3O^+][OAc^-]}{[HOAc]}$$

TABLE 15.3

Common Acids and Their Acid-Dissociation Equilibrium Constants

Strong Acids	K_a	Intermediate Acids	K_a
hydroiodic acid (HI)	3×10^9	chlorous acid ($HClO_2$)	1.1×10^{-2}
hydrobromic acid (HBr)	1×10^9	phosphoric acid (H_3PO_4)	7.1×10^{-3}
hydrochloric acid (HCl)	1×10^6	Weak Acids	K_a
sulfuric acid (H_2SO_4)	1×10^3		
perchloric acid ($HClO_4$)	1×10^3	citric acid ($C_6H_7O_8$)	7.5×10^{-4}
nitric acid (HNO_3)	28	hydrofluoric acid (HF)	7.2×10^{-4}
chromic acid (H_2CrO_4)	10	formic acid (HCO_2H)	1.8×10^{-4}
		acetic acid (CH_3CO_2H)	1.8×10^{-5}
		boric acid (H_3BO_3)	5.8×10^{-10}

Values of K_a can be used to estimate the relative strengths of acids — the larger the value of K_a, the stronger the acid. By definition, a compound is classified as a strong acid when its value of K_a is larger than 1. Weak acids usually have K_a values that are very much smaller than 1. A list of common acids and their acid-dissociation equilibrium constants is given in Table 15.3. A more extensive list can be found in Table A-9 in the appendix.

There is a 10^{58}-fold difference between the values of K_a for the strongest (HI: $K_a = 3 \times 10^9$) and weakest ($CH_4: K_a = 10^{-49}$) Brønsted acids. To put this factor of 10^{58} into perspective, we might note that the radius of a typical atom (10^{-8} centimeter) and the distance from the earth to the portion of the universe that can be seen with the strongest telescopes (about 10 billion light-years) are separated by a factor of only 10^{36}. It shouldn't be surprising to find that we will have to develop slightly different techniques for handling calculations with strong acids such as hydrochloric acid, weak acids such as vinegar, and very weak acids such as dilute solutions of boric acid.

15.5 STRONG ACIDS

The simplest acid-base equilibria are those in which a strong acid (or base) is dissolved in water.

Calculate the pH of the solution formed when a single drop of 2 M hydrochloric acid is added to 100 milliliters of water.

The key to this calculation is remembering that HCl is a strong acid ($K_a = 1 \times 10^6$) and that strong acids can be assumed to dissociate completely in water.

$$HCl(aq) + H_2O(l) \longrightarrow H_3O^+(aq) + Cl^-(aq)$$

The H_3O^+ ion concentration at equilibrium is therefore essentially equal to the initial concentration of the acid.

A useful rule of thumb says that there are roughly 20 drops in each milliliter. One drop of 2 M HCl therefore has a volume of 5×10^{-5} liters.

One drop = 1/20 mL = 0.05 mL = 5×10^{-5} L

The number of moles of HCl added to the water can be calculated from the volume and concentration of the hydrochloric acid.

$$\frac{2 \text{ mol HCl}}{\cancel{L}} \times 5 \times 10^{-5} \cancel{L} = 1 \times 10^{-4} \text{ mol HCl}$$

The initial concentration of HCl is equal to the number of moles of HCl added to the beaker divided by the volume of water in the beaker.

$$\frac{1 \times 10^{-4} \text{ mol HCl}}{0.100 \text{ L}} = 1 \times 10^{-3} \text{ M HCl}$$

According to this calculation, the initial concentration of HCl is 1×10^{-3} M. If we assume that this acid dissociates completely, the H_3O^+ ion concentration at equilibrium is also 1×10^{-3} M.

$$HCl(aq) \ + H_2O(l) \longrightarrow \ H_3O^+(aq) \ + \ Cl^-(aq)$$

Initial:	1×10^{-3} M			
Equilibrium:	0		1×10^{-3} M	1×10^{-3} M

If the H_3O^+ ion concentration in this solution is 1×10^{-3} M, the pH of the solution is 3.

$$pH = -\log [H_3O^+]$$
$$= -\log [1 \times 10^{-3}] = -(-3) = 3$$

15.6 WEAK ACIDS

Equilibrium problems involving weak acids can usually be solved by applying the techniques developed in Chapter 14.

Calculate the H_3O^+, OAc$^-$, and HOAc concentrations at equilibrium in an 0.10 M solution of acetic acid in water.

We can start this calculation as we did the gas-phase equilibrium problems, by building a representation of what we know about the reaction.

$$HOAc(aq) + H_2O(l) \rightleftharpoons H_3O^+(aq) + OAc^-(aq) \quad K_a = 1.8 \times 10^{-5}$$

Initial:	0.10 M	0	0
Equilibrium:	?	?	?

We can then compare the initial reaction quotient (Q_a) with the equilibrium constant (K_a) for the reaction and conclude that the reaction must shift to the right to reach equilibrium.

$$Q_a = \frac{(H_3O^+)(OAc^-)}{(HOAc)} = \frac{(0)(0)}{(0.10)} = 0 < K_a$$

Recognizing that we get one H_3O^+ ion and one OAc$^-$ ion each time an HOAc dissociates allows us to write equations for the equilibrium concentrations of the three components of the reaction.

$$HOAc(aq) + H_2O(l) \rightleftharpoons H_3O^+(aq) + OAc^-(aq) \quad K_a = 1.8 \times 10^{-5}$$

Initial:	0.10 M	0	0
Equilibrium:	$0.10 - \Delta$	Δ	Δ

Substituting what we know into the K_a expression gives the following equation.

$$K_a = \frac{[H_3O^+][OAc^-]}{[HOAc]} = \frac{[\Delta][\Delta]}{[0.10 - \Delta]} = 1.8 \times 10^{-5}$$

Although we could rearrange this equation and then solve it with the quadratic formula, it is tempting first to test the assumption that Δ is small compared with the initial concentration of acetic acid.

$$\frac{[\Delta][\Delta]}{[0.10]} \cong 1.8 \times 10^{-5}$$

Solving this approximate equation gives the following value for Δ.

$$\Delta \cong 0.0013 \ M$$

Is Δ small enough to be ignored in this problem? Yes — it is less than 5% of the initial concentration of acetic acid.

$$\frac{0.0013}{0.10} \times 100\% = \mathbf{1.3\%}$$

We can therefore use this value of Δ to calculate the equilibrium concentrations of H_3O^+, OAc^-, and HOAc.

$$[HOAc] = 0.10 - \Delta = 0.10 \ M$$
$$[H_3O^+] = \Delta = 0.0013 \ M$$
$$[OAc^-] = \Delta = 0.0013 \ M$$

We can test these results by substituting them back into the expression for K_a.

$$K_a = \frac{[H_3O^+][OAc^-]}{[HOAc]} = \frac{[0.0013][0.0013]}{[0.10]} = \mathbf{1.7 \times 10^{-5}}$$

Within experimental error, the results of this calculation agree with the value of K_a for acetic acid.

15.7 HIDDEN ASSUMPTIONS IN WEAK-ACID CALCULATIONS

How many assumptions were made in the calculation in Section 15.6? It may have appeared that we made one assumption — that Δ is small compared with the initial concentration of HOAc. In fact, two assumptions were made. The other assumption was hidden in the way the problem was set up.

$$HOAc(aq) + H_2O(l) \rightleftharpoons H_3O^+(aq) + OAc^-(aq) \quad K_a = 1.8 \times 10^{-5}$$

Initial:	0.10 M	0	0
Equilibrium:	0.10 − Δ	Δ	Δ

The amount of H_3O^+ ion in water is so small that it is common to assume that the initial concentration of this ion in an acid-dissociation equilibrium calculation is zero, which isn't quite true.

There are actually two sources of the H_3O^+ ion in this solution. We get H_3O^+ ions from the dissociation of acetic acid.

$$HOAc(aq) + H_2O(l) \rightleftharpoons H_3O^+(aq) + OAc^-(aq)$$

But we also get some H_3O^+ ions from the dissociation of water.

$$2 \ H_2O(l) \rightleftharpoons H_3O^+(aq) + OH^-(aq)$$

The initial concentration of the H_3O^+ ion is therefore not quite zero — it is $1.0 \times 10^{-7} \ M$ at room temperature.

Before we can trust the results of the calculation in Section 15.6, we have to check both of the assumptions made in this calculation.

1. **The assumption that the amount of acid that dissociates is small compared with the initial concentration of the acid.**
2. **The assumption that enough acid dissociates to allow us to ignore the dissociation of water.**

We have already confirmed the validity of the first assumption. Only 1.3% of the acetic acid molecules dissociate in this solution.

$$\frac{[OAc^-]}{[HOAc]} \times 100\% = \frac{[0.0013]}{[0.10]} \times 100\% = 1.3\%$$

We can now check the second assumption. According to the calculation in Section 15.6, the concentration of the H_3O^+ ion from the dissociation of acetic acid is 0.0013 M. The maximum concentration of the H_3O^+ ion from the dissociation of water is 1.0×10^{-7} M. The H_3O^+ ion concentration from dissociation of HOAc is therefore 13,000 times the H_3O^+ ion concentration in pure water.

$$\frac{0.0013}{1.0 \times 10^{-7}} = 13,000$$

The H_3O^+ ion concentration from the dissociation of water in this solution is actually smaller than the H_3O^+ ion concentration in pure water because of LeChatelier's principle. As noted in Section 15.2, the H_3O^+ ion concentration generated by the dissociation of acetic acid suppresses the dissociation of water. Thus, the second assumption is valid in this calculation. As a later section will show, this assumption only fails for dilute solutions of very weak acids.

IMPLICATIONS OF THE ASSUMPTIONS
15.8 IN WEAK-ACID CALCULATIONS

The two assumptions that are made in weak-acid equilibrium problems can be restated as follows.

1. **The amount of dissociation of the acid is small enough that the change in the concentration of the acid as the reaction comes to equilibrium can be ignored.**
2. **The amount of dissociation of the acid is large enough that the H_3O^+ ion concentration from the dissociation of water can be ignored.**

In other words, the acid must be *weak* enough that Δ is small compared with the initial concentration of the acid. But it must also be *strong* enough that the H_3O^+ ions from the acid overwhelm the contribution to the H_3O^+ ion concentration from the dissociation of water. This invokes an image out of the fairy tale concerning Goldilocks and the three bears. In order for the approach taken to the calculation for acetic acid to work, the acid has to be "just right." If it's too strong, Δ won't be small enough to be ignored. If it's too weak, the dissociation of water will have to be included in the calculation.

Fortunately, many acids are "just right." To illustrate this point, the next section will use both assumptions in a series of calculations designed to identify the factors that influence the H_3O^+ ion concentration in aqueous solutions of weak acids.

15.9 FACTORS THAT INFLUENCE THE H$_3$O$^+$ ION CONCENTRATION IN WEAK-ACID SOLUTIONS

It seems reasonable to expect that the H$_3$O$^+$ ion concentration depends on the value of the acid-dissociation equilibrium constant for the acid.

Exercise 15.3

Calculate the H$_3$O$^+$ ion concentration and the pH of 0.10 M solutions of the following acids.

    (a) Hypochlorous acid, HOCl $K_a = 2.9 \times 10^{-8}$
    (b) Hypobromous acid, HOBr $K_a = 2.4 \times 10^{-9}$
    (c) Hypoiodous acid, HOI $K_a = 2.3 \times 10^{-11}$

Solution

The first calculation can be set up as follows.

$$HOCl(aq) + H_2O(l) \rightleftharpoons H_3O^+(aq) + OCl^-(aq) \quad K_a = 2.9 \times 10^{-8}$$

Initial: 0.10 M 0   0
Equilibrium: 0.10 $-$ Δ Δ Δ

Substituting this information into the K_a expression gives the following equation.

$$K_a = \frac{[H_3O^+][OCl^-]}{[HOCl]} = \frac{[\Delta][\Delta]}{[0.10 - \Delta]} = 2.9 \times 10^{-8}$$

We can now try the assumption that Δ is small compared with the initial concentration of the acid.

$$\frac{[\Delta][\Delta]}{[0.10]} \cong 2.9 \times 10^{-8}$$

Solving this equation for Δ gives the following result.

$$\Delta \cong 0.000054\ M = 5.4 \times 10^{-5}\ M$$

Both assumptions made in this calculation are legitimate: Δ is small compared with the initial concentration of HOCl, but it is more than 500 times the H$_3$O$^+$ ion concentration in pure water. We can therefore use this value of Δ to calculate the pH of the solution.

$$[H_3O^+] = \Delta = 5.4 \times 10^{-5}\ M$$
$$pH = -\log [H_3O^+] = -\log [5.4 \times 10^{-5}] = 4.27$$

    Combining the results of this calculation with similar calculations for hypobromous and hypoiodous acid gives the following results.

        HOCl: $[H_3O^+] = 5.4 \times 10^{-5}\ M$ pH $= 4.27$
        HOBr: $[H_3O^+] = 1.5 \times 10^{-5}\ M$ pH $= 4.82$
        HOI: $[H_3O^+] = 1.5 \times 10^{-6}\ M$ pH $= 5.82$

    As expected, the H$_3$O$^+$ ion concentration at equilibrium—and therefore the pH of the solution—depends on the value of K_a for the acid. The H$_3$O$^+$ ion concentration becomes smaller and the pH of the solution becomes larger as the value of K_a becomes smaller. The next exercise shows that the H_3O^+ ion concentration at equilibrium also depends on the initial concentration of the acid.

Exercise 15.4

Calculate the H_3O^+ ion concentration and the pH of acetic acid solutions with the following concentrations.

(a) $0.10\ M$ (b) $0.010\ M$ (c) $0.0010\ M$ (d) $0.00010\ M$

Solution

Any one of these calculations can be set up as follows.

$$\text{HOAc}(aq) + H_2O(l)\ \ H_3O^+(aq) + \text{OAc}^-(aq)\qquad K_a = 1.8 \times 10^{-5}$$

Initial:	$0.10\ M$	0	0
Equilibrium:	$0.10 - \Delta$	Δ	Δ

Substituting this information into the K_a expression gives the following equation.

$$K_a = \frac{[H_3O^+][\text{OAc}^-]}{[\text{HOAc}]} = \frac{[\Delta][\Delta]}{[0.10 - \Delta]} = 1.8 \times 10^{-5}$$

Having performed this calculation earlier, we know that Δ is small compared with the initial concentration of acetic acid.

$$\frac{[\Delta][\Delta]}{[0.10]} \cong 1.8 \times 10^{-5}$$

Solving this equation for Δ gives the following result.

$$\Delta \cong 0.0013\ M = 1.3 \times 10^{-3}\ M$$

Since both assumptions are legitimate for this calculation, we can use this value of Δ to calculate the pH of the solution.

$$[H_3O^+] = \Delta = 1.3 \times 10^{-3}\ M$$
$$\text{pH} = -\log [H_3O^+] = \mathbf{2.89}$$

Repeating this calculation for the different initial concentrations gives the following results.

$0.10\ M$ HOAc:	$[H_3O^+] = 1.3 \times 10^{-3}\ M$	pH $= 2.9$
$0.010\ M$ HOAc:	$[H_3O^+] = 4.2 \times 10^{-4}\ M$	pH $= 3.4$
$0.0010\ M$ HOAc:	$[H_3O^+] = 1.3 \times 10^{-4}\ M$	pH $= 3.9$
$0.00010\ M$ HOAc:	$[H_3O^+] = 4.2 \times 10^{-5}\ M$	pH $= 4.4$

The concentration of the H_3O^+ ion in an aqueous solution gradually becomes smaller and the pH of the solution becomes larger as the solution becomes more dilute. The results of Exercises 15.3 and 15.4 provide a basis for constructing a model that allows us to predict whether we can legitimately ignore the dissociation of water when we work equilibrium problems involving weak acids. Two factors must be built into this model: (1) the strength of the acid as reflected by the value of K_a and (2) the strength of the solution as reflected by the initial concentration of the acid.

15.10 WHEN CAN THE DISSOCIATION OF WATER BE IGNORED?

In Chapter 14 we introduced the following rule of thumb.

The value of Δ can be ignored when it is less than 5% of the initial concentration to which it is added or from which it is subtracted.

The same rule of thumb can be applied to the question of whether the dissociation of water can be legitimately ignored.

The dissociation of water can be ignored when the H_3O^+ ion concentration from this reaction is less than 5% of the total H_3O^+ concentration.

At first glance this might lead us to question the validity of ignoring the dissociation of water in the last calculation in Exercise 15.3. In that problem, the H_3O^+ ion concentration from the dissociation of the acid was found to be 1.5×10^{-6} M. If the H_3O^+ ion concentration from the dissociation of water is 1.0×10^{-7} M, it might seem that water contributes more than 5% to the total H_3O^+ ion concentration.

$$\frac{1.0 \times 10^{-7}}{1.5 \times 10^{-6}} \times 100\% = 6.7\%$$

This percentage, however, fails to take into account LeChatelier's principle.

Remember that there are two sources of the H_3O^+ ion in this solution.

$$HOI(aq) + H_2O(l) \rightleftharpoons H_3O^+(aq) + OI^-(aq)$$
$$2\,H_2O(l) \rightleftharpoons H_3O^+(aq) + OH^-(aq)$$

LeChatelier's principle suggests that adding an acid to water drives the equilibrium between water and its ions back toward neutral H_2O molecules. In other words, adding an acid reduces the amount of water that dissociates.

If we look at the reaction for the dissociation of water

$$2\,H_2O(l) \rightleftharpoons H_3O^+(aq) + OH^-(aq)$$

it becomes clear that the amount of H_3O^+ ion from the dissociation of water is always equal to the amount of OH^- ion from this reaction. We can represent this statement with the following equation.

$$[H_3O^+]_w = [OH^-]_w$$

The total H_3O^+ ion concentration in an acid solution is equal to the sum of the H_3O^+ ion concentrations from the two sources of this ion.

$$[H_3O^+]_t = [H_3O^+]_a + [H_3O^+]_w$$

As long as the H_3O^+ ion from the dissociation of water is less than 5% of the total, it can be ignored. It is therefore useful to examine a solution in which the dissociation of water contributes exactly 5% to the total H_3O^+ ion concentration.

$$[H_3O^+]_w = 0.05\,[H_3O^+]_t$$

Remember that this solution contains two sources of the H_3O^+ ion but only one source of the OH^- ion — water. The product of the total H_3O^+ ion concentration times the OH^- ion concentration from the dissociation of water is therefore equal to K_w.

$$[H_3O^+]_t[OH^-]_w = 1.0 \times 10^{-14}$$

But the H_3O^+ and OH^- ion concentrations from the dissociation of water are equal.

$$[H_3O^+]_w = [OH^-]_w$$

The OH^- ion concentration from water is therefore also equal to 5% of the total H_3O^+ ion concentration.

$$[OH^-]_w = 0.05 \, [H_3O^+]_t$$

Substituting this equation into the K_w expression gives the following result.

$$[H_3O^+]_t (0.05 \, [H_3O^+]_t) = 1.0 \times 10^{-14}$$

This equation can now be solved for the total H_3O^+ ion concentration.

$$[H_3O^+]_t = 4.5 \times 10^{-7} \, M$$

According to this calculation, the dissociation of water can be ignored whenever the total H_3O^+ ion concentration is larger than 4.5×10^{-7} M. In the calculation for hypoiodous acid in Exercise 15.3, the concentration of the H_3O^+ ion from dissociation of HOI was 1.5×10^{-6}. The total H_3O^+ ion concentration is therefore large enough that the dissociation of water can be ignored. Thus, the results of this calculation are legitimate.

We can summarize what has been said about weak acids in water in the following instructions.

1. **Perform the calculations as illustrated above, assuming that Δ is much less than the initial concentration of the acid and that the amount of H_3O^+ from the dissociation of water can be ignored.**

2. **Conclude that the calculation is correct if Δ is less than 5% of the initial concentration of the acid and the H_3O^+ ion concentration is larger than 4.5×10^{-7} M.**

Calculations for weak acids when both assumptions are valid can be generalized as follows. Assume that we have a generic acid — HA — for which the initial concentration is C_a.

$$HA(aq) + H_2O(l) \rightleftharpoons H_3O^+(aq) + A^-(aq)$$

Initial:	C_a	0	0
Equilibrium:	$C_a - \Delta$	Δ	Δ

The equilibrium constant expression for this reaction is written as follows.

$$K_a = \frac{[H_3O^+][A^-]}{[HA]} = \frac{[\Delta][\Delta]}{[C_a - \Delta]}$$

If we assume that Δ is small compared with the initial concentration of the acid, we get the following approximate equation,

$$K_a \cong \frac{[\Delta][\Delta]}{[C_a]}$$

which can be rearranged as follows.

$$\Delta \cong \sqrt{K_a C_a}$$

To use this equation to repeat the first calculation in Exercise 15.4, we substitute the values of K_a and the initial concentration of acid.

$$\Delta \cong \sqrt{K_a C_a} = \sqrt{(1.8 \times 10^{-5})(0.10)} = 1.3 \times 10^{-3}$$

Remembering that the H_3O^+ ion concentration was defined as Δ gives a result that is identical to the one obtained in Exercise 15.4.

$$[H_3O^+] = 1.3 \times 10^{-3}$$

We now have techniques for solving equilibrium problems that involve strong acids or that involve weak acids for which both our assumptions are legitimate. Next, we need to develop techniques to handle problems for which one or the other of our assumptions is not valid.

15.11 NOT-SO-WEAK ACIDS

Let's start with acid solutions that aren't weak enough to allow us to assume that Δ is negligible.

Calculate the H_3O^+, $HClO_2$, and ClO_2^- concentrations at equlibrium in a 0.100 M solution of chlorous acid ($K_a = 1.1 \times 10^{-2}$).

$$HClO_2(aq) + H_2O(l) \rightleftharpoons H_3O^+(aq) + ClO_2^-(aq)$$

The first step involves building a model of the problem.

$$HClO_2(aq) + H_2O(l) \rightleftharpoons H_3O^+(aq) + ClO_2^-(aq) \quad K_a = 1.1 \times 10^{-2}$$

Initial: 0.100 M 0   0
Equilibrium: 0.100 − Δ Δ Δ

We then substitute this information into the K_a expression.

$$K_a = \frac{[H_3O^+][ClO_2^-]}{[HClO_2]} = \frac{[\Delta][\Delta]}{[0.100 - \Delta]} = 1.1 \times 10^{-2}$$

There is nothing wrong with trying the assumption that Δ is small compared with the initial concentration of the acid—even if we suspect it is wrong.

$$\frac{[\Delta][\Delta]}{[0.100]} \cong 1.1 \times 10^{-2}$$

Solving this approximate equation gives a value for Δ that is 33% of the initial concentration of chlorous acid.

$$\Delta \cong 0.033 \ M$$

The assumption that Δ is small therefore fails miserably.

There are two ways out of this difficulty. We can expand the original equation and solve it by the quadratic formula.

$$\frac{[\Delta][\Delta]}{[0.100 - \Delta]} = 1.1 \times 10^{-2}$$

Or we can plug the first approximation of the value of Δ into this equation and use successive approximations (see Section 14.13) to solve the problem.

Either technique gives the following value of Δ.

$$\Delta = 0.028 \ M$$

Using this value of Δ gives the following results.

$$[HClO_2] = 0.100 - \Delta = 0.072 \ M$$
$$[ClO_2^-] = \Delta = 0.028 \ M$$
$$[H_3O^+] = \Delta = 0.028 \ M$$

Substituting these values back into the equilibrium expression gives a value of K_a that agrees with the value given in the problem, within experimental error.

$$K_a = \frac{[H_3O^+][ClO_2^-]}{[HClO_2]} = \frac{[0.028][0.028]}{[0.072]} = 1.1 \times 10^{-2}$$

Chlorous acid does not belong among the class of strong acids that dissociate more or less completely. Nor does it really fit in the category of weak acids, which dissociate only to a negligible extent. Since the amount of dissociation in this solution is about 28%, it might be classified as a not-so-weak acid.

15.12 VERY WEAK ACIDS

It is more difficult to solve equilibrium problems when the acid is too weak to allow us to ignore the dissociation of water. Deriving an equation that can be used to solve this class of problems is therefore easier than solving them one at a time. To derive such an equation, we start by assuming that we have a generic acid, HA, that dissolves in water. We have two sources of the H_3O^+ ion in this solution.

$$HA(aq) + H_2O(l) \rightleftharpoons H_3O^+(aq) + A^-(aq)$$
$$2 \ H_2O(l) \rightleftharpoons H_3O^+(aq) \times OH^-(aq)$$

Thus, the total H_3O^+ ion concentration is the sum of the H_3O^+ ion concentrations from the acid and the water.

$$[H_3O^+]_t = [H_3O^+]_a + [H_3O^+]_w$$

The only source of the OH^- ion in this solution is the dissociation of water. The product of the total H_3O^+ ion concentration times the OH^- ion concentration from the dissociation of water is therefore equal to K_w.

$$[H_3O^+]_t[OH^-]_w = 1.0 \times 10^{-14}$$

Because the dissociation reactions produce an H_3O^+ ion for each A^- ion and an H_3O^+ ion for each OH^- ion, the total H_3O^+ ion concentration is equal to the sum of the A^- and OH^- ion concentrations.

$$[H_3O^+]_t = [A^-] + [OH^-]_w$$

This last equation simply states that the sum of the positive ions formed by the dissociation of the acid and water is equal to the sum of the negative ions produced by these reactions.

Let's rearrange this last equation.

$$[A^-] = [H_3O^+]_t - [OH^-]_w$$

We can now substitute the relationship between the OH^- and total H_3O^+ ion concentrations

$$[OH^-]_w = \frac{K_w}{[H_3O^+]_t}$$

for the last term in this equation.

$$[A^-] = [H_3O^+]_t - \frac{K_w}{[H_3O^+]_t}$$

The acid-dissociation equilibrium constant expression for the acid is written as follows.

$$K_a = \frac{[H_3O^+]_t[A^-]}{[HA]}$$

This equation reminds us that the dissociation of the acid is influenced by the total H_3O^+ ion concentration, not just the H_3O^+ ion from the acid. Let's now substitute the equation for the A^- ion concentration into this K_a expression.

$$K_a = \frac{[H_3O^+]_t([H_3O^+]_t - K_w/[H_3O^+]_t)}{[HA]}$$

Combining terms gives the following form of the equation.

$$K_a[HA] = [H_3O^+]_t^2 - K_w$$

Rearranging this equation and taking the square root of both sides gives the following result.

$$[H_3O^+]_t = \sqrt{K_a[HA] + K_w}$$

We can generate a more useful version of this equation by introducing an assumption. Let's assume that HA is a weak acid—so weak that Δ is small compared with the initial concentration of the acid. This is a reasonable assumption, because we are trying to generate an equation that can be used to solve equilibrium problems that involve acids that are so weak we can't ignore the dissociation of water.

By convention, the symbol used to represent the initial concentration of the acid is C_a. If Δ is small compared with the initial concentration of the acid, then the concentration of HA when this reaction reaches equilibrium will be virtually the same as the initial concentration.

$$[HA] \cong C_a$$

Substituting this approximation into the equation derived in this section gives the following result.

$$[H_3O^+]_t = \sqrt{K_aC_a + K_w}$$

Exercise 15.5

Calculate the H_3O^+ concentration in a 0.0001 M solution of hydrocyanic acid (HCN).

$$HCN(aq) + H_2O(l) \rightleftharpoons H_3O^+(aq) + CN^-(aq) \qquad K_a = 6 \times 10^{-10}$$

Solution

This is a dilute solution of a fairly weak acid. It is therefore the kind of solution for which the dissociation of water is likely to make an important contribution to the total H_3O^+ concentration. We know K_a for the acid, as well as the initial concentration. We can therefore use the equation derived earlier to calculate the total H_3O^+ ion concentration from both the acid and water.

$$[H_3O^+]_t = \sqrt{K_a C_a + K_w}$$
$$= \sqrt{(6 \times 10^{-10})(1 \times 10^{-4}) + (1 \times 10^{-14})}$$
$$= 3 \times 10^{-7}\ M$$

15.13 SUMMARIZING THE CHEMISTRY OF WEAK ACIDS

This section compares the way in which the H_3O^+ concentration is calculated for pure water, a weak acid, a not-so-weak acid, and a very weak acid.

PURE WATER

The product of the concentrations of the H_3O^+ and OH^- ions in pure water is equal to K_w.

$$[H_3O^+][OH^-] = K_w$$

But the H_3O^+ and OH^- ion concentrations in pure water are the same.

$$[H_3O^+] = [OH^-]$$

Substituting the second equation into the first gives the following result.

$$[H_3O^+]^2 = K_w$$

The H_3O^+ ion concentration in pure water is therefore equal to the square root of K_w.

$$\mathbf{[H_3O^+] = \sqrt{K_w}}$$

WEAK ACIDS

The generic equilibrium constant expression for a weak acid is written as follows.

$$K_a = \frac{[H_3O^+][A^-]}{[HA]}$$

If the acid is strong enough to allow us to ignore the dissociation of water, the H_3O^+ ion and A^- ion concentrations in this solution are about equal.

$$[H_3O^+] \cong [A^-]$$

Substituting this information into the acid-dissociation equilibrium constant expression gives the following result.

$$K_a = \frac{[H_3O^+]^2}{[HA]}$$

The concentration of HA molecules at equilibrium is equal to the initial concentration of the acid minus the amount that dissociates — Δ.

$$K_a = \frac{[H_3O^+]^2}{[C_a - \Delta]}$$

If the acid is weak enough that Δ is small compared with the initial concentration of the acid, we get the following approximate equation.

$$K_a = \frac{[H_3O^+]^2}{C_a}$$

Rearranging this equation and taking the square root of both sides gives the following result.

$$[H_3O^+] = \sqrt{K_a C_a}$$

NOT-SO-WEAK ACIDS

When a not-so-weak acid dissociates in water, the change in the concentration of the HA molecules as the reaction comes to equilibrium is too large compared with the initial concentration of the acid to be ignored. For these acids, the following equation must be solved with the quadratic formula or by successive approximations.

$$K_a = \frac{[H_3O^+]^2}{[C_a - \Delta]}$$

VERY WEAK ACIDS

When the acid is so weak that we can't ignore the dissociation of water, we use the following equation to calculate the concentration of the H$_3$O$^+$ ion at equilibrium.

$$[H_3O^+] = \sqrt{K_a C_a + K_w}$$

It is useful to compare the four equations used to calculate the H$_3$O$^+$ ion concentration in these solutions.

Pure water: $\quad[H_3O^+] = \sqrt{K_w}$

Weak acid: $\quad[H_3O^+] = \sqrt{K_a C_a}$

Not-so-weak acid: $\quad[H_3O^+] = \sqrt{K_a(C_a - \Delta)}$

Very weak acid: $\quad[H_3O^+] = \sqrt{K_a C_a + K_w}$

The first three equations are special cases of the fourth. When we can ignore the dissociation of the acid — because there is no acid in the solution — we get the first equation. When we can ignore the dissociation of water, we get the second equation. When we can ignore the dissociation of water but not the amount of acid that dissociates, we get the third equation. When we can't ignore the dissociation of either the acid or water, we have to use the last equation.

This discussion gives us a basis for deciding when we can ignore the dissociation of water. Remember our rule of thumb — we can ignore anything that makes a contribution of less than 5% to the total. Now compare the most inclusive equation for the H$_3$O$^+$ ion concentration

$$[H_3O^+] = \sqrt{K_a C_a + K_w}$$

with the equation that assumes the dissociation of water can be ignored.

$$[H_3O^+] = \sqrt{K_a C_a}$$

The only difference is the K_w term, which is under the square root sign. If the H$_3$O$^+$ ion concentration were directly proportional to the sum of $K_a C_a$ plus K_w, we could assume that the error would be less than 5% when $K_a C_a$ was more than 20 times larger than K_w. This is not a linear relationship, however. Because the H$_3$O$^+$ ion concentration is proportional to the square root of $K_a C_a$ plus K_w, it can be shown that $K_a C_a$ only has to be 10 times larger than K_w to give an error of less than 5% in the concentration of the H$_3$O$^+$ ion. Stated differently, this means that as long as

K_aC_a is at least 10 times larger than K_w, the dissociation of water can be ignored. This is the basis for the following rule.

> When K_aC_a for a weak acid is *larger* than 1.0×10^{-13}, we can ignore the dissociation of water. When K_aC_a is *smaller* than 1.0×10^{-13}, the dissociation of water must be included in the calculation.

Exercise 15.6

Calculate the pH of 0.023 M solution of saccharin (HSc, or C_7HNO_3S), assuming that K_a is 2.1×10^{-12} for this artificial sweetener.

Solution

Can we ignore the dissociation of water in this solution?

$$K_aC_a = (2.1 \times 10^{-12})(0.023) = 4.8 \times 10^{-14} < 1 \times 10^{-13}$$

No. The product of K_a times the initial concentration is too small for the dissociation of water to be ignored. The H_3O^+ ion concentration in this solution has to be calculated from the following equation.

$$[H_3O^+] = \sqrt{K_aC_a + K_w}$$
$$= \sqrt{(2.1 \times 10^{-12})(0.023) + 1.0 \times 10^{-14}} = 2.4 \times 10^{-7} \, M$$

The pH of the solution is therefore 6.62.

$$pH = -\log[H_3O^+] = 6.62$$

15.14 BASES

With a few slight modifications, the techniques applied to equilibrium calculations for acids are also valid for solutions of bases in water.

> **Calculate the pH of a 0.100 M aqueous solution of ammonia, assuming that NH_3 acts as a base toward water.**

We start by writing an equation for the reaction between ammonia and water.

$$NH_3(aq) + H_2O(l) \rightleftharpoons NH_4^+(aq) + OH^-(aq)$$

Strict adherence to the rules for writing equilibrium constant expressions leads to the following equation for this reaction.

$$K_c = \frac{[NH_4^+][OH^-]}{[NH_3][H_2O]}$$

But, taking a lesson from our experience with acid-dissociation equilibria, we can assume that the concentration of water in this solution is more or less constant and build the $[H_2O]$ term into the value of the equilibrium constant. Reactions between a base and water are described in terms of a *base-ionization equilibrium constant, K_b*.

$$K_b = \frac{[NH_4^+][OH^-]}{[NH_3]} = K_c \times [H_2O]$$

The values of K_b for a limited number of bases are given in Table A-10 in the appendix. For NH_3, K_b is 1.8×10^{-5}.

We can organize what we know about this equilibrium by using the format we used for equilibria involving acids.

$$NH_3(aq) + H_2O(l) \rightleftharpoons NH_4^+(aq) + OH^-(aq) \quad K_b = 1.8 \times 10^{-5}$$

Initial:	0.10 M	0	0
Equilibrium:	0.10 − Δ	Δ	Δ

Substituting this information into the equilibrium constant expression gives the following equation.

$$K_b = \frac{[NH_4^+][OH^-]}{[NH_3]} = \frac{[\Delta][\Delta]}{[0.10 - \Delta]} = 1.8 \times 10^{-5}$$

K_b for ammonia is small enough to allow us to consider the following approximation.

$$\frac{[\Delta][\Delta]}{[0.10]} \cong 1.8 \times 10^{-5}$$

Solving this approximate equation gives the following result.

$$\Delta \cong 1.3 \times 10^{-3}$$

This value of Δ is small enough compared with the initial concentration of NH_3 to be ignored and yet large enough compared with the OH^- ion concentration in water for the dissociation of water to be ignored. It can therefore be used to calculate the pOH of the solution.

$$pOH = -\log (1.3 \times 10^{-3}) = \mathbf{2.89}$$

This in turn can be used to calculate the pH of the solution.

$$pH = 14 - pOH = \mathbf{11.11}$$

Equilibrium problems involving bases are relatively easy to solve if we know the value of K_b for the base. Values of K_b are listed in Table A-10 for only a limited number of compounds, however. The first step in many base equilibrium calculations therefore involves determining the value of K_b for the reaction from the value of K_a for the conjugate acid (see Section 11.8). Consider the following problem, for example.

Calculate the pH of a 0.030 M solution of sodium benzoate ($C_6H_5CO_2Na$) in water if the value of K_a for benzoic acid ($C_6H_5CO_2H$) is 6.3×10^{-5}.

Benzoic acid and sodium benzoate are members of a family of food preservatives whose ability to retard the rate at which food spoils has helped produce a 10-fold reduction in the incidence of stomach cancer. To save time and space, we'll abbreviate benzoic acid as HOBz and sodium benzoate as NaOBz. Benzoic acid, as its name implies, is an acid. Sodium benzoate is a salt of the conjugate base, the OBz^- or benzoate ion.

Salts dissociate into their ions when they dissolve in water.

$$NaOBz(s) \xrightarrow{H_2O} Na^+(aq) + OBz^-(aq)$$

The benzoate ion produced in this reaction acts as a base toward water, picking up a proton to form the conjugate acid and a hydroxide ion.

$$OBz^-(aq) + H_2O(l) \rightleftharpoons HOBz(aq) + OH^-(aq)$$

The base-ionization equilibrium constant expression for this reaction is therefore written as follows.

$$K_b = \frac{[\text{HOBz}][\text{OH}^-]}{[\text{OBz}^-]}$$

Although the value of K_b for this base isn't given in the statement of the problem, we are given the value of K_a for benzoic acid. The next step in solving the problem therefore involves calculating the value of K_b for the OBz$^-$ ion from the value of K_a for HOBz.

Compare the K_a and K_b expressions for benzoic acid and its conjugate base.

$$K_a = \frac{[\text{H}_3\text{O}^+][\text{OBz}^-]}{[\text{HOBz}]} \qquad K_b = \frac{[\text{HOBz}][\text{OH}^-]}{[\text{OBz}^-]}$$

Both equations contain the ratio of the equilibrium concentrations of the acid and its conjugate base. K_a is proportional to [OBz$^-$] divided by [HOBz], and K_b is proportional to [HOBz] divided by [OBz$^-$].

It is therefore possible to calculate K_b from K_a, or vice versa. If we start with K_a, our goals are simple—remove the [H$_3$O$^+$] term and introduce the [OH$^-$] term. We can do this by multiplying the top and bottom of the K_a expression by the OH$^-$ ion concentration.

$$K_a = \frac{[\text{H}_3\text{O}^+][\text{OBz}^-]}{[\text{HOBz}]} \times \frac{[\text{OH}^-]}{[\text{OH}^-]}$$

Rearranging this equation gives the following result.

$$K_a = \frac{[\text{OBz}^-]}{[\text{HOBz}][\text{OH}^-]} \times [\text{H}_3\text{O}^+][\text{OH}^-]$$

The two terms in this equation should look familiar. The first is the inverse of the K_b expression, and the second is the expression for K_w.

$$K_a = \frac{1}{K_b} \times K_w$$

Multiplying both sides of this equation by K_b gives the following result.

$$K_a K_b = K_w$$

The value of K_b for the reaction between the benzoate ion and water can therefore be calculated from K_a for benzoic acid.

$$K_b = \frac{K_w}{K_a} = \frac{1.0 \times 10^{-14}}{6.3 \times 10^{-5}} = 1.6 \times 10^{-10}$$

Now that we know K_b for the benzoate ion, we can solve this problem with the techniques used to handle weak-acid equilibria. We start by building a representation for the problem.

$$\text{OBz}^-(aq) + \text{H}_2\text{O}(l) \rightleftharpoons \text{HOBz}(aq) + \text{OH}^-(aq) \quad K_b = 1.6 \times 10^{-10}$$

Initial:	0.030 M	0	0
Equilibrium:	0.030 $- \Delta$	Δ	Δ

We then substitute this information into the K_b expression.

$$K_b = \frac{[\text{HOBz}][\text{OH}^-]}{[\text{OBz}^-]} = \frac{[\Delta][\Delta]}{[0.030 - \Delta]} = 1.6 \times 10^{-10}$$

K_b is relatively small, so we can assume that very little of the benzoate ion will actually react with water to form benzoic acid. We can therefore assume that Δ is small compared with 0.030.

$$\frac{[\Delta][\Delta]}{[0.030]} \cong 1.6 \times 10^{-10}$$

Solving this approximate equation for Δ gives the following result.

$$\Delta \cong 2.2 \times 10^{-6} \, M$$

It's obvious that the assumption that Δ is small is valid. We can therefore use Δ to calculate the pOH of the solution.

$$[OH^-] = \Delta = 2.2 \times 10^{-6} \, M$$
$$pOH = -\log [OH^-] = 5.66$$

The problem asked for the pH of the solution, however, so we use the relationship between pH and pOH to calculate the pH.

$$pH = 14 - pOH = 8.34$$

We can generalize the results of this calculation by looking at the equation solved to obtain the value of Δ.

$$\frac{[\Delta][\Delta]}{[0.030]} \cong 1.6 \times 10^{-10}$$

Rearranging this equation gives the following result.

$$\Delta \cong \sqrt{(1.6 \times 10^{-10})(0.030)}$$

The first term in this equation is the value of K_b for the benzoate ion, and the second term is the initial concentration of this base. Thus, in general, the following equation can be used to calculate the value of Δ.

$$\Delta \cong \sqrt{K_b C_b}$$

This equation is completely analogous to the one used to solve weak-acid equilibrium problems. It is important to remember, however, that in this case, Δ is equal to the OH^- ion concentration at equilibrium.

We made two assumptions in this calculation.

1. We assumed that Δ was small enough compared with the initial concentration of the base so that it could be ignored in the $[0.030 - \Delta]$ term.

2. We assumed that all of the OH^- ion at equilibrium came from the reaction between the benzoate ion and water. (We ignored the contribution from the dissociation of water.)

We have already confirmed the validity of the first assumption. What about the second? The OH^- ion concentration obtained from this calculation is $2.1 \times 10^{-6} \, M$, which is 21 times the OH^- ion concentration in pure water. According to LeChatelier's principle, however, the addition of a base suppresses the dissociation of water. This means that the dissociation of water makes a contribution of significantly less than 5% to the total OH^- ion concentration in this solution. It can therefore be ignored.

Section 15.9 stated that two factors influence the H_3O^+ ion concentration in an acid solution—K_a and C_a. Solutions of bases dissolved in water give analogous results. Two factors affect the OH^- ion concentration in aqueous solutions of bases—K_b and C_b. In pure water, the OH^- ion concentration is equal to the square root of K_w.

$$\text{Pure water:} \qquad [OH^-] = \sqrt{K_w}$$

In solutions of weak bases in water, the OH^- ion concentration is the square root of the product of K_b times the initial concentration of the base.

$$\text{Weak bases:} \qquad [OH^-] = \sqrt{K_b C_b}$$

In solutions of bases that are either so weak or so dilute that we can't ignore the dissociation of water, the OH^- ion concentration is the square root of the sum of these terms.

$$\text{Very weak bases:} \qquad [OH^-] = \sqrt{K_b C_b + K_w}$$

By analogy with our discussion for weak acids in Section 15.13, we can ignore the dissociation of water when K_w is less than 10% of the product of K_b times C_b. We can turn this into the following general rule.

When $K_b C_b$ for a weak base is *larger* than 1.0×10^{-13}, we can ignore the dissociation of water. When $K_b C_b$ is *smaller* than 2.0×10^{-13}, we have to include the dissociation of water in our calculations.

Exercise 15.7

Calculate the HOAc, OAc^-, and OH^- concentrations at equilibrium in a solution of 0.0500 moles of sodium acetate (NaOAc) dissolved in enough water to give 500 milliliters of solution. (HOAc: $K_a = 1.8 \times 10^{-5}$)

Solution

Sodium acetate dissolves in water to give an aqueous solution of the Na^+ and OAc^- ions.

$$NaOAc(s) \xrightarrow{H_2O} Na^+(aq) + OAc^-(aq)$$

The acetate ion then acts as a base toward water, picking up a proton to give acetic acid and the hydroxide ion.

$$OAc^-(aq) + H_2O(l) \rightleftharpoons HOAc(aq) + OH^-(aq)$$

The first step in solving this problem involves calculating the initial concentration of the acetate ion.

$$\frac{0.0500 \text{ mol } OAc^-}{0.500 \text{ L}} = 0.100 \; M \; OAc^-$$

The problem we are trying to solve can be summarized as follows.

$$OAc^-(aq) + H_2O(l) \rightleftharpoons HOAc(aq) + OH^-(aq) \qquad K_b = ?$$

Initial:	0.100 M	0	0
Equilibrium:	$0.100 - \Delta$	Δ	Δ

The only other information we need to solve this problem is the value of K_b for the reaction between the acetate ion and water. We can obtain this from the K_a for the conjugate acid.

$$K_b = \frac{K_w}{K_a} = \frac{1.0 \times 10^{-14}}{1.8 \times 10^{-5}} = 5.6 \times 10^{-10}$$

At this point we test whether the solution is basic enough to allow us to ignore the dissociation of water.

$$K_b C_b = (5.6 \times 10^{-10})(0.100) = 5.6 \times 10^{-11} > 1.0 \times 10^{-13}$$

The product of K_b times the initial concentration of the base is large enough to allow us to legitimately ignore the dissociation of water.

We can therefore set up the calculation as follows.

$$K_b = \frac{[HOAc][OH^-]}{[OAc^-]} = \frac{[\Delta][\Delta]}{[0.100 - \Delta]} = 5.6 \times 10^{-10}$$

The value of K_b for this base is relatively small, so we can feel reasonably confident that Δ is small compared with the initial concentration of base.

$$\frac{[\Delta][\Delta]}{[0.100]} \cong 5.6 \times 10^{-10}$$

Solving this approximate equation for Δ gives the following result.

$$\Delta \cong 7.5 \times 10^{-6} \, M$$

We have already confirmed the validity of the assumption that the dissociation of water can be ignored in this problem. We now test the assumption that Δ is small compared with the initial concentration of the base and find this assumption is also valid. Thus, we can use this value of Δ to calculate the equilibrium concentrations of HOAc, OAc$^-$, and OH$^-$.

$$[OAc^-] = 0.10 - \Delta = 0.10 \, M$$
$$[HOAc] = [OH^-] = \Delta = 7.5 \times 10^{-6} \, M$$

Exercise 15.8

Salts of the fluoride ion are routinely used to fluoridate water to help prevent tooth decay. Calculate the pH of an 0.0010 M solution of NaF in water. (HF: $K_a = 7.2 \times 10^{-4}$)

Solution

The only way to decide whether we can ignore the dissociation of water in this solution is to calculate the product of K_b for this reaction times the initial concentration of the fluoride ion. To do this we need to calculate K_b for the F$^-$ ion from K_a for hydrofluoric acid.

$$K_b = \frac{K_w}{K_a} = \frac{1.0 \times 10^{-14}}{7.2 \times 10^{-4}} = 1.4 \times 10^{-11}$$

The product of K_b times the initial concentration of the F$^-$ ion can then be compared with 1.0×10^{-13}.

$$K_b C_b = (1.4 \times 10^{-11})(0.0010) = 1.4 \times 10^{-14} < 1.0 \times 10^{-13}$$

The F$^-$ ion is too weak a base, and the solution is too dilute, for the dissociation of water to be ignored. The OH$^-$ ion concentration in this solution is therefore calculated from the following equation.

$$[OH^-] = \sqrt{K_b C_b + K_w}$$
$$= \sqrt{(1.4 \times 10^{-11})(0.0010) + 1.0 \times 10^{-14}} = 1.5 \times 10^{-7} \, M$$

The pOH of this solution is 6.82

$$pOH = -\log[OH^-] = 6.82$$

and the pH is 7.18.

$$pH = 14 - pOH = 7.18$$

Dilute solutions of NaF in water are such weak bases that NaF is often described as a neutral salt.

15.15 MIXTURES OF ACIDS AND BASES

In Exercise 5.4, we calculated the H_3O^+ ion concentration in a 0.100 M acetic acid solution.

$$0.100 \; M \; \text{HOAc:} \qquad [H_3O^+] = 1.3 \times 10^{-3} \; M$$

In Exercise 5.7, we calculated the OH^- ion concentration in a 0.100 M solution of sodium acetate in water.

$$0.100 \; M \; \text{NaOAc:} \qquad [OH^-] = 7.5 \times 10^{-6} \; M$$

Converting these results into the pH of the solutions gives the following results.

$$0.100 \; M \; \text{HOAc:} \qquad pH = 2.89$$
$$0.100 \; M \; \text{NaOAc:} \qquad pH = 8.88$$

What would happen if we added enough sodium acetate to an acetic acid solution so that the solution was 0.100 M in both HOAc and NaOAc? Would this solution be acidic, basic, or neutral?

Think about the difference between pure water and 0.100 M solutions of HOAc and NaOAc.

$$H_2O \; (pH = 7)$$

0.100 M HOAc   0.100 M NaOAc

(pH = 2.89)   (pH = 8.88)

Adding HOAc ($K_a = 1.8 \times 10^{-5}$) to water results in a larger change in the pH of the solution than adding an equivalent amount of NaOAc ($K_b = 5.6 \times 10^{-10}$). We might therefore expect the solution to be acidic, but not as acidic as HOAc by itself.

Why shouldn't the mixture be as acidic as HOAc by itself? There are two sources of the OAc^- ion in this solution. One is the dissociation of acetic acid.

$$HOAc(aq) + H_2O(l) \rightleftharpoons H_3O^+(aq) + OAc^-(aq)$$

The other is the dissociation of sodium acetate.

$$NaOAc(s) \xrightarrow{H_2O} Na^+(aq) + OAc^-(aq)$$

These reactions share a common ion — the OAc^- ion. LeChatelier's principle leads us to predict that adding an excess of the OAc^- ion to an HOAc solution should shift the equilibrium between HOAc and the H_3O^+ and OAc^- ions to the left. Adding NaOAc should therefore reduce the extent to which HOAc dissociates.

We can quantitate this discussion by doing the following calculation.

**Calculate the pH of a solution that is 0.100 M in both HOAc and NaOAc.
(HOAc: $K_a = 1.8 \times 10^{-5}$)**

We can set up this problem as follows.

$$HOAc(aq) + H_2O(l) \rightleftharpoons H_3O^+(aq) + OAc^-(aq) \quad K_a = 1.8 \times 10^{-5}$$

Initial:	0.100 M	0	0.100 M
Equilibrium:	?	?	?

We haven't described any rules we can use to predict whether the dissociation of
water can be ignored in this calculation, so let's make this assumption and check its
validity later.

If all of the H_3O^+ ion concentration at equilibrium comes from the dissociation
of the acetic acid, the reaction has to shift to the right to reach equilibrium.

$$HOAc(aq) + H_2O(l) \rightleftharpoons H_3O^+(aq) + OAc^-(aq) \quad K_a = 1.8 \times 10^{-5}$$

Initial:	0.100 M	0	0.100 M
Equilibrium:	0.100 − ΔM	Δ	0.100 + Δ

Substituting this information into the K_a expression gives the following result.

$$K_a = \frac{[H_3O^+][OAc^-]}{[HOAc]} = \frac{[\Delta][0.100 + \Delta]}{[0.100 - \Delta]} = 1.8 \times 10^{-5}$$

By now, the next step should be automatic — assume that Δ is small compared with
the initial concentrations of HOAc and OAc⁻.

$$\frac{[\Delta][0.100]}{[0.100]} \cong 1.8 \times 10^{-5}$$

Solving this approximate equation for Δ gives the following result.

$$\Delta \cong 1.8 \times 10^{-5} \, M$$

Are our two assumptions legitimate? Is Δ small enough compared with 0.100 to be
ignored? Is Δ large enough so that the dissociation of water can be ignored? The
answer to both questions is yes. Thus, we can use this value of Δ to calculate the
H_3O^+ ion concentration at equilibrium and therefore the pH of the solution.

$$[H_3O^+] = \Delta = 1.8 \times 10^{-5} \, M$$
$$pH = -\log [H_3O^+] = 4.74$$

We predicted that this solution would be acidic, but less acidic than HOAc by itself
and that is exactly what we find.

15.16 BUFFERS

Mixtures of a weak acid and its conjugate base, such as HOAc and the OAc⁻ ion,
have a number of interesting properties. By carefully selecting the ratio of the
concentrations of the acid and its conjugate base, for example, we can prepare
solutions of known pH that can be used for tasks such as calibrating pH meters.

Exercise 15.9

Calculate the initial concentration of sodium acetate that would have to be present
in a 0.100 M HOAc solution to give a pH of 5.00.

Solution

We can set up this problem in terms of two unknowns — the amount of HOAc that dissociates, Δ, and the initial concentration of NaOAc, C_b.

$$\text{HOAc}(aq) + \text{H}_2\text{O}(l) \rightleftharpoons \text{H}_3\text{O}^+(aq) + \text{OAc}^-(aq) \quad K_a = 1.8 \times 10^{-5}$$

Initial:	0.100 M	0	C_b
Equilibrium:	$0.100 - \Delta$	Δ	$C_b + \Delta$

Note that just as we used C_a for the initial concentration of an acid, we use C_b for the initial concentration of a base.

We have two unknowns and only one equation, so we need to find a way to remove one of the unknowns. What else do we know about the problem? We know that the pH of the solution is 5.00. That means we can calculate the H_3O^+ ion concentration at equilibrium.

$$\text{pH} = -\log[\text{H}_3\text{O}^+]$$
$$[\text{H}_3\text{O}^+] = 10^{-\text{pH}} = 10^{-5.00} = 1.00 \times 10^{-5}$$

If we assume that essentially all of the H_3O^+ ion at equilibrium comes from the dissociation of HOAc, we conclude that Δ is equal to 1.0×10^{-5} M.

$$\Delta \cong 1.0 \times 10^{-5}$$

The problem now has only one unknown.

$$\text{HOAc}(aq) + \text{H}_2\text{O}(l) \rightleftharpoons \text{H}_3\text{O}^+(aq) + \text{OAc}^-(aq) \quad K_a = 1.8 \times 10^{-5}$$

Initial:	0.100 M	0	C_b
Equilibrium:	$0.100 - (1.0 \times 10^{-5})$ M	1.0×10^{-5} M	$C_b + (1.0 \times 10^{-5})$ M

Substituting this information into the K_a expression gives the following equation.

$$K_a = \frac{[\text{H}_3\text{O}^+][\text{OAc}^-]}{[\text{HOAc}]} = \frac{[1.0 \times 10^{-5}][C_b + (1.0 \times 10^{-5})]}{[0.100]} = 1.8 \times 10^{-5}$$

This equation can be solved for C_b. But the mathematics is much easier if we assume that C_b is larger than 1.0×10^{-5}. This assumption gives us the following approximate equation.

$$\frac{[1.0 \times 10^{-5}][C_b]}{[0.100]} \cong 1.8 \times 10^{-5}$$

Solving for C_b gives the following result.

$$C_b \cong 0.18 \ M$$

Our assumption that C_b is larger than 1.0×10^{-5} is valid, and we can conclude that 0.18 moles of NaOAc must be added to each liter of 0.100 M HOAc to give a solution that has a pH of 5.000.

Mixtures of a weak acid and its conjugate base are called buffers. The term *buffer* usually means "to lessen or absorb shock." These mixtures are buffers because they lessen or absorb the drastic change in pH that occurs when small amounts of acids or bases are added to water.

In Section 15.5 we concluded that adding a single drop of 2 M hydrochloric acid to 100 milliliters of water changes the pH from 7 to 3. In other words, a single drop of 2 M HCl in 100 milliliters of water raises the H_3O^+ ion concentration in this solution by a factor of 10,000. Let's see what happens when we add a drop of 2 M HCl to 100 milliliters of a buffer solution.

Exercise 15.10

Calculate the effect of adding one drop of 2 M HCl to 100 mL of a buffer solution that is 0.100 M in both acetic acid and sodium acetate.

Solution

Before the acid is added, the concentrations of HOAc and the OAc^- ions in this solution are the same.

$$K_a = \frac{[H_3O^+][OAc^-]}{[HOAc]} = \frac{[H_3O^+][0.100]}{[0.100]} = 1.8 \times 10^{-5}$$

The H_3O^+ ion concentration is therefore equal to K_a for the acid.

$$[H_3O^+] = 1.8 \times 10^{-5}$$

Accordingly, the pH of the solution is 4.74.

$$pH = -\log [H_3O^+] = \mathbf{4.74}$$

In Section 15.5 we concluded that adding a single drop of 2 M HCl to 100 milliliters of water was equivalent to adding 1×10^{-3} moles of HCl per liter of solution. Adding this acid upsets the equilibrium between HOAc and the H_3O^+ and OAc^- ions—the H_3O^+ ion concentration is now too large for the solution to be at equilibrium.

$$Q_a = \frac{(H_3O^+)(OAc^-)}{(HOAc)} = \frac{(1.0 \times 10^{-3})(0.100)}{(0.100)} = 1.0 \times 10^{-3} > K_a$$

LeChatelier's principle suggests that the reaction will have to shift toward HOAc to reach equilibrium.

For the sake of argument, let's assume that all of the H_3O^+ ion from the hydrochloric acid is consumed to convert OAc^- ions into HOAc.

$$OAc^-(aq) + H_3O^+(aq) \longrightarrow HOAc(aq) + H_2O(l)$$

Because we added 0.001 M HCl to this solution, this reaction would lead to the following intermediate conditions.

$$HOAc(aq) + H_2O(l) \rightleftharpoons H_3O^+(aq) + OAc^-(aq) \quad K_a = 1.8 \times 10^{-5}$$

Initial:	0.100 M	0.100 M
Intermediate:	0.101 M	0.099 M

In order for the reaction to come to equilibrium, some of the acetic acid must dissociate to form H_3O^+ and OAc^- ions.

$$HOAc(aq) + H_2O(l) \quad H_3O^+(aq) + OAc^-(aq) \quad K_a = 1.8 \times 10^{-5}$$

Intermediate:	0.101 M		0.099 M
Equilibrium:	$0.101 - \Delta$	Δ	$0.099 + \Delta$

We can substitute this information into the K_a expression

$$K_a = \frac{[H_3O^+][OAc^-]}{[HOAc]} = \frac{[\Delta][0.099 + \Delta]}{[0.101 - \Delta]} = 1.8 \times 10^{-5}$$

Assuming that Δ is relatively small gives the following approximate equation.

$$\frac{[\Delta][0.099]}{[0.101]} \cong 1.8 \times 10^{-5}$$

When we solve this equation for Δ, we find that there has been no significant change in the H_3O^+ ion concentration.

$$\Delta \cong 1.8 \times 10^{-5}$$

As a result, there is no significant change in the pH of this solution.

$$pH = -\log [H_3O^+] = 4.74$$

Although a single drop of 2 M HCl reduces the pH of 100 milliliters of water by 4 pH units (from 7 to 3), there is effectively no change in the pH of the buffer when a drop of 2 M HCl is added to 100 milliliters of this solution. The pH of the solution is indeed buffered against the effect of small amounts of acid or base.

Buffers can be made from a weak acid and its conjugate base, such as acetic acid and a salt of the acetate ion.

$$\underset{\text{weak acid}}{HOAc(aq)} + H_2O(l) \rightleftharpoons H_3O^+(aq) + \underset{\text{conjugate base}}{OAc^-(aq)}$$

Buffers can also be made from a weak base and its conjugate acid, such as ammonia and a salt of the ammonium ion.

$$\underset{\text{conjugate acid}}{NH_4^+(aq)} + H_2O(aq) \rightleftharpoons H_3O^+(aq) + \underset{\text{weak base}}{NH_3(aq)}$$

There is an important difference between these buffers, however. Mixtures of HOAc and the OAc^- ion form an *acidic* buffer, with a pH below 7. Mixtures of NH_3 and the NH_4^+ ion form a *basic* buffer, with a pH above 7.

We can predict whether a buffer will be acidic or basic by comparing the values of K_a and K_b for the conjugate acid-base pair. K_a for acetic acid is significantly larger than K_b for the acetate ion. We therefore expect mixtures of this conjugate acid-base pair to be acidic.

$$HOAc \ (K_a = 1.8 \times 10^{-5}) \qquad OAc^- \ (K_b = 5.6 \times 10^{-10})$$

K_b for ammonia, on the other hand, is much larger than K_a for the ammonium ion. Mixtures of this acid-base pair are therefore basic.

$$NH_3 \ (K_b = 1.8 \times 10^{-5}) \qquad NH_4^+ \ (K_a = 5.6 \times 10^{-10})$$

In general, if K_a for the acid is larger than 1×10^{-7}, the buffer prepared from the acid and its conjugate base will be acidic. If K_b is larger than 1×10^{-7}, the buffer prepared with that base and its conjugate acid will be basic.

15.17 BUFFERING CAPACITY AND pH TITRATION CURVES

The most important property of a buffer is its ability to resist changes in pH when small quantities of acid or base are added to the solution. We can develop a model for the capacity of a buffer to absorb acid or base by looking at how the buffer resists changes in pH. Consider an HOAc/OAc$^-$ buffer, for example.

$$K_a = \frac{[H_3O^+][OAc^-]}{[HOAc]} = 1.8 \times 10^{-5}$$

TABLE 15.4

*The Effect of Adding Hydrochloric Acid
to Water and Two Buffer Solutions*

Mol HCl / L solution	pH of water	pH of Buffer 1 (0.100 M HOAc/ 0.100 M OAc⁻)	pH of Buffer 2 (0.200 M HOAc/ 0.200 M OAc⁻)
0	7	4.74	4.74
0.000001	6	4.74	4.74
0.00001	5	4.74	4.74
0.0001	4	4.74	4.74
0.001	3	4.74	4.74
0.01	2	4.66	4.70
0.1	1	2.72	4.27

Rearranging the K_a expression for this buffer gives the following equation.

$$[H_3O^+] = K_a \times \frac{[HOAc]}{[OAc^-]}$$

According to this equation, the H_3O^+ ion concentration, and therefore the pH of the solution, will remain essentially constant as long as the ratio of the concentrations of HOAc and the OAc⁻ ion is more or less constant.

Suppose we start with a solution that contains equal amounts of HOAc and the OAc⁻ ion. When we add small amounts of acid to this solution, some of the acetate ions are converted into acetic acid.

$$OAc^-(aq) + H_3O^+(aq) \longrightarrow HOAc(aq) + H_2O(l)$$

But the ratio of these concentrations doesn't change by much, so the pH stays essentially the same. When we add small amounts of base to the solution, some of the acetic acid is converted into acetate ions.

$$HOAc(aq) + OH^-(aq) \longrightarrow OAc^-(aq) + H_2O(l)$$

Once again, the ratio of these concentrations stays more or less the same, and so does the pH.

As long as the concentrations of HOAc and the OAc⁻ ion in the buffer are larger than the amount of acid or base added to the solution, the pH remains constant. Table 15.4 compares the effects of adding various amounts of hydrochloric acid to water and to a pair of buffer solutions. The first column in the table gives the number of moles of HCl added per liter. The second column shows the effect of adding this much HCl to water, and the third column shows the effect on a buffer solution that contains 0.100 M HOAc and 0.100 M NaOAc. Even when as much as 0.01 moles per liter of acid are added to the buffer, there is very little change in the pH of this solution, as shown in Figure 15.8.

The capacity of a buffer can be defined as follows.

Buffering capacity: **The amount of acid or base that can be added before the pH of the solution changes significantly.**

The fourth column in Table 15.4 shows the results of adding hydrochloric acid to a buffer solution that contains twice as much acetic acid and twice as much sodium acetate as the solution in the third column. Note that the buffering capacity of this solution is even greater than that of the solution in the third column. This buffer protects against quantities of acid or base as large as 0.1 moles per liter.

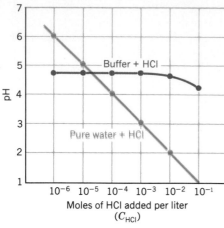

FIG. 15.8 This figure compares the effects of adding hydrochloric acid to water and to a buffer that consists of 0.100 M HOAc and 0.100 M NaOAc. The pH of water is very sensitive to hydrochloric acid. The pH of the buffer, however, remains remarkably constant until relatively large amounts of hydrochloric acid have been added, at which point the buffer becomes exhausted.

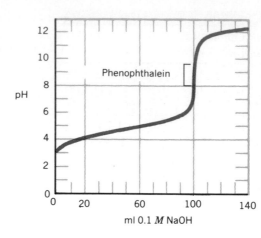

FIG. 15.9 This figure shows what happens to the pH of 0.100 M solution of acetic acid as it is titrated with 0.100 M sodium hydroxide. The pH increases at first, as some of the HOAc is converted into OAc⁻ ions. But this creates a buffer, which resists changes in pH until it has become exhausted. At that point, the pH rises rapidly, then slowly levels off as the solution begins to behave more like 0.100 M NaOH.

The chemistry of buffers and the concept of buffering capacity have a lot to do with the shape of the **titration curve** in Figure 15.9, which shows a graph of the pH of a solution of acetic acid as it is titrated with a strong base, sodium hydroxide. Four points (*A*, *B*, *C*, and *D*) on this curve will be discussed in some detail.

Point *A* represents the pH of the acetic acid solution at the start of the titration. The pH at this point is 2.9—the pH of a 0.10 M HOAc solution. As NaOH is added to this solution, the HOAc molecules are converted into the conjugate base—the OAc⁻ ions.

$$HOAc(aq) + OH^-(aq) \longrightarrow OAc^-(aq) + H_2O(l)$$

Point *B* is the point at which exactly half of the HOAc molecules have been converted to OAc⁻ ions. At this point, the concentration of the HOAc molecules is equal to the concentration of the OAc⁻ ions in the solution. The H_3O^+ ion concentration is proportional to the ratio of the concentrations of HOAc and OAc⁻, however.

$$[H_3O^+] = K_a \times \frac{[HOAc]}{[OAc^-]}$$

If the ratio of the HOAc and OAc⁻ concentrations is 1 : 1, the concentration of the H_3O^+ ion at Point *B* is equal to the acid-dissociation equilibrium constant for the acid.

$$\text{Point } B: \qquad [H_3O^+] = K_a$$

This provides us with a way of measuring K_a for an acid. We can titrate a sample of the acid with a strong base, plot the titration curve, and then find the point along this curve at which exactly half of the acid has been consumed. The H_3O^+ ion concentration at this point will be equal to the value of K_a for the acid.

Point *C* is the **equivalence point** for the titration—the point at which the number of moles of base that have been added to the solution is exactly equivalent to the number of moles of acid present at the start of the titration. The goal of an acid-base titration is to find the equivalence point. What is actually observed is the **endpoint** of the titration—Point *D*—the point at which an indicator changes color.

Every effort is made to bring the endpoint as close as possible to the equivalence point of a titration. Suppose that phenolphthalein is used as the indicator for this titration. Phenolphthalein turns from colorless to pink as the pH of the solution changes from 8 to 10. If we wait until the phenolphthalein permanently turns pink, the endpoint will fall beyond the equivalence point for the titration. So we try to find the point at which adding one drop of base turns the entire solution to a pink color that fades in about 10 seconds while the solution is stirred. In theory, this is one drop before the endpoint of the titration and closer to the equivalence point.

Note the shape of the pH titration curve in Figure 15.9. The pH rises rapidly at first, because we are adding a strong base to a weak acid and the base neutralizes some of the acid. The curve then levels off, and the pH remains more or less constant as we add base, because some of the HOAc present initially is converted into OAc⁻ ions to form a buffer solution. The pH of this buffer solution stays relatively constant as we add more base to the system until most of the acid has been converted to its conjugate base. At that point the pH rises rapidly, indicating that essentially all of the HOAc in the system has been converted into OAc⁻ ions. The pH then gradually levels off as the solution begins to look like a 0.10 M NaOH solution.

The titration curve for a weak base, such as ammonia, titrated with a strong acid, such as hydrochloric acid, would be analogous to the curve in Figure 15.9. The principal difference is that the pH value is high at first and decreases as acid is added.

15.18 DIPROTIC ACIDS

The acid equilibrium problems discussed so far have focused on a family of compounds known as **monoprotic acids.** Each of these acids has a single H^+ ion, or proton, which it can donate when it acts as a Brønsted acid. Hydrochloric acid (HCl), acetic acid (CH_3CO_2H or HOAc), nitric acid (HNO_3), and benzoic acid ($C_6H_5CO_2H$) are all monoprotic acids.

$$HCL(aq) + H_2O(l) \longrightarrow H_3O^+(aq) + Cl^-(aq)$$
$$CH_3CO_2H(aq) + H_2O(l) \rightleftharpoons H_3O^+(aq) + CH_3CO_2^-(aq)$$
$$HNO_3(aq) + H_2O(l) \longrightarrow H_3O^+(aq) + NO_3^-(aq)$$
$$C_6H_5CO_2H(aq) + H_2O(l) \rightleftharpoons H_3O^+(aq) + C_6H_5CO_2^-(aq)$$

Several important acids can be classified as **polyprotic acids,** which can lose more than one H^+ ion when they act as Brønsted acids. **Diprotic acids,** such as sulfuric acid (H_2SO_4), carbonic acid (H_2CO_3), hydrogen sulfide (H_2S), chromic acid (H_2CrO_4), oxalic acid ($H_2C_2O_4$), and amino acids such as glycine ($C_2H_6NO_2$), have two acidic hydrogen atoms. **Triprotic acids,** such as phosphoric acid (H_3PO_4) and citric acid ($C_6H_8O_7$), have three.

There is often a very large difference in the ease with which these acids lose the first and second or second and third protons. When sulfuric acid is classified as a strong acid, students often assume that it loses both of its protons when it reacts with water. That isn't a legitimate assumption. Sulfuric acid is a strong acid because K_a for the loss of the first proton is much larger than 1. We can therefore assume that essentially all of the H_2SO_4 molecules in an aqueous solution lose the first proton to form the HSO_4^-, or hydrogen sulfate, ion.

$$H_2SO_4(aq) + H_2O(l) \longrightarrow H_3O^+(aq) + HSO_4^-(aq) \qquad K_{a1} = 1 \times 10^3$$

But K_a for the loss of the second proton is only 10^{-2}, which suggests that only 10% of the H_2SO_4 molecules in a 1 M solution lose a second proton.

$$HSO_4^-(aq) + H_2O(l) \rightleftharpoons H_3O^+(aq) + SO_4^{2-}(aq) \qquad K_{a2} = 1.2 \times 10^{-2}$$

Why do students so often believe that H_2SO_4 molecules give up both protons when they react with water? It may be the result of their having been told that H_2SO_4 loses both of its protons when it reacts with a base. Sulfuric acid reacts with ammonia, for example, to form an aqueous solution of ammonium sulfate.

$$H_2SO_4(aq) + 2\ NH_3(aq) \longrightarrow 2\ NH_4^+(aq) + SO_4^{2-}(aq)$$

The point being made here is subtle, but important. When H_2SO_4 reacts with a strong base such as NaOH — or even a moderately strong base such as NH_3 — it loses both protons. But water is a weak base, and H_2SO_4 loses an average of only 1.1 protons when it reacts with water.

Table 15.5 gives values of K_a for some common polyprotic acids. The large difference between the values of K_a for the loss of the first and second protons by a polyprotic acid is important because it means we can assume that these acids dissociate one step at a time.

Let's look at the consequence of this assumption by examining the chemistry of a 0.10 M solution of H_2S in water. Hydrogen sulfide is the foul-smelling gas that gives rotten eggs their unpleasant odor. It is an excellent source of the S^{2-} ion, however, and is therefore commonly used in introductory chemistry laboratories. H_2S is a weak acid that dissociates in steps. Some of the H_2S molecules lose a proton in the first step to form the HS^-, or hydrogen sulfide, ion.

First step: $$H_2S(aq) + H_2O(l) \rightleftharpoons H_3O^+(aq) + HS^-(aq)$$

A small fraction of the HS^- ions formed in this reaction then go on to lose another H^+ ion in a second step.

Second step: $$HS^-(aq) + H_2O(l) \rightleftharpoons H_3O^+(aq) + S^{2-}(aq)$$

Since there are two distinct steps in this reaction, we can write two distinct equilibrium constant expressions.

$$K_{a1} = \frac{[H_3O^+][HS^-]}{[H_2S]} = 1.0 \times 10^{-7}$$

$$K_{a2} = \frac{[H_3O^+][S^{2-}]}{[HS^-]} = 1.3 \times 10^{-13}$$

Although each of these equations contains three terms, there are only four unknowns — $[H_3O^+]$, $[H_2S]$, $[HS^-]$, and $[S^{2-}]$ — because the $[H_3O^+]$ and $[HS^-]$ terms appear in both equations. The $[H_3O^+]$ term represents the total H_3O^+ ion concentration from both steps and therefore must have the same value in both equations. Similarly, the $[HS^-]$ term, which describes the balance between the HS^-

TABLE 15.5

Acid-Dissociation Equilibrium Constants for Common Polyprotic Acids

Acid	K_{a1}	K_{a2}	K_{a3}
sulfuric acid (H_2SO_4)	1.0×10^3	1.2×10^{-2}	
chromic acid (H_2CrO_4)	9.6	3.2×10^{-7}	
oxalic acid ($H_2C_2O_4$)	5.4×10^{-2}	5.4×10^{-5}	
sulfurous acid (H_2SO_3)	1.7×10^{-2}	6.4×10^{-8}	
phosphoric acid (H_3PO_4)	7.1×10^{-3}	6.3×10^{-8}	4.2×10^{-13}
glycine ($C_2H_6NO_2$)	4.5×10^{-3}	2.5×10^{-10}	
citric acid ($C_6H_8O_7$)	7.5×10^{-4}	1.7×10^{-5}	4.0×10^{-7}
carbonic acid (H_2CO_3)	4.5×10^{-7}	4.7×10^{-11}	
hydrogen sulfide (H_2S)	1.0×10^{-7}	1.3×10^{-13}	

ions formed in the first step and the HS⁻ ions consumed in the second step, must have the same value for both equations.

Four equations are needed to solve for four unknowns. We already have two equations—the K_{a1} and K_{a2} expressions. We are going to have to find either two more equations or a pair of assumptions that can generate two equations.

We can base one assumption on the fact that the value of K_{a1} for this acid is almost a million times the value of K_{a2}.

$$K_{a1} = 1.0 \times 10^{-7} \qquad K_{a2} = 1.3 \times 10^{-13}$$

This means that only a negligibly small fraction of the HS⁻ ions formed in the first step go on to dissociate in the second step. Most of the H_3O^+ ions in this solution come from the dissociation of H_2S, and most of the HS⁻ ions formed in this reaction remain in solution. As a result, we can assume that the H_3O^+ and HS⁻ ion concentrations are more or less equal.

First assumption: $[H_3O^+] \cong [HS^-]$

We need one more equation—and therefore one more assumption. Note that H_2S is a weak acid ($K_{a1} = 1.0 \times 10^{-7}$, $K_{a2} = 1.3 \times 10^{-13}$). Thus, we can assume that most of the H_2S molecules present initially will still be present when the solution reaches equilibrium. In other words, we can assume that the equilibrium concentration of H_2S is approximately equal to the initial concentration.

Second assumption: $[H_2S] \cong C_{H_2S}$

We now have four equations in four unknowns.

$$K_{a1} = \frac{[H_3O^+][HS^-]}{[H_2S]} = 1.0 \times 10^{-7}$$

$$K_{a2} = \frac{[H_3O^+][S^{2-}]}{[HS^-]} = 1.3 \times 10^{-13}$$

$$[H_3O^+] \cong [HS^-]$$

$$[H_2S] \cong C_{H_2S}$$

Since there is always a unique solution to a problem with four equations in four unknowns, we are now ready to try the following calculation.

Calculate the H_3O^+, H_2S, HS^-, and S^{2-} concentrations at equilibrium in a 0.10 M solution of H_2S in water.

Because K_{a1} is so much larger than K_{a2} for this acid, we can work with the equilibrium expression for the first step without worrying about the second step for the moment.

We therefore start with the expression for K_{a1} for this acid.

$$K_{a1} = \frac{[H_3O^+][HS^-]}{[H_2S]} = 1.0 \times 10^{-7}$$

We then invoke one of our assumptions.

$$[H_2S] = C_{H_2S} = 0.10 \ M$$

Substituting this approximation into the K_{a1} expression gives the following equation.

$$K_{a1} = \frac{[H_3O^+][HS^-]}{[0.10]} = 1.0 \times 10^{-7}$$

We now invoke our other assumption.

$$[H_3O^+] \cong [HS^-] = \Delta$$

Substituting this approximation into the K_{a1} expression gives the following result.

$$K_{a1} = \frac{[\Delta][\Delta]}{[0.10]} \cong 1.0 \times 10^{-7}$$

We now solve this approximate equation for Δ.

$$\Delta \cong 1.0 \times 10^{-4}\ M$$

Note that we are three-fourths of the way to our goal. We know the H_2S, H_3O^+, and HS^- concentrations — if our two assumptions are valid.

$$[H_2S] \cong 0.10\ M$$
$$[H_3O^+] \cong 1.0 \times 10^{-4}\ M$$
$$[HS^-] \cong 1.0 \times 10^{-4}\ M$$

Having extracted the values of three unknowns from the first equilibrium expression, we turn to the second equilibrium expression.

$$K_{a2} = \frac{[H_3O^+][S^{2-}]}{[HS^-]} = 1.3 \times 10^{-13}$$

Substituting the known values of the H_3O^+ and HS^- ion concentrations into this expression gives the following equation.

$$\frac{[\cancel{1.0 \times 10^{-4}}][S^{2-}]}{[\cancel{1.0 \times 10^{-4}}]} = 1.3 \times 10^{-13}$$

Because the equilibrium concentrations of the H_3O^+ and HS^- ions are more or less the same in this solution, the S^{2-} ion concentration at equilibrium is equal to the value of K_{a2} for this acid.

$$[S^{2-}] \cong 1.3 \times 10^{-13}\ M$$

It is now time to check our assumptions. Is the dissociation of H_2S small compared with the initial concentration? Yes. The HS^- and H_3O^+ ion concentrations obtained from this calculation are $1.0 \times 10^{-4}\ M$, which is 0.1% of the initial concentration of H_2S. The following assumption is therefore valid.

$$[H_2S] = C_{H_2S} = 0.10\ M$$

Is the difference between the S^{2-} and HS^- ion concentrations large enough to allow us to assume that essentially all of the H_3O^+ ions at equilibrium are formed in the first step and that essentially all of the HS^- ions formed in this step remain in solution? Yes. The S^{2-} ion concentration obtained from this calculation is 10^9 times smaller than the HS^- ion concentration. Our other assumption is therefore also valid.

$$[H_3O^+] = [HS^-]$$

We can summarize the concentrations of the various components of this equilibrium as follows.

$$[H_2S] = 0.10\ M$$
$$[H_3O^+] = [HS^-] = 1.0 \times 10^{-4}\ M$$
$$[S^{2-}] = 1.3 \times 10^{-13}\ M$$

15.19 TRIPROTIC ACIDS

Our technique for working equilibrium problems involving diprotic acids was based on two assumptions.

1. **The difference between K_{a1} and K_{a2} is large enough to allow us to work the problem one step at a time.**
2. **The acid is weak enough so that the equilibrium concentration is approximately equal to the initial concentration.**

There are two potential sources of trouble with this approach to polyprotic acids — one of the two assumptions may fail. Let's look at a polyprotic acid for which the second assumption fails, to show how this difficulty can be overcome.

Calculate the H_3O^+, H_3PO_4, $H_2PO_4^-$, HPO_4^{2-}, and PO_4^{3-} concentrations at equilibrium in a 0.100 M phosphoric acid solution. (H_3PO_4: $K_{a1} = 7.1 \times 10^{-3}$, $K_{a2} = 6.3 \times 10^{-8}$, $K_{a3} = 4.2 \times 10^{-13}$)

The difference between K_{a1}, K_{a2}, and K_{a3} for this acid is large enough to allow us to do this calculation one step at a time. But K_{a1} is a little too big to allow us to assume that Δ is small compared with the initial concentration of phosphoric acid.

Let's assume that this acid dissociates by steps and analyze the first step — the most extensive reaction.

$$K_{a1} = \frac{[H_3O^+][H_2PO_4^-]}{[H_3PO_4]} = 7.1 \times 10^{-3}$$

We now assume that the difference between K_{a1} and K_{a2} is large enough that most of the H_3O^+ ions come from this step and most of the $H_2PO_4^-$ ions formed in this step remain in solution.

$$[H_3O^+] \cong [H_2PO_4^-] = \Delta$$

Substituting this into the K_{a1} expression gives the following equation.

$$\frac{[\Delta][\Delta]}{[0.100 - \Delta]} = 7.1 \times 10^{-3}$$

Although we can't assume that Δ is small compared with 0.100, we don't really need this assumption, because we can use the quadratic formula to solve the equation. Or we can make the assumption, obtain an approximate value of Δ, and substitute Δ back into the problem until successive approximations converges on the correct value of Δ. Either way, we obtain the same answer.

$$\Delta = 0.023 \, M$$

We can then use this value of Δ to calculate the following concentrations.

$$[H_3PO_4] = 0.100 - \Delta = 0.077 \, M$$
$$[H_3O^+] = [H_2PO_4^-] = 0.023 \, M$$

We now turn to the second strongest acid in this solution.

$$K_{a2} = \frac{[H_3O^+][HPO_4^{2-}]}{[H_2PO_4^-]} = 6.3 \times 10^{-8}$$

Substituting what we know about the H_3O^+ and $H_2PO_4^-$ ion concentrations into this expression gives the following equation.

$$\frac{[\text{0.023}][\text{HPO}_4{}^{2-}]}{[\text{0.023}]} = 6.3 \times 10^{-8}$$

The $\text{HPO}_4{}^{2-}$ ion concentration in this solution is therefore equal to K_{a2}.

$$[\text{HPO}_4{}^{2-}] = 6.3 \times 10^{-8} \, M$$

We have only one more equation, the equilibrium expression for the weakest acid in the solution.

$$K_{a3} = \frac{[\text{H}_3\text{O}^+][\text{PO}_4{}^{3-}]}{[\text{HPO}_4{}^{2-}]} = 4.2 \times 10^{-13}$$

Substituting what we know about the concentrations of the H_3O^+ and $\text{HPO}_4{}^{2-}$ ions into this expression gives the following equation.

$$\frac{[0.023][\text{PO}_4{}^{3-}]}{[6.3 \times 10^{-8}]} = 4.2 \times 10^{-13}$$

This equation can be solved for the phosphate ion concentration.

$$[\text{PO}_4{}^{3-}] = 1.2 \times 10^{-18} \, M$$

Summarizing the results of these calculations helps us check the assumptions made along the way.

$$[\text{H}_3\text{PO}_4] = 0.077 \, M$$
$$[\text{H}_3\text{O}^+] = [\text{H}_2\text{PO}_4{}^-] = 0.023 \, M$$
$$[\text{HPO}_4{}^{2-}] = 6.3 \times 10^{-8} \, M$$
$$[\text{PO}_4{}^{3-}] = 1.2 \times 10^{-18} \, M$$

The only approximation used in working this problem was the assumption that the acid dissociates one step at a time. Is the difference between the concentrations of the $\text{H}_2\text{PO}_4{}^-$ and $\text{HPO}_4{}^{2-}$ ions large enough to justify the assumption that essentially all of the H_3O^+ ions come from the first step? Yes. Is it large enough to justify the assumption that essentially all of the $\text{H}_2\text{PO}_4{}^-$ formed in the first step remains in solution? Yes.

You may never encounter an example of the second source of trouble for polyprotic acid calculations—acids for which the differences between successive values of K_a are too small to allow us to assume stepwise dissociation. This assumption works even when we might expect it to fail.

Exercise 15.11

Calculate the H_3O^+, H_3Cit, H_2Cit^-, HCit^{2-}, and Cit^{3-} concentrations in a $1.00 \, M$ solution of citric acid. (H_3Cit: $K_{a1} = 7.5 \times 10^{-4}$, $K_{a2} = 1.7 \times 10^{-5}$, $K_{a3} = 4.0 \times 10^{-7}$)

Solution

At first glance, K_{a1}, K_{a2}, and K_{a3} for this acid might not seem to differ enough to allow us to assume stepwise dissociation. The only way to tell for sure is to try the assumption and see if it works.

We start with the expression for the first step.

$$K_{a1} = \frac{[\text{H}_3\text{O}^+][\text{H}_2\text{Cit}^-]}{[\text{H}_3\text{Cit}]} = 7.5 \times 10^{-4}$$

Even though citric acid is a weak acid, the dissociation of water can be ignored in this calculation.

$$K_{a1}C_a = (7.5 \times 10^{-4})(1.00) = 7.5 \times 10^{-4} \gg 1 \times 10^{-13}$$

We can therefore assume that most of the H_3O^+ ions at equilibrium come from the citric acid. Furthermore, we can assume that most of the H_2Cit^- ions formed in the first step remain in solution. In other words, we can assume that the concentrations of the H_3O^+ and H_2Cit^- ions are equal.

$$[H_3O^+] \cong [H_2Cit^-] = \Delta$$

Substituting this assumption into the K_{a1} expression gives the following equation.

$$\frac{[\Delta][\Delta]}{[H_3Cit]} \cong 7.5 \times 10^{-4}$$

Citric acid is weak enough that the equilibrium concentration of the undissociated H_3Cit molecules is roughly equal to the initial concentration of this acid.

$$[H_3Cit] \cong C_{H_3Cit}$$

Substituting this assumption into the K_{a1} expression gives the following result.

$$\frac{[\Delta][\Delta]}{[1.00]} \cong 7.5 \times 10^{-4}$$

We can now solve this approximate equation for Δ.

$$\Delta \cong 0.027 \ M$$

If our assumptions are valid, this value of Δ can be used to calculate the following equilibrium concentrations.

$$[H_3Cit] = 0.100 \ M$$
$$[H_3O^+] = [H_2Cit^-] = 0.027 \ M$$

We can now turn to the K_{a2} expression.

$$K_{a2} = \frac{[H_3O^+][HCit^{2-}]}{[H_2Cit^-]} = 1.7 \times 10^{-5}$$

Substituting what we know about the H_3O^+ and H_2Cit^- concentrations into this expression gives the following equation.

$$\frac{[\cancel{0.027}][HCit^{2-}]}{[\cancel{0.027}]} = 1.7 \times 10^{-5}$$

The $HCit^{2-}$ ion concentration is therefore equal to K_{a2} for this acid.

$$[HCit^{2-}] = 1.7 \times 10^{-5} \ M$$

We can now turn to the K_{a3} expression.

$$K_{a3} = \frac{[H_3O^+][Cit^{3-}]}{[HCit^{2-}]} = 4.0 \times 10^{-7}$$

Substituting the known values for the H_3O^+ and $HCit^{2-}$ ion concentrations into this expression gives the following equation.

$$\frac{[0.027][Cit^{3-}]}{[1.7 \times 10^{-5}]} = 4.0 \times 10^{-7}$$

Solving for the Cit^{3-} ion concentration gives the following result.

$$[Cit^{3-}] = 2.5 \times 10^{-10} \, M$$

Summarizing the results of this calculation allows us to test the assumptions made in deriving them.

$$[H_3Cit] = 0.100 \, M$$
$$[H_3O^+] = [H_2Cit^-] = 0.027 \, M$$
$$[HCit^{2-}] = 1.7 \times 10^{-5} \, M$$
$$[Cit^{3-}] = 2.5 \times 10^{-10} \, M$$

Is it legitimate to assume that the amount of H_3Cit that dissociates in the first step is small compared with the initial concentration of this acid? Yes. Is it legitimate to assume that most of the H_3O^+ ion comes from the first step? Yes. Is it legitimate to assume that most of the H_2Cit^- ion formed in the first step remains in solution? Yes. Since all of the assumptions made in this calculation are valid, the results are also valid.

15.20 DIPROTIC BASES

The techniques we have been using to work with polyprotic acids can be extended to polyprotic bases. The only challenge is calculating the values of K_b for the base.

Exercise 15.12

Calculate the H_2CO_3, HCO_3^-, CO_3^{2-}, and OH^- concentrations at equilibrium in a solution that was initially 0.100 M in Na_2CO_3. (H_2CO_3: $K_{a1} = 4.5 \times 10^{-7}$; $K_{a2} = 4.7 \times 10^{-11}$).

Solution

Sodium carbonate dissociates into its ions when it dissolves in water.

$$Na_2CO_3(aq) \xrightarrow{\text{H}_2\text{O}} 2 \, Na^+(aq) + CO_3^{2-}(aq)$$

The carbonate ion now acts as a base toward water, picking up a pair of protons (one at a time) to form the bicarbonate ion, HCO_3^-, and then eventually carbonic acid, H_2CO_3.

$$CO_3^{2-}(aq) + H_2O(l) \rightleftharpoons HCO_3^-(aq) + OH^-(aq) \qquad K_{b1} = ?$$
$$HCO_3^-(aq) + H_2O(l) \rightleftharpoons H_2CO_3(aq) + OH^-(aq) \qquad K_{b2} = ?$$

The first step in solving this problem involves determining the values of K_{b1} and K_{b2} for the carbonate ion. We can start by writing equilibrium constant expressions for these reactions and comparing them with the K_{a1} and K_{a2} expressions for carbonic acid.

$$K_{b1} = \frac{[HCO_3^-][OH^-]}{[CO_3^{2-}]} \qquad K_{a2} = \frac{[H_3O^+][CO_3^{2-}]}{[HCO_3^-]}$$

$$K_{b2} = \frac{[H_2CO_3][OH^-]}{[HCO_3^-]} \qquad K_{a1} = \frac{[H_3O^+][HCO_3^-]}{[H_2CO_3]}$$

The expressions for K_{b1} and K_{a2} depend on the concentrations of the HCO_3^- and

CO_3^{2-} ions. The expressions for K_{b2} and K_{a1} depend on the HCO_3^- and H_2CO_3 concentrations. We can therefore calculate K_{b1} from K_{a2} and K_{b2} from K_{a1}.

We can start by multiplying the top and bottom of the K_{a1} expression by the OH^- ion concentration to introduce the $[OH^-]$ term.

$$K_{a1} = \frac{[H_3O^+][HCO_3^-]}{[H_2CO_3]} \times \frac{[OH^-]}{[OH^-]}$$

One of our goals is removing the $[H_3O^+]$ term from this equation, so we'll group terms as follows.

$$K_{a1} = \frac{[HCO_3^-]}{[H_2CO_3][OH^-]} \times [H_3O^+][OH^-]$$

The first term in this equation is the inverse of the K_{b2} expression, and the second term is the K_w expression.

$$K_{a1} = \frac{1}{K_{b2}} \times K_w$$

Rearranging this equation gives the following result.

$$K_{a1}K_{b2} = K_w$$

Similarly, we can multiply the top and bottom of the K_{a2} expression by the OH^- ion concentration.

$$K_{a2} = \frac{[H_3O^+][CO_3^{2-}]}{[HCO_3^-]} \times \frac{[OH^-]}{[OH^-]}$$

Collecting terms gives the following result.

$$K_{a2} = \frac{[CO_3^{2-}]}{[HCO_3^-][OH^-]} \times [H_3O^+][OH^-]$$

The first term in this equation is the inverse of K_{b1}, and the second term is K_w.

$$K_{a2} = \frac{1}{K_{b1}} \times K_w$$

This equation can therefore be rearranged as follows.

$$K_{a2}K_{b1} = K_w$$

We can now calculate the values of K_{b1} and K_{b2} for the carbonate ion.

$$K_{b1} = \frac{K_w}{K_{a2}} = \frac{1.0 \times 10^{-14}}{4.7 \times 10^{-11}} = 2.1 \times 10^{-4}$$

$$K_{b2} = \frac{K_w}{K_{a1}} = \frac{1.0 \times 10^{-14}}{4.5 \times 10^{-7}} = 2.2 \times 10^{-8}$$

We are finally ready to do the calculations. We start with the K_{b1} expression, because the CO_3^{2-} ion is the strongest base in this solution and therefore the strongest source of the OH^- ion.

$$K_{b1} = \frac{[HCO_3^-][OH^-]}{[CO_3^{2-}]}$$

The difference between K_{b1} and K_{b2} for the carbonate ion is large enough to suggest that most of the OH^- ions come from this step and most of the HCO_3^-

formed in this reaction remains in solution.

$$[OH^-] \cong [HCO_3^-] = \Delta$$

The value of K_{b1} is small enough to allow us to assume that Δ is small compared with the initial concentration of the carbonate ion. If this is true, the concentration of the CO_3^{2-} ion at equilibrium will be roughly equal to the initial concentration of Na_2CO_3.

$$[CO_3^{2-}] \cong C_{Na_2CO_3}$$

Substituting this information into the K_{b1} expression gives the following result.

$$K_{b1} = \frac{[HCO_3^-][OH^-]}{[CO_3^{2-}]} = \frac{[\Delta][\Delta]}{[0.100]} \cong 2.1 \times 10^{-4}$$

This approximate equation can now be solved for Δ.

$$\Delta \cong 0.0046 \, M$$

We can use this value of Δ to calculate the equilibrium concentrations of OH^-, HCO_3^-, and CO_3^{2-}.

$$[CO_3^{2-}] = 0.100 - \Delta \cong 0.095 \, M$$
$$[OH^-] = [HCO_3^-] = \Delta \cong 0.0046 \, M$$

We now turn to the K_{b2} expression.

$$K_{b2} = \frac{[H_2CO_3][OH^-]}{[HCO_3^-]} = 2.2 \times 10^{-8}$$

Substituting what we know about the OH^- and HCO_3^- ion concentrations into this equation gives the following result.

$$\frac{[H_2CO_3][0.0046]}{[0.0046]} = 2.2 \times 10^{-8}$$

According to this equation, the H_2CO_3 concentration at equilibrium is equal to K_{b2} for the carbonate ion.

$$[H_2CO_3] = 2.2 \times 10^{-8} \, M$$

Summarizing the results of our calculation, once again, allows us to test the assumptions made in generating these results.

$$[CO_3^{2-}] = 0.095 \, M$$
$$[OH^-] = [HCO_3^-] = 4.6 \times 10^{-3} \, M$$
$$[H_2CO_3] = 2.2 \times 10^{-8} \, M$$

All of our assumptions are valid. The extent of the reaction between the CO_3^{2-} ion and water to give the HCO_3^- ion is less than 5% of the initial concentration of Na_2CO_3. Most of the OH^- ion comes from the first step, and most of the HCO_3^- ion formed in this step remains in solution.

15.21 CALCULATIONS FOR COMPOUNDS THAT COULD BE EITHER ACIDS OR BASES

Sometimes the hardest part of a calculation is deciding whether the compound is an acid or a base. Look at the following problem, for example.

Calculate the pH of a 0.100 *M* NaHCO$_3$ solution.

Sodium bicarbonate dissolves in water to give the bicarbonate ion.

$$NaHCO_3(s) \xrightarrow{H_2O} Na^+(aq) + HCO_3^-(aq)$$

In theory, the bicarbonate ion can act as both a Brønsted acid and a Brønsted base toward water.

$$HCO_3^-(aq) + H_2O(l) \rightleftharpoons H_3O^+(aq) + CO_3^{2-}(aq)$$
$$HCO_3^-(aq) + H_2O(l) \rightleftharpoons H_2CO_3(aq) + OH^-(aq)$$

Which reaction predominates? Is the HCO$_3^-$ ion more likely to act as an acid or as a base?

We can answer this question by comparing the equilibrium constants for these reactions. The equilibrium in which the HCO$_3^-$ ion acts as a Brønsted acid is described by K_{a2} for carbonic acid.

$$K_{a2} = \frac{[H_3O^+][CO_3^{2-}]}{[HCO_3^-]} = 4.7 \times 10^{-11}$$

The equilibrium in which the HCO$_3^-$ ion acts as a Brønsted base is described by K_{b2} for the carbonate ion.

$$K_{b2} = \frac{[H_2CO_3][OH^-]}{[HCO_3^-]} = 2.2 \times 10^{-8}$$

Since K_{b2} is larger than K_{a2}, the HCO$_3^-$ ion is more strongly basic than acidic. Solutions of the HCO$_3^-$ ion in water can therefore be handled as monoprotic bases.

Exercise 15.13

Phosphoric acid, H$_3$PO$_4$, is obviously an acid. The phosphate ion, PO$_4^{3-}$, is a base. Predict whether solutions of the H$_2$PO$_4^-$ and HPO$_4^{2-}$ ions should be acidic or basic.

Solution

Let's start by looking at the stepwise dissociation of phosphoric acid.

$$H_3PO_4(aq) + H_2O(l) \rightleftharpoons H_3O^+(aq) + H_2PO_4^-(aq) \qquad K_{a1} = 7.1 \times 10^{-3}$$
$$H_2PO_4^-(aq) + H_2O(l) \rightleftharpoons H_3O^+(aq) + HPO_4^{2-}(aq) \qquad K_{a2} = 6.3 \times 10^{-8}$$
$$HPO_4^{2-}(aq) + H_2O(l) \rightleftharpoons H_3O^+(aq) + PO_4^{3-}(aq) \qquad K_{a3} = 4.2 \times 10^{-13}$$

We can then look at the steps by which the PO$_4^{3-}$ ion picks up protons from water to form phosphoric acid.

$$PO_4^{3-}(aq) + H_2O(l) \rightleftharpoons HPO_4^{2-}(aq) + OH^-(aq) \qquad K_{b1} = ?$$
$$HPO_4^{2-}(aq) + H_2O(l) \rightleftharpoons H_2PO_4^-(aq) + OH^-(aq) \qquad K_{b2} = ?$$
$$H_2PO_4^-(aq) + H_2O(l) \rightleftharpoons H_3PO_4(aq) + OH^-(aq) \qquad K_{b3} = ?$$

Applying the procedure used for sodium carbonate in Exercise 15.12 gives the following relationships among the values of these six equilibrium constants.

$$K_{a1}K_{b3} = K_w \qquad K_{a2}K_{b2} = K_w \qquad K_{a3}K_{b1} = K_w$$

Note that since K_w is a constant, we have to multiply the largest value of K_a times the smallest value of K_b, and vice versa.

We can predict whether the $H_2PO_4^-$ ion should be an acid or a base by looking at the values of the K_a and K_b constants that characterize its reactions with water.

$$H_2PO_4^-(aq) + H_2O(l) \rightleftharpoons H_3O^+(aq) + HPO_4^{2-}(aq) \qquad K_{a2} = 6.3 \times 10^{-8}$$

$$H_2PO_4^-(aq) + H_2O(l) \rightleftharpoons H_3PO_4(aq) + OH^-(aq) \qquad K_{b3} = 1.4 \times 10^{-12}$$

According to these values, solutions of the $H_2PO_4^-$ ion in water should be acidic.

We can use the same method to decide whether the HPO_4^{2-} ion is an acid or a base when dissolved in water. We start by looking at the values of K_a and K_b that characterize its reactions with water.

$$HPO_4^{2-}(aq) + H_2O(l) \rightleftharpoons H_3O^+(aq) + PO_4^{3-}(aq) \qquad K_{a3} = 4.2 \times 10^{-13}$$

$$HPO_4^{2-}(aq) + H_2O(l) \rightleftharpoons H_2PO_4^-(aq) + OH^-(aq) \qquad K_{b2} = 1.6 \times 10^{-7}$$

Solutions of the HPO_4^{2-} ion in water should be basic.

Measurements of the pH of 0.100 M solutions of these compounds yield the following results.

H_3PO_4	$H_2PO_4^-$	HPO_4^-	PO_4^{3-}
pH = 1.6	pH = 4.1	pH = 10.1	pH = 12.7

These experimental results are consistent with the predictions of this exercise.

SUMMARY

The chemistry of water is dominated by the following equilibrium.

$$2\,H_2O(aq) \rightleftharpoons H_3O^+(aq) + OH^-(aq)$$

Because an H_3O^+ ion is created for each OH^- ion when water dissociates, the concentrations of these ions in pure water are the same: $[H_3O^+] = [OH^-] = 1.0 \times 10^{-14}\,M$. When an acid is added to water, the concentration of the H_3O^+ ion becomes larger. As expected from LeChatelier's principle, this suppresses the dissociation of water, and the OH^- ion concentration becomes smaller. The opposite occurs when a base dissolves in water.

The tendency of an acid to dissociate in water is often expressed in terms of the acid-dissociation equilibrium constant, K_a. The K_a expression for a generic acid — HA — is written as follows.

$$K_a = \frac{[H_3O^+][A^-]}{[HA]}$$

Because strong acids dissociate more or less completely, the pH of these solutions is usually equal to the concentration of the acid. For strong acids, the dissociation of water can be ignored for all but the most dilute solutions.

For weak acids, we often assume that the amount of dissociation is small compared with the initial concentra-

tion of the acid but large enough to allow us to ignore the dissociation of water.

$$\Delta = \sqrt{K_a C_a}$$

For very weak acids, we have to include the dissociation of water in our calculations.

$$\Delta = \sqrt{K_a C_a + K_w}$$

When the acid is too strong to assume that Δ is small, but not strong enough to be included among the strong acids, we solve the following equation using either the quadratic formula or successive approximations.

$$K_a = \frac{[\Delta][\Delta]}{[C_a - \Delta]}$$

The reaction between a base and water is described in terms of a base-ionization equilibrium constant expression.

$$K_b = \frac{[BH^+][OH^-]}{[B]}$$

When the value of K_b is known, these equilibria are handled with the same techniques as solutions of acids in water. When K_b is not known, it can be calculated from the value of K_a for the conjugate acid.

$$K_a K_b = K_w$$

Mixtures of weak acids and their conjugate bases, or weak bases and their conjugate acids, are known as buffers. They have the remarkable ability to resist changes in pH when small amounts of acid or base are added to the solution. Much of the shape of a pH titration curve is dictated by the chemistry of buffers. When a weak acid is titrated with a strong base, for example, the pH initially rises as some of the acid is neutralized. The net effect of this reaction, however, is the creation of a buffer solution that resists further changes in pH until essentially all of the acid has been consumed. At that point, the pH rises rapidly. This rapid change in pH near the equivalence point of the titration indicates the point at which equivalent amounts of acid and base have been added to the solution.

Calculations involving diprotic and triprotic acids are greatly simplified by the fact that there is usually a large difference between the magnitude of K_a for the loss of the first and second or second and third protons. The equilibria in which these acids lose protons can therefore be handled one at a time, starting with the loss of the first proton. Calculations for polyprotic bases are very similar.

Some compounds, such as $NaHCO_3$, $NaHS$, NaH_2PO_4, and Na_2HPO_4, can act as both Brønsted acids and Brønsted bases. The easiest way to decide which of these reactions dominates is to evaluate the relative magnitude of the equilibrium constants for the reactions. The $H_2PO_4^-$ ion, for example, is more strongly acidic, whereas the HPO_4^{2-} ion is more strongly basic.

PROBLEMS

The Acid-Base Chemistry of Water

15-1 Define the following terms; *acid, base, monoprotic, diprotic, triprotic, weak acid, strong acid, weak base, strong base.*

15-2 Describe the difference between strong acids, such as hydrochloric acid, and weak acids, such as acetic acid, and the difference between strong bases, such as sodium hydroxide, and weak bases, such as ammonia.

15-3 Explain why the concentrations of the H_3O^+ ion and the OH^- ion in pure water are the same.

15-4 The dissociation of water is an endothermic reaction.

$$2\,H_2O(aq) \rightleftharpoons H_3O^+(aq) + OH^-(aq) \qquad \Delta H° = 55.84\ kJ$$

Use the discussion of LeChatelier's principle in Chapter 14 to predict what should happen to the fraction of water molecules that dissociate into ions as the temperature of water increases.

15-5 Explain why adding a strong acid to water suppresses the dissociation of water.

15-6 Explain why it is impossible for water to be at equilibrium when it contains large quantities of both the H_3O^+ and OH^- ions.

pH and pOH

15-7 Calculate the number of H_3O^+ and OH^- ions in 1.00 milliliter of pure water.

15-8 Calculate the pH and pOH of a 0.035 M HCl solution.

15-9 Calculate the pH and pOH of a solution that contains 0.568 grams of HCl per 250 milliliters of solution.

15-10 Calculate the pH and pOH of a solution that contains 5×10^{-8} moles of HCl per liter of solution.

15-11 Explain why you don't have to worry about the dissociation of water in Problem 15-9 but you do have to worry about the dissociation of water in Problem 15-10.

15-12 Calculate the H_3O^+ and OH^- ion concentrations in a solution that has a pH of 3.72.

15-13 Explain why the pH of a solution becomes smaller as the H_3O^+ ion concentration becomes larger.

Acid-Dissociation Equilibrium Constants

15-14 Which of the following factors influences the value of K_a for the dissociation of formic acid?

$$HCO_2H(aq) + H_2O(aq) \rightleftharpoons HCO_2^-(aq) + H_3O^+(aq)$$

(a) temperature (b) pressure (c) pH (d) the concentration of HCO_2H (e) the concentration of the HCO_2^- ion

15-15 Which of the following solutions is the most acidic?

(a) 0.10 M CH_3CO_2H: $K_a = 1.8 \times 10^{-5}$
(b) 0.10 M HCO_2H: $K_a = 1.8 \times 10^{-4}$
(c) 0.10 M $ClCH_2CO_2H$: $K_a = 1.4 \times 10^{-3}$
(d) 0.10 M Cl_2CHCO_2H: $K_a = 5.1 \times 10^{-2}$

15-16 Which of the following compounds is the strongest base?

(a) $CH_3CO_2^-$ (CH_3CO_2H: $K_a = 1.8 \times 10^{-5}$)
(b) HCO_2^- (HCO_2H: $K_a = 1.8 \times 10^{-4}$)
(c) $ClCH_3CO_2^-$ ($ClCH_2CO_2H$: $K_a = 1.4 \times 10^{-3}$)
(d) $Cl_2CHCO_2^-$ (Cl_2CHCO_2H: $K_a = 7.8 \times 10^{-3}$)

15-17 List acetic acid, chlorous acid, hydrofluoric acid, and nitrous acid in order of increasing strength if 0.10 M solutions of these acids contain the following equilibrium concentrations.

Acid	HA	A⁻	H₃O⁺
HOAc	0.099M	0.0013 M	0.0013 M
HOClO	0.050 M	0.050 M	0.050 M
HF	0.092 M	0.0081 M	0.0081 M
HNO₂	0.093 M	0.0069 M	0.0069 M

Strong Acids

15-18 Explain why the H_3O^+ ion concentration in a strong acid solution depends on the concentration of the solution but not the value of K_a for the acid.

15-19 Calculate the pH of a 0.056 M solution of hydrochloric acid.

15-20 Nitric acid is often grouped with sulfuric acid and hydrochloric acid as one of the strong acids. Calculate the pH of 0.10 M nitric acid assuming that it is a strong acid that dissociates completely. Calculate the pH of this solution using the value of K_a for the acid. (HNO_3: $K_a = 28$)

Hidden Assumptions in Weak-Acid Calculations

15-21 Describe the two assumptions that are commonly made in weak-acid equilibrium problems. Describe how you can test whether these assumptions are valid for a particular calculation.

15-22 Explain why the techniques used to calculate the equilibrium concentrations of the components of a weak acid solution can't be used for either strong acids or very weak acids.

Factors That Influence the H₃O⁺ Ion Concentration in Weak Acid Solutions

15-23 Explain why the H_3O^+ ion concentration in a weak acid solution depends on both the value of K_a for the acid and the concentration of the acid.

15-24 Which of the following solutions has the largest H_3O^+ ion concentration?

(a) 0.10 M HOAc (b) 0.010 M HOAc
(c) 0.0010 M HOAc

Weak Acids

15-25 Calculate the pH and the percent of HOAc molecules that dissociate in 0.10 M, 0.010 M, and 0.0010 M solutions of acetic acid. What happens to the percent ionization as the solution becomes more dilute? What happens to the pH? (HOAc: $K_a = 1.8 \times 10^{-4}$)

15-26 Formic acid, HCO_2H, was first isolated by the destructive distillation of ants. In fact, the name comes from the Latin word for ants, *formi*. Calculate the HCO_2H, HCO_2^-, and H_3O^+ concentrations in an 0.010 M solution of formic acid in water. (HCO_2H: $K_a = 1.8 \times 10^{-4}$)

15-27 Hydrogen cyanide, HCN, is a gas that dissolves in water to form hydrocyanic acid. Calculate the H_3O^+, HCN, and CN^- concentrations in a 0.174 M solution of hydrocyanic acid. (HCN: $K_a = 6 \times 10^{-10}$)

15-28 Calculate the molarity of an HCN solution that is 0.01% ionized at equilibrium. (HCN: $K_a = 6 \times 10^{-10}$)

15-29 The first disinfectant used by Joseph Lister was called carbolic acid. This substance is now known as phenol. Calculate the H_3O^+ ion concentration in a 0.0167 M solution of phenol, PhOH. (PhOH: $K_a = 1.0 \times 10^{-10}$)

15-30 Calculate the concentration of acetic acid that would give an H_3O^+ ion concentration of 2.0×10^{-3} M. (HOAc: $K_a = 1.8 \times 10^{-5}$)

15-31 Calculate the value of K_a for ascorbic acid (vitamin C) assuming that 2.8% of the ascorbic acid molecules in a 0.10 M solution dissociate.

15-32 Calculate the value of K_a for nitrous acid, HNO_2, assuming that a 0.10 M solution is 7.1% dissociated at equilibrium.

Not-So-Weak Acids

15-33 Calculate the H_3O^+ ion concentration in 0.10 M solutions of acetic acid (CH_3CO_2H: $K_a = 1.8 \times 10^{-5}$), chloroacetic acid ($ClCH_2CO_2H$: $K_a = 1.4 \times 10^{-3}$), and dichloroacetic acid (Cl_2CHCO_2H: $K_a = 5.1 \times 10^{-2}$). What happens to the H_3O^+ ion concentration as K_a becomes larger?

15-34 Trichloroacetic acid (Cl_3CCO_2H: $K_a = 0.22$) is a much stronger acid than acetic acid. Calculate the concentrations of Cl_3CO_2H, $Cl_3CCO_2^-$, and H_3O^+ in a 0.250 M solution of trichloroacetic acid.

When Can the Dissociation of Water Be Ignored?

15-35 After working a weak-acid equilibrium problem, we can use a rule of thumb to help us decide whether it was legitimate to ignore the dissociation of water in our calculations. Describe this rule of thumb.

15-36 Explain why the H_3O^+ ion concentration from the dissociation of a weak acid does not have to be 20 times larger than the H_3O^+ ion concentration in pure water for the dissociation of water to be ignored.

Very Weak Acids

15-37 Under which of the following conditions can we legitimately ignore the contribution to the total H_3O^+ ion concentration from the dissociation of water?

(a) When $K_aC_a < 1.0 \times 10^{-13}$ (b) When $K_aC_a > 1.0 \times 10^{-13}$ (c) When $K_aC_a = 1.0 \times 10^{-13}$

15-38 Calculate the H_3O^+ ion concentration in a solution that is 1×10^{-4} M in hydrocyanic acid. (HCN: $K_a = 6 \times 10^{-10}$)

15-39 Explain why you can ignore the dissociation of water in Problem 15-28 but not in Problem 15-41.

Summarizng the Chemistry of Weak Acids

15-40 Boric acid, H_3BO_3, is a weak acid found in many first-aid products, such as eye drops. Calculate the H_3O^+ and OH^- ion concentrations in a 0.0024 M solution of boric acid in water. Clearly state and justify any assumptions you make. (H_3BO_3: $K_a = 5.8 \times 10^{-10}$)

Bases

15-41 Which of the following equations correctly describe the relationship between K_b for the formate ion, HCO_2^-, K_a for formic acid, HCO_2H?

(a) $K_b = K_w \times K_a$ (b) $K_b = K_a/K_w$ (c) $K_b = K_w/K_a$
(d) $K_b = K_w + K_a$ (e) $K_b = K_w - K_a$

15-42 Use the relationship between K_a for an acid and K_b for its conjugate base to explain why strong acids have weak conjugate bases and weak acids have strong conjugate bases.

15-43 Calculate the HCO_2H, OH^-, and HCO_2^- ion concentrations in a solution that contains 0.020 moles of sodium formate ($NaHCO_2$) in 250 milliliters of solution. (HCO_2H: $K_a = 1.8 \times 10^{-4}$)

15-44 Calculate the OH^-, $HOBr$, and OBr^- ion concentrations in a solution that contains 0.050 moles of sodium hypobromite ($NaOBr$) in 500 milliliters of solution. ($HOBr$: $K_a = 2.4 \times 10^{-9}$)

15-45 Calculate the pH of a 0.756 M solution of NaOAc. ($HOAc$: $K_a = 1.8 \times 10^{-5}$)

15-46 A solution of NH_3 dissolved in water is known as both aqueous ammonia and ammonium hydroxide. Use the value of K_b for the following reaction to explain why aqueous ammonia is the better name.

$$NH_3(aq) + H_2O \rightleftharpoons NH_4^+(aq) + OH^-(aq) \quad K_b = 1.8 \times 10^{-5}$$

15-47 At 25°C, a 0.10 M aqueous solution of methylamine, CH_3NH_2, is 6.9% ionized.

$$CH_3NH_2(aq) + H_2O \rightleftharpoons CH_3NH_3^+(aq) + OH^-(aq)$$

Calculate K_b for methylamine. Is methylamine a stronger base or a weaker base than ammonia?

15-48 Explain why a 0.10 M aqueous solution of NaF is slightly basic (pH = 8) whereas a 0.10 M aqueous solution of NaCl is neutral. (HF: $K_a = 7.2 \times 10^{-4}$, HCl: $K_a = 1 \times 10^6$)

15-49 What is the molarity of an aqueous ammonia solution that has an OH^- ion concentration of 1.0×10^{-3} M?

15-50 Proteins contain nitrogen. When they are digested, they are converted to carbohydrates, which do not contain nitrogen. The nitrogen in proteins therefore has to be excreted from the organism. Mammals excrete this nitrogen as urea, H_2NCONH_2. Calculate the pH of a 0.10 M solution of urea. (H_2NCONH_2: $K_b = 1.5 \times 10^{-14}$)

15-51 Calculate K_b for hydrazine, H_2NNH_2, if the pH of a 0.10 M aqueous solution of this rocket fuel is 10.53.

15-52 Calculate the pH of a 0.015 M aqueous solution of calcium acetate, $Ca(OAc)_2$. ($HOAc$: $K_a = 1.8 \times 10^5$)

Buffers, Buffering Capacity, and pH Titration Curves

15-53 Explain how buffers resist changes in pH.

15-54 Explain why a mixture of HOAc and NaOAc is an acidic buffer but a mixture of NH_3 and NH_4Cl is a basic buffer. ($HOAc$: $K = 1.8 \times 10^{-5}$, NH_4^+: $K_a = 5.6 \times 10^{-10}$)

15-55 Which of the following mixtures would make the best buffer?

(a) HCl and NaCl   (b) NaOAc and NH_3 (c) HOAc and NH_4Cl (d) NaOAc and NH_4Cl (e) NH_3 and NH_4Cl

15-56 Which of the following solutions is an acidic buffer?

(a) 0.10 M HCl and 0.10 M NaOH   (b) 0.10 M HCl and 0.10 M NaCl   (c) 0.10 M HCO_2H and 0.10 M $NaHCO_2$ (d) 0.10 M NH_3 and 0.10 M NH_4Cl

15-57 Which of the following solutions is a basic buffer?

(a) 0.10 M HCl and 0.10 M NaOH   (b) 0.10 M HCl and 0.10 M NaCl   (c) 0.10 M HCO_2H and 0.10 M $NaHCO_2$ (d) 0.10 M NH_3 and 0.10 M NH_4Cl

15-58 What is the best way of increasing the buffering capacity of a buffer made from $NaHCO_2$ dissolved in an aqueous solution of HCO_2H?

(a) Increase the concentration of HCO_2H (b) Increase the concentration of $NaHCO_2$ (c) Increase the concentrations of both HCO_2H and $NaHCO_2$ (d) Increase the ratio of the concentration of HCO_2H to the concentration of $NaHCO_2$ (e) Increase the ratio of the concentration of $NaHCO_2$ to the concentration of HCO_2H

15-59 Describe in detail the experiment you would use to measure the value of K_a for formic acid, HCO_2H.

15-60 Sketch a titration curve for a weak acid reacting with a strong base. Label the equivalence point, the endpoint, and the point at which the H_3O^+ ion concentration is equal to K_a for the acid.

Buffer Calculations

15-61 Calculate the pH of a solution that contains 0.010 moles per liter of proprionic acid (HOPr) and 0.080 moles per liter of potassium proprionate (KOPr). (HOPr: $K_a = 1.3 \times 10^{-5}$)

15-62 Calculate the pH of a solution prepared from 0.040 moles of sodium nitrite, $NaNO_2$, dissolved in 200 milliliters of 0.10 M nitrous acid. (HNO_2: $K_a = 5.1 \times 10^{-4}$)

15-63 Calculate the H_3O^+ ion concentration and the pH of a buffer solution prepared from 0.218 moles of sodium acetate dissolved in 500 milliliters of 0.100 M acetic acid. ($HOAc$: $K_a = 1.8 \times 10^{-5}$)

15-64 Calculate the H_3O^+ ion concentration in a solution that is 0.050 M in acetic acid and 0.10 M in sodium acetate. Calculate what happens when 1×10^{-3} moles of HCl are added to this solution. ($HOAc$: $K_a = 1.8 \times 10^{-5}$)

15-65 Calculate how many moles of sodium formate ($NaHCO_2$) would have to be added to 500 milliliters of 0.100 M formic acid (HCO_2H) to give a solution buffered at pH 4.11. (HCO_2H: $K_a = 1.8 \times 10^{-4}$)

15-66 Calculate the ratio of concentrations of hypochlorous acid, HOCl, and sodium hypochlorite, NaOCl, needed to produce a buffer solution with a pH of 7.60. (HOCl: $K_a = 2.9 \times 10^{-8}$)

15-67 Calculate what weight of NaOH must be added to 500 milliliters of 0.100 M NaH$_2$PO$_4$ to give a buffer with a pH of 8.10. (H$_2$PO$_4^-$: $K_a = 6.3 \times 10^{-8}$)

15-68 How much NaHCO$_2$ would you have to add to 0.10 M HCO$_2$H to get a buffer solution with a pH of 3.4? (HCO$_2$H: $K_a = 1.8 \times 10^{-4}$)

15-69 Most bacteria will not grow in solutions that are more acidic than pH = 4.5. What ratio of concentrations of HOAc and OAc$^-$ would you need to prepare a buffer with a pH of 4.5?

15-70 What ratio of concentrations of HPO$_4^{2-}$ and H$_2$PO$_4^-$ would you need to prepare a buffer with a pH of 7.00? (H$_2$PO$_4^-$: $K_a = 6.3 \times 10^{-8}$)

15-71 Calculate what weight of NH$_4$Cl must be added to 100 milliliters of 0.300 M NH$_3$ to obtain a solution with a pH of 9.00.

Polyprotic Acids

15-72 Calculate the H$_3$O$^+$, CO$_3^{2-}$, HCO$_3^-$, and H$_2$CO$_3$ concentrations at equilibrium in a solution that initially contained 0.10 moles of carbonic acid per liter. (H$_2$CO$_3$: $K_{a1} = 4.5 \times 10^{-7}$, $K_{a2} = 4.7 \times 10^{-11}$.

15-73 Calculate the equilibrium concentrations of all of the important components of a 0.25 M malonic acid, HO$_2$CCH$_2$CO$_2$H, solution. Use the symbol H$_2M$ as an abbreviation for malonic acid and assume stepwise dissociation of this acid. (H$_2M$: $K_{a1} = 1.4 \times 10^{-5}$, $K_{a2} = 2.1 \times 10^{-6}$)

15-74 Check the validity of assuming in Problem 15-79 that malonic acid dissociates in a stepwise fashion by comparing the concentrations of the HM^- and M^{2-} ions. Is this assumption valid?

15-75 Glycine, the simplest of the amino acids found in proteins, is a diprotic acid with the formula HO$_2$CCH$_2$NH$_3^+$. If we symbolize glycine as H$_2Gly^+$, we can write the following equations for the stepwise dissociation of this amino acid.

$$H_2Gly^+(aq) + H_2O \rightleftharpoons$$
$$HGly(aq) + H_3O^+(aq) \qquad K_{a1} = 4.5 \times 10^{-3}$$
$$HGly(aq) + H_2O \rightleftharpoons Gly^-(aq) + H_3O^+ \qquad K_{a2} = 2.5 \times 10^{-10}$$

Calculate the concentrations of H$_3$O$^+$, Gly^-, HGly, and H$_2Gly^+$ in a 2.0 M solution of glycine in water.

15-76 Which of the following equations accurately describes a 2.0 M solution of glycine?

(a) $[H_3O^+] \cong [H_2Gly^+]$ (b) $[H_3O^+] < [H_2Gly^+]$
(c) $[H_3O^+] > [HGly]$ (d) $[H_3O^+] \cong [HGly]$ (e) $[H_3O^+] < [HGly]$

15-77 Which of the following equations results from the fact that glycine is a weak diprotic acid?

(a) $[Gly^-] \cong C_{H_2Gly^+}$ (b) $[Gly^-] \cong K_{a2}$ (c) $[Gly^-] \cong [HGly]$
(d) $[Gly^-] > [HGly]$ (e) $[Gly^-] \cong [H_3O^+]$

15-78 Oxalic acid (H$_2$C$_2$O$_4$) has been implicated in diseases such as gout and kidney stones. Calculate the H$_3$O$^+$, H$_2$C$_2$O$_4$,

HC$_2$O$_4^-$, and C$_2$O$_4^{2-}$ concentrations in a 1.25 M solution of oxalic acid.

$$H_2C_2O_4(aq) + H_2O \rightleftharpoons$$
$$H_3O^+(aq) + HC_2O_4^-(aq) \qquad K_{a1} = 5.4 \times 10^{-2}$$
$$HC_2O_4^-(aq) + H_2O \rightleftharpoons$$
$$H_3O^+(aq) + C_2O_4^{2-}(aq) \qquad K_{a2} = 5.4 \times 10^{-5}$$

Polyprotic Bases

15-79 Which of the following sets of equations can be used to calculate K_{b1} and K_{b2} for sodium oxalate, Na$_2$C$_2$O$_4$, from K_{a1} and K_{a2} for oxalic acid, H$_2$C$_2$O$_4$?

(a) $K_{b1} = K_w \times K_{a1}$ and $K_{b2} = K_w \times K_{a2}$ (b) $K_{b1} = K_w/K_{a1}$ and $K_{b2} = K_w/K_{a2}$ (c) $K_{b1} = K_{a1}/K_w$ and $K_{b2} = K_{a2}/K_w$
(d) $K_{b1} = K_w \times K_{a2}$ and $K_{b2} = K_w \times K_{a1}$ (e) $K_{b1} = K_w/K_{a2}$ and $K_{b2} = K_w/K_{a1}$

15-80 Calculate the pH of a 0.028 M solution of sodium oxalate, Na$_2$C$_2$O$_4$, in water.

$$H_2C_2O_4(aq) + H_2O \rightleftharpoons$$
$$H_3O^+(aq) + HC_2O_4^-(aq) \qquad K_{a1} = 5.4 \times 10^{-2}$$
$$HC_2O_4^-(aq) + H_2O \rightleftharpoons$$
$$H_3O^+(aq) + C_2O_4^{2-}(aq) \qquad K_{a2} = 5.4 \times 10^{-5}$$

15-81 Calculate the H$_3$O$^+$, OH$^-$, H$_2$CO$_3$, HCO$_3^-$, and CO$_3^{2-}$ concentrations in a 0.150 M solution of sodium carbonate, Na$_2$CO$_3$, in water. (H$_2$CO$_3$: $K_{a1} = 4.5 \times 10^{-7}$, $K_{a2} = 4.7 \times 10^{-11}$)

Compounds That Could Be Either Acids or Bases

15-82 Use K_{a1} and K_{a2} for carbonic acid as the basis for calculating the pH of a 0.10 M NaHCO$_3$ solution.

$$H_2CO_3(aq) + H_2O \rightleftharpoons$$
$$H_3O^+(aq) + HCO_3^-(aq) \qquad K_{a1} = 4.5 \times 10^{-7}$$
$$HCO_3^-(aq) + H_2O \rightleftharpoons$$
$$H_3O^+(aq) + CO_3^{2-}(aq) \qquad K_{a2} = 4.7 \times 10^{-11}$$

15-83 Which of the following statements concerning sodium hydrogen sulfide, NaHS, is correct? (H$_2$S: $K_{a1} = 1.0 \times 10^{-7}$, $K_{a2} = 1.3 \times 10^{-13}$)

(a) NaHS is an acid because K_{a1} for H$_2$S is much larger than K_{a2}. (b) NaHS is an acid because K_{a1} for H$_2$S is smaller than K_{b1} for Na$_2$S. (c) NaHS is a base because K_{b1} for Na$_2$S is much larger than K_{b2}. (d) NaHS is a base because K_{b2} for Na$_2$S is larger than K_{a2} for H$_2$S.

15-84 Calculate the pH of 0.10 M solutions of H$_3$PO$_4$, NaH$_2$PO$_4$, Na$_2$HPO$_4$, and Na$_3$PO$_4$. (H$_3$PO$_4$: $K_{a1} = 7.1 \times 10^{-3}$, $K_{a2} = 6.3 \times 10^{-8}$, $K_{a3} = 4.2 \times 10^{-13}$)

15-85 Predict whether an aqueous solution of sodium hydrogen sulfate, NaHSO$_4$, should be acidic, basic, or neutral. (H$_2$SO$_4$: $K_{a1} = 1 \times 10^3$, $K_{a2} = 1.2 \times 10^{-2}$)

15-86 Predict whether an aqueous solution of sodium hydrogen sulfite, NaHSO$_3$, should be acidic, basic, or neutral. (H$_2$SO$_3$: $K_{a1} = 1.7 \times 10^{-2}$, $K_{a2} = 6.4 \times 10^{-8}$)

SOLUBILITY PRODUCT EQUILIBRIA

16.1 WHAT HAPPENS WHEN SOLIDS DISSOLVE IN WATER?

If asked to describe what happens when a teaspoon of sugar is stirred into a cup of coffee, you might answer as follows.

> The sugar initially settles to the bottom of the cup. When the coffee is stirred, it dissolves to produce a sweeter cup of coffee.

Most people hesistate when asked to extend this description to the molecular level. If they are chemistry students, like yourself, they often ask to be reminded of the molecular formula for sugar. When told that the sugar used in cooking is $C_{12}H_{22}O_{11}$, they write equations such as the following.

$$C_{12}H_{22}O_{11}(s) \xrightarrow{\text{H}_2\text{O}} C_{12}H_{22}O_{11}(aq)$$

Although this equation is true, it doesn't provide any hints about why some solids don't dissolve in water. To describe what happens at the molecular level when a solid dissolves we have to refer back to the distinctions introduced in Chapter 12 among molecular solids, covalent solids, and ionic solids.

Molecular solids are covalent compounds that exist as molecules, which are held together in the solid state by relatively weak intermolecular forces. The sugar we use to sweeten coffee or tea is an example of a molecular solid. It is known by the common name *sucrose* and consists of molecules with the formula $C_{12}H_{22}O_{11}$. When sucrose dissolves in water, the weak bonds between sucrose molecules are broken, and the individual $C_{12}H_{22}O_{11}$ molecules are released into solution.

It takes energy to break the bonds between the $C_{12}H_{22}O_{11}$ molecules in sucrose. It also takes energy to break the hydrogen bonds in water that must be disrupted to insert one of these molecules into solution. Sugar dissolves in water because energy is given off when the slightly polar sucrose molecules form intermolecular bonds with the polar water molecules. The weak bonds that form between the solute and the solvent compensate for the energy needed to disrupt the structure of both the pure solute and the solvent. In the case of sugar and water, this process works so well that up to 1800 grams of sucrose can dissolve in a liter of water.

Some elements and compounds that are held together by covalent bonds do not form molecules. They form *covalent solids,* which consist of extended arrays of covalent bonds. Two examples of covalent solids are diamond and silicon carbide (SiC). It takes much more energy to break a covalent bond than it does to break the weak bonds between molecules. As a result, covalent solids are essentially insoluble in water.

Ionic solids (or salts) are compounds that consist of positive and negative ions. These ions are held together by the strong force of attraction between particles with opposite charges. When one of these solids dissolves in water, the particles that form the solid are released into solution, where they become associated with the polar solvent molecules. This dissociation can be symbolized as follows.

$$NaCl(s) \xrightarrow{\text{H}_2\text{O}} Na^+(aq) + Cl^-(aq)$$

As a rule, we can assume that salts dissociate into their ions when they dissolve in water. An ionic compound will dissolve in water when the energy given off as the ions interact with water molecules compensates for the energy needed to break the ionic bonds in the solid and the energy required to separate the water molecules so that the ions can be inserted into solution.

As noted in Chapter 7, it takes an enormous amount of energy to rip apart an ionic crystal. The lattice energy of sodium chloride, for example, is 787.3 kJ/mol.

This means that 787.3 kilojoules of energy is released when a mole of Na^+ and Cl^- ions in the gas phase come together to form solid NaCl. But it also means that 787.3 kJ/mol of energy is needed to transform solid NaCl into isolated Na^+ and Cl^- ions in the gas phase.

$$NaCl(s) \longrightarrow Na^+(g) + Cl^-(g) \qquad \Delta H^\circ = 787.3 \text{ kJ/mol}$$

It takes so much energy to separate the Na^+ and Cl^- ions in NaCl that we might not expect this compound to dissolve in water.

The force of attraction between Na^+ ions and water molecules is so large, however, that 783.5 kJ/mol of energy is released when the Na^+ and Cl^- ions interact with water molecules.

$$Na^+(g) + Cl^-(g) \xrightarrow{H_2O} Na^+(aq) + Cl^-(aq) \qquad \Delta H^\circ = -783.5 \text{ kJ/mol}$$

It takes a great deal of energy to take apart a sodium chloride crystal, but a similar amount of energy is released when the ions interact with water molecules to form an aqueous solution. The overall enthalpy of reaction for the process in which solid NaCl dissolves in water is therefore very small.

$$NaCl(s) \longrightarrow Na^+(g) + Cl^-(g) \qquad\qquad \Delta H_1^\circ = 787.3 \text{ kJ/mol}$$

$$\underline{Na^+(g) + Cl^-(g) \xrightarrow{H_2O} Na^+(aq) + Cl^-(aq) \qquad \Delta H_2^\circ = -783.5 \text{ kJ/mol}}$$

$$NaCl(s) \xrightarrow{H_2O} Na^+(aq) + Cl^-(aq) \qquad\qquad \Delta H^\circ = 3.8 \text{ kJ/mol}$$

A large positive value for the overall enthalpy of reaction would have suggested that the reactants were more stable than the products of this reaction. A large negative value for ΔH° would have suggested that the products were more stable than the reactants. Because ΔH° for this reaction is small, we can only conclude that thermodynamics favors neither the reactants nor the products. We might therefore expect NaCl to be moderately soluble in water. In fact, up to 360 grams of NaCl will dissolve in a liter of water at room temperature, to give a solution with a concentration of about 6 M.

Now that we have some idea of why NaCl dissolves in water we can understand why other ionic compounds, such as silver chloride, do not. The lattice energy for AgCl is very large — 915.7 kJ/mol.

$$AgCl(s) \longrightarrow Ag^+(g) + Cl^-(g) \qquad \Delta H^\circ = 915.7 \text{ kJ/mol}$$

The energy released when the Ag^+ and Cl^- ions interact with water is also large.

$$Ag^+(g) + Cl^-(g) \xrightarrow{H_2O} Ag^+(aq) + Cl^-(aq) \qquad \Delta H^\circ = -850.2 \text{ kJ/mol}$$

But it isn't large enough to compensate for the energy needed to separate the ions in this crystal. The overall enthalpy of reaction is therefore unfavorable — the products of the reaction are less stable than the reactants.

$$AgCl(s) \longrightarrow Ag^+(g) + Cl^-(g) \qquad\qquad \Delta H_1^\circ = 915.7 \text{ kJ/mol}$$

$$\underline{Ag^+(g) + Cl^-(g) \xrightarrow{H_2O} Ag^+(aq) + Cl^-(aq) \qquad \Delta H_2^\circ = -850.2 \text{ kJ/mol}}$$

$$AgCl(s) \xrightarrow{H_2O} Ag^+(aq) + Cl^-(aq) \qquad\qquad \Delta H^\circ = 65.5 \text{ kJ/mol}$$

We would therefore expect little AgCl to dissolve in water. In fact, less than 0.002 gram of AgCl will dissolve in a liter of water at room temperature. The solubility of silver chloride in water is so small that it is often said to be "insoluble" in water, even though this term is misleading.

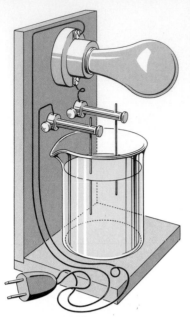

FIG. 16.1 This conductivity apparatus can be used to demonstrate the difference between aqueous solutions of ionic and covalent compounds. Ionic compounds dissociate into ions when they dissolve in water. These ions are attracted toward the metal wires, which carry an electric charge. Reactions that occur when the ions collide with the wires complete the electric circuit, so the light bulb glows.

16.2 SOLUBILITY EQUILIBRIA

What evidence do we have that NaCl and AgCl behave the way the thermodynamic calculations in the preceding section predict they should? Experimental evidence to support these conclusions is based on the following principle, which is one of the basic assumptions of solubility equilibria.

When solids dissolve in water, they dissociate to give the elementary particles from which they are formed.

Thus, molecular solids dissociate to give individual molecules

$$C_{12}H_{22}O_{11}(s) \xrightarrow{H_2O} C_{12}H_{22}O_{11}(aq)$$

and ionic solids dissociate to give solutions of the positive and negative ions they contain.

$$NaCl(s) \xrightarrow{H_2O} Na^+(aq) + Cl^-(aq)$$

We can detect the presence of ions in an aqueous solution with the conductivity apparatus shown in Figure 16.1. This apparatus consists of a light bulb connected to a pair of metal wires that can be immersed in a beaker of water. The circuit is not complete. In order for the light bulb to glow when the apparatus is plugged into an electrical outlet, there must be a way for electrical charge to flow through the solution from one of these metal wires to the other.

When an ionic solid dissolves in water, the positive and negative ions are free to flow through this aqueous solution. The positive ions flow toward one of the metal wires, which carries a net negative charge. The negative ions flow toward the other metal wire, which has a positive charge. Oxidation-reduction reactions occur when these ions collide with the metal wires. The negative ions lose electrons to the positively charged metal wire, and the positive ions gain electrons from the negatively charged wire. The net result is a transfer of electrons that completes the circuit and lets the light bulb glow.

As we might expect, the brightness of the bulb is proportional to the concentration of the ions in the solution. Slightly soluble salts such as calcium sulfate make the light bulb glow dimly. When the metal wires are immersed in a solution of a very soluble salt, such as NaCl, the light bulb glows brightly.

The conductivity apparatus in Figure 16.1 only gives us qualitative information about the relative concentrations of the ions in different solutions. It is possible to build a more sophisticated instrument that gives quantitative measurements of the conductivity of a solution, which is directly proportional to the concentration of the ions in the solution. Figure 16.2 shows what happens to the conductivity of water as solid silver chloride is added to form an aqueous solution.

The system conducts an electric current even before any AgCl is added because of the H_3O^+ and OH^- ions in water. The solution becomes a slightly better conductor when AgCl is added, because some of this salt dissolves to give Ag^+ and Cl^- ions, which can carry an electric current through the solution. The conductivity continues to increase as more AgCl is added, until about 0.002 gram of this salt has dissolved per liter of solution.

The fact that the conductivity does not increase after the solution has reached this concentration tells us that there is a limit on the solubility of this salt in water. Once the solution reaches this limit, no more AgCl dissolves, regardless of how much solid we add to the system. This is exactly what we should expect if the solubility of AgCl is controlled by an equilibrium. Once the solution reaches equilibrium, the rate at which AgCl dissolves to form Ag^+ and Cl^- ions is equal to the

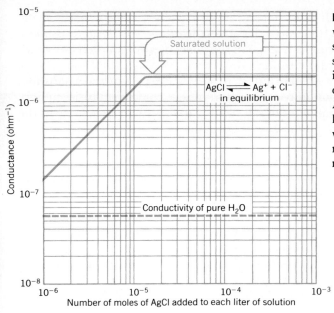

FIG. 16.2 The conductivity of water depends on the amount of solid AgCl added per liter of solution. The conductivity increases at first as the AgCl dissolves and dissociates into Ag^+ and Cl^- ions. Once the solution has become saturated with AgCl, the conductivity remains the same no matter how much solid is added.

rate at which these ions recombine to form AgCl.

$$AgCl(s) \underset{}{\overset{H_2O}{\rightleftharpoons}} Ag^+(aq) + Cl^-(aq)$$

Evidence to support this conclusion comes from conductivity measurements, such as the one illustrated in Figure 16.3. When the salt is first added, it dissolves and dissociates rapidly.

$$AgCl(s) \xrightarrow[\text{dissociate}]{\text{dissolve}} Ag^+(aq) + Cl^-(aq)$$

The conductivity of the solution therefore rapidly increases.

Eventually, the concentrations of these ions become large enough that the reverse reaction starts to compete with the forward reaction.

$$Ag^+(aq) + Cl^-(aq) \xrightarrow[\text{precipitate}]{\text{associate}} AgCl(s)$$

This leads to a decrease in the rate at which Ag^+ and Cl^- ions enter the solution.

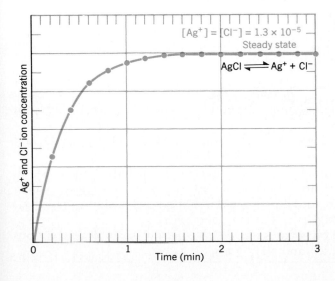

FIG. 16.3 The concentration of Ag^+ and Cl^- ions versus time as a sample of AgCl dissolves in water.

Eventually, the Ag^+ and Cl^- ion concentrations become large enough that the rate at which precipitation occurs exactly balances the rate at which AgCl dissolves. Once that happens, there is no net change in the concentration of these ions with time, and the reaction is at equilibrium.

When this system reaches equilibrium it is often called a **saturated solution,** because it contains the maximum concentration of ions that can exist in equilibrium with the solid salt. The amount of salt needed to form a given volume of the saturated solution is called the **solubility** of the salt.

16.3 SOLUBILITY RULES

The data needed to do the thermodynamic calculations in Section 16.1 are available for few compounds. These calculations are therefore more useful in explaining the results of measurements of the solubility of salts than at predicting what happens when a salt is added to water.

There are a number of obvious patterns in the data obtained from measuring the solubility of different salts. These patterns form the basis for the rules outlined in Table 16.1, which can be used to predict whether a given salt will dissolve in water. The rules are based on the following set of definitions of the terms **soluble, insoluble,** and **slightly soluble.**

> **A salt is soluble if it dissolves in water to give a solution with a concentration of at least 0.1 moles per liter at room temperature.**

> **A salt is insoluble if the concentration of an aqueous solution is less than 0.001 M at room temperature.**

TABLE 16.1

Solubility Rules for Ionic Compounds in Water

Soluble Salts

1. The Na^+, K^+, and NH_4^+ ions form soluble salts. Thus, NaCl, KNO_3, $(NH_4)_2SO_4$, Na_2S, and $(NH_4)_2CO_3$ are soluble.
2. The nitrate (NO_3^-) ion forms soluble salts. Thus, $Cu(NO_3)_2$ and $Fe(NO_3)_3$ are soluble.
3. The chloride (Cl^-), bromide (Br^-), and iodide (I^-) ions generally form soluble salts. Exceptions to this rule include salts of the Pb^{2+}, Hg_2^{2+}, Ag^+, and Cu^+ ions. $ZnCl_2$ is soluble, but CuBr is not.
4. The sulfate (SO_4^{2-}) ion generally forms soluble salts. Exceptions include $BaSO_4$, $SrSO_4$, and $PbSO_4$, which are insoluble, and Ag_2SO_4, $CaSO_4$, and Hg_2SO_4, which are only slightly soluble.

Insoluble Salts

1. Sulfides (S^{2-}) are usually insoluble. Exceptions include Na_2S, K_2S, $(NH_4)_2S$, MgS, CaS, SrS, and BaS.
2. Oxides (O^{2-}) are usually insoluble. Exceptions include Na_2O, K_2O, SrO, and BaO, which are soluble, and CaO, which is slightly soluble.
3. Hydroxides (OH^-) are usually insoluble. Exceptions include NaOH, KOH, $Sr(OH)_2$, $Ba(OH)_2$, which are soluble, and $Ca(OH)_2$, which is slightly soluble.
4. Chromates (CrO_4^{2-}) are usually insoluble. Exceptions include Na_2CrO_4, K_2CrO_4, $(NH_4)_2CrO_4$, $MgCrO_4$, $CaCrO_4$, and $SrCrO_4$.
5. Phosphates (PO_4^{3-}) and carbonates (CO_3^{2-}) are usually insoluble. Exceptions include salts of the Na^+, K^+, and NH_4^+ ions.

Slightly soluble salts give solutions with concentrations that fall between these extremes.

Exercise 16.1

Which of the following salts are soluble in water?

(a) $PbSO_4$ (d) $Ba(OH)_2$
(b) $Mg(NO_3)_2$ (e) $(NH_4)_2CrO_4$
(c) ZnO (f) $CuCl_2$

Solution

The salts (b), (d), (e), and (f) are soluble in water. Note that (f) is $CuCl_2$, *not* CuCl. CuCl contains copper in the $+1$ oxidation state, and the Cu^+ ion forms an insoluble chloride. $CuCl_2$ contains copper in the $+2$ oxidation state, and $CuCl_2$ is soluble.

16.4 THE SOLUBILITY PRODUCT EXPRESSION

Silver chloride is so insoluble in water that a saturated solution contains only about 1.3×10^{-5} moles of AgCl per liter of water.

$$AgCl(s) \xrightleftharpoons{H_2O} Ag^+(aq) + Cl^-(aq)$$

Strict adherence to the rules for writing equilibrium constant expressions for this reaction gives the following result.

$$K_c = \frac{[Ag^+][Cl^-]}{[AgCl]}$$

Water is not included in the equilibrium constant expression because it is neither consumed nor produced in this reaction, even though it is a vital component of the system.

Two of the terms in this expression are easy to interpret. The $[Ag^+]$ and $[Cl^-]$ terms represent the concentrations of the Ag^+ and Cl^- ions in moles per liter when this reaction is at equilibrium — when the solution is saturated with AgCl.

The third term — [AgCl] — is more ambiguous. It doesn't represent the concentration of AgCl dissolved in water, because there is no AgCl dissolved in water. (AgCl dissociates into Ag^+ ions and Cl^- ions when it dissolves in water.) The term can't represent the amount of solid AgCl in the system, because the equilibrium is not affected by the amount of excess solid added to the system, as shown by Figure 16.2. The [AgCl] term has to be translated quite literally as the concentration of AgCl — the number of moles of AgCl in a liter of solid AgCl.

The concentration of AgCl in solid AgCl can be calculated from its density and molecular weight.

$$5.56 \; \frac{g\ AgCl}{cm^3} \times 1 \; \frac{cm^3}{mL} \times 1000 \; \frac{mL}{L} \times \frac{1\ mol\ AgCl}{143.34\ g} = 38.8 \; mol\ AgCl/L$$

This quantity is a constant. The number of moles per liter in solid AgCl is the same at the start of the reaction as it is when the reaction reaches equilibrium.

Since the [AgCl] term is a constant, it can be built into the equilibrium constant for the reaction.

$$[Ag^+][Cl^-] = K_c \times [AgCl]$$

This equation suggests that the product of the equilibrium concentrations of the Ag^+ and Cl^- ions in this solution is equal to a constant. Since this constant is proportional to the solubility of the salt, it is called the *solubility product equilibrium constant* for the reaction, or K_{sp}.

$$K_{sp} = [\text{Ag}^+][\text{Cl}^-]$$

The K_{sp} expression for a salt is the product of the concentrations of the ions, with each concentration raised to a power equal to the coefficient of that ion in the balanced equation for the solubility equilibrium. Solubility product constants for a number of so-called insoluble salts are given in Table A-11 in the appendix.

Exercise 16.2

Write K_{sp} expressions for saturated solutions of the following salts.

(a) CaF_2 (b) Bi_2S_3

Solution

(a) We start with a balanced equation for the equilibrium we want to describe.

$$\text{CaF}_2(s) \xrightleftharpoons{\text{H}_2\text{O}} \text{Ca}^{2+}(aq) + 2\ \text{F}^-(aq)$$

The solubility product expression is therefore written as follows.

$$K_{sp} = [\text{Ca}^{2+}][\text{F}^-]^2$$

(b) Once again, we start with a balanced equation for the equilibrium.

$$\text{Bi}_2\text{S}_3(s) \xrightleftharpoons{\text{H}_2\text{O}} 2\ \text{Bi}^{3+}(aq) + 3\ \text{S}^{2-}(aq)$$

Five ions are produced when this salt dissolves in water, so the K_{sp} expression contains a total of five terms.

$$K_{sp} = [\text{Bi}^{3+}]^2[\text{S}^{2-}]^3$$

16.5 THE RELATIONSHIP BETWEEN K_{sp} AND THE SOLUBILITY OF A SALT

K_{sp} is called the solubility product because it is literally the product of the solubilities of the ions in moles per liter. The solubility product of a salt can therefore be calculated from its solubility, or vice versa.

Exercise 16.3

All photographic films are based on the sensitivity of silver bromide to light. When light hits crystals of AgBr, a small fraction of the Ag^+ ions are reduced to silver metal. The rest of the Ag^+ ions in these crystals are reduced to silver metal when the film is developed. AgBr crystals that did not absorb light then have to be removed from the film in order to "fix" the image. Calculate the solubility of AgBr in water in grams per liter, assuming that K_{sp} for AgBr is equal to 5.0×10^{-13}.

Solution

We start with a balanced equation for the equilibrium.

$$\text{AgBr}(s) \xrightleftharpoons{\text{H}_2\text{O}} \text{Ag}^+(aq) + \text{Br}^-(aq)$$

We then write the solubility product expression for this reaction.

$$K_{sp} = [Ag^+][Br^-] = 5.0 \times 10^{-13}$$

One equation can't be solved for two unknowns — the Ag^+ and Br^- ion concentrations. We can generate a second equation, however, by noting that one Ag^+ ion is released for every Br^- ion. Because there is no other source of either ion in this solution, the concentrations of these ions at equilibrium must be the same.

$$[Ag^+] = [Br^-]$$

Substituting this equation into the K_{sp} expression

$$[Ag^+][Br^-] = 5.0 \times 10^{-13}$$

gives the following result.

$$[Ag^+]^2 = 5.0 \times 10^{-13}$$

Taking the square root of both sides of this equation gives the equilibrium concentrations of the Ag^+ and Br^- ions.

$$[Ag^+] = [Br^-] = 7.1 \times 10^{-7} \, M$$

Once we know how many moles of AgBr dissolve in a liter of water, we can calculate the number of grams per liter.

$$7.1 \times 10^{-7} \, \frac{\text{mol AgBr}}{\text{L}} \times 187.8 \, \frac{\text{g AgBr}}{\text{mol}} = 1.3 \times 10^{-4} \, \frac{\text{g AgBr}}{\text{L}}$$

The solubility of AgBr in water is only 0.00013 gram per liter. It therefore isn't practical to try to wash the unexposed AgBr off photographic film with water. The technique used to remove the AgBr will be described in the next chapter.

Solubility product calculations with 1 : 1 salts such as AgBr are relatively easy to perform. Because there are just as many Ag^+ ions as Br^- ions in AgBr, the concentrations of these ions at equilibrium must be the same. In order to extend such calculations to compounds with more complex formulas we need to understand the relationship between the solubility of a salt — S — and the concentrations of its ions at equilibrium.

Exercise 16.4

Several compounds were studied as possible sources of the fluoride ion for use in toothpaste. Write equations that describe the relationship between the solubility of CaF_2 and the equilibrium concentrations of the Ca^{2+} and F^- ions as a first step toward evaluating its use as a fluoridating agent.

Solution

As always, we start with the balanced equation.

$$CaF_2(s) \xrightleftharpoons{\text{H}_2\text{O}} Ca^{2+}(aq) + 2 \, F^-(aq)$$

Salts dissociate when they dissolve in water. For every mole of CaF_2 that dissolves, we get a mole of Ca^{2+} ions. The equilibrium concentration of the Ca^{2+} ion is therefore equal to the solubility of this compound in moles per liter.

$$[Ca^{2+}] = S$$

For every mole of CaF_2 that dissolves, we get twice as many moles of F^- ions. The F^- ion concentration at equilibrium is therefore equal to twice the solubility of the compound in units of moles per liter.

$$[F^-] = 2\,S$$

Exercise 16.5

Calculate the solubility of calcium fluoride in grams per liter and comment on its potential as a fluoridating agent. (CaF_2: $K_{sp} = 4.0 \times 10^{-11}$)

Solution

According to Exercise 16.2, the following is the solubility product expression for CaF_2.

$$K_{sp} = [Ca^{2+}][F^-]^2$$

Exercise 16.4 gave us the following equations for the relationship between the solubility of this salt and the concentrations of the Ca^{2+} and F^- ions.

$$[Ca^{2+}] = S$$
$$[F^-] = 2\,S$$

Substituting this information into the K_{sp} expression gives the following equation.

$$[S][2\,S]^2 = 4.0 \times 10^{-11}$$

Combining terms gives us a simpler version.

$$4\,S^3 = 4.0 \times 10^{-11}$$

This equation can be solved for the solubility of CaF_2 in units of moles per liter.

$$S = 2.2 \times 10^{-4}\,M$$

Once we know how many moles of CaF_2 dissolve in a liter, we can calculate the number of grams per liter.

$$2.2 \times 10^{-4}\,\frac{\text{mol } CaF_2}{L} \times 78.1\,\frac{\text{g } CaF_2}{\text{mol}} = 0.017\,\frac{\text{g } CaF_2}{L}$$

The solubility of calcium fluoride is fairly small — 0.017 grams per liter. Stannous fluoride, tin (II) fluoride, is over 10,000 times as soluble, so SnF_2 was chosen as the first compound used in fluoride toothpastes.

The techniques used in the preceding exercise are equally valid for salts that contain more positive ions than negative ions.

Exercise 16.6

Calculate the solubility in grams per liter of silver sulfide in order to decide whether it is accurately labeled when described as an insoluble salt. (Ag_2S: $K_{sp} = 6.3 \times 10^{-50}$)

Solution

Silver sulfide is another example of a 2 : 1 salt.

$$Ag_2S(s) \xrightleftharpoons{H_2O} 2\,Ag^+(aq) + S^{2-}(aq)$$

The concentration of the S^{2-} ion in a saturated solution is equal to the solubility of this salt.

$$[S^{2-}] = S$$

The equilibrium concentration of the Ag^+ ion is twice as large.

$$[Ag^+] = 2\,S$$

Substituting this information into the K_{sp} expression for Ag_2S

$$K_{sp} = [Ag^+]^2[S^{2-}]$$

gives the following equation.

$$[2\,S]^2[S] = 6.3 \times 10^{-50}$$
$$4\,S^3 = 6.3 \times 10^{-50}$$

This equation can be solved for the solubility of Ag_2S in moles per liter.

$$S = 2.5 \times 10^{-17}\,M$$

The weight of Ag_2S that dissolves in a liter of water can be calculated from the solubility in moles per liter.

$$2.5 \times 10^{-17}\,\frac{\text{mol }Ag_2S}{L} \times 247.8\,\frac{\text{g }Ag_2S}{\text{mol}} = 6.2 \times 10^{-15}\,\frac{\text{g }Ag_2S}{L}$$

This calculation suggests that 6.2 femtograms of Ag_2S dissolve in a liter of water. This amount is at least 10^{10} times smaller than the smallest sample that can be weighed on an analytical balance. The only reasonable way of describing this salt is as "insoluble" in water.

16.6 COMMON MISCONCEPTIONS ABOUT SOLUBILITY PRODUCT CALCULATIONS

Let's focus on one step in the preceding exercise. We started with the solubility product expression for Ag_2S.

$$K_{sp} = [Ag^+]^2[S^{2-}]$$

We then substituted the relationship between the concentrations of these ions and the solubility of the salt into this equation.

$$[2\,S]^2[S] = 6.3 \times 10^{-50}$$

When similar problems are done in chemistry classrooms across the United States, someone inevitably asks the following questions: "Why did you double the Ag^+ ion concentration and then square it?" "Aren't you counting this term twice?"

This is such a common question that it deserves special attention. The question results from confusion about the symbols used in the calculation. Remember that the symbol S in this equation stands for the solubility of Ag_2S. Furthermore, remember that we get two Ag^+ ions for each Ag_2S unit that dissolves in water. The Ag^+ ion concentration at equilibrium is therefore twice the solubility of the salt — or $2\,S$.

We square the Ag^+ ion concentration term because the equilibrium constant expression for this reaction is proportional to the product of the concentrations of the three products of the reaction.

$$K_{sp} = [Ag^+][Ag^+][S^{2-}]$$

It is just more convenient to write this equation in the condensed form.

$$K_{sp} = [Ag^+]^2[S^{2-}]$$

Another common mistake in solubility product calculations occurs when students are asked to write an equation that describes the relationship between the concentrations of the Ag^+ and S^{2-} ions in a saturated Ag_2S solution. It is all too easy to look at the formula for this compound — Ag_2S — and then write the following equation.

$$[S^{2-}] = 2\,[Ag^+]$$

This seems reasonable to some, who argue that there are twice as many Ag^+ ions as S^{2-} ions in the compound. But the equation is wrong. Because two Ag^+ ions are produced for each S^{2-} ion, there are twice as many silver ions as sulfide ions in this solution. This situation is correctly described by the following equation.

$$[Ag^+] = 2\,[S^{2-}]$$

How can you avoid making this mistake? After you write the equation that you think describes the relationship between the concentrations of the ions, try it to see if it works. Suppose two units of Ag_2S dissolve in water. How many S^{2-} ions would you get? Two. How many Ag^+ ions would you get? Four. Can you substitute this concrete example into your equation and get it to work?

$$[Ag^+] = 2\,[S^{2-}]$$
$$4 = 2 \times 2$$

Yes, if the equation is written correctly.

16.7 USING K_{sp} AS A MEASURE OF THE SOLUBILITY OF A SALT

The value of K_a for the dissociation of an acid is directly proportional to the strength of the acid.

$$K_a = \frac{[H_3O^+][A^-]}{[HA]}$$

If we find the following K_a values in a table, we can immediately conclude that formic acid is a stronger acid than acetic acid.

> Formic acid (HCO_2H): $K_a = 1.8 \times 10^{-4}$
> Acetic acid (CH_3CO_2H): $K_a = 1.8 \times 10^{-5}$

The same can be said about values of K_b.

$$K_b = \frac{[HB^+][OH^-]}{[B]}$$

The following base-ionization equilibrium constants immediately allow us to conclude that methylamine is a stronger base than ammonia.

> Methylamine (CH_3NH_2): $K_b = 4.8 \times 10^{-4}$
> Ammonia (NH_3): $K_b = 1.8 \times 10^{-5}$

Unfortunately, there is no simple way of predicting the relative solubilities of salts from their K_{sp} if the salts produce different numbers of positive and negative ions when they dissolve in water.

Exercise 16.7

Which salt — $CaCO_3$ or Ag_2CO_3 — is more soluble in water in units of moles per liter? ($CaCO_3$: $K_{sp} = 2.8 \times 10^{-9}$, Ag_2CO_3: $K_{sp} = 8.1 \times 10^{-12}$)

Solution

We might expect $CaCO_3$ to be more soluble than Ag_2CO_3, because it has a larger K_{sp}. The only way to test this prediction is to calculate the solubilities of both compounds.

The solubility product expression for $CaCO_3$ has the following form.

$$K_{sp} = [Ca^{2+}][CO_3{}^{2-}]$$

The concentrations of both the Ca^{2+} and $CO_3{}^{2-}$ ions in a saturated solution of this salt are equal to the solubility of the salt: $[Ca^{2+}] = [CO_3{}^{2-}] = S$.

$$[S][S] = 2.8 \times 10^{-9}$$

Taking the square root of both sides therefore gives the solubility of this salt.

$$S = 5.3 \times 10^{-5} \, M$$

Ag_2CO_3 is a 2:1 salt, for which the solubility product expression is written as follows.

$$K_{sp} = [Ag^+]^2[CO_3{}^{2-}]$$

The $CO_3{}^{2-}$ ion concentration is equal to the solubility of the salt, but the Ag^+ ion concentration is twice as large.

$$[CO_3{}^{2-}] = S$$
$$[Ag^+] = 2\,S$$

Substituting this information into the K_{sp} expression gives the following equation.

$$[2\,S]^2[S] = 8.1 \times 10^{-12}$$
$$4\,S^3 = 8.1 \times 10^{-12}$$

This equation can be solved for the solubility of Ag_2CO_3.

$$S = 1.3 \times 10^{-4} \, M$$

In spite of the fact that K_{sp} for $CaCO_3$ is larger than K_{sp} for Ag_2CO_3, $CaCO_3$ is less soluble than Ag_2CO_3.

$$\mathbf{Ag_2CO_3:} \qquad S = 0.00013 \, M$$
$$\mathbf{CaCO_3:} \qquad S = 0.000053 \, M$$

THE ROLE OF THE ION PRODUCT (Q_{sp})
16.8 IN SOLUBILITY CALCULATIONS

Consider a saturated solution of AgCl in water.

$$AgCl(s) \xrightleftharpoons{\text{H}_2\text{O}} Ag^+(aq) + Cl^-(aq)$$

Because AgCl is a 1:1 salt, which dissociates in water to give an Ag^+ ion for every Cl^- ion, the concentrations of the Ag^+ and Cl^- ions in this solution are equal.

Saturated solution of AgCl in water: $\qquad [Ag^+] = [Cl^-]$

Imagine what happens when a few crystals of solid $AgNO_3$ are added to this saturated solution of AgCl in water. According to the rules in Table 16.1, silver nitrate is a soluble salt. It therefore dissolves and dissociates into Ag^+ and NO_3^- ions. As a result, there are two sources of the Ag^+ ion in this solution.

$$AgNO_3(s) \xrightarrow{H_2O} \mathbf{Ag^+(aq)} + NO_3^-(aq)$$

$$AgCl(s) \xrightleftharpoons{H_2O} \mathbf{Ag^+(aq)} + Cl^-(aq)$$

The immediate result of adding $AgNO_3$ to the solution is therefore an increase in the Ag^+ ion concentration.

When this happens, the solution can no longer be at equilibrium. The product of the concentrations of the Ag^+ and Cl^- ions is too large. In more formal terms, we can argue that the *ion product* (Q_{sp}) for the solution is larger than the solubility product (K_{sp}).

$$Q_{sp} = (Ag^+)(Cl^-) > K_{sp}$$

The ion product is literally the product of the concentrations of the ions at any moment in time. When it is equal to the solubility product for the salt, the system is at equilibrium.

The reaction eventually comes back to equilibrium after the excess ions have precipitated from solution as solid AgCl. When equilibrium is reestablished, however, the values of the Ag^+ and Cl^- ion concentrations will have changed. Because there are two sources of the Ag^+ ion in this solution, the $[Ag^+]$ term is larger than the $[Cl^-]$ term.

Saturated solution of AgCl to which $AgNO_3$ has been added: $[Ag^+] > [Cl^-]$

Now imagine what happens when a few crystals of NaCl are added to a saturated solution of AgCl in water. There are two sources of the chloride ion in this solution.

$$NaCl(s) \xrightarrow{H_2O} Na^+(aq) + \mathbf{Cl^-(aq)}$$

$$AgCl(s) \xrightleftharpoons{H_2O} Ag^+(aq) + \mathbf{Cl^-(aq)}$$

Once again, the ion product is larger than the solubility product.

$$Q_{sp} = (Ag^+)(Cl^-) > K_{sp}$$

Furthermore, when the reaction comes back to equilibrium, the $[Cl^-]$ term is larger than the $[Ag^+]$ term.

Saturated solution of AgCl to which NaCl has been added: $[Ag^+] < [Cl^-]$

Figure 16.4 shows a small portion of the graph that could be constructed of the infinite number of combinations of Ag^+ and Cl^- ion concentrations. Any point along the curved line in this graph corresponds to a system at equilibrium. For any point along this line, the product of the Ag^+ and Cl^- ion concentrations is equal to the K_{sp} for AgCl.

Point *A* represents a solution of two sources of the Ag^+ ion—such as $AgNO_3$ and AgCl—dissolved in water that has come to equilibrium. Point *B* represents a saturated solution of AgCl in pure water, in which the $[Ag^+]$ and $[Cl^-]$ terms are equal. The solution corresponding to Point *C* results when two sources of the Cl^- ion—such as NaCl and AgCl—are added to water and the solution is allowed to come to equilibrium.

All other points in this diagram represent solutions that aren't at equilibrium.

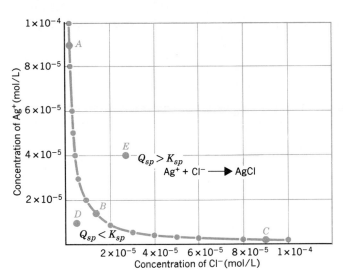

FIG. 16.4 The inverse relationship between the Ag^+ and Cl^- ion concentrations in a saturated solution of AgCl. The solid line is called the saturation curve for AgCl. Every point along this line represents a solution at equilibrium. Any point below the saturation curve represents a solution for which Q_{sp} is smaller than K_{sp}. Any point above the saturation curve describes a solution for which Q_{sp} is larger than K_{sp}; AgCl will have to precipitate from this solution before it can reach equilibrium.

Any point *below* the solid line (such as Point D) represents a solution for which the ion product is smaller than the solubility product.

$$\text{Point } D: \qquad Q_{sp} < K_{sp}$$

If more AgCl were added to the solution at Point D, it would dissolve.

$$\text{If } Q_{sp} < K_{sp}: \qquad AgCl(s) \longrightarrow Ag^+(aq) + Cl^-(aq)$$

Points *above* the solid line (such as Point E) represent solutions for which the ion product is larger than the solubility product.

$$\text{Point } E: \qquad Q_{sp} > K_{sp}$$

The solution described by Point E will eventually come to equilibrium after enough solid AgCl has precipitated.

$$\text{If } Q_{sp} > K_{sp}: \qquad Ag^+(aq) + Cl^-(aq) \longrightarrow AgCl(s)$$

16.9 THE COMMON ION EFFECT

When $AgNO_3$ is added to a saturated solution of AgCl, it is often described as a source of a **common ion**—the Ag^+ ion.

Common ion: An ion that comes from two different salts in a solution.

Solutions to which both NaCl and AgCl have been added also contain a common ion—the Cl^- ion. This section focuses on the effect of common ions on solubility product equilibria.

Exercise 16.8

Calculate the solubility of AgCl in pure water. (AgCl: $K_{sp} = 1.8 \times 10^{-10}$)

Solution

There is only one source of the Ag^+ and Cl^- ions in this solution. At equilibrium, the concentrations of these ions must be the same.

$$[Ag^+] = [Cl^-]$$

Substituting this information into the solubility product expression

$$K_{sp} = [Ag^+][Cl^-]$$

leads to the conclusion that the concentration of either ion at equilibrium is equal to the square root of the solubility product.

$$[Ag^+] = [Cl^-] = \sqrt{K_{sp}} = 1.3 \times 10^{-5} \, M$$

What happens to the solubility of AgCl when we try to dissolve this salt in a solution that is already 0.10 M in NaCl? As a rule, we can assume that salts dissociate when they dissolve. A 0.10 M NaCl solution therefore contains 0.10 moles per liter of the Cl^- ion. What effect does the presence of this common ion have on the solubility of AgCl in the NaCl solution?

The Cl^- ion is one of the products of the solubility equilibrium.

$$AgCl(s) \xrightleftharpoons{H_2O} Ag^+(aq) + Cl^-(aq)$$

LeChatelier's principle leads us to expect that AgCl will be even less soluble in an 0.10 M Cl^- solution than it is in pure water.

Exercise 16.9

Calculate the solubility in moles per liter of AgCl in 0.10 M NaCl. (AgCl: $K_{sp} = 1.8 \times 10^{-10}$)

Solution

What do we know about this system? We can no longer assume that the Ag^+ and Cl^- ion concentrations at equilibrium will be equal, because there are two sources of the Cl^- ion in this solution — AgCl and NaCl.

$$[Ag^+] \neq [Cl^-]$$

The best way of setting up this problem is to revert to the format used in Chapters 14 and 15.

$$AgCl(s) \rightleftharpoons Ag^+(aq) + Cl^-(aq) \qquad K_{sp} = 1.8 \times 10^{-10}$$

Initial:	0	0.10 M
Equilibrium:	S	$0.10 + S$

Initially, there is no Ag^+ ion in the solution, but the Cl^- ion concentration is 0.10 M. As the solution comes to equilibrium, some of the AgCl will dissolve, and the concentrations of both the Ag^+ and Cl^- ions will increase. These concentrations will increase by an amount equal to the solubility of AgCl in this solution — S.

We can now write the solubility product expression for this reaction.

$$K_{sp} = [Ag^+][Cl^-]$$

And we can substitute what we know about the equilibrium concentrations of the Ag^+ and Cl^- ions into this equation.

$$[S][0.10 + S] = 1.8 \times 10^{-10}$$

We could expand the equation and solve it with the quadratic formula. But that would involve a lot of work. Let's see if we can find an assumption that makes the calculation easier.

What do we know about the value of S? In pure water, the solubility of AgCl is only 0.000013 M. In this solution, we expect it to be even smaller. It therefore seems reasonable to expect that S should be small compared with the initial concentration of the Cl^- ion.

$$[S][0.10] \cong 1.8 \times 10^{-10}$$

Solving this approximate equation for S gives the following result.

$$S \cong 1.8 \times 10^{-9} \, M$$

Note that the assumption used to generate the approximate equation is valid. The Cl^- ion concentration from the dissociation of AgCl is about 50 million times smaller than the initial Cl^- ion concentration. This assumption works very well with common ion problems involving insoluble salts because the K_{sp} values for these salts are so small.

Let's compare the results of Exercises 16.8 and 16.9.

In pure water: $\qquad S = 1.3 \times 10^{-5} \, M$
In 0.10 M NaCl: $\qquad S = 1.8 \times 10^{-9} \, M$

The difference between the solubility of AgCl in these two solutions is due to the *common ion effect*.

Common ion effect: The decrease in the solubility of a salt when it is dissolved in a solution that already contains one of its ions.

Exercise 16.10

Compare the solubility of Ag_2S in pure water (Exercise 16.6) with the solubility of Ag_2S in a solution that is 0.050 M in Na_2S. (Ag_2S: $K_{sp} = 6.3 \times 10^{-50}$)

Solution

The key to this problem is recognizing that Na_2S is a soluble salt, which dissociates when it dissolves in water. The initial concentration of the S^{2-} ion is therefore 0.050 M.

$$Ag_2S(s) \rightleftharpoons 2 \, Ag^+(aq) + S^{2-}(aq) \qquad K_{sp} = (6.3 \times 10^{-50})$$

Initial:	0	0.050 M
Equilibrium:	2 S	0.10 + S

When Ag_2S dissolves in this solution we get a little more S^{2-} ion and twice as much Ag^+ ion. Substituting this information into the K_{sp} expression

$$K_{sp} = [Ag^+]^2[S^{2-}]$$

gives the following result.

$$[2 \, S]^2[0.050 + S] = 6.3 \times 10^{-50}$$

This equation is hard to solve by brute force, but we know that the solubility of Ag_2S in water is very low. We can therefore assume that essentially all of the S^{2-} ion in this solution comes from the Na_2S. Assuming that S is negligibly small compared with 0.050 gives the following approximate equation.

$$[2 \, S]^2[0.050] \cong 6.3 \times 10^{-50}$$

The equation can be solved for the solubility of Ag_2S.

$$S \cong 5.6 \times 10^{-25}\ M$$

Does the common ion effect make any difference in the solubility of Ag_2S?

In pure water: $S = 2.5 \times 10^{-17}\ M$
In 0.050 M Na_2S: $S = 5.6 \times 10^{-25}\ M$

Yes! Consider the magnitude of the last number. According to our calculations, it takes about 3.0 liters of 0.050 M Na_2S to dissolve enough Ag_2S to give a single S^{2-} ion and a pair of Ag^+ ions.

16.10 HOW TO KEEP A SALT FROM DISSOLVING

The common ion effect can make an "insoluble" salt even less soluble in water.

Exercise 16.11

Gold(III) chloride is an exception to the rule that chlorides are soluble in water. Calculate the solubility of $AuCl_3$ in grams per liter in the following solutions. ($AuCl_3$: $K_{sp} = 3.2 \times 10^{-25}$)

(a) Pure water
(b) A solution of 0.014 grams of $ZnCl_2$ dissolved in a liter of water

Solution

(a) The solubility product expression for this compound is written as follows.

$$K_{sp} = [Au^{3+}][Cl^-]^3$$

The Au^{3+} ion concentration at equilibrium is equal to the solubility of the salt, and the concentration of the Cl^- ion is three times as large.

$$[S][3\ S]^3 = 3.2 \times 10^{-25}$$

Combining terms in this equation gives the following result.

$$27\ S^4 = 3.2 \times 10^{-25}$$

This equation can be solved for the solubility of $AuCl_3$ in moles per liter.

$$S = 3.3 \times 10^{-7}\ M$$

The resulting information can be used to calculate the solubility of $AuCl_3$ in grams per liter.

$$3.3 \times 10^{-7}\ \frac{\text{mol } AuCl_3}{L} \times 303.3\ \frac{\text{g } AuCl_3}{\text{mol}} = 1.0 \times 10^{-4}\ \frac{\text{g } AuCl_3}{L}$$

(b) We can start by calculating the concentration of $ZnCl_2$ in this solution.

$$0.014\ \frac{\text{g } ZnCl_2}{L} \times \frac{1\ \text{mol } ZnCl_2}{136.3\ \text{g}} = 1.0 \times 10^{-4}\ M$$

When this compound dissolves in water we get twice as many Cl^- ions. The problem can therefore be set up as follows.

$$AuCl_3(s) \rightleftharpoons Au^{3+}(aq) + \quad 3\ Cl^-(aq) \qquad K_{sp} = 3.2 \times 10^{-25}$$

| Initial: | 0 | 0.00020 M |
| Equilibrium: | S | $0.00020 \times 3\,S$ |

Substituting this information into the K_{sp} expression gives the following equation.

$$[S][0.00020 + 3\,S]^3 = 3.2 \times 10^{-25}$$

Although the amount of $ZnCl_2$ in the solution is small, the amount of Cl^- ions from $ZnCl_2$ is still several hundred times larger than the amount of Cl^- ions given off when $AuCl_3$ dissolves in pure water. That means we can try assuming that the Cl^- ion given off when $AuCl_3$ dissolves in this solution is negligibly small compared with the initial concentration of the Cl^- ion. This assumption generates the following approximate equation.

$$[S][0.00020]^3 \cong 3.2 \times 10^{-25}$$

This equation can be solved for an approximate value of the solubility of $AuCl_3$.

$$S \cong 4.0 \times 10^{-14}\ M$$

The assumption that S is small compared with the initial concentration of the Cl^- ion is valid. Much less Cl^- ion is contributed to the solution by $AuCl_3$ than is initially present from $ZnCl_2$. Converting the solubility into units of grams per liter gives the following result.

$$4.0 \times 10^{-14}\ \frac{\text{mol } AuCl_3}{\text{L}} \times 303.3\ \frac{\text{g } AuCl_3}{\text{mol}} = 1.2 \times 10^{-11}\ \frac{\text{g } AuCl_3}{\text{L}}$$

According to these calculations, adding a minuscule amount of $ZnCl_2$ to water decreases the solubility of $AuCl_3$ by a factor of about 8,000,000, from 0.1 milligrams per liter to 12 picograms per liter — from the smallest weight that an analytical balance can measure to a level that tests the limits of the best instruments chemists have developed so far.

16.11 HOW TO KEEP A SALT FROM PRECIPITATING

The preceding sections showed how the common ion effect can keep a salt from dissolving. This section focuses on the opposite effect — preventing a salt from precipitating from solution.

Exercise 16.12

Use the solubility product of $Cr(OH)_3$ to describe why it is impossible to prepare a 0.10 M Cr^{3+} solution at neutral pH. [$Cr(OH)_3$: $K_{sp} = 6.3 \times 10^{-31}$].

Solution

At neutral pH, the H_3O^+ and OH^- ion concentrations in water are $1.0 \times 10^{-7}\ M$. To decide whether $Cr(OH)_3$ would precipitate if the Cr^{3+} ion concentration was 0.10 M, we need to compare the ion product for this solution with the value of K_{sp} for the compound.

$$Q_{sp} = (Cr^{3+})(OH^-)^3$$
$$= (0.10)(1.0 \times 10^{-7})^3 = 1.0 \times 10^{-22} \gg 6.3 \times 10^{-31}$$

The value of Q_{sp} for the solution is very much larger than the K_{sp} for $Cr(OH)_3$, which means that $Cr(OH)_3$ would precipitate from a neutral solution long before the Cr^{3+} ion concentration could reach 0.10 M.

Exercise 16.13

Calculate the pH at which $Cr(OH)_3$ just starts to precipitate from a solution that is 0.10 M in the Cr^{3+} ion.

Solution

Chromium(III) hydroxide starts to precipitate from this solution when the ion product is equal to the solubility product. We can therefore begin by writing the K_{sp} expression for this salt.

$$K_{sp} = [Cr^{3+}][OH^-]^3$$

Next, we substitute the concentration of the Cr^{3+} ion into this equation

$$[0.10][OH^-]^3 = 6.3 \times 10^{-31}$$

and solve the equation for the OH^- ion concentration at which $Cr(OH)_3$ just starts to precipitate.

$$[OH^-] = 1.8 \times 10^{-10} \, M$$

We can then calculate the pOH of the solution

$$pOH = -\log [OH^-] = 9.74$$

and use this information to calculate the pH.

$$pH = 14 - pOH = \mathbf{4.26}$$

According to this calculation, the OH^- ion concentration is too small for $Cr(OH)_3$ to precipitate from an 0.10 M Cr^{3+} solution if the pH of the solution is kept below 4.26.

The results of similar calculations for a number of different Cr^{3+} ion concentrations are given in Table 16.2. Only a small fraction of these data will fit on a normal graph, such as Figure 16.5. We can overcome this problem by plotting the log of the Cr^{3+} ion concentrations versus the log of the OH^- ion concentrations. The log $[Cr^{3+}]$ term for the data in Table 16.2 ranges from 0 to -7, and the log $[OH^-]$ term ranges from -7.7 to -10.1.

Another way to overcome the problem involves plotting the Cr^{3+} and OH^- concentrations on log-log graph paper, which is specifically designed to condense

TABLE 16.2

The Inverse Relationship Between the Equilibrium Concentrations of the Cr^{3+} and OH^- Ions

$[Cr^{3+}]$ (mol/L)	$[OH^-]$ (mol/L)
1.0	8.6×10^{-11}
1×10^{-1}	1.8×10^{-10}
1×10^{-2}	4.0×10^{-10}
1×10^{-3}	8.6×10^{-10}
1×10^{-4}	1.8×10^{-9}
1×10^{-5}	4.0×10^{-9}
1×10^{-6}	8.6×10^{-9}
1×10^{-7}	1.8×10^{-8}

FIG. 16.5 The saturation curve for $Cr(OH)_3$ over a narrow range of Cr^{3+} and OH^- concentrations. The solid line describes the combinations of Cr^{3+} and OH^- concentrations for which the solution is at equilibrium.

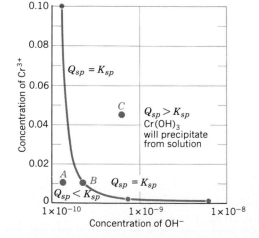

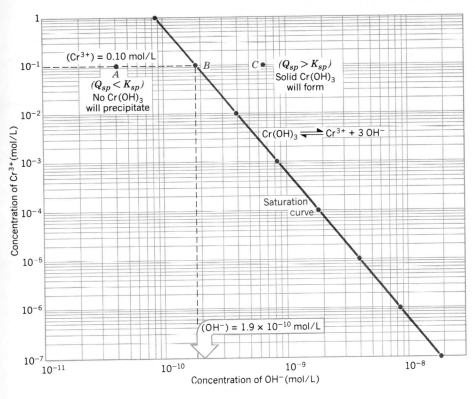

FIG. 16.6 Log-log graph paper can be used to plot the saturation curve for $Cr(OH)_3$ over the entire range of concentrations in Table 16.2. The horizontal dashed line at the upper left represents a Cr^{3+} ion concentration of 0.10 M. This line intersects the saturation curve at an OH^- ion concentration of 1.9×10^{-10} M. As long as we keep the OH^- ion concentration *below* 1.9×10^{-10} M, no $Cr(OH)_3$ will precipitate from a 0.10 M Cr^{3+} ion solution.

several orders of magnitude onto a single scale. Figure 16.6 shows a log-log plot of the data in Table 16.2.

The points along the solid line in Figure 16.6 represent combinations of Cr^{3+} and OH^- ion concentrations at which the solution is saturated — in other words, at equilibrium. Point A represents a solution in which the OH^- ion concentration is too small for $Cr(OH)_3$ to precipitate when the Cr^{3+} ion concentration is 0.10 M. Point B corresponds to the set of conditions calculated in Exercise 16.13. Point C describes a pair of Cr^{3+} and OH^- ion concentrations for which the ion product is too large; $Cr(OH)_3$ would precipitate from any solution that momentarily contained the Cr^{3+} and OH^- ion concentrations that correspond to Point C.

16.12 HOW TO SEPARATE IONS BY SELECTIVE PRECIPITATION

The last few sections have shown how the concentration of an ion can be controlled to either prevent a solid from dissolving or keep it in solution. By combining these processes we should be able to separate a mixture of two or more ions.

The technique used in this section is known as *selective precipitation.* It involves adding a reagent that selectively brings one of the ions out of solution as a precipitate while it leaves other ions in the solution. The solid precipitate can be collected by filtration. Or the sample can be spun in a centrifuge until the precipitate collects at the bottom. The solution can then be decanted, or poured off.

Some mixtures can be separated on the basis of the solubility rules outlined in Table 16.1. For example, we can separate the Ag^+ ion from a solution that contains the Cu^{2+} ion by adding a source of the Cl^- ion to this solution. Silver chloride is an insoluble salt ($K_{sp} = 1.8 \times 10^{-10}$), which will precipitate from solution. Copper(II) chloride is soluble in water — up to 73 grams of $CuCl_2$ will dissolve in 100 milliliters of water.

Selective precipitation becomes more of a challenge when the ions to be separated form salts with similar solubilities. Mn^{2+} and Ni^{2+} ions, for example, both form insoluble sulfides.

$$MnS: \qquad K_{sp} = 3 \times 10^{-13}$$
$$NiS: \qquad K_{sp} = 3.2 \times 10^{-19}$$

The 10^{-6}-fold difference in the solubility products of these salts raises an interesting question. Is this difference large enough to allow us to selectively precipitate the less soluble NiS salt from a mixture of these ions, without precipitating MnS as well?

Exercise 16.14

Describe the conditions under which Ni^{2+} ions can be precipitated as NiS from a solution that is 0.10 M in the Ni^{2+} and Mn^{2+} ions, while the Mn^{2+} ions are left in solution.

Solution

We can precipitate Ni^{2+} from solution as NiS by adding a source of the S^{2-} ion to this solution. The S^{2-} ion concentration must be carefully adjusted, however, to meet the following criteria.

1. The S^{2-} ion concentration must be large enough to precipitate as much of the Ni^{2+} as possible.
2. The S^{2-} ion concentration must be small enough so that no MnS precipitates from solution.

The second of these criteria is easier to test than the first. It is relatively easy to determine when the S^{2-} ion concentration is too large.

Exercise 16.15

Calculate the S^{2-} ion concentration at which MnS will begin to precipitate from a solution that is 0.10 M in Mn^{2+} ions. (MnS: $K_{sp} = 3 \times 10^{-13}$)

Solution

We can start with the solubility product expression for MnS.

$$K_{sp} = [Mn^{2+}][S^{2-}]$$

We then substitute what we know about the Mn^{2+} ion concentration into this equation

$$[0.10][S^{2-}] = 3 \times 10^{-13}$$

and calculate the S^{2-} ion concentration at which MnS just starts to precipitate.

$$[S^{2-}] = 3 \times 10^{-12} M$$

According to Exercise 16.15, we can keep Mn^{2+} ions from precipitating from a 0.10 M solution if we can keep the S^{2-} ion concentration smaller than $3 \times 10^{-12} M$. Is this S^{2-} ion concentration large enough to effectively remove Ni^{2+} ions from the mixture?

Exercise 16.16

Calculate the Ni^{2+} ion concentration in a solution to which enough S^{2-} ion has been added to raise the concentration of this ion to $3 \times 10^{-12} M$. (NiS: $K_{sp} = 3.2 \times 10^{-19}$)

Solution

Initially, all of the S^{2-} ion added to this solution combines with Ni^{2+} ions to form NiS, which precipitates from the solution. Eventually, the Ni^{2+} ion concentration becomes so small that S^{2-} ion starts to accumulate in the solution. At what point does the Ni^{2+} ion concentration become small enough so that the S^{2-} ion concentration at equilibrium is $3 \times 10^{-12} M$?

We start with the solubility product expression for NiS.

$$K_{sp} = [Ni^{2+}][S^{2-}]$$

We then substitute the known S^{2-} ion concentration into this equation.

$$[Ni^{2+}][3 \times 10^{-12}] = 3.2 \times 10^{-19}$$

Solving for the Ni^{2+} ion concentration gives the following result.

$$[Ni^{2+}] = 1 \times 10^{-7} M$$

In other words, the S^{2-} concentration at equilibrium can't get as large as $3 \times 10^{-12} M$ until the Ni^{2+} ion concentration has been reduced by a factor of 10^6, from $0.10 M$ to $1 \times 10^{-7} M$.

The 10^6-fold difference between the solubility products for MnS and NiS is large enough to allow us to separate the Mn^{2+} and Ni^{2+} ions in a mixture that is initially $0.10 M$ in both ions. We can separate the ions by carefully controlling the S^{2-} ion concentration in the solution so that it remains just below $3 \times 10^{-12} M$. Through this process, we can precipitate enough NiS to reduce the Ni^{2+} ion concentration by a factor of 1,000,000, without allowing any MnS to precipitate.

Would this technique work equally well if the task involved separating the Mn^{2+} and Ni^{2+} ions when both concentrations were $0.010 M$? Or when the Ni^{2+} ion was present at a $0.10 M$ concentration but the Mn^{2+} concentration was only $0.0010 M$? Instead of repeating the calculations for each set of initial concentrations, we can construct a graph that allows us to answer this question for almost any combination of Mn^{2+} and Ni^{2+} ion concentrations.

Figure 16.7 shows the saturation curves for NiS and MnS plotted on the same piece of log-log graph paper. The solid line at the left describes pairs of Ni^{2+} and S^{2-} ion concentrations at which NiS is in equilibrium with these ions. The solid line at the right does the same for the equilibrium between MnS and the Mn^{2+} and S^{2-} ions.

The horizontal line at the bottom of the graph represents an Ni^{2+} ion concentration of $1 \times 10^{-6} M$. This line intersects the saturation curve for NiS at an S^{2-} ion concentration of $3.2 \times 10^{-13} M$. The vertical line that represents this S^{2-} ion concentration doesn't intersect the MnS saturation curve until the Mn^{2+} ion concentration is $1 M$.

What does this graph tell us? As long as the initial concentrations of the Ni^{2+} and Mn^{2+} ions are less than $1 M$, we can reduce the Ni^{2+} ion concentration to $1 \times 10^{-6} M$ by adding S^{2-} ions without precipitating MnS.

FIG. 16.7 The saturation curves for NiS and MnS plotted on log-log graph paper. The horizontal dashed line at the bottom of this figure represents an Ni^{2+} ion concentration of $1 \times 10^{-6}\,M$. This line intersects the saturation curve for NiS at an S^{2-} ion concentration of $3.2 \times 10^{-13}\,M$. A vertical line corresponding to this S^{2-} ion concentration intersects the MnS saturation curve at a Mn^{2+} ion concentration of about 1 M. This means that we can precipitate Ni^{2+} from solution as NiS without precipitating MnS as long as the Ni^{2+} ion concentration is more than $1 \times 10^{-6}\,M$ and the Mn^{2+} ion concentration is less than 1 M.

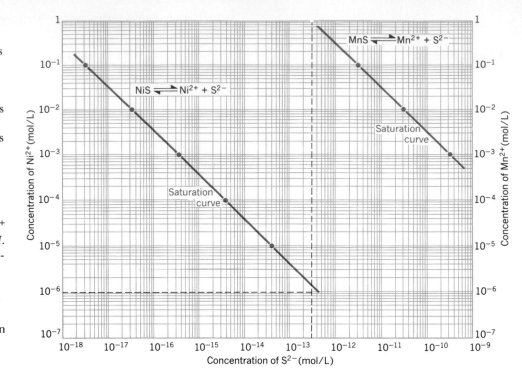

16.13 HOW TO ADJUST THE CONCENTRATION OF AN ION

The preceding section leaves an important question unanswered: How do we adjust the S^{2-} ion concentration in a solution so that it approaches but does not exceed $3 \times 10^{-12}\,M$?

Chapter 15 laid the foundation for controlling the concentration of ions — such as the S^{2-} ion — that are conjugate bases of weak acids. We start with an aqueous solution of the weak acid — in this case, hydrogen sulfide dissolved in water.

$$H_2S(aq) + H_2O(l) \rightleftharpoons H_3O^+(aq) + HS^-(aq) \qquad K_{a1} = 1.0 \times 10^{-7}$$
$$HS^-(aq) + H_2O(l) \rightleftharpoons H_3O^+(aq) + S^{2-}(aq) \qquad K_{a2} = 1.3 \times 10^{-13}$$

At room temperature, hydrogen sulfide is a gas that is only marginally soluble in water. A saturated solution of this gas has a concentration of about 0.10 M at room temperature. Using the techniques introduced in Section 15.18, we can calculate the S^{2-} ion concentration in a 0.10 M H_2S solution.

$$[S^{2-}] = K_{a2} = 1.3 \times 10^{-13}\,M$$

The S^{2-} ion concentration in this solution is close to $3 \times 10^{-12}\,M$. How could we bring it even closer?

LeChatelier's principle suggests that we should be able to either increase or decrease the amount of dissociation of this acid by adjusting the pH of the solution. If we add a strong acid to an aqueous solution, the amount of S^{2-} ion should decrease as the equilibria described earlier in this section are driven toward the left. If we add a strong base, the S^{2-} ion concentration should increase as the equilibria are pulled to the right.

In pure water, H_2S dissociates by losing one proton at a time. But when we add either a strong acid or a strong base to this solution, the equilibria shift so much in one direction or the other that we can treat H_2S as if it dissociated by losing two protons in a single step.

$$H_2S(aq) + 2\,H_2O(l) \rightleftharpoons 2\,H_3O^+(aq) + S^{2-}(aq)$$

The equilibrium constant expression for the overall reaction is equal to the product of the equilibrium constant expressions for the individual steps.

$$\frac{[H_3O^+][HS^-]}{[H_2S]} \times \frac{[H_3O^+][S^{2-}]}{[HS^-]} = \frac{[H_3O^+]^2[S^{2-}]}{[H_2S]}$$

The equilibrium constant for the overall reaction is therefore equal to the product of K_{a1} times K_{a2}.

$$\frac{[H_3O^+]^2[S^{2-}]}{[H_2S]} = K_{a1} \times K_{a2} = 1.3 \times 10^{-20}$$

A saturated solution of H_2S has an initial concentration of 0.10 M. Because H_2S is a weak acid, we can assume that the concentration of this acid at equilibrium is approximately equal to its initial concentration.

$$\frac{[H_3O^+]^2[S^{2-}]}{[0.10]} = 1.3 \times 10^{-20}$$

This equation can be rearranged as follows.

$$[H_3O^+]^2 = \frac{1.3 \times 10^{-20} \times [0.10]}{[S^{2-}]}$$

Taking the square root of both sides gives the following result.

$$[H_3O^+] = \sqrt{\frac{1.3 \times 10^{-21}}{[S^{2-}]}}$$

Exercise 16.17

Calculate the pH to which a saturated solution of H_2S in water must be adjusted for the S^{2-} ion concentration to be 3×10^{-12} M.

Solution

The key to this problem is recognizing the relationship between the S^{2-} and H_3O^+ ion concentrations in a saturated H_2S solution to which either an acid or a base has been added.

$$[H_3O^+] = \sqrt{\frac{1.3 \times 10^{-21}}{[S^{2-}]}}$$

Once we have this equation, we can calculate what H_3O^+ ion concentration would be required to force the S^{2-} ion concentration to take on a particular value. For example, the H_3O^+ ion concentration must be 2×10^{-5} M if we want the S^{2-} ion concentration to be equal to 3×10^{-12} M.

$$[H_3O^+] = \sqrt{\frac{1.3 \times 10^{-21}}{3 \times 10^{-12}}} = 2 \times 10^{-5} \ M$$

Once we know the H_3O^+ ion concentration needed, we can calculate the pH of the solution.

$$pH = -\log [H_3O^+] = 4.7$$

Combining the results of this exercise with the calculations in the preceding section leads us to conclude that we can separate the Ni^{2+} ion from the Mn^{2+} ion by adding a saturated solution of H_2S that has been buffered at a pH of about 4.7. Most of the Ni^{2+} ion will precipitate as NiS, which can be collected by filtration. All of the Mn^{2+} ion will remain in solution.

SUMMARY

There is a limit on how much of any solid will dissolve in a liter of water. A solution is saturated when this limit is reached. In a saturated solution, an equilibrium exists between the solid and the particles released into the solution when the solid dissolves—the solid dissolves and dissociates at the same rate at which these particles come together to form additional solid, which precipitates from solution.

This chapter focused on the solubility of salts in aqueous solutions. One of the basic assumptions behind this discussion was the notion that salts dissociate into their ions when they dissolve in water. The equilibrium between a salt and its ions in a saturated solution can be described with a solubility product equilibrium constant expression, K_{sp}. The concentration of water is not included in this expression, because water is neither consumed nor produced in these reactions. The concentration of the solid is also left out of the expression, because it is a constant. The K_{sp} for a salt is equal to the product of the concentrations of the ions formed when the salt dissolves in water, with each concentration raised to a power equal to the coefficient of that ion in a balanced equation for the reaction.

The common ion effect can decrease the solubility of a salt. Adding a source of a common ion to a saturated solution of a salt produces a situation in which the ion product (Q_{sp}) for this solution is larger than the solubility product (K_{sp}). As expected from LeChatelier's principle, adding a common ion to a solubility product equilibrium shifts this equilibrium toward the left. Some of the ions in the solution come together to form the salt, which precipitates from solution.

LeChatelier's principle also suggests a way of increasing the solubility of a salt. We can achieve this goal by adding a reagent to the solution that decreases the concentration of one of the ions formed when the salt dissolves. Adding an acid, for example, can reduce the OH^- ion concentration in water to a point at which insoluble hydroxides such as $Cr(OH)_3$ no longer precipitate from solution.

Carefully controlling the concentration of one of the ions in a solution makes it possible to precipitate one component of a solution without precipitating any others. This technique, known as selective precipitation, offers an excellent way of separating one metal ion from another.

PROBLEMS

What Happens When Solids Dissolve in Water?

16-1 Give at least one example of each of the following kinds of solids.

(a) molecular (b) covalent (c) ionic

16-2 Both molecular and covalent compounds contain covalent bonds. Explain why molecular compounds are often soluble in water but covalent compounds are not.

16-3 Explain why NaCl is soluble in water but AgCl is not.

16-4 Use an example to explain what is meant by the general rule that salts dissociate when they dissolve in water.

Solubility Equilibria

16-5 Explain why the bulb in the conductivity apparatus in Figure 16.1 glows more brightly when the wires are immersed in a solution of NaCl than when the wires are immersed in tap water.

16-6 Explain why the addition of a few small crystals of silver chloride makes water a better conductor of electricity. Explain why the conductivity gradually increases as more AgCl is added, until it eventually reaches a maximum. Describe what is happening in the solution when its conductivity reaches the maximum.

Solubility Rules

16-7 Describe the difference between *soluble, insoluble,* and *slightly soluble* salts.

16-8 Summarize the solubility rules for ionic compounds.

16-9 Which of the following salts is insoluble in water?

(a) $Ba(NO_3)_2$ (b) $BaCl_2$ (c) $BaCO_3$ (d) BaS (e) $Ba(OAc)_2$

16-10 Which of the following salts is insoluble in water?

(a) $(NH_4)_2SO_4$ (b) K_2CrO_4 (c) Na_2S (d) $Pb(NO_3)_2$
(e) $Cr(OH)_3$

16-11 Which of the following salts is soluble in water?

(a) PbS (b) PbO (c) $PbCrO_4$ (d) $PbCO_3$ (e) $Pb(NO_3)_2$

The Solubility Product Expression

16-12 Explain why the $[Ag^+]$ and $[Cl^-]$ terms are variables but the [AgCl] term is a constant no matter how much AgCl is added to a saturated solution of silver chloride in water.

16-13 Why isn't the concentration of solid AgCl included in the equilibrium constant expression for the following reaction?

$$AgCl(s) \xrightarrow{H_2O} Ag^+(aq) + Cl^-(aq)$$

16-14 What is the correct solubility product expression for the following reaction?

$$Ca_3(PO_4)_2(s) \xrightarrow{H_2O} 3\ Ca^{2+}(aq) + 2\ PO_4^{3-}(aq)$$

(a) $K_{sp} = \dfrac{[Ca^{2+}][PO_4^{3-}]}{[Ca_3(PO_4)_2]}$

(b) $K_{sp} = \dfrac{[Ca^{2+}]^3[PO_4^{3-}]^2}{[Ca_3(PO_4)_2]}$

(c) $K_{sp} = [Ca^{2+}][PO_4^{3-}]$

(d) $K_{sp} = [Ca^{2+}]^3[PO_4^{3-}]^2$

(e) $K_{sp} = [Ca^{2+}]^2[PO_4^{3-}]^3$

16-15 Which of the following is the correct solubility product expression for $Al_2(SO_4)_3$?

(a) $K_{sp} = [Al^{3+}][SO_4^{2-}]$ (b) $K_{sp} = [2\ Al^{3+}][3\ SO_4^{2-}]$
(c) $K_{sp} = [Al^{3+}]^2[SO_4^{2-}]^3$ (d) $K_{sp} = [2\ Al^{3+}]^2[3\ SO_4^{2-}]^3$

16-16 Write the solubility product expression for each of the following salts.

(a) $BaCrO_4$ (b) $CaCO_3$ (c) PbF_2 (d) Ag_2S

16-17 Write the solubility product expression for each of the following salts.

(a) Ag_3PO_4 (b) $Fe(OH)_3$ (c) Bi_2S_3 (d) $Pb_3(PO_4)_2$

16-18 Describe the experiment you would use to measure K_{sp} for $PbCrO_4$.

The Relationship Between K_{sp} and the Solubility of a Salt

16-19 Write an equation that describes the relationship between the concentrations of the Mg^{2+} and F^- ions in a saturated solution of magnesium fluoride.

16-20 Write an equation that describes the relationship between the concentrations of the Ag^+ and CrO_4^{2-} ions in a saturated solution of Ag_2CrO_4.

16-21 Write an equation that describes the relationship between the concentrations of the Bi^{3+} and S^{2-} ions in a saturated solution of Bi_2S_3.

16-22 Which of the following equations describes the relationship between the solubility product for MgF_2 and the solubility of this compound?

(a) $K_{sp} = 2\ S$ (b) $K_{sp} = S^2$ (c) $K_{sp} = 2\ S^2$ (d) $K_{sp} = S^3$
(e) $K_{sp} = 4\ S^3$

16-23 Hg_2Cl_2 contains the Hg_2^{2+} and Cl^- ions. Which of the following equations describes the relationship between the solubility product and the solubility of this compound?

(a) $K_{sp} = S^3$ (b) $K_{sp} = 4\ S^3$ (c) $K_{sp} = S^4$ (d) $K_{sp} = 16\ S^4$

16-24 Which is more soluble, Ag_2S or HgS? (Ag_2S: $K_{sp} = 6.3 \times 10^{-50}$, HgS: $K_{sp} = 4 \times 10^{-53}$)

16-25 Which is more soluble, $PbSO_4$ or PbI_2? ($PbSO_4$: $K_{sp} = 1.6 \times 10^{-8}$, PbI_2: $K_{sp} = 7.1 \times 10^{-9}$)

16-26 Mercury forms salts that contain either the Hg^{2+} ion or the Hg_2^{2+} ion. Which is more soluble, HgS or Hg_2S? (HgS: $K_{sp} = 4 \times 10^{-53}$, Hg_2S: $K_{sp} = 1.0 \times 10^{-47}$)

Solubility Product Calculations

16-27 What is the concentration of the CrO_4^{2-} ion in a saturated solution of barium chromate dissolved in water if the Ba^{2+} ion concentration is 1.1×10^{-5} M?

16-28 What is the concentration of the CrO_4^{2-} ion in a saturated solution of silver chromate dissolved in water if the Ag^+ ion concentration is 1.3×10^{-4} M?

16-29 What is the concentration of the CN^- ion in a saturated solution of zinc cyanide dissolved in water if the Zn^{2+} ion concentration is 4.0×10^{-5} M?

16-30 What is the solubility product for silver bromide if the solubility of AgBr in water is 1.3×10^{-5} grams per liter?

16-31 What is the solubility product for strontium fluoride if the solubility of SrF_2 in water is 0.107 grams per liter?

16-32 Silver acetate, $Ag(OAc)_2$, is marginally soluble in water. What is the solubility product for silver acetate if 1.190 grams of $Ag(OAc)_2$ dissolves in 99.40 milliliters of water?

16-33 Lithium salts, such as lithium carbonate, are used to treat manic-depressives. What is the solubility product for lithium carbonate if 1.36 grams of Li_2CO_3 dissolve in 100 milliliters of water?

16-34 What is the solubility product for $Mg(OH)_2$ if a saturated solution in water has a concentration of 0.000165 M?

16-35 People who have had the misfortune of going through a series of x-rays of the gastrointestinal tract have been given a suspension of solid barium sulfate in water to drink. $BaSO_4$ is used instead of other Ba^{2+} salts, which also reflect x-rays, because it is so remarkably insoluble in water. (Thus the patient is exposed to the minimum amount of toxic Ba^{2+} ion.) What is the solubility product for barium sulfate if 1 gram of $BaSO_4$ dissolves in 400,000 grams of water?

16-36 What is the solubility product for calcium hydroxide if a saturated solution of $Ca(OH)_2$ has a pH of 12.35 at room temperature?

16-37 What is the solubility product for $Ca(OH)_2$ if it takes 22.24 milliliters of a 0.100 M HCl solution to neutralize 100 milliliters of saturated $Ca(OH)_2$?

16-38 What is the solubility of silver iodide in water in grams per liter if the solubility product for AgI is 8.3×10^{-17}?

16-39 What is the solubility of silver sulfide in water in grams per 100 milliliters if the solubility product for Ag_2S is 6.3×10^{-50}?

16-40 What is the solubility in water in grams per 100 milliliters for each of the following salts?

(a) BaC_2O_4 ($K_{sp} = 2.3 \times 10^{-8}$) (b) $CaCO_3$ ($K_{sp} = 2.8 \times 10^{-9}$) (c) PbF_2 ($K_{sp} = 2.7 \times 10^{-8}$) (d) Hg_2S ($K_{sp} = 1.0 \times 10^{-47}$) (e) HgS ($K_{sp} = 4 \times 10^{-53}$)

16-41 What is the solubility in water in grams per 100 milliliters for each of the following salts?

(a) Hg_2CrO_4 ($K_{sp} = 2.0 \times 10^{-9}$) (b) Bi_2S_3 ($K_{sp} = 1 \times 10^{-97}$) (c) $Ca_3(PO_4)_2$ ($K_{sp} = 2.0 \times 10^{-29}$) (d) $Pb_3(PO_4)_2$ ($K_{sp} = 8.0 \times 10^{-43}$) (e) Ag_3PO_4 ($K_{sp} = 1.4 \times 10^{-16}$)

16-42 List the following salts in order of increasing solubility in water.

(a) Ag_2S ($K_{sp} = 6.3 \times 10^{-50}$) (b) Bi_2S_3 ($K_{sp} = 1 \times 10^{-97}$) (c) CuS ($K_{sp} = 6.3 \times 10^{-36}$) (d) HgS ($K_{sp} = 4 \times 10^{-53}$)

The Common Ion Effect

16-43 Define the term *common ion effect*.

16-44 Describe how LeChatelier's principle can be used to explain the common ion effect.

16-45 Describe what happens to the equilibrium concentrations of the Ag^+ and Cl^- ions when 10 grams of NaCl is added to a liter of a saturated solution of silver chloride in water.

16-46 Which of the following statements is true?

(a) MgF_2 is more soluble in 0.100 M NaF than in pure water. (b) MgF_2 is less soluble in 0.100 M NaF than in pure water. (c) MgF_2 is just as soluble in 0.100 M NaF as in pure water.

16-47 In which of the following solutions would Ag_2S be least soluble? (Ag_2S: $K_{sp} = 6.3 \times 10^{-50}$, H_2S: $K_{a1} = 1.0 \times 10^{-7}$, $K_{a2} = 1.3 \times 10^{-13}$)

(a) pure water (b) 0.0010 M Na_2S (c) 0.10 M H_2S

16-48 Calculate the solubility of $Al(OH)_3$ in moles per liter in a solution buffered at pH 9.31. [$Al(OH)_3$: $K_{sp} = 1.3 \times 10^{-33}$]

How to Keep a Salt From Dissolving or Precipitating

16-49 Calculate the equilibrium concentration of the Ag^+ ion in a solution prepared by dissolving 3.21 grams of potassium iodide in 350 milliliters of water and then adding silver iodide until the solution is saturated with AgI. (AgI: $K_{sp} = 8.3 \times 10^{-17}$)

16-50 How many grams of AgI will dissolve in 100 milliliters of the solution described in Problem 16-49?

16-51 How many grams of silver sulfide will dissolve in 500 milliliters of an 0.050 M S^{2-} solution? (Ag_2S: $K_{sp} = 6.3 \times 10^{-50}$)

16-52 Calculate the solubility of Bi_2S_3 in each of the following solutions. (Bi_2S_3: $K_{sp} = 1 \times 10^{-97}$)

(a) pure water (b) a solution in which [S^{2-}] is 5×10^{-8} M (c) a pH 1 solution that is 0.10 M in H_2S

16-53 In which of the following solutions is $Pb(OH)_2$ most soluble? [$Pb(OH)_2$: $K_{sp} = 1.2 \times 10^{-15}$]

(a) pure water (b) 0.010 M NaOH (c) 0.010 M HCl

16-54 Calculate the solubility of $Co(OH)_3$ in both pure water and a pH 9 solution. [$Co(OH)_3$: $K_{sp} = 1.6 \times 10^{-44}$]

16-55 At what pH does $Cr(OH)_3$ just start to precipitate from a solution that is 0.025 M in $Cr(NO_3)_3$? [$Cr(OH)_3$: $K_{sp} = 6.3 \times 10^{-31}$]

How to Separate Ions by Selective Precipitation

16-56 Assume that you are given a solution that contains 5.31 grams of $BaCl_2$ in 135 milliliters of water, and assume that you can add a source of the SO_4^{2-} ion without significantly changing the volume of the solution. What is the sulfate ion concentration when 99.9% of the Ba^{2+} has been precipitated as $BaSO_4$? ($BaSO_4$: $K_{sp} = 1.1 \times 10^{-10}$)

16-57 Start with a solution that contains 7.33 grams of K_2CrO_4 per 500 milliliters. Assume that you can add Ag^+ ions to this solution without changing its volume significantly. What is the Ag^+ ion concentration when 99.9% of the CrO_4^{2-} has been precipitated as Ag_2CrO_4? (Ag_2CrO_4: $K_{sp} = 1.1 \times 10^{-12}$)

16-58 Both barium sulfite, $BaSO_3$, and barium sulfate, $BaSO_4$, are insoluble in water. $BaSO_3$ dissolves in acid, however, whereas $BaSO_4$ does not. Explain why. ($BaSO_3$: $K_{sp} = 8 \times 10^{-7}$; $BaSO_4$: $K_{sp} = 1.1 \times 10^{-10}$; H_2SO_3: $K_{a1} = 1.7 \times 10^{-2}$, $K_{a2} = 6.4 \times 10^{-8}$; H_2SO_4: $K_{a1} = 10^3$, $K_{a2} = 1.2 \times 10^{-2}$)

16-59 Ag^+ forms a strong complex with ammonia.

$$Ag^+(aq) + 2\ NH_3(aq) \rightleftharpoons Ag(NH_3)_2^+(aq) \qquad K_c = 1.1 \times 10^7$$

Should AgCl be more soluble or less soluble in a solution containing ammonia than in pure water? Explain your answer using chemical equations to illustrate your arguments. (AgCl: $K_{sp} = 1.8 \times 10^{-10}$)

16-60 Which sulfide, PbS or ZnS, precipitates first when we slowly add S^{2-} ions to a solution that is 1.0×10^{-4} M in both Pb^{2+} and Zn^{2+} ions? What is the concentration of this ion when the second ion starts to precipitate? (PbS: $K_{sp} = 8.0 \times 10^{-28}$, ZnS: $K_{sp} = 1.6 \times 10^{-24}$)

16-61 Which halide, AgCl or AgI, precipitates first when we slowly add Ag^+ ions to a solution that is 1.0×10^{-5} M in both Cl^- and I^- ions? What is the concentration of this ion when the second ion starts to precipitate? (AgCl: $K_{sp} = 1.8 \times 10^{-10}$, AgI: $K_{sp} = 8.3 \times 10^{-17}$)

16-62 Which of the following statements is true? (CdS: $K_{sp} = 8 \times 10^{-27}$, CuS: $K_{sp} = 6.3 \times 10^{-36}$, Ag_2S: $K_{sp} = 6.3 \times 10^{-50}$)

(a) CdS is less soluble than CuS. (b) CdS is about 10^9 times more soluble than CuS. (c) Ag_2S is less soluble than either CdS or CuS. (d) Ag_2S is about 10 times more soluble than CuS.

16-63 What happens when H_2S is added very slowly to a solution that contains 0.10 M concentrations of the Cu^{2+} and Cd^{2+} ions? (CdS: $K_{sp} = 8 \times 10^{-27}$, CuS: $K_{sp} = 6.3 \times 10^{-36}$, Ag_2S: $K_{sp} = 6.3 \times 10^{-50}$)

(a) CdS will precipitate first.   (b) CuS will precipitate first.   (c) Most of the Cu^{2+} will still be in solution when most of the Cd^{2+} has been precipitated.   (d) Both CdS and CuS will precipitate at the same time.   (e) Neither CdS or CuS will precipitate.

16-64 A solution is prepared from small amounts of CdS and CuS added to pure water. What happens when a strong acid such as HCl is added to this solution? (CdS: $K_{sp} = 8 \times 10^{-27}$, CuS: $K_{sp} = 6.3 \times 10^{-36}$, Ag_2S: $K_{sp} = 6.3 \times 10^{-50}$)

(a) Only CdS will dissolve at first.   (b) Both CdS and CuS will dissolve at the same time.   (c) Some CdS will dissolve, but more CuS will precipitate.   (d) More of both CdS and CuS will precipitate.

16-65 Describe how you would separate Pb^{2+} from Hg^{2+} in a solution that was 0.10 M in both ions. (*Hint:* PbS: $K_{sp} = 8.0 \times 10^{-28}$, HgS: $K_{sp} = 4 \times 10^{-53}$)

How to Adjust the Concentration of an Ion

16-66 Hydrogen sulfide, H_2S, is a weak acid in water.

$$H_2S(aq) + H_2O \rightleftharpoons H_3O^+(aq) + HS^-(aq) \quad K_{a1} = 1.0 \times 10^{-7}$$
$$HS^-(aq) + H_2O \rightleftharpoons H_3O^+(aq) + S^{2-}(aq) \quad K_{a2} = 1.3 \times 10^{-13}$$

If your goal was to dissolve a ZnS precipitate, would it be better to add a strong acid or a strong base? Explain your answer using chemical equations to illustrate your reasoning. (ZnS: $K_{sp} = 1.6 \times 10^{-24}$)

16-67 Calculate the maximum concentrations of Zn^{2+}, Hg^{2+}, and Bi^{3+} that can exist in the presence of 1.0×10^{-20} M S^{2-} ion. (ZnS: $K_{sp} = 1.6 \times 10^{-24}$, HgS: $K_{sp} = 4 \times 10^{-53}$, Bi_2S_3: $K_{sp} = 1 \times 10^{-97}$)

16-68 Explain why iron(III) hydroxide, $Fe(OH)_3$, precipitates when aqueous solutions of iron(III) chloride, $FeCl_3$, and sodium carbonate, Na_2CO_3, are mixed. [$Fe(OH)_3$: $K_{sp} = 4 \times 10^{-38}$]

16-69 What will precipitate from a solution that is 0.10 M in Cd^{2+}, Fe^{3+}, HCl, and H_2S?

16-70 Which of the following cations will precipitate as the sulfide from a solution that is 0.10 M in H_2S, 0.3 M in HCl, and 10^{-3} M in each of the following ions?

(a) Fe^{3+} (b) Ag^+ (c) Cu^{2+} (d) Ni^{2+}

COMPLEX ION AND COMBINED EQUILIBRIA

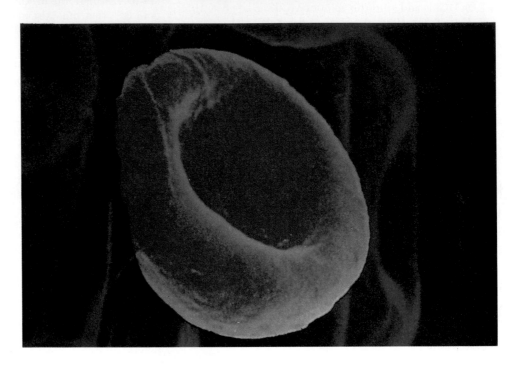

17.1 THE CHEMISTRY OF AQUEOUS SOLUTIONS OF THE Cu²⁺ ION

The basic assumption behind the discussion of solubility equilibria in Chapter 16 was the idea that salts dissociate into their ions when they dissolve in water. Copper sulfate, for example, is assumed to dissociate into the Cu^{2+} and SO_4^{2-} ions in water, as shown in Figure 17.1.

$$CuSO_4(s) \xrightarrow{H_2O} Cu^{2+}(aq) + SO_4^{2-}(aq)$$

Figure 17.2 shows what happens when a few drops of 2 M NH₃ are added to a 0.10 M CuSO₄ solution. The first thing we notice is the formation of a light blue, almost bluish-white, precipitate. We can explain this observation by combining what we know about acid-base equilibria and solubility equilibria. Ammonia acts as a base toward water to form a mixture of the ammonium and hydroxide ions.

$$NH_3(aq) + H_2O(l) \rightleftharpoons NH_4^+(aq) + OH^-(aq) \qquad K_b = 1.8 \times 10^{-5}$$

The OH^- ions formed in this reaction combine with Cu^{2+} ions in the solution to form a $Cu(OH)_2$ precipitate.

$$Cu^{2+}(aq) + 2\ OH^-(aq) \rightleftharpoons Cu(OH)_2(s) \qquad K_{sp} = 2.2 \times 10^{-20}$$

In theory, the OH^- ion concentration should become larger when more base is added to the solution. As a result, more $Cu(OH)_2$ should precipitate from the solution, which is what we see here. Figure 17.3, however, shows the effect of adding an excess of 2 M NH₃ to the solution. In the presence of excess ammonia, the $Cu(OH)_2$ precipitate dissolves, and the solution turns deep blue.

This demonstration raises several important questions. "Why does the $Cu(OH)_2$ precipitate dissolve in excess ammonia?" "Why does the color of the solution become more intense and why does the color shift to a distinctly deeper blue when this happens?"

17.2 COMPLEX IONS

The first step toward answering these questions involves writing the electron configuration of copper metal and its Cu^{2+} ions. In Section 5.20, we noted that the experimental electron configuration for an isolated copper atom in the gas phase differs slightly from the predictions of the Aufbau principle.

$$Cu = [Ar]\ 4s^1\ 3d^{10}$$

We explained this by invoking the general principle that there is something inherently stable about electron configurations in which the subshells are either full ($3d^{10}$), half-full ($3d^5$), or empty ($3d^0$).

When a copper atom loses a single electron to form a Cu^+ ion, it achieves the following electron configuration.

$$Cu^+ = [Ar]\ 3d^{10}$$

If it loses a second electron to form a Cu^{2+} ion, this electron must come from the $3d$ subshell.

$$Cu^{2+} = [Ar]\ 3d^9$$

It is sometimes useful to think about the electron configuration of the Cu^{2+} ion in terms of the entire set of valence-shell orbitals. In addition to the nine electrons

FIG. 17.1 The first thing that happens when 2 M NH₃ is added to an aqueous solution of the Cu^{2+} ion is the formation of a pale blue, almost bluish white, $Cu(OH)_2$ precipitate.

FIG. 17.2 LeChatelier's principle suggests that adding more ammonia to the solution in Figure 17.1 should increase the amount of precipitate, which is what is observed at first.

FIG. 17.3 In the presence of excess ammonia, the $Cu(OH)_2$ precipitate dissolves to form a deep blue solution of the $Cu(NH_3)_4^{2+}$ complex ion.

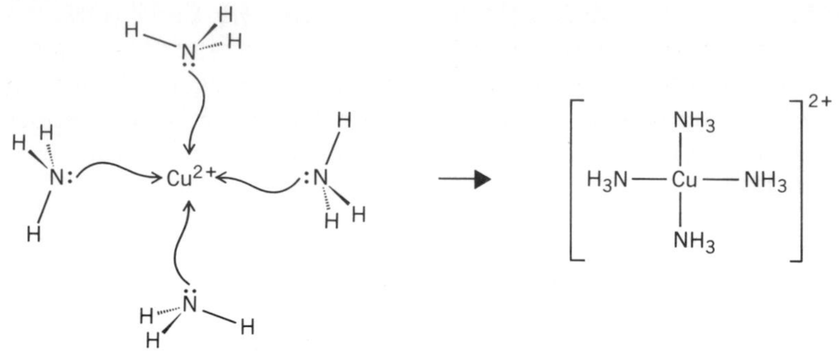

FIG. 17.4 The Cu^{2+} ion has four empty valence-shell orbitals— $4s$, $4p_x$, $4p_y$, and $4p_z$—that can accept pairs of nonbonding electrons from NH_3 molecules to form covalent Cu—N bonds.

in the $3d$ subshell, this ion also has an empty $4s$ orbital and a set of three empty $4p$ orbitals.

$$Cu^{2+} = [Ar]\ 4s^0\ 3d^9\ 4p^0$$

The Cu^{2+} ion can therefore pick up pairs of nonbonding electrons from four NH_3 molecules to form covalent Cu—N bonds, as shown in Figure 17.4.

$$Cu^{2+}(aq) + 4\ NH_3(aq) \rightleftharpoons Cu(NH_3)_4{}^{2+}$$

G. N. Lewis was the first to recognize the similarity between this reaction and the prototypical acid-base reaction, in which an H^+ ion combines with an OH^- ion to form a neutral water molecule.

$$H^+ \quad :\ddot{O}-H^- \longrightarrow H-\ddot{O}-H$$

Both reactions involve the transfer of a pair of nonbonding electrons from one atom to an empty orbital on another atom to form a covalent bond. Both reactions can therefore be interpreted in terms of an electron-pair acceptor combining with an electron-pair donor.

Lewis suggested that we could expand our definition of acids by assuming that an acid is any atom, ion, or molecule that acts like the H^+ ion to accept a pair of nonbonding electrons.

Lewis acid: An electron-pair acceptor.

A Lewis base, on the other hand, is any atom, ion, or molecule that acts like the OH^- ion to donate a pair of nonbonding electrons.

Lewis base: An electron-pair donor.

The product of the reaction of a Lewis acid with a Lewis base is often called an *acid-base complex.* When the Cu^{2+} ion reacts with four NH_3 molecules, the product of this reaction is called a *complex ion.*

$$Cu^{2+}(aq) + 4\ NH_3(aq) \rightleftharpoons Cu(NH_3)_4{}^{2+}(aq)$$

electron-pair acceptor (Lewis acid)	electron-pair donor (Lewis base)	acid-base complex, or complex ion

Any atom, ion, or molecule with one or more empty valence-shell orbitals can be a Lewis acid.

Exercise 17.1

Is boron trifluoride a potential Lewis acid?

Solution

The only way to answer this question is to decide whether there are empty valence-shell orbitals on the molecule. The best way of doing this is to write the Lewis structure. The rules outlined in Chapter 8 give the following Lewis structure for BF_3.

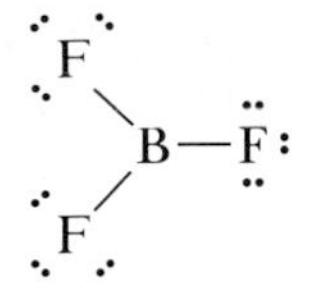

$$\overset{\cdot\,\cdot}{\underset{\overset{\cdot}{\underset{\cdot\,\cdot}{F}}}{\overset{\overset{\cdot}{\underset{\cdot}{F}}}{\diagdown}}} \text{B} \text{—} \ddot{\text{F}}\text{:}$$

All of the fluorine atoms have an octet of valence electrons. But there are only six valence electrons on the boron atom. Since all of the electrons on the boron atom are paired, this atom has an empty valence-shell orbital, which could accept a pair of electrons. Thus BF_3 has the potential to be a Lewis acid.

Boron trifluoride is stable in the sense that it does not decompose at room temperature, but it is highly reactive, combining almost instantly with Lewis bases such as NH_3 to form an acid-base complex that contains a covalent bond between the boron and nitrogen atoms.

$$BF_3(g) + NH_3(g) \longrightarrow F_3B\text{—}NH_3(s)$$

Any atom, ion, or molecule that contains pairs of nonbonding electrons is a Lewis base. The Lewis structures in Figure 17.5, for example, suggest that all of these substances can act as Lewis bases to form complex ions.

Exercise 17.2

Predict which of the following compounds can act as a Lewis base.

(a) CH_3^+ (b) CH_3 (c) CH_3^-

Solution

The only way to decide whether a compound can be a Lewis base is to look for pairs of valence-shell nonbonding electrons. We start by writing the Lewis structures for these three compounds.

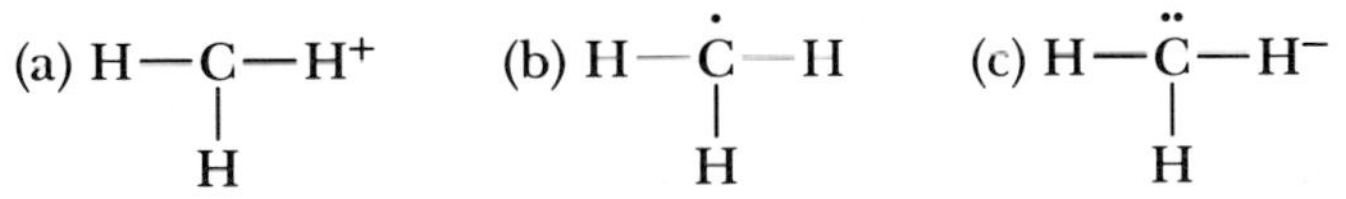

(a) $\underset{\overset{|}{H}}{H\text{—}C\text{—}H^+}$ (b) $\underset{\overset{|}{H}}{H\text{—}\overset{\cdot}{C}\text{—}H}$ (c) $\underset{\overset{|}{H}}{H\text{—}\overset{\cdot\,\cdot}{C}\text{—}H^-}$

Only the CH_3^- ion has a pair of nonbonding electrons that it could donate to a Lewis acid, or electron-pair acceptor. The CH_3^- ion is therefore the only one of these compounds that has the potential to be a Lewis base.

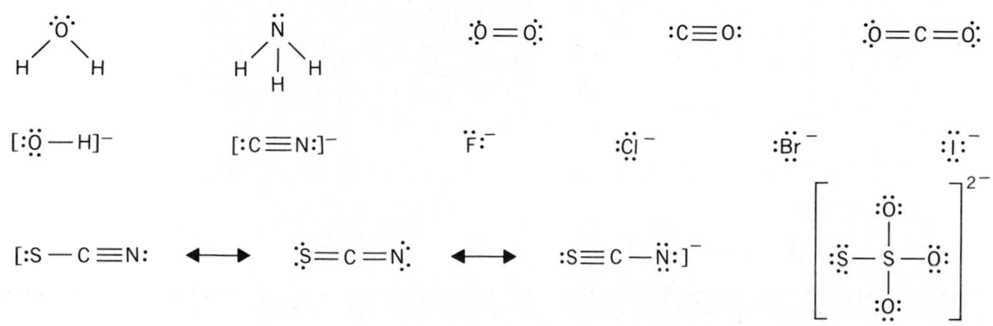

FIG. 17.5 Lewis structures of some potential Lewis bases.

17.3 COMPLEX ION COORDINATION NUMBERS

Complex ions, such as the $Cu(NH_3)_4^{2+}$ ion, are also known as *coordination complexes,* because they contain neutral molecules or anions linked, or coordinated, to a metal ion. The number of Lewis bases, or complexing agents, that can bind to the metal ion is called the *coordination number.*

The Fe^{3+} ion combines with the thiocyanate ion in aqueous solution to form a deep red $Fe(SCN)_2^+$ complex ion, as shown in Figure 17.6.

$$Fe^{3+}(aq) + 2\ SCN^-(aq) \rightleftharpoons Fe(SCN)_2^+(aq)$$

Because this complex contains two SCN^- ions, the coordination number is 2. The Ag^+ ion combines with ammonia in aqueous solution to form another two-coordinate complex ion.

$$Ag^+(aq) + 2\ NH_3(aq) \rightleftharpoons Ag(NH_3)_2^+(aq)$$

As we have already seen, Cu^{2+} ions combine with excess ammonia to form a deep blue $Cu(NH_3)_4^{2+}$ complex ion with a coordination number of 4.

$$Cu^{2+}(aq) + 4\ NH_3(aq) \rightleftharpoons Cu(NH_3)_4^{2+}(aq)$$

Another example of a four-coordinate complex ion is formed when Co^{2+} ions react with thiocyanate ions.

$$Co^{2+}(aq) + 4\ SCN^-(aq) \rightleftharpoons Co(SCN)_4^{2-}(aq)$$

Both 2 and 4 are common coordination numbers for transition-metal complex ions. Another is 6. Co^{3+} and Cr^{3+} ions, for example, form six-coordinate complexes with ammonia.

$$Cr^{3+}(aq) + 6\ NH_3(aq) \rightleftharpoons Cr(NH_3)_6^{3+}(aq)$$
$$Co^{3+}(aq) + 6\ NH_3(aq) \rightleftharpoons Co(NH_3)_6^{3+}(aq)$$

Iron forms six-coordinate complexes with the cyanide ion in which the iron atom is formally in the $+2$ and $+3$ oxidation states.

$$Fe^{2+}(aq) + 6\ CN^-(aq) \rightleftharpoons Fe(CN)_6^{4-}(aq)$$
$$Fe^{3+}(aq) + 6\ CN^-(aq) \rightleftharpoons Fe(CN)_6^{3-}(aq)$$

FIG. 17.6 Dilute solutions of the Fe^{3+} ion are essentially colorless. When the SCN^- ion is added to these solutions, however, the $Fe(SCN)^{2+}$ and $Fe(SCN)_2^+$ complex ions are formed, which gives the solutions a blood-red color.

Exercise 17.3

Predict the charge on the complex ion formed in each of the following reactions.

(a) $Mn^{2+}(aq) + 6 H_2O(l) \rightarrow$
(b) $Zn^{2+}(aq) + 4 CN^-(aq) \rightarrow$
(c) $Ag^+(aq) + 2 S_2O_3^{2-}(aq) \rightarrow$

Solution

Charge is conserved in all chemical reactions. Thus, the sum of the charges on the transition metal ion and the Lewis bases that coordinate to this ion must be equal to the charge on the complex ion. The following complex ions are formed in these reactions: $Mn(H_2O)_6^{2+}$, $Zn(CN)_4^{2-}$, and $Ag(S_2O_3)_2^{3-}$. The charges on these complexes are $+2$, -2, and -3, respectively.

A given metal ion often forms complexes with the same coordination number. The Ag^+ ion, for example, forms two-coordinate complexes with Cl^-, Br^-, I^-, CN^-, SCN^-, NH_3, and $S_2O_3^{2-}$. The Cd^{2+}, Cu^{2+}, Hg^{2+}, Pb^{2+}, and Zn^{2+} ions tend to form four-coordinate complexes such as $Cu(NH_3)_4^{2+}$ and $Zn(CN)_4^{2-}$. A detailed discussion of coordination numbers will be found in Chapter 22. For now, we'll simply note that the coordination number often becomes larger as the charge on the metal ion increases. The Cu^+ ion, for example, forms a two-coordinate $Cu(NH_3)_2^+$ ion, whereas the Cu^{2+} ion forms a four-coordinate $Cu(NH_3)_4^+$ complex ion.

17.4 COMPLEX IONS IN AQUEOUS SOLUTION

Section 16.11 discussed how to prepare a $0.10\ M\ Cr^{3+}$ ion solution by adjusting the OH^- ion concentration in water to make the ion product for the Cr^{3+} and OH^- ions smaller than the solubility product for $Cr(OH)_3$. An equation that describes what happens when $Cr(NO_3)_3$ is used to prepare a $0.10\ M\ Cr^{3+}$ solution might look like this.

$$Cr(NO_3)_3(s) \xrightarrow{H_2O} Cr^{3+}(aq) + 3\ NO_3^-(aq)$$

What does the term $Cr^{3+}(aq)$ mean? The aq symbol is used to remind us that the Cr^{3+} ions are tightly bound to neighboring water molecules. The electron configuration for the Cr^{3+} ion indicates the presence of six empty valence-shell orbitals on this atom — the $4s$, three $4p$, and two $3d$ orbitals.

$$Cr^{3+} = [Ar]\ 4s^0\ 3d^3\ 4p^0$$

The Cr^{3+} ion can therefore accept pairs of nonbonding electrons from six neighboring water molecules to form a $Cr(H_2O)_6^{3+}$ complex ion, as shown in Figure 17.7.

This behavior is not restricted to the Cr^{3+} ion. In aqueous solution, all of the first-row transition metal ions are present as six-coordinate complex ions.

$Sc(H_2O)_6^{3+}$	$Fe(H_2O)_6^{3+}$
$Ti(H_2O)_6^{2+}$	$Co(H_2O)_6^{3+}$
$V(H_2O)_6^{2+}$	$Ni(H_2O)_6^{2+}$
$Cr(H_2O)_6^{3+}$	$Cu(H_2O)_6^{2+}$
$Mn(H_2O)_6^{2+}$	$Zn(H_2O)_6^{2+}$

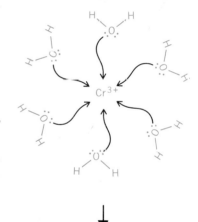

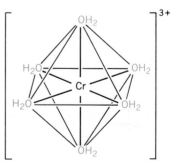

FIG. 17.7 The Cr^{3+} ion has six empty valence-shell orbitals — $4s$, $4p_x$, $4p_y$, $4p_z$, $3d_{x^2-y^2}$, and $3d_{z^2}$ — that can accept pairs of nonbonding electrons from water molecules to form covalent Cr—O bonds.

A similar behavior is observed for second- and third-row transition metals.

This raises an interesting point about the difference between symbols such as $Fe^{3+}(aq)$ and symbols such as $Cr(H_2O)_6{}^{3+}$. By convention, the first type of symbol is used to represent solutions that contain complex ions—such as the $Fe(H_2O)_6{}^{3+}$ ion—that are difficult, if not impossible, to isolate. This complex ion is formed when iron(III) salts dissolve in water.

$$Fe(NO_3)_3(s) + 6\ H_2O(l) \longrightarrow Fe(H_2O)_6{}^{3+}(aq) + 3\ NO_3{}^-(aq)$$

But when the water is allowed to evaporate, $Fe(NO_3)_3$ is isolated, not $[Fe(H_2O)_6](NO_3)_3$.

Once again by convention, the second type of symbol is used to identify compounds that can be isolated. Under the proper conditions, for example, it is possible to isolate a six-coordinate complex of Cr^{3+} with water molecules, such as $[Cr(H_2O)_6]Br_3$.

17.5 THE STEPWISE FORMATION OF COMPLEX IONS

When a transition metal ion binds Lewis bases to form a coordination complex, or complex ion, it picks up these bases one at a time. The Ag^+ ion, for example, combines with NH_3 in a two-step reaction. It first picks up one NH_3 molecule to form a one-coordinate complex.

$$Ag^+(aq) + NH_3(aq) \rightleftharpoons Ag(NH_3)^+(aq)$$

This intermediate then picks up a second NH_3 molecule in a separate step.

$$Ag(NH_3)^+(aq) + NH_3(aq) \rightleftharpoons Ag(NH_3)_2{}^+(aq)$$

It is possible to write equilibrium constant expressions for each step in a complex-ion formation reaction. The equilibrium constants for these reactions are known as **complex formation equilibrium constants, K_f.** The equilibrium constant expressions for the two steps in the formation of the $Ag(NH_3)_2{}^+$ complex ion are written as follows.

$$K_{f1} = \frac{[Ag(NH_3)^+]}{[Ag^+][NH_3]}$$

$$K_{f2} = \frac{[Ag(NH_3)_2{}^+]}{[Ag(NH_3)^+][NH_3]}$$

When polyprotic acids were discussed in Chapter 15, we noted that the difference between the ease with which these acids lose the first and second protons is very large.

$$H_2S(aq) + H_2O(l) \rightleftharpoons H_3O^+(aq) + HS^-(aq) \qquad K_{a1} = 1.0 \times 10^{-7}$$
$$HS^-(aq) + H_2O(l) \rightleftharpoons H_3O^+(aq) + S^{2-}(aq) \qquad K_{a1} = 1.3 \times 10^{-13}$$

As a result, when we perform calculations with polyprotic acids we can assume that the reactions occur in distinct steps.

This assumption is harder to justify for complex formation equilibria. K_{f1} and K_{f2} for the complex between Ag^+ and ammonia, for example, differ by a factor of only 4.

$$Ag^+(aq) + NH_3(aq) \rightleftharpoons Ag(NH_3)^+(aq) \qquad K_{f1} = 1.7 \times 10^3$$
$$Ag(NH_3)^+(aq) + NH_3(aq) \rightleftharpoons Ag(NH_3)_2{}^+(aq) \qquad K_{f2} = 6.5 \times 10^3$$

TABLE 17.1

The Effect of Changes in the NH_3 Concentration on the Fraction of Silver Present as the Ag^+, $Ag(NH_3)^+$, or $Ag(NH_3)_2^+$ Ions

$[NH_3]$	Ag^+ (%)	$Ag(NH_3)^+$ (%)	$Ag(NH_3)_2^+$ (%)
10^{-6}	99.8	0.2	0.001
10^{-5}	98.2	1.7	0.1
10^{-4}	78.1	13.3	8.6
10^{-3}	7.3	12.4	80.4
10^{-2}	0.09	1.5	98.4
10^{-1}	0.0009	0.2	99.8

This means that most of the Ag^+ ions that pick up one NH_3 molecule to form the $Ag(NH_3)^+$ complex ion are likely to pick up another NH_3 to form the two-coordinate $Ag(NH_3)_2^+$ complex ion. Table 17.1 summarizes the concentrations of the Ag^+, $Ag(NH_3)^+$, and $Ag(NH_3)_2^+$ ions over a range of concentrations of NH_3. These data are plotted in Figure 17.8.

Essentially all of the silver is present as the Ag^+ ion at very low concentrations of NH_3. As the NH_3 concentration becomes larger, the amount of Ag^+ ion rapidly becomes smaller. Even at NH_3 concentrations as small as 0.010 M, the two-coordinate $Ag(NH_3)_2^+$ complex ion is the dominant species in this solution.

The concentration of the one-coordinate $Ag(NH_3)^+$ intermediate is never very large. Either the NH_3 concentration is so small that most of the silver is present as the Ag^+ ion, or it is large enough that essentially all of the silver is present as the two-coordinate $Ag(NH_3)_2^+$ complex ion.

If the only important components of this equilibrium are the free Ag^+ ion (at low NH_3 concentrations) and the two-coordinate $Ag(NH_3)_2^+$ complex ion (at high concentrations of NH_3), we can best handle the complex formation equilibria for this reaction by collapsing the individual steps in the reaction into an overall equation such as the following.

$$Ag^+(aq) + 2\,NH_3(aq) \rightleftharpoons Ag(NH_3)_2^+(aq)$$

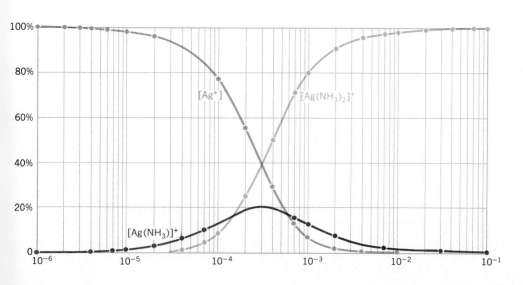

FIG. 17.8 The effect of changes in the NH_3 concentration on the proportion of the total silver ion concentration present as Ag^+, $Ag(NH_3)^+$, and $Ag(NH_3)_2^+$ ions. At NH_3 concentrations as low as 3.2×10^{-3} M, more than 95% of the silver ions are present as two-coordinate $Ag(NH_3)_2^+$ complex ions.

The overall complex formation equilibrium constant expression for this reaction is written as follows.

$$K_f = \frac{[Ag(NH_3)_2^+]}{[Ag^+][NH_3]^2}$$

This expression is equal to the product of the equilibrium constant expressions for the individual steps in the reaction.

$$\frac{[Ag(NH_3)_2^+]}{[Ag^+][NH_3]^2} = \frac{[\cancel{Ag(NH_3)^+}]}{[Ag^+][NH_3]} \times \frac{[Ag(NH_3)_2^+]}{[\cancel{Ag(NH_3)^+}][NH_3]}$$

The overall complex formation equilibrium constant is therefore equal to the product of the K_f values for the individual steps.

$$K_f = K_{f1} \times K_{f2} = 1.1 \times 10^7$$

Overall complex formation equilibrium constants for common complex ions can be found in Table A-12 in the appendix.

Exercise 17.4

Calculate the overall complex formation equilibrium constant for the two-coordinate $Fe(SCN)_2^+$ complex ion from the following data.

$$Fe^{3+}(aq) + SCN^-(aq) \rightleftharpoons Fe(SCN)^{2+}(aq) \qquad K_{f1} = 890$$
$$Fe(SCN)^{2+}(aq) + SCN^-(aq) \rightleftharpoons Fe(SCN)_2^+(aq) \qquad K_{f2} = 2.6$$

Solution

The difference between the stepwise formation equilibrium constants for this complex is relatively small.

$$K_{f1} = \frac{[Fe(SCN)^{2+}]}{[Fe^{3+}][SCN^-]} = 890$$

$$K_{f2} = \frac{[Fe(SCN)_2^+]}{[Fe(SCN)^{2+}][SCN^-]} = 2.6$$

Solutions of these complex ions are therefore best described in terms of an overall complex formation equilibrium.

$$Fe^{3+}(aq) + 2\ SCN^-(aq) \rightleftharpoons Fe(SCN)_2^+(aq)$$

The equilibrium constant expression for the overall reaction is equal to the product of the expressions for the individual steps in the reaction.

$$\frac{[Fe(SCN)_2^+]}{[Fe^{3+}][SCN^-]^2} = \frac{[\cancel{Fe(SCN)^{2+}}]}{[Fe^{3+}][SCN^-]} \times \frac{[Fe(SCN)_2^+]}{[\cancel{Fe(SCN)^{2+}}][SCN^-]}$$

The overall equilibrium constant is therefore the product of K_{f1} and K_{f2}.

$$K_f = K_{f1} \times K_{f2} = 2.3 \times 10^3$$

17.6 COMPLEX DISSOCIATION EQUILIBRIUM CONSTANTS

Complex ions can also be described in terms of complex dissociation equilibria. We can start by assuming, for example, that most of the silver ions in an aqueous solution are present as two-coordinate $Ag(NH_3)_2^+$ complex ions. We can then assume that some of these ions dissociate to form $Ag(NH_3)^+$ complex ions.

$$Ag(NH_3)_2^+(aq) \rightleftharpoons Ag(NH_3)^+(aq) + NH_3(aq)$$

Some of this intermediate then dissociates into NH_3 and the free Ag^+ ion.

$$Ag(NH_3)^+(aq) \rightleftharpoons Ag^+(aq) + NH_3(aq)$$

Complex dissociation equilibrium constants (K_d) can be written for each of these reactions.

$$K_{d1} = \frac{[Ag(NH_3)^+][NH_3]}{[Ag(NH_3)_2^+]}$$

$$K_{d2} = \frac{[Ag^+][NH_3]}{[Ag(NH_3)^+]}$$

Alternatively, the individual steps in the reaction can be collapsed into an overall equation.

$$Ag(NH_3)_2^+(aq) \rightleftharpoons Ag^+(aq) + 2\ NH_3(aq)$$

This equation can be described by the following overall complex dissociation equilibrium constant expression.

$$K_d = \frac{[Ag^+][NH_3]^2}{[Ag(NH_3)_2^+]} = K_{d1} \times K_{d2}$$

Exercise 17.5

Calculate the overall complex dissociation equilibrium constant for the $Cu(NH_3)_4^{2+}$ ion if the overall K_f for this complex is 2.1×10^{13}.

Solution

The overall complex formation equilibrium constant describes the following reaction.

$$Cu^{2+}(aq) + 4\ NH_3(aq) \rightleftharpoons Cu(NH_3)_4^{2+}(aq)$$

The equilibrium constant expression for this reaction is therefore written as follows.

$$K_f = \frac{[Cu(NH_3)_4^{2+}]}{[Cu^{2+}][NH_3]^4}$$

The overall complex dissociation equilibrium constant describes the reaction written in the opposite direction.

$$Cu(NH_3)_4^{2+}(aq) \rightleftharpoons Cu^{2+}(aq) + 4\ NH_3(aq)$$

The equilibrium constant expression for this reaction is therefore the inverse of the complex formation equilibrium expression

$$K_d = \frac{[Cu^{2+}][NH_3]^4}{[Cu(NH_3)_4^{2+}]}$$

and the value of K_d is equal to the inverse of K_f.

$$K_d = 1/K_f = 4.8 \times 10^{-14}$$

If you have any doubt about the validity of this equation, look at the product of the K_d and K_f expressions.

$$\frac{[Cu^{2+}][NH_3]^4}{[Cu(NH_3)_4^{2+}]} \times \frac{[Cu(NH_3)_4^{2+}]}{[Cu^{2+}][NH_3]^4} = 1$$

Thus, the product of K_d times K_f must be equal to 1.

$$K_d \times K_f = 1$$

Exercise 17.6

Derive equations that describe the relationships among the values of K_{f1}, K_{f2}, K_{d1}, and K_{d2} for the $Ag(NH_3)_2^+$ complex ion.

Solution

Perhaps the best way of starting this derivation involves grouping equilibrium expressions that contain similar terms.

$$K_{f1} = \frac{[Ag(NH_3)^+]}{[Ag^+][NH_3]} \qquad\qquad K_{d2} = \frac{[Ag^+][NH_3]}{[Ag(NH_3)^+]}$$

$$K_{f2} = \frac{[Ag(NH_3)_2^+]}{[Ag(NH_3)^+][NH_3]} \qquad K_{d1} = \frac{[Ag(NH_3)^+][NH_3]}{[Ag(NH_3)_2^+]}$$

K_{f1} is the inverse of K_{d2}, and K_{f2} is the inverse of K_{d1}. These equilibrium constants are therefore related by the following equations.

$$K_{f1} = 1/K_{d2} \qquad K_{f2} = 1/K_{d1}$$

These equations can also be written as follows.

$$K_{f1} \times K_{d2} = 1 \qquad K_{f2} \times K_{d1} = 1$$

If you have any doubt about the validity of these equations, test them. Look at the product of the K_{f1} and K_{d2} equilibrium constant expressions, for example.

$$\frac{[Ag(NH_3)^+]}{[Ag^+][NH_3]} \times \frac{[Ag^+][NH_3]}{[Ag(NH_3)^+]} = 1$$

Or look at the product of the K_{f2} and K_{d1} equilibrium constant expressions.

$$\frac{[Ag(NH_3)_2^+]}{[Ag(NH_3)^+][NH_3]} \times \frac{[Ag(NH_3)^+][NH_3]}{[Ag(NH_3)_2^+]} = 1$$

17.7 APPROXIMATE COMPLEX ION CALCULATIONS

There is no fundamental difference between the approaches used to solve complex ion equilibrium problems and those introduced in the discussion of gas-phase equilibria in Chapter 14.

Exercise 17.7

Calculate the equilibrium concentration of the Fe^{3+} ion in a solution that is initially 0.10 M Fe^{3+} and 1.0 M SCN^-. [$Fe(SCN)_2^+$: $K_f = 2.3 \times 10^3$]

Solution

It might seem tempting to set up this problem as follows.

$$Fe^{3+}(aq) + 2\ SCN^-(aq) \rightleftharpoons Fe(SCN)_2^+(aq) \qquad K_f = 2.3 \times 10^3$$

Initial:	0.10 M	1.0 M	0
Equilibrium:	0.10 − Δ	1.0 − 2Δ	Δ

This format assumes that some of the Fe^{3+} ions in the solution combine with exactly twice as many SCN^- ions to form $Fe(SCN)_2^+$ complex ions. Substituting this information into the equilibrium constant expression for this reaction gives the following equation.

$$K_f = \frac{[Fe(SCN)_2^+]}{[Fe^{3+}][SCN^-]^2} = \frac{[\Delta]}{[0.10 - \Delta][1.0 - 2\Delta]^2} = 2.3 \times 10^3$$

One way to solve this equation is to hope that Δ is small compared with the initial concentrations of the Fe^{3+} and SCN^- ions.

$$\frac{[\Delta]}{[0.10][1.0]^2} \cong 2.3 \times 10^3$$

Solving this approximate equation gives the following result.

$$\Delta \cong 230$$

This answer obviously violates the assumption that Δ is small. We should have known from the beginning that this assumption was questionable, because the initial reaction quotient is very far from equilibrium.

$$Q_f = \frac{Fe(SCN)_2^+}{(Fe^{3+})(SCN^-)^2} = \frac{0}{(0.10)(1.0)^2} = 0 \ll 2.3 \times 10^3$$

Section 14.11 introduced a way around this problem. We define a set of intermediate conditions in which we drive the reaction as far as possible toward the right.

$$Fe^{3+}(aq) + 2\ SCN^-(aq) \rightleftharpoons Fe(SCN)_2^+(aq) \quad K_f = 2.3 \times 10^3$$

Initial:	0.10 M	1.0 M	0
Change:	− 0.10 M	− 2(0.10) M	+ 0.10 M
Intermediate:	0	0.8 M	0.10 M

The reaction then comes to equilibrium from these intermediate conditions.

$$Fe^{3+}(aq) + 2\ SCN^-(aq) \rightleftharpoons Fe(SCN)_2^+(aq) \qquad K_f = 2.3 \times 10^3$$

Intermediate:	0	0.8 M	0.10 M
Equilibrium:	Δ	0.8 + 2Δ	0.10 − Δ

Substituting this information into the equilibrium constant expression gives the following equation.

$$\frac{[0.10 - \Delta]}{[\Delta][0.8 + 2\Delta]^2} = 2.3 \times 10^3$$

We now assume, once again, that Δ is relatively small.

$$\frac{[0.10]}{[\Delta][0.8]^2} \cong 2.3 \times 10^3$$

Solving this approximate equation gives the following result.

$$\Delta \cong 6.8 \times 10^{-5}$$

The assumption is valid this time, because the problem was defined so that it is valid. We can now use this value of Δ to answer the original question.

$$[Fe^{3+}] = 6.8 \times 10^{-5} \, M$$

Even though the complex formation equilibrium constant is not very large, essentially all of the Fe^{3+} ion in this solution is complexed. This is the result of the combined effects of a large value of K_f and a relatively large concentration of the SCN^- ion.

Exercise 17.8

Calculate the concentration of the Cu^{2+} ion in a solution that is initially 0.10 M Cu^{2+} and 1.0 M NH_3. [$Cu(NH_3)_4{}^{2+}$: $K_f = 2.1 \times 10^{13}$]

Solution

We can set up this calculation as follows.

$$Cu^{2+}(aq) + 4 \, NH_3(aq) \rightleftharpoons Cu(NH_3)_4{}^{2+}(aq) \quad K_f = 2.1 \times 10^{13}$$

Initial:	0.10 M	1.0 M	0
Intermediate:	0	0.6 M	0.10 M
Equilibrium:	Δ	$0.6 + 4\Delta$	$0.10 - \Delta$

Because K_f for this complex is very large, essentially all of the Cu^{2+} ions will be tied up as $Cu(NH_3)_4{}^{2+}$ ions at equilibrium. We therefore define a set of intermediate conditions in which the reaction is pushed as far as possible to the right — that is, in which all of the Cu^{2+} is converted into $Cu(NH_3)_4{}^{2+}$ complex ions. We then let the reaction come to equilibrium from these intermediate conditions.

Substituting this information into the equilibrium constant expression for this complex

$$K_f = \frac{[Cu(NH_3)_4{}^{2+}]}{[Cu^{2+}][NH_3]^4}$$

gives the following equation.

$$\frac{[0.10 - \Delta]}{[\Delta][0.6 + 4\Delta]^4} = 2.1 \times 10^{13}$$

To solve this equation, we need to make an approximation. The value of K_f is so large we can assume that very little $Cu(NH_3)_4{}^{2+}$ complex ion dissociates as the reaction comes to equilibrium.

$$\frac{[0.10]}{[\Delta][0.6]^4} \cong 2.1 \times 10^{13}$$

Solving this approximate equation gives the following result.

$$\Delta \cong 3.7 \times 10^{-14}$$

The assumption that Δ is small is valid, and we can use the results of the calculation to determine the concentration of the free Cu^{2+} ion in this solution.

$$[Cu^{2+}] = 3.7 \times 10^{-14} \, M$$

Exercise 17.9

Use the results of the preceding exercise to explain why $Cu(OH)_2$ dissolves in excess ammonia. [$Cu(OH)_2$: $K_{sp} = 2.2 \times 10^{-20}$, NH_3: $K_b = 1.8 \times 10^{-5}$]

Solution

To explain why $Cu(OH)_2$ dissolves, we might first consider why it precipitates in the first place. Ammonia acts as a base toward water.

$$NH_3(aq) + H_2O(l) \rightleftharpoons NH_4^+(aq) + OH^-(aq)$$

$$K_b = \frac{[NH_4^+][OH^-]}{[NH_3]} = 1.8 \times 10^{-5}$$

Even fairly dilute solutions—such as $0.0010\ M\ NH_3$—have significant concentrations of the OH^- ion.

$$\frac{[\Delta][\Delta]}{[0.0010 - \Delta]} = 1.8 \times 10^{-5}$$

Solving this equation for Δ gives the following result.

$$[OH^-] = 1.3 \times 10^{-4}\ M$$

Even in very dilute NH_3 solution, the OH^- ion concentration is therefore large enough to precipitate $Cu(OH)_2$ from a $0.10\ M\ Cu^{2+}$ ion solution.

$$Q_{sp} = (Cu^{2+})(OH^-)^2$$
$$= (0.10)(1.3 \times 10^{-4})^2 = 1.7 \times 10^{-9} \gg K_{sp}$$

As the amount of NH_3 added to the solution becomes larger, the concentration of the OH^- ion also becomes larger. But it doesn't become very much larger. A 1000-fold increase in the NH_3 concentration—to $1.0\ M$—only results in a 30-fold increase in the OH^- ion concentration—to $4.2 \times 10^{-3}\ M$.

As the amount of NH_3 added to the solution becomes larger, the concentration of free Cu^{2+} ions rapidly becomes smaller, because these ions are tied up as $Cu(NH_3)_4^{2+}$ complex ions. According to the preceding exercise, the Cu^{2+} ion concentration is only $3.7 \times 10^{-14}\ M$ in $1.0\ M\ NH_3$. The ion product for $Cu(OH)_2$ under these conditions is about the same size as the solubility product for this compound.

$$Q_{sp} = (Cu^{2+})(OH^-)^2$$
$$= (3.7 \times 10^{-14})(4.2 \times 10^{-3})^2 = 6.5 \times 10^{-19} \approx K_{sp}$$

As soon as the NH_3 concentration exceeds $1\ M$, the Cu^{2+} ion concentration becomes so small that the ion product for $Cu(OH)_2$ is smaller than K_{sp}, and the $Cu(OH)_2$ precipitate dissolves.

17.8 USING COMPLEX ION EQUILIBRIA TO DISSOLVE AN INSOLUBLE SALT

Section 17.7 provided an example of how complex ion equilibria can be used to dissolve an insoluble salt. The key to achieving this goal can be stated simply.

Choose a complex for which K_f is large enough to ensure that the concentration of the uncomplexed metal ion is too small for the ion product to exceed the solubility product.

To show how this can be done, let's return to a problem encountered in Exercise 16.3. An essential step in processing photographic film involves removing the AgBr crystals that did not capture light. Because this step permanently fixes the image onto the film, the reagent used to achieve it is called a fixer. Exercises 17.10 and 17.11 test potential fixers.

Exercise 17.10

Calculate the solubility of AgBr in a solution that is 4 M in NH_3 at equilibrium. (AgBr: $K_{sp} = 5 \times 10^{-13}$, $Ag(NH_3)_2^+$: $K_f = 1.1 \times 10^7$)

Solution

Two equilibria must be considered in this problem: the solubility product for AgBr

$$AgBr(s) \xrightleftharpoons{H_2O} Ag^+(aq) + Br^-(aq)$$

and the complex formation equilibrium for the $Ag(NH_3)_2^+$ complex ion.

$$Ag^+(aq) + 2\ NH_3(aq) \rightleftharpoons Ag(NH_3)_2^+$$

AgBr will dissolve if the $Ag(NH_3)_2^+$ complex is strong enough to reduce the Ag^+ ion concentration to the point at which the product of the Ag^+ and Br^- ion concentrations at equilibrium is less than the K_{sp} of AgBr.

One way of deciding whether K_f for this complex ion is large enough to overcome the solubility product equilibrium involves combining the equilibria to give the following overall equation.

$$AgBr(s) + 2\ NH_3(aq) \rightleftharpoons Ag(NH_3)_2^+(aq) + Br^-(aq)$$

The equilibrium constant for this reaction is the product of K_f and K_{sp}.

$$K = K_f \times K_{sp} = (1.1 \times 10^7)(5 \times 10^{-13}) = 5.5 \times 10^{-6}$$

The equilibrium constant is much smaller than 1, so we might not expect very much AgBr to dissolve. The equilibrium constant expression for this reaction is written as follows.

$$\frac{[Ag(NH_3)_2^+][Br^-]}{[NH_3]^2} = 5.5 \times 10^{-6}$$

If we assume that most of the Ag^+ that goes into solution is complexed, we can calculate the solubility of AgBr in this solution as follows.

$$\frac{[S][S]}{[4]^2} = 5.5 \times 10^{-6}$$

Solving for S gives the following result.

$$S = 9.4 \times 10^{-3}$$

AgBr is therefore only marginally soluble in 4 M NH_3.

Exercise 17.11

Predict the solubility of AgBr in a 1.0 M solution of the $S_2O_3^{2-}$ ion. (AgBr: $K_{sp} = 5 \times 10^{-13}$, $Ag(S_2O_3)_2^{3-}$: $K_f = 2.9 \times 10^{13}$)

Solution

The Ag^+ ion forms a two-coordinate complex with the thiosulfate ion.

$$Ag^+(aq) + S_2O_3{}^{2-}(aq) \rightleftharpoons Ag(S_2O_3)_2{}^{3-}(aq)$$

Adding this equilibrium to the solubility product equilibrium for AgBr gives the following result.

$$AgBr(s) + S_2O_3{}^{2-}(aq) \rightleftharpoons Ag(S_2O_3)_2{}^{3-}(aq) + Br^-(aq)$$

The equilibrium constant for this reaction is the product of K_f times K_{sp}.

$$K = K_f \times K_{sp} = (2.9 \times 10^{13})(5 \times 10^{-13}) = 15$$

The equilibrium constant is now larger than 1, which leads us to expect that a significant amount of AgBr will dissolve in this solution.

The equilibrium constant expression for the reaction is written as follows.

$$\frac{[Ag(S_2O_3)_2{}^{3-}][Br^-]}{[S_2O_3{}^{2-}]^2} = 15$$

Because the equilibrium constant is larger than 1, we can't ignore the amount of the initial concentration of $S_2O_3{}^{2-}$ consumed as this reaction comes to equilibrium.

$$\frac{[S][S]}{[1 - 2\,S]^2} = 15$$

Solving this equation with the quadratic formula gives the following result.

$$S = 0.44\ M$$

The $Ag(S_2O_3)_2{}^{3-}$ complex ion is strong enough to dissolve a considerable amount of AgBr—up to 0.44 moles per liter of solution. It therefore isn't surprising to find that the thiosulfate ion is used as the fixer in the processing of virtually all commercial photographic films.

USING COMPLEX ION EQUILIBRIA
17.9 TO KEEP A SALT FROM PRECIPITATING

Complex ion equilibria can also be used to adjust the concentration of an ion to keep an insoluble salt from precipitating from solution.

Exercise 17.12

The Hg^{2+} ion concentration in a neutral solution must be kept below $1 \times 10^{-12}\ M$ to keep $Hg(OH)_2$ from precipitating. Calculate how much NaCl must be added to a liter of a $0.10\ M\ Hg^{2+}$ solution to reduce the Hg^{2+} ion concentration to $1 \times 10^{-12}\ M$. ($HgCl_4{}^{2-}$: $K_f = 1 \times 10^{16}$)

Solution

Often, one of the most difficult tasks in problem solving is recognizing the problem to be solved. What information does this problem contain that gives us a hint about how it can be set up? We can start by noting that it contains a complex formation equilibrium constant. We can therefore start with a balanced equation for the equilibrium reaction.

$$Hg^{2+}(aq) + 4\ Cl^-(aq) \rightleftharpoons HgCl_4{}^{2-}(aq)$$

We can then search for information about either the initial concentration or the equilibrium concentration of these ions.

We know the initial concentration of the Hg^{2+} ion — 0.10 M — and the initial concentration of the $HgCl_4{}^{2-}$ complex ion — 0 M. What we don't know is the initial concentration of the Cl^- ion — x.

$$Hg^{2+}(aq) + 4\ Cl^-(aq) \qquad HgCl_4{}^{2-}(aq) \qquad K_f = 1 \times 10^{16}$$

Initial:	0.10 M	x	0

The initial reaction quotient ($Q_f = 0$) is very far from equilibrium ($K_f = 1 \times 10^{16}$). We therefore define an intermediate set of conditions in which we push the reaction as far as we can to the right.

$$Hg^{2+}(aq) + \quad 4\ Cl^-(aq) \qquad HgCl_4{}^{2-}(aq) \qquad K_f = 1 \times 10^{16}$$

Initial:	0.10 M	x	0
Intermediate:	0	$x - 0.40\ M$	0.10 M

We then let the reaction come to equilibrium from these intermediate conditions.

$$Hg^{2+}(aq) + \quad 4\ Cl^-(aq) \qquad HgCl_4{}^{2-}(aq) \quad K_f = 1 \times 10^{16}$$

Intermediate:	0	$x - 0.40\ M$	0.10 M
Equilibrium:	Δ	$x - 0.40 + 4\Delta$	$0.10 - \Delta$

This problem seems to contain two unknowns — Δ and x — which is unfortunate, because we can't solve the equilibrium constant expression for two unknowns.

We therefore search the problem for a way of reducing the number of unknowns. We've already abstracted all of the information about the initial conditions. But we haven't used all of the information about the equilibrium conditions. Our goal is to reduce the Hg^{2+} ion concentration at equilibrium to $1 \times 10^{-12}\ M$. If the equilibrium concentration of this ion is $1 \times 10^{-12}\ M$, then Δ is no longer an unknown — it's equal to 1×10^{-12}. Plugging this value of Δ into the definition of the problem gives the following result.

$$Hg^{2+}(aq) \quad + \quad 4\ Cl^-(aq) \rightleftharpoons HgCl_4{}^{2-}(aq) \quad K_f = 1 \times 10^{16}$$

Intermediate:	0	$x - 0.40\ M$	0.10 M
Equilibrium:	$1 \times 10^{-12}\ M$	$x - 0.40\ M$	0.10 M

We now have only one unknown, so we need only one equation.

$$K_f = \frac{[HgCl_4{}^{2-}]}{[Hg^{2+}][Cl^-]^4}$$

Substituting what we know about the equilibrium concentrations into this equation gives the following result.

$$\frac{[0.10]}{[1 \times 10^{-12}][x - 0.40]^4} = 1 \times 10^{16}$$

We can solve this equation by rearranging it as follows.

$$[x - 0.40]^4 = 1 \times 10^{-5}$$

We can then take the fourth root of both sides of the equation

$$[x - 0.40] = (1 \times 10^{-5})^{1/4} = 0.056$$

and solve for the value of x.

$$x = 0.46$$

According to this calculation, we can reduce the Hg^{2+} ion concentration in the solution to 1×10^{-12} M by adding 0.46 moles of Cl^- ion to each liter of solution.

17.10 THE RIGOROUS APPROACH TO COMPLEX ION CALCULATIONS

The approach to complex equilibrium problems in Sections 17.7 through 17.9 is based on the following assumptions.

1. **The difference between the stepwise formation constants for the complex (K_{f1}, K_{f2}, and so on) is small. We can therefore collapse the individual steps into a single overall reaction.**

2. **The overall formation constant is large enough to allow us to assume that essentially all of the metal ions at equilibrium are present as complex ions.**

At times, we want to know the concentrations of the intermediate complex ions — no matter how small they might be. We may also encounter complex ion equilibria for which the overall K_f is not very large or for which the concentration of the complexing agent is small. In such cases, we need a rigorous approach to complex ion calculations. To illustrate such an approach, we'll calculate one of the sets of data in Table 17.1.

The data in Table 17.1 are based on the following complex equilibria.

$$Ag^+(aq) + NH_3(aq) \rightleftharpoons Ag(NH_3)^+(aq) \qquad K_{f1} = 1.7 \times 10^3$$
$$Ag(NH_3)^+(aq) + NH_3(aq) \rightleftharpoons Ag(NH_3)_2^+(aq) \qquad K_{f2} = 6.5 \times 10^3$$

Let's start with the equilibrium constant expression for the first step.

$$K_{f1} = \frac{[Ag(NH_3)^+]}{[Ag^+][NH_3]}$$

This equation can be solved for the concentration of the one-coordinate $Ag(NH)3)^+$ complex ion as follows.

$$[Ag(NH_3)^+] = K_{f1}[Ag^+][NH_3]$$

We now turn to the equilibrium constant expression for the second step

$$K_{f2} = \frac{[Ag(NH_3)_2^+]}{[Ag(NH_3)^+][NH_3]}$$

and solve for the concentration of the two-coordinate $Ag(NH_3)_2^+$ complex ion.

$$[Ag(NH_3)_2^+] = K_{f2}[Ag(NH_3)^+][NH_3]$$

We now substitute the first equation into the second.

$$[Ag(NH_3)_2^+] = K_{f2}(K_{f1}[Ag^+][NH_3])[NH_3]$$

The net result of this derivation is a set of equations that relate the concentrations of the $Ag(NH_3)^+$ and $Ag(NH_3)_2^+$ complex ions to a pair of constants—K_{f1} and K_{f2}—and a pair of variables—$[Ag^+]$ and $[NH_3]$.

$$[Ag(NH_3)^+] = K_{f1}[Ag^+][NH_3]$$
$$[Ag(NH_3)_2^+] = K_{f1}K_{f2}[Ag^+][NH_3]^2$$

We'll now write an equation that describes the relationship between the initial concentration of the silver ion—C_{Ag^+}—and the concentrations of the components of the solution that contain the silver ion at equilibrium.

$$C_{Ag^+} = [Ag^+] + [Ag(NH_3)^+] + [Ag(NH_3)_2^+]$$

We then substitute the equations for the concentrations of the $Ag(NH_3)^+$ and $Ag(NH_3)_2^+$ complex ions into this relationship.

$$C_{Ag^+} = [Ag^+] + K_{f1}[Ag^+][NH_3] + K_{f1}K_{f2}[Ag^+][NH_3]^2$$

We can then factor out the $[Ag^+]$ term to obtain the following equation.

$$C_{Ag^+} = [Ag^+](1 + K_{f1}[NH_3] + K_{f1}K_{f2}[NH_3]^2)$$

If we define a new term—D—as follows

$$D = 1 + K_{f1}[NH_3] + K_{f1}K_{f2}[NH_3]^2$$

we can simplify our statement of the initial Ag^+ ion concentration.

$$C_{Ag^+} = [Ag^+] \times D$$

This equation can be rearranged as follows.

$$\frac{[Ag^+]}{C_{Ag^+}} = \frac{1}{D}$$

The percent of the initial Ag^+ ion concentration that is present as the one-coordinate $Ag(NH_3)^+$ ion at equilibrium can be calculated from the following equation.

$$\% [Ag(NH_3)^+] = \frac{[Ag(NH_3)^+]}{C_{Ag^+}} \times 100$$

Substituting the equation we derived for the $[Ag(NH_3)_2^+]$ term into this equation gives the following result.

$$\% [Ag(NH_3)^+] = \frac{K_{f1}[Ag^+][NH_3]}{C_{Ag^+}} \times 100$$

We can now use the relationship among C_{Ag^+}, $[Ag^+]$, and D to obtain the following form of this equation.

$$\% [Ag(NH_3)^+] = \frac{K_{f1}[NH_3]}{D} \times 100$$

Now we'll derive a similar equation for the two-coordinate $Ag(NH_3)_2^+$ complex ion. We start with the equation that defines the percent of the initial Ag^+ ion concentration that is present as the two-coordinate $Ag(NH_3)_2^+$ complex ion at equilibrium.

$$\% [Ag(NH_3)_2^+] = \frac{[Ag(NH_3)_2^+]}{C_{Ag^+}} \times 100\%$$

Substituting the relationship we derived for the concentration of the $Ag(NH_3)_2^+$ complex ion into the numerator of this equation gives the following result.

$$\% \,[Ag(NH_3)_2^+] = \frac{K_{f1}K_{f2}[Ag^+][NH_3]^2}{C_{Ag^+}} \times 100$$

Again, we can use the relationship among C_{Ag^+}, $[Ag^+]$, and D to obtain the following form of this equation.

$$\% \,[\mathbf{Ag(NH_3)_2^+}] = \frac{K_{f1}K_{f2}[\mathbf{NH_3}]^2}{D} \times 100$$

Exercise 17.13

Calculate the fraction of the initial Ag^+ ion concentration present as the Ag^+, $Ag(NH_3)^+$, and $Ag(NH_3)_2^+$ ions at equilibrium in a solution that is 0.0010 M in NH_3 at equilibrium. ($K_{f1} = 1.7 \times 10^3$, $K_{f2} = 6.5 \times 10^3$)

Solution

We can start by calculating the D—or denominator—term for this system.

$$\begin{aligned}
D &= 1 + K_{f1}[NH_3] + K_{f1}K_{f2}[NH_3]^2 \\
&= 1 + (1.7 \times 10^3)(0.0010) + (1.7 \times 10^3)(6.5 \times 10^3)(0.0010)^2 \\
&= 13.75
\end{aligned}$$

This value can be substituted into the equation for the percent of the initial concentration of Ag^+ that is present as Ag^+ ions at equilibrium.

$$\% \,[Ag^+] = \frac{1}{D} \times 100 = \frac{1}{13.75} \times 100 = \mathbf{7.3\%}$$

The values of D, K_{f1}, and the NH_3 concentration at equilibrium can be substituted into the equation for the concentration of the one-coordinate complex ion.

$$\% \,[Ag(NH_3)^+] = \frac{K_{f1}[NH_3]}{D} \times 100 = \frac{(1.7 \times 10^3)(0.0010)}{13.75} \times 100 = \mathbf{12.4\%}$$

Finally, the values of D, K_{f1}, K_{f2}, and the NH_3 concentration can be substituted into the equation for the concentration of the two-coordinate complex ion.

$$\begin{aligned}
\% \,[Ag(NH_3)_2^+] &= \frac{K_{f1}K_{f2}[NH_3]^2}{D} \times 100 \\
&= \frac{(1.7 \times 10^3)(6.5 \times 10^3)(0.0010)^2}{13.75} \times 100 = \mathbf{80.4\%}
\end{aligned}$$

According to this calculation, 7.3% of the Ag^+ ions added to this solution will be present as free, or uncomplexed, Ag^+ ions in a 0.0010 M NH_3 solution. Another 12.4% will be present as one-coordinate $Ag(NH_3)^+$ complex ions. The vast majority of the Ag^+ ions added to this solution—80.4%—will be present at equilibrium as two-coordinate $Ag(NH_3)_2^+$ ions.

Exercise 17.14

Calculate the fraction of the initial Cu^{2+} ion concentration present as four-coordinate $Cu(NH_3)_4^{2+}$ complex ions in a solution that is found experimentally to be 0.10 M NH$_3$ at equilibrium. ($K_{f1} = 2.0 \times 10^4$, $K_{f2} = 4.7 \times 10^3$, $K_{f3} = 1.1 \times 10^3$, $K_{f4} = 2.0 \times 10^2$)

Solution

A logical extension of the procedures used to derive the equations in the preceding exercise gives the following equations for the five components of this solution that contain Cu^{2+} ions.

$$\% \,[Cu^{2+}] = \frac{1}{D} \times 100$$

$$\% \,[Cu(NH_3)^{2+}] = \frac{K_{f1}[NH_3]}{D} \times 100$$

$$\% \,[Cu(NH_3)_2{}^{2+}] = \frac{K_{f1}K_{f2}[NH_3]^2}{D} \times 100$$

$$\% \,[Cu(NH_3)_3{}^{2+}] = \frac{K_{f1}K_{f2}K_{f3}[NH_3]^3}{D} \times 100$$

$$\% \,[Cu(NH_3)_4{}^{2+}] = \frac{K_{f1}K_{f2}K_{f3}K_{f4}[NH_3]^4}{D} \times 100$$

The denominator term in these equations is defined as follows.

$$D = 1 + K_{f1}[NH_3] + K_{f1}K_{f2}[NH_3]^2 + K_{f1}K_{f2}K_{f3}[NH_3]^3 + K_{f1}K_{f2}K_{f3}K_{f4}[NH_3]^4$$

These equations may at first appear very complex. They contain a clear pattern, however, which should make them relatively easy to extend to any series of coordination numbers.

The notation for calculations of this nature can be simplified by using the following convention.

$$\beta_0 = 1$$
$$\beta_1 = K_{f1} = 2.0 \times 10^4$$
$$\beta_2 = K_{f1}K_{f2} = (2.0 \times 10^4)(4.7 \times 10^3)$$
$$\beta_3 = K_{f1}K_{f2}K_{f3} = (2.0 \times 10^4)(4.7 \times 10^3)(1.1 \times 10^3)$$
$$\beta_4 = K_{f1}K_{f2}K_{f3}K_{f4} = (2.0 \times 10^4)(4.7 \times 10^3)(1.1 \times 10^3)(2.0 \times 10^2)$$

When this convention is used, the term in which we are interested — $\% \,[Cu(NH_3)_4{}^{2+}]$ — is defined by the following equation.

$$\% \,[Cu(NH_3)_4{}^{2+}] = \frac{\beta_4[NH_3]^4}{D} \times 100\%$$

D is now defined as follows.

$$\begin{aligned} D &= \beta_0 + \beta_1[NH_3] + \beta_2[NH_3]^2 + \beta_3[NH_3]^3 + \beta_4[NH_3]^4 \\ &= 1 + (2.0 \times 10^4)(0.10) + (9.4 \times 10^7)(0.10)^2 \\ &\quad + (1.0 \times 10^{11})(0.10)^3 + (2.1 \times 10^{13})(0.10)^4 \\ &= 2.2 \times 10^9 \end{aligned}$$

The values of β_4, the NH$_3$ concentration at equilibrium, and D are then substituted into the equation for the percent of the initial Cu^{2+} ion concentration present as four-coordinate Cu(NH$_3$)$_4{}^{2+}$ complex ions at equilibrium.

$$\% \,[Cu(NH_3)_4{}^{2+}] = \frac{\beta_4[NH_3]^4}{D} \times 100 = \frac{(2.1 \times 10^{13})(0.10)^4}{(2.2 \times 10^9)} \times 100 = \mathbf{95\%}$$

Even when the NH_3 concentration is as small as 0.10 M, 95% of the Cu^{2+} ion concentration is present as four-coordinate $Cu(NH_3)_4^{2+}$ complex ions.

17.11 A QUALITATIVE VIEW OF COMBINED EQUILIBRIA

Most of the discussion so far has focused on individual equilibria. There are several reasons for this.

1. **It is much easier to handle these reactions one at a time.**
2. **The results obtained when the reactions are handled one at a time are useful. They are in very close agreement with experimental measurements.**

This section will examine how LeChatelier's principle can be applied to systems in which many chemical equilibria exist simultaneously.

Consider what happens when solid $CuSO_4$ dissolves in an aqueous solution of NH_3. If we start with enough copper sulfate, we can observe a solubility product equilibrium.

$$CuSO_4(s) \rightleftharpoons Cu^{2+}(aq) + SO_4^{2-}(aq)$$

But this is not the only reaction that occurs in this solution.

The SO_4^{2-} ion is a weak Brønsted base that can pick up an H^+ ion from water to form the hydrogen sulfate and hydroxide ions.

$$SO_4^{2-}(aq) + H_2O(l) \rightleftharpoons HSO_4^-(aq) + OH^-(aq)$$

There are other sources of the OH^- ion in this solution. Water, of course, dissociates to some extent to form the OH^- ion.

$$2 H_2O(l) \rightleftharpoons H_3O^+(aq) + OH^-(aq)$$

Ammonia also reacts with water, to some extent, to form the NH_4^+ and OH^- ions.

$$NH_3(aq) + H_2O(l) \rightleftharpoons NH_4^+(aq) + OH^-(aq)$$

The Cu^{2+} ion released into solution when $CuSO_4$ dissolves reacts with ammonia to form a series of four different complex ions.

$$Cu^{2+}(aq) + NH_3(aq) \rightleftharpoons Cu(NH_3)^{2+}(aq)$$
$$Cu(NH_3)^{2+}(aq) + NH_3(aq) \rightleftharpoons Cu(NH_3)_2^{2+}(aq)$$
$$Cu(NH_3)_2^{2+}(aq) + NH_3(aq) \rightleftharpoons Cu(NH_3)_3^{2+}(aq)$$
$$Cu(NH_3)_3^{2+}(aq) + NH_3(aq) \rightleftharpoons Cu(NH_3)_4^{2+}(aq)$$

If the concentration of the OH^- ion in this solution gets too large, the Cu^{2+} ion precipitates as $Cu(OH)_2$, which gives us a second solubility product equilibrium.

$$Cu(OH)_2(s) \rightleftharpoons Cu^{2+}(aq) + 2 OH^-(aq)$$

As we can see, the simple process of dissolving copper(II) sulfate in aqueous ammonia can involve at least nine different equilibria. We must consider each of these reactions if we want to predict what will happen under a particular set of experimental conditions.

Based on the calculations in Section 17.10, we can make at least one simplifying assumption. We can assume that the complex ion equilibria in this sytem can be represented by a single equation in which the Cu^{2+} ion combines with four NH_3 molecules to form the four-coordinate $Cu(NH_3)_4^{2+}$ ion. We can therefore con-

FIG. 17.9 The relationships among the equilibria that exist in a saturated solution of $CuSO_4$ in ammonia.

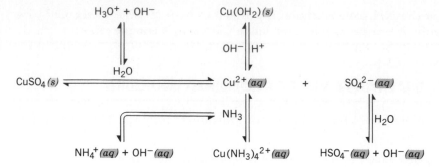

struct a fairly complete model of what happens in this solution if we take into account the following equilibria, which are summarized in Figure 17.9.

$$CuSO_4(s) \rightleftharpoons Cu^{2+}(aq) + SO_4^{2-}(aq)$$
$$SO_4^{2-}(aq) + H_2O(l) \rightleftharpoons HSO_4^-(aq) + OH^-(aq)$$
$$2\,H_2O(l) \rightleftharpoons H_3O^+(aq) + OH^-(aq)$$
$$NH_3(aq) + H_2O(l) \rightleftharpoons NH_4^+(aq) + OH^-(aq)$$
$$Cu^{2+}(aq) + 4\,NH_3(aq) \rightleftharpoons Cu(NH_3)_4^{2+}(aq)$$
$$Cu(OH)_2(s) \rightleftharpoons Cu^{2+}(aq) + 2\,OH^-(aq)$$

Exercise 17.15

Predict the effect of adding each of the following substances to a saturated solution of $CuSO_4$ in aqueous NH_3.

(a) $CuSO_4(s)$ (c) $NaOH(aq)$ (e) $NH_3(aq)$
(b) $HNO_3(aq)$ (d) $Na_2SO_4(aq)$

Solution

We can predict the effect of adding each of these reagents by applying LeChatelier's principle to Figure 17.9.

(a) Adding excess $CuSO_4$ to a saturated solution has no effect on any of the equilibria shown in the figure.

(b) Nitric acid is a strong acid that should convert most of the NH_3 into NH_4^+ ions. Anything that removes NH_3 from the solution tends to destroy the $Cu(NH_3)_4^{2+}$ complex ion. Adding nitric acid therefore tends to increase the Cu^{2+} ion concentration, which should cause $CuSO_4$ to precipitate from solution. No $Cu(OH)_2$ precipitates, however, because the concentration of the OH^- ion in this solution is too small.

(c) Sodium hydroxide is a strong base. In theory, it should react with the NH_4^+ ion in this solution to form more NH_3. In practice, there isn't very much NH_4^+ ion in the solution to begin with, so adding NaOH has little effect on most of the equilibria in the solution. The presence of excess OH^- ion, however, causes some of the Cu^{2+} ion in solution to precipitate as $Cu(OH)_2$.

(d) Sodium sulfate is a source of the SO_4^{2-} ion. The common ion effect therefore predicts that $CuSO_4$ will precipitate from solution if Na_2SO_4 is added.

(e) As the NH_3 concentration becomes larger, more of the Cu^{2+} ion is tied up as the $Cu(NH_3)_4^{2+}$ complex ion. This reduces the concentration of the free Cu^{2+} ion, which makes the ion product for $CuSO_4$ smaller than the solubility product. As a result, adding NH_3 can cause more $CuSO_4$ to dissolve.

Another example of combined equilibria revolves around the aqueous chemistry of the Fe^{3+} ion. Dilute solutions of this ion are essentially colorless. In the presence of the thiocyanate ion, however, the blood-red solution shown in Figure 17.6 is formed. This solution consists of a pair of complex ions.

$$Fe^{3+}(aq) + SCN^-(aq) \rightleftharpoons Fe(SCN)^{2+}(aq) \qquad K_f = 890$$
$$Fe(SCN)^{2+}(aq) + SCN^-(aq) \rightleftharpoons Fe(SCN)_2^+(aq) \qquad K_f = 2.6$$

The Fe^{3+} ion also forms a complex with the citrate — or Cit^{3-} — ion.

$$Fe^{3+}(aq) + Cit^{3-}(aq) \rightleftharpoons Fe(Cit)(aq) \qquad K_f = 6.3 \times 10^{11}$$

This neutral coordination complex has a pale yellow color.

Exercise 17.16

Figure 17.10 shows photographs of the following aqueous solutions.

(a) A dilute aqueous solution of the Fe^{3+} ion.

(b) A mixture of the $Fe^{3+}(aq)$ and $SCN^-(aq)$ ions.

(c) A mixture of the $Fe^{3+}(aq)$ and $SCN^-(aq)$ ions to which a strong acid has been added.

(d) A mixture of the $Fe^{3+}(aq)$ and $Cit^{3-}(aq)$ ions.

(e) A mixture of the $Fe^{3+}(aq)$ and $Cit^{3-}(aq)$ ions to which a strong acid has been added.

(f) A mixture of the $Fe^{3+}(aq)$, $SCN^-(aq)$, and $Cit^{3-}(aq)$ ions.

(g) A mixture of the $Fe^{3+}(aq)$, $SCN^-(aq)$, and $Cit^{3-}(aq)$ ions to which a strong acid has been added.

Explain the color (or lack of color) of each of these solutions.

Solution

(a) The color of dilute solutions of the Fe^{3+} ion is so weak they are essentially colorless.

(b) When the SCN^- ion is added to an aqueous solution of the Fe^{3+} ion, the $Fe(SCN)^{2+}$ and $Fe(SCN)_2^+$ complex ions are formed, and the solution turns a blood-red color.

(c) Nothing seems to happen when a strong acid is added to a mixture of the Fe^{3+} and SCN^- ions. This tells us something about the strength of the conjugate acid of the SCN^- ion — thiocyanic acid, HSCN. If HSCN were a weak acid, adding a strong acid to the solution would tie up the SCN^- ion as HSCN. The SCN^- ion would no longer be free to form complexes with the Fe^{3+} ion. This would decrease the amount of the $Fe(SCN)^{2+}$ and $Fe(SCN)_2^+$ complex ions in solution, thereby decreasing the intensity of the color of this

FIG. 17.10 The petri dishes in this photograph show the results of mixing the reagents identified in Exercise 17.16.

solution. Since this is not observed, we conclude that HSCN must be a relatively strong acid. This is consistent with the K_a for thiocyanic acid found in Table A-9 in the appendix.

(d) Mixing the $Fe^{3+}(aq)$ and $Cit^{3-}(aq)$ ions forms the Fe(Cit) complex, which has a pale yellow color.

(e) Adding a strong acid destroys the Fe(Cit) complex by converting the citrate ion into its conjugate acid, citric acid, H_3Cit. This observation suggests that citric acid is relatively weak, which is consistent with the K_a data in the appendix.

(f) The K_f for the Fe(Cit) complex is much larger than the overall K_f for the $Fe(SCN)_2^+$ complex ion. Given a choice between SCN^- and Cit^{3-} ions, more Fe^{3+} ions form complexes with Cit^{3-} ions. A solution containing a mixture of the three ions therefore has a yellow color.

(g) When a strong acid is added to a mixture of the Fe^{3+}, Cit^{3-}, and SCN^- ions, the Cit^{3-} ions are converted into citric acid — H_3Cit. When the Cit^{3-} ions are removed from solution, the only complexing agents left in the solution are SCN^- ions. The solution therefore turns the blood-red color of the $Fe(SCN)_2^+$ complex ion.

17.12 A QUANTITATIVE VIEW OF COMBINED EQUILIBRIA

We can perform quantitative calculations for solutions of combined equilibria if we keep in mind how these equilibria are coupled.

Exercise 17.17

Predict whether AgOH will precipitate from a solution buffered at pH 9.12 that is initially 0.00010 M $AgNO_3$ and 0.10 M in NH_4NO_3. (AgOH: $K_{sp} = 2.0 \times 10^{-8}$, NH_3: $K_b = 1.8 \times 10^{-5}$, $Ag(NH_3)_2^+$: $K_f = 1.1 \times 10^7$, H_2O: $K_w = 1.0 \times 10^{-14}$)

Solution

We can start by listing the important reactions in this solution. Silver nitrate and ammonium nitrate are both soluble salts.

$$AgNO_3(s) \xrightarrow{H_2O} Ag^+(aq) + NO_3^-(aq)$$

$$NH_4NO_3(s) \xrightarrow{H_2O} NH_4^+(aq) + NO_3^-(aq)$$

Because the reactions occur in water, we have to consider the dissociation of water in any list of relevant equilibria.

$$2\,H_2O(l) \rightleftharpoons H_3O^+(aq) + OH^-(aq) \qquad K_w = 1.0 \times 10^{-14}$$

Some of the NH_4^+ ions will react with water to give NH_3. Since the problem gives us the value of K_b for NH_3, not K_a for the NH_4^+ ion, let's look at this reaction in terms of the following equation.

$$NH_3(aq) + H_2O(l) \rightleftharpoons NH_4^+(aq) + OH^-(aq) \qquad K_b = 1.8 \times 10^{-5}$$

The Ag^+ ion formed when silver nitrate dissolves in water can combine with the ammonia to form a complex ion.

$$Ag^+(aq) + 2\,NH_3(aq) \rightleftharpoons Ag(NH_3)_2^+(aq) \qquad K_a = 1.1 \times 10^7$$

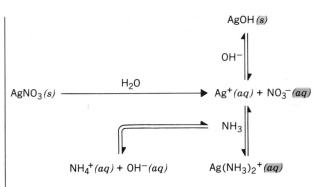

FIG. 17.11 The relationships among the equilibria that must be considered to solve the problem in Exercise 17.17.

It can also combine with the OH^- ion to form an AgOH precipitate.

$$AgOH(s) \xrightleftharpoons{H_2O} Ag^+(aq) + OH^-(aq) \qquad K_{sp} = 2.0 \times 10^{-8}$$

The relationships among these equilibria are shown in Figure 17.11.

We can solve some problems by looking at the initial conditions and working toward the final answer. Others, such as this, are so complex it is useful to look at the goal and then work backwards. The goal in this problem is to decide whether AgOH precipitates from a given solution. In order to make this decision, we need two pieces of information—the Ag^+ and OH^- ion concentrations at equilibrium. We can therefore divide the problem into two parts whose individual subgoals consist of determining the values of the $[Ag^+]$ and $[OH^-]$ terms.

It doesn't matter which subgoal we start with, so we'll choose the one that should be the easiest to reach: finding the OH^- ion concentration. The problem gives the pH of the buffer solution—9.12. We can calculate the H_3O^+ ion concentration at equilibrium as follows.

$$[H_3O^+] = 10^{-pH} = 10^{-9.12} - 7.6 \times 10^{-10} \ M$$

We can then use this H_3O^+ concentration to calculate the OH^- ion concentration.

$$[OH^-] = \frac{K_w}{[H_3O^+]} = \frac{1.0 \times 10^{-14}}{7.6 \times 10^{-10}} = 1.3 \times 10^{-5} \ M$$

We can now turn to our second subgoal—finding the concentration of the Ag^+ ion at equilibrium. We know that this ion is in equilibrium with the $Ag(NH_3)_2^+$ complex ion.

$$K_f = \frac{[Ag(NH_3)_2^+]}{[Ag^+][NH_3]^2}$$

If we knew the concentrations of NH_3 and the $Ag(NH_3)_2^+$ complex ion, we could calculate the Ag^+ ion concentration at equilibrium. This gives us two new subgoals: determining the $Ag(NH_3)_2^+$ and NH_3 concentrations.

Determining the $Ag(NH_3)_2^+$ ion concentration is relatively easy. We know that the initial concentration of the Ag^+ ion in this solution is 0.00010 M. We also know that this ion forms a strong complex with NH_3. We can therefore assume that essentially all of the silver ions in this solution will be present as $Ag(NH_3)_2^+$ complex ions.

$$[Ag(NH_3)_2^+] = 0.00010 \ M$$

Now all we need is the NH_3 concentration. We know that ammonia is in equilibrium with the NH_4^+ ion.

$$K_b = \frac{[NH_4^+][OH^-]}{[NH_3]} = 1.8 \times 10^{-10}$$

And we already know the OH^- ion concentration in this solution.

$$\frac{[NH_4^+][1.3 \times 10^{-5}]}{[NH_3]} = 1.8 \times 10^{-5}$$

Rearranging this equation gives the following result.

$$\frac{[NH_4^+]}{[NH_3]} = 1.4$$

The concentration of the NH_4^+ ion at equilibrium is 1.4 times the concentration of NH_3.

$$[NH_4^+] = 1.4\,[NH_3]$$

We also know that the solution was initially 0.10 M in NH_4^+ ions. These ions are now present as NH_3 molecules, as NH_4^+ ions, or as part of $Ag(NH_3)_2^+$ complex ions.

$$[NH_3] + [NH_4^+] + 2\,[Ag(NH_3)_2^+] = 0.10\ M$$

We have already determined that the concentration of the $Ag(NH_3)_2^+$ complex ion in this solution is very low — 0.00010 M — so we can ignore the last term in this equation. This gives us two equations in two unknowns.

$$[NH_4^+] = 1.4\,[NH_3]$$
$$[NH_3] + [NH_4^+] = 0.10\ M$$

These equations can be solved for the NH_3 and NH_4^+ concentrations at equilibrium.

$$[NH_3] = 0.042\ M$$
$$[NH_4^+] = 0.058\ M$$

We can now return to the complex formation equilibrium

$$K_f = \frac{[Ag(NH_3)_2^+]}{[Ag^+][NH_3]^2}$$

and substitute into this equation the equilibrium concentrations of NH_3 and the $Ag(NH_3)_2^+$ complex ion.

$$\frac{[0.00010]}{[Ag^+][0.042]^2} = 1.1 \times 10^7$$

We can then solve this equation for the Ag^+ ion concentration at equilibrium.

$$[Ag^+] = 5.2 \times 10^{-9}\ M$$

We can now look at the product of the Ag^+ and OH^- ion concentrations.

$$[Ag^+][OH^-] = [5.2 \times 10^{-9}][1.3 \times 10^{-5}] = 6.8 \times 10^{-14} \ll K_{sp}$$

The ion product for this solution is very much smaller than the solubility product for AgOH, which means that AgOH won't precipitate from solution.

Problems such as Exercise 17.17 require an organized approach. We must divide these problems into small steps, decide the order in which the steps should be carried out, apply what we know about aqueous equilibria to each step, and never lose track of information obtained in previous steps. There is no magic formula that can help us either divide these problems into steps or decide the order in which steps should be handled. In general, though, we follow this sequence.

1. **Identify the equilibria that must be included in the model of the solution.**
2. **Draw a figure that shows how these equilibria are coupled.**
3. **Find the simplest equilibrium—the one for which all of the needed data are available—and solve this part of the problem.**
4. **Ask yourself: "Where did this get me?" "What can I do with this information?"**
5. **Use the results of one step to solve another until all of the equilibria have been solved.**

Exercise 17.18

Calculate the solubility of zinc sulfide in a solution that is initially 0.10 M in H_2S and that is buffered at pH 3.00. Repeat this calculation for a solution buffered at pH 9.00.

Solution

We can start by listing the equilibria that might be relevant in this problem and looking up their equilibrium constants in the appendix.

$$ZnS(s) \xrightleftharpoons{H_2O} Zn^{2+}(aq) + S^{2-}(aq) \qquad K_{sp} = 1.6 \times 10^{-24}$$
$$2\,H_2O(l) \rightleftharpoons H_3O^+(aq) + OH^-(aq) \qquad K_w = 1.0 \times 10^{-14}$$
$$H_2S(aq) + H_2O(l) \rightleftharpoons H_3O^+(aq) + HS^-(aq) \qquad K_{a1} = 1.0 \times 10^{-7}$$
$$HS^-(aq) + H_2O(l) \rightleftharpoons H_3O^+(aq) + S^{2-}(aq) \qquad K_{a2} = 1.3 \times 10^{-13}$$

What next? We might calculate the H_3O^+ ion concentration in a pH 3.00 solution.

$$[H_3O^+] = 10^{-pH} = 10^{-3.00} = 1.0 \times 10^{-3}\,M$$

We can treat the dissociation of H_2S in the presence of excess acid (or base) as if it occurred in a single step.

$$H_2S(aq) + 2\,H_2O(l) \rightleftharpoons 2\,H_3O^+(aq) + S^{2-}(aq) \qquad K_a = 1.3 \times 10^{-20}$$

Because H_2S is a relatively weak acid, we can assume that the equilibrium concentration of the undissociated H_2S molecules is approximately equal to the initial concentration of this acid.

$$[H_2S] \cong 0.10\,M$$

We know the H_2S and H_3O^+ concentrations at equilibrium.

$$\frac{[H_3O^+]^2[S^-]}{[H_2S]} = \frac{[1.0 \times 10^{-3}]^2[S^{2-}]}{[0.10]} = 1.3 \times 10^{-20}$$

Thus, we can calculate the concentration of the S^{2-} ion at equilibrium before any ZnS dissolves.

$$[S^{2-}] = 1.3 \times 10^{-15}\,M$$

When ZnS dissolves, we get more S^{2-} ion. The amount of ZnS that dissolves can be calculated from the following equation, where S is the solubility of ZnS.

$$[Zn^{2+}][S^{2-}] = K_{sp}$$
$$[S][S + 1.3 \times 10^{-15}] = 1.6 \times 10^{-24}$$

We can solve this equation with the quadratic formula. Alternatively, we can assume that the S^{2-} ion concentration from the dissociation of H_2S is negligibly small compared with the S^{2-} ion concentration from ZnS. This assumption gives us the following approximate equation.

$$[S]^2 \cong 1.6 \times 10^{-24}$$

Either way, we get the same answer.

$$[S] \cong 1.3 \times 10^{-12} \, M$$

When we repeat this calculation for a pH 9.00 solution, we find that the S^{2-} ion from the dissociation of H_2S is now $1.3 \times 10^{-3} \, M$. The S^{2-} ion concentration at equilibrium is equal to the solubility of ZnS plus $1.3 \times 10^{-3} \, M$.

$$[Zn^{2+}][S^{2-}] = K_{sp}$$
$$[S][S + 1.3 \times 10^{-3}] = 1.6 \times 10^{-24}$$

There is little doubt that the S^{2-} ion concentration from the dissociation of H_2S is very much larger than the S^{2-} ion concentration from ZnS.

$$[S][1.3 \times 10^{-3}] \cong 1.6 \times 10^{-24}$$

We can then solve this approximate equation for the solubility of ZnS in a pH 9.00 solution.

$$S \cong 1.2 \times 10^{-21} \, M$$

The solubility of ZnS is very much smaller in this solution than in the pH 3.00 buffer. It has decreased to a point at which fewer than 740 Zn^{2+} ions dissolve in a liter of solution. In general, ZnS becomes significantly less soluble as the solution becomes more basic.

$$\text{pH 3.00:} \quad S = 1.3 \times 10^{-12} \, M$$
$$\text{pH 9.00:} \quad S = 1.2 \times 10^{-21} \, M$$

17.13 AN INTRODUCTION TO QUALITATIVE ANALYSIS

Qualitative analysis brings together all of the techniques introduced to handle equilibria in the four chapters on this topic. The term is used to describe a procedure in which the atoms, ions, or molecules in a sample are identified. The rest of this chapter is devoted to a discussion of a qualitative analysis, or "qual," scheme that can be used to unambiguously determine the presence or absence of the following ions in an unknown solution.

Ag^+, Hg_2^{2+}, Hg^{2+}, Pb^{2+}, Cu^{2+}, Bi^{3+}, Sb^{3+}, Co^{2+}, Mn^{2+}, Ni^{2+}, Cr^{3+}, Ba^{2+}, Ca^{2+}, Na^+, K^+, NH_4^+

Similar schemes can be developed to test for almost any combination of ions.

It is relatively easy to develop tests for individual ions. We can test for the presence of the Cu^{2+} ion in a solution by noting whether the characteristic deep blue of the $Cu(NH_3)_4^{2+}$ complex ion appears when ammonia is added to the solution. We can detect the Fe^{3+} ion by the blood-red color formed when SCN^- ion is added to the sample. To test for the Ni^{2+} ion, we can add a complexing agent known as dimethylglyoxime (DMG), which forms a red $Ni(DMG)_2$ complex.

Unfortunately, many of these tests can't be done in the presence of other ions.

The Ni^{2+} ion, for example, is not the only ion that forms complexes with DMG, nor is Fe^{3+} the only ion that forms colored complexes with SCN^-. Every qualitative analysis scheme is therefore divided into two kinds of procedures. Separatory steps are used to separate small groups of ions or individual ions from the solution. Confirmatory steps are then used to confirm or deny the presence of a given ion in the sample.

The qual scheme discussed here revolves around two processes: selective precipitation and selective dissolution. Selective precipitation, for example, is used to divide the original unknown into five groups of ions, as shown in Figure 17.12. Each group is isolated by adding a reagent that causes the ions in this group to precipitate from solution. The resulting solid is then separated from the solution, which is then treated with the next reagent.

The three ions in Group I — Ag^+, Hg_2^{2+}, and Pb^{2+} — are precipitated by adding hydrochloric acid to the original unknown. After the solid precipitate has been isolated, the solution is treated with H_2S in the presence of a strong acid. This precipitates the ions in Group II — Pb^{2+}, Hg^{2+}, Cu^{2+}, Sb^{3+}, and Bi^{3+}. This solid is then isolated, and the solution is treated with H_2S in the presence of a base, which precipitates the ions in Group III — Co^{2+}, Ni^{2+}, Mn^{2+}, and Cr^{3+}. Once again, the solid is separated from the solution, which is treated with a source of the CO_3^{2-} ion. This treatment precipitates the ions in Group IV — Ba^{2+} and Ca^{2+}. After this solid has been separated, the solution that remains contains the three ions in Group V — Na^+, K^+, and NH_4^+.

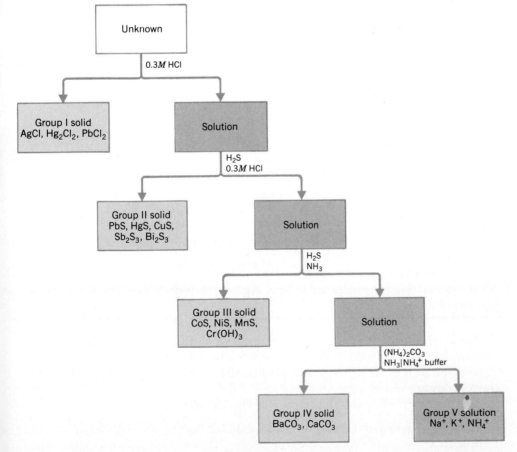

FIG. 17.12 Because it is difficult to test for individual ions in a complex mixture, the overall structure of qualitative analysis involves separating the ions in the mixture into groups of ions that can be handled separately.

This procedure divides the original mixture into four solids and a solution, each of which contains between two and five ions. Each group is then separated into individual ions, which are subjected to confirmatory tests.

17.14 GROUP I: Ag^+, Hg_2^{2+}, AND Pb^{2+}

The ions in Group I in this qual scheme are Ag^+, Hg_2^{2+}, and Pb^{2+}. The Hg_2^{2+} ion is the only common polyatomic metal ion. We can understand why mercury forms Hg_2^{2+} ions by looking at the electron configuration of a neutral mercury atom.

$$Hg = [Xe]\, 6s^2\, 4f^{14}\, 5d^{10}$$

There are only two valence electrons on a neutral mercury atom (see Section 8.1). When one of these valence electrons is removed to form an Hg^+ ion, this ion has an unpaired electron. Pairs of Hg^1 ions therefore combine to form Hg_2^{2+} molecules for exactly the same reasons that hydrogen atoms combine to form H_2 molecules.

$$Hg_2^{2+} = [Hg - Hg]^{2+}$$

The Group I ions are precipitated as their chloride salts. The solubility products for AgCl and Hg_2Cl_2 are very small; the K_{sp} for $PbCl_2$ is somewhat larger.

$$PbCl_2: \qquad K_{sp} = 1.6 \times 10^{-5}$$
$$AgCl: \qquad K_{sp} = 1.8 \times 10^{-10}$$
$$Hg_2Cl_2: \qquad K_{sp} = 1.3 \times 10^{-18}$$

If enough hydrochloric acid is added to a sample of these ions to raise the Cl^- concentration to 0.10 M, very few Ag^+ or Hg_2^{2+} ions will remain in solution.

Exercise 17.19

Calculate the residual Ag^+ ion concentration when the Cl^- ion concentration is 0.10 M.

Solution

We can start with the solubility product expression for AgCl.

$$K_{sp} = [Ag^+][Cl^-]$$

Substituting what we know into this equation gives the following result.

$$[Ag^+][0.10] = 1.8 \times 10^{-10}$$
$$[Ag^+] = 1.8 \times 10^{-9}\ M$$

If the original solution contained 0.10 M Ag^+, 99.99999982% of the Ag^+ has been precipitated from solution!

Those who are new to qualitative analysis sometimes believe in the following philosophy: "If one drop is good, two drops are better!" Why stop when the Cl^- ion concentration is 0.10 M? Why not precipitate even more AgCl by making this concentration even larger?

Ag^+ ions precipitate in the presence of the Cl^- ion.

$$Ag^+(aq) + Cl^-(aq) \rightleftharpoons AgCl(s) \qquad K_{sp} = 1.8 \times 10^{-10}$$

But they also form a soluble complex ion with the chloride ion.

$$Ag^+(aq) + 2\ Cl^-(aq) \rightleftharpoons AgCl_2^-(aq) \qquad K_f = 1.1 \times 10^5$$

Fortunately, the K_{sp} for AgCl is very small, and K_f for the AgCl$_2^-$ complex ion is not very large. In the presence of reasonable amounts of Cl$^-$, AgCl precipitates from this solution. In the presence of a large excess of the Cl$^-$ ion, however, some of the precipitate will dissolve. The qual scheme represents a delicate balance between many such competing processes and must be followed carefully if errors are to be avoided.

Once Group I has been precipitated from the original unknown, we need a way to subdivide the ions in this group so that we can test for each ion individually, as shown in Figure 17.13. Separating Pb^{2+} from Ag$^+$ and Hg$_2$$^{2+}$ ions is relatively easy. This separation is based on two facts. PbCl$_2$ is much more soluble than either AgCl or Hg$_2$Cl$_2$, and most salts become more soluble as the solution is heated.

Thus, to selectively dissolve Pb^{2+} ions we treat the Group I precipitate with boiling water. If we treat this solid with boiling water, stir, centrifuge the hot solution, and decant the liquid while the solution is still hot, we should be able to separate the Pb^{2+} ions from the AgCl and Hg$_2$Cl$_2$ precipitate, which remains behind. We can then add chromate ion to the solution to test for Pb^{2+} ions. If Pb^{2+} ions are present, a yellow PbCrO$_4$ precipitate should form. If a precipitate doesn't form, we conclude that this solution does not contain Pb^{2+} ions.

We now need to separate AgCl from Hg$_2$Cl$_2$. As we have already seen, the Ag$^+$ ion forms a complex with ammonia.

$$Ag^+(aq) + 2\ NH_3(aq) \rightleftharpoons Ag(NH_3)_2^+(aq) \qquad K_f = 1.1 \times 10^7$$

This complex is strong enough to reduce the Ag$^+$ ion concentration to a point at which the product of the Ag$^+$ and Cl$^-$ ion concentrations is smaller than the K_{sp} for AgCl. Any AgCl in the sample therefore dissolves.

$$AgCl(s) + 2\ NH_3(aq) \rightleftharpoons Ag(NH_3)_2^+(aq) + 2\ Cl^-(aq)$$

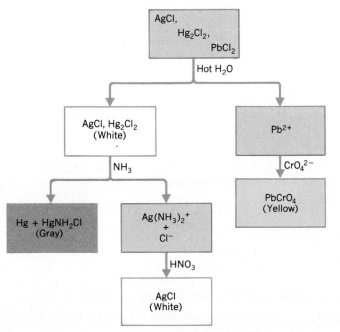

FIG. 17.13 The procedures used to separate and then identify the ions in Group I.

The Hg_2^{2+} ion does not form a complex with ammonia. It reacts with NH_3 to form a mixture of finely divided mercury metal and an Hg^{2+} salt that contains the NH_2^- and Cl^- ions.

$$Hg_2Cl_2(s) + 2\ NH_3(aq) \longrightarrow Hg(s) + HgNH_2Cl(s) + NH_4^+(aq) + Cl^-(aq)$$

Finely divided mercury metal is a black powder, and $HgNH_2Cl$ is a white solid. If the original sample contains Hg_2^{2+} ions, the white precipitate obtained when hydrochloric acid is added to the unknown will turn gray in the presence of ammonia.

We can now test for the Ag^+ ion by adding nitric acid to the solution that may contain the $Ag(NH_3)_2^+$ complex ion. The nitric acid converts NH_3 into NH_4^+ ions, thereby destroying the complex ion. This increases the concentration of the free Ag^+ ion in this solution until it once again precipitates as $AgCl$.

17.15 SEPARATING GROUPS II AND III FROM EACH OTHER

Both Group II (Pb^{2+}, Hg^{2+}, Cu^{2+}, Sb^{3+}, and Bi^{3+}) and Group III (Co^{2+}, Mn^{2+}, Ni^{2+}, and Cu^{2+}) are precipitated from solution by adding a source of the sulfide ion. The Group II ions precipitate when the unknown is treated with H_2S in the presence of acid. The Group III ions precipitate when H_2S is added to the unknown in the presence of a base.

The key to separating the ions in Group II from those in Group III is the large difference between the solubility products of their sulfides.

	Group II			Group III
PbS	$K_{sp} = 8 \times 10^{-28}$		CoS	$K_{sp} = 4 \times 10^{-21}$
CuS	$K_{sp} = 6.3 \times 10^{-36}$		MnS	$K_{sp} = 3 \times 10^{-13}$
HgS	$K_{sp} = 4 \times 10^{-53}$		NiS	$K_{sp} = 3.2 \times 10^{-19}$
Bi$_2$S$_3$	$K_{sp} = 1 \times 10^{-97}$			

Section 16.12 showed that a 10^6-fold difference between the solubility products of MnS and NiS is large enough to allow one of these ions to be separated from the other. If we can separate MnS and NiS, we should have no difficulty in separating the ions in Group II from those in Group III, because the difference between their solubility products is even larger.

Section 16.13 provided a way of adjusting the S^{2-} ion concentration in an aqueous solution to make it large enough to precipitate the highly insoluble Group II sulfides without precipitating any of the more soluble Group III sulfides. We start with a weak source of the S^{2-} ion — $H_2S(aq)$ — and then adjust the pH of the solution.

The dissociation of hydrogen sulfide in the presence of excess acid or base is best described in terms of the following overall equation.

$$H_2S(aq) + 2\ H_2O(l) \rightleftharpoons 2\ H_3O^+(aq) + S^{2-}(aq) \qquad K_a = 1.3 \times 10^{-20}$$

The equilibrium constant expression for this reaction is written as follows.

$$K_a = \frac{[H_3O^+]^2[S^{2-}]}{[H_2S]}$$

Solving for the S^{2-} ion concentration gives the following equation.

$$[S^{2-}] = K_a \times \frac{[H_2S]}{[H_3O^+]^2}$$

A saturated solution of H_2S in water has a concentration of about 0.10 M at room temperature. Since this is a weak acid ($K_a = 1.3 \times 10^{-20}$), we can assume that most of the H_2S is still present as undissociated acid molecules at equilibrium.

$$[S^{2-}] = 1.3 \times 10^{-20} \times \frac{[0.10]}{[H_3O^+]^2}$$

This gives us a simple relationship between the concentrations of the S^{2-} and H_3O^+ ions in this solution.

$$[S^{2-}] = \frac{1.3 \times 10^{-21}}{[H_3O^+]^2}$$

Table 17.2 summarizes the results of calculations of the S^{2-} ion concentration versus the pH of a saturated solution of H_2S in water.

If we bubble H_2S directly into solution, we may produce local concentrations of the S^{2-} ion that are too large. It should also be noted that H_2S is extremely toxic. Hydrogen sulfide is easy to detect at first, because it has the odor of rotten eggs. Our sense of smell soon tires, however, and eventually we can't smell this compound any more. We should therefore take all possible precautions to minimize exposure to this gas.

The easiest way to ensure that the S^{2-} ion concentration does not become too large, and to minimize exposure to H_2S gas, is to generate H_2S in the test tube in which the reaction is run. We do this by using thioacetamide as the source of H_2S. Thioacetamide reacts with water when heated gently to give the acetate ion, the ammonium ion, and H_2S gas.

$$CH_3\!-\!\overset{\overset{\textstyle S}{\|}}{C}\!-\!NH_2(aq) + 2\ H_2O(aq) \longrightarrow CH_3CO_2^-(aq) + NH_4^+(aq) + H_2S(g)$$

TABLE 17.2

The Dependence of the S^{2-} Ion Concentration in a Saturated H_2S Solution on the pH of the Solution

pH	$[S^{2-}]$ (moles per liter)
1	1.3×10^{-19}
3	1.3×10^{-15}
5	1.3×10^{-11}
7	1.3×10^{-7}
10	0.13

Exercise 17.20

Calculate the optimum S^{2-} ion concentration to separate the ions in Group II from those in Group III, starting with a solution in which all of the ions are present at a concentration of 0.10 M.

Solution

The ions in these two groups that are the hardest to separate are the Pb^{2+} and Co^{2+} ions, because PbS is the most soluble of the Group II sulfides, and CoS is the least soluble of the Group III sulfides. If we can find an S^{2-} ion concentration that can separate these two ions, we can separate any Group II ion from any Group III ion. Our goal is therefore to find an S^{2-} ion concentration large enough to precipitate PbS from solution but small enough that no CoS precipitates.

It is easy to calculate the S^{2-} ion concentration at which CoS starts to precipitate from an 0.10 M Co^{2+} solution. We start with the solubility product expression for this salt.

$$K_{sp} = [Co^{2+}][S^{2-}]$$

We then substitute the Co^{2+} ion concentration into this equation.

$$[0.10][S^{2-}] = 4 \times 10^{-21}$$
$$[S^{2-}] = 4 \times 10^{-20}\ M$$

As long as we keep the S^{2-} ion concentration below $4 \times 10^{-20}\ M$, no CoS precipitates.

What fraction of the Pb^{2+} ion remains in solution when the S^{2-} ion concentration is 4×10^{-20} M? We start, once again, with the solubility product expression for the salt.

$$K_{sp} = [Pb^{2+}][S^{2-}]$$

We then substitute the known S^{2-} ion concentration into this equation.

$$[Pb^{2+}][4 \times 10^{-20}] = 8 \times 10^{-28}$$
$$[Pb^{2+}] = 2 \times 10^{-8} \, M$$

If the solution was initially 0.10 M in Pb^{2+}, only 0.00003% of the Pb^{2+} remains in solution when the S^{2-} ion reaches this concentration. In other words, 99.99997% of the Pb^{2+} can be removed from solution before any Co^{2+} starts to precipitate.

At what pH does a saturated solution of H_2S have an S^{2-} ion concentration of 4×10^{-20} M? Rearranging the equilibrium constant expression for the dissociation of H_2S gives the following equation.

$$[H_3O^+]^2 = K_a \times \frac{[H_2S]}{[S^{2-}]}$$

Substituting what we know about this solution gives the following result.

$$[H_3O^+]^2 = \frac{1.3 \times 10^{-20}[0.10]}{4 \times 10^{-20}}$$

Taking the square root of both sides gives the H_3O^+ ion concentration

$$[H_3O^+] = 0.18 \, M$$

which corresponds to a pH of 0.7.

$$pH = -\log [H_3O^+] = 0.7$$

We can precipitate the Group II ions without precipitating any of the Group III ions by adjusting the pH of the solution to 0.7.

17.16 GROUP II: Pb^{2+}, Hg^{2+}, Cu^{2+}, Sb^{3+}, AND Bi^{3+}

The first step in the identification of the ions in Group II involves dividing this group into two subgroups, which are easier to handle. This separation is based on the fact that two of these ions form soluble complex ions with the S^{2-} ion.

$$HgS(s) + S^{2-}(aq) \rightleftharpoons HgS_2{}^{2-}(aq)$$
$$Sb_2S_3(s) + 3 \, S^{2-}(aq) \rightleftharpoons 2 \, SbS_3{}^{3-}(aq)$$

These ions therefore dissolve when excess S^{2-} ion is added to the Group II precipitate, as shown in Figure 17.14.

FIG. 17.14 The first step in the analysis of Group II involves dividing the ions into two smaller subgroups, which are then separated into individual ions.

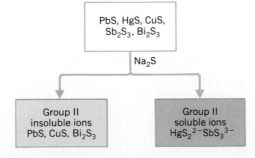

THE Hg^{2+} AND Sb^{3+} SUBGROUP

Before we can test for the Hg^{2+} or Sb^{3+} ion we need to separate these ions from each other. We start by adding hydrochloric acid to the solution (see Figure 17.15). This drives the following equilibrium toward H_2S, which bubbles out of solution.

$$S^{2-}(aq) + 2\ H^+(aq) \longrightarrow H_2S(g)$$

This process destroys the HgS_2^{2-} and SbS_3^{3-} complex ions and allows HgS and Sb_2S_3 to precipitate from solution once again.

The Hg^{2+} and Sb^{3+} ions both form complexes with Cl^- ions.

$$Hg^{2+}(aq) + 4\ Cl^-(aq) \rightleftharpoons HgCl_4^{2-}(aq) \qquad K_f = 1.2 \times 10^{15}$$
$$Sb^{3+}(aq) + 4\ Cl^-(aq) \rightleftharpoons SbCl_4^-(aq) \qquad K_f = 5.2 \times 10^4$$

But the K_{sp} for HgS is so small that the extremely small Hg^{2+} ion concentration in a solution that contains excess Cl^- ion is still too large to allow HgS to dissolve. The K_{sp} for antimony sulfide has not been measured. Experimentally, however, we find that this compound is soluble enough to dissolve in the presence of excess Cl^-.

$$Sb_2S_3(s) + 8\ Cl^-(aq) \rightleftharpoons 2\ SbCl_4^-(aq) + 2\ S^{2-}(aq)$$

Adding excess hydrochloric acid to a mixture of HgS and Sb_2S_3 gives a solution that contains the soluble $SbCl_4^-$ complex ion and a black precipitate of HgS. We can test for the Sb^{3+} ion by treating the $SbCl_4^-$ complex ion with H_2S and neutralizing the excess acid with ammonia. In a basic H_2S solution, the S^{2-} ion concentration is large enough to allow the $SbCl_4^-$ complex ion to be converted back to the insoluble Sb_2S_3 precipitate, but not large enough to form the soluble SbS_3^{3-} complex ion. There is no doubt that the precipitate formed at this stage is Sb_2S_3, because this salt has a characteristic orange color.

Any precipitate that does not dissolve in excess hydrochloric acid should be HgS. We can confirm the presence of the Hg^{2+} ion by adding a solution of the hypochlorite ion in hydrochloric acid to this solid. The OCl^- ion is an oxidizing agent, which oxidizes the S^{2-} ion to elemental sulfur.

$$S^{2-}(aq) + OCl^-(aq) + 2\ H^+(aq) \longrightarrow S(s) + Cl^-(aq) + H_2O(l)$$

This process removes essentially all of the S^{2-} ion from the solution, so the HgS precipitate has to dissolve. We can confirm the fact that this precipitate is HgS by treating the solution with a source of the Sn^{2+} ion in strong acid. The Sn^{2+} ion is a reducing agent that reduces the Hg^{2+} ions to mercury metal, which settles out of solution as a finely divided black powder.

$$Hg^{2+}(aq) + SnCl_3^-(aq) + 3\ Cl^-(aq) \longrightarrow Hg(s) + SnCl_6^{2-}(aq)$$

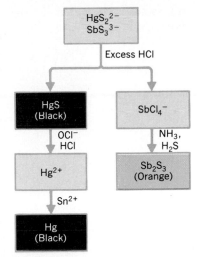

FIG. 17.15 The procedures used to identify the ions in the Hg^{2+} and Sb^{3+} subgroup.

THE Pb^{2+}, Cu^{2+}, AND Bi^{3+} SUBGROUP

The identification of the ions in the other subgroup in Group II starts by treating this solid with nitric acid (see Figure 17.16). Nitric acid is a strong acid ($K_a = 28$). It is also a strong oxidizing agent, and it oxidizes the S^{2-} ion in this solution to elemental sulfur.

$$3\ S^{2-}(aq) + 8\ H^+(aq) + 2\ NO_3^-(aq) \longrightarrow 3\ S(s) + 2\ NO(g) + 4\ H_2O(l)$$

By removing essentially all of the S^{2-} ion from the solution, nitric acid causes the precipitate that contains PbS, CuS, and Bi_2S_3 to dissolve.

FIG. 17.16 The flowchart for the Pb^{2+}, Cu^{2+}, and Bi^{3+} subgroup.

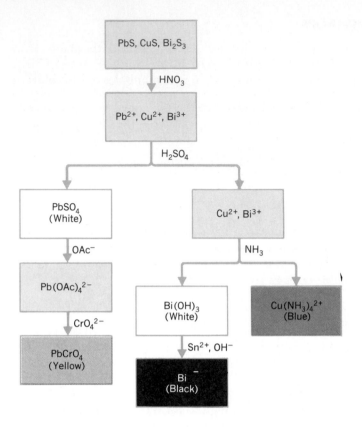

The Pb^{2+} ion can be separated from the Cu^{2+} and Bi^{3+} ions by adding sulfuric acid, which causes $PbSO_4$ to precipitate from solution. We can confirm that the solid that forms is $PbSO_4$ by adding the acetate ion to this moderately insoluble salt ($K_{sp} = 1.6 \times 10^{-8}$) to form the soluble $Pb(OAc)_4^{2-}$ complex ion ($K_f = 3 \times 10^8$).

$$PbSO_4(s) + 4\ OAc^-(aq) \rightleftharpoons Pb(OAc)_4^{2-}(aq) + SO_4^{2-}(aq)$$

If we add a source of the CrO_4^{2-} ion to this solution, however, lead chromate precipitates, because $PbCrO_4$ is much less soluble in water ($K_{sp} = 2.8 \times 10^{-13}$).

$$Pb(OAc)_4^{2-}(aq) + CrO_4^{2-}(aq) \rightleftharpoons PbCrO_4(s) + 4\ OAc^-(aq)$$

Students often ask: "Why do we find Pb^{2+} ions in Group II if we precipitate this ion in Group I?" The answer is that $PbCl_2$ is too soluble for all of the Pb^{2+} to precipitate in Group I. Furthermore, if we add too much hydrochloric acid when Group I is precipitated, some of the $PbCl_2$ redissolves to form the soluble $PbCl_4^{2-}$ complex ion. Some Pb^{2+} ion is therefore usually carried over into Group II.

We now turn to the solution that contains the Cu^{2+} and Bi^{3+} ions. Ammonia is added to this solution to neutralize the acid used to generate the solution. The Cu^{2+} ions form a characteristic blue complex with NH_3, which confirms the presence of these ions. The Bi^{3+} ions don't form complexes with ammonia. Instead, they precipitate from solution as $Bi(OH)_3$. We can confirm that the precipitate is $Bi(OH)_3$ by adding a source of the Sn^{2+} ion in basic solution. In this solution, the tin(II) ion is present as the $Sn(OH)_3^-$ complex ion. This ion is a strong enough reducing agent to reduce the Bi^{3+} ion to bismuth metal, which appears as a finely divided black powder.

17.17 GROUP III: Co²⁺, Ni²⁺, Mn²⁺, AND Cr³⁺

The Group III ions are precipitated by addition of a source of H_2S in the presence of ammonia to the solution left over after Groups I and II have been removed. Three of the Group III ions precipitate as sulfides: CoS, NiS, and MnS. The fourth ion comes out of solution as an insoluble hydroxide: $Cr(OH)_3$. (This is the first time a base has been added to the original unknown. It is therefore the first point at which insoluble hydroxides can precipitate.)

There are too many ions in this group to be handled simultaneously. We therefore split the ions into two subgroups, each containing two ions. This separation is based on the fact that MnS and $Cr(OH)_3$ dissolve in dilute acid but CoS and NiS do not (see Figure 17.17).

It is easy to understand why $Cr(OH)_3$ dissolves in dilute acid. The acid lowers the OH^- ion concentration enough to make the ion product for this salt smaller than the solubility product. It is also easy to understand why MnS dissolves in dilute acid. This salt has the largest solubility product of any of the sulfides in the qual scheme. Dilute acid drives the following equilibrium so far to the right that the ion product for MnS is smaller than the solubility product.

$$S^{2-}(aq) + 2\,H^+(aq) \longrightarrow H_2S(l)$$

CoS ($K_{sp} = 4 \times 10^{-21}$) and NiS ($K_{sp} = 3.2 \times 10^{-19}$) are much less soluble than MnS

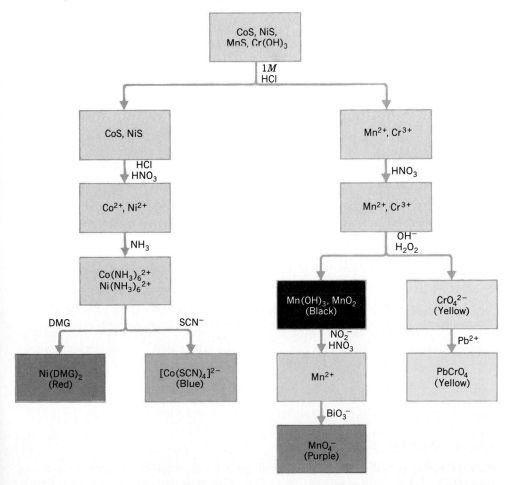

FIG. 17.17 The first step in the analysis of Group III involves dividing the ions into smaller subgroups, which can be handled one at a time.

($K_{sp} = 3 \times 10^{-13}$). If we avoid using excess acid, and if we perform the separation fast enough, we can selectively dissolve MnS without dissolving either CoS or NiS.

THE Mn^{2+} AND Cr^{3+} SUBGROUP

The first step in the identification of the ions in the subgroup that contains Mn^{2+} and Cr^{3+} involves adding nitric acid to remove all of the S^{2-} ion. Some of the S^{2-} ion is converted into H_2S, which bubbles out of solution.

$$S^{2-}(aq) + 2 H^+(aq) \longrightarrow H_2S(s)$$

The rest is oxidized to elemental sulfur.

$$3 S^{2-}(aq) + 8 H^+(aq) + 2 NO_3^-(aq) \longrightarrow 3 S(s) + 2 NO(g) + 4 H_2O(l)$$

A solution of hydrogen peroxide in base is then added to the Mn^{2+} and Cr^{3+} ions. Both ions form insoluble hydroxides.

$$Mn^{2+}(aq) + 2 OH^-(aq) \longrightarrow Mn(OH)_2(s)$$
$$Cr^{3+}(aq) + 3 OH^-(aq) \longrightarrow Cr(OH)_3(s)$$

In excess base, some of the Cr^{3+} forms a soluble complex ion.

$$Cr(OH)_3(s) + OH^-(aq) \rightleftharpoons Cr(OH)_4^-(aq)$$

However, H_2O_2 is a strong enough oxidizing agent to oxidize the Mn^{2+} ions to either the $+3$ or $+4$ oxidation state.

$$2 Mn(OH)_2(s) + H_2O_2(aq) \longrightarrow 2 Mn(OH)_3(s)$$
$$Mn(OH)_2(s) + H_2O_2(aq) \longrightarrow MnO_2(s) + 2 H_2O(l)$$

It is also strong enough to oxidize the Cr^{3+} ions to the $+6$ oxidation state.

$$Cr(OH)_3(s) + H_2O_2(aq) + 3 OH^-(aq) \longrightarrow CrO_4^{2-}(aq) + 4 H_2O(l)$$

The net effect of these reactions is to precipitate the manganese as a mixture of $Mn(OH)_3$ and MnO_2 and leave the chromium in solution as the CrO_4^{2-} ion.

The presence of the CrO_4^{2-} ion should be obvious from its distinctive yellow-orange color. It can be confirmed, however, by adding the soluble lead acetate complex to the solution.

$$Pb(OAc)_4^{2-}(aq) + CrO_4^{2-}(aq) \rightleftharpoons PbCrO_4(s) + 4 OAc^-(aq)$$

A solution of the nitrite ion (NO_2^-) in nitric acid is now used to reduce the $Mn(OH)_3/MnO_2$ precipitate back to Mn^{2+} ions.

$$2 Mn(OH)_3(s) + NO_2^-(aq) + 4 H^+(aq) \longrightarrow 2 Mn^{2+}(aq) + NO_3^-(aq) + 5 H_2O(l)$$
$$MnO_2(s) + NO_2^-(aq) + 2 H^+(aq) \longrightarrow Mn^{2+}(aq) + NO_3^-(aq) + H_2O(l)$$

We can confirm the presence of the Mn^{2+} ion by adding a very powerful oxidizing agent known as the bismuthate ion. This ion oxidizes Mn^{2+} to the permanganate ion, MnO_4^-, in which the oxidation number of manganese is $+7$.

$$2 Mn^{2+}(aq) + 5 BiO_3^-(aq) + 14 H^+(aq) \longrightarrow 2 MnO_4^-(aq) + 5 Bi^{3+}(aq) + 7 H_2O(l)$$

The permanganate ion formed in this reaction has a characteristic purple color.

THE Co^{2+} AND Ni^{2+} SUBGROUP

CoS and NiS do not dissolve in dilute acid, but they dissolve in a mixture of a strong acid (HCl) and an oxidizing agent (HNO_3). This mixture of acids not only drives the S^{2-} ion toward H_2S, which bubbles out of solution, but it oxidizes the remaining S^{2-} ion to elemental sulfur.

The resulting solution, which contains the Co^{2+} and Ni^{2+} ions, is then neutralized with ammonia to form the following six-coordinate complex ions.

$$Co^{2+}(aq) + 6\ NH_3(aq) \rightleftharpoons Co(NH_3)_6{}^{2+}(aq)$$

$$Ni^{2+}(aq) + 6\ NH_3(aq) \rightleftharpoons Ni(NH_3)_6{}^{2+}(aq)$$

The chemistry of these ions is different enough to allow us to test for Ni^{2+} in the presence of Co^{2+}, and vice versa.

We treat half of the solution with dimethylglyoxime and the other half with the thiocyanate ion in the presence of acetone. If the Ni^{2+} ion is present, it forms a bright-red $Ni(DMG)_2$ complex, which is insoluble in water. If Co^{2+} ions are present, they form a characteristic blue $Co(SCN)_4{}^{2-}$ complex ion.

17.18 GROUP IV: Ba²⁺ AND Ca²⁺

The ions in Group IV are precipitated from solution as insoluble carbonates.

$$Ba^{2+}(aq) + CO_3{}^{2-}(aq) \rightleftharpoons BaCO_3(s)$$

$$Ca^{2+}(aq) + CO_3{}^{2-}(aq) \rightleftharpoons CaCO_3(s)$$

What is the best source of the $CO_3{}^{2-}$ ion to add to the original unknown to precipitate the Group IV ions? We don't want to add Na_2CO_3, because that would contaminate the solution with Na^+ ions, which we must test for in Group V. A similar argument eliminates K_2CO_3 as a reagent for this step.

In theory, we could use carbonic acid — $H_2CO_3(aq)$ — but solutions of carbonic acid are notoriously unstable.

$$H_2CO_3(aq) \longrightarrow H_2O(l) + CO_2(g)$$

Let's look at the reagents used to precipitate Groups I, II, and III.

Group I: HCl
Group II: H_2S in HCl
Group III: H_2S in NH_3

We have already added NH_3 to this solution, which therefore contains at least some $NH_4{}^+$ ions.

$$NH_3(aq) + H_2O(l) \rightleftharpoons NH_4{}^+(aq) + OH^-(aq)$$

The reagent of choice for precipitating Group IV is therefore the ammonium salt of the carbonate ion — $(NH_4)_2CO_3$. Ammonia is added at the same time as $(NH_4)_2CO_3$ to adjust the pH of the solution so that the $CO_3{}^{2-}$ ion concentration is large enough to precipitate the Group IV ions.

The first step in separating Ba^{2+} ions from Ca^{2+} involves dissolving the carbonate precipitate in acetic acid, as shown in Figure 17.18. The acid reacts with the $CO_3{}^{2-}$ ion to form carbonic acid

$$CO_3{}^{2-}(aq) + 2\ H^+(aq) \longrightarrow H_2CO_3(aq)$$

which decomposes to carbon dioxide and water.

$$H_2CO_3(aq) \longrightarrow H_2O(l) + CO_2(g)$$

We then add chromate ion, which precipitates Ba^{2+} but not Ca^{2+}.

$$Ba^{2+}(aq) + CrO_4{}^{2-}(aq) \rightleftharpoons BaCrO_4(s)$$

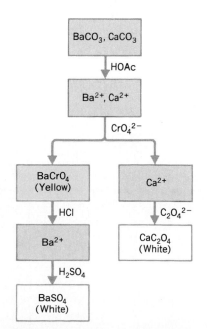

FIG. 17.18 The procedures used to identify the ions in Group IV.

To confirm the presence of Ba^{2+} ions, we dissolve this precipitate in hydrochloric acid and then watch $BaSO_4$ precipitate when sulfuric acid is added to the solution.

The presence of Ca^{2+} ions is confirmed by adding the oxalate ion, $C_2O_4^{2-}$, which forms an insoluble precipitate.

$$Ca^{2+}(aq) + C_2O_4^{2-}(aq) \rightleftharpoons CaC_2O_4(s)$$

Exercise 17.21

The chemistry of the oxalate — $C_2O_4^{2-}$ — and carbonate — CO_3^{2-} — ions is often confused. Draw Lewis structures to illustrate the difference between these ions.

Solution

Strict adherence to the rules outlined in Chapter 8 gives the following Lewis structures for these ions.

17.19 GROUP V: Na^+, K^+, AND NH_4^+

To remove ions from solution in the qual scheme, we start with the ions that are easiest to precipitate. Each subsequent group is a little harder to precipitate. Eventually, we are left with a solution that contains the ions that are the hardest to precipitate — Na^+, K^+, and NH_4^+. We leave these ions in solution.

The test for Na^+ is based on the fact that this ion forms a mixed salt with Zn^{2+}, acetate (OAc^-), and uranyl (UO_2^{2+}) ions.

$$NaZn(UO_2)_3(OAc)_9 \cdot 6\,H_2O$$

Note that the positive charge on one Na^+ ion, one Zn^{2+} ion, and three UO_2^{2+} ions balances the charge on the nine acetate ions.

The test for K^+ is based on the fact that the potassium and sodium salts of the $Co(NO_2)_6^{3-}$ complex ion are soluble in water but the mixed salt $NaK_2Co(NO_2)_6$ is insoluble in water. Thus, if a yellow precipitate forms when $Na_3Co(NO_2)_6$ is added to the solution, it must contain some K^+ ion.

When we finally test for the NH_4^+ ion, we have to start with a fresh sample of the unknown solution. Because both NH_3 and the NH_4^+ ion were added in the procedure that divided the unknown into groups of ions, the solution that contains Group V will give a positive test for the NH_4^+ ion, even if this ion was not present in the original unknown.

We can test for the NH_4^+ ion by adding a strong base to a sample of the original unknown. If this ion is present, the base should convert at least some of it to NH_3, which can bubble out of solution when it is heated.

$$NH_4^+(aq) + OH^-(aq) \rightleftharpoons NH_3(aq) + H_2O(l)$$

The ammonia given off in this reaction can be detected on the basis of either its odor or its reaction with a piece of wet litmus paper held above the solution.

SUMMARY

Any ion or molecule that can donate a pair of nonbonding electrons is a Lewis base. An ion or molecule with one or more empty valence-shell orbitals that can accept a pair of nonbonding electrons is a Lewis acid. Lewis acids combine with Lewis bases to form acid-base complexes held together by the sharing of a pair of electrons in a covalent bond. When transition-metal ions act as Lewis acids, they tend to form acid-base complexes, or coordination complexes, that carry a net charge. These coordination complexes are often known as complex ions.

Because the values of K_f for the individual steps in the formation of a complex ion are often about the same, complex ion equilibria are typically collapsed into a single overall equation. In calculations involving complex ion equilibria, reactions are driven as far as possible toward the complex, and a small amount of the complex ion is then allowed to dissociate.

Complex ion equilibria are particularly important in solutions in which more than one equilibrium occurs at the same time. They are therefore an essential part of any discussion of combined equilibria. A qualitative understanding of combined equilibria can be obtained by application of LeChatelier's principle. More quantitative results come from solving combined equilibrium problems one step at a time.

Complex ion equilibria and combined equilibria play an important role in qualitative analysis, the set of procedures that has been developed for identifying the ions in an unknown solution.

PROBLEMS

17-1 Define the following terms: *Lewis acid, Lewis base, complex ion,* and *coordination number*

17-2 Explain why main-group elements form simple ions, such as the Al^{3+} and O^{2-} ions, but transition metals form complex ions, such as the $Fe(SCN)_2^+$ and $Cu(NH_3)_4^{2+}$ ions.

17-3 Explain the difference between Lewis acids and Brønsted acids and give examples of each. Explain the difference between Lewis bases and Brønsted bases and give examples of each.

17-4 Give an example of a compound that is a Lewis acid but not a Brønsted acid.

17-5 Which of the following is a Lewis acid?

   (a) CO   (b) C_2H_2 (c) BeF_2 (d) CH_4 (e) NF_3

17-6 Which of the following is a Lewis acid?

   (a) CH_3^+ (b) CH_4 (c) NH_3 (d) BF_4^- (e) O^{2-}

17-7 Which of the following is not a Lewis acid?

   (a) H^+ (b) BF_3 (c) CO   (d) Cu^{2+} (e) Fe^{3+}

17-8 Which of the following is not a Lewis base?

   (a) NH_4^+ (b) OH^- (c) Cl^- (d) O_2 (e) SCN^-

17-9 Which of the following Lewis acids is not a Brønsted acid?

   (a) HF   (b) HOAc   (c) H_3PO_4 (d) NH_3 (e) BF_3

17-10 Which of the following is not a Lewis acid-base reaction?

   (a) $BF_3 + KF \rightarrow KBF_4$ (b) $H_2 + Cl_2 \rightarrow 2\ HCl$ (c) $CaO + CO_2 \rightleftharpoons CaCO_3$ (d) $SiF_4 + 2\ F^- \rightleftharpoons SiF_6^{2-}$ (e) $Ag^+ + 2\ NH_3 \rightleftharpoons Ag(NH_3)_2^+$

17-11 In which of the following reactions is water a Lewis acid?

   (a) $H_2O + O^{2-} \rightarrow 2\ OH^-$ (b) $H_2O + NaCl \rightarrow NaCl(aq)$
   (c) $H_3O^+ + OH^- \rightleftharpoons 2\ H_2O$ (d) $Fe^{3+} + H_2O \rightleftharpoons Fe(OH)_3 + 3\ H^+$ (e) $CO_2 + H_2O \rightleftharpoons H_2CO_3$

17-12 In which of the following reactions is water a Lewis base?

   (a) $H_2O + O^2 \rightarrow 2\ OH^-$ (b) $H_2O + NaCl \rightarrow NaCl(aq)$
   (c) $H_3O^+ + OH^- \rightleftharpoons 2\ H_2O$ (d) $Fe^{3+} + H_2O \rightleftharpoons Fe(OH)_3 + 3\ H^+$ (e) $CO_2 + H_2O \rightleftharpoons H_2CO_3$

17-13 Explain why the charges on the $Fe(SCN)^{2+}$ and $Fe(SCN)_2^+$ ions are $+2$ and $+1$, respectively.

17-14 Calculate the oxidation numbers of the transition metal ions in the $Zn(NH_3)_4^{2+}$, $Fe(SCN)_2^+$, $Sn(OH)_3^-$, $Co(SCN)_4^{2-}$, and $Ag(S_2O_3)_2^{3-}$ complex ions.

The Stepwise Formation of Complex Ions

17-15 Explain the difference between polyprotic acids and complex ions that allows us to assume that polyprotic acids, dissociate in steps, whereas the dissociation of complex ions, can be collapsed into a single overall reaction.

17-16 Describe the conditions under which the Fe^{3+} ion is most likely to be the dominant species (type of ion) in a mixture of Fe^{3+} and SCN^- ions.

17-17 Describe the conditions under which the dominant species in a mixture of the Fe^{3+} and SCN^- ions is most likely to be the two-coordinate $Fe(SCN)_2^+$ ion.

17-18 Cu^{2+} forms a four-coordinate complex with ammonia.

$$Cu^{2+}(aq) + 4\ NH_3(aq) \rightleftharpoons Cu(NH_3)_4^{2+}(aq)$$

What is the relationship between the overall complex formation equilibrium constant for this reaction, K_f, and the stepwise formation constants, K_{f1}, K_{f2}, K_{f3} and K_{f4}?

17-19 Calculate the complex formation equilibrium constant, K_f, for the following overall reaction

$$Ag^+(aq) + 2 \, S_2O_3^{2-}(aq) \rightleftharpoons Ag(S_2O_3)_2^{3-}(aq)$$

from the values of the stepwise formation constants.

$Ag^+(aq) + S_2O_3^{2-}(aq) \rightleftharpoons Ag(S_2O_3)^-(aq) \quad K_{f1} = 6.6 \times 10^8$

$Ag(S_2O_3)^-(aq) + S_2O_3^{2-}(aq) \rightleftharpoons Ag(S_2O_3)_2^{3-}(aq)$
$\qquad\qquad\qquad\qquad\qquad\qquad\qquad K_{f2} = 4.4 \times 10^4$

17-20 Calculate the complex formation equilibrium constant, K_f, for the following overall reaction

$$Cd^{2+}(aq) + 4 \, CN^-(aq) \rightleftharpoons Cd(CN)_4^{2-}(aq)$$

from the values of the stepwise formation constants.

$Cd^{2+}(aq) + CN^-(aq) \rightleftharpoons Cd(CN)^+(aq) \quad K_{f1} = 3.0 \times 10^5$

$Cd(CN)^+(aq) + CN^-(aq) \rightleftharpoons Cd(CN)_2(aq) \quad K_{f2} = 1.3 \times 10^5$

$Cd(CN)_2(aq) + CN^-(aq) \rightleftharpoons Cd(CN)_3^-(aq) \quad K_{f3} = 4.3 \times 10^4$

$Cd(CN)_3^-(aq) + CN^-(aq) \rightleftharpoons Cd(CN)_4^{2-}(aq) \quad K_{f4} = 3.5 \times 10^3$

17-21 Which of the following solutions has the smallest concentration of the Ag^+ ion?

(a) $0.10 \, M \, Ag^+$ and $1.0 \, M \, Cl^-$ $(AgCl_2^-: K_f = 1.1 \times 10^5)$
(b) $0.10 \, M \, Ag^+$ and $1.0 \, M \, NH_3$ $[Ag(NH_3)_2^+: K_f = 1.1 \times 10^7]$ (c) $0.10 \, M \, Ag^+$ and $1.0 \, M \, S_2O_3^{2-}$ $[Ag(S_2O_3)_2^{3-}: K_f = 2.9 \times 10^{13}]$ (d) $0.10 \, M \, Ag^+$ and $1.0 \, M \, CN^-$ $[Ag(CN)_2^-: K_f = 1.3 \times 10^{21}]$

17-22 Which of the following solutions has the largest concentration of the Hg^{2+} ion?

(a) $0.10 \, M \, Hg^{2+}$ and $1.0 \, M \, Cl^-$ $(HgCl_4^{2-}: K_f = 1.2 \times 10^{15})$ (b) $0.10 \, M \, Hg^{2+}$ and $1.0 \, M \, Br^-$ $(HgBr_4^{2-}: K_f = 1 \times 10^{21})$ (c) $0.10 \, M \, Hg^{2+}$ and $1.0 \, M \, I^-$ $(HgI_4^{2-}: K_f = 6.8 \times 10^{29})$ (d) $0.10 \, M \, Hg^{2+}$ and $1.0 \, M \, CN^-$ $[Hg(CN)_4^{2-}: K_f = 3 \times 10^{41}]$

Complex Dissociation Equilibrium Constants

17-23 Which of the following equations describes the relationship between the complex formation equilibrium constant, K_f, and the complex dissociation equilibrium constant, K_d, for the $Fe(CN)_6^{3-}$ complex ion.

(a) $K_f = K_d$ (b) $K_f = K_w/K_d$ (c) $K_f = K_d/K_w$ (d) $K_f = K_w \times K_d$ (e) $K_f = 1/K_d$

17-24 Derive the relationships among K_{f1}, K_{f2}, K_{d1}, and K_{d2} for the $Fe(SCN)^{2+}$ and $Fe(SCN)_2^+$ complex ions.

Approximate Complex Ion Calculations

17-25 Calculate the Fe^{3+} ion concentration at equilibrium in a solution prepared by adding 0.100 moles of SCN^- 250 milliliters of $0.0010 \, M \, Fe(NO_3)_3$.

$Fe^{3+}(aq) + 2 \, SCN^-(aq) \rightleftharpoons Fe(SCN)_2^+(aq) \quad K_f = 2.3 \times 10^3$

17-26 Calculate the Cu^{2+} ion concentration at equilibrium in a solution that is initially $0.10 \, M$ in Cu^{2+} and $4 \, M$ in NH_3. $[Cu(NH_3)_4^{2+}: K_f = 2.1 \times 10^{13}]$

17-27 Calculate the Zn^{2+} ion concentration at equilibrium in a solution prepared by dissolving 0.220 moles of $ZnCl_2$ in 500 milliliters of $2.0 \, M$ ammonia. $[Zn(NH_3)_4^{2+}: K_f = 2.9 \times 10^9]$

17-28 Calculate the Sb^{3+} ion concentration at equilibrium in a $0.10 \, M \, Sb^{3+}$ ion solution that has been buffered at pH 8.00. $[Sb(OH)_4^-: K_f = 2 \times 10^{38}]$

17-29 Calculate the Sb^{3+} ion concentration at equilibrium in a solution that is initially $0.10 \, M$ in Sb^{3+} and $6 \, M$ in HCl. $(SbCl_4^-: K_f = 5.2 \times 10^4)$

17-30 Calculate the Co^{3+} ion concentration at equilibrium in a solution that is initially $0.10 \, M$ in Co^{3+} and $1.0 \, M$ in SCN^-. $[Co(SCN)_4^-: K_f = 1 \times 10^3]$

17-31 Calculate the Cd^{2+} ion concentration at equilibrium in a solution we prepare by adding 10.0 milliliters of $15 \, M$ aqueous ammonia to 100 milliliters of a solution of 7.00×10^{-3} grams of $CdCl_2$ in water. $[Cd(NH_3)_4^{2+}: K_f = 1.3 \times 10^7]$

Using Complex Ion Equilibria to Dissolve an Insoluble Salt

17-32 We can separate AgCl from Hg_2Cl_2 by adding $4 \, M \, NH_3$. How many moles of AgCl will dissolve in 1.00 liter of $4.00 \, M$ NH_3? $[AgCl: K_{sp} = 1.8 \times 10^{-10}, Ag(NH_3)_2^+: K_f = 1.2 \times 10^7]$

17-33 How many grams of AgBr will dissolve in 250 milliliters of $6.0 \, M \, NH_3$? $[Ag(NH_3)_2^+: K_f = 1.1 \times 10^7, AgBr: K_{sp} = 5.0 \times 10^{-13}]$

17-34 K_f for the $Ag(NH_3)_2^+$ complex ion is not large enough to allow silver bromide to dissolve in $4 \, M \, NH_3$. Will 0.0001 moles of AgBr dissolve in $15 \, M \, NH_3$? $[AgBr: K_{sp} = 5.0 \times 10^{-13}, Ag(NH_3)_2^+: K_f = 1.1 \times 10^7]$

17-35 K_f for the thiosulfate complex is not large enough to allow silver iodide to dissolve in $S_2O_3^{2-}$ solutions, but the value of K_f for the cyanide complex ion, $Ag(CN)_2^-$, is much larger. Will 0.0001 moles of AgI dissolve in a liter of $4 \, M \, CN^-$? $[AgI: K_{sp} = 8.3 \times 10^{-17}, Ag(CN)_2^-: K_f = 1 \times 10^{21}]$

Using Complex Ion Equilibria to Keep a Salt From Precipitating

17-36 What concentration of NH_3 must be present in a $0.10 \, M$ $AgNO_3$ solution to prevent AgCl from precipitating when 4.0 grams of sodium chloride is added to 250 milliliters of this solution? $[Ag(NH_3)_2^+: K_f = 1.1 \times 10^7, AgCl: K_{sp} = 1.8 \times 10^{-10}]$

17-37 Will $Co(OH)_3$ precipitate from a solution that is initially $0.10 \, M$ in Co^{3+} and $1.0 \, M$ in SCN^- if this solution is buffered at pH 7.00? $[Co(OH)_3: K_{sp} = 1.6 \times 10^{-44}, Co(SCN)_4^-: K_f = 1 \times 10^3]$

17-38 It is possible to keep $Co(OH)_3$ from precipitating from a $0.010 \, M \, CoCl_3$ solution by buffering the solution at pH 9.10 with a buffer that contains NH_3 and the NH_4^+ ion. How much

6 M NH_3 and 6 M HCl must be added per liter of this solution to prevent $Co(OH)_3$ from precipitating? [NH_3: $K_b = 1.8 \times 10^{-5}$, $Co(OH)_3$: $K_{sp} = 1.0 \times 10^{-43}$, $Co(NH_3)_6^{3+}$: $K_f = 2 \times 10^{35}$]

The Rigorous Approach to Complex Ion Calculations

17-39 Write the equation used to calculate the percent of the total iron concentration present as the one-coordinate $Fe(SCN)^{2+}$ ion in a solution that contains both the Fe^{3+} and SCN^- ions.

17-40 Calculate the percent of the total iron concentration present as the $Fe(SCN)^{2+}$ ion in a 0.10 M Fe^{3+} ion solution to which enough SCN^- has been added to raise the concentration of this ion to 1.0 M. [$Fe(SCN)_2^+$: $K_{f1} = 890$, $K_{f2} = 2.6$]

17-41 Calculate the percent of the total iron concentration present as the two-coordinate $Fe(SCN)_2^+$ ion in a 0.10 M Fe^{3+} ion solution to which enough SCN^- has been added to raise the concentration of this ion to 1.0 M. [$Fe(SCN)_2^+$: $K_{f1} = 890$, $K_{f2} = 2.6$]

17-42 Calculate the percent of the total iron concentration present as the Fe^{3+} ion in a 0.10 M Fe^{3+} ion solution to which enough F^- ion has been added to raise the concentration of this ion to 1.0 M. (FeF_3: $K_{f1} = 1.9 \times 10^5$, $K_{f2} = 1.1 \times 10^4$, $K_{f3} = 5.8 \times 10^2$)

17-43 Calculate the percent of the total cadmium ion concentration present as the $Cd(CN)_4^{2-}$ complex ion in a 0.10 M Cd^{2+} ion solution to which enough CN^- ion has been added to raise the concentration of this ion to 1.0 M. [$Cd(CN)_4^{2-}$: $K_{f1} = 3.0 \times 10^5$, $K_{f2} = 1.3 \times 10^5$, $K_{f3} = 4.3 \times 10^4$, $K_{f4} = 3.5 \times 10^3$]

17-44 Calculate the percent of the total copper ion concentration present as the Cu^{2+}, $Cu(NH_3)^{2+}$, $Cu(NH_3)_2^{2+}$, and $Cu(NH_3)_3^{2+}$ ions in a solution to which 0.010 moles of Cu^{2+} has been added per liter of 1.0 M NH_3 solution.

A Qualitative View of Combined Equilibria

17-45 Fe^{3+} forms blood-red complexes with the thiocyanate ion, SCN^-. What is the best way to increase the concentration of these complex ions in a solution that contains the following equilibria?

$$2\ H_2O(aq) \rightleftharpoons H_3O^+(aq) + OH^-(aq) \qquad K_w = 1.0 \times 10^{-14}$$
$$Fe(OH)_3(s) \rightleftharpoons Fe^{3+}(aq) + 3\ OH^-(aq) \qquad K_{sp} = 4 \times 10^{-38}$$
$$Fe^{3+} + SCN^-(aq) \rightleftharpoons Fe(SCN)^{2+}(aq) \qquad K_f = 890$$
$$Fe(SCN)^{2+}(aq) + SCN^-(aq) \rightleftharpoons Fe(SCN)_2^+(aq) \qquad K_f = 2.6$$
$$SCN^-(aq) + H_2O(aq) \rightleftharpoons HSCN(aq) + OH^-(aq) \qquad K_a = 71$$

(a) Add HNO_3. (b) Add NaOH. (c) Add NaSCN. (d) Add $Fe(OH)_3$.

17-46 What is the effect of adding a strong acid such as HCl to a solution that contains the $Zn(CN)_4^{2-}$ complex ion if HCN is a weak acid ($K_a = 6 \times 10^{-10}$)?

(a) The Zn^{2+} ion concentration increases. (b) The Zn^{2+} ion concentration decreases. (c) The Zn^{2+} ion concentration remains the same. (d) The Zn^{2+} and CN^- ion

concentrations both increase. (e) There is no way of predicting what will happen to the Zn^{2+} ion concentration.

17-47 What is one way to increase the concentration of the Cu^{2+} ion in a saturated solution of $CuSO_4$ in ammonia in which the following equilibria are possible?

$$CuSO_4(s) \rightleftharpoons Cu^{2+}(aq) + SO_4^{2-}(aq)$$
$$Cu^{2+}(aq) + 4\ NH_3(aq) \rightleftharpoons Cu(NH_3)_4^{2+}(aq)$$
$$Cu^{2+}(aq) + 2\ OH^-(aq) \rightleftharpoons Cu(OH)_2(s)$$
$$NH_3(aq) + H_2O \rightleftharpoons NH_4^+(aq) + OH^-(aq)$$
$$2\ H_2O(aq) \rightleftharpoons H_3O^+(aq) + OH^-(aq)$$
$$SO_4^{2-}(aq) + H_2O \rightleftharpoons HSO_4^-(aq) + OH^-(aq)$$

(a) Add an acid, such as HNO_3. (b) Add a base, such as NaOH. (c) Increase the ammonia concentration. (d) Add more $CuSO_4$. (e) None of these increases the Cu^{2+} ion concentration in this solution.

Qualitative Analysis: Group I

17-48 Give at least five examples of separatory steps from the qualitative analysis scheme described in this chapter. Give at least five examples of confirmatory steps.

17-49 Identify the ions in the five groups of this qual scheme.

17-50 Explain why the +1 oxidation number of mercury corresponds with the Hg_2^{2+} ion.

17-51 Explain why Pb^{2+} is found in both Group I and Group II of this qual scheme.

17-52 Explain why $PbCl_2$ dissolves when the Group I precipitate is treated with hot water, whereas AgCl and Hg_2Cl_2 do not.

17-53 Explain why it is a mistake to add too much Cl^- when precipitating the Group I ions.

Qualitative Analysis: Group II

17-54 Describe why the ions in Group II are precipitated from solution before the ions in Group III.

17-55 Explain what happens if the S^{2-} ion concentration is not carefully controlled when the Group II ions are precipitated.

17-56 Explain how Hg^{2+} and Sb^{3+} can be separated from the other ions in Group II.

17-57 Explain why HgS doesn't dissolve in acid but does dissolve when treated with the hypochlorite (OCl^-) ion.

17-58 The Cu^{2+} and Bi^{3+} ions both form insoluble hydroxides. Explain why $Bi(OH)_3$ precipitates but $Cu(OH)_2$ does not when an aqueous solution of NH_3 is added to these ions.

Qualitative Analysis: Group III

17-59 Explain why $Cr(OH)_3$ precipitates along with CoS, NiS, and MnS in Group III.

17-60 Explain why $Cr(OH)_3$ and MnS dissolve when 1 M HCl is added to the Group III precipitate, whereas CoS and NiS do not.

17-61 Explain why CoS and NiS dissolve in a mixture of hydrochloric acid and nitric acid but not in 1 M hydrochloric acid.

17-62 Describe how hydrogen peroxide is used to separate the Mn^{2+} and Cr^{3+} ions in this qual scheme.

Qualitative Analysis: Group IV

17-63 Explain why adding either Na_2CO_3 or K_2CO_3 to precipitate the ions in Group IV would be a mistake.

17-64 Explain why $BaCO_3$ and $CaCO_3$ dissolve when treated with acetic acid.

17-65 Describe the difference between the carbonate (CO_3^{2-}) ion and the oxalate ($C_2O_4^{2-}$) ion.

Qualitative Analysis: Group V

17-66 Explain why the ions in Group V include Na^+, K^+, and NH_4^+.

17-67 Explain why we have to go back to a fresh sample of the unknown to test for the NH_4^+ ion.

17-68 Calculate the oxidation number of the uranium in the uranyl ion, UO_2^{2+}. Use the electron configuration of uranium given in Section 5.19 to explain why uranium forms an ion with this oxidation number.

Qualitative Analysis Calculations

17-69 When we test for the Ag^+ ion in the qual scheme, we dilute 10 drops of the unknown with 10 drops of water and 2 drops of 6 M HCl. Calculate the Cl^- ion concentration in this solution, assuming that there are 20 drops in one milliliter.

Calculate the minimum Ag^+ ion concentration necessary to give a positive test, assuming that K_{sp} for AgCl is 1.8×10^{-10}.

17-70 The qual scheme is based on the assumption that ions in Group I are completely precipitated before Group II is analyzed. Assume that the unknown contains 0.10 M Hg_2^{2+} ion. Calculate the fraction of this ion that remains in solution after the Group I ions have been precipitated.

17-71 Assume that the solid AgCl precipitate formed when 10 drops of unknown are diluted with 10 drops of water and 2 drops of 6 M Cl is then added is collected. Should all of this AgCl dissolve when the precipitate is mixed with 10 drops of 4 M NH_3? [$Ag(NH_3)_2^+$: $K_f = 1.2 \times 10^7$]

17-72 The Group II cations precipitate from a 0.10 M H_2S solution in the presence of 0.3 M HCl. Calculate the S^{2-} ion concentration in this solution (H_2S: $K_{a1} = 1.0 \times 10^{-7}$, $K_{a2} = 1.3 \times 10^{-13}$). Predict which sulfides will precipitate from this solution if the solution is also 0.10 M in the following ions: Hg^{2+}, Cu^{2+}, Pb^{2+}, Co^{2+}, Ni^{2+}, and Mn^{2+}.

$$K_{sp}: HgS = 4 \times 10^{-53} \qquad CoS = 4 \times 10^{-21}$$
$$CuS = 6.3 \times 10^{-36} \qquad NiS = 3.2 \times 10^{-19}$$
$$PbS = 8 \times 10^{-28} \qquad MnS = 3 \times 10^{-13}$$

17-73 Calculate the CO_3^{2-} ion concentration in a 0.10 M HCO_3^- solution buffered with equal numbers of moles of NH_3 and NH_4^+. Is this CO_3^{2-} concentration large enough to precipitate $BaCO_3$ when the solution is mixed with an equal volume of a 0.10 M Ba^{2+} ion solution? ($BaCO_3$: $K_{sp} = 5.1 \times 10^{-9}$; H_2CO_3: $K_{a1} = 4.5 \times 10^{-7}$, $K_{a2} = 4.7 \times 10^{-11}$; NH_3: $K_b = 1.8 \times 10^{-5}$)

OXIDATION-REDUCTION REACTIONS

EVOLUTION OF THE THEORY
18.1 OF OXIDATION-REDUCTION REACTIONS

The first step toward a theory of chemical reactions was taken by Georg Ernst Stahl in 1697 when he proposed the *phlogiston* theory, which was based on the following observations.

1. Metals share many properties.
2. Metals often produce salts, or "calxes," when heated.
3. These salts are not as dense as the metals, in much the same way that the ash formed when wood burns is not as dense as the wood.
4. Some of these salts form metals when heated with charcoal.
5. With only a few exceptions, the salts are found in nature, not the metals.

These observations led Stahl to the following conclusions.

1. Phlogiston (from the Greek *phlogistos*, "to burn") is given off whenever something burns.
2. Wood and charcoal are particularly rich in phlogiston, because they leave very little ash when they burn. (Candles must be almost pure phlogiston, because they leave no ash when they burn.)
3. Because they are found in nature, salts must be simpler than metals.
4. Metals form salts by giving off phlogiston.

$$\text{Metal} \longrightarrow \text{salt} + \text{phlogiston}$$

5. Metals can be made by adding phlogiston to salts.

$$\text{Salt} + \text{phlogiston} \longrightarrow \text{metal}$$

6. Because charcoal is rich in phlogiston, heating calxes in the presence of charcoal produces metals.

This model was remarkably successful. It explained why metals have similar properties—they all contained phlogiston. It explained the relationship between metals and their salts—they were related by the gain or loss of phlogiston. It explained why a candle goes out when placed in a bell jar—the air eventually becomes saturated with phlogiston.

There was only one problem with the phlogiston theory. As early as 1630, Jean Rey had noted that tin gains weight when it forms a salt. From our point of view, this seems to be a fatal flaw: If phlogiston is given off when a metal forms a salt, why does the salt weigh more than the metal? It didn't bother chemists at the time, however. Stahl explained this observation by suggesting that the weight increased because air entered the metal to fill the vacuum left after the phlogiston escaped.

The phlogiston theory was the basis for research in chemistry for most of the 18th century. It was not until 1772 that Antoine Lavoisier noted that nonmetals gain large amounts of weight when burned in air. The weight of phosphorus, for example, increases by a factor of about 2.3. The magnitude of this change led Lavoisier to conclude that phosphorus must combine with something in air when it burns. This conclusion was reinforced by the observation that when phosphorus burns in a limited amount of air, the volume of the air decreases by a factor of 1/5.

Lavoisier proposed the name *oxygene* (literally, "acid-former") for the substance absorbed from air when a compound burns, because the products of the combustion of nonmetals such as phosphorus are acids.

$$P_4(s) + 5\ O_2(g) \longrightarrow P_4O_{10}(s)$$
$$P_4O_{10}(s) + 6\ H_2O(aq) \longrightarrow 4\ H_3PO_4(aq)$$

Lavoisier's oxygen theory of combustion was eventually accepted, and chemists began to describe any reaction between an element or compound and oxygen as *oxidation*. The reaction between magnesium metal and oxygen, for example, involves the oxidation of magnesium.

$$2\ Mg(s) + O_2(g) \longrightarrow 2\ MgO(s)$$

By the turn of the 20th century, it seemed that all oxidation reactions had one thing in common—oxidation always seemed to involve the loss of electrons. Chemists therefore developed a model for these reactions that focused on the transfer of electrons. Magnesium metal, for example, was thought to lose electrons to form Mg^{2+} ions when it reacted with oxygen. By convention, the element or compound that gained these electrons was said to undergo *reduction*. In this case, O_2 molecules were said to be reduced to form O^{2-} ions.

A classic demonstration of oxidation-reduction reactions involves placing a piece of copper wire into an aqueous solution of the Ag^+ ion, as shown in Figure 18.1. The reaction involves the net transfer of electrons from copper metal to Ag^+ ions to produce silver metal and Cu^{2+} ions.

$$Cu(s) + 2\ Ag^+(aq) \longrightarrow Cu^{2+}(aq) + 2\ Ag(s)$$

The Cu^{2+} ions formed in this reaction are responsible for the light blue color of the solution. Their presence can be confirmed by adding ammonia to this solution to form the deep blue $Cu(NH_3)_4^{2+}$ complex ion, as shown in Figure 18.2.

Chemists eventually recognized that oxidation-reduction reactions don't always involve the transfer of electrons. The product of the reaction between carbon and oxygen, for example, is a covalent molecule—CO_2—not an ionic compound composed of C^{4+} and O^{2-} ions.

$$C(s) + O_2(g) \longrightarrow CO_2(g)$$

They therefore developed the concept of oxidation number to extend the idea of oxidation and reduction to reactions in which electrons are not really gained or lost.

The best model of oxidation-reduction reactions is based on the following definitions, first introduced in Section 7.11.

> **Oxidation** occurs when the oxidation number of an atom becomes **larger**.
>
> **Reduction** occurs when the oxidation number of an atom becomes **smaller**.

According to this model, carbon is oxidized when it combines with oxygen, because the oxidation number of the carbon increases from 0 to +4. Oxygen is reduced in this reaction, because its oxidation number decreases from 0 to -2.

$$C(s) + O_2(g) \longrightarrow CO_2(g)$$

FIG. 18.1 Oxidation-reduction reactions can involve the direct transfer of electrons from one atom, ion, or molecule to another. In this case, Ag^+ ions come in contact with the surface of a piece of copper metal and are reduced to silver metal, which plates out of the solution. At the same time, the copper metal is oxidized to form Cu^{2+} ions, which are released into the solution.

FIG. 18.2 We can demonstrate the fact that Cu^{2+} ions are released when Ag^+ ions react with copper metal by adding ammonia to this solution to form the characteristic deep blue $Cu(NH_3)_4^{2+}$ complex ion.

The burning of natural gas in a petroleum flare stack is an example of an oxidation-reduction reaction.

18.2 OXIDATION-REDUCTION REACTIONS

We find examples of oxidation-reduction, or ***redox,*** reactions almost every time we analyze the chemical reactions used as sources of either heat or work. On the macroscopic scale, the burning of natural gas is an example of an oxidation-reduction reaction.

$$CH_4(g) + 2 O_2(g) \longrightarrow CO_2(g) + 2 H_2O(g)$$

At the microscopic level, oxidation-reduction is involved in the sequence of reactions our bodies use to burn sugars, such as glucose.

$$C_6H_{12}O_6(aq) + 6 O_2(g) \longrightarrow 6 CO_2(g) + 6 H_2O(l)$$

and the fatty acids in the fats we eat.

$$CH_3(CH_2)_{16}CO_2H + 26 O_2(g) \longrightarrow 18 CO_2(g) + 18 H_2O(g)$$

We don't have to restrict ourselves to reactions that give off energy to find examples of oxidation-reduction. Silver metal tarnishes when it is oxidized by trace quantities of H_2S or SO_2 in the atmosphere, or when it comes in contact with foods, such as eggs, that are rich in sulfur compounds.

$$4 Ag(s) + 2 H_2S(g) + O_2(g) \longrightarrow 2 Ag_2S(s) + 2 H_2O(g)$$

Fortunately, the film of Ag_2S that collects on the metal surface forms a protective coating that slows down further oxidation of the silver metal.

There are two ways of reversing the tarnishing process. We can polish the silver with a complexing agent, such as thiourea (H_2NCSNH_2). This reagent forms a soluble complex ion with the Ag^+ ion, which allows the Ag_2S film to be washed off the surface of the metal. Alternatively, we can reduce the Ag^+ ions back to silver metal. One way of doing this is to wrap the object in aluminum foil and immerse it in a salt solution. According to the table of relative reducing strength in Section 7.14, aluminum is a much better reducing agent than silver. Aluminum therefore reduces the Ag^+ ions in the Ag_2S to silver metal.

$$Al(s) + 3 Ag^+(aq) \longrightarrow Al^{3+}(aq) + 3 Ag(s)$$

The tarnishing of silver is just one example of a broad class of oxidation-reduction reactions of metals that fall under the general heading of ***corrosion.*** Another example is the series of reactions that occur when iron or steel rusts. In theory, iron should react with oxygen to form a mixture of iron(II) and iron(III) oxides.

$$2 Fe(s) + O_2(g) \longrightarrow 2 FeO(s)$$
$$2 Fe(s) + 3 O_2(g) \longrightarrow Fe_2O_3(s)$$

In theory, iron is a better reducing agent than water, and it should reduce water to form an aqueous solution of Fe^{2+} ions and H_2 gas.

$$Fe(s) + 2 H_2O(l) \longrightarrow Fe^{2+}(aq) + 2 OH^-(aq) + H_2(g)$$

In practice, these reactions are so slow they can be ignored.

In the presence of both oxygen and water, however, iron is oxidized to give a hydrated form of iron(II) oxide, as shown in Figure 18.3.

$$2 Fe(s) + O_2(aq) + 2 H_2O(l) \longrightarrow FeO \cdot H_2O(s)$$

Because this compound has the same empirical formula as $Fe(OH)_2$, it is often

The tarnish that forms on silver is also an example of an oxidation-reduction reaction.

mistakenly called iron(II), or ferrous, hydroxide. The $FeO \cdot H_2O$ formed in this reaction is further oxidized by O_2 dissolved in water to give a hydrated form of iron(III), or ferric, oxide.

$$4\ FeO \cdot H_2O(s) + O_2(aq) + 2\ H_2O(l) \longrightarrow 2\ Fe_2O_3 \cdot 3\ H_2O(s)$$

To further complicate matters, $FeO \cdot H_2O$ formed at the metal surface combines with $Fe_2O_3 \cdot 3\ H_2O$ to give a hydrated form of magnetic iron oxide.

$$FeO \cdot H_2O(s) + Fe_2O_3 \cdot 3\ H_2O(s) \longrightarrow Fe_3O_4 \cdot n\ H_2O(s)$$

Because these reactions occur only in the presence of both water and oxygen, cars tend to rust where water collects. Furthermore, because the simplest way of preventing iron from rusting is to coat the metal so that it doesn't come in contact with water, cars were originally painted for only one reason — to slow down the formation of rust.

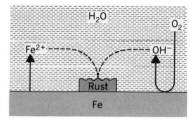

Rust forms on iron metal that is exposed to both oxygen and water.

18.3 ASSIGNING OXIDATION NUMBERS

The key to recognizing oxidation-reduction reactions is the ability to assign oxidation numbers. It is therefore a good idea to take another look at the rules for assigning oxidation numbers introduced in Section 2.13. The concept of *oxidation number* was defined in Section 2.13 as follows.

> **The oxidation number of an atom is equal to the charge that would be present on the atom if the compound was composed of ions.**

If NaCl is assumed to consist of Na^+ and Cl^- ions, the oxidation numbers of the sodium and chlorine atoms are $+1$ and -1, respectively. If CH_4 is assumed to contain C^{4-} and H^+ ions, the oxidation numbers of the carbon and hydrogen atoms are -4 and $+1$.

It doesn't matter whether the compound actually contains ions or not. The oxidation number is the charge an atom would have *if* the compound was ionic. The concept of oxidation number is nothing more than a bookkeeping system used to keep track of electrons in chemical reactions. This system is based on a few rules, summarized in Table 18.1.

Any set of rules, no matter how good, will only get you so far. From that point on, you have to rely on a combination of common sense and prior knowledge. Questions to keep in mind while assigning oxidation numbers include the following. Are there any recognizable ions hidden in the molecule? Does the oxidation number make sense in terms of the known electron configuration of the atom?

FIG. 18.3 The reaction that occurs when iron rusts is not a direct reaction between iron metal and oxygen gas to form Fe_2O_3. At the metal surface, the iron reacts with oxygen dissolved in water to give a hydrated form of iron(II) oxide, $FeO \cdot H_2O$. This material migrates across the metal surface until it comes in contact with OH^- ions produced by the reduction of the O_2 gas dissolved in the water. At this point, it is oxidized further to give a hydrated form of iron(III) oxide, $Fe_2O_3 \cdot 3\ H_2O$. The iron(II) oxide and iron(III) oxide can combine to give a hydrated form of magnetic iron oxide, $Fe_3O_4 \cdot n\ H_2O$.

Exercise 18.1

Determine the oxidation number of each element in the following compounds.

(a) BaO_2 (b) $(NH_4)_2MoO_4$ (c) $HgS_2{}^{2-}$ (d) $Na_3Co(NO_2)_6$ (e) CS_2

Solution

(a) If the oxidation number of the oxygen in BaO_2 were -2, the oxidation number of the barium would have to be $+4$. But elements in Group IIA can't form $+4$ ions. This compound must therefore be barium peroxide, $[Ba^{2+}][O_2{}^{2-}]$. Barium is therefore $+2$, and oxygen is -1.

(b) $(NH_4)_2MoO_4$ contains the NH_4^+ ion, in which hydrogen is $+1$ and nitrogen is -3. Because there are two NH_4^+ ions, the other half of the compound must be an MoO_4^{2-} ion, in which molybdenum is $+6$ and oxygen is -2.

(c) Sulfur, in Group VIA, tends to form -2 ions. The HgS_2^{2-} complex ion therefore contains mercury in the $+2$ oxidation state bound to a pair of sulfur ions with oxidation numbers of -2.

(d) Sodium is in the $+1$ oxidation state in all of its compounds. This compound therefore contains the $[Co(NO_2)_6]^{3-}$ complex ion. Those familiar with the common ions introduced in Chapter 2 should recognize the NO_2^-, or nitrite, ion, in which nitrogen is $+3$ and oxygen is -2. The oxidation state of the cobalt atom is therefore $+3$.

(e) The most electronegative element in a compound always has the negative oxidation number. Since sulfur tends to form -2 ions, the oxidation number of the sulfur in CS_2 is -2, and that of the carbon is $+4$.

TABLE 18.1

Rules for Assigning Oxidation Numbers

1. The oxidation number of an atom is zero in any neutral substance that contains atoms of only one element. The oxygen atoms in O_2 and O_3, the phosphorus atoms in P_4, and the sulfur atoms in S_8, for example, all have oxidation numbers of zero.
2. The oxidation number of a positive or negative ion that contains only a single atom is equal to the charge on the ion. For example, $Na^+ = +1$, $Mg^{2+} = +2$, $Cl^- = -1$, $O^{2-} = -2$, and so on.
3. When hydrogen is combined with another nonmetal, its oxidation number is $+1$. Hydrogen is therefore $+1$ in the following compounds: CH_4, NH_3, PH_3, H_2O, H_2S, HF, HCl, and HBr.
4. When hydrogen is combined with a metal, its oxidation number is -1. Hydrogen is therefore -1 in the following compounds: LiH, NaH, CaH_2, and $LiAlH_4$.
5. The metals in Group IA (Li, Na, K, Rb, Cs, and Fr) always form compounds in the $+1$ oxidation state.
6. The metals in Group IIA (Be, Mg, Ca, Sr, Ba, and Ra) always form compounds in the $+2$ oxidation state.
7. Oxygen almost always has an oxidation number of -2. Exceptions include molecules that contain oxygen—oxygen bonds, such as O_2 and O_3, in which the oxidation number of oxygen is zero, and H_2O_2 and the O_2^{2-} ion, in which the oxidation number of oxygen is -1.
8. The nonmetals in Group VIIA (F, Cl, Br, I, and At) usually form compounds in the -1 oxidation state.
9. The sum of the oxidation numbers of the atoms in a molecule or ion is equal to the charge on the molecule or ion.
10. When in doubt about which element in a compound is positive and which is negative, assign the negative oxidation number to the most electronegative atom. Thus, in SO_2, sulfur is $+4$ and oxygen is -2.

18.4 RECOGNIZING OXIDATION-REDUCTION REACTIONS

Chemical reactions, no matter how simple or complex, can usually be divided into two classes: oxidation-reduction or acid-base. Acid-base reactions can involve the transfer of an H^+ ion from a Brønsted acid to a Brønsted base.

$$CH_3CO_2H(aq) + OH^-(aq) \rightleftharpoons CH_3CO_2^-(aq) + H_2O(l)$$

Brønsted acid   Brønsted base

They can also involve the sharing of a pair of electrons by an electron-pair donor (Lewis base) and an electron-pair acceptor (Lewis acid).

$$Co^{3+}(aq) + 6\ NO_2^-(aq) \rightleftharpoons Co(NO_2)_6^{3-}(aq)$$

Lewis acid   Lewis base

Oxidation-reduction reactions can involve the transfer of one or more electrons.

$$Al(s) + 3\ Ag^+(aq) \longrightarrow Al^{3+}(aq) + 3\ Ag(s)$$

They can also involve the transfer of oxygen, hydrogen, or even halogen atoms.

$$2\ Mg(s) + O_2(g) \longrightarrow 2\ MgO(s)$$
$$CO_2(g) + H_2(g) \rightleftharpoons CO(g) + H_2O(g)$$
$$SF_4(g) + F_2(g) \longrightarrow SF_6(g)$$

Fortunately, there is an almost foolproof method of distinguishing between acid-base and redox reactions.

Reactions in which none of the atoms undergoes a change in oxidation number are acid-base, or *metathesis*, reactions.

Reactions in which at least one atom undergoes a change in oxidation number are oxidation-reduction reactions.

There is no change in the oxidation number of any atom in either of the foregoing examples of acid-base reactions.

$$CH_3CO_2H + OH^- \rightleftharpoons CH_3CO_2^- + H_2O$$

$$Co^{3+} + 6\ NO_2^- \rightleftharpoons Co(NO_2)_6^{3-}$$

The word *metathesis* literally means "interchange" or "transposition," and it is often used to describe changes that occur in the order of letters or sounds in a word as a language develops—for example, the transposition, or metathesis, that transformed the Old English word *brid* into the modern word *bird*. In chemistry, metathesis is used to describe reactions that interchange atoms or groups of atoms between molecules.

At least one atom undergoes a change in oxidation number in every oxidation-reduction reaction.

$$Al(s) + 3\ Ag^+(aq) \longrightarrow Al^{3+}(aq) + 3\ Ag(s)$$

$$\begin{array}{cccc} 0 & +1 & +3 & 0 \end{array}$$

$$2\ Mg(s) + O_2(g) \longrightarrow 2\ MgO(s)$$

$$\begin{array}{cccc} 0 & 0 & +2 & -2 \end{array}$$

$$CO_2(g) + H_2(g) \longrightarrow CO(g) + H_2O(g)$$

$$\begin{array}{cccc} +4 & 0 & +2 & +1 \end{array}$$

$$SF_4(g) + F_2(g) \longrightarrow SF_6(g)$$

$$\begin{array}{cccc} +4 & 0 & +6 & -1 \end{array}$$

Exercise 18.2

Classify each of the following as either an acid-base or an oxidation-reduction reaction.

(a) $Hg_2^{2+}(aq) + 2\ OH^-(aq) \longrightarrow Hg_2O(s) + H_2O(l)$

(b) $Hg_2^{2+}(aq) + Sn^{2+}(aq) \longrightarrow 2\ Hg(s) + Sn^{4+}(aq)$

(c) $Hg_2^{2+}(aq) + H_2S(aq) \longrightarrow Hg(s) + HgS(s) + 2\ H^+(aq)$

(d) $Hg_2CrO_4(s) + 2\ OH^-(aq) \longrightarrow Hg_2O(s) + CrO_4^{2-}(aq) + H_2O(l)$

(e) $HgS(s) + 2\ S_2^{2-}(aq) \rightleftharpoons HgS_2^{2-}(aq) + S_3^{2-}(aq)$

Solution

(a) Acid-base. Mercury is in the $+1$ oxidation state in both the Hg_2^{2+} ion and in $[Hg_2^{2+}][O^{2-}]$.

(b) Oxidation-reduction. Mercury is reduced from the $+1$ to the 0 oxidation state, while tin is oxidized from $+2$ to $+4$.

(c) Oxidation-reduction. This is a disproportionation reaction (see Section 10.8) in which mercury is simultaneously reduced from $+1$ to 0 and oxidized from $+1$ to $+2$.

(d) Acid-base. Mercury is in the $+1$ oxidation state in both $[Hg_2^{2+}][CrO_4^{2-}]$ and $[Hg_2^{2+}][O^{2-}]$.

(e) Oxidation-reduction. Mercury is the $+2$ oxidation state in both $[Hg^{2+}][S^{2-}]$ and the $([Hg^{2+}][S^{2-}]_2)^{2-}$ complex ion. But part of the sulfur in the S_2^{2-} ion is reduced to S^{2-}, and part is oxidized to S_3^{2-}.

$$HgS(s) + 2\ S_2^{2-}(aq) \rightleftharpoons HgS_2^{2-}(aq) + S_3^{2-}(aq)$$

$$\begin{array}{ccc} -1 & -2 & -\tfrac{2}{3} \end{array}$$

18.5 BALANCING OXIDATION-REDUCTION EQUATIONS

A trial-and-error approach to balancing chemical equations was introduced in Section 3.13. This approach involves playing with the equation—adjusting the ratio of the reactants and products—until the following goals have been achieved.

GOALS FOR BALANCING CHEMICAL EQUATIONS

1. The same number of atoms of each element is found on both sides of the equation. In other words, mass is conserved.

2. The sum of the positive and negative charges is the same on both sides of the equation. Electrons are neither created nor destroyed, and therefore charge is conserved.

We can illustrate this approach by applying it to the first step in the Ostwald process for converting ammonia into nitric acid (see Section 10.6). The following is the skeleton equation for this reaction.

$$\text{\_\_\_\_ } NH_3(g) + \text{\_\_\_\_ } O_2(g) \longrightarrow \text{\_\_\_\_ } NO(g) + \text{\_\_\_\_ } H_2O(g)$$

We might start by noting that there are an odd number of hydrogen atoms on the left and an even number on the right side of this equation. We can therefore try to balance the hydrogen atoms, as follows.

$$2\ NH_3(g) + \text{\_\_\_\_ } O_2(g) \longrightarrow \text{\_\_\_\_ } NO(g) + 3\ H_2O(g)$$

But this upsets the balance of nitrogen atoms, so we double the amount of NO produced.

$$2\ NH_3(g) + \text{\_\_\_\_ } O_2(g) \longrightarrow 2\ NO(g) + 3\ H_2O(g)$$

The products contain an odd number of oxygen atoms, so we double all of the coefficients as a first step toward balancing the number of oxygen atoms.

$$4\ NH_3(g) + \text{\_\_\_\_ } O_2(g) \longrightarrow 4\ NO(g) + 6\ H_2O(g)$$

We then add a total of 10 oxygen atoms to the reactants to balance the 10 oxygen atoms among the products.

$$4\ NH_3(g) + 5\ O_2(g) \longrightarrow 4\ NO(g) + 6\ H_2O(g)$$

Finally, we check to see if the equation is truly balanced. Both sides of the equation have the same number of atoms of each element, and both sides have the same net electric charge.

There are two situations in which relying on trial and error can get you into trouble. Sometimes the equation is too complex to be solved by trial and error within a reasonable amount of time. Look at the following equations, for example.

$$3\ Cu(s) + 8\ HNO_3(aq) \longrightarrow 3\ Cu^{2+}(aq) + 2\ NO(g) + 6\ NO_3^-(aq) + 4\ H_2O(l)$$
$$2\ Mn^{2+}(aq) + 5\ BiO_3^-(aq) + 14\ H^+(aq) \longrightarrow 2\ MnO_4^-(aq) + 5\ Bi^{3+}(aq) + 7\ H_2O(l)$$

Other times, more than one balanced chemical equation can be written. The following are just a few of the balanced equations that can be written for the reaction between the permanganate ion and hydrogen peroxide, for example.

$$2\ MnO_4^-(aq) + H_2O_2(aq) + 6\ H^+(aq) \longrightarrow 2\ Mn^{2+}(aq) + 3\ O_2(g) + 4\ H_2O(l)$$
$$2\ MnO_4^-(aq) + 3\ H_2O_2(aq) + 6\ H^+(aq) \longrightarrow 2\ Mn^{2+}(aq) + 4\ O_2(g) + 6\ H_2O(l)$$

$$2\ MnO_4^-(aq) + 5\ H_2O_2(aq) + 6\ H^+(aq) \longrightarrow 2\ Mn^{2+}(aq) + 5\ O_2(g) + 8\ H_2O(l)$$
$$2\ MnO_4^-(aq) + 7\ H_2O_2(aq) + 6\ H^+(aq) \longrightarrow 2\ Mn^{2+}(aq) + 6\ O_2(g) + 10\ H_2O(l)$$

Equations such as these have to be balanced by a more systematic approach than trial and error. Two common alternatives are the oxidation number and half-reaction methods.

18.6 THE OXIDATION NUMBER METHOD OF BALANCING REDOX EQUATIONS

The *oxidation number method* of balancing redox equations is based on the fact that electrons are neither created nor destroyed in a chemical reaction. Therefore, the total change in the oxidation numbers of the atoms oxidized in the reaction must balance the total change in the oxidation numbers of the atoms reduced. This method involves the following steps.

STEP 1: *Write a skeleton equation for the reaction.* The skeleton equation for the reaction between ammonia and chlorine to form nitrogen and hydrogen chloride, for example, is written as follows.

$$NH_3 + Cl_2 \longrightarrow N_2 + HCl$$

(The symbols for the states of the reactants and products are often ignored until the final equation is written.)

STEP 2: *Assign oxidation numbers to atoms on both sides of the equation.*

$$\underset{-3\ +1}{NH_3} + \underset{0}{Cl_2} \longrightarrow \underset{0}{N_2} + \underset{+1\ -1}{HCl}$$

STEP 3: *Determine which atoms are oxidized and which are reduced.*

$$\underset{-3\ +1}{NH_3} + \underset{0}{Cl_2} \longrightarrow \underset{0}{N_2} + \underset{+1\ -1}{HCl}$$

oxidation

reduction

STEP 4: *Calculate the total change in oxidation number for the atoms oxidized and the total change for the atoms reduced.* Nitrogen is oxidized from the -3 to the 0 oxidation state in this reaction. Because two molecules of NH_3 are consumed for every N_2 molecule produced, the change in the oxidation number of the nitrogen atoms is 2×3, or 6.

Oxidation: $\quad \underset{-3}{2\ NH_3} \longrightarrow \underset{0}{N_2} \quad$ change in oxidation number
$$2 \times 3 = 6$$

Chlorine is reduced from the 0 to the -1 oxidation state in the reaction. Because two molecules of HCl are produced for each molecule of Cl_2 consumed, the change in the oxidation number of the chlorine atoms is 2×-1, or -2.

Reduction: $\quad \underset{0}{Cl_2} \longrightarrow \underset{-1}{2\ HCl} \quad$ change in oxidation number
$$2 \times -1 = -2$$

STEP 5: *Adjust the coefficients of the equation so that the change in oxidation number* *during oxidation balances the change during reduction.* If the total change in oxidation number when two NH_3 molecules are oxidized is $+6$, the total change in oxidation number when Cl_2 is reduced must be -6. To achieve this balance, we can assume that three Cl_2 molecules are reduced every time two NH_3 molecules are oxidized.

$$2\ NH_3 + 3\ Cl_2 \longrightarrow N_2 + 6\ HCl$$

STEP 6: *Balance the remainder of the equation by inspection, if necessary.* The equation generated in Step 5 is balanced. Both sides contain the same number of nitrogen, hydrogen, and chlorine atoms, and both sides have the same charge. The following is therefore the balanced equation for this reaction.

$$2\ NH_3(g) + 3\ Cl_2(g) \longrightarrow N_2(g) + 6\ HCl(g)$$

When the reaction occurs in water, however, it may be necessary to add or subtract H_2O molecules, H^+ ions, or OH^- ions from one side of the equation or the other to obtain a balanced equation. Exercise 18.3 gives an example of this procedure.

Exercise 18.3

Use the oxidation number method to write a balanced equation for the reaction between bromine and base to form the bromide and bromate ions.

Solution

STEP 1: *Write a skeleton equation for the reaction.*

$$Br_2 + OH^- \longrightarrow Br^- + BrO_3^-$$

STEP 2: *Assign oxidation numbers to atoms on both sides of the equation.*

$$Br_2 + OH^- \longrightarrow Br^- + BrO_3^-$$
$$0 \quad\ -2\ +1 \quad\ -1 \quad\ +5\ -2$$

STEP 3: *Determine which atoms are oxidized and which are reduced.*

$$Br_2 + OH^- \longrightarrow Br^- + BrO_3^-$$
$$0 \quad\ -2\ +1 \quad\ -1 \quad\ +5\ -2$$

reduction

oxidation

STEP 4: *Calculate the total change in oxidation number for the atoms oxidized and the total change for the atoms reduced.* Two BrO_3^- ions are produced for each molecule of Br_2 oxidized. Because each bromine atom is oxidized from 0 to $+5$, the total change in oxidation number is 2×5, or 10.

Oxidation: $Br_2 \longrightarrow 2\ BrO_3^-$ change in oxidation number
$$0 +5 2 \times 5 = 10$$

Two Br^- ions are produced for each Br_2 molecule reduced. Because each bromine atom is reduced from 0 to -1, the total change in oxidation number is 2×-1, or -2.

Reduction: $Br_2 \longrightarrow 2\ Br^-$ change in oxidation number
$$0 -1 2 \times -1 = -2$$

STEP 5: *Adjust the coefficients so that the change in oxidation number during oxidation*

An aqueous solution of bromine dissolved in water has the characteristic color of elemental bromine.

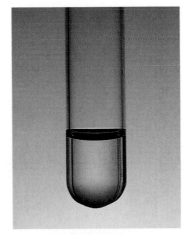

In the presence of strong base, the characteristic color of an aqueous solution of Br_2 fades as the bromine disproportionates to form the Br^- and BrO_3^- ions.

balances the change during reduction. The total change in oxidation number when a Br_2 molecule is oxidized to a pair of BrO_3^- ions is $+10$. The total change in oxidation number when Br_2 molecules are reduced to Br^- ions must therefore be -10. Thus, five Br_2 molecules must be reduced (5×-2) for every Br_2 molecule oxidized ($+10$).

$$\begin{array}{r} 5(Br_2 \longrightarrow 2\ Br^-) \\ + \underline{1(Br_2 \longrightarrow 2\ BrO_3^-)} \\ 6\ Br_2 \longrightarrow 2\ BrO_3^- + 10\ Br^- \end{array}$$

STEP 6: *Balance the remainder of the equation by inspection, if necessary.* The equation generated in Step 5 is not balanced.

$$6\ Br_2 \longrightarrow 2\ BrO_3^- + 10\ Br^-$$

The reaction occurs in a basic solution, however, which means that we can add OH^- ions and H_2O molecules to either side of the equation until it is balanced in terms of both charge and mass. The trick is deciding which side of the equation gets the OH^- ions and which gets the water molecules. One way to make this decison is to remember that we need to balance both the number of atoms and the charge. As it now stands, the right side of the equation has a net charge of -12, and the left side has no net charge. One way to balance the charge is to add 12 OH^- ions to the reactants.

$$6\ Br_2 + \mathbf{12\ OH^-} \longrightarrow 2\ BrO_3^- + 10\ Br^-$$

Six of the oxygen atoms in the OH^- ions end up in the BrO_3^- ions. What happens to the other 6 oxygen atoms and the 12 hydrogen atoms? The most reasonable answer is that they combine to form 6 H_2O molecules.

$$6\ Br_2 + 12\ OH^- \longrightarrow 2\ BrO_3^- + 10\ Br^- + \mathbf{6\ H_2O}$$

Although this equation appears balanced in terms of both charge and mass, it is not quite correct. The best answer to this exercise would be the equation with the smallest possible coefficients, so we divide all of the coefficients by 2.

$$3\ Br_2 + 6\ OH^- \longrightarrow BrO_3^- + 5\ Br^- + 3\ H_2O$$

We can now complete the exercise by identifying the states of the reactants and products, as follows.

$$3\ Br_2(aq) + 6\ OH^-(aq) \longrightarrow BrO_3^-(aq) + 5\ Br^-(aq) + 3\ H_2O(l)$$

18.7 THE HALF-REACTION METHOD OF BALANCING REDOX EQUATIONS

The second approach to balancing oxidation-reduction reactions is based on the assumption that we can divide these reactions into separate oxidation and reduction half-reactions. We balance these half-reactions individually and then combine them by assuming, once again, that electrons are neither created nor destroyed in a chemical reaction.

When we used the oxidation number method, we combined oxidation and reduction so that electrons were conserved and then balanced the equation. When we use the ***half-reaction method,*** we balance the two halves of the equation and then combine these half-reactions so that electrons are conserved. The first steps of both approaches are the same.

STEP 1: *Write a skeleton equation for the reaction.* The skeleton equation for the reaction between the Cu^{2+} and I^- ions to form CuI and the I_3^- ion, for example, is written as follows.

$$Cu^{2+} + I^- \longrightarrow CuI + I_3^-$$

STEP 2: *Assign oxidation numbers to atoms on both sides of the equation.*

$$Cu^{2+} + I^- \longrightarrow CuI + I_3^-$$
$$+2 \quad\; -1 \qquad +1 \; -1 \quad\; -\tfrac{1}{3}$$

STEP 3: *Determine which atoms are oxidized and which are reduced.*

$$Cu^{2+} + I^- \longrightarrow CuI + I_3^-$$
$$+2 \quad\; -1 \qquad +1 \; -1 \quad\; -\tfrac{1}{3}$$

reduction

oxidation

STEP 4: *Divide the reaction into oxidation and reduction half-reactions, and balance these half-reactions one at a time.* This reaction can be arbitrarily divided into two half-reactions. One half-reaction describes what happens during oxidation.

$$\text{Oxidation:} \qquad I^- \longrightarrow I_3^-$$

The other describes the reduction half of the reaction.

$$\text{Reduction:} \qquad Cu^{2+} \longrightarrow CuI$$

It doesn't matter which half-reaction we balance first. Let's start with oxidation. Our goal is to balance this half-reaction in terms of both charge and mass. We start by balancing the number of iodine atoms on both sides of the equation.

$$\text{Oxidation:} \qquad 3\, I^- \longrightarrow I_3^-$$

We then balance the charge by noting that two electrons must be removed from three I^- ions to produce an I_3^- ion.

$$\text{Oxidation:} \qquad 3\, I^- \longrightarrow I_3^- + 2\, e^-$$

We can now turn to the reduction half-reaction. To balance the number of atoms on both sides of this equation, we need to add an iodide ion to the reactants.

$$\text{Reduction:} \qquad Cu^{2+} + I^- \longrightarrow CuI$$

We can now balance the charge by noting that it takes one electron to reduce copper from the $+2$ to the $+1$ oxidation state.

$$\text{Reduction:} \qquad Cu^{2+} + I^- + e^- \longrightarrow CuI$$

STEP 5: *Combine the two half-reactions so that electrons are neither created nor destroyed.* Two electrons are given off in the oxidation half-reaction.

$$\text{Oxidation:} \qquad 3\, I^- \longrightarrow I_3^- + 2\, e^-$$

One electron is consumed in the reduction half-reaction.

$$\text{Reduction:} \qquad Cu^{2+} + I^- + e^- \longrightarrow CuI$$

We can combine these half-reactions so that electrons are conserved by multiplying the reduction half-reaction by 2.

$$\frac{(3\ I^- \longrightarrow I_3^- + 2\ e^-)}{+\ 2(Cu^{2+} + I^- + e^- \longrightarrow CuI)}$$
$$2\ Cu^{2+} + 5\ I^- \longrightarrow 2\ CuI + I_3^-$$

STEP 6: *Balance the remainder of the equation by inspection, if necessary.* Step 5 produces an equation that is balanced in terms of both charge and mass.

$$2\ Cu^{2+}(aq) + 5\ I^-(aq) \longrightarrow 2\ CuI(s) + I_3^-(aq)$$

As you will see in the next section, we sometimes have to fine-tune the balanced equation by adding H_2O molecules, H^+ ions, or OH^- ions to one side or the other.

Exercise 18.4

The amount of Cu^{2+} ion in a solution can be determined as follows. Any Cu^{2+} ion in the solution is first allowed to react with excess I^- to form a mixture of CuI and I_3^-.

$$2\ Cu^{2+}(aq) + 5\ I^-(aq) \longrightarrow 2\ CuI(s) + \mathbf{I_3^-(aq)}$$

The triiodide ion produced in this reaction is then titrated with thiosulfate. Use the half-reaction method to write a balanced equation for the reaction between I_3^- and $S_2O_3^{2-}$.

$$\mathbf{I_3^-(aq)} + S_2O_3^{2-}(aq) \longrightarrow I^-(aq) + S_4O_6^{2-}(aq)$$

Solution

STEP 1: *Write a skeleton equation for the reaction.*

$$I_3^- + S_2O_3^{2-} \longrightarrow I^- + S_4O_6^{2-}$$

STEP 2: *Assign oxidation numbers to atoms on both sides of the equation.*

$$I_3^- + S_2O_3^{2-} \longrightarrow I^- + S_4O_6^{2-}$$
$$-\tfrac{1}{3} \quad +2\ -2 \qquad -1 \quad +2\tfrac{1}{2}\ -2$$

STEP 3: *Determine which atoms are oxidized and which are reduced.*

$$I_3^- + S_2O_3^{2-} \longrightarrow I^- + S_4O_6^{2-}$$
$$-\tfrac{1}{3} \quad +2\ -2 \qquad -1 \quad +2\tfrac{1}{2}\ -2$$

reduction

oxidation

STEP 4: *Divide the reaction into oxidation and reduction half-reactions, and balance these half-reactions one at a time.* This reaction can be divided into the following half-reactions.

Oxidation: $S_2O_3^{2-} \longrightarrow S_4O_6^{2-}$

$+2 \qquad\qquad +2\frac{1}{2}$

Reduction: $I_3^- \longrightarrow I^-$

$-\frac{1}{3} \qquad\qquad -1$

The reduction half-reaction is the reverse of the half-reaction seen earlier in this section.

Reduction: $I_3^- + 2\,e^- \longrightarrow 3\,I^-$

To balance the oxidation half-reaction, we note that combining a pair of $S_2O_3^{2-}$ ions would produce an ion with a -4 charge. To get an $S_4O_6^{2-}$ ion, we have to remove two electrons.

Oxidation: $2\,S_2O_3^{2-} \longrightarrow S_4O_6^{2-} + 2\,e^-$

STEP 5: *Combine these half-reactions so that electrons are neither created nor destroyed.* Two electrons are given off in the oxidation half-reaction, and two electrons are picked up in the reduction half-reaction. We can therefore obtain a balanced chemical equation by simply combining these half-reactions.

$$(2\,S_2O_3^{2-} \longrightarrow S_4O_6^{2-} + 2\,e^-)$$
$$+ (I_3^- + 2\,e^- \longrightarrow 3\,I^-)$$
$$\overline{I_3^- + 2\,S_2O_3^{2-} \longrightarrow 3\,I^- + S_4O_6^{2-}}$$

STEP 6: *Balance the remainder of the equation by inspection, if necessary.* Since the overall equation is already balanced in terms of both charge and mass, we simply reintroduce the symbols describing the states of the reactants and products.

$$I_3^-(aq) + 2\,S_2O_3^{2-}(aq) \longrightarrow 3\,I^-(aq) + S_4O_6^{2-}(aq)$$

18.8 REDOX REACTIONS IN ACIDIC SOLUTIONS

Some might argue that we don't need to use oxidation numbers or half-reactions to balance equations such as those in Section 18.7, because they can be balanced by trial and error.

$$2\,Cu^{2+}(aq) + 5\,I^-(aq) \longrightarrow 2\,CuI(s) + I_3^-(aq)$$
$$I_3^-(aq) + 2\,S_2O_3^{2-}(aq) \longrightarrow 3\,I^-(aq) + S_4O_6^{2-}(aq)$$

But the oxidation number and half-reaction techniques become indispensable in balancing reactions such as the oxidation of sulfur dioxide by the dichromate ion in acidic solution.

$$SO_2(aq) + Cr_2O_7^{2-}(aq) \xrightarrow{\ H^+\ } SO_4^{2-}(aq) + 2\,Cr^{3+}(aq)$$

The reason why this equation is inherently more difficult to balance has nothing to do with the ratio of moles of SO_2 to moles of $Cr_2O_7^{2-}$. It results from the fact that the solvent takes an active role in both half-reactions.

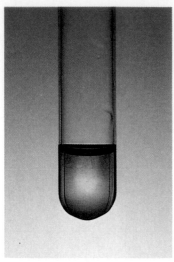

Aqueous solutions of the $Cr_2O_7^{2-}$ ion have a characteristic orange color. When SO_2 gas is bubbled through the solution, this color fades as this ion is reduced to Cr^{3+}.

Exercise 18.5

Use half-reactions to balance the equation for the reaction between sulfur dioxide and the dichromate ion in acidic solution.

$$SO_2(aq) + Cr_2O_7^{2-}(aq) \xrightarrow{H^+} SO_4^{2-}(aq) + 2\ Cr^{3+}(aq)$$

Solution

STEP 1: *Write a skeleton equation for the reaction.*

$$SO_2 + Cr_2O_7^{2-} \longrightarrow SO_4^{2-} + 2\ Cr^{3+}$$

STEP 2: *Assign oxidation numbers to atoms on both sides of the equation.*

$$SO_2\ +\ Cr_2O_7^{2-} \longrightarrow SO_4^{2-} + 2\ Cr^{3+}$$
$$\ +4\ -2\quad +6\quad -2\qquad\qquad +6\ -2\qquad +3$$

STEP 3: *Determine which atoms are oxidized and which are reduced.*

$$SO_2 + Cr_2O_7^{2-} \longrightarrow SO_4^{2-} + 2\ Cr^{3+}$$
$$+4\ -2\quad +6\ -2\qquad\qquad +6\ -2\qquad +3$$

oxidation

reduction

STEP 4: *Divide the reaction into oxidation and reduction half-reactions, and balance these half-reactions.* This reaction can be divided into the following half-reactions.

Oxidation: $$SO_2 \longrightarrow SO_4^{2-}$$
$$\quad +4\qquad\qquad +6$$

Reduction: $$Cr_2O_7^{2-} \longrightarrow Cr^{3+}$$
$$\qquad +6\qquad\qquad +3$$

It doesn't matter which half-reaction we balance first, so let's start with reduction. Because two chromium atoms are reduced from the +6 to the +3 oxidation state, six electrons are consumed in this half-reaction.

Reduction: $$Cr_2O_7^{2-} + 6\ e^- \longrightarrow 2\ Cr^{3+}$$
$$\qquad +6\qquad\qquad\qquad +3$$

What happens to the oxygen atoms when the chromium atoms are reduced? The seven oxygen atoms in the $Cr_2O_7^{2-}$ ions are formally in the -2 oxidation state. If the chromium is reduced to Cr^{3+}, and the oxygen atoms are released into solution in the -2 oxidation state, the oxygen atoms will be present as O^{2-} ions. But it doesn't make sense to write this half-reaction as follows.

Reduction: $$Cr_2O_7^{2-} + 6\ e^- \longrightarrow 2\ Cr^{3+} + 7\ O^{2-}$$

The reaction is being run in an acidic solution, and the O^{2-} ions should immediately react with the H^+ ions in this solution to form water. The following is therefore a more realistic equation for this half-reaction.

Reduction: $$Cr_2O_7^{2-} + 14\ H^+ + 6\ e^- \longrightarrow 2\ Cr^{3+} + 7\ H_2O$$

We can now turn to the oxidation half-reaction and start by noting that two electrons are given off when sulfur is oxidized from the +4 to the +6 oxidation state.

$$\text{Oxidation:} \quad SO_2 \longrightarrow SO_4^{2-} + 2\ e^-$$
$$\phantom{\text{Oxidation:} \quad SO_2 \longrightarrow } {}_{+4} {}_{+6}$$

How can we balance the charge on both sides of this equation? The key to answering this question is remembering that the reaction is being run in acid. Thus, we can add H^+ ions or H_2O molecules to either side of the equation. In this case, we need to add four H^+ ions to the products to give an equation that has the same net charge on both sides.

$$\text{Oxidation:} \quad SO_2 \longrightarrow SO_4^{2-} + 2\ e^- + \mathbf{4\ H^+}$$

We can balance the number of hydrogen and oxygen atoms on both sides of this equation by adding a pair of H_2O molecules to the reactants.

$$\text{Oxidation:} \quad SO_2 + \mathbf{2\ H_2O} \longrightarrow SO_4^{2-} + 2\ e^- + 4\ H^+$$

STEP 5: *Combine the two half-reactions so that electrons are neither created nor destroyed.* Six electrons are consumed in the reduction half-reaction, and two electrons are given off in the oxidation half-reaction. We can combine these half-reactions so that electrons are conserved by multiplying the reduction half-reaction by 3.

$$(Cr_2O_7^{2-} + 14\ H^+ + 6\ e^- \longrightarrow 2\ Cr^{3+} + 7\ H_2O)$$
$$+\ 3(SO_2 + 2\ H_2O \longrightarrow SO_4^{2-} + 2\ e^- + 4\ H^+)$$
$$\overline{Cr_2O_7^{2-} + 3\ SO_2 + 14\ H^+ + 6\ H_2O \longrightarrow}$$
$$2\ Cr^{3+} + 3\ SO_4^{2-} + 12\ H^+ + 7\ H_2O$$

STEP 6: *Balance the remainder of the equation by inspection, if necessary.* Although the equation appears balanced, we are not quite finished with it. We can simplify the equation by subtracting 12 H^+ ions and 6 H_2O molecules from each side. The result is the following balanced equation.

$$Cr_2O_7^{2-}(aq) + 3\ SO_2(aq) + 2\ H^+(aq) \longrightarrow 2\ Cr^{3+}(aq) + 3\ SO_4^{2-}(aq) + H_2O(l)$$

Exercise 18.6

We can determine the molarity of a permanganate ion solution of unknown concentration by titrating this solution with a known amount of oxalic acid until the purple color of the MnO_4^- ion just disappears.

$$H_2C_2O_4(aq) + MnO_4^-(aq) + H^+(aq) \longrightarrow CO_2(g) + Mn^{2+}(aq)$$

Use the half-reaction method to write a balanced equation for this reaction.

Solution
STEP 1: *Write a skeleton equation for the reaction.*

$$H_2C_2O_4 + MnO_4^- + H^+ \longrightarrow CO_2 + Mn^{2+}$$

STEP 2: *Assign oxidation numbers to atoms on both sides of the equation.*

$$H_2C_2O_4 + MnO_4^- + H^+ \longrightarrow CO_2 + Mn^{2+}$$
$${}_{+1\ +3\ -2} {}_{+7\ -2} {}_{+1} {}_{+4\ -2} {}_{+2}$$

Solutions of the MnO_4^- ion have a deep purple color.

STEP 3: *Determine which atoms are oxidized and which are reduced.*

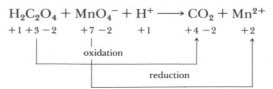

STEP 4: *Divide the reaction into oxidation and reduction half-reactions, and balance these half-reactions.* This reaction can be divided into the following half-reactions.

$$\text{Oxidation:} \qquad \underset{+3}{H_2C_2O_4} \longrightarrow \underset{+4}{CO_2}$$

$$\text{Reduction:} \qquad \underset{+7}{MnO_4^-} \longrightarrow \underset{+2}{Mn^{2+}}$$

We'll balance the reduction half-reaction first. It takes five electrons to reduce manganese from $+7$ to $+2$.

$$\text{Reduction:} \qquad MnO_4^- + 5\ e^- \longrightarrow Mn^{2+}$$

Because the reaction is run in acid, we can add H^+ ions or H_2O molecules to either side of the equation. To balance the charge, we need to add eight H^+ ions to the reactants.

$$\text{Reduction:} \qquad MnO_4^- + 8\ H^+ + 5\ e^- \longrightarrow Mn^{2+}$$

To balance the number of hydrogen and oxygen atoms, we need to add four H_2O molecules to the products.

$$\text{Reduction:} \qquad MnO_4^- + 8\ H^+ + 5\ e^- \longrightarrow Mn^{2+} + 4\ H_2O$$

We can now turn to the oxidation half-reaction. This is a two-electron oxidation, because two carbon atoms are oxidized from the $+3$ to the $+4$ oxidation state.

$$\text{Oxidation:} \qquad \underset{+3}{H_2C_2O_4} \longrightarrow \underset{+4}{2\ CO_2} + 2\ e^-$$

We can balance both charge and mass by noting that two H^+ ions are given off when oxalic acid is oxidized to carbon dioxide.

$$\text{Oxidation:} \qquad H_2C_2O_4 \longrightarrow 2\ CO_2 + 2\ e^- + 2\ H^+$$

STEP 5: *Combine the two half-reactions so that electrons are neither created nor destroyed.* Five electrons are consumed in the reduction half-reaction, and two electrons are given off in the oxidation half-reaction. We can combine these half-reactions so that electrons are conserved by using the lowest common multiple of 5 and 2.

$$2(MnO_4^- + 8\ H^+ + 5\ e^- \longrightarrow Mn^{2+} + 4\ H_2O)$$
$$+ 5(H_2C_2O_4 \longrightarrow 2\ CO_2 + 2\ e^- + 2\ H^+)$$
$$\overline{\begin{array}{l} 2\ MnO_4^- + 16\ H^+ + 5\ H_2C_2O_4 \longrightarrow \\ \qquad 10\ CO_2 + 2\ Mn^{2+} + 8\ H_2O + 10\ H^+ \end{array}}$$

STEP 6: *Balance the remainder of the equation by inspection, if necessary.* We can generate the simplest balanced equation for this reaction by subtracting 10 H^+ ions from both sides of the equation derived in Step 5.

$$2\ MnO_4^-(aq) + 5\ H_2C_2O_4(aq) + 6\ H^+(aq) \longrightarrow 10\ CO_2(g) + 2\ Mn^{2+}(aq) + 8\ H_2O(l)$$

Exercise 18.7

Use the half-reaction method to write a balanced chemical equation for the reaction between iodine and nitric acid to form the iodate ion and nitrogen dioxide.

$$I_2(s) + HNO_3(aq) \longrightarrow IO_3^-(aq) + NO_2(g)$$

Solution

STEP 1: *Write a skeleton equation for the reaction.*

$$I_2 + HNO_3 \longrightarrow IO_3^- + NO_2$$

STEP 2: *Assign oxidation numbers to atoms on both sides of the equation.*

$$I_2 + HNO_3 \longrightarrow IO_3^- + NO_2$$
$$\quad 0 \quad\; +1\,+5\,-2 \quad\;\; +5\,-2 \quad +4\,-2$$

STEP 3: *Determine which atoms are oxidized and which are reduced.*

$$I_2 + HNO_3 \longrightarrow IO_3^- + NO_2$$
$$\quad 0 \quad\; +1\,+5\,-2 \quad\;\; +5\,-2 \quad +4\,-2$$

oxidation

reduction

STEP 4: *Divide the reaction into oxidation and reduction half-reactions, and balance these half-reactions.* The reaction can be divided into the following half-reactions.

$$\text{Oxidation:} \qquad I_2 \longrightarrow IO_3^-$$
$$\qquad\qquad\qquad\quad 0 \qquad\;\; +5$$

$$\text{Reduction:} \qquad HNO_3 \longrightarrow NO_2$$
$$\qquad\qquad\qquad\quad +5 \qquad\;\; +4$$

Each iodine atom loses five electrons when it is oxidized from the 0 to the $+5$ oxidation state. This half-reaction is therefore a 10-electron oxidation.

$$\text{Oxidation:} \qquad I_2 \longrightarrow 2\,IO_3^- + 10\,e^-$$
$$\qquad\qquad\qquad\quad 0 \qquad\;\; +5$$

Because the reaction is run in nitric acid, we can balance this half-reaction by adding H^+ ions or H_2O molecules to either side of the equation. Let's start by adding enough H^+ ions to the products to balance the charge.

$$\text{Oxidation:} \qquad I_2 \longrightarrow 2\,IO_3^- + 10\,e^- + 12\,H^+$$

We can now balance the number of hydrogen and oxygen atoms by adding six H_2O molecules to the reactants.

$$\text{Oxidation:} \qquad I_2 + 6\,H_2O \longrightarrow 2\,IO_3^- + 10\,e^- + 12\,H^+$$

It takes one electron to reduce nitrogen from the $+5$ to the $+4$ oxidation state.

$$\text{Reduction:} \qquad HNO_3 + e^- \longrightarrow NO_2$$

Because this reaction is run in nitric acid, we can add a single H^+ ion to the reactants to balance the charge.

$$\text{Reduction:} \qquad HNO_3 + H^+ + e^- \longrightarrow NO_2$$

We can balance the number of hydrogen and oxygen atoms by adding a water molecule to the products.

Reduction: $HNO_3 + H^+ + e^- \longrightarrow NO_2 + H_2O$

STEP 5: *Combine the two half-reactions so that electrons are neither created nor destroyed.* Ten electrons are given off in the oxidation half-reaction, and only one electron is consumed during reduction. We can combine these half-reactions so that electrons are conserved by multiplying the reduction half-reaction by 10.

$$(I_2 + 6\ H_2O \longrightarrow 2\ IO_3^- + 10\ e^- + 12\ H^+)$$
$$+\ 10(HNO_3 + H^+ + e^- \longrightarrow NO_2 + H_2O)$$
$$\overline{\qquad\qquad\qquad\qquad\qquad\qquad\qquad\qquad\qquad\qquad}$$
$$I_2 + 10\ HNO_3 + 6\ H_2O + 10\ H^+ \longrightarrow$$
$$2\ IO_3^- + 10\ NO_2 + 12\ H^+ + 10\ H_2O$$

STEP 6: *Balance the remainder of the equation by inspection, if necessary.* We have to subtract 6 H_2O molecules and 10 H^+ ions from each side to get the simplest balanced equation for this reaction.

$$I_2(aq) + 10\ HNO_3(aq) \longrightarrow 2\ IO_3^-(aq) + 10\ NO_2(g) + 2\ H^+(aq) + 4\ H_2O(l)$$

Exercise 18.8

An endless number of balanced equations can be written for the reaction between the permanganate ion and hydrogen peroxide in acidic solution to form the manganese(II) ion and oxygen.

$$MnO_4^-(aq) + H_2O_2(aq) + H^+(aq) \longrightarrow Mn^{2+}(aq) + O_2(g)$$

Use the half-reaction method to determine the correct stoichiometry for this reaction.

Solution

STEP 1: *Write a skeleton equation for the reaction.*

$$MnO_4^- + H_2O_2 + H^+ \longrightarrow Mn^{2+} + O_2$$

STEP 2: *Assign oxidation numbers to atoms on both sides of the equation.*

$$MnO_4^- + H_2O_2 + H^+ \longrightarrow Mn^{2+} + O_2$$
$$\begin{array}{ccccc} +7\ -2 & +1\ -1 & +1 & +2 & 0 \end{array}$$

STEP 3: *Determine which atoms are oxidized and which are reduced.*

$$MnO_4^- + H_2O_2 + H^+ \longrightarrow Mn^{2+} + O_2$$
$$\begin{array}{ccccc} +7\ -2 & +1\ -1 & +1 & +2 & 0 \end{array}$$

reduction

oxidation

STEP 4: *Divide the reaction into oxidation and reduction half-reactions, and balance these half-reactions.* We've already balanced the reduction half-reaction (see Exercise 18.6).

Reduction: $MnO_4^- + 8\ H^+ + 5\ e^- \longrightarrow Mn^{2+} + 4\ H_2O$

To balance the oxidation half-reaction, we have to remove two electrons from a pair of oxygen atoms in the -1 oxidation state to form a neutral O_2 molecule.

Oxidation: $H_2O_2 \longrightarrow O_2 + 2\ e^-$

We can then add a pair of H^+ ions to the products to balance both charge and mass in this half-reaction.

Oxidation: $H_2O_2 \longrightarrow O_2 + 2\ H^+ + 2\ e^-$

STEP 5: *Combine the two half-reactions so that electrons are neither created nor destroyed.* Two electrons are given off during oxidation, and five electrons are consumed during reduction. We can combine these half-reactions so that electrons are conserved by using the lowest common multiple of 5 and 2.

$$2(MnO_4^- + 8\ H^+ + 5\ e^- \longrightarrow Mn^{2+} + 4\ H_2O)$$
$$+\ 5(H_2O_2 \longrightarrow O_2 + 2\ H^+ + 2\ e^-)$$
$$\overline{\begin{array}{l} 2\ MnO_4^- + 5\ H_2O_2 + 16\ H^+ \longrightarrow \\ \qquad\qquad 2\ Mn^{2+} + 5\ O_2 + 10\ H^+ + 8\ H_2O \end{array}}$$

STEP 6: *Balance the remainder of the equation by inspection, if necessary.* The simplest balanced equation for this reaction is obtained when 10 H^+ ions are subtracted from each side of the equation derived in Step 5.

$$2\ MnO_4^-(aq) + 5\ H_2O_2(aq) + 6\ H^+(aq) \longrightarrow 2\ Mn^{2+}(aq) + 5\ O_2(g) + 8\ H_2O(l)$$

18.9 REDOX REACTIONS IN BASIC SOLUTIONS

The oxidation-number and half-reaction techniques are also valuable for balancing reactions in basic solutions. The key to success with these reactions is recognizing that basic solutions contain water and hydroxide ions. We can therefore add H_2O molecules or OH^- ions as needed to either side of the equation.

In Section 18.8 we obtained the following equation for the reaction between the permanganate ion and hydrogen peroxide in an acidic solution.

$$2\ MnO_4^-(aq) + 5\ H_2O_2(aq) + 6\ H^+(aq) \longrightarrow 2\ Mn^{2+}(aq) + 5\ O_2(g) + 8\ H_2O$$

It might be interesting to see whether the ratio of moles of MnO_4^- to moles of H_2O_2 consumed in this reaction changes when the reaction occurs in a basic solution.

Exercise 18.9

Use half-reactions to write a balanced equation for the reaction between the permanganate ion and hydrogen peroxide in a basic solution to form manganese dioxide and oxygen.

$$MnO_4^-(aq) + H_2O_2(aq) \xrightarrow{\ OH^-\ } MnO_2(s) + O_2(g)$$

Solution

STEP 1: *Write a skeleton equation for the reaction.*

$$MnO_4^- + H_2O_2 \longrightarrow MnO_2 + O_2$$

STEP 2: *Assign oxidation numbers to atoms on both sides of the equation.*

$$MnO_4^- + H_2O_2 \longrightarrow MnO_2 + O_2$$
$${+7\ -2}\quad\ {+1\ -1}\quad\ \ {+4\ -2}\quad\ \ 0$$

When H_2O_2 is added to a basic solution of the MnO_4^- ion it is oxidized to form O_2 gas, which bubbles out of solution, and the MnO_4^- ion is reduced to MnO_2.

STEP 3: *Determine which atoms are oxidized and which are reduced.*

$$MnO_4^- + H_2O_2 \longrightarrow MnO_2 + O_2$$

$$\underset{+7\ -2}{} \quad \underset{+1\ -1}{} \quad \underset{+4\ -2}{} \quad \underset{0}{}$$

reduction

oxidation

STEP 4: *Divide the reaction into oxidation and reduction half-reactions, and balance these half-reactions.* This reaction can be divided into the following half-reactions.

Reduction: $MnO_4^- \longrightarrow MnO_2$
$$\qquad\qquad\qquad +7 \qquad\quad +4$$

Oxidation: $H_2O_2 \longrightarrow O_2$
$$\qquad\qquad\qquad -1 \qquad\quad 0$$

Let's start by balancing the reduction half-reaction. It takes three electrons to reduce manganese from the $+7$ to the $+4$ oxidation state.

Reduction: $MnO_4^- + 3\ e^- \longrightarrow MnO_2$

We can try to balance either the number of atoms or the charge on both sides of the equation. One way of balancing the net charge of -4 on the left side of the equation is to add four OH^- ions to the products.

Reduction: $MnO_4^- + 3\ e^- \longrightarrow MnO_2 + 4\ OH^-$

We can now balance the number of hydrogen and oxygen atoms by adding two H_2O molecules to the reactants.

Reduction: $MnO_4^- + 3\ e^- + 2\ H_2O \longrightarrow MnO_2 + 4\ OH^-$

Next, we turn to the oxidation half-reaction. Two electrons are lost when hydrogen peroxide is oxidized to form O_2 molecules.

Oxidation: $H_2O_2 \longrightarrow O_2 + 2\ e^-$

We can balance the charge on this half-reaction by adding a pair of OH^- ions to the reactants.

Oxidation: $H_2O_2 + 2\ OH^- \longrightarrow O_2 + 2\ e^-$

The only way to balance the number of hydrogen and oxygen atoms is to add two H_2O molecules to the products.

Oxidation: $H_2O_2 + 2\ OH^- \longrightarrow O_2 + 2\ H_2O + 2\ e^-$

STEP 5: *Combine the two half-reactions so that electrons are neither created nor destroyed.* Two electrons are given off during the oxidation half-reaction, and three electrons are consumed in the reduction half-reaction. We can combine these half-reactions by using the lowest common multiple of 2 and 3.

$$2(MnO_4^- + 3\ e^- + 2\ H_2O \longrightarrow MnO_2 + 4\ OH^-)$$
$$+\ 3(H_2O_2 + 2\ OH^- \longrightarrow O_2 + 2\ H_2O + 2\ e^-)$$
$$\overline{\begin{array}{l} 2\ MnO_4^- + 3\ H_2O_2 + 4\ H_2O + 6\ OH^- \\ \qquad\quad 2\ MnO_2 + 3\ O_2 + 8\ OH^- + 6\ H_2O \end{array}}$$

STEP 6: *Balance the remainder of the equation by inspection, if necessary.* The simplest balanced equation is obtained when four H_2O molecules and six OH^- ions are subtracted from each side of the equation derived in Step 5.

$$2 \ MnO_4^-(aq) + 3 \ H_2O_2(aq) \longrightarrow 2 \ MnO_2(s) + 3 \ O_2(g) + 2 \ OH^-(aq) + 2 \ H_2O(l)$$

Note that the ratio of moles of MnO_4^- to moles of H_2O_2 consumed in this reaction is different in acidic and basic solutions. This difference results from the fact that MnO_4^- is reduced all the way to Mn^{2+} in acid but the reaction stops at MnO_2 in basc.

18.10 COMMON OXIDIZING AGENTS AND REDUCING AGENTS

Another way of looking at oxidation-reduction reactions was introduced in Section 7.12. It focuses on the role played by a particular reactant in a chemical reaction. What is the role of the nitric acid in the following reaction, for example?

$$I_2(s) + 10 \ HNO_3(aq) \longrightarrow 2 \ IO_3^-(aq) + 10 \ NO_2(g) + 2 \ H^+(aq) + 4 \ H_2O(l)$$

According to Exercise 18.7, iodine is oxidized to the iodate ion in this reaction and nitric acid is reduced to nitrogen dioxide.

$$\text{Oxidation:} \qquad \underset{0}{I_2} \longrightarrow \underset{+5}{IO_3^-}$$

$$\text{Reduction:} \qquad \underset{+5}{HNO_3} \longrightarrow \underset{+4}{NO_2}$$

Nitric acid removes electrons from iodine molecules and thereby oxidizes the iodine. Thus, nitric acid acts as an *oxidizing agent* in this reaction.

> **Oxidizing agent: An atom, ion, or molecule that gains electrons in a chemical reaction and thereby oxidizes the substance it reacts with.**

Iodine, on the other hand, is a *reducing agent* in this reaction. By giving up electrons, it reduces the nitric acid.

> **Reducing agent: An atom, ion, or molecule that loses electrons in a chemical reaction and thereby reduces the substance it reacts with.**

Atoms, ions, and molecules that have an unusually large affinity for electrons tend to be good oxidizing agents. Elemental fluorine, for example, is the strongest common oxidizing agent. F_2 is such a good oxidizing agent that metals, quartz, asbestos, and even water burst into flame in its presence. Other good oxidizing agents include O_2, O_3, and Cl_2, which are the elemental forms of the second and third most electronegative elements, respectively.

Another place to look for good oxidizing agents is among compounds that contain elements with unusually large oxidation numbers. Examples include the permanganate (MnO_4^-), chromate (CrO_4^{2-}), and dichromate ($Cr_2O_7^{2-}$) ions, as well as nitric acid (HNO_3), perchloric acid ($HClO_4$), and sulfuric acid (H_2SO_4). These compounds are strong oxidizing agents because elements become more electronegative as the oxidation states of their atoms become larger, as first noted in Section 8.6.

Good reducing agents include the active metals, such as sodium, magnesium, aluminum, and zinc, which have relatively small ionization energies and low electronegativities. Metal hydrides, such as NaH, CaH_2, and $LiAlH_4$, which formally contain the H^- ion, are also good reducing agents.

TABLE 18.2

Half-Reactions for Some Common Oxidizing Agents and Reducing Agents

Half-Reactions for Common Oxidizing Agents	*Half-Reactions for Common Reducing Agents*
$F_2 + 2\,e^- \rightleftharpoons 2\,F^-$	$Na \rightleftharpoons Na^+ + e^-$
$Cl_2 + 2\,e^- \rightleftharpoons 2\,Cl^-$	$Mg \rightleftharpoons Mg^{2+} + 2\,e^-$
$O_2 + 4\,H^+ + 4\,e^- \rightleftharpoons 2\,H_2O$	$Al \rightleftharpoons Al^{3+} + 3\,e^-$
$2\,H^+ + 2\,e^- \rightleftharpoons H_2$	$Zn \rightleftharpoons Zn^{2+} + 2\,e^-$
$MnO_4^- + 8\,H^+ + 5\,e^- \rightleftharpoons Mn^{2+} + 4\,H_2O$	$2\,H^- \rightleftharpoons H_2 + 2\,e^-$
$MnO_4^- + 2\,H_2O + 3\,e^- \rightleftharpoons MnO_2 + 4\,OH^-$	
$Cr_2O_7^{2-} + 14\,H^+ + 6\,e^- \rightleftharpoons 2\,Cr^{3+} + 7\,H_2O$	
$CrO_4^{2-} + 8\,H^+ + 3\,e^- \rightleftharpoons Cr^{3+} + 4\,H_2O$	
$HNO_3 + H^+ + e^- \rightleftharpoons NO_2 + H_2O$	
$HNO_3 + 3\,H^+ + 3\,e^- \rightleftharpoons NO + 2\,H_2O$	
$H_2SO_4 + 2\,H^+ + 2\,e^- \rightleftharpoons SO_2 + 2\,H_2O$	

There is a class of compounds that can act as either oxidizing agents or reducing agents. One example is hydrogen gas, which acts as an oxidizing agent when it combines with metals

$$2\,Na(s) + H_2(g) \longrightarrow 2\,NaH(s)$$

and as a reducing agent when it reacts with nonmetals.

$$H_2(g) + Cl_2(g) \longrightarrow 2\,HCl(g)$$

Another example is hydrogen peroxide, in which the oxygen atom is in the -1 oxidation state. Because this oxidation state lies between the extremes of the more common 0 and -2 oxidation states of oxygen, H_2O_2 can act as either an oxidizing agent or a reducing agent.

A summary of half-reactions for some common oxidizing agents and reducing agents can be found in Table 18.2.

18.11 THE RELATIVE STRENGTHS OF OXIDIZING AND REDUCING AGENTS

The permanganate ion is an excellent oxidizing agent. As we've seen, it oxidizes oxalic acid to CO_2 and hydrogen peroxide to oxygen.

$$2\,MnO_4^-(aq) + 5\,H_2C_2O_4(aq) + 6\,H^+(aq) \longrightarrow$$
$$2\,Mn^{2+}(aq) + 10\,CO_2(g) + 8\,H_2O(l)$$

$$2\,MnO_4^-(aq) + 5\,H_2O_2(aq) + 6\,H^+(aq) \longrightarrow 2\,Mn^{2+}(aq) + 5\,O_2(g) + 8\,H_2O(l)$$

It can also oxidize the Cl^- ion to Cl_2, Fe^{2+} ions to Fe^{3+} ions, and SO_2 to the SO_4^{2-} ion.

$$2\,MnO_4^-(aq) + 10\,Cl^-(aq) + 16\,H^+(aq) \longrightarrow 2\,Mn^{2+}(aq) + 5\,Cl_2(g) + 8\,H_2O(l)$$

$$MnO_4^-(aq) + 5\,Fe^{2+}(aq) + 8\,H^+(aq) \longrightarrow Mn^{2+}(aq) + 5\,Fe^{3+}(aq) + 4\,H_2O(l)$$

$$2\,MnO_4^-(aq) + 5\,SO_2(aq) + 2\,H_2O(l) \longrightarrow 2\,Mn^{2+}(aq) + 5\,SO_4^{2-}(aq) + 4\,H^+(aq)$$

This raises an interesting question. If the MnO_4^- ion is such a good oxidizing agent, how do we make this ion from Mn^{2+}? In theory, all we have to do is find an oxidizing agent that is even stronger. In practice, that can be difficult. There are only a few compounds that are stronger oxidizing agents than the MnO_4^- ion. These compounds include the periodate (IO_4^-), bismuthate (BiO_3^-), and peroxydisulfate ($S_2O_8^{2-}$) ions. In the qualitative analysis scheme described in Chapter 17, the bismuthate ion was used to generate permanganate ions.

$$2\ Mn^{2+}(aq) + 5\ BiO_3^-(aq) + 14\ H^+(aq) \longrightarrow 2\ MnO_4^-(aq) + 5\ Bi^{3+}(aq) + 7\ H_2O(l)$$

This equation is a perfect example of a phenomenon first discussed in Section 7.14. Spontaneous oxidation-reduction reactions convert the stronger of a pair of oxidizing agents and the stronger of a pair of reducing agents into a weaker oxidizing agent and a weaker reducing agent. The fact that the following reaction occurs as written suggests that BiO_3^- is a stronger oxidizing agent than MnO_4^- and that Mn^{2+} is a stronger reducing agent than Bi^{3+}.

$$\underset{\substack{\text{stronger} \\ \text{reducing agent}}}{2\ Mn^{2+}} + \underset{\substack{\text{stronger} \\ \text{oxidizing agent}}}{5\ BiO_3^-} + 14\ H^+ \longrightarrow \underset{\substack{\text{weaker} \\ \text{oxidizing agent}}}{2\ MnO_4^-} + \underset{\substack{\text{weaker} \\ \text{reducing agent}}}{5\ Bi^{3+}} + 7\ H_2O$$

On the basis of many experiments, the common oxidation-reduction half-reactions have been organized into a table in which the strongest reducing agents are at one end and the strongest oxidizing agents are at the other, as shown in Table 18.3. By convention, all of the half-reactions are written in the direction of reduction. Furthermore, by convention, the strongest reducing agents are usually found at the top of the table.

Fortunately, you don't have to memorize these conventions. This table is an extension of Table 7.3. All you have to do is remember that the active metals, such as sodium and potassium, are excellent reducing agents, and look for these entries in the table. The strongest reducing agents will be found at the corner of the table where sodium and potassium metal are listed.

Exercise 18.10

Use Table 18.3 to organize the following oxidizing and reducing agents in order of increasing strength.

Reducing agents: Cl^-, Cu, H_2, H^-, HF, Pb, and Zn
Oxidizing agents: Cr^{3+}, $Cr_2O_7^{2-}$, Cu^{2+}, H^+, O_2, O_3, and Na^+

Solution

According to Table 18.3, the reducing agents given here become stronger in the following order.

$$HF < Cl^- < Cu < H_2 < Pb < Zn < H^-$$

The oxidizing agents become stronger in the following order.

$$Na^+ < Cr^{3+} < H^+ < Cu^{2+} < O_2 < Cr_2O_7^{2-} < O_3$$

Exercise 18.11

Use Table 18.3 to predict whether the following oxidation-reduction reactions should occur as written.

TABLE 18.3

The Relative Strengths of Common Oxidizing Agents and Reducing Agents

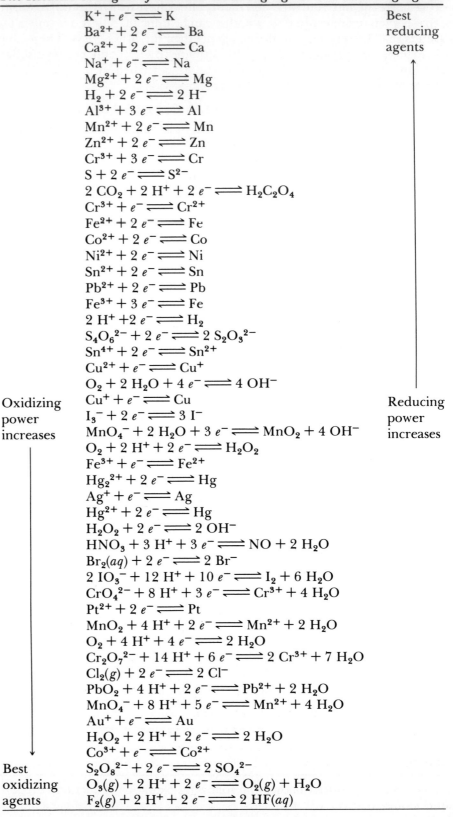

$K^+ + e^- \rightleftharpoons K$	Best reducing agents
$Ba^{2+} + 2\,e^- \rightleftharpoons Ba$	
$Ca^{2+} + 2\,e^- \rightleftharpoons Ca$	
$Na^+ + e^- \rightleftharpoons Na$	
$Mg^{2+} + 2\,e^- \rightleftharpoons Mg$	
$H_2 + 2\,e^- \rightleftharpoons 2\,H^-$	
$Al^{3+} + 3\,e^- \rightleftharpoons Al$	
$Mn^{2+} + 2\,e^- \rightleftharpoons Mn$	
$Zn^{2+} + 2\,e^- \rightleftharpoons Zn$	
$Cr^{3+} + 3\,e^- \rightleftharpoons Cr$	
$S + 2\,e^- \rightleftharpoons S^{2-}$	
$2\,CO_2 + 2\,H^+ + 2\,e^- \rightleftharpoons H_2C_2O_4$	
$Cr^{3+} + e^- \rightleftharpoons Cr^{2+}$	
$Fe^{2+} + 2\,e^- \rightleftharpoons Fe$	
$Co^{2+} + 2\,e^- \rightleftharpoons Co$	
$Ni^{2+} + 2\,e^- \rightleftharpoons Ni$	
$Sn^{2+} + 2\,e^- \rightleftharpoons Sn$	
$Pb^{2+} + 2\,e^- \rightleftharpoons Pb$	
$Fe^{3+} + 3\,e^- \rightleftharpoons Fe$	
$2\,H^+ + 2\,e^- \rightleftharpoons H_2$	
$S_4O_6^{2-} + 2\,e^- \rightleftharpoons 2\,S_2O_3^{2-}$	
$Sn^{4+} + 2\,e^- \rightleftharpoons Sn^{2+}$	
$Cu^{2+} + e^- \rightleftharpoons Cu^+$	
$O_2 + 2\,H_2O + 4\,e^- \rightleftharpoons 4\,OH^-$	

Oxidizing power increases — Reducing power increases

$Cu^+ + e^- \rightleftharpoons Cu$
$I_3^- + 2\,e^- \rightleftharpoons 3\,I^-$
$MnO_4^- + 2\,H_2O + 3\,e^- \rightleftharpoons MnO_2 + 4\,OH^-$
$O_2 + 2\,H^+ + 2\,e^- \rightleftharpoons H_2O_2$
$Fe^{3+} + e^- \rightleftharpoons Fe^{2+}$
$Hg_2^{2+} + 2\,e^- \rightleftharpoons Hg$
$Ag^+ + e^- \rightleftharpoons Ag$
$Hg^{2+} + 2\,e^- \rightleftharpoons Hg$
$H_2O_2 + 2\,e^- \rightleftharpoons 2\,OH^-$
$HNO_3 + 3\,H^+ + 3\,e^- \rightleftharpoons NO + 2\,H_2O$
$Br_2(aq) + 2\,e^- \rightleftharpoons 2\,Br^-$
$2\,IO_3^- + 12\,H^+ + 10\,e^- \rightleftharpoons I_2 + 6\,H_2O$
$CrO_4^{2-} + 8\,H^+ + 3\,e^- \rightleftharpoons Cr^{3+} + 4\,H_2O$
$Pt^{2+} + 2\,e^- \rightleftharpoons Pt$
$MnO_2 + 4\,H^+ + 2\,e^- \rightleftharpoons Mn^{2+} + 2\,H_2O$
$O_2 + 4\,H^+ + 4\,e^- \rightleftharpoons 2\,H_2O$
$Cr_2O_7^{2-} + 14\,H^+ + 6\,e^- \rightleftharpoons 2\,Cr^{3+} + 7\,H_2O$
$Cl_2(g) + 2\,e^- \rightleftharpoons 2\,Cl^-$
$PbO_2 + 4\,H^+ + 2\,e^- \rightleftharpoons Pb^{2+} + 2\,H_2O$
$MnO_4^- + 8\,H^+ + 5\,e^- \rightleftharpoons Mn^{2+} + 4\,H_2O$
$Au^+ + e^- \rightleftharpoons Au$
$H_2O_2 + 2\,H^+ + 2\,e^- \rightleftharpoons 2\,H_2O$
$Co^{3+} + e^- \rightleftharpoons Co^{2+}$

Best oxidizing agents

$S_2O_8^{2-} + 2\,e^- \rightleftharpoons 2\,SO_4^{2-}$
$O_3(g) + 2\,H^+ + 2\,e^- \rightleftharpoons O_2(g) + H_2O$
$F_2(g) + 2\,H^+ + 2\,e^- \rightleftharpoons 2\,HF(aq)$

(a) $Zn(s) + 2 H^+(aq) \longrightarrow Zn^{2+}(aq) + H_2(g)$

(b) $2 Ag(s) + S(s) \longrightarrow Ag_2S(s)$

(c) $2 Ag(s) + Cu^{2+}(aq) \longrightarrow 2 Ag^+(aq) + Cu(s)$

(d) $MnO_4^-(aq) + 3 Fe^{2+}(aq) + OH^-(aq) \longrightarrow MnO_2(s) + 3 Fe^{3+}(aq)$

(e) $MnO_4^-(aq) + 3 Fe^{2+}(aq) + H^+(aq) \longrightarrow Mn^{2+}(aq) + 3 Fe^{3+}(aq)$

Solution

(a) Yes. Zn is a better reducing agent than H_2, and H^+ is a better oxidizing agent than Zn^{2+}. Zn should therefore dissolve in acid.

(b) No. S^{2-} is a better reducing agent than Ag, and Ag^+ is a better oxidizing agent than S. Silver doesn't tarnish because it reduces sulfur; it tarnishes because it reacts with sulfur compounds in the presence of oxygen, and the oxygen is reduced.

(c) No. Cu is a better reducing agent than Ag, and Ag^+ is a better oxidizing agent than Cu^{2+}.

(d) No. MnO_4^- is not a strong enough oxidizing agent in base to oxidize Fe^{2+} to Fe^{3+}.

(e) Yes. MnO_4^- is a strong enough oxidizing agent in acid to oxidize Fe^{2+} to Fe^{3+}.

Exercise 18.12

Which of the following pairs of ions cannot exist simultaneously in aqueous solutions.

(a) Cu^+ and Fe^{3+} (b) Co^{3+} and Fe^{2+} (c) Fe^{3+} and I^- (d) Al^{3+} and Co^{2+}

Solution

(a) According to Table 18.3, the Cu^+ ion is a significantly better reducing agent than the Fe^{2+} ion, and the Fe^{3+} ion is a better oxidizing agent than Cu^{2+}. Solutions that contain both Cu^+ and Fe^{3+} should undergo a spontaneous oxidation-reduction reaction, so these ions can't exist simultaneously in the same solution.

$$Cu^+(aq) + Fe^{3+}(aq) \longrightarrow Cu^{2+}(aq) + Fe^{2+}(aq)$$

(b) The Co^{3+} ion is one of the strongest oxidizing agents in Table 18.3, more than strong enough to oxidize Fe^{2+} ions to Fe^{3+} ions.

$$Co^{3+}(aq) + Fe^{2+}(aq) \longrightarrow Co^{2+}(aq) + Fe^{3+}(aq)$$

The Co^{3+} and Fe^{2+} ions therefore cannot coexist in aqueous solution.

(c) According to Table 18.3, the Fe^{3+} ion is strong enough to oxidize I^- ions to I_2.

$$2 Fe^{3+}(aq) + 2 I^-(aq) \longrightarrow Fe^{2+}(aq) + I_2(aq)$$

These ions cannot exist in the same solution.

(d) The Al^{3+} ion is one of the weakest oxidizing agents in Table 18.3, whereas the Co^{2+} ion is one of the weakest reducing agents. These ions can't undergo an oxidation-reduction reaction with each other and can therefore coexist in aqueous solution.

$$Al^{3+}(aq) + Co^{2+}(aq) \longrightarrow\!\!\!\!\!/$$

SUMMARY

Oxidation-reduction reactions can be systematically balanced by two techniques: the oxidation number and half-reaction methods. The oxidation number approach combines oxidation and reduction so that electrons are conserved and then balances the equation. The half-reaction method balances the two halves of the reaction and then combines them so that electrons are conserved.

Spontaneous oxidation-reduction reactions convert the stronger of a pair of oxidizing agent and the stronger

of a pair of reducing agent into a weaker oxidizing agent and a weaker reducing agent. The fact that sodium metal can reduce aluminum chloride to aluminum metal, for example, suggests that sodium is a stronger reducing agent than aluminum. The results of a large number of observations of this nature are summarized in a table of the relative strengths of common oxidizing agents and reducing agents. This table can be used to predict whether oxidation-reduction should occur.

PROBLEMS

Evolution of the Theory of Oxidation-Reduction Reactions

18-1 Describe the phlogiston theory of chemical reactions.

18-2 Summarize the fatal flaw in the phlogiston theory, which inevitably led to its collapse.

18-3 Define *oxidation* and *reduction* in terms of changes in oxidation number.

18-4 Define *oxidation* and *reduction* in terms of the transfer of electrons.

18-5 Inorganic chemists often think about oxidation in terms of the loss of electrons. Magnesium metal is oxidized, for example, when magnesium atoms lose two electrons to form Mg^{2+} ions. Organic chemists often think about oxidation in terms of the loss of a pair of hydrogen atoms. Ethyl alcohol, CH_3CH_2OH, is oxidized to acetaldehyde, CH_3CHO, for example, by the removal of a pair of hydrogen atoms. Show that both reactions obey the general rule that oxidation is any process in which the oxidation number of an atom becomes larger.

18-6 Inorganic chemists often think about reduction in terms of the gain of electrons. O_2 is reduced when it picks up four electrons to form a pair of O^{2-} ions. Organic chemists often think about reduction in terms of the gain of a pair of hydrogen atoms. Ethylene, C_2H_4, is reduced when it reacts with H_2 to form ethane, C_2H_6. Show that both reactions obey the general rule that reduction is any process in which the oxidation number of an atom becomes smaller.

Oxidation-Reduction Reactions

18-7 Our bodies convert the carbohydrates, fats, and proteins we consume into simple sugars such as glucose, $C_6H_{12}O_6$, which we then burn to form CO_2 and H_2O. Show that the combustion of glucose is an example of an oxidation-reduction reaction.

18-8 Organisms that can live in the absence of oxygen are

called anaerobic. They obtain their energy by converting glucose, $C_6H_{12}O_6$, into compounds such as lactic acid, $C_3H_6O_3$. Show that these organisms capture energy from their environment with no net oxidation or reduction.

18-9 Write chemical equations for the corrosion of silver and iron and show that corrosion results from oxidation-reduction reactions.

Assigning Oxidation Numbers

18-10 An area of active research in recent years has involved compounds, such as $Re_2Cl_8^{2-}$, $Cr_2Cl_9^{3-}$, and $Mo_2Cl_8^{4-}$, that contain metal—metal bonds. Calculate the oxidation number of the metal atom in each of these compounds.

18-11 Calculate the oxidation number of the aluminum atom in the following compounds.

(a) AlH_3 (b) $LiAlH_4$ (c) $Al(H_2O)_6^{3+}$ (d) Al_2O_3
(e) $Al(OH)_4^-$ (f) Li_3AlH_6 (g) Al_4C_3 (h) $NaAl(OH)_2CO_3$
(the active ingredient in Rolaids)

18-12 Which of the following compounds contain hydrogen in a negative oxidation state?

(a) H_2S (b) H_2O (c) NH_3 (d) PH_4^+ (e) $LiAlH_4$
(f) HF (g) CaH_2 (h) CH_4

18-13 Arrange the following compounds in order of increasing oxidation number of the carbon atom.

(a) C (b) CO (c) CO_2 (d) H_2CO (e) CH_3OH (f) CH_4

18-14 Carbon can have any oxidation number between -4 and $+4$. Calculate the oxidation number of carbon in the following compounds.

(a) CF_4 (b) COF_2 (c) CO (d) CO_2 (e) C_3O_2 (f) CS_2
(g) CH_3Li (h) CH_3^+ (i) CH_4 (j) CH_3^- (k) H_2CO
(l) CO_3^{2-} (m) H_2CO_3 (n) HCO_2H

18-15 Sulfur can have any oxidation number between $+6$ and

−2. Calculate the oxidation number of sulfur in the following compounds.

(a) S_8 (b) H_2S (c) ZnS (d) Na_2S_2 (e) SF_4 (f) SF_5^-
(g) SF_6 (h) SF_3^+ (i) SO_2 (j) SO_3 (k) SO_3^{2-} (l) SO_4^{2-}
(m) $S_2O_3^{2-}$ (n) $S_2O_6^{2-}$ (o) $S_2O_8^{2-}$ (p) H_2SO_3
(q) H_2SO_4 (r) $H_2S_2O_3$

18-16 Calculate the oxidation number of manganese in the following compounds.

(a) MnO (b) Mn_2O_3 (c) MnO_2 (d) MnO_3 (e) Mn_2O_7
(f) $Mn(OH)_2$ (g) $Mn(OH)_3$ (h) H_2MnO_3 (i) H_2MnO_4
(j) $HMnO_4$ (k) $CaMnO_3$ (l) $MnSO_4$ (m) $Mn(CO)_6^-$

18-17 Prussian blue is a pigment with the formula $Fe_4[Fe(CN)_6]_3$. If this compound contains the $Fe(CN)_6^{4-}$ ion, what is the oxidation state of the other four iron atoms? Turnbull's blue is a pigment with the formula $Fe_3[Fe(CN)_6]_2$. This compound contains the $Fe(CN)_6^{3-}$ ion. What is the oxidation state of the other three iron atoms?

Recognizing Oxidation-Reduction Reactions

18-18 Describe the difference between metathesis reactions and oxidation-reduction reactions. Give examples of both.

18-19 We can remove the Ag_2S that forms when silver tarnishes by polishing the object with something containing cyanide ions or by wrapping the object in aluminum foil and immersing it in salt water.

$$Ag_2S(s) + 4\ CN^-(aq) \longrightarrow 2\ Ag(CN)_2^-(aq) + S^{2-}(aq)$$
$$3\ Ag_2S(s) + 2\ Al(s) \longrightarrow 6\ Ag(s) + Al_2S_3(s)$$

Which of these reactions involves oxidation-reduction?

18-20 Decide whether each of the following reactions involves oxidation-reduction. If it does, identify what is oxidized and what is reduced.

(a) $CO_2(g) + H_2O(l) \rightleftharpoons H_2CO_3(aq)$
(b) $Fe_2O_3(s) + 3\ CO(g) \rightarrow 2\ Fe(l) + 3\ CO_2(g)$
(c) $SiO_2(s) + 3\ C(s) \rightarrow SiC(s) + 2\ CO(g)$
(d) $CO_2(g) + H_2(g) \rightleftharpoons CO(g) + H_2O(g)$
(e) $CO(g) + 2\ H_2(g) \rightarrow CH_3OH(l)$

18-21 Decide whether each of the following reactions involves oxidation-reduction. If it does, identify what is oxidized and what is reduced.

(a) $Mg(s) + 2\ HCl(aq) \rightarrow MgCl_2(aq) + H_2(g)$
(b) $I_2(s) + 3\ Cl_2(g) \rightarrow 2\ ICl_3(l)$
(c) $HCl(aq) + NaOH(aq) \rightarrow NaCl(aq) + H_2O(l)$
(d) $2\ Na(s) + 2\ H_2O(l) \rightarrow 2\ NaOH(aq) + H_2(g)$
(e) $2\ H_2(g) + O_2(g) \rightarrow 2\ H_2O(g)$

18-22 Decide whether each of the following reactions involves oxidation-reduction. If it does, identify what is oxidized and what is reduced.

(a) $Ca_3P_2(s) + 6\ H_2O(l) \rightarrow 3\ Ca(OH)_2(aq) + 2\ PH_3(g)$
(b) $2\ PH_3(g) + 4\ O_2(g) \rightarrow H_3PO_4(s)$
(c) $PH_3(g) + HCl(g) \rightarrow PH_4Cl(s)$
(d) $P_4(s) + 5\ O_2(g) \rightarrow P_4O_{10}(s)$

The Oxidation Number Method of Balancing Redox Equations

18-23 Use the oxidation number method to balance the following oxidation-reduction equation.

$$CuO(s) + NH_3(g) \longrightarrow Cu(s) + N_2(g) + H_2O(l)$$

18-24 Use the oxidation number method to balance the following oxidation-reduction equations.

(a) $CS_2(g) + O_2(g) \rightarrow CO_2(g) + SO_2(g)$
(b) $CH_4(g) + O_2(g) \rightarrow CO_2(g) + H_2O(g)$

18-25 Use the oxidation number method to balance the following oxidation-reduction equations.

(a) $SO_2(g) + H_2S(g) \rightarrow S_8(s) + H_2O(l)$
(b) $H_2S(g) + O_2(g) \rightarrow H_2O(g) + S_8(s)$

18-26 Use the oxidation number method to balance the following oxidation-reduction equations.

(a) $Fe_2O_3(s) + CO(g) \rightarrow Fe(l) + CO_2(g)$
(b) $Cu_2S(s) + O_2(g) \rightarrow Cu_2O(s) + SO_2(g)$

18-27 Use the oxidation number method to balance the following oxidation-reduction equations.

(a) $PbS(s) + O_2(g) \rightarrow PbO(s) + SO_2(g)$
(b) $FeS_2(s) + O_2(g) \rightarrow Fe_2O_3(s) + SO_2(g)$

18-28 Use the oxidation number method to balance the following oxidation-reduction equations.

(a) $NH_3(g) + O_2(g) \rightarrow NO(g) + H_2O(g)$
(b) $NO_2(g) + H_2O(l) \rightarrow HNO_3(aq) + NO(g)$
(c) $NH_4NO_3(s) \rightarrow N_2(g) + O_2(g) + H_2O(g)$
(d) $HNO_3(aq) + SO_2(g) \rightarrow H_2SO_4(aq) + 2\ NO_2(g)$

18-29 Use the oxidation number method to balance the following oxidation-reduction equations.

(a) $HCl(aq) + MnO_2(s) \rightarrow Cl_2(g) + Mn^{2+}(aq)$
(b) $HI(aq) + Fe^{3+}(aq) \rightarrow I_2(aq) + Fe^{2+}(aq)$
(c) $Cl_2(g) + OH^-(aq) \rightarrow Cl^-(aq) + OCl^-(aq)$
(d) $Br_2(aq) + SO_2(g) \rightarrow H_2SO_4(aq) + HBr(aq)$

18-30 Use the oxidation number method to balance the following oxidation-reduction equations.

(a) $HIO_3(aq) + HI(aq) \rightarrow I_2(aq)$
(b) $HNO_2(aq) + HI(aq) \rightarrow I_2(aq) + NO(g)$
(c) $H_2SO_4(aq) + HI(aq) \rightarrow I_2(aq) + H_2S(g)$
(d) $I^-(aq) + IO_3^-(aq) + H^+(aq) \rightarrow I_2(aq)$

18-31 Use the oxidation number method to balance the following oxidation-reduction equations.

(a) $(NH_4)_2Cr_2O_7(s) \rightarrow N_2(g) + Cr_2O_3(s) + H_2O(g)$
(b) $PbO_2(s) + H_2S(g) \rightarrow PbS(s) + SO_2(g)$
(c) $NH_3(aq) + Cl_2(g) \rightarrow N_2(g) + NH_4Cl(aq)$
(d) $H_2S(g) + HNO_3(aq) \rightarrow H_2SO_4(aq) + NO(g)$

18-32 Some of the earliest matches consisted of white phosphorus and potassium chlorate glued to wooden sticks. Write a balanced equation for the oxidation-reduction reaction that gave off enough energy to ignite the stick.

$$P_4(s) + KClO_3(s) \rightarrow P_4O_{10}(s) + KCl(s)$$

The Half-Reaction Method of Balancing Redox Equations

18-33 Use half-reactions to balance the following oxidation-reduction equations.

(a) $HCl(aq) + HNO_3(aq) \rightarrow NO(g) + Cl_2(g)$
(b) $HBr(aq) + H_2SO_4(aq) \rightarrow SO_2(g) + Br_2(aq)$
(c) $HCl(aq) + MnO_4^-(aq) \rightarrow Cl_2(g) + Mn^{2+}(aq)$
(d) $I_3^-(aq) + S_2O_3^{2-}(aq) \rightarrow I^-(aq) + S_4O_6^{2-}(aq)$

18-34 Use half-reactions to balance the following oxidation-reduction equations.

(a) $Cu(s) + HNO_3(aq) \rightarrow Cu^{2+}(aq) + NO(g)$
(b) $Cu(s) + HNO_3(aq) \rightarrow Cu^{2+}(aq) + NO_2(g)$
(c) $CuO(s) + NH_3(g) \rightarrow Cu(s) + N_2(g)$

18-35 Use half-reactions to balance the following oxidation-reduction equations.

(a) $HCl(aq) + MnO_2(s) \rightarrow Cl_2(g) + Mn^{2+}(aq)$
(b) $HI(aq) + Fe^{3+}(aq) \rightarrow I_2(aq) + Fe^{2+}(aq)$
(c) $Cl_2(g) + OH^-(aq) \rightarrow Cl^-(aq) + OCl^-(aq)$
(d) $Br_2(aq) + SO_2(g) \rightarrow H_2SO_4(aq) + HBr(aq)$

18-36 Use half-reactions to balance the following oxidation-reduction equations.

(a) $HIO_3(aq) + HI(aq) \rightarrow I_2(aq)$
(b) $HNO_2(aq) + HI(aq) \rightarrow I_2(aq) + NO(g)$
(c) $H_2SO_4(aq) + HI(aq) \rightarrow I_2(aq) + H_2S(g)$
(d) $I^-(aq) + IO_3^-(aq) + H^+(aq) \rightarrow I_2(aq)$

18-37 Use half-reactions to balance the following oxidation-reduction equations.

(a) $H_2S(g) + H_2O_2(aq) \rightarrow S_8(s)$
(b) $Fe^{3+}(aq) + SH^-(aq) \rightarrow Fe^{2+}(aq) + S_8(s) + H^+(aq)$
(c) $HgS(s) + HNO_3(aq) \rightarrow Hg^{2+}(aq) + S_8(s) + NO(g)$
(d) $I_2(aq) + S_2O_3^{2-}(aq) \rightarrow I^-(aq) + S_4O_6^{2-}(aq)$

Redox Reactions in Acidic Solutions

18-38 Use half-reactions to balance the following oxidation-reduction equations for reactions that occur in acidic solution.

(a) $I^-(aq) + CrO_4^{2-}(aq) \rightarrow I_2(aq) + Cr^{3+}(aq)$
(b) $Fe^{2+}(aq) + Cr_2O_7^{2-}(aq) \rightarrow Fe^{3+}(aq) + Cr^{3+}(aq)$
(c) $H_2S(g) + CrO_4^{2-}(aq) \rightarrow S_8(s) + Cr^{3+}(aq)$

18-39 Use half-reactions to balance the following oxidation-reduction equations for reactions that occur in acidic solution.

(a) $S_2O_3^{2-}(aq) + MnO_4^-(aq) \rightarrow SO_4^{2-}(aq) + Mn^{2+}(aq)$
(b) $H_2C_2O_4(aq) + MnO_4^-(aq) \rightarrow CO_2(g) + Mn^{2+}(aq)$
(c) $Mn^{2+}(aq) + PbO_2(s) \rightarrow MnO_4^-(aq) + Pb^{2+}(aq)$

18-40 Balance the following oxidation-reduction equations that occur in acidic solution.

(a) $Ag(s) + MnO_4^-(aq) \rightarrow Ag^+(aq) + Mn^{2+}(aq)$
(b) $Fe^{2+}(aq) + MnO_4^-(aq) \rightarrow Fe^{3+}(aq) + Mn^{2+}(aq)$
(c) $SO_2(g) + MnO_4^-(aq) \rightarrow H_2SO_4(aq) + Mn^{2+}(aq)$
(d) $H_2O_2(aq) + MnO_4^-(aq) \rightarrow O_2(g) + Mn^{2+}(aq)$

18-41 Balance the following oxidation-reduction equations that occur in acidic solution.

(a) $Cr(s) + O_2(g) + H^+(aq) \rightarrow Cr^{3+}(aq)$
(b) $Fe^{2+}(aq) + Cr_2O_7^{2-}(aq) \rightarrow Fe^{3+}(aq) + Cr^{3+}(aq)$
(c) $BrO_3^-(aq) + Cr^{3+}(aq) \rightarrow Br_2(aq) + HCrO_4^-(aq)$

Redox Reactions in Basic Solutions

18-42 Balance the following oxidation-reduction equations that occur in basic solution.

(a) $I^-(aq) + MnO_4^-(aq) \rightarrow IO_3^-(aq) + MnO_2(s)$
(b) $NO(g) + MnO_4^-(aq) \rightarrow NO_3^-(aq) + MnO_2(s)$
(c) $NH_3(aq) + MnO_4^-(aq) \rightarrow N_2(g) + MnO_2(s)$
(d) $CH_3OH(aq) + MnO_4^-(aq) \rightarrow H_2CO(aq) + MnO_2(s)$

18-43 Balance the following oxidation-reduction equations that occur in basic solution.

(a) $Cr(s) + OH^-(aq) \rightarrow Cr(OH)_4^-(aq) + H_2(g)$
(b) $Cr(OH)_3(s) + H_2O_2(aq) \rightarrow CrO_4^{2-}(aq)$
(c) $SO_3^{2-}(aq) + O_2(g) \rightarrow SO_4^{2-}(aq)$

18-44 Compounds are said to undergo disproportionation when they are simultaneously oxidized and reduced. Use half-reactions to balance the following disproportionation equations.

(a) $Cl_2(g) + OH^-(aq) \rightarrow Cl^-(aq) + OCl^-(aq)$
(b) $Cl_2(g) + OH^-(aq) \rightarrow Cl^-(aq) + ClO_3^-(aq)$
(c) $P_4(s) + OH^-(aq) \rightarrow PH_3(g) + H_2PO_2^-(aq)$
(d) $H_2O_2(aq) \rightarrow O_2(g) + H_2O(aq)$

Industrial Redox Reactions

18-45 Write a balanced equation for the oxidation-reduction reaction used to extract gold from its ores.

$$Au(s) + CN^-(aq) + O_2(g) \longrightarrow Au(CN)_2^-(aq) + OH^-(aq)$$

18-46 Write a balanced equation for the Raschig process, in which ammonia reacts with the hypochlorite ion in a basic solution to form hydrazine.

$$NH_3(aq) + OCl^-(aq) \longrightarrow N_2H_4(aq) + Cl^-(aq)$$

Common Oxidizing Agents and Reducing Agents

18-47 What chemical and physical properties make Cl_2, O_2, CrO_4^{2-}, and MnO_4^- good oxidizing agents?

18-48 What chemical and physical properties make Na, Al, and Zn good reducing agents?

18-49 Which of the following can't be an oxidizing agent?

(a) Cl^- (b) Br_2 (c) Fe^{3+} (d) Zn (e) CaH_2

18-50 Which of the following can't be a reducing agent?

(a) H_2 (b) Cl_2 (c) Fe^{3+} (d) Al (e) LiH

18-51 Which of the following can be both an oxidizing agent and a reducing agent?

(a) H_2 (b) I_2 (c) H_2O_2 (d) P_4 (e) S_8

The Relative Strengths of Oxidizing and Reducing Agents

18-52 Use Table 18.3 to determine which of the following transition metals is the strongest reducing agent.

(a) Cr (b) Mn (c) Fe (d) Co (e) Ni

18-53 Use Table 18.3 to determine which of the following transition metal oxides is the strongest oxidizing agent.

(a) MnO_4^- in acid (b) MnO_4^- in base (c) MnO_2
(d) CrO_4^{2-} (e) $Cr_2O_7^{2-}$

18-54 Use Table 18.3 to predict which of the following oxidation-reduction reactions should occur.

(a) $Co(s) + Cr^{3+}(aq) \rightarrow Co^{2+}(aq) + Cr^{3+}(aq)$
(b) $H_2(g) + Zn^{2+}(aq) \rightarrow Zn(s) + 2 H^+(aq)$
(c) $Sn(s) + H^+(aq) \rightarrow Sn^{2+}(aq) + H_2(g)$
(d) $3 Na(l) + AlCl_3(l) \rightarrow 3 NaCl(l) + Al(l)$

18-55 Use Table 18.3 to predict which of the following oxidation-reduction reactions should occur.

(a) $2 S_2O_3^{2-}(aq) + I_3^-(aq) \rightarrow S_4O_6^{2-}(aq) + 3 I^-(aq)$
(b) $PbO_2(s) + Mn^{2+}(aq) \rightarrow Pb^{2+}(aq) + MnO_2(s)$
(c) $CrO_4^{2-}(aq) + Mn^{2+}(aq) \rightarrow Cr^{3+}(aq) + MnO_4^-(aq)$
(d) $2 H_2O_2(aq) \rightarrow 2 H_2O(aq) + O_2(g)$

18-56 Use Table 18.3 to predict the products of the following oxidation-reduction reactions.

(a) $HNO_3(aq) + Mn(s) \rightarrow$
(b) $Mg(s) + CrCl_3(s) \rightarrow$
(c) $Cr^{3+}(aq) + O_2(aq) + H^+(aq) \rightarrow$
(d) $F_2(g) + Au(s) \rightarrow$
(e) $2 H_2O_2(aq) \rightarrow$

18-57 Which of the following pairs of ions can't coexist in aqueous solution?

(a) Cr^{2+} and MnO_4^- (b) Fe^{3+} and $Cr_2O_7^{2-}$ (c) Cr^{2+} and I^- (d) Mn^{2+} and Cl^-

18-58 Which of the following pairs of ions can't coexist in aqueous solution?

(a) Na^+ and $S_2O_8^{2-}$ (b) Hg^{2+} and Cl^- (c) Cr^{2+} and I_3^- (d) Cu^{2+} and I^-

18-59 Solutions of the Sn^{2+} ion are unstable because they are slowly oxidized by air to the Sn^{4+} ion. Explain why adding small pieces of tin metal to such a solution helps keep the concentration of the Sn^{2+} ion high enough to allow us to use the solution as a reagent in the qualitative analysis scheme.

18-60 On an exam, students were asked to use half-reactions to write a balanced equation for the following reaction.

$$SO_2(g) + Cr_2O_7^{2-}(aq) \xrightarrow{H^+} SO_4^{2-}(aq) + Cr^{3+}(aq)$$

One student wrote the following half-reactions.

Oxidation: $SO_2 + 4 OH^- \longrightarrow SO_4^{2-} + H_2O + 2 e^-$
Reduction: $Cr_2O_7^{2-} + 7 H_2O + 6 e^- \longrightarrow 2 Cr^{3+} + 14 OH^-$

The student then wrote the following overall equation.

$$3 SO_2(g) + Cr_2O_7^{2-}(aq) + 4 H_2O(aq) \longrightarrow$$
$$3 SO_4^{2-}(aq) + 2 Cr^{3+}(aq) + 2 OH^-(aq)$$

If this question was worth 10 points, how much partial credit would you give the student? Explain why.

Stoichiometry Problems Involving Oxidation-Reduction Reactions

18-61 Oxalic acid reacts with the chromate ion in acidic solution as follows.

$$3 H_2C_2O_4(aq) + 2CrO_4^{2-}(aq) + 10 H^+(aq) \longrightarrow$$
$$6 CO_2(g) + 2 Cr^{3+}(aq) + 8 H_2O(aq)$$

If 10.0 milliliters of oxalic acid consumes 40.0 milliliters of 0.0250 M CrO_4^{2-} solution, what is the molarity of the oxalic acid solution?

18-62 A 2.50-gram sample of bronze was dissolved in sulfuric acid.

$$Cu(s) + 2 H_2SO_4(aq) \longrightarrow CuSO_4(aq) + SO_2(g) + 2 H_2O(aq)$$

The $CuSO_4$ formed in this reaction was mixed with KI to form CuI.

$$2 CuSO_4(aq) + 5 I^-(aq) \longrightarrow 2 CuI(s) + I_3^-(aq) + 2 SO_4^{2-}(aq)$$

The I_3^- formed in this reaction was then titrated with $S_2O_3^{2-}$.

$$I_3^-(aq) + 2 S_2O_3^{2-}(aq) \longrightarrow 3 I^-(aq) + S_4O_6^{2-}(aq)$$

If 31.5 milliliters of 1.00 M $S_2O_3^{2-}$ was consumed in this titration, what was the percent by weight of copper in the original sample of bronze?

CHAPTER 19

ELECTROCHEMISTRY

CHAPTER CONTENTS

19.1 CORROSION

The Statue of Liberty, standing more than 300 feet tall from the base of her pedestal to the tip of her torch, was a gift to the United States from the people of France. Designed and built in Paris by the sculptor Frederic Bartholdi, she consists of 300 thin copper plates weighing 80 tons supported by an iron skeleton designed by Alexandre Eiffel.

The iron skeleton consists of more than 1300 iron bars that were bent to follow the contours of the copper skin and then connected to the skin with copper bands, or saddles, as shown in Figure 19.1. Because the iron bars are flat, they are free to move through the saddles, thereby allowing the skin to expand or contract as the temperature changes. Asbestos strips were originally inserted between the iron bars and the copper saddles to keep the two metals from touching. Over the years, these insulating strips wore away, and the two metals came into contact.

As soon as the iron bars and copper plate came into contact, one of these metals started to corrode. Which one? Was it the copper skin or the iron skeleton? There are several ways of finding the answer to this question. We can try to answer it on the basis of observation. Several years ago, when the media were saturated with photographs of Liberty undergoing repair, did you notice any obvious damage to the outer skin? Were there numerous holes, for example? No; which means the damage must have been to the iron skeleton that supports the skin.

We can also apply the theory of oxidation-reduction reactions to this question. According to the table of common oxidizing agents and reducing agents given in Chapter 18, iron metal is a better reducing agent than copper metal. We might therefore expect the iron skeleton to oxidize, or corrode, before the copper skin.

We can confirm this prediction with a simple experiment. We start by scraping the sides of a copper penny and a short length of iron wire until both metals are shiny. We then wrap the iron wire around the copper penny. To simulate the effect of the ocean environment, we soak a paper towel in water and sprinkle salt on the towel. To simulate the effect of pollutants in the atmosphere, we add a few drops of vinegar to the towel. We then place the damp towel at the bottom of a petri dish, place the copper penny on top of the towel, and cover the dish with clear plastic wrap. Over a period of days, it becomes obvious that the iron wire, not the copper penny, corrodes.

It isn't surprising that iron corrodes, or rusts, when exposed to salt water. What is surprising is the fact that iron corrodes as much as 1000 times as fast when it comes in contact with copper metal. Some of the iron bars in contact with the copper skin in the Statue of Liberty lost two-thirds of their cross-sectional area to corrosion in less than 100 years.

To understand why iron corrodes so much faster when it comes in contact with copper we have to build a model for ***electrochemical reactions*** — chemical reactions that involve the flow of electrons. Once established, this model will not only explain the corrosion of metals but also the electrochemical reactions that fuel the batteries in digital watches and cameras, the fuel cells that provide energy for satellites, and the lead storage batteries in cars. It will also provide a basis for understanding electrolysis as a technique for preparing strong oxidizing agents (such as F_2 and Cl_2) and strong reducing agents (such as Na, K, and Al).

FIG. 19.1 The copper plates that form the skin of the Statue of Liberty are supported by an iron skeleton. Asbestos was originally placed between the two metals to keep them from making electrical contact. Because copper and iron do not expand and contract at the same rate when the temperature changes, this asbestos was gradually worn away by friction. Once the two metals came into contact, the iron began to corrode.

Exercise 19.1

Predict what will happen if builders use iron nails to attach a copper roof to a building.

Solution

Iron nails corrode so fast when in contact with copper metal that the roof falls off, often within six months to two years.

19.2 ELECTRICAL WORK FROM SPONTANEOUS OXIDATION-REDUCTION REACTIONS

Section 18.11 introduced the following rule, which can be used to predict whether an oxidation-reduction reaction should occur.

> **Spontaneous oxidation-reduction reactions convert the stronger of a pair of oxidizing agents and the stronger of a pair of reducing agents into a weaker oxidizing agent and a weaker reducing agent.**

Exercise 19.2

Use Table 18.3 to predict whether zinc metal should dissolve in acid.

Solution

Zinc will dissolve in acid if the metal is a strong enough reducing agent to reduce the H^+ ions in this solution to H_2 gas.

$$Zn(s) + 2\ H^+(aq) \longrightarrow Zn^{2+}(aq) + H_2(g)$$

According to Table 18.3, zinc is a stronger reducing agent than H_2, and the H^+ ion is a stronger oxidizing agent than the Zn^{2+} ion.

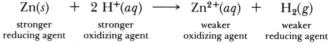

$$\underset{\substack{\text{stronger} \\ \text{reducing agent}}}{Zn(s)} + \underset{\substack{\text{stronger} \\ \text{oxidizing agent}}}{2\ H^+(aq)} \longrightarrow \underset{\substack{\text{weaker} \\ \text{oxidizing agent}}}{Zn^{2+}(aq)} + \underset{\substack{\text{weaker} \\ \text{reducing agent}}}{H_2(g)}$$

Zinc should therefore dissolve in acid.

We can test this prediction by adding a few chunks of mossy zinc to a beaker of concentrated hydrochloric acid, as shown in Figure 19.2. Within a few minutes, the zinc metal dissolves completely, and significant amounts of hydrogen gas are liberated.

This is a perfect example of a spontaneous oxidation-reduction reaction.

1. It is exothermic, in this case giving off 153.89 kilojoules per mole of zinc consumed.

2. The equilibrium constant for the reaction is very large ($K_c = 6 \times 10^{25}$), and we can therefore write the equation for the reaction as if essentially all of the reactants were converted to products.

$$Zn(s) + 2\ H^+(aq) \longrightarrow Zn^{2+}(aq) + H_2(g)$$

3. It can be formally divided into separate oxidation and reduction half-reactions.

$$\text{Oxidation:} \qquad Zn \longrightarrow Zn^{2+} + 2\ e^-$$
$$\text{Reduction:} \qquad 2\ H^+ + 2\ e^- \longrightarrow H_2$$

4. The energy given off by this reaction can be captured and made to work.

FIG. 19.2 Zinc metal reacts spontaneously with strong acids, such as concentrated hydrochloric acid, to form H_2 gas and an aqueous solution of Zn^{2+} ions.

$$Zn(s) + 2\ H^+(aq) \longrightarrow$$
$$Zn^{2+}(aq) + H_2(g).$$

The reaction gives off enough heat to boil HCl gas out of the solution to form the fumes shown in this photograph.

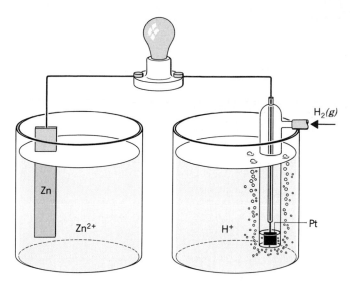

FIG. 19.3 In theory, the reaction between zinc metal and hydrochloric acid will do work if the electrons transferred in this reaction are forced to flow through an external wire. In practice, connecting a strip of zinc metal immersed in $Zn^{2+}(aq)$ solution to an electrode immersed in hydrochloric acid is not enough to make the light bulb glow.

According to the first law of thermodynamics, introduced in Section 9.8, the energy given off in a chemical reaction can be converted into heat, work, or a mixture of heat and work. When we drop pieces of zinc metal into an acidic solution, most of the energy produced is given off as heat, although some work is done to expand the hydrogen gas formed in this reaction. By running the half-reactions in separate containers, however, we can force the electrons to flow from the oxidation to the reduction half-reaction through an external wire. This allows us to capture the energy given off in the reaction as electrical work.

We can start by immersing a thin strip of zinc metal in a 1.00 M aqueous solution of Zn^{2+} ions, as shown in Figure 19.3. We then immerse a short piece of platinum wire in a second beaker filled with 1.00 M hydrochloric acid and bubble H_2 gas over the Pt wire. Finally, we connect the zinc metal and platinum wire to an electric circuit that contains a light bulb.

We've now made a system in which electrons can flow from one half-reaction, or **half-cell,** to another. The same driving force that makes zinc metal react with acid when the two are in contact should operate in this system. Zinc atoms on the metal surface lose electrons to form Zn^{2+} ions, which go into solution.

$$\text{Oxidation:} \qquad Zn \longrightarrow Zn^{2+} + 2\ e^-$$

The electrons given off in this half-reaction flow through the circuit and eventually accumulate on the platinum wire to give this wire a net negative charge. The H^+ ions from the hydrochloric acid are attracted to this negative charge and migrate toward the platinum wire. When the H^+ ions touch the platinum metal surface, they pick up electrons to form hydrogen atoms, which immediately combine to form H_2 molecules.

$$\text{Reduction:} \qquad 2\ H^+ + 2\ e^- \longrightarrow H_2$$

The flow of electrons through the external wire should do enough work to light the light bulb. Nothing happens, however, when we set up the cell as shown in Figure 19.3. The zinc metal doesn't dissolve, H_2 gas isn't given off at the platinum wire, and the light bulb doesn't glow.

In order to get the light bulb to glow, we have to complete the circuit. As it now stands, electrons are transferred from one half-cell to the other. The Zn/Zn^{2+} half-cell loses electrons and thereby picks up a positive charge that interferes with

FIG. 19.4 Adding a salt bridge to the apparatus in Figure 19.3 completes the electrical circuit. The flow of electrons through the external wire now makes the light bulb glow.

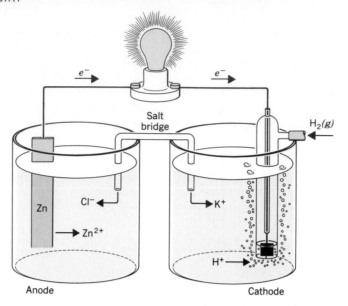

the transfer of more electrons. The reduction of H^+ ions to H_2 molecules in the H_2/H^+ half-cell leads to a buildup of negative charge in this half-cell as the positively charged H^+ ions are removed from the solution. This negative charge also interferes with the transfer of more electrons.

We can complete the circuit with a U-tube filled with a saturated solution of KCl, as shown in Figure 19.4. Negatively charged Cl^- ions flow out of one end of the U-tube to balance the positive charge on the Zn^{2+} ions created in the half-cell on the left. Positively charged K^+ ions flow out of the other end of the tube to replace the H^+ ions consumed in the half-cell on the right. The U-tube is called a *salt bridge*, because it contains a solution of an ionic compound, or salt, that serves as a bridge to complete the electric circuit. Now there is nothing to interfere with the flow of electrons from the oxidation half-cell to the reduction half-cell. As soon as the salt bridge is inserted into the circuit, the light bulb begins to glow.

19.3 VOLTAIC CELLS FROM SPONTANEOUS OXIDATION-REDUCTION REACTIONS

Almost exactly 200 years have passed since the first step was taken to capture chemical energy to do electrical work. In 1791, an Italian professor of anatomy named Luigi Galvani noticed that the leg muscles of a freshly dissected frog contracted when they were connected to one of the frog's nerves with a circuit composed of two metals. This work soon came to the attention of an Italian professor of physics, Alessandro Volta, who eventually concluded that the frog's leg was a detector of the electric current generated by the bimetallic circuit.

Further experiments led Volta to distinguish between two classes of materials that conducted electric currents. One class contains metals, such as copper, zinc, and silver, that generate an electric current when brought into contact. The other class contains liquids, such as aqueous salt solutions, that carry the electric current. By assembling a cell consisting of alternating silver and zinc metal discs separated by moist cardboard, Volta was able to construct the first continuous source of electric current.

In recognition of the contributions of Luigi Galvani and Alessandro Volta, electrochemical cells that use a spontaneous oxidation-reduction reaction to generate an electric current are known as either *galvanic cells* or *voltaic cells.* For reasons that will become apparent in the next section, we will refer to these cells from now on as *voltaic cells.*

Let's take another look at the voltaic cell in Figure 19.4. Within each half-cell, reaction occurs on the surface of the metal electrode. At the zinc electrode, zinc atoms lose electrons and go into solution as Zn^{2+} ions. The electrons that accumulate on the zinc metal flow through the metal until they reach the wire that connects the zinc electrode with the platinum wire. They then flow down the platinum wire, where they eventually come in contact with an H^+ ion in the neighboring solution and reduce this ion to a hydrogen atom, which combines with another hydrogen atom to form an H_2 molecule.

The electrode at which oxidation takes place in a voltaic cell is called the *anode.* The electrode at which reduction occurs is called the *cathode.* There are several ways of remembering these definitions. For example, you can use the following mnemonic device: Note that the first letter of *c*athode appears in redu*c*tion, but not oxidation, while the first letter of *a*node appears in oxid*a*tion, but not reduction.

You can also remember the identity of the cathode and the anode by recognizing that positive ions, or *cations,* flow toward the *cathode,* while negative ions, or *anions,* flow toward the *anode.* In the Zn/H^+ voltaic cell, H^+ ions flow toward the cathode, where they are reduced to H_2 gas. On the other side of the cell, Cl^- ions are released from the salt bridge and flow toward the anode, where the zinc metal is oxidized.

Exercise 19.3

Draw a picture of a voltaic cell based on the following reaction.

$$Zn(s) + Cu^{2+}(aq) \longrightarrow Zn^{2+}(aq) + Cu(s)$$

Label the cathode and the anode, show the direction in which the electrons flow through the external circuit, and identify the half-reaction that occurs at each electrode.

Solution

This cell is based on the following half-reactions.

$$\text{Oxidation:} \qquad Zn \longrightarrow Zn^{2+} + 2\ e^-$$
$$\text{Reduction:} \qquad Cu^{2+} + 2\ e^- \longrightarrow Cu$$

Oxidation always occurs at the anode, and reduction always occurs at the cathode. Electrons flow from the anode (where they are given off) to the cathode (where they are consumed). This cell can be represented by the drawing in Figure 19.5.

19.4 STANDARD-STATE CELL POTENTIALS FOR VOLTAIC CELLS

The voltaic cells described in the preceding section do electrical work by driving a electric current through a wire. The potential of a cell to do work is known as the *cell potential.* Cell potentials are measured with a voltmeter, which replaces the light bulb in the external circuit, as shown in Figure 19.5. The cell potential is

FIG. 19.5 The voltaic cell described in Exercise 19.3 can be represented by the drawing shown here. Zinc metal is oxidized at the anode to form Zn^{2+} ions, which go into solution, and Cu^{2+} ions are reduced at the cathode to form copper metal. Electrons flow from the anode to the cathode through an external wire. The potential of this cell for doing work can be measured with a voltmeter.

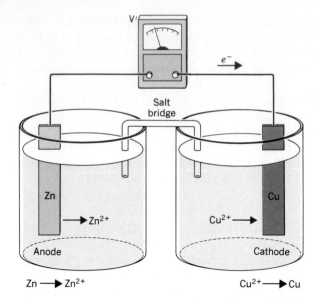

reported in units of *volts*. (You can remember that *volt*aic cells use a spontaneous chemical reaction to do electrical work by noting that they produce a cell potential measured in *volts*.)

The potential of a voltaic cell depends on the concentrations of the ions and the partial pressures of any gases involved in the reaction and on the temperature at which the reaction is run. The following set of ***standard-state conditions*** for electrochemical measurements has been defined.

1. **All solutions are 1 *M*.**
2. **All gases have a partial pressure of 1 atm.**
3. **All measurements are done at 25°C.**

Cell potentials measured under standard-state conditions are represented by the symbol $E°$.

Standard-state cell potentials are given in units of volts. By definition, it takes 1 joule of energy to transport 1 coulomb of electrical charge across a potential of 1 volt.

$$1 \text{ V} = \frac{1 \text{ J}}{1 \text{ C}}$$

Exercise 19.4

Calculate the potential of a cell that consumes 19.3 kilojoules of energy to transport the charge on a mole of electrons.

Solution

According to the values of the fundamental constants in Table A-2, a mole contains 6.022×10^{23} particles, and the charge on an electron is 1.602×10^{-19} coulomb. A mole of electrons therefore carries a charge of 96,480 coulombs (to four significant figures).

$$6.022 \times 10^{23} \, \frac{e^-}{\text{mol}} \times 1.602 \times 10^{-19} \, \frac{\text{C}}{e^-} = \textbf{96,480 C}$$

According to this equation, the cell potential is equal to the amount of energy required to move the electrons divided by the magnitude of the charge on the electrons. This cell therefore has a potential of 0.200 volt.

$$\frac{19,300 \text{ J}}{96,480 \text{ C}} = 0.200 \text{ V}$$

The standard-state cell potential, $E°$, measures the strength of the driving force behind the chemical reaction. The larger the difference between the oxidizing and reducing strengths of the reactants and products, the larger the cell potential. To obtain relatively large cell potentials, we have to react a strong reducing agent with a strong oxidizing agent.

The experimental value for the standard-state cell potential for the reaction between zinc metal and acid is 0.76 volt.

$$Zn(s) + 2 \text{ H}^+(aq) \longrightarrow Zn^{2+}(aq) + H_2(g) \qquad E° = 0.76 \text{ V}$$

The fact that the cell potential depends on the difference between the oxidizing strength and the reducing strength of the reactants and products of a reaction has an important implication. The cell potential for this reaction measures the relative reducing power of zinc metal compared with hydrogen gas, but it doesn't give us any information about the absolute value of the reducing power for either zinc metal or H_2.

We encountered a similar problem with enthalpies of reaction in Chapter 9. There, we solved the problem by arbitrarily defining the enthalpy of formation of an element in its most stable state at 25°C and 1 atm as exactly 0.000... kJ/mol. We then defined all other enthalpies of reaction relative to this arbitrary point.

We can do the same thing with cell potentials. We will arbitrarily define the standard-state potential for the reduction of H^+ ions to H_2 gas as exactly zero volts.

$$2 \text{ H}^+ + 2 \ e^- \longrightarrow H_2 \qquad E° = 0.000... \text{ V}$$

We will then use this arbitrary reference point to calibrate the potential of any other half-reaction.

The standard-state cell potential for a reaction is the sum of the standard-state potentials for the oxidation and reduction half-reactions.

$$E° = E°_{ox} + E°_{red}$$

As noted above, the standard-state cell potential for the reaction between zinc metal and hydrochloric acid is 0.76 volt.

$$0.76 \text{ V} = E°_{ox} + E°_{red}$$

If $E°_{red}$ for the reduction of H^+ ions to H_2 molecules is defined as exactly 0.000... volt, then $E°_{ox}$ for the oxidation of zinc metal to Zn^{2+} ions must be 0.76 volt.

$$0.76 \text{ V} = E°_{ox} + 0.000 \text{ V}$$
$$E°_{ox} = 0.76 \text{ V}$$

We can come to the same conclusion by arguing that the standard-state cell potential for the reaction between zinc and acid can be 0.76 volt only if the standard-state half-cell potential for the oxidation of zinc metal is 0.76 volt.

$$
\begin{array}{ll}
Zn \longrightarrow Zn^{2+} + 2 \ e^- & E°_{ox} = 0.76 \text{ V} \\
\underline{+2 \text{ H}^+ + 2 \ e^- \longrightarrow H_2} & \underline{E°_{red} = 0.00 \text{ V}} \\
Zn + 2 \text{ H}^+ \longrightarrow Zn^{2+} + H_2 & E° = E°_{ox} + E°_{red} = 0.76 \text{ V}
\end{array}
$$

19.5 PREDICTING SPONTANEOUS OXIDATION-REDUCTION REACTIONS FROM THE SIGN OF $E°$

In the preceding section, we stated that the size, or magnitude, of the cell potential is a measure of the driving force behind the reaction. The larger the difference between the oxidizing and reducing strengths of the reactants and products, the larger the cell potential. The sign of the cell potential tells us the direction in which the reaction should occur.

> **Oxidation-reduction reactions that have a positive overall cell potential are spontaneous.**

Zinc metal reacts with acid, for example, because the overall cell potential for this reaction is positive.

$$Zn(s) + 2 H^+(aq) \longrightarrow Zn^{2+}(aq) + H_2(g) \qquad E° = 0.76 \text{ V}$$

Exercise 19.5

Use the overall standard-state cell potential for the following reactions to predict which reactions occur spontaneously.

(a) $Cu(s) + 2 Ag^+(aq) \longrightarrow Cu^{2+}(aq) + 2 Ag(s)$ $\qquad E° = 0.46 \text{ V}$
(b) $2 Fe^{3+}(aq) + 2 Cl^-(aq) \longrightarrow 2 Fe^{2+}(aq) + Cl_2(g)$ $\qquad E° = -0.59 \text{ V}$
(c) $2 Fe^{3+}(aq) + 2 I^-(aq) \longrightarrow 2 Fe^{2+}(aq) + I_2(aq)$ $\qquad E° = 0.24 \text{ V}$
(d) $2 H_2O_2(aq) \longrightarrow 2 H_2O(l) + O_2(aq)$ $\qquad E° = 1.09 \text{ V}$

Solution

Any reaction for which the overall cell potential is positive will occur spontaneously. Reactions (a), (c), and (d) are therefore spontaneous.

What happens to the cell potential when we reverse the direction in which the reaction is written?

Exercise 19.6

Use the standard-state cell potential for the following reaction

$$Cu(s) + 2 H^+(aq) \longrightarrow Cu^{2+}(aq) + H_2(g) \qquad E° = -0.34 \text{ V}$$

to predict the standard-state cell potential for the opposite reaction.

$$Cu^{2+}(aq) + H_2(g) \longrightarrow Cu(s) + 2 H^+(aq) \qquad E° = ?$$

Solution

Turning the reaction around doesn't change the relative strengths of Cu^{2+} and H^+ ions as oxidizing agents or of copper metal and H_2 as reducing agents. The size, or magnitude, of the potential remains the same. But turning the equation around changes the sign of the cell potential and can therefore turn an unfavorable reaction into one that is spontaneous or vice versa.

The standard-state cell potential for the reduction of Cu^{2+} ions by H_2 gas is therefore $+0.34$ V.

$$Cu^{2+}(aq) + H_2(g) \longrightarrow Cu(s) + 2 H^+(aq) \qquad E° = -(-0.34 \text{ V}) = +0.34 \text{ V}$$

According to this exercise, copper should not dissolve in 1 M acid, because the standard-state potential for the following reaction is negative.

$$Cu(s) + 2 H^+(aq) \longrightarrow Cu^{2+}(aq) + H_2(g) \qquad E° = -0.34 \text{ V}$$

When the reaction is reversed, however, the potential becomes positive and the reaction becomes favorable.

$$Cu^{2+}(aq) + H_2(g) \longrightarrow Cu(s) + 2 H^+(aq) \qquad E° = ?$$

We therefore expect that hydrogen is a strong enough reducing agent to reduce Cu^{2+} ions to copper metal. This is consistent with an observation made in Section 7.11, where we noted that hydrogen can reduce copper(II) oxide to copper metal.

$$CuO(s) + H_2(g) \longrightarrow Cu(s) + H_2O(g)$$

19.6 STANDARD-STATE REDUCTION HALF-CELL POTENTIALS

The standard-state cell potentials for some common half-reactions are given in Table 19.1; a more complete table can be found in the appendix. These tables are based on the following conventions.

1. **All half-reactions are written as reduction half-reactions.**
2. **All half-reactions that produce a better reducing agent than H_2 are given negative half-cell reduction potentials.**
3. **The most negative half-cell potentials are placed at the top of the list of half-reactions.**

This is not the first time you have seen a table like this. Ever since oxidation and reduction were introduced in Chapter 7, you have encountered tables of relative oxidizing and reducing power that list the strongest reducing agents in the upper right-hand corner and the strongest oxidizing agents in the lower left-hand corner. The beauty of this particular version is that it gives a quantitative measure of the difference between the strengths of a pair of reducing or oxidizing agents.

There is no need to remember that the strength of the reducing agents increases toward the upper right-hand part of this table or that the strength of the oxidizing agents increases toward the bottom left hand corner. You can recreate this pattern each time you encounter the table by remembering some of the basic chemistry of the elements developed in earlier chapters of this text. Take a look at the half-reaction at the top of the table.

$$K^+ + e^- \rightleftharpoons K \qquad E°_{red} = -2.924 \text{ V}$$

What do we know about potassium metal? Potassium is one of the most reactive metals — it bursts into flame when added to water, for example. Furthermore, we know that metals are reducing agents in all of their chemical reactions. When we find potassium in this table, we can therefore conclude that it is listed among the strongest reducing agents.

Conversely, look at the last reaction in the table.

$$F_2 + 2 e^- \rightleftharpoons 2 F^- \qquad E°_{red} = 3.03 \text{ V}$$

We know that nonmetals tend to be oxidizing agents. We also know that fluorine is the most electronegative element in the periodic table. It shouldn't be surprising to find that F_2 is the strongest oxidizing agent in Table 19.1.

TABLE 19.1

Standard-State Reduction Potentials, E°_{red}

Half-Reaction	E°_{red}	
$K^+ + e^- \rightleftharpoons K$	-2.924	Best
$Ba^{2+} + 2\,e^- \rightleftharpoons Ba$	-2.90	reducing
$Ca^{2+} + 2\,e^- \rightleftharpoons Ca$	-2.76	agents
$Na^+ + e^- \rightleftharpoons Na$	-2.7109	
$Mg^{2+} + 2\,e^- \rightleftharpoons Mg$	-2.375	
$H_2 + 2\,e^- \rightleftharpoons 2\,H^-$	-2.23	
$Al^{3+} + 3\,e^- \rightleftharpoons Al$	-1.706	
$Mn^{2+} + 2\,e^- \rightleftharpoons Mn$	-1.04	
$Zn^{2+} + 2\,e^- \rightleftharpoons Zn$	-0.7628	
$Cr^{3+} + 3\,e^- \rightleftharpoons Cr$	-0.74	
$S + 2\,e^- \rightleftharpoons S^{2-}$	-0.508	
$2\,CO_2 + 2\,H^+ + 2\,e^- \rightleftharpoons H_2C_2O_4$	-0.49	
$Cr^{3+} + e^- \rightleftharpoons Cr^{2+}$	-0.41	
$Fe^{2+} + 2\,e^- \rightleftharpoons Fe$	-0.409	
$Co^{2+} + 2\,e^- \rightleftharpoons Co$	-0.28	
$Ni^{2+} + 2\,e^- \rightleftharpoons Ni$	-0.23	
$Sn^{2+} + 2\,e^- \rightleftharpoons Sn$	-0.1364	
$Pb^{2+} + 2\,e^- \rightleftharpoons Pb$	-0.1263	
$Fe^{3+} + 3\,e^- \rightleftharpoons Fe$	-0.036	
$2\,H^+ + 2\,e^- \rightleftharpoons H_2$	$0.000...$	
$S_4O_6{}^{2-} + 2\,e^- \rightleftharpoons 2\,S_2O_3{}^{2-}$	0.0895	
$Sn^{4+} + 2\,e^- \rightleftharpoons Sn^{2+}$	0.15	
$Cu^{2+} + e^- \rightleftharpoons Cu^+$	0.158	
$Cu^{2+} + 2\,e^- \rightleftharpoons Cu$	0.3402	Reducing
$O_2 + 2\,H_2O + 4\,e^- \rightleftharpoons 4\,OH^-$	0.401	strength
$Cu^+ + e^- \rightleftharpoons Cu$	0.522	increases
$I_3^- + 2\,e^- \rightleftharpoons 3\,I^-$	0.5338	
$MnO_4^- + 2\,H_2O + 3\,e^- \rightleftharpoons MnO_2 + 4\,OH^-$	0.588	
$O_2 + 2\,H^+ + 2\,e^- \rightleftharpoons H_2O_2$	0.682	
$Fe^{3+} + e^- \rightleftharpoons Fe^{2+}$	0.770	
$Hg_2{}^{2+} + 2\,e^- \rightleftharpoons Hg$	0.7961	
$Ag^+ + e^- \rightleftharpoons Ag$	0.7996	
$Hg^{2+} + 2\,e^- \rightleftharpoons Hg$	0.851	
$H_2O_2 + 2\,e^- \rightleftharpoons 2\,OH^-$	0.88	
$HNO_3 + 3\,H^+ + 3\,e^- \rightleftharpoons NO + 2\,H_2O$	0.96	
$Br_2(aq) + 2\,e^- \rightleftharpoons 2\,Br^-$	1.087	
$2\,IO_3^- + 12\,H^+ + 10\,e^- \rightleftharpoons I_2 + 6\,H_2O$	1.19	
$CrO_4{}^{2-} + 8\,H^+ + 3\,e^- \rightleftharpoons Cr^{3+} + 4\,H_2O$	1.195	
$MnO_2 + 4\,H^+ + 2\,e^- \rightleftharpoons Mn^{2+} + 2\,H_2O$	1.208	
$O_2 + 4\,H^+ + 4\,e^- \rightleftharpoons 2\,H_2O$	1.229	
$Cr_2O_7{}^{2-} + 14\,H^+ + 6\,e^- \rightleftharpoons 2\,Cr^{3+} + 7\,H_2O$	1.33	
$Cl_2(g) + 2\,e^- \rightleftharpoons 2\,Cl^-$	1.3583	
$PbO_2 + 4\,H^+ + 2\,e^- \rightleftharpoons Pb^{2+} + 2\,H_2O$	1.467	
$MnO_4^- + 8\,H^+ + 5\,e^- \rightleftharpoons Mn^{2+} + 4\,H_2O$	1.491	
$Au^+ + e^- \rightleftharpoons Au$	1.68	
$H_2O_2 + 2\,H^+ + 2\,e^- \rightleftharpoons 2\,H_2O$	1.776	
$S_2O_8{}^{2-} + 2\,e^- \rightleftharpoons 2\,SO_4{}^{2-}$	2.05	
$O_3(g) + 2\,H^+ + 2\,e^- \rightleftharpoons O_2(g) + H_2O$	2.07	
$F_2(g) + 2\,H^+ + 2\,e^- \rightleftharpoons 2\,HF(aq)$	3.03	

Oxidizing strength increases

Best oxidizing agents

Referring to either end of this table can also help you remember the sign convention for cell potentials. Section 19.5 introduced the following rule: Oxidation-reduction reactions that have a positive overall cell potential are spontaneous. This is consistent with the data in Table 19.1. We know that fluorine likes to gain electrons to form fluoride ions, and the half-cell potential for this reaction is positive.

$$F_2 + 2\ e^- \rightleftharpoons 2\ F^- \qquad E^\circ_{red} = 3.03\ \text{V}$$

We also know that potassium is an excellent reducing agent. Thus, the potential for the reduction of K^+ ions to potassium metal is negative

$$K^+ + e^- \rightleftharpoons K \qquad E^\circ_{red} = -2.924\ \text{V}$$

but the potential for the oxidation of potassium to K^+ ions is positive.

$$K \rightleftharpoons K^+ + e^- \qquad E^\circ_{ox} = -(-2.924\ \text{V}) = 2.924\ \text{V}$$

Exercise 19.7

Which of the following is the strongest oxidizing agent?

(a) H_2O_2 in acid
(b) H_2O_2 in base
(c) MnO_4^- in acid
(d) MnO_4^- in base
(e) CrO_4^{2-} in acid

Solution

We can start by searching Table 19.1 for the standard-state reduction half-cell potential for each of these solutions.

(a) $H_2O_2 + 2\ H^+ + 2\ e^- \rightleftharpoons 2\ H_2O$ $E^\circ_{red} = 1.776\ \text{V}$
(b) $H_2O_2 + 2\ e^- \rightleftharpoons 2\ OH^-$ $E^\circ_{red} = 0.88\ \text{V}$
(c) $MnO_4^- + 8\ H^+ + 5\ e^- \rightleftharpoons Mn^{2+} + 4\ H_2O$ $E^\circ_{red} = 1.491\ \text{V}$
(d) $MnO_4^- + 2\ H_2O + 3\ e^- \rightleftharpoons MnO_2 + 4\ OH^-$ $E^\circ_{red} = 0.588\ \text{V}$
(e) $CrO_4^{2-} + 8\ H^+ + 3\ e^- \rightleftharpoons Cr^{3+} + 4\ H_2O$ $E^\circ_{red} = 1.195$

According to these data, the strongest oxidizing agent is H_2O_2 in acid.

19.7 PREDICTING STANDARD-STATE CELL POTENTIALS

In 1836, a chemistry professor from London named John Frederic Daniell developed the first voltaic cell stable enough to be used as a battery. This so-called Daniell cell consisted of a zinc rod coated with mercury immersed in dilute sulfuric acid inside a porous cup, which was in turn immersed in a solution of copper(II) sulfate in contact with a copper container (see Figure 19.6).

For our purposes, it is easier to work with the idealized Daniell cell shown in Figure 19.7. We can use the known values of the standard-state reduction potentials for the Cu/Cu^{2+} and Zn/Zn^{2+} half-cells to predict the standard-state cell

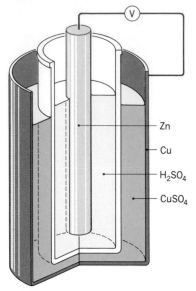

FIG. 19.6 The first stable battery was a voltaic cell developed by John Frederic Daniell. The Daniell cell consisted of a zinc rod coated with mercury immersed in dilute H_2SO_4 in a porous cup, which was placed in a solution of $CuSO_4$ in a copper metal container.

potential for the Daniell cell and to determine which electrode is the anode and which is the cathode.

We can start by writing a balanced chemical equation for the reaction that occurs in this cell. Table 19.1 suggests that zinc is a better reducing agent than copper and that the Cu^{2+} ion is a better oxidizing agent than the Zn^{2+} ion. The overall reaction must therefore involve the reduction of Cu^{2+} ions by zinc metal.

$$Zn(s) + Cu^{2+}(aq) \longrightarrow Zn^{2+}(aq) + Cu(s)$$

We can then divide this reaction into separate oxidation and reduction half-reactions.

$$\text{Reduction:} \qquad Cu^{2+} + 2\ e^- \longrightarrow Cu$$
$$\text{Oxidation:} \qquad Zn \longrightarrow Zn^{2+} + 2\ e^-$$

The potential for the reduction of Cu^{2+} ions to copper metal can be found in Table 19.1. To find the potential for the oxidation of zinc metal, we have to reverse the sign on the potential for the Zn/Zn^{2+} couple in this table.

$$\text{Reduction:} \qquad Cu^{2+} + 2\ e^- \longrightarrow Cu \qquad E°_{red} = 0.34\ V$$
$$\text{Oxidation:} \qquad Zn \longrightarrow Zn^{2+} + 2\ e^- \qquad E°_{ox} = -(-0.76\ V)$$

The overall standard-state cell potential for this cell is the sum of the standard-state potentials for the two half-cells — 1.10 V.

$$
\begin{array}{ll}
Cu^{2+} + 2\ e^- \longrightarrow Cu & E°_{red} = 0.34\ V \\
+\ Zn \longrightarrow Zn^{2+} + 2\ e^- & E°_{ox} = 0.76\ V \\
\hline
Zn + Cu^{2+} \longrightarrow Zn^{2+} + Cu & E° = E°_{red} + E°_{ox} = 1.10\ V
\end{array}
$$

We can then remember that oxidation always occurs at the anode and reduction always occurs at the cathode. The Zn/Zn^{2+} half-cell is therefore the anode, and the Cu^{2+}/Cu half-cell is the cathode, as shown in Figure 19.7.

FIG. 19.7 An idealized line drawing of a Daniell cell. Zinc metal is oxidized to Zn^{2+} ions at the anode, and Cu^{2+} ions are reduced to copper metal at the cathode. K^+ ions flow from the salt bridge into the cathode half-cell to compensate for the Cu^{2+} ions consumed in this reaction, and Cl^- ions flow from the salt bridge into the anode half-cell to balance the charge on the Zn^{2+} ions produced at this electrode.

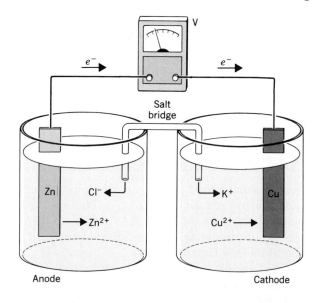

Exercise 19.8

Use the data in Table 19.1 to explain why copper metal does not dissolve in a typical strong acid, such as hydrochloric acid,

$$Cu(s) + 2\ H^+(aq) \longrightarrow Cu^{2+}(aq) + H_2(g)$$

but will dissolve in dilute nitric acid.

$$3\ Cu(s) + 2\ HNO_3(aq) + 6\ H^+(aq) \longrightarrow 3\ Cu^{2+}(aq) + 2\ NO(g) + 4\ H_2O(l)$$

Solution

Copper does not dissolve in a typical strong acid because the overall cell potential for the oxidation of copper metal to Cu^{2+} ions coupled with the reduction of H^+ ions to H_2 is negative.

$$
\begin{array}{ll}
Cu \longrightarrow Cu^{2+} + 2\ e^- & E°_{ox} = -(0.34\ V) \\
\underline{+\ 2\ H^+ + 2\ e^- \longrightarrow H_2} & \underline{E°_{red} = 0.000...\ V} \\
Cu + 2\ H^+ \longrightarrow Cu^{2+} + H_2 & E° = E°_{ox} + E°_{red} = -\mathbf{0.34\ V}
\end{array}
$$

Copper dissolves in nitric acid because the half-reaction at the cathode now involves the reduction of nitric acid to NO gas, and the potential for this half-reaction is strong enough to overcome the unfavorable half-cell potential for oxidation of copper metal to Cu^{2+} ions.

$$
\begin{array}{ll}
3(Cu \longrightarrow Cu^{2+} + 2\ e^-) & E°_{ox} = -(0.34\ V) \\
\underline{+\ 2(HNO_3 + 3\ H^+ + 3\ e^- \longrightarrow NO + 2\ H_2O)} & \underline{E°_{red} = 0.96\ V} \\
3\ Cu + 2\ HNO_3 + 6\ H^+ \longrightarrow & E° = E°_{ox} + E°_{red} = \mathbf{0.62\ V} \\
\qquad\qquad 3\ Cu^{2+} + 2\ NO + 4\ H_2O &
\end{array}
$$

The last step in Exercise 19.8 illustrates an important point. The units of half-cell potentials are volts, not volts per mole or volts per electron. All we have to do when combining half-reactions is add the two half-cell potentials. We do not multiply these potentials by the integers used to balance the number of electrons transferred in the reaction. In Exercise 19.8, for example, we added the half-cell potential for oxidation to the half-cell potential for reduction without multiplying these potentials by the coefficients used to balance the number of electrons given off and consumed in the reaction.

19.8 LINE NOTATION FOR VOLTAIC CELLS

Voltaic cells can be described by a standard line notation based on the following conventions.

1. **A single vertical line indicates a change in state or phase.**
2. **Within a half-cell, the reactants are always listed before the products.**
3. **Concentrations of aqueous solutions are written in parentheses after the symbol for the ion or molecule.**
4. **A double vertical line is used to indicate a salt bridge.**
5. **The line notation for the anode (oxidation) is always written before the line notation for the cathode (reduction).**

The line notation for a standard-state Daniell cell is written as follows.

$$Zn|Zn^{2+}(1.0 \ M)\|Cu^{2+}(1.0 \ M)|Cu$$

anode   cathode
(oxidation)   (reduction)

Electrons flow from the anode to the cathode in a voltaic cell. (They flow from the electrode at which they are given off to the electrode at which they are consumed.) Reading from left to right, this line notation therefore corresponds to the direction in which electrons flow.

Exercise 19.9

Write the line notation for the cell shown in Figure 19.4.

Solution

The cell in Figure 19.4 is based on the following half-reactions.

Oxidation: $Zn \longrightarrow Zn^{2+} + 2 \ e^-$

Reduction: $2 \ H^+ + 2 \ e^- \longrightarrow H_2$

Since we list the reactants before the products in a half-cell, we write the following line notation for the anode.

$$Zn|Zn^{2+}(1.0 \ M)$$

The line notation for the cathode has to indicate that H^+ ions are reduced to H_2 gas on a platinum metal surface. This can be done as follows.

$$H^+(1.0 \ M)|H_2 \ (1 \ atm)|Pt$$

Since line notation is read in the direction in which electrons flow, we place the notation for the anode before the notation for the cathode, as follows.

$$Zn|Zn^{2+}(1.0 \ M)\|H^+(1.0 \ M)|H_2 \ (1 \ atm)|Pt$$

anode   cathode

19.9 USING STANDARD-STATE HALF-CELL POTENTIALS TO UNDERSTAND CHEMICAL REACTIONS

As noted in Section 19.5, the sign of the overall standard-state cell potential for a chemical reaction can be used to predict whether the reaction will occur spontaneously. The same information can be obtained from a table of standard-state half-cell reduction potentials.

Exercise 19.10

Which of the following pairs of ions cannot coexist in aqueous solution.

   (a) Cu^{2+} and Zn^{2+} (b) Hg^{2+} and Sn^{2+} (c) Fe^{3+} and Cr^{2+}

Solution

To answer this question, we have to search the table of standard-state reduction potentials to determine what oxidation-reduction reactions might occur when these ions are mixed.

(a) Cu^{2+} and Zn^{2+} are both oxidizing agents. Neither Table 19.1 nor the expanded table in the appendix contains a half-reaction in which either ion is a reducing agent. Thus, Cu^{2+} and Zn^{2+} ions can't undergo a redox reaction, and both ions can exist in aqueous solution at the same time.

(b) According to the data in Table 19.1, the Hg^{2+} ion is a strong enough oxidizing agent to oxidize Sn^{2+} to Sn^{4+}.

$$
\begin{array}{ll}
Hg^{2+} + 2\ e^- \longrightarrow Hg & E^\circ_{red} = 0.851\ V \\
+\ Sn^{2+} \longrightarrow Sn^{4+} + 2\ e^- & E^\circ_{ox} = -(0.14\ V) \\
\hline
Hg^{2+} + Sn^{2+} \longrightarrow Hg + Sn^{4+} & E^\circ = E^\circ_{red} + E^\circ_{ox} = 0.70\ V
\end{array}
$$

Therefore, these ions cannot coexist in aqueous solution.

(c) According to Table 19.1, the Fe^{3+} ion is a strong enough oxidizing agent to oxidize Cr^{2+} to Cr^{3+}.

$$
\begin{array}{ll}
Fe^{3+} + e^- \longrightarrow Fe^{2+} & E^\circ_{red} = 0.770\ V \\
+\ Cr^{2+} \longrightarrow Cr^{3+} + e^- & E^\circ_{ox} = -(-0.41\ V) \\
\hline
Fe^{3+} + Cr^{2+} \longrightarrow Fe^{2+} + Cr^{3+} & E^\circ = E^\circ_{red} + E^\circ_{ox} = 1.18\ V
\end{array}
$$

Therefore, these ions cannot coexist in aqueous solution.

Exercise 19.11

Nearly every form of life contains an enzyme known as catalase, which decomposes hydrogen peroxide to oxygen and water.

$$2\ H_2O_2(aq) \longrightarrow O_2(g) + 2\ H_2O(l)$$

Explain why hydrogen peroxide is unstable relative to oxygen and water.

Solution

Because the oxygen atoms in hydrogen peroxide are in the -1 oxidation state, H_2O_2 can be either oxidized to O_2 or reduced to water.

$$
\begin{array}{ll}
H_2O_2 + 2\ H^+ + 2\ e^- \longrightarrow 2\ H_2O & E^\circ_{red} = 1.776\ V \\
+\ H_2O_2 \longrightarrow O_2 + 2\ H^+ + 2\ e^- & E^\circ_{ox} = -(0.682\ V) \\
\hline
2\ H_2O_2 \longrightarrow O_2 + 2\ H_2O & E^\circ = E^\circ_{red} + E^\circ_{ox} = 1.094\ V
\end{array}
$$

According to the data in Table 19.1, H_2O_2 is strong enough to oxidize itself. The enzyme just increases the rate at which this reaction occurs.

19.10 THE NERNST EQUATION

Think about what happens when the Daniell cell in Figure 19.7 is used to do electrical work.

1. The zinc electrode becomes lighter as zinc atoms are oxidized to Zn^{2+} ions, which go into solution.

2. The copper electrode becomes heavier as Cu^{2+} ions in the solution are reduced to copper metal.

3. The concentration of Zn^{2+} ions at the anode increases while the concentration of the Cu^{2+} ions at the cathode decreases.

4. Negative ions flow from the salt bridge into the anode to balance the charge on the Zn^{2+} ions produced at this electrode. Positive ions flow from the salt bridge into the cathode to compensate for the Cu^{2+} ions consumed in this reaction.

There is one thing missing from this list. Over a period of time, the cell loses its charge. It runs down, and eventually it has to be replaced.

Why does a battery run down? What happens when it loses its charge? We can answer these questions by thinking about the changes that occur in a voltaic cell when it does electrical work. Let's assume that our cell is initially a standard-state Daniell cell in which the concentrations of the Zn^{2+} and Cu^{2+} ions are both exactly 1 molar.

$$Zn(s) + Cu^{2+}(aq) \longrightarrow Zn^{2+}(aq) + Cu(s)$$

As the cell does work, the Zn^{2+} and Cu^{2+} ion concentrations change. This change must affect the potential of the battery. In Section 19.4, the cell potential was said to be a measure of the driving force behind the reaction. As the reaction goes forward—as zinc metal is consumed and copper metal is produced—the driving force behind the reaction must become weaker. Therefore, the cell potential must become smaller.

According to the calculations in Section 19.7, the standard-state cell potential for a Daniell cell is 1.10 volts.

$$Zn(s) + Cu^{2+}(aq) \longrightarrow Zn^{2+}(aq) + Cu(s) \qquad E° = 1.10 \text{ V}$$

As the cell does work, however, the potential gradually decreases. This raises an interesting question: When does the cell potential become zero? It might be tempting to assume that the cell potential disappears when we run out of the limiting reagent—when the last piece of zinc metal disappears, or when the last few copper ions have been reduced. This assumption would be wrong, however. The cell potential becomes zero when the driving force behind the reaction disappears—in other words, when the reaction reaches equilibrium.

When the reaction is at equilibrium, there is no net change in the amount of zinc metal or copper ions in the system, so no electrons flow from the anode to the cathode. If there is no longer a net flow of electrons, the cell can no longer do electrical work. Its potential for doing work must be zero.

We need at least three terms to write an equation that describes what happens to the potential of a battery as it runs down.

1. A term for the cell potential at any moment in time, E.

2. A term for the cell potential when the system is at standard-state conditions, $E°$.

3. A term that describes how close the system is to equilibrium, X.

If the cell potential gradually decreases as the cell comes to equilibrium, we would expect this equation to take the following form.

$$E = E° - X$$

In 1889, Hermann Walther Nernst showed that the X term in this equation could be written as follows.

$$E = E° - \frac{RT}{nF} \ln Q$$

In this equation, known as the *Nernst equation,* ln indicates a natural or Naperian logarithm, which is a log to the base e, where e is an irrational number equal to 2.71828.... The other terms have the following meanings.

E is the cell potential at any moment in time.

$E°$ is the cell potential when the reaction is at standard-state conditions — when all concentrations are 1 M, all partial pressures are 1 atm, and the temperature is 25°C.

R is the ideal gas constant in units of joules per mole.

T is the temperature in kelvins.

n is the number of electrons transferred in the balanced chemical equation.

F is a constant known as a faraday, which is equal to the charge on a mole of electrons.

Q is the reaction quotient, or the ratio of the concentrations of the products to the concentrations of the reactants at that moment in time.

Before we can use the Nernst equation to calculate the potential for the Daniell cell when it is no longer at standard-state conditions, we need to determine the values of $E°$, R, n, F, and Q for this cell. As mentioned, the standard-state potential for the cell is 1.10 volts. The ideal gas constant is 8.314 J/mol. Two electrons are transferred from zinc metal to Cu^{2+} ions in this reaction, so n is 2 for this cell. The charge on a mole of electrons was calculated as part of Exercise 19.4. Carrying this calculation to as many significant figures as possible gives the following result.

$$6.022045 \times 10^{23}\, \frac{e^-}{mol} \times 1.6021892 \times 10^{-19}\, \frac{C}{e^-} = 96{,}484.56\ C/mol$$

To find the reaction quotient, Q, we divide the concentrations of the products of the reaction by the product of the concentrations of the reactants.

$$Zn(s) + Cu^{2+}(aq) \longrightarrow Zn^{2+}(aq) + Cu(s)$$

Since we never include the concentrations of solids in reaction quotients or equilibrium expressions, Q for this reaction is equal to the concentration of the Zn^{2+} ion divided by the concentration of the Cu^{2+} ion.

$$Q = \frac{(Zn^{2+})}{(Cu^{2+})}$$

For the sake of argument, we'll assume that the temperature is 25°C. We'll also replace the natural, or Naperian, logarithm to the base e with a log to the base 10 by taking advantage of the following approximation.

$$\ln X \cong 2.303 \log_{10} X$$

When this is done, the Nernst equation takes on the following form.

$$E = E° - 2.303\, \frac{RT}{nF} \log_{10} Q$$

Substituting what we know about the Daniell cell into this equation gives the following result.

$$E = E° - \frac{2.303(8.314\ J/mol)(298\ K)}{(2)(96{,}480\ C/mol)} \log \left[\frac{(Zn^{2+})}{(Cu^{2+})} \right]$$

Simplifying this equation yields the following result, which represents the cell potential of the Daniell cell at any moment in time.

$$E = E° - \frac{0.0591}{2} \log \left[\frac{(Zn^{2+})}{(Cu^{2+})} \right]$$

Exercise 19.12

Calculate the potential in the following cell at 25°C when exactly half of the Cu^{2+} ions that were present initially have been consumed.

$$Zn|Zn^{2+}(1.00\ M)||Cu^{2+}(1.00\ M)|Cu$$

Solution

At 25°C, the Nernst equation for this cell can be written as follows.

$$E = E° - \frac{0.0591}{2} \log \left[\frac{(Zn^{2+})}{(Cu^{2+})} \right]$$

The initial concentrations of the Zn^{2+} and Cu^{2+} ions were both 1.00 M. When half of the Cu^{2+} ions have been consumed, the Zn^{2+} concentration will be 1.50 M and the Cu^{2+} concentration will be 0.50 M. The cell potential at that moment in time will have dropped from 1.10 volts to 1.08 volts.

$$E = 1.10\ V - \frac{0.0591}{2} \log \left[\frac{(1.50)}{(0.50)} \right]$$
$$E = 1.10\ V - 0.0154\ V = \mathbf{1.08\ V}$$

Exercise 19.12 raises an important point. The value of E depends on the logarithm of the ratio of the concentrations of the products and the reactants. As a result, the potential of a cell or battery is more or less constant until virtually all of the reactants have been converted into products, as can be seen in the following exercise.

Exercise 19.13

Calculate the potential in the following cell when 99.9999% of the Cu^{2+} ions have been consumed.

$$Zn|Zn^{2+}(1.00\ M)||Cu^{2+}(1.00\ M)|Cu$$

Solution

When 99.9999% of the Cu^{2+} ions have been consumed, the concentration of the Cu^{2+} ion will have decreased to only 0.000001 M, while the Zn^{2+} ion concentration will have increased to 1.999999 M. Even though only 0.0001% of the limiting reactant is left, the cell potential is still 0.91 volt compared with a standard-state value of 1.10 volts.

$$E = 1.10\ V - \frac{0.0591}{2} \log \left[\frac{(1.999999)}{(1 \times 10^{-6})} \right]$$
$$E = 1.10\ V - 0.186\ V = \mathbf{0.91\ V}$$

The Nernst equation can be used to calculate the potential of a half-cell or a cell that is operating at non-standard-state conditions.

Exercise 19.14

Calculate the potential at 25°C for the following cell.

$$Cu|Cu^{2+}(0.024\ M)||Ag^+(0.0048\ M)|Ag$$

Solution

We can start by translating the line notation into an equation for the reaction.

$$Cu(s) + 2\ Ag^+(aq) \longrightarrow Cu^{2+}(aq) + 2\ Ag(s)$$

We can then look up the standard-state potentials for the two half-reactions and calculate the standard-state cell potential.

Oxidation: $\quad Cu \longrightarrow Cu^{2+} + 2\ e^- \qquad E^\circ_{ox} = -(0.3402\ V)$

Reduction: $\quad Ag^+ + e^- \longrightarrow Ag \qquad \underline{E^\circ_{red} = 0.7996\ V}$

$$E^\circ = E^\circ_{red} + E^\circ_{ox} = 0.4594\ V$$

We now set up the Nernst equation for this cell, noting that n is 2 because two electrons are transferred in the balanced equation for the reaction.

$$E = E^\circ - \frac{0.0591}{2} \log\left[\frac{(Cu^{2+})}{(Ag^+)^2}\right]$$

Substituting the concentrations of the Cu^{2+} and Ag^+ ions into this equation gives the value for the cell potential.

$$E = 0.4594\ V - \frac{0.0591}{2} \log\left[\frac{(0.024)}{(0.0048)^2}\right]$$

$$E = 0.4594\ V - 0.0892\ V = \textbf{0.3702 V}$$

The Nernst equation can also help explain the effect of changes in the pH of electrochemical reactions. Figure 19.8 shows the results of a demonstration in which glucose, $C_6H_{12}O_6$, is used to reduce Ag^+ ions to silver metal. The standard-state half-cell potential for the reduction of Ag^+ ions is approximately 0.800 volt.

$$Ag^+ + e^- \rightleftharpoons Ag \qquad E^\circ = 0.800\ V$$

The standard-state half-cell potential for the oxidation of glucose is -0.050 V.

$$C_6H_{12}O_6 + H_2O \rightleftharpoons C_6H_{12}O_7 + 2\ H^+ + 2\ e^- \qquad E^\circ = -0.050\ V$$

The overall standard-state cell potential for this reaction is therefore favorable.

$$2\ (Ag^+ + e^- \longrightarrow Ag) \qquad\qquad E^\circ = 0.800\ V$$
$$\underline{C_6H_{12}O_6 + H_2O \longrightarrow C_6H_{12}O_7 + 2\ H^+ + 2\ e^- \qquad E^\circ = -0.050\ V}$$
$$2\ Ag^+ + C_6H_{12}O_6 + H_2O \longrightarrow 2\ Ag + C_6H_{12}O_7 + 2\ H^+ \qquad E^\circ = \textbf{0.750 V}$$

The reaction isn't run under standard-state conditions in this demonstration, however. It usually takes place in a solution to which aqueous ammonia has been added. As we should expect from the discussion in Chapter 17, the concentration of the Ag^+ ion in this solution is negligibly small — most of the silver is present as the $Ag(NH_3)_2^+$ complex ion. The standard-state half-cell potential for the reduc-

FIG. 19.8 This photograph shows the silver mirror that can be plated out on the inside of a flask by reduction of Ag^+ ions with glucose in an aqueous ammonia solution.

tion of this complex is considerably smaller than the standard-state potential for the reduction of the Ag^+ ion.

$$Ag(NH_3)_2^+ + e^- \rightleftharpoons Ag + 2\ NH_3 \qquad E° = 0.373\ V$$

This leads to a significant decrease in the overall standard-state cell potential for the reaction, because the $Ag(NH_3)_2^+$ ion is a much weaker oxidizing agent than the Ag^+ ion.

However, the half-reaction for the oxidation of glucose contains a pair of H^+ ions.

$$C_6H_{12}O_6 + H_2O \rightleftharpoons C_6H_{12}O_7 + 2\ H^+ + 2\ e^-$$

The half-cell potential for this reaction therefore depends on the pH of the solution. Because two H^+ ions are given off when glucose is oxidized, the reaction quotient for this reaction depends on the square of the H^+ ion concentration. A change in this solution from standard-state conditions (pH 1) to the pH of an aqueous ammonia solution (pH ~ 11) therefore results in an increase of 0.651 volt in the half-cell potential for this reaction.

The increase in the reducing strength of glucose when the reaction is run at pH 11 more than compensates for the decrease in oxidizing strength that results from the predominance of the $Ag(NH_3)_2^+$ complex ion. Thus, the overall cell potential for the reduction of silver ions to silver metal is more favorable in aqueous ammonia than under standard-state conditions.

$$
\begin{array}{ll}
Ag(NH_3)_2^+ + e^- \longrightarrow Ag + 2\ NH_3 & E = 0.373\ V \\
\underline{C_6H_{12}O_6 + H_2O \longrightarrow C_6H_{12}O_7 + 2\ H^+ + 2\ e^-} & \underline{E° = 0.601\ V} \\
2\ Ag^+ + C_6H_{12}O_6 + H_2O \longrightarrow 2\ Ag + C_6H_{12}O_7 + 2\ H^+ & \mathbf{E° = 0.974\ V}
\end{array}
$$

19.11 USING THE NERNST EQUATION TO MEASURE EQUILIBRIUM CONSTANTS

The Nernst equation can also be used to measure the equilibrium constant for a reaction. To understand the relationship between cell potentials and equilibrium constants, we have to recognize what happens to the cell potential as an oxidation-reduction reaction comes to equilibrium. As the reaction approaches equilibrium, the driving force behind the reaction decreases, and the cell potential approaches zero. The cell potential reaches zero when the reaction reaches equilibrium.

$$\text{At equilibrium:} \qquad E = 0$$

What implications does this have for the Nernst equation?

$$E = E° - \frac{2.303\ RT}{nF} \log Q$$

At equilibrium, the reaction quotient is equal to the equilibrium constant.

$$\text{At equilibrium:} \qquad E = E° - \frac{2.303\ RT}{nF} \log K$$

But the overall cell potential for a reaction at equilibrium is zero.

$$\text{At equilibrium:} \qquad 0 = E° - \frac{2.303\ RT}{nF} \log K$$

Solving for the standard-state cell potential gives the following equation.

At equilibrium:
$$E° = \frac{2.303\, RT}{nF} \log K$$

Solving this equation for the logarithm of the equilibrium constant gives the following result.

At equilibrium:
$$\log K = \frac{nFE°}{2.303\, RT}$$

This equation suggests that we can calculate the equilibrium constant for any redox reaction from its standard-state cell potential.

Exercise 19.15

Calculate the equilibrium constant at 25 °C for the reaction between zinc metal and acid.

$$Zn(s) + 2\, H^+(aq) \longrightarrow Zn^{2+}(aq) + H_2(g)$$

Solution

We can start by calculatiing the overall standard-state cell potential for this reaction.

$$Zn \longrightarrow Zn^{2+} + 2\, e^- \qquad\qquad E°_{ox} = -(-0.7628\text{ V})$$
$$\underline{+\, 2\, H^+ + 2\, e^- \longrightarrow H_2} \qquad\qquad \underline{E°_{red} = 0.0000\text{ V}}$$
$$Zn(s) + 2\, H^+(aq) \longrightarrow Zn^{2+}(aq) + H_2(g) \qquad E° = 0.7628\text{ V}$$

We then write the Nernst equation

$$E = E° - \frac{2.303\, RT}{nF} \log Q$$

and substitute into this equation the implications of the fact that the reaction is at equilibrium.

$$0 = E° - \frac{2.303\, RT}{nF} \log K$$

This equation can be rearranged as follows.

$$E° = \frac{2.303\, RT}{nF} \log K$$

It can then be solved for the logarithm of the equilibrium constant.

$$\log K = \frac{nFE°}{2.303\, RT}$$

Substituting what we know about the reaction into this equation gives the following result.

$$\log K = \frac{(2)(96{,}480\text{ C})(0.7628\text{ V})}{(2.303)(8.314\text{ J/mol})(298\text{ K})} = \mathbf{25.8}$$

If the logarithm of the equilibrium constant is equal to 25.8, then the equilibrium constant for this reaction is equal to 10 raised to the 25.8 power.

$$K = 10^{25.8} = 6.3 \times 10^{25}$$

This is a very large equilibrium constant; equilibrium obviously lies heavily on the side of the products. The equilibrium constant is so large that the equation for this reaction is written as if it proceeds to completion.

$$Zn(s) + 2\ H^+(aq) \longrightarrow Zn^{2+}(aq) + H_2(g)$$

We can even use this technique to calculate equilibrium constants for reactions that don't seem to involve oxidation-reduction.

Exercise 19.16

Calculate the complex formation equilibrium constant for the $Zn(NH_3)_4{}^{2+}$ complex ion.

Solution

The table of standard-state reduction potentials in the appendix contains the following reduction half-reactions.

$$Zn(NH_3)_4{}^{2+} + 2\ e^- \rightleftharpoons Zn + 4\ NH_3 \qquad E^\circ_{red} = -1.04\ V$$
$$Zn^{2+} + 2\ e^- \rightleftharpoons Zn \qquad E^\circ_{red} = -0.7628\ V$$

By reversing the first half-reaction and adding it to the second, we can obtain an overall equation that corresponds to the complex formation equilibrium.

$$\begin{array}{ll} Zn + 4\ NH_3 \longrightarrow Zn(NH_3)_4{}^{2+} + 2\ e^- & E^\circ_{ox} = 1.04\ V \\ + Zn^{2+} + 2\ e^- \longrightarrow Zn & E^\circ_{red} = -0.7628\ V \\ \hline Zn^{2+} + 4\ NH_3 \longrightarrow Zn(NH_3)_4{}^{2+} & E^\circ = 0.28\ V \end{array}$$

We can then start with the Nernst equation and assume that the reaction is at equilibrium. In this case, the equilibrium constant for the reaction is equal to the complex formation constant for the $Zn(NH_3)_4{}^{2+}$ complex ion—K_f.

$$0 = E^\circ - \frac{2.303\ RT}{nF} \log K_f$$

We can rearrange this equation as follows

$$\log K_f = \frac{nFE^\circ}{2.303\ RT}$$

and substitute what we know about the reaction into it.

$$\log K_f = \frac{(2)(96{,}480\ C)(0.28\ V)}{(2.303)(8.314\ J/mol)(298\ K)} = 9.47$$

The complex formation equilibrium constant for this reaction is equal to 10 raised to the 9.47 power.

$$K_f = 10^{9.47} = 2.9 \times 10^9$$

This value of K_f for the $Zn(NH_3)_4{}^{2+}$ complex ion is the same as the one given in Table A-12 in the appendix. That isn't surprising, because many experimental values for equilibrium constants are the results of measurements such as these.

Exercise 19.17

Calculate the solubility product at 25°C for $Mg(OH)_2$.

Solution

The table of standard-state reduction potentials in the appendix contains the following reduction half-reactions.

$$Mg(OH)_2 + 2\ e^- \rightleftharpoons Mg + 2\ OH^- \qquad E^\circ_{red} = -2.69\ V$$
$$Mg^{2+} + 2\ e^- \rightleftharpoons Mg \qquad E^\circ_{red} = -2.375\ V$$

This time we have to reverse the second half-reaction to obtain an overall equation that corresponds to the solubility product equilibrium.

$$Mg(OH)_2 + 2\ e^- \longrightarrow Mg + 2\ OH^- \qquad E^\circ_{red} = -2.69\ V$$
$$\underline{+\ Mg \longrightarrow Mg^{2+} + 2\ e^- \qquad\qquad E^\circ_{ox} = 2.375\ V}$$
$$Mg(OH)_2 \longrightarrow Mg^{2+} + 2\ OH^- \qquad E^\circ = -0.32\ V$$

Once again, we start by solving the Nernst equation for the relationship between the equilibrium constant and the cell potential for the reaction

$$\log K_{sp} = \frac{nFE^\circ}{2.303\ RT}$$

and then substitute what we know about the reaction into this equation.

$$\log K_{sp} = \frac{(2)(96,480\ C)(-0.32\ V)}{(2.303)(8.314\ J/mol)(298\ K)} = -10.82$$

Solving for the equilibrium constant gives a value of 1.5×10^{-11}.

$$K_{sp} = 1.5 \times 10^{-11}$$

This value matches the one given in Table A-11 of the appendix, within experimental error.

19.12 ELECTROLYTIC CELLS

The cells discussed so far have one thing in common. They all use a spontaneous chemical reaction to drive an electric current through an external circuit.

These voltaic cells are important because they are the basis for the batteries that fuel modern society. But they are not the only kind of electrochemical cell. It is also possible to construct a cell that does work on a chemical system by driving an electric current through it. These cells are called *electrolytic cells.*

> **Electrolytic cells use an electric current to drive oxidation-reduction reactions that would not occur spontaneously.**

The process by which an electric current is used to drive oxidation-reduction reactions is called *electrolysis.*

19.13 ELECTROLYSIS AS A MEANS OF PREPARING ELEMENTS

Tables of standard-state reduction potentials can help us understand the chemistry behind the discovery of a number of elements. Chlorine, for example, was first prepared in 1774 by reacting hydrochloric acid with manganese dioxide.

$$2 \text{ HCl}(aq) + \text{MnO}_2(s) + 2 \text{ H}^+(aq) \qquad \text{Cl}_2(g) + \text{Mn}^{2+}(aq) + 2 \text{ H}_2\text{O}(l)$$

MnO_2 in acid ($E^\circ_{red} = 1.208$ V) was the first oxidizing agent available to chemists that was almost as strong as Cl_2 ($E^\circ_{red} = 1.358$ V). It was therefore the first oxidizing agent strong enough to drive this reaction to a point at which appreciable amounts of Cl_2 could be isolated. Today, chlorine is prepared in the lab by reacting hydrochloric acid with potassium permanganate

$$10 \text{ HCl}(aq) + 2 \text{ MnO}_4^-(aq) + 6 \text{ H}^+(aq) \longrightarrow 5 \text{ Cl}_2(g) + 2 \text{ Mn}^{2+}(aq) + 8 \text{ H}_2\text{O}(l)$$

because the MnO_4^- ion in acid ($E^\circ_{red} = 1.491$ V) is one of the few common oxidizing reagents that is actually stronger than Cl_2.

Crude samples of aluminum were first prepared in 1825 by reaction of aluminum chloride with potassium metal dissolved in mercury at temperatures high enough to melt the reactants, and the first commercial preparation of aluminum in 1854 was based on the reaction between sodium metal and aluminum chloride.

$$3 \text{ K}(l) + \text{AlCl}_3(l) \longrightarrow \text{Al}(l) + 3 \text{ KCl}(l)$$
$$3 \text{ Na}(l) + \text{AlCl}_3(l) \longrightarrow \text{Al}(l) + 3 \text{ NaCl}(l)$$

This approach had the disadvantage of coupling the cost of aluminum metal with the cost of sodium or potassium, which made aluminum in the 1850s more expensive than silver. It was used because sodium ($E^\circ_{red} = -2.71$ V) and potassium ($E^\circ_{red} = -2.92$ V) are among the handful of common reducing agents that are stronger than aluminum ($E^\circ_{red} = -1.71$ V) and thus strong enough to reduce Al^{3+} ions to aluminum metal.

In theory, we should be able to make sodium metal by reducing one of its salts, such as sodium chloride. Unfortunately, the potential for reducing Na^+ ions to sodium metal is unusually large, and this reaction is very unfavorable.

$$\text{Na}^+ + e^- \longrightarrow \text{Na} \qquad E^\circ = -2.71 \text{ V}$$

To drive the reaction to completion, we need a reducing agent with a potential of at least 3 volts. Sodium was therefore prepared for the first time by electrolysis with a series of batteries that produced an electric current with enough potential to reduce the Na^+ ions to sodium metal.

19.14 THE ELECTROLYSIS OF MOLTEN NaCl

Let's take a closer look at what happens when an electric current is used to reduce the Na^+ ions in sodium chloride to sodium metal. The first question is: How do we get the electric current into the sample? The answer is to connect the battery to inert electrodes that can conduct an electric current and then insert the electrodes into the sample, as shown in Figure 19.9.

One of the electrodes picks up a net negative charge as the battery pushes electrons onto this electrode. The other electrode picks up a positive charge as the battery draws electrons away. The positively charged Na^+ ions are attracted toward the negative electrode, and the negatively charged Cl^- ions are attracted toward the positive electrode. Unfortunately, nothing happens if we start with sodium chloride at room temperature, because the Na^+ and Cl^- ions are trapped in the crystal lattice. None of the ions are free to move through the sample.

One way of solving this problem is to melt the sodium chloride (MP = 801°C).

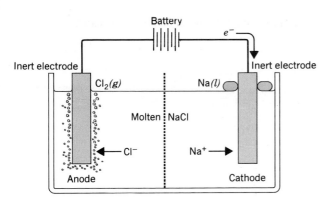

FIG. 19.9 Electrolysis of molten sodium chloride involves using an electric current to reduce Na^+ ions to sodium metal at the cathode and to oxidize Cl^- ions to Cl_2 gas at the anode. Because sodium metal has a low melting point and a relatively low density, it collects as a liquid above the molten sodium chloride near the cathode.

The Na^+ and Cl^- ions are then free to move through the molten salt, so they migrate toward the electrode with the opposite charge.

What happens when one of the Na^+ ions collides with the negative electrode? The battery carries a potential large enough to force the Na^+ ions to pick up electrons to form sodium metal, which collects as a liquid floating at the top of the molten salt. The sodium produced in this reaction is a liquid because sodium melts at relatively low temperatures (MP = 97.8°C). It floats above the molten salt because its density (0.968 g/cm³) is much smaller than that of sodium chloride (2.164 g/cm³). The overall reaction at the negative electrode can be written as follows.

$$\text{Negative electrode (cathode):} \qquad Na^+ + e^- \longrightarrow Na$$

Since reduction always takes place at the cathode, the negative electrode must be the cathode of this cell.

What happens when one of the Cl^- ions collides with the positive electrode? The standard reduction potential for chlorine is about 1.36 volts.

$$Cl_2 + 2\,e^- \longrightarrow 2\,Cl^- \qquad E^\circ_{red} = 1.3583 \text{ V}$$

This means that oxidizing chloride ions to chlorine gas is an uphill battle.

$$2\,Cl^- \longrightarrow Cl_2 + 2\,e^- \qquad E^\circ_{ox} = -(1.3583 \text{ V})$$

But the battery carries enough potential to oxidize the Cl^- ions to Cl_2 gas, which bubbles off at the positive electrode of this cell. The overall reaction at the positive electrode can be written as follows.

$$\text{Positive electrode (anode):} \qquad 2\,Cl^- \longrightarrow Cl_2 + 2\,e^-$$

Since oxidation always takes place at the anode, the positive electrode in this cell must be the anode.

Students often ask: "What is the meaning of the dotted vertical line in drawings of electrolytic cells?" One way to answer this question is to think about the conditions under which the cell is operating. We know that it has to operate at high temperatures in order to keep the sodium chloride from crystallizing. We also know that molten sodium metal collects at one end of the cell and chlorine gas collects at the other.

What would happen if the molten sodium metal came in contact with Cl_2 gas at 800°C? Sodium reacts violently with chlorine at room temperature. At 800°C, the reaction would be awesome. The dotted line in Figure 19.9 symbolizes a diaphragm that keeps the Cl_2 gas from coming into contact with the sodium metal, while allowing ions to move from one side of the cell to the other.

The net effect of passing an electric current through molten sodium chloride is to decompose this compound into its elements, sodium metal and chlorine gas.

Electrolysis of NaCl:

Cathode (−): $Na^+ + e^- \longrightarrow Na$

Anode (+): $2\ Cl^- \longrightarrow Cl_2 + 2\ e^-$

The potential required to oxidize Cl^- ions to Cl_2 is -1.36 volts, and the potential needed to reduce Na^+ ions to sodium metal is -2.71 volts. The battery used to drive this reaction must therefore have a potential of at least 4.07 volts.

This example explains why the process is called electrolysis. The suffix *-lysis* comes from the Greek stem meaning to loosen or split up. Electrolysis literally uses an electric current to split a compound into its elements.

$$2\ NaCl(l) \xrightarrow{\text{electrolysis}} 2\ Na(l) + Cl_2(g)$$

This example also illustrates the difference between voltaic cells and electrolytic cells. Voltaic cells use the energy given off in a spontaneous reaction to do electrical work. Electrolytic cells use electrical work as source of energy to drive the reaction in the opposite direction.

19.15 THE ELECTROLYSIS OF AQUEOUS NaCl

What would happen if we dissolved sodium chloride in water and then tried to pass an electric current through this solution? Sodium chloride dissociates in water to give an aqueous solution of the Na^+ and Cl^- ions.

$$NaCl(s) \xrightarrow{H_2O} Na^+(aq) + Cl^-(aq)$$

When we connect a pair of inert electrodes to a battery and immerse them in this solution, we get the electrolysis cell shown in Figure 19.10.

One of the electrodes picks up a net negative charge as it gains electrons from the battery. The other picks up a positive charge as it loses electrons to the battery. The Na^+ ions will migrate toward the negative electrode, and the Cl^- ions will migrate toward the positive electrode.

Everything so far is the same as in the electrolysis of molten sodium chloride. But when the time comes to predict what will happen when these ions come into contact with the electrodes, this system is more complex. There are now two substances that can be reduced at the cathode, Na^+ ions and water molecules.

FIG. 19.10 Electrolysis of aqueous sodium chloride results in reduction of water to form H_2 gas at the cathode and oxidation of Cl^- ions to form Cl_2 gas at the anode.

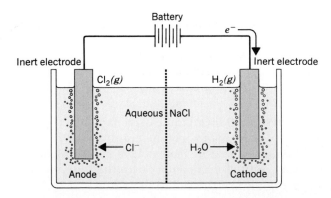

Cathode ($-$):

$$Na^+ + e^- \longrightarrow Na \qquad\qquad E^\circ_{red} = -2.71 \text{ V}$$

$$2\ H_2O + 2\ e^- \longrightarrow H_2 + 2\ OH^- \qquad E^\circ_{red} = -0.83 \text{ V}$$

There are also two substances that can be oxidized at the anode, Cl^- ions and water molecules.

Anode ($+$):

$$2\ Cl^- \longrightarrow Cl_2 + 2\ e^- \qquad\qquad E^\circ_{ox} = -1.36 \text{ V}$$

$$2\ H_2O \longrightarrow O_2 + 4\ H^+ + 4\ e^- \qquad E^\circ_{ox} = -1.23 \text{ V}$$

This doesn't mean that we should expect sodium metal to accumulate at the cathode of this cell. Any sodium formed at the cathode would immediately react with water to form Na^+ ions and H_2 gas.

$$2\ Na(s) + 2\ H_2O(l) \longrightarrow 2\ Na^+(aq) + 2\ OH^-(aq) + H_2(g)$$

The only product formed at the cathode is hydrogen gas.

Cathode ($-$): $\qquad 2\ H_2O(l) + 2\ e^- \longrightarrow H_2(g) + 2\ OH^-(aq)$

What should we expect to see at the anode? There are two possible oxidation half-reactions in this cell.

$$2\ Cl^- \longrightarrow Cl_2 + 2\ e^- \qquad\qquad E^\circ_{ox} = -1.36 \text{ V}$$

$$2\ H_2O \longrightarrow O_2 + 4\ H^+ + 4\ e^- \qquad E^\circ_{ox} = -1.23 \text{ V}$$

The standard-state reduction potentials for these half-reactions are so close to each other that we might expect to see a mixture of Cl_2 and O_2 gas collect at the anode. In practice, the only product of this reaction is Cl_2.

Why does the reaction produce only Cl_2? At first glance, it would seem easier to oxidize water ($E^\circ_{ox} = -1.23$ volts) to O_2 than to oxidize Cl^- ions ($E^\circ_{ox} = -1.36$ volts) to Cl_2. Several factors must be kept in mind, however. First, the reaction is never allowed to approach standard-state conditions. The concentration of the Cl^- ion is kept very large, which influences the reduction potential for the chloride ion. (The solution is typically 25% NaCl by weight.) The pH is also kept very high, which influences the reduction potential for water.

The deciding factor is a phenomenon known as **overvoltage.**

Overvoltage: The difference between the theoretical and experimental voltages required for the oxidation or reduction of a substance.

Under ideal conditions, a potential of 1.23 volts is large enough to oxidize water to O_2 gas. Under real conditions, however, it usually takes a much larger voltage to initiate this reaction. The overvoltage for the oxidation of water can be as large as 1 volt. By carefully choosing the electrode to maximize the overvoltage for the oxidation of water to O_2 gas and then carefully controlling the potential at which the cell operates, we can ensure that only chlorine is produced in this reaction.

In summary, electrolysis of aqueous solutions of sodium chloride does not give the same products as electrolysis of molten sodium chloride. Electrolysis of molten NaCl decomposes this compound into its elements.

$$2\ NaCl(l) \xrightarrow{\ electrolysis\ } 2\ Na(l) + Cl_2(g)$$

Electrolysis of aqueous NaCl solutions gives a mixture of hydrogen and chlorine gas and an aqueous sodium hydroxide solution.

$$2\ NaCl(aq) + 2\ H_2O(l) \xrightarrow{\text{electrolysis}} 2\ Na^+(aq) + 2\ OH^-(aq) + H_2(g) + Cl_2(g)$$

If our goal is to make sodium metal, we have to electrolyze molten sodium chloride. If our goal is to make chlorine, it is easier to run the reaction in aqueous solution. Since the demand for chlorine is much larger than the demand for sodium, electrolysis of aqueous sodium chloride is a more important process commercially. Electrolysis of aqueous NaCl solution has two other advantages. It produces H_2 gas at the cathode, which can be collected and sold. It also produces NaOH, which can be drained from the bottom of the electrolytic cell and sold.

19.16 ELECTROLYSIS OF WATER

A standard apparatus for the electrolysis of water is shown in Figure 19.11.

$$2\ H_2O(l) \xrightarrow{\text{electrolysis}} 2\ H_2(g) + O_2(g)$$

A pair of inert electrodes are sealed in opposite ends of a container designed to collect the H_2 and O_2 gas given off in this reaction. The electrodes are then connected to a battery or another source of electric current.

By itself, water is a very poor conductor of electricity. Thus, we have to add an electrolyte to the water to provide ions that can flow through the solution, thereby completing the electric circuit. The electrolyte must be soluble in water. It should also be relatively inexpensive. Most importantly, it must contain ions that are harder to oxidize or reduce than water.

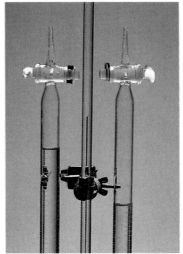

FIG. 19.11 Electrolysis of an aqueous Na_2SO_4 solution results in the reduction of water to form H_2 gas at the cathode, and oxidation of water to form O_2 gas at the anode.

$$2\ H_2O + 2\ e^- \longrightarrow H_2 + 2\ OH^- \qquad E^\circ_{red} = -0.83\ V$$
$$2\ H_2O \longrightarrow O_2 + 4\ H^+ + 4\ e^- \qquad E^\circ_{ox} = -1.23\ V$$

According to Table 19.1, the following cations are harder to reduce than water: Li^+, Rb^+, K^+, Cs^+, Ba^{2+}, Sr^{2+}, Ca^{2+}, Na^+, and Mg^{2+}. Two of these cations are more likely candidates than the others because they form soluble salts: Na^+ and K^+.

What is the best anion to use? Table 19.1 suggests the SO_4^{2-} ion, because it is the most difficult anion to oxidize. The potential for oxidation of this ion to the peroxydisulfate ion is -2.05 volts.

$$2\ SO_4^{2-} \longrightarrow S_2O_8^{2-} + 2\ e^- \qquad E^\circ_{ox} = -2.05\ V$$

When an aqueous solution of either Na_2SO_4 or K_2SO_4 is electrolyzed in the apparatus shown in Figure 19.11, H_2 gas collects at one electrode and O_2 gas collects at the other.

What would happen if we added an indicator such as bromothymol blue to this apparatus? Bromothymol blue turns yellow in acidic solutions (pH < 6) and blue in basic solutions (pH > 7.6). According to the equations for the two half-reactions, the indicator should turn yellow at the anode and blue at the cathode.

$$\text{Cathode } (-): \qquad 2\ H_2O + 2\ e^- \longrightarrow H_2 + 2\ OH^-$$
$$\text{Anode } (+): \qquad 2\ H_2O \longrightarrow O_2 + 4\ H^+ + 4\ e^-$$

The solution turns blue at the cathode because OH^- ions accumulate at this electrode when water is reduced to H_2 gas. It turns yellow at the anode because H^+ ions are formed when water is oxidized to O_2 gas.

19.17 FARADAY'S LAW

Humphry Davy's experiments with electrochemistry led to his discovery of sodium, potassium, magnesium, calcium, barium, and strontium. It has been said, however, that his greatest discovery was Michael Faraday. Faraday was apprenticed to a bookbinder in London at the age of 13. His exposure to works brought into the shop for binding excited his interest in science, and he started attending lectures on science and doing experiments in chemistry and electricity on his own. He was eventually hired by Davy as a secretary and scientific assistant and thereby began a life-long commitment to research in both chemistry and physics.

Faraday's early research on electrolysis led him to propose a relationship between the amount of current passed through a solution and the weight of the substance decomposed or produced by this current. *Faraday's law of electrolysis* can be expressed as follows.

> **The amount of a substance either decomposed or produced at one of the electrodes in an electrolytic cell is directly proportional to the amount of electricity that passes through the cell.**

We can use Faraday's law to calculate the weight of sodium produced in a given amount of time when molten sodium chloride is electrolyzed or the volume of hydrogen gas produced in a given amount of time when water is electrolyzed. First, however, we need to understand the relationship between the amount of electric charge that flows through the cell and the rate at which this charge is transferred.

The amount of electric charge that flows through the cell is measured in units of coulombs (C). The rate at which this charge flows is known as the current, which is measured in amperes, or amps. By definition, 1 coulomb of charge is transferred when a current of 1 amp flows for 1 second.

$$1 \text{ C} = 1 \text{ amp} \times 1 \text{ sec} = 1 \text{ amp-sec}$$

Exercise 19.18

Calculate the electrical charge that flows through an electrolysis cell when a current of 10.0 amperes is run for a period of 4.00 hours.

Solution

A 10.0-amp current flowing for a period of 4.00 hours will transfer coulombs of electric charge.

$$10.0 \text{ amp} \times 4.00 \text{ hr} \times 60 \frac{\text{min}}{\text{hr}} \times 60 \frac{\text{sec}}{\text{min}} = 144{,}000 \text{ amp-sec} = \textbf{144,000 C}$$

Exercise 19.19

Calculate the weight of sodium metal that will be deposited at the cathode when a 10.0-amp current is passed through an electrolysis cell containing molten sodium chloride for a period of 4.00 hours.

Solution

According to the preceding exercise, 144,000 coulombs of electric charge will pass through the electrolysis cell during this period. We can convert this quantity of charge into moles of electrons by taking advantage of a calculation first performed in Exercise 19.4 and then repeated in Section 19.10. By multiplying the number of electrons in a mole by the charge on an electron, we can calculate the charge on a mole of electrons.

$$6.022045 \times 10^{23} \, \frac{e^-}{mol} \times 1.6021892 \times 10^{-19} \, \frac{C}{e^-} = 96{,}484.56 \, C/mol$$

Section 19.10 defined the charge on a mole of electrons as a **faraday,** F.

$$1 \, F = 96{,}484.56 \, \frac{C}{mol}$$

Using this definition, we can calculate the number of moles of electrons transferred when 144,000 coulombs of electric charge flow through the cell.

$$144{,}000 \, C \times \frac{1 \, mol \, e^-}{96{,}480 \, C} = \textbf{1.49 mol } e^-$$

According to the balanced equation for the reaction that occurs at the cathode of this cell, we get 1 mole of sodium for every mole of electrons.

$$\text{Cathode } (-): \qquad Na^+ + e^- \longrightarrow Na$$

Thus, we get 1.49 moles, or 34.3 grams, of sodium in 4.00 hours.

$$1.49 \, mol \, Na \times 22.99 \, \frac{g \, Na}{mol} = \textbf{34.3 g Na}$$

We would have to run this electrolysis for more than two days to prepare a pound of sodium.

Exercise 19.20

Calculate the volume of H_2 gas at 25°C and 1.00 atm that will collect at the cathode when an aqueous solution of Na_2SO_4 is electrolyzed for 2.00 hours with a 10.0-amp current.

Solution

We can start by calculating the amount of electrical charge that passes through the solution.

$$10.0 \, amp \times 2.00 \, hr \times 60 \, \frac{min}{hr} \times 60 \, \frac{sec}{min} = 72{,}000 \, amp\text{-}sec = \textbf{72{,}000 C}$$

We can then calculate how many moles of electrons carry this charge.

$$72{,}000 \, C \times 1 \, \frac{mol \, e^-}{96{,}480 \, C} = \textbf{0.746 mol } e^-$$

The equation for the reaction that produces H_2 gas at the cathode indicates that we get a mole of H_2 gas for every 2 moles of electrons.

$$\text{Cathode } (-): \qquad 2 \, H_2O + 2 \, e^- \longrightarrow H_2(g) + 2 \, OH^-$$

Half as many moles of H_2 gas are therefore produced at the cathode.

$$0.746 \; \cancel{\text{mol} \; e^-} \times \frac{1 \; \text{mol} \; H_2}{2 \; \cancel{\text{mol} \; e^-}} = 0.373 \; \text{mol} \; H_2$$

We can now turn to the ideal gas equation and calculate the volume of this gas at 25°C and 1 atm.

$$V = \frac{nRT}{P} = \frac{(0.373 \; \cancel{\text{mol}})(0.08206 \; \text{L-atm}/\cancel{\text{mol-K}})(298 \; \cancel{K})}{(1 \; \cancel{\text{atm}})} = 9.12 \; L$$

We can extend the general pattern outlined in Exercises 19.18 through 19.20 to answer questions that might seem impossible at first glance.

Exercise 19.21

Determine the oxidation number of the chromium in an unknown salt if electrolysis of a sample of this salt for 1.50 hours with a 10.0-amp current deposits 9.71 grams of chromium metal at the cathode.

Solution

We can start by calculating the amount of electric charge that passed through the cell during the electrolysis.

$$10.0 \; \text{amp} \times 1.50 \; \cancel{\text{hr}} \times 60 \; \frac{\cancel{\text{min}}}{\cancel{\text{hr}}} \times 60 \; \frac{\text{sec}}{\cancel{\text{min}}} = 54{,}000 \; \text{amp-sec} = \mathbf{54{,}000 \; C}$$

We can then calculate the number of moles of electrons that passed through the cell.

$$54{,}000 \; \cancel{C} \times 1 \; \frac{\text{mol} \; e^-}{96{,}484 \; \cancel{C}} = 0.560 \; \text{mol} \; e^-$$

We can't pursue this line of thought any further unless we know the balanced equation for the reaction at the cathode. So let's go back to the original statement of the question and see what else can be done.

The problem tells us the weight of chromium deposited at the cathode. It might be useful to calculate the number of moles of chromium generated.

$$9.71 \; \cancel{g} \; Cr \times \frac{1 \; \text{mol} \; Cr}{52.0 \cancel{g}} = 0.187 \; \text{mol} \; Cr$$

We now know the number of moles of chromium metal produced and the number of moles of electrons it took to produce this metal. What is the relationship between the moles of electrons consumed in this reaction and the moles of chromium produced?

$$\frac{0.560 \; \text{mol} \; e^-}{0.187 \; \text{mol} \; Cr} = \frac{3}{1}$$

Three moles of electrons are consumed for every mole of chromium metal produced. We can explain this by assuming that the net reaction at the cathode involves reduction of Cr^{3+} ions to chromium metal.

$$\text{Cathode} \; (-): \qquad Cr^{3+} + 3 \; e^- \longrightarrow Cr$$

Thus, the oxidation number of chromium in the unknown salt must be +3.

19.18 GALVANIC CORROSION AND CATHODIC PROTECTION

The following behavior is observed when the corrosion of metals is studied.

1. Iron metal dissolves in strong acids.
2. Iron rusts very slowly in contact with dry air.
3. Iron rusts very slowly in contact with water that does not contain dissolved O_2 gas.
4. Iron rusts more rapidly when in contact with water that contains dissolved O_2.
5. Iron rusts much more rapidly when it comes in contact with copper metal.
6. Iron does not rust when it is in contact with zinc or magnesium metal.

The model for electrochemical reactions developed in this chapter can be used to explain these observations.

Iron atoms on the surface of the metal slowly react with oxygen in the atmosphere to form a mixture of oxides.

$$2\ Fe(s) + O_2(g) \longrightarrow 2\ FeO(s)$$
$$2\ Fe(s) + 3\ O_2(g) \longrightarrow Fe_2O_3(s)$$
$$3\ Fe(s) + 2\ O_2(g) \longrightarrow Fe_3O_4(s)$$

This reaction is very uneven, however, and there are a number of holes in the oxide layer that forms on the surface of the metal. These holes allow oxygen atoms to migrate toward the metal surface below the oxide layer. They also allow iron atoms to migrate toward the atmosphere above the oxide layer. Corrosion therefore continues slowly but surely. At sites where the metal was deformed when it was worked or shaped, the iron oxide often breaks off, exposing a fresh metal surface to further corrosion and pitting.

Iron metal dissolves in acid, as would be expected from its position in Table 19.1. If the acid is treated to remove any dissolved O_2 gas, it reacts with iron to form solutions of the Fe^{2+} ion, because the overall potential for reaction between iron metal and acid to form Fe^{2+} ions is larger than the potential for the reaction to form Fe^{3+} ions.

Fe^{2+}:

oxidation:	$Fe \longrightarrow Fe^{2+} + 2\ e^-$	$E_{ox}^{\circ} = -(-0.409\ V)$
reduction:	$2\ H^+ + 2\ e^- \longrightarrow H_2$	$E_{red}^{\circ} = 0.000\ V$
overall:	$Fe + 2\ H^+ \longrightarrow Fe^{2+} + H_2$	$E^{\circ} = \mathbf{0.409\ V}$

Fe^{3+}:

oxidation:	$Fe \longrightarrow Fe^{3+} + 3\ e^-$	$E_{ox}^{\circ} = -(-0.036\ V)$
reduction:	$2\ H^+ + 2\ e^- \longrightarrow H_2$	$E_{red}^{\circ} = 0.000$
overall:	$2\ Fe + 6\ H^+ \longrightarrow 2\ Fe^{3+} + 3\ H_2$	$E^{\circ} = \mathbf{0.036\ V}$

In the absence of dissolved O_2, iron metal does not react with water, because the concentration of the H^+ ion in water is far from the standard state of 1 M. Substituting the $1 \times 10^{-7}\ M\ H^+$ ion concentration in water into the Nernst equation for the reaction between iron and H^+ ions to form Fe^{2+} lowers the cell potential for this reaction by 0.414 V. This is a large enough change to make the reaction slightly unfavorable.

When both water and air are present, iron corrodes more rapidly than in dry air because the oxidation and reduction reactions no longer have to occur at the same point on the metal surface. The iron metal now simultaneously acts as the anode, cathode, and circuit through which electrons flow from the anode to the cathode, as shown in Figure 19.12.

At one point on the metal surface, the iron is oxidized to Fe^{2+} ions.

$$\text{Anode:} \qquad F \longrightarrow Fe^{2+} + 2\,e^- \qquad E^{\circ}_{ox} = 0.409 \text{ V}$$

The electrons released in this reaction can flow through the metal to another point, at which O_2 dissolved in the water is reduced.

$$\text{Cathode:} \qquad O_2 + 2\,H_2O + 4\,e^- \longrightarrow 4\,OH^- \qquad E^{\circ}_{red} = 0.401 \text{ V}$$

The overall reaction for this step is written as follows.

$$Fe(s) + O_2(g) + 2\,H_2O \longrightarrow Fe^{2+}(aq) + 4\,OH^-(aq) \qquad E^{\circ} = \mathbf{0.810 \text{ V}}$$

The Fe^{2+} and OH^- ions produced in these half-reactions diffuse toward each other and eventually combine to form a hydrated iron(II) oxide that undergoes further oxidation to form rust, $Fe_2O_3 \cdot 3\,H_2O$.

This reaction is greatly facilitated when the iron metal comes in contact with copper metal. A voltaic cell is created at the point at which the two metal touch, as shown in Figure 19.13. The iron metal acts as the anode, giving up electrons to form Fe^{2+} ions.

$$\text{Anode:} \qquad Fe \longrightarrow Fe^{2+} + 2\,e^- \qquad E^{\circ}_{ox} = 0.409 \text{ V}$$

These electrons flow to the copper metal, which becomes the cathode at which O_2 dissolved in water is reduced.

$$\text{Cathode:} \qquad O_2 + 2\,H_2O + 4\,e^- \longrightarrow 4\,OH^- \qquad E^{\circ}_{red} = 0.401 \text{ V}$$

Once again, the Fe^{2+} ions produced at the anode diffuse toward the OH^- ions given off at the cathode and eventually combine with these OH^- ions to form rust. This process is called **galvanic corrosion** because the reaction is similar to what happens in a galvanic, or voltaic, cell.

Copper metal doesn't corrode when it touches iron, because copper metal is a weaker reducing agent than iron. We can see this by imagining the voltaic cell that would be created if copper acted as the anode and the electrons flowed to iron, which would act as the cathode at which O_2 was reduced.

$$\text{Anode:} \qquad Cu \longrightarrow Cu^{2+} + 2\,e^- \qquad E^{\circ}_{ox} = -0.340 \text{ V}$$
$$\text{Cathode:} \qquad O_2 + 2\,H_2O + 4\,e^- \longrightarrow 4\,OH^- \qquad E^{\circ}_{red} = 0.401 \text{ V}$$

The overall potential for this reaction is much smaller than the potential for the reaction that occurs when iron acts as the anode.

$$Cu(s) + O_2(g) + 2\,H_2O \longrightarrow Cu^{2+}(aq) + 4\,OH^-(aq) \qquad E^{\circ} = 0.061 \text{ V}$$
$$Fe(s) + O_2(g) + 2\,H_2O \longrightarrow Fe^{2+}(aq) + 4\,OH^-(aq) \qquad E^{\circ} = 0.810 \text{ V}$$

The iron therefore corrodes before the copper.

Comparing Figures 19.12 and 19.13 raises a more subtle question. The overall reaction for the corrosion of iron is the same, regardless of whether copper metal is in contact with the iron metal. Thus, the overall standard-state potential for these reactions must be the same. Why, then, does iron corrode so much faster when it comes in contact with copper metal? The answer seems to rest with the concept of

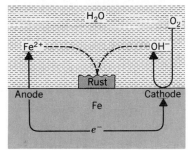

FIG. 19.12 When iron metal is immersed in water that contains dissolved oxygen, oxidation of the iron and reduction of O_2 can occur at different points on the metal surface. The net effect of this reaction is the formation of Fe^{2+} ions and OH^- ions, which diffuse toward each other to form a complex that is further oxidized to form rust, $Fe_2O_3 \cdot 3\,H_2O$.

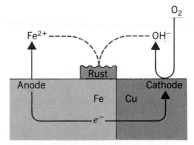

FIG. 19.13 When iron comes in contact with copper metal, a galvanic cell is created in which the iron in the presence of water and oxygen is oxidized to Fe^{2+} ions. The electrons given off in this process are transferred to the copper metal, which uses them to reduce O_2 to OH^- ions.

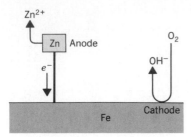

FIG. 19.14 The corrosion of iron can be prevented by cathodic protection. The iron is either coated with zinc to form galvanized iron or connected to a piece of zinc or magnesium metal. Zinc and magnesium are both stronger reducing agents than iron, so they are oxidized instead of the iron. When zinc or magnesium acts as the anode of this galvanic cell, the iron must become the cathode at which O_2 is reduced. Making iron the cathode of this cell protects it from corrosion.

overvoltage introduced in Section 19.15. The actual potential for the reduction of O_2 on a metal surface is usually larger than the value reported in tables of standard-state potentials. The overvoltage for the reduction of O_2 on copper metal, however, is significantly smaller than the overvoltage on iron. Since the overvoltage is defined as the extra voltage that must be applied to a reaction to get it to occur at the same rate, the reaction depicted in Figure 19.13 for iron metal in contact with copper should occur more rapidly than the corresponding reaction on an iron metal surface.

Iron can be protected from corrosion if it is allowed to stay in contact with a metal that is a better reducing agent, as shown in Figure 19.14. This process is called ***cathodic protection,*** because it involves making iron the cathode of a voltaic cell.

The zinc metal acts as the anode of a voltaic cell.

$$\text{Anode:} \qquad Zn \longrightarrow Zn^{2+} + 2\,e^- \qquad E_{ox}^\circ = 0.763 \text{ V}$$

The electrons given off in this reaction flow through the wire to the iron metal, which acts as the cathode at which O_2 is reduced.

$$\text{Cathode:} \qquad O_2 + 2\,H_2O + 4\,e^- \longrightarrow 4\,OH^- \qquad E_{red}^\circ = 0.401 \text{ V}$$

The overall potential for this reaction is much larger than the potential for the reaction that would occur if iron was the anode.

$$Zn(s) + O_2(g) + 2\,H_2O \longrightarrow Zn^{2+}(aq) + 4\,OH^-(aq) \qquad E^\circ = 1.164 \text{ V}$$

Thus, oxidation of the iron won't occur until the anode is either completely consumed or the electrical contact between the iron and zinc metal is broken. Cathodic protection is the theory behind so-called galvanized iron, which is iron metal coated with a thin layer of zinc. Until virtually all of the zinc has corroded, the iron metal that provides the structure for this material remains intact.

SUMMARY

By physically separating the oxidation and reduction halves of a reaction it is possible to force the electrons that flow from one of these half-reactions to the other to pass through an external wire. The net result is an electrochemical cell that can do electrical work. Cells of this sort are commonly known as either galvanic or voltaic cells.

The cell potential for a spontaneous oxidation-reduction reaction is a measure of the driving force behind the reaction. A large cell potential is observed when a strong reducing agent reacts with a strong oxidizing agent, because the driving force behind this reaction is relatively large. By convention, the sign of the cell potential is always positive for a spontaneous oxidation-reduction reaction.

The magnitude of the potential for an oxidation-reduction reaction depends on the conditions under which the measurement is made. Under standard-state conditions, all solutions have a concentration of 1 M, all gases have a partial pressure of 1 atm, and the temperature is 25°C. The standard-state cell potential for an oxidation-reduction reaction is represented by the symbol E°.

The standard-state potential for the reduction of H^+ ions to H_2 gas is arbitrarily defined as exactly zero volts. This convention allows us to use measurements of the cell potentials of oxidation-reduction reactions to generate a table of standard-state half-cell potentials. By convention, these half-cell potentials are written in the direction of reduction; half-reactions that give a better reducing agent than H_2 are given negative half-cell reduction potentials; and the most negative half-cell potentials are written at the top of the table. The standard-state cell potential for an oxidation-reduction reaction is the sum of the half-cell potentials for the oxidation and reduction halves of the reaction.

The Nernst equation can be used to calculate the potential for a half-cell or a cell that isn't at standard-state conditions. When an oxidation-reduction reaction is at equilibrium, there is no longer any driving force pushing the reaction forward. In other words, E must be zero. Rewriting the Nernst equation to reflect this condition allows us to calculate the equilibrium constant for an oxidation-reduction reaction from the value of $E°$ for the reaction. The larger the value of $E°$, the larger the equilibrium constant, because the standard-state conditions for reactions that have large values of $E°$ are very far from equilibrium.

Voltaic cells use a spontaneous oxidation-reduction reaction to drive an electric current through a wire. Electrolytic cells use an electric current to drive an oxidation-reduction reaction in the opposite direction. We can calculate the amount of material deposited at the anode or the cathode of an electrolytic cell by noting that a 1-amp current must flow for 1 second to transfer 1 coulomb of electric charge and that the charge on a mole of electrons is 96,484.56 coulombs.

PROBLEMS

Electrical Work from Spontaneous Oxidation-Reduction Reactions

19-1 Define the terms *coulomb, volt, ampere,* and *faraday.*

19-2 Describe an experiment that would allow you to determine the relative strengths of zinc, copper, silver, and iron metal as reducing agents.

19-3 Explain why oxidation and reduction half-reactions have to be physically separated for an oxidation-reduction reaction to do work.

19-4 Describe the function of a salt bridge in an electrochemical cell. Explain what happens when the salt bridge is removed from the system and why.

19-5 Humphrey Davy described a "living salt bridge," which consists of the first and second fingers of the hand inserted into the two half-cells of a voltaic cell. Explain how this accomplishes the same function as a U-tube filled with a saturated KCl solution.

Voltaic Cells from Spontaneous Oxidation-Reduction Reactions

19-6 Describe the relationships among the following terms: *cathode, anode, cation,* and *anion.*

19-7 Explain why cations flow towards the cathode and anions flow towards the anode of an electrochemical cell, regardless of whether it is a voltaic or electrolytic cell.

19-8 Explain why oxidation occurs at the anode and reduction occurs at the cathode of both voltaic and electrolytic cells.

Standard-State Cell Potentials for Voltaic Cells

19-9 Describe the difference between E and $E°$ for a cell.

19-10 Describe the conditions under which $E°$ is measured.

19-11 Describe what the magnitude of $E°$ for an oxidation-reduction reaction indicates.

19-12 Explain why cell potentials measure the relative strengths of a pair of oxidizing agents and the relative strengths of a pair of reducing agents, but not their absolute strengths.

19-13 Describe the arbitrary convention that is used to turn measurements of the standard-state cell potential for a reaction into a table of standard-state half-cell reduction potentials.

Predicting Spontaneous Oxidation-Reduction Reactions from the Sign of $E°$

19-14 Describe what the sign of $E°$ for an oxidation-reduction reaction indicates.

19-15 Describe what happens to the sign of $E°$ for an oxidation-reduction reaction when the direction in which the reaction is written is reversed.

19-16 Which of the following oxidation-reduction reactions should occur as written when run under standard-state conditions?

(a) $Al(s) + Cr^{3+}(aq) \rightarrow Al^{3+}(aq) + Cr(s)$ $E° = 0.966$ V
(b) $3\ Cr^{2+}(aq) \rightarrow Cr(s) + 2\ Cr^{3+}(aq)$ $E° = 0.33$ V
(c) $Fe(s) + Cr^{3+}(aq) \rightarrow Cr(s) + Fe^{3+}(aq)$ $E° = -0.70$ V
(d) $3\ H_2(g) + 2\ Cr^{3+}(aq) \rightarrow 2\ Cr(s) + 6\ H^+(aq)$ $E° = -0.74$

19-17 Which of the following oxidation-reduction reactions should occur as written when run under standard-state conditions?

(a) $2\ Fe^{2+}(aq) + H_2O_2(aq) \rightarrow 2\ Fe^{3+}(aq) + 2\ OH^-(aq)$ $E° = 0.11$ V   (b) $2\ Fe^{2+}(aq) + Cl_2(aq) \rightarrow 2\ Fe^{3+}(aq) + 2\ Cl^-(aq)$ $E° = 0.588$ V   (c) $2\ Fe^{2+}(aq) + Br_2(aq) \rightarrow 2\ Fe^{3+}(aq) + 2\ Br^-(aq)$ $E° = 0.317$ V   (d) $2\ Fe^{2+}(aq) + I_2(aq) \rightarrow 2\ Fe^{3+}(aq) + 2\ I^-(aq)$ $E° = -0.235$ V

19-18 Use the results of Problem 19-17 to determine the relative strengths of H_2O_2, Cl_2, Br_2, and I_2 as oxidizing agents.

Standard-State Reduction Half-Cell Potentials

19-19 What do the following standard-state half-cell reduction potentials tell us about the relative strengths of zinc and copper metal as reducing agents? What do they tell us about the relative strengths of Zn^{2+} and Cu^{2+} ions as oxidizing agents?

$$Zn^{2+} + 2\,e^- \rightleftharpoons Zn \qquad E° = -0.7628 \text{ V}$$
$$Cu^{2+} + 2\,e^- \rightleftharpoons Cu \qquad E° = 0.3402 \text{ V}$$

Predicting Standard-State Cell Potentials

19-20 Describe what happens inside a Daniell cell when it does work.

$$Zn(s) + Cu^{2+}(aq) \longrightarrow Zn^{2+}(aq) + Cu(s)$$

19-21 Calculate $E°$ for the following reaction and predict whether the reaction should occur spontaneously as written when run under standard-state conditions.

$$Al^{3+}(aq) + Fe(s) \longrightarrow Fe^{3+}(aq) + Al(s)$$

19-22 Calculate $E°$ for the following reaction and predict whether the reaction should occur spontaneously as written when run under standard-state conditions.

$$Ca(s) + 2\,H_2O(l) \longrightarrow Ca^{2+}(aq) + 2\,OH^-(aq) + H_2(g)$$

19-23 Calculate $E°$ for the following reaction and predict whether the reaction should occur spontaneously as written when run under standard-state conditions.

$$6\,Fe^{2+}(aq) + Cr_2O_7^{2-}(aq) + 14\,H^+(aq) \longrightarrow$$
$$6\,Fe^{3+}(aq) + 2\,Cr^{3+}(aq) + 7\,H_2O(l)$$

19-24 Calculate $E°$ for the following reaction and predict whether the reaction should occur spontaneously as written when run under standard-state conditions.

$$2\,Cu^+(aq) \longrightarrow Cu(s) + Cu^{2+}(aq)$$

19-25 Use standard-state half-cell reduction potentials to predict whether a $1.00\ M$ $Fe^{3+}(aq)$ solution should react with a $1.00\ M$ $H^-(aq)$ solution to produce $Fe^{2+}(aq)$ and $H_2(g)$.

19-26 Predict which of the following reactions should occur spontaneously as written when run under standard-state conditions.

(a) $Zn(s) + 2\,H^+(aq) \rightarrow Zn^{2+}(aq) + H_2(g)$
(b) $Cr(s) + 3\,Fe^{3+}(aq) \rightarrow Cr^{3+}(aq) + 3\,Fe^{2+}(aq)$
(c) $Mn(s) + Mg^{2+}(aq) \rightarrow Mn^{2+}(aq) + Mg(s)$

19-27 Predict which of the following reactions should occur spontaneously as written when run under standard-state conditions.

(a) $NO_2^-(aq) + ClO^-(aq) \rightarrow NO_3^-(aq) + Cl^-(aq)$
(b) $2\,ClO_2^-(aq) \rightarrow ClO^-(aq) + ClO_3^-(aq)$
(c) $3\,Cu(s) + 2\,HNO_3(aq) + 6\,H^+(aq) \rightarrow 3\,Cu^{2+}(aq)$
$+ 2\,NO(g) + 4\,H_2O(l)$

Line Notation for Voltaic Cells

19-28 Write the line notation for the following voltaic cells under standard-state conditions.

$$Zn(s) + Cu^{2+}(aq) \longrightarrow Cu(s) + Zn^{2+}(aq)$$
$$2\,Ag^+(aq) + Cu(s) \longrightarrow Cu^{2+}(aq) + 2\,Ag(s)$$

19-29 Write the line notation for the voltaic cell in which AgCl is reduced at a silver metal electrode and copper metal is oxi-

dized at the other electrode to form Cu^{2+} ions under standard-state conditions.

$$2\,AgCl(s) + Cu(s) \longrightarrow Ag(s) + Cu^{2+}(aq) + 2\,Cl^-(aq)$$

19-30 Determine $E°$ for the following cells and decide in which direction these reactions are spontaneous.

$$Zn|Zn^{2+}\ (1.0\ M)||Fe^{2+}\ (1.0\ M)|Fe$$
$$Pt|Mn^{2+}\ (1.0\ M)|MnO_4^-\ (1.0\ M)||Fe^{2+}\ (1.0\ M)|Fe$$

19-31 Draw a diagram of the following cell, label the anode and cathode, and calculate $E°$ for the cell.

$$Zn|Zn^{2+}\ (1.0\ M)||H^+\ (1.0\ M)|H_2\ (1\ atm)|Pt$$

19-32 Draw a diagram of the following cell, label the anode and cathode, and calculate $E°$ for the cell.

$$Zn|Zn^{2+}\ (1.0\ M)||MnO_4^-\ (1.0\ M)|Mn^{2+}\ (1.0\ M)|Pt$$

Using Standard-State Half-Cell Potentials to Understand Chemical Reactions

19-33 Which of the following pairs of ions can't coexist in aqueous solution?

(a) Na^+, S^{2+} (b) Zn^{2+}, I^- (c) Hg_2^{2+}, F^- (d) Fe^{2+}, Hg^{2+}
(e) Ag^+, Hg^{2+}

19-34 Which of the following pairs of ions cannot coexist in aqueous solution?

(a) Sn^{2+}, Fe^{2+} (b) Au^+, Br^- (c) Fe^{3+}, CrO_4^{2-} (d) Fe^{3+}, SO_4^{2-} (e) MnO_4^-, I^-

19-35 Explain why copper, gold, mercury, platinum, and silver can be found in their metallic state in nature.

19-36 Which of the following is the strongest reducing agent?

(a) Zn (b) Fe (c) H_2 (d) Cu (e) Ag

19-37 Which of the following is the strongest oxidizing agent?

(a) H_2O_2 in OH^- (b) H_2O_2 in H^+ (c) Na (d) O_2 in H^+
(e) Al

19-38 Which of the following is the strongest reducing agent?

(a) H^+ (b) H_2 (c) H^- (d) H_2O (e) O_2

19-39 Which is the better oxidizing agent?

(a) H^- or K (b) Sn or Fe^{2+} (c) Ag^+ or Au^+

The Nernst Equation

19-40 Explain why there is only one value of $E°$ for a cell but many different values of E.

19-41 Explain why the cell potential becomes smaller as the cell comes closer to equilibrium.

19-42 What does it mean when E for a cell is equal to zero? What does it mean when $E°$ is equal to zero?

19-43 Which of the following describes an oxidation-reduction reaction at equilibrium?

(a) $E = 0$ (b) $E° = 0$ (c) $E = E°$ (d) $Q = 0$ (e) $\ln K = 0$

19-44 Write the Nernst equation for the following half-cell and calculate the half-cell potential, assuming that the H$^+$ ion concentration is $10^{-7}\ M$.

$$2\ H^+(aq) + 2\ e^- \rightleftharpoons H_2(g)$$

19-45 Describe how increasing the pH of the following half-reaction affects the half-cell reduction potential.

$$MnO_4^-(aq) + 8\ H^+(aq) + 5\ e^- \rightleftharpoons Mn^{2+}(aq)$$

Does the permanganate ion become a stronger oxidizing agent or a weaker oxidizing agent as the solution becomes more basic?

19-46 Write the Nernst equation for the following reaction and calculate the cell potential, assuming that the Al^{3+} ion concentration is $1.2\ M$ and the Fe^{3+} ion concentration is $2.5\ M$.

$$Al^{3+}(aq) + Fe(s) \longrightarrow Al(s) + Fe^{3+}(aq)$$

19-47 Calculate $E°$ for the following reaction.

$$2\ Fe^{2+}(aq) + H_2O_2(aq) \longrightarrow 2\ Fe^{3+}(aq) + 2\ OH^-(aq)$$

Write the Nernst equation for this reaction and calculate the cell potential for a system initially at the standard state that has reached a pH of 10.

19-48 Assume that we start with a Daniell cell at standard-state conditions.

$$Zn|Zn^{2+}(1\ M)||Cu^{2+}(1\ M)|Cu$$

Calculate the cell potential under the following sets of conditions.

(a) Ninety-nine percent of the zinc metal and Cu^{2+} ions have been consumed. (b) The reaction has reached 99.99% completion. (c) The reaction has reached 99.9999% completion. (d) The Cu^{2+} ion concentration is only $1 \times 10^{-8}\ M$. (e) The reaction has reached equilibrium.

19-49 Calculate the standard-state cell potential for the voltaic cell built around the following oxidation-reduction reaction.

$$2\ MnO_4^-(aq) + 5\ H_2O_2(aq) + 6\ H^+(aq) \longrightarrow$$
$$2\ Mn^{2+}(aq) + 8\ H_2O(l) + 5\ O_2(g)$$

Use the Nernst equation to predict the effect on the cell potential of an increase in the pH of the solution. Predict the effect of an increase in the H$_2$O$_2$ concentration.

19-50 The standard-state cell potential, $E°$, is zero for any cell in which the reaction at the cathode is the opposite of the reaction at the anode.

$$Cu|Cu^{2+}(1.0\ M)||Cu^{2+}(1.0\ M)|Cu$$

A small voltage can be obtained, hoever, from a cell in which the Cu^{2+} ion concentration is different in the two half-cells. Calculate E for the following concentration cell.

$$Cu|Cu^{2+}(2.50\ M)||Cu^{2+}(0.18\ M)|Cu$$

Using the Nernst Equation to Measure Equilibrium Constants

19-51 Calculate the equilibrium constant at 25°C for the Daniell cell.

$$Zn(s) + Cu^{2+}(aq) \longrightarrow Zn^{2+}(aq) + Cu(s)$$

19-52 Use the standard-state reduction potentials for the following half-cells to calculate the solubility product for AgI.

$$Ag^+ + e^- \rightleftharpoons Ag \qquad E° = 0.7996\ V$$
$$AgI + e^- \rightleftharpoons Ag + I^- \qquad E° = -0.164\ V$$

19-53 Calculate the complex formation equilibrium constant for the Ag(NH$_3$)$_2^+$ complex ion from the following data.

$$Ag^+ + e^- \rightleftharpoons Ag \qquad E° = 0.7996\ V$$
$$Ag(NH_3)_2^+ + e^- \rightleftharpoons Ag + 2\ NH_3 \qquad E° = 0.373\ V$$

19-54 Use the data in Table A-13 to calculate the solubility product of AgCl.

19-55 Use the data in Table A-13 to calculate the solubility product of Hg$_2$Cl$_2$.

19-56 Use the data in Table A-13 to calculate the complex formation constants for the Zn(CN)$_4^{2-}$ complex ion.

19-57 Use the data in Table A-13 to calculate the complex formation constants for the Ni(NH$_3$)$_6^{2+}$ and Co(NH$_3$)$_6^{2+}$ complex ions.

19-58 Calculate $E°$ for the following reaction.

$$Cu(s) + 2\ Ag^+(aq) \longrightarrow Cu^{2+}(aq) + Ag(s)$$

What would happen to the cell potential if enough NaCl was added to increase the Cl$^-$ concentration to $1.0\ M$? (AgCl: $K_{sp} = 1.8 \times 10^{-10}$)

Electrolytic Cells

19-59 Explain why electrolysis of aqueous NaCl produces H$_2$ gas and not sodium metal at the cathode and Cl$_2$ gas and not O$_2$ gas at the anode.

19-60 Calculate the standard-state cell potential necessary to electrolyze molten NaCl to form sodium metal and chlorine gas.

19-61 Calculate the standard-state cell potential necessary to electrolyze an aqueous solution of NaCl.

19-62 Which of the following statements best describes what happens when MgCl$_2$ is electrolyzed?

(a) Mg metal forms at the anode. (b) Mg^{2+} ions are oxidized at the cathode. (c) Cl$_2$ gas is formed at the anode by the oxidation of Cl$^-$. (d) Cl$^-$ ions flow toward the cathode.

19-63 Explain why an electrolyte, such as Na$_2$SO$_4$, has to be added to water before the water can be electrolyzed to H$_2$ and O$_2$.

19-64 Describe what would happen if NiSO$_4$ instead of Na$_2$SO$_4$ was added to the water to be electrolyzed.

19-65 Why does the solution around the cathode become basic when an aqueous solution of Na$_2$SO$_4$ is electrolyzed?

19-66 What are the most likely products of the electrolysis of an aqueous solution of KBr?

(a) $K^+(aq)$ and $Br^-(aq)$ (b) $K(s) + Br_2(l)$
(c) $K^+(aq) + OH^-(aq) + H_2(g) + Br_2(l)$
(d) $K^+(aq) + OH^-(aq) + O_2(g) + Br_2(l)$

19-67 Describe what happens during the electrolysis of the following solutions.

(a) $FeCl_3(aq)$ (b) $NaI(aq)$ (c) $K_2SO_4(aq)$ (d) $H_2SO_4(aq)$

19-68 Why can't aluminum metal be prepared by the electrolysis of an aqueous solution of the Al^{3+} ion?

19-69 Which of the following reactions occurs at the anode during electrolysis of an aqueous Na_2SO_4 solution?

(a) $SO_4^{2-}(aq) \rightarrow SO_2(g) + O_2(g) + 2\ e^-$
(b) $2\ H_2O(l) \rightarrow O_2(g) + 4\ H^+(aq) + 4\ e^-$
(c) $2\ H_2O(l) + O_2(g) + 2\ e^- \rightarrow 4\ OH^-(aq)$
(d) $SO_4^{2-}(aq) + 4\ H^+(aq) + 2\ e^- \rightarrow SO_2(g) + H_2O(l)$

Faraday's Law

19-70 Describe an experiment that would use Faraday's law to determine the strength of an electric current.

19-71 Calculate the amount of silver metal that can be prepared with 1 coulomb of electric charge.

19-72 The rate at which electric power is delivered is measured in units of kilowatts. One watt is the power delivered by a 1-amp current at a potential of 1 volt. Calculate the number of kilowatt-hours of electricity it takes to prepare a mole of sodium metal by electrolysis of molten NaCl.

19-73 If a 5-amp current was used, how long would it take to prepare a mole of sodium metal? Of magnesium metal? Of aluminum metal?

19-74 Predict the products of the electrolysis of an aqueous solution of LiBr, and calculate the weight of each product formed by electrolysis of an $LiBr(aq)$ solution for 1.0 hours with a 2.5-amp current.

19-75 Calculate the amount of electric current necessary to produce a metric ton (1000 kilograms) of Cl_2 gas.

19-76 Calculate the molarity of 1.00 liter of 2.0 M $CuSO_4$ after the solution has been electrolyzed for 2.5 hours with a 4.5-amp current.

19-77 Calculate the ratio of the weight of O_2 to the weight of H_2 produced when an aqueous Na_2SO_4 solution is electrolyzed for 2.56 hours with a 1.34-amp current.

19-78 Under a particular set of conditions, electrolysis of an aqueous $AgNO_3$ solution generated 1.00 gram of silver metal. What weight of I_2 would be produced by the electrolysis of an aqueous solution of NaI under the same condition?

19-79 Calculate the ratio of the weight of Br_2 that collects at the anode to the weight of aluminum metal plated out at the cathode when molten $AlBr_3$ is electrolyzed for 10.5 hours with a 20.0-amp current.

19-80 Assume that aqueous solutions of the following salts are electrolyzed for 20.0 minutes with a 10.0-amp current. Which solution will deposit the most grams of metal at the cathode?

(a) $ZnCl_2$ (b) $ZnBr_2$ (c) WCl_6 (d) $ScBr_3$ (e) $HfCl_4$

19-81 Which of the following compounds will give the most grams of metal when a 10.0-amp current is passed through molten samples of these salts for 2.0 hours?

(a) KCl   (b) $CaCl_2$ (c) $ScCl_3$

19-82 An electric current is passed through a series of three cells filled with aqueous solutions of $AgNO_3$, $Cu(NO_3)_2$, and $Fe(NO_3)_3$. What weight of each metal will be deposited by a 1.58-amp current flowing for 3.54 hours?

19-83 What are the values of x and y in the formula Ce_xCl_y if electrolysis of an aqueous solution of this salt for 16.5 hours with a 1.00-amp current deposits 21.6 grams of cerium metal?

19-84 What is the oxidation number of the manganese in an unknown salt if electrolysis of an aqueous solution of this salt for 30 minutes with a 10.0-amp current generates 2.56 grams of manganese metal?

19-85 Gold forms compounds in the $+1$ and $+3$ oxidation states. What is the oxidation number of gold in a compound that deposits 1.53 grams of gold metal when electrolyzed for 15 minutes with a 2.50-amp current?

Galvanic Corrosion and Cathodic Protection

19-86 Use cell potentials to explain why iron slowly oxidizes to form Fe_2O_3.

19-87 Use cell potentials to explain why iron dissolves in strong acids.

19-88 Use the Nernst equation to explain why the H^+ ion concentration in water is too small for iron to dissolve in water from which all oxygen has been removed.

19-89 Use cell potentials to explain why iron very slowly dissolves in water that contains oxygen.

19-90 Explain why it is a mistake to describe rust as iron(III) oxide — Fe_2O_3.

19-91 When iron rusts, the metal surface acts as the anode of a voltaic cell. Explain why the iron metal becomes the cathode of a voltaic cell when it is connected to a piece of magnesium or zinc.

19-92 Explain why galvanized iron, which contains a thin coating of zinc metal, corrodes much more slowly than iron by itself.

CHEMICAL THERMODYNAMICS

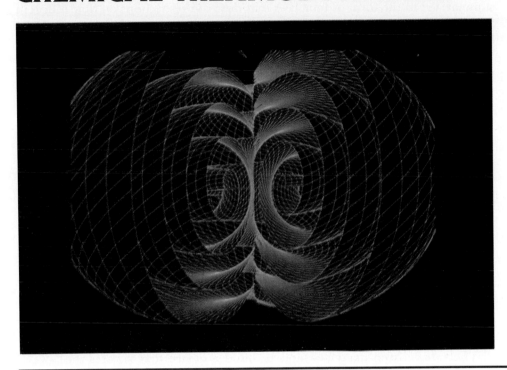

20.1 CHEMICAL THERMODYNAMICS

Chapter 9 defined *thermodynamics* as the branch of science that deals with the relationship between heat and other forms of energy, such as work. It is frequently summarized in the form of four laws that describe restrictions on how different forms of energy can be interconverted. Chapter 9 introduced the zeroth and first laws of thermodynamics.

Zeroth law:
Two objects that are in thermal equilibrium with a third object are in thermal equilibrium with each other.

First law:
Energy is conserved; it can be neither created nor destroyed. Thus, the change in the internal energy of a system is equal to the sum of the heat gained (or lost) by the system and the work done by (or on) the system.

$$\Delta E = q + w$$

This chapter briefly reviews the discussion in Chapter 9 and then introduces the second and third laws of thermodynamics.

Second law:
Spontaneous natural processes are accompanied by an increase in the entropy of the universe.

Third law:
The entropy of a perfect crystal is zero when the temperature of the crystal is at absolute zero (0 K).

There have been many attempts to build a device that violates the laws of thermodynamics. All have failed. Thermodynamics is one of the few areas of science in which there are no exceptions. At a time when all other areas of physics seemed open to question, Einstein wrote about thermodynamics as follows.

Thermodynamics is the only science about which I am firmly convinced that, within the framework of the applicability of its basic principles, it will never be overthrown.

The birth of thermodynamics is often traced to the work of a French scientist, Sadi Carnot, who first analyzed the factors that control the amount of work that can be done by a steam engine in his book *Reflections on the Motive Power of Fire*, published in 1824. Much of thermodynamics is still more important to physicists and engineers than chemists. This chapter will focus on *chemical thermodynamics*, the portion of thermodynamics that pertains to chemical reactions.

20.2 THE FIRST LAW OF THERMODYNAMICS

One of the basic assumptions of thermodynamics is the idea that we can arbitrarily divide the universe into a system and its surroundings, as shown in Figure 20.1. As noted in Section 9.5, the boundary between the system and its surroundings can be either real or imaginary. It can consist of the walls of a beaker that separates a solution from the rest of the universe (as in Figure 20.2). Or it can be an imaginary set of points that divide the air just above the surface of a metal from the rest of the atmosphere (as in Figure 20.3).

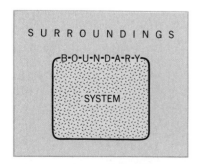

FIG. 20.1 Thermodynamics assumes that the universe can be divided into a system and its surroundings, which are separated by a boundary.

One of the fundamental properties of a chemical system is its **internal energy, E,** which is proportional to the sum of the kinetic and potential energies of the particles that form the system. As noted in Section 9.8, the internal energy of an ideal gas is directly proportional to its temperature, because the potential energy of an ideal gas is zero and the kinetic energy of the gas particles depends on the temperature of the gas.

$$E = \frac{3}{2} RT$$

In this equation, R is the ideal gas constant in units of joules per mole kelvin (J/mol-K) and T is the temperature in units of kelvin. The internal energy of systems that are more complex than an ideal gas can't be measured directly. But the internal energy of the system is still directly proportional to its temperature. This means that we can monitor changes in the internal energy of any system by watching what happens to the temperature of the system.

The internal energy of a system is a state function, as defined in Section 9.10. This means that the internal energy of a system at any time depends only on the state of the system, not the path used to get the system to that state. To illustrate, we can conduct a simple thought experiment that involves a beaker of water on a hot plate. Assume that a thermometer immersed in the water reads 73.5°C. This measurement can only describe the state of the system at that moment in time. It can't tell us whether the water was heated directly from room temperature to 73.5°C or heated from room temperature to 100°C and then allowed to cool. Temperature — and the internal energy of the system to which it is proportional — is therefore a state function. Because the internal energy of the system is a state function, any change in this quantity is equal to the difference between its initial and final values.

$$\Delta E = E_f - E_i$$

The first law of thermodynamics in its simplest form states that energy is conserved. It can be transferred from the system to its surroundings, or vice versa, but it can't be created or destroyed. A more useful form of the first law describes how energy is conserved.

The change in the internal energy of a system is equal to the sum of the heat gained (or lost) by the system and the work done by (or on) the system.

Mathematically, the first law can be written as follows.

$$\Delta E = q + w$$

In this equation, ΔE is the change in the internal energy of the system, q is the heat gained or lost by the system, and w is the work done by the system on its surroundings, or vice versa.

You can remember the sign convention for the relationship between the internal energy of a system and the heat gained or lost by the system by thinking about a concrete example, such as a beaker of water. When the hot plate is turned on, the system gains heat. As a result, both the temperature and the internal energy of the system increase, and ΔE is *positive*. When the hot plate is turned off, the water loses heat to its surroundings as it cools to room temperature, and ΔE is *negative*.

The sign convention for the relationship between internal energy and work can also be remembered by referring to a concrete example, such as a light bulb. When

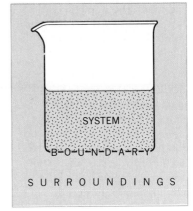

FIG. 20.2 The boundary between the system and its surroundings can be as real as the walls of a beaker that separates a solution (the system) from the rest of the universe (the surroundings).

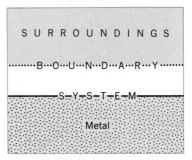

FIG. 20.3 The boundary between the system and its surroundings can be as imaginary as an arbitrary set of points that separate the air just above a metal surface (the system) from the rest of the atmosphere (the surroundings).

FIG. 20.4 The sign conventions for heat and work are similar when the first law of thermodynamics is written as follows.

$$\Delta E = q + w$$

The internal energy of the system decreases when the system either loses heat to its surroundings or does work on the surroundings. Conversely, the internal energy increases when the system gains heat from its surroundings or when work is done on the system.

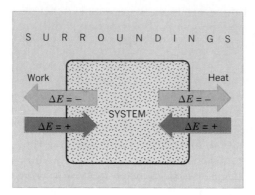

work is done on this system as an electric current is driven through the tungsten wire, the system becomes hotter — in other words, ΔE is *positive*. Conversely, ΔE is *negative* when the system does work on its surroundings.

The sign conventions for heat, work, and internal energy are summarized in Figure 20.4. The internal energy and temperature of a system decrease ($\Delta E < 0$) when the system either loses heat or does work on its surroundings. Conversely, the internal energy and temperature increase ($\Delta E > 0$) when the system gains heat from its surroundings or when the surroundings do work on the system.

20.3 ENTHALPY VERSUS INTERNAL ENERGY

The first law of thermodynamics can be used to measure the change in the internal energy of a system that occurs during a chemical reaction.

$$\Delta E = q + w$$

In this case, the system is the chemical reaction and the boundary is the container in which the reaction is run. In the course of the reaction, heat is either given off or absorbed. If we run the reaction under conditions in which no work is done by the system on its surroundings, or vice versa, the heat given off or absorbed will be equal to the change in the internal energy of the system.

$$\Delta E = q \qquad \text{(if and only if } w = 0)$$

Two kinds of work are normally associated with a chemical reaction: **electrical work** and **work of expansion.** Chapter 19 showed how chemical reactions can do work by driving an electric current through an external wire. This requires running the reaction under special conditions, however, so it is easy to prevent a chemical reaction from doing electrical work. Chemical reactions can also do work of expansion on their surroundings, by increasing the volume of the system. Section 9.7 showed that work of expansion is proportional to the pressure against which the system expands times the change in the volume of the system.

$$w = -P\Delta V$$

The sign convention for this equation reflects the fact that the internal energy of the system decreases ($\Delta E < 0$) when the system does work on its surroundings.

When a chemical reaction is run under conditions of constant volume, no work of expansion is possible. The heat given off or absorbed by a reaction run at constant volume is therefore equal to the change in the internal energy that occurs during the reaction.

$$\Delta E = q_v$$

Although it is possible to run reactions in sealed containers at constant volume, most chemical reactions are run in open flasks and beakers. Here, volume is not constant, because gas can either enter or leave the container during the reaction. The system is at constant pressure, however, because the total pressure inside the container is always equal to atmospheric pressure.

If a gas is driven out of the flask during the reaction, the system does work on its surroundings. If the net effect of the reaction is to pull a gas into the flask, the surroundings do work on the system. We can still measure the amount of heat given off or absorbed during the reaction, but it is no longer equal to the change in the internal energy of the system, because some of the heat has been converted into work.

$$q_P = \Delta E - w$$

We can get around this problem by introducing the concept of **enthalpy (H)**, the sum of the internal energy of the system plus the product of the pressure of the gas in the system times the volume of the system.

$$H = E + PV$$

The change in the enthalpy of the system during a chemical reaction is equal to the change in its internal energy plus the change in the product of the pressure times the volume of the system.

$$\Delta H = \Delta E + \Delta(PV)$$

If the reaction is run at constant pressure, this equation takes the following form.

$$\Delta H = \Delta E + P\Delta V$$

Substituting the first law into this equation gives the following.

$$\Delta H = (q_P + w) + P\Delta V$$

Assuming that the only work done by the reaction is work of expansion gives an equation in which the $P\Delta V$ terms cancel.

$$\Delta H = (q_P - \cancel{P\Delta V}) + \cancel{P\Delta V}$$

Thus, the heat given off or absorbed during a chemical reaction run at constant pressure is equal to the change in the enthalpy of the system.

$$\Delta H = q_P$$

As first noted in Section 9.12, the relationship between the change in the internal energy of the system during a chemical reaction and the enthalpy of reaction can be summarized as follows.

1. The heat given off or absorbed when a reaction is run at constant *volume* is equal to the change in the internal energy of the system.

$$\Delta E = q_v$$

2. The heat given off or absorbed when a reaction is run at constant *pressure* is equal to the change in the enthalpy of the system.

$$\Delta H = q_P$$

3. The change in the enthalpy of the system during a chemical reaction is equal to the change in the internal energy plus the change in the product of the pressure of the gas in the system and its volume.

$$\Delta H = \Delta E + \Delta(PV)$$

4. The difference between ΔE and ΔH is small for reactions that involve only liquids and solids, because there is little, if any, change in the volume of the system during the reaction. The difference can be relatively large, however, for reactions that involve gases if the number of moles of gas changes in the course of the reaction.

Exercise 20.1

Which of the following processes are run at constant volume and which are run at constant pressure?

(a) Titration of a strong acid with a strong base
(b) Decomposition of $CaCO_3$ by heating limestone in a crucible with a bunsen burner
(c) Reaction between zinc metal and an aqueous solution of Cu^{2+} ions to form copper metal and Zn^{2+} ions
(d) Measurement of the calories in a 1-ounce serving of a breakfast cereal made when the cereal is burned in a bomb calorimeter

Solution

Processes (a), (b), and (c) are all run under conditions of constant pressure. Only (d) is run at constant volume.

FIG. 20.5 There is a preferred direction to most chemical reactions. It isn't surprising to find that magnesium metal dissolves in concentrated hydrochloric acid to form H_2 gas, which bubbles out of solution. But it would be surprising to find bubbles of H_2 appearing at the surface of the solution and then sinking through the solution until they vanished, while a piece of magnesium metal mysteriously appeared.

20.4 SPONTANEOUS CHEMICAL REACTIONS

The first law of thermodynamics is often described as suggesting that we can't get something for nothing. It allows us to build an apparatus that does work, but it places important restrictions on that apparatus. It says that we have to be willing to pay a price for work in terms of a loss of either heat or internal energy. It also puts a limit on the amount of work we can get for a given investment of either heat or internal energy.

The first law allows us to convert heat into work or work into heat. It also allows us to change the internal energy of a system by transferring either heat or work between the system and its surroundings. But it doesn't tell us whether one of these changes is more easily achieved than another. Our experiences, however, tell us that there is a preferred direction to many natural processes. We aren't surprised when a cup of coffee gradually cools during the course of a meal, for example, or when the ice in a glass of lemonade gradually melts. But we would be surprised if a cup of coffee grew hotter until it boiled or if the water in a glass of lemonade spontaneously froze on a hot summer day, although neither process violates the first law of thermodynamics.

Similarly, we aren't surprised to see a piece of zinc metal dissolve in a strong acid to give bubbles of hydrogen gas, as shown in Figure 20.5.

$$Zn(s) + 2\ H^+(aq) \longrightarrow Zn^{2+}(aq) + H_2(g)$$

But if we saw a film in which H_2 bubbles formed on the surface of a solution and then sank through the solution until they disappeared, while a strip of zinc metal formed in the middle of the solution, we would conclude that the film was being run backward.

Many chemical and physical processes are reversible and yet tend to proceed in one direction—a direction in which they are said to be *spontaneous*. This raises an important question: What makes a reaction spontaneous; what drives the reaction in one direction and not the other? So many spontaneous reactions are exothermic that it is tempting to argue that spontaneous chemical reactions are those that give off energy. The reaction between aluminum and bromine, for example, is exothermic.

$$2 \text{ Al}(s) + 3 \text{ Br}(l) \longrightarrow 2 \text{ AlBr}_3(s) \qquad \Delta H° = -511 \text{ kJ/mol AlBr}_3$$

So is the reaction between hydrogen and oxygen

$$2 \text{ H}_2(g) + \text{O}_2(g) \longrightarrow 2 \text{ H}_2\text{O}(g) \qquad \Delta H° = -241.83 \text{ kJ/mol H}_2\text{O}$$

and the reaction between phosphorus and oxygen.

$$\text{P}_4(s) + 5 \text{ O}_2(g) \longrightarrow \text{P}_4\text{O}_{10}(s) \qquad \Delta H° = -2985 \text{ kJ/mol P}_4\text{O}_{10}$$

We might therefore draw the following conclusion.

One of the driving forces behind a chemical reaction is a tendency to give off energy.

There are also spontaneous reactions, however, that absorb energy from their surroundings. At 100°C, water boils spontaneously even though the reaction is endothermic and the system must absorb energy from its surroundings.

$$\text{H}_2\text{O}(l) \longrightarrow \text{H}_2\text{O}(g) \qquad \Delta H° = 40.88 \text{ kJ/mol}$$

Ammonium nitrate dissolves spontaneously in water, even though energy is absorbed when this reaction takes place.

$$\text{NH}_4\text{NO}_3(s) \xrightarrow{\text{H}_2\text{O}} \text{NH}_4^+(aq) + \text{NO}_3^-(aq) \qquad \Delta H° = 25.7 \text{ kJ/mol}$$

The tendency of a spontaneous reaction to give off energy can't be the only driving force behind a chemical reaction. There must be another factor that helps determine whether a reaction is spontaneous. This factor, known as *entropy*, is a measure of the disorder of the system.

20.5 ENTROPY AS A MEASURE OF DISORDER

Perhaps the best way to understand entropy as a driving force in nature is to conduct a simple thought experiment. Start with a new deck of cards. Open the deck, remove the jokers, and then turn the deck so that you can read the cards. The top card will be the ace of spades, followed by the two, three, and four of spades, and so on, as shown in Figure 20.6. Now divide the cards in half, shuffle the deck, and notice what happens: The deck becomes more disordered. The more often the deck is shuffled, the more disordered it becomes. What makes a deck of cards become more random, or more disordered, when shuffled?

In 1877, Ludwig Boltzmann provided us with a basis for answering this question when he showed that the entropy of a system is a measure of the amount of disorder in the system. A deck of cards fresh from the manufacturer is perfectly ordered and

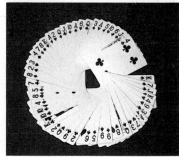

FIG. 20.6 A deck of cards, fresh from the manufacturer, is a perfectly ordered system. When the cards are shuffled, they become more and more disordered. The driving force behind this process is a natural tendency for systems to move toward greater disorder.

TABLE 20.1

Number of Equivalent Combinations (W)
for Various Types of Poker Hands

Hand	W	ln W
Royal flush (AKQJ10 in one suit)	4	1.39
Straight flush (five cards in sequence in one suit)	36	3.58
Four of a kind	624	6.44
Full house (three of a kind plus a pair)	3,744	8.23
Flush (five cards in the same suit)	5,108	8.54
Straight (five cards in sequence)	10,200	9.23
Three of a kind	54,912	10.91
Two pairs	123,552	11.72
One pair	1,098,240	13.91
No pairs	1,302,540	14.08
Total	2,598,960	

the entropy of this system is zero. When the deck is shuffled, however, the entropy of the system increases as the deck becomes more and more disordered. There are 8.066×10^{67} different ways of organizing a deck of cards. The probability of obtaining any particular order of cards when the deck is shuffled is therefore 1 part in 8.066×10^{67}. In theory, it is possible to shuffle a deck of cards over and over again until the cards fall into perfect order. But it isn't very likely!

Boltzmann proposed the following equation to describe the relationship between entropy and the amount of disorder in a system.

$$S = K \ln W$$

In this equation, S is the entropy of the system, k is a proportionality constant equal to the ideal gas constant divided by Avogadro's constant, ln symbolizes a log to the base e, and W is the number of equivalent ways of describing the state of the system. According to this equation, the entropy of a system increases as the number of equivalent ways of describing the state of the system increases.

The relationship between the number of equivalent ways of describing the state of a system and the amount of disorder in the system can be demonstrated with another analogy based on a deck of cards. There are 2,598,960 different hands that could be dealt in a game of five-card poker. More than half of these hands are essentially worthless. Winning hands are much rarer. Only 3,744 combinations correspond to a "full house," for example. Table 20.1 gives the number of equivalent combinations of cards for each category of poker hand, which is the value of W for this category. As the hand becomes more disordered, the value of W becomes larger, and the hand becomes intrinsically less valuable.

20.6 ENTROPY AND THE SECOND LAW OF THERMODYNAMICS

The second law of thermodynamics describes the relationship between entropy and the spontaneity of natural processes.

Natural processes that are accompanied by an increase in entropy tend to be spontaneous.

We can apply this principle to chemical reactions by noting that entropy is a state function that is proportional to the disorder of the system.

$\Delta S > 0$ implies that the system becomes more *disordered*.

$\Delta S < 0$ implies that the system becomes more *ordered*.

Entropy favors any process that leads to an increase in the disorder of the system.

The following generalizations can help us decide when a system becomes more disordered during a chemical reaction.

1. Solids have a much more regular structure than liquids and are therefore more ordered than liquids.
2. The particles in a gas are in a state of constant, random motion. Gases are therefore less regular — more disordered — than the corresponding liquids.
3. Any process that increases the number of particles in the system increases the amount of disorder.

Exercise 20.2

Predict which of the following processes will lead to an increase in the entropy of the system.

(a) $N_2(g) + 3\ H_2(g) \longrightarrow 2\ NH_3(g)$

(b) $H_2O(l) \longrightarrow H_2O(g)$

(c) $CaCO_3(s) \longrightarrow CaO(s) + CO_2(g)$

(d) $NH_4NO_3(s) \xrightarrow{H_2O} NH_4^+(aq) + NO_3^-(aq)$

Solution

(a) The total number of molecules decreases in this reaction, which means the entropy of the system decreases as it becomes more ordered.

(b) Gases are much more disordered than the corresponding liquids, so the entropy of the system increases.

(c) Reactions in which a compound decomposes into two products lead to an increase in entropy, because the system becomes more disordered. The increase in entropy in this reaction is even larger because the starting material is a solid and one of the products is a gas.

(d) Isolated NH_4^+ and NO_3^- ions in aqueous solutions are more disordered than solid NH_4NO_3, so the entropy of the system increases in this reaction.

The two driving forces that control the direction in which a chemical reaction occurs are the changes in the enthalpy and the entropy of the system that occur in the course of the reaction.

Chemical reactions that are exothermic — that give off energy — are often driven to completion by this factor.

Chemical reactions that lead to an increase in the disorder of a system — and therefore its entropy — are often driven to completion by this factor.

Exercise 20.3

Decide whether enthalpy or entropy is the driving force behind the following reaction, discussed in Section 10.6.

$$2 \, NO_2(g) \rightleftharpoons N_2O_4(g)$$

Solution

Section 10.6 noted that NO_2 is unusual because it is a stable molecule with an unpaired electron in its Lewis structure.

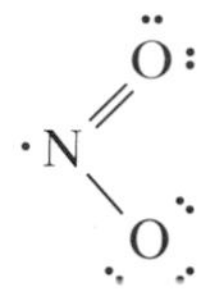

When a pair of NO_2 molecules collide in the proper orientation, a covalent bond can form between the nitrogen atoms to produce the dimer, N_2O_4.

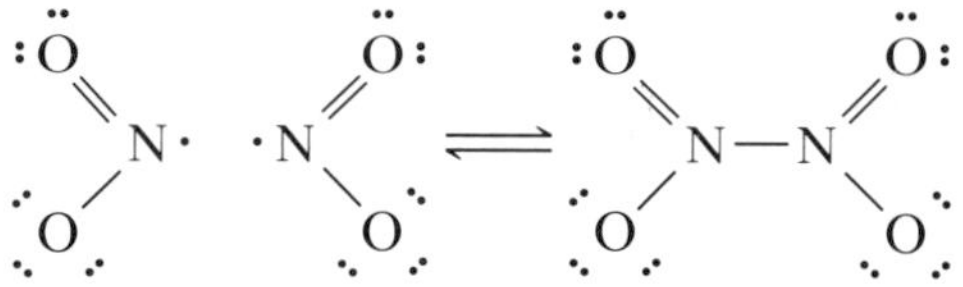

This reaction is favored by enthalpy, because it forms a new bond and is therefore exothermic. It is not favored by entropy, because it leads to a decrease in the disorder of the system. Enthalpy is therefore the driving force behind this reaction.

20.7 THE THIRD LAW OF THERMODYNAMICS

The third law of thermodynamics defines absolute zero on the entropy scale.

The entropy of a perfect crystal is zero when the temperature of the crystal is at absolute zero (0 K).

The crystal must be perfect, or else there will be some inherent disorder. It also must be at 0 K; otherwise there will be thermal motion within the crystal, which leads to disorder.

As the crystal warms to temperatures above 0 K, the particles in the crystal start to move, generating some disorder. The entropy of the crystal gradually increases with temperature as the average kinetic energy of the particles increases. At the melting point, the entropy of the system increases abruptly as the compound is transformed into a liquid, which is not as well ordered as the solid. The entropy of the liquid gradually increases as the liquid becomes warmer, because of the gradual increase in the vibrational, rotational, and translational motion of the particles. At the boiling point, there is another abrupt increase in the entropy of the substance as it is transformed into a random, chaotic gas.

The data for sulfur trioxide in Table 20.2 provide an example of what happens to the entropy of a substance as it changes from a solid to a liquid and then a gas. Note that the units of entropy are joules per mole kelvin (J/mol-K).

TABLE 20.2

Standard-State Entropies for the Solid, Liquid, and Gaseous Forms of Sulfur Trioxide

Compound	$S°$ ($J/mol\text{-}K$)
$SO_3(s)$	70.7
$SO_3(l)$	113.8
$SO_3(g)$	256.7

20.8 STANDARD-STATE ENTROPIES OF REACTION

The difference between the sum of the entropies of the reactants and products of a chemical reaction is known as the *entropy of reaction, ΔS*. When this difference is measured under standard-state conditions, the result is the *standard-state entropy of reaction, ΔS°*.

Standard-state Conditions:

All solutions have concentrations of 1 *M*.

All gases have partial pressures of 1 atm.

The temperature is 25°C.

The change in any state function is calculated by subtracting the value for the initial state from the value for the final state. The standard-state entropy of reaction is therefore equal to the sum of the standard-state entropies of the products minus the sum of the standard-state entropies of the reactants.

$$\Delta S° = \sum S° \text{ (products)} - \sum S° \text{ (reactants)}$$

Standard-state entropies for the reactants and products of some common chemical reactions are given in Table A-15 in the appendix.

Exercise 20.4

Calculate the standard-state entropy of reaction for the following equations and explain the sign of $\Delta S°$ for each.

(a) $Hg(l) \rightleftharpoons Hg(g)$

(b) $NaCl(s) \xrightarrow{H_2O} Na^+(aq) + Cl^-(aq)$

(c) $2\ NO_2(g) \rightleftharpoons N_2O_4(g)$

(d) $N_2(g) + O_2(g) \longrightarrow 2\ NO(g)$

Solution

(a) Table A-15 contains the following standard-state entropy data for this reaction.

Compound	$S°$ (J/mol-K)
$Hg(l)$	76.02
$Hg(g)$	174.9

The balanced equation for the process states that 1 mole of mercury vapor is produced for each mole of liquid mercury that boils. The standard-state entropy of reaction is therefore calculated as follows.

$$\Delta S° = \sum S° \text{ (products)} - \sum S° \text{ (reactants)}$$
$$= [1 \text{ mol } Hg(g) \times 174.9 \text{ J/mol-K}] - [1 \text{ mol } Hg(l) \times 76.02 \text{ J/mol-K}]$$
$$= 98.9 \text{ J/K}$$

The sign of $\Delta S°$ is positive, because this process transforms a liquid into a gas, which is inherently more disordered.

(b) Table A-15 contains the following standard-state entropy data for this reaction.

Compound	$S°$ (J/mol-K)
$NaCl(s)$	72.13
$Na^+(aq)$	59.0
$Cl^-(aq)$	56.5

In this reaction, 1 mole of Na^+ ions and 1 mole of Cl^- ions are produced for each mole of $NaCl$ that dissolves. The standard-state entropy of reaction is therefore calculated as follows.

$$\Delta S° = \sum S° \text{ (products)} - \sum S° \text{ (reactants)}$$
$$= (1 \text{ mol } Na^+ \times 59.0 \text{ J/mol-K} + 1 \text{ mol } Cl^- \times 56.5 \text{ J/mol-K})$$
$$- (1 \text{ mol } NaCl \times 72.13 \text{ J/mol-K})$$
$$= 43.4 \text{ J/K}$$

The sign of $\Delta S°$ for this reaction is positive because Na^+ and Cl^- ions in aqueous solution are free to move independently of each other and are therefore more disordered than solid $NaCl$.

(c) Table A-15 contains the following standard-state entropy data for this reaction.

Compound	$S°$ (J/mol-K)
NO_2	240.0
N_2O_4	304.2

In this equation, 1 mole of N_2O_4 is formed for every 2 moles of NO_2 consumed, and the value of $\Delta S°$ is calculated as follows.

$$\Delta S° = \sum S° \text{ (products)} - \sum S° \text{ (reactants)}$$
$$= (1 \text{ mol } N_2O_4 \times 304.2 \text{ J/mol-K}) - (2 \text{ mol } NO_2 \times 240.0 \text{ J/mol-K})$$
$$= -175.8 \text{ J/K}$$

The sign of $\Delta S°$ is negative, because two molecules come together in this reaction to form a larger, more ordered product.

(d) Table A-15 contains the following standard-state entropy data for this reaction.

Compound	$S°$ (J/mol-K)
$NO(g)$	210.7
$N_2(g)$	191.5
$O_2(g)$	205.03

The balanced equation for this reaction indicates that 2 moles of NO are produced when 1 mole of N_2 reacts with 1 mole of O_2. Thus, the standard-state entropy of reaction is calculated as follows.

$$\Delta S° = \sum S° \text{ (products)} - \sum S° \text{ (reactants)}$$
$$= (2 \text{ mol } NO \times 210.7 \text{ J/mol-K})$$
$$- (1 \text{ mol } N_2 \times 191.5 \text{ J/mol-K} + 1 \text{ mol } O_2 \times 205.03 \text{ J/mol-K})$$
$$= 24.9 \text{ J/K}$$

$\Delta S°$ for this reaction is small but positive, because the product of the reaction (NO) is slightly more disordered than the reactants (N_2 and O_2).

20.9 THE DIFFERENCE BETWEEN ENTHALPY OF REACTION AND ENTROPY OF REACTION CALCULATIONS

At first glance, tables of thermodynamic data seem inconsistent. Consider, for example, the data in Table 20.3, taken from Table A-15 in the appendix.

The enthapy data in this table are given in terms of the enthalpy of formation of each substance, $\Delta H_f°$. This quantity is the heat given off or absorbed when the substance is made from its elements in their most stable state at 25°C and 1 atm. The enthalpy of formation of AlF_3, for example, is the heat given off in the following reaction.

$$2\ Al(s) + 3\ F_2(g) \longrightarrow 2\ AlF_3(s)$$

The enthalpy data in this table are relative numbers, which compare each compound with its elements. The entropy data, on the other hand, are given as absolute numbers, $S°$.

Enthalpy data are listed as relative measurements because there is no absolute zero on the enthalpy scale. All we can measure is the heat given off or absorbed by a chemical reaction. Thus, all we can determine is the difference between the enthalpies of the reactants and the products of a reaction. We therefore define the enthalpy of formation of the elements in their most stable states at 25°C and 1 atm as zero and report all compounds as either more or less stable than their elements.

Entropy data are different. The third law of thermodynamics provides an absolute zero on the entropy scale. The absolute entropy of any element or compound at 25°C can be measured by comparing it with a perfect crystal at absolute zero.

20.10 GIBBS FREE ENERGY

We have stated that two forces determine the direction in which a reaction is spontaneous—the enthalpy of reaction and the entropy of reaction. Some reactions are spontaneous because they give off energy in the form of heat ($\Delta H < 0$). Others are spontaneous because they lead to an increase in the disorder of the system ($\Delta S > 0$). The following exercise shows how calculations of $\Delta H°$ and $\Delta S°$ can be used to identify the driving force behind a chemical reaction.

TABLE 20.3

Thermodynamic Data for Aluminum and Its Compounds

Substance	$\Delta H_f°(kJ/mol)$	$S°\ (J/mol\text{-}K)$
Al(s)	0	28.3
Al(g)	326	164.4
$Al_2O_3(s)$	−1676	50.92
$AlBr_3(s)$	−511	180
$AlCl_3(s)$	−704.2	110.7
$AlF_3(s)$	−1504	66.44

Exercise 20.5

Calculate $\Delta H°$ and $\Delta S°$ for the following reaction and decide in which direction each of these factors will drive the reaction.

$$N_2(g) + 3 H_2(g) \rightleftharpoons 2 NH_3(g)$$

Solution

Table A-15 contains the following data for this reaction.

Compound	$\Delta H_f°(kJ/mol)$	$S°$ $(J/mol\text{-}K)$
$N_2(g)$	0	191.5
$H_2(g)$	0	130.57
$NH_3(g)$	-46.1	192.3

The reaction is exothermic ($\Delta H° < 0$), which means that it is favored by the enthalpy of reaction.

$$\begin{aligned}
\Delta H° &= \sum \Delta H_f°(\text{products}) - \sum \Delta H_f°(\text{reactants}) \\
&= (2 \text{ mol } NH_3 \times -46.1 \text{ kJ/mol}) \\
&\quad - (1 \text{ mol } N_2 \times 0 \text{ kJ/mol} + 3 \text{ mol } H_2 \times 0 \text{ kJ/mol}) \\
&= -92.2 \text{ kJ}
\end{aligned}$$

There is a significant increase in the order of the system ($\Delta S° < 0$), however, when N_2 and H_2 combine to form NH_3. The reaction is therefore unfavorable in terms of the entropy of reaction.

$$\begin{aligned}
\Delta S° &= \sum S° (\text{products}) - \sum S° (\text{reactants}) \\
&= (2 \text{ mol } NH_3 \times 192.3 \text{ J/mol-K}) \\
&\quad - (1 \text{ mol } N_2 \times 191.5 \text{ J/mol-K} + 3 \text{ mol } H_2 \times 130.57 \text{ J/mol-K}) \\
&= -198.6 \text{ J/K}
\end{aligned}$$

This exercise raises an important question: What happens when one of the potential driving forces behind a chemical reaction is favorable and the other is not? We can answer this question by defining a new quantity known as the ***Gibbs free energy (G)*** of the system, which reflects the balance between these forces. The name of this quantity recognizes the contributions of J. Willard Gibbs, a professor of mathematical physics at Yale from 1871 until the turn of the century, who is considered by some to be the greatest scientist yet produced by the United States.

The Gibbs free energy of a system is a measure of the energy available to do work under conditions of constant pressure. G for a system at any moment in time is defined as the enthalpy (H) of the system minus the product of the temperature (T) times the entropy (S) of the system.

$$G = H - TS$$

The Gibbs free energy of the system is a state function, because it is defined in terms of other thermodynamic properties of the system that are state functions. The change in the Gibbs free energy of the system that occurs during a chemical reaction is given by the following equation.

$$\Delta G = \Delta H - \Delta TS$$

If the reaction is run at constant temperature, this equation can be written as follows.

$$\Delta G = \Delta H - T\Delta S$$

If the data are collected under standard-state conditions, the result is the ***standard-state free energy of reaction (ΔG°).***

$$\Delta G° = \Delta H° - T\Delta S°$$

The entropy term is subtracted from the enthalpy term in these equations because these terms have opposite sign conventions. According to the foregoing discussion, reactions are spontaneous when ΔH is negative and ΔS is positive.

$$\begin{matrix} \Delta H < 0 \\ \Delta S > 0 \end{matrix} \quad \text{favorable, or spontaneous reaction}$$

The opposite is true when $\Delta H°$ is positive and $\Delta S°$ is negative.

$$\begin{matrix} \Delta H > 0 \\ \Delta S < 0 \end{matrix} \quad \text{unfavorable, or not spontaneous reaction}$$

Substituting these terms into the equation that defines the free energy of the system suggests that any reaction for which $\Delta G°$ is negative will be favorable (or spontaneous) and that any reaction for which $\Delta G°$ is positive will be unfavorable.

$$\Delta G < 0 \quad \text{favorable, or spontaneous reaction}$$
$$\Delta G > 0 \quad \text{unfavorable, or not spontaneous reaction}$$

Why do we multiply the $\Delta S°$ term, but not the $\Delta H°$ term, in the following equation by the temperature of the system?

$$\Delta G° = \Delta H° - T\Delta S°$$

Remember that the enthalpy of reaction is stated in units of kilojoules per mole (kJ/mol) but the entropy of reaction is stated in units of joules per mole kelvin (J/mol-K). We therefore have to multiply the entropy term by the temperature in kelvin so that the two terms have the same units.

When a reaction is favored by both enthalpy ($\Delta H < 0$) and entropy ($\Delta S > 0$), there is no need to calculate the value of ΔG to decide whether the reaction should proceed. The same can be said for reactions favored by neither enthalpy ($\Delta H > 0$) nor entropy ($\Delta S < 0$). Free energy calculations become important, however, for reactions favored by only one of these factors.

Exercise 20.6

Use calculations of $\Delta H°$, $\Delta S°$, and $\Delta G°$ for the following reaction to explain why NH_4NO_3 spontaneously dissolves in water.

$$NH_4NO_3(s) \xrightarrow{H_2O} NH_4^+(aq) + NO_3^-(aq)$$

Solution

Table A-15 contains the following data for this reaction.

Compound	$\Delta H_f°(kJ/mol)$	$S° (J/mol\text{-}K)$
$NH_4NO_3(s)$	-365.6	151.1
$NH_4^+(aq)$	-132.5	113
$NO_3^-(aq)$	-207.4	146

This reaction is endothermic, and therefore unfavorable in terms of the enthalpy of reaction.

$$\Delta H° = \sum \Delta H_f°(\text{products}) - \sum \Delta H_f°(\text{reactants})$$
$$= (1 \text{ mol } NH_4^+ \times -132.5 \text{ kJ/mol} + 1 \text{ mol } NO_3^- \times -207.4 \text{ kJ/mol})$$
$$- (1 \text{ mol } NH_4NO_3 \times -365.6 \text{ kJ/mol})$$
$$= 25.6 \text{ kJ}$$

The reaction leads to a significant increase in the disorder of the system, however, and is therefore favored by the entropy of reaction.

$$\Delta S° = \sum S°(\text{products}) - \sum S°(\text{reactants})$$
$$= (1 \text{ mol } NH_4^+ \times 113 \text{ J/mol-K} + 1 \text{ mol } NO_3^- \times 146 \text{ J/mol-K})$$
$$- (1 \text{ mol } NH_4NO_3 \times 151.1 \text{ J/mol-K})$$
$$= 108 \text{ J/K}$$

To decide whether NH_4NO_3 should dissolve in water we have to compare the $\Delta H°$ and $T\Delta S°$ terms to see which is larger. Before we can do this, however, we have to convert the temperature from celsius to kelvin.

$$25°C + 273 = 298 \text{ K}$$

We then recognize the difference in the units in which enthalpy of reaction and entropy of reaction data are reported. The units of $\Delta H°$ in this calculation are kilojoules, whereas the units of $\Delta S°$ are joules per kelvin. Before we can compare the magnitudes of the $\Delta H°$ and $T\Delta S°$ terms we have to convert them to a consistent set of units. Perhaps the easiest way of doing this is to convert the units of $\Delta H°$ to joules. We then multiply the entropy term by the absolute temperature and subtract this quantity from the enthalpy term.

$$\Delta G° = \Delta H° - T\Delta S°$$
$$= 25,600 \text{ J} - (298 \text{ K} \times 108 \text{ J/K})$$
$$= 25,600 \text{ J} - 32,000 \text{ J} = -6600 \text{ J}$$

At 25°C, the standard-state free energy for this reaction is negative because the entropy term is larger than the enthalpy term.

$$G° = -6.6 \text{ kJ}$$

The reaction is therefore spontaneous at this temperature.

20.11 THE EFFECT OF TEMPERATURE ON THE FREE ENERGY OF CHEMICAL REACTIONS

The balance between the contributions from the enthalpy and entropy terms to the free energy of a chemical reaction depends on the temperature at which the reaction is run.

Exercise 20.7

Predict whether the following reaction is spontaneous at 25°C.

$$N_2(g) + 3 H_2(g) \rightleftharpoons 2 NH_3(g)$$

Solution

The values of $\Delta H°$ and $\Delta S°$ for this reaction were calculated in Exercise 20.6. According to that calculation, this reaction is favored by enthalpy but not by entropy.

$$\Delta H° = -92.2 \text{ kJ} \qquad \text{(favorable)}$$
$$\Delta S° = -198.6 \text{ J/K} \qquad \text{(unfavorable)}$$

Before we can compare these terms to see which is larger, we have to incorporate into our calculation the temperature at which the reaction is run. We can start by calculating the temperature in units of kelvin.

$$25°C + 273 = 298 \text{ K}$$

We then multiply the entropy of reaction by the absolute temperature.

$$\begin{aligned}\Delta G° &= \Delta H° - T\Delta S° \\ &= (-92,200 \text{ J}) - (298 \text{ K} \times -198.6 \text{ J/K}) \\ &= (-92,200 \text{ J}) + 59,200 \text{ J} \\ &= -33,000 \text{ J}\end{aligned}$$

According to this calculation, the reaction is spontaneous at 25°C.

$$\Delta G° = -33.0 \text{ kJ}$$

What happens as we raise the temperature of the reaction? The equation used to define free energy suggests that the entropy term will become more important as the temperature increases. Since the entropy term is unfavorable, the reaction should become less favorable as the temperature increases.

Exercise 20.8

Predict whether the following reaction is still spontaneous at 500°C.

$$N_2(g) + 3 H_2(g) \rightleftharpoons 2 NH_3(g)$$

(Assume that the values of $\Delta H°$ and $\Delta S°$ used in Exercise 20.7 are still valid at this temperature.)

Solution

To decide whether the reaction is still spontaneous, we need to calculate the temperature on the kelvin scale.

$$500°C + 273 = 773 \text{ K}$$

We then multiply the entropy term by this temperature and subtract the resulting term from the value of $\Delta H°$ for the reaction.

$$\begin{aligned}\Delta G° &= \Delta H° - T\Delta S° \\ &= (-92,200 \text{ J}) - (773 \text{ K} \times -198.6 \text{ J/K}) = 61,300 \text{ J} \\ &= 61.3 \text{ kJ}\end{aligned}$$

Because the unfavorable entropy term becomes larger as the temperature of the reaction increases, the reaction changes from one which is favorable at low temperatures to one that is unfavorable at high temperatures.

20.12 BEWARE OF OVERSIMPLIFICATIONS

The calculation in Exercise 20.7 of $\Delta G°$ for the following reaction at 25°C is perfectly legitimate.

$$N_2(g) + 3 H_2(g) \rightleftharpoons 2 NH_3(g)$$

The enthalpy and entropy data used in this calculation were standard-state measurements defined in terms of concentrations of 1 M, partial pressures of 1 atm, and a temperature of 25°C. It is therefore legitimate to use these data to predict the standard-state free energy for this reaction at 25°C. The calculation of $\Delta G°$ for this reaction at 500°C in Exercise 20.8, however, was based on the assumption that the enthalpy and entropy data defined at 25°C are still valid at 500°C.

This is a useful assumption, but it may not be a valid one. Small but significant changes in $\Delta H°$ and $\Delta S°$ occur as the temperature of the reaction changes. These changes are often small enough over moderate temperature ranges, however. Thus, the calculation in Exercise 20.8 gives results that are reasonably close to those obtained experimentally.

20.13 STANDARD-STATE FREE ENERGIES OF REACTION

Section 20.10 showed how to calculate the standard-state free energy of reaction from the standard-state enthalpy and entropy of reaction.

$$\Delta G° = \Delta H° - T\Delta S°$$

It is also possible to calculate $\Delta G°$ from tabulated standard-state free energy data. Since there is no absolute zero on the free-energy scale, the easiest way to tabulate such data is in terms of *standard-state free energies of formation*, $\Delta G_f°$.

Standard-state free energy of formation, $\Delta G_f°$: The difference between the free energy of a substance and the free energies of its elements in their most stable states at 25°C and 1 atm, all measurements being made under standard-state conditions.

Exercise 20.9

Explain why $\Delta G_f°$ for $N_2(g)$ is 0 kJ/mol.

Solution

The standard-state free energy of formation for any element in its most stable state at 25°C and 1 atm is 0 kJ/mol, because there is no difference between the starting materials and the products in the equation that would be written to describe the formation of this substance from its elements in their most stable states at 25°C and 1 atm.

$$N_2(g) \longrightarrow N_2(g) \qquad \Delta G_f° = 0 \text{ kJ/mol}$$

In Chapter 9, we noted that the enthalpy of formation of an element in its most stable state at 25°C and 1 atm pressure is equal to zero. For the same reasons, the free energy of formation of an element in its most stable state at 25°C and 1 atm pressure is also zero.

$\Delta G°$ for any reaction can be calculated from $\Delta G_f°$ for the reactants and products in much the same way that $\Delta H°$ for a reaction is calculated from $\Delta H_f°$ data. The standard-state free energy of reaction is equal to the sum of the free energies of formation of the products of the reaction minus the sum of the free energies of formation of the reactants.

$$\Delta G° = \sum \Delta G_f°(\text{products}) - \sum \Delta G_f°(\text{reactants})$$

Exercise 20.10

Calculate $\Delta H°$ and $\Delta G°$ for the following reaction.

$$Ba(OH)_2 \cdot 8 \ H_2O(s) + 2 \ NH_4NO_3(s) \longrightarrow Ba(NO_3)_2(aq) + 2 \ NH_3(aq) + 10 \ H_2O(l)$$

Use these results to explain why stirring a mixture of $Ba(OH)_2 \cdot 8 \ H_2O$ and NH_4NO_3 can cause the mixture to absorb enough energy from its surroundings to freeze a beaker containing these solids to a wooden board, as shown in Section 9.13.

Solution

$Ba(OH)_2$ is a base and NH_4NO_3 is an acid. When they react, the products of this reaction dissolve in the water given off in the reaction. Table A-15 contains the following data for this reaction.

Compound	$\Delta H_f^\circ(kJ/mol)$	$\Delta G_f^\circ(kJ/mol)$
$Ba(OH)_2 \cdot 8 \ H_2O(s)$	-3342.2	-2792.8
$Ba(NO_3)_2(aq)$	-952.36	-783.28
$NH_4NO_3(s)$	-365.6	-184.0
$NH_3(aq)$	-80.29	-26.50
$H_2O(l)$	-285.83	-237.18

The value of ΔH° for the reaction is equal to the sum of the enthalpies of formation of the products minus the sum of the enthalpies of formation of the reactants.

$$\begin{aligned} \Delta H^\circ &= \sum \Delta H_f^\circ(\text{products}) - \sum \Delta H_f^\circ(\text{reactants}) \\ &= [1 \ \text{mol } Ba(NO_3)_2 \times -952.36 \ kJ/\text{mol} + 2 \ \text{mol } NH_3 \times -80.29 \ kJ/\text{mol} \\ &\quad + 10 \ \text{mol } H_2O \times -285.83 \ kJ/\text{mol}] - [1 \ \text{mol } Ba(OH)_2 \\ &\quad \times -3342.2 \ kJ/\text{mol} + 2 \ \text{mol } NH_4NO_3 \times -365.6 \ kJ/\text{mol}] \\ &= \mathbf{102.2 \ kJ} \end{aligned}$$

The standard-state enthalpy of reaction is positive, and the reaction is therefore endothermic—it absorbs heat from its surroundings.

The value of ΔG° for the reaction is equal to the sum of the free energies of formation of the products minus the sum of the free energies of formation of the reactants.

$$\begin{aligned} \Delta G^\circ &= \sum \Delta G_f^\circ(\text{products}) - \sum \Delta G_f^\circ(\text{reactants}) \\ &= [1 \ \text{mol } Ba(NO_3)_2 \times -783.28 \ kJ/\text{mol} + 2 \ \text{mol } NH_3 \times -26.50 \ kJ/\text{mol} \\ &\quad + 10 \ \text{mol } H_2O \times -237.18 \ kJ/\text{mol}] - [1 \ \text{mol } Ba(OH)_2 \\ &\quad \times -2792.8 \ kJ/\text{mol} + 2 \ \text{mol } NH_4NO_3 \times -184.0 \ kJ/\text{mol}] \\ &= \mathbf{-47.3 \ kJ} \end{aligned}$$

Although the reaction is not favored by the enthalpy of reaction, it is apparently favored enough by the entropy of reaction to make the standard-state free energy of reaction slightly negative. The reaction is therefore spontaneous, even though it absorbs heat from its surroundings when it occurs.

20.14 INTERPRETING STANDARD-STATE FREE ENERGY OF REACTION DATA

The value of ΔG° for a reaction measures the difference between the sum of the free energies of the reactants and products when all reactants and products are present at standard-state conditions. Consider the following reaction, for example.

$$N_2(g) + 3 \ H_2(g) \rightleftharpoons 2 \ NH_3(g)$$

The standard state for this reaction assumes that all reactants and products are present at partial pressures of 1 atm and a temperature of 25°C. The standard state for this reaction is therefore described by the following reaction quotient expression.

Standard state:

$$Q_p = \frac{P_{NH_3}{}^2}{P_{N_2} P_{H_2}{}^3} = 1$$

According to Exercise 20.7, $\Delta G°$ for this reaction is -33.0 kilojoules at 25°C. What does this imply about the reaction? The *sign* of $\Delta G°$ tells us the direction in which the reaction has to shift to come to equilibrium. The fact that $\Delta G°$ is negative means that, starting from standard-state conditions, the reaction must shift to the right, converting some of the reactants into products, in order to reach equilibrium. The *magnitude* of $\Delta G°$ tells us how far the standard state is from equilibrium. The larger the value of $\Delta G°$, the further the reaction has to go to get from the standard-state conditions to equilibrium.

20.15 THE RELATIONSHIP BETWEEN FREE ENERGY AND EQUILIBRIUM CONSTANTS

When a reaction leaves the standard state because of a change in the ratio of the products to the reactants, we have to describe the system in terms of non-standard-state free energies of reaction.

$$\Delta G = \Delta H - T\Delta S$$

The difference between $\Delta G°$ and ΔG for a reaction is important. There is only one possible value of $\Delta G°$ for a reaction at a given temperature, but there are an infinite number of possible values of ΔG.

To examine the implications of this fact, let's turn once more to the following reaction, for which $\Delta G°$ is -33.0 kilojoules at 25°C.

$$N_2(g) + 3\ H_2(g) \rightleftharpoons 2\ NH(g)$$

The sign of $\Delta G°$ tells us the direction in which the reaction must shift to convert a system under standard-state conditions to one at equilibrium. Since $\Delta G°$ for this reaction is negative, the reaction has to shift to the right to reach equilibrium.

When the reaction is at standard-state conditions, the free energy of reaction is -33.0 kilojoules. As it comes to equilibrium, the driving force behind the reaction decreases. In other words, as the reaction approaches equilibrium, ΔG becomes smaller. Eventually, the reaction will reach equilibrium, and ΔG will be equal to 0—there will no longer be any driving force behind the reaction.

At equilibrium: **$\Delta G = 0$**

The larger the value of $\Delta G°$, the farther the reaction has to go before it reaches equilibrium.

The relationship between the free energy of reaction at any moment in time (ΔG) and the standard-state free energy of reaction ($\Delta G°$) is described by the following equation.

$$\Delta G = \Delta G° + RT \ln Q$$

In this equation, R is the ideal gas constant in units of J/mol-K, T is the temperature in kelvin, ln indicates a logarithm to the base e, and Q is the reaction quotient at that moment in time.

Exercise 20.11

Show that the free energy of reaction at any moment in time (ΔG) for the following reaction becomes equal to the standard-state free energy of reaction ($\Delta G°$) when the reaction is at standard-state conditions.

$$N_2(g) + 3\ H_2(g) \rightleftharpoons 2\ NH_3(g)$$

Solution

The relationship between the free energy of reaction under any set of conditions (ΔG) and the standard-state free energy of reaction ($\Delta G°$) is given by the following equation.

$$\Delta G = \Delta G° + RT \ln Q$$

When the reaction is at standard-state conditions, the reaction quotient is equal to 1.

$$Q_p = \frac{P_{NH_3}^{2}}{P_{N_2} P_{H_2}^{3}} = 1$$

The logarithm of 1 to any base is zero. Thus, when the reaction is at standard-state conditions, the $RT \ln Q$ term is zero, and ΔG is equal to $\Delta G°$.

$$\Delta G = \Delta G° \qquad (\text{when } Q = 1)$$

Exercise 20.12

Prove the validity of the following equation linking the standard-state free energy of a reaction and the equilibrium constant for the reaction.

$$\Delta G° = -RT \ln K$$

Solution

We can start with the equation that describes the relationship between ΔG and $\Delta G°$.

$$\Delta G = \Delta G° + RT \ln Q$$

We can then note that the driving force behind a chemical reaction is zero when the reaction is at equilibrium.

$$0 = \Delta G° + RT \ln K$$

Rearranging this equation gives the following result.

$$\boldsymbol{\Delta G° = -RT \ln K}$$

The relationship between the standard-state free energy of reaction, $\Delta G°$, and the equilibrium constant for a reaction, K, can be written in terms of a logarithm to the base e.

$$\Delta G° = -RT \ln K$$

It can also be written in terms of a log to the base 10, as follows.

$$\Delta G° = -2.303 \, RT \log K$$

This relationship allows us to calculate the equilibrium constant for any reaction from the standard-state free energy of reaction, or vice versa.

Students often appreciate the power of this equation to do calculations but fail to understand where it comes from. Why are the standard-state free energy of a reaction and the equilibrium constant for the reaction related? It seems at first glance as if these two quantities should be very different. One describes the system when the concentrations of all reagents in solution are 1 M and the partial pressures of all gases are 1 atm. The other describes the system when it lies at equilibrium, which can be very far from standard-state conditions.

The key to understanding the relationship between $\Delta G°$ and K is to appreciate the fact that the magnitude of $\Delta G°$ tells us how far the standard state is from equilibrium. The smaller the value of $\Delta G°$, the closer the standard state is to equilibrium. The larger the value of $\Delta G°$, the further the reaction has to go to reach equilibrium.

Exercise 20.13

Calculate the equilibrium constant at 25°C for the following reaction.

$$N_2(g) + 3 \, H_2(g) \rightleftharpoons 2 \, NH_3(g)$$

Solution

Table A-15 contains the following information for this reaction.

Compound	$\Delta H_f°(kJ/mol)$	$S°\ (J/mol\text{-}K)$
$N_2(g)$	0	191.5
$H_2(g)$	0	130.57
$NH_3(g)$	-46.1	192.3

According to these data, $\Delta H°$ for this reaction is -92.2 kJ and $\Delta S°$ is -198.6 J/K. At 25°C, $\Delta G°$ is therefore -33.0 kJ.

$$\begin{aligned} \Delta G° &= \Delta H° - T\Delta S° \\ &= (-92.2 \text{ kJ}) - (298.15 \text{ K} \times -198.6 \text{ J/K}) = \mathbf{-33.0 \text{ kJ}} \end{aligned}$$

But $\Delta G°$ is proportional to the logarithm of the equilibrium constant for this reaction.

$$\Delta G° = -2.303 \, RT \log K$$

Solving for the log of the equilibrium constant gives the following equation.

$$\log K = -\frac{\Delta G°}{2.303 \, RT}$$

Substituting the known value of $\Delta G°$, R, and T into this equation gives the following result.

$$\log K = -\frac{-33,000 \, \cancel{J}}{(2.303)(8.314 \, \cancel{J}/\text{mol-}\cancel{K})(298 \, \cancel{K})} = \mathbf{5.78}$$

The equilibrium constant for this reaction at 25°C is therefore 6.0×10^5.

$$K = 10^{5.78} = \mathbf{6.0 \times 10^5}$$

The calculation in Exercise 20.13 raises several important points. It is important to remember that enthalpies of reaction are usually given in kilojoules, whereas entropies of reaction are given in joules. Thus, a factor of 1000 must be incorporated into the calculation before the $T\Delta S°$ term is subtracted from the $\Delta H°$ term. It is also a good idea to check the final answer to see whether it makes sense, because it is easy to make mistakes when handling the sign of the relationship between $\Delta G°$ and K. For example, $\Delta G°$ for the reaction in Exercise 20.13 is negative. The reaction is therefore spontaneous, and the equilibrium should lie heavily on the side of the products. The equilibrium constant should therefore be much larger than 1, which it is.

The nature of the equilibrium constant obtained in this calculation should also be clarified. As we saw in Chapter 14, the equilibrium constant for a reaction can be expressed in two ways — K_c and K_p. We can write equilibrium constant expressions in terms of the partial pressures of the reactants and products, if the reactants and products are gases.

$$K_p = \frac{P_{NH_3}^2}{P_{N_2}P_{H_2}^3}$$

We can also write these expressions in terms of molar concentrations.

$$K_c = \frac{[NH_3]^2}{[N_2][H_2]^3}$$

For gas-phase reactions, such as the one in Exercise 20.13, the equilibrium constant obtained by this calculation is based on the partial pressures of the gases. The equilibrium constant that comes from this calculation is K_p, not K_c. For reactions in solution, the equilibrium constant that comes from the calculation is based on concentrations. It involves K_c, not K_p.

Exercise 20.14

Use the standard-state free energy of formation data in Table A-15 to calculate the acid-dissociation equilibrium constant (K_a) at 25°C for formic acid.

$$HCO_2H(aq) \rightleftharpoons H_3O^+(aq) + HCO_2^-(aq)$$

Solution

Table A-15 contains the following information for this reaction.

Compound	$\Delta G_f°(kJ/mol)$
$HCO_2H(aq)$	-372.3
$H_3O^+(aq)$	0.00
$HCO_2^-(aq)$	-351.0

According to these data, $\Delta G°$ for this reaction is 21.3 kJ.

$$\begin{aligned}\Delta G° &= \sum \Delta G_f°(\text{products}) - \sum \Delta G_f°(\text{reactants})\\ &= (1 \text{ mol } H_3O^+ \times 0.00 \text{ kJ/mol} + 1 \text{ mol } HCO_2^- \times -351.0 \text{ kJ/mol})\\ &\quad - (1 \text{ mol } HCO_2H \times -372.3 \text{ kJ/mol})\\ &= \textbf{21.3 kJ}\end{aligned}$$

We can now turn to the relationship between $\Delta G°$ and the equilibrium constant for the reaction

$$\Delta G° = -2.303 \, RT \log K$$

and solve for the logarithm of the equilibrium constant.

$$\log K = -\frac{\Delta G°}{2.303 \, RT}$$

Substituting the known values of $\Delta G°$, R, and T into this equation gives the following result.

$$\log K = -\frac{21,300 \, \cancel{J}}{(2.303)(8.341 \, \cancel{J}/\text{mol-}\cancel{K})(298 \, \cancel{K})} = -3.73$$

We can now calculate the value of the equilibrium constant.

$$K_a = 10^{-3.73} = 1.9 \times 10^{-4}$$

The value of K_a obtained in this calculation agrees with the value for formic acid given in Table A-9, within experimental error.

20.16 THE TEMPERATURE DEPENDENCE OF EQUILIBRIUM CONSTANTS

When equilibrium constants were introduced in Chapter 14, we noted that they are not strictly constant, because they change with temperature. We are now ready to understand why.

. The standard-state free energy of reaction is a measure of how far the standard state is from equilibrium.

$$\Delta G° = -2.303 \, RT \log K$$

But the magnitude of $\Delta G°$ depends on the temperature of the reaction.

$$\Delta G° = \Delta H° - T\Delta S°$$

As a result, the equilibrium constant must depend on the temperature of the reaction.

A good example of this phenomenon is the reaction in which NO_2 dimerizes to form N_2O_4.

$$2 \, NO_2(g) \rightleftharpoons N_2O_4(g)$$

In Exercise 20.3 we stated that this reaction is favored by enthalpy because it forms a new bond, which makes the system more stable. We also stated that the reaction isn't favored by entropy, because it leads to a decrease in the disorder of the system.

NO_2 is a brown gas, but N_2O_4 is colorless. We can therefore monitor the extent to which NO_2 dimerizes to form N_2O_4 by determining the intensity of the brown color in a sealed tube of this gas. What should happen to the equilibrium between NO_2 and N_2O_4 as the temperature is lowered?

For the sake of argument, let's assume that there is no significant change in either $\Delta H°$ or $\Delta S°$ as the system is cooled. The contribution to the free energy of the reaction from the enthalpy term is therefore constant, but the contribution from the entropy term becomes smaller as the temperature is lowered.

$$\Delta G° = \Delta H° - T\Delta S°$$

As the tube is cooled and the entropy term becomes less important, the net effect is a shift in the equilibrium toward the right. Figure 20.7 shows what happens to the intensity of the brown color when a sealed tube containing NO_2 gas is immersed in liquid nitrogen. The amount of NO_2 in the tube decreases as the tube is cooled.

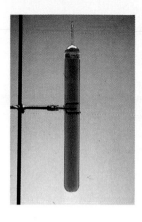

FIG. 20.7 The equilibrium between NO_2 and N_2O_4 is influenced by two factors. $\Delta H°$ for this reaction favors N_2O_4. But $\Delta S°$ favors NO_2, which is more disordered. The contribution to $\Delta G°$ from the enthalpy term is more or less constant regardless of the temperature of the system. But the contribution from the entropy term is very sensitive to temperature. At low temperatures, the enthalpy term is larger than the entropy term, and the equilibrium strongly favors N_2O_4. As the temperature increases, the equilibrium shifts towards NO_2. Thus, the characteristic brown color of NO_2 gas disappears as a tube containing NO_2 is moved into a liquid nitrogen bath ($-196°C$).

Exercise 20.15

Calculate the equilibrium constant for the following reaction at the temperature of boiling water (100°C), ice (0°C), a dry-ice/acetone bath ($-78°C$), and liquid nitrogen ($-196°C$).

$$2\ NO_2(g) \rightleftharpoons N_2O_4(g)$$

Solution

Table A-15 contains the following information for this reaction.

Compound	$\Delta H_f°(kJ/mol)$	$S°\ (J/mol\text{-}K)$
$NO_2(g)$	33.2	240.0
$N_2O_4(g)$	9.17	304.2

According to these data, the reaction is favored by enthalpy.

$$\Delta H° = (1\ \text{mol}\ N_2O_4 \times 9.16\ kJ/\text{mol}) - (2\ \text{mol}\ NO_2 \times 33.2\ kJ/\text{mol})$$
$$= -57.2\ kJ$$

But it is not favored by entropy.

$$\Delta S° = (1\ \text{mol}\ N_2O_4 \times 304.2\ J/\text{mol-K}) - (2\ \text{mol}\ NO_2 \times 240.0\ J/\text{mol-K})$$
$$= -175.8\ J/K$$

If we assume that these values of $\Delta H°$ and $\Delta S°$ are still valid at 100°C, the value of $\Delta G°$ at this temperature is 8400 joules.

$$\Delta G° = \Delta H° - T\Delta S°$$
$$= -57,200\ J - (373\ K)(-175.8\ J/K) = 8400\ J$$

Repeating this calculation at the other temperatures gives the following results.

$$100°C: \qquad \Delta G° = 8.4\ kJ$$
$$0°C: \qquad \Delta G° = -9.2\ kJ$$
$$-78°C: \qquad \Delta G° = -22.9\ kJ$$
$$-196°C: \qquad \Delta G° = -43.6\ kJ$$

We can now take advantage of the relationship between $\Delta G°$ and the equilibrium constant for the reaction.

$$\Delta G° = -2.303\ RT \log K_p$$

Solving for the logarithm of the equilibrium constant gives the following equation.

$$\log K_p = -\frac{\Delta G°}{2.303\ RT}$$

Let's start by calculating the value of $\log K_p$ when the reaction is at 100°C.

$$\log K_p = -\frac{8{,}400\ \cancel{J}}{(2.303)(8.314\ \cancel{J}/\text{mol-}\cancel{K})(373\ \cancel{K})} = -1.18$$

The equilibrium constant at this temperature is therefore 0.066.

$$K_p = 10^{-1.18} = \mathbf{0.066}$$

Repeating this calculation at the other temperatures gives the following results.

$$
\begin{array}{ll}
100°\text{C:} & K_p = 0.066 \\
0°\text{C:} & K_p = 58 \\
-78°\text{C:} & K_p = 1.3 \times 10^6 \\
-196°\text{C:} & K_p = 3.7 \times 10^{29}
\end{array}
$$

At 100°C, the unfavorable entropy term is relatively important, and the equilibrium lies on the side of NO_2. As the reaction is cooled, the entropy term becomes less important, and the equilibrium shifts toward N_2O_4. At the temperature of liquid nitrogen, essentially all of the NO_2 condenses to form N_2O_4.

20.17 THE RELATIONSHIP BETWEEN FREE ENERGY AND CELL POTENTIALS

The value of ΔG for a reaction at any moment in time tells us two things. The sign of ΔG tells us in what direction the reaction has to shift to reach equilibrium. The magnitude of ΔG tells us how far the reaction is from equilibrium at that moment in time.

In Chapter 19, we saw that the potential of an electrochemical cell is a measure of how far an oxidation-reduction reaction is from equilibrium. In that chapter, we used the Nernst equation to describe the relationship between the cell potential at any moment in time and the standard-state cell potential.

$$E = E° - \frac{RT}{nF}\ln Q$$

Let's rearrange this equation as follows.

$$nFE = nFE° - RT\ln Q$$

We can now compare it with the equation used in this chapter to describe the relationship between the free energy of reaction at any moment in time and the standard-state free energy of reaction.

$$\Delta G = \Delta G° + RT\ln Q$$

These equations are similar because the Nernst equation is a special case of the more general free energy relationship. We can convert one of these equations to the other by taking advantage of the following relationships between the free energy of a reaction and the cell potential of the reaction when it is run as an electrochemical cell.

$$\Delta G = -nFE$$
$$\Delta G° = -nFE°$$

Exercise 20.16

Calculate the standard-state free energy of reaction and the standard-state cell potential for the following reaction.

$$Zn(s) + Cu^{2+}(aq) \longrightarrow Zn^{2+}(aq) + Cu(s)$$

Use these data to test the relationship $\Delta G° = -nFE°$

Solution

The following standard-state free energy of formation data can be found in Table A-15.

Compound	$\Delta G_f°$ (kJ/mol)
$Zn(s)$	0
$Zn^{2+}(aq)$	−147.06
$Cu(s)$	0
$Cu^{2+}(aq)$	65.49

According to these data, $\Delta G°$ for the reaction is -212.55 kJ/mol.

$$\Delta G° = \sum \Delta G_f°(\text{products}) - \sum \Delta G_f°(\text{reactants})$$
$$= (1 \text{ mol } Zn^{2+} \times -147.06 \text{ kJ/mol} + 1 \text{ mol } Cu \times 0 \text{ kJ/mol})$$
$$- (1 \text{ mol } Zn \times 0 \text{ kJ/mol} + 1 \text{ mol } Cu^{2+} \times 65.49 \text{ kJ/mol})$$
$$= -212.55 \text{ kJ}$$

According to the data in Table A-13, the standard-state cell potential for this cell is 1.1030 volts.

$Zn \longrightarrow Zn^{2+} + 2\,e^-$		$E_{ox}° = +0.7628$ V
$Cu^+ + 2\,e^- \longrightarrow Cu$		$E_{red}° = 0.3402$ V
$Zn + Cu^{2+} \longrightarrow Zn^{2+} + Cu$		$E° = E_{ox}° + E_{red}° = 1.1030$ V

Before we can use these data to test the following equation, we must note that a potential of 1 volt is equivalent to 1 joule per coulomb (J/C).

$$\Delta G° = -nFE°$$

Thus, a cell potential of 1.1030 volts is equivalent to 1.1030 J/C. We now note that two electrons are transferred from zinc metal to copper ions in the balanced equation for this reaction, and so n is equal to 2. Substituting the value of n, Faraday's constant, and the cell potential in joules per coulomb into this equation gives the following result.

$$\Delta G° = -nFE° = (2)(96,484 \text{ C})(1.1030 \text{ J/C}) = -212.83 \text{ kJ}$$

The standard-state free energy of reaction calculated from the cell potential is essentially the same as the value obtained from the table of standard-state free energies of formation. Thus, the following relationship holds, within experimental error.

$$\Delta G^{\circ} = -nFE^{\circ}$$

Exercise 20.20

Derive the relationship between the standard-state cell potential for an electrochemical cell and the equilibrium constant for the reaction.

Solution

We can start with the equation that describes the relationship between the standard-state free energy of reaction and the standard-state cell potential for an electrochemical cell.

$$\Delta G^{\circ} = -nFE^{\circ}$$

We can then write the relationship between the standard-state free energy of reaction and the equilibrium constant for the reaction.

$$\Delta G^{\circ} = -2.303 \, RT \log K$$

Substituting one of these equations into the other gives the following result.

$$-nFE^{\circ} = -2.303 \, RT \log K$$

Multiplying both sides of this equation by -1 gives the following equation.

$$nFE^{\circ} = 2.303 \, RT \log K$$

Solving for the log K term gives the equation used in Chapter 19 to calculate the equilibrium constants for electrochemical reactions from their standard-state cell potentials.

$$\log K = \frac{nFE^{\circ}}{2.303 \, RT}$$

SUMMARY

The first law of thermodynamics describes the relationship between heat (q), work (w), and the internal energy of a system (E). According to the first law, the change in the internal energy of the system is equal to the sum of the heat transferred from the system to its surroundings (or vice versa) and the work done by the system on its surroundings (or vice versa).

$$\Delta E = q + w$$

When a chemical reaction is run in a sealed container, at constant volume, the change in the internal energy of the system is equal to the heat given off or absorbed during the reaction.

$$\Delta E = q_V$$

When the reaction is run in an open container, at constant pressure, the heat of reaction is equal to the change in the enthalpy of the system.

$$\Delta H = q_P$$

Two factors determine the direction in which a chemical reaction will occur — the change in the enthalpy of the system and the change in the entropy of the system that occur during the reaction. Reactions that give off energy tend to occur spontaneously. Reactions that lead to an increase in the disorder of the system — and therefore an increase in its entropy — also tend to occur spontaneously.

The change in the Gibbs free energy of a system during a chemical reaction is a measure of the balance between the two forces that drive the reaction. The Gibbs free energy (G) is the sum of the enthalpy of the system (H) plus the product of the absolute temperature (T)

times the entropy of the system (S).

$$G = H - TS$$

At constant temperature, the change in the Gibbs free energy of the system during a reaction is given by the following equation.

$$\Delta G = \Delta H - T\Delta S$$

Any reaction for which ΔG is negative occurs spontaneously.

The sign of ΔG for a reaction indicates the direction in which the reaction has to proceed to reach equilibrium. When ΔG is negative, the reaction has to shift to the right — toward the products — to reach equilibrium. When ΔG is positive, the reaction has to shift to the left, toward the reactants. The magnitude of ΔG indicates how far the reaction is from equilibrium. There are an infinite number of possible values of ΔG for a reaction but only one value of $\Delta G°$, which represents measurements made under standard-state conditions.

The value of $\Delta G°$ for a reaction indicates how far the standard-state conditions are from equilibrium. As a reaction moves from standard-state conditions toward equilibrium, the value of ΔG for the reaction gradually decreases until it eventually reaches zero when the reaction is at equilibrium. The relationship between the free energy of reaction at any moment in time (ΔG) and the standard-state free energy of reaction ($\Delta G°$) is given by the following equation.

$$\Delta G = \Delta G° + RT \ln Q$$

This equation can also be written as follows.

$$\Delta G = \Delta G° + 2.303\, RT \log Q$$

When the reaction is at equilibrium, ΔG is zero and the reaction quotient is equal to the equilibrium constant.

$$0 = \Delta G° + 2.303\, RT \log K$$

Solving for $\Delta G°$ gives the following equation.

$$\Delta G° = -2.303\, RT \log K$$

This equation shows how the equilibrium constant for a reaction can be calculated from the standard-state free energy of reaction (or vice versa).

The value of $\Delta G°$ for a reaction depends on the temperature of the reaction.

$$\Delta G° = \Delta H° - T\,\Delta S°$$

Since the equilibrium constant depends on the value of $\Delta G°$, the equilibrium constant, too, depends on the temperature at which the reaction is run.

The value of $\Delta G°$ for an oxidation-reduction reaction is proportional to the overall cell potential for the reaction.

$$\Delta G° = -nFE°$$

This means that the standard-state free energy of reaction for an oxidation-reduction can be calculated from the cell potential, or vice versa.

PROBLEMS

The First Law of Thermodynamics

20-1 Define the following terms: *heat, work, internal energy, enthalpy, entropy,* and *free energy.*

20-2 Describe the first law of thermodynamics in terms of both a verbal definition and a mathematical equation.

20-3 What will happen to the internal energy of the system when an exothermic reaction is run under conditions of constant volume?

20-4 Calculate the amount of heat absorbed by 3 moles of an ideal gas if the temperature of the gas increases by 3.5°C.

20-5 Calculate the temperature of 3 moles of an ideal gas after it expands from a volume of 1 liter to a volume of 5 liters at a constant pressure of 1 atm.

20-6 Calculate the change in the internal energy of a system that does 1246 joules of work on its surroundings and at the same time absorbs 456 joules of heat from its surroundings.

Predict whether the temperature of the system will become larger or smaller.

20-7 Calculate the amount of heat that must be added to an ideal gas that expands from 1.25 liters to 8.00 liters against a constant pressure of 5.00 atm in order to keep the temperature of the gas constant.

20-8 Work of expansion can be expressed in units of liter-atmospheres (L-atm). For example, a piston with a diameter of 25.0 centimeters that moves 15.0 centimeters under a constant pressure of 9.00 atm does 66.3 L-atm worth of work. Use the two values of the ideal gas constant ($R = 0.08206$ L-atm/mol-K and $R = 8.314$ J/mol-K) to convert 66.3 L-atm of work into joules.

Enthalpy versus Internal Energy

20-9 Describe the difference between ΔE and ΔH for a chemical reaction.

20-10 Under what conditions is the heat given off or absorbed in a chemical reaction equal to the change in the internal energy of the system? Under what conditions is it equal to the enthalpy of reaction?

20-11 For which of the following reactions is ΔE roughly equal to ΔH?

(a) $2 H_2(g) + O_2(g) \rightarrow 2 H_2O(l)$
(b) $Zn(s) + Cu^{2+}(aq) \rightarrow Zn^{2+}(aq) + Cu(s)$
(c) $Zn(s) + 2 H^+(aq) \rightarrow Zn^{2+}(aq) + H_2(g)$
(d) $NH_3(g) + HCl(g) \rightarrow NH_4Cl(s)$
(e) $HCl(aq) + NaOAc(aq) \rightarrow HOAc(aq) + NaCl(aq)$

20-12 Explain why there is only one possible value of $\Delta H°$ for a reaction, whereas there are many values of ΔH.

Entropy as a Measure of Disorder

20-13 Describe the relationship between the entropy of a system and the number of equivalent ways in which the system can be described.

20-14 Explain why the entropy of a poker hand is inversely proportional to its ability to win a game of poker.

20-15 At first glance, the number of possible poker hands might be expected to be $52 \times 51 \times 50 \times 49 \times 48$, or 311,875,200. Why is the actual number of different hands only 2,598,960, which is $5 \times 4 \times 3 \times 2 \times 1$ (or 5!) times smaller?

20-16 Describe entropy as a driving force that can determine the direction in which a natural process occurs.

20-17 Give examples of natural processes that spontaneously lead to an increase in the entropy (or disorder) of the system.

Entropy and the Second Law of Thermodynamics

20-18 In which of the following reactions will the entropy of the system become larger?

(a) $H_2(g) + Cl_2(g) \rightarrow 2 HCl(g)$
(b) $2 NO_2(g) \rightleftharpoons N_2O_4(g)$
(c) $CaO(s) + CO_2(g) \rightleftharpoons CaCO_3(s)$
(d) $2 H_2(g) + O_2(g) \rightarrow 2 H_2O(g)$
(e) $2 NH_3(g) \rightleftharpoons N_2(g) + 3 H_2(g)$

20-19 In which of the following processes is $\Delta S°$ negative?

(a) $2 H_2O_2(aq) \rightarrow 2 H_2O(aq) + O_2(g)$
(b) $CO_2(s) \rightleftharpoons CO_2(g)$
(c) $H_2O(l) \rightleftharpoons H_2O(g)$
(d) $4 Al(s) + 3 O_2(g) \rightarrow$
$2 Al_2O_3(s)$ (e) $NaCl(s) \xrightarrow{H_2O} Na^+(aq) + Cl^-(aq)$

20-20 In which of the following processes is $\Delta S°$ positive?

(a) $2 NO(g) + Cl_2(g) \rightleftharpoons 2 NOCl(g)$
(b) $NaCl(s) \rightleftharpoons NaCl(l)$
(c) $3 O_2(g) \rightleftharpoons 2 O_3(g)$
(d) $C_2H_4(g) + H_2(g) \rightarrow C_2H_6(g)$

20-21 Which of the following processes should have the most positive value of $\Delta S°$?

(a) $N_2(g) + O_2(g) \rightarrow 2 NO(g)$
(b) $3 C_2H_2(g) \rightarrow C_6H_6(l)$
(c) $H_2O(l) \rightleftharpoons H_2O(g)$
(d) $4 Al(s) + 3 O_2(g) \rightarrow 2 Al_2O_3(s)$

20-22 Which of the following processes should have the most positive value of $\Delta S°$?

(a) $H_2O(l) \rightleftharpoons H_2O(s)$

(b) $NaNO_3(s) \xrightarrow{H_2O} Na^+(aq) + NO_3^-(aq)$
(c) $2 HCl(g) \rightarrow H_2(g) + Cl_2(g)$
(d) $2 H_2(g) + O_2(g) \rightarrow 2 H_2O(g)$

The Third Law of Thermodynamics

20-23 Explain why the standard-state entropy of a solid increases with temperature.

20-24 Explain why the standard-state entropy of a compound increases as the compound is converted from a solid to a liquid or from a liquid to a gas.

Standard-State Entropies of Reaction

20-25 Calculate $\Delta S°$ for the following reaction, in which a mixture of CO_2 and H_2 gas is converted into methanol.

$$CO(g) + 2 H_2(g) \longrightarrow CH_3OH(l)$$

20-26 Calculate $\Delta S°$ for the following reaction.

$$P_4O_{10}(s) + 6 H_2O(l) \longrightarrow 4 H_3PO_4(aq)$$

20-27 Calculate $\Delta S°$ for the following reaction.

$$NH_4NO_2(s) \longrightarrow N_2(g) + 2 H_2O(g)$$

Comment on both the sign and the magnitude of $\Delta S°$.

20-28 Show how $\Delta S°$ for the following reaction can be used to convince students that O_2 is more disordered than O_3, even though $S°$ for O_3 is larger than $S°$ for O_2.

$$3 O_2(g) \rightleftharpoons 2 O_3(g)$$

20-29 Compare $S°$ for the various forms of elemental phosphorus.

Compound	$S° (J/mol\text{-}K)$
$P(s)$	41.09
$P(g)$	163.193
$P_2(g)$	218.129
$P_4(g)$	279.98

Explain why $S°$ increases in the order $P < P_2 < P_4$, even though the system becomes more ordered.

20-30 Look up $S°$ data for aqueous solutions of the following ions: Li^+, Na^+, K^+, Fe^{2+}, Fe^{3+}, Mg^{2+}, Ba^{2+}, and Al^{3+}. Use these data to determine two factors that lead to an increase in the standard-state entropy of an aqueous solution of an ion.

20-31 The $S°$ data for ions in aqueous solutions are slightly misleading. For all other substances, the zero point on the entropy scale is a perfect crystal at 0 K. For ionic solutions, however, the zero point is defined as the entropy of a 1 M H^+ ion solution. Explain why the existence of two entropy scales with different reference points does not influence the value of $\Delta S°$ for the following reaction, which involves solids and gases as well as aqueous solutions.

$$Zn(s) + 2 H^+(aq) \longrightarrow Zn^{2+}(aq) + H_2(g)$$

20-32 Explain why enthalpy and free energy data for compounds are given as enthalpies of formation ($\Delta H_f°$) and free energies of formation ($\Delta G_f°$), whereas entropy data are not given in this form.

Gibbs Free Energy

20-33 Explain the difference between ΔG and $\Delta G°$ for a chemical reaction.

20-34 What does it mean when ΔG for a reaction is zero? What does it mean when $\Delta G°$ is zero?

20-35 Which of the following combinations of $\Delta H°$ and $\Delta S°$ always indicates a spontaneous reaction?

  (a) $\Delta H° > 0, \Delta S° < 0$ (b) $\Delta H° < 0, \Delta S° > 0$
  (c) $\Delta H° > 0, \Delta S° > 0$ (d) $\Delta H° < 0, \Delta S° < 0$
  (e) $\Delta H° = 0, \Delta S° = 0$

20-36 Predict the signs of $\Delta H°$ and $\Delta S°$ for the following reactions without referring to a table of thermodynamic data, and explain your predictions.

  (a) $2 H_2(g) + O_2(g) \rightarrow 2 H_2O(g)$
  (b) $2 Na(s) + Cl_2(g) \rightarrow 2 NaCl(s)$
  (c) $N_2(g) + 3 H_2(g) \rightleftharpoons 2 NH_3(g)$
  (d) $2 Cu(NO_3)_2(s) \rightarrow 2 CuO(s) + 4 NO_2(g) + O_2(g)$

20-37 Predict the signs of $\Delta H°$ and $\Delta S°$ for the following reaction without referring to a table of thermodynamic data, and explain your predictions.

$$NH_3(g) \xrightarrow{\text{H}_2\text{O}} NH_3(aq)$$

Explain why the odor of NH_3 gas that collects above this solution becomes more intense as the temperature increases.

20-38 Calculate $\Delta H°$ and $\Delta S°$ for the decomposition of limestone when heated to form lime and carbon dioxide.

$$CaCO_3(s) \longrightarrow CaO(s) + CO_2(g)$$

Identify the driving force behind the reaction.

20-39 Calculate $\Delta H°$ and $\Delta S°$ for the thermite reaction.

$$Fe_2O_3(s) + 2 Al(s) \longrightarrow Al_2O_3(s) + 2 Fe(s)$$

Identify the driving force behind the reaction.

20-40 The first step in extracting iron ore from pyrite, FeS_2, involves roasting the ore in the presence of oxygen to form iron(III) oxide and sulfur dioxide.

$$4 FeS_2(s) + 11 O_2(g) \longrightarrow 2 Fe_2O_3(s) + 8 SO_2(g)$$

Calculate $\Delta H°$ and $\Delta S°$ for this reaction and determine the driving force behind the reaction.

20-41 Calculate $\Delta H°$ and $\Delta S°$ for the following reaction.

$$2 KMnO_4(s) + 5 H_2O_2(aq) + 6 H^+(aq) \longrightarrow$$
$$2 K^+(aq) + 2 Mn^{2+}(aq) + 5 O_2(g) + 8 H_2O(l)$$

What is the major driving force behind this reaction?

20-42 It is possible to make hydrogen chloride by reacting phosphorus pentachloride with water and boiling the HCl out of this solution.

$$PCl_5(g) + 5 H_2O(aq) \longrightarrow H_3PO_4(aq) + 5 HCl(g)$$

Calculate $\Delta H°$ and $\Delta S°$ for this reaction. What is the driving force behind the reaction: the enthalpy of reaction, the entropy of reaction, or LeChatelier's principle?

20-43 Determine which of the following processes is spontaneous at 25°C.

  (a) $CH_3OH(l) \rightleftharpoons CH_3OH(g)$
  (b) $CH_3OH(l) \rightarrow HCHO(g) + H_2(g)$
  (c) $2 CH_3OH(l) \rightarrow CH_4(g) + O_2(g)$
  (d) $CH_3OH(l) \rightarrow CO(g) + 2 H_2(g)$

20-44 Calculate $\Delta H°$ and $\Delta S°$ for the following reaction.

$$3 Fe(s) + 4 H_2O(aq) \longrightarrow Fe_3O_4(s) + 4 H_2(g)$$

Explain why it is a mistake to try to use water to put out a fire that contains white-hot iron metal.

20-45 Use thermodynamic data to explain why sodium reacts with water but silver metal does not.

$$2 Na(s) + 2 H_2O(l) \longrightarrow 2 Na^+(aq) + 2 OH^-(aq) + H_2(g)$$
$$2 Ag(s) + 2 H_2O(l) \nrightarrow 2 Ag^+(aq) + 2 OH^-(aq) + H_2(g)$$

20-46 Use thermodynamic data to explain why zinc reacts with 1 M acid but not with water.

$$Zn(s) + 2 H^+(aq) \longrightarrow Zn^{2+}(aq) + H_2(g)$$
$$Zn(s) + 2 H_2O(l) \nrightarrow Zn^{2+}(aq) + 2 OH^-(aq) + H_2(g)$$

20-47 Calculate $\Delta H°$ and $\Delta S°$ for the following reaction.

$$2 NH_4NO_3(s) \longrightarrow 2 N_2(g) + O_2(g) + 2 H_2(g)$$

Which of these factors explains why ammonium nitrate is a potential explosive?

20-48 Silane, SiH_4, decomposes to form elemental silicon and hydrogen.

$$SiH_4(s) \longrightarrow Si(s) + 2 H_2(g)$$

Calculate $\Delta H°$ and $\Delta S°$ for this reaction and predict the effect on the reaction of an increase in the temperature at which it is run.

20-49 For which of the following reactions would you expect $\Delta H°$ and $\Delta G°$ to be about the same?

1906.(a) $4 Fe(s) + 3 O_2(g) \rightarrow 2 Fe_2O_3(s)$
(b) $2 Na(s) + 2 H_2O(aq) \rightarrow 2 Na^+(aq) + 2 OH^-(aq) + H_2(g)$
(c) $Fe_2O_3(s) + 2 Al(s) \rightarrow Al_2O_3(s) + 2 Fe(s)$
(d) $N_2O_4(g) \rightleftharpoons 2 NO_2(g)$
(e) $CaC_2(s) + 2 H_2O(aq) \rightarrow Ca^{2+}(aq) + 2 OH^-(aq) + C_2H_2(g)$

The Effect of Temperature on the Free Energy of Chemical Reactions

20-50 For which of the following reactions do conditions change from unfavorable to favorable as the temperature increases?

(a) $2 CO(g) + O_2(g) \rightarrow 2 CO_2(g)$
$\Delta H^\circ = -565.97$ kJ/mol, $\Delta S^\circ = -173.00$ J/mol-K

(b) $2 H_2O(g) \rightarrow 2 H_2(g) + O_2(g)$
$\Delta H^\circ = 483.64$ kJ/mol, $\Delta S^\circ = 90.01$ J/mol-K

(c) $2 N_2O(g) \rightarrow 2 N_2(g) + O_2(g)$
$\Delta H^\circ = -164.1$ kJ/mol, $\Delta S^\circ = 148.66$ J/mol-K

(d) $PbCl_2(s) \xrightarrow{H_2O} Pb^{2+}(aq) + 2 Cl^-(aq)$
$\Delta H^\circ = 23.39$ kJ/mol, $\Delta S^\circ = -12.5$ J/mol-K

20-51 Explain why the equilibrium constant for the following reaction decreases as the temperature increases.

$$N_2(g) + 3 H_2(g) \rightleftharpoons 2 NH_3(g)$$

20-52 Which of the following diagrams best describes the relationship between ΔG° and temperature for the following reaction?

$$2 H_2(g) + O_2(g) \longrightarrow 2 H_2O(g)?$$

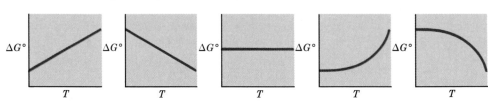

Standard-State Free Energies of Reaction

20-53 For which of the following substances is the standard-state free energy of formation equal to zero?

(a) $H(g)$ (b) $Hg(g)$ (c) $H_2O(l)$ (d) $O_2(g)$ (e) $O_3(g)$

20-54 Calculate ΔG° for the following reaction from ΔH° and ΔS° data for this reaction. Compare the result of this calculation with the value obtained from free energy of formation data.

$$CS_2(l) + 3 O_2(g) \longrightarrow CO_2(g) + 2 SO_2(g)$$

20-55 Adding salt to water does not change ΔH° for the following process.

$$H_2O(s) \rightleftharpoons H_2O(l)$$

But it increases ΔS° for this process, because it increases the entropy of the solution without changing the entropy of the solid. Show how this can be used to explain the fact that adding salt to water lowers its melting point.

20-56 Adding salt to water does not change ΔH° for the following process.

$$H_2O(l) \rightleftharpoons H_2O(g)$$

But it decreases ΔS° for this process because it increases the entropy of the liquid without changing the entropy of the gas.

Show how this can be used to explain the fact that adding salt to water raises its boiling point.

20-57 Use the concept of standard-state free energies of reaction, ΔG°, to explain why the vapor pressure of water becomes larger as the temperature of the water increases.

The Relationship Between Free Energy and Equilibrium Constants

20-58 Which of the following correctly describes a reaction at equilibrium?

(a) $\Delta G = 0$ (b) $\Delta G^\circ = 0$ (c) $\Delta G = \Delta G^\circ$ (d) $Q = 0$
(e) $\ln K = 0$

20-59 Calculate the equilibrium constant at 25°C for the following reaction from the standard-state free energies of formation of the reactants and products.

$$CO(g) + 2 H_2(g) \rightleftharpoons CH_3OH(g)$$

20-60 Calculate K_p at 1000 K for the following reaction.

$$2 CH_4(g) \rightleftharpoons C_2H_6(g) + H_2(g)$$

Describe the assumptions you had to make to perform this calculation.

20-61 Calculate the equilibrium constant at both 25°C and 450°C for the following reaction.

$$N_2(g) + 3 H_2(g) \rightleftharpoons 2 NH_3(g)$$

20-62 Calculate the equilibrium constant at 25°C for the following reaction.

$$2 HI(g) + Cl_2(g) \rightleftharpoons 2 HCl(g) + I_2(s)$$

20-63 Calculate ΔH°, ΔS°, and ΔG° for the following reactions.

$$HCl(aq) \rightleftharpoons H^+(aq) + Cl^-(aq)$$
$$CH_3CO_2H(aq) \rightleftharpoons H^+(aq) + CH_3CO_2^-(aq)$$

Use these data to calculate the values of K_a for hydrochloric and acetic acid. Compare the results of these calculations with the data in Table A-9.

20-64 Calculate ΔH°, ΔS°, and ΔG° for the following reaction.

$$NH_3(aq) + H_2O(aq) \rightleftharpoons NH_4^+(aq) + OH^-(aq)$$

Use these data to calculate the value of K_b for ammonia. Compare the result of this calculation with the value of K_b in Table A-10.

20-65 Calculate ΔH°, ΔS°, and ΔG° for the following reactions.

$$H_2CO_3(aq) \rightleftharpoons H^+(aq) + HCO_3^-(aq)$$
$$HCO_3^-(aq) \rightleftharpoons H^+(aq) + CO_3^{2-}(aq)$$

Use these data to calculate values of K_{a1} and K_{a2} for carbonic acid. Compare the results of these calculations with the data in Table A-9.

20-66 Calculate $\Delta H°$, $\Delta S°$, and $\Delta G°$ for the following reaction.

$$HgS(s) \xrightarrow{H_2O} Hg^{2+}(aq) + S^{2-}(aq)$$

Use these data to calculate the solubility product constant for HgS. Compare the result of this calculation with the data in Table A-11.

20-67 Calculate $\Delta H°$, $\Delta S°$, and $\Delta G°$ at 25°C for the following reaction.

$$AgCl(s) \xrightarrow{H_2O} Ag^+(aq) + Cl^-(aq)$$

Use these data to calculate the solubility product for silver chloride. Compare the result of this calculation with the data in Table A-11.

20-68 Use the data in Table A-15 to predict which of the following salts should be less soluble in water, PbS or $PbSO_4$.

20-69 Use thermodynamic data to calculate the solubility of LiF in water. Compare the result of this calculation with the solubility reported in the *Handbook of Chemistry and Physics*, 0.12 grams of LiF per 100 grams of water.

20-70 Use $\Delta G_f°$ for the $Ag^+(aq)$ and $S^{2-}(aq)$ ions and the solubility product constant for Ag_2S to estimate $\Delta G_f°$ for $Ag_2S(s)$.

20-71 Calculate K_f for the $Zn(NH_3)_4^{2+}$ complex ion, assuming that $\Delta G°$ is -54.0 kJ/mol for the following reaction.

$$Zn^{2+}(aq) + 4 NH_3(aq) \rightleftharpoons Zn(NH_3)_4^{2+}(aq)$$

20-72 Use $\Delta G_f°$ for the $Fe^{3+}(aq)$ and $SCN^-(aq)$ ions and the complex formation constant for the following reaction to estimate $\Delta G_f°$ for the $Fe(SCN)^{2+}$ complex ion.

$$Fe^{3+}(aq) + SCN^-(aq) \rightleftharpoons Fe(SCN)^{2+} \qquad K_f = 8.9 \times 10^2$$

The Temperature Dependence of Equilibrium Constants

20-73 Synthesis gas can be made by reaction of red-hot coal with steam.

$$C(s) + H_2O(g) \rightleftharpoons CO(g) + H_2(g)$$

Calculate the temperature at which the equilibrium constant for this reaction is equal to 1.

20-74 Calculate $\Delta H°$, $\Delta S°$, $\Delta G°$, and the equilibrium constant at 25°C for the following reaction.

$$CO_2(g) + H_2(g) \rightleftharpoons CO(g) + H_2O(g)$$

Predict the effect of an increase in the temperature of the system on the equilibrium constant for this reaction.

20-75 What happens to the equilibrium constant for the following reaction as the temperature increases?

$$PCl_5(g) \rightleftharpoons PCl_3(g) + Cl_2(g)$$

20-76 At approximately what temperature does the equilibrium constant for the following reaction become larger than 1?

$$PCl_5(g) \rightleftharpoons PCl_3(g) + Cl_2(g)$$

20-77 For which of the following reactions would the equilibrium constant be expected to increase with increasing temperature?

(a) $2 H_2(g) + O_2(g) \rightleftharpoons 2 H_2O(g)$
(b) $2 HCl(g) \rightleftharpoons H_2(g) + Cl_2(g)$
(c) $2 NH_3(g) \rightleftharpoons N_2(g) + 3 H_2(g)$

20-78 Calculate the equilibrium constant for the following reaction at 25°C, 200°C, 400°C, and 600°C.

$$2 SO_3(g) \rightleftharpoons 2 SO_2(g) + O_2(g)$$

20-79 Calculate $\Delta H°$ and $\Delta S°$ for the following reaction.

$$Cu(s) + 2 H^+(aq) \rightleftharpoons Cu^{2+}(aq) + H_2(g)$$

Predict what will happen to the equilibrium constant of this reaction as the temperature increases.

The Relationship Between Free Energy and Cell Potentials

20-80 If the following reaction is spontaneous as written, which of the following statements is true?

$$Zn(s) + Cu^{2+}(aq) \longrightarrow Cu(s) + Zn^{2+}$$

(a) $K_{eq} > 1$, $\Delta G° < 0$, and $E° < 0$
(b) $K_{eq} > 1$, $\Delta G° > 0$, and $E° < 0$
(c) $K_{eq} > 1$, $\Delta G° < 0$, and $E° > 0$
(d) $K_{eq} < 1$, $\Delta G° < 0$, and $E° > 0$
(e) $K_{eq} < 1$, $\Delta G° > 0$, and $E° > 0$

20-81 Calculate $\Delta G°$ for a standard-state Daniell cell, assuming that the standard-state cell potential is 1.103 volts.

$$Zn(s) + Cu^{2+}(aq) \longrightarrow Zn^{2+}(aq) + Cu(s)$$

20-82 Calculate $\Delta G°$ for the following reaction, assuming that the standard-state cell potential for the reaction is 0.4594 volt.

$$Cu(s) + 2 Ag^+(aq) \longrightarrow Cu^{2+}(aq) + 2 Ag(s)$$

Is this reaction spontaneous at room temperature?

20-83 Calculate $\Delta G°$ for the following reaction from the standard-state cell potential for the reaction, $E° = -0.3402$ volt.

$$Cu(s) + 2 H^+(aq) \longrightarrow Cu^{2+}(aq) + H_2(g)$$

Is this reaction spontaneous at room temperature?

20-84 Calculate the equilibrium constant for the following reaction from the standard-state cell potential for the reaction, $E° = -0.3402$ volt.

$$Cu(s) + 2 H^+(aq) \longrightarrow Cu^{2+}(aq) + H_2(g)$$

CHAPTER 21

KINETICS

CHAPTER CONTENTS

21.1 THERMODYNAMIC VERSUS KINETIC CONTROL

Thermodynamics is a powerful tool for predicting what should (or should not) happen in a chemical reaction. It is ideally suited for answering questions that begin, "What if . . . ?" Thermodynamics predicts, for example, that hydrogen should react with oxygen to form water.

$$2\ H_2(g) + O_2(g) \longrightarrow 2\ H_2O(g) \qquad \Delta G° = -228.60\ \text{kJ/mol}\ H_2O$$

It also predicts that iron(III) oxide should react with aluminum metal to form aluminum oxide and iron metal.

$$Fe_2O_3(s) + 2\ Al(s) \longrightarrow Al_2O_3(s) + 2\ Fe(s) \qquad \Delta G° = -840.\ \text{kJ/mol}\ Fe_2O_3$$

The thermite reaction used to weld iron rails while building railroads involves the reaction between powdered aluminum and iron oxide to form aluminum oxide and molten iron metal.

Both of these reactions can, in fact, occur. We can experience the first by touching a balloon filled with H_2 gas with a candle tied to the end of a meter stick. We can demonstrate the other by adding a few drops of glycerin to a small quantity of barium peroxide placed on top of a mixture of Fe_2O_3 and powdered aluminum.

This doesn't mean that H_2 and O_2 burst spontaneously into flame the instant these gases are mixed. In the absence of a spark, mixtures of H_2 and O_2 can be stored for years with no detectable change. Nor does it mean that scrupulous attention must be paid to prevent Fe_2O_3 from ever coming in contact with aluminum metal, lest some violent reaction take place. In the absence of a source of heat, such as the reaction between glycerin and barium peroxide, mixtures of Fe_2O_3 and powdered aluminum are so stable they are sold in 50-pound lots under the trade name "Thermite."

It is a mistake to lose sight of the fact that thermodynamics can never tell us what will happen. It can only tell us what should or might happen. Reactions that behave as thermodynamics predicts they should are said to be under ***thermodynamic control***. It's not surprising from a thermodynamic point of view, for example, that potassium metal bursts into flame when it comes in contact with water.

$$2\ K(s) + 2\ H_2O(l) \longrightarrow 2\ K^+(aq) + 2\ OH^-(aq) + H_2(g)$$
$$\Delta G° = -406.8\ \text{kJ/mol}\ H_2$$

Other reactions are said to be under ***kinetic control***. Thermodynamics predicts that these reactions should occur, but the rate of the reactions is negligibly slow. Examples of reactions under kinetic control include the reactions in which elemental sulfur and its compounds burn. Thermodynamics predicts that these reactions should form SO_3 ($\Delta G_f° = -371.1$ kJ/mol), not SO_2 ($\Delta G_f° = -300.2$ kJ/mol), because SO_3 is more stable than SO_2. Under normal conditions, however, these reactions invariably stop at SO_2, because the reaction between SO_2 and O_2 to form SO_3 is extremely slow.

$$S_8(s) + 8\ O_2(g) \longrightarrow 8\ SO_2(g)$$
$$CS_2(g) + 3\ O_2(g) \longrightarrow CO_2(g) + 2\ SO_3(g)$$
$$2\ H_2S(g) + 3\ O_2(g) \longrightarrow 2\ H_2O(g) + 2\ SO_2(g)$$

In order to fully understand chemical reactions, we need to combine the predictions of thermodynamics with studies of the factors that influence the rates of chemical reactions. These factors fall under the general heading of ***chemical kinetics***.

21.2 CHEMICAL KINETICS DEFINED

Chemical kinetics can be defined as the study of the following.

1. The rate at which reactants are converted into the products of a chemical reaction.
2. The factors, such as temperature, pressure, concentration, and catalysts, that influence the rate of a reaction.
3. The sequence of steps, or the *mechanism*, by which the reactants are converted into products.

Before we can study the rate of a chemical reaction, we have to find a way to measure the rate at which one of the reactants is consumed or one of the products of the reaction is produced.

One of the first kinetic experiments was done in 1850 by Ludwig Wilhemy, a lecturer in physics at the University of Heidelberg. Wilhemy studied the rate at which sucrose, or cane sugar, reacts with acid to form a mixture of fructose and glucose known as invert sugar.

$$\text{Sucrose} + H_3O^+ \longrightarrow \underbrace{\text{glucose} + \text{fructose}}_{\text{invert sugar}}$$

Wilhemy found that the rate at which sucrose is consumed in this reaction is directly proportional to its concentration when the pH and temperature of the solution are held constant. The rate of this reaction at any moment in time is therefore given by the following equation, in which k is a proportionality constant and *(sucrose)* is the concentration of sucrose in units of moles per liter at that moment in time.

$$\textbf{Rate} = k\textbf{(sucrose)}$$

Because the sucrose concentration becomes smaller as sucrose is consumed in the reaction, the rate of the reaction gradually decreases.

Another factor that influences the rate of a chemical reaction is temperature, as can be seen in Figure 21.1. This figure shows a graph of the concentration of N_2O_5

FIG. 21.1 One of the factors that influence the rate of a reaction is the temperature at which the reaction is run. This graph shows the change in the concentration of N_2O_5 versus time when this compound decomposes to NO_2 and O_2 at 35°C and 45°C. These data show that the rate of the reaction increases with temperature.

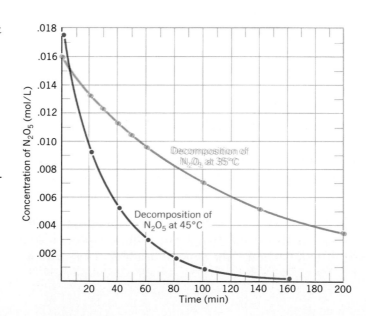

versus time as this compound decomposes to NO_2 and O_2.

$$2\ N_2O_5(g) \longrightarrow 4\ NO_2(g) + O_2(g)$$

Note that the concentration of N_2O_5 falls much more rapidly when the reaction is run at 45°C than when the reaction is run at 35°C, which shows that the rate of this reaction increases with temperature.

A third factor that can influence the rate of a chemical reaction is the presence of a *catalyst.*

> **Catalyst: A substance that alters the rate of a chemical reaction without being consumed in the reaction.**

The remainder of this chapter will be devoted to building a model that explains how concentration, temperature, and catalysts influence the rates of chemical reactions.

21.3 RATE OF A CHEMICAL REACTION DEFINED

What does the term *rate of reaction* mean? The word *rate* is used to indicate how rapidly (or slowly) something happens. Rate is usually expressed as a ratio of the number of events per unit of time. For example, the rate at which cars are sold is equal to the total number of cars that change hands divided by the period of time during which this number is counted.

Speed and velocity are special cases of rates. Speed describes how fast an object is moving. To calculate speed, we divide the distance traveled by the time needed to cover this distance.

$$\text{Speed} = \frac{\text{distance}}{\text{time}}$$

By convention, the term *velocity* is assumed to indicate both the speed and the direction in which the object is moving.

The rate of a chemical reaction is the "speed" at which the reactants are consumed or the products are produced. In this case, the distance traveled is the change in the concentration, $\Delta(X)$, that occurs in a given amount of time, Δt.

$$\text{Rate of reaction} = \frac{\Delta(X)}{\Delta t}$$

21.4 IS THE RATE OF A CHEMICAL REACTION CONSTANT?

To determine the rate of a chemical reaction, we measure the change in the concentration of one of the reactants or products with time. Table 21.1 contains the data used to construct the graph of concentration versus time for the decomposition of N_2O_5 at 45°C shown in Figure 21.1.

To find the rate of reaction over any period of time, we divide the change in the concentration of N_2O_5 by the time over which this change occurred. Because N_2O_5 is consumed in this reaction, the change in the N_2O_5 concentration — $\Delta(N_2O_5)$ — is a negative number. (The final concentration of N_2O_5 is smaller than the initial concentration of N_2O_5.) By convention, rate data are reported as positive num-

TABLE 21.1

Concentration of N_2O_5 versus Time for the Decomposition of N_2O_5 at 45°C

Time (min)	(N_2O_5) (mol/L)
0	0.01756
20	0.00933
40	0.00531
60	0.00295
80	0.00167
100	0.00094
160	0.00014

FIG. 21.2 The instantaneous rate of reaction at any moment in time can be calculated from the slope of a tangent to a graph of concentration versus time. At a point 20 minutes after the beginning of this reaction, the rate of reaction is 2.4×10^{-4} moles per liter per minute. After 80 minutes, the rate of reaction has decreased to 5.0×10^{-5} moles per liter per minute.

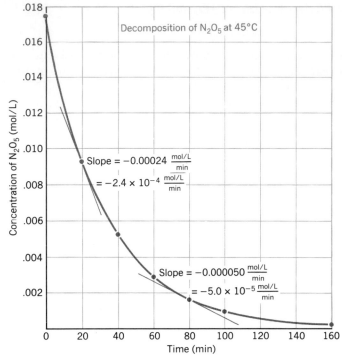

bers. The rate of this reaction is therefore defined by the following equation.

$$\text{Rate} = -\frac{\Delta(N_2O_5)}{\Delta t}$$

According to the data in Table 21.1, the rate of reaction over the first 20-minute period is 8.31×10^{-4} moles per liter per minute.

$$\text{Rate} = -\frac{\Delta(N_2O_5)}{\Delta t} = -\frac{(0.00933 - 0.01756) \text{ mol/L}}{(20 - 0) \text{ min}} = 8.31 \times 10^{-4} \frac{\text{mol/L}}{\text{min}}$$

Exercise 21.1

Use the data in Table 21.1 to determine whether the rate of the decomposition of N_2O_5 is constant.

Solution

To see whether the rate of reaction is constant, we can repeat the calculation of the rate of reaction over the second, third, fourth, fifth, and sixth time intervals in this table.

$$\text{2nd: Rate} = -\frac{\Delta(N_2O_5)}{\Delta t} = -\frac{(0.00531 - 0.00933) \text{ mol/L}}{(40 - 20) \text{ min}} = 2.01 \times 10^{-4} \frac{\text{mol/L}}{\text{min}}$$

$$\text{3rd: Rate} = -\frac{\Delta(N_2O_5)}{\Delta t} = -\frac{(0.00295 - 0.00531) \text{ mol/L}}{(60 - 40) \text{ min}} = 1.18 \times 10^{-4} \frac{\text{mol/L}}{\text{min}}$$

$$\text{4th: Rate} = -\frac{\Delta(N_2O_5)}{\Delta t} = -\frac{(0.00167 - 0.00295) \text{ mol/L}}{(80 - 60) \text{ min}} = 0.64 \times 10^{-4} \frac{\text{mol/L}}{\text{min}}$$

$$\text{5th: Rate} = -\frac{\Delta(N_2O_5)}{\Delta t} = -\frac{(0.00094 - 0.00167) \text{ mol/L}}{(100 - 80) \text{ min}} = 0.37 \times 10^{-4} \frac{\text{mol/L}}{\text{min}}$$

6th: $\text{Rate} = -\dfrac{\Delta(N_2O_5)}{\Delta t} = -\dfrac{(0.00014 - 0.00094)\ \text{mol/L}}{(160 - 100)\ \text{min}} = 0.13 \times 10^{-4}\ \dfrac{\text{mol/L}}{\text{min}}$

The rate of decomposition of N_2O_5 is not constant. It decreases by a factor of 15 between the first and sixth time intervals.

The calculations in Exercise 21.1 illustrate an important point.

The rates of most chemical reactions are not constant. Most reactions slow down as the reactants are converted into products.

21.5 INSTANTANEOUS RATES OF REACTION AND THE INITIAL RATE OF REACTION

The procedure used to calculate the rate of reaction in the preceding section was based on the assumption that the rate of reaction is constant during the period between measurements. The results of these calculations show that this assumption isn't valid. The rate of reaction not only changes from one time period to another, it is constantly changing. This fact has three important consequences.

1. **The rate of reaction changes while it is being measured.**
2. **All that can be measured is the *average* rate of reaction, which is the rate of reaction averaged over the time period between concentration measurements.**
3. **It is therefore a good idea to make rate measurements over periods of time that are short compared with the time it takes for the reaction to occur.**

Let's assume that we can find a way to measure the infinitesimally small change in the concentration of one the reactants or products—$d(X)$—that occurs during an infinitesimally short period of time—dt. The ratio of these quantities is the *instantaneous rate of reaction.*

$$\text{Instantaneous rate of reaction} = \frac{d(X)}{dt}$$

Instantaneous rates of reaction are inherently better than measurements taken over longer periods of time, because the rate of reaction doesn't change during the measurement.

The instantaneous rate of reaction can be calculated from a graph of the concentration of the reactant (or product) versus time. Figure 21.2 shows how the instantaneous rate of reaction for the decomposition of N_2O_5 at 45°C can be calculated. The quantity $d(N_2O_5)/dt$ is equal to the slope of a tangent to the plot of the concentration of N_2O_5 versus time at any moment in time.

We can study the effects of different factors on the rate of a chemical reaction by comparing the results of different experiments. The easiest way to do this is to compare instantaneous rates of reaction that are calculated at the same moment in time. In theory, we can make this comparison at any time during the course of the reaction. We could study the effect of temperature on the rate of decomposition of N_2O_5, for example, by comparing instantaneous rates of reaction at 20 minutes for reactions run at 35°C and 45°C. By convention, chemists often choose to compare the results of kinetic experiments by extrapolating to the *initial instantaneous rate of reaction.*

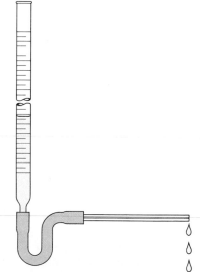

FIG. 21.3 We can build a physical model of a kinetic system by connecting a buret that has no stopcock to a piece of capillary tubing. The rate at which water flows through the capillary tubing is proportional to the force with which it is pushed down the tube.

Initial instantaneous rate of reaction: The instantaneous rate of reaction
extrapolated to the instant when the reagents were first mixed.

21.6 A PHYSICAL ANALOG OF KINETIC SYSTEMS

TABLE 21.2	
Volume of Water in Buret versus Time	
Volume of water in buret (mL)	Time (sec)
50	0
40	19
30	42
20	72
10	116
0	203

A physical model that has many of the properties of chemical kinetics can be
constructed from a 50-milliliter buret and a 1-foot length of capillary tubing, as
shown in Figure 21.3. The buret is filled with water, and the volume of water in the
buret is monitored versus time as the water gradually flows through the capillary
tubing. A typical set of data obtained with this apparatus is given in Table 21.2 and
plotted in Figure 21.4.

The plot of the volume of water in the buret versus time in Figure 21.4 is similar
to the plots of the concentration of N_2O_5 versus time in Figures 21.1 and 21.2. The
average rate at which the first 10 milliliters of water drains from the buret is 0.53
milliliters per second.

$$\text{First 10 mL:} \qquad \text{Rate} = -\frac{(40 - 50)\ \text{mL}}{(19 - 0)\ \text{s}} = 0.53\ \text{mL/s}$$

But the average rate steadily becomes smaller for the second, third, fourth, and
fifth 10-milliliter portions.

$$\text{Second 10 mL:} \qquad \text{Rate} = -\frac{(30 - 40)\ \text{mL}}{(42 - 19)\ \text{s}} = 0.43\ \text{mL/s}$$

$$\text{Third 10 mL:} \qquad \text{Rate} = -\frac{(20 - 30)\ \text{mL}}{(72 - 42)\ \text{s}} = 0.33\ \text{mL/s}$$

$$\text{Fourth 10 mL:} \qquad \text{Rate} = -\frac{(10 - 20)\ \text{mL}}{(116 - 72)\ \text{s}} = 0.23\ \text{mL/s}$$

$$\text{Fifth 10 mL:} \qquad \text{Rate} = -\frac{(0 - 10)\ \text{mL}}{(203 - 116)\ \text{s}} = 0.11\ \text{mL/s}$$

If the rate at which water escaped from the buret were constant, we could
calculate this rate at any moment in time and get the same answer. But the rate at
which water escapes from the buret depends on the pressure pushing the water
through the tube. As the height of the column of water in the buret decreases, less

FIG. 21.4 A graph of the volume
of water in the buret versus
time shows that the rate at which
water flows through the tube
becomes smaller as the volume
of water in the buret becomes
smaller.

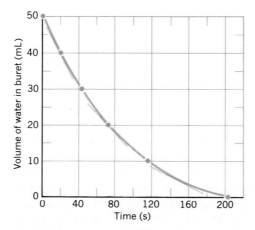

pressure is exerted on the water in the capillary tube, and the water flows through the tube at a slower rate.

The best rate measurements are therefore those taken over the shortest period of time. Instead of calculating the rate averaged over a 20-second or 80-second period, we would do better to plot these data, draw a smooth curve through the points, and then calculate the instantaneous rate by determining the slope of a tangent to the curve at that moment in time.

If we want to compare the results of this experiment with data for a 2-foot or 4-foot length of capillary tubing, for example, we need to compare measurements taken at similar points along the curve. One way of doing this is to extrapolate the data to the instantaneous rate at the beginning of the process and compare initial instantaneous rates for different lengths of tubing.

21.7 RATE LAWS AND RATE CONSTANTS

The data in Table 21.1 summarize experimental measurements of the concentration of N_2O_5 versus time when this compound decomposes into a mixture of NO_2 and O_2.

$$2 \, N_2O_5(g) \longrightarrow 4 \, NO_2(g) + O_2(g)$$

These data agree with the predictions of the following equation for the rate of this reaction.

$$\text{Rate} = -\frac{d(N_2O_5)}{dt} = k(N_2O_5)$$

This equation is called the **rate law** for the reaction. The proportionality constant, k, is known as the **rate constant.**

Exercise 21.2

Calculate the rate constant for the decomposition of N_2O_5 at 45°C if the instantaneous rate of decomposition after 20 minutes is 2.4×10^{-4} moles per liter per minute and the concentration of N_2O_5 at that moment is 0.00933 moles per liter.

Solution

We can start with the rate law for the reaction.

$$\text{Rate} = -\frac{d(N_2O_5)}{dt} = k(N_2O_5)$$

We can then substitute the known concentration of N_2O_5 and the known rate of reaction into this equation.

$$2.4 \times 10^{-4} \, \frac{\text{mol/L}}{\text{min}} = k(0.00933 \text{ mol/L})$$

Solving for the rate constant gives the following result.

$$k = 0.026 \, \frac{1}{\text{min}} = \textbf{0.026 min}^{-1}$$

The rate of this reaction at any moment in time can be predicted from the concentration of N_2O_5 at that moment, the rate law for the reaction, and the rate constant for the reaction.

Exercise 21.3

Predict the initial instantaneous rate of reaction for the decomposition of N_2O_5 at $45°C$ if the initial concentration of N_2O_5 is 0.01756 moles per liter.

Solution

Once again, we can start with the rate law for the reaction.

$$\text{Rate} = -\frac{d(N_2O_5)}{dt} = k(N_2O_5)$$

We know the rate constant for the reaction from the preceding exercise, and we know the initial concentration of N_2O_5. We should therefore be able to calculate the initial instantaneous rate of reaction.

$$\text{Rate} = (0.026 \text{ min}^{-1})(0.01756 \text{ mol/L}) = 4.6 \times 10^{-4} \frac{\text{mol/L}}{\text{min}}$$

The rate of reaction is equal to the change in the concentration of N_2O_5, in moles per liter, divided by the time over which this change occurs, in minutes. The rate of this reaction is therefore reported in units of moles per liter per minute. Because the number of moles of N_2O_5 per liter is equal to the molarity, the rate can also be reported in terms of the change in molarity per minute — M/min.

21.8 THE RATE LAW VERSUS THE STOICHIOMETRY OF A REACTION

The rate laws for some reactions are consistent with the stoichiometry of the reactions. Nitrogen oxide reacts with oxygen, for example, according to the following equation, which indicates that two NO molecules are consumed for every molecule of O_2.

$$2 \text{ NO}(g) + O_2(g) \longrightarrow 2 \text{ NO}_2(g)$$

The rate law for this reaction seems to agree with this stoichiometry, because it depends on the square of the NO concentration times the concentration of O_2.

$$\text{Rate} = k(NO)^2(O_2)$$

The rate law for the decomposition of N_2O_5, however, illustrates an important point.

The rate law for a chemical reaction must be determined experimentally; it cannot be predicted from the stoichiometry of the reaction.

The balanced equation for the decomposition of N_2O_5 contains two molecules of N_2O_5.

$$2 \text{ N}_2O_5(g) \longrightarrow 4 \text{ NO}_2(g) + O_2(g)$$

But the rate law for this reaction contains the concentration of N_2O_5 raised only to the first power.

$$\text{Rate} = k(N_2O_5)$$

There are many examples of reactions for which the rate law is very different from what might be expected on the basis of the stoichiometry of the reaction. Sulfur dioxide reacts with oxygen on a platinum metal surface, for example, according to a balanced equation that is similar to the equation for the reaction between NO and O_2.

$$2\ SO_2(g) + O_2(g) \longrightarrow 2\ SO_3(g)$$

But when the kinetics of this reaction were studied, the rate of reaction was found to obey the following rate law.

$$\text{Rate} = k\frac{(SO_2)}{(SO_3)^{1/2}}$$

There is only one way to obtain the rate law for a reaction — it must be determined by experiment.

21.9 DIFFERENT WAYS OF EXPRESSING THE RATE OF REACTION

There is usually more than one way to measure the rate of a chemical reaction. We can study the decomposition of hydrogen iodide, for example, by measuring the rate at which either H_2 or I_2 is formed in the following reaction.

$$2\ HI(g) \longrightarrow H_2(g) + I_2(g)$$

Alternatively, we can study the rate at which HI is consumed.

Let's begin by measuring the rate at which H_2 or I_2 is formed. The stoichiometry of the reaction suggests that the rate at which H_2 and I_2 are formed must be the same. The rate law for this reaction can be written as follows.

$$\frac{d(I_2)}{dt} = \frac{d(H_2)}{dt} = k(HI)^2$$

What would happen if we studied the rate at which HI is consumed? Because HI is consumed in the reaction, the change in its concentration must be a negative number, as noted in Section 21.4. Since the rate of a chemical reaction is always reported as a positive number, we must change the sign before reporting the rate of reaction for a reactant that is being consumed.

$$\text{Rate} = -\frac{d(HI)}{dt} = k'(HI)^2$$

The negative sign does two things. Mathematically, it converts a negative change in the concentration of HI into a positive rate. Physically, it reminds us that the concentration of the reactant decreases with time.

What is the relationship between the rate of reaction we obtained by monitoring the formation of H_2 or I_2 and the rate we obtained by watching HI disappear? The stoichiometry of the reaction says that two HI molecules are consumed for every molecule of H_2 or I_2 produced. This means that the rate of decomposition of HI is twice as fast as the rate at which H_2 and I_2 are formed.

We can translate this relationship into a mathematical equation as follows.

$$-\frac{d(HI)}{dt} = 2\frac{d(H_2)}{dt} = 2\frac{d(I_2)}{dt}$$

This equation can also be written in the following form.

$$\frac{d(I_2)}{dt} = \frac{d(H_2)}{dt} = -\frac{1}{2}\left[\frac{d(HI)}{dt}\right]$$

The rate constant obtained from the rate at which H_2 and I_2 are formed in this reaction (k) is not the same as the rate constant obtained from the rate at which HI is consumed (k').

Exercise 21.4

Calculate the rate at which HI disapears in the following reaction at a moment in time when I_2 is being formed at a rate of 1.8×10^{-6} moles per liter per second.

$$2\ HI(g) \longrightarrow H_2(g) + I_2(g)$$

Solution

The balanced equation for the reaction shows that 2 moles of HI disappear for every mole of I_2 formed. Thus, HI is consumed in this reaction twice as fast as I_2 is formed.

$$-\frac{d(HI)}{dt} = 2\frac{d(I_2)}{dt} = 2 \times 1.8 \times 10^{-6}\ \frac{mol/L}{s} = 3.6 \times 10^{-6}\ M/s$$

Students sometimes get the wrong answer to this exercise because they become confused about whether the equation for the calculation should be written as follows

$$-\frac{d(HI)}{dt} = 2\frac{d(I_2)}{dt}$$

or as the following alternative equation.

$$-\frac{d(HI)}{dt} = \frac{1}{2}\left[\frac{d(I_2)}{dt}\right]$$

You can avoid mistakes by checking to see whether the answer makes sense.

The balanced equation for the reaction says that 2 moles of HI are consumed for every mole of I_2 produced. HI should therefore disappear (3.6×10^{-6} M/sec) twice as fast as I_2 is formed (1.8×10^{-6} M/sec). This agrees with the relative magnitudes of the rates of reaction calculated in this exercise.

21.10 ORDER AND MOLECULARITY

Some chemical reactions occur in a single step. The transfer of a chlorine atom from $ClNO_2$ to NO, for example, is a one-step reaction.

$$ClNO_2(g) + NO(g) \longrightarrow ClNO(g) + NO_2(g)$$

Many chemical reactions, however, occur in a series of steps. The decomposition of N_2O_5, for example, occurs in three steps. In the first step, N_2O_5 decomposes into NO_2 and NO_3.

$$N_2O_5(g) \rightleftharpoons NO_2(g) + NO_3(g)$$

Some of the products of the first step then combine to form a mixture of NO_2, NO, and O_2.

$$NO_2(g) + NO_3(g) \longrightarrow NO_2(g) + NO(g) + O_2(g)$$

The NO produced in this reaction then combines with the rest of the NO_3 produced in the first step to form more NO_2.

$$NO(g) + NO_3(g) \longrightarrow 2\ NO_2(g)$$

When these individual steps are combined in the correct proportion, the result is the overall equation for the decomposition of N_2O_5 to NO_2 and O_2.

$$2\ (N_2O_5 \longrightarrow NO_2 + NO_3)$$
$$NO_2 + NO_3 \longrightarrow NO_2 + O_2 + NO$$
$$\underline{NO + NO_3 \longrightarrow 2\ NO_2}$$
$$2\ N_2O_5 \longrightarrow 4\ NO_2 + O_2$$

It is often useful to classify the steps in a reaction in terms of *molecularity*—that is, in terms of whether one, two, or three molecules are consumed. When a single molecule is consumed, the step is called *unimolecular.* When two molecules are consumed, it is *bimolecular.* A few reactions are *trimolecular.* The best-known example is the reaction between NO and O_2 to form NO_2, which requires that all three molecules react simultaneously.

$$2\ NO(g) + O_2(g) \longrightarrow 2\ NO_2(g)$$

Exercise 21.5

Calculate the molecularity of the three steps in the decomposition of N_2O_5.

Solution

All we have to do is count the number of reactant molecules consumed in each of these steps to decide that the first step is unimolecular and the other two steps are bimolecular.

Step 1:	$N_2O_5 \longrightarrow NO_2 + NO_3$	(unimolecular)
Step 2:	$NO_2 + NO_3 \longrightarrow NO_2 + O_2 + NO$	(bimolecular)
Step 3:	$NO + NO_3 \longrightarrow 2\ NO_2$	(bimolecular)

It is also useful to classify reactions in terms of their *order.* The decomposition of N_2O_5 is a *first-order reaction,* because the rate of reaction depends on the concentration of N_2O_5 raised to the first power.

$$\text{Rate} = k(N_2O_5)$$

The decomposition of HI

$$2\ HI \longrightarrow H_2 + I_2$$

is a *second-order reaction,* because the rate of reaction depends on the concentration of HI raised to the second power.

$$\text{Rate} = k(HI)^2$$

When the rate of a reaction depends on more than one reagent, we classify the reaction in terms of the order of each reagent.

Exercise 21.6

Classify the order of the reaction between NO and O_2 to form NO_2.

$$2\, NO(g) + O_2(g) \longrightarrow 2\, NO_2(g)$$

Assume that the rate law for this reaction is as follows.

$$Rate = k(NO)^2(O_2)$$

Solution

This reaction is first-order in O_2, second-order in NO, and third-order overall.

What is the difference between the molecularity and the order of a chemical reaction? The molecularity of a reaction, or of a step within a reaction, describes what happens in the reaction on the molecular level. The order of a reaction describes what happens on the macroscopic scale. We can determine the order experimentally by watching the products appear or the reactants disappear. The molecularity of the reaction is something we deduce to explain these experimental results.

Exercise 21.7

Calculate the units of the rate constants for first-order and second-order reactions in which the rate of reaction is measured in moles per liter per second.

Solution

The rate law for a first-order reaction depends on the concentration of the reactant to the first power.

$$Rate = k(X)$$

If the rate of the reaction has units of moles per liter per second, and the concentration of X is given in units of moles per liter, the units of k for this reaction must be the reciprocal of seconds, $1/s$ or s^{-1}.

$$\frac{mol/L}{s} = \frac{1}{s} \times mol/L$$

The rate law for a second-order reaction depends on the concentration of the reactant to the second power.

$$Rate = k(X)^2$$

If the rate of the reaction has units of moles per liter, the units of k must now be the reciprocal of moles per liter times the reciprocal of seconds.

$$\frac{mol/L}{s} = \left(\frac{1}{mol/L} \times \frac{1}{s}\right) \times (mol/L)^2$$

The number of moles of reactant per liter is equal to the molarity of the reactant. The units of second-order rate constants are therefore often written as follows— $M^{-1}\, s^{-1}$.

21.11 DETERMINING THE ORDER OF A REACTION FROM INITIAL RATES OF REACTION

The only way to find the rate law for a reaction is to determine it by experiment. One way of achieving this goal is to measure the initial instantaneous rate of reaction at different initial concentrations of the reactants. We'll use this approach to determine the rate law for the decomposition of hydrogen peroxide in the presence of the iodide ion.

$$2 H_2O_2(aq) \xrightarrow{I^-} 2 H_2O(l) + O_2(g)$$

Data on initial instantaneous rates of reaction for five experiments run at different initial concentrations of H_2O_2 and I^- are summarized in Table 21.3.

The initial I^- ion concentration is the same in the first three trials. The only difference in these experiments is the initial concentration of H_2O_2. Let's compare the first three experiments two at a time. The difference between Trial 1 and Trial 2 is an increase by a factor of 2 in the initial H_2O_2 concentration, which leads to an increase by a factor of 2 in the initial instantaneous rate of reaction.

$$\frac{\text{Rate for Trial 2}}{\text{Rate for Trial 1}} = \frac{4.6 \times 10^{-7} \ M/s}{2.3 \times 10^{-7} \ M/s} = 2$$

The only difference between Trial 1 and Trial 3 is a factor-of-3 increase in the initial H_2O_2 concentration, which produces a factor-of-3 increase in the initial instantaneous rate of reaction.

$$\frac{\text{Rate for Trial 3}}{\text{Rate for Trial 1}} = \frac{6.9 \times 10^{-7} \ M/s}{2.3 \times 10^{-7} \ M/s} = 3$$

We can therefore conclude that the initial rate of reaction is directly proportional to the initial H_2O_2 concentration.

Experiments 1, 4, and 5 were run at the same initial concentration of H_2O_2 but different initial concentrations of the I^- ion. When we compare Trials 1 and 4 we see that doubling the initial I^- concentration leads to a twofold increase in the rate of reaction.

$$\frac{\text{Rate for Trial 4}}{\text{Rate for Trial 1}} = \frac{4.6 \times 10^{-7} \ M/s}{2.3 \times 10^{-7} \ M/s} = 2$$

Trials 1 and 5 show that tripling the initial I^- concentration leads to a threefold

TABLE 21.3

Rate of Reaction Data for the Decomposition of H_2O_2 in the Presence of the I^- Ion

	Initial (H_2O_2) (M)	Initial (I^-) (M)	Initial Instantaneous Rate of Reaction (M/s)
Trial 1:	1.0×10^{-2}	2.0×10^{-3}	2.3×10^{-7}
Trial 2:	2.0×10^{-2}	2.0×10^{-3}	4.6×10^{-7}
Trial 3:	3.0×10^{-2}	2.0×10^{-3}	6.9×10^{-7}
Trial 4:	1.0×10^{-2}	4.0×10^{-3}	4.6×10^{-7}
Trial 5:	1.0×10^{-2}	6.0×10^{-3}	6.9×10^{-7}

increase in the initial rate of reaction. We therefore conclude that the initial rate of the reaction is directly proportional to the initial concentration of the I^- ion.

The results of these experiments are consistent with a rate law for this reaction that is first-order in both H_2O_2 and I^-.

$$\text{Rate} = k(H_2O_2)(I^-)$$

Exercise 21.8

Hydrogen iodide decomposes to give a mixture of hydrogen and iodine.

$$2\,HI(g) \longrightarrow H_2(g) + I_2(g)$$

Use the following data on initial instantaneous rates of reaction to determine whether the decomposition of HI in the gas phase is first-order or second-order in hydrogen iodide.

	Initial (HI)	Initial Instantaneous Rate of Reaction (M/s)
Trial 1:	1.0×10^{-2}	4.0×10^{-6}
Trial 2:	2.0×10^{-2}	1.6×10^{-5}
Trial 3:	3.0×10^{-2}	3.6×10^{-5}

Solution

We can start by comparing Trials 1 and 2. When the initial concentration of HI is doubled, the initial rate of reaction increases by a factor of 4.

$$\frac{\text{Rate for Trial 2}}{\text{Rate for Trial 1}} = \frac{1.6 \times 10^{-5}\,M/s}{4.0 \times 10^{-6}\,M/s} = 4$$

Let's now compare Trials 1 and 3. When the initial concentration of HI is tripled, the initial rate increases by a factor of 9.

$$\frac{\text{Rate for Trial 3}}{\text{Rate for Trial 1}} = \frac{3.6 \times 10^{-5}\,M/s}{4.0 \times 10^{-6}\,M/s} = 9$$

The only way to explain this observation is to assume that the rate of reaction is proportional to the square of the HI concentration. The reaction is therefore second-order in HI.

$$\text{Rate} = k(HI)^2$$

21.12 THE INTEGRATED FORM OF THE FIRST-ORDER RATE LAW

So far, we have focused our attention on the rate of a chemical reaction at a particular moment in time. Exercise 21.3 showed how we can calculate the rate of reaction from the rate law for the reaction, the rate constant for the reaction, and the concentrations of the reactants at any moment in time. The rate law introduced in Section 21.7 isn't as useful for predicting the concentrations of the reactants that remain in solution or the products formed at a given moment in time. For these calculations, we use the **integrated form** of the rate law.

We can start with the rate law for a reaction that is first-order in the disappearance of one reactant, X.

$$-\frac{d(X)}{dt} = k(X)$$

If you haven't been exposed to calculus, you will simply have to accept the fact that this equation can be integrated to give the following result.

$$\log\left[\frac{(X)}{(X)_0}\right] = -\frac{kt}{2.303}$$ integrated form of the first-order rate law

In this equation, (X) represents the concentration of the reactant at any moment in time, $(X)_0$ is the initial concentration of the reactant, k is the rate constant for the reaction, and t is the time since the reaction started.

If you have been exposed to calculus, you should understand the steps by which the integrated form of this equation is derived. We start with the first-order rate law equation.

$$-\frac{d(X)}{dt} = k(X)$$

We then rearrange the equation as follows.

$$\frac{1}{(X)}d(X) = -kdt$$

Our goal is to integrate both sides of this equation. Mathematically, integrating an equation is equivalent to finding the area under the curve that would be produced if this function was graphed. This process is indicated with integral signs, as follows.

$$\int \frac{1}{(X)}d(X) - \int -kdt$$

We are interested in the area under this curve between the time when the reaction first starts ($t = 0$) and some later time (t).

$$\int_0^x \frac{1}{(X)}d(X) = \int_0^t -kdt$$

By convention, the integral of this equation has the following form.

$$\ln\left[\frac{(X)}{(X)_0}\right] = -kt$$

Logarithms to the base e can be replaced by logarithms to the base 10 by use of the following approximation.

$$\ln X \cong 2.303 \log_{10} X$$

The integrated form of the first-order rate law can therefore be written as follows.

$$2.303 \log\left[\frac{(X)}{(X)_0}\right] = -kt$$

Or it can be written in the following form.

$$\log\left[\frac{(X)}{(X)_0}\right] = -\frac{kt}{2.303}$$ integrated form of the first-order rate law

When using this equation, remember that (X) is the concentration of the reactant at any moment in time, $(X)_0$ is the initial concentration of the reactant, k is the rate constant for the reaction, and t is the time since the reaction started.

Exercise 21.9

Write the integrated form of the first-order rate law for a reaction that is first-order in the formation of one product, X.

$$\frac{d(X)}{dt} = k(X)$$

Solution

The integrated form of this equation will look similar to the one derived above, but it will have the opposite sign convention.

$$\log\left[\frac{(X)}{(X)_0}\right] = \frac{kt}{2.303}$$

The integrated form of the first-order rate equation is useful for calculating the concentration of X at any moment in time from the initial concentration, the rate constant for the reaction, and the amount of time since the reaction started.

Exercise 21.10

Calculate how long it would take for the ^{14}C in a piece of charcoal to decay to half of its original concentration, assuming that ^{14}C decays to ^{14}N by first-order kinetics with a rate constant of 1.21×10^{-4} yr^{-1}.

$$^{14}C \longrightarrow {}^{14}N + e^-$$

Solution

If ^{14}C decays by first-order kinetics, the rate law for this reaction can be written as follows.

$$\text{Rate} = -\frac{d(^{14}C)}{dt} = k(^{14}C)$$

The following is the integrated form of this rate law.

$$\log\left[\frac{(^{14}C)}{(^{14}C)_0}\right] = -\frac{kt}{2.303}$$

The question asks us to calculate the amount of time it would take for the ^{14}C to decay to half of its original value. We are therefore interested in the moment when the concentration of ^{14}C in the charcoal — (^{14}C) — has half of its initial value — $(^{14}C)_0$.

$$(^{14}C) = \frac{1}{2}(^{14}C)_0$$

Substituting this relationship into the integrated form of the rate law gives the following equation.

$$\log\left[\frac{1/2\,(^{14}C)_0}{(^{14}C)_0}\right] = -\frac{kt}{2.303}$$

This equation can be simplified as follows.

$$\log (1/2) = -kt/2.303$$

Solving for t gives the following result.

$$t = \frac{-2.303 \log (1/2)}{k} = \frac{-2.303 \log (1/2)}{1.21 \times 10^{-4} \ yr^{-1}} = 5730 \ years$$

According to Exercise 21.10, it takes 5730 years for half of the ^{14}C in the sample to decay. This length of time is often called the **half-life** of ^{14}C. In general, the half-life for a first-order kinetic process can be calculated from the rate constant as follows.

$$t_{1/2} = \frac{-2.303 \log(1/2)}{k} = \frac{0.693}{k}$$

21.13 THE INTEGRATED FORM OF THE SECOND-ORDER RATE LAW

The following is the rate law for a reaction that is second-order in reactant X.

$$-\frac{d(X)}{dt} = k(X)^2$$

To derive the integrated form of this rate law, we start by rearranging the equation as follows.

$$-\frac{1}{(X)^2} \, d(X) = kt$$

We then integrate both sides of this equation.

$$\int_0^x -\frac{1}{(X)^2} \, d(X) = \int_0^t kt$$

The following is the result of this integration.

$$\frac{1}{(X)} - \frac{1}{(X)_0} = kt \qquad \text{integrated form of the second-order rate law}$$

Once again, the (X) term is the concentration of X at any moment in time, $(X)_0$ is the initial concentration of X, k is the rate constant for the reaction, and t is the time since the reaction started.

The integrated form of the second-order rate law shows how the concentration of X at any moment in time can be calculated from the initial concentration, the rate constant for the reaction, and the amount of time since the reaction began.

Exercise 21.11

Calculate the amount of time it would take for acetaldehyde, CH_3CHO, to decompose from an initial concentration of 0.00750 M to a concentration only 20% as large, assuming that acetaldehyde decomposes by second-order kinetics with a rate constant of 0.334 $M^{-1} \ s^{-1}$.

Solution

The rate law for the decomposition of acetaldehyde by second-order kinetics is written as follows.

$$-\frac{d(CH_3CHO)}{dt} = k(CH_3CHO)^2$$

The following is the integrated form of this rate law.

$$\frac{1}{(CH_3CHO)} - \frac{1}{(CH_3CHO)_0} = kt$$

The initial concentration of acetaldehyde is 0.00750 M, and the final concentration is 20% as large, or 0.00150 M.

$$\frac{1}{(0.00150)} - \frac{1}{(0.00750)} = kt$$

Substituting the rate constant for this reaction ($k = 0.334$) into this equation and then solving for t gives the following result.

$$t = 1600 \text{ seconds}$$

Exercise 21.12

Derive the formula for calculating the half-life of a second-order reaction.

Solution

We start with the integrated form of the second-order rate law.

$$\frac{1}{(X)} - \frac{1}{(X)_0} = kt$$

We then ask: "How long it would take for the concentration of X to decay from its initial value, $(X)_0$, to a value half as large?"

$$\frac{1}{1/2\,(X)_0} - \frac{1}{(X)_0} = kt_{1/2}$$

The top and bottom halves of the first terms of the equation can be multiplied by 2.

$$\frac{2}{(X)_0} - \frac{1}{(X)_0} = kt_{1/2}$$

Subtracting as indicated gives the following equation.

$$\frac{1}{(X)_0} = kt_{1/2}$$

We can then solve this equation to find the half-life.

$$t_{1/2} = \frac{1}{k(X)_0}$$

Note the difference between the equations for calculating the half-life of first-order and second-order reactions. The half-life of a first-order reaction is a constant, which is proportional to the rate constant for the reaction.

$$t_{1/2} = \frac{0.693}{k} \qquad \text{first-order reaction}$$

The half-life for a second-order reaction is inversely proportional to both the rate constant for the reaction and the initial concentration of the reactant that is consumed in the reaction.

$$t_{1/2} = \frac{1}{k(X)_0} \qquad \text{second-order reaction}$$

21.14 DETERMINING THE ORDER OF A REACTION WITH THE INTEGRATED FORMS OF RATE LAWS

The integrated forms of the rate laws for first-order and second-order reactions provide another way of determining the order of a reaction. We start by assuming, for the sake of argument, that the reaction is first-order in reactant X.

$$-\frac{d(X)}{dt} = k(X)$$

We then test this assumption by checking data on concentration versus time for the reaction to see whether they fit the first-order rate law.

$$\log\left[\frac{(X)}{(X)_0}\right] = -\frac{kt}{2.303} \qquad \text{first-order reaction}$$

The integrated form of the first-order rate law can be written as follows.

$$\log (X) - \log (X)_0 = -kt/2.303$$

Rearranging this equation gives the following result.

$$\log (X) = \log (X)_0 - kt/2.303$$

For a given set of data, this equation contains two variables — $\log (X)$ and t — and two constants — $\log (X)_0$ and $k/2.303$. The equation can therefore be set up in terms of the equation for a straight line.

$$y = mx + b$$

$$\log (X) = \left(\frac{k}{2.303}\right) t + \log (X)_0$$

If the reaction is first-order in X, a plot of the log of the concentration of X versus time will be a straight line with a slope equal to $-k/2.303$, as shown in Figure 21.5.

If the plot of $\log (X)$ versus time is not a straight line, the reaction can't be first-order in X. We therefore assume, for the sake of argument, that it is second-order in X.

$$-\frac{d(X)}{dt} = k(X)^2$$

We can test this assumption by checking whether the experimental data fit the integrated form of the second-order rate law.

$$\frac{1}{(X)} - \frac{1}{(X)_0} = kt$$

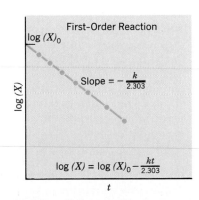

First-Order Reaction

$$\log (X) = \log (X)_0 - \frac{kt}{2.303}$$

FIG. 21.5 If the rate law for a reaction is first-order in X, and in nothing else, a plot of the log of the concentration of X at any moment in time versus the amount of time since the reaction started will be a straight line with a slope equal to the negative of the rate constant divided by 2.303 — $-k/2.303$.

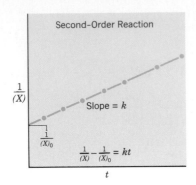

FIG. 21.6 If the rate law for a reaction is second-order in X, and in nothing else, a plot of the inverse of the concentration of X at any moment in time versus the amount of time since the reaction started will be a straight line with a slope equal to the rate constant, k.

For a given set of data, this equation contains two variables — (X) and t — and two constants — $(X)_0$ and k. Therefore, it can also be set up in terms of the equation for a straight line.

$$y = mx + b$$

$$\frac{1}{(X)} = kt + \frac{1}{(X)_0}$$

If the reaction is second-order in X, a plot of the reciprocal of the concentration of X versus time will be a straight line with a slope equal to k, as shown in Figure 21.6.

Exercise 21.13

Use the data in Table 21.2 to determine whether the rate at which water flows from a buret through a capillary tube is first-order or second-order.

Solution

We start by calculating the log of the volume of water and the reciprocal of the volume in the buret at each point at which measurements were taken.

Volume of Water in Buret (mL)	Time (s)	log V	1/V
50	0	1.699	0.020
40	19	1.602	0.025
30	42	1.477	0.033
20	72	1.301	0.050
10	116	1.000	0.10
0	203	—	—

A plot of log V versus t (Figure 21.7) gives a straight line, within experimental error. But when $1/V$ versus t is plotted (Figure 21.8), the points deviate significantly from a straight line. We therefore conclude that these data best fit first-order kinetics.

FIG. 21.7 A plot of the log of the volume of water in the buret versus the amount of time it takes for the water to flow through a capillary tube is a straight line, within experiment error, which shows that this is a first-order process.

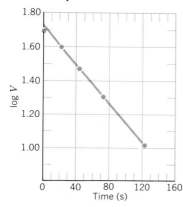

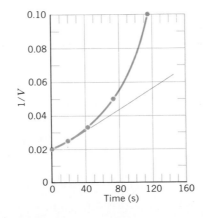

FIG. 21.8 A plot of the reciprocal of the volume of water in the buret versus the amount of time it takes for the water to flow through a capillary tube isn't a straight line, which shows that this is not a second-order process.

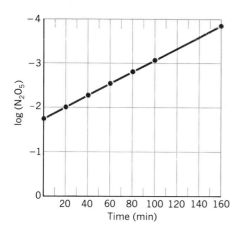

FIG. 21.9 A plot of the log of the concentration of N_2O_5 versus time for the decomposition of N_2O_5 is a straight line, which shows that this reaction is first-order in N_2O_5.

FIG. 21.10 A plot of the reciprocal of the concentration of N_2O_5 versus time for the decomposition of N_2O_5 isn't a straight line, which shows that this reaction is not second-order in N_2O_5.

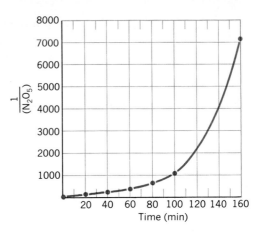

Exercise 21.14

Use the data in Table 21.1 to determine whether the decomposition of N_2O_5 is a first-order or second-order reaction.

Solution

The first step in solving this problem involves calculating the log of the N_2O_5 concentration and the reciprocal of the concentration for each point at which a measurement was taken.

Time (min)	(N_2O_5) (mol/L)	$\log(N_2O_5)$	$1/(N_2O_5)$
0	0.01756	−1.755	56.95
20	0.00933	−2.030	107
40	0.00531	−2.275	188
60	0.00295	−2.530	339
80	0.00167	−2.777	599
100	0.00094	−3.027	1060
160	0.00014	−3.854	7100

We then construct graphs of $\log(N_2O_5)$ versus t (Figure 21.9) and $1/(N_2O_5)$ versus t (Figure 21.10). Only one of these graphs, Figure 21.9, gives a straight line. We therefore conclude that these data best fit a first-order kinetic equation.

$$-\frac{d(N_2O_5)}{dt} = k(N_2O_5)$$

21.15 REACTIONS THAT ARE FIRST-ORDER IN TWO REACTANTS

If a reaction is first-order in reactant X, the rate law for the reaction is written as follows.

$$-\frac{d(X)}{dt} = k(X)$$

The integrated form of this equation implies that a plot of log (X) versus time gives a straight line.

$$\log\left[\frac{(X)}{(X)_0}\right] = -\frac{kt}{2.303}$$

If the reaction is second-order in X, the rate law is written as follows.

$$-\frac{d(X)}{dt} = k(X)^2$$

The integrated form of this equation implies that a plot of $1/(X)$ versus time gives a straight line.

$$\frac{1}{(X)} - \frac{1}{(X)_0} = kt$$

What about reactions that are first-order in two reactants, X and Y, and therefore second-order overall?

$$-\frac{d(X)}{dt} = k(X)(Y)$$

A plot of $1/(X)$ versus time won't give a straight line, because the reaction is not second-order in X. Unfortunately, neither will a plot of log (X) versus time, because the reaction is no longer strictly first-order in X. It is first-order in both X and Y.

One way around this problem is to turn the reaction into one that is *pseudo-first-order* by making the concentration of one of the reactants so large that it is effectively constant. The rate law for the reaction is still first-order in both reactants.

$$-\frac{d(X)}{dt} = k(X)(Y)$$

But if the initial concentration of one reactant is very much larger than that of the other, the concentration of the excess reactant will be essentially constant during the course of the reaction. The rate of reaction will therefore appear to be sensitive only to changes in the concentration of the other reactant.

If the reaction is studied under conditions for which there is a large excess of Y, then, the concentration of Y will remain essentially constant during the reaction. Thus, the reaction will appear to be first-order in X. A plot of log (X) versus time will therefore give a straight line.

$$-\frac{d(X)}{dt} = [k(Y)](X) = k'(X)$$

If there is a large excess of X, the reaction will appear to be first-order in Y. Under these conditions, a plot of log (Y) versus time will be linear.

$$-\frac{d(X)}{dt} = [k(X)](Y) = k'(Y)$$

The value of the rate constant obtained from this equation—k'—won't be the actual rate constant for the reaction. It will be the product of the rate constant for the reaction times the concentration of the reagent that is present in excess.

21.16 A COLLISION THEORY MODEL OF CHEMICAL REACTIONS

The reaction between $ClNO_2$ and NO is an example of a single-step reaction.

$$ClNO_2(g) + NO(g) \longrightarrow NO_2(g) + ClNO(g)$$

The rate of this reaction is first-order in both $ClNO_2$ and NO and second-order overall. The rate of formation of NO_2, for example, is given by the following rate law.

$$\frac{d(NO_2)}{dt} = k(ClNO_2)(NO)$$

This rate law is easy to explain if we assume that molecules must collide in order to react. Before a chlorine atom can be transferred from $ClNO_2$ to NO, for example, the two molecules must collide.

Figure 21.11 shows a small portion of a system in which the reaction between $ClNO_2$ and NO takes place. What factors influence the rate at which the molecules in this container collide? In Section 4.15, we concluded that the number of collisions per second in a gas depends on the number of molecules per liter. Doubling the number of $ClNO_2$ or NO molecules per liter of gas should double the rate at which these molecules collide. The rate of this reaction is therefore directly proportional to the concentration of both $ClNO_2$ and NO.

$$\frac{d(NO_2)}{dt} = k(ClNO_2)(NO)$$

Collisions between reactant molecules are necessary in order for a reaction to take place. But they aren't sufficient. One reason why collisions do not always convert the reactants into products is the orientation of the molecules when they collide. Consider the Lewis structures of the reactants and products for the reaction between $ClNO_2$ and NO.

$$ClNO_2(g) + NO(g) \longrightarrow NO_2(g) + ClNO$$

In the course of this reaction, a chlorine atom is transferred from one nitrogen atom to another. In order for the reaction to occur, the nitrogen atom in NO must collide with the chlorine atom in $ClNO_2$.

Reaction won't occur if the oxygen end of the NO molecule collides with the chlorine atom on $ClNO_2$.

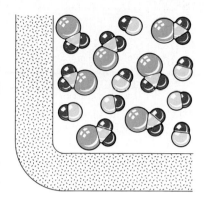

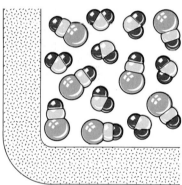

FIG. 21.11 This figure represents a snapshot of a small portion of a container in which $ClNO_2$ reacts with NO to form NO_2 and ClNO. What factors control the rate at which this reaction occurs? The collision theory model of chemical reactions assumes that molecules must collide in order to react. Anything that increases the frequency of collisions increases the rate of reaction, so the rate of reaction must be proportional to the concentration of either reactant.

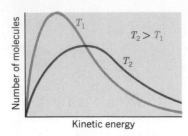

FIG. 21.12 The kinetic molecular theory states that the average kinetic energy of a gas is proportional to the temperature of the gas, and nothing else. At any given temperature, however, some of the gas particles are moving faster than others.

Nor will it occur if one of the oxygen atoms on $ClNO_2$ collides with the nitrogen atom on NO.

Another factor that influences whether reaction will occur is the energy the molecules carry when they collide. Not all of the molecules have the same kinetic energy, as shown in Figure 21.12. The kinetic energy molecules carry when they collide is important because it is the principal source of the energy that must be invested in a reaction to get it started.

The overall standard-state free energy for the reaction between $ClNO_2$ and NO is favorable.

$$ClNO_2(g) + NO(g) \longrightarrow NO_2(g) + ClNO(g) \qquad \Delta G° = -23.6 \text{ kJ/mol}$$

But before the reactants can be converted into products, the free energy of the system must climb a barrier. This barrier is the **activation energy** for the reaction, as shown in Figure 21.13. The vertical axis in this diagram represents the free energy of the system at any moment in time. The horizontal axis represents the **reaction coordinate,** which summarizes the infinitesimally small steps that must be taken in order for the reactants to be converted into the products of this reaction.

Why does a reaction have an activation energy? Think about what has to happen in order for $ClNO_2$ to react with NO. First, and foremost, these two molecules have to collide, thereby organizing the system. Not only do they have to be brought together but they have to be held in exactly the right orientation relative to each other to ensure that reaction can occur. Both of these factors raise the free energy of the system by lowering the entropy. Some energy also must be invested in the system to begin breaking the $Cl-NO_2$ bond before the $Cl-NO$ bond can begin to form.

NO and $ClNO_2$ molecules that collide in the correct orientation, with enough kinetic energy to climb the activation energy barrier, can react to form NO_2 and ClNO. As the temperature of the system increases, the number of molecules that carry enough energy to react when they collide also increases. The rate of reaction therefore increases with temperature. As a general rule, the rate of a chemical reaction doubles for every 10°C increase in the temperature of the system.

21.17 THE MECHANISMS OF CHEMICAL REACTIONS

FIG. 21.13 A useful model for thinking about the rate of chemical reactions is a plot of the free energy of the system versus the reaction coordinate. The overall free energy of reaction, $\Delta G°$, is equal to the difference between the free energies of the reactants and products. The activation energy for the reaction, E_a, is the difference between the free energy of the reactants and the top of an energy barrier that must be overcome before the reaction can occur.

The rate laws for chemical reactions can be explained by the following general rules.

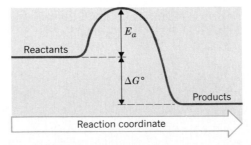

1. **The rate of any step in a reaction is directly proportional to the concentrations of the reagents consumed in that step.**
2. **The overall rate law for a reaction is determined by the sequence of steps, or the mechanism, by which the reactants are converted into the products of the reaction.**
3. **The overall rate law for a reaction is dominated by the rate law for the slowest step in the reaction.**

Some reactions, such as the transfer of a chlorine atom from $ClNO_2$ to NO, occur in a single step.

$$ClNO_2(g) + NO(g) \longrightarrow NO_2(g) + ClNO(g)$$

The rate of this reaction depends on the frequency of collisions between $ClNO_2$ and NO molecules. Anything that changes the concentration of either reactant will change the frequency with which collisions occur and the rate of reaction. The overall rate of a one-step reaction is therefore proportional to the product of the concentrations of the reactants consumed in the reaction.

$$\frac{d(NO_2)}{dt} = k(NO)(ClNO_2)$$

Many reactions, such as the decomposition of N_2O_5, occur in two or more steps. The first step in this case is a relatively fast reaction in which N_2O_5 decomposes to form NO_2 and NO_3. The products of this step then undergo a much slower reaction to form NO, NO_2, and O_2. The NO produced in the second step then reacts with NO_3 in a fast step to form more NO_2.

Step 1:	$N_2O_5 \rightleftharpoons NO_2 + NO_3$	(fast step)
Step 2:	$NO_2 + NO_3 \longrightarrow NO + NO_2 + O_2$	(slow step)
Step 3:	$NO + NO_3 \longrightarrow 2\ NO_2$	(fast step)

Combining these reactions in the proper ratio yields the overall equation for the decomposition of N_2O_5.

$$2\ N_2O_5 \rightleftharpoons 2\ NO_2 + 2\ NO_3$$
$$NO_2 + NO_3 \longrightarrow NO + NO_2 + O_2$$
$$\underline{NO + NO_3 \longrightarrow 2\ NO_2}$$
$$2\ N_2O_5 \longrightarrow 4\ NO_2 + O_2$$

The overall rate of reaction can be no faster than the rate of the slowest step in the mechanism. The slowest step is called the *rate-limiting step,* because it places a limit on the rate at which the overall reaction can occur. In the decomposition of N_2O_5, the second step is the rate-limiting step. No matter how fast the first and third steps take place, the decomposition of N_2O_5 cannot proceed any faster than the rate-limiting second step.

The rate of any step in a reaction is directly proportional to the concentrations of the reactants consumed in that step. The following is the rate law for the second step in the decomposition of N_2O_5.

Step 2: $\quad$ rate $= k(NO_2)(NO_3)$

If the other steps in the reaction are much faster, the overall rate of reaction is more or less equal to the rate of this rate-limiting step.

$$-\frac{d(N_2O_5)}{dt} = k(NO_2)(NO_3)$$

This equation is not very useful, because it is difficult to measure the concentrations of intermediates, such as NO_3, that are simultaneously formed and consumed in the reaction. It would be better to have an equation that related the overall rate of reaction to the concentrations of the original reactants.

Let's take advantage of the fact that the first step in this reaction is reversible.

$$\text{Step 1:} \qquad N_2O_5 \rightleftharpoons NO_2 + NO_3$$

Since the rate of any step in a reaction is directly proportional to the concentration of the reagents consumed in that step, the rate of the forward reaction depends on the concentration of N_2O_5.

$$\text{Step 1:} \qquad \text{rate}_{forward} = k_f(N_2O_5)$$

The rate of the reverse reaction in this step depends on the concentrations of both NO_2 and NO_3.

$$\text{Step 1:} \qquad \text{rate}_{reverse} = k_r(NO_2)(NO_3)$$

Because the first step in this reaction is very much faster than the second, the first step should have enough time to come to equilibrium. When that happens, the rate of the forward and reverse reactions for the first step are equal.

$$\text{Step 1:} \qquad \text{rate}_{forward} = \text{rate}_{reverse}$$
$$k_f(N_2O_5) = k_r(NO_2)(NO_3)$$

This equation can be rearranged as follows.

$$(NO_2)(NO_3) = \frac{k_f}{k_r}(N_2O_5)$$

Substituting this equation into the overall rate law for the reaction

$$-\frac{d(N_2O_5)}{dt} = k(NO_2)(NO_3)$$

gives the following result.

$$-\frac{d(N_2O_5)}{dt} = k\left(\frac{k_f}{k_r}\right)(N_2O_5)$$

Since k, k_f, and k_r are all constants, they can be replaced by a single constant, k', to give the experimental rate law for this reaction.

$$-\frac{d(N_2O_5)}{dt} = k'(N_2O_5)$$

This analysis of the decomposition of N_2O_5 is useful because it helps us understand the relationship between the experimental rate law for a reaction and the mechanism of the reaction. But it can be misleading. The usual goal is not to use a mechanism to explain the experimental rate law, but to use the experimental rate law to help determine the mechanism of the reaction.

Exercise 21.15

Propose mechanisms for the following reactions that are consistent with the experimentally determined rate law for each reaction.

$$CH_3Br + OH^- \longrightarrow CH_3OH + Br^- \qquad \text{rate} = k(CH_3Br)(OH^-)$$

$$(CH_3)_3CBr + OH^- \longrightarrow (CH_3)_3COH + Br^- \qquad \text{rate} = k((CH_3)_3CBr)$$

Solution

The rate law for the first reaction is consistent with a single-step, bimolecular reaction in which OH^- ions collide with CH_3Br to form a C—OH bond and displace a Br^- ion.

$$HO^- + CH_3—Br \longrightarrow CH_3—OH + Br^-$$

The second reaction must occur by a different mechanism, because the rate of this reaction is independent of the OH^- ion concentration. The rate law for this reaction is consistent with a mechanism that involves two or more steps, of which the slowest, or rate-limiting, step, is a unimolecular reaction that involves only $(CH_3)_3CBr$. One of the products of this reaction is the Br^- ion. The rate-limiting step might therefore involve dissociation of $(CH_3)_3CBr$ into $(CH_3)_3C^+$ and Br^- ions. The $(CH_3)_3C^+$ ion might then rapidly combine with OH^- ions to form the other product of the reaction, $(CH_3)_3COH$.

$$(CH_3)_3CBr \longrightarrow (CH_3)_3C^+ + Br^- \qquad \text{slow step}$$

$$(CH_3)_3C^+ + OH^- \longrightarrow (CH_3)_3COH \qquad \text{fast step}$$

21.18 CATALYSTS AND THE RATES OF CHEMICAL REACTIONS

Aqueous solutions of hydrogen peroxide are stable — until we add a few drops of blood, a freshly cut slice of turnip or horseradish, or a small quantity of the I^- ion, at which point the hydrogen peroxide rapidly decomposes (see Figure 21.14).

$$2 H_2O_2(aq) \longrightarrow 2 H_2O(l) + O_2(g)$$

This demonstration provides an excellent example of a catalyst, which was defined in Section 21.2 as a substance that alters the rate of a chemical reaction without being consumed in the reaction.

Four criteria must be satisfied in order for something to be classified as a catalyst.

1. Catalysts increase the rate of reaction.
2. Catalysts are not consumed by the reaction.
3. A small quantity of catalyst should be able to affect the rate of reaction for a large amount of reactant.
4. Catalysts do not change the equilibrium constant for the reaction.

The first criterion provides the basis for defining a catalyst as something that increases the rate of a chemical reaction. The second reflects the fact that anything consumed in the reaction is a reactant, not a catalyst. The third criterion is a consequence of the second — because catalysts are not consumed in the reaction, they can catalyze the reaction over and over again. The fourth criterion results

FIG. 21.14 A 30% aqueous solution of hydrogen peroxide, H_2O_2, is stable by itself. But when a drop of blood, a piece of platinum metal, an $I^-(aq)$ solution, or a slice of freshly cut turnip or horseradish that contains the enzyme catalase is added to the solution, it rapidly decomposes.

FIG. 21.15 A catalyst increases the rate of a chemical reaction by providing an alternative mechanism that has a smaller activation energy.

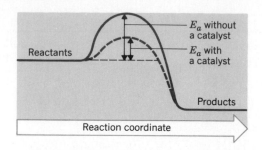

TABLE 21.4

Activation Energies for the Decomposition of Hydrogen Peroxide

Catalyst	Activation Energy (kJ/mol)	Relative Rate of Reaction
none	75.3	1
I^-	56.5	2.0×10^3
Pt	49.0	4.1×10^4
catalase	8	6.3×10^{11}

from the fact that catalysts speed up the rates of the forward and reverse reactions equally, so the equilibrium constant for the reaction remains the same.

Catalysts increase the rates of chemical reactions by providing a new mechanism that has a smaller activation energy, as shown in Figure 21.15. A larger proportion of the collisions that occur between reactants now have enough energy to overcome the activation energy for the reaction. As a result, the rate of reaction increases. The effect of several catalysts on the activation energy for the decomposition of aqueous solutions of hydrogen peroxide is summarized in Table 21.4.

To illustrate how a catalyst can decrease the activation energy for a reaction by providing another pathway for the reaction, let's look at the mechanism for the decomposition of hydrogen peroxide catalyzed by the I^- ion. In the presence of this ion, the decomposition of H_2O_2 doesn't have to occur in a single step. It can occur in two steps, both of which are easier and therefore faster. In the first step, the I^- ion is oxidized by H_2O_2 to OI^-.

$$H_2O_2(aq) + I^-(aq) \longrightarrow H_2O(l) + OI^-(aq)$$

In the second step, the OI^- ion is reduced to I^- by H_2O_2.

$$OI^-(aq) + H_2O_2(aq) \longrightarrow H_2O(l) + O_2(g) + I^-(aq)$$

There is no net change in the concentration of the I^- ion as a result of these reactions, and the I^- ion therefore satisfies the criteria for a catalyst.

THE RELATIONSHIP BETWEEN THE RATE CONSTANTS
21.19 AND THE EQUILIBRIUM CONSTANT FOR A REACTION

There is a simple relationship between the equilibrium constant for a reversible reaction and the rate constants for the forward and reverse reactions, if the mechanism for the reaction involves only a single step. To understand this relationship, let's turn once more to a reversible reaction that we know occurs by a simple, one-step mechanism.

$$ClNO_2(g) + NO(g) \longrightarrow NO_2(g) + ClNO(g)$$

The rate of the forward reaction is equal to a rate constant for this reaction, k_f, times the concentrations of the reactants, $ClNO_2$ and NO.

$$Rate_{forward} = k_f(ClNO_2)(NO)$$

The rate of the reverse reaction is equal to a second rate constant, k_r, times the concentrations of the products, NO_2 and ClNO.

$$Rate_{reverse} = k_r(NO_2)(ClNO)$$

This system will reach equilibrium when the rate of the forward reaction is equal to the rate of the reverse reaction.

$$Rate_{forward} = rate_{reverse}$$

Substituting the rate laws for the forward and reverse reactions at equilibrium into this equation gives the following result.

$$k_f[NO][NO_2Cl] = k_r[NOCl][NO_2]$$

This equation can be rearranged to give the equilibrium constant expression for the reaction.

$$\frac{k_f}{k_r} = \frac{[NOCl][NO_2]}{[NO][NO_2Cl]} = K_c$$

Thus, the equilibrium constant for a one-step reaction is equal to the forward rate constant divided by the reverse rate constant.

$$K_c = \frac{k_f}{k_r}$$

21.20 DETERMINING THE ACTIVATION ENERGY OF A REACTION

The collision theory model of chemical kinetics predicts that the rate of a chemical reaction should be proportional to the temperature at which the reaction is run. As the temperature is increased, the molecules move faster and therefore collide more often. They also carry more kinetic energy. Thus, the proportion of collisions that carry enough energy to overcome the activation energy for the reaction increases with temperature.

How do we bridge the gap between this prediction of what should happen on the molecular scale and the experimental rate law for chemical reactions on the macroscopic scale? Let's return, one more time, to the rate law for the decomposition of N_2O_5.

$$-\frac{d(N_2O_5)}{dt} = k(N_2O_5)$$

We know that the rate of the reaction increases with temperature. The only way to explain this observation is to assume that the rate constant for a reaction depends on the temperature at which the reaction is run.

In 1889, Svante Arrhenius derived the following equation for the dependence of the rate constant on the temperature of the reaction.

$$k = Ze^{-E_a/RT}$$

In this equation, k is the rate constant for the reaction, Z is a proportionality

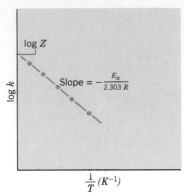

FIG. 21.16 The Arrhenius equation can be used to determine the activation energy for a reaction because a plot of the log of the rate constant for the reaction at different temperatures versus the inverse of the temperature in kelvin is a straight line with a slope equal to $-E_a/2.303\ R$.

constant characteristic of the reaction, E_a is the activation energy for the reaction, R is the ideal gas constant in joules per mole kelvin, and T is the temperature in kelvin.

The Arrhenius equation can be used to determine the activation energy for a reaction. We start by taking the logarithm to the base e of both sides of the equation.

$$\ln k = \ln (Ze^{-E_a/RT})$$
$$= \ln Z + \ln e^{-E_a/RT}$$
$$= \ln Z - E_a/RT$$

For the sake of convenience, we then introduce the following approximation.

$$\ln X \cong 2.303 \log_{10} X$$

This gives the following form of the Arrhenius equation.

$$\log k = \log Z - \frac{E_a}{2.303\ RT}$$

According to this equation, a plot of $\log k$ versus $1/T$ should give a straight line with a slope of $-E_a/2.303\ R$, as shown in Figure 21.16.

Exercise 21.16

Determine the activation energy for the decomposition of HI from the following data for the dependence of the rate constant on temperature.

Temperature (°C)	Rate Constant (M/s)
300	2.91×10^{-6}
400	8.38×10^{-4}
500	7.65×10^{-2}

Solution

We can determine the activation energy of any reaction from a plot of the logarithm of the rate constants versus the reciprocal of the absolute temperature. We therefore start by converting these temperatures to the kelvin scale and then calculating $1/T$. We then calculate the log of the rate constants and construct a graph of these data similar to the one shown in Figure 21.16.

When this is done, we get a straight line with a slope of -9875 K. The slope of this line is equal to $-E_a/2.303\ R$.

$$-9875\ \text{K} = -\frac{E_a}{2.303(8.314\ \text{J/mol-K})}$$

Solving for E_a gives the activation energy for this reaction.

$$E_a = \textbf{189 kJ/mol}$$

The Arrhenius equation can also be used to predict what will happen to the rate constant for the reaction when the temperature of the reaction changes. By paying strict attention to the mathematics of logarithms, it is possible to show that the Arrhenius equation

$$k = Ze^{-E_a/RT}$$

can be used to derive the following relationship.

$$\log\left(\frac{k_1}{k_2}\right) = \frac{E_a}{2.303\ R}\left(\frac{1}{T_2} - \frac{1}{T_1}\right)$$

Exercise 21.17

Calculate the rate of decomposition of HI at 600°C.

Solution

We start with the following form of the Arrhenius equation.

$$\log\left(\frac{k_1}{k_2}\right) = \frac{E_a}{2.303\ R}\left(\frac{1}{T_2} - \frac{1}{T_1}\right)$$

We then pick any one of the three data points used in the preceding exercise as T_1 and allow the value of T_2 to be 600°C.

$$T_1 = 300°C \qquad k_1 = 2.91 \times 10^{-6}\ M/s$$
$$T_2 = 600°C \qquad k_2 = ?$$

We then convert the temperatures to kelvin.

$$T_1 = 300°C + 273 = 573\ K$$
$$T_2 = 600°C + 273 = 873\ K$$

Substituting what we know about the system into the equation given above gives the following result.

$$\log\left[\frac{(2.91 \times 10^{-6})}{k_2}\right] = \frac{189,000\ J/mol}{2.303 \times 8.314\ J/mol\text{-}K}\left(\frac{1}{873\ K} - \frac{1}{573\ K}\right)$$

We can simplify the right-hand side of this equation as follows.

$$\log\left[\frac{(2.91 \times 10^{-6})}{k_2}\right] = -5.920$$

We then take the antilog of both sides of the equation.

$$\frac{(2.91 \times 10^{-6})}{k_2} = 10^{-5.920} = 1.20 \times 10^{-6}$$

Solving for k_2 gives the rate constant for this reaction at 600°C.

$$k_2 = 2.4\ M/s$$

According to this calculation, increasing the temperature of the reaction from 300°C to 600°C increases the rate constant for the reaction by a factor of almost a million. If all other factors were held constant, the rate of the reaction would increase by the same amount.

The Arrhenius equation can also be used to calculate what happens to the rate of a reaction when a catalyst lowers the activation energy.

Exercise 21.18

Calculate the relative values of the rate constants for the decomposition of H_2O_2 at 25°C when this reaction is run in the absence of a catalyst ($E_a = 75.3$ kJ/mol) and when the reaction is catalyzed by I^- ions ($E_a = 56.5$ kJ/mol).

Solution

We can start by substituting what we know about the reaction in the absence of a catalyst into the Arrhenius equation.

$$k = Ze^{-E_a/RT}$$
$$= Ze^{(-75,300\,J/mol)/(8.314\,J/mol\text{-}K)(298\,K)}$$
$$= Ze^{-30.4}$$

We can then repeat this calculation for the reaction in the presence of I^- ions.

$$k_{I^-} = Ze^{-E_a/RT}$$
$$= Ze^{(-56,500\,J/mol)/(8.314\,J/mol\text{-}K)(298\,K)}$$
$$= Ze^{-22.8}$$

The relative rate constants can be calculated by comparing these two equations.

$$\frac{k_{I^-}}{k} = \frac{Ze^{-22.8}}{Ze^{-30.4}} = \frac{2000}{1}$$

If all other factors are held constant, the relative rates of the reaction will differ by the same amount.

21.21 THE KINETICS OF ENZYME-CATALYZED REACTIONS

Enzymes are proteins that catalyze chemical reactions in living systems. In 1913, Lenor Michaelis and his student M. L. Menton studied the rate at which an enzyme isolated from yeast catalyzed the hydrolysis of sucrose, or cane sugar, into fructose and glucose.

$$\text{Sucrose} + H_2O \longrightarrow \text{fructose} + \text{glucose}$$

At first glance, this reaction seems similar to the reaction between sucrose and acid, which had been studied by Ludwig Wilhemy 60 years earlier, as noted in Section 21.2.

$$\text{Sucrose} + H_3O^+ \longrightarrow \text{fructose} + \text{glucose}$$

Wilhemy found that the reaction between sucrose and acid was first-order in sucrose at constant pH.

$$\text{Rate} = k(\text{sucrose})$$

Michaelis and Menton, however, found that the initial rate of the enzyme-catalyzed reaction was first-order in sucrose only at low concentrations of sucrose. At high concentrations, the initial rate of reaction was independent of the sucrose concentration. In other words, at high sucrose concentrations, the rate of reaction became constant. A reaction that does not depend on the concentration of one of the reactants is said to be *zero-order* in that reactant.

The difference between what Michaelis and Menton found and what was expected for this reaction is shown in Figure 21.17. There is a maximum initial rate of reaction, rate$_{max}$, for the enzyme-catalyzed reaction that can't be exceeded no matter how much sucrose we add to the solution. Adding more enzyme to the solution, however, does increase rate$_{max}$.

Michaelis and Menton suggested that this behavior could be explained by assuming that the reaction proceeds through a two-step mechanism. In the first step, the enzyme combines with sucrose to form a complex, *ES*.

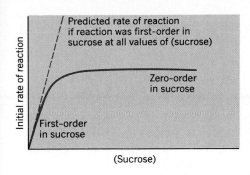

FIG. 21.17 If the reaction in which an enzyme hydrolyzes sucrose to fructose and glucose was first-order in the concentration of sucrose, the initial rate of reaction should be directly proportional to the initial concentration of sucrose. Michaelis and Menton, however, showed that this reaction is first-order in sucrose at low concentrations of sucrose but zero-order in sucrose at high concentrations.

$$E + S \underset{k_2}{\overset{k_1}{\rightleftharpoons}} ES$$

The enzyme then converts sucrose into the products of the reaction, which are released in a second step that regenerates the enzyme.

$$ES \underset{k_4}{\overset{k_3}{\rightleftharpoons}} E + P$$

If the first reaction is faster than the second, the overall rate of this reaction is more or less equal to the rate of the second step, which is proportional to the concentration of the enzyme-sucrose complex.

$$\text{Rate} = k_3(ES)$$

The maximum rate of reaction will be seen when essentially all of the enzyme is tied up as the *ES* complex.

$$\text{Rate}_{max} = k_3(E_t)$$

In this equation, E_t is the sum of the concentrations of the free enzyme, E, and the enzyme-sucrose complex, ES.

There is a limit to the rate at which the enzyme can consume sucrose. When there is a large excess of sucrose in this solution, every time the enzyme operates on a molecule of sucrose, it will immediately pick up another. No matter how much more sucrose is added to the solution, the reaction can't occur any faster. The reaction no longer depends on the sucrose concentration and is therefore zero-order in sucrose.

$$\text{Rate} = k(\text{sucrose})^0 = k$$

SUMMARY

Chemical kinetics involves the study of the rates of chemical reactions and the factors that influence these rates. It is complicated by the fact that the rate of each individual step in a reaction is proportional to the concentrations of the reactants consumed in that step. As a result, the rates of chemical reactions aren't constant; they change with time.

We determine the rate of a chemical reaction by measuring the change in the concentration of one of the reactants — $\Delta(X)$ — that occurs in a given period of time — Δt. Because the rate of reaction changes as it is being measured, the best data are obtained when an infinitesimally small change in concentration — $d(X)$ — is measured over an infinitesimally small length of time — dt. This measurement is known as the instantaneous rate of reaction — $d(X)/dt$. Because the rate of reaction changes with time, we often extrapolate these data to the initial instantaneous rate of reaction.

The rate of a chemical reaction, or of any individual step in a reaction, can be described by a rate law. For one-step reactions, the rate law is consistent with the stoichiometry of the reaction. For reactions that occur in more than one step, however, the rate laws may differ significantly from what we would expect on the basis of the stoichiometry of the reaction. As a result, the rate law for a reaction must be determined experimentally.

The individual steps in a chemical reaction can be classified in terms of the *molecularity*. Steps may be unimolecular, bimolecular, or, occasionally, trimolecular. The rate law for the reaction is classified in terms of the order of the reaction. The molecularity of the steps in a reaction can be deduced from the mechanism proposed for the reaction; the order of the reaction is determined experimentally.

We can determine the order of a reaction by studying how the initial instantaneous rate of reaction changes when the initial concentrations of the reactants are varied. We can also deduce it by comparing the kinetic data with the predictions of the integrated forms of the rate laws.

$$\log\left[\frac{(X)}{(X)_0}\right] = -\frac{kt}{2.303}$$ integrated form of the first-order rate law

$$\frac{1}{(X)} - \frac{1}{(X)_0} = kt$$ integrated form of the second-order rate law

The collision theory model of chemical kinetics suggests that the rate of a chemical reaction depends on the activation energy for the reaction. As the temperature of the reaction increases, more of the collisions between reactant molecules carry enough kinetic energy to cross the activation energy barrier for the reaction. Thus, the rate of reaction increases with temperature. The Arrhenius equation describes the relationship between the rate constant for the reaction at a given temperature and the activation energy for the reaction.

$$\log k = \log Z - \frac{E_a}{2.303\,RT}$$

Another way of increasing the rate of a chemical reaction is to add a catalyst, which decreases the activation energy by providing an alternative mechanism for the reaction.

PROBLEMS

Chemical Kinetics Defined

21-1 Define the terms *thermodynamic control* and *kinetic control* and give an example of each.

21-2 Describe how the terms *rate of reaction, rate law,* and *rate constant* differ. Give an example of each.

21-3 Describe the difference between the *rate of reaction* measured over a finite period of time and the *instantaneous rate of reaction.* What is the advantage of determining the instantaneous rate of reaction?

21-4 Explain why the rate of each step in a chemical reaction is proportional to the concentrations of the reactants consumed in that step.

21-5 Explain why the rates of many chemical reactions increase as the temperature increases.

21-6 Explain why the rate of a chemical reaction changes during the reaction.

21-7 Which of the following graphs best describes the relationship between the rate of a chemical reaction and the temperature of the reaction?

21-8 Describe what happens to the rate at which a reactant is consumed during the course of a reaction. Describe what happens to the rate at which the products are produced.

21-9 Explain why mixtures of H_2 and O_2 gas do not react when stored at room temperature for several years, whereas the reaction is complete within a few days at 300°C, within a few hours at 500°C, and almost instantaneously at 700°C.

Rate Laws and Rate Constants

21-10 What are the units of the rate constant for the following reaction

$$N_2O_4(g) \longrightarrow 2\,NO_2(g)$$

if the following is the rate law for this reaction?

$$-d(N_2O_4)/dt = k(N_2O_4)$$

21-11 What are the units of the rate constant for the following reaction

$$2\,NO_2(g) \longrightarrow N_2O_4(g)$$

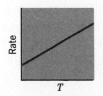

if the following is the rate law for this reaction?

$$-d(NO_2)/dt = k(NO_2)^2$$

21-12 What are the units of the rate constant for the following reaction

$$2 Br^-(aq) + H_2O_2(aq) + 2 H^+(aq) \longrightarrow Br_2(aq) + 2 H_2O(aq)$$

if the following is the rate law for this reaction?

$$-d(H_2O_2)/dt = k(Br^-)(H_2O_2)(H^+)?$$

21-13 The rate of a reaction is said to be diffusion-controlled when the reaction occurs as fast as the reactants diffuse through the solution to collide with each other. The following is a good example of a diffusion-controlled reaction.

$$H_3O^+(aq) + OH^-(aq) \longrightarrow 2 H_2O(aq)$$

Assuming that the rate constant for this reaction is 1.4×10^{11} L/mol-s at $25\,^{\circ}C$ and that the reaction obeys the following rate law, calculate the rate of reaction in a neutral solution (pH = 7.00).

$$Rate = k(H_3O^+)(OH^-)$$

The Rate Law versus the Stoichiometry of a Reaction

21-14 Describe the difference between the stoichiometry and the mechanism of a reaction. Give an example.

21-15 Describe the conditions under which the rate law for the reaction is most likely to reflect the stoichiometry of the reaction.

21-16 Describe one or more factors that can make the rate law for the reaction differ from what the stoichiometry of the reaction would lead us to expect.

Order and Molecularity

21-17 Describe the difference between unimolecular and bimolecular reactions. Explain how the terms *unimolecular* and *bimolecular* differ in meaning from the terms *first-order* and *second-order*.

21-18 Which of the following graphs best describes the rate at which N_2O_4 decomposes to NO_2 if the following reaction is first-order in N_2O_4?

$$N_2O_4(g) \longrightarrow 2 NO_2(g)$$

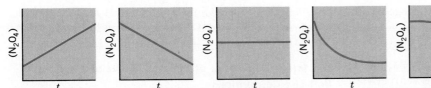

Determining the Order of a Reaction from Initial Rates of Reaction

21-19 Nitrogen oxide reacts with chlorine to form nitrosyl chloride.

$$2 NO(g) + Cl_2(g) \longrightarrow 2 NOCl(g)$$

Determine the rate law for this reaction from the following data on initial instantaneous rates of reaction.

Initial (NO) (M)	Initial (Cl_2) (M)	Initial Rate of Reaction (M/s)
0.10	0.10	0.117
0.20	0.10	0.468
0.30	0.10	1.054
0.30	0.20	2.107
0.30	0.30	3.161

21-20 Use the results of the preceding problem to determine the rate constant for this reaction. Predict the initial instantaneous rate of reaction when the initial NO and Cl_2 concentrations are both 0.50 M.

21-21 Nitrogen oxide reacts with oxygen to form nitrogen dioxide.

$$2 NO(g) + O_2(g) \longrightarrow 2 NO_2(g)$$

Determine the rate law for this reaction from the following data on initial instantaneous rates of reaction.

Initial Pressure (NO) (mmHg)	Initial Pressure O_2 (mmHg)	Initial Rate of Reaction (mmHg/s)
100	100	0.355
150	100	0.800
250	100	2.22
150	130	1.04
150	180	1.44

21-22 Use the results of the preceding problem to determine the rate constant for this reaction. Predict the initial rate of reaction when the initial pressures for both NO and O_2 are 250 mmHg.

21-23 Methyl iodide, CH_3I, reacts with the OH^- ion in aqueous solution to form methanol, CH_3OH, and the iodide ion.

$$CH_3I(aq) + OH^-(aq) \longrightarrow CH_3OH(aq) + I^-(aq)$$

Determine the rate law for this reaction from the following data on initial instantaneous rates of reaction.

Initial (CH_3I) (M)	Initial (OH^-) (M)	Initial Rate of Reaction (M/s)
1.35	0.10	8.78×10^{-6}
1.00	0.10	6.50×10^{-6}
0.85	0.10	5.53×10^{-6}
0.85	0.15	8.29×10^{-6}
0.85	0.25	1.38×10^{-5}

21-24 Use the results of the preceding question to determine the rate constant for this reaction. Predict the initial instantaneous rate of reaction when the initial CH_3I concentration is 0.10 M and the initial OH^- concentration is 0.050 M.

Different Ways of Expressing the Rate of Reaction

21-25 Which reactant or product in the following reaction changes most in concentration over time?

$$6\ Fe^{2+}(aq) + Cr_2O_7{}^{2-}(aq) + 14\ H^+(aq) \longrightarrow$$
$$6\ Fe^{3+}(aq) + 2\ Cr^{3+}(aq) + 7\ H_2O(aq)$$

21-26 Which equation describes the relationship between the rates at which Cl_2 and F_2 are consumed in the following reaction?

$$Cl_2(g) + 3\ F_2(g) \longrightarrow 2\ ClF_3(g)$$

(a) $-d(Cl_2)/dt = -d(F_2)/dt$
(b) $-d(Cl_2)/dt = 3[-d(F_2)/dt]$
(c) $3[-d(Cl_2)/dt] = -d(F_2)/dt$

21-27 Which equation describes the relationship between the rates at which Cl_2 is consumed and ClF_3 is produced in the following reaction.

$$Cl_2(g) + 3\ F_2(g) \longrightarrow 2\ ClF_3(g)$$

(a) $-d(Cl_2)/dt = -d(ClF_3)/dt$
(b) $-d(Cl_2)/dt = d(ClF_3)/dt$
(c) $-d(Cl_2)/dt = 2[-d(ClF_3)/dt]$
(d) $2[-d(Cl_2)/dt] = -d(ClF_3)/dt$
(e) $-d(Cl_2)/dt = 2[d(ClF_3)/dt]$
(f) $2[-d(Cl_2)/dt] = d(ClF_3)/dt$

21-28 Assume that the instantaneous rate at which NO_2 is consumed in the following reaction at some moment in time is 0.0592 M/s. Calculate the instantaneous rate of formation of N_2O_4 at this moment in time.

$$2\ NO_2(g) \longrightarrow N_2O_4(g)$$

21-29 The rate constant for the following reaction is 0.039 $M^{-1}\,K^{-1}$ when the rate at which HI is consumed is measured at 500°C.

$$2\ HI(g) \longrightarrow H_2(g) + I_2(g)$$

What is the rate constant for the formation of I_2 if the following is the rate law for this reaction?

$$-d(HI)/dt = k(HI)^2$$

21-30 Ammonia burns in the gas phase to form nitrogen oxide and water.

$$4\ NH_3(g) + 5\ O_2(g) \longrightarrow 4\ NO(g) + 6\ H_2O(g)$$

Derive the relationship between the rates at which NH_3 and O_2 are consumed in this reaction. Derive the relationship between the rates at which O_2 is consumed and H_2O is produced.

21-31 Calculate the instantaneous rate of formation of NO and the instantaneous rate of disappearance of NH_3 at some moment in time for the following reaction if the instantaneous rate of formation of water is 0.040 M/s.

$$4\ NH_3(g) + 5\ O_2(g) \longrightarrow 4\ NO(g) + 6\ H_2O(g)$$

21-32 The instantaneous rate of disappearance of $MnO_4{}^-$ in the following reaction is 4.56×10^{-3} M/s at some moment in time.

$$10\ I^- + 2\ MnO_4{}^- + 16\ H^+ \longrightarrow 2\ Mn^{2+} + 5\ I_2 + 8\ H_2O$$

What is the rate of appearance of I_2 at the same moment?

The Integrated Forms of the Rate Laws

21-33 Describe the kinds of problems that are best solved by using the rate law for a chemical reaction, such as the following.

$$-d(N_2O_5)/dt = k(N_2O_5)$$

Describe the kinds of problems that are best solved with the integrated form of the rate law.

$$\log\left[\frac{(N_2O_5)}{(N_2O_5)_0}\right] = -\frac{kt}{2.303}$$

21-34 Water was heated in a test tube to 75.0°C and then allowed to cool to room temperature. If this process followed first-order kinetics with a rate constant of 8.0×10^{-4} s^{-1}, what was the temperature of the water after 400 seconds?

21-35 In an acceleration test for a BMW 325es, the following speed-versus-time data were collected after the car shifted into fourth gear.

Speed (mph):	80.4	83.9	87.5	91.2	95.1	99.3
Time (s):	16	18	20	22	24	26

Are these data consistent with first-order kinetics? Predict the time at which the speed should reach 100 mph.

21-36 Calculate the rate constant for the following reaction

$$2\ NO_2(g) \longrightarrow N_2O_4(g)$$

assuming that it takes 0.005 seconds for the initial concentration of NO_2 to decrease from 0.50 M to 0.25 M and that the rate law for this reaction is the following.

$$-d(N_2O_4) = k(NO_2)^2$$

21-37 For the decomposition of hydrogen peroxide

$$2\ H_2O_2(aq) \longrightarrow 2\ H_2O(aq) + O_2(g)$$

the reaction is first-order in H_2O_2.

$$-d(H_2O_2)/dt = k(H_2O_2)$$

How long would it take for half of the H_2O_2 in a 10-gallon sample to be consumed if the rate constant for this reaction is 5.6×10^{-2} s^{-1}?

The Integrated Forms of the Rate Laws and Half-Life Calculations

21-38 Calculate the rate constant for the following reaction

$$NH_4{}^+(aq) + H_2O(aq) \longrightarrow NH_3(aq) + H_3O^+(aq)$$

if the half-life for this reaction is 0.0282 seconds at 25°C and the rate law for the reaction is the following.

$$-d(NH_4{}^+)/dt = k(NH_4{}^+)$$

21-39 What would happen to the half-life for the following reaction if the initial concentration of NO_2 was doubled?

$$2 \, NO_2(g) \longrightarrow N_2O_4(g)$$

Assume this reaction is second-order in NO_2.

$$-d(NO_2) = k(NO_2)^2$$

21-40 The following reaction is first-order in both reactants and therefore second-order overall.

$$CH_3I(aq) + OH^-(aq) \longrightarrow CH_3OH(aq) + I^-(aq)$$

However, when the reaction is run in a buffer solution, in which the OH^- ion concentration is constant, it is pseudo–first-order in CH_3I.

$$-d(CH_3I)/dt = k(CH_3I)$$

What is the half-life of this reaction in a pH 10.00 buffer if the rate constant for this pseudo–first-order reaction is $6.5 \times 10^{-9} \, s^{-1}$?

21-41 The age of a rock can be estimated by measuring the amount of ^{40}Ar trapped inside. This calculation is based on the fact that ^{40}K decays to ^{40}Ar by a first-order rate process. (It also assumes that none of the ^{40}Ar produced by this reaction has escaped from the rock since the rock was formed.)

$$^{40}K + e^- \longrightarrow {}^{40}Ar \qquad k = 5.81 \times 10^{-11} \, yr^{-1}$$

Calculate the half-life of this radioactive decay.

21-42 Another way of determining the age of a rock involves measuring the extent to which the ^{87}Rb in the rock has decayed to ^{87}Sr.

$$^{87}Rb \longrightarrow {}^{87}Sr + e^- \qquad k = 1.42 \times 10^{-11} \, yr^{-1}$$

What fraction of the ^{87}Rb would still remain in a rock when half of the ^{40}K has decayed?

21-43 ^{14}C measurements on the linen wrappings from the Book of Isaiah in the Dead Sea scrolls suggest that the scrolls contain about 79.5% of the ^{14}C expected in living tissue. How old are these scrolls, if the half-life for the decay of ^{14}C is 5730 years?

21-44 The Lascaux cave near Montignac in France contains a series of cave paintings. Radiocarbon dating of charcoal taken from this site suggests an age of 15,520 years. What fraction of the ^{14}C present in living tissue is still present in this sample? (^{14}C: $t_{1/2} = 5730$ yr)

21-45 A skull fragment found in 1936 at Baldwin Hills, California, was dated by ^{14}C analysis. Approximately 100 grams of bone was cleaned and treated with $1 \, M \, HCl(aq)$ to destroy the mineral content of the bone. The bone protein was then collected, dried, and pyrolyzed. The CO_2 produced was collected and purified, and the ratio of ^{14}C to ^{12}C was measured. If this sample contained roughly 5.7% of the ^{14}C present in living tissue, how old was the skeleton? (^{14}C: $t_{1/2} = 5730$ yr)

21-46 Charcoal samples from Stonehenge in England emitted 62.3% of the disintegrations per gram of carbon per minute expected for living tissue. What is the age of this sample? (^{14}C: $t_{1/2} = 5730$ yr)

21-47 A lump of beeswax was excavated in England near a collection of Bronze Age objects roughly 2500 to 3000 years of age. Radiocarbon analysis of the beeswax suggests an activity equal to roughly 90.3% of the activity observed for living tissue. Was this beeswax part of the hoard of Bronze Age objects or not?

21-48 The activity of the ^{14}C in living tissue is 15.3 disintegrations per minute per gram of carbon. The limit for reliable determination of ^{14}C ages is 0.10 disintegration per minute per gram of carbon. Calculate the maximum age of a sample that can be dated accurately by radiocarbon dating. Assume a half-life of 5730 years.

Determining the Order of a Reaction with the Integrated Forms of Rate Laws

21-49 Determine the rate law for the following reaction

$$2 \, N_2O(g) \longrightarrow 2 \, N_2(g) + O_2(g)$$

from the following data, which show what happens to the concentration of N_2O as this compound decomposes into N_2 and O_2.

(N_2O) (M):	0.100	0.086	0.079	0.075	0.066	0.059	0.049
Time (s):	0	80	120	160	240	320	480

21-50 Use the results of the preceding problem to calculate the rate constant for this reaction. Predict the concentration of N_2O that would remain after 900 seconds.

21-51 Determine the rate law for the following reaction

$$BH_4^-(aq) + 4 \, H_2O(aq) \longrightarrow B(OH)_4^-(aq) + 4 \, H_2(g)$$

from the following experimental data.

(BH_4^-) (M):	0.100	0.088	0.077	0.068	0.060	0.052	0.046
Time (hr):	0	24	48	72	96	120	144

21-52 Use the results of the preceding question to calculate the half-life for the reaction.

21-53 Triphenylphosphine, PPh_3, reacts with nickel tetracarbonyl, $Ni(CO_4)$, to displace a molecule of carbon monoxide.

$$Ni(CO)_4 + PPh_3 \longrightarrow Ph_3PNi(CO)_3 + CO$$

The following data were obtained when this reaction was run at 25°C with a constant PPh_3 concentration.

$(Ni(CO)_4)$ (M):	10.0	8.6	5.8	4.4	3.3	2.5
Time (s):	0	40	80	120	160	200

Use these data to determine whether the reaction is first-order or second-order in $Ni(CO)_4$.

21-54 The rate of the reaction in the preceding problem does not depend on the concentration of PPh_3. Combine this fact with the results of the preceding problem to determine whether the rate law for this reaction is consistent with the following mechanism.

$$Ni(CO)_4 \longrightarrow Ni(CO)_3 + CO \qquad \text{(slow step)}$$
$$Ni(CO)_3 + PPh_3 \longrightarrow Ph_3PNi(CO)_3 \qquad \text{(fast step)}$$

21-55 $Cr(NH_3)_5Cl^{2+}$ reacts with the OH^- ion in aqueous solution to displace Cl^- from the complex ion.

$$Cr(NH_3)_5Cl^{2+}(aq) + OH^-(aq) \longrightarrow$$
$$Cr(NH_3)_5OH^{2+}(aq) + Cl^-(aq)$$

The following data were obtained when this reaction was run at 25°C in a buffer solution at constant pH.

$(Cr(NH_3)_5Cl^{2+})$ (M):	1.00	0.81	0.66	0.54	0.44	0.35
Time (min):	0	3	6	9	12	15

Use these data to determine whether this reaction is first-order or second-order in $Cr(NH_3)_5Cl^{2+}$.

21-56 The rate of the reaction in the preceding problem is proportional to the pH of the buffer solution in which the reaction is run. Each time the buffer is changed so that the OH^- ion concentration is doubled, the rate of reaction increases by a factor of 2. Combine this observation with the results of the calculation in the preceding problem to determine the rate law for this reaction.

21-57 Show that the rate law derived in the preceding problem is consistent with the following mechanism.

$$Cr(NH_3)_5Cl^{2+} + OH^- \longrightarrow$$
$$Cr(NH_3)_4(NH_2)(Cl)^+ + H_2O \quad \text{(slow step)}$$
$$Cr(NH_3)_4(NH_2)(Cl)^+ \longrightarrow Cr(NH_3)_4(NH_2)^{2+} + Cl^- \quad \text{(fast step)}$$
$$Cr(NH_3)_4(NH_2)^{2+} + H_2O \longrightarrow Cr(NH_3)_5(OH)^{2+} \quad \text{(fast step)}$$

21-58 Dimethyl ether, CH_3OCH_3, decomposes at high temperatures (500°C) as shown in the following equation.

$$CH_3OCH_3(g) \longrightarrow CH_4(g) + H_2(g) + CO(g)$$

The following data show how the pressure of CH_3OCH_3 changes with time. Use these data to determine the order of this reaction.

$P_{CH_3OCH_3}$ (mmHg):	312	280	253	229	156
Time (s):	0	390	777	1195	3155

Reactions That are First-Order in Two Reactants

21-59 The following reaction is first-order in both CH_3I and OH^-.

$$CH_3I(aq) + OH^-(aq) \longrightarrow CH_3OH(aq) + I^-(aq)$$

Describe how to turn this reaction into one that is pseudo-first-order in CH_3I.

A Collision Theory Model of Chemical Reactions

21-60 Describe the factors that determine whether a collision between two molecules will lead to a chemical reaction.

21-61 Describe the relationship between the rate of a chemical reaction and the activation energy for the reaction.

The Mechanisms of Chemical Reactions

21-62 Each of the following reactions was found experimentally to be second-order overall. Which of them is most likely to be an elementary reaction that occurs in a single step?

(a) $2 NO_2(g) + Cl_2(g) \rightarrow 2 NO_2Cl(g)$
(b) $2 Br^-(aq) + H_2O_2(aq) + H^+(aq) \rightarrow Br_2(aq) + 2 H_2O(aq)$
(c) $N_2O_3(g) \rightarrow NO(g) + NO_2(g)$
(d) $3 O_2(g) \rightarrow 2 O_3(g)$
(e) $2 NO(g) \rightarrow N_2(g) + O_2(g)$

21-63 The rate law for the reaction

$$2 NO(g) + O_2(g) \longrightarrow 2 NO_2(g)$$

is first-order in O_2, second-order in NO, and third-order overall.

$$\text{Rate} = k(NO)^2(O_2)$$

Show how this rate law is consistent with the following two-step mechanism for this reaction.

$$2 NO \rightleftharpoons N_2O_2 \quad \text{(fast step)}$$
$$N_2O_2 + O_2 \longrightarrow 2 NO_2 \quad \text{(slow step)}$$

21-64 NO reacts with H_2 according to the overall equation

$$2 NO(g) + 2 H_2(g) \longrightarrow N_2(g) + 2 H_2O(g)$$

and the following mechanism.

$$2 NO + H_2 \longrightarrow N_2 + H_2O_2 \quad \text{(slow step)}$$
$$H_2O_2 + H_2 \longrightarrow 2 H_2O \quad \text{(fast step)}$$

What is the experimental rate law for this reaction?

21-65 The following reaction

$$2 NO_2(g) + F_2(g) \longrightarrow 2 NO_2F(g)$$

has the following experimental rate law.

$$\text{Rate} = k(NO_2)(F_2)$$

This rate law is consistent with which of the following mechanisms?

(a) $2 NO_2 + F_2 \rightarrow 2 NO_2F$
(b) $NO_2 + F_2 \rightarrow NO_2F + F$ (fast step)
    $NO_2 + F \rightarrow NO_2F$ (slow step)
(c) $NO_2 + F_2 \rightarrow NO_2F + F$ (slow step)
    $NO_2 + F \rightarrow NO_2F$ (fast step)
(d) $F_2 \rightleftharpoons 2 F$ (slow step)
    $2 NO_2 + 2 F \rightarrow 2 NO_2F$ (fast step)

21-66 The following reaction

$$3 NO(g) \longrightarrow N_2O(g) + NO_2(g)$$

has the following experimental rate law.

$$\text{Rate} = k(NO)^3$$

Which of the following mechanisms provides the best explanation for this rate law?

(a) $NO + NO + NO \rightarrow N_2O + NO_2$ (one-step reaction)
(b) $2 NO \rightarrow N_2O_2$ (slow step)
    $N_2O_2 + NO \rightarrow N_2O + NO_2$ (fast step)
(c) $2 NO \rightleftharpoons N_2O_2$ (fast step)
    $N_2O_2 + NO \rightleftharpoons N_2O + NO_2$ (slow step)

21-67 Predict the rate law for the oxidation of the iodide ion by hypochlorite

$$I^-(aq) + OCl^-(aq) \longrightarrow Cl^-(aq) + OI^-(aq)$$

if the reaction proceeds by the following mechanism.

$$OCl^- + H_2O \rightleftharpoons HOCl + OH^- \qquad \text{(fast step)}$$
$$I^- + HOCl \longrightarrow HOI + Cl^- \qquad \text{(slow step)}$$
$$HOI + OH^- \longrightarrow OI^- + H_2O \qquad \text{(fast step)}$$

21-68 The following is the experimental rate law for the reaction between hydrogen and bromine to form hydrogen bromide, HBr.

$$\text{Rate} = k(H_2)(Br_2)^{1/2}$$

Explain how the following mechanism is consistent with this rate law.

$$Br_2 \rightleftharpoons 2\ Br \qquad \text{(fast step)}$$
$$Br + H_2 \longrightarrow HBr + H \qquad \text{(slow step)}$$
$$H + Br_2 \longrightarrow HBr + Br \qquad \text{(fast step)}$$
$$II + HBr \longrightarrow H_2 + Br_2 \qquad \text{(fast step)}$$

Catalysts and the Rate of Chemical Reactions

21-69 Describe the four properties of a catalyst. Give an example of a catalyzed reaction and show how the catalyst meets these criteria.

The Relationship Between the Rate Constants and the Equilibrium Constant for a Reaction

21-70 Describe the relationship between the forward and reverse rate constants and the equilibrium constant for a one-step reaction.

21-71 The following is a one-step reaction.

$$CH_3Cl(aq) + I^-(aq) \longrightarrow CH_3I(aq) + Cl\ (aq)$$

What is the equilibrium constant for this reaction if the rate constant for the forward reaction is 5.2×10^{-7} L/mol-K and the rate constant for the reverse reaction is 1.5×10^{-11} L/mol-K?

Determining the Activation Energy of a Reaction

21-72 Assume that a catalyst decreases the activation energy for the forward reaction from 120 kJ/mol to 90 kJ/mol. What will happen to the activation energy for the reverse reaction if

E_a for the reverse reaction is 185 kJ/mol in the absence of the catalyst?

21-73 The rate constant for the decomposition of N_2O_5 increases from 1.52×10^{-5} s^{-1} at 25°C to 3.83×10^{-3} s^{-1} at 45°C. Calculate the activation energy for this reaction.

21-74 Calculate the activation energy for the following reaction, if the rate constant for this reaction increases from 8.71×10^1 M^{-1} s^{-1} at 500 K to 1.53×10^3 M^{-1} s^{-1} at 650 K.

$$2\ NO_2(g) \longrightarrow 2\ NO(g) + O_2(g)$$

21-75 Calculate the activation energy for the decomposition of NO_2

$$2\ NO_2(g) \longrightarrow N_2(g) + 2\ O_2(g)$$

from the temperature dependence of the rate constant for this reaction.

Temperature (°C):	319	330	354	378	383
k (M^{-1} s^{-1}):	0.522	0.755	1.70	4.02	5.03

21-76 Calculate the rate constant at 780 K for the following reaction, assuming that the rate constant for the reaction is 3.5×10^{-7} M^{-1} s^{-1} at 550 K and the activation energy is 188 kJ/mol.

$$2\ HI(g) \longrightarrow H_2(g) + I_2(g)$$

21-77 Calculate the rate constant at 75°C for the following reaction, assuming that the rate constant for this reaction is 6.5×10^{-5} M^{-1} s^{-1} at 25°C and the activation energy is 92.9 kJ/mol.

$$CH_3I(aq) + OH^-(aq) \longrightarrow CH_3OH(aq) + I^-(aq)$$

The Kinetics of Enzyme-Catalyzed Reactions

21-78 Explain why the rate of the enzyme-catalyzed hydrolysis of sucrose is first-order in sucrose at low concentrations of this substance.

21-79 Explain how the fact that only a limited amount of a given enzyme is present in living tissue results in the fact that the rate of enzyme-catalyzed reactions becomes zero-order in substrate at very high concentrations of substrate.

TRANSITION METAL COMPLEXES

CHAPTER CONTENTS

Main-group Elements | Transition Metals | Main-group Elements

H																	H	He
Li	Be											B	C	N	O	F	Ne
Na	Mg												Al	Si	P	S	Cl	Ar
K	Ca	Sc	Ti	V	Cr	Mn	Fe	Co	Ni	Cu	Zn	Ga	Ge	As	Se	Br	Kr
Rb	Sr	Y	Zr	Nb	Mo	Tc	Ru	Rh	Pb	Ag	Cd	In	Sn	Sb	Te	I	Xe
Cs	Ba	La	Hf	Ta	W	Re	Os	Ir	Pt	Au	Hg	Tl	Pb	Bi	Po	At	Rn
Fr	Ra	Ac	104	105	106	107		109									

Lanthanides	Ce	Pr	Nd	Pm	Sm	Eu	Gd	Tb	Dy	Ho	Er	Tm	Yb	Lu
Actinides	Th	Pa	U	Np	Pu	Am	Cm	Bk	Cf	Es	Fm	Md	No	Lr

FIG. 22.1 The elements in the periodic table are divided into four categories: (1) main-group elements, (2) transition metals, (3) lanthanides, and (4) actinides.

22.1 TRANSITION METALS AND COORDINATION COMPLEXES

The elements in the periodic table are often divided into four classes, or categories: (1) main-group elements, (2) transition metals, (3) lanthanides, and (4) actinides (see Figure 22.1). The main-group elements include the active metals in the two columns on the extreme left of the periodic table and the metals, semimetals, and nonmetals in the six columns on the far right. The transition metals are metallic elements that serve as a bridge, or transition, between the two sides of the table. The lanthanides and actinides are also called the inner transition metals, because they belong between the first and second elements in the last two rows of transition metals.

There are several ways of dividing the elements between the categories of main-group elements and transition metals (see Figure 22.2). One approach is based on the pattern of electron configurations in the periodic table. This approach notes that s orbitals are typically filled among the two columns of main-group elements on the left side of the table, p orbitals are filled in the six columns on the right side of the table, and d orbitals are filled among the transition metals. This system classifies zinc (Zn), cadmium (Cd), and mercury (Hg) as transition metals, because they are found among the elements where the d orbitals are filled.

Another approach classifies the elements in terms of their chemical properties. This system notes that zinc, cadmium, and mercury have filled subshells of d orbitals. The Zn^{2+}, Cd^{2+}, and Hg^{2+} ions therefore have more in common with main-group ions, such as the Ga^{3+}, Sn^{4+}, and Pb^{4+} ions, than with transition-metal ions, such as Fe^{3+} and Cr^{3+}.

$$Zn^{2+} = [Ar]\, 3d^{10} \qquad Ga^{3+} = [Ar]\, 3d^{10}$$
$$Cd^{2+} = [Kr]\, 4d^{10} \qquad Sn^{4+} = [Kr]\, 4d^{10}$$
$$Hg^{2+} = [Xe]\, 5d^{10}\, 4f^{14} \qquad Pb^{4+} = [Xe]\, 5d^{10}\, 4f^{14}$$

This convention classifies zinc, cadmium, and mercury as main-group elements instead of transition metals.

Which system is better? The answer isn't obvious. This text uses the first approach — it includes zinc, cadmium, and mercury among the transition metals. The basis for this decision is simple. The noble gases (He, Ne, Ar, Kr, Xe, and Rn) are classified as main-group elements even though they have filled s and p subshells.

$$Kr = [Ar]\, 4s^2\, 3d^{10}\, 4p^6$$

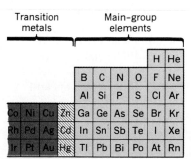

FIG. 22.2 There is some dispute about whether zinc, cadmium, and mercury should be classified as transition metals or main-group metals. We will classify them as transition metals, because they belong in the part of the periodic table where the d subshell is filled.

FIG. 22.3 The transition metals have many of the same physical properties as the main-group metals, including the tendency to look like metals. These photographs show samples of Zinc, Cobalt, Platinum, and Rhodium.

Lutetium (Lu) and lawrencium (Lr) are usually listed among the lanthanides and actinides, even though they have filled subshells of f orbitals.

$$Lu = [Xe]\ 6s^2\ 5d^1\ 4f^{14}$$
$$Lr = [Rn]\ 7s^2\ 6d^1\ 5f^{14}$$

The authors therefore include zinc, cadmium, and mercury among the transition metals, even though they have filled d subshells.

$$Zn = [Ar]\ 4s^2\ 3d^{10}$$
$$Cd = [Kr]\ 5s^2\ 4d^{10}$$
$$Hg = [Xe]\ 6s^2\ 4f^{14}\ 5d^{10}$$

The fact that chemists disagree about whether zinc, cadmium, and mercury should be classified as main-group elements or transition metals suggests that the differences between these classes of elements are not always clear. Transition metals are like main-group metals in many ways. They all look like metals (see Figure 22.3); they are malleable and ductile; they conduct heat and electricity; and they all form positive ions. You can appreciate the extent to which the physical properties of main-group metals and transition metals overlap by noting that the two best conductors of electricity are a transition metal (copper) and a main-group metal (aluminum).

There are also ways in which these metals differ. The transition metals tend to be more electronegative than the main-group metals. They are therefore more likely to form covalent compounds.

Exercise 22.1

Use electronegativities to classify the following compounds as either ionic, polar, or covalent.

(a) NaCl   (b) $CrCl_3$ (c) $HgCl_2$

Solution

(a) Sodium chloride is classified as an ionic compound because of the large difference between the electronegativities of the two elements.

$$NaCl: \quad Cl \quad EN = 3.16$$
$$Na \quad \underline{EN = 0.93}$$
$$\Delta EN = 2.23$$

The ionic character of this compound is responsible for its high melting point (801°C) and boiling point (1465°C).

(b) The bonds in chromium(III) chloride are best described as polar.

$$CrCl_3: \quad Cl \quad EN = 3.16$$
$$Cr \quad \underline{EN = 1.66}$$
$$\Delta EN = 1.50$$

The fact that this compound is less ionic than NaCl explains why it forms a gas at much lower temperatures (947°C).

(c) Although $HgCl_2$ dissociates to form Hg^{2+} and Cl^- ions when it dissolves in water, the compound by itself is best described as covalent.

$$HgCl_2: \quad \begin{array}{ll} Cl & EN = 3.16 \\ Hg & \underline{EN = 2.00} \\ & \Delta EN = 1.16 \end{array}$$

The covalent nature of this compound explains its relatively low melting point (277°C) and boiling point (304°C).

There are other differences between main-group metals and transition metals. The main-group metals form salts, such as NaCl, Mg_3N_2, and CaS, in which there are just enough negative ions to balance the charge on the positive ions. The transition metals form similar compounds, such as $FeCl_3$, HgI_2, or $Cd(OH)_2$, in which there are just enough negative ions to balance the charge on the metal ions. But transition metals are more likely than main-group metals to form complexes, such as the $FeCl_4^-$, HgI_4^{2-}, and $Cd(OH)_4^{2-}$ ions, that have an excess number of negative ions.

Another difference between main-group and transition metal ions is the ease with which they form stable compounds with neutral molecules, such as water and ammonia. Salts of main-group metal ions dissolve in water to form aqueous solutions.

$$NaCl(s) \xrightarrow{H_2O} Na^+(aq) + Cl^-(aq)$$

When we let the water evaporate, however, we get back the original starting material — NaCl(s). Salts of transition-metal ions can display very different behavior. Chromium(III) chloride is a violet compound, which dissolves in liquid ammonia to form a yellow compound with the formula $CrCl_3 \cdot 6 NH_3$ that can be isolated when the excess ammonia is allowed to evaporate (see Figure 22.4).

$$CrCl_3(s) + 6 NH_3(l) \longrightarrow CrCl_3 \cdot 6 NH_3(s)$$

Compounds such as the $FeCl_4^-$ ion and $CrCl_3 \cdot 6 NH_3$ have been given many names. Some call them *coordination compounds,* because they contain ions or molecules linked, or coordinated, to a transition metal. Others call them *complex ions* or *coordination complexes* (the terms we used in Chapter 17), because they are Lewis acid-base complexes. The ions or molecules that bind to transition metal ions to form these complexes are called ***ligands*** (from Latin, "to tie or bind"). The number of ligands bound to the transition-metal ion is called the ***coordination number.***

Although complexes are particularly important in the chemistry of the transition metals, it should be noted that some main-group elements also form complexes. Aluminum, tin, and lead, for example, form complexes such as the AlF_6^{3-}, $SnCl_4^{2-}$, and PbI_4^{2-} ions.

FIG. 22.4 One way transition metals differ from main-group metals is their ability to bind neutral molecules to form compounds with different chemical and physical properties. $CrCl_3$, shown on the left, dissolves in liquid ammonia to form the compound shown on the right, which has the formula $CrCl_3 \cdot 6 NH_3$.

22.2 WERNER'S MODEL OF COORDINATION COMPLEXES

In 1893, Alfred Werner proposed a model for coordination complexes that still serves as the basis for work in this field. Werner first became interested in coordination compounds while preparing lectures for a course on atomic theory in the summer of 1892. He announced his model for coordination compounds before the end of the same year and spent the remainder of his life collecting evidence to support his theory.

Before we introduce Werner's model of coordination complexes, it might be useful to look at some of the evidence that was available at the time.

1. At least three different cobalt(III) complexes can be isolated when $CoCl_2$ is dissolved in aqueous ammonia and then oxidized by air to the $+3$ oxidation state. A fourth complex can be made by slightly different techniques. These four complexes have the following empirical formulas.

$CoCl_3 \cdot 6\ NH_3$	orange-yellow
$CoCl_3 \cdot 5\ NH_3 \cdot H_2O$	red
$CoCl_3 \cdot 5\ NH_3$	purple
$CoCl_3 \cdot 4\ NH_3$	green

2. These complexes all have one thing in common — the reactivity of the ammonia has been drastically reduced. By itself, ammonia reacts rapidly with hydrochloric acid to form ammonium chloride.

$$NH_3(aq) + HCl(aq) \longrightarrow NH_4^+(aq) + Cl^-(aq)$$

But these complexes don't react with hydrochloric acid, even when the reaction is run at $100°C$.

$$CoCl_3 \cdot 6\ NH_3(aq) + HCl(aq) \xrightarrow{\ 100°C\ }\!\!\!\!\!/\!\!\!/$$

3. All three of the chloride ions in $CoCl_3 \cdot 6\ NH_3$ and $CoCl_3 \cdot 5\ NH_3 \cdot H_2O$ precipitate when Ag^+ ions are added to aqueous solutions of these compounds. Only two of the Cl^- ions in the $CoCl_3 \cdot 5\ NH_3$ complex and only one of the Cl^- ions in $CoCl_3 \cdot 4\ NH_3$ can be precipitated with Ag^+ ions.

4. Measurements of the conductivity of aqueous solutions of these complexes suggest that the $CoCl_3 \cdot 6\ NH_3$ and $CoCl_3 \cdot 5\ NH_3 \cdot H_2O$ complexes dissociate in water to give a total of four ions. $CoCl_3 \cdot 5\ NH_3$ dissociates to give three ions, and $CoCl_3 \cdot 4\ NH_3$ dissociates to give only two ions.

Werner explained these observations by suggesting that transition metal ions such as the Co^{3+} ion have a *primary valence* and a *secondary valence.*

Primary valence: The number of negative ions needed to satisfy the charge on the metal ion.

The primary valence of a metal ion is the same in all of its complexes. Furthermore, the primary valence of a metal ion can only be satisfied by negative ions, such as the Cl^- ion. In each of the cobalt(III) complexes discussed in this section, it takes three Cl^- ions to satisfy the primary valence of the Co^{3+} ion.

Secondary valence: The number of ions or molecules physically bound to the transition metal ion.

As we would now describe it, the secondary valence is equal to the coordination number of the metal ion.

Werner assumed that the secondary valence of the transition metal in the four cobalt(III) complexes discussed in this section is 6. Thus, he concluded that their formulas could be written as follows.

$[Co(NH_3)_6^{3+}][Cl^-]_3$	orange-yellow
$[Co(NH_3)_5(H_2O)^{3+}][Cl^-]_3$	red
$[Co(NH_3)_5(Cl)^{2+}][Cl^-]_2$	purple
$[Co(NH_3)_4(Cl)_2^+][Cl^-]$	green

Exercise 22.2

Describe how Werner's formulas explain both the number of ions formed when the following compounds dissolve in water and the number of chloride ions that precipitate when these solutions are treated with Ag^+ ions.

   (a) $[Co(NH_3)_6]Cl_3$ (c) $[Co(NH_3)_5(Cl)]Cl_2$

   (b) $[Co(NH_3)_5(H_2O)]Cl_3$ (d) $[Co(NH_3)_4(Cl)_2]Cl$

Solution

The secondary valence of cobalt is 6 in each of these complexes. Thus, the cobalt is coordinated to a total of six ligands in each complex. Each complex also has a total of three chloride ions. Some of the Cl^- ions are free to dissociate when the complex dissolves in water. But others are bound to the Co^{3+} ion and neither dissociate nor react with Ag^+.

   (a) The three chloride ions in $[Co(NH_3)_6]Cl_3$ are free to dissociate when this complex dissolves in water. The complex therefore dissociates in water to give a total of four ions.

$$[Co(NH_3)_6]Cl_3(s) \xrightarrow{H_2O} [Co(NH_3)_6{}^{3+}](aq) + 3\ Cl^-(aq)$$

All three Cl^- ions are free to precipitate with Ag^+.

   (b) The three Cl^- ions are free to dissociate when $[Co(NH_3)_5(H_2O)]Cl_3$ dissolves in water and therefore precipitate when Ag^+ ions are added to the solution.

$$[Co(NH_3)_5(H_2O)Cl_3(s) \xrightarrow{H_2O} [Co(NH_3)_5(H_2O)^{3+}](aq) + 3\ Cl^-(aq)$$

   (c) One of the chloride ions is bound to the cobalt in the $[Co(NH_3)(Cl)]Cl_2$ complex. Only three ions are formed when this compound dissolves in water, and only two Cl^- ions are free to precipitate with Ag^+ ions.

$$[Co(NH_3)_5(Cl)][Cl]_2(s) \xrightarrow{H_2O} [Co(NH_3)_5(Cl)^{2+}](aq) + 2\ Cl^-(aq)$$

   (d) Two of the chloride ions are bound to the cobalt in $[Co(NH_3)_6Cl_2]Cl$. Only two ions are formed when this compound dissolves in water, and only one Cl^- ion is free to precipitate with Ag^+ ions.

$$[Co(NH_3)_4(Cl)_2][Cl](s) \xrightarrow{H_2O} [Co(NH_3)_4(Cl)_2{}^+](aq) + Cl^-(aq)$$

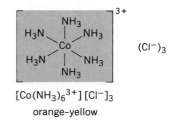

$[Co(NH_3)_6{}^{3+}]\ [Cl^-]_3$
orange-yellow

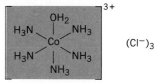

$[Co(NH_3)_5(H_2O)^{3+}]\ [Cl^-]_3$
red

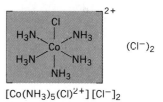

$[Co(NH_3)_5(Cl)^{2+}]\ [Cl^-]_2$
purple

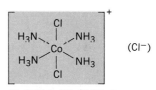

$[Co(NH_3)_4(Cl)_2{}^+]\ [Cl^-]$
green

FIG. 22.5 The structures of the four cobalt(III) complexes that played a major role in the development of Werner's theory of coordination complexes.

Werner also assumed that transition metal complexes had definite shapes. According to his theory, the ligands in six-coordinate cobalt(III) complexes are oriented toward the corners of an octahedron, as shown in Figure 22.5.

22.3 TYPICAL COORDINATION NUMBERS

Transition metal complexes have been characterized with coordination numbers that range from 1 to 12. The most common coordination numbers, however, are 2, 4, and 6. Examples of complexes with these coordination numbers are given in Table 22.1.

TABLE 22.1

Examples of Common Coordination Numbers

Metal Ion	Ligand	Complex	Coordination Number
Ag^+	$+ \ 2 \ NH_3$	$\rightleftharpoons Ag(NH_3)_2{}^+$	2
Cu^+	$+ \ 2 \ NH_3$	$\rightleftharpoons Cu(NH_3)_2{}^+$	2
Ag^+	$+ \ 2 \ S_2O_3{}^{2-}$	$\rightleftharpoons Ag(S_2O_3)_2{}^{3-}$	2
Ag^+	$+ \ 2 \ Cl^-$	$\rightleftharpoons AgCl_2{}^-$	2
Pb^{2+}	$+ \ 2 \ OAc^-$	$\rightleftharpoons Pb(OAc)_2$	2
Fe^{3+}	$+ \ 2 \ SCN^-$	$\rightleftharpoons Fe(SCN)_2{}^+$	2
Zn^{2+}	$+ \ 4 \ CN^-$	$\rightleftharpoons Zn(CN)_4{}^{2-}$	4
Cu^{2+}	$+ \ 4 \ NH_3$	$\rightleftharpoons Cu(NH_3)_4{}^{2+}$	4
Hg^{2+}	$+ \ 4 \ I^-$	$\rightleftharpoons HgI_4{}^{2-}$	4
Co^{2+}	$+ \ 4 \ SCN^-$	$\rightleftharpoons Co(SCN)_4{}^{2-}$	4
Ni^{2+}	$+ \ 4 \ CN^-$	$\rightleftharpoons Ni(CN)_4{}^{2-}$	4
Sb^{3+}	$+ \ 4 \ Cl^-$	$\rightleftharpoons SbCl_4{}^-$	4
Sn^{2+}	$+ \ 4 \ OH^-$	$\rightleftharpoons Sn(OH)_4{}^{2-}$	4
Fe^{2+}	$+ \ 6 \ H_2O$	$\rightleftharpoons Fe(H_2O)_6{}^{2+}$	6
Fe^{3+}	$+ \ 6 \ H_2O$	$\rightleftharpoons Fe(H_2O)_6{}^{3+}$	6
Co^{3+}	$+ \ 6 \ NH_3$	$\rightleftharpoons Co(NH_3)_6{}^{3+}$	6
Ni^{2+}	$+ \ 6 \ NH_3$	$\rightleftharpoons Ni(NH_3)_6{}^{2+}$	6
Fe^{2+}	$+ \ 6 \ CN^-$	$\rightleftharpoons Fe(CN)_6{}^{4-}$	6
Al^{3+}	$+ \ 6 \ F^-$	$\rightleftharpoons AlF_6{}^{3-}$	6
Sn^{4+}	$+ \ 6 \ OH^-$	$\rightleftharpoons Sn(OH)_6{}^{2-}$	6

Note that the charge on the complex is always the sum of the charges on the ions or molecules that form the complex.

$$Cu^{2+} + 4 \ NH_3 \rightleftharpoons Cu(NH_3)_4{}^{2+}$$

$$Pb^{2+} + 2 \ OAc^- \rightleftharpoons Pb(OAc)_2$$

$$Fe^{2+} + 6 \ CN^- \rightleftharpoons Fe(CN)_6{}^{4-}$$

Note also that the coordination number of a complex often increases as the charge on the metal ion becomes larger. Cu^+ ions form complexes with a coordination number of 2, whereas Cu^{2+} ions have a coordination number of 4.

$$Cu^+ + 2 \ NH_3 \rightleftharpoons Cu(NH_3)_2{}^+$$

$$Cu^{2+} + 4 \ NH_3 \rightleftharpoons Cu(NH_3)_4{}^{2+}$$

Similarly, Sn^{2+} ions tend to form four-coordinate complexes, whereas Sn^{4+} ions are more likely to form six-coordinate complexes.

$$Sn^{2+} + 4 \ OH^- \rightleftharpoons Sn(OH)_4{}^{2-}$$

$$Sn^{4+} + 6 \ OH^- \rightleftharpoons Sn(OH)_6{}^{2-}$$

Exercise 22.3

Calculate the charge on the transition metal ion in each of the following complexes.

(a) $Na_2Co(SCN)_4$ (b) $Ni(NH_3)_6(NO_3)_2$ (c) K_2PtCl_6

Solution

(a) This complex contains Na^+ and $Co(SCN)_4{}^{2-}$ ions. Each thiocyanate (SCN^-) ion carries a charge of -1. Since the net charge on the complex is -2, the cobalt ion must carry a charge of $+2$.

(b) This complex contains the $Ni(NH_3)_6{}^{2+}$ and $NO_3{}^-$ ions. Since ammonia is a neutral molecule, the nickel must carry a charge of $+2$.

(c) This complex contains the K^+ and $PtCl_6{}^{2-}$ ions. Each chloride ion carries a charge of -1. Since the net charge on the $PtCl_6{}^{2-}$ ion is -2, the platinum must carry a charge of $+4$.

22.4 THE ELECTRON CONFIGURATION OF TRANSITION METAL IONS

The relationship between the electron configurations of main-group elements and their ions was described in Section 5.22. There, we suggested that aluminum loses its outermost, or valence, electrons when it forms Al^{3+} ions.

$$Al = [Ne]\ 3s^2\ 3p^1 \qquad Al^{3+} = [Ne]$$

Oxygen, on the other hand, gains enough electrons to fill its valence shell.

$$O = [He]\ 2s^2\ 2p^4 \qquad O^{2-} = [He]\ 2s^2\ 2p^6 = [Ne]$$

The relationship between the electron configurations of transition metal elements and their ions is more complex. To understand the problem, we'll look at some of the chemistry of cobalt. Cobalt forms both $+2$ and $+3$ ions. The Co^{2+} ion is found in the $Co(SCN)_4{}^{2-}$ complex ion, and the Co^{3+} ion forms complexes such as the $Co(NH_3)_6{}^{3+}$ ion.

Which valence electrons are removed from a neutral cobalt atom to form the Co^{2+} and Co^{3+} ions? The following is the electron configuration of a neutral cobalt atom.

$$Co = [Ar]\ 4s^2\ 3d^7$$

The rules used to predict the configurations of neutral atoms in Section 5.19 suggest that the $4s$ orbital has a lower energy than the $3d$ orbitals. We might therefore expect cobalt to lose electrons from the higher-energy $3d$ orbitals. That is not what is observed, however. Experiment shows that the Co^{2+} and Co^{3+} ions have the following electron configurations.

$$Co^{2+} = [Ar]\ 3d^7$$
$$Co^{3+} = [Ar]\ 3d^6$$

In general, electrons are removed from s orbitals before they are removed from d orbitals when transition metals are ionized.

Students often ask, "Why are electrons removed from $4s$ orbitals before $3d$ orbitals if they are placed in $4s$ orbitals before $3d$ orbitals?" There are many ways of answering this question. We can start by noting that the difference between the energies of the $3d$ and $4s$ orbitals is very small. It is for this reason that chromium and copper have electron configurations that differ slightly from the predictions of the aufbau principle (see Section 5.20).

$$Cr = [Ar]\ 4s^1\ 3d^5$$
$$Cu = [Ar]\ 4s^1\ 3d^{10}$$

Native copper.

It should also be noted that the relative energies of the $4s$ and $3d$ orbitals used during the aufbau process holds true for neutral atoms. When transition metals form positive ions, the $3d$ orbitals become more stable than the $4s$ orbitals. As a result, electrons are removed from the $4s$ orbital before the $3d$ orbitals.

Exercise 22.4

Predict the electron configurations of the following ions.

(a) Sc^{3+} (b) Cr^{3+} (c) Fe^{3+}

Solution

(a) We start with the configuration of the neutral atom and then remove three electrons.

$$Sc = [Ar]\ 4s^2\ 3d^1$$
$$Sc^{3+} = [Ar]$$

Although it is a transition metal, scandium is very similar to main-group metals such as magnesium.

(b) We start, once again, with the configuration of the neutral atom.

$$Cr = [Ar]\ 4s^1\ 3d^5$$

Three valence electrons are lost when Cr^{3+} ions are formed. One comes from the $4s$ orbital, and the others come from the $3d$ orbitals.

$$Cr^{3+} = [Ar]\ 3d^3$$

(c) Since iron is the sixth element in the first series of transition metals, it has six electrons in the d subshell.

$$Fe = [Ar]\ 4s^2\ 3d^6$$

Two of the three electrons that must be removed to form an Fe^{3+} ion come from the $4s$ orbital. The other comes from a $3d$ orbital.

$$Fe^{3+} = [Ar]\ 3d^5$$

LEWIS ACID – LEWIS BASE APPROACH
22.5 TO BONDING IN COMPLEXES

As noted in earlier chapters, G. N. Lewis was the first to recognize that the reaction between a transition metal ion (such as the Co^{3+} ion) and ligands (such as ammonia) to form a coordination complex was analogous to the reaction between the H^+ and OH^- ions to form water.

The reaction between H^+ and OH^- ions involves the donation of a pair of electrons from the OH^- ion to the H^+ ion to form a covalent bond.

$$H - \overset{..}{\underset{..}{O}} :^- \quad H^+ \longrightarrow H - \overset{..}{\underset{..}{O}} - H$$

The H^+ ion can therefore be described as an electron-pair acceptor. The OH^- ion, on the other hand, is an electron-pair donor. Lewis proposed that any ion or molecule that behaves like the H^+ ion should be an acid.

Lewis acid: Any ion or molecule that can accept a pair of electrons—an electron-pair acceptor.

Any ion or molecule that behaves like the OH^- ion is a Lewis base.

Lewis base: Any ion or molecule that can donate a pair of electrons—an electron-pair donor.

When cobalt(III) ions react with ammonia, the metal ion accepts pairs of non-bonding electrons from the ligands to form covalent cobalt—nitrogen bonds (see Figure 22.6). The metal ion is therefore a Lewis acid, and the ligands coordinated to this metal ion are Lewis bases.

$$Co^{3+} \quad + \quad 6\ NH_3 \quad \longrightarrow \quad Co(NH_3)_6^{3+}$$

electron-pair acceptor (Lewis acid) \quad electron-pair donor (Lewis base) \quad acid-base complex

The Co^{3+} ion is an electron-pair acceptor, or Lewis acid, because it has empty valence-shell orbitals that can be used to hold pairs of electrons. To emphasize these empty valence orbitals, we can write the configuration of the Co^{3+} ion as follows.

$$Co^{3+} = [Ar]\ 4s^0\ 3d^6\ 4p^0$$

There is room in the valence shell of this ion for 12 more electrons. (The $4s$ orbital can hold 2 electrons, 4 more electrons can be added to the $3d$ subshell, and the $4p$ subshell can hold a total of 6 electrons.)

The NH_3 molecule is an electron-pair donor, or Lewis base, because it has a pair of nonbonding electrons on the nitrogen atom. The $Co(NH_3)_6^{3+}$ complex is formed when pairs of electrons from six ammonia molecules are donated to the Co^{3+} ion to form covalent metal—nitrogen bonds.

Co^{3+} ⇅ ⇅ ⇅ _ _ _ _ _ _ _   empty atomic orbitals
              $3d$ $4s$ $4p$

$Co(NH_3)_6^{3+}$ ⇅ ⇅ ⇅ ⇅ ⇅   ⇅   ⇅ ⇅ ⇅
               $3d$ $4s$ $4p$

This model explains why transition metal ions form coordination complexes. Transition metal ions have empty valence-shell orbitals that can accept pairs of electrons from a Lewis base. The model also predicts which ions or molecules can act as ligands to form coordination complexes. According to this model, ligands must be Lewis bases. They must contain at least one pair of nonbonding electrons that can be donated to a metal ion.

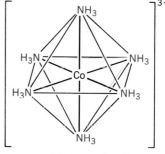

FIG. 22.6 The reaction between Co^{3+} ions and ammonia to form the $Co(NH_3)_6^{3+}$ ion is an example of a Lewis acid–Lewis base reaction. Each of the six NH_3 molecules acts as a Lewis base, donating a pair of nonbonding electrons to the Co^{3+} ion. The Co^{3+} ion acts as a Lewis acid, using empty valence orbitals to pick up pairs of nonbonding electrons. The net result of this reaction is the formation of six covalent cobalt—nitrogen bonds.

Exercise 22.5

Which of the following ions or molecules are Lewis acids?

(a) Cu^{2+} (b) Cl^- (c) CO

Solution

(a) The Cu^{2+} ion is a Lewis acid, because it contains empty valence-shell orbitals.

$$Cu^{2+} = [Ar]\ 4s^0\ 3d^9\ 4p^0$$

(b) The Cl⁻ ion is not a Lewis acid, because the valence-shell orbitals are filled.

$$Cl^- = [Ne]\, 3s^2\, 3p^6 = [Ar]$$

(c) The only way to decide whether a molecule is a Lewis acid is to draw its Lewis structure.

$$:C\equiv O:$$

CO is not a Lewis acid, because its valence orbitals are all filled.

Exercise 22.6

Which of the following ions or molecules are Lewis bases?

(a) Ca^{2+} (b) Cl^- (c) CO

Solution

(a) The Ca^{2+} is not a Lewis base, because it doesn't have nonbonding pairs of electrons in its valence shell.

$$Ca^{2+} = [Ar]$$

(b) The Lewis structure for the Cl⁻ ion is written as follows.

$$:\overset{..}{\underset{..}{Cl}}:^-$$

This ion can be an electron-pair donor and is therefore a Lewis base.

(c) CO also contains nonbonding pairs of electrons, so it is a Lewis base.

$$:C\equiv O:$$

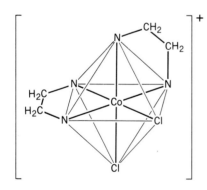

$[Co(en)_2Cl_2]^+$

FIG. 22.7 Ethylenediamine is an example of a bidentate (literally, two-toothed) ligand, which "bites" the metal in two places. Because multidentate ligands hold the metal as if they were claws, these ligands are also known as chelating ligands (from the Greek name for claw).

22.6 TYPICAL LIGANDS

Any ion or molecule with a pair of nonbonding electrons can be a ligand. Many of these ligands can be described as **monodentate** (literally, "one-toothed"), because they "bite" the metal in only one place. Examples of typical monodentate ligands are given in Table 22.2.

Other ligands can attach to the metal more than once. Ethylenediamine is a typical **bidentate ligand.**

$$H_2N\underset{\cdot\cdot}{\overset{CH_2CH_2}{\diagup\diagdown}}NH_2$$
ethylenediamine
(en)

Each end of this molecule contains a pair of nonbonding electrons that can form a covalent bond to a metal ion. Ethylenediamine is also an example of a **chelating ligand.** The term *chelate* comes from a Greek stem meaning "claw." It is used to describe ligands that can grab the metal in two or more places, as a claw would. A typical ethylenediamine complex is shown in Figure 22.7.

TABLE 22.2

Typical Monodentate Ligands

$:\!\overset{..}{\underset{..}{F}}\!:^-$ $\quad:\!\overset{..}{\underset{..}{Cl}}\!:^-$ $\quad:\!\overset{..}{\underset{..}{Br}}\!:^-$ $\quad:\!\overset{..}{\underset{..}{I}}\!:^-$ $\quad:\!\overset{..}{\underset{..}{O}}\!:^{2-}$ $\quad:\!\overset{..}{\underset{..}{O}}\!-H^-$

(water: O bonded to H and H)   (ammonia: N bonded to three H) $:\overset{..}{O}\!=\!\overset{..}{O}:$ $:C\!\equiv\!O:$ $:\overset{..}{O}\!=\!C\!=\!\overset{..}{O}:$

$:C\!\equiv\!N\!:^-$ $\qquad:\!\overset{..}{\underset{..}{S}}\!-\!C\!\equiv\!N\!:^-$ $\qquad:\!\overset{..}{\underset{..}{S}}\!-\!\overset{\overset{\displaystyle :\!O\!:}{|}}{\underset{\underset{\displaystyle :\!O\!:}{|}}{S}}\!-\!\overset{..}{\underset{..}{O}}\!:^{2-}$

By linking ethylenediamine fragments it is possible to make *tridentate ligands*

$$\text{H}_2\text{N}\overset{\text{CH}_2\text{CH}_2}{\diagup}\overset{}{\underset{..}{\text{NH}}}\overset{\text{CH}_2\text{CH}_2}{\diagup}\text{NH}_2$$

diethylenetriamine (dien)

and *tetradentate ligands.*

$$\text{H}_2\text{N}\overset{\text{CH}_2\text{CH}_2}{\diagup}\overset{}{\underset{..}{\text{NH}}}\overset{\text{CH}_2\text{CH}_2}{\diagup}\overset{}{\underset{..}{\text{NH}}}\overset{\text{CH}_2\text{CH}_2}{\diagup}\text{NH}_2$$

triethylenetetramine (trien)

By adding $-\text{CH}_2\text{CO}_2^-$ groups to an ethylenediamine framework, it is possible to form a *hexadentate ligand,* which can single-handedly satisfy the secondary valence of a transition metal ion (see Figure 22.8).

$$
\begin{array}{ccc}
\overset{\displaystyle \overset{..}{O}}{\underset{}{\|}} & & \overset{\displaystyle \overset{..}{O}}{\underset{}{\|}} \\
^-\!:\!\overset{..}{\underset{..}{O}}\!-\!\text{CCH}_2 & & \text{CH}_2\text{C}\!-\!\overset{..}{\underset{..}{O}}\!:^- \\
 & :\!\text{NCH}_2\text{CH}_2\text{N}\!: & \\
^-\!:\!\overset{..}{\underset{..}{O}}\!-\!\text{CCH}_2 & & \text{CH}_2\text{C}\!-\!\overset{..}{\underset{..}{O}}\!:^- \\
\overset{}{\underset{\displaystyle \overset{..}{\underset{..}{O}}}{\|}} & & \overset{}{\underset{\displaystyle \overset{..}{\underset{..}{O}}}{\|}}
\end{array}
$$

ethylenediaminetetraacetate (EDTA)

A number of multidentate ligands are shown in Table 22.3. These ligands form complexes with transition metal ions that are more stable than analogous monodentate ligands. This extra stability results from a decrease in the unfavorable entropy factor associated with the formation of complexes. When one end of a bidentate ligand binds to a transition metal, the other end is so close to the metal that the unfavorable contribution from entropy associated with closing the chelate ring is very much smaller than the entropy associated with binding a second monodentate ligand to the metal ion.

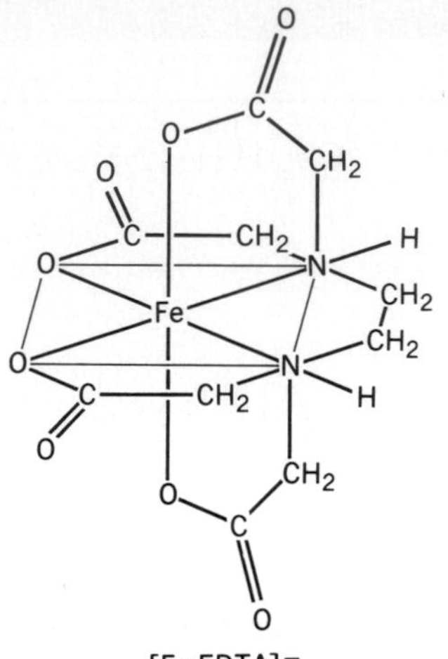

FIG. 22.8 Ethylenediamine-tetraacetate, or EDTA, is a hex-adentate ligand that can single-handedly satisfy the secondary valence of transition metal ions with a coordination number of 6. EDTA forms very strong complexes with many transition metal ions. K_f for the FeEDTA complex shown here is 1.7×10^{24}.

[Fe EDTA]$^-$

TABLE 22.3

Typical Multidentate Ligands

Bidentate Ligands

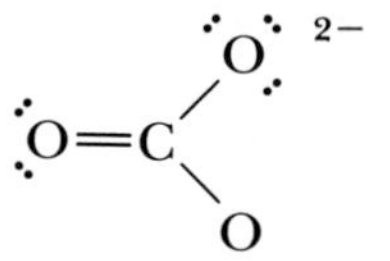

carbonate

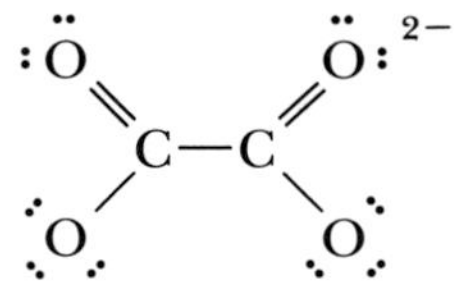

oxalate (ox)

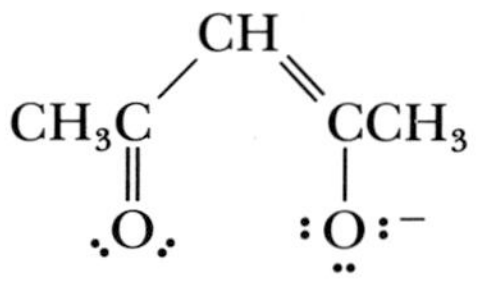

acetylacetonate (acac)

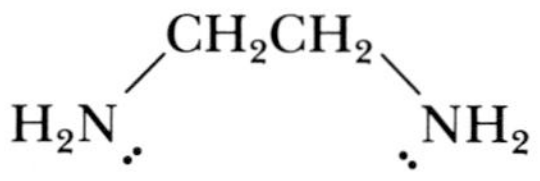

ethylenediamine (en)

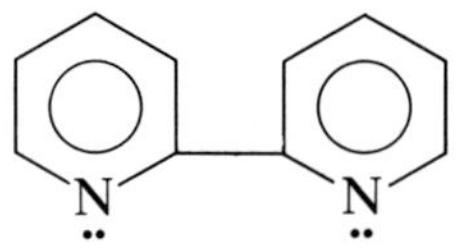

2,2'-bipyridyl (bipy)

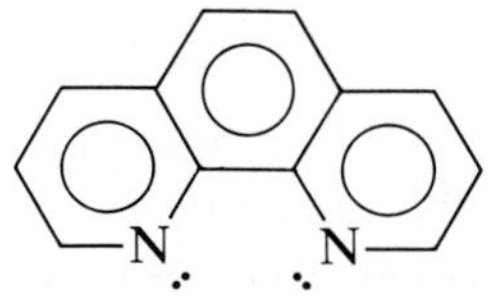

1,10-phenanthroline (phen)

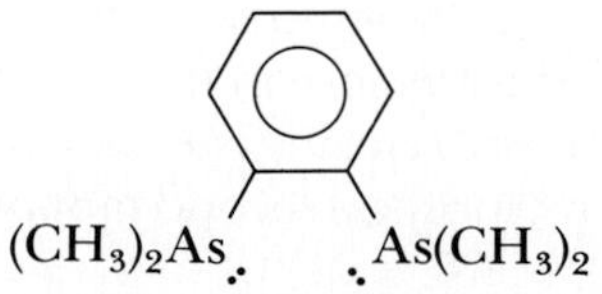

o-phenylenebisdimethylarsine (diars)

Tridentate Ligands

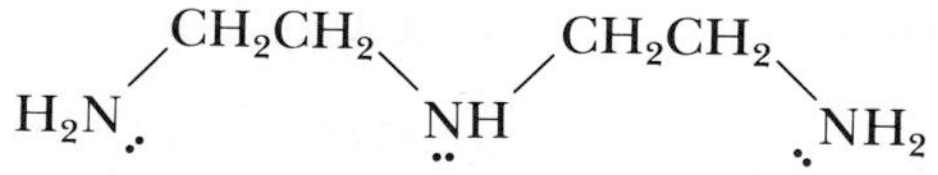

$$CH_2CH_2 \qquad\qquad CH_2CH_2$$
$$H_2N \qquad\qquad NH \qquad\qquad NH_2$$

diethylenetriamine (dien)

Tetradentate Ligands

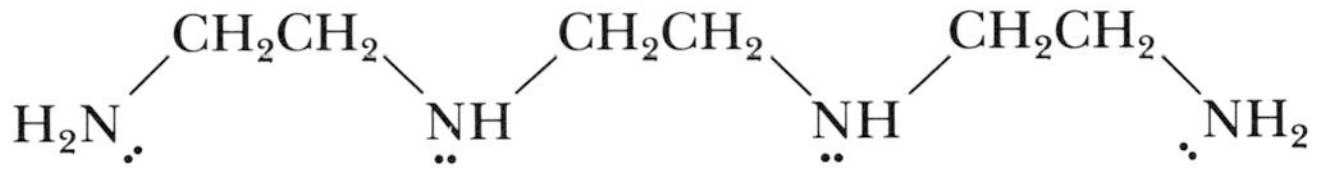

$$CH_2CH_2 \qquad CH_2CH_2 \qquad CH_2CH_2$$
$$H_2N \qquad NH \qquad NH \qquad NH_2$$

triethylenetetraamine (trien)

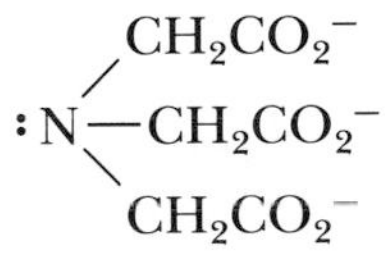

$$CH_2CO_2^-$$
$$:N—CH_2CO_2^-$$
$$CH_2CO_2^-$$

nitrilotriacetate (NTA)

Hexadentate Ligands

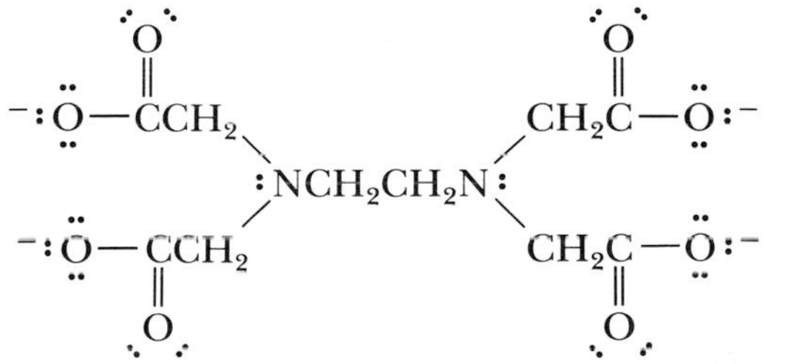

$$
\begin{array}{cc}
\overset{..}{\overset{..}{O}} & \overset{..}{\overset{..}{O}} \\
\| & \| \\
^-:\overset{..}{\overset{..}{O}}—CCH_2 & CH_2C—\overset{..}{O}:^- \\
& :NCH_2CH_2N: \\
^-:\overset{..}{\overset{..}{O}}—CCH_2 & CH_2C—\overset{..}{O}:^- \\
\| & \| \\
\overset{..}{\overset{..}{O}} & \overset{..}{\overset{..}{O}}
\end{array}
$$

ethylenediaminetetraacetate (EDTA)

22.7 COORDINATION COMPLEXES IN THE LABORATORY, IN INDUSTRY, AND IN NATURE

THE LABORATORY

Coordination complexes play a vital role in controlling the solubility of transition metal compounds. As noted in Chapter 17, the $Ag(NH_3)_2^+$ complex ion serves as the basis for dissolving and then reprecipitating AgCl during qualitative analysis.

$$AgCl(s) + 2\ NH_3(aq) \rightleftharpoons Ag(NH_3)_2^+(aq) + Cl^-(aq)$$
$$Ag(NH_3)_2^+(aq) + Cl^-(aq) + 2\ H^+(aq) \rightleftharpoons AgCl(s) + 2\ NH_4^+(aq)$$

Coordination complexes also serve as the basis for spot tests used to detect transition metal ions. Cu^{2+} ions can be detected from the dark blue complex that forms when aqueous ammonia is added to the solution.

$$Cu^{2+}(aq) + 4\ NH_3(aq) \rightleftharpoons Cu(NH_3)_4^{2+}(aq)$$

Thiocyanate forms a blue complex in the presence of the Co^{2+} ion

$$Co^{2+}(aq) + 4\ SCN^-(aq) \rightleftharpoons Co(SCN)_4^{2-}(aq)$$

and blood-red complexes with the Fe^{3+} ion.

$$Fe^{3+}(aq) + SCN^-(aq) \rightleftharpoons Fe(SCN)^{2+}(aq)$$
$$Fe(SCN)^{2+}(aq) + SCN^- \rightleftharpoons Fe(SCN)_2^+(aq)$$

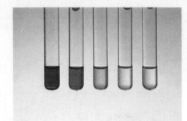

FIG. 22.9 Coordination complexes are often used to measure the concentrations of transition metal ion solutions. The intensity of the blue color of solutions containing the $Cu(NH_3)_4^{2+}$ ion, for example, is directly proportional to the concentration of this complex ion.

The characteristic red precipitate formed when dimethylglyoxime (DMG) is added to a solution can be used to test for the presence of Ni^{2+} ion.

$$Ni^{2+}(aq) + 2\ DMG^-(aq) \rightleftharpoons Ni(DMG)_2(s)$$

Complex ions such as the $Cu(NH_3)_4^{2+}$, $Co(SCN)_4^{2-}$, and $Fe(SCN)_2^+$ ions are also used to measure the amount of Cu^{2+}, Co^{2+}, or Fe^{3+} ion in a solution. In each case, the intensity of the light passing through the solution is inversely proportional to the concentration of the complex ion (see Figure 22.9).

INDUSTRY

The entire photographic industry is based on complex-ion chemistry. All commercial film, regardless of whether it is black and white or color, uses one of the silver halides (AgCl, AgBr, or AgI) to capture the photographic image. When the photons in light strike a silver halide crystal, a few of the Ag^+ ions pick up electrons from neighboring molecules to form silver atoms.

Before the latent image captured by this process can be seen, the film has to be developed. The developer reduces the remaining Ag^+ ions in the crystals that have been partially reduced by light. In order to capture, or fix, the image permanently, it is necessary to remove the AgX crystals that did not react with light. This is done by adding a complexing agent, such as the thiosulfate ion, which reacts with these salts to form a soluble complex ion.

$$AgX(s) + 2\ S_2O_3^{2-}(aq) \rightleftharpoons Ag(S_2O_3)_2^{3-}(aq) + X^-(aq)$$

Complex ion chemistry also plays an important role in separating metal ions as a first step in preparing metals from their ores. For centuries, gold could be isolated only from relatively rich ores. In 1890, Macarthur and Forest introduced the cyanide process, which provided a way to isolate gold from low-grade ores. This process is based on the fact that gold metal reacts with the cyanide ion in the presence of air to form a soluble $Au(CN)_2^-$ complex ion, which can be reduced to gold metal.

$$4\ Au(s) + 8\ CN^-(aq) + O_2(g) + 2\ H_2O(aq) \longrightarrow 4\ Au(CN)_2^-(aq) + 4\ OH^-(aq)$$

NATURE

Complex ions play a central role in the chemistry of living systems. As evidence of this, we'll look at three biologically important complexes: vitamin B_{12}, chlorophyll *a*, and the heme found in the oxygen-carrying proteins hemoglobin and myoglobin.

As early as 1926, it was known that patients who suffered from pernicious anemia became better when they ate liver. It took more than 20 years, however, to determine that the vitamin B_{12} in liver was one of the active factors; and it took another 10 years to determine the structure of vitamin B_{12} shown in Figure 22.10*a*. Vitamin B_{12} is a six-coordinate Co^{3+} complex. The Co^{3+} ion is coordinated to four nitrogen atoms that lie in a planar molecule known as a ***corrin ring***. It is also coordinated to a fifth nitrogen atom and to a sixth group, labeled *R* in the figure. The *R* group may be a CN^-, NO_2^-, SO_3^{2-}, or OH^- ion, depending on the source of the vitamin B_{12}.

Every carbon atom in our bodies can be traced to a reaction in which plants use the energy in sunlight to make glucose ($C_6H_{12}O_6$) from CO_2 and water.

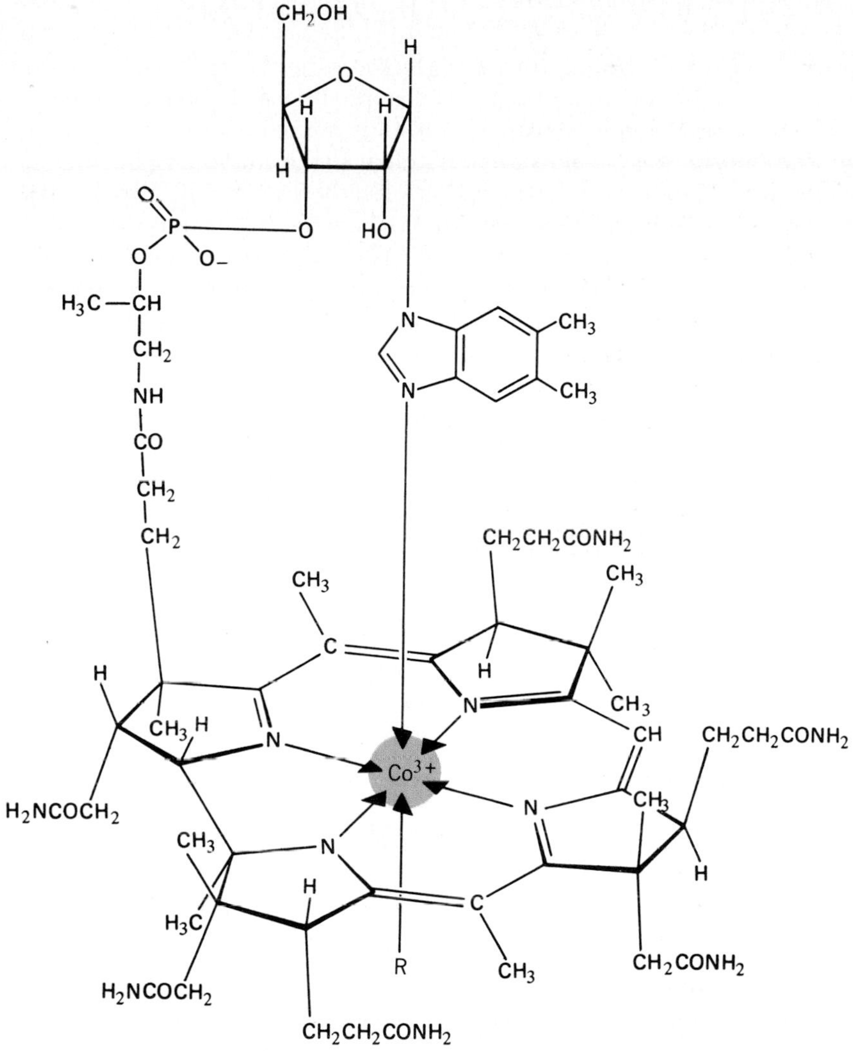

Vitamin B_{12}

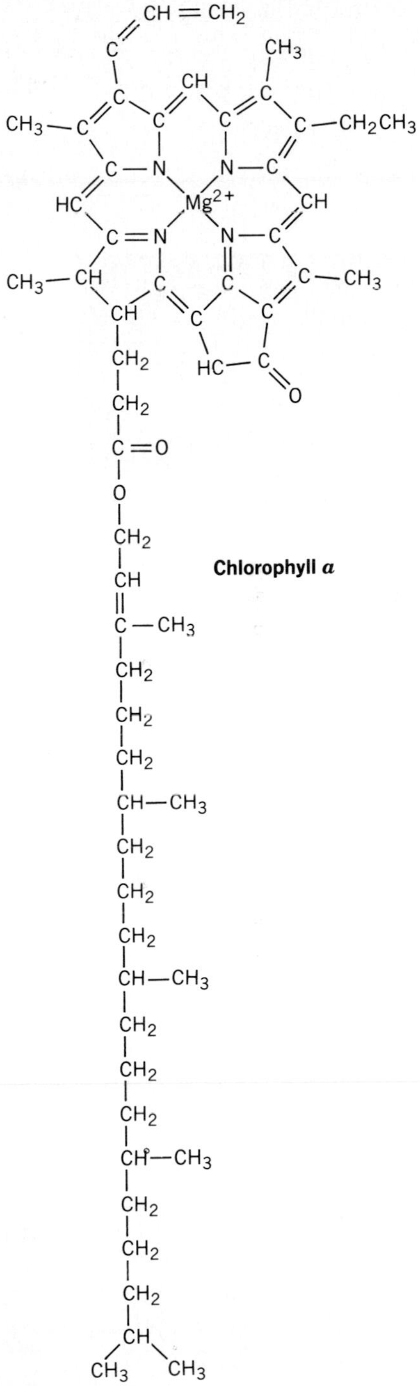

Chlorophyll *a*

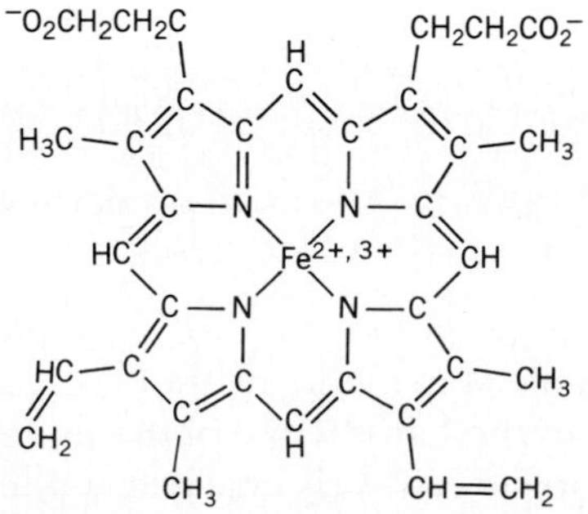

Heme

FIG. 22.10 The structures of the biologically important coordination complexes known as vitamin B_{12}, chlorophyll *a*, and heme.

$$6 \ CO_2(g) + 6 \ H_2O(aq) \xrightarrow[\text{chlorophyll}]{\text{light}} C_6H_{12}O_6(aq) + 6 \ O_2(g)$$

The fundamental step in the photosynthesis of glucose is the absorption of sunlight by the chlorophyll in higher plants and algae. Chlorophyll a (see Figure 22.10b) is a complex in which an Mg^{2+} ion is coordinated to four nitrogen atoms in a planar **chlorin ring**. The chlorin ring in chlorophyll is similar to the corrin ring in vitamin B_{12}, but not identical. The minor differences between these rings adjust their sizes so that each ring is exactly the right size to hold the correct transition metal ion.

The coordination complex shown in Figure 22.10c is known as a **heme**. A heme contains an Fe^{2+} or Fe^{3+} ion coordinated to four nitrogen atoms in a **porphyrin ring**. These complexes are found in myoglobin and hemoglobin, which carry oxygen from the lungs to the muscles. The porphyrin ring in a heme is slightly larger than the corrin ring in vitamin B_{12} and slightly smaller than the chlorin ring in chlorophyll a, so that it is just the right size to hold the Fe^{2+} or Fe^{3+} ion.

A computer model of the structure of hemoglobin, the oxygen-carrying protein in blood. Oxygen forms a Lewis acid–base complex with the iron atoms represented by the yellow balls in this structure.

22.8 NOMENCLATURE OF COMPLEXES

The official rules for naming chemical compounds are established by nomenclature committees of the International Union of Pure and Applied Chemistry (IUPAC). The IUPAC nomenclature of coordination complexes is based on the following rules.

RULES FOR NAMING COORDINATION COMPLEXES

1. The name of the positive ion is always written before the name of the negative ion.

2. The name of the ligand is written before the name of the metal to which it is coordinated.

3. The Greek prefixes *mono, di-, tri-, tetra-, penta-, hexa-,* and so on are used to indicate the number of ligands when the ligands are relatively simple. The Greek prefixes *bis-, tris-,* and *tetrakis-* are used with more complicated ligands, particularly those whose names already contain Greek prefixes, such as ethylene*di*amine.

4. The names of negative ligands always end in *o* — for example, *fluoro* (F^-), *chloro* (Cl^-), *bromo* (Br^-), *iodo* (I^-), *oxo* (O^{2-}), *hydroxo* (OH^-), *cyano* (CN^-), and so on.

5. A handful of neutral ligands are given common names — for example, **aquo** (H_2O), **ammine** (NH_3), and **carbonyl** (CO).

6. Ligands are listed in the following order: negative ions, neutral molecules, and positive ions. Ligands with the same charge are listed in alphabetical order.

7. The oxidation number of the metal atom is indicated by a Roman numeral in parentheses after the name of the metal atom.

8. The names of complexes with a net negative charge end in *-ate.* For example, $Co(SCN)_4^{2-}$ is the tetrathiocyanatocobaltate(II) ion. When the symbol for the metal is derived from its Latin name, *-ate* is added to the Latin name of the metal. Thus, negatively charged iron complexes are ferrates, negatively charged copper complexes are cuprates, and so on.

Exercise 22.7

Name the following coordination complexes.

(a) $K_4Fe(CN)_6$ (d) $[Cr(NH_3)_5(H_2O)][(NO_3)_3]$
(b) $Fe(acac)_3$ (e) $[Cr(NH_3)_4Cl_2]Cl$
(c) $[Cr(en)_3]Cl_3$

Solution

(a) This salt contains the K^+ and $Fe(CN)_6^{4-}$ ions. It is therefore potassium hexacyanoferrate(II).

(b) Tris(acetylacetonato)iron(III)

(c) Tris(ethylenediamine)chromium(III) chloride

(d) Pentaammineaquochromium(III) nitrate

(e) Dichlorotetraamminechromium(III) chloride

22.9 ISOMERS

GEOMETRIC *(CIS/TRANS)* ISOMERS

An important test of Werner's theory of coordination complexes revolved around whether these complexes formed *isomers* (literally, "equal parts"). The term *isomer*, you may recall, describes compounds with the same chemical formula but different structures. There are two possible isomers for the $Co(NH_3)_4Cl_2^+$ complex ion, for example, as shown in Figure 22.11. These structures differ in the orientation of the two chloride ions around the Co^{3+} ion. In the *trans* isomer, the chlorides occupy positions across from each other in the octahedron. In the *cis* isomer, they occupy adjacent positions. You can remember the difference between *cis* and *trans* isomers by noting that the prefix *trans* is commonly used to describe things that are on opposite sides, or across from each other, as in *transatlantic* or *transcontinental*.

At the time Werner proposed his theory, only one isomer of the $[Co(NH_3)_4Cl_2]Cl$ complex was known, the green complex described in Section 22.2. Werner predicted that a second isomer should exist, and his discovery in 1907 of a purple compound with the same chemical formula was a key step in convincing scientists who were still critical of his model.

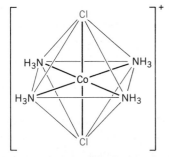

Trans-$[Co(NH_3)_4Cl_2]^+$

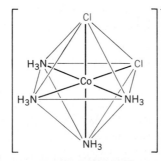

Cis-$[Co(NH_3)_4Cl_2]^+$

FIG. 22.11 There are two possible structures for the $Co(NH_3)_4Cl_2^+$ complex ion. In the *trans* isomer, the two chlorides lie across from each other. In the *cis* isomer, the chlorides are in adjacent positions in the octahedron.

FIG. 22.12 Geometric isomers can also be found for square planar complexes, such as $Pt(NH_3)_2Cl_2$. In the *trans* isomer, the two Cl^- ligands are across from each other (as are the two NH_3 ligands). In the *cis* isomer, the Cl^- ligands are in adjacent positions (as are the NH_3 ligands).

The *cis*- and *trans*-$[Co(NH_3)_4Cl_2]Cl$ complexes are examples of **geometric isomers**—they contain the same ligands, but they differ in their three-dimensional structure, or geometry. Geometric isomers are also possible in four-coordinate complexes that have a square planar geometry. Figure 22.12 shows the structures of the *cis* and *trans* isomers of dichlorodiammineplatinum(II). The *cis* isomer is being used as a drug to treat brain tumors, under the trade name *cisplatin*. This planar complex inserts itself into the grooves in the double helical structure of the DNA in cells. This inhibits the replication of DNA, thereby slowing down the rate at which the tumor grows, which allows the body's natural defense mechanisms to act on the tumor.

LINKAGE ISOMERS

Ligands that have nonbonding electrons on opposite ends, such as the SCN^- ion, sometimes bind to transition metals in more than one way. The following compounds are examples of **linkage isomers.** They differ only in the way the SCN^- ion is linked to the metal ion.

$(bipy)_2Pd(NCS)_2$

$(bipy)_2Pd(SCN)_2$

CHIRAL ISOMERS

Complexes such as Co(en) and $Co(acac)_3$ are **chiral** (from the Greek *cheir*, "hand"). They exist as pairs of isomers that are mirror images of each other, much as the right and left hands are mirror images of each other (see Figure 22.13). These optical isomers, or **stereoisomers,** have almost identical physical properties. They have the same melting point, boiling point, density, and color. They differ only in the way they interact with plane-polarized light.

Light consists of electric and magnetic fields that oscillate equally in all directions perpendicular to the path of the light ray. When light is passed through a polarizer, such as a lens in a pair of Polaroid sunglasses, these oscillations are confined to a single plane.

Compounds that can rotate plane-polarized light are said to be **optically active.** Those that rotate the plane of polarization to the right (clockwise) are said to be **dextrorotatory** (from the Latin *dexter*, "right"). Those that rotate the plane to the left (counterclockwise) are **levorotatory** (from the Latin *laevus*, "left"). All optically active compounds are chiral. They exist as pairs of stereoisomers, which are mirror images of each other.

The effect of optically active compounds on light can be demonstrated by filling a large tube with a saturated solution of cane sugar in water and then shining polarized light through the tube to form a spiral of light of different colors along the length of the tube.

22.10 COMMON OXIDATION STATES OF TRANSITION METALS

We can explain the chemistry of the transition metals by assuming that electrons are removed from the valence-shell s orbitals on the metals before the valence-shell d orbitals. Vanadium and iron, for example, form $+2$ ions by losing the two electrons in the $4s$ valence orbital.

$$V = [Ar]\ 4s^2\ 3d^3 \qquad V^{2+} = [Ar]\ 3d^3$$
$$Fe = [Ar]\ 4s^2\ 3d^6 \qquad Fe^{2+} = [Ar]\ 3d^6$$

Iridium forms a $+3$ ion by losing two electrons from the $6s$ orbital and one electron from a $5d$ orbital.

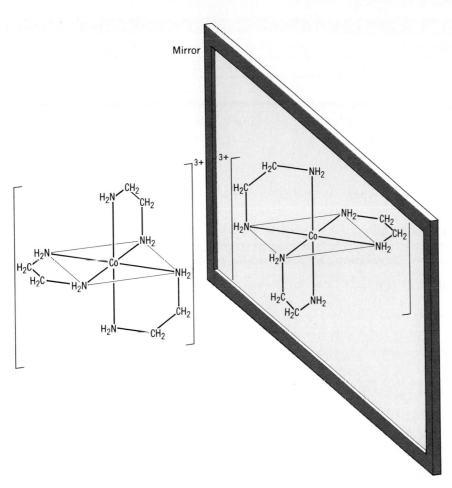

FIG. 22.10 The Co(en)$_3^{3+}$ complex ion is an example of a compound that exists as a pair of optical isomers, or stereo-isomers, which are mirror images of each other.

$$\text{Ir} = [\text{Xe}]\ 6s^2\ 4f^{14}\ 5d^7 \qquad \text{Ir}^{3+} = [\text{Xe}]\ 4f^{14}\ 5d^6$$

Because the valence electrons in these ions are concentrated in d orbitals, these oxidation states are often described as d^3 or d^6 configurations.

The elements in the first column of transition metals (Sc, Y, La, and Ac) act very much like main-group metals. They invariably lose all of their valence electrons to form $+3$ ions that have a d^0 configuration.

$$\text{Sc} = [\text{Ar}]\ 4s^2\ 3d^1 \qquad \text{Sc}^{3+} = [\text{Ar}]$$

Most of the other transition metals form two or more oxidation states. Titanium, for example, forms compounds in the $+2$ (TiO) and $+4$ (TiO$_2$) oxidation states. Vanadium can be isolated in any oxidation state ranging from $+2$ to $+5$. Manganese is even more versatile; it forms compounds in any oxidation state from -1 to $+7$.

Not all of these oxidation states are equally stable. The most stable, and therefore the most common, oxidation states of the first-row transition metals are given in Table 22.4. There are clear patterns in these oxidation states.

1. **The most common oxidation states correspond to a d^0, d^3, d^5, d^6, d^8, or d^{10} configuration.**

2. **There are no examples of common oxidation states corresponding to a d^1, d^2, d^4 configuration.**

3. **There is only one example each of a d^7 and a d^9 configuration.**

TABLE 22.4										

Common Oxidation States of the First-Row Transition Metals

	Sc	Ti	V	Cr	Mn	Fe	Co	Ni	Cu	Zn
+1									d^{10}	
+2			d^3		d^5	d^6	d^7	d^8	d^9	d^{10}
+3	d^0			d^3		d^5	d^6			
+4		d^0			d^3					
+5			d^0							
+6				d^0						
+7					d^0					

Exercise 22.8

Which of the following represent stable oxidation states of the transition metals?

(a) V^{3+} (b) Cr^{3+} (c) Mn^{3+} (d) Co^{3+} (e) Ni^{3+}

Solution

In order to predict whether the oxidation states of these ions are stable, we need to determine the electron configuration of each ion.

$$V^{3+} = [Ar]\, 3d^2$$
$$Cr^{3+} = [Ar]\, 3d^3$$
$$Mn^{3+} = [Ar]\, 3d^4$$
$$Co^{3+} = [Ar]\, 3d^6$$
$$Ni^{3+} = [Ar]\, 3d^7$$

Only the Cr^{3+} and Co^{3+} ions have electron configurations that tend to be stable — d^3 and d^6. It is possible to form the V^{3+}, Mn^{3+}, or Ni^{3+} ions, but there is no reason to expect them to be stable.

Another point about common oxidation states of transition metal ions is worth mentioning. Transition metal ions can be found in aqueous solution with $+1$, $+2$, or $+3$ charges. Among the first series of transition metals, the following ions correspond to stable oxidation states.

+1:	Cu^+
+2:	V^{2+}, Mn^{2+}, Fe^{2+}, Co^{2+}, Ni^{2+}, Cu^{2+}, Zn^{2+}
+3:	Sc^{3+}, Cr^{3+}, Fe^{3+}, Co^{3+}

None of these transition metals can be found in aqueous solution with a charge larger than $+3$.

In Section 8.6, we concluded that atoms become more electronegative as the positive charge on the atom increases. By the time these transition metals reach oxidation states of $+4$ (or larger), they are so electronegative they form covalent bonds with oxygen atoms. As soon as one of the transition metals is oxidized beyond the $+3$ oxidation state, it reacts with water to form a covalent oxide. Note, for example, that the half-reaction for the oxidation of Mn^{2+} ions to the $+7$ oxidation state is written as follows.

$$Mn^{2+}(aq) + 4\, H_2O(aq) \longrightarrow MnO_4^-(aq) + 8\, H^+(aq) + 5\, e^-$$

Manganese in a +7 oxidation state has such a high affinity for electrons that it rips water molecules apart to steal negatively charged O^{2-} ions and release H^+ ions into the solution.

It is useful to have a way of distinguishing between the charge on a transition metal ion and the oxidation state of the transition metal. By convention, symbols such as Mn^{2+} refer to ions that carry a +2 charge. Symbols such as Mn(VII) are used to describe compounds in which manganese is in the +7 oxidation state.

Mn(VII) is not the only example of an oxidation state powerful enough to decompose water. As soon as Mn^{2+} is oxidized to the Mn(IV) oxidation state, it reacts with water to form MnO_2. The Cr^{3+} ion can be found in aqueous solution. But once this ion is oxidized to Cr(VI), it reacts with water to form the CrO_4^{2-} or $Cr_2O_7^{2-}$ ion. Vanadium exists in aqueous solutions as the V^{2+} ion. But once it is oxidized to the +4 or +5 oxidation state, it reacts with water to form the VO^{2+} or VO_2^+ ion.

22.11 THE VALENCE BOND APPROACH TO BONDING IN COMPLEXES

The idea that atoms are held together in covalent bonds by sharing pairs of electrons was first proposed by G. N. Lewis in 1902. It was not until 1927, however, that Walter Heitler and Fritz London showed how the sharing of pairs of electrons holds a covalent molecule together.

The Heitler-London model of covalent bonds, which was outlined briefly in Section 8.3, was the basis of a theory known as the *valence bond model*. This theory assumes that covalent bonds result from the sharing of pairs of valence-shell electrons. The last major step in the evolution of the valence bond theory was the suggestion by Linus Pauling that atomic orbitals mix to form hybrid orbitals, such as the sp, sp^2, sp^3, dsp^3, and d^2sp^3 orbitals discussed in Section 8.15.

It is easy to apply the valence bond theory to some coordination complexes—such as the $Co(NH_3)_6^{3+}$ ion. Here, we start with the electron configuration of the Co^{3+} ion.

$$Co^{3+} = [Ar]\, 3d^6$$

We then look at the valence-shell orbitals and note that the $4s$ and $4p$ orbitals are empty.

$$Co^{3+} = [Ar]\, 4s^0\, 3d^6\, 4p^0$$

By concentrating the $3d^6$ electrons in three of the orbitals in this subshell, it is possible to create an electron configuration that can be written as follows.

The empty $3d_{x^2-y^2}$, $3d_{z^2}$, $4s$, $4p_x$, $4p_y$, and $4p_z$ orbitals are then mixed to form a set of empty d^2sp^3 orbitals that point toward the corners of an octahedron. Each of these orbitals can accept a pair of nonbonding electrons from a neutral NH_3 molecule to form a complex in which the cobalt atom has a filled shell of valence electrons.

Exercise 22.9

Use the valence bond model to explain why Fe^{2+} ions combine with cyanide ligands to form the $Fe(CN)_6^{4-}$ complex ion.

Solution

We start with the electron configuration of the transition metal ion.

$Fe^{2+} = [Ar]\ 4s^0\ 3d^6\ 4p^0$

By investing a little energy in the system, we can pair the six $3d$ electrons, thereby creating empty $3d_{x^2-y^2}$ and $3d_{z^2}$ orbitals.

$$Fe^{3+} \quad \underset{3d}{\underline{\text{↿⇂}}\ \underline{\text{↿⇂}}\ \underline{\text{↿⇂}}\ \underline{\quad}\ \underline{\quad}} \quad \underset{4s}{\underline{\quad}} \quad \underset{4p}{\underline{\quad}\ \underline{\quad}\ \underline{\quad}}$$

These orbitals can then be mixed with the empty $4s$ and $4p$ orbitals to form a set of six empty d^2sp^3 hybrid orbitals, which can accept pairs of nonbonding electrons from the CN^- ligands to form an octahedral complex in which the metal atom has a filled shell of valence electrons.

$$Fe(CN)_6^{4-} \quad \underset{3d}{\underline{\text{↿⇂}}\ \underline{\text{↿⇂}}\ \underline{\text{↿⇂}}\ \underline{\text{↿⇂}}\ \underline{\text{↿⇂}}} \quad \underset{4s}{\underline{\text{↿⇂}}} \quad \underset{4p}{\underline{\text{↿⇂}}\ \underline{\text{↿⇂}}\ \underline{\text{↿⇂}}}$$

Some complexes, such as the $Ni(NH_3)_6^{2+}$ ion, are a bit harder to explain with the valence bond theory. The first step in building a model of this compound involves writing the configuration of the Ni^{2+} ion.

$Ni = [Ar]\ 4s^2\ 3d^8 \qquad Ni^{2+} = [Ar]\ 3d^8$

This configuration creates a problem. There are eight electrons in the $3d$ orbitals. Even if we invest the energy necessary to pair all of the $3d$ electrons, we can't find a pair of empty $3d$ orbitals to use to form a set of d^2sp^3 hybrids.

$$Ni^{2+} \quad \underset{3d}{\underline{\text{↿⇂}}\ \underline{\text{↿⇂}}\ \underline{\text{↿⇂}}\ \underline{\text{↿⇂}}\ \underline{\quad}} \quad \underset{4s}{\underline{\quad}} \quad \underset{4p}{\underline{\quad}\ \underline{\quad}\ \underline{\quad}}$$

There is a way to overcome this problem, however. The five $4d$ orbitals on nickel are all empty. We can therefore form a set of empty sp^3d^2 hybrid orbitals by mixing the $4d_{x^2-y^2}$, $4d_{z^2}$, $4s$, $4p_x$, $4p_y$, and $4p_z$ orbitals. These hybrid orbitals can then accept pairs of nonbonding electrons from six ammonia molecules to form a complex ion.

$$[Ni(NH_3)_6]^{2+} \quad \underset{3d}{\underline{\text{↿⇂}}\ \underline{\text{↿⇂}}\ \underline{\text{↿⇂}}\ \underline{\text{↿⇂}}\ \underline{\quad}} \quad \underset{4s}{\underline{\text{↿⇂}}} \quad \underset{4p}{\underline{\text{↿⇂}}\ \underline{\text{↿⇂}}\ \underline{\text{↿⇂}}} \quad \underset{4d}{\underline{\text{↿⇂}}\ \underline{\text{↿⇂}}\ \underline{\quad}\ \underline{\quad}}$$

22.12 THE EFFECTIVE ATOMIC NUMBER (EAN) RULE

In the same year that Heitler and London proposed the valence bond theory, Nevil Sidgwick introduced the concept of *effective atomic number*. This concept assumes that transition metal ions pick up just enough ligands to form complexes in which the valence-shell orbitals are filled. The effective atomic number (EAN) of a complex is calculated as follows.

1. Start with the number of electrons on a neutral metal atom. Iron, for

example, has an atomic number of 26. A neutral iron atom therefore has 26 electrons.

2. **Add or subtract enough electrons to account for the charge on the metal atom.** The $Fe(NH_3)_6^{2+}$ complex ion, for example, formally contains the Fe^{2+} ion, so we subtract two electrons.

3. **Add two electrons for every ligand atom bound to the metal atom.**

According to these rules, the effective atomic number for the $Fe(NH_3)_6^{2+}$ complex ion is 36.

$$Fe(NH_3)_6^{2+}: \qquad EAN = 26 - 2 + 6(2) = 36$$

How many electrons does an iron atom need to fill its valence shell? A glance at the periodic table tells us that iron is in the same row as the noble gas krypton, which has 36 valence electrons. Because the effective atomic number for this complex ion is 36, the complex is said to obey the EAN rule.

Exercise 22.10

Use the effective atomic number rule to predict the number of ligands in each of the following complexes.

(a) $Cr(CO)_x$ (b) $Fe(H_2O)_x^{2+}$ (c) $Ir(NH_3)_x^{3+}$

Solution

(a) Chromium needs 36 electrons to fill its valence shell. There are 24 electrons on a neutral chromium atom. Since this complex has no charge, 12 electrons are needed from the ligands to reach an EAN of 36. Thus, x must be 6 in this complex.

$$Cr(CO)_6: \qquad EAN = 24 + 0 + 2(6) = 36$$

(b) Iron also needs a total of 36 electrons to fill its valence shell. There are 26 electrons on a neutral iron atom. The complex has a charge of $+2$, however, so the value of x, again, is 6.

$$Fe(H_2O)_6^{2+}: \qquad EAN = 26 - 2 + 2(6) = 36$$

(c) Iridium needs 86 electrons to fill its valence shell. There are 77 electrons on a neutral iridium atom. Since the complex has a charge of $+3$, the value of x is 6.

$$Ir(NH_3)_6^{3+}: \qquad EAN = 77 - 3 + 2(6) = 86$$

There are classes of transition metal complexes that obey the EAN rule time after time. Unfortunately, there are also many exceptions to this rule. Table 22.5 lists examples of complexes that obey the EAN rule, as well as a number of complexes that do not.

22.13 CRYSTAL FIELD THEORY

At almost exactly the same time that chemists such as Sidgwick were developing the valence bond model for coordination complexes, physicists such as Hans Bethe, John Van Vleck, and Leslie Orgel were developing an alternative model known as

TABLE 22.5

The Effective Atomic Numbers of Common Transition Metal Complexes

Complex	EAN of Complex	EAN of Corresponding Noble Gas
$Cr(CO)_6$	$24 + 0 + 6(2) = 36$	36
$Fe(CO)_4{}^{2-}$	$26 + 2 + 4(2) = 36$	36
$Fe(H_2O)_6{}^{2+}$	$26 - 2 + 6(2) = 36$	36
$Fe(CN)_6{}^{4-}$	$26 - 2 + 6(2) = 36$	36
$Co(NH_3)_6{}^{3+}$	$27 - 3 + 6(2) = 36$	36
$Ni(CO)_4$	$28 + 0 + 4(2) = 36$	36
$Zn(NH_3)_4{}^{2+}$	$30 - 2 + 4(2) = 36$	36
$V(H_2O)_6{}^{2+}$	$23 - 2 + 6(2) = 33$	36
$Cr(NH_3)_6{}^{3+}$	$24 - 3 + 6(2) = 33$	36
$MnF_6{}^{4-}$	$25 - 2 + 6(2) = 35$	36
$Fe(H_2O)_6{}^{3+}$	$26 - 3 + 6(2) = 35$	36
$Co(SCN)_4{}^{2-}$	$27 - 2 + 4(2) = 33$	36
$Ni(CN)_4{}^{2-}$	$28 - 2 + 4(2) = 34$	36
$Cu(NH_3)_4{}^{2+}$	$29 - 2 + 4(2) = 35$	36

crystal field theory. The crystal field model tried to describe what happens to the energies of the valence orbitals of an atom or ion in a crystal. This model is based on the assumption that the bonds that hold the crystal together are ionic.

OCTAHEDRAL CRYSTAL FIELDS

The structure of manganese(II) oxide consists of a closest-packed array of O^{2-} ions with Mn^{2+} ions in all of the octahedral holes (see Section 12.9). Each Mn^{2+} ion is surrounded by six O^{2-} ions arranged toward the corners of an octahedron of neighboring O^{2-} ions, and vice versa. MnO is therefore analogous to a transition metal complex in which a central Mn^{2+} ion is coordinated to six ligands, as shown in Figure 22.14. Such a transition metal complex is called an ***octahedral complex.***

What is the relationship between the $4s$ and $4p$ orbitals on an isolated Mn^{2+} ion and those on an Mn^{2+} ion buried in an MnO crystal? The energies of the orbitals on the ion in the crystal are higher. They have been raised by the repulsion between the electrons in these orbitals and the electrons on the six O^{2-} ions that surround the metal ion. The three $4p$ orbitals are still degenerate, however. All these orbitals are still equal in energy, because each $4p$ orbital points toward two O^{2-} ions at the corners of the octahedron.

Repulsion between electrons on the O^{2-} ions and electrons in the $3d$ orbitals on the Mn^{2+} in MnO also increases the energy of these orbitals. But the five $3d$ orbitals on the Mn^{2+} ion are no longer degenerate. Two of these orbitals ($d_{x^2-y^2}$ and d_{z^2}) point directly toward the six O^{2-} ions, as shown in Figure 22.15. The other three orbitals (d_{xy}, d_{xz}, and d_{yz}) lie between the O^{2-} ions.

The energy of the $3d$ orbitals increases when the six O^{2-} ions are brought close to the Mn^{2+} ion. However, the energy of two of these orbitals ($d_{x^2-y^2}$ and d_{z^2}) increases much more than the energy of the other three (d_{xy}, d_{xz}, and d_{yz}), as shown in Figure 22.16.

The five $3d$ orbitals in an isolated Mn^{2+} ion are degenerate — they have exactly the same energy. The only difference in these orbitals is their orientation in space,

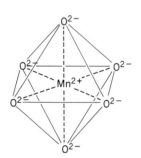

FIG. 22.14 Crystal field theory tries to predict what will happen to the energies of the orbitals on a transition metal ion when the ion is trapped in a crystal lattice. MnO is an example of a compound in which each Mn^{2+} ion is affected by an octahedral crystal field from the six O^{2-} ions that form the octahedral hole in which the Mn^{2+} ion is trapped. The results of this model can be extended to octahedral transition metal complexes, such as the $Mn(H_2O)_6{}^{2+}$ ion.

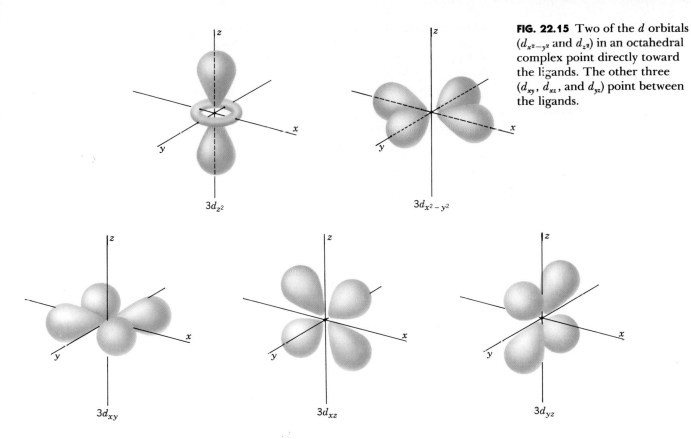

FIG. 22.15 Two of the d orbitals ($d_{x^2-y^2}$ and d_{z^2}) in an octahedral complex point directly toward the ligands. The other three (d_{xy}, d_{xz}, and d_{yz}) point between the ligands.

which has no effect on the energy of an orbital in an isolated atom or ion. The crystal field of the six O^{2-} ions in MnO, however, splits this degeneracy. Three of the orbitals are now lower in energy than the other two.

For reasons that are neither obvious nor important, the d_{xy}, d_{xz}, and d_{yz} orbitals are called the t_{2g} orbitals. By convention, the $d_{x^2-y^2}$ and d_{z^2} orbitals are called the e_g orbitals. The easiest way of remembering this convention is to note that there are three orbitals in the t_{2g} set.

$$t_{2g} = d_{xy}, d_{xz}, \text{ and } d_{yz}$$
$$e_g \;= d_{x^2-y^2} \text{ and } d_{z^2}$$

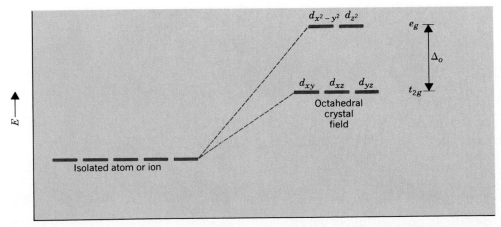

FIG. 22.16 The two d orbitals that point toward the ligands in an octahedral complex are higher in energy than the three d orbitals that point between the ligands. The five d orbitals are said to be split into t_{2g} and e_g sets of orbitals. The difference between the energies of the t_{2g} and e_g orbitals in an octahedral complex is represented by the symbol Δ_o.

The difference between the energies of the t_{2g} and e_g orbitals in an octahedral transition metal complex is represented by the symbol Δ_o. The effect of this splitting of energy is not trivial—Δ_o is 242 kJ/mol for the $Ti(H_2O)_6{}^{3+}$ ion. The magnitude of the effect changes from complex to complex, however. It depends on the identity of the metal ion, the charge on this ion, and the nature of the ligand.

TETRAHEDRAL CRYSTAL FIELDS

Copper(I) chloride crystallizes as a closest-packed array of Cl^- ions with Cu^+ ions in half of the tetrahedral holes (see Section 12.9). Each Cu^+ ion is surrounded by a tetrahedron of neighboring Cl^- ions. CuCl is therefore analogous to a transition metal complex in which the metal ion is coordinated to four ligands arranged toward the corners of a tetrahedron, as shown in Figure 22.17. Such a transition metal complex is called a **tetrahedral complex.**

Once again, the crystal field of the neighboring negative ions splits the degeneracy of the five $3d$ atomic orbitals on the transition metal ion. The tetrahedral crystal field splits these orbitals into the same t_{2g} and e_g sets of orbitals as the octahedral crystal field.

$$t_{2g} = d_{xy}, d_{xz}, \text{ and } d_{yz}$$
$$e_g = d_{x^2-y^2} \text{ and } d_{z^2}$$

But the two orbitals in the e_g set are lower in energy than the three orbitals in the t_{2g} set, as shown in Figure 22.18.

To understand the splitting of d orbitals in a tetrahedral crystal field, imagine four ligands lying at alternating corners of a cube to form a tetrahedral geometry, as shown in Figure 22.19. The $d_{x^2-y^2}$ and d_{z^2} orbitals on the metal ion at the center of the cube lie between the ligands, and the d_{xy}, d_{xy}, and d_{yz} orbitals point toward the ligands. As a result, the splitting observed in a tetrahedral crystal field is the opposite of the splitting in an octahedral complex.

Because there are fewer ligands in a tetrahedral complex, the magnitude of the splitting is smaller. The difference between the energies of the t_{2g} and e_g orbitals in a tetrahedral complex is slightly less than half as large as the splitting in analogous octahedral complexes.

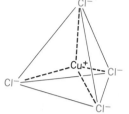

FIG. 22.17 CuCl is an example of a compound in which each Cu^+ ion is affected by a tetrahedral crystal field from the four Cl^- ions that form the hole in which the Cu^+ ion is trapped. The results of this model can be extended to tetrahedral transition metal complexes, such as the $Co(SCN)_4{}^{2-}$ ion.

FIG. 22.18 In a tetrahedral complex, the e_g orbitals are lower in energy than the t_{2g} orbitals. The difference between the energies of the e_g and t_{2g} sets of orbitals in a tetrahedral complex is smaller than the difference in an equivalent octahedral complex: $\Delta_t = \frac{4}{9}\Delta_o$.

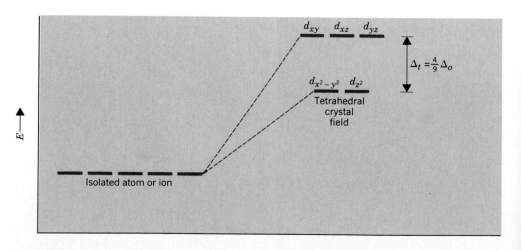

SQUARE PLANAR COMPLEXES

The principles of the crystal field theory can be extended to square planar complexes, such as $Pt(NH_3)_2Cl_2$. The results of such an analysis are shown in Figure 22.20.

22.14 CRYSTAL FIELD THEORY AND COMMON TRANSITION METAL OXIDATION STATES

We don't need crystal field theory to explain why so many transition metal ions have oxidation states that correspond to d^0, d^5, or d^{10} electron configurations. Time after time, we have seen that configurations in which valence subshells are either full, half full, or empty are inherently stable.

Crystal field theory is useful, however, in explaining some of the other common configurations outlined in Table 22.4. The stability of complexes such as the $V(H_2O)_6^{2+}$ and $Cr(NH_3)_6^{3+}$ ions can be attributed to the fact that a d^3 configuration in an octahedral crystal field corresponds to a half-filled subshell.

Octahedral d^3 complexes:

$$\underline{\uparrow} \quad \underline{\uparrow} \quad \underline{\uparrow}$$

The same argument can be used to explain why octahedral d^6 complexes such as $Fe(H_2O)_6^{2+}$ and $Co(NH_3)_6^{3+}$ are stable.

Octahedral d^6 complexes:

$$\underline{\uparrow\downarrow} \quad \underline{\uparrow\downarrow} \quad \underline{\uparrow\downarrow}$$

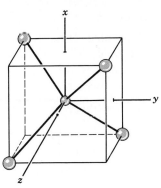

FIG. 22.19 The splitting of d orbitals in a tetrahedral complex is the opposite of the splitting in octahedral complexes. In a tetrahedral complex, the orbitals that lie along the x, y, and z axes of the coordinate system point between the ligands. The orbitals that lie in the xy, xz, and yz planes, however, point toward the ligands.

Exercise 22.11

Use crystal field theory to explain why Co^{2+} ions tend to form tetrahedral complexes, such as the $Co(SCN)_4^{2-}$ ion, instead of octahedral complexes, such as the Co^{3+} ion.

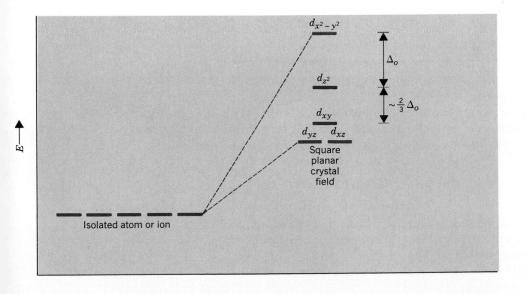

FIG. 22.20 The splitting of d orbitals in square planar transition metal complexes.

Solution

We have already seen why Co^{3+} ions form octahedral complexes. In an octahedral crystal field a d^6 configuration implies that the t_{2g} orbitals are totally filled.

To understand why Co^{2+} ions form tetrahedral complexes, we have to recognize that this ion has a d^7 configuration.

$$Co^{2+} = [Ar]\ 3d^7$$

Crystal field theory suggests that tetrahedral d^7 complexes should be more stable than octahedral d^7 complexes. Only tetrahedral d^7 complexes have a configuration in which the t_{2g} and e_g sets of orbitals are either full or half full.

Tetrahedral d^7 complexes:

$$\underline{\uparrow}\ \underline{\uparrow}\ \underline{\uparrow}$$
$$\underline{\uparrow\downarrow}\ \underline{\uparrow\downarrow}$$

Octahedral d^7 complexes:

$$\underline{\uparrow}\ \underline{}$$
$$\underline{\uparrow\downarrow}\ \underline{\uparrow\downarrow}\ \underline{\uparrow\downarrow}$$

It is not surprising that ions with a d^7 configuration tend to form tetrahedral complexes, such as the $Co(SCN)_4{}^{2-}$ complex ion.

Crystal field theory also explains why d^8 metal ions (such as Ni^{2+}, Pd^{2+}, and Pt^{2+}) and d^9 metal ions (such as Cu^{2+}) form square planar complexes. In this geometry, all of the orbitals are either filled, half filled, or empty.

square planar d^8 complexes:

$$\underline{}$$
$$\underline{\uparrow\downarrow}$$
$$\underline{\uparrow\downarrow}$$
$$\underline{\uparrow\downarrow}\ \underline{\uparrow\downarrow}$$

square planar d^9 complexes:

$$\underline{\uparrow}$$
$$\underline{\uparrow\downarrow}$$
$$\underline{\uparrow\downarrow}$$
$$\underline{\uparrow\downarrow}\ \underline{\uparrow\downarrow}$$

22.15 THE SPECTROCHEMICAL SERIES

The splitting of d orbitals in the crystal field model obviously depends on the geometry of the complex. It also depends on the nature of the metal ion, the charge on this ion, and the ligands that surround the metal. Studies of the absorption of light by transition metal complexes have provided measurements of Δ for a large number of complexes. For a given ligand, the value of Δ decreases in the following order.

$$Pt^{4+} > Ir^{3+} > Rh^{3+} > Co^{3+} > Cr^{3+} > Fe^{3+} > Fe^{2+} > Co^{2+} > Ni^{2+} > Mn^{2+}$$

strong-field ions   weak-field ions

Metal ions at one end of this continuum are called strong-field ions because the splitting due to the crystal field is unusually strong. Conversely, ions at the other end of the spectrum are known as weak-field ions.

For a given metal, the value of Δ decreases in the following order.

$$CO \sim CN^- > NO_2^- > NH_3 > -NCS^- > H_2O > OH^- > F^- > -SCN^- \sim Cl^- > Br^- > I^-$$

strong-field ligands   weak-field ligands

Ligands that give rise to large differences between the energies of the t_{2g} and e_g orbitals are called strong-field ligands. At the opposite extreme are the weak-field ligands.

Because they result from studies of the absorption spectra of transition metal complexes, these generalizations are known as the **spectrochemical series.** The range of values of Δ for a given geometry is remarkably large. The value of Δ_o, for example, is as small as 100 kJ/mol in the $Ni(H_2O)_6^{2+}$ ion and as large as 520 kJ/mol in the $Rh(CN)_6^{3-}$ ion.

22.16 HIGH-SPIN VERSUS LOW-SPIN COMPLEXES

In Section 5.19, we suggested that degenerate orbitals are filled according to Hund's rules.

1. One electron is added to each of the degenerate orbitals in a subshell before a second electron is added to any orbital in the subhsell.
2. Electrons added to a subshell have the same value for the spin quantum number until each orbital in the subshell has at least one electron.

Octahedral transition metal ions with d^1, d^2, or d^3 configurations are therefore written as follows.

(octahedral crystal field)

As soon as we try to add a fourth electron, however, we are faced with a problem. We have to decide whether the fourth electron should be used to pair one of the electrons in the lower-energy (t_{2g}) set of orbitals or whether it should be placed in one of the higher-energy (e_g) orbitals.

low-spin d^4 high-spin d^4

One of these configurations is called **high-spin** because it contains four unpaired electrons with the same spin. The other is called **low-spin** because it contains only two unpaired electrons. The same problem occurs with octahedral d^5, d^6, ad d^7 complexes, as shown in Figure 22.21.

For octahedral d^8, d^9, and d^{10} complexes, there is only one way to write satisfactory configurations, as shown in Figure 22.22. Thus, we only have to worry about high-spin versus low-spin octahedral complexes when there are four, five, six, or seven electrons in the d orbitals.

In theory, the choice between high-spin and low-spin configurations for octahedral d^4, d^5, d^6, or d^7 complexes is easy. All we have to do is compare the energy it takes to pair electrons with the energy it takes to excite an electron to the higher-energy (e_g) orbitals. If it takes less energy to pair the electrons, the complex is low-spin. If it takes less energy to excite the electron, the complex is high-spin.

For the sake of argument, let's assume that it takes the same energy to pair electrons in the t_{2g} orbitals of an octahedral complex regardless of the identity of

FIG. 22.21 In an octahedral crystal field, high-spin and low-spin configurations can be written for d^4, d^5, d^6, and d^7 complexes.

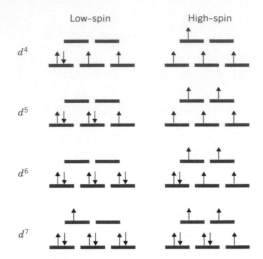

the metal or ligand. The amount of energy needed to excite an electron into the higher-energy (e_g) orbitals, however, depends on the value of Δ_o for the complex.

As a result, we expect to find low-spin complexes among metal ions and/or ligands that lie toward the high-field end of the spectrochemical series. High-spin complexes are expected among metals and/or ligands that lie toward the low-field end of these series.

$$\text{low-spin } d^6 \qquad \text{high-spin } d^6$$

Compounds in which all of the electrons are paired are **_diamagnetic_**—they are repelled by both poles of a magnet. Compounds that contain one unpaired electron or more are **_paramagnetic_**—they are attracted to the poles of a magnet. The force of attraction between paramagnetic complexes and a magnetic field is proportional to the number of unpaired electrons in the complex. We can therefore determine whether a complex is high-spin or low-spin by measuring the strength of the interaction between the complex and a magnetic field.

Exercise 22.12

Explain why the $Co(NH_3)_6^{3+}$ ion is a diamagnetic, low-spin complex but the CoF_6^{3-} ion is a paramagnetic, high-spin complex.

Solution

Both complexes contain the Co^{3+} ion, which lies toward the strong-field end of the spectrochemical series. NH_3 also lies toward the strong-field end of this series. As a result, Δ_o for the $Co(NH_3)_6^{3+}$ ion is relatively large. When the splitting is large, it takes less energy to pair the electrons in the t_{2g} orbitals than it does to excite the electrons to the e_g orbitals. The $Co(NH_3)_6^{3+}$ ion is therefore a low-spin d^6 complex in which the electrons are all paired, and the ion is diamagnetic.

The F^- ion lies toward the low-field end of the spectrochemical series. As a result, Δ_o for the CoF_6^{3-} is much smaller. Here, it takes less energy to excite electrons into the e_g orbitals than to pair them in the t_{2g} orbitals. The CoF_6^{3-} ion is therefore a high-spin d^6 complex, which contains unpaired electrons and is paramagnetic.

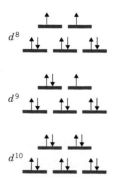

FIG. 22.22 There are no high-spin and low-spin configurations for octahedral d^8, d^9, and d^{10} complexes, because there is only one way of writing these configurations for an octahedral crystal field.

22.17 LIGAND FIELD THEORY

The valence bond model described in Section 22.11 and the crystal field theory described in Section 22.13 explain some aspects of the chemistry of the transition metals, but neither model is good at predicting all of the properties of transition metal complexes. A third model, based on molecular orbital theory, was therefore developed. This so-called *ligand field theory* is more powerful than either the valence bond or crystal field theories. Unfortunately, it is also more abstract.

The ligand field model for an octahedral transition metal complex such as the $Co(NH_3)_6^{3+}$ ion assumes that the $3d$, $4s$, and $4p$ orbitals on the metal overlap with one orbital on each of the six ligands to form a total of 15 molecular orbitals, as shown in Figure 22.23. Six of these are *bonding molecular orbitals*, whose energies are much lower than those of the original atomic orbitals. Another six are *antibonding molecular orbitals*, whose energies are higher than those of the original atomic orbitals. Three are best described as *nonbonding molecular orbitals*, because they have essentially the same energy as the $3d$ atomic orbitals on the metal.

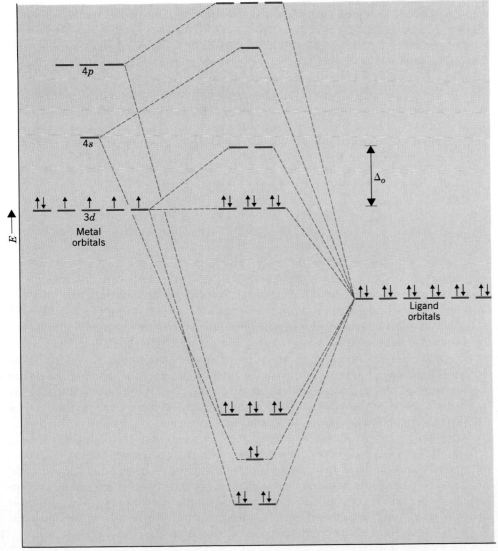

FIG. 22.23 Ligand field theory is a molecular orbital alternative to the valence bond and crystal field models of coordination complexes. The molecular orbitals in this diagram are generated when the $3d$, $4s$, and $4p$ orbitals on the transition metal overlap with an orbital on each of the six ligands that contribute nonbonding electrons, to form an octahedral coordination complex.

Ligand field theory contains elements of both the valence bond and crystal field models. It enables the $3d$, $4s$, and $4p$ orbitals on the metal to overlap with orbitals on the ligand to form the octahedral covalent bond skeleton that holds this complex together. At the same time, this model generates a set of five orbitals in the center of the diagram that are split into t_{2g} and e_g subshells, as predicted by the crystal field theory.

As a result, we don't have to worry about "inner-shell" versus "outer-shell" metal complexes. In effect, we can use the $3d$ orbitals in two different ways. We can use them to form the covalent bond skeleton and then use them again to form the orbitals that hold the electrons that were originally in the $3d$ orbitals of the transition metal.

22.18 THE COLORS OF TRANSITION METAL COMPLEXES

TABLE 22.6

Characteristic Colors of Some Common Transition Metal Ions

Cr^{2+}	blue
Cr^{3+}	blue-violet
Co^{2+}	pinkish red
Cu^{2+}	light blue
Fe^{3+}	yellow to reddish brown
Mn^{2+}	faint pink
Ni^{2+}	green
Zn^{2+}	colorless
CrO_4^{2-}	yellow
$Cr_2O_7^{2-}$	orange
MnO_4^-	deep purple

One of the joys of working with transition metals is the extraordinary range of colors exhibited by their compounds. By making small changes in the combination of ammonia, water, and chloride ligands on a Co^{3+} ion, for example, we can vary the color of the complex over the entire visible spectrum, from red to violet, as shown in Section 22.2.

We can also vary the color of the complex by keeping the ligand constant and changing the metal ion. The characteristic colors of aqueous solutions of some common transition metal ions are given in Table 22.6.

The color of these complexes also changes with the oxidation state of the metal atom. By passing a solution of the VO_2^+ ion down a column packed with zinc metal it is possible to change the color of the solution from yellow to blue to green and then to violet as the vanadium is reduced, one electron at a time, from the $+5$ to the $+2$ oxidation state (see Figure 22.24.)

VO_2^+	yellow
VO^{2+}	blue
V^{3+}	green
V^{2+}	violet

These observations inevitably lead to the question of why so many transition metal complexes are colored and why the color is so sensitive to changes in the metal, ligand, or oxidation state. The first step in explaining the color of these complexes is to understand the physics of color.

There are two ways of producing the sensation of color. We can add color where none exists or subtract it from white light. The three primary *additive colors* are red, green, and blue. When all three are present at the same intensity, we get white light. The three primary *subtractive colors* are cyan, magenta, and yellow. When light enters an object that absorbs these three colors with the same intensity, the light is absorbed. If the object absorbs light strongly enough, or if the intensity of the light is dim enough, all of the light can be absorbed.

The additive and subtractive colors are complementary, as shown in Figure 22.25. If we subtract yellow from white light we get a mixture of cyan and magenta—and the light looks blue. If we subtract blue—a mixture of cyan and magenta—from white light, the light looks yellow. The CrO_4^{2-} ion appears yellow because it absorbs blue light. The $Cu(NH_3)_4^{2+}$ complex ion has a blue color because it absorbs light in the yellow portion of the spectrum. Table 22.7 describes what happens when light in different portions of the visible spectrum is absorbed.

FIG. 22.24 The color of a transition metal complex depends on the oxidation state of the metal ion as well as the ligands bound to the metal ion. This photograph shows the step-by-step reduction of the VO_2^+ ion to the V^{2+} ion with zinc metal.

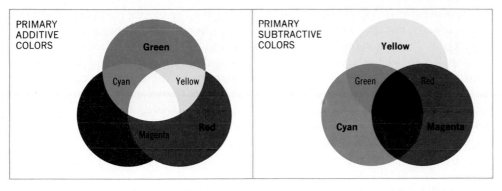

FIG. 22.25 The three primary additive colors are green, blue, and red. The three primary subtractive colors are cyan, magenta, and yellow. Each of the additive colors can be prepared from two of the subtractive colors. Mixing yellow and cyan gives green; mixing yellow and magenta gives red; mixing cyan and magenta gives blue. Each of the subtractive colors can be prepared from two of the additive colors. Green and red give yellow; red and blue give magenta; blue and green give cyan.

What happens to the energy of the light absorbed by a compound? Light is absorbed when it carries just enough energy to excite an electron from one orbital to another. Compounds therefore absorb light when the difference between the energy of an orbital that contains an electron and the energy of an empty orbital corresponds to a wavelength in the narrow band of the electromagnetic spectrum that is visible to the naked eye.

How much energy is associated with the absorption of a typical photon, such as a photon with a wavelength of 480 nanometers? As we can see from Table 22.7, a photon with a wavelength of 480 nanometers falls in the blue portion of the electromagnetic spectrum. If we know the wavelength of this photon, we can calculate its frequency.

$$\nu = \frac{c}{\lambda} = \frac{2.998 \times 10^8 \text{ m/s}}{480 \times 10^{-9} \text{ m}} = 6.246 \times 10^{14} \text{ s}^{-1}$$

From the frequency and Planck's constant, we can calculate its energy.

$$E = h\nu = (6.626 \times 10^{-34} \text{ J-s})(6.246 \times 10^{14} \text{ s}^{-1})$$
$$= 4.138 \times 10^{-19} \text{ J}$$

But this is the energy associated with only a single photon. Multiplying by Avogadro's constant gives us the energy of the photon in units of kilojoules per mole.

$$4.138 \times 10^{-19} \frac{\text{J}}{\cancel{\text{photon}}} \times 6.022 \times 10^{23} \frac{\cancel{\text{photons}}}{\text{mol}} = 249.2 \frac{\text{kJ}}{\text{mol}}$$

TABLE 22.7

The Relationship between the Color of Transition Metal Complexes and the Wavelength of Light Absorbed

Wavelength Absorbed (nm)	Color of Light Absorbed	Color of Complex
410	violet	lemon yellow
430	indigo	yellow
480	blue	orange
500	blue-green	red
530	green	purple
560	lemon yellow	violet
580	yellow	indigo
610	orange	blue
680	red	blue-green

Repeating this calculation for all frequencies in the visible portion of the spectrum reveals that the energy associated with a photon of light ranges from 160 to 300 kJ/mol. If we refer back to the discussion of the spectrochemical series in Section 22.15, we find that these energies fall in the range of values of Δ_o. Anything that changes the difference between the energies of the t_{2g} and e_g orbitals in an octahedral or tetrahedral transition metal complex therefore influences the color of the light absorbed by this complex. It is not surprising that the color of the complex is sensitive to factors such as the identity of the metal, the nature of the ligand, and the oxidation state of the metal.

SUMMARY

The transition metals are the elements that serve as a bridge, or transition, between the main-group elements on either side of the periodic table. Although they have many of the same chemical and physical properties as the main-group metals, they differ in two ways: (1) they tend to be more electronegative than the main-group metals and therefore more likely to form covalent compounds, and (2) they are more likely to form coordination complexes, such as the $Cu(NH_3)_4^{2+}$ and $Co(NH_3)_6^{3+}$ ions.

These coordination complexes result from reactions between a transition metal ion acting as a Lewis acid and one or more ligands acting as Lewis bases. In each case, a pair of nonbonding electrons on the ligand—the Lewis base—is donated to an empty orbital on the transition metal—the Lewis acid—to form a covalent bond.

The coordination number of a transition metal complex is equal to the number of ligands bound to the transition metal. The most common coordination numbers are 4 and 6. Four-coordinate complexes usually have structures that are either tetrahedral or square planar. Six-coordinate complexes tend to have an octahedral structure.

Coordination complexes often form isomers—two or more compounds with the same empirical formula but different structures. Kinds of isomers include geometric (cis/trans) isomers, linkage isomers, and chiral isomers.

The common oxidation numbers for transition-metal ions correspond to d^0, d^3, d^5, d^6, d^8, or d^{10} electron configurations. Transition metal ions that have a charge of $+1$, $+2$, or $+3$ can exist in water. Ions that have a larger charge are so electronegative that they form covalent bonds to O^{2-} ions they remove from neighboring water molecules.

The valence bond theory can be applied to coordination complexes. When there are enough empty valence-shell d orbitals—such as the $3d$ orbitals—the transition metal ion is assumed to form a set of empty dsp^2 or d^2sp^3 hybrid orbitals that can accept pairs of nonbonding electrons from the ligands. When there aren't enough empty valence-shell d orbitals, the transition metal is assumed to use the empty d orbitals in the next shell—such as the $4d$ orbitals—to form sets of sp^2d or sp^3d^2 hybrid orbitals that can accept pairs of electrons from the ligands.

The crystal field theory can be used to explain why common oxidation numbers for transition metal ions tend to correspond to d^0, d^3, d^5, d^6, d^8, or d^{10} electron configurations. This model was originally developed to explain how the energies of the orbitals in an ion are affected when the ion is trapped in the holes of an ionic crystal. When the model is extended to transition metal complexes, it suggests that the five valence-shell d orbitals on the transition metal are no longer degenerate. In an octahedral complex, they are split into a set of t_{2g} orbitals with relatively low energies and a set of e_g orbitals with higher energies. In a tetrahedral complex, the d orbitals are again split into a t_{2g} and an e_g set, but here the e_g orbitals have lower energies than the t_{2g} orbitals.

When the difference between the energies of the t_{2g} and e_g sets of orbitals is relatively small, electrons are distributed among these orbitals as expected from Hund's rules. The net result is a complex that is described as high-spin. When the difference between the energies of the t_{2g} and e_g sets of orbitals is relatively large, it often takes less energy to pair electrons in the lower-energy set of orbitals than to satisfy Hund's rules. The net result is a complex that is best described as low-spin.

Elements of the valence bond and crystal field theories are combined in the ligand field theory. In effect, this theory allows us to use $3d$ orbitals to form the covalent bond skeleton and also to form the orbitals that hold the electrons that were originally in the $3d$ orbitals of the transition metal.

PROBLEMS

Transition Metals and Coordination Complexes

22-1 List the transition metals in the fourth row of the periodic table. Identify the subshell of atomic orbitals filled among this series.

22-2 Describe the arguments for including zinc, cadmium, and mercury among the transition metals. Describe the arguments for grouping these metals with the main-group elements. What do you believe is the best way of classifying these elements?

22-3 Describe some of the ways in which transition metals are different from main-group metals such as aluminum, tin, and lead. Describe ways in which they are similar.

Werner's Model of Coordination Complexes

22-4 Werner wrote the formula of one of his coordination complexes as $CoCl_3 \cdot 6\,NH_3$. Today, we write this compound as $[Co(NH_3)_6]Cl_3$ to indicate the presence of $Co(NH_3)_6^{3+}$ and Cl^- ions. Write modern formulas for the compounds Werner described as $CoCl_3 \cdot 5\,NH_3$, $CoCl_3 \cdot 4\,NH_3$, and $CoCl_3 \cdot 5\,NH_3 \cdot H_2O$.

22-5 Describe the difference between the primary and secondary valence of the Co^{3+} ion in $[Co(NH_3)_6]Cl_3$ and $[Co(NH_3)_4Cl_2]Cl$.

22-6 Explain why aqueous solutions of $CoCl_3 \cdot 6\,NH_3$ are better conductors of electricity than aqueous solutions of $CoCl_3 \cdot 4\,NH_3$.

22-7 Explain why Ag^+ ions precipitate three chloride ions from an aqueous solution of $CoCl_3 \cdot 6\,NH_3$ but only one chloride ion from an aqueous solution of $CoCl_3 \cdot 4\,NH_3$.

22-8 Define the terms *coordination number* and *ligand*.

22-9 Use the examples in Table 22.1 to identify one of the factors that influence the coordination number of a transition metal ion.

22-10 Determine both the coordination number and the charge on the transition metal ion in each of the following complexes.

(a) CuF_4^{2-} (b) $Cr(CO)_6$ (c) $Fe(CN)_6^{4-}$ (d) $Pt(NH_3)_2Cl_2$

22-11 Determine both the coordination number and the charge on the transition metal ion in each of the following complexes.

(a) $Co(SCN)_4^{2-}$ (b) $Fe(acac)_3$ (c) $Ni(en)_2(H_2O)_2^{2+}$ (d) $Co(NH_3)_5(H_2O)^{3+}$

The Electron Configuration of Transition Metal Ions

22-12 Write the electron configuration of the following transition metal ions.

(a) V^{2+} (b) Cr^{2+} (c) Mn^{2+} (d) Fe^{2+} (e) Ni^+ (f) Cu^{2+}

22-13 Explain why the Co^{2+} ion can be described as a d^7 ion.

22-14 Which of the following ions can be described as d^5?

(a) Cr^{2+} (b) Mn^{2+} (c) Fe^{3+} (d) Co^{3+} (e) Cu^+

Lewis Acid – Lewis Base Approach to Bonding in Complexes

22-15 Describe what to look for in deciding whether an ion or molecule is a Lewis acid.

22-16 Explain why Lewis acids, such as the Co^{3+} ion, pick up Lewis bases, or ligands, to form coordination complexes.

22-17 Which of the following are Lewis acids?

(a) Fe^{3+} (b) BF_3 (c) H^+ (d) Ag^+ (e) Cu^{2+}

22-18 Which of the following are Lewis bases?

(a) CO (b) O_2 (c) Cl^- (d) N_2 (e) NH_3

22-19 Which of the following are Lewis bases?

(a) CN^- (b) SCN^- (c) CO_3^{2-} (d) NO^+ (e) $S_2O_3^{2-}$

Typical Ligands

22-20 Define the terms *monodentate ligand*, *bidentate ligand*, *tridentate ligand*, and *tetradentate ligand*. Give an example of each.

22-21 Draw the structures of the following coordination complexes.

(a) $Fe(acac)_3$ (b) $Co(en)_3^{3+}$ (c) $Fe(EDTA)^-$ (d) $Fe(CN)_6^{4-}$

Nomenclature of Complexes

22-22 Name the following complexes.

(a) $Cu(NH_3)_4^{2+}$ (b) $Mn(H_2O)_6^{2+}$ (c) $Fe(CN)_6^{4-}$ (d) $Ni(en)_3^{2+}$ (e) $Cr(acac)_3$

22-23 Name the following complexes.

(a) $Pt(NH_3)_2Cl_2$ (b) $Ni(CO)_4$ (c) $Co(en)_3^{3+}$ (d) $Na[Mn(CO)_5]$

22-24 Name the following complexes.

(a) $Na_3[Co(NO_2)_6]$ (b) $Na_2[Zn(CN)_4]$ (c) $[Co(NH_3)_4Cl_2]Cl$ (d) $[Ag(NH_3)_2]Cl$

22-25 Write the formulas for the following compounds.

(a) hexamminechromium(III) chloride (b) chloropentamminechromium(III) chloride (c) triethylenediamminecobalt(III) chloride (d) potassium tetranitritodiamminecobaltate(III)

Isomers

22-26 Which of the following square planar complexes can form *cis* / *trans* isomers?

(a) $Cu(NH_3)_4^{2+}$ (b) $Pt(NH_3)_2Cl_2$ (c) $RhCl_3(CO)$ (d) $IrCl(CO)(PH_3)_2$

22-27 Which of the following octahedral complexes can form *cis/trans* isomers?

(a) $Co(NH_3)_6^{3+}$ (b) $Co(NH_3)_5Cl^{2+}$ (c) $Co(NH_3)_5(H_2O)^{3+}$
(d) $Co(NH_3)_4Cl_2^+$ (e) $Co(NH_3)_4(H_2O)_2^{3+}$

22-28 The SCN^- ion forms linkage isomers in which either the nitrogen or the sulfur is bound to a transition metal ion. Which of the following ligands can also form linkage isomers?

(a) CO_2 (b) CN^- (c) $S_2O_3^{2-}$ (d) CO (e) OH^-

22-29 Compounds are optically active when the mirror image of the compound cannot be superimposed upon itself. Draw the mirror images of the following complex ions and determine which of these ions exist as chiral isomers.

(a) $Cu(NH_3)_4^{2+}$ (a square planar complex) (b) $Co(NH_3)_6^{2+}$ (an octahedral complex) (c) $Ag(NH_3)_2^+$ (a linear complex) (d) $Cr(en)_3^{3+}$ (an octahedral complex)

Common Oxidation States of Transition Metals

22-30 Which of the following electron configurations do not correspond to a common oxidation number of transition metal ions?

(a) d^0 (b) d^3 (c) d^4 (d) d^5 (e) d^6

22-31 Which of the following transition metal ions are in an oxidation state in which the electron configuration of the metal is formally d^3?

(a) V^{2+} (b) Cr^{3+} (c) MnO_2 (d) Fe^{2+} (e) Co^{3+}

22-32 Which of the following transition metal ions are in an oxidation state in which the electron configuration of the metal is either d^5 or d^6?

(a) Mn^{2+} (b) Fe^{2+} (c) Fe^{3+} (d) Co^{3+} (e) Ni^{2+}

22-33 Which of the following transition metal ions are in an oxidation state in which the electron configuration of the metal is not formally d^0?

(a) Sc^{3+} (b) Ti^{4+} (c) VO_2^+ (d) CrO_4^{2-} (e) Fe^{3+}

22-34 Use the general trends in electron configurations of transition metal ions to predict the common oxidation states of the following elements.

(a) Sc (b) Ti (c) V (d) Cr (e) Mn

22-35 Use the general trends in electron configurations of transition metal ions to predict the common oxidation states of the following elements.

(a) Fe (b) Co (c) Ni (d) Cu (e) Zn

22-36 Give examples of transition metal ions in oxidation states from $+1$ through $+7$. Examples of monatomic metal ions with charges of $+1$, $+2$, and $+3$ are easy to find. Can you find an example of a monatomic transition metal ion with a charge larger than $+3$?

The Valence Bond Approach to Bonding in Complexes

22-37 Describe the difference between the valence bond model for the $Co(NH_3)_6^{3+}$ complex ion and the valence bond model for the $Ni(NH_3)_6^{2+}$ complex ion.

22-38 Describe the difference between inner-shell and outer-shell hybrid orbitals in the valence bond model of transition metal complexes.

22-39 Apply the valence bond model of bonding in transition metal complexes to $Cr(CO)_6$ and $Fe(CO)_5$.

22-40 Apply the valence bond model of bonding in transition metal complexes to the $Zn(NH_3)_4^{2+}$ and $Fe(H_2O)_6^{3+}$ complex ions.

The Effective Atomic Number (EAN) Rule

22-41 Use the effective atomic number rule to predict the charge on the $Mn(CO)_5^{x-}$ ion.

22-42 Use the effective atomic number rule to predict the charge on the HgI_4^{x-} ion.

22-43 Use the effective atomic number rule to predict the coordination number of the Fe^{2+} ion in the $Fe(NH_3)_x^{2+}$ ion.

22-44 Use the effective atomic number rule to predict the coordination number of the Cd^{2+} ion in the $Cd(OH)_x^{2-}$ ion.

22-45 Use the effective atomic number rule to predict the coordination number of the Ni atom in $Ni(CO)_x$.

Crystal Field Theory

22-46 Describe what happens to the energies of the $3d$ atomic orbitals in an octahedral crystal field.

22-47 Describe what happens to the energies of the $3d$ atomic orbitals in a tetrahedral crystal field.

22-48 Which of the $3d$ atomic orbitals in an octahedral crystal field belong to the t_{2g} set of orbitals? Which belong to the e_g set?

22-49 What do the orbitals in a t_{2g} set have in common? What do the orbitals in the e_g set have in common?

22-50 The $3d$ orbitals are split into t_{2g} and e_g sets in both octahedral and tetrahedral crystal fields. Is there any difference between the orbitals that go into the t_{2g} set in octahedral and in tetrahedral crystal fields?

22-51 Describe the two major differences between the splitting of the five $3d$ atomic orbitals in octahedral and in tetrahedral crystal fields.

22-52 Describe why we don't need crystal field theory to explain why so many transition metal ions have common oxidation states corresponding to d^0, d^5, or d^{10} electron configurations.

22-53 Use the splitting of the $3d$ atomic orbitals in an octahedral crystal field to explain the stability of oxidation states corresponding to d^3 and d^6 electron configurations in the $Cr(NH_3)_6^{3+}$ and $Fe(H_2O)_6^{2+}$ complex ions.

22-54 Use the splitting of the $3d$ atomic orbitals in a tetrahedral crystal field to explain the stability of oxidation states corresponding to the d^7 electron configuration in the $Co(SCN)_4{}^{2-}$ complex ion.

22-55 Use the splitting of the $3d$ atomic orbitals in a square planar crystal field to explain the stability of oxidation states corresponding to d^8 and d^9 electron configurations in the $Pt(NH_3)_2Cl_2$ and $Cu(NH_3)_4{}^{2+}$ complexes.

22-56 The difference between the energies of the t_{2g} and e_g sets of atomic orbitals in an octahedral or tetrahedral crystal field depends on both the metal atom and the ligands that form the complex. Which of the following metal ions would give the largest difference?

(a) Rh^{3+} (b) Cr^{3+} (c) Fe^{3+} (d) Co^{2+} (e) Mn^{2+}

22-57 Use the crystal field model to explain why Co^{3+} forms octahedral complexes, such as the $Co(NH_3)_6{}^{3+}$ ion, but Co^{2+} forms tetrahedral complexes, such as the $Co(SCN)_4{}^{2-}$ ion.

22-58 Use the crystal field model to explain why the d^8 Ni^{2+} ion forms octahedral complexes, such as the $Ni(NH_3)_6{}^{2+}$ ion, or square planar complexes, such as the $Ni(CN)_4{}^{2-}$ ion, but a neutral nickel atom that can be thought of as d^{10} forms tetrahedral $Ni(CO)_4$ complexes.

High-Spin Versus Low-Spin Complexes

22-59 Which of the following ligands gives the largest splitting of the t_{2g} and e_g sets of orbitals in an octahedral complex?

(a) CO   (b) NH_3 (c) H_2O (d) OH^- (e) I^-

22-60 Describe the difference between a high-spin and a low-spin d^6 complex.

22-61 What factors determine whether a complex is high-spin or low-spin?

22-62 Explain why the $Co(NH_3)_6{}^{3+}$ ion is a high-spin d^6 complex but $CoF_6{}^{3-}$ is a low-spin d^6 complex.

22-63 One of the $Fe(H_2O)_6{}^{2+}$ and $Fe(CN)_6{}^{4-}$ complex ions is high-spin and the other is low-spin. Which is which?

Ligand Field Theory

22-64 Describe how ligand field theory eliminates the difference between inner-shell complexes, such as the $Co(NH_3)_6{}^{3+}$ ion, and outer-shell complexes, such as the $Ni(NH_3)_6{}^{2+}$ ion.

22-65 Explain how ligand field theory allows the valence-shell d orbitals on the transition metal to be used simultaneously to form the skeleton structure of the complex and to hold the electrons that were originally in the d orbitals on the transition metal.

The Colors of Transition Metal Complexes

22-66 Describe the characteristic colors of aqueous solutions of the following transition metal ions.

(a) Cu^{2+} (b) Fe^{3+} (c) Ni^{2+} (d) $CrO_4{}^{2-}$ (e) $MnO_4{}^-$

22-67 Explain why so many of the pigments used in oil paints, such as vermilion (HgS), cadmium red (CdS), cobalt yellow [$K_3Co(NO_2)_6$], chrome yellow ($PbCrO_4$), Prussian blue ($Fe_4[Fe(CN)_6]_3$), and cobalt blue ($CoO \cdot Al_2O_3$), contain transition metal ions.

22-68 Explain why $Cu(NH_3)_4{}^{2+}$ complexes have a deep blue color if they do not absorb blue light. What frequencies of light do they absorb?

22-69 $CrO_4{}^{2-}$ ions are bright yellow. In what portion of the visible spectrum do these ions absorb light?

22-70 When $CrO_4{}^{2-}$ reacts with acid to form $Cr_2O_7{}^{2-}$ ions, the color shifts from bright yellow to orange. Does this mean that the frequencies of light absorbed shift toward a higher or a lower frequency?

22-71 Ni^{2+} forms a complex with the dimethylglyoxime (DMG) ligand that absorbs light in the blue-green portion of the spectrum. What is the color of this $Ni(DMG)_2$ complex?

22-72 Explain why a white piece of paper looks as if it has a faint pink color to a person who has been working for several hours at a computer terminal that has a "green" screen.

CHAPTER 23

NUCLEAR CHEMISTRY

CHAPTER CONTENTS

23.1 THE DISCOVERY OF RADIOACTIVITY

John Dalton proposed his atomic theory in 1803, and a table of atomic weights was available in 1865 when Dmitri Mendeléeff developed the periodic table. It may therefore seem hard to believe that the existence of atoms was still debated well into the 20th century. The concept of the atom was accepted as a useful tool, but there was no conclusive evidence that atoms really existed and no information about the structure of the atom until about the turn of the century.

In 1895, William Conrad Roentgen announced the discovery of x-rays, which were produced when cathode rays struck the anode in a cathode-ray tube (see Section 5.3). Roentgen noted that the glass walls of the tube emitted light, or *fluoresced,* when they were struck by these x-rays. More importantly, he found that x-rays passed through what had always been assumed to be solid matter (see Figure 23.1).

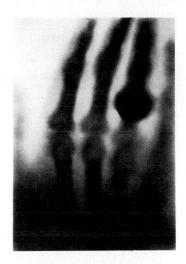

FIG. 23.1 One of the earliest x-ray images. There is reason to believe this is an x-ray of Roentgen's wife's hand.

The discovery of a form of radiation that could pass through solids fired the imagination of a generation of scientists, who rushed to study the new phenomenon. Henri Becquerel decided to investigate the connection between x-rays and fluorescence. He knew that salts of uranium, such as potassium uranyl sulfate, fluoresce when exposed to sunlight. He therefore wrapped a photographic plate in black paper, placed crystals of $K_2UO_2(SO_4) \cdot 2\ H_2O$ on top of the plate, and exposed the crystals to sunlight. When he developed the plates, black spots were found beneath the places where the crystals had been. This suggested that the uranium salts had emitted some form of radiation, which had passed through the paper and fogged the photographic plate.

Becquerel found the same results, however, when the crystals and photographic plate were prepared and kept in the dark. Furthermore, much better images were obtained with pure uranium metal, which did not fluoresce when exposed to sunlight. It soon became evident that a new form of radiation had been discovered, which eventually became known as ***radioactivity.*** By 1898, Marie Curie had found that compounds of thorium were also radioactive. After painstaking effort she eventually isolated two more radioactive elements, polonium and radium, from uranium-containing ores.

In 1899, Ernest Rutherford studied the absorption of radioactivity by thin sheets of metal foil and found two components—***alpha (α) particles,*** which were absorbed by metal foil that was only a few thousandths of a centimeter thick, and ***beta (β) particles,*** which could pass through 100 times as much foil before they were absorbed. Shortly thereafter, a third form of radiation, ***gamma (γ) rays,*** was discovered that could penetrate as much as several centimeters of lead.

These three forms of radiation differ in how they are affected by electric and magnetic fields, as shown in Figure 23.2. Analysis of the effect of these fields on α-particles suggested that an α-particle had the same charge-to-mass ratio as an He^{2+} ion. The equivalence of α-particles and He^{2+} ions was eventually confirmed by Rutherford through a beautiful experiment. Rutherford built an apparatus that allowed α-particles to pass through a very thin glass wall into an evacuated flask. After a few days, he was able to detect the helium gas that accumulated in this flask on the basis of the characteristic absorption spectrum of the element.

Early experiments with electric and magnetic fields suggested that β-particles had the same charge-to-mass ratio as electrons. Further experiments have shown no detectable difference between β-particles and electrons. The only reason to retain the name β-particle is to emphasize the fact that these particles are ejected from the nucleus of an atom when it undergoes radioactive decay.

FIG. 23.2 The effect of an electric field on α-, β-, and γ-radiation. Alpha particles are attracted toward the negative electrode and are therefore positively charged. Beta particles are attracted toward the positive electrode and are therefore negatively charged. Gamma rays are not affected by an electric field and are therefore electrically neutral.

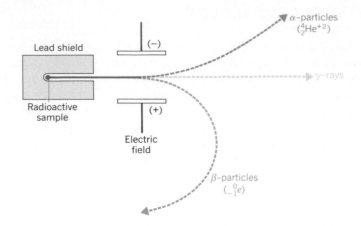

The fact that γ-rays are not deflected by either electric or magnetic fields suggests that these rays don't carry an electric charge. Since they travel at the speed of light, they are classified as a separate form of electromagnetic radiation (see Section 5.11), which carry even more energy than x-rays.

At the turn of the century, when radioactivity was discovered, atoms were assumed to retain their identities throughout all chemical and physical processes. Ernest Rutherford and Frederick Soddy found, however, that radioactive substances became less active with time. More importantly, they noted that radioactivity was accompanied by the release of atoms of a different element. By 1903, they had concluded that radioactivity was accompanied by a change in the structure of the atom and therefore proposed a model that assumed that radiation was emitted when an element decayed into a different kind of atom.

By 1910, an obvious problem had to be faced. At least 40 radioactive elements could be isolated when uranium decayed to lead, but there was space for only 11 elements between lead and uranium. In 1913, Kasimir Fajans and Frederick Soddy provided an explanation for these results based on the following rules.

1. α-particles were emitted when an element decayed to form an element that was two places to the left in the periodic table.
2. β-particles were emitted when the element decayed to form a new element that belonged one place to the right.
3. Radioactive elements that fall in the same place in the periodic table are different forms of the same element.

Soddy proposed the name *isotope* to describe different radioactive atoms that are in the same position in the periodic table. J. J. Thomson and Francis Aston then used the results of experiments done with a mass spectrometer (see Section 3.1) to show that isotopes are atoms of the same element that have different atomic masses.

23.2 THE STRUCTURE OF THE ATOM

The discovery of the electron in 1897 by J. J. Thomson suggested that there was an internal structure to the "indivisible" building blocks of matter known as atoms. Thomson addressed the question: How many electrons does an atom contain? By studying the scattering of light, x-rays, and α-particles by matter, he concluded that the number of electrons in an atom was between 0.2 and 2 times the weight of the atom. In Thomson's model of the atom, electrons were imbedded in a sea of positive charge distributed uniformly over the volume of the atom.

By 1911, Rutherford had found that the scattering of α-particles by thin pieces of metal foil could only be explained by assuming that all of the positive charge and most of the mass of the atom were concentrated in an infinitesimally small fraction of the total volume, known as the nucleus. These experiments also provided an estimate of the charge on the nucleus. Rutherford found, for example, that the nucleus of a gold atom had a charge about 80 times the charge on an electron.

Shortly before World War I, H. G. J. Moseley reported measurements of the frequencies of the x-rays emitted by cathode-ray tubes when different metals were used as the anode. Moseley interpreted these results in terms of a model in which the number of positively charged particles in the nucleus of an atom was equal to the atomic number. The nucleus of a gold atom, for example, has a charge of $+79$, because the atomic number of this element is 79.

The discovery of the neutron in 1932 explained the discrepancy between the charge on the nucleus and the weight of an atom. A gold atom that weighs 197 amu consists of a nucleus that contains 79 protons and 118 neutrons surrounded by 79 electrons. By convention, the numbers of protons and neutrons in an atom are represented by the following symbol, where E is the element, Z is the atomic number, and M is the mass number of the atom.

$$_Z^M E$$

The symbol for the only naturally occurring isotope of gold, for example, is written as follows.

$$_{79}^{197} Au$$

The number of protons in the nucleus is equal to the atomic number, Z. The number of neutrons is equal to the difference between the mass number and the number of protons ($M - Z$).

Exercise 23.1

The convention just described can also be applied to subatomic particles. The only difference is the use of lowercase letters for the symbol used to identify the particle. Write the symbols for electrons, protons, and neutrons.

Solution

In order to write the symbols for these particles we need to know the charge and mass number of each particle. Electrons carry a charge of -1, protons have a charge of $+1$, and neutrons are electrically neutral. According to the data in Table A-2 in the appendix, protons and neutrons have a mass of about 1 amu, and the mass of the electron is negligibly small. These particles are therefore described by the following symbols.

$$_{-1}^{0} e \qquad _1^1 p \qquad _0^1 n$$

electron   proton   neutron

At times, it is useful to indicate the number of neutrons in an atom to facilitate comparisons between atoms. We do that by adding the neutron number as a subscript in the lower right-hand corner of the symbol for the atom. The symbol for gold, for example, can be written as follows.

$$^{197}_{79}\text{Au}_{118}$$

The upper right-hand corner of the symbol is reserved for the charge on the atom. The number of electrons in a neutral atom is equal to the number of protons. If the atom gains or loses electrons, however, it picks up a net charge. A polonium atom that contains 84 protons, 82 electrons, and 130 neutrons is identified by the following symbol.

$$^{214}_{84}\text{Po}^{2+}_{130}$$

Because anyone with access to a periodic table can find the atomic number of an element, a shorthand notation is often used that reports only the mass number of the atom and the symbol of the element. The shorthand notation for the only naturally occurring isotope of gold is ^{197}Au.

Exercise 23.2

Determine the number of protons, neutrons, and electrons in a $^{210}\text{Pb}^{2+}$ ion.

Solution

The atomic number of lead is 82, so this ion contains 82 protons. If the ion has a net charge of $+2$, it must have two more protons than electrons — thus, it has a total of 80 electrons. Because neutrons and protons both weigh about 1 amu, the mass number is equal to the sum of the number of protons and neutrons. The difference between the mass number (210) and the atomic number (82) is therefore equal to the number of neutrons in the nucleus of the atom. This ion contains 128 neutrons.

A particular combination of protons and neutrons is called a **nuclide.** ^{12}C, for example, is a nuclide that contains six protons and six neutrons. Nuclides with the same number of protons are called **isotopes. Isobars** are nuclides with the same mass number. Nuclides with the same number of neutrons are **isotones.**

Exercise 23.3

The following sets of nuclides contain examples of isotopes, isobars, and isotones. Classify each set of nuclides into one of these categories.

(a) ^{12}C, ^{13}C, and ^{14}C (b) ^{40}Ar, ^{40}K, and ^{40}Ca (c) ^{14}C, ^{15}N, and ^{18}O

Solution

(a) ^{12}C, ^{13}C, and ^{14}C all contain six protons. They are therefore examples of isotopes.
(b) ^{40}Ar, ^{40}K, and ^{40}Ca all have the same mass number, so they are isobars.
(c) ^{14}C, ^{15}N, and ^{18}O can't be isotopes, because they contain different numbers of protons. They can't be isobars, because they have different mass numbers. They all contain eight neutrons, however, so they are examples of isotones.

23.3 MODES OF RADIOACTIVE DECAY

Early studies of radioactivity indicated that three different kinds of radiation were emitted, which were symbolized by the first three letters of the Greek alphabet: α, β, and γ.

ALPHA DECAY

Alpha decay is usually restricted to the heavier elements in the periodic table. Only a handful of nuclei with atomic numbers less than 83 decay by the emission of an α-particle. To predict the product of α-decay, we start by assuming that both mass number and charge are conserved in a nuclear reaction. Alpha decay of the ^{238}U "parent" nuclide, for example, produces ^{234}Th as the "daughter" nuclide.

$$^{238}_{92}U \longrightarrow \, ^{234}_{90}Th + \, ^{4}_{2}He \qquad (\alpha\text{-decay})$$

The sum of the mass numbers of the products $(234 + 4)$ is equal to the mass number of the reactant (238). The sum of the charges on the nuclei of the products $(90 + 2)$ is equal to the charge on the nucleus of the parent nuclide (92).

BETA DECAY

Three different modes of radioactive decay fall into the category of beta decay: (1) electron, or β^-, emission, (2) electron capture, and (3) positron, or β^+, emission.

In **electron (β^-) emission,** an electron is ejected from the nucleus, and the charge on the nucleus increases by 1. Electron, or β^-, emitters are found throughout the periodic table, from the lightest elements (^3H) to the heaviest (^{255}Es). The product of β^- emission can be predicted by assuming that mass number and charge are conserved in nuclear reactions. If ^{40}K is a β^- emitter, for example, the product of this reaction must be ^{40}Ca.

$$^{40}_{19}K \longrightarrow \, ^{40}_{20}Ca + \, ^{0}_{-1}e \qquad (\text{electron, or } \beta^-, \text{ emission})$$

Once again, the sum of the mass numbers of the products $(40 + 0)$ is equal to the mass number of the parent nuclide (40), and the sum of the charge on the products $(20 - 1)$ is equal to the charge on the parent nuclide (19).

The energy of the electron emitted by ^{40}K covers a continuous spectrum from slightly more than 1000 kJ/mol — 1 MJ/mol — to a maximum of 127.8 MJ/mol. In 1930, Wolfgang Pauli explained this by suggesting that the energy given off during this reaction is divided between two particles emitted during β-decay. The second particle is an **antineutrino** ($\bar{\nu}$), the antimatter equivalent to the **neutrino** (from the Italian meaning "a little neutral particle"). The decay of a nuclide by electron emission should therefore be written as follows.

$$^{14}_{6}C \longrightarrow \, ^{14}_{7}N + \, ^{0}_{-1}e + \bar{\nu} \qquad (\text{electron emission})$$

Nuclei can also decay by capturing one of the electrons that surround the nucleus. **Electron capture** leads to a decrease of 1 in the charge on the nucleus. Like all other forms of radioactive decay, this reaction gives off a considerable amount of energy. The energy given off is carried by an x-ray photon, which is often represented by the symbol $h\nu$, where h is Planck's constant and ν is the frequency of the x-ray. The product of the reaction can be predicted, once again, by assuming that mass number and charge are conserved.

$$^{40}_{19}K + \, ^{0}_{-1}e \longrightarrow \, ^{40}_{18}Ar + h\nu \qquad (\text{electron capture})$$

The electron captured by the nucleus in this reaction is usually a $1s$ electron, because that is the electron closest to the nucleus.

A third form of beta decay is called **positron (β^+) emission.** The **positron** is the antimatter equivalent of an electron — it has the same mass as an electron, but the opposite charge. Positron, or β^+, decay leads to the creation of a daughter nuclide with one less positive charge on the nucleus than the parent. It is accompanied by

$$\ce{^{40}_{19}K ->[] ^{40}_{20}Ca + ^{0}_{-1}e + \bar{\nu}}$$

$$\ce{^{40}_{19}K -> ^{40}_{18}Ar + ^{0}_{+1}e + \nu}$$

$$\ce{^{40}_{19}K + ^{0}_{-1}e -> ^{40}_{18}Ar + h\nu}$$

FIG. 23.3 ^{40}K is an unusual nuclide because it simultaneously decays by all three forms of β-decay—electron emission, electron capture, and positron emission. The two most common modes of decay for this isotope are electron emission, which increases the atomic number of the nuclide, and electron capture, which decreases the atomic number.

the release of a neutrino (ν).

$$\ce{^{40}_{19}K -> ^{40}_{18}Ar + ^{0}_{+1}e + \nu} \qquad \text{(positron emission)}$$

Positrons have a very short life. They rapidly lose their kinetic energy as they pass through matter. As soon as they come to rest, they combine with an electron to form two γ-ray photons.

$$\ce{^{0}_{+1}e + ^{0}_{-1}e -> 2\,\gamma}$$

The three forms of β-decay for the ^{40}K nuclide are summarized in Figure 23.3. Note that the mass number of the parent and daughter nuclides are the same for electron emission, electron capture, and position emission. All three forms of β-decay therefore interconvert isobars.

GAMMA EMISSION

Daughter nuclides produced by α-decay or β-decay are often emitted in an excited state. The excess energy associated with this excited state is released when the nucleus emits a photon in the γ-ray portion of the electromagnetic spectrum. Most of the time, the γ-ray is emitted within 10^{-12} seconds after the α- or β-particle. At times, however, gamma decay is delayed, and a short-lived, or **metastable,** nuclide is formed. These metastable nuclides are identified by a small letter m written after the mass number. ^{60m}Co, for example, is produced by the electron emission of ^{60}Fe.

$$\ce{^{60}_{26}Fe -> ^{60m}_{27}Co + ^{0}_{-1}e + \bar{\nu}}$$

The metastable ^{60m}Co nuclide has a half-life of 10.5 minutes. Since a photon of electromagnetic radiation carries neither charge nor mass, the product of the γ-ray emission by ^{60m}Co is ^{60}Co.

$$\ce{^{60m}_{27}Co -> ^{60}_{27}Co + \gamma} \qquad \text{(γ-ray emission)}$$

SPONTANEOUS FISSION

Nuclides become less stable as the charge on the nucleus increases. Nuclides with atomic numbers of 90 or more undergo a form of radioactive decay known as **spontaneous fission.** In this reaction, the parent nucleus splits into smaller nuclei. The reaction is accompanied by the ejection of one or more neutrons.

$$\ce{^{252}_{98}Cf -> ^{140}_{54}Xe + ^{108}_{44}Ru + 4\,^{1}_{0}n}$$

For all but the very heaviest isotopes, spontaneous fission is a very slow reaction. The rate of α-decay of ^{238}U, for example, is almost two million times faster than the rate of spontaneous fission of this nuclide.

Exercise 23.4

Predict the products of the following nuclear reactions.

(a) Electron emission by ^{14}C
(b) Positron emission by ^{8}B
(c) Electron capture by ^{22}Na
(d) Alpha emission by ^{204}Rn
(e) Gamma-ray emission by ^{56m}Ni

Solution

We can predict the product of each reaction by writing an equation in which both mass number and charge are conserved.

(a) $^{14}_{6}C \longrightarrow ^{14}_{7}N + ^{0}_{-1}e + \bar{\nu}$ (d) $^{204}_{86}Rn \longrightarrow ^{200}_{84}Po + ^{4}_{2}He$

(b) $^{8}_{5}B \longrightarrow ^{8}_{4}Be + ^{0}_{+1}e + \nu$ (e) $^{56m}_{28}Ni \longrightarrow ^{56}_{28}Ni + \gamma$

(c) $^{22}_{11}Na + ^{0}_{-1}e \longrightarrow ^{22}_{10}Ne + h\nu$

23.4 NEUTRON-RICH VERSUS NEUTRON-POOR NUCLIDES

In 1934, Enrico Fermi proposed a theory to explain beta decay. In this model, neutrons were assumed to decay to form protons by the emission of an electron and an antineutrino.

$$^{1}_{0}n \longrightarrow ^{1}_{1}p + ^{0}_{-1}e + \bar{\nu} \qquad (\beta^{-}\text{-emission})$$

Electron, or β^{-}, emission therefore results in an *increase* in the atomic number of the nucleus.

$$^{14}_{6}C \longrightarrow ^{14}_{7}N + ^{0}_{-1}e + \bar{\nu}$$

Free neutrons have in fact been observed to undergo this reaction, and the half-life of an isolated neutron is only about 11 minutes.

According to this theory, electron capture transforms a proton into a neutron.

$$^{1}_{1}p + ^{0}_{-1}e \longrightarrow ^{1}_{0}n + h\nu \qquad (\text{electron capture})$$

Electron capture therefore *decreases* the atomic number of the nucleus.

$$^{7}_{4}Be + ^{0}_{-1}e \longrightarrow ^{7}_{3}Li + h\nu$$

Positron emission also transforms a proton into a neutron.

$$^{1}_{1}p \longrightarrow ^{1}_{0}n + ^{0}_{+1}e + \nu \qquad (\beta^{+}\text{-emission})$$

As a result, it also *decreases* the atomic number of the nucleus.

$$^{11}_{6}C \longrightarrow ^{11}_{5}B + ^{0}_{+1}e + \nu$$

A plot of the number of neutrons versus the number of protons for all *stable* naturally occurring isotopes is shown in Figure 23.4. Several conclusions can be drawn from this figure.

1. Only a very narrow band of neutron-to-proton ratios result in a stable nuclide, which does not undergo radioactive decay.
2. The ratio of neutrons to protons in stable nuclides gradually increases as the number of protons in the nucleus increases.
3. Light nuclides, such as ^{12}C, contain about the same number of neutrons and protons. Heavy nuclides, such as ^{238}U, contain up to 1.6 times as many neutrons as protons.
4. There are no stable nuclides with atomic numbers larger than 83.
5. This narrow band of stable nuclei is surrounded by a sea of instability.
6. Nuclei that lie above this line have too many neutrons and are said to be **neutron-rich**.
7. Nuclei that lie below this line have too few neutrons and are said to be **neutron-poor**.

The most likely mode of decay for a neutron-rich nucleus is one that converts a neutron into a proton. Neutron-rich nuclei therefore tend to decay by electron emission. Every radioactive isotope with an atomic number smaller than 83 that can be considered neutron-rich decays by electron, or β^-, emission. ^{14}C, ^{32}P, and ^{35}S, for example, are all neutron-rich nuclei that decay by the emission of an electron.

$$^{35}_{16}S \longrightarrow ^{35}_{17}Cl + ^{0}_{-1}e + \bar{v} \qquad (\beta^- \text{ emission})$$

Neutron-poor nuclei should decay by modes that convert a proton into a neutron. Neutron-poor nuclides with atomic numbers less than 83 tend to decay by either electron capture or positron emission. Many of these nuclides decay by both routes, but positron emission is more often observed in the lighter nuclides, such as ^{22}Na.

$$^{22}_{11}Na \longrightarrow ^{22}_{10}Ne + ^{0}_{+1}e + v \qquad (\beta^+ \text{ emission})$$

FIG. 23.4 A graph of the number of neutrons versus the number of protons for all stable naturally occurring nuclei. Nuclei that lie to the right of this band of stability are neutron-poor. Nuclei to the left of the band are neutron-rich.

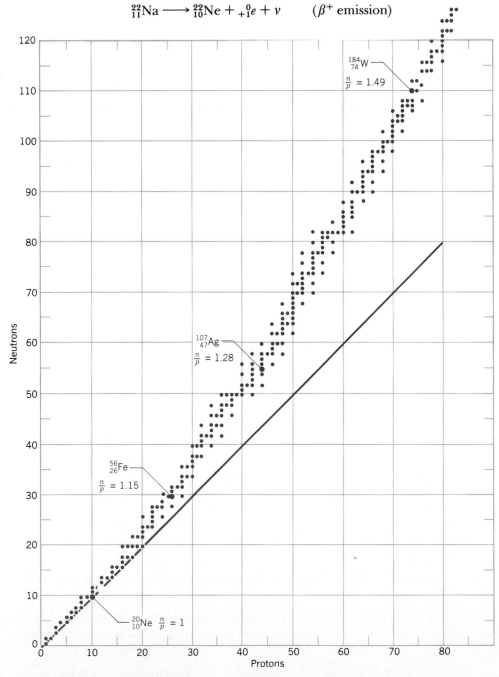

Electron capture is more common among heavier nuclides, such as ^{125}I, because the 1s electrons are held closer to the nucleus as the charge on the nucleus becomes larger.

$$^{125}_{53}I + {_{-1}^{0}e} \longrightarrow {^{125}_{52}Te} + h\nu \qquad \text{(electron capture)}$$

A third mode of decay is observed in neutron-poor nuclides that have atomic numbers larger than 83. Although it may not be obvious at first, α-decay also increases the ratio of neutrons to protons.

$$^{238}_{92}U \longrightarrow {^{234}_{90}Th} + {^{4}_{2}He} \qquad (\alpha\text{-decay})$$

Exercise 23.5

Explain why neutron-poor nuclides such as ^{238}U decay by emitting α-particles.

Solution

The parent nuclide (^{238}U) in this reaction has 92 protons and 146 neutrons, which means that the neutron-to-proton ratio is 1.587. The daughter nuclide (^{234}Th) has 90 protons and 144 neutrons, so its neutron-to-proton ratio is 1.600. Because an α-particle has the same number of protons and neutrons (two of each), the net result of α-particle decay is the formation of a nuclide that has a larger neutron-to-proton ratio. The daughter nuclide is therefore less likely to be neutron-poor.

Exercise 23.6

Predict the most likely modes of decay and the products of decay of the following nuclides.

(a) ^{17}F (b) ^{105}Ag (c) ^{185}Ta

Solution

Before we can predict how these nuclides decay, we have to decide whether each nuclide is neutron-rich or neutron-poor. We can do this by comparing the nuclides with stable isotopes of the elements.

(a) The atomic weight of fluorine is 18.998 amu. The ^{17}F nuclide weighs less than the average fluorine atom, which means it contains fewer neutrons. Thus, it is likely to be neutron-poor. Because it is a relatively light nuclide, ^{17}F might be expected to decay by positron emission.

$$^{17}_{9}F \longrightarrow {^{17}_{8}O} + {_{+1}^{0}e} + \nu \qquad (\beta^{+}\text{-emission})$$

(b) The atomic weight of silver is 107.87 amu. The ^{105}Ag nuclide weighs less than the average silver atom and therefore contains fewer neutrons than stable isotopes of silver. Since it is a relatively heavy neutron-poor nuclide, we might expect it to decay by electron capture.

$$^{105}_{47}Ag + {_{-1}^{0}e} \longrightarrow {^{105}_{46}Pd} + h\nu \qquad \text{(electron capture)}$$

(c) The atomic weight of tantalum is 180.9479 amu. The ^{185}Ta isotope therefore contains more neutrons than the average tantalum atom. If this nuclide is neutron-rich, it will decay by electron emission.

$$^{185}_{73}Ta \longrightarrow {^{185}_{74}W} + {_{-1}^{0}e} + \bar{\nu} \qquad (\beta^{-}\text{-emission})$$

23.5 BINDING ENERGY CALCULATIONS

We should be able to predict the mass of an atom from the number of electrons, protons, and neutrons it contains. We can predict the mass of a helium atom, for example, from the masses of these three subatomic particles.

$$\text{Neutron} = 1.0086650 \text{ amu}$$
$$\text{Proton} = 1.0072765 \text{ amu}$$
$$\text{Electron} = 5.4858026 \times 10^{-4} \text{ amu}$$

A helium atom contains two protons, two neutrons, and two electrons. Its predicted mass is therefore 4.0329802 amu.

$$2(1.0072765) \text{ amu} = 2.0145530 \text{ amu}$$
$$2(1.0086650) \text{ amu} = 2.0173300 \text{ amu}$$
$$2(0.0005486) \text{ amu} = \underline{0.0010972 \text{ amu}}$$
$$\text{Total mass} = 4.0329802 \text{ amu}$$

The experimental mass of a helium atom — 4.0026033 amu — is not the same as the predicted mass. The helium atom weighs 0.0303769 amu less than the sum of its parts.

$$\text{Predicted mass} = 4.0329802 \text{ amu}$$
$$\text{Observed mass} = \underline{4.0026033 \text{ amu}}$$
$$\text{Mass defect} = 0.0303769 \text{ amu}$$

The difference between the mass of an atom and the sum of the masses of its protons, neutrons, and electrons is called the *mass defect.*

The mass defect of an atom reflects the stability of the nucleus. It is equal to the energy released when the nucleus is formed from its constituent protons and neutrons. Accordingly, the mass defect is also known as the *binding energy* of the nucleus.

The binding energy serves the same function for nuclear reactions as $\Delta H°$ serves for a chemical reaction — it measures the difference between the stability of the products of the reaction and the starting materials. The larger the binding energy, the larger the amount of energy released when the nucleus is formed. Thus, the larger the binding energy, the more stable the nucleus. The binding energy can also be viewed as the amount of energy that must be added to the system to take the nucleus apart and form isolated neutrons and protons. It is therefore quite literally the energy that binds together the neutrons and protons in the nucleus.

The binding energy of a nuclide can be calculated from its mass defect by use of Einstein's equation that relates mass and energy.

$$E = mc^2$$

To obtain the binding energy in units of joules, we first convert the mass defect from atomic mass units into kilograms.

$$0.0303769 \frac{\text{amu}}{\text{atom}} \times 1.6605655 \times 10^{-24} \frac{\text{g}}{\text{amu}} \times \frac{1 \text{ kg}}{1000 \text{ g}} = 5.04428 \times 10^{-29} \frac{\text{kg}}{\text{atom}}$$

Substituting this information into Einstein's equation gives a binding energy for the helium atom of 4.53358×10^{-12} joules.

$$E = (5.04428 \times 10^{-29} \text{ kg/atom})(2.9979246 \times 10^8 \text{ m/s})^2$$
$$= 4.53358 \times 10^{-12} \text{ J/atom}$$

Multiplying by Avogadro's number gives a binding energy for helium of 2.730×10^{12} joules per mole, or 2.730 billion kilojoules per mole—an enormous amount of energy.

$$4.53358 \times 10^{-12} \ \frac{\text{J}}{\text{atom}} \times 6.022 \times 10^{23} \ \frac{\text{atoms}}{\text{mol}} = 2.730 \times 10^{12} \text{ J/mol}$$

This calculation helps us understand some of the fascination of nuclear reactions. The energy released when natural gas is burned is about 800 kJ/mol. The synthesis of a mole of helium releases 3.4 million times as much energy.

Since most nuclear reactions are carried out on very small samples of material, the mole is not a reasonable basis of measurement. Binding energies are usually expressed in units of electron volts or million electron volts per atom. The binding energy for helium is 28.3×10^6 eV/atom, or 28.3 MeV/atom.

$$4.53358 \times 10^{-12} \ \frac{\text{J}}{\text{atom}} \times \frac{1 \text{ eV}}{1.6021892 \times 10^{-19} \text{ J}} = 28.30 \times 10^6 \text{ eV/atom}$$

To reemphasize the difference between the energies of nuclear and chemical reactions, note that the binding energy for the helium nucleus is more than a million times larger than the first ionization energy of any element in the periodic table.

Calculations of the binding energy can be simplified by using the following conversion factor between the mass defect in atomic mass units and the binding energy in million electron volts.

$$1.000... \text{ amu} = 931.5016 \text{ MeV}$$

Exercise 23.7

Calculate the binding energy of ^{235}U if this atom weighs 235.0349 amu.

Solution

A neutral ^{235}U atom contains 92 protons, 92 electrons, and 143 neutrons. The predicted mass of a ^{235}U atom is therefore 236.9600 amu. (This calculation uses four decimal places because the experimental mass of the nuclide is known to only four decimal places.)

$$92(1.00728) \text{ amu} = 92.6698 \text{ amu}$$
$$143(1.00867) \text{ amu} = 144.2398 \text{ amu}$$
$$\underline{92(0.0005486) \text{ amu} = \ \ \ 0.0505 \text{ amu}}$$
$$\text{Total mass} = 236.9601 \text{ amu}$$

To calculate the mass defect for this nucleus, we subtract the observed mass from the predicted mass.

$$\text{Predicted mass} = 236.9600 \text{ amu}$$
$$\underline{\text{Observed mass} = 235.0349 \text{ amu}}$$
$$\text{Mass defect} = \ \ \ 1.925 \text{ amu}$$

Using the conversion factor that relates the binding energy to the mass defect, we obtain a binding energy for ^{235}U of 1793.1 MeV per atom.

$$1.925 \; \frac{\text{amu}}{\text{atom}} \times 931.5016 \; \frac{\text{MeV}}{\text{amu}} = 1793.1 \; \text{MeV/atom}$$

Binding energies gradually become larger with atomic number, although they tend to level off near the end of the periodic table. A more useful quantity is the *binding energy per nucleon,* a measure of the stability of the nucleus. (A ***nucleon*** is a proton or neutron in an atom's nucleus.) We calculate this quantity by dividing the binding energy for a nuclide by the number of protons and neutrons in the nuclide.

The binding energy per nucleon ranges from about 5 to 8 MeV for most nuclei. It reaches a maximum, however, at an atomic mass of about 60 amu (see Figure 23.5). The largest binding energy per nucleon is observed for ^{56}Fe, which is the most stable nuclide in the periodic table.

The graph of binding energy per nucleon versus atomic mass explains why energy is released when relatively small nuclei combine, or fuse, to form larger nuclei in ***fusion reactions.***

$$^{12}_{6}\text{C} + ^{12}_{6}\text{C} \longrightarrow ^{24}_{12}\text{Mg} \qquad \text{(fusion)}$$

It also explains why energy is released when relatively heavy nuclei split apart in ***fission*** (literally, "to split or cleave") ***reactions.***

$$^{235}_{92}\text{U} \longrightarrow ^{139}_{56}\text{Ba} + ^{94}_{36}\text{Kr} + 2 \, ^{1}_{0}n \qquad \text{(fission)}$$

There are a number of small irregularities in the binding energy curve at the low end of the mass spectrum, as shown in Figure 23.6. The ^{4}He nucleus, for example, is much more stable than its nearest neighbors. The unusual stability of the ^{4}He nucleus may explain why α-particle decay is so much faster than spontaneous

FIG. 23.5 The binding energy of a nucleus gradually increases with atomic number from one end of the periodic table to the other. The binding energy per nucleon, however, reaches a maximum at about ^{56}Fe. In theory, nuclei that are lighter than ^{56}Fe can become more stable by fusing together, and nuclei that are significantly heavier than ^{56}Fe can become more stable by splitting apart.

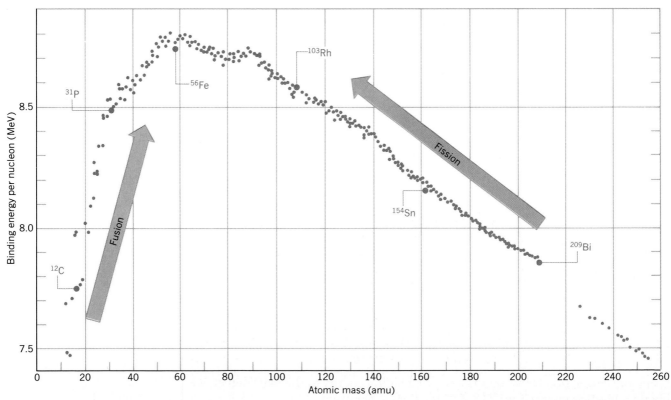

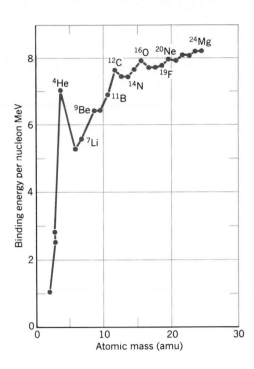

FIG. 23.6 The binding energy per nucleon for most stable nuclei is between about 7.8 and 8.8 MeV per nucleon. There is more variability in the binding energy per nucleon for the very light nuclei, however. The binding energy per nucleon for ^{4}He is particularly large. The unusual stability of this nuclide explains why the ejection of a helium nucleus, or α-particle, is observed when some radioactive elements decay.

fission. It also explains the enormous energy released when helium is synthesized from hydrogen atoms in the sun.

23.6 THE KINETICS OF RADIOACTIVE DECAY

Radioactive nuclei decay by first-order kinetics. The rate of decay is the product of a rate constant (k) times the number of atoms of isotope in the sample (N).

$$\text{Rate} = -\frac{dN}{dt} = kN$$

This equation is similar to the rate law for a first-order chemical reaction introduced in Section 21.7. Radioactive decay differs from the behavior of a first-order reaction in one way — the rate of radioactive decay does not depend on the temperature of the system. The rate of decay is also independent of chemical changes. The rate of decay of ^{238}U, for example, is exactly the same in uranium metal as it is in uranium hexafluoride (UF_6) or any other compound of this element.

The rate at which a radioactive isotope decays is called the *activity* of the isotope. The most common unit of activity is the *curie* (Ci), which was originally defined as the number of disintegrations per second of 1 gram of ^{226}Ra but is now defined as an activity of 3.700×10^{10} disintegrations per second.

Exercise 23.8

The most abundant isotope of uranium is ^{238}U — 99.275% of the atoms in a sample of uranium are ^{238}U. Calculate the activity of the ^{238}U in 1 liter of a 1.00 M solution of the uranyl ion, $UO_2{}^{2+}$, assuming that the rate constant for the decay of this isotope is 4.87×10^{-18} disintegrations per second.

Solution

A 1.00 M UO_2^{2+} solution contains 6.02×10^{23} uranium atoms per liter. The rate of this reaction depends on the constant — k — and the number of uranium atoms in the sample — N.

$$\frac{-dN}{dt} = kN$$

$$= (4.87 \times 10^{-18} \text{ s}^{-1})(6.02 \times 10^{23} \text{ atoms}) = 2.93 \times 10^6 \text{ atoms/s}$$

To calculate the activity of this sample, we convert from disintegrations per second (atom/s) to curies.

$$2.93 \times 10^6 \text{ atoms/s} \times \frac{1 \text{ Ci}}{3.700 \times 10^{10} \text{ atoms/s}} = 7.92 \times 10^{-5} \text{ Ci}$$

The curie is a very large unit of measurement. Activities of samples handled in the laboratory are therefore often reported in millicuries or microcuries. This sample has an activity of 79.2 microcuries — 79.6 μCi.

The rates at which different radioactive nuclei decay can be compared in terms of either the rate constants for the decay or the half-lives of the nuclei. We can conclude that ^{14}C decays more rapidly than ^{238}U, for example, by noting that the rate constant for the decay of ^{14}C is much larger than that for ^{238}U.

$$^{14}\text{C}: \quad k = 1.210 \times 10^{-4} \text{ yr}^{-1}$$
$$^{238}\text{U}: \quad k = 1.53 \times 10^{-10} \text{ yr}^{-1}$$

We can reach the same conclusion by noting that the half-life for the decay of ^{14}C is much shorter than that for ^{235}U.

$$^{14}\text{C}: \quad t_{1/2} = 5730 \text{ yr}$$
$$^{235}\text{U}: \quad t_{1/2} = 4.51 \times 10^9 \text{ yr}$$

The *half-life* for the decay of a radioactive nuclide is the length of time it takes for exactly half of the nuclei in the sample to decay. In Section 21.12, we concluded that the half-life of a first-order process is inversely proportional to the rate constant for the process.

$$t_{1/2} = \frac{\ln 2}{k} = \frac{0.693}{k}$$

Exercise 23.9

Calculate the half-life of ^{226}Ra from the definitions of the curie.

Solution

A curie was originally defined as the activity of 1 gram of ^{226}Ra. We might therefore start this problem by calculating the number of ^{226}Ra atoms in a 1-gram sample.

$$1.00 \text{ g Ra} \times \frac{1 \text{ mol Ra}}{226.0 \text{ g}} = 4.42 \times 10^{-3} \text{ mol Ra}$$

$$4.42 \times 10^{-3} \text{ mol Ra} \times 6.02 \times 10^{23} \frac{\text{atoms}}{\text{mol}} = 2.66 \times 10^{21} \text{ Ra atoms}$$

A curie is currently defined as an activity of 3.700×10^{10} disintegrations per second.

$$\text{Rate} = 3.700 \times 10^{10} \frac{\text{atoms}}{\text{s}}$$

This rate is equal to the product of the rate constant for the reaction times the number of ^{226}Ra atoms in the sample.

$$\text{Rate} = \frac{-dN}{dt} = kN$$

Substituting one of the equations that defines the rate of reaction into the other gives the following result.

$$3.700 \times 10^{10} \frac{\text{atoms}}{\text{s}} = kN$$

Thus, the rate constant for this decay is the rate of reaction divided by the number of atoms in the sample.

$$k = \frac{3.700 \times 10^{10} \text{ atoms/s}}{2.66 \times 10^{21} \text{ atoms}} = 1.39 \times 10^{-11} \text{ s}^{-1}$$

We can now obtain the half-life of this decay from its rate constant.

$$t_{1/2} = \frac{0.693}{k} = \frac{(0.693)}{(1.39 \times 10^{-11} \text{ s}^{-1})} = 5.0 \times 10^{10} \text{ s}$$

Converting seconds into years, we find that the half-life of ^{226}Ra is roughly 1600 years.

The half-life of a nuclide can be used in simple calculations.

Exercise 23.10

Calculate how much ^{14}C would be left in a sample after eight half-lives.

Solution

Half of the ^{14}C present initially decays during the first half-life, half of what is left decays during the second half-life, and so on. The ^{14}C left after eight half-lives is equal to one-half raised to the eighth power.

$$(1/2)^8 = 0.00391$$

According to this calculation, less than 0.4% of the original ^{14}C is left.

For more complex calculations, it is easier to convert the half-life of the nuclide into a rate constant and then use the integrated form of the first-order rate law described in Section 21.12.

Exercise 23.11

Calculate how long it would take for a sample of ^{222}Rn that weighs 0.750 grams to decay to 0.100 grams. Assume a half-life of 3.823 days.

Solution

We can calculate the rate constant for this decay from the half-life.

$$k = \frac{\ln 2}{t_{1/2}} = \frac{0.6931}{3.823 \text{ d}} = 0.1813 \text{ d}^{-1}$$

We now turn to the integrated form of the first-order rate law.

$$\log\left[\frac{(N)}{(N_0)}\right] = -\frac{kt}{2.303}$$

The ratio of the number of atoms (N) in the sample when it weighs 0.100 grams to the number of atoms (N_0) when it weighs 0.750 grams is the same as the ratio of grams at the end of the time period to the number of grams present initially.

$$\log\left[\frac{(0.100)}{(0.750)}\right] = -\frac{(0.1813 \text{ d}^{-1})(t)}{2.303}$$

Solving for t, we find that it takes 11.1 days for 0.750 grams of ^{222}Rn to decay to 0.100 grams of this nuclide.

$$t = 11.1 \text{ d}$$

23.7 DATING BY RADIOACTIVE DECAY

The earth is constantly bombarded by cosmic rays emitted by the sun. The total energy received in the form of cosmic rays is relatively small—roughly equal to the energy received by the planet from starlight. The energy of a single cosmic ray is very high, however—on the order of several billion electron volts. These highly energetic rays react with atoms in the atmosphere to produce neutrons, which then react with nitrogen atoms in the atmosphere to produce ^{14}C.

$$^{14}_{7}\text{N} + ^{1}_{0}n \longrightarrow ^{14}_{6}\text{C} + ^{1}_{1}\text{H}$$

^{14}C is a neutron-rich nuclide that decays by electron emission with a half-life of 5730 years.

$$^{14}_{6}\text{C} \longrightarrow ^{14}_{7}\text{N} + ^{0}_{-1}e + \bar{\nu}$$

Just after World War II, Willard F. Libby proposed a way to use these reactions to estimate the age of carbon-containing substances. The ^{14}C dating technique for which Libby received the Nobel prize was based on the following assumptions.

1. ^{14}C is produced in the atmosphere at a more or less constant rate.
2. Carbon atoms circulate between the atmosphere, the oceans, and living organisms at a rate very much faster than they decay. As a result, there is a constant concentration of ^{14}C in all living things.
3. After death, organisms no longer pick up ^{14}C. By comparing the activity of a sample with the activity of living tissue we can estimate how long it has been since the organism died.

The natural abundance of ^{14}C is about 1 part in 10^{12}, and the average activity of living tissue is 15.3 disintegrations per minute per gram of carbon. Typical samples used for radiocarbon dating include charcoal, wood, cloth, paper, sea shells, limestone, flesh, hair, soil, peat, and bone. Since most iron samples also contain carbon, it is possible to estimate the time since iron was last fired by analyzing for ^{14}C.

Exercise 23.12

The skin, bone, and clothing of "Whiskey Lil," an adult female mummy discovered in Chimney Cave, Lake Winnemucca, Nevada, were dated by radiocarbon analysis. How old is this mummy if the sample retained 73.9% of the activity of living tissue? (^{14}C: $t_{1/2} = 5730$ years)

Solution

Because ^{14}C decays by first-order kinetics, the log of the ratio of the ^{14}C in the sample today (N) to the amount that would be present if it was still alive (N_0) is proportional to the rate constant for this decay and the time since death.

$$\log\left[\frac{(N)}{(N_0)}\right] = -\frac{kt}{2.303}$$

The rate constant for this reaction can be calculated from its half-life.

$$k = \frac{\ln 2}{t_{1/2}} = \frac{0.6931}{5730 \text{ yr}} = 1.210 \times 10^{-4} \text{ yr}^{-1}$$

If the sample has retained 73.9% of its activity, the ratio of the activity today (N) to the original activity (N_0) is 0.739. We can now substitute what we know into the integrated form of the first-order rate law.

$$\log(0.739) = -\frac{(1.210 \times 10^{-4} \text{ yr}^{-1})(t)}{2.303}$$

Solving this equation gives the following results,

$$t = 2500 \text{ yr}$$

This skull taken from Rancho La Brea has been dated by ^{14}C as 9000 years old.

One of Libby's assumptions is questionable. The amount of ^{14}C in the atmosphere has not been constant with time. Because of changes in solar activity and the earth's magnetic field, it has varied by as much as $\pm 5\%$. More recently, contamination from the burning of fossil fuels and the testing of nuclear weapons has caused significant changes in the amount of radioactive carbon in the atmosphere. Radiocarbon dates are therefore reported in years before the present era—before 1950.

Studies of bristlecone pines allow us to correct for changes in the abundance of ^{14}C with time. The bristlecone pine, a remarkable tree that grows in the White Mountains of California, can live for up to five thousand years. By studying the ^{14}C activity of samples taken from the annual growth rings in these trees, researchers have developed a calibration curve for ^{14}C dates from the present back to at least 5145 B.C.

After roughly 45,000 years (eight half-lives), a sample retains only 0.4% of the activity of living tissue. At that point it becomes too old to date by radiocarbon techniques. Other radioactive isotopes, however, can be used to date rocks, soils, or archaeological objects that are much older. Potassium-argon dating, for example, has been used to date samples ranging from 2500 to 4.3 billion years old.

Naturally occurring potassium contains 0.0118% by weight of the radioactive ^{40}K isotope. This isotope decays to ^{40}Ar with a half-life of 1.3 billion years. The ^{40}Ar produced after a rock crystallizes is trapped in the crystal lattice. It can be released, however, when the rock is melted at temperatures up to 2000°C. By measuring the amount of ^{40}Ar released when the rock is melted and comparing it with the amount of potassium in the sample, scientists can determine the time since the rock crystallized.

23.8 IONIZING VERSUS NON-IONIZING RADIATION

We live in a sea of radiation. We are exposed to infrared radiation, ultraviolet radiation, visible rays, and cosmic rays from the sun. We are subjected to radio waves from local radio and television transmitters, microwaves that leak from microwave ovens, and x-rays produced by the cathode-ray tubes in our television sets. We are also exposed to both natural and human-made sources of radioactivity, including γ-rays from soils and rocks and α-particles from trace contaminants in brick and clay.

In recent years, people have learned to fear the effects of this radiation. They don't want to live near nuclear reactors. They are frightened by reports of links between excess exposure to sunlight and skin cancer. They are afraid of the danger of leakage from microwave ovens or the radiation produced by their television sets.

It is easy to understand this fear. Several factors combine to heighten the public's anxiety about the short-range and long-range effects of radiation. Perhaps the most important source of fear is the fact that radiation can't be detected by the average person—it is both tasteless and odorless. Furthermore, the effects of exposure to radiation might not appear for months or years.

To understand the biological effects of radiation we must first understand the difference between ionizing and non-ionizing radiation. The two most common processes that arise when radiation is absorbed by matter are excitation and ionization. Excitation occurs when the energy of the radiation excites the motion of the atoms or molecules or excites an electron from an occupied orbital to an empty, higher-energy orbital. Ionization occurs when enough energy is absorbed to remove an electron from the atom or molecule.

When living tissue, which is 70 to 90% water, is irradiated, most of the energy is absorbed by water molecules. When we calculate the dividing line between radiation that excites electrons and radiation that forms ions we therefore use the ionization energy of water, which is 1216 kJ/mol, in our calculation. Radiation that carries less energy can only excite the water molecule; it is therefore called *non-ionizing radiation.* Radiation that carries more energy, however, can remove an electron from a water molecule to form an ion and is therefore called *ionizing radiation.*

Table 23.1 contains estimates of the energies of various kinds of radiation. Radio waves, microwaves, infrared radiation, and visible light are all forms of non-ionizing radiation. High-energy ultraviolet rays, x-rays, gamma rays, and α- and β^--particles are forms of ionizing radiation.

When ionizing radiation passes through living tissue, electrons are removed from neutral water molecules to produce H_2O^+ ions. Between three and four water molecules are ionized for every 100 eV of energy absorbed in the form of ionizing radiation.

$$H_2O \longrightarrow H_2O^+ + e^-$$

The electrons produced in this reaction are picked up by neutral water molecules to form H_2O^- ions. The net reaction therefore involves the transfer of an electron from one water molecule to another.

$$2\ H_2O \longrightarrow H_2O^+ + H_2O^-$$

TABLE 23.1

Energies of Ionizing and Non-Ionizing Forms of Radiation

Radiation	Typical Frequency	Typical Energy	
Nuclear Radiation			
(from ^{238}U)		4.1×10^8 kJ/mol	
(from ^{14}C)		1.5×10^7 kJ/mol	
Electromagnetic Radiation			Ionizing Radiation
cosmic rays	6×10^{21} s^{-1}	2.4×10^9 kJ/mol	
gamma rays	3×10^{20} s^{-1}	1.2×10^8 kJ/mol	
x-rays	3×10^{17} s^{-1}	1.2×10^5 kJ/mol	
ultraviolet	3×10^{15} s^{-1}	1200 kJ/mol	
visible	5×10^{14} s^{-1}	200 kJ/mol	Non-Ionizing radiation
infrared	3×10^{13} s^{-1}	12 kJ/mol	
microwaves	3×10^9 s^{-1}	1.2×10^{-3} kJ/mol	
radio waves	3×10^7 s^{-1}	1.2×10^{-5} kJ/mol	

These ions should not be confused with the H_3O^+ and OH^- ions produced when acids and bases dissolve in water. The H_2O^+ and H_2O^- ions are examples of *free radicals* — atoms, molecules, or ions with an unpaired electron in the outermost or valence, shell (see Figure 23.7). Free radicals are highly reactive. They combine with other radicals to pair the odd electrons, or they act as either electron donors (reducing agents) or electron acceptors (oxidizing agents) to get rid of the unpaired electrons.

The radicals formed when ionizing radiation passes through water are among the strongest oxidizing agents that can exist in aqueous solution. At the molecular level, these oxidizing agents destroy biologically active molecules by either removing electrons or removing hydrogen atoms. These reactions lead to damage to the cell membrane, nucleus, chromosomes, or the mitochondria. The damage can inhibit cell division, result in cell death, or produce a malignant cell.

FIG. 23.7 The Lewis structures of the H_2O^+ and OH^+ ions. Note that both of these ions contain an unpaired electrons, which makes them extremely reactive.

23.9 BIOLOGICAL EFFECTS OF IONIZING RADIATION

From the time that radioactivity was discovered, it was obvious that it caused damage. Glass containers used to store radium compounds, for example, turned a rich purple and eventually cracked because of radiation damage. As early as 1901, Pierre Curie discovered that a sample of radium placed on his skin produced wounds that were very slow to heal. What some find surprising is the magnitude of the difference between the biological effects of non-ionizing radiation, such as light and microwaves, and ionizing radiation, such as high-energy ultraviolet radiation, x-rays, γ-rays, and α- and β^--particles.

The principal effect of energy absorbed from non-ionizing radiation is an increase in the temperature of the system. Radiation at the low-energy end of the electromagnetic spectrum excites the movement of atoms and molecules, which is equivalent to heating the sample. Radiation near the visible portion of the spectrum excites electrons into higher-energy orbitals. When the electron eventually falls back to a lower-energy state, the excess energy is given off to neighboring molecules in the form of heat.

Biological systems are sensitive to heat. We experience this each time we cook

This dessicator was used to store radium compounds. Radiation emitted by these compounds eventually caused enough damage to the structure of the glass to turn it a deep purple color.

with a microwave oven or spend an afternoon in the sun. But it takes a great deal of non-ionizing radiation to reach dangerous levels. We can assume, for example, that absorption of enough radiation to produce an increase of 8° to 10°C in body temperature would be fatal. Since the average 70-kilogram human is 80% water by weight, we can use the heat capacity of water—75.37 J/mol-K—to calculate that it would take about 3 million joules of non-ionizing radiation to kill the average human. If this energy was carried by visible light with a frequency of 6×10^{15} s^{-1}, it would correspond to absorption of slightly more than a mole of photons.

Ionizing radiation is between 10,000 and 10 billion times as dangerous. A dose of only 300 joules of ionizing radiation is fatal for the average human, even though this radiation raises the temperature of the body by only 0.001°C. Whereas it takes almost a mole of photons of visible light to produce a fatal dose of non-ionizing radiation, absorption of only 7×10^{-10} moles of α-particles emitted by ^{238}U is fatal.

There are three ways of measuring ionizing radiation.

1. Measure the *activity* of the source, in units of disintegrations per second or curies, using a Geiger counter.
2. Measure the radiation to which an object is *exposed*, in units of roentgens, by measuring the amount of ionization produced when this radiation passes through a sample of air.
3. Measure the radiation *absorbed* by the object, in units of radiation absorbed doses, or rads.

The easiest measurement to make is the activity of the source. The most useful measurement is the radiation absorbed by the object, which is the hardest to obtain.

One **radiation absorbed dose,** or **rad,** corresponds to the absorption of 10^{-5} joules of energy per gram of body weight, or 0.01 J/kg. One rad therefore produces an increase in body temperature of about 2×10^{-6}°C. At first glance, the rad may seem to be a negligibly small unit of measurement. The destructive power of the radicals produced when water is ionized is so large, however, that cells are inactivated at a dose of 100 rads, and a dose of 400 to 450 rads is fatal for the average human.

Not all forms of radiation have the same efficiency for damaging biological organisms. The faster energy is lost as the radiation passes through the tissue, the more damage it does. To correct for the differences in **radiation biological effectiveness (RBE)** among various forms of radiation, a second unit of absorbed dose has been defined. The **roentgen equivalent man,** or **rem,** is the product of the absorbed dose in rads times the biological effectiveness of the radiation.

$$\textbf{rems} = \textbf{rads} \times \textbf{RBE}$$

Typical values for the RBE of several forms of radiation are given in Table 23.2.

Estimates of the per capita exposure to radiation in the United States are summarized in Table 23.3. These estimates include both external and internal sources of natural background radiation. External sources include cosmic rays from the sun and α-particles and γ-rays emitted from rocks and soil. Internal sources include nuclides that enter the body when we breathe (^{14}C, ^{85}Kr, ^{220}Rn, and ^{222}Rn) and through the food chain (^{3}H, ^{14}C, ^{40}K, ^{90}Sr, ^{131}I, and ^{137}Cs). The actual dose from natural radiation depends on where one lives. People who live in the Rocky Mountains, for example, receive twice as much radiation as the national average because there is less atmosphere to filter out the cosmic rays from the sun.

TABLE 23.2

The Radiation Biological Effectiveness
of Various Forms of Radiation

Radiation	RBE
x-rays and γ-rays	1
β^--particles with energies larger than 0.03 MeV	1
β^--particles with energies less than 0.3 MeV	1.7
thermal (slow-moving) neutrons	3
fast-moving neutrons or protons	10
α-particles of heavy ions	20

TABLE 23.3

Average Whole-Body Exposure Levels
for Sources of Ionizing Radiation

Source	Per Capita Dose (rems/yr)
natural background	0.082
medical x-rays	0.077
nuclear test fallout	0.005
consumer and industrial products	0.005
nuclear power industry	0.001

The average dose from medical x-rays has decreased in recent years because of advances in the sensitivity of the photographic film used for x-rays. Radiation from nuclear test fallout has also decreased as a result of the atmospheric nuclear test ban. Not all nations have signed this treaty, however, and the fallout from a Chinese atmospheric test in 1976 led to the ^{131}I contamination of milk in the Harrisburg, Pennsylvania, vicinity at a level of 300 pCi (3.00×10^{-10} Ci) per liter, which was about eight times the level of contamination (41 pCi per liter) that resulted from the accident at Three Mile Island.

The contribution to the radiation absorbed dose from consumer and industrial products includes radiation from construction materials, x-rays emitted by television sets, and inhaled tobacco smoke. The most recent estimate of the total radiation emitted from the mining and milling of uranium, the fabrication of reactor fuels, the storage of radioactive wastes, and the operation of nuclear reactors is less than 0.001 rem per year.

The total dose from ionizing radiation for the average American is about 0.170 rem per year. The Committee on Biological Effects of Ionizing Radiation of the National Academy of Sciences recently estimated that an increase in this dose to a level of 1 rem per year would result in 169 additional deaths from cancer per million people exposed. This can be compared with the 170,000 cancer deaths that would normally occur in a population this size that was not exposed to this level of radiation.

Exercise 23.13

Calculate the rems of radiation absorbed by the average person from the ^{14}C in his or her body. Assume the activity of the ^{14}C in the average body is 0.08 μCi, the energy of the β^--particles emitted when ^{14}C decays is 0.156 MeV, and about one-third of this energy is captured by the body.

Solution

We can calculate the number of β^- particles emitted by ^{14}C from the activity of this isotope.

$$0.08 \, \mu\text{Ci} \times \frac{1 \, \text{Ci}}{10^6 \, \mu\text{Ci}} \times \frac{3.700 \times 10^{10} \, \text{atoms/s}}{1 \, \text{Ci}} = 3000 \, \text{atoms/s}$$

To calculate the effect of this radiation over a year, we have to calculate how many ^{14}C nuclei decay during this time.

$$3000 \, \frac{\text{atoms}}{\text{s}} \times 60 \, \frac{\text{s}}{\text{min}} \times 60 \, \frac{\text{min}}{\text{hr}} \times 24 \, \frac{\text{hr}}{\text{d}} \times 365 \, \frac{\text{d}}{\text{yr}} = 9.5 \times 10^{10} \, \text{atoms/yr}$$

We can calculate the amount of energy absorbed from the number of atoms that disintegrate, the energy per disintegration, and the fraction of this energy absorbed by the body.

$$9.5 \times 10^{10} \, \frac{\text{atoms}}{\text{yr}} \times 0.156 \, \frac{\text{MeV}}{\text{atom}} \times \frac{1}{3} = 4.9 \times 10^9 \, \frac{\text{MeV}}{\text{yr}}$$

This is equal to 7.8×10^{-4} joules per year.

$$4.9 \times 10^{15} \, \frac{\text{eV}}{\text{yr}} \times \frac{1.60 \times 10^{-19} \, \text{J}}{1 \, \text{eV}} = 7.8 \times 10^{-4} \, \text{J/yr}$$

Averaged over a 70-kilogram body weight, this amounts to about 1×10^{-5} joules per gram of body weight per year. If 1 rad is equal to 0.01 joules per kilogram, the radiation absorbed dose from ^{14}C decay is 0.001 rad per year. Using an RBE for β^--decay of 1, this corresponds to 0.001 rems, or 1 millirem, per year.

The principal effect of low doses of ionizing radiation is to induce cancers, which may take up to 20 years to develop. What is the effect of high doses of ionizing radiation? Cells that are actively dividing are more sensitive to radiation than cells that are not. Thus, cells in the liver, kidney, muscle, brain, and bone are more resistant to radiation than the cells of bone marrow, the reproductive organs, the epithelium of the intestine, and the skin, which suffer the most damage from radiation. Damage to the bone marrow is the main cause of death at moderately high levels of exposure (200 to 1000 rads). Damage to the gastrointestinal tract is the major cause of death for exposures on the order of 100 to 10,000 rads. Massive damage to the central nervous system is the cause of death from extremely high exposures (over 10,000 rads).

23.10 NATURAL VERSUS INDUCED RADIOACTIVITY

The vast majority of the nuclides found in nature are stable. If, as estimated, our planet is 4.6 billion years old, the only radioactive isotopes that should remain are members of three classes.

1. Isotopes such as ^{238}U with half-lives on the order of 10^9 years.
2. Daughter nuclides such as ^{234}Th ($t_{1/2} = 24.1$ days) produced when long-lived radioactive nuclides (radionuclides) decay.
3. Nuclides such as ^{14}C that are still being synthesized.

Half of the elements in the periodic table have an odd number of protons, because atomic numbers that are odd are just as likely to occur as those that are even. However, about 80% of stable nuclides have an even number of protons. Very few elements with an odd atomic number have more than one stable isotope. Stable isotopes abound, however, among elements with even atomic numbers. Ten stable isotopes are known for tin ($Z = 50$), for example. It is also interesting to note that 91% of the stable isotopes of elements with an odd number of protons have an even number of neutrons. These observations suggest that certain combinations of protons and neutrons are particularly stable.

In Chapter 5 we said that there are magic numbers of electrons—electron configurations with 2, 10, 18, 36, 54, and 86 electrons are unusually stable. There also seem to be magic numbers of neutrons and protons. Nuclei with 2, 8, 20, 28, 50, 82, or 126 protons or neutrons are unusually stable. This observation explains the anomalously large binding energies observed in Figure 23.6 for $_2^4$He, $_6^{12}$C, $_8^{16}$O, and $_{10}^{20}$Ne. In each case, the nucleus has an even number of both protons and neutrons. $_{10}^{20}$Ne has a magic number of nucleons when both protons and neutrons are counted; ^{4}He and ^{16}O have magic numbers of both protons and neutrons.

If nuclei tend to be more stable when they have even numbers of protons and neutrons, it isn't surprising that nuclides with an odd number of both protons and neutrons are unstable. ^{40}K is one of only five naturally occurring nuclides that contain both an odd number of protons and an odd number of neutrons. This nuclide simultaneously undergoes the electron capture and positron emission expected for neutron-poor nuclides and the electron emission observed with neutron-rich nuclides, as shown in Figure 23.3.

Only 18 radioactive isotopes with atomic numbers of 80 or less can be found in nature. With the exception of ^{14}C, which is continuously synthesized in the atmosphere, all these elements have lifetimes longer than 10^9 years. Although these isotopes all undergo radioactive decay, they decay so slowly that reasonable quantities are still present today, 4.6 billion years after the planet was formed.

Another 45 radioactive isotopes found in nature have atomic numbers larger than 80. These nuclides fall into three families, one of which is shown in Figure 23.8. The parent nuclide is ^{232}Th, which undergoes α-decay to form ^{228}Ra. The

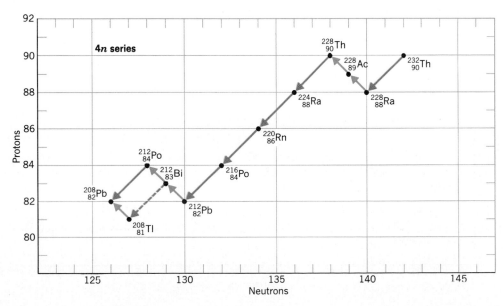

FIG. 23.8 The naturally occurring radioactive isotopes with atomic numbers of 80 or more fall into three series. The $4n$ series starts with ^{232}Th and eventually decays to ^{208}Pb. The mass number of every member of this series is equal to some integer times 4.

product of this reaction decays by β^--emission to form ^{228}Ac, which decays to ^{228}Th, and so on, until the stable ^{208}Pb isotope is formed. This family of radionuclides is called the $4n$ series, because all its members have a mass number that can be divided by 4.

A second family of radioactive nuclei starts with ^{238}U and decays to form the stable ^{206} isotope, as shown in Figure 23.9. Every member of this series has a mass number that fits the equation $4n + 2$. The third family, known as the $4n + 3$ series, starts with ^{235}U and decays to ^{207}Pb, as shown in Figure 23.10.

There is evidence to suggest that a $4n + 1$ series once existed, which started with ^{237}Np and decayed to form the only stable isotope of bismuth, ^{209}Bi, as shown in Figure 23.11. The half-life of every member of this series is less than 2×10^6 years, however, so none of the nuclides produced by the decay of neptunium remain in detectable quantities on the earth.

In 1934, Irene Curie, the daughter of Pierre and Marie Curie, and her husband, Frédéric Joliot, announced the first synthesis of artificial radioactive isotopes. They bombarded a thin foil of aluminum metal with α-particles produced by the decay of polonium and found that the aluminum target became radioactive. Chemical analysis showed that the product of this reaction was an isotope of phosphorus.

$$^{27}_{13}\text{Al} + {}^4_2\text{He} \longrightarrow {}^{30}_{15}\text{P} + {}^1_0n$$

In the next 50 years, more than 2000 other artificial radionuclides were synthesized.

A shorthand notation has been developed for nuclear reactions such as the reaction discovered by Curie and Joliot.

$$^{27}_{13}\text{Al}(\alpha, n)^{30}_{15}\text{P}$$

The parent (or target) nuclide and the daughter nuclide are separated by parentheses, which contain the symbols for the particle that hits the target and the particle (or particles) released in this reaction.

The nuclear reactions used to synthesize artificial radionuclides are characterized by enormous activation energies. Two devices are used to obtain this energy —linear accelerators, or cyclotrons, and nuclear reactors. Linear accelerators or

FIG. 23.9 A second naturally occurring series, known as the $4n + 2$ sequence, starts with ^{238}U and ends when the decay reaches the stable ^{206}Pb nuclide.

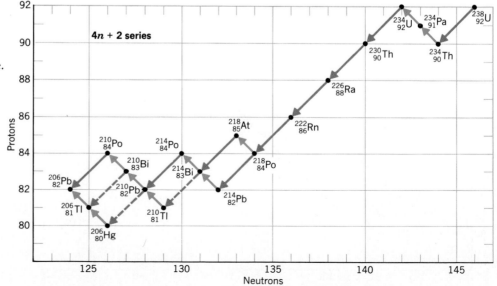

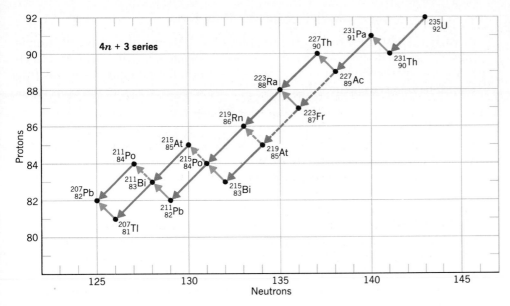

FIG. 23.10 The third naturally occurring decay sequence, known as the $4n + 3$ sequence, starts with ^{235}U and ends with the stable ^{207}Pb nuclide.

cyclotrons can be used to excite charged particles such as protons, electrons, α-particles, or even heavier ions, which are then focused on a stationary target.

The capture of a positively charged particle usually produces a neutron-poor isotope. The following is an example of a reaction that can be induced by a cyclotron or linear accelerator.

$$^{24}_{12}\text{Mg} + ^{2}_{1}\text{H} \longrightarrow ^{22}_{11}\text{Na} + ^{4}_{2}\text{He}$$

Artificial radionuclides are also synthesized in nuclear reactors, which are excellent sources of slow-moving, or **thermal**, neutrons. The absorption of a neutron usually results in a neutron-rich isotope. The following neutron absorption reaction occurs in the cooling systems of nuclear reactors cooled with liquid sodium metal.

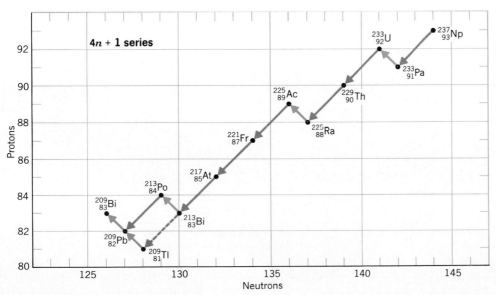

FIG. 23.11 There is evidence to suggest that a $4n + 1$ series, which started with ^{237}Np and decayed to ^{209}Bi, once existed. The half-life of every radioactive member of this series is so short, however, that none of these nuclides remain in detectable amounts on the earth.

$$^{23}_{11}Na + ^{1}_{0}n \longrightarrow ^{24}_{11}Na + \gamma$$

As early as 1940, these techniques were used to synthesize elements with atomic numbers larger than the heaviest naturally occurring element, uranium. McMillan and Abelson synthesized neptunium and plutonium, for example, by irradiating ^{238}U with neutrons to form ^{239}U,

$$^{238}_{92}U + ^{1}_{0}n \longrightarrow ^{239}_{92}U + \gamma$$

which undergoes spontaneous β^--decay to form ^{239}Np and then ^{239}Pu.

$$^{239}_{92}U \longrightarrow ^{239}_{93}Np + ^{0}_{-1}e + \bar{\nu}$$
$$^{239}_{93}Np \longrightarrow ^{239}_{94}Pu + ^{0}_{-1}e + \bar{\nu}$$

Larger bombarding particles were eventually used to produce even heavier transuranium elements.

$$^{253}_{99}Es + ^{4}_{2}He \longrightarrow ^{256}_{101}Md + ^{1}_{0}n$$
$$^{246}_{96}Cm + ^{12}_{6}C \longrightarrow ^{254}_{102}No + 4^{1}_{0}n$$

The half-lives for α-decay and spontaneous fission become shorter as the atomic number of the element becomes larger. Element 104, for example, has a half-life for spontaneous fission of 0.3 seconds. Elements therefore become harder to characterize as the atomic number increases.

Recent theoretical work has predicted that a magic number of protons might exist at $Z = 114$. This work suggests that there is an island of stability in the sea of unstable nuclides, as illustrated in Figure 23.12. If this theory is correct, superheavy elements could be formed if we could find a way to cross the gap between elements $Z = 109$ through $Z = 114$.

There is some debate about the number of neutrons needed to overcome the

FIG. 23.12 The stable nuclei fall within a narrow band of neutron-to-proton ratios. Glenn Seaborg visualizes this as a ridge of stability in a sea of instability. Theory predicts, however, that an island of stability will exist for superheavy elements with atomic numbers around 114 and atomic masses close to 300 amu. Synthesis of the superheavy elements requires that we jump across the sea of instability in one step.

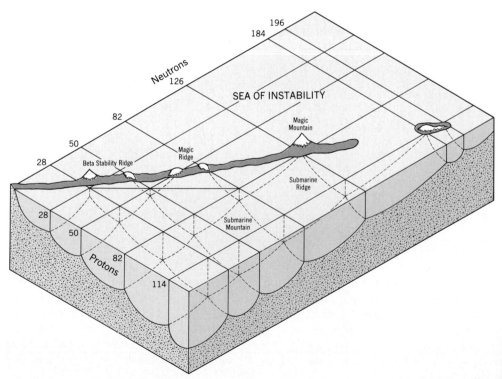

proton-proton repulsion in a nucleus with 114 protons. The best estimates suggest that at least 184 neutrons, and perhaps as many as 196, would be needed. It is not an easy task to bring together two particles that give both the correct number of total protons and the necessary neutrons to produce a nuclide with a half-life long enough to be detected.

If we start with a relatively long-lived parent nuclide, such as ^{251}Cf ($t_{1/2} = 800$ yr) and bombard this nucleus with a heavy ion, such as ^{32}S, we can envision producing a daughter nuclide with the correct atomic number, but the mass number would be too small by at least 16 amu.

$$^{251}_{98}Cm + {}^{32}_{16}S \longrightarrow {}^{282}_{114}X + {}^{1}_{0}n$$

Exercise 23.14

When a new radionuclide is synthesized, it has to be separated from the starting material. This means that the chemistry of this element must be predicted on the basis of the behavior of elements that belong to the same group in the periodic table. Predict the group in which element 114 belongs. What oxidation states are expected for this element?

Solution

We can use Figure 5.22 and the aufbau principle (see Section 5.19) to predict the following electron configuration for element 114.

$1s^2\ 2s^2\ 2p^6\ 3s^2\ 3p^6\ 4s^2\ 3d^{10}\ 4p^6\ 5s^2\ 4d^{10}\ 5p^6\ 6s^2\ 4f^{14}\ 5d^{10}\ 6p^6\ 7s^2\ 5f^{14}\ 6d^{10}\ 7p^2$

This configuration can also be written as follows.

$$X = [Rn]\ 7s^2\ 5f^{14}\ 6d^{10}\ 7p^2$$

In Section 8.1, we concluded that filled d and f subshells are ignored when valence electrons on an atom are counted. Thus, the valence electrons for this element are the $7s^2$ and $7p^2$ electrons. The element must belong in Group IVA, directly below Pb. Possible oxidation states are $+2$ and $+4$.

An expanded periodic table for elements up to $Z = 168$ is shown in Figure 23.13. Elements 104 through 112 are transition metals that fill the $6d$ orbitals. Elements 113 through 120 are main-group elements in which the $7p$ and $8s$ orbitals are filled. The next subshell is the $5g$ atomic orbital, which can hold up to 18 electrons. There is reason to believe that the $5g$ and $6f$ orbitals will be filled at the same time. The next 32 elements are therefore grouped into a so-called superactinide series.

23.11 NUCLEAR FISSION

The graph of binding energy per nucleon in Figure 23.5 suggests that nuclides with masses larger than about 130 amu should spontaneously split apart to form lighter, more stable, nuclides. Experimentally, we find that spontaneous fission reactions occur for only the very heaviest nuclides—those with mass numbers of 230 or more. When they do occur, these reactions are often very slow. The half-life for the spontaneous fission of ^{238}U, for example, is 10^{16} years, or about two million times longer than the age of our planet. This observation can be explained with the following model.

FIG. 23.13 An extended version of the periodic table developed by Glenn Seaborg that predicts the positions for all elements up to atomic number 153.

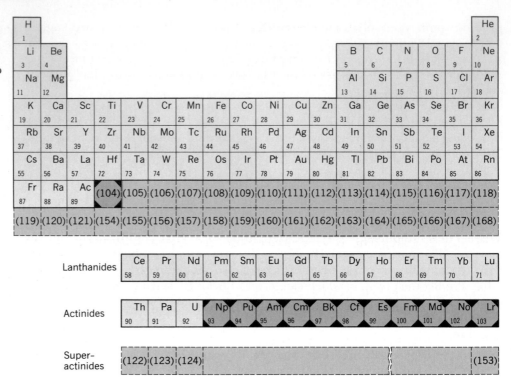

The Rutherford-Geiger-Marsden experiment, which provided the first evidence for the existence of nuclei, was introduced in Section 5.7. At that time, we concluded that the deflection of the α-particles could be explained by the force of repulsion between the positively charged α-particle and the positively charged nucleus of the atom. Because it occurs between the nucleus and a particle that is outside the nucleus, this is an *external* force of repulsion.

Theory suggests that there is also an *internal* barrier, which gives rise to an activation energy that opposes the ejection of a positively charged particle from the nucleus of an atom. As long as this activation energy is large, the spontaneous fission of nuclei such as ^{238}U will be slow.

The force of repulsion between particles of the same charge depends on the charges (Z_1 and Z_2) and the distance between the particles (r).

$$F = \frac{Z_1 Z_2}{r^2}$$

It is therefore easier for a nucleus such as ^{238}U to eject an α-particle ($Z_1 = +2$ and $Z_2 = +90$) than it is for this nucleus to split in half ($Z_1 = +46$ and $Z_2 = +46$) in a spontaneous fission reaction. This may help explain why so many of the heavier nuclei that cannot exhibit spontaneous fission can undergo alpha decay.

There is no force of repulsion between a positively charged nucleus and neutral particles such as neutrons. As a result, nuclei that repel positively charged α-particles are able to absorb slow-moving, thermal neutrons, which can induce fission reactions.

When ^{235}U absorbs a thermal neutron, for example, it splits into two particles of uneven mass and releases an average of 2.5 neutrons. The following equation gives just one example of the many reactions that occur during the induced fission of ^{235}U.

$$^{235}_{92}U + ^{1}_{0}n \longrightarrow ^{139}_{56}Ba + ^{94}_{36}Kr + 3\,^{1}_{0}n$$

More than 370 daughter nuclides with atomic masses between 72 and 161 amu are formed in the thermal-neutron-induced fission of ^{235}U. A plot of the relative frequency versus atomic mass of the daughter nuclides produced in this reaction is shown in Figure 23.14.

Several isotopes of uranium undergo induced fission. But the only naturally occurring isotope in which we can induce fission with thermal neutrons is ^{235}U, which is present at an abundance of only 0.72%. The induced fission of this isotope releases an average of 200 MeV per atom, or 80 million kilojoules per gram of ^{235}U. Comparing this figure with the 50 kJ/g released when natural gas is burned reveals the attraction of nuclear fission as a source of power.

The first artificial nuclear reactor was built by Enrico Fermi and co-workers beneath the University of Chicago's football stadium and brought on line on December 2, 1942. This reactor, which produced several kilowatts of power, consisted of a pile of graphite blocks weighing 385 tons stacked in layers around a cubical array of 40 tons of uranium metal and uranium oxide.

Spontaneous fission of ^{238}U or ^{235}U in this reactor produced a very small number of neutrons. But enough uranium was present so that one of these neutrons induced the fission of a ^{235}U nucleus, thereby releasing an average of 2.5 neutrons, which catalyzed the fission of additional ^{235}U nuclei in a chain reaction. The amount of fissionable material necessary for the chain reaction to sustain itself is called the ***critical mass.***

Pitchblende and yellow cake, which are about 75% U_3O_8.

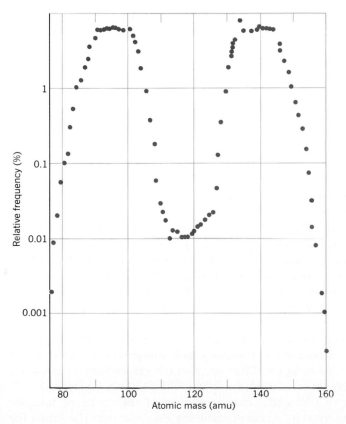

FIG. 23.14 If the fission of ^{235}U produced two fragments with roughly the same mass, we would expect that the largest number of daughter nuclides for the thermal-neutron-induced fission of this isotope would have a mass of about 120 amu. As shown by this graph of the relative frequency versus the mass number of the daughter nuclei, few products with this mass are in fact formed when ^{235}U undergoes thermal-neutron-induced fission. For reasons that are not quite understood, fission reactions tend to produce one fragment that is roughly 1.4 times heavier than the other.

The Fermi reactor at Chicago served as a prototype of larger reactors constructed in 1943 at Oak Ridge, Tennessee, and Hanford, Washington, to produce ^{239}Pu for one of the atomic bombs dropped on Japan at the end of World War II. Some of the neutrons released in the chain reaction are absorbed by ^{238}U to form ^{239}U.

$$^{238}_{92}U + {}^{1}_{0}n \longrightarrow {}^{239}_{92}U + \gamma \qquad \text{(neutron capture)}$$

The product of this reaction undergoes decay by the successive loss of two β^--particles to form ^{239}Pu.

$$^{239}_{92}U \longrightarrow {}^{239}_{93}Np + {}^{0}_{-1}e + \bar{\nu} \qquad (\beta^- \text{ decay})$$
$$^{239}_{93}Np \longrightarrow {}^{239}_{94}Pu + {}^{0}_{-1}e + \bar{\nu} \qquad (\beta^- \text{ decay})$$

^{238}U is an example of a *fertile nuclide*. It does not undergo fission with thermal neutrons, but it can be converted to ^{239}Pu, which does undergo thermal-neutron-induced fission. ^{232}Th is another fertile nuclide — it captures a neutron to form ^{233}Th, which undergoes β^--decay to form ^{233}U, which is also capable of induced fission.

$$^{232}_{90}Th(n, \beta^-)^{233}_{90}Th \xrightarrow{\beta^-} {}^{233}_{91}Pa \xrightarrow{\beta^-} {}^{233}_{92}U$$

Fission reactors can be designed to handle naturally abundant ^{235}U, as well as fuels described as slightly enriched (2–5% ^{235}U), highly enriched (20–30% ^{235}U), or fully enriched (more than 90% ^{235}U). Heat generated in the reactor core is transferred to a cooling agent in a closed system. The cooling agent is passed through a series of heat exchangers in which water is heated to steam. The steam produced in these exchangers then drives a turbine that generates electrical power. There are two ways of specifying the power of such a plant — the thermal energy produced by the reactor or the electrical energy generated by the turbines. The electrical capacity of the plant is usually about one-third of the thermal power.

It takes 10^{11} fissions per second to produce 1 watt of electrical power. Thus, about 1 gram of fuel is consumed per day per megawatt of electrical energy produced. This means that 1 gram of waste products is produced per megawatt per day — including 0.5 grams of ^{239}Pu. These waste products must be either reprocessed to generate more fuel or stored for the tens of thousands of years it takes for the level of radiation to reach a safe limit.

If we could design a reactor in which the ratio of the ^{239}Pu or ^{233}U produced to the ^{235}U consumed was greater than 1, the reactor would generate more fuel than it consumed. Such reactors are known as *breeders*, and the first commercial breeder reactors are now operating in France.

The key to an efficient breeder reactor is a fuel that gives the largest possible number of neutrons released per neutron absorbed. The breeder reactors being built today use a mixture of PuO_2 and UO_2 as the fuel and fast neutrons to activate fission. Fast neutrons carry an energy of at least several KeV and therefore travel 10,000 or more times faster than thermal neutrons. ^{239}Pu in the fuel assembly absorbs one of these fast neutrons and undergoes fission with the release of three neutrons. ^{238}U in the fuel then captures one of these neutrons to produce additional ^{239}Pu.

The advantage of breeder reactors is obvious — they mean a limitless supply of fuel for nuclear reactors. There are significant disadvantages, however. Breeder reactors are more expensive to build. They are also useless without a subsidiary industry to collect the fuel, process it, and ship the ^{239}Pu to new reactors.

It is the reprocessing of ^{239}Pu that concerns most of the critics of breeder reactors. ^{239}Pu is so dangerous as a cancer-inducing material that the limits for exposure in the nuclear industry allow workers to inhale no more than 0.2 micrograms of plutonium over their lifetimes. There is also concern that the ^{239}Pu produced by these reactors might be stolen and assembled into bombs by terrorist organizations.

The fate of breeder reactors in the United States is linked to economic considerations. Because of the costs of building these reactors and safely reprocessing the ^{239}Pu produced, the breeder reactor becomes economical only when the scarcity of uranium drives its price so high that the breeder reactor becomes cost effective by comparison. If nuclear energy is to play a dominant role in the generation of electrical energy in the 21st century, however, breeder reactors may be essential.

The "pile" Fermi constructed at the University of Chicago in 1942 was the first artificial nuclear reactor, but not the first fission reactor known. In 1972, a group of French scientists discovered that uranium ore from a deposit in the Oklo mine in Gabon, West Africa, contained 0.4% ^{235}U instead of the 0.72% abundance found in all other sources of this ore. Analysis of the trace elements in the ore lead to the conclusion that five natural fission reactors operated in this deposit for a period of 600,000 to 800,000 years about 2 billion years ago.

23.12 NUCLEAR FUSION

The graph of binding energy per nucleon in Figure 23.5 suggests another way of obtaining useful energy from nuclear reactions. Fusing two light nuclei can liberate as much energy as is released in the fission of ^{235}U and ^{239}Pu. The fusion of four protons to form a helium nucleus, two positrons, and two neutrinos, for example, generates 24.7 MeV of energy.

$$4\,^{1}_{1}\text{H} \longrightarrow \,^{4}_{2}\text{He} + 2\,^{0}_{+1}e + 2$$

This fusion reaction can be observed in nature. Our sun produces the vast majority of the energy radiated from its surface by the fusion of protons to form helium atoms within its core. Fusion reactions have been duplicated in human devices. The enormous destructive power of the ^{235}U-fueled atomic bomb dropped on Hiroshima on August 6, 1945, which killed 75,000 people, and the ^{239}Pu-fueled bomb dropped on Nagasaki three days later touched off a violent debate after World War II about the building of the next superweapon—a fusion, or "hydrogen," bomb. Alumni of the Manhattan project, the project that had led to the development of the first atomic bomb, were divided on the issue. Ernest Lawrence and Edward Teller fought for the construction of the fusion device. J. Robert Oppenheimer and Enrico Fermi argued against it. The decision was made to develop the weapon, and the first artificial fusion reaction occurred when the hydrogen bomb was tested in November 1952.

The history of fusion research is therefore quite different from that of fission research. With fission, the reactor came first, and then the bomb was built. With fusion, the bomb was built long before any progress was made toward the development of a controlled thermonuclear fusion reactor. More than 30 years after the first hydrogen bomb was exploded, the feasibility of controlled fusion reactions is still open to debate. The reaction that is most likely to fuel the first thermonuclear fusion reactor is the *d-t*, or deuterium-tritium, reaction. This reaction fuses two isotopes of hydrogen, deuterium (^{2}H) and tritium (^{3}H), to form helium and a neutron.

$$^{2}_{1}\text{H} + \,^{3}_{1}\text{H} \longrightarrow \,^{4}_{2}\text{He} + \,^{1}_{0}n$$

Computer graphics of a plasma in hydrogen gas.

FIG. 23.15 This ZT-40M Reversed Field Pinch fusion device at the Los Alamos National Laboratory was able to sustain a plasma of 8 million °C for 8 milliseconds in January 1986.

An extremely high-speed x-ray pinhole photo of energy released from a laser fusion target.

If we consider the consequences of this reaction we can begin to understand why it is called a ***thermonuclear reaction*** and why it is so difficult to produce in a controlled manner. The *d-t* reaction requires that we fuse two positively charged particles. This means that we must provide enough energy to overcome the force of repulsion between these particles before fusion can occur. To produce a self-sustaining reaction, we have to provide the particles with enough *thermal* energy so that they can fuse when they collide.

Each fusion reaction is therefore characterized by a specific ***ignition temperature***, which must be surpassed before the reaction can occur. The *d-t* reaction has an ignition temperature above 10^8 K. In a hydrogen bomb, a fission reaction produced by a small atomic bomb is used to heat the contents to the temperature required to initiate fusion. Obtaining the same result in a controlled reaction is much more challenging.

Any substance at temperatures approaching 10^8K will exist as a completely ionized gas, or ***plasma.*** The goals of fusion research at present include the following.

1. To achieve the required temperature to ignite the fusion reaction.
2. To keep the plasma together at this temperature long enough to get useful amounts of energy out of the thermonuclear fusion reactions.
3. To obtain more energy from the thermonuclear reactions than is used to heat the plasma to the ignition temperature.

These are not trivial goals. The only reasonable container for a plasma at 10^8 K is a magnetic field. Both doughnut-shaped (toroidal) and linear magnetic bottles have been proposed as fusion reactors. A typical toroidal fusion reactor design is shown in Figure 23.15. But further problems result from the fact that the reactors that produce high enough temperatures for ignition are not the same as the reactors that have produced long enough confinement times for the plasma to provide useful amounts of energy.

A second approach to a controlled fusion reactor is based on the use of lasers. In theory, fuel pellets containing the proper reagents for the thermonuclear reaction could be hit by pulsed beams of laser power. If enough power was delivered, the fuel pellets would collapse upon themselves, or implode, to reach densities several orders of magnitude greater than normal. This could produce a plasma both hot enough and dense enough to initiate fusion reactions.

23.13 NUCLEAR SYNTHESIS

From the most primitive societies to the most complex, people have strived to explain how the world was created. Within the last 60 years, scientists have generated a new story of creation that offers a model for understanding how nuclei are synthesized.

In 1929, Edwin Hubble first provided evidence to suggest that our universe is expanding. Between 1946 and 1948, George Gamow and co-workers generated a model that assumed that the primordial substance, or ***ylem,*** from which all other matter was created was an extraordinarily hot, dense singularity, which exploded in a "Big Bang" and has been expanding ever since. This model assumes that neutrons in the ylem were transformed into protons by β^--decay. Neutrons and protons then combined to form ^{4}He atoms before the temperature and pressure of the fireball decayed to the point at which no further nuclear reactions were possible.

The first-generation stars condensed out of this cloud of hydrogen and helium. As the gas condensed by gravitational attraction, it became warmer. Eventually, the temperature reached 10^7 K, and first-generation, main sequence stars were born. The temperatures at the cores of these stars were high enough to ignite the following thermonuclear reactions, which transform hydrogen to helium.

$$^1_1H + {}^1_1H \longrightarrow {}^2_1H + {}^0_{+1}e + \nu$$

$$^1_1H + {}^2_1H \longrightarrow {}^3_2He$$

$$^3_2He + {}^3_2He \longrightarrow {}^4_2He + 2\,{}^1_1H$$

The net result of these reactions was the formation of a helium atom from four protons.

$$4\,{}^1_1H \longrightarrow {}^4_2He + 2\,{}^0_{+1}e + 2\nu$$

Eventually the heat generated in this reaction was enough to halt the gravitational collapse of the star, which then entered a stable period during which the energy given off by this reaction balanced the energy radiated at the surface.

The hydrogen-burning reactions in a main-sequence star are concentrated in the core. When enough hydrogen has been consumed, the core begins to collapse, and the temperature of the core rises above 10^8 K. (The larger the star, the more rapidly it radiates energy from its surface, and the more rapidly it consumes the hydrogen in the core.) As the core collapses, the hydrogen-containing outer shell expands, and the surface of the star cools. Stars that have reached this point in their evolution include the so-called red giants, which are no longer considered to be main-sequence stars.

Temperatures in the core of these red giants are high enough to ignite further thermonuclear fusion reactions, such as the following.

$$3\,{}^4_2He \longrightarrow {}^{12}_6C$$

This reaction becomes the principal source of energy in a red giant, although there is undoubtedly some burning of hydrogen to helium in the outer shell of the star. As the amount of ^{12}C in the core increases, further reactions occur to form ^{16}O and ^{20}Ne.

$$^{12}_6C + {}^4_2He \longrightarrow {}^{16}_8O + \gamma$$

$$^{16}_8O + {}^4_2He \longrightarrow {}^{20}_{10}Ne + \gamma$$

Eventually, the helium in the core is exhausted, and the core collapses further, reaching temperatures of $6-7 \times 10^8$ K. At this point, more complex reactions take place that produce nuclides such as ^{28}Si and ^{32}S.

$$^{12}_6C + {}^{16}_8O \longrightarrow {}^{28}_{14}Si + \gamma$$

$$^{16}_8O + {}^{16}_8O \longrightarrow {}^{32}_{16}S + \gamma$$

Further gravitational collapse heats the core to temperatures above 10^9 K, and a complex sequence of reactions takes place to synthesize the nuclei with the highest binding energies, such as Fe and Ni. If the star explodes in a supernova, its contents are ejected across space. Second-generation stars that condense in this region contain not only hydrogen and helium but elements with higher atomic number.

The best estimates of the age of the Milky Way suggest that our galaxy is about 15 billion years old. Our sun and its planets, however, are only 4.65 billion years old. This suggests that the sun is a second-generation star. In such stars, the transformation of hydrogen to helium can be catalyzed by ^{12}C, as shown in Figure 23.16.

$$^{12}_6C + {}^1_1H \longrightarrow {}^{13}_7N + \gamma$$

$$^{13}_7N \longrightarrow {}^{13}_6C + {}^0_{+1}e + \nu$$

$$^{13}_6C + {}^1_1H \longrightarrow {}^{14}_7N + \gamma$$

$$^{14}_7N + {}^1_1H \longrightarrow {}^{15}_8O + \gamma$$

$$^{15}_8O \longrightarrow {}^{15}_7N + {}^0_{+1}e + \nu$$

$$^{15}_7N + {}^1_1H \longrightarrow {}^{12}_6C + {}^4_2He$$

FIG. 23.16 Second-generation stars, such as our sun, contain nuclei synthesized by their predecessors. These stars do not use the same mechanism to synthesize helium as first-generation stars but use a sequence of reactions, such as those shown here, that are catalyzed by isotopes heavier than helium.

Two processes can synthesize elements with atomic numbers larger than that of iron. One of them is relatively slow *(s-process),* and the other is very rapid *(r-process).* Since the only way to synthesize nuclei with atomic numbers larger than iron's is by neutron absorption, both the *s*-process and the *r*-process result from (n,γ) reactions.

In the *s*-process, neutrons are captured one at a time to form a neutron-rich nuclide, which has enough time to undergo α- or β^--decay before another neutron can be absorbed. An example of an *s*-process sequence of reactions starts with ^{120}Sn. The capture of a neutron produces ^{121}Sn, which undergoes β^- decay. If β^- decay occurs before this nuclide captures another neutron, ^{121}Sb, a stable isotope of antimony, is formed. Eventually, ^{121}Sb captures a neutron to produce ^{122}Sb, which is transformed into ^{122}Te by β^--decay. ^{122}Te can undergo β^--decay to form ^{122}I, or it can capture a neutron to form ^{123}Te. With ^{123}Te, we encounter a series of stable isotopes of tellurium. Thus, neutrons are slowly absorbed, one at a time, until we reach ^{127}Te, which decays to ^{127}I, the most abundant isotope of iodine.

$$^{120}_{50}\text{Sn}(n,\ \gamma)^{121}_{50}\text{Sn} \xrightarrow{\beta^-} {}^{121}_{51}\text{Sb}(n,\ \gamma)^{122}_{51}\text{Sb} \xrightarrow{\beta^-} {}^{122}_{52}\text{Te}(5n,\ 5\gamma)^{127}_{52}\text{Te} \xrightarrow{\beta^-} {}^{127}_{53}\text{I}$$

This slow process can't account for very heavy nuclides, such as ^{232}Th and ^{238}U, because the lifetimes of the intermediate nuclei with atomic numbers between 83 and 90 are too short for this step-by-step absorption of neutrons to proceed. Synthesizing appreciable quantities of uranium and thorium requires a rapid process. In the *r*-process, a number of neutrons are captured in rapid succession, before there is time for α- or β^--decay to take place. Achieving an *r*-process reaction, however, requires a very high neutron flux. (These reactions occur during nuclear explosions, for example.) The neutron flux needed to fuel such reactions is not likely to occur in a normal star. During the moment when a star explodes as a supernova, however, the conditions are ripe for *r*-process reactions. The heavier elements on this planet, then, were produced in a series of supernova explosions that occurred in this portion of the galaxy before our solar system condensed.

SUMMARY

We can predict the products of nuclear reactions by assuming that both mass number and charge are conserved. If ^{238}U undergoes α-particle decay, the product of this reaction must be ^{234}Th.

$$^{238}_{92}\text{U} \longrightarrow {}^{234}_{90}\text{Th} + {}^4_2\text{He} \qquad \text{(alpha decay)}$$

There are three forms of beta decay. One corresponds to the emission of an electron.

$$^{40}_{19}\text{K} \longrightarrow {}^{40}_{20}\text{Ca} + {}^{\ 0}_{-1}e + \bar{\nu} \qquad (\beta^- \text{ emission})$$

One involves the emission of a positron.

$$^{40}_{19}\text{K} \longrightarrow {}^{40}_{18}\text{Ar} + {}^0_{+1}e + \nu \qquad (\beta^+ \text{ emission})$$

And one involves the capture of an electron.

$$^{40}_{19}\text{K} + {}^{\ 0}_{-1}e \longrightarrow {}^{40}_{18}\text{Ar} + h\nu \qquad \text{(electron capture)}$$

Alpha and beta decay often produce nuclides in an excited state. The daughter nuclide emits its excess energy in the form of a γ-ray.

$$^{60m}_{27}\text{Co} \longrightarrow {}^{60}_{27}\text{Co} + \gamma \qquad (\gamma\text{-ray emission})$$

Alpha decay produces a nuclide with a larger neutron-to-proton ratio. Beta decay interconverts nuclides that have the same mass number—in other words, isobars. Gamma ray emission doesn't change either the mass number or the charge on the nuclide.

Stable, naturally occurring nuclides only exist within

a narrow band of neutron-to-proton ratios. When this ratio is too large, the nuclide is said to be neutron-rich. When the ratio is too small, it is neutron-poor. Neutron-rich nuclides decay by β^- emission. Neutron-poor nuclides decay by β^+ emission, electron capture, of α-particle emission.

Nuclei can also undergo decay by spontaneous fission. The rate at which a nuclide splits into two unequal parts increases with the atomic number of the nuclide. For most nuclides, the rate of spontaneous fission is very slow. The rate of this reaction can be enhanced, however, by irradiation of the nuclide with neutrons. The capture of a neutron lowers the activation energy associated with the fission of the nuclide, thereby significantly increasing the rate at which the nuclide undergoes fission.

The difference between the mass of an atom and the sum of the masses of the electrons, protons, and neutrons in the atom is known as the mass defect. The mass defect is a direct measure of the binding energy that holds the nuclide together. The binding energy of nuclides increases with atomic number. But when the binding energy per nucleon is calculated, nuclides are found to be most stable at or near the ^{56}Fe isotope. Nuclides that are heavier than ^{56}Fe can become more stable by undergoing fission reactions. Nuclides that are lighter than ^{56}Fe can become more stable by undergoing fusion reactions.

PROBLEMS

The Discovery of Radioactivity

23-1 Identify the particle given off when a nucleus undergoes α-decay. Identify the particle given off during β-decay. Describe the relationship between γ-rays and other forms of electromagnetic radiation.

23-2 Describe an experiment that could be used to determine whether an atom emits α-particles, β-particles, or γ-rays.

The Structure of the Atom

23-3 Define the terms *atomic number, mass number, nuclide, nucleon,* and *nucleus.*

23-4 Define the terms *isotopes, isobars,* and *isotones.* Give examples of each.

23-5 Calculate the number of electrons, protons, and neutrons in neutral atoms of the following nuclides.

(a) ^{14}C (b) ^{75}As (c) ^{90}Sr (d) ^{184}W (e) ^{239}Pu

23-6 Calculate the number of electrons, protons, and neutrons in the following ions.

(a) ^{40}K$^+$ (b) ^{79}Se^{2-} (c) ^{103}Rh^{3+} (d) ^{127}I$^-$ (e) ^{209}Bi^{3+}

Modes of Radioactive Decay

23-7 Explain why the three forms of β-decay interconvert isobars.

23-8 Describe how electron emission, electron capture, and positron emission differ. Give an example of each.

23-9 Which of the following reactions interconvert isotopes? Which interconvert isobars? Which interconvert isotones?

(a) electron emission (b) electron capture (c) positron emission (d) α-emission (e) neutron emission (f) neutron absorption (g) α-emission followed by two β^- decays

23-10 Identify the missing particle in each of the following equations and name the form of radioactive decay.

(a) $^{125}_{53}$I $+ ^{0}_{-1}e \rightarrow X$ (b) $^{235}_{92}$U $\rightarrow X + ^{140}_{54}$Xe $+ 5\ ^{1}_{0}n$
(c) $^{90}_{38}$Sr $\rightarrow X + ^{0}_{-1}e + \bar{\nu}$ (d) $^{40}_{19}$K $\rightarrow X + ^{0}_{+1}e + \nu$
(e) $^{228}_{90}$Th $\rightarrow X + ^{4}_{2}$He

23-11 Write balanced equations for the β^--particle decay of the following nuclides.

(a) ^{14}C (b) ^{35}S (c) ^{52}V (d) ^{99}Mo (e) ^{241}Pu

23-12 Write balanced equations for the α-particle decay of the following nuclides.

(a) ^{188}Pt (b) ^{204}At (c) ^{224}Ra (d) ^{237}Np (e) ^{248}Cf

23-13 Write balanced equations for the electron capture reactions of the following nuclides.

(a) ^{7}Be (b) ^{22}Na (c) ^{37}Ar (d) ^{62}Cu (e) ^{128}Ba

23-14 Write balanced equations for the positron emission reactions of the following nuclides.

(a) ^{18}F (b) ^{34}Cl (c) ^{45}Ti (d) ^{73}Se (e) ^{90}Mo

23-15 Predict the products of the following nuclear reactions.

(a) Electron emission by ^{32}P (b) Positron emission by ^{11}C (c) α-decay by ^{212}Rn (d) Electron capture by ^{125}Xe

23-16 Predict the products of the following nuclear reactions.

(a) $^{208}_{82}$Pb(^{2}H, n) (b) $^{252}_{98}$Cf($^{13}_{6}$C, $8n$) (c) $^{95}_{42}$Mo(n, γ)
(d) $^{202}_{84}$Po($^{22}_{10}$Ne, $4n$) (e) $^{14}_{7}$N(n, p) (f) $^{63}_{29}$Cu($p, 2n$)

23-17 Identify the missing particle or particles in the following reactions and write a balanced equation for each reaction.

(a) $^{238}_{92}$U(_____, $3\ n$)$^{239}_{94}$Pu (b) _____(α, n)$^{242}_{96}$Cm
(c) $^{250}_{98}$Cf($^{11}_{5}$B, _____)$^{257}_{103}$Lr (d) $^{249}_{98}$Cf(_____, $4\ n$)$^{257}_{104}$Unq

Neutron-Rich Versus Neutron-Poor Nuclides

23-18 Explain why neutron-rich nuclides decay by electron (β^-) emission.

23-19 Explain why neutron-poor nuclides decay by either electron capture, positron (β^+) emission, the emission of an alpha particle, or spontaneous fission.

23-20 Which of the following nuclides are most likely to be neutron-rich?

(a) ^{14}C (b) ^{24}Na (c) ^{26}Si (d) ^{27}Al (e) ^{31}P

23-21 Which of the following nuclides are most likely to be neutron-poor?

(a) ^3H (b) ^{11}C (c) ^{14}N (d) ^{40}K (e) ^{61}Cu

23-22 Explain why ^{17}Ne, ^{18}Ne, and ^{19}Ne decay by positron emission but ^{23}Ne and ^{24}Ne decay by electron emission.

23-23 Which isotope of carbon is most likely to decay by positron emission?

(a) ^{11}C (b) ^{12}C (c) ^{13}C (d) ^{14}C

23-24 Which isotope of carbon is most likely to decay by electron emission?

(a) ^{11}C (b) ^{12}C (c) ^{13}C (d) ^{14}C

Binding Energy Calculations

23-25 Define the terms *mass defect* and *binding energy*.

23-26 Calculate the binding energy of ^6Li in million electron volts per atom if the exact mass of this nuclide is 6.01512 amu. Calculate the binding energy per nucleon.

23-27 Calculate the binding energy of ^{60}Ni in million electron volts per atom if the exact mass is 59.9332 amu. Calculate the binding energy per nucleon.

23-28 Calculate the exact mass of ^{238}U if the binding energy per nucleon is 7.570198 MeV.

23-29 Which nuclide in problems 23-26 through 23-28 has the largest binding energy? Which has the largest binding energy per nucleon?

23-30 Calculate the energy released in the following reaction.

$$^{24}Mg(^2H, p)^{25}Mg$$

Use the following data for the masses of the particles involved in the reaction: ^{24}Mg, 23.98504 amu; ^{25}Mg, 24.98584 amu; ^2H, 2.0140 amu.

23-31 Calculate the energy released in the following reaction.

$$^{10}B(n, \gamma)^7Li$$

Use the following data for the masses of the particles involved in the reaction: ^{10}B, 10.0129 amu; ^7Li, 7.01600 amu; ^4He, 4.00260 amu.

The Kinetics of Radioactive Decay

23-32 The half-life of ^{32}P is 14.3 days. Calculate how long it would take for a 1-gram sample of ^{32}P to decay to each of the following quantities of ^{32}P.

(a) 0.500 gram (b) 0.250 gram (c) 0.125 gram
(d) 0.0625 gram

23-33 Calculate the half-life for the decay of ^{39}Cl if a 1-gram sample decays to 0.125 gram in 165 minutes.

23-34 Calculate the rate constants (in s^{-1}) for the decay of the following nuclides from their half-lives.

(a) ^{18}F, 110 minutes (b) ^{54}Mn, 312 days (c) ^3H, 12.26 years (d) ^{14}C, 5730 years (e) ^{129}I, 1.6×10^7 years

23-35 A 1-gram sample of ^{22}Na decays to 0.20 gram in 6.04 years. Calculate the half-life for this decay, the rate constant, and the time it would take for this sample to decay to 0.075 gram.

23-36 Calculate the time required for a 2.50-gram sample of ^{51}Cr to decay to 1.00 gram, assuming that the half-life is 27.8 days.

23-37 A sample of ^{210}Po initially weighed 2.000 grams. After 25 days, 0.125 gram of ^{210}Po remained, the rest of the sample having decayed to the stable ^{206}Pb isotope. Calculate the half-life of ^{210}Po and the weight of ^{206}Pb formed.

23-38 Forgeries that had been accepted by art authorities as paintings by the Dutch artist Vermeer (1632 – 1675) have been detected by measuring of the activity of the ^{210}Pb isotope in the lead paints. When lead is extracted from its ores, it is separated from the ^{226}Ra, which is the source of the ^{210}Pb isotope. If the half-life for the decay of ^{210}Pb is 21 years, what fraction of the ^{210}Pb would be present in a 300-year-old painting? What fraction would remain in a 10-year-old forgery?

23-39 Use the following data for ^{48}Cr to calculate the rate constant and the half-life for the decay of this isotope by electron capture.

mass (g)	100	80	60	40	20
time (min)	0	444	1017	1825	3205

Units of Activity for Radioactive Decay

23-40 The threat to people's health from radon in the air trapped in their houses has received attention in recent years. If the average level of radon in a house is approximately 1 picocurie (pCi) per liter of air, how many radon atoms are there per liter? Assume that ^{222}Rn is the principal source of this activity and that the half-life for the decay of this nuclide is 3.823 days.

23-41 Calculate the number of disintegrations per minute in a 1.00-milligram sample of ^{238}U, assuming that the half-life is 4.47×10^9 years.

23-42 Calculate the activity, in disintegrations per second, for a 1.00-milligram sample of each of the following isotopes of argon.

(a) ^{35}Ar, $t_{1/2} = 1.85$ (b) ^{41}Ar, $t_{1/2} = 1.83$ h (c) ^{37}Ar, $t_{1/2} = 35.1$ d (d) ^{39}Ar, $t_{1/2} = 270$ yr

23-43 Calculate the activity in curies for a 1.00-milligram sample of the following isotopes of uranium.

(a) ^{228}U, $t_{1/2} = 9.3$ min (b) ^{230}U, $t_{1/2} = 20.8$ d (c) ^{232}U, $t_{1/2} = 72$ yr (d) ^{236}U, $t_{1/2} = 2.39 \times 10^7$ yr

23-44 Calculate the half-life of ^{227}Ac, assuming that a 0.100-milligram sample has an activity of 2.75×10^8 disintegrations per second.

23-45 Calculate the weight of 1.00 millicurie (mCi) of ^{14}C, assuming that the half-life of this nuclide is 5730 years.

Dating by Radioactive Decay

23-46 The ^{14}C in living matter has an activity of 15.3 disintegrations, or "counts," per minute (cpm). What is the age of an artifact that has an activity of 4 cpm? (^{14}C: $t_{1/2}$ = 5730 yr)

23-47 A skull fragment found in 1936 at Baldwin Hills, California, was dated by radiocarbon analysis. Approximately 100 grams of bone was cleaned and treated with 1 M hydrochloric acid to destroy the mineral content of the bone. The bone protein was collected, dried, and pyrolyzed. The CO_2 produced was collected and purified and the ratio of ^{14}C to ^{12}C was measured. If this sample contained roughly 5.7% of the ^{14}C present in living tissue, how old is the skeleton? (^{14}C: $t_{1/2}$ = 5730 yr)

23-48 Measurements on the linen wrappings from the Book of Isaiah in the Dead Sea scrolls suggest that the scrolls contain about 79.5% of the ^{14}C expected in living tissue. How old are these scrolls? (^{14}C: $t_{1/2}$ = 5730 yr)

23-49 The Lascaux cave near Montignac in France contains a series of remarkable cave paintings. Radiocarbon dating of charcoal taken from this site suggests an age of 15,520 years. What fraction of the ^{14}C present in living tissue is still present in this sample? (^{14}C: $t_{1/2}$ = 5730 yr)

23-50 Charcoal samples from Stonehenge in England emit 62.3% of the disintegrations per gram of carbon per minute expected for living tissue. What is the age of this charcoal? (^{14}C: $t_{1/2}$ = 5730 yr)

23-51 A lump of beeswax was excavated in England near a collection of Bronze Age objects that are between 2500 to 3000 years old. Radiocarbon analysis of the beeswax suggests an activity roughly 90.3% of the activity observed for living tissue. Was this beeswax part of the hoard of Bronze Age objects, or did it date from another period?

23-52 The activity of the ^{14}C in living tissue is 15.3 disintegrations per minute per gram of carbon, and the limit for reliable determination of ^{14}C ages is 0.10 disintegration per minute per gram of carbon. Calculate the maximum age of a sample that can be dated accurately by radiocarbon dating, assuming that the half-life for the decay of ^{14}C is 5730 years.

Ionizing versus Non-Ionizing Radiation

23-53 Use the relationship between the energy and the frequency of a photon (see Section 5.13) to calculate the energy in kilojoules per mole of a photon of blue light that has a frequency of 6.5×10^{14} s^{-1}. How do the results of this calculation compare with the ionization energy of water—1216 kJ/mol?

23-54 Calculate the energy in kilojoules per mole for an x-ray that has a frequency of 3×10^{17} s^{-1}. How do the results of this calculation compare with the ionization energy of water?

23-55 Explain why ionizing radiation is so much more dangerous than non-ionizing radiation.

23-56 Describe the difference between an H_2O^+ ion and an H_3O^+ ion. Describe the difference between an H_2O^- ion and an OH^- ion. If a free radical is an ion or molecule that contains one or more unpaired electrons, which of these ions are free radicals?

Biological Effects of Ionizing Radiation

23-57 Radioactivity is measured in units that describe the amount of radiation given off, the amount of radiation to which an object is exposed, the amount of radiation absorbed, or the toxicity of the radiation to biological systems. Sort the following units into these categories.

(a) curies (b) rads (c) rems (d) roentgens

23-58 Explain why sources of α-particles are intrinsically more dangerous than sources of β^- particles.

Natural Versus Induced Radioactivity

23-59 Describe the difference between natural and induced radioactivity. Give examples of both processes.

23-60 The first artificial radioactive elements were synthesized by Irene Curie and Frédéric Joliot, who bombarded ^{10}B and ^{27}Al with α-particles to form ^{13}N and ^{30}P. Write balanced equations for these reactions, identify the particle ejected in each reaction, and predict the mode of decay expected for the products of these reactions.

23-61 Russell, Soddy, and Fajans predicted that the emission of one α- and two β^--particles by a nuclide would produce an isotope of that nuclide. Which isotope of ^{216}Po is produced by such decay? What intermediate nuclides are formed?

23-62 In the first synthesis of an isotope of mendelevium, element 101, Ghiorso and co-workers bombarded ^{253}Es with α-particles. Starting with less than 10^{-12} gram of einsteinium, they isolated one atom of mendelvium after a period of a few hours. If a neutron was emitted in this reaction, what isotope of Md was produced? Another isotope of mendelevium was produced by bombardment of ^{238}U with ^{19}F atoms. If five neutrons were ejected in this reaction, what isotope of Md was produced?

23-63 ^{256}Lr is produced when ^{243}Am is bombarded with ^{18}O. How many neutrons are emitted in this reaction? ^{256}Lr decays by both electron capture and α-particle emission. What are the daughter nuclides produced in these reactions?

23-64 $^{238}_{92}U$ decays by the emission of eight α-particles and six β^--particles. What stable nuclide is formed at the end of this decay chain?

23-65 How many alpha and beta particles are emitted when $^{232}_{90}Th$ decays to $^{208}_{82}Pb$?

Nuclear Fission and Nuclear Fusion

23-66 Describe the difference between fission and fusion reactions. Give examples of both processes.

23-67 Explain why relatively light nuclides give off energy when they fuse to form heavier nuclides, whereas relatively heavy nuclides give off energy when they undergo fission.

23-68 Describe the difference between spontaneous and induced fission reaction. Explain why nuclei undergoing induced fission reactions have much shorter half-lives.

23-69 Describe the advantages and disadvantages of fusion reactors versus fission reactors.

Nuclear Synthesis

23-70 Describe the changes in the nuclear reactions that fuel stars as the stars become older.

23-71 What evidence do we have that the sun is a second-generation star?

23-72 Describe the difference between the s-process and r-process for the synthesis of nuclides. Explain why the s-process can't synthesize relatively heavy naturally occurring nuclides, such as ^{238}U.

CHAPTER 24

THE ORGANIC CHEMISTRY OF CARBON

CHAPTER CONTENTS

24.1 THE SPECIAL ROLE OF CARBON

FIG. 24.1 The German chemist Friedrich Wöhler (1800–1882) was a professor of chemistry at the University of Göttingen. He was an outstanding teacher who formed life-long ties with his students and contributed to many different phases of research in chemistry during the 19th century. He is remembered today for his synthesis of an organic compound (urea) from inorganic starting materials.

For more than 200 years, chemists have divided compounds into two classes. Compounds that were isolated from plants and animals were said to be **organic,** whereas those extracted from ores and minerals were **inorganic.** At one time, chemists believed that organic compounds — such as formic acid (HCO_2H), which was first obtained by the distillation of ants — were fundamentally different from inorganic compounds. Organic compounds were thought to contain a *vital force* that was only found in living systems.

The synthesis of urea from inorganic starting materials by Friederich Wöhler marked the first step in the decline of the vital force theory. Wöhler (see Figure 24.1) was interested in the chemistry of cyanate compounds, which we now know to be compounds that contain the OCN^-, or cyanate, ion. In 1828, he tried to synthesize ammonium cyanate (NH_4OCN) from silver cyanate ($AgOCN$) and ammonium chloride (NH_4Cl). What he expected can be described by the following equation.

$$AgOCN(s) + NH_4Cl(aq) \longrightarrow AgCl(s) + NH_4OCN(aq)$$

The product he isolated from this reaction, however, had none of the properties of cyanate compounds. It was a white, crystalline material that was identical to urea (H_2NCONH_2), which had been isolated from urine.

Neither Wöhler nor his contemporaries claimed that his results disproved the vital force theory. But these results set in motion a series of experiments that eventually led to the synthesis of a number of organic compounds from inorganic starting materials. This in turn inevitably led to the removal of vitalism from the list of theories that had any relevance to chemistry, although it did not lead to the death of the theory, which still had proponents more than 90 years later.

If the difference between organic and inorganic compounds is not the presence of some mysterious vital force required for their synthesis, what is the basis for distinguishing between these classes of compounds? Most compounds extracted from living organisms contain carbon. It is therefore common to identify organic chemistry as the chemistry of carbon. But this definition includes compounds such as calcium carbonate ($CaCO_3$), as well as the elemental forms of carbon such as diamond and graphite, which are clearly inorganic (see Section 10.10). Perhaps the best definition of *organic chemistry* identifies it as the chemistry of compounds that contain both carbon and hydrogen. Such compounds will be the subject of the remainder of this chapter.

Even though organic chemistry focuses on compounds that contain carbon and hydrogen, while inorganic chemistry includes all of the compounds of the other 107 elements, over 95% of the compounds chemists have isolated from natural sources or synthesized in the laboratory are *organic* compounds. Carbon obviously plays a special role in the chemistry of the elements. This role is the result of a combination of factors, including the number of valence electrons on a neutral carbon atom, the electronegativity of carbon, and the atomic radius of carbon atoms (see Table 24.1).

Carbon has four valence electrons — $2s^2 2p^2$ — and it must either gain four electrons or lose four electrons to reach a rare-gas configuration. The electronegativity of carbon, however, is too small for carbon to gain electrons from most elements to form C^{4-} ions and too large for carbon to lose electrons to form C^{4+} ions. Carbon therefore forms covalent bonds with a large number of other elements, including the hydrogen, nitrogen, oxygen, phosphorus, and sulfur found in living systems.

TABLE 24.1	
The Physical Properties of Carbon	
electronic configuration	$1s^2 2s^2 2p^2$
melting point	
graphite	sublimes between 3652 and 3697°C
diamond	above 3550°C
boiling point	
graphite	4827°C
diamond	4827°C
density	
graphite	2.25 g/mL
diamond	3.514 g/mL
1st ionization energy	1086.4 kJ/mol
electron affinity	122.3 kJ/mol
electronegativity	2.55
C—C single bond length	
diamond	0.1544 nm
C_2H_6	0.1533 nm
covalent radius	0.077 nm
C—C bond energy	330 kJ/mol
energy to transform C(s) into C(g)	716.68 kJ/mol
enthalpy of formation	
graphite	0.000 kJ/mol
diamond	2.425 kJ/mol

Carbon atoms are relatively small—the covalent radius is only 0.077 nanometers. As a result, carbon atoms can come close enough together to form strong C=C double bonds or even C≡C triple bonds. Carbon also forms strong double and triple bonds to nitrogen and oxygen. It can even form strong double bonds to elements such as phosphorus or sulfur that don't form strong double bonds to themselves.

Several years ago, an unmanned Viking spacecraft carried out experiments designed to search for evidence of life on Mars. These experiments were based on the assumption that living systems contain carbon, and the absence of any evidence for carbon-based life on that planet was assumed to mean that no life existed. Several factors make carbon essential to life: (1) the ease with which it forms bonds to itself, (2) its tendency to form multiple bonds to C, N, O, P, and S atoms, and (3) the strength of these covalent bonds. These factors provide an almost infinite variety of potential structures for organic compounds, such as vitamin C, shown in Figure 24.2. No other element can provide the variety of combinations and permutations of compounds necessary for life to exist.

Vitamin C

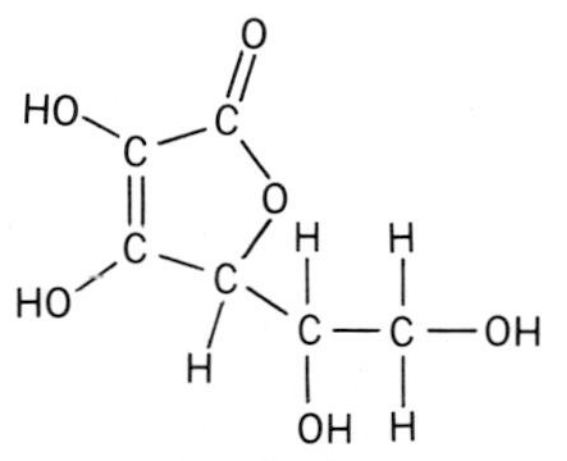

FIG. 24.2 Carbon is the only element that provides the diverse structures necessary for life to exist. Vitamin C is a relatively simple example of the complexity that organic compounds can achieve.

THE SATURATED HYDROCARBONS:
24.2 ALKANES AND CYCLOALKANES

Compounds that contain only hydrogen and carbon are known as ***hydrocarbons.*** Hydrocarbons that contain as many hydrogen atoms as possible are said to be ***saturated;*** saturated hydrocarbons are also known as ***alkanes.*** The carbon atoms in an alkane are all sp^3 hybridized.

The simplest saturated hydrocarbon, or alkane, is methane — CH_4. To understand the Lewis structure of methane, note the presence of four electrons in the valence shell of a neutral carbon atom. This suggests that carbon can combine wih four hydrogen atoms to form a compound in which it shares a total of eight valence electrons.

Methane is an example of a general rule that carbon is **tetravalent** — it tends to form a total of four bonds in its compounds.

Exercise 24.1

Predict the formula of ethane, the saturated hydrocarbon (or alkane) that contains two carbon atoms.

Solution

As a rule, compounds that contain more than one carbon atom are held together by C—C bonds. If we assume that carbon is tetravalent, the formula of this compound must be C_2H_6.

The alkane that contains three carbon atoms is known as propane. Propane has the formula C_3H_8 and the following skeleton structure.

The four-carbon alkane is butane, with the formula C_4H_{10}.

The alkanes listed in Table 24.2 all have the same general formula. For every n carbon atoms, they have $2n + 2$ hydrogen atoms. Alkanes therefore have the generic formula C_nH_{2n+2}.

The boiling points of the alkanes gradually increase as the molecular weight of the compound increases. At room temperature, the lighter alkanes are gases; the midweight alkanes are liquids; and the heavier alkanes are solids, or tars.

The alkanes in Table 24.2 are all **straight-chain hydrocarbons,** in which the carbon atoms form a chain that runs from one end of the molecule to the other. Alkanes also form **branched** structures. The smallest hydrocarbon in which a branch can occur has four carbon atoms.

TABLE 24.2

The Saturated Hydrocarbons, or Alkanes

Name	Molecular Formula	Melting Point (°C)	Boiling Point (°C)	State at 25°C
methane	CH_4	−182.48	−164	gas
ethane	C_2H_6	−183.3	−88.63	gas
propane	C_3H_8	−189.69	−42.07	gas
butane	C_4H_{10}	−138.35	−0.5	gas
pentane	C_5H_{12}	−129.72	36.07	liquid
hexane	C_6H_{14}	−95	68.95	liquid
heptane	C_7H_{16}	−90.61	98.42	liquid
octane	C_8H_{18}	−56.79	124.66	liquid
nonane	C_9H_{20}	−51	150.798	liquid
decane	$C_{10}H_{22}$	−29.7	174.1	liquid
undecane	$C_{11}H_{24}$	−24.59	195.9	liquid
dodecane	$C_{12}H_{26}$	−9.6	216.3	liquid
tridecane	$C_{13}H_{28}$	−5.5	235.4	liquid
eicosane	$C_{20}H_{42}$	36.8	343	solid
triacontane	$C_{30}H_{62}$	65.8	449.7	solid

$$CH_3-\overset{\overset{\displaystyle CH_3}{|}}{CH}-CH_3 \qquad \text{isobutane}$$

This compound has the same molecular formula as butane (C_4H_8), but a different structure.

As noted in earlier chapters, compounds with the same formula and different structures are known as *isomers* (from the Greek *isos,* "equal," and *meros,* "parts"). The difference between the structures of butane and isobutane is shown in Figure 24.3. These compounds are examples of *structural isomers,* which differ in the sequence in which their atoms are bonded together.

There are three structural isomers of pentane, C_5H_{12}. The first is "normal" pentane, or *n*-pentane.

$$CH_3-CH_2-CH_2-CH_2-CH_3 \qquad \text{n-pentane}$$

A branched isomer is also possible; it was originally named isopentane.

$$CH_3-\overset{\overset{\displaystyle CH_3}{|}}{CH}-CH_2CH_3 \qquad \text{isopentane}$$

FIG. 24.3 Two isomers with the formula C_4H_{10} are possible; they are known as butane and isobutane. Butane is an example of a straight-chain hydrocarbon, because the four carbon atoms form a continuous chain. Isobutane is an example of a branched hydrocarbon. The carbon atoms in these saturated hydrocarbons, or alkanes, are all sp^3 hybridized.

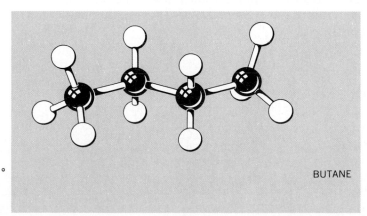

BUTANE

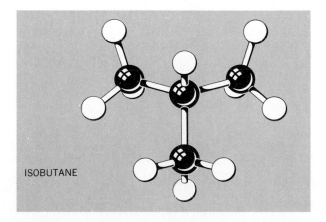

ISOBUTANE

When a more highly branched isomer was discovered, it was named neopentane (literally, the new isomer of pentane).

$$CH_3 - \underset{\underset{\displaystyle CH_3}{|}}{\overset{\overset{\displaystyle CH_3}{|}}{C}} - CH_3 \qquad \text{neopentane}$$

Exercise 24.2

Determine the number of structural isomers of hexane, C_6H_{14}.

Solution

There are five structural isomers. There is a straight-chain, or normal, isomer.

$$CH_3 - CH_2 - CH_2 - CH_2 - CH_2 - CH_3$$

There are two isomers with a single carbon branch.

$$\underset{\displaystyle CH_3 - \overset{\overset{\displaystyle CH_3}{|}}{CH} - CH_2 - CH_2 - CH_3}{} \qquad CH_3 - CH_2 - \overset{\overset{\displaystyle CH_3}{|}}{CH} - CH_2 - CH_3$$

And there are two isomers with two branches.

$$CH_3 - \overset{\overset{\displaystyle CH_3}{|}}{CH} - \overset{\overset{\displaystyle CH_3}{|}}{CH} - CH_3 \qquad CH_3 - \underset{\underset{\displaystyle CH_3}{|}}{\overset{\overset{\displaystyle CH_3}{|}}{C}} - CH_2 - CH_3$$

The number of isomers of a compound increases rapidly with the number of carbon atoms. There are over four billion isomers of $C_{30}H_{62}$, for example.

If the carbon chain that forms the backbone of a straight-chain hydrocarbon is long enough, we can envision the two ends coming together to form a *cyclic hydrocarbon,* or *cycloalkane.* One hydrogen atom has to be removed from each end of the hydrocarbon chain to form the C—C bond that closes the ring. Cycloalkanes therefore have two fewer hydrogen atoms than the parent alkane and a generic formula of C_nH_{2n}. Cyclohexane, for example, has the formula C_6H_{12}.

A systematic approach to the naming of alkanes and cycloalkanes consists of the following steps.

1. Find the longest continuous chain of carbon atoms in the skeleton structure of the compound. Name the compound as a derivative of the alkane with this number of carbon atoms.

This compound is a derivative of heptane because the longest chain contains a total of seven carbon atoms.

2. Name the substituents on the chain—the groups attached to the longest chain of carbon atoms. Substituents derived from alkanes are named with the ending *-yl* replacing the ending *-ane*. This compound contains methyl (CH_3—) and ethyl (CH_3CH_2—) substituents.

3. Number the chain starting at the end nearest the first substituent.

4. Use the results of the third step to specify the carbon atoms on which the substituents are located. For example, this compound is 4-ethyl-2-methyl-heptane.

5. Use the prefixes *di-*, *tri-*, and *tetra-* to describe substituents that are found two, three, or four times on the same chain of carbon atoms.

Exercise 24.3

Name the following compound.

Solution

The longest continuous chain in this compound contains five carbon atoms, so the compound is a derivative of pentane. There are three identical CH_3—substituents on the backbone. Two of these methyl groups are on the second carbon, and one is on the fourth carbon. This compound is therefore *2,2,4-trimethylpentane*. Since the compound contains a total of eight carbon atoms, it is also known by the common name *isoctane*.

24.3 THE UNSATURATED HYDROCARBONS: ALKENES AND ALKYNES

Carbon not only forms alkanes with long chains of strong C—C single bonds, it also forms strong C=C double bonds. Compounds that contain C=C double bonds were once known as *olefins* (literally, "to make an oil"), because they were hard to crystallize — they tend to remain oily liquids when cooled. These compounds are now called *alkenes* to emphasize the fact that they are derivatives of alkanes from which two hydrogen atoms have been removed. (The generic formula for an alkane is C_nH_{2n+2}; the generic formula for an alkene with one C=C bond has two fewer hydrogen atoms, C_nH_{2n}.) Alkenes are examples of *unsaturated hydrocarbons,* because they have fewer hydrogen atoms than the corresponding alkanes. The carbon atoms in C=C double bonds are sp^2 hybridized.

Many alkenes have common names, such as ethylene and propylene.

$$H_2C=CH_2 \qquad \text{ethylene}$$
$$H_2C=CH—CH_3 \qquad \text{propylene}$$

But the systematic nomenclature for alkenes names these compounds as derivatives of the parent alkanes. The presence of the C=C double bond is indicated by a change in the *-ane* ending on the name of the alkane to *-ene*.

$$\begin{array}{ll} CH_3—CH_3 & CH_2=CH_2 \\ \text{ethane} & \text{ethene} \\ CH_3—CH_2—CH_3 & CH_2=CH—CH_3 \\ \text{propane} & \text{propene} \end{array}$$

The location of the C=C double bond in the skeleton structure of the compound is indicated by specifying the number of the carbon atom at which the C=C bond starts. The following compounds, for example, are named 1-butene and 2-butene.

$$CH_2=CH—CH_2—CH_3 \qquad CH_3—CH=CH—CH_3$$
$$\text{1-butene} \qquad\qquad \text{2-butene}$$

The names of substituents are added as prefixes to the name of the alkane.

Exercise 24.4

Name the following compound.

$$\begin{array}{c} \qquad\qquad CH_3 \\ \qquad\qquad | \\ CH_3—CH=C—CH_2—CH—CH_3 \\ \qquad\qquad\qquad\qquad | \\ \qquad\qquad\qquad\qquad CH_3 \end{array}$$

Solution

This compound is a derivative of hexane, because the longest carbon chain contains six carbon atoms. Because it contains a C=C double bond, it is a hex*ene*. Because the double bond links the second and third carbon atoms, it is a 2-hexene. The $CH_3—$, or methyl, substituents are on the second and fifth carbon atom, so the compound is *3,5-dimethyl-2-hexene*.

The geometry of alkenes differs from that of the parent alkanes. According to the VSEPR theory described in Section 8.12, the three regions where electrons are found in the valence shell of the carbon atoms in a C=C double bond are arranged

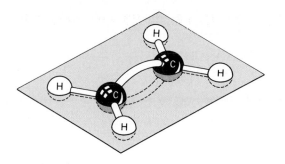

FIG. 24.4 The six atoms that form a $C=C$ double bond and the bond's nearest neighbors all lie in the same plane, in accord with the predictions of the VSEPR theory that the geometry around each carbon atom should be trigonal planar. The carbon atoms in the $C=C$ double bond are both sp^3 hybridized.

toward the corners of an equilateral triangle.

$$\underset{H}{\overset{H}{\diagdown}}C=C\underset{H}{\overset{H}{\diagup}}$$

This prediction is confirmed by experiment, which suggests that the six atoms that form the $C=C$ double bond and its nearest neighbors all lie in the same plane, as shown in Figure 24.4.

The presence of a $C=C$ double bond greatly increases the number of isomers of a compound. There are two isomers of butane and three isomers of pentane, for example. But there are four isomers of butene and seven isomers of pentene.

Alkenes form structural and geometric isomers. Structural isomers differ in the location of the $C=C$ bond. Butene, for example, has a structural isomer in which the $C=C$ bond is located between the first and second carbon atoms and a structural isomer in which the double bond is located between the second and third carbon atoms.

$$CH_2=CH-CH_2-CH_3 \qquad CH_3-CH=CH-CH_3$$
$$\text{1-butene} \hspace{5em} \text{2-butene}$$

Geometric isomers differ in the way substitutents are arranged around the $C=C$ double bond. The isomer with similar substituents on the same side of the double bond is called *cis*, a Latin stem meaning "on this side." The isomer in which similar substituents are on opposite sides of the double bond, or across from each other, is called *trans*, a Latin stem meaning "across." This *cis* isomer of 2-butene, for example, has both CH_3- groups on one side of the double bond. In the *trans* isomer the CH_3- groups are on opposite sides of the double bond.

$$\underset{II}{\overset{CH_3}{\diagdown}}C=C\underset{H}{\overset{CH_3}{\diagup}} \qquad \underset{H}{\overset{CH_3}{\diagdown}}C=C\underset{CH_3}{\overset{H}{\diagup}}$$
$$\textit{cis-2-butene} \hspace{5em} \textit{trans-2-butene}$$

Exercise 24.5

Name the geometric and structural straight-chain isomers of pentene (C_5H_{10}).

Solution

There are two structural isomers of pentene.

$$CH_2=CHCH_2CH_2CH_3 \qquad CH_3CH=CHCH_2CH_3$$
$$\text{1-pentene} \hspace{5em} \text{2-pentene}$$

There are no *cis/trans* isomers of 1-pentene, because there is only one way of arranging the substituents around the double bond.

$$
\begin{array}{c}
\text{H} \qquad\qquad \text{CH}_2\text{CH}_2\text{CH}_3 \\
\diagdown\;\;\diagup \\
\text{C}=\text{C} \\
\diagup\;\;\diagdown \\
\text{H} \qquad\qquad \text{H}
\end{array}
$$

Cis and *trans* geometric isomers are possible for 2-pentene, however.

$$
\begin{array}{cc}
\begin{array}{c}
\text{CH}_3 \qquad\qquad \text{CH}_2\text{CH}_3 \\
\diagdown\qquad\qquad\diagup \\
\text{CH}=\text{CH} \\
\diagup\qquad\qquad\diagdown \\
\text{H} \qquad\qquad\qquad \text{H}
\end{array}
&
\begin{array}{c}
\text{CH}_3 \qquad\qquad \text{H} \\
\diagdown\qquad\qquad\diagup \\
\text{CH}=\text{CH} \\
\diagup\qquad\qquad\diagdown \\
\text{H} \qquad\qquad\qquad \text{CH}_2\text{CH}_3
\end{array} \\
\textit{cis}\text{-2-pentene} & \textit{trans}\text{-2-pentene}
\end{array}
$$

Compounds that contain C≡C triple bonds are called **alkynes.** These compounds have four fewer hydrogen atoms than the parent alkanes, so the generic formula for an alkyne with one C≡C triple bond is C_nH_{2n-2}. The carbon atoms in C≡C triple bonds are *sp* hybridized. The simplest alkyne has the formula C_2H_2 and is known by the common name *acetylene*.

$$\text{H}-\text{C}\equiv\text{C}-\text{H}$$
acetylene

The systematic nomenclature names alkynes as derivatives of the parent alkane, with the ending *-yne* replacing *-ane*, as in the following examples.

$$\text{CH}_3-\text{C}\equiv\text{C}-\text{CH}_2-\text{CH}_3 \qquad\qquad \text{HC}\equiv\text{C}-\text{CH}_2-\text{CH}_3$$
2-pentyne   1-butyne

$$
\begin{array}{c}
\qquad\qquad\qquad \text{CH}_3 \\
\qquad\qquad\qquad | \\
\text{CH}_3-\text{C}\equiv\text{C}-\text{CH}_2-\text{CH}-\text{CH}_2\text{CH}_3
\end{array}
$$
5-methyl-2-heptyne

In addition to compounds that contain one double bond (alkenes) or one triple bond (alkynes), we can also envision compounds with two double bonds *(dienes)*, three double bonds *(trienes)*, or a combination of double and triple bonds.

$$\text{CH}_3-\text{CH}=\text{CH}-\text{CH}_2-\text{C}\equiv\text{CH}$$
2-hexen-5-yne

$$
\begin{array}{c}
\qquad\text{CH}-\text{CH} \\
\diagup\qquad\qquad\diagdown \\
\text{H}_2\text{C} \qquad\qquad \text{CH}_2
\end{array}
$$
1,3-butadiene

24.4 THE REACTIONS OF ALKANES, ALKENES, AND ALKYNES

Saturated hydrocarbons such as methane (CH_4), propane (C_3H_8), and butane (C_4H_{10}) react with oxygen at high temperatures to form CO_2 and H_2O.

$$C_3H_8(g) + 5\ O_2(g) \longrightarrow 3\ CO_2(g) + 4\ H_2O(l)$$

At room temperature, however, alkanes are generally inert to chemical reactions.

Unsaturated hydrocarbons such as alkenes and alkynes are much more reactive. They react rapidly with bromine, for example, to add a Br_2 molecule across the $C\!=\!C$ double bond.

$$CH_3-CH\!=\!CH-CH_3 + Br_2 \longrightarrow CH_3-\underset{\underset{Br}{|}}{\overset{\overset{Br}{|}}{CH}}-CH-CH_3$$

2-butene   2,3-dibromobutane

This reaction provides one way to test for alkenes or alkynes. Solutions of bromine dissolved in CCl_4 have an intense red-orange color. When such a solution is mixed with a sample of an alkane, no change is observed. When it is mixed with an alkene or alkyne, however, the color of Br_2 rapidly disappears.

The reaction between 2-butene and bromine to form 2,3-dibromobutane is just one example of an **addition reaction** of alkenes and alkynes — in the course of this reaction, a molecule of bromine adds across the $C\!=\!C$ double bond. Bromine isn't the only reagent that undergoes addition reactions with alkenes and alkynes. Hydrogen bromide (HBr) adds across a $C\!=\!C$ double bond to form the corresponding alkyl bromide, in which the hydrogen ends up on the carbon atom that had more hydrogen atoms to begin with. Addition of HBr to propene, for example, gives 2-bromopentane.

$$CH_3CH\!=\!CH_2 + HBr \longrightarrow CH_3CHBrCH_3$$

In the presence of a suitable catalyst, H_2 adds across the double or triple bond to convert an alkene or alkyne to the corresponding alkane.

$$CH_3CH\!=\!CHCH_3 + H_2 \xrightarrow{Pt} CH_3CH_2CH_2CH_3$$

In the presence of an acid catalyst, such as sulfuric acid, it is also possible to add a water molecule across a $C\!=\!C$ double bond.

$$CH_3CH\!=\!CHCH_3 + H_2O \xrightarrow{H_2SO_4} CH_3\underset{\overset{|}{OH}}{\overset{\overset{OH}{|}}{CH}}CH_2CH_3$$

Addition reactions provide a way to add new substituents to a hydrocarbon chain and thereby produce new derivatives of the parent alkanes.

24.5 NATURALLY OCCURRING HYDROCARBONS AND THEIR DERIVATIVES

It is easy to draw out a discussion of the properties of simple hydrocarbons, such as methane, ethane, acetylene, and so on, which are not intrinsically interesting. When this happens, we lose sight of one of the fascinating aspects of organic chemistry — the myriad ways in which nature uses the chemistry of carbon to produce a seemingly endless variety of structures and functions. It might therefore be useful to look at some naturally occurring hydrocarbons and their derivatives, so that we can begin to appreciate the complexity of their chemistry.

Complex hydrocarbons and their derivatives are found throughout nature. Natural rubber, for example, is a hydrocarbon that contains long chains of alternating $C\!=\!C$ double bonds and $C\!-\!C$ single bonds.

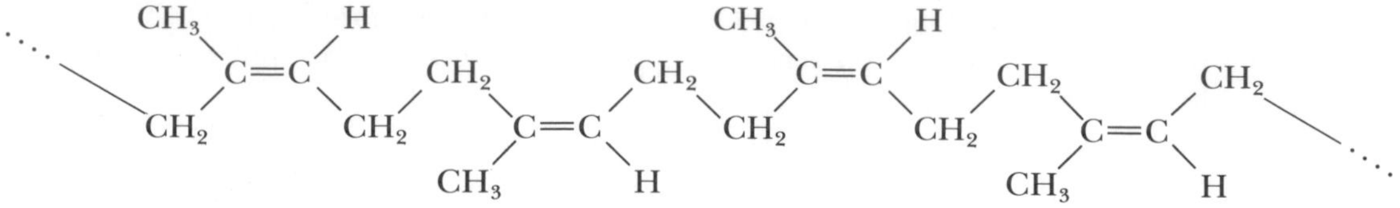

natural rubber

Long hydrocarbon chains are particularly common in **lipids,** the class of biological compounds that includes animal fats and vegetable oils. The first soaps were obtained by reacting animal fat with NaOH (see Section 13.13) to form compounds such as sodium stearate, which consists of a long, nonpolar hydrocarbon chain attached to a polar CO_2^- head.

$$CH_3CH_2CH_2CH_2CH_2CH_2CH_2CH_2CH_2CH_2CH_2CH_2CH_2CH_2CH_2CH_2CH_2CO_2^-Na^+$$

sodium stearate

Another important class of naturally occurring hydrocarbons are the **terpenes,** which can be distilled from plants. Terpenes often have very distinctive odors. β-Pinene (Figure 24.5a), for example, is responsible for the characteristic odor of turpentine. The terpenes also include derivatives of hydrocarbons such as citronellal (Figure 24.5b), which is one of the compounds that gives rise to the odor of lemons.

Vitamins (literally, "vital amines") are important components of nutrition that cannot be made by the body and must therefore be obtained from the diet. The so-called fat-soluble vitamins, such as vitamin A_1, D_2, and E, have structures based on complex hydrocarbon skeletons (Figure 24.6a). The **steroids** (such as cholesterol, shown in Figure 24.6b), which are important regulators of biological activity, also have structures based on a complex hydrocarbon skeleton.

24.6 THE AROMATIC HYDROCARBONS AND THEIR DERIVATIVES

At the turn of the 19th century, one of the signs of living the good life was having gas lines connected to your house, so that you could use gas lanterns to light the house after dark. The gas burned in these lanterns was called coal gas, because it was produced by the heating of coal in the absence of air. The principal component of coal gas was methane, CH_4.

In 1825, Michael Faraday determined the empirical formula of an oily liquid with a distinct odor that collected in tanks used to store coal gas at high pressures and found that this compound had the same number of carbon and hydrogen atoms. Ten years later, Eilhardt Mitscherlich produced the same material by heating benzoic acid with lime. Mitscherlich named this substance *benzin,* which became *benzene* when translated into English, and determined the molecular formula of the compound — C_6H_6.

Benzene is obviously an unsaturated hydrocarbon, because it has far less hydrogen than the equivalent saturated hydrocarbon — C_6H_{14}. At first glance, it is tempting to classify benzene as an alkene with four C=C double bonds, or perhaps an alkyne with two C≡C triple bonds. But benzene is too stable to be an alkene or alkyne. Alkenes and alkynes react rapidly with potassium permanganate ($KMnO_4$), but benzene does not. Alkenes also rapidly add Br_2 to the C=C double bond. Benzene does not react with bromine, by itself. Furthermore, when it does react with bromine — in the presence of $FeBr_3$ — the product of this reaction is a com-

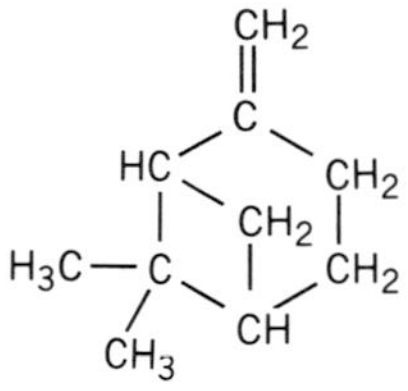

β-Pinene
(turpentine)

(a)

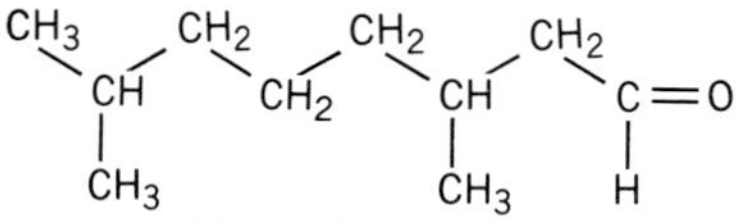

Citronellal
(oil of lemon)

(b)

FIG. 24.5 Complex hydrocarbons and their derivatives are found throughout nature. This figure shows the structures of a pair of compounds that can be isolated from plants. (a) β-Pinene is responsible for the characteristic odor of turpentine. (b) Citronellal is also known as oil of lemon, because it gives rise to the characteristic odor of lemons.

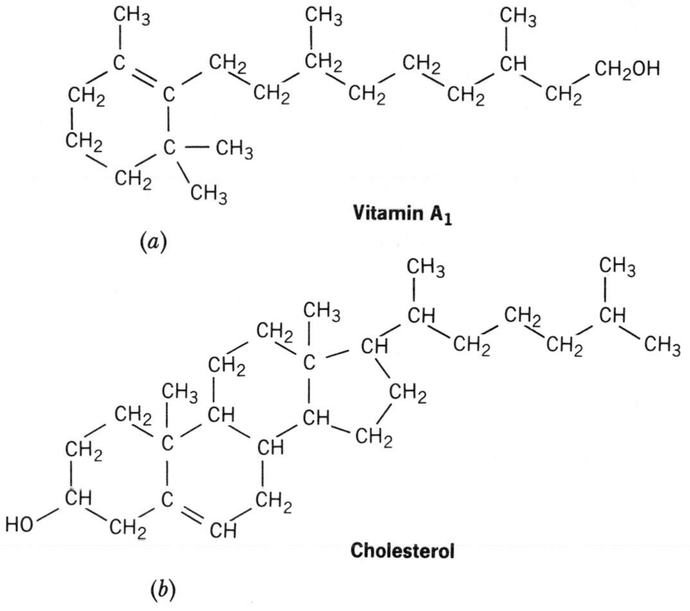

Vitamin A₁

(a)

Cholesterol

(b)

FIG. 24.6 (*a*) Vitamin A₁ is a derivative of a complex hydrocarbon and is a vital amine involved in the process by which our eyes capture images. (*b*) Cholesterol, another complex hydrocarbon derivative, is an important component of cell membranes, and it is the biosynthetic precursor for many of the other steroids our bodies produce.

pound in which a bromine atom has been substituted for a hydrogen atom, not added to the compound.

$$C_6H_6 + Br_2 \xrightarrow{FeBr_3} C_6H_5Br + HBr$$

As the number of compounds isolated from coal increased, it became obvious that benzene was not unique. An entire class of compounds was discovered whose molecular formulas suggested the presence of multiple C=C bonds, yet these compounds were too stable to be alkenes. Since they often had a distinct odor, or aroma, they became known as ***aromatic compounds.***

The structure of benzene was a recurring problem throughout most of the 19th century. The first solution to this problem was advanced by Friedrich August Kekulé in 1865. (Kekulé's interest in the structure of organic compounds may have resulted from the fact that he first enrolled at the University of Giessen as a student of architecture.) One day, while dozing before a fire, Kekulé dreamed of long rows of atoms twisting in a snakelike motion until one of the snakes seized hold of its own tail. This dream led Kekulé to propose that benzene consists of a ring of six carbon atoms with alternating C—C single bonds and C=C double bonds (see Figure 24.7). Since there are two ways in which these bonds can alternate, Kekulé proposed that benzene was a mixture of two compounds in equilibrium.

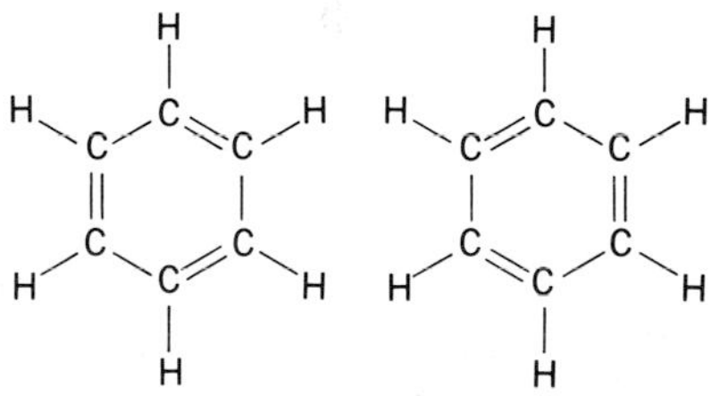

Kekulé's structure explained the molecular formula of benzene, but it did not explain why benzene failed to behave like an alkene. The unusual stability of benzene was not understood until the development of the theory of resonance described in Section 8.9. This theory states that molecules for which two or more satisfactory Lewis structures can be drawn are an average, or hybrid, of these structures. Benzene, for example, is a resonance hybrid of the two Kekulé structures.

This model is consistent with the crystal structure of benzene, which shows a planar six-membered ring in which all six carbon—carbon bonds are the same length (0.139 nm). It also explains why the length of these bonds is intermediate between the length of C—C single bonds (0.154 nm) and C=C double bonds (0.133 nm).

Kekulé structures
for benzene

FIG. 24.7 Kekulé's solution to the problem of explaining why the formula of benzene is C₆H₆ involved assuming that this compound consisted of a six-membered ring of carbon atoms with alternating C—C single bonds and C=C double bonds. Because there were two ways of writing the structure of this molecule, Kekulé assumed that benzene was a mixture of two isomers that were in equilibrium with each other.

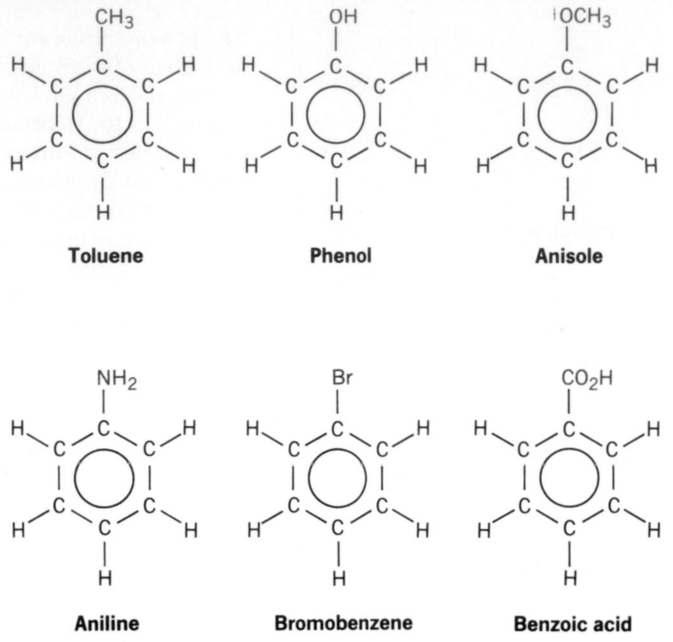

FIG. 24.8 Benzene is not the only aromatic compound. Other compounds in this class include toluene, phenol, anisole, aniline, bromobenzene, and benzoic acid.

Toluene   Phenol   Anisole

Aniline   Bromobenzene   Benzoic acid

According to resonance theory, molecules that are hybrids of two or more Lewis structures are more stable than those that aren't. It is this extra stability that makes benzene and other aromatic derivatives less reactive than normal alkenes. To emphasize the difference between benzene and a simple alkene we will replace the Kekulé structure for benzene with the aromatic ring shown in Figure 24.8. The circle in the center of the aromatic ring indicates that the electrons in the ring are not localized — they are not restricted between individual pairs of atoms. They are delocalized — they are free to move around the ring.

There are three ways in which a pair of substituents can be placed on an aromatic ring. In the *ortho* (*o*) isomer, the substituents are in adjacent positions on the ring. In the *meta* (*m*) isomer, they are separated by one carbon atom. In the *para* (*p*) isomer, they are on opposite ends of the ring. The three isomers of dimethylbenzene, or xylene, are shown in Figure 24.9.

Exercise 24.6

Predict the structure of *ortho*-dichlorobenzene, which is one of the active ingredients in moth balls.

Solution

Ortho isomers of benzene contain two substituents at adjacent positions in the six-membered ring. Thus, *ortho*-dichlorobenzene has the following structure.

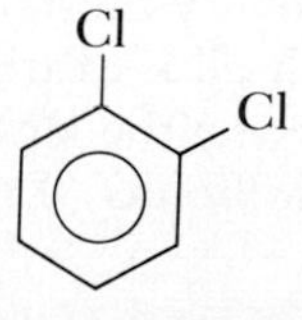

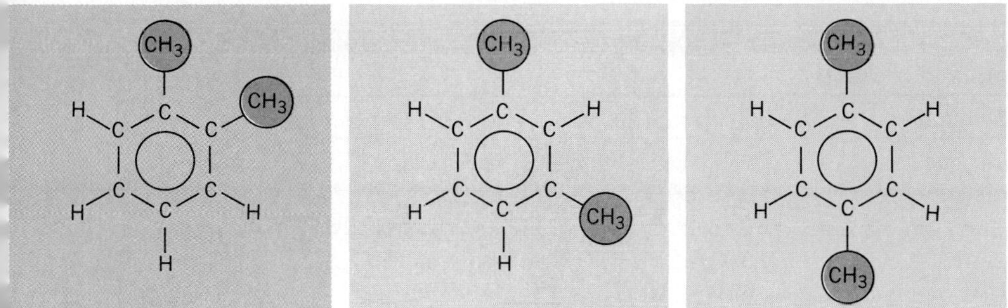

FIG. 24.9 The three isomers of xylene, or dimethylbenzene. The isomer in which the two methyl groups are on adjacent carbon atoms is called the *ortho* isomer. The *para* isomer contains methyl groups on opposite ends of the benzene ring. The structure of the *meta* isomer lies between these extremes.

Aromatic compounds can contain more than one six-membered ring. Naphthalene, anthracene, and phenanthrene (see Figure 24.10) are examples of aromatic compounds that contain two or more fused benzene rings.

24.7 THE CHEMISTRY OF PETROLEUM PRODUCTS

The term **petroleum** comes from the Latin stems *petra*, "rock," and *oleum*, "oil." It is used to describe a broad range of hydrocarbons that are found as gases, liquids, or solids beneath the surface of the earth. The two most common forms of petroleum are natural gas and crude oil.

Natural gas is a mixture of lightweight alkanes. The composition of natural gas depends on the source, but a typical sample might contain 80% methane (CH_4), 7% ethane (C_2H_6), 6% propane (C_3H_8), 4% butane and isobutane (C_4H_{10}), and 3% pentanes (C_5H_{12}). The C_3, C_4, and C_5 hydrocarbons are usually removed before the gas is sold; commercial natural gas is therefore primarily a mixture of methane and ethane. The propane and butanes removed from natural gas are usually liquified under pressure and sold as liquified petroleum gases (LPG).

Natural gas was known in England as early as 1659. But it did not replace coal gas as an important source of energy in the United States until after World War II, when a network of gas pipelines was constructed. By 1980, annual consumption of natural gas had grown to more than 55,000 billion cubic feet, which represented almost 30% of total U.S. energy consumption.

The first oil well — drilled by Edwin Drake in 1859, in Titusville, Pennsylvania — produced up to 800 gallons per day. By 1980, consumption of oil had reached 2500 million gallons per day. About 225 billion barrels of oil were produced by the petroleum industry between 1859 and 1970. Another 200 billion barrels were produced between 1970 and 1980. The total proven world reserves of crude oil in 1970 were estimated at 546 billion barrels, with perhaps another 800 to 900 billion barrels of oil that remained to be found. It took 500 million years for the petroleum beneath the earth's crust to accumulate. At the present rate of consumption, we might be able to exhaust the supply of petroleum by the 200th anniversary of the first oil well.

Crude oil is a complex mixture that is between 50 and 95% hydrocarbon by weight. The first step in refining crude oil involves separating the oil into different hydrocarbon fractions by distillation (see Section 13.19). A typical set of petroleum fractions is given in Table 24.3. Since there are a number of factors that influence the boiling point of a hydrocarbon, these petroleum fractions are complex mixtures. More than 500 different hydrocarbons have been identified in the gasoline fraction, for example.

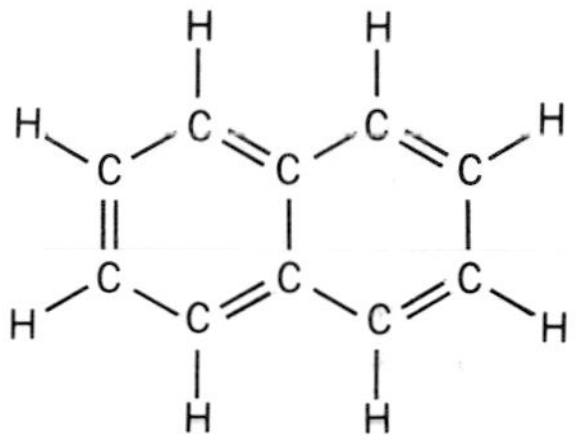

Naphthalene ($C_{10}H_8$)

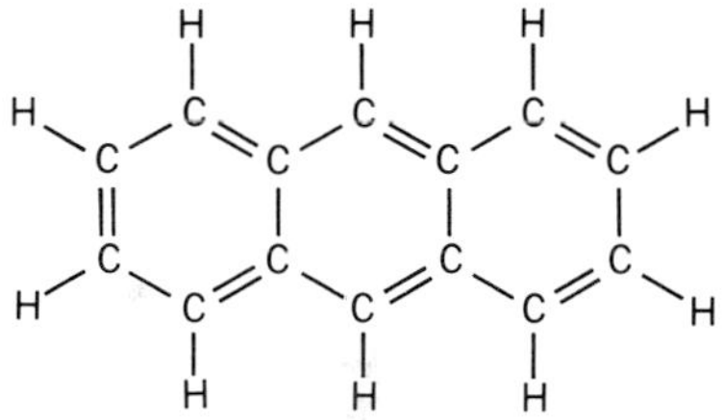

Anthracene ($C_{14}H_{10}$)

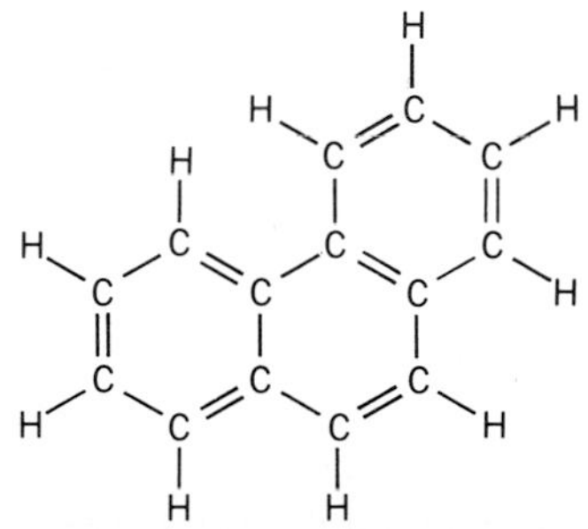

Phenanthrene ($C_{14}H_{10}$)

FIG. 24.10 More complex aromatic compounds can be obtained by fusing two or more six-membered rings. The examples shown here include compounds that have been given the common names *naphthalene*, *anthracene*, and *phenanthrene*.

A plume of natural gas being burned at the Brooklyn Union Gas Company in Brooklyn, New York.

TABLE 24.3

Petroleum Fractions

Fraction	Boiling Range (°C)	Number of Carbon Atoms
natural gas	below 20	C_1 to C_4
petroleum ether	20 to 60	C_5 to C_6
gasoline	40 to 200	C_5 to C_{12}, mainly C_6 to C_8
kerosene	150 to 260	mostly C_{12} to C_{13}
fuel oils	above 260	C_{14} and higher
lubricants	above 400	C_{20} and above
asphalt or coke	residue	polycyclic

About 10% of the product of the distillation of crude oil is a fraction known as *straight-run gasoline,* which served as a satisfactory fuel during the early days of the internal combustion engine. As the automobile engine developed, however, it was made more powerful by increasing the **compression ratio**—the extent to which the gasoline/air mixture is compressed by the piston before it is ignited by the spark plug. Straight-run gasoline burns unevenly in high-compression engines, which produces a shock wave that causes the engine to "knock," or "ping." As the petroleum industry matured, it faced two problems—increasing the yield of gasoline from each barrel of crude oil and decreasing the tendency of gasoline to knock when it burned.

Knocking is related to the structure of the hydrocarbons in gasoline.

1. Branched alkanes and cycloalkanes burn more evenly than straight-chain alkanes.
2. Short alkanes (C_4H_{10}) burn more evenly than long alkanes (C_7H_{16}).
3. Alkenes burn more evenly than alkanes.
4. Aromatic hydrocarbons burn more evenly than cycloalkanes.

The most commonly used measure of a gasoline's ability to burn without knocking is its **octane number.** Octane numbers compare a gasoline's tendency to knock against the tendency of a blend of two hydrocarbons—heptane and 2,2,4-trimethylpentane, or isooctane—to knock. Heptane (C_7H_{16}) is a long, straight-chain alkane, which burns unevenly and produces a great deal of knocking. Highly branched alkanes such as 2,2,4-trimethylpentane are more resistant to knocking. Gasolines that match a blend of 87% isooctane and 13% heptane are given an octane number of 87.

Natural gas can be transported by cooling it until it condenses to form a liquid.

TABLE 24.4

Hydrocarbon Octane Numbers

Hydrocarbon	Road Index Octane Number
heptane	0
2-methylheptane	23
hexane	25
2-methylhexane	44
1-heptene	60
pentane	62
1-pentene	84
butane	91
cyclohexane	97
2,2,4-trimethylpentane (isooctane)	100
benzene	101
toluene	112

There are three ways of reporting octane numbers. Measurements made under conditions of high speed and high temperature are reported as *motor octane numbers*. Measurements taken under relatively mild engine conditions are known as *research octane numbers*. The *road-index octane numbers* reported on gasoline pumps are an average of these two. Road index octane numbers for a few pure hydrocarbons are given in Table 24.4.

The gradual increase in the number of diesel-powered automobiles has sparked an interest in the properties of diesel fuel. Diesel fuel is essentially the same as the fuel oil burned to heat homes. It contains C_{14} or higher-number hydrocarbons, and its quality is reported in terms of a **cetane number,** which is based on the way two isomers of $C_{16}H_{34}$ burn in a diesel engine. *Cetane,* the straight-chain, or *n*-hexadecane, isomer, is assigned a cetane number of 100, and the highly branched isomer 2,3,4,5,6,7,8-heptamethylnonane is assigned a cetane number of 15.

By 1922, a number of compounds had been discovered that could increase the octane number of gasoline. Adding as little as 6 milliliters of tetraethyllead $Pb(CH_2CH_3)_4$ to a gallon of gasoline, for example, can increase the octane number by 15 to 20 units. This discovery gave rise to the first "ethyl" gasoline and enabled the petroleum industry to produce aviation gasolines with octane numbers greater than 100.

Another way to increase the octane number is called **thermal reforming.** At high temperatures (500–600°C) and high pressures (25–50 atm), straight-chain alkanes isomerize to branched alkanes and cycloalkanes, thereby increasing the octane number of the gasoline. Running this reaction in the presence of hydrogen and a catalyst, such as a mixture of silica (SiO_2) and alumina (Al_2O_3), results in **catalytic reforming,** which can produce a gasoline with even higher octane numbers. Thermal or catalytic reforming and gasoline additives such as tetraethyllead increase the octane number of the straight-run gasoline obtained from the distillation of crude oil, but neither process increases the yield of gasoline from a barrel of oil.

The data in Table 24.3 suggest that we could increase the yield of gasoline by "cracking" the hydrocarbons that end up in the kerosene or fuel oil fractions into smaller pieces. **Thermal cracking** was discovered as early as the 1860s. At high temperatures (500°C) and high pressures (25 atm), long-chain hydrocarbons break into smaller pieces. A saturated C_{12} hydrocarbon in kerosene, for example,

A petroleum refinery.

might break into two C_6 fragments. The ratio of H to C atoms in the starting material requires that one of the products of this reaction must contain a $C=C$ double bond.

$$C_{12}H_{26} \longrightarrow C_6H_{14} + C_6H_{12}$$

$$CH_3(CH_2)_{10}CH_3 \longrightarrow CH_3CH_2CH_2CH_2CH_2CH_3 + CH_2=CHCH_2CH_2CH_2CH_3$$

The presence of alkenes in thermally cracked gasolines increases the octane number (70) relative to that of straight-run gasoline (60), but it also makes thermally cracked gasoline less stable for long-term storage. Thermal cracking has therefore been replaced by *catalytic cracking,* which uses catalysts instead of high temperatures and pressures to crack long-chain hydrocarbons into smaller fragments for use in gasoline.

About 87% of the crude oil refined in 1980 went into the production of fuels such as gasoline, kerosene, and fuel oil. The remaining 13% went for nonfuel uses, such as petroleum solvents, industrial greases and waxes, or as starting materials for the synthesis of *petrochemicals.* Petroleum products are used to produce synthetic fibers such as nylon, orlon, and dacron and other polymers such as polystyrene, polyethylene, and synthetic rubber. They also serve as raw materials in the production of refrigerants, aerosols, antifreeze, detergents, dyes, adhesives, alcohols, explosives, weed killers, insecticides, and insect repellents. The H_2 given off when alkanes are converted to alkenes or when cycloalkanes are converted to aromatic hydrocarbons can be used to produce a number of inorganic petrochemicals, such as ammonia, ammonium nitrate, and nitric acid. As a result, most fertilizers as well as other agricultural chemicals are also petrochemicals.

24.8 THE CHEMISTRY OF COAL

Coal can be defined as a sedimentary rock that burns. It was formed by the decomposition of plant matter and is a complex substance that can be found in many forms. Coal is often subdivided into four classes: anthracite, bituminous, sub-bituminous, and lignite. Elemental analysis gives empirical formulas such as $C_{135}H_{97}O_9NS$ for bituminous coal and $C_{240}H_{90}O_4NS$ for high-grade anthracite. A typical structure for coal is shown in Figure 24.11.

Anthracite coal is a dense, hard rock with a jet-black color and a metallic luster.

Petroleum products.

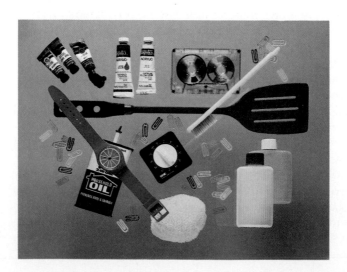

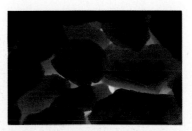

FIG. 24.11 A model for one portion of the extended structure of coal.

It contains between 86% and 98% C by weight, and it burns slowly, with a pale blue flame and very little smoke. *Bituminous coal,* or soft coal, contains between 69% and 86% C by weight and is the most abundant form of coal. *Sub-bituminous coal* contains less carbon and more water and is therefore a less efficient source of heat. *Lignite coal,* or brown coal, is a very soft coal that can be as much as 70% water by weight.

The total energy consumption in the United States for 1980 was equivalent to 84×10^{15} kilojoules. Of this total, 41% came from oil, 29% from natural gas, and 24% from coal. Coal is unique as a source of energy in the United States, however, because none of the 8234 billion pounds used in 1980 was imported. Furthermore, the proven reserves are so large we can continue using coal at this level of consumption for at least 2000 years.

At the time this text was written, coal was the most cost-efficient fuel for heating. The cost of coal delivered to the Purdue University physical plant was $1.41 per million kilojoules of heating energy. The equivalent cost for natural gas would have been $5.22 and number 2 fuel oil would have cost $7.34. Although coal is cheaper than natural gas and oil, it is considerably more difficult to handle. As a result, there has been a long history of efforts to turn coal into either a gaseous or a liquid fuel.

Anthracite coal burning.

COAL GASIFICATION

As early as 1800, coal gas was made from coal heated in the absence of air. Coal gas is rich in CH_4 and gives off up to 20.5 kilojoules per liter of gas burned. Coal gas — or town gas, as it was also known — became so popular that most major cities and many small towns had a local gashouse that produced it, and gas burners were adjusted to burn a fuel that produced 20.5 kJ/L. Gas lanterns, of course, were eventually replaced by electric lights. But coal gas was still used for cooking and heating until the more efficient natural gas (38.3 kJ/L) became readily available.

A slightly less efficient fuel known as water gas can be made by reacting the carbon in coal with steam.

$$C(s) + H_2O(g) \longrightarrow CO(g) + H_2(g) \qquad \Delta H° = 118 \text{ kJ/mol C}$$

Water gas burns to give CO_2 and H_2O, releasing roughly 11.2 kilojoules per liter of gas consumed. Note that the enthalpy of reaction for the preparation of water gas is positive, which means that this reaction is endothermic. As a result, the preparation of water gas typically involves alternating blasts of steam and either air or oxygen through a bed of white-hot coal. The reactions between coal and oxygen to produce CO and CO_2 are exothermic and provide enough energy to drive the reaction between steam and coal.

Water gas formed by the reaction of coal with oxygen and steam is a mixture of CO, CO_2, and H_2. The ratio of H_2 to CO can be increased if more water is added to this mixture, to take advantage of a reaction known as the water-gas shift reaction.

$$CO(g) + H_2O(g) \longrightarrow CO_2(g) + H_2(g) \qquad \Delta H° = -41 \text{ kJ/mol CO}$$

The concentration of CO_2 can be decreased if CO_2 is allowed to react with coal at high temperatures to form CO.

$$C(s) + CO_2(g) \longrightarrow 2\ CO(g) \qquad \Delta H° = +160 \text{ kJ/mol C}$$

Water gas from which the CO_2 has been removed is called synthesis gas, because it can be used as a starting material for a variety of organic and inorganic compounds. It can be used as the source of H_2 for the synthesis of ammonia, for example.

$$N_2(g) + 3\ H_2(g) \longrightarrow 2\ NH_3(g)$$

It can also be used to make methyl alcohol, or methanol, which is a starting material for the synthesis of alkenes and aromatic compounds, acetic acid, formaldehyde, and other alcohols, such as ethyl alcohol, or ethanol, C_2H_5OH.

$$CO(g) + 2\ H_2(g) \longrightarrow CH_3OH(l)$$

Synthesis gas can also be used to produce methane, or synthetic natural gas (SNG).

$$CO(g) + 3\ H_2(g) \longrightarrow CH_4(g) + H_2O(g)$$
$$2\ CO(g) + 2\ H_2(g) \longrightarrow CH_4(g) + CO_2(g)$$

COAL LIQUEFACTION

The first step toward making liquid fuels from coal involves the manufacture of synthesis gas (CO and H_2) from coal. In 1925, Franz Fischer and Hans Tropsch developed a catalyst that converted CO and H_2 at 1 atm and 250 to 300°C into liquid hydrocarbons. By 1941, Fischer-Tropsch plants produced 740,000 tons of petroleum products per year in Germany.

Fischer-Tropsch technology is based on a complex series of reactions, which use

H_2 to reduce CO to $—CH_2—$ groups linked to form long-chain hydrocarbons.

$$CO(g) + 2\ H_2(g) \longrightarrow (—CH_2—)_n(l) + H_2O(g) \qquad \Delta H° = -165\ \text{kJ/mol CO}$$

The water produced in this reaction combines with CO in the water-gas shift reaction to form H_2 and CO_2.

$$CO(g) + H_2O(g) \longrightarrow CO_2(g) + H_2(g) \qquad \Delta H° = -41\ \text{kJ/mol CO}$$

The following is therefore the overall Fischer-Tropsch reaction.

$$2\ CO(g) + H_2(g) \longrightarrow (—CH_2—)_n(l) + CO_2(g) \qquad \Delta H° = -124\ \text{kJ/mol CO}$$

At the end of World War II, Fischer-Tropsch technology was under study in most industrial nations. The low cost and high availability of crude oil, however, led to a decline in interest in liquid fuels made from coal. The only commercial plants using this technology today are in the Sasol complex in South Africa, which uses 30.3 million tons of coal per year.

Another approach to liquid fuels is based on the reaction between CO and H_2 to form methanol, CH_3OH.

$$CO(g) + 2\ H_2(g) \longrightarrow CH_3OH(l)$$

Methanol can be used directly as a fuel, or it can be converted into gasoline by a catalyst recently developed by Mobil Oil Company.

As the supply of petroleum becomes smaller and its cost continues to rise, a gradual shift may be observed toward liquid fuels made from coal. Whether this takes the form of a return to a modified Fischer-Tropsch technology, the conversion of methanol to gasoline, or other alternatives, only time will tell.

24.9 FUNCTIONAL GROUPS

Sodium metal reacts with water to give an aqueous solution of NaOH and H_2 gas.

$$2\ Na(s) + 2\ H_2O(l) \longrightarrow 2\ Na^+(aq) + 2\ OH^-(aq) + H_2(g)$$

Sodium also reacts with methanol to give H_2 gas and a solution of Na^+ and CH_3O^- ions dissolved in the alcohol.

$$2\ Na(s) + 2\ CH_3OH(l) \longrightarrow 2\ Na^+ + 2\ CH_3O^- + H_2(g)$$

Instead of trying to memorize both reactions, we can build a general rule of which these reactions are examples.

Sodium metal reacts with compounds that contain the $—OH$ substituent, or functional group, to give sodium salts of the conjugate base and H_2 gas.

Functional groups are atoms or groups of atoms in organic compounds that give the compounds some of their characteristic properties. They provide a way of integrating large quantities of information about the chemistry of organic compounds. We don't have to worry about the difference between 1-butene and 2-methyl-3-hexene, for example. We can focus on the fact that both compounds are alkenes that add Br_2 across the $C=C$ double bond.

$$H_2C=CHCH_2CH_3 + Br_2 \longrightarrow CH_2BrCHBrCH_2CH_3$$

$$\underset{\underset{CH_3CHCH=CHCH_2CH_3}{|}}{CH_3} + Br_2 \longrightarrow \underset{\underset{CH_3CHCHBrCHBrCH_2CH_3}{|}}{CH_3}$$

Some of the important functional groups in organic chemistry are given in Table 24.5.

Alkyl halides are derivatives of an alkane in which one or more halogens (F, Cl, Br, or I) have been substituted for a hydrogen. They are named as derivatives of the parent alkane, although they often have common names as well.

CH_3Cl	chloromethane, or methyl chloride
CH_2Cl_2	dichloromethane, or methylene chloride
$CHCl_3$	trichloromethane, or chloroform
CCl_4	tetrachloromethane, or carbon tetrachloride

An *alcohol* is a compound (such as CH_3OH) that contains an —OH group. Alcohols are therefore derivatives of water in which one of the hydrogen atoms has been replaced by an alkyl group. *Ethers* are derivatives of water in which both hydrogen atoms have been replaced by alkyl groups. There are two structural isomers of C_2H_6O, for example. One is an alcohol; the other is an ether.

$$C_2H_6O:$$
$$CH_3—CH_2—OH \qquad CH_3—O—CH_3$$
$$\text{alcohol} \qquad\qquad\qquad \text{ether}$$

Amines are derivatives of ammonia in which the hydrogen atoms have been replaced by one or more alkyl groups.

$$(CH_3)_2CH—NH_2 \qquad \overset{\displaystyle CH_3}{\underset{\displaystyle CH_3}{\diagup\!\!\diagup}}NH \qquad CH_3—\overset{\displaystyle |}{\underset{\displaystyle |}{N}}—CH_3$$
$$\underset{\displaystyle CH_3}{}$$
$$\text{amines}$$

A number of functional groups are based on the C=O double bond, which is known as a *carbonyl group;* these include aldehydes, ketones, carboxylic acids, esters, and amides. When one of the substituents on the carbonyl group is a hydrogen atom, the compound is an *aldehyde.*

$$\overset{\displaystyle O}{\overset{\displaystyle \|}{H—C—H}} \qquad \overset{\displaystyle O}{\overset{\displaystyle \|}{CH_3—C—H}}$$
$$\text{aldehydes}$$

When both of the substituents are alkyl groups, it is a *ketone.*

$$\overset{\displaystyle O}{\overset{\displaystyle \|}{CH_3—C—CH_3}} \qquad \overset{\displaystyle O}{\overset{\displaystyle \|}{(CH_3)_2CH—C—CH_3}}$$
$$\text{ketones}$$

When one of the substituents on the carbonyl group is an —OH group, the compound is a *carboxylic acid.* Acetic acid, CH_3CO_2H, is an example of a carboxylic acid. So is formic acid, HCO_2H.

$$\overset{\displaystyle O}{\overset{\displaystyle \|}{CH_3—C—OH}} \qquad \overset{\displaystyle O}{\overset{\displaystyle \|}{H—C—OH}}$$
$$\text{carboxylic acids}$$

When the hydrogen atom in a carboxylic acid is replaced by an alkyl group, the result is an *ester.*

TADLE 24.5

Functional Group	Name	Example (common name)
$-C-H$	alkane	$CH_3-CH_2-CH_3$ (propane)
$C=C$	alkene	$CH_3-CH=CH_2$ (propene)
$-C\equiv C-$	alkyne	$H-C\equiv C-H$ (acetylene)
$-F, -Cl, -Br, -I$	alkyl halide	CH_3-Cl (methyl chloride)
$-OH$	alcohol	CH_3-OH (methyl alcohol)
$-O-$	ether	CH_3-O-CH_3 (dimethyl ether)
$-NH_2$	amine	CH_3-NH_2 (methylamine)
$-\overset{\displaystyle O}{\overset{\|}{C}}-H$	aldehyde	$CH_3-\overset{\displaystyle O}{\overset{\|}{C}}-H$ (acetaldehyde)
$-\overset{\displaystyle O}{\overset{\|}{C}}-$	ketone	$CH_3-\overset{\displaystyle O}{\overset{\|}{C}}-CH_3$ (acetone)
$-\overset{\displaystyle O}{\overset{\|}{C}}-OH$	carboxylic acid	$CH_3-\overset{\displaystyle O}{\overset{\|}{C}}-OH$ (acetic acid)
$-\overset{\displaystyle O}{\overset{\|}{C}}-O-$	ester	$CH_3-\overset{\displaystyle O}{\overset{\|}{C}}-O-CH_3$ (methyl acetate)
$-\overset{\displaystyle O}{\overset{\|}{C}}-NH_2$	amide	$CH_3-\overset{\displaystyle O}{\overset{\|}{C}}-NH_2$ (acetamide)

$$CH_3\overset{\displaystyle O}{\overset{\|}{C}}-O-CH_2CH_3$$
ester

If one of the substituents on the carbonyl is an amino group ($-NH_2$), the compound is an **amide**.

$$CH_3-\overset{\displaystyle O}{\overset{\|}{C}}-NH_2 \qquad CH_3-\overset{\displaystyle O}{\overset{\|}{C}}-NH-CH_3$$
amides

Exercise 24.7

Identify the functional groups in the following compounds.

(a) Vitamin C (Figure 24.2)
(b) β-Pinene (Figure 24.6a)
(c) Citronellal (Figure 24.6b)
(d) Vitamin A$_1$ (Figure 24.7a)
(e) Cholesterol (Figure 24.7b)

Solution

(a) Vitamin C is an alcohol, because it contains $-OH$ groups. It is also an alkene, because it has a $C=C$ double bond. The most difficult functional group to recognize in this molecule is the ester linkage ($-CO-O-$).

(b) β-Pinene is an alkene, because it contains a $C{=}C$ double bond.
(c) Citronellal is an aldehyde, because it contains a $-CHO$ group.
(d) Vitamin A_1 is both an alcohol ($-OH$) and an alkene ($C{=}C$).
(e) Cholesterol is also both an alcohol ($-OH$) and an alkene ($C{=}C$).

24.10 ALKYL HALIDES

Alkanes are inert to most chemical reactions, with three major exceptions. As already noted, they can be isomerized and cracked. They also burn to form carbon dioxide and water.

$$CH_4(g) + 2\,O_2(g) \longrightarrow CO_2(g) + 2\,H_2O(g)$$

And they react with halogens to form alkyl halides.

$$CH_4(g) + Cl_2(g) \longrightarrow CH_3Cl(g) + HCl(g)$$

Chlorination— or substitution of chlorine for hydrogen atoms— can continue, to form dichloromethane (CH_2Cl_2), chloroform ($CHCl_3$), and carbon tetrachloride (CCl_4).

Alkyl halides enter into more chemical reactions than the parent alkanes. Chloroform ($CHCl_3$) and carbon tetrachloride (CCl_4) react with hydrogen fluoride to form a mixture of chlorofluorocarbons, such as $CHCl_2F$, $CHClF_2$, CHF_3, CCl_3F, CCl_2F_2, and $CClF_3$, which are sold under trade names such as Freon and Genetron. The freons are inert gases with high densities, low boiling points, low toxicities, and no odor. As a result, they have found extensive use as propellants in antiperspirants and hair sprays. Controversy over the role of chlorofluorocarbons in the depletion of the earth's ozone layer led the U.S. Environmental Protection Agency to ban the use of CCl_2F_2 and CCl_3F in aerosols in 1978. CCl_2F_2, CCl_3F, and $CHFCl_2$ are still used as refrigerants in the air-conditioning industry, however.

THE MECHANISM OF THE CHLORINATION REACTION

The reaction between CH_4 and Cl_2 has the following properties.

1. The reaction does not take place in the dark or at low temperatures.
2. The reaction occurs in the presence of ultraviolet light or at temperatures above 250°C.
3. Once the reaction gets started, it continues even after the light is turned off.
4. The products of the reaction include CH_2Cl_2, $CHCl_3$, and CCl_4, as well as CH_3Cl.
5. The reaction also produces some C_2H_6.

These facts are consistent with a chain-reaction mechanism that involves three processes: chain initiation, chain propagation, and chain termination.

CHAIN INITIATION A Cl_2 molecule can dissociate into a pair of chlorine atoms by absorbing energy from either ultraviolet light or heat.

$$Cl_2 + \text{energy} \longrightarrow 2\,Cl\cdot \qquad \Delta H° = 121.7 \text{ kJ/mol } Cl\cdot$$

The Cl atom produced in this reaction is a *free radical*— an atom, ion, or molecule with one unpaired electron or more. The unpaired electron is symbolized by a single dot— for example, $Cl\cdot$.

CHAIN PROPAGATION Free radicals, such as the Cl· atom, are extremely reactive. The Cl· atom can remove a hydrogen atom from CH_4 to form HCl and a CH_3· radical. The CH_3· radical then removes a chlorine atom from a Cl_2 molecule to form CH_3Cl and a new Cl· radical.

$$CH_4 + Cl\cdot \longrightarrow CH_3\cdot + HCl \qquad \Delta H^\circ = -16 \text{ kJ/mol Cl}$$
$$CH_3\cdot + Cl_2 \longrightarrow CH_3Cl + Cl\cdot \qquad \Delta H^\circ = -87 \text{ kJ/mol } CH_3$$

Since a Cl· atom is generated in the second reaction for every Cl· atom consumed in the first, this reaction continues in a chain-like fashion until the radicals involved in these chain-propagation steps are destroyed.

CHAIN TERMINATION The radicals that keep the reaction going eventually combine in a chain-terminating step. Chain termination can occur in three ways.

$$Cl\cdot + Cl\cdot \longrightarrow Cl_2 \qquad \Delta H^\circ = -243.4 \text{ kJ/mol } Cl_2$$
$$CH_3\cdot + Cl\cdot \longrightarrow CH_3Cl \qquad \Delta H^\circ = -330 \text{ kJ/mol } CH_3Cl$$
$$CH_3\cdot + CH_3\cdot \longrightarrow CH_3CH_3 \qquad \Delta H^\circ = -350 \text{ kJ/mol } C_2H_6$$

Exercise 24.8

Use the free-radical chain-reaction mechanism to explain the five observations listed for the reaction between CH_4 and Cl_2.

Solution

1. The reaction does not occur in the dark or at low temperatures because energy must be absorbed to generate the free radicals that carry the reaction.
2. The reaction occurs in the presence of ultraviolet light because the photon has enough energy to dissociate a Cl_2 molecule to a pair of Cl· atoms. The reaction occurs at high temperatures because thermal energy can also be absorbed to dissociate Cl_2 molecules to form Cl· atoms.
3. The reaction continues after the light has been turned off because light is only needed to generate the Cl· atoms that start the reaction.
4. Reaction doesn't stop at CH_3Cl because the Cl· atom is so reactive it can abstract additional hydrogen atoms to form CH_2Cl, $CHCl_3$, and eventually CCl_4.
5. The formation of C_2H_6 is a clear indication that the reaction proceeds through a free-radical mechanism. When two CH_3· radicals collide, they combine to form a CH_3CH_3, or C_2H_6, molecule.

24.11 ALCOHOLS AND ETHERS

An alcohol is a compound with an —OH group attached to a saturated carbon atom. Alcohols have common names based on the name of the alkyl group.

CH_3OH	methyl alcohol
CH_3CH_2OH	ethyl alcohol
$\begin{array}{c} CH_3 \\ \diagdown \\ \qquad CHOH \\ \diagup \\ CH_3 \end{array}$	isopropyl alcohol

The systematic nomenclature uses the ending *-ol* to indicate an alcohol and a number to identify the carbon that carries the —OH group. Isopropyl alcohol, for example, is also known as 2-propanol. In the name of an alcohol, the carbon atom with the —OH substituent must be a member of the longest chain.

Exercise 24.9

Use the systematic nomenclature to name the following alcohol.

$$\underset{\text{CH}_3-\text{CH}_2-\text{CH}_2-\overset{\displaystyle \overset{\text{CH}_2-\text{CH}_3}{|}}{\text{CH}}-\text{CH}_2-\text{OH}}{}$$

Solution

The longest chain of carbon atoms in this compound contains six atoms. It might therefore seem tempting to name this compound as a derivative of hexane.

$$\underset{\text{CH}_3-\text{CH}_2-\text{CH}_2-\overset{\displaystyle \overset{\text{CH}_2-\text{CH}_3}{|}}{\text{CH}}-\text{CH}_2-\text{OH}}{}$$

But the longest chain that carries the —OH substituent is only five carbon atoms long. The compound is therefore named as a derivative of pentane.

$$\underset{\text{CH}_3-\text{CH}_2-\text{CH}_2-\overset{\displaystyle \overset{\text{CH}_2-\text{CH}_3}{|}}{\text{CH}}-\text{CH}_2-\text{OH}}{} \qquad \text{2-ethyl-1-pentanol}$$

Methanol, or methyl alcohol, is also known as wood alcohol, because people originally made it by heating wood until a liquid distilled. Methanol is highly toxic, and many people have become blind or died from drinking it. Ethanol, or ethyl alcohol, is the alcohol associated with "alcoholic" beverages. It has been made for at least 6000 years by the addition of yeast to solutions that are rich in either sugars or starches. The yeast cells obtain energy from enzyme-catalyzed reactions that convert these sugars or starches to ethanol and CO_2.

$$C_6H_{12}O_6(aq) \longrightarrow 2\ CH_3CH_2OH(aq) + 2\ CO_2(g)$$

When the alcohol reaches a concentration of 10 to 12% by volume, the yeast cells die. Brandy, rum, gin, and the various whiskeys that have a higher concentration of alcohol are prepared by distillation of the alcohol produced by the fermentation reaction just described.

Ethanol is not as toxic as methanol, but it is still dangerous. Blood alcohol levels of 1.5 to 3 grams per liter result in intoxication, and levels of 4 to 6 grams per liter can lead to coma or death. The breath analyzer used to detect drunken drivers operates on the basis of the fact that persons who have been drinking exhale some alcohol when they breathe. The alcohol in the breath is collected and allowed to react with the dichromate ($Cr_2O_7{}^{2-}$) ion in acid solution.

$$3\ CH_3CH_2OH(aq) + 2\ Cr_2O_7{}^{2-}(aq) + 16\ H^+(aq) \longrightarrow$$
$$3\ CH_3CO_2(aq) + 4\ Cr^{3+}(aq) + 11\ H_2O(l)$$

The $Cr_2O_7{}^{2-}$ ion is orange and the Cr^{3+} ion is green, so the extent of this reaction can be followed by monitoring the color of the solution.

Ethanol is oxidized to CO_2 and H_2O by alcohol dehydrogenase enzymes in the body. This reaction gives off 30 kilojoules of energy per gram, which makes ethanol a better source of energy than carbohydrates (17 kJ/g) and almost as good a source of energy as fat (38 kJ/g). An ounce of 80-proof liquor can provide as much as 3% of the average daily caloric intake, and drinking alcohol can contribute to obesity. Many alcoholics are malnourished, however, because of the lack of vitamins in the calories they obtain from alcoholic beverages.

It is often useful to classify alcohols on the basis of their structure as either *primary* (1°), *secondary* (2°), or *tertiary* (3°). Ethanol is a primary alcohol, because there is only one alkyl group attached to the carbon that carries the —OH substituent. The structure of a primary alcohol can be abbreviated as RCH_2OH, where R stands for an alkyl group. Isopropyl alcohol, or rubbing alcohol, is an example of a secondary alcohol, which has two alkyl groups on the carbon atom with the —OH substituent (R_2CHOH). An example of a tertiary alcohol (R_3COH) is *t*-butyl alcohol, or 2-methyl-2-propanol.

$$CH_3CH_2OH \qquad CH_3\overset{\displaystyle OH}{\underset{\displaystyle |}{C}}HCH_3 \qquad CH_3\overset{\displaystyle OH}{\underset{\displaystyle |}{\underset{\displaystyle |}{\underset{\displaystyle CH_3}{C}}}}CH_3$$

primary alcohol   secondary alcohol   tertiary alcohol

Another class of alcohols are the **phenols,** in which an —OH group is attached to one of the carbon atoms in a benzene ring, as was shown in Figure 24.8. Phenols are potent disinfectants. When antiseptic techniques were first introduced to surgery in the 1860s by Joseph Lister, it was phenol (or carbolic acid, as it was then known) that was used. Phenol derivatives are still used in commercial disinfectants such as Lysol.

The hydrogen bonds between H_2O molecules in water explain the unusually high boiling point of water (see Section 13.9). Alcohols can form similar hydrogen bonds (see Figure 24.12). As a result, alcohols have boiling points that are much higher than those of alkanes with similar molecular weights. The boiling point of ethanol, for example, is 78.5°C, whereas propane, with about the same molecular weight, boils at −42.1°C.

Alcohols are Brønsted acids in aqueous solution.

$$CH_3CH_2OH(aq) + H_2O(l) \rightleftharpoons H_3O^+(aq) + CH_3CH_2O^-(aq)$$

The acidity of methanol and ethanol is about the same as that of water; alcohols with higher molecular weights tend to be slightly less acidic than water. Because of

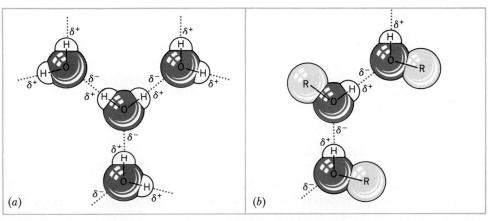

(a)   (b)

FIG. 24.12 Hydrogen bonding can be found in both water (*a*) and alcohols (*b*).

their acidity, alcohols react with sodium metal to produce sodium salts of the corresponding conjugate base.

$$2\ Na(s) + 2\ CH_3CH_2OH(l) \longrightarrow 2\ [Na^+][CH_3CH_2O^-] + H_2(g)$$

The conjugate base of an alcohol is known as an *alkoxide.*

$$[Na^+][CH_3O^-] \qquad [Na^+][CH_3CH_2O^-]$$

sodium methoxide   sodium ethoxide

Alcohols can be prepared by adding water to an alkene in the presence of a strong acid, such as concentrated sulfuric acid.

$$CH_3CH{=}CH_2 + H_2O \xrightarrow{H_2SO_4} CH_3\overset{\overset{\displaystyle OH}{|}}{CH}CH_3$$

Alcohols can also be prepared by substitution reactions between an alkyl halide and the OH^- ion.

$$CH_3CH_2CH_2Br(aq) + OH^-(aq) \longrightarrow CH_3CH_2CH_2OH(aq) + Br^-(aq)$$

An ether $(R{-}O{-}R)$ is a compound in which an oxygen atom bridges two alkyl groups. Its name includes the word *ether* attached to the names of the alkyl groups.

$$CH_3{-}O{-}CH_3 \qquad\qquad\text{dimethyl ether}$$
$$CH_3CH_2{-}O{-}CH_2CH_3 \qquad\text{diethyl ether}$$

$$CH_3O{-}\bigcirc\!\!\!\!\!\!\!\bigcirc \qquad\qquad\text{methyl phenyl ether}$$

Diethyl ether, often known by the generic name "ether," was once used extensively as an anesthetic. Mixtures of diethyl ether and air explode in the presence of a spark, however, and ether has been replaced by safer anesthetics.

Ethers boil at lower temperatures than alcohols that have the same molecular formula. Diethyl ether ($CH_3CH_2OCH_2CH_3$) boils at $34.5°C$, whereas 1-butanol ($CH_3CH_2CH_2CH_2OH$) boils at $118°C$. This can be explained by the fact that there are no hydrogen bonds between ether molecules. Ethers can form hydrogen bonds with water molecules, however, so their solubility in water falls between those of the nonpolar alkanes and the more polar alcohols. Diethyl ether dissolves in water to form solutions that are as much as 10% ether by weight.

Ethers can be synthesized by splitting out a molecule of water between two alcohols in the presence of heat and concentrated sulfuric acid.

$$CH_3CH_2OH + HOCH_2CH_3 \xrightarrow{conc\ H_2SO_4,\ 140°C} CH_3CH_2OCH_2CH_3 + H_2O$$

Once formed, ethers are essentially inert to chemical reactions. They don't react with most oxidizing or reducing agents, and they are stable to most acids and bases except at high temperatures. They are therefore frequently used as solvents for chemical reactions.

24.12 SUBSTITUTION REACTIONS

The preceding section noted that alcohols could be made by reacting alkyl halides with the OH^- ion. Both CH_3Br and $(CH_3)_3CBr$, for example, react with solutions of the hydroxide ion to form the analogous alcohol.

$$CH_3-Br + OH^- \longrightarrow CH_3-OH + Br^-$$

$$CH_3-\underset{\underset{CH_3}{|}}{\overset{\overset{CH_3}{|}}{C}}-Br + OH^- \longrightarrow CH_3-\underset{\underset{CH_3}{|}}{\overset{\overset{CH_3}{|}}{C}}-OH + Br^-$$

As noted in Exercise 21.15, however, the kinetics of these reactions are different. The rate of the reaction between CH_3Br and the OH^- ion is proportional to the concentrations of both reactants.

$$Rate = k(CH_3Br)(OH^-)$$

But the rate of the reaction between $(CH_3)_3CBr$ and the OH^- ion is only proportional to the concentration of the alkyl halide.

$$Rate = k[(CH_3)_3CBr]$$

The fact that the rate laws for these reactions are different means that the mechanisms of the reactions must also be different.

The reaction between CH_3Br and the OH^- ion involves attack by the OH^- ion on the carbon atom. Through this mechanism, a new carbon-oxygen bond is formed at the same time that the carbon—bromine bond is broken.

$$HO^- + CH_3Br \longrightarrow HO \cdots \overset{\overset{\displaystyle H \quad H}{\diagdown \quad \diagup}}{\underset{\underset{\displaystyle H}{|}}{C}} \cdots Br \longrightarrow HO-CH_3 + Br^-$$

The OH^- ion acts as a Lewis base in this reaction — it donates a pair of nonbonding electrons to the carbon atom to form a new covalent bond. Organic chemists prefer to call the OH^- ion a **nucleophile** (literally, something that loves nuclei). But there is no difference between what organic chemists call a nucleophile and what other chemists call a Lewis base.

The reaction between CH_3Br and the OH^- ion is an example of a **nucleophilic substitution reaction.** It is obviously a substitution reaction, because an OH^- ion is substituted for a Br^- ion. It is a nucleophilic substitution reaction because it involves attack on an organic compound (CH_3Br) by a nucleophile (the OH^- ion). Since the rate-limiting step in this reaction involves both the CH_3Br and OH^- molecules, the reaction is bimolecular. It is therefore called a *bimolecular nucleophilic substitution,* or S_N2, reaction.

The observed rate law for the reaction between $(CH_3)_3CBr$ and the OH^- ion is consistent with a two-step mechanism. The first step is a slow, rate-determining reaction in which the carbon—bromine bond is broken to form an intermediate. The second step is a much faster reaction between this intermediate and the OH^- ion to form the alcohol.

In theory, there are three ways in which the C—Br bond could break. The pair of electrons could be divided equally between the two atoms that formed this bond to generate a pair of free radicals.

$$CH_3-\underset{\underset{CH_3}{|}}{\overset{\overset{CH_3}{|}}{C}}-Br \longrightarrow CH_3-\underset{\underset{CH_3}{|}}{\overset{\overset{CH_3}{|}}{C}}\cdot + Br\cdot$$

Both electrons could leave with the bromine atom to form a positively charged *carbonium* ion and a negatively charged Br^- ion.

$$CH_3-\underset{\underset{CH_3}{|}}{\overset{\overset{CH_3}{|}}{C}}-Br \longrightarrow CH_3-\underset{\underset{CH_3}{|}}{\overset{\overset{CH_3}{|}}{C}}^+ + Br^-$$

Or, both electrons could stay on the carbon atom to form a negatively charged *carbanion* and a positively charged Br^+ ion.

$$CH_3-\underset{\underset{CH_3}{|}}{\overset{\overset{CH_3}{|}}{C}}-Br \longrightarrow CH_3-\underset{\underset{CH_3}{|}}{\overset{\overset{CH_3}{|}}{C}}^- + Br^+$$

Only one of these alternatives provides a useful intermediate. The mechanism that involves a carbonium ion produces a Lewis acid, which can combine with a Lewis base such as the OH^- ion to form an alcohol.

$$CH_3-\underset{\underset{CH_3}{|}}{\overset{\overset{CH_3}{|}}{C}}^+ + OH^- \longrightarrow CH_3-\underset{\underset{CH_3}{|}}{\overset{\overset{CH_3}{|}}{C}}-OH$$

The observed rate law for the reaction between $(CH_3)_3CBr$ and OH^- is therefore consistent with the following mechanism.

$$(CH_3)_3CBr \longrightarrow (CH_3)_3C^+ + Br^- \qquad \text{slow step}$$
$$(CH_3)_3C^+ + OH^- \longrightarrow (CH_3)_3C-OH \qquad \text{fast step}$$

Since the rate-determining step for this reaction involves the dissociation of the $(CH_3)_3CBr$ molecule, the reaction is known as a *unimolecular nucleophilic substitution,* or S_N1, reaction.

Why does $(CH_3)_3CBr$ react with the OH^- ion by the S_N1 mechanism when CH_3Br does not? The S_N1 mechanism proceeds through a carbonium ion intermediate. Experimental evidence suggests that the stability of carbonium ions decreases in the following order.

$$CH_3-\underset{\underset{CH_3}{|}}{\overset{\overset{CH_3}{|}}{C}}^+ > CH_3-\overset{\overset{CH_3}{|}}{C}H^+ > CH_3-CH_2^+ > CH_3^+$$

Tertiary carbonium ions are as much as 290 kJ/mol more stable than primary carbonium ions. Thus, it is much easier for $(CH_3)_3Br$ to form a carbonium ion intermediate than it is for CH_3Br to undergo this reaction.

Why does CH_3Br react with the OH^- ion by the S_N2 mechanism when $(CH_3)_3CBr$ does not? The S_N2 mechanism requires that the OH^- ion attack the carbon atom. The OH^- ion can get past the small hydrogen atoms in CH_3Br much more easily than it can get past the bulkier CH_3^- groups in $(CH_3)_3CBr$.

$$HO^- \overset{H}{\underset{H}{\overset{\diagdown}{\diagup}}}C-Br \longrightarrow HO\cdots\overset{H \quad H}{\underset{H}{C}}\cdots Br \longrightarrow HO-\overset{H \quad H}{\underset{H}{C}} + Br^-$$

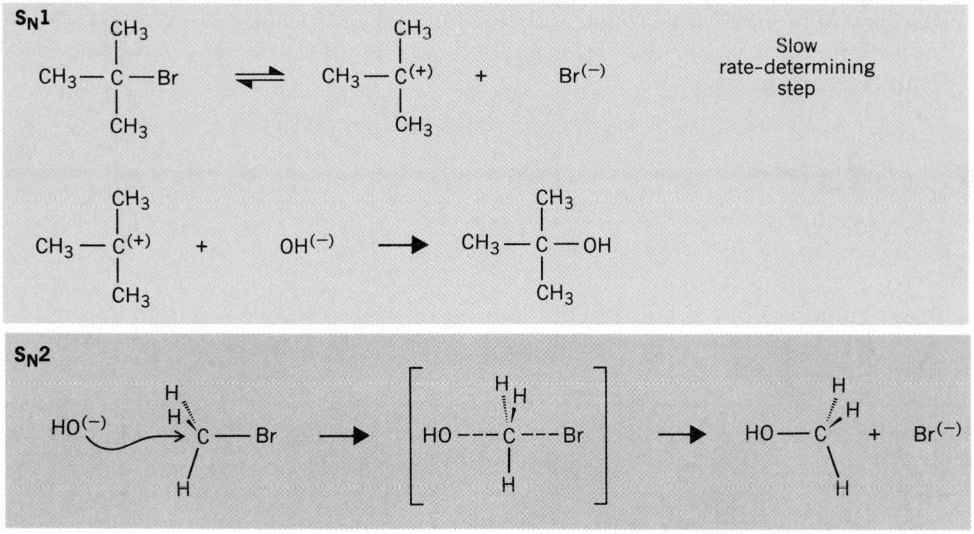

S$_N$1

Slow
rate-determining
step

S$_N$2

FIG. 24.13 Two distinct mechanisms have been discovered for nucleophilic substitution reactions such as the reaction between CH_3Br or $(CH_3)_3CBr$ and the OH^- ion. The S_N1 mechanism is a two-step reaction in which the carbon—bromine bond breaks to form a positively charged carbonium ion and a negatively charged Br^- ion. A rapid reaction follows in which the $(CH_3)_3C^+$ ion combines with OH^- to form the alcohol. The S_N2 mechanism involves attack on the carbon atom by the OH$^-$ ion to form a carbon—oxygen bond as the carbon—bromine bond is broken.

The S$_N$1 and S$_N$2 mechanisms for the reaction of alkyl halides with OH$^-$ are summarized in Figure 24.13.

24.13 OXIDATION-REDUCTION REACTIONS

Organic reactions can be classified as either oxidation-reduction or metathesis reactions. If there is a change in the oxidation number of any of the atoms in the reaction, it is an oxidation-reduction reaction. If there is no change in the oxidation numbers it is a metathesis, or acid-base, reaction. By assuming that hydrogen is in the $+1$ oxidation state when bound to a nonmetal, we can assign oxidation numbers to the carbon atoms in the following simple examples of alkanes, alkenes, and alkynes.

alkane	CH_3-CH_3	$C = -3$
alkene	$H_2C=CH_2$	$C = -2$
alkyne	$HC\equiv CH$	$C = -1$

Oxidation-reduction reactions include the addition of H_2 to an alkene to form an alkane.

$$H_2C=CH_2 + H_2 \longrightarrow H_3C-CH_3$$
$$+1\ -2 \qquad\qquad 0 \qquad +1\ -3$$

reduction

oxidation

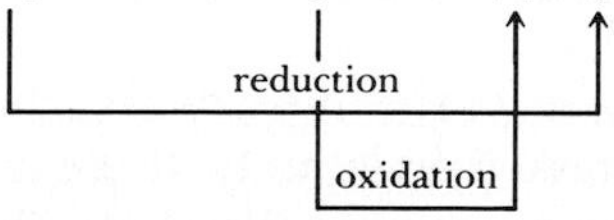

H_2 acts as a reducing agent in this reaction — the carbon atoms in the C=C double bond are reduced, and the hydrogen atoms in H_2 are oxidized.

The oxidation numbers of the compounds in Table 24.6 can be used to classify other organic reactions as either oxidation-reduction or metathesis reactions. Metathesis reactions include nucleophilic substitution reactions such as the synthesis of an alcohol from an alkyl halide.

$$CH_3Cl + OH^- \longrightarrow CH_3OH + Cl^-$$
$$-2\ +1\ -1 \quad -2\ +1 \qquad -2\ +1\ -2 \quad\ -1$$

The reaction between methane and chlorine, on the other hand, is an oxidation-reduction reaction in which Cl_2 is an oxidizing agent that removes two electrons from the carbon atom.

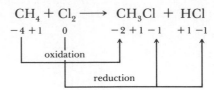

Other examples of oxidation-reduction reactions are shown in Figure 24.14.

Exercise 24.10

Use the oxidation numbers of the carbon atoms in italics to classify the following reactions as either oxidation-reduction or metathesis reactions.

(a) $2\ CH_3OH \xrightarrow{H_2SO_4} CH_3OCH_3 + H_2O$

(b) $HCO_2H + CH_3OH \longrightarrow HCO_2CH_3 + H_2O$

(c) $CO + 2\ H_2 \longrightarrow CH_3OH$

Solution

(a) There is no change in the oxidation number of the carbon atom when an alcohol is converted to an ether, so this is a metathesis reaction.

$$2\ \underset{-2}{CH_3}OH \longrightarrow \underset{-2}{CH_3}OCH_3 + H_2O$$

(b) There is no change in the oxidation number of the carbon atom when formic acid reacts with an alcohol to form an ester. Thus, this reaction is also an example of metathesis.

$$\underset{+2}{HCO_2}H + CH_3OH \longrightarrow \underset{+2}{HCO_2}CH_3 + H_2O$$

(c) The carbon atom in carbon monoxide is reduced from $+2$ to -2 when CO combines with H_2 to form methanol, so this is an oxidation-reduction reaction.

$$\underset{+2}{CO} + 2\ H_2 \longrightarrow \underset{-2}{CH_3}OH$$

Assigning oxidation numbers to the individual carbon atoms in complex organic molecules can be a difficult task. Fortunately, there is another way of recognizing oxidation-reduction reactions in organic chemistry. The three reactions in Figure 24.14 all have one thing in common. In each case, the carbon atom is oxidized when hydrogen atoms are removed from that atom. This observation can be generalized as follows.

Oxidation occurs whenever hydrogen atoms are removed from a carbon atom.

How do we recognize when reduction takes place? All we have to do is reverse the rule just given.

Reduction occurs whenever hydrogen atoms are added to a carbon atom.

TABLE 24.6

Common Oxidation Numbers of Carbon

Functional Group	Example	Oxidation Number of Carbon
alkane	CH_4	-4
alkene	$H_2C{=}CH_2$	-2
alcohol	CH_3OH	-2
ether	CH_3OCH_3	-2
alkyl halide	CH_3Cl	-2
amine	CH_3NH_2	-2
alkyne	$HC{\equiv}CH$	-1
aldehyde	H_2CO	0
acid	HCO_2H	$+2$
	CO_2	$+4$

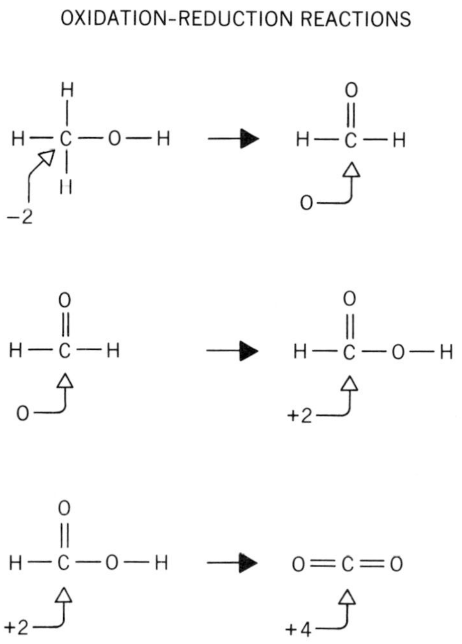

OXIDATION-REDUCTION REACTIONS

FIG. 24.14 Each of the transformations shown here involves the oxidation of a carbon atom. In the first reaction, the carbon atom is oxidized from the -2 to the 0 oxidation state. In the second reaction, it is oxidized to the $+2$ state, and the third reaction involves oxidation of the carbon atom to the $+4$ oxidation state.

Exercise 24.11

Determine whether the carbon atoms highlighted with color in Figure 24.15 are oxidized or reduced in the reactions shown in the figure.

Solution

(a) The carbon atoms are reduced in this reaction, because they pick up hydrogen atoms.

(b) The carbon atom is oxidized in this reaction, because it loses hydrogen atoms.

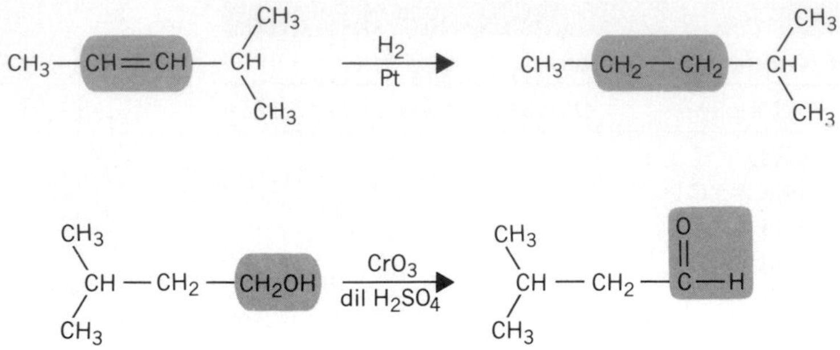

FIG. 24.15 Since it is difficult to assign oxidation numbers to the carbon atoms in complex organic molecules, we often use the rule that adding hydrogen atoms to a carbon atom results in the reduction of that atom, whereas removing hydrogen atoms from a carbon atom results in the oxidation of that atom. (See Exercise 24.11.)

$$CH_3-CH{=}CH-\underset{\underset{CH_3}{|}}{\overset{\overset{CH_3}{|}}{CH}} \quad \xrightarrow[\text{Pt}]{H_2} \quad CH_3-CH_2-CH_2-\underset{\underset{CH_3}{|}}{\overset{\overset{CH_3}{|}}{CH}}$$

$$\underset{\underset{CH_3}{|}}{\overset{\overset{CH_3}{|}}{CH}}-CH_2-CH_2OH \quad \xrightarrow[\text{dil } H_2SO_4]{CrO_3} \quad \underset{\underset{CH_3}{|}}{\overset{\overset{CH_3}{|}}{CH}}-CH_2-\overset{\overset{\displaystyle O}{\|}}{C}-H$$

24.14 ALDEHYDES AND KETONES

As noted earlier, the $C{=}O$ double bond is known as a carbonyl group. When there is at least one hydrogen atom attached to the carbonyl carbon, the compound is an aldehyde. Aldehydes have common names derived from the names of the corresponding carboxylic acids.

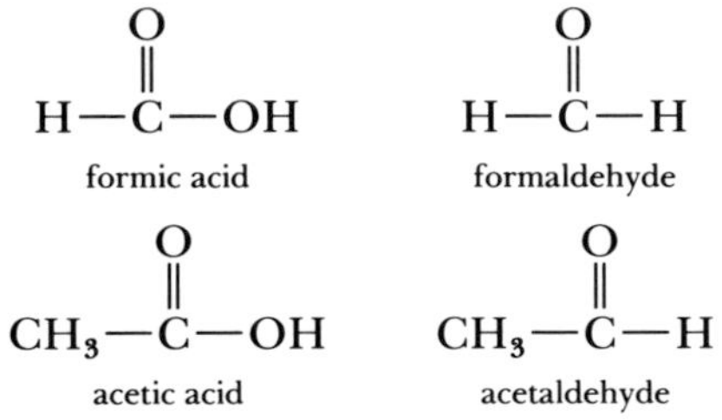

$$\overset{\overset{\displaystyle O}{\|}}{H-C-OH} \qquad \overset{\overset{\displaystyle O}{\|}}{H-C-H}$$
$$\text{formic acid} \qquad\qquad \text{formaldehyde}$$

$$\overset{\overset{\displaystyle O}{\|}}{CH_3-C-OH} \qquad \overset{\overset{\displaystyle O}{\|}}{CH_3-C-H}$$
$$\text{acetic acid} \qquad\qquad \text{acetaldehyde}$$

The systematic name for an aldehyde is obtained by adding -*al* to the name of the parent alkane.

$$\overset{\overset{\displaystyle O}{\|}}{H-C-H} \qquad \overset{\overset{\displaystyle O}{\|}}{CH_3-C-H}$$
$$\text{methanal} \qquad\qquad \text{ethanal}$$

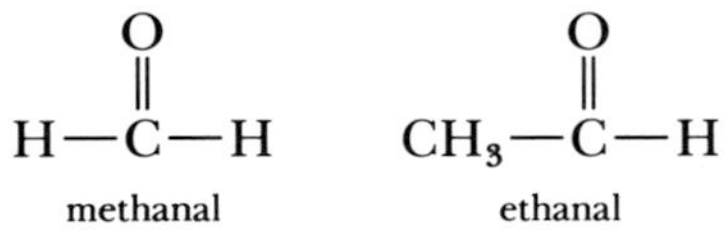

When both substituents on the carbonyl carbon are alkyl groups, the compound is a ketone. The common name for a ketone includes the word *ketone* and the names of the two alkyl groups.

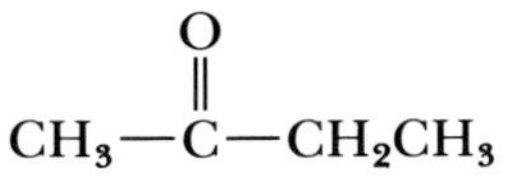

$$\overset{\overset{\displaystyle O}{\|}}{CH_3-C-CH_2CH_3} \qquad \text{methyl ethyl ketone}$$

The systematic name is obtained by adding -*one* to the name of the parent alkane. Numbers are used to indicate the location of the $C{=}O$ group. Methyl ethyl ketone, for example, is also known as 2-butanone.

Exercise 24.12

The simplest ketone is often known by the name *acetone*.

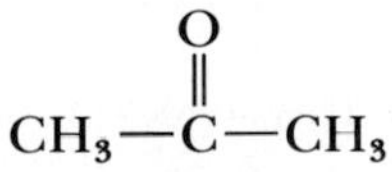

$$\overset{\overset{\displaystyle O}{\|}}{CH_3-C-CH_3} \qquad \text{acetone}$$

Determine the common name and the systematic name for this compound, which is frequently used as a solvent for organic reactions.

Solution

All we have to do to determine the common name of this compound is identify the alkyl groups. Since the compound contains two CH_3—, or methyl, groups, it has the common name *dimethyl ketone*. The systematic name is based on the fact that the compound can be thought of as a derivative of propane with a carbonyl (C=O) group on the second carbon atom. It is therefore also known as *2-propanone*.

Both aldehydes and ketones can be reduced by H_2 in the presence of a catalyst, such as platinum metal. When the starting material is an aldehyde, the product is a primary alcohol.

aldehyde   primary alcohol

When the starting material is a ketone, the product is a secondary alcohol.

ketone   secondary alcohol

This explains why it is important to distinguish between primary, secondary, and tertiary alcohols. It also provides a hint as to how aldehydes and ketones can be synthesized. Aldehydes can be prepared by oxidization of primary alcohols; ketones result from the oxidation of secondary alcohols.

Chromium trioxide (CrO_3) in concentrated sulfuric acid is often used to oxidize secondary alcohols to ketones.

Unfortunately, CrO_3 in H_2SO_4 is such a good oxidizing agent that it not only oxidizes a primary alcohol to an aldehyde but also oxidizes the aldehyde to the corresponding carboxylic acid.

Weaker oxidizing agents, such as CrO_3 in dilute sulfuric acid, must be used to prepare aldehydes from primary alcohols.

It is convenient to write the carbonyl group as a covalent C=O double bond, but it is also somewhat misleading. The difference between the electronegativities of carbon and oxygen is 0.89, and so the C=O bond is moderately polar. The bonding in the C=O bond can be thought of as a hybrid of the following Lewis structures.

To represent the polar nature of this hybrid, we indicate the presence of a slight negative charge (δ^-) on the oxygen and a slight positive charge (δ^+) on the carbon of the C=O double bond.

$$\delta^+ \ \diagdown C = O^{\delta^-}$$

Reagents that attack the electron-rich δ^- end of this bond are called *electrophiles* (literally, "lovers of electrons"). Electrophiles include ions (such as the H^+ and Fe^{3+} ions) and neutral molecules (such as $AlCl_3$ and BF_3) that are Lewis acids, or electron-pair acceptors. Reagents that attack the electron-poor δ^+ end of the C=O bond are called nucleophiles (literally, "lovers of nuclei," as noted earlier). Nucleophiles are Lewis bases such as NH_3 or the OH^- ion.

electrophile (H^+, Fe^{3+}, BF_3, etc.)

nucleophile
(OH^-, NH_3, etc.)

We can use this model to explain the reactions of carbonyl compounds. Aldehydes and ketones, for example, react with a number of hydride ion (H^-) donors. The H^- ion is a Lewis base, or nucleophile, that attacks the δ^+ end of the C=O bond. When this happens, the two valence electrons on the H^- ion form a covalent bond to the carbon atom. Since carbon is tetravalent, one pair of electrons in the C=O bond is displaced onto the oxygen to form an intermediate with a formal negative charge on the oxygen atom.

This intermediate ion can then remove an H^+ ion from water to form an alcohol.

The most common sources of the H^- ion for this reaction are metal hydrides, such as lithium aluminum hydride, $LiAlH_4$.

The polar nature of the C=O double bond also explains what happens when carbon dioxide reacts with water to form carbonic acid.

$$CO_2(g) + H_2O(l) \longrightarrow H_2CO_3(aq)$$

CO_2 is a linear molecule that contains two C=O double bonds. Water acts as a nucleophile, attacking the δ^+ end of the C=O bond and thereby forming a new carbon—oxygen single bond.

A hydrogen atom then migrates from the positively charged oxygen atom to the negatively charged oxygen atom to form carbonic acid, H_2CO_3.

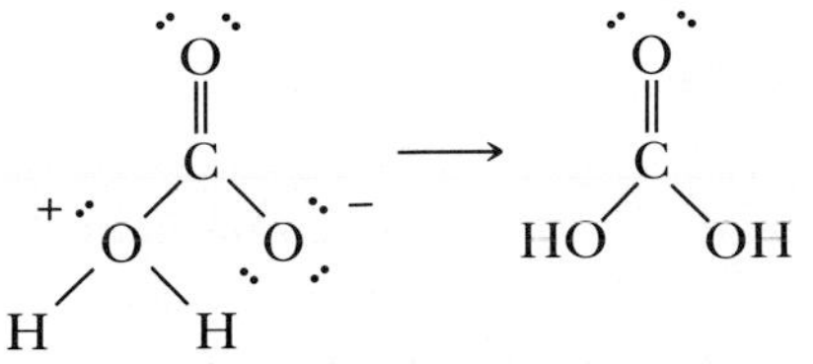

24.15 CARBOXYLIC ACIDS AND CARBOXYLATE IONS

When one of the substituents on a carbonyl group is an —OH group, the compound is a carboxylic acid with the generic formula RCO_2H. The **carboxylate ion** (RCO_2^-) is formed by the loss of an H^+ ion from a carboxylic acid. This ion is a hybrid of two resonance structures.

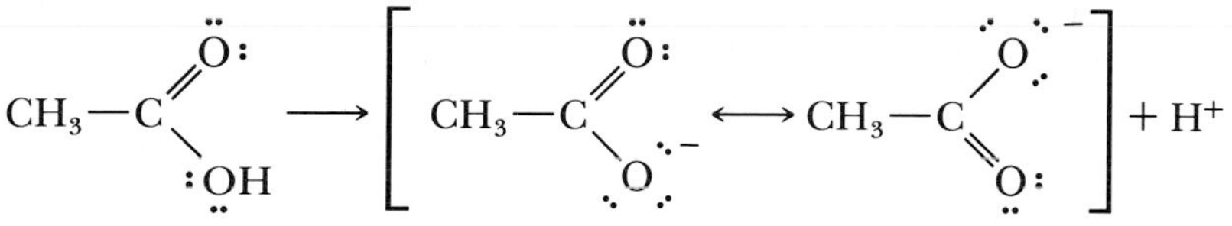

carboxylic
acid

carboxylate ion

The net effect of resonance is to delocalize the negative charge on the molecule over the entire —CO_2^- group. This tends to stabilize the carboxylate ion and makes carboxylic acids much stronger acids than are analogous alcohols. K_a for carboxylic acids such as formic (HCO_2H) and acetic (CH_3CO_2H) acid is on the order of 10^{-4} to 10^{-5}, whereas alcohols have values of K_a of only 10^{-16}.

Carboxylic acids were among the first organic compounds to be discovered. They therefore have well-entrenched common names that are often derived from the Latin stems of their sources in nature. Formic acid (Latin *formica*, "an ant"), for example, and acetic acid (Latin *acetum*, "vinegar") were first isolated by distillation from ants and vinegar, respectively. Butyric acid (Latin *butyrum*, "butter") is found in rancid butter; and caproic, caprylic, and capric acids (Latin *caper*, "goat") are all obtained from goat fat. We give systematic names to simple carboxylic acids by adding *-oic acid* to the name of the parent alkane. A list of common carboxylic acids is given in Table 24.7.

Formic acid and acetic acid have a sharp, pungent odor. As the length of the alkyl chain increases, the odor of carboxylic acids becomes more unpleasant. Butyric acid, for example, is found in sweat, and the odor of rancid meat is due in part to carboxylic acids released as the meat spoils.

The data in Table 24.7 show that the solubility of carboxylic acids in water also changes with the length of the alkyl chain. The —CO_2H end of this molecule is polar and therefore soluble in water. As the alkyl chain gets longer, the molecule becomes more nonpolar and less soluble in water. The —CO_2H "head" of long-chain carboxylic acids is hydrophilic (water-loving), and the hydrocarbon "tail" is hydrophobic (water-hating).

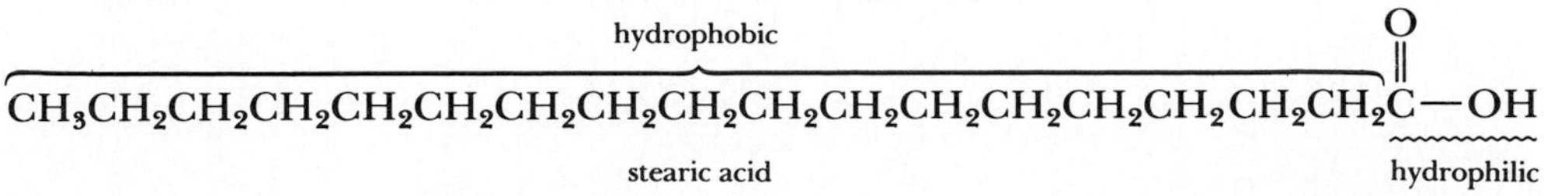

stearic acid

TABLE 24.7

Common Carboxylic Acids

Compound	Common Name	Systematic Name	Solubility in H_2O (g / 100 mL)
Saturated carboxylic acids and fatty acids			
HCO_2H	formic acid	methanoic acid	
CH_3CO_2H	acetic acid	ethanoic acid	
$CH_3CH_2CO_2H$	propionic acid	propanoic acid	
$CH_3CH_2CH_2CO_2H$	butyric acid	butanoic acid	
$CH_3(CH_2)_3CO_2H$	valeric acid	pentanoic acid	4.97
$CH_3(CH_2)_4CO_2H$	caproic acid	hexanoic acid	0.968
$CH_3(CH_2)_6CO_2H$	caprylic acid	octanoic acid	0.068
$CH_3(CH_2)_8CO_2H$	capric acid	decanoic acid	0.015
$CH_3(CH_2)_{10}CO_2H$	lauric acid	dodecanoic acid	0.0055
$CH_3(CH_2)_{12}CO_2H$	myristic acid	tetradecanoic acid	0.0020
$CH_3(CH_2)_{14}CO_2H$	palmitic acid	hexadecanoic acid	0.00072
$CH_3(CH_2)_{16}CO_2H$	stearic acid	octadecanoic acid	0.00029
Unsaturated fatty acids			
$CH_3(CH_2)_7CH\!=\!CH(CH_2)_7CO_2H$	oleic acid		
$CH_3(CH_2)_4CH\!=\!CHCH_2CH\!=\!$ $CH(CH_2)_7CO_2H$	linoleic acid		
$CH_3CH_2CH\!=\!CH_2CH_2CH\!=\!$ $CHCH_2CH\!=\!CH(CH_2)_7CO_2H$	linolenic acid		

Since long-chain carboxylic acids are obtained from animal fat, they are often called *fatty acids.*

Fatty acids react with NaOH to form salts of the corresponding carboxylate ion.

$$\overbrace{CH_3CH_2CH_2CH_2CH_2CH_2CH_2CH_2CH_2CH_2CH_2CH_2CH_2CH_2CH_2CH_2CH_2}^{\text{hydrophobic}}\underset{\text{hydrophilic}}{\underbrace{\overset{\displaystyle O}{\overset{\|}{C}}\!-\!O^-}}\ Na^+$$

sodium stearate

These salts are more polar than carboxylic acids and therefore more soluble in water than all but the simplest carboxylic acids. Fatty carboxylate ion salts make excellent soaps (see Section 13.13).

Compounds that contain two $-CO_2H$ functional groups are known as dicarboxylic acids. A number of dicarboxylic acids (see Table 24.8) can be isolated from natural sources. Tartaric acid, for example, is a by-product of the fermentation of wine, and succinic, fumaric, malic, and oxaloacetic acid are intermediates in the metabolism of sugars to CO_2 and H_2O.

Several tricarboxylic acids are associated with the metabolism of sugar to CO_2 and H_2O, as well. The most important example of this class of compounds is citric acid.

$$\begin{array}{c} CH_2CO_2H \\ | \\ HO\!-\!C\!-\!CO_2H \\ | \\ CH_2CO_2H \end{array}$$

citric acid

TABLE 24.8

Common Dicarboxylic Acids

Compound	Common Name
$HO_2C—CO_2H$	oxalic acid
$HO_2C—CH_2—CO_2H$	malonic acid
$HO_2C—CH_2—CH_2—CO_2H$	succinic acid
cis-$HO_2C—CH=CH—CO_2H$	maleic acid
trans-$HO_2C—CH=CH—CO_2H$	fumaric acid
$HO_2C—CH_2—\overset{\displaystyle OH}{\underset{\displaystyle \vert}{CH}}—CO_2H$	malic acid
$HO_2C—\overset{\displaystyle OH}{\underset{\displaystyle \vert}{CH}}—\overset{\displaystyle OH}{\underset{\displaystyle \vert}{CH}}—CO_2H$	tartaric acid
$HO_2C—CH_2—\overset{\displaystyle O}{\overset{\displaystyle \|}{C}}—CO_2H$	oxaloacetic acid
$HO_2C—(CH_2)_4—CO_2H$	adipic acid

24.16 CARBOXYLIC ACID ESTERS

Carboxylic acids ($—CO_2H$) react with alcohols (ROH) in the presence of either acid or base to form **carboxylic acid esters** ($—CO_2R$). Acetic acid, for example, reacts with methanol to form methyl acetate and water.

$$CH_3—\overset{O}{\overset{\|}{C}}\underset{OH} + HOCH_3 \longrightarrow CH_3—\overset{O}{\overset{\|}{C}}\underset{O—CH_3} + HOH$$

acetic acid   methanol   methyl acetate

To name these esters, we identify the alkyl group in the alcohol and then add *-ate* to the stem of the name of the carboxylic acid.

$$CH_3CO_2CH_3 \qquad\qquad CH_3CH_2CH_2CO_2CH_2CH_3$$

methyl acetate or methyl ethanoate   ethyl butyrate or ethyl butanoate

The term *ester* is commonly used to describe the product of the reaction of any strong acid with an alcohol. For example, sulfuric acid reacts with methanol to form a diester known as dimethyl sulfate.

$$HO—\overset{O}{\underset{O}{\overset{\|}{\underset{\|}{S}}}}—OH + 2\ CH_3OH \longrightarrow CH_3O—\overset{O}{\underset{O}{\overset{\|}{\underset{\|}{S}}}}—OCH_3$$

Phosphoric acid reacts with alcohols to form triesters such as triethyl phosphate.

$$HO—\overset{O}{\underset{OH}{\overset{\|}{\underset{\vert}{P}}}}—OH + 3\ CH_3CH_2OH \longrightarrow CH_3CH_2O—\overset{O}{\underset{OCH_2CH_3}{\overset{\|}{\underset{\vert}{P}}}}—OCH_2CH_3$$

CARBOXYLIC ACID ESTER

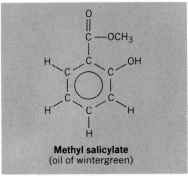

Methyl salicylate
(oil of wintergreen)

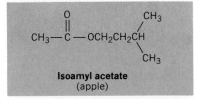

Isoamyl acetate
(apple)

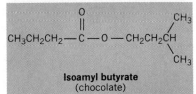

Isoamyl butyrate
(chocolate)

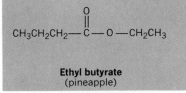

Ethyl butyrate
(pineapple)

FIG. 24.16 Carboxylic acid esters often have pleasant odors or flavors. Among this class of compounds are methyl salicylate, isoamyl acetate, isoamyl butyrate, and ethyl butyrate.

Although compounds of the type $-CO_2R$ should be called *carboxylic acid esters,* the term *ester* is commonly used instead.

Carboxylic acid esters with low molecular weights are colorless, volatile liquids that often have a pleasant odor. They are important components of both natural and synthetic flavors (see Figure 24.16). Methyl salicylate, for example, is an ester found in oil of wintergreen. Synthetic flavors include esters such as isoamyl acetate (apple), ethyl butyrate (pineapple), and isoamyl butyrate (chocolate).

Carboxylic acid esters made from long-chain fatty acids are found in animal fats and vegetable oils. Fats (such as butter and lard) and oils (such as olive oil, corn oil, an peanut oil) are esters of an alcohol known as glycerol that contain as many as three long-chain fatty acids, such as oleic acid, palmitic acid, and stearic acid.

$$
\begin{array}{ll}
\underset{\displaystyle \text{glycerol}}{\begin{array}{l}
\text{CH}_2\!-\!\text{OH} \\[4pt]
\text{CH}\!-\!\text{OH} \\[4pt]
\text{CH}_2\!-\!\text{OH}
\end{array}}
& +
\underset{\displaystyle \text{fatty acids}}{\begin{array}{l}
\overset{\text{O}}{\overset{\|}{\text{HOC}}}(\text{CH}_2)_7\text{CH}\!=\!\text{CH}(\text{CH}_2)_7\text{CH}_3 \\[4pt]
\overset{\text{O}}{\overset{\|}{\text{HOC}}}(\text{CH}_2)_{14}\text{CH}_3 \\[4pt]
\overset{\text{O}}{\overset{\|}{\text{HOC}}}(\text{CH}_2)_{16}\text{CH}_3
\end{array}}
\end{array}
\longrightarrow
\underset{\displaystyle \text{triacylglycerols or triglycerides}}{\begin{array}{l}
\text{CH}_2\!-\!\text{O}\!-\!\overset{\text{O}}{\overset{\|}{\text{C}}}\!-\!(\text{CH}_2)_7\text{CH}\!=\!\text{CH}(\text{CH}_2)_7\text{CH}_3 \\[4pt]
\text{CH}\!-\!\text{O}\!-\!\overset{\text{O}}{\overset{\|}{\text{C}}}\!-\!(\text{CH}_2)_{14}\text{CH}_3 \\[4pt]
\text{CH}_2\!-\!\text{O}\!-\!\overset{\text{O}}{\overset{\|}{\text{C}}}\!-\!(\text{CH}_2)_{16}\text{CH}_3
\end{array}}
$$

Animal fats are solids at room temperature. They contain roughly equal amounts of saturated fatty acids (such as palmitic and stearic acid) and unsaturated fatty acids (such as oleic and linoleic acid). Vegetable oils, on the other hand, are liquids at room temperature, and they contain as much as 80% unsaturated fatty acids. Fats and oils are the most efficient way for plants and animals to store energy. The energy released when fats and oils are consumed is more than twice the energy released by sugars, starches, or proteins.

Other naturally occurring fatty acid esters include the waxes that provide a protective coating for leaves, feathers, and fur. Beeswax, for example, is a mixture of esters formed from long-chain alcohols and fatty acids.

$$
\text{CH}_3(\text{CH}_2)_{24}\!-\!\overset{\text{O}}{\overset{\|}{\text{C}}}\!-\!\text{O}\!-\!(\text{CH}_2)_{30}\text{CH}_3
$$
a typical component of beeswax

Fats and oils undergo a reaction called ***saponification*** (literally, "the making of soap"). They react with either NaOH or KOH to form glycerol and the sodium or potassium salts of the carboxylate ion. This reaction provides us with both a source of carboxylic acids from natural sources and a way to make soaps.

$$
\begin{array}{l}
\text{CH}_2\text{O}\overset{\text{O}}{\overset{\|}{\text{C}}}(\text{CH}_2)_{16}\text{CH}_3 \\[6pt]
\text{CHO}\overset{\text{O}}{\overset{\|}{\text{C}}}(\text{CH}_2)_{16}\text{CH}_3 \;+\; 3\,\text{NaOH} \\[6pt]
\text{CH}_2\text{O}\overset{\text{O}}{\overset{\|}{\text{C}}}(\text{CH}_2)_{16}\text{CH}_3
\end{array}
\longrightarrow
\begin{array}{l}
\text{CH}_2\text{OH} \\[4pt]
\text{CHOH} \;+\; 3\,[\text{CH}_3(\text{CH}_2)_{16}\text{CO}_2^-][\text{Na}^+] \\[4pt]
\text{CH}_2\text{OH}
\end{array}
$$

24.17 AMINES, ALKALOIDS, AND AMIDES

Amines are derivatives of ammonia in which one or more of the hydrogen atoms have been replaced by alkyl groups. We indicate the degree of substitution by labeling the amine as either primary (RNH_2), secondary (R_2NH), or tertiary (R_3N). The common names of these compounds are derived from the names of the alkyl groups.

$$(CH_3)_2CHNH_2 \qquad CH_3NHCH_2CH_3 \qquad (CH_3)_3N$$

isopropylamine ethylmethylamine trimethylamine

The systematic names are derived from the names of the parent alkanes by adding the prefix -*amino* and a number specifying the carbon that carries the —NH_2 group.

$$CH_3-CH=CH-CH_2-\overset{\overset{\displaystyle NH_2}{|}}{CH}-CH_3 \qquad \text{5-amino-2-hexene}$$

The chemistry of amines mirrors the chemistry of ammonia. Amines are weak bases that pick up a proton to form ammonium salts. Trimethylamine, for example, picks up a proton to form the trimethylammonium ion.

$$(CH_3)_3N + H^+ \longrightarrow (CH_3)_3NH^+$$

These salts are more soluble in water than the corresponding amines, and this reaction can be used to dissolve otherwise insoluble amines in aqueous solution.

Amines that can be isolated from plants are known as **alkaloids.** They include poisons such as nicotine, coniine, and strychnine (shown in Figure 24.17). The

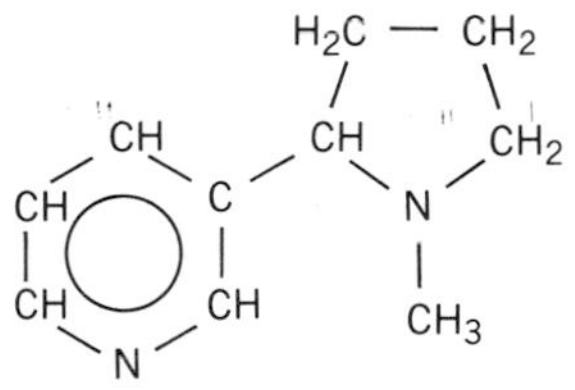

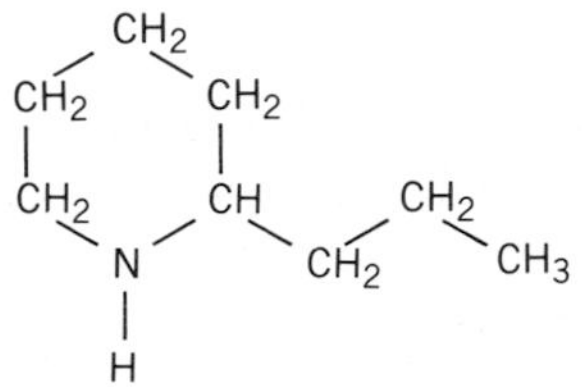

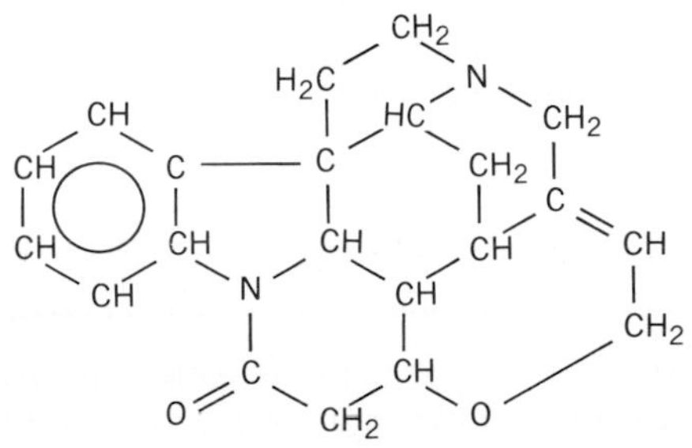

FIG. 24.17 Complex amines isolated from plants are known as alkaloids. Nicotine, an alkaloid from the tobacco plant, has pleasant, invigorating effects when taken in minuscule quantities but is extremely toxic in larger amounts. Coniine is the active ingredient in hemlock, a poison that has been used since the time of Socrates. Strychnine, another toxic alkaloid, has been a popular poison in murder mysteries.

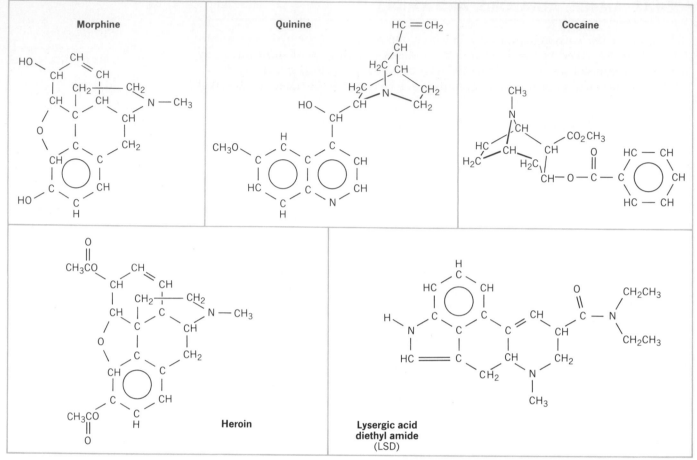

FIG. 24.18 Morphine, quinine, and cocaine are naturally occurring alkaloids. Morphine can be isolated from poppies, quinine is obtained from the bark of the cinchona tree, and cocaine is obtained from coca leaves. Heroin is a synthetic derivative of morphine that is both more powerful and more dangerous. LSD is a synthetic derivative of an alkaloid that can be isolated from a mold that grows on rye; it was first synthesized in a search for new analgesics.

alkaloids also include a number of drugs, such as morphine, quinine, and cocaine, that can be isolated from plants, as well as synthetic analogs such as heroin and LSD (see Figure 24.18).

Carboxylic acids react with alcohols to form esters. They can also react with amines to form amides. When the acid reacts with ammonia, the product of this reaction has the generic formula $RCONH_2$.

$$CH_3-\overset{O}{\underset{\|}{C}}-OH + NH_3 \longrightarrow CH_3-\overset{O}{\underset{\|}{C}}-NH_2 + H_2O$$

NH_3 acts as a nucleophile in this reaction, attacking the δ^+ end of the C=O double bond to form a covalent carbon—nitrogen bond, as shown in Figure 24.19. If an H^+ ion migrates from the ammonium ion ($-NH_3^+$) end of this molecule to the alkoxide ($-O^-$) end, the net result is the addition of an NH_3 molecule across the C=O double bond. This intermediate can then split out a water molecule to form the amide.

24.18 GRIGNARD REAGENTS

Our discussion so far has allowed us to build a small repertoire of reactions that can be used to convert one functional group to another. For example, we have briefly

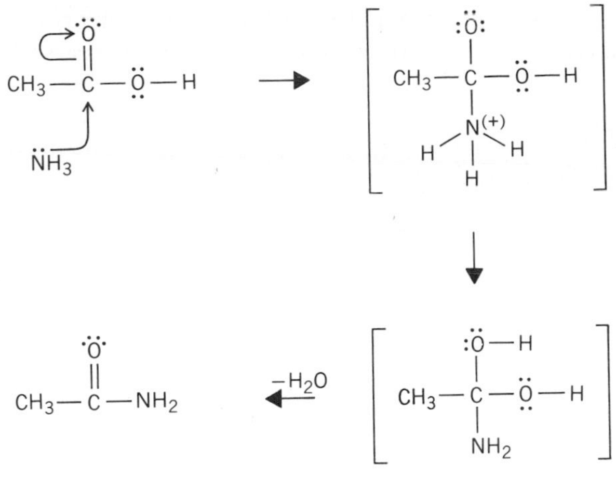

FIG. 24.19 The reaction between ammonia and a carboxylic acid to form an amide is very similar to the reaction between a carboxylic acid and an alcohol to form an ester. Ammonia acts as a Lewis base, or nucleophile, which attacks the δ^+ end of the C=O double bond to form a carbon—nitrogen covalent bond. Migration of an H^+ ion from the positive end of the product of this reaction to the negative end of the molecule, followed by the elimination of water, results in the formation of the amide.

discussed converting alkenes or alkynes to alkanes, alkanes to alkyl halides, alkyl halides to alcohols, alcohols to either aldehydes or ketones, and aldehydes to carboxylic acids. We have also shown how carboxylic acids are converted into esters and amides. We have yet to encounter, however, a reaction that addresses a very basic question: How do we make C—C bonds? One answer resulted from the work that François Auguste Victor Grignard started as part of his Ph.D. research at the turn of the century.

Grignard noted that alkyl halides react with magnesium metal in diethyl ether (Et$_2$O) to form compounds that contain a metal—carbon bond. Methyl bromide, for example, forms methylmagnesium bromide.

$$CH_3Br + Mg \xrightarrow{Et_2O} CH_3MgBr$$

Carbon is more electronegative than magnesium ($\Delta EN = 1.24$), and the metal—carbon bond has a significant amount of ionic character. **Grignard reagents** such as CH$_3$MgBr are best thought of as hybrids of ionic and covalent Lewis structures.

$$\left[CH_3-Mg-\ddot{B}r: \longleftrightarrow [CH_3:^-][Mg^{2+}][:\ddot{B}r:^-] \right]$$

Grignard reagents are therefore our first source of carbanions (literally, "anions of carbon"). The Lewis structure of the CH$_3^-$ ion suggests that these carbanions are Lewis bases, or electron-pair donors.

Grignard reagents such as methylmagnesium bromide are therefore sources of a nucleophile that can attack the δ^+ end of the C=O double bond in aldehydes and ketones.

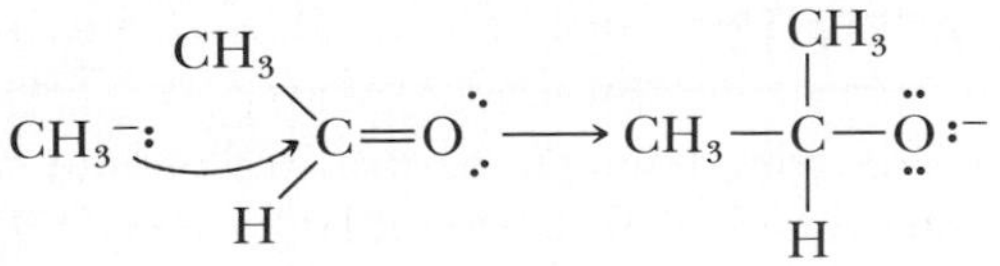

If we treat this intermediate with water, we get an alcohol.

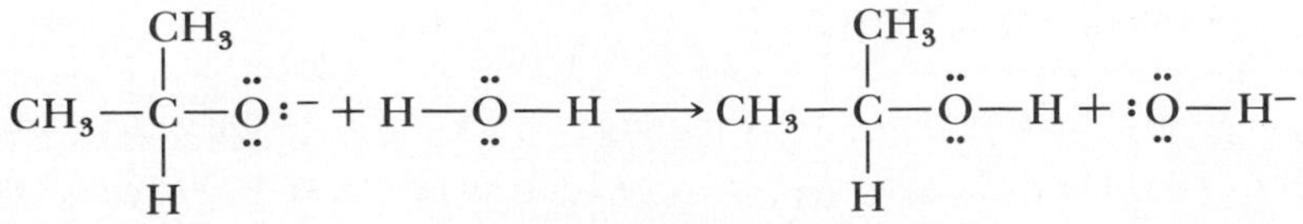

The net effect of these two steps is the formation of a carbon—carbon bond.

By reacting a Grignard reagent with formaldehyde we can add a single carbon atom to form a primary alcohol.

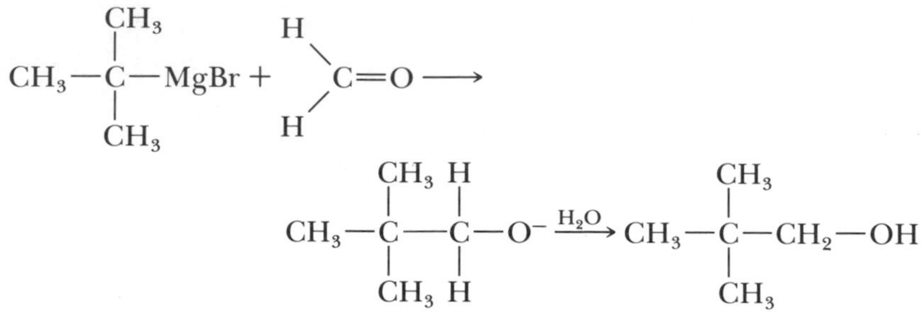

$$CH_3-\underset{\underset{CH_3}{|}}{\overset{\overset{CH_3}{|}}{C}}-MgBr + \overset{H}{\underset{H}{}}C=O \longrightarrow$$

$$CH_3-\underset{\underset{CH_3}{|}}{\overset{\overset{CH_3}{|}}{C}}-\underset{\underset{H}{|}}{\overset{\overset{H}{|}}{C}}-O^- \xrightarrow{H_2O} CH_3-\underset{\underset{CH_3}{|}}{\overset{\overset{CH_3}{|}}{C}}-CH_2-OH$$

This alcohol can then be oxidized to the corresponding aldehyde.

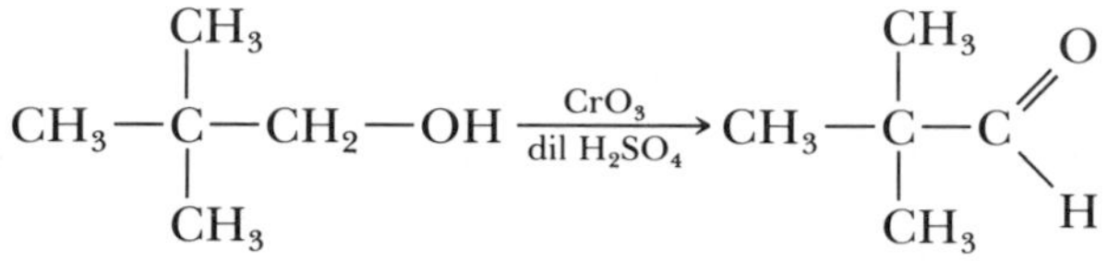

$$CH_3-\underset{\underset{CH_3}{|}}{\overset{\overset{CH_3}{|}}{C}}-CH_2-OH \xrightarrow[\text{dil } H_2SO_4]{CrO_3} CH_3-\underset{\underset{CH_3}{|}}{\overset{\overset{CH_3}{|}}{C}}-C\overset{O}{\underset{H}{}}$$

The Grignard reagent therefore provides us with a way of performing the following transformation.

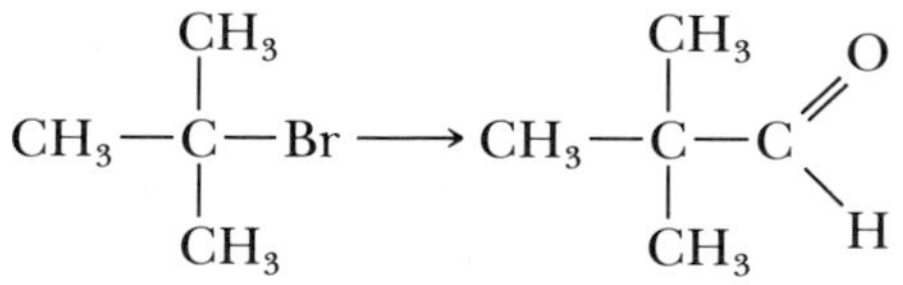

$$CH_3-\underset{\underset{CH_3}{|}}{\overset{\overset{CH_3}{|}}{C}}-Br \longrightarrow CH_3-\underset{\underset{CH_3}{|}}{\overset{\overset{CH_3}{|}}{C}}-C\overset{O}{\underset{H}{}}$$

A single carbon atom can also be added if the Grignard reagent is allowed to react with CO_2 to form a carboxylic acid.

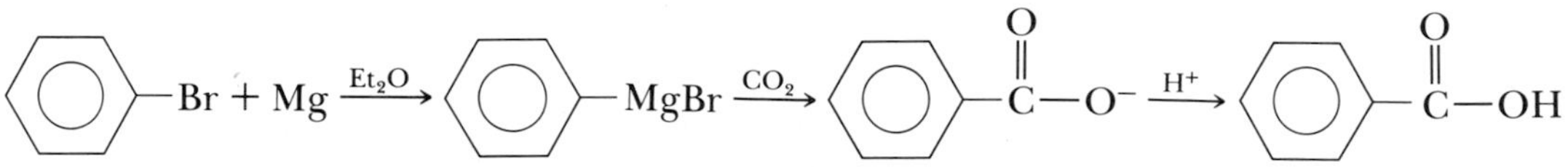

$$\text{C}_6\text{H}_5-Br + Mg \xrightarrow{Et_2O} \text{C}_6\text{H}_5-MgBr \xrightarrow{CO_2} \text{C}_6\text{H}_5-\overset{O}{\underset{}{C}}-O^- \xrightarrow{H^+} \text{C}_6\text{H}_5-\overset{O}{\underset{}{C}}-OH$$

The most important aspect of the chemistry of Grignard reagents is the ease with which this reaction allows us to couple alkyl chains. Isopropylmagnesium bromide, for example, can be used to graft an isopropyl group onto the hydrocarbon chain of an appropriate ketone, as shown in Figure 24.20.

24.19 OPTICAL ACTIVITY

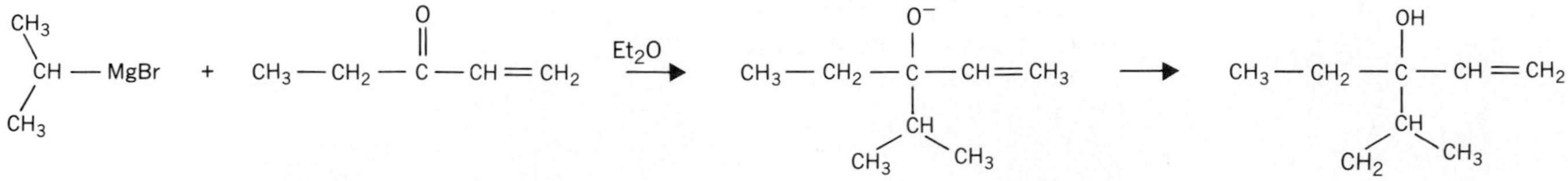

FIG. 24.20 Grignard reagents provide a powerful way of introducing alkyl groups onto a hydrocarbon chain.

Sections 24.2 and 24.3 described some structural and geometric isomers of organic compounds. Structural isomers, such as ethanol (CH_3CH_2OH) and dimethyl ether (CH_3OCH_3), have the same formula but differ in the sequence in which their atoms

$$\underset{\underset{CH_3}{|}}{\overset{\overset{CH_3}{|}}{CH}}-MgBr + CH_3-CH_2-\overset{O}{\underset{}{C}}-CH=CH_2 \xrightarrow{Et_2O} CH_3-CH_2-\underset{\underset{CH}{|}}{\overset{\overset{O^-}{|}}{C}}-CH=CH_3 \longrightarrow CH_3-CH_2-\underset{\underset{CH}{|}}{\overset{\overset{OH}{|}}{C}}-CH=CH_2$$

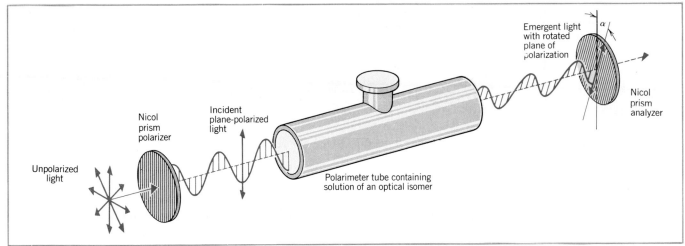

are joined. Geometric isomers, such as *cis-* and *trans-*2-butene, have similar structures but differ in the ways in which their substituents are arranged around a particular functional group. Structural isomers have very different chemical and physical properties. Ethanol reacts with sodium, but dimethyl ether does not. Ethanol boils at 78.4°C, whereas dimethyl ether boils at −23.7°C. Geometric isomers usually have similar chemical properties but different physical properties. *Cis-*2-butene, for example, freezes at −138.91°C, and *trans-*2-butene freezes at −105.55°C.

A third category of isomers includes **optical isomers,** or **stereoisomers.** Light consists of magnetic and electric fields that oscillate equally in all directions perpendicular to the path of the light ray, as shown in Figure 24.21. When light is passed through a polarizer, however, the oscillations are confined to a single plane. In 1813, Jean Baptiste Biot noticed that the plane in which polarized light oscillates was rotated either to the right or the left when this light was passed through single crystals of quartz or aqueous solutions of tartaric acid or sugars.

Compounds that can rotate plane-polarized light are said to be **optically active.** Those that rotate the plane clockwise (to the right) are said to be **dextrorotatory** (from the Latin *dexter,* "right"). Those that rotate the plane counterclockwise (to the left) are called **levorotatory** (from the Latin *laevus,* "left").

In 1848, Louis Pasteur observed that sodium ammonium tartrate forms two different kinds of crystals, which are mirror images of each other, much as the right hand is a mirror image of the left hand (see Figure 24.22). By separating one type of crystal from the other with a pair of forceps he was able to prepare two samples of this compound. One sample was dextrorotatory when dissolved in aqueous solution, and the other was levorotatory. Since the optical activity remained after the compound had been dissolved in water, it could not be the result of macroscopic

FIG. 24.21 Light consists of magnetic and electric fields that oscillate equally in all directions perpendicular to the path of the light ray. When light is passed through a polarizer, such as the lenses of Polaroid sunglasses, these oscillations are confined to a single plane. Optically active compounds rotate the plane of this plane-polarized light either to the left or to the right.

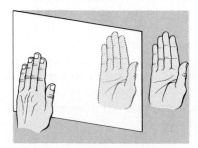

FIG. 24.22 The right hand is the mirror image of the left hand. Neither hand is superimposable on the other. Molecules that have this quality are said to be *chiral,* or handed, and they exist as pairs of optical isomers, or stereoisomers.

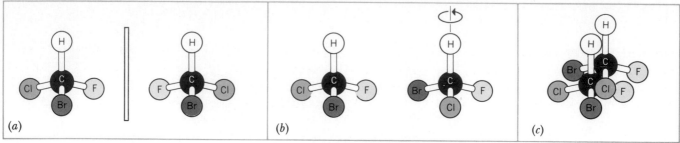

FIG. 24.23 (a) All molecules have a mirror image. This figure shows the structure of the CHFClBr molecule and its mirror image. (b) To determine whether a molecule is optically active, we have to decide whether the mirror image can be superimposed on the original molecule. We could start by rotating the mirror image 180° around the C—H bond. (c) Note that these structures are not the same—they are not superimposable. This compound is therefore chiral. One of these stereoisomers will rotate the plane of plane-polarized light to the right, and the other will rotate it to the left.

properties of the crystals. He therefore suggested that there must be some asymmetry to the structure of this molecule that allowed it to exist in two forms.

In 1874, Jacobus Henricus van't Hoff and Joseph Achille Le Bel recognized that a tetrahedral carbon atom with four different substituents could exist in two forms that were mirror images of each other. According to van't Hoff and Le Bel, CHFClBr should be optically active, because it contains four different substituents on a tetrahedral carbon atom. Figure 24.23a shows one possible arrangement of these substituents and the mirror image of this structure. If we rotate one of these molecules by 180° around the C—H bond we get the structures shown in Figure 24.23b. As Figure 24.23c shows, these structures are different—they can't be superimposed on each other.

Molecules such as CHFClBr are often described as being **_chiral_** (from the Greek _cheir_, "hand"). An object is chiral if it is not superimposable on its mirror image. Hands are chiral—there is no way to superimpose one hand on the other. Gloves are also chiral—right-hand gloves do not fit on the left hand and vice versa. Feet and shoes are both chiral, but socks are not.

Compounds that are chiral are optically active. Chirality is a property of a molecule that results from its structure. Optical activity is a macroscopic property of a collection of these molecules that arises from the way they interact with light. To decide whether a compound should be optically active, we look for centers of chirality—carbon atoms that have four different substituents.

SUMMARY

Organic chemistry is best defined as the chemistry of compounds that contain both carbon and hydrogen. The chemistry of these compounds is organized around functional groups, such as alkanes, alkenes, alkynes, alkyl halides, alcohols, ethers, aldehydes, ketones, carboxylic acids, esters, and amides. Alkanes contain only C—C and C—H single bonds—they are often described as saturated hydrocarbons because they contain as many hydrogen atoms as possible. Unsaturated hydrocarbons include alkenes (which contain C=C double bonds), alkynes (with C≡C triple bonds), and aromatic compounds (which contain one or more benzene ring).

Structural isomers, such as 1-butene and 2-butene, contain the same functional groups but differ in the location of these groups. Geometric isomers, such as _cis_-2-butene and _trans_-2-butene, differ in the geometries around the functional groups. When a compound can't be superimposed on its mirror image, it is said to be chiral, or optically active; it exists as a pair of stereoisomers.

Organic reactions can be divided into two categories, metathesis reactions (such as nucleophilic substitution reactions) and oxidation-reduction reactions. By definition, oxidation-reduction reactions involve a change in the oxidation number of one or more atoms. Since it is difficult to assign oxidation numbers to the carbon atoms in complex organic compounds, it is useful to remember the general rule that adding hydrogen atoms to a compound corresponds to the reduction of the carbon atoms, whereas removing hydrogen atoms from a compound involves oxidation.

PROBLEMS

The Special Role of Carbon

24-1 Describe how the properties of carbon explain the fact that over 95% of the compounds chemists have isolated from natural sources or synthesized in their labs are organic compounds.

24-2 Explain why carbon forms covalent bonds with so many other elements.

24-3 Explain why carbon forms relatively strong double bonds, not only with itself, but with other nonmetals such as nitrogen, oxygen, phosphorus, and sulfur.

The Saturated Hydrocarbons: Alkanes and Cycloalkanes

24-4 Define the following terms: *alkane, hydrocarbon, saturated hydrocarbon, straight-chain hydrocarbon, branched hydrocarbon,* and *cyclic hydrocarbon.* Give an example of each.

24-5 Explain why alkanes have the generic formula C_nH_{2n+2} and write the generic formulas for cycloalkanes, alkenes, and alkynes.

24-6 Describe the difference between *n*-pentane, isopentane, and neopentane. Classify these compounds as either structural isomers, geometric isomers, or optical isomers.

24-7 Predict the number of structural isomers of heptane, C_7H_{16}.

24-8 Explain why the melting point and boiling point of hydrocarbons seem to increase with their molecular weights.

24-9 Write the molecular formula for the saturated hydrocarbon that has the following carbon skeleton, and name this compound.

$$\begin{array}{ccccc} C & - & C & - & C & - & C \\ & & | & & | \\ & & C & & C \\ & & & & | \\ & & & & C \end{array}$$

24-10 Write the molecular formula for the saturated hydrocarbon that has the following carbon skeleton, and name this compound.

$$\begin{array}{c} C \quad\quad C-C \\ | \quad\quad\quad | \\ C \quad\quad\quad C \\ | \quad\quad\quad | \\ C-C-C-C-C \\ | \\ C \\ | \\ C \end{array}$$

24-11 Write the molecular formula for the compound that has the following skeleton structure, and name this compound.

$$CH_3 \quad Cl$$

The Unsaturated Hydrocarbons: Alkenes and Alkynes

24-12 Define the following terms: *alkene, alkyne, olefin,* and *unsaturated hydrocarbon.*

24-13 Draw the structures of the alkenes that have the formula C_6H_{12}, and name these compounds.

24-14 Draw the structures of the alkynes that have the formula C_5H_8, and name these compounds.

24-15 Use the VSEPR theory to predict the shape of the ethylene (C_2H_4) molecule.

24-16 Explain the difference between the structural and geometric isomers of butene (C_4H_8).

24-17 Explain why alkynes can have structural isomers but not geometric isomers.

24-18 Describe the hybridization of each of the carbon atoms in the following compound.

$$CH_3-CH=CH-CH_2-C\equiv CH$$

24-19 Which of the following compounds does not have the same molecular formula as the others?

(a) cyclopentane (b) methylcyclobutane (c) 1,1-dimethylcyclopropane (d) 1-pentene (e) pentane

24-20 Which of the following compounds have *cis* and *trans* isomers?

(a) $CHCl_3$ (b) $F_2C=CF_2$ (c) $Cl_2C=CH_2CH_3$
(d) $FClC=CFCl$ (e) $H_2C=CHF$

24-21 Which of the following compounds have *cis* and *trans* isomers?

(a) 1-pentene (b) 2-pentene (c) 2-methyl-2-butene
(d) 2-chloro-2-butene (e) 2-pentyne

24-22 Write the molecular formula for the compound that has the following skeleton structure, and name this compound.

$$\begin{array}{c} C-C=C-C-C \\ \quad | \quad | \\ \quad C \quad C \end{array}$$

24-23 Write the molecular formula for the compound that has the following skeleton structure, and name this compound.

$$\begin{array}{c} C-C-C \quad C \\ \quad\quad | \quad | \\ \quad\quad C=C \\ \quad\quad | \\ \quad\quad C-C \end{array}$$

24-24 Write the molecular formula for the compound that has the following skeleton structure, and name this compound.

$$\begin{array}{c} C-C-C \\ | \\ C\equiv C-C-C-C \\ \quad\quad\quad | \\ \quad\quad\quad C \end{array}$$

The Reactions of Alkanes, Alkenes, and Alkynes

24-25 Write a balanced equation for the reaction in which isoctane, C_8H_{18}, burns in oxygen to form CO_2 and H_2O vapor.

24-26 Define the term *addition reaction,* and give examples of an addition reaction of an alkene.

24-27 Predict the product of the addition reaction between Br_2 and 3-pentene.

24-28 Predict the product of the addition reaction between water and 2-butene in the presence of sulfuric acid.

24-29 Predict the product of the addition reaction between H_2 and 3-pentyne in the presence of a platinum metal catalyst.

24-30 Which of the following is a product of the reaction between chlorine and ethylene?

(a) CH_3CHCl_2 (b) CH_3CH_2Cl (c) $ClCH_2CH_2Cl$
(d) $Cl_2CHCHCl_2$

24-31 Which of the following does not react with Br_2 dissolved in CCl_4?

(a) pentane (b) 1-pentene (c) 2-pentene (d) 1-pentyne
(e) 2-pentyne

The Aromatic Hydrocarbons and their Derivatives

24-32 Draw the structures of the following aromatic compounds: aniline, anisole, benzene, and benzoic acid.

24-33 Draw the structures of the ortho-, meta-, and para-isomers of bromotoluene.

24-34 TNT is an abbreviation for 2,4,6-trinitrotoluene. Toluene is a derivative of benzene in which a methyl (CH_3-) group is substituted for one of the hydrogen atoms. Trinitrotoluene is a derivative of toluene in which NO_2 groups have replaced three more hydrogen atoms on the benzene ring. Draw the structures of all the possible isomers of trinitrotoluene. Label the isomer that is 2,4,6-trinitrotoluene.

24-35 On an exam, a student described benzene as a mixture of two structures that are rapidly being converted from one to the other. What is wrong with this student's answer?

The Chemistry of Petroleum Products and Coal

24-36 Explain why a mixture of CO and H_2 can be used as a fuel.

24-37 Natural gas, petroleum ether, gasoline, kerosene, and asphalt are all different forms of hydrocarbons that give off energy when burned. Describe how these substances differ. What happens to the boiling points of these mixtures as the average length of the hydrocarbon chain increases?

24-38 Which of the following compounds has the largest octane number?

(a) *n*-butane (b) *n*-pentane (c) *n*-hexane (d) *n*-octane

24-39 Which of the following won't increase the octane number of gasoline?

(a) Increasing the concentration of branched-chain alkanes (b) Increasing the concentration of cycloalkanes (c) Increasing the concentration of aromatic hydrocarbons (d) Increasing the average length of the hydrocarbon chains

24-40 Describe how thermal cracking and catalytic reforming increase the amount of high-octane gasoline that can be obtained from a barrel of oil.

24-41 Coal gas can be obtained when coal is heated in the absence of air. Water gas can be obtained when coal reacts with steam. Describe the difference between these gases, and explain why much more water gas can be extracted from a ton of coal.

24-42 Give examples of compounds that contain each of the following functional groups.

(a) an alcohol (b) an aldehyde (c) an amine (d) an amide (e) an alkyl halide (f) an alkene (g) an alkyne

24-43 Describe the difference between the members of each of the following pairs of functional groups.

(a) An alkyl halide and an alcohol (b) An alcohol and an aldehyde (c) An amine and an amide (d) An ether and an ester (e) An aldehyde and a ketone

24-44 Classify each of the following compounds as an alcohol, an aldehyde, an ether, or a ketone.

(a) $CH_3CH_2\overset{\displaystyle O}{\overset{\|}{C}}H$ (b) CH_3CH_2OH

(c) $CH_3CH_2OCH_2CH_3$ (d) $CH_3CH_2\overset{\displaystyle O}{\overset{\|}{C}}CH_3$

24-45 Classify each of the following compounds as a primary, a secondary, or a tertiary alcohol.

(a) CH_3CH_2OH (b) $CH_3CH_2\overset{\displaystyle CH_3}{\overset{|}{C}}HOH$ (c) $CH_3CH_2\overset{\displaystyle CH_3}{\underset{\displaystyle CH_3}{\overset{|}{\underset{|}{C}}}OH}$

(d) $CH_3CH_2\overset{\displaystyle OH}{\overset{|}{C}}HCH_2CH_3$

24-46 Organic compounds often contain more than one functional group. Identify all of the functional groups in the strychnine molecule shown in Figure 24.17.

24-47 Identify all of the functional groups in the heroin molecule shown in Figure 24.18.

24-48 Identify all of the functional groups in the LSD molecule shown in Figure 24.18.

Alkyl Halides

24-49 Describe the free-radical chain-reaction mechanism involved in the chlorination of methane to form methyl chloride.

Clearly label the chain-initiation, chain-propagation, and chain-termination steps.

24-50 Use Lewis structures to describe the difference between a CH_3^+ ion, a $CH_3 \cdot$ free radical, and a CH_3^- ion.

Alcohols and Ethers

24-51 Describe the differences in the structures of water, methyl alcohol, and dimethyl ether.

24-52 Ethyl alcohol and dimethyl ether have the same chemical formula, C_2H_6O. Explain why one compound reacts rapidly with sodium metal but the other does not.

24-53 Predict the product of the dehydration of ethyl alcohol with sulfuric acid at 130°C.

24-54 Describe the differences among primary alcohols, secondary alcohols, tertiary alcohols, and phenols. Give an example of each.

24-55 Explain why alcohols are Brønsted acids in water.

24-56 Write the structures of the conjugate bases, formed when each of the following alcohols acts as a Brønsted acid.

(a) CH_3CH_2OH (ethyl alcohol) (b) C_6H_5OH (phenol)
(c) $(CH_3)_2CHOH$ (isopropyl alcohol)

24-57 Use the fact that there are no hydrogen bonds between ether molecules to explain why diethyl ether ($C_4H_{10}O$) boils at 34.5°C, whereas 1-butanol ($C_4H_{10}O$) boils at 118°C.

Substitution Reactions

24-58 Define the term *substitution reaction* and describe the difference between unimolecular (S_N1) and bimolecular (S_N2) substitution reactions.

24-59 Write the mechanism of an S_N2 reaction and show why there is no difference between what organic chemists call a *nucleophile* and what inorganic chemists call a *Lewis base*.

Oxidation-Reduction Reactions

24-60 Arrange the following in order of increasing oxidation number of the carbon atom.

(a) C (b) HCHO (c) HCO_2H (d) CO (e) CO_2
(f) CH_4 (g) CH_3OH

24-61 Organic reactions can be divided into two classes: metathesis, or acid-base, reactions and oxidation-reduction reactions. Sort the following reactions into these categories.

(a) $CH_4 + Cl_2 \rightarrow CH_3Cl + HCl$ (b) $CH_3OH \rightarrow HCHO$
(c) $HCHO \rightarrow HCO_2H$ (d) $CH_3OH + HBr \rightarrow CH_3Br + H_2O$ (e)

$$\underset{\text{CH}_3\text{CCH}_3}{\overset{\text{O}}{\parallel}} \rightarrow \underset{\text{CH}_3\text{CHCH}_3}{\overset{\text{OH}}{\vert}}$$

24-62 Organic compounds can be oxidized by removing one or more hydrogen atoms from a carbon atom. Use this rule to label the following transformations as either oxidation or reduction.

(a) $CH_3CH_2OH \rightarrow CH_3CHO$ (b) $CH_3CHO \rightarrow CH_3CO_2H$ (c) $(CH_3)_2C{=}O \rightarrow (CH_3)_2CHOH$
(d) $CH_3CH{=}CHCH_3 \rightarrow CH_3CH_2CH_2CH_3$
(e) $(CH_3)_2CHC{\equiv}CH \rightarrow (CH_3)_2CHCH_2CH_3$

24-63 Which of the following can be oxidized to form an aldehyde?

(a) CH_3CH_2OH (b) $CH_3CHOHCH_3$ (c) CH_3OCH_3
(d) $(CH_3)_2C{=}O$

24-64 Which of the following should be most difficult to oxidize?

(a) CH_3CH_2OH (b) $(CH_3)_2CHOH$ (c) $(CH_3)_3COH$
(d) CH_3CHO

24-65 Which of the following can be prepared by reduction of CH_3CHO?

(a) CH_3CH_3 (b) CH_3CH_2OH (c) CH_3CHO
(d) CH_3CO_2H (e) $CH_3CH_2CO_2H$

24-66 Which of the following can be prepared by oxidation of CH_3CHO?

(a) CH_3CH_3 (b) CH_3CH_2OH (c) CH_3CHO
(d) CH_3CO_2H (e) $CH_3CH_2CO_2H$

24-67 Predict the product of the oxidation of 2-methyl-3-pentanol.

Aldehydes and Ketones

24-68 Compounds that react with the electron-rich δ^- end of a carbonyl group are called electrophiles. Use Lewis structures to show that there is no difference between an electrophile and a Lewis acid.

24-69 At which end of a carbonyl group will the following substances react?

(a) H^+ (b) H^- (c) OH^- (d) NH_3 (e) BF_3

24-70 Explain why mild oxidation of a primary alcohol gives an aldehyde, whereas oxidation of a secondary alcohol gives a ketone.

24-71 Explain why strong oxidizing agents can't be used to convert a primary alcohol to an aldehyde.

Carboxylic Acids, Carboxylate Ions, and Carboxylic Acid Esters

24-72 Explain how a carboxylic acid, a carboxylate ion, and a carboxylic acid ester differ.

24-73 What major differences between carboxylic acids (such as butyric acid, $CH_3CH_2CH_2CO_2H$) and esters (such as ethyl butyrate, $CH_3CH_2CH_2CO_2CH_2CH_3$) help explain why butyric acid gives rise to the odor of rotten meat but ethyl butyrate gives rise to the pleasant odor of pineapple?

24-74 Explain why fats, which are esters of long-chain carboxylic acids, are insoluble in water. Explain what happens during saponification that makes the resulting compounds marginally soluble in water.

Amines, Alkaloids, and Amides

24-75 Classify each of the compounds in Figure 24.17 as either a primary amine, a secondary amine, a tertiary amine, or an amide.

24-76 Classify each of the compounds in Figure 24.18 as a primary amine, a secondary amine, a tertiary amine, and/or an amide.

Grignard Reagents

24-77 Use Lewis structures to decide whether a Grignard reagent is a nucleophile (Lewis base) or an electrophile (Lewis acid). Predict which end of a C=O bond the Grignard reagent will attack.

24-78 If CH_3CH_2MgBr can be thought of as containing the $CH_3CH_2^-$, Mg^{2+}, and Br^- ions, what should be the product of the reaction between this Grignard reagent and dilute aqueous hydrochloric acid?

24-79 Write a step-by-step mechanism for the reaction between CH_3MgBr and $H_2C=O$ to form CH_3CH_2OH.

24-80 Predict the product of the reaction between $(CH_3CH_2)_2CHMgBr$ and $CH_3CH_2COCH_3$.

24-81 If CH_3CH_2OH is used as the only source of carbon, which of the following can be synthesized by a Grignard reaction?

(a) $CH_3CH_2CH_2OH$ (b) CH_3CH_2CHO
(c) $CH_3CH_2OCH_2CH_3$ (d) $CH_3CO_2CH_2CH_3$
(e) $CH_3CH_2CH_2CO_2CH_2CH_3$

24-82 Which of the following reagents is required to synthesize the following compound?

$$CH_3-\overset{\overset{\displaystyle OH}{|}}{CH}-\overset{\overset{\displaystyle |}{CH}}{\underset{\underset{\displaystyle CH_3}{|}}{CH}}-CH_3$$

(a) $(CH_3)_2CHMgBr$ (b) CH_3CHO (c) CH_3MgBr
(d) $CH_3COCH_2CH_3$ (e) CH_3COCH_3 (f) CH_3CH_2MgBr
(g) CH_3COCH_3

Optical Activity

24-83 Describe how structural isomers, geometric isomers, and optical isomers, or stereoisomers, differ.

24-84 Objects that cannot be superimposed on their mirror images are said to be *chiral,* and chiral molecules are optically active. Which of the following molecules are optically active?

(a) CH_4 (b) CH_3Cl (c) $CHCl_3$ (d) $CHFCl_2$ (e) $CHFClBr$

24-85 Which of the following molecules are optically active?

(a) C_2H_4 (b) C_6H_6 (c) $C_6H_4Cl_2$ (d) $CH_3CH_2\overset{\overset{\displaystyle OH}{|}}{CH}CH_3$

Qualitative Organic Analysis

24-86 Describe a way of determining whether a compound is an alkane or an alkene.

24-87 Describe a way of determining whether a compound is a carboxylic acid or an ester.

24-88 Describe a way of determining whether a compound is an alcohol or an ether.

24-89 Describe a way of determining whether a compound is a primary or a tertiary alcohol.

24-90 Describe a way of determining whether a compound is an alcohol or an alkyl halide.

24-91 Describe a way of determining whether a compound is a carboxylic acid or an amide.

24-92 Describe a way of determining whether a compound is an alcohol or an amine.

24-93 Describe a way of determining whether a compound is an aldehyde or a ketone.

POLYMERS: SYNTHETIC AND NATURAL

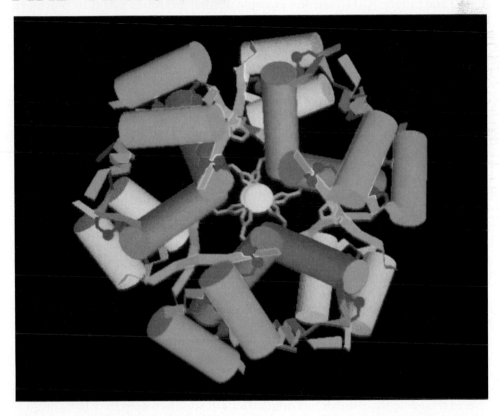

25.1 POLYMERS, PLASTICS, AND ELASTOMERS

In 1833, Jons Jacob Berzelius proposed that compounds with the same formula but different structures should be called *isomers* (literally, "equal parts"). At the same time, he suggested that compounds with the same empirical formula but different molecular weights should be called **polymers** (literally, "many parts"). Ethylene (C_2H_4) and butene (C_4H_8) are one example of what Berzelius meant by polymers, acetylene (C_2H_2) and benzene (C_6H_6) are another.

The term *polymer* eventually came to mean compounds that had very large molecular weights, such as cellulose and natural rubber. In 1920, Hermann Staudinger proposed the first explanation of why polymer molecules are so heavy. He discovered that polymers are long chains of relatively simple repeating units. Cellulose, for example, is a polymer of $—C_6H_{10}O_5—$ units, and natural rubber contains large numbers of $—CH_2C(CH_3)=CHCH_2—$ units.

The term *rubber* was first used in 1770 by Joseph Priestley to describe the gum from a South American tree that could erase pencil marks. Natural rubber has only limited uses, however, because it is tacky, strong-smelling, perishable, too soft when warm, and too hard when cold.

Shortly after natural rubber was first brought to England, it was discovered that cloth could be treated with rubber to produce a waterproof fabric, which unfortunately suffered from all the disadvantages of natural rubber. In 1823, Charles Macintosh found that he could overcome these disadvantages by dissolving natural rubber in a petroleum solvent and placing this solution between two pieces of cloth—thereby producing the first "macintoshes."

Nathaniel Hayward was the first to note that rubber loses some of its sticky properties when treated with sulfur. It was Charles Goodyear, however, who in 1839 accidentally dropped a mixture of natural rubber and sulfur onto a hot stove and discovered "vulcanized" rubber, which is stable over a wide range of temperatures and far more durable than natural rubber.

Today, rubber is just one example of a family of polymers known as **elastomers.** To qualify as an elastomer, a polymer must be both flexible and elastic. It must be able to stretch to at least twice its original length and then snap back to its original shape.

Cellulose, a polymer of $—C_6H_{10}O_5$ units, forms the cell walls of plants. Wood is about 50% cellulose by weight, whereas cotton contains as much as 90% cellulose. Cellulose has been used for centuries to make paper from wood pulp and cloth from cotton. In the last hundred years, it has also served as the starting material for the synthesis of the first plastics and the first synthetic fibers.

Each $—C_6H_{10}O_5—$ repeating unit in cellulose contains three $—OH$ groups that can react with nitric acid to form nitrate esters. In 1869, John Wesley Hyatt found that mixtures of cellulose nitrate and camphor dissolve in alcohol to produce a plastic substance he named celluloid.

> **Plastics** are polymers that flow and can therefore take on complex shapes when heated, molded, or milled.

Cellulose nitrate, or celluloid, was first used as a substitute for ivory in the manufacture of a variety of items ranging from combs to billiard balls. Cellulose nitrate is extremely flammable, however (see Figure 25.1) and it has been replaced by other plastics for almost all uses except the manufacture of ping-pong balls. No other plastic has been found that has quite the same bounce as celluloid.

Cellulose was also the source of the first synthetic fiber—rayon. Cellulose from

FIG. 25.1 The cellulose nitrate that is still used today for ping-pong balls is flammable.

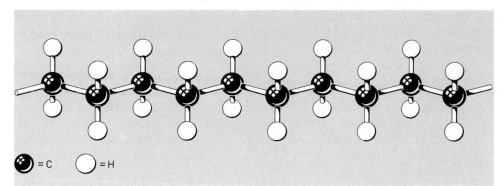

FIG. 25.2 Each carbon atom in a linear polymer has a tetrahedral geometry.

= C   = H

wood pulp can't be used to make fibers because it contains too many impurities. It can be purified, however, by treatment with sodium hydroxide and then carbon disulfide (CS_2) to produce a viscous solution. When this solution is forced through tiny holes in a nozzle into an acid bath, the cellulose fiber known as rayon is regenerated. A similar process is used to make a thin film of regenerated cellulose known as cellophane.

25.2 DEFINITIONS OF TERMS

LINEAR, BRANCHED, AND CROSS-LINKED POLYMERS

The term *polymer* is used in this chapter to describe compounds with relatively large molecular weights that consist of molecules formed by the linking together of many small *monomers.* The simplest example of such a polymer is polyethylene, which is formed by the polymerization of ethylene molecules.

$$CH_2 = CH_2 \longrightarrow [... - CH_2 - CH_2 - CH_2 - CH_2 - CH_2 - CH_2 - ...]$$

ethylene   polyethylene

Polyethylene is often called a *linear,* or *straight-chain, polymer* because it consists of a long string of carbon—carbon bonds. This term is somewhat misleading, however, because the geometry around each carbon atom is tetrahedral and the chain is not quite linear, as shown in Figure 25.2. Furthermore, as the chain grows, it folds back on itself in a random fashion to form the structure shown in Figure 25.3.

Polymers with branches at irregular intervals along the polymer chain are called *branched polymers* (see Figure 25.4). The branches make it more difficult for these polymer molecules to pack in a regular array and therefore decrease the crystallinity of the polymer.

Cross-linked polymers contain branches that connect polymer chains, as shown in Figure 25.5. At first, adding cross-links between polymer chains makes the polymer more elastic. The vulcanization of rubber, for example, results from the introduction of short chains of sulfur atoms that link the polymer chains in the natural rubber. When the number of cross-links is relatively large, however, the polymer becomes more rigid.

Linear and branched polymers are found in a class of materials known as *thermoplastics.* These materials flow when heated and can be molded into a variety of shapes, which they retain when they cool. Lightly cross-linked polymers are found in elastomers — materials that return to their original shapes after they have been

Linear polymer

FIG. 25.3 Linear, or straight-chain, polymers can fold back upon themselves in a random fashion.

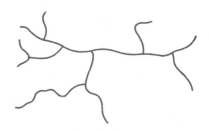

Branched polymer

FIG. 25.4 Branched polymers contain short side chains that extend from the main backbone of the polymer.

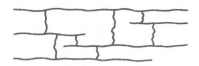

Cross-linked polymer

FIG. 25.5 Cross-linked polymers have branches that connect polymer chains.

stretched. Heavy cross-linking produces materials known as ***thermoset plastics.*** Once the cross-links form, these polymers take on a shape that cannot be changed without destruction of the plastic.

Exercise 25.1

Polyethylene can be obtained in two different forms. High-density polyethylene (0.94 g/cm³) is a linear polymer. Low-density polyethylene (0.92 g/cm³) is a branched polymer with short side-chains on up to 3% of the atoms along the polymer chain. Explain how the structure of these polymers gives rise to the difference in their densities.

Solution

Linear polymers are more regular than branched polymers. This means that linear polymers can pack more tightly, with less wasted space. As a result, linear polymers are slightly more dense than branched polymers.

HOMOPOLYMERS AND COPOLYMERS

Polyethylene is an example of a ***homopolymer,*** which is formed by the polymerization of a single monomer. Other polymers, such as synthetic rubber, are ***copolymers*** formed by polymerizing two or more different monomers. Ethylene ($CH_2=CH_2$) and propylene ($CH_2=CH-CH_3$) can be copolymerized, for example, to produce a polymer that has two kinds of repeating units.

$$CH_2=CH_2 + CH_2=CH-CH_3 \longrightarrow [...(-CH_2CH_2-)_x(-CH_2-\overset{\overset{\displaystyle CH_3}{|}}{CH}-)_y...]$$

Copolymers are classified on the basis of the way the monomers are arranged along the polymer chain, as shown in Figure 25.6. Random copolymers contain repeating units arranged in a purely random fashion. Regular copolymers contain

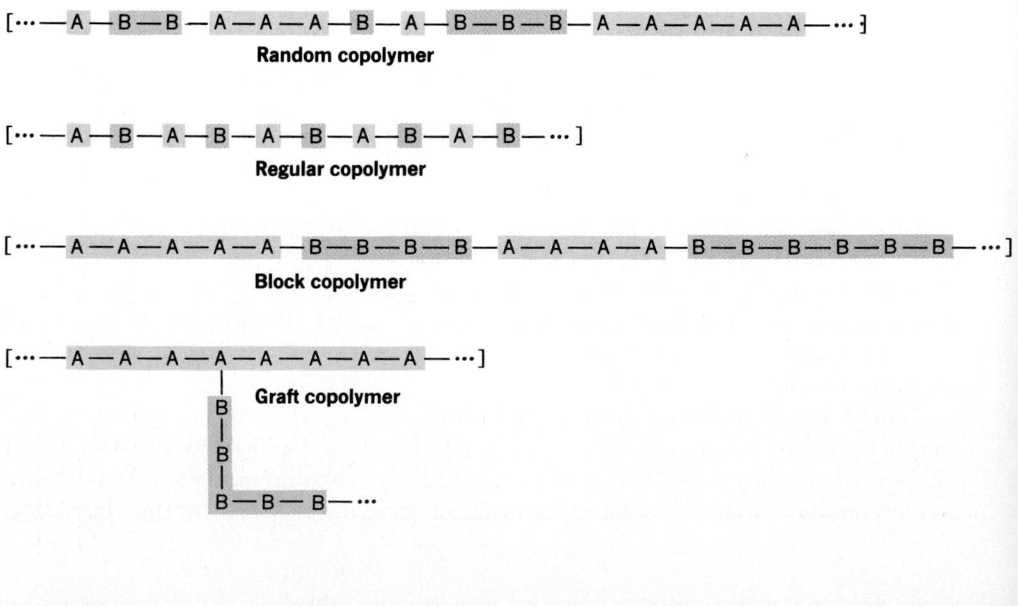

FIG. 25.6 Random, regular, block, and graft copolymers differ in how the monomers that form the polymer chain alternate.

[···—A—B—B—A—A—A—B—A—B—B—B—A—A—A—A—A—···]
Random copolymer

[···—A—B—A—B—A—B—A—B—A—B—···]
Regular copolymer

[···—A—A—A—A—A—B—B—B—B—A—A—A—A—B—B—B—B—B—B—···]
Block copolymer

[···—A—A—A—A—A—A—A—A—···]
Graft copolymer
B
B
B—B—B—···

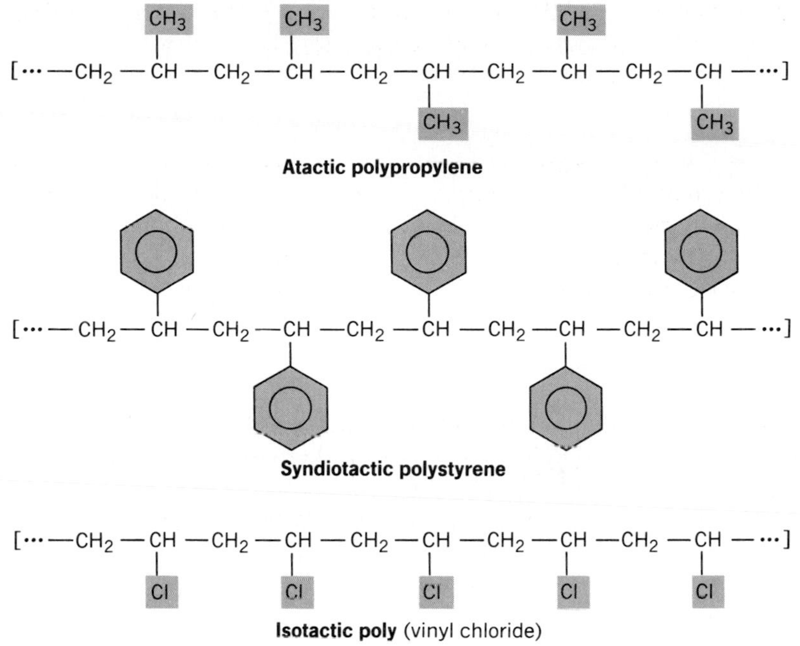

FIG. 25.7 Polymers that contain side chains can be classified as atactic, syndiotactic, or isotactic on the basis of the way the side chains are arranged with respect to the polymer backbone.

a sequence of regularly alternating repeating units. The repeating units in block copolymers occur in blocks of different lengths. Graft copolymers have a chain of one repeating unit grafted onto the backbone of another.

TACTICITY

Some monomers, such as propylene (CH_2=CH—CH_3), form polymers with regular substituents, or side chains, on the polymer chain. These polymers possess a property known as **tacticity** (from the Latin *tacticus*, "fit for arranging"). Tacticity results from the different ways in which these substituents can be arranged on the polymer backbone.

If the substituents are arranged in an irregular, random fashion, the polymer is **atactic** (literally, "no arrangement"). An atactic polypropylene, for example, might have the structure shown in Figure 25.7. If the substituents alternate regularly from one side of the chain to the other, the polymer is **syndiotactic.** When all of the substituents are on the same side of the chain, the polymer is **isotactic** (literally, "the same arrangement").

Exercise 25.2

Atactic polypropylene is a soft, rubbery material with no commercial value. The isotactic polymer is a rigid substance with an excellent resistance to mechanical stress. Explain the difference between these two forms of polypropylene.

Solution

Isotactic polypropylene is easier to pack in a regular fashion than the atactic form of this polymer. The net result is an increase in the crystallinity of the solid, which in turn makes the solid more rigid.

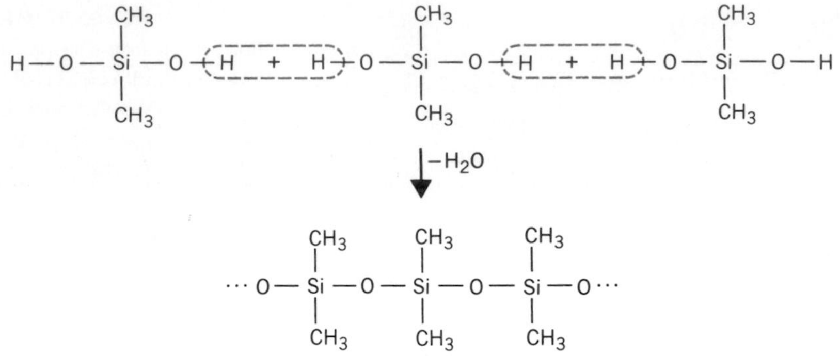

FIG. 25.8 A silicone is an example of a condensation polymer — each time a bond is formed between the monomers that make up the polymer chain, a molecule of water is eliminated.

Silicone

ADDITION VERSUS CONDENSATION POLYMERS

Polyethylene, polypropylene, and poly (vinyl chloride) are examples of **addition polymers,** which are formed by adding monomers to a growing polymer chain. The repeating unit in addition polymers has the same formula as the monomer from which the polymer is formed.

$$CH_2{=}CH_2 \longrightarrow [...(-CH_2-CH_2-)_n ...] \text{ polyethylene}$$
$$CH_2{=}CHCH_3 \longrightarrow [...(-CH_2-\underset{\underset{CH_3}{|}}{CH}-)_n ...] \text{ polypropylene}$$
$$CH_2{=}CHCl \longrightarrow [...(-CH_2-\underset{\underset{Cl}{|}}{CH}-)_n ...] \text{ poly(vinyl chloride)}$$

To *condense* means to make something more dense, or compact — to express an idea in fewer words, for example. A polymer formed when a small molecule is condensed out of two reactants is called a **condensation polymer.** $(CH_3)_2Si(OH)_2$, for example, rapidly polymerizes to form a condensation polymer known as a silicone, as shown in Figure 25.8. Note that the repeating unit in a condensation polymer is smaller than the monomer from which it is made.

Exercise 25.3

Classify the products of the following polymerization reactions as either addition or condensation polymers.

(a) Poly(methyl methacrylate), sold as Lucite or Plexiglass

$$CH_2{=}\underset{\underset{CH_3}{|}}{\overset{\overset{CH_3}{|}}{C}}-CO_2CH_3 \longrightarrow [...(-CH_2-\underset{\underset{CO_2CH_3}{|}}{\overset{\overset{CH_3}{|}}{C}}-)_n ...]$$

(b) Nylon 6

$$H_2N-(CH_2)_5-CO_2H \longrightarrow [...(-HN-(CH_2)_5-\overset{\overset{O}{\|}}{C}-)_n ...]$$

Solution

(a) Poly(methyl methacrylate) is an example of an addition polymer. Every atom in the monomer ends up in the repeating unit of the polymer.

(b) Nylon 6 is an example of a condensation polymer. A molecule of water condenses from this polymer each time a new repeating unit is added.

25.3 ELASTOMERS

Elastomers are polymers with the properties of rubber — they are highly flexible and elastic. To be elastic, a polymer needs the following properties.

1. It must contain long, flexible molecules that are coiled in the natural state and can be stretched without breaking, as shown in Figure 25.9.

2. It must contain a few cross-links between polymer chains so that one molecule does not slip past another when stretched.

3. It cannot contain too many cross-links, however, or else it would be too rigid to stretch.

4. The intermolecular force of attraction (see Section 12.2) between chains must be relatively small, so that the polymer can curl back into its coiled shape after it has been stretched.

We can understand these requirements by taking a closer look at the chemistry of natural rubber, which is a polymer of a C_5H_8 hydrocarbon known as isoprene.

$$CH_2{=}\overset{\overset{\textstyle CH_3}{|}}{C}{-}CH{=}CH_2 \longrightarrow [...({-}CH_2{-}\overset{\overset{\textstyle CH_3}{|}}{C}{=}CH{-}CH_2{-})_n ...]$$

isoprene   natural rubber

The double bonds in natural rubber are all in the *cis* form, as shown in Figure 25.10. This structure gives rise to long, flexible molecules. The force of attraction between polymer chains is relatively small, so the polymer can curl back into its original shape after the molecules have been oriented by stretching. By treating natural rubber with sulfur it is possible to induce a small number of cross-links to form between these polymer chains. This greatly increases the elastic properties of the rubber, as shown in Figure 25.11.

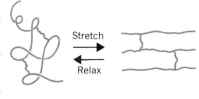

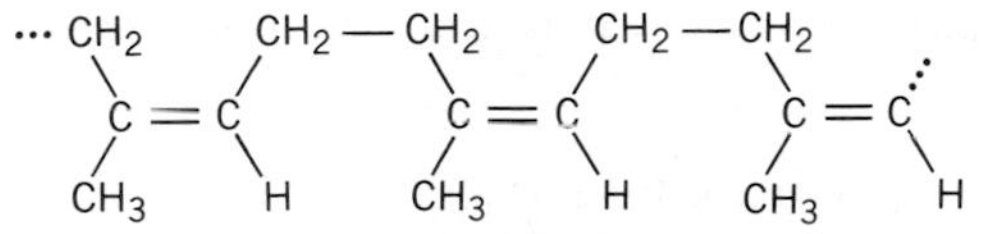

FIG. 25.9 The polymer chains become oriented when an elastomer is stretched. The presence of a few cross-links between polymer chains keeps the chains from slipping past each other when this happens. If the force of attraction between the polymer chains is relatively small, they will coil back into a random shape when the tension is released.

FIG. 25.10 Natural rubber is an addition polymer formed from the *cis* isomer of isoprene.

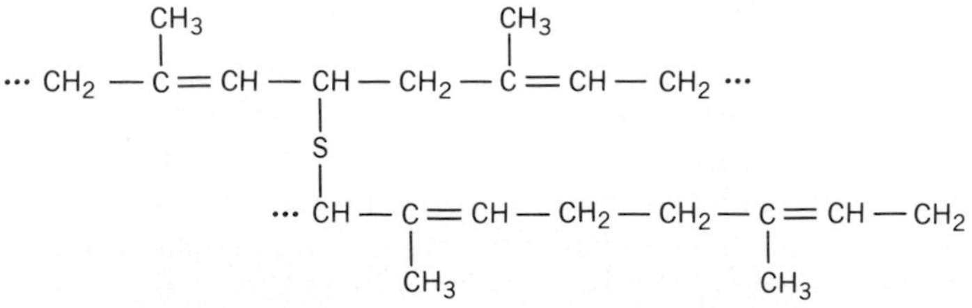

cis-1, 4-Polyisoprene (natural rubber)

FIG. 25.11 The elastic properties of natural rubber can be greatly increased when this substance is treated with a little sulfur, which forms a small number of cross-links between the polymer chains.

At first glance, it might seem easy to make synthetic rubber. All we have to do is synthesize isoprene and then find a suitable catalyst that can polymerize this monomer. The task is made more difficult by the fact that the *cis* isomer of isoprene rearranges into the *trans* isomer during polymerization and the fact that the *trans* isomer of polyisoprene, which is known as gutta percha, isn't elastic. It is therefore important to control the geometry around the $C=C$ double bond during polymerization to make sure that as few of these bonds as possible are converted to the *trans* geometry. Until recently, this wasn't possible, and other approaches to making synthetic rubber were necessary.

One solution to the problem involved polymerizing derivatives of butadiene, because the geometry around the $C=C$ double bond in these polymers is not as important. This was done when the first major synthetic rubber, neoprene, was synthesized from chloroprene, or 2-chloro-1,3-butadiene.

$$CH_2=\overset{\overset{\displaystyle Cl}{\displaystyle |}}{C}-CH=CH_2 \longrightarrow [...(-CH_2-\overset{\overset{\displaystyle Cl}{\displaystyle |}}{C}=CH-CH_2-)_n...]$$

chloroprene   polychloroprene, or neoprene

Another approach involves polymerizing isobutylene to form butyl rubber.

$$CH_2=C\underset{\diagdown CH_3}{\overset{\diagup CH_3}{}} \longrightarrow [...(-CH_2-\overset{\overset{\displaystyle CH_3}{\displaystyle |}}{\underset{\underset{\displaystyle CH_3}{\displaystyle |}}{C}}-)_n...]$$

isobutylene   polyisobutylene, or butyl rubber

The most important of the synthetic rubbers, however, is a copolymer of 75% butadiene and 25% styrene known as styrene-butadiene rubber (SBR). Roughly 40% of the rubber used in the world today is SBR; another 35% is natural rubber that has been treated with sulfur.

Recent advances in the design of catalysts for addition polymerization reactions have made it possible to polymerize isoprene to form a synthetic rubber that is virtually identical to natural rubber. The cost of isoprene, however, is large enough to make synthetic polyisoprene more expensive than natural rubber. As a result, less than 3% of the total demand for rubber is now met by this synthetic form of natural rubber.

25.4 FREE-RADICAL POLYMERIZATION REACTIONS

It isn't difficult to form addition polymers from monomers containing $C=C$ double bonds — many of these compounds polymerize spontaneously during storage unless polymerization is actively inhibited. One of the reasons why thermally cracked gasoline was not as good as straight-run gasoline in the early years of the petroleum industry, for example, was the tendency of the alkenes in this gasoline to polymerize.

Many addition polymers are made by a free-radical polymerization reaction. We have defined a free radical as a neutral atom or molecule with one or more unpaired electrons and suggested that free-radical reactions proceed by a chain-reaction mechanism that contains three steps: chain initiation, chain propagation, and chain termination.

CHAIN INITIATION

A source of free radicals is needed to initiate the chain reaction. Benzoyl peroxide is an example of a free-radical initiator that decomposes on heating or in the presence of ultraviolet radiation to form free radicals.

CHAIN PROPAGATION

The free radical ($R\cdot$) produced in the chain-initiation step adds to an alkene ($C\!=\!C$) to form a new free radical, which then reacts with additional monomers in a chain reaction.

$$R\cdot + H_2C\!=\!CH_2 \longrightarrow RCH_2CH_2\cdot$$
$$RCH_2CH_2\cdot + CH_2\!=\!CH_2 \longrightarrow RCH_2CH_2CH_2CH_2\cdot$$
$$RCH_2CH_2CH_2CH_2\cdot + CH_2\!=\!CH_2 \longrightarrow RCH_2CH_2CH_2CH_2CH_2CH_2\cdot$$

CHAIN TERMINATION

Whenever pairs of radicals combine to form a covalent bond, the chain reactions carried by these radicals are terminated.

$$R(CH_2CH_2)_n CH_2CH_2\cdot + R\cdot \longrightarrow R(CH_2CH_2)_{n+1}R$$
$$R(CH_2CH_2)_n CH_2CH_2\cdot + \cdot CH_2CH_2(CH_2CH_2)_m R \longrightarrow$$
$$R(CH_2CH_2)_n(CH_2CH_2)_2(CH_2CH_2)_m R$$

Exercise 25.4

Which of the following atoms or molecules is a free radical?

(a) CH_3 (b) OH   (c) Cl   (d) NH_3

Solution

The Lewis structures for (a), (b), and (c) all contain an unpaired electron, which means that they are all free radicals. Only the last molecule has a Lewis structure in which all the electrons are paired.

25.5 IONIC AND COORDINATION POLYMERIZATION REACTIONS

Addition polymers can also be made by chain reactions that proceed through ionic intermediates or by chain reactions in which the alkene is coordinated to a transition metal catalyst.

ANIONIC POLYMERIZATION

Anionic polymerization, like free-radical polymerization, takes place by three steps: chain initiation, chain propagation, and chain termination. However, the intermediate that carries the chain reaction is a negatively charged ion. We stated that Grignard reagents are hybrids of ionic and covalent Lewis structures.

$$\left[CH_3-Mg-\ddot{B}r\!: \longleftrightarrow [CH_3^-\!:][Mg^{2+}][:\ddot{B}r\!:^-] \right]$$

Methylmagnesium bromide is therefore a good source of the CH_3^- ion. Other sources of negatively charged carbon atoms, or carbanions, include alkyllithium and trialkylaluminum compounds. These compounds can also be viewed as hybrids of resonance structures that include carbanions.

$$\left[CH_3-Li \longleftrightarrow [CH_3^-][Li^+] \right]$$
methyl lithium

$$\left[CH_3-Al-CH_3 \longleftrightarrow [Al^{3+}][CH_3^-]_3 \right]$$
$$CH_3$$
trimethyl aluminum

The CH_3^- ion from one of these metal alkyls can attack an alkene to form a carbon—carbon bond, as shown in Figure 25.12. The product of this reaction is a new carbanion that can attack another alkene in a chain-propagation step. The chain reaction is terminated when the carbanion eventually reacts with water.

CATIONIC POLYMERIZATION

The intermediate that carries the chain reaction in polymerization reactions can also be a positive ion, or cation. In this case, the chain reaction is initiated by adding a strong acid to an alkene to form a carbonium ion, as shown in Figure 25.13. The ion produced in this reaction combines with additional monomers to produce a growing polymer chain. The chain reaction is terminated when the carbonium ion reacts with water.

COORDINATION POLYMERIZATION

In 1963, Karl Ziegler and Giulio Natta received the Nobel prize in chemistry for their work with coordination compound catalysts for addition polymerization

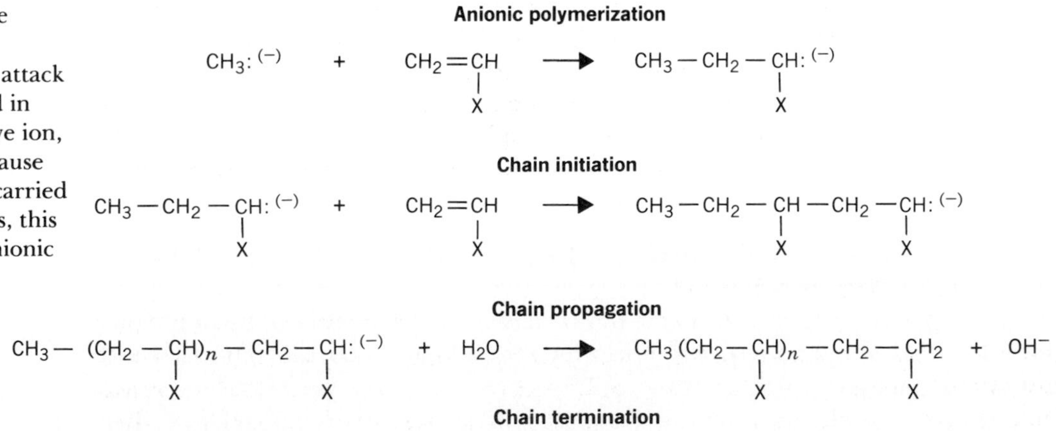

FIG. 25.12 Addition polymers result from chain-reaction mechanisms that consist of three steps: chain initiation, chain propagation, and chain termination. This figure shows the mechanism for the chain reaction that results from attack on the C=C double bond in the monomer by a negative ion, such as the CH_3^- ion. Because the chain is initiated and carried by negatively charged ions, this mechanism is known as anionic polymerization.

Anionic polymerization

$$CH_3\!:^{(-)} \quad + \quad CH_2{=}CH \longrightarrow CH_3-CH_2-CH\!:^{(-)}$$
$$\qquad\qquad\qquad\qquad X \qquad\qquad\qquad\qquad\qquad X$$

Chain initiation

$$CH_3-CH_2-CH\!:^{(-)} \quad + \quad CH_2{=}CH \longrightarrow CH_3-CH_2-CH-CH_2-CH\!:^{(-)}$$
$$\qquad\quad X \qquad\qquad\qquad\qquad X \qquad\qquad\qquad\qquad X \qquad\quad X$$

Chain propagation

$$CH_3-(CH_2-CH)_n-CH_2-CH\!:^{(-)} \quad + \quad H_2O \longrightarrow CH_3(CH_2{-}CH)_n-CH_2-CH_2 \quad + \quad OH^-$$
$$\qquad\qquad X \qquad\qquad\qquad X \qquad\qquad\qquad\qquad\qquad\qquad X \qquad\qquad X$$

Chain termination

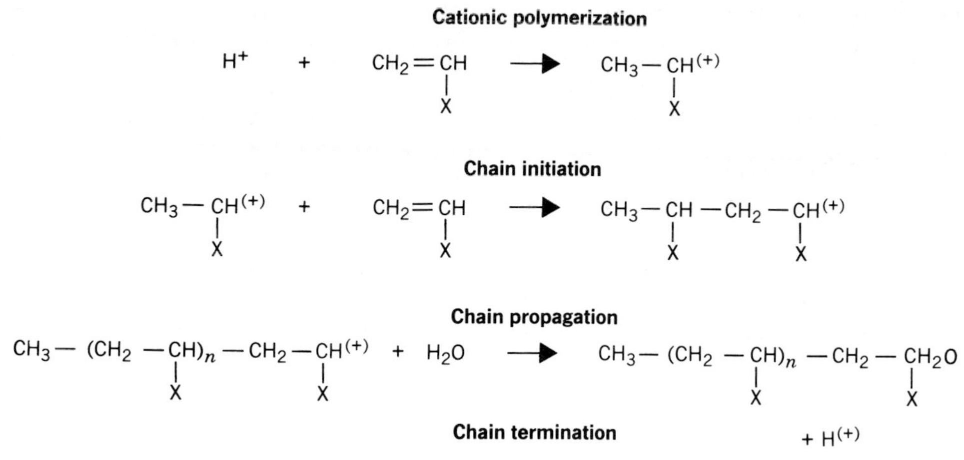

Cationic polymerization

$$H^+ \quad + \quad CH_2{=}\underset{X}{CH} \quad \longrightarrow \quad CH_3{-}\underset{X}{CH}^{(+)}$$

Chain initiation

$$CH_3{-}\underset{X}{CH}^{(+)} \quad + \quad CH_2{=}\underset{X}{CH} \quad \longrightarrow \quad CH_3{-}\underset{X}{CH}{-}CH_2{-}\underset{X}{CH}^{(+)}$$

Chain propagation

$$CH_3{-}(CH_2{-}\underset{X}{CH})_n{-}CH_2{-}\underset{X}{CH}^{(+)} \quad + \quad H_2O \quad \longrightarrow \quad CH_3{-}(CH_2{-}\underset{X}{CH})_n{-}CH_2{-}\underset{X}{CH_2O}$$

Chain termination $+ H^{(+)}$

FIG. 25.13 Addition polymers can also result from a cationic polymerization mechanism in which the chain is initiated and carried by positively charged ions.

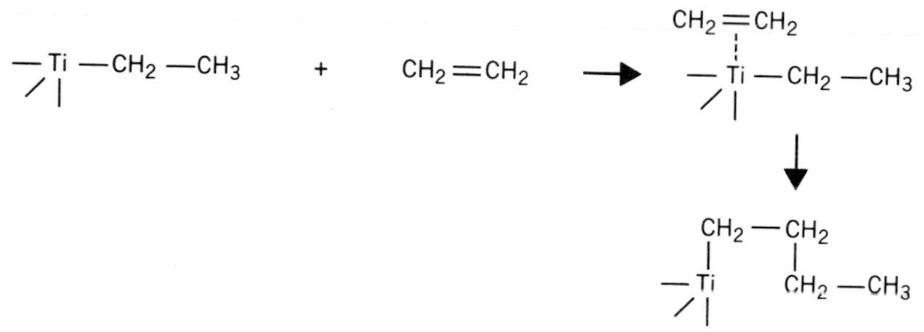

$$-\underset{|}{\overset{\diagup}{Ti}}{-}CH_2{-}CH_3 \quad + \quad CH_2{=}CH_2 \quad \longrightarrow \quad \begin{array}{c} CH_2{=}CH_2 \\ \vdots \\ -\underset{|}{\overset{\diagup}{Ti}}{-}CH_2{-}CH_3 \end{array}$$

$$\downarrow$$

$$-\underset{|}{\overset{\diagup}{Ti}}\begin{array}{c} CH_2{-}CH_2 \\ | \quad \quad | \\ CH_2{-}CH_3 \end{array}$$

FIG. 25.14 Ziegler-Natta catalysts provide another mechanism for making addition polymers. This reaction proceeds through an intermediate in which the alkene is coordinated to a transition metal. Because the polymer chain is formed when one ligand coordinated to the titanium atom attacks another ligand bound to this metal atom, the mechanism produces a polymer in which both the linearity and the tacticity can be carefully controlled.

reactions. These so-called Ziegler-Natta catalysts provide an unprecedented opportunity to control the linearity and tacticity of the polymer.

Free-radical polymerization of ethylene produces a low-density, branched polymer with side chains of one to five carbon atoms on up to 3% of the atoms along the polymer chain. Ziegler-Natta catalysts produce a more linear polymer, which is more rigid, with a higher density and a higher tensile strength. Polypropylene produced by free-radical reactions, for example, is a soft, rubbery, atactic polymer with no commercial value. Ziegler-Natta catalysts provide an isotactic polypropylene, which is harder, tougher, and more crystalline.

Typical Ziegler-Natta catalysts are mixtures of titanium(III) chloride ($TiCl_3$) and triethylaluminum [$Al(CH_2CH_3)_3$]. The first step in this reaction involves the transfer of an ethyl group from aluminum to titanium to produce a Lewis acid. An alkene then acts as a Lewis base, or electron-pair donor, to form a transition metal complex, as shown in Figure 25.14. The Ti—CH_2CH_3 bond serves as a source of a carbanion, which attacks the alkene to form a carbon—carbon bond. A new alkene then coordinates to the titanium atom, and the reaction continues. In essence, the titanium atom in this reaction provides a template on which the alkene and the carbanion combine to give a linear polymer with carefully controlled stereochemistry.

25.6 ADDITION POLYMERS

Addition polymers, such as polyethylene, polypropylene, poiy(vinyl chloride), and polystyrene, tend to be either linear or branched polymers with little or no cross-linking. As a result, they are thermoplastic materials, which flow easily when heated and can be molded into a variety of shapes. The structures, names, and trade names of some common addition polymers are given in Table 25.1.

POLYETHYLENE

There are two principal routes to producing polyethylene (see Figure 25.15). Low-density polyethylene is produced by free-radical polymerization at high temperatures (200°C) and high pressures (above 1000 atm). The high-density polymer results from Ziegler-Natta catalysis at temperatures below 100°C and pressures less than 100 atm.

More polyethylene is produced each year than any other plastic—about 7800 million pounds of low-density and 4400 million pounds of high-density polyethylene were sold in 1980. Polyethylene has no taste or odor and is lightweight, nontoxic, and relatively inexpensive. It is used as a film for packaging food, clothing, and hardware; for example, most commercial trash bags, sandwich bags, and plastic wrapping are made from polyethylene films. Polyethylene is also used for everything from seat covers to milk cartons, pails, pans, and dishes.

POLYPROPYLENE

The isotactic polypropylene from Ziegler-Natta–catalyzed polymerization is a rigid, thermally stable polymer with an excellent resistance to stress, cracking, and chemical reaction. Although it costs more per pound than polyethylene, it is much stronger. Thus, bottles made from polypropylene can be thinner, contain less polymer, and often cost less than conventional polyethylene products. Polypropylene's most important impact on today's college student undoubtedly takes the form of the plastic stackable chairs and other furniture that abound on college campuses.

POLY(TETRAFLUOROETHYLENE)

Tetrafluoroethylene (CF_2=CF_2) is a gas that boils at -76°C and is therefore stored in cylinders at high pressure. In 1938, Roy Plunkett received a cylinder of tetrafluoroethylene that did not deliver as much gas as it should have. Instead of returning the cylinder, he cut it open with a hacksaw and discovered a white, waxy powder that was the first polytetrafluoroethylene polymer. After considerable effort, a less fortuitous route to obtaining this polymer was discovered, and polytetrafluoroethylene, or Teflon, became commercially available.

Teflon is a remarkable substance. It has the best resistance to chemical attack of any polymer, and it can be used at any temperature between -73°C and 260°C with no effect on its properties. It also has a very low coefficient of friction. In simpler terms, it has a waxy or slippery touch. Even materials such as crude rubber, adhesives, bread dough, and candy can't stick to a Teflon-coated surface.

POLY(VINYL CHLORIDE) AND POLY(VINYLIDENE CHLORIDE)

Chlorine is one of the top ten industrial chemicals in the United States—more than 20 billion pounds are produced annually. About 20% of this is used to make the vinyl chloride monomer (CH_2=$CHCl$) for the production of poly(vinyl chloride), or PVC. The Cl substituents on the polymer chain make PVC more fire-resistant than polyethylene or polypropylene. They also increase the force of attrac-

FIG. 25.15 Linear polyethylene is slightly more dense than the branched polymer. Samples of the low-density (0.92 g/cm³) polyethylene can be distinguished from the high-density (0.94 g/cm³) polymer by dropping them into a mixture of solvents that has an intermediate density. The low-density polymer floats to the top, while the high-density polymer sinks to the bottom.

TABLE 25.1

Addition Polymers

Structure	Name	Trademark or Common Name
$(-CH_2-CH_2-)_n$	polyethylene	
$(-CF_2-CF_2-)_n$	poly(tetrafluoroethylene)	Teflon
$(-CH_2-\underset{\underset{CH_3}{\mid}}{CH}-)_n$	polypropylene	Herculon
$(-CH_2-\underset{\underset{CH_3}{\mid}}{\overset{\overset{CH_3}{\mid}}{C}}-)_n$	polyisobutylene	butyl rubber
$(-CH_2-CH-)_n$ (phenyl)	polystyrene	
$(-CH_2-\underset{\underset{CN}{\mid}}{CH}-)_n$	polyacrylonitrile	Orlon
$(-CH_2-\underset{\underset{Cl}{\mid}}{CH}-)_n$	poly(vinyl chloride)	PVC
$(-CH_2-\underset{\underset{Cl}{\mid}}{\overset{\overset{Cl}{\mid}}{C}}-)_n$	poly(vinylidene chloride)	Saran
$(-CH_2-\underset{\underset{CO_2CH_3}{\mid}}{CH}-)_n$	poly(methyl acrylate)	
$(-CH_2-\underset{\underset{CO_2CH_3}{\mid}}{\overset{\overset{CH_3}{\mid}}{C}}-)_n$	poly(methyl methacrylate)	Plexiglass, Lucite
$(-CH_2-\overset{\overset{H}{\mid}}{C}=\overset{\overset{H}{\mid}}{C}-CH_2-)_n$	polybutadiene	
$(-CH_2-\overset{\overset{Cl}{\mid}}{C}=CH-CH_2-)_n$	polychloroprene	neoprene
$(-CH_2-\overset{\overset{H}{\mid}}{C}=\overset{\overset{CH_3}{\mid}}{C}-CH_2-)_n$	poly(cis-1,4-isoprene)	natural rubber
$(-CH_2-\overset{\overset{H}{\mid}}{C}=\underset{\underset{CH_3}{\mid}}{C}-CH_2-)_n$	poly(trans-1,4-isoprene)	gutta percha

tion between polymer chains, which increases the hardness of the plastic. The properties of PVC can be varied over a wide range by addition of plasticizers, stabilizers, fillers, and dyes, making PVC a remarkably versatile plastic.

A copolymer of vinyl chloride CH_2=$CHCl$ and vinylidene chloride CH_2=CCl_2 is sold under the trade name Saran. The same increase in the force of attraction between polymer chains that makes PVC harder than polyethylene gives thin films of Saran a tendency to cling to itself.

Exercise 25.5

Use the discussion of intermolecular forces of attraction in Section 12.2 to explain why Saran Wrap clings to itself, whereas plastic wraps made out of polyethylene do not.

Solution

Section 12.2 introduced three intermolecular forces of attraction grouped under the heading of van der Waals forces: dipole-dipole, dipole–induced dipole, and induced dipole–induced dipole forces.

Dipole-dipole interactions in polyethylene are relatively weak, because the electrons in the C—C and C—H bonds in this polymer are shared equally by the atoms in each bond. These interactions are much stronger in Saran Wrap, because the C—Cl bonds are more polar.

Dipole–induced dipole and induced dipole–induced dipole interactions are also stronger in Saran Wrap. As noted in Section 12.2, atoms become more polarizable as they become larger, because there are more electrons to be polarized.

The net increase in the force of attraction between polymer molecules results in an increase in the tendency of the plastic to cling to itself.

ACRYLICS

Acrylic acid is the common name for 2-propenoic acid.

$$CH_2\!=\!CH\!-\!\overset{\displaystyle O}{\overset{\displaystyle \|}{C}}\!-\!OH \qquad \text{acrylic acid}$$

Acrylic fibers such as Orlon are made from a derivative of acrylic acid known as acrylonitrile.

$$CH_2\!=\!CH\!-\!CN \longrightarrow (-CH_2\!-\!\underset{\displaystyle CN}{\overset{\displaystyle |}{CH}}\!-\!)_n$$

acrylonitrile

polyacrylonitrile

Acrylic polymers are formed by polymerization of one of the esters of this acid, such as methyl acrylate.

$$CH_2\!=\!CH\!-\!\overset{\displaystyle O}{\overset{\displaystyle \|}{C}}\!-\!OCH_3 \longrightarrow (-CH_2\!-\!\underset{\displaystyle CO_2CH_3}{\overset{\displaystyle |}{CH}}\!-\!)_n$$

methyl acrylate

poly(methyl acrylate)

One of the most important acrylic polymers is poly(methyl methacrylate), or PMMA, which is sold under the trade names Lucite and Plexiglass.

$$CH_2=\overset{\overset{\displaystyle CH_3}{|}}{C}-CO_2CH_3 \longrightarrow (-CH_2-\overset{\overset{\displaystyle CH_3}{|}}{\underset{\underset{\displaystyle CO_2CH_3}{|}}{C}}-)_n$$

methyl methacrylate

poly(methyl methacrylate)

PMMA is a lightweight, crystal-clear, glass-like polymer used in airplane windows, taillight lenses, and light fixtures. Because it is hard, stable to sunlight, and extremely durable, PMMA is also used to make the reflectors embedded between lanes of interstate highways.

The unusual transparency of PMMA makes this polymer ideal for hard contact lenses. Unfortunately, PMMA is impermeable to oxygen and water. Oxygen must therefore be transported to the cornea of the eye in the tears and then passed under the contact lens each time the eye blinks. Soft plastic lenses that pass both oxygen and water are made from poly(2-hydroxyethyl methacrylate), which is cross-linked with ethylene glycol dimethacrylate.

$$(-CH_2-\overset{\overset{\displaystyle CH_2OH}{\overset{|}{\overset{\displaystyle CH_2}{|}}}}{\underset{\underset{\displaystyle CO_2CH_3}{|}}{C}}-)_n$$

poly(2-hydroxyethyl methacrylate)

$$\begin{array}{cc} CH_2 & CH_2 \\ \| & \| \\ CH_3-C & C-CH_3 \\ | & | \\ O=C & C=O \\ | & | \\ O & O \\ | & | \\ CH_2 & -CH_2 \end{array}$$

ethylene glycol dimethacrylate

25.7 CONDENSATION POLYMERS

The first synthetic plastic was bakelite, which was developed by Leo Baekland between 1905 and 1914. The synthesis of bakelite starts with the reaction between formaldehyde (H_2CO) and phenol (C_6H_5OH) to form a mixture of *ortho-* and *para-*substituted phenols, as shown in Figure 25.16. At temperatures above 100°C, these phenols condense to form a polymer in which the aromatic rings are bridged by either $-CH_2-O-CH_2-$ or $-CH_2-$ linkages. The cross-linking in this polymer is so extensive that it is a thermoset plastic.

Research on condensation polymers started by Wallace Carothers and co-workers at duPont in the 1920s and 1930s eventually led to the discovery of the substances shown in Table 25.2.

In Section 24.17, we noted that carboxylic acids react with amines to form amides.

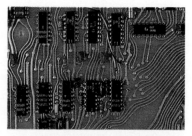

Circuit board with microchips encapsulated with bakelite.

$$R-\overset{\overset{\displaystyle O}{\|}}{C}-OH + H_2NR \longrightarrow R-\overset{\overset{\displaystyle O}{\|}}{C}-NHR + H_2O$$

carboxylic acid · amine · amide

When Carothers and his co-workers reacted dicarboxylic acids with diamines, as

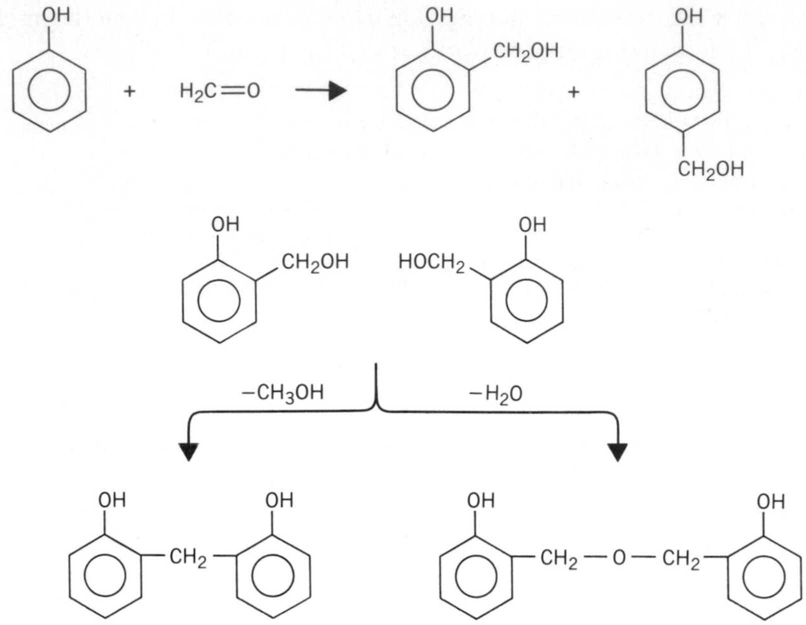

FIG. 25.16 One of the first commercial plastics was bakelite, a highly cross-linked polymer formed by condensation of the products of the reaction between phenol and formaldehyde. This polymer contains so many cross-links it is a thermoset plastic—once it sets, there is no way to change its shape without destroying the plastic.

shown in Figure 25.17, they obtained a *polyamide* with the following repeating structure.

$$[-NH-(CH_2)_x-NH-\overset{O}{\overset{\|}{C}}-(CH_2)_y-\overset{O}{\overset{\|}{C}}-]_n$$
polyamide

Unfortunately, these polyamides didn't seem to have any worthwhile properties. Work therefore turned to a second class of condensation polymers known as *polyesters*, which were made by reacting dicarboxylic acids with dialcohols, as shown in Figure 25.18.

$$[-O-(CH_2)_x-O-\overset{O}{\overset{\|}{C}}-(CH_2)_y-\overset{O}{\overset{\|}{C}}-]_n$$
polyester

While studying these polyesters, Julian Hill found that he could wind a small amount of this polymer on the end of a stirring rod and draw it slowly out of a solution as a silky fiber. One day, when Carothers wasn't in the lab, Hill and his colleagues tried to see how long a fiber they could make by stretching a sample of

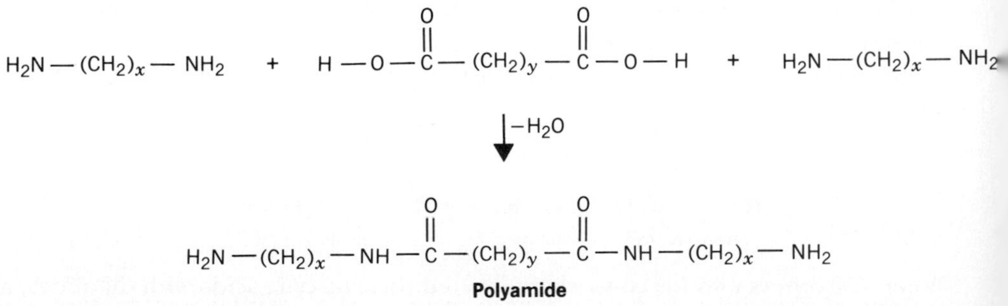

FIG. 25.17 Polyamides are condensation polymers usually formed by reaction of a dicarboxylic acid with a diamine.

$$H_2N-(CH_2)_x-NH_2 \quad + \quad H-O-\overset{O}{\overset{\|}{C}}-(CH_2)_y-\overset{O}{\overset{\|}{C}}-O-H \quad + \quad H_2N-(CH_2)_x-NH_2$$

$$\downarrow -H_2O$$

$$H_2N-(CH_2)_x-NH-\overset{O}{\overset{\|}{C}}-(CH_2)_y-\overset{O}{\overset{\|}{C}}-NH-(CH_2)_x-NH_2$$
Polyamide

TABLE 25.2

Condensation Polymers

Structure	Trademark or Common Name

Polyamides

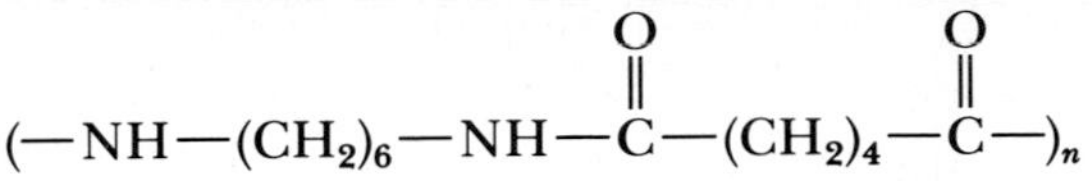

$(-NH-(CH_2)_6-NH-\overset{\overset{O}{\|}}{C}-(CH_2)_4-\overset{\overset{O}{\|}}{C}-)_n$ — nylon 6,6

$(-NH-(CH_2)_6-NH-\overset{\overset{O}{\|}}{C}-(CH_2)_8-\overset{\overset{O}{\|}}{C}-)_n$ — nylon 6,10

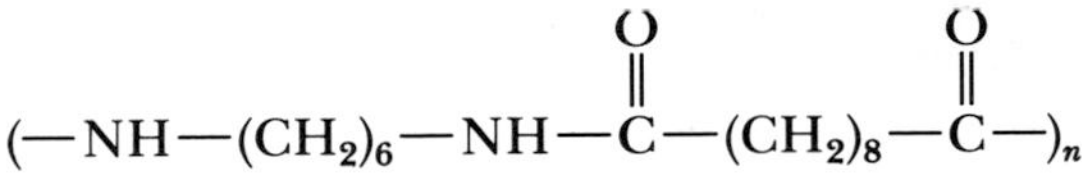

$(-NH-(CH_2)_5-\overset{\overset{O}{\|}}{C}-)_n$ — nylon 6

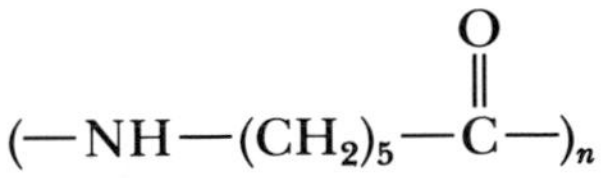

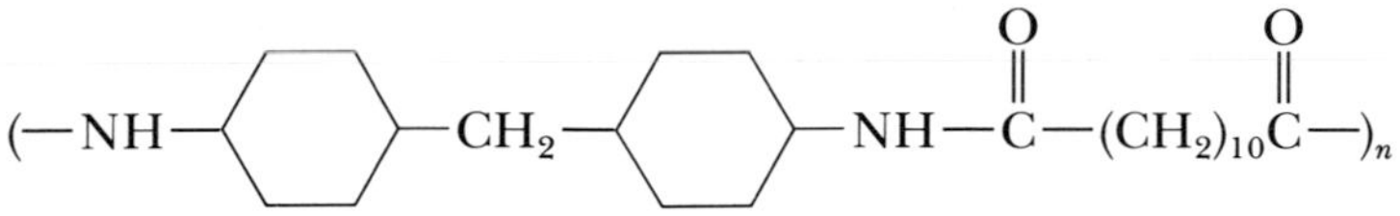

— Qiana

Polyaramides

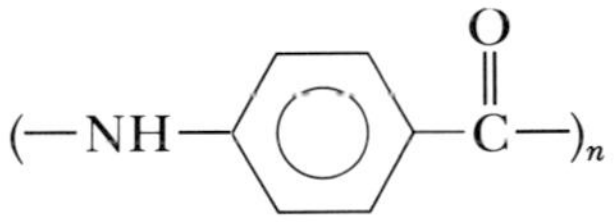

— Kevlar

Polyesters

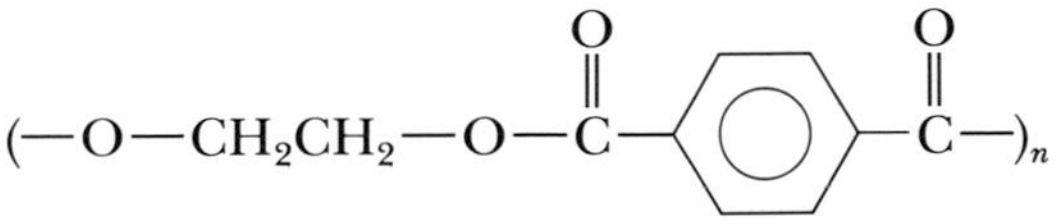

— Dacron, Mylar

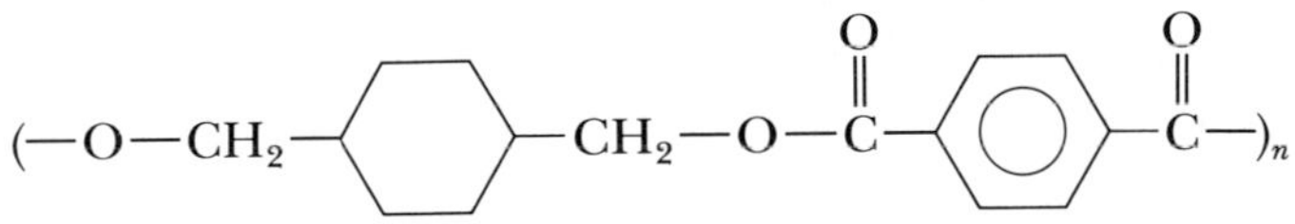

— Kodel

Polycarbonates

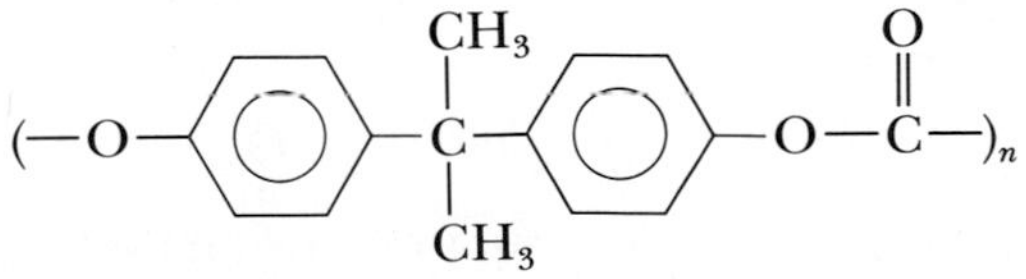

— Lexan

Silicones

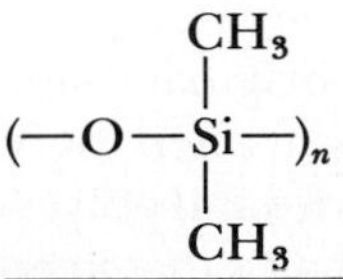

$(-O-\underset{\underset{CH_3}{|}}{\overset{\overset{CH_3}{|}}{Si}}-)_n$ — silicone rubber

FIG. 25.18 Polyesters are condensation polymers formed by reaction of dicarboxylic acids with dialcohols.

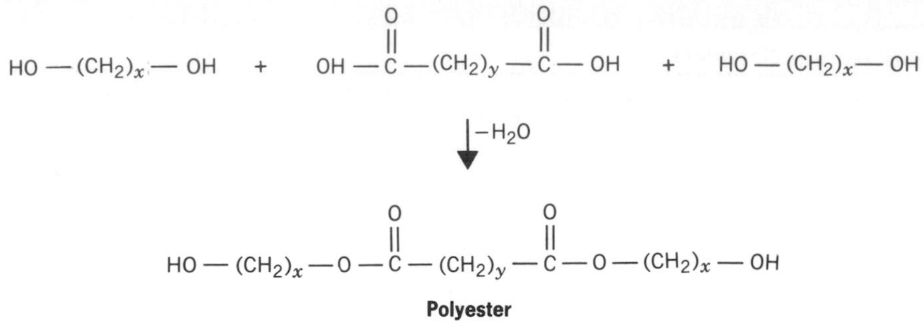

$$HO-(CH_2)_x-OH \quad + \quad OH-\overset{\overset{\displaystyle O}{\|}}{C}-(CH_2)_y-\overset{\overset{\displaystyle O}{\|}}{C}-OH \quad + \quad HO-(CH_2)_x-OH$$

$$\Big\downarrow -H_2O$$

$$HO-(CH_2)_x-O-\overset{\overset{\displaystyle O}{\|}}{C}-(CH_2)_y-\overset{\overset{\displaystyle O}{\|}}{C}-O-(CH_2)_x-OH$$

Polyester

this polymer as they ran down the hall. They soon realized that this playful exercise had oriented the polymer molecules in two dimensions and produced a new material with superior properties.

They immediately tried the same thing with one of the polyamides and produced a sample of the first synthetic fiber — **nylon.** We can demonstrate this process by pouring a solution of adipoyl chloride in CH_2Cl_2 into a beaker and then carefully adding a solution of hexamethylenediamine in water, as shown in Figure 25.19.

$$\underset{\text{hexamethylene diamine}}{H_2N-(CH_2)_6-NH_2} + \underset{\text{adipoyl chloride}}{Cl-\overset{\overset{\displaystyle O}{\|}}{C}-(CH_2)_4\overset{\overset{\displaystyle O}{\|}}{C}-Cl} \longrightarrow \underset{\text{nylon 6,6}}{(-NH-(CH_2)_6-NH-\overset{\overset{\displaystyle O}{\|}}{C}-(CH_2)_4-\overset{\overset{\displaystyle O}{\|}}{C}-)_n}$$

A thin film of polymer forms at the interface between these two phases. If we grasp this film with a pair of tweezers, we can draw a continuous string of nylon from the solution. This particular polyamide is known as nylon 6,6, because the polymer is formed from a diamine that has six carbon atoms and a derivative of a dicarboxylic acid that also has six carbon atoms.

Exercise 25.6

A synthetic fiber known as nylon 6 has the following structure.

$$(-NH-(CH_2)_5-CO-)_n$$

Explain how this polymer is made.

Solution

This particular polyamide can be made from a monomer that contains both an $-NH_2$ and a $-CO_2H$ functional group. In this case, the polymer is made by repeating the following condensation reaction.

$$H_2N(CH_2)_5\overset{\overset{\displaystyle O}{\|}}{C}OH + H_2N(CH_2)_5\overset{\overset{\displaystyle O}{\|}}{C}OH \longrightarrow H_2N(CH_2)_5\overset{\overset{\displaystyle O}{\|}}{C}-NH(CH_2)_5\overset{\overset{\displaystyle O}{\|}}{C}OH + H_2O$$

FIG. 25.19 A filament of Nylon 6,6 can be prepared by using tweezers to pull on the thin film that forms at the interface between a solution of adipoyl chloride in a nonpolar solvent and hexamethylenediamine dissolved in water.

The effect of pulling on the polymer with the tweezers in Figure 25.19 is much like that of stretching an elastomer — the polymer molecules become oriented in two dimensions. Why don't the polymer molecules return to their original shape when we stop pulling? Section 25.3 suggested that polymers are elastic when there is no strong force of attraction between the polymer chains. Polyamides and polyesters, however, form strong hydrogen bonds between the polymer chains that keep the polymer molecules oriented as shown in Figure 25.20.

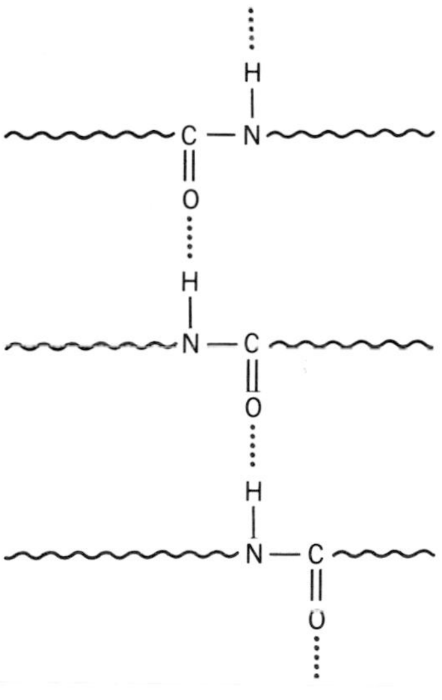

FIG. 25.20 Elastomers return to their original structure when stretched because the force of attraction between adjacent polymer chains is very weak. The hydrogen bonds that form between the polymer chains when polyamides and polyesters are stretched help to keep the chains oriented in a two-dimensional fiber.

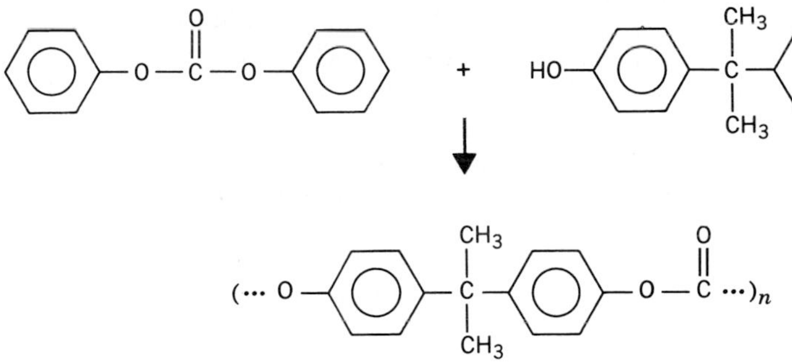

FIG. 25.21 Polycarbonates are condensation polymers formed when a diester of carbonic acid is allowed to react with a dialcohol.

The first synthetic polyester fibers were produced from the reaction between ethylene glycol and terephthalic acid or one of its esters to give poly(ethylene terephthalate).

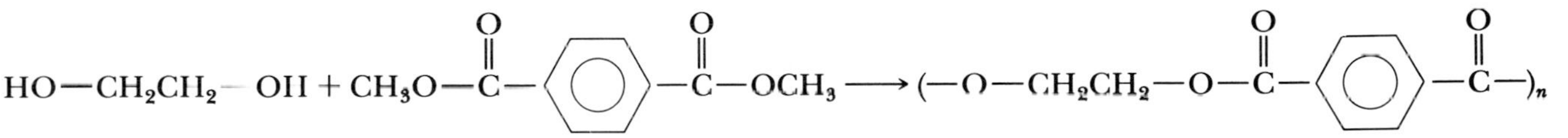

This polymer is still used to make thin films (Mylar) as well as textile fibers (Dacron and Fortrel).

Carbonic acid (H_2CO_3) is too unstable to form esters directly, but its acid chloride, $COCl_2$, reacts with alcohols to form carbonate esters analogous to those formed when carboxylic acids react with alcohols to form esters.

$$\underset{}{R{-}OH} + \underset{\text{carbonic acid}}{HO{-}\overset{\overset{\displaystyle O}{\|}}{C}{-}OH} + HO{-}R \longrightarrow \underset{\text{carbonate ester}}{R{-}O{-}\overset{\overset{\displaystyle O}{\|}}{C}{-}O{-}R} + 2\,H_2O$$

Polycarbonates are produced when one of these carbonate esters reacts with an appropriate alcohol, as shown in Figure 25.21. The polycarbonate shown in the figure is known as Lexan. It has a very high resistance to impact and is used in safety glass, bullet-proof windows, and motorcycle helmets.

25.8 NATURAL POLYMERS: PROTEINS, CARBOHYDRATES, AND NUCLEIC ACIDS

In their search for new synthetic polymers, chemists have often been guided by the compounds and reactions they found in nature. This chapter approaches this relationship from the other direction. The second half of the chapter applies the structure and properties of synthetic polymers to the more diverse world of natural polymers.

Biochemistry can be defined as the chemistry of living organisms. It is often convenient to subdivide biochemistry into the study of four classes of biomolecules: proteins, carbohydrates, nucleic acids, and lipids.

The name ***protein,*** which comes from the Greek stem for "first," indicates the relative importance of this class of biomolecules. About half of the dry weight of an organism is due to proteins, and they exhibit the most diverse array of functions of any class of biomolecules. Among the functions of proteins we can include the following.

Structure: The actin and myosin in muscles, the collagen in skin and bone, and the keratins in hair, horn, and hoof are all examples of proteins whose primary function is to produce the structure of the organism.

Catalysis: Most of the chemical reactions in living systems are catalyzed by enzymes, which are proteins.

Control: Many proteins regulate or control biological activity. Insulin, for example, controls the rate at which sugar is metabolized.

Energy: Many proteins, including the seed proteins, the casein in milk, and the albumin in eggs are used primarily as a way of storing food energy.

Transport: O_2 is carried through the bloodstream by the proteins hemoglobin and myoglobin. Other proteins are involved in the transport of sugars, amino acids, and ions across the cell membranes.

Protection: The first line of defense against viruses and bacteria are the antibodies produced by the immune system, which are based on proteins.

The name ***carbohydrate*** reflects the fact that many of the compounds in this class have the empirical formula CH_2O — they are literally hydrates of carbon. Carbohydrates are the primary source of food energy for most living systems. They include sugars, such as glucose ($C_6H_{12}O_6$) and sucrose ($C_{12}H_{22}O_{11}$), as well as polymers of sugars, such as starch, glycogen, and cellulose. Carbohydrates are produced from CO_2 and H_2O during photosynthesis and are therefore the end product of the process by which plants capture the energy in sunlight.

$$6\ CO_2(g) + 6\ H_2O(l) \xrightarrow{\text{photosynthesis}} C_6H_{12}O_6(aq) + 6\ O_2(g)$$

The name ***nucleic acid*** was given to a class of relatively strong acids that were found in the nuclei of cells. As monomers, nucleic acids such as adenosine triphosphate (ATP) are involved in the process by which cells capture food energy and make it available for fuel. As polymers, they serve as the repository of genetic information that allows an organism to grow and eventually reproduce.

Proteins, carbohydrates, and nucleic acids are all grouped on the basis of structure. The fourth class of biomolecules, the ***lipids,*** is organized on the basis of properties. Any molecule in a biological system that is soluble in a nonpolar solvent is classified as a lipid (from the Greek *lipos,* "fat"). Although they cluster to form the membranes that divide a cell into compartments and separate the cell from its environment, lipids don't form polymers like proteins, carbohydrates, and nucleic acids.

25.9 THE AMINO ACIDS

Proteins are formed by polymerization of a family of monomers known as the ***amino acids.*** These monomers are called amino acids because they contain both an amino ($-NH_2$) and a carboxylic acid ($-CO_2H$) functional group. With only one exception, the amino acids found in proteins are primary amines in which the

—NH_2 substituent is on the carbon atom adjacent to the —CO_2H group. They therefore have the following generic formula.

$$R \\ | \\ H_2N-CH-CO_2H$$

an amino acid

The chemistry of amino acids is complicated by the fact that the —NH_2 group is a base and the —CO_2H group is an acid. In aqueous solution, an H^+ ion is transferred from one end of the molecule to the other to form a ***zwitterion*** (from the German meaning "mongrel ion," or hybrid ion).

$$R \\ | \\ H_2N-CH-CO_2(H) \longrightarrow \quad H_3N^+-CH-CO_2^-$$

Zwitterions are simultaneously electrically charged and electrically neutral—they contain positive and negative charges, but the net charge on the molecule is zero.

Exercise 25.7

Amino acids in aqueous solution at a neutral pH are present as zwitterions with the following generic formula.

$$R \\ | \\ H_3N^+-CH-CO_2^-$$

Predict what should happen when one of these amino acids is dissolved in a strong acid or a strong base.

Solution

In a strongly acidic solution, the —CO_2^- end of this molecule picks up an H^+ ion, or proton, to form a molecule with a net positive charge.

$$R \\ | \\ H_3N^+-CH-CO_2H$$

In the presence of a strong base, the —NH_3^+ end of the molecule loses an H^+ ion, or proton, to form a molecule with a net negative charge.

$$R \\ | \\ H_2N-CH-CO_2^-$$

Twenty common amino acids can be found in proteins (see Table 25.3). Most of these amino acids differ only in the nature of the R-group substituent. The common amino acids are therefore divided into four categories on the basis of these R-groups. Amino acids with nonpolar substituents are said to be ***hydrophobic*** (water-hating). Another class of amino acids have polar R-groups, which can form hydrogen bonds to water and are therefore called ***hydrophilic*** (water-loving). The remaining amino acids have substituents that carry either negative or positive charges when the amino acids are dissolved in aqueous solution at neutral pH. These last two classes are strongly hydrophilic.

TABLE 25.3

The Twenty Common Amino Acids

Name	Structure (at neutral pH)	Name	Structure (at neutral pH)
Nonpolar (Hydrophobic) R-Groups		*Nonpolar (Hydrophobic) R-Groups*	

Glycine (GLY)

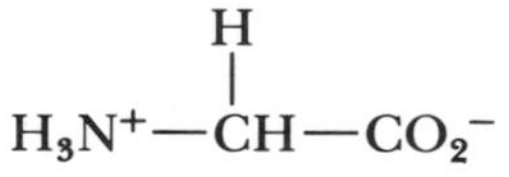

$$\overset{\overset{\displaystyle H}{|}}{H_3N^+—CH—CO_2^-}$$

Alanine (ALA)

$$\overset{\overset{\displaystyle CH_3}{|}}{H_3N^+—CH—CO_2^-}$$

Valine (VAL)

$$H_3N^+—CH—CO_2^- \text{ with } CH(CH_3)(CH_3)$$

Leucine (LEU)

$$H_3N^+—CH—CO_2^- \text{ with } CH_2—CH(CH_3)(CH_3)$$

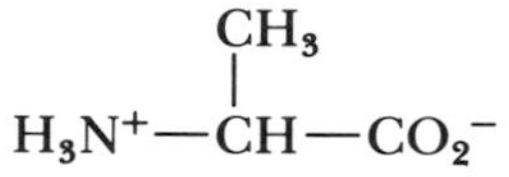

Isoleucine (ILE)

$$H_3N^+—CH—CO_2^- \text{ with } CH(CH_3)(CH_2CH_3)$$

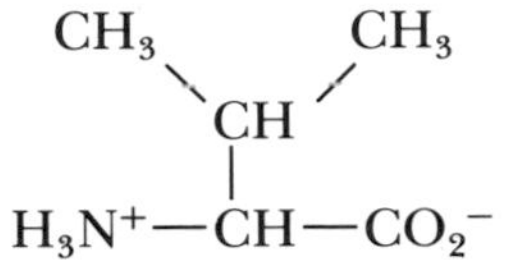

Proline (PRO)

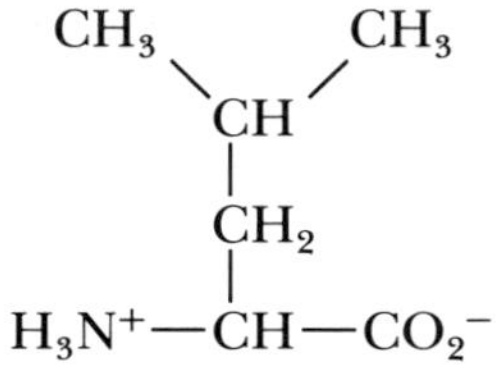

$$\begin{array}{c} H_2C—CH_2 \\ H_2C \quad CH—CO_2^- \\ N^+ \\ H \quad H \end{array}$$

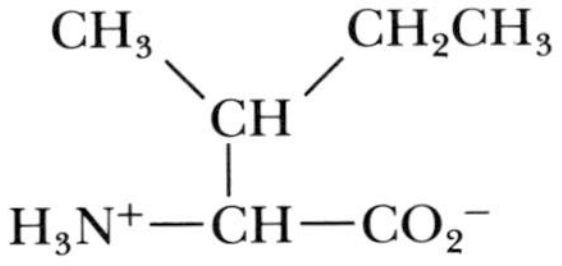

Methionine (MET)

$$H_3N^+—CH—CO_2^- \text{ with } CH_2—CH_2—S—CH_3$$

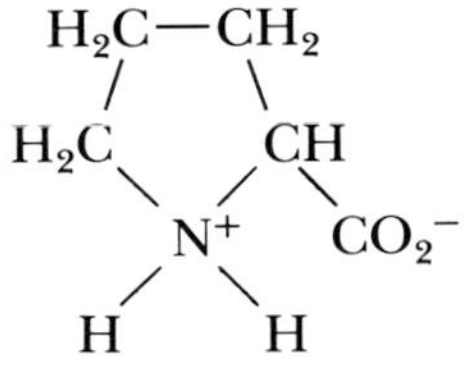

Phenylalanine (PHE)

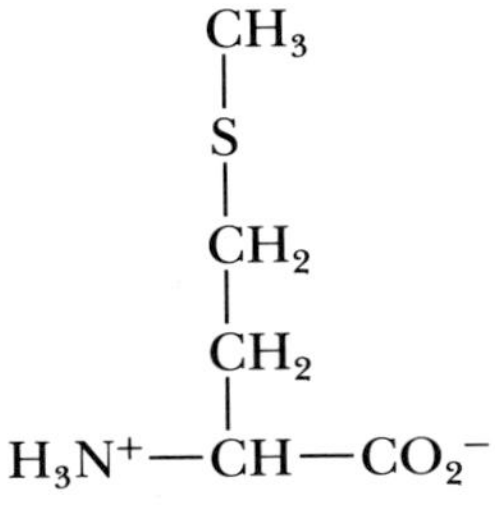

$$H_3N^+—CH—CO_2^- \text{ with } CH_2—C_6H_5$$

Tryptophan (TRP)

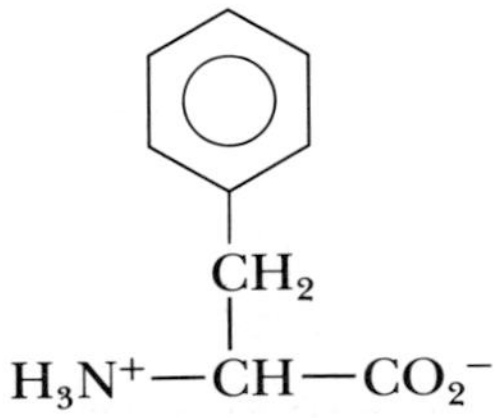

$$H_3N^+—CH—CO_2^- \text{ with } CH_2—\text{(indole ring, N—H)}$$

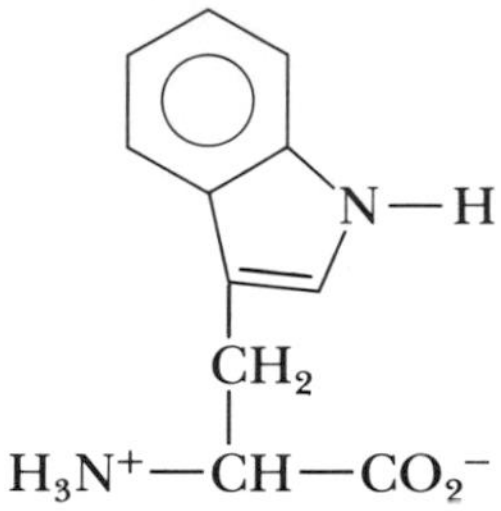

Polar (Hydrophilic) R-Groups

Serine (SER)

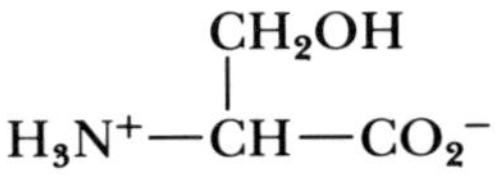

$$\overset{\overset{\displaystyle CH_2OH}{|}}{H_3N^+—CH—CO_2^-}$$

Threonine (THR)

$$H_3N^+—CH—CO_2^- \text{ with } CH(CH_3)(OH)$$

Tyrosine (TYR)

$$H_3N^+—CH—CO_2^- \text{ with } CH_2—C_6H_4—OH$$

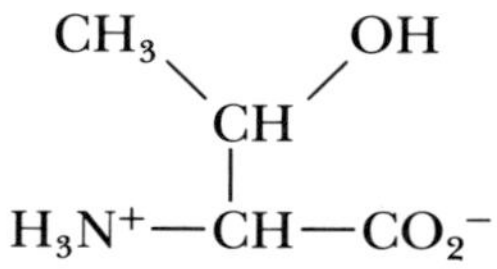

Cysteine (CYS)

$$\overset{\overset{\displaystyle CH_2SH}{|}}{H_3N^+—CH—CO_2^-}$$

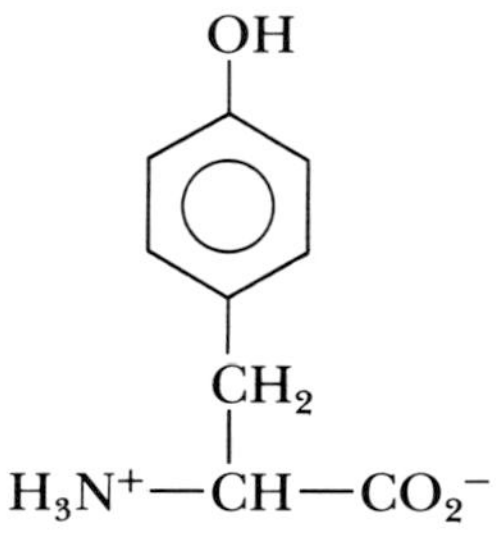

Asparagine (ASN)

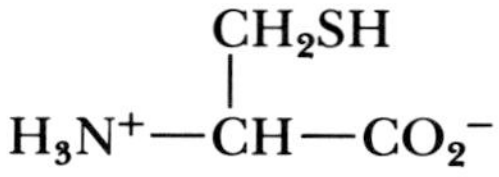

$$H_3N^+—CH—CO_2^- \text{ with } CH_2—\overset{\overset{\displaystyle O}{||}}{C}—NH_2$$

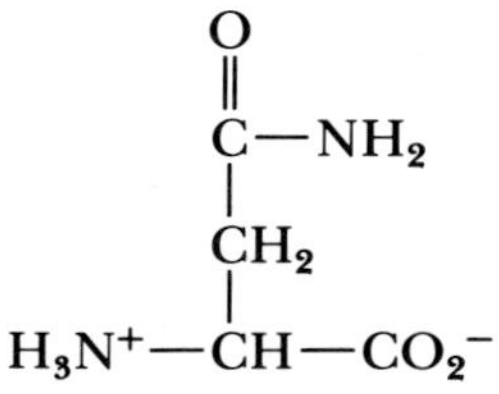

Glutamine (GLN)

$$H_3N^+—CH—CO_2^- \text{ with } CH_2—CH_2—\overset{\overset{\displaystyle O}{||}}{C}—NH_2$$

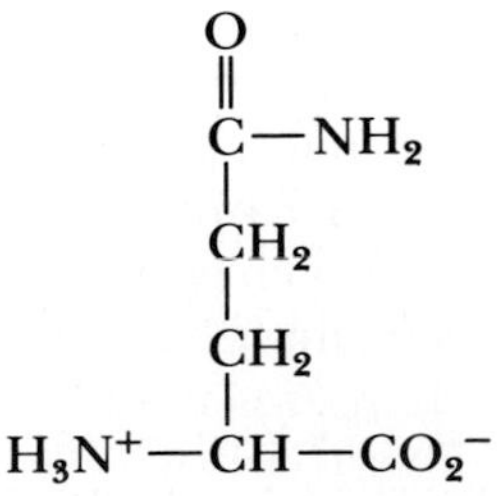

TABLE 25.3 continued

The Twenty Common Amino Acids

Name	Structure (at neutral pH)	Name	Structure (at neutral pH)
Negatively Charged R-Groups		*Positively Charged R-Groups*	

Aspartic acid (ASP)

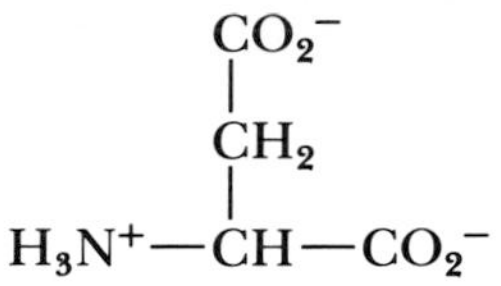

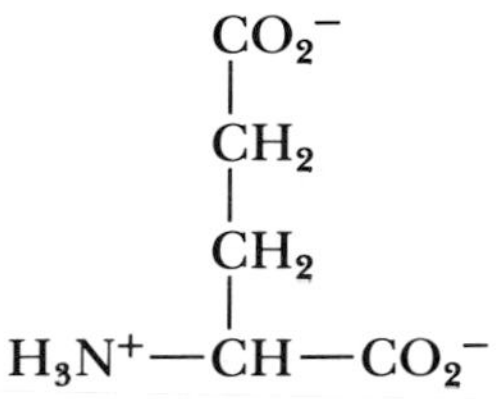

$$CO_2^-$$
$$|$$
$$CH_2$$
$$|$$
$$H_3N^+-CH-CO_2^-$$

Glutamic acid (GLU)

$$CO_2^-$$
$$|$$
$$CH_2$$
$$|$$
$$CH_2$$
$$|$$
$$H_3N^+-CH-CO_2^-$$

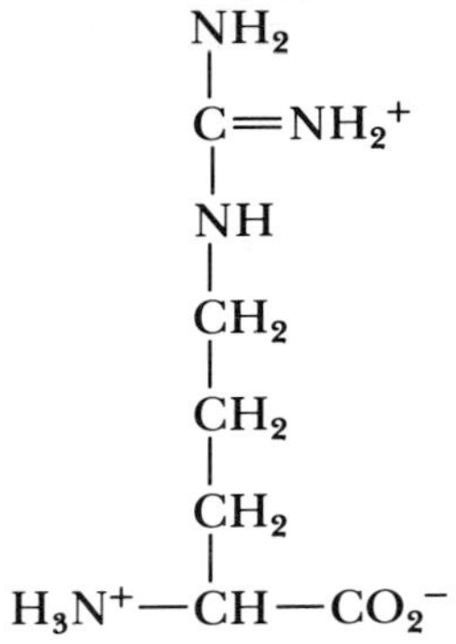

Positively Charged R-Groups

Lysine (LYS)

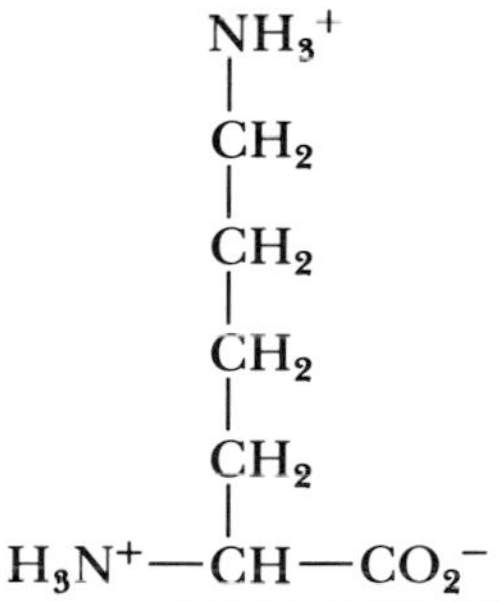

$$NH_3^+$$
$$|$$
$$CH_2$$
$$|$$
$$CH_2$$
$$|$$
$$CH_2$$
$$|$$
$$CH_2$$
$$|$$
$$H_3N^+-CH-CO_2^-$$

Arginine (ARG)

$$NH_2$$
$$|$$
$$C=NH_2^+$$
$$|$$
$$NH$$
$$|$$
$$CH_2$$
$$|$$
$$CH_2$$
$$|$$
$$CH_2$$
$$|$$
$$H_3N^+-CH-CO_2^-$$

Histidine (HIS)

$$H_3N^+-CH-CO_2^-$$

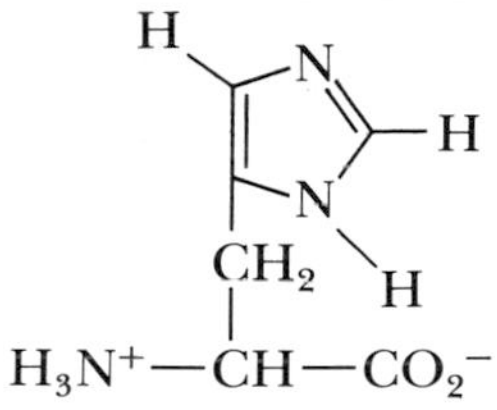

Exercise 25.8

Classify the following amino acids.

(a) Valine: $R = -CH(CH_3)_2$
(b) Serine: $R = -CH_2OH$
(c) Aspartic acid: $R = -CH_2CO_2H$
(d) Lysine: $R = -(CH_2)_4NH_2$

Solution

(a) Valine is grouped among the amino acids with nonpolar, or hydrophobic, R-groups.

(b) Serine is a member of the class of amino acids with polar, or hydrophilic, R-groups.

(c) Aspartic acid is classified as a hydrophilic amino acid with a negatively charged R-group, because this substituent is present as a $-CH_2CO_2^-$ ion in aqueous solution at neutral pH.

(d) Lysine is categorized as a hydrophilic amino acid with a positively charged R-group, because this substituent is present as a $-(CH_2)_4NH_3^+$ ion in aqueous solution at neutral pH.

FIG. 25.22 Amino acids, peptides, and proteins are usually optically active compounds that can exist in the form of two isomers. This figure shows the D and L isomers of the amino acid alanine.

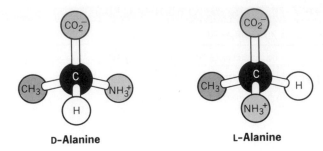

We concluded earlier that carbon atoms with four different substituents are chiral and therefore optically active. With the exception of glycine, all of the common amino acids contain at least one chiral carbon atom. These amino acids therefore exist as pairs of stereoisomers. The isomer that rotates plane-polarized light to the right is called the D, or *dextrorotatory*, isomer (from the Latin meaning "turn to the right"). The isomer that rotates light to the left is called the L, or *levorotatory*, isomer (from the Latin meaning "turn to the left").

The structures of the D and L isomers of alanine are shown in Figure 25.22. Although D-amino acids can be found in nature, they are never used to make proteins. All of the amino acids in proteins are L isomers.

25.10 PEPTIDES AND PROTEINS

Proteins are linear polymers of between 40 and 10,000 (or more) amino acids. The average molecular weight of an amino acid is about 110 amu. A modestly sized protein with only 300 amino acids therefore has a molecular weight of 33,000 grams per mole, and very large proteins can have molecular weights as high as 1,000,000 grams per mole.

Proteins are formed by the reaction of the $—CO_2H$ end of one amino acid with the $—NH_2$ end of another to form an amide. The $—CO—NH—$ amide bond that links amino acids is also known as a *peptide bond*, because relatively short polymers of the amino acids are known as *peptides*.

$$
\underset{\substack{| \\ R \\ |}}{H_3N^+}-CH-\overset{O}{\overset{\|}{C}O^- + H_3N^+-CH-\overset{O}{\overset{\|}{C}O^-} \longrightarrow H_3N^+-\underset{R}{CH}-\overset{O}{\overset{\|}{C}}-NH-\underset{R}{CH}-CO_2^- + H_2O
$$

a dipeptide

The same $—CO—NH—$ bond forms the backbone of both proteins and synthetic fibers such as nylon. This raises an interesting question: How do we explain the enormous range of structures and functions of proteins when nylon has such monotonously regular properties?

The monomers used to produce nylon are symmetrical. As a result, there is only one way in which a diamine can react with a dicarboxylic acid. The nylons therefore have a regular structure that repeats monotonously from one end of the polymer chain to the other.

The two ends of the amino acid molecule are different — each monomer has both an $—NH_2$ head and a $—CO_2H$ tail. Four different dipeptides can therefore be formed from only two amino acids. Aspartic acid (ASP) could react with phenylalanine (PHE), for example, to give two symmetrical dipeptides—PHE—PHE and ASP—ASP—and two unsymmetrical dipeptides—PHE—ASP and ASP—PHE—as shown in Figure 25.23. When the full range of amino acids is brought into this calculation, we find that it is possible to make 400 different dipeptides, 64 million different hexapeptides, and 10^{52} different peptides that contain 40 amino acids.

Differences between the structures of even such closely related dipeptides as ASP—PHE and PHE—ASP give rise to significant differences in their properties as well. The methyl ester of ASP—PHE, for example, has a very sweet taste and is sold as an artificial sweetener under the trade name *aspartame.*

aspartame

The ester of the dipeptide with the opposite arrangement of amino acids, PHE—ASP, does not taste sweet and has no commercial value. As the length of the polymer chain increases and the number of possible combinations of R-groups increases, polymer chains with an almost infinite variety of structures and properties are produced.

In recent years, a group of naturally occurring peptides that mimic pain-killing drugs such as morphine has been discovered in human brain cells. These *enkephalins* (from the Greek meaning "in the head") hold the promise of a synthetic pain-killer that is both safe and nonaddictive.

One of the enkephalins is a pentapeptide that contains four different amino acids—tyrosine (TYR), glycine (GLY), phenylalanine (PHE), and methionine (MET). The first step in describing the structure of this peptide is to list the amino acids in the order in which they are found on the peptide chain: TYR—GLY—GLY—PHE—MET. We then have to identify the amino acid at the —CO_2H end of the chain and the amino acid at the —NH_2 end. By convention, proteins are listed from the N-terminal amino acid residue toward the C-terminal end. The structure of this enkephalin is shown in Figure 25.24.

The structure of this pentapeptide illustrates the perils that face anyone who tries to synthesize peptides, or proteins, from amino acids. In order to make this

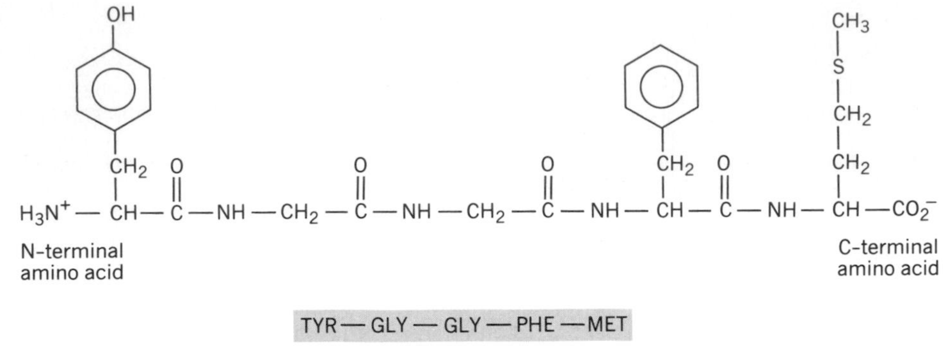

FIG. 25.24 The structure of a naturally occurring pentapeptide that belongs to the family of enkephalins found to bind to the same sites in the brain as synthetic pain-killers such as morphine.

TYR — GLY — GLY — PHE — MET

enkephalin in large quantities, chemists would have to overcome the following problems.

1. Peptide bond formation is an uphill process in terms of energy ($\Delta G° = +17$ kJ/mol), so a way must be found to drive this reaction forward.

2. The amino acids must be added to the chain one at a time, in a very carefully controlled fashion.

3. The *R*-groups of certain amino acids must be protected during polymerization so that no reactions take place on these side-chains.

Solid-phase techniques have been developed for making synthetic polypeptides. Recent developments in molecular biology, however, suggest that it might be easier to use genetic engineering to teach bacteria to produce polypeptides and proteins on an industrial scale. This technique is already being used to produce synthetic insulin—a protein that contains 51 amino acids.

25.11 THE STRUCTURE OF PROTEINS

THE PRIMARY STRUCTURE OF PROTEINS

The primary structure of a protein is nothing more than its sequence of amino acids, read off one at a time, as if printed on ticker-tape. Insulin obtained from cows, for example, consists of two chains (*A* and *B*) with the primary structures shown in Figure 25.25. It has long been known, however, that there is more to the

FIG. 25.25 Insulin is a relatively small protein that consists of two chains held together by —S—S— bonds between side chains of cysteine residues that become adjacent to each other when the protein folds.

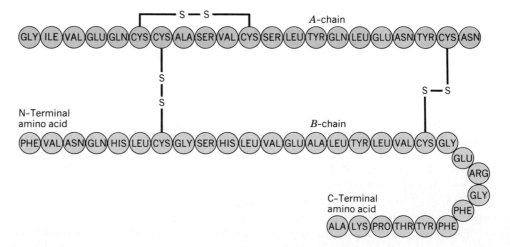

structure of a protein than just the sequence of amino acids. The polypeptide chain folds back on itself to form a secondary structure. Interactions between amino acid side chains then produces a tertiary structure. For some proteins, interactions between individual polypeptide chains give rise to a quaternary structure.

THE SECONDARY STRUCTURE OF PROTEINS

The peptide bond in proteins is a resonance hybrid of two Lewis structures.

The Lewis structure on the left implies that the geometry around the carbon atom should be trigonal planar and that the carbon atom and its three nearest neighbors should all lie in the same plane. The Lewis structure on the right suggests a trigonal planar geometry for the nitrogen atom as well. Because the peptide bond is a hybrid of these resonance forms, all six of these atoms must lie in the same plane.

Since the N—H and C=O bonds are relatively polar, hydrogen bonds can form between peptide bonds.

The two-dimensional nature of the peptide bond limits the number of ways in which the polypeptide chain can be arranged in space. By building models of polypeptides, Linus Pauling and Robert Corey discovered two ways in which an ideal polypeptide chain could fold back on itself to maximize the number of hydrogen bonds between peptides. In one of these structures, the polypeptide chain forms the right-handed α-helix shown in Figure 25.26. The other structure suggested by Pauling and Corey for an ideal polypeptide chain is the β-pleated sheet shown in Figure 25.27.

THE TERTIARY STRUCTURE OF PROTEINS

Many proteins, such as ribonuclease, shown at the beginning of this chapter, have structures that lie between the extremes of ideal α-helixes and β-pleated sheets.

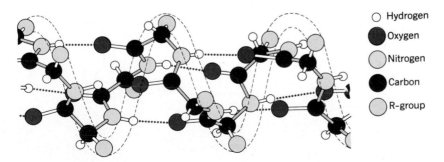

○ Hydrogen
● Oxygen
◔ Nitrogen
● Carbon
◯ R-group

FIG. 25.26 The hydrogen bonds between adjacent peptides enable the polypeptide chain to coil to form a right-handed α-helix. One 360° turn along this helix contains 3.6 amino acid residues, and the distance between adjacent coils is 0.54 nanometer.

FIG. 25.27 In another polypeptide structure, the polymer chain folds back on itself to form a structure that looks something like a pleated sheet. Two different β-pleated sheet structures are found in nature. They differ in whether the adjacent polypeptide chains run in the same (parallel) or opposite (anti-parallel) directions. This drawing shows the anti-parallel β-structure in silk.

○ Hydrogen
● Oxygen
○ Nitrogen
● Carbon
○ R-group

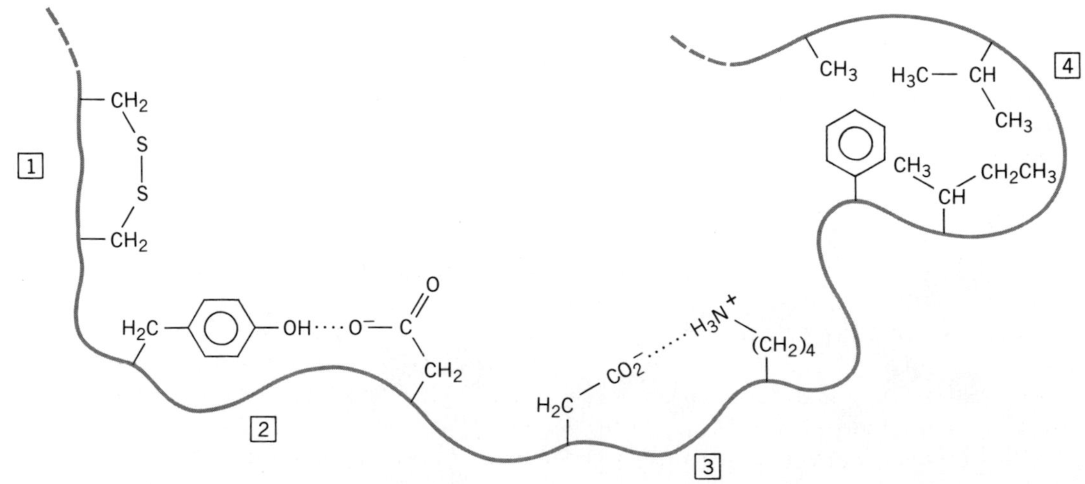

FIG. 25.28 Four factors are primarily responsible for the tertiary structure of proteins: (1) —S—S—, or disulfide, bonds between adjacent cysteine residues, (2) hydrogen bonds between amino acid side chains, (3) ionic bonds between positively and negatively charged side chains, and (4) hydrophobic interactions between nonpolar side chains.

This results from the fact that other factors besides the hydrogen bonds in the secondary structure influence the way proteins fold to form three-dimensional structures.

The tertiary structure of a protein results from four factors related to the side chains of the amino acids that form the backbone of the protein, as shown in Figure 25.28.

Disulfide (S—S) linkages. If the folding of a protein brings two cysteine residues together, the two —SH side chains can undergo oxidation to form a covalent —S—S— bond that either cross-links the polypeptide chain or holds two peptide chains together.

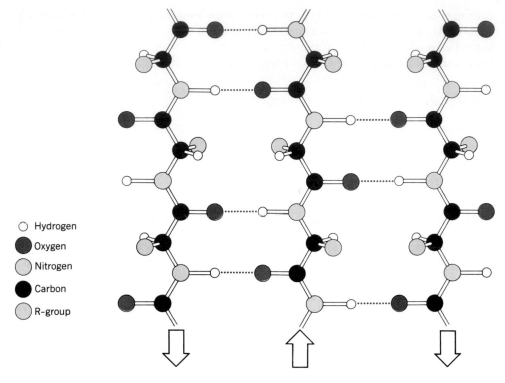

Hydrogen bonding. In addition to the hydrogen bonding between peptide bonds that gives rise to the secondary structure of the protein, hydrogen bonds can also form between amino acid side chains.

Ionic bonding. The structure of a protein can be stabilized by the force of attraction between amino acid side chains of opposite charge, such as the $-NH_3^+$ side chain of LEU and the $-CO_2^-$ side chain of ASP.

Hydrophobic interactions. The side chains of the amino acids GLY, ALA, VAL, LEU, ILE, PRO, MET, PHE, and TRP are hydrophobic. Proteins often fold so that these water-hating substituents are buried within the protein, where they can interact to form hydrophobic pockets that stabilize this structure of the protein.

Exercise 25.9

Hair and wool contain proteins, known as the soft keratins, that are rich in the amino acid cysteine. Horn and hoofs contain the hard keratins. Predict whether hard keratins should contain more or less cysteine than their soft analogs.

Solution

The $-S-S-$ bonds between cysteine side chains in a protein can be used to cross-link individual protein molecules. Anything that increases the number of cross-links should make the cell that contains these proteins more rigid. We would therefore predict that the hard keratins in horn and hoofs should contain more cysteine than the soft keratins in hair and wool.

Exercise 25.10

Human hair contains proteins that are about 14% cysteine. Use this fact to explain what happens when someone gets a "permanent."

Solution

Hair curls as it grows because of the disulfide ($-S-S-$) links between cysteine residues on adjacent protein molecules. It takes three steps to change the way hair curls. The first step involves shaping the hair to our satisfaction and then locking it into place with curlers. The hair is then treated with a mild reducing agent that reduces the $-S-S-$ bonds to give a pair of $-SH$ groups. This relaxes the structure of the proteins in the hair, allowing them to pick up the structure dictated by the curlers. The $-SH$ side chains on cysteine residues that are now adjacent to each other are oxidized by the O_2 in air. New $-S-S-$ linkages form, locking the hair "permanently" in place—at least, until new hair grows.

THE QUATERNARY STRUCTURE OF PROTEINS

Hemoglobin is the protein that absorbs O_2 in the lungs and carries it through the bloodstream to the muscles, where it is used. Each of these proteins consists of four polypeptide chains—two α-chains that contain 141 amino acids each and two β-chains that contain 146 amino acids each. Hemoglobin is therefore an example of a protein that has a quaternary structure. It consists of four polymer chains, which must be assembled to form the complete protein.

The polymer chains in a quaternary protein are not covalently linked like the two polypeptide chains in insulin. The force of attraction between the α- and β-chains in hemoglobin results from interactions between hydrophobic substituents on these polymer chains. In other quaternary proteins, hydrogen bonding or ionic bonding interactions between amino acid side chains on the outer surfaces of adjacent polymer chains are responsible for assembling the polymer chains.

THE DENATURATION OF PROTEINS

Proteins are fragile molecules that are remarkably sensitive to changes in structure. For example, the replacement of a single glutamic acid by a nonpolar valine at the sixth position on the β-chains of hemoglobin causes the disease known as sickle-cell anemia. The introduction of a hydrophobic VAL residue at this position changes the quaternary structure of hemoglobin. The "sticky," nonpolar side chain on a valine residue at this position causes hemoglobin molecules to cluster together in an abnormal fashion, interfering with their function as oxygen-carrying proteins.

Sickle-cell anemia is the result of a change in the way the protein is assembled from amino acids. The structure of a protein can also be changed after it has been made. Anything that causes a protein to leave its normal, or natural, structure is said to **denature** the protein. Factors that can lead to denaturation include the following.

1. Heating, which causes one portion of the protein molecule to both vibrate and rotate more rapidly than the rest of the protein. This disrupts the secondary and tertiary structure of the protein, which inevitably leads to denaturation. The changes we observe when we fry an egg result from denaturation caused by heating.

2. Changes in pH that interfere with ionic bonding between amino acid side chains.

3. Detergents, which make nonpolar amino acid side chains soluble and thereby destroy the hydrophobic interactions that give rise to the tertiary and quaternary structure of the protein.

4. Oxidizing or reducing agents that either create or destroy S—S bonds.

5. Reagents such as urea (H_2NCONH_2), which disrupt the hydrogen bonds that form the secondary structure of the protein.

25.12 CARBOHYDRATES AND THE MONOSACCHARIDES

The family of compounds known as carbohydrates was once limited to compounds that had the empirical formula CH_2O — literally, hydrates of carbon. In recent years, carbohydrates have been classified on the basis of their structures, not their formulas. They are now defined more broadly as polyhydroxy aldehydes and ketones. Among the compounds that belong to this family are cellulose, starch, glycogen, and most sugars.

Exercise 25.11

The following compounds are now considered to be carbohydrates because of their structures. Which of these compounds satisfy the original definition of carbohydrates?

(a) glucose, $C_6H_{12}O_6$ (c) cellulose, $(-C_6H_{10}O_5-)_n$
(b) sucrose, $C_{12}H_{22}O_{11}$ (d) ribose, $C_5H_{10}O_5$

Solution

Only glucose and ribose have the empirical formula CH_2O. The molecular formula of sucrose suggests that this compound is formed by condensation of a pair of $C_6H_{12}O_6$ monomers by the elimination of a molecule of water. The formula for the repeating unit in cellulose suggests that this polymer is similarly formed.

We are so familiar with the food we call "sugar" that we often forget that this substance is actually just one member of a class of carbohydrates that taste sweet. The sugar we cook with is more precisely known as **sucrose** or cane sugar, $C_{12}H_{22}O_{11}$. Two other common sugars have the same formula but different structures—**maltose,** or malt sugar, and **lactose,** or milk sugar. There are also a number of sugars with the formula $C_6H_{12}O_6$, including both **glucose** and **fructose.** Glucose is often known as blood sugar, because it is the sugar with the highest concentration in the bloodstream. Fructose is also known as fruit sugar, because it is found in fruit and honey.

Carbohydrates serve a variety of functions.

1. They are important intermediates in the series of reactions that cells use to break down, or digest, complex molecules.
2. They serve as starting materials for the synthesis of proteins, nucleic acids, and lipids.
3. They are synthesized by plants from CO_2 and H_2O during photosynthesis and are therefore the first step in the process by which plants absorb energy from the sun.
4. They are used by both plants and animals to store food energy.
5. They are important in the structure of plant cells.

There are three important classes of carbohydrates: (1) monosaccharides, such as glucose and fructose, (2) disaccharides, such as sucrose and lactose, and (3) polysaccharides, such as cellulose and starch.

The **monosaccharides** are white, crystalline solids that contain a single aldehyde or ketone functional group. They are divided into two classes—**aldoses** and **ketoses**—on the basis of whether they contain an *ald*ehyde or a *ket*one. Each is also classified as a triose, tetrose, pentose, hexose, or heptose, depending on whether it contains three, four, five, six, or seven carbon atoms.

Almost without exception, the monosaccharides are optically active compounds. Although both D and L isomers are possible, most of the monosaccharides found in nature are in the D configuration. Structures for the D and L isomer of the simplest aldose—glyceraldehyde—are shown below.

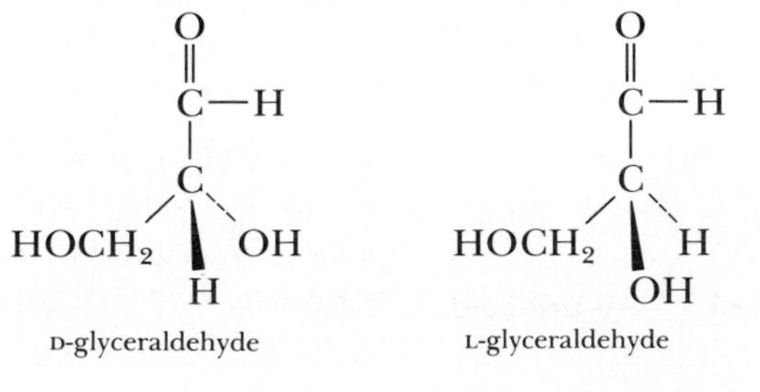

D-glyceraldehyde   L-glyceraldehyde

The structures of many monosaccharides were first determined by Emil Fischer in the 1880s and 1890s and are still written according to a convention he developed. The Fischer projection, as it is called, attempts to represent what the molecule would look like if its three-dimensional structure were projected onto a piece of paper. By convention, Fischer projections are written vertically, with the aldehyde or ketone group at the top. The —OH group on the second-to-last carbon atom is written on the right side of the skeleton structure for the D isomer and on the left for the L isomer. Fischer projections for the two isomers of glyceraldehyde are shown below.

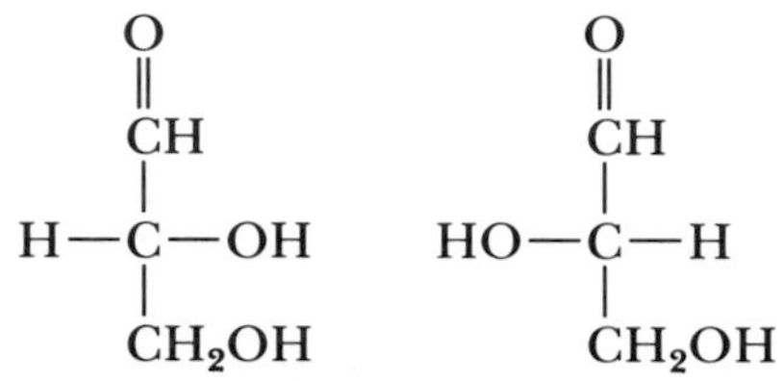

D-glyceraldehyde   L-glyceraldehyde

Fischer projections for some of the more common monosaccharides are given in Figure 25.29.

Exercise 25.12

Use the Fischer projections in Figure 25.29 to explain the difference between the structures of glucose and fructose.

Solution

Both compounds satisfy the definition of a carbohydrate as a polyhydroxy aldehyde or ketone. Both have the same molecular formula — $C_6H_{12}O_6$. Both have the same structure for the third, fourth, fifth, and sixth carbon atoms. The difference between these compounds is the fact that one of them is an aldehyde (glucose) and the other is a ketone (fructose).

FIG. 25.29 Fischer projections for some of the common monosaccharides.

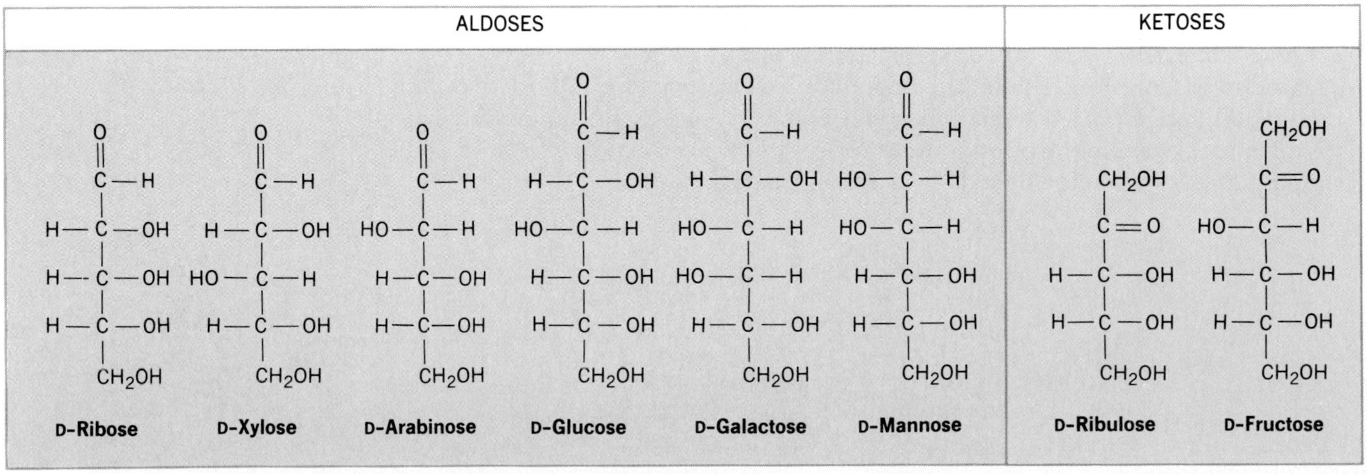

D-glucose

α-D-glucopyranose

β-D-glucopyranose

FIG. 25.30 The —OH group near one end of a carbohydrate can attack the C=O group at the other end to form a six-membered cyclic compound known as a pyranose. This figure shows the α- and β-isomers of the pyranose that results when glucose forms a cyclic compound.

If the carbon chain is long enough, the alcohol at one end of a monosaccharide can attack the carbonyl group at the other end to form a cyclic compound. When a six-membered ring is formed, the structure is called a *pyranose* (see Figure 25.30). When a five-membered ring is formed, it is called a *furanose* (see Figure 25.31). There are two possible structures for the pyranose and furanose forms of a monosaccharide—the so-called α- and β-isomers.

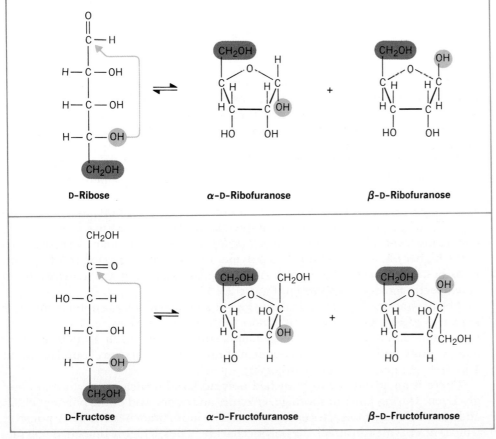

D-Ribose α-D-Ribofuranose β-D-Ribofuranose

D-Fructose α-D-Fructofuranose β-D-Fructofuranose

FIG. 25.31 When a carbohydrate forms a five-membered ring, it is known as a furanose. This figure shows the α- and β-isomers of the furanose that result when D-ribose and D-fructose form cyclic compounds.

25.13 CARBOHYDRATES: THE DISACCHARIDES AND POLYSACCHARIDES

Disaccharides are formed by the condensation of a pair of monosaccharides. The structures of three important disaccharides with the formula $C_{12}H_{22}O_{11}$ are shown in Figure 25.32. Maltose, or malt sugar, which forms when starch breaks down, is an important component of the barley malt used to brew beer. Lactose, or milk sugar, is a disaccharide found in milk. Very young children have a special enzyme known as *lactase* that helps digest lactose. As they grow older, many people, especially people of Asian or African descent, lose the ability to digest lactose and cannot tolerate milk. Because human milk has twice as much lactose as milk from cows, young children who develop lactose intolerance while they are being breast-fed are often helped by a switch to cows' milk or a synthetic formula based on sucrose.

The substance most people refer to as sugar is the disaccharide sucrose, which is extracted from either sugar cane or beets. Sucrose is the sweetest of the disaccharides — it is roughly three times as sweet as maltose and six times as sweet as lactose. In recent years, however, the cost of sucrose has steadily increased. It has been replaced in many commercial products by corn syrup, which is obtained when the polysaccharides in cornstarch are broken down. Corn syrup is primarily D-glucose. Unfortunately, glucose is only about 70% as sweet as sucrose. Fructose, however, is about two and a half times as sweet as glucose. A commercial process has been developed that uses an isomerase enzyme to convert about half of the glucose in corn syrup into fructose. This high-fructose corn sweetener is just as sweet as sucrose and has found extensive use in soft drinks.

Monosaccharides and disaccharides represent only a small fraction of the total amount of carbohydrates in the natural world. The great bulk of the carbohydrates in nature are present as *polysaccharides,* which have relatively large molecular weights. The polysaccharides serve two principal functions: They are used by both plants and animals to store glucose as a source of future food energy, and they provide some of the mechanical structure of cells.

Very few forms of life receive a constant supply of energy from their environments. In order to survive, plant and animal cells had to develop a way of storing energy during times of plenty in order to survive the times of shortage that follow.

Plants store food energy as polysaccharides such as *starch.* There are two basic kinds of starch — amylose and amylopectin. *Amylose* is found in algae and other lower forms of plants. It is a linear polymer of approximately 600 glucose residues whose structure we can predict by simply extending the structure of maltose by adding more α-D-glucopyranose rings. *Amylopectin* is the dominant form of starch in the higher plants. It is a branched polymer of about 6000 glucose residues with branches on 1 in every 24 glucose rings. A small portion of the structure of an amylopectin molecule is shown in Figure 25.33.

The polysaccharide that animals use for the short-term storage of food energy is known as *glycogen.* Glycogen has almost the same structure as amylopectin, with two minor differences. The glycogen molecule is larger, with 10,000 or more glucose units, and it contains twice as many branches. Branches are found on about 1 in every 12 glucose rings in glycogen.

There is an advantage to branched polysaccharides such as amylopectin and glycogen. During times of shortage, enzymes attack one end of the polymer chain and cut off glucose molecules one at a time. The more branches, the more points at which the enzyme can attack the polysaccharide. Thus, a highly branched polysac-

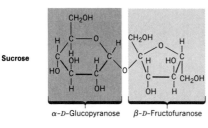

α-Maltose

α-D-Glucopyranose

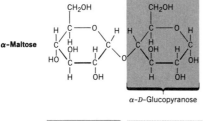

Sucrose

α-D-Glucopyranose β-D-Fructofuranose

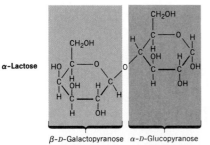

α-Lactose

β-D-Galactopyranose α-D-Glucopyranose

FIG. 25.32 The structures of three common disaccharides — maltose, sucrose, and lactose.

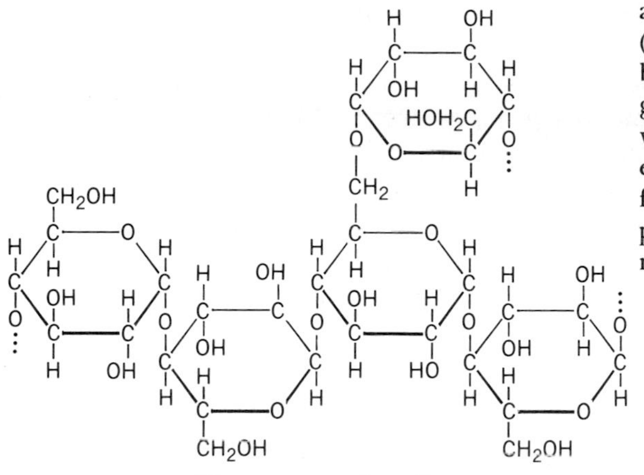

FIG. 25.33 The glycogen in animals and the amylopectin (starch) in plants are both branched polymers of α-D-glucopyranose monomers, which are used to store food energy. Amylose is another form of starch that is a linear polymer of α-D-glucopyranose monomers.

Glycogen or amylopectin

charide is better suited for the rapid release of glucose to the cell than a linear polymer.

Polysaccharides are also used to form the walls of plant and bacterial cells. Cells that don't have a cell wall often break open in solutions whose salt concentrations are either too low (hypotonic) or too high (hypertonic). If the concentration of ions in the solution is much smaller than the concentration inside the cell, osmotic pressure forces water into the cell to bring the system into balance, which causes the cell to burst. If the ionic concentration in the solution is too high, osmotic pressure forces water out of the cell, and the cell breaks open as it shrinks. The cell wall provides the mechanical strength that helps protect plant cells that live in fresh-water ponds (too little salt) or seawater (too much salt) from osmotic shock. The cell wall also provides the mechanical strength that allows plant cells to support the weight of other cells.

The most abundant structural polysaccharide is *cellulose.* There is so much cellulose in the cell walls of plants that it is the most abundant of all biological molecules. Cellulose is a linear polymer of glucose molecules, with a structure that resembles that of amylose more closely than that of amylopectin. There is an important difference between cellulose and amylose, however, which you can see by comparing Figures 25.33 and 25.34. Cellulose is formed by the linking of β-glucopyranose rings, instead of the α-glucopyranose isomers used in starch and glycogen.

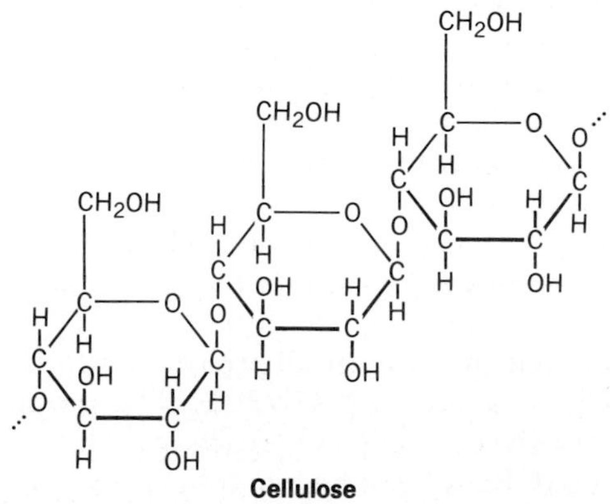

FIG. 25.34 Cellulose is a structural polysaccharide that is also a linear polymer of D-glucopyranose monomers. Cellulose differs from amylose, however, because it is a polymer of the β-isomer of D-glucopyranose.

Cellulose

Cellulose and starch provide an excellent example of the importance of structure in determining the function of biomolecules. At the turn of the century, Emil Fischer proposed that the structure of an enzyme is carefully matched to the substance on which it acts, in much the same way that a lock and key are matched. As a result, the α-amylase enzyme in saliva that breaks down the α-linkages between glucose molecules in starch cannot act on the β-linkages in cellulose.

Most animals can't digest cellulose because they don't have an enzyme that can cleave β-linkages between glucose molecules. Cellulose in their diet serves only as fiber, or roughage. The digestive tracts of some animals, such as cows, horses, sheep, and goats, contain bacteria that have enzymes that cleave these β-linkages; so these animals can digest cellulose.

Exercise 25.13

Termites provide another example of the symbiotic relationship between bacteria and higher organisms. Termites cannot digest the cellulose in the wood they eat, but their digestive tracts are infested with bacteria that can. Propose a simple way of ridding a house from termites, without killing other insects that might be beneficial.

Solution

Killing termites is not a difficult task. There are a number of poisons that do an excellent job. Unfortunately, these compounds may also poison other insects or even small animals or children. The simplest way of getting rid of termites is to treat the wood with a poison that kills the bacteria that digest the cellulose in wood, so that the termites eventually starve to death.

25.14 LIPIDS

Proteins, carbohydrates, and nucleic acids are classified on the basis of similarities in their structure and function. The lipids, however, are grouped on the basis of one of their physical properties—their tendency to dissolve in nonpolar solvents such as chloroform ($CHCl_3$), benzene (C_6H_6), and diethyl ether ($CH_3CH_2OCH_2CH_3$). Because they are soluble in nonpolar solvents, lipids tend to be insoluble in water, and they often feel oily or greasy to the touch.

NEUTRAL FATS AND OILS

In Section 24.14, we stated that carboxylic acids become less soluble in water as the length of the alkyl chain increases and that the long-chain carboxylic acids are called *fatty acids* because they can be obtained from animal fats. The names and structures of some of the fatty acids are given in Table 25.4. *Saturated fatty acids* are long-chain carboxylic acids that contain no C=C double bonds; *unsaturated fatty acids* are long-chain carboxylic acids that contain one or more C=C double bonds.

Free fatty acids are almost never found in nature. Almost all are tied up with alcohols or amines to form esters (RCO_2R) or amides ($RCONHR$). The most abundant lipids are the triesters formed when a glycerol molecule reacts with three fatty acids, as shown in Figure 25.35. Such lipids have been known by a variety of

TABLE 25.4

Fatty Acids

		Saturated Fatty Acids	
Compound	*Common Name*	*Systematic Name*	
$CH_3(CH_2)_4CO_2H$	caproic acid	hexanoic acid	
$CH_3(CH_2)_6CO_2H$	caprylic acid	octanoic acid	
$CH_3(CH_2)_8CO_2H$	capric acid	decanoic acid	
$CH_3(CH_2)_{10}CO_2H$	lauric acid	dodecanoic acid	
$CH_3(CH_2)_{12}CO_2H$	myristic acid	tetradecanoic acid	
$CH_3(CH_2)_{14}CO_2H$	palmitic acid[a]	hexadecanoic acid	
$CH_3(CH_2)_{16}CO_2H$	stearic acid[a]	octadecanoic acid	
$CH_3(CH_2)_{18}CO_2H$	arachidic acid	eicosanoic acid	

Unsaturated Fatty Acids	
Compound	*Common Name*
$CH_3(CH_2)_5CH{=}CH(CH_2)_7CO_2H$	palmitoleic acid[b]
$CH_3(CH_2)_7CH{=}CH(CH_2)_7CO_2H$	oleic acid[b]
$CH_3(CH_2)_4CH{=}CHCH_2CH{=}CH(CH_2)_7CO_2H$	linoleic acid[b]
$CH_3CH_2CH{=}CH_2CH_2CH{=}CHCH_2CH{=}CH(CH_2)_7CO_2H$	linolenic acid
$CH_3(CH_2)_4CH{=}CHCH_2CH{=}CHCH_2CH{=}CHCH_2CH{=}CH(CH_2)_3CO_2H$	arachidonic acid[b]

[a] The most abundant saturated fatty acids in animals.   [b] The most abundant unsaturated fatty acids in animals.

names, including *fat, neutral fat, glyceride, triglyceride,* and, most recently, *triacylglycerol.*

Most fats are complex mixtures of different triacylglycerols. As the percentage of unsaturated fatty acids in these fats becomes larger, the fat melts at lower temperatures. The principal difference between vegetable oils and animal fats is the degree of unsaturation of the fatty acids. Beef fat, which is roughly 50% unsaturated fatty acids, is a solid. Olive oil, which is roughly 80% unsaturated, is a liquid. A major industry has evolved around the reduction of the unsaturated fatty acids in vegetable oils to form solid fats.

Fats and oils are used by living cells for only one purpose—to store energy. They are a far more efficient storage system than glycogen or starch because they give off between two and three times as much energy when they are burned. This explains why the seeds of many plants are relatively rich in oils, which provide the energy the seed needs to grow until the leaves can begin to produce energy by photosynthesis.

Roughly 15% of the weight of the average human is fat. Although the average human has about 2500 kilojoules of energy available in the form of glycogen, fat stored in the body represents more than 400,000 kilojoules. If all of this energy were stored as glycogen, the human body would have to weigh at least one-third more than it does now.

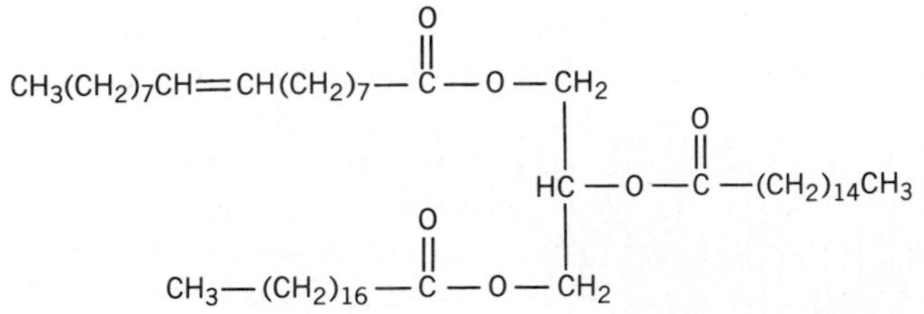

FIG. 25.35 Neutral fats and oils are triesters of glycerol ($HOCH_2CHOHCH_2OH$) and three fatty acids.

FIG. 25.36 When one of the fatty acids in a neutral fat or oil is replaced with a phosphate, the result is a polar lipid molecule that contains two nonpolar, hydrophobic tails, and a polar, hydrophilic, phosphate-ion head. This compound is just one member of a class of molecules known as the phospholipids.

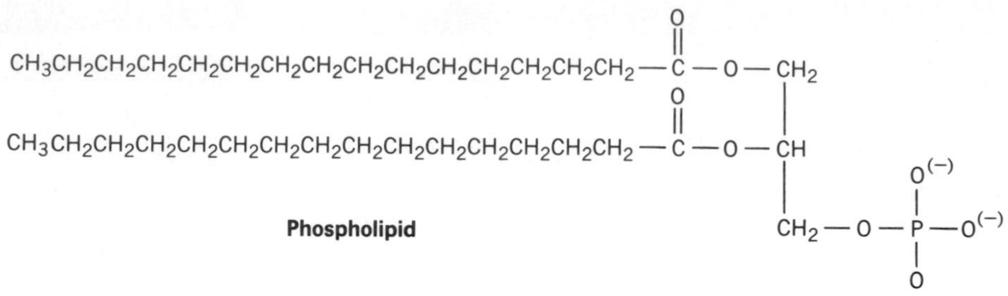

$$CH_3CH_2CH_2CH_2CH_2CH_2CH_2CH_2CH_2CH_2CH_2CH_2CH_2CH_2CH_2 - \overset{\overset{O}{\|}}{C} - O - CH_2$$

$$CH_3CH_2CH_2CH_2CH_2CH_2CH_2CH_2CH_2CH_2CH_2CH_2CH_2CH_2CH_2 - \overset{\overset{O}{\|}}{C} - O - CH$$

$$CH_2 - O - \overset{\overset{O^{(-)}}{|}}{\underset{O}{\overset{}{P}}} - O^{(-)}$$

Phospholipid

POLAR LIPIDS

Fats and oils are neutral compounds. If one of the fatty acids in a triacylglycerol is replaced by a phosphate group, however, the result is a **phospholipid** that has two nonpolar hydrophobic tails and a charged hydrophilic head, as shown in Figure 25.36. These phospholipids spontaneously associate to form the membranes (see Figure 25.37) that surround living cells and help subdivide cells into compartments. Although lipids technically don't form polymers like proteins or carbohydrates, they do associate to form structures of macromolecular size and therefore belong in any discussion of polymers.

FATTY ALCOHOLS, WAXES, TERPENES, AND STEROIDS

A number of lipids can be classified as neither neutral fats and oils nor as polar lipids. **Waxes,** for example, are esters of long-chain fatty acids and long-chain fatty alcohols.

$$CH_3(CH_2)_{24} - \overset{\overset{O}{\|}}{C} - O - (CH_2)_{30}CH_3$$

a typical component of beeswax

FIG. 25.37 Phospholipids spontaneously cluster to form the membranes that separate cells from each other and divide, or compartmentalize, each cell.

They have one principal function—to make the skin, hair, wool, and fur of animals, the feathers of birds, and the leaves of plants water-resistant, which reduces the amount of water these organisms lose through evaporation.

The **terpenes,** which were introduced in Section 24.5, are also lipids, because they are naturally occurring substances that are soluble in nonpolar solvents. The function of terpenes in plants is not completely understood. Some terpenes act as

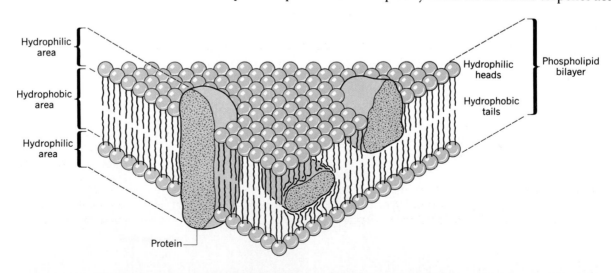

Steroids:

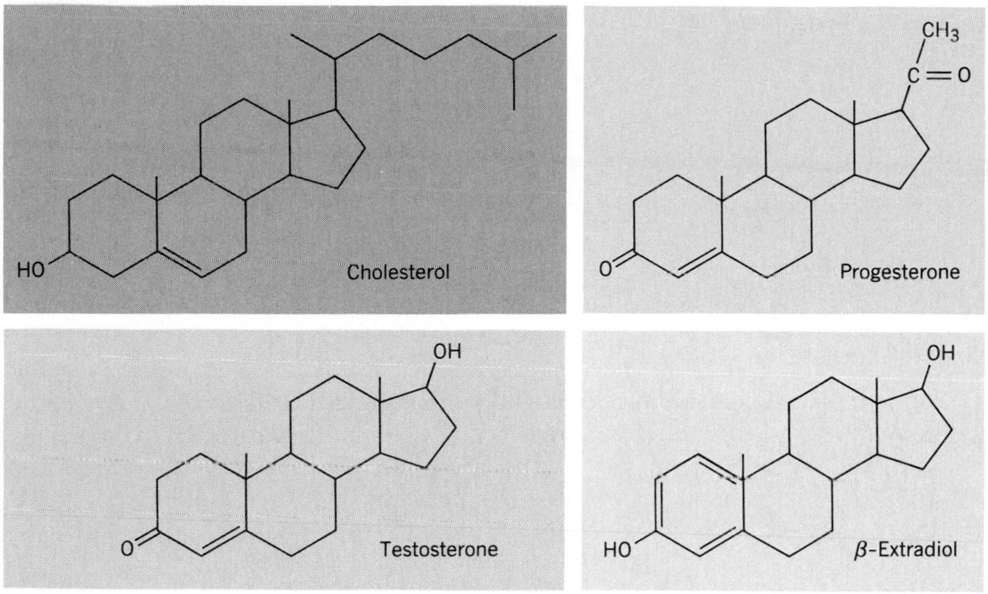

FIG. 25.38 The family of compounds known as lipids consists of biologically important compounds that are soluble in nonpolar solvents. Steroids such as cholesterol, progesterone, testosterone, and β-estradiol are lipids.

Cholesterol

Progesterone

Testosterone

β-Extradiol

pigments—the β-carotenes, for example, are responsible for the color of carrots. Many of the terpenes have very distinctive odors and are therefore thought to be involved in the processes by which plants attract certain species of insects and repel others.

Terpenes also serve as the biosynthetic precursors of the **steroids**, such as cholesterol, progesterone, testosterone, and β-estradiol, shown in Figure 25.38. Cholesterol is the most abundant of these steroids. It helps to stabilize the structure of membranes and is found in almost all animal tissue. The disease atherosclerosis, or hardening of the arteries, occurs when fat tissue containing cholesterol is deposited on the walls of the major blood vessels. Although the disease is not necessarily caused by high levels of cholesterol in the blood, the two are closely related.

25.15 NUCLEIC ACIDS

Nucleic acids were first isolated in 1869, shortly after the Civil War. Their name reflects the fact that they are relatively strong acids that were isolated from the nuclei of cells. The nucleic acids are polymers with molecular weights as high as 100,000,000 grams per mole. They can be broken down, or digested, to form monomers known as nucleotides.

Each **nucleotide** is composed of three units—a sugar, an amine, and a phosphate ion—as shown in Figure 25.39. Nucleic acids are divided into two classes, DNA and RNA, on the basis of the structure of the sugar used to form its nucleotides. **Ribonucleic acid**, or **RNA**, is built on a β-D-ribofuranose ring. **Deoxyribonucleic acid**, or **DNA**, contains a modified ribofuranose in which the —OH group on the second carbon atom has been removed, as shown in Figure 25.40.

The five amines that form nucleic acids fall into two categories, *purines* and *pyrimidines* (see Figure 25.41). There are three pyrimidines—cytosine, thymine, and uracil—and two purines—adenine and guanine. DNA and RNA each contain four nucleotides. Both contain the same purines—adenine and guanine—and both also contain the pyrimidine cytosine. But the fourth nucleotide in DNA is thymine, whereas RNA uses uracil to complete its quartet of nucleotides.

Cytidine monophosphate

FIG. 25.39 Nucleotides contain a furanose sugar, an amine, and a phosphate ion. In this example, the amine is cytosine and the nucleotide is known as cytidine monophosphate.

FIG. 25.40 One of the differences between DNA and RNA is the sugar on which the nucleotide is built. RNA uses β-ᴅ-ribofuranose. DNA uses a similar sugar in which the —OH group on the second carbon atom has been replaced by a hydrogen. Since this sugar is *deoxy*ribofuranose, DNA is known as deoxyribonucleic acid.

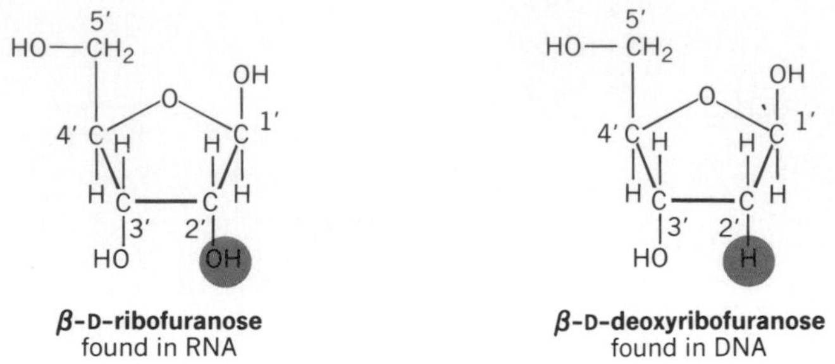

β-ᴅ-ribofuranose
found in RNA

β-ᴅ-deoxyribofuranose
found in DNA

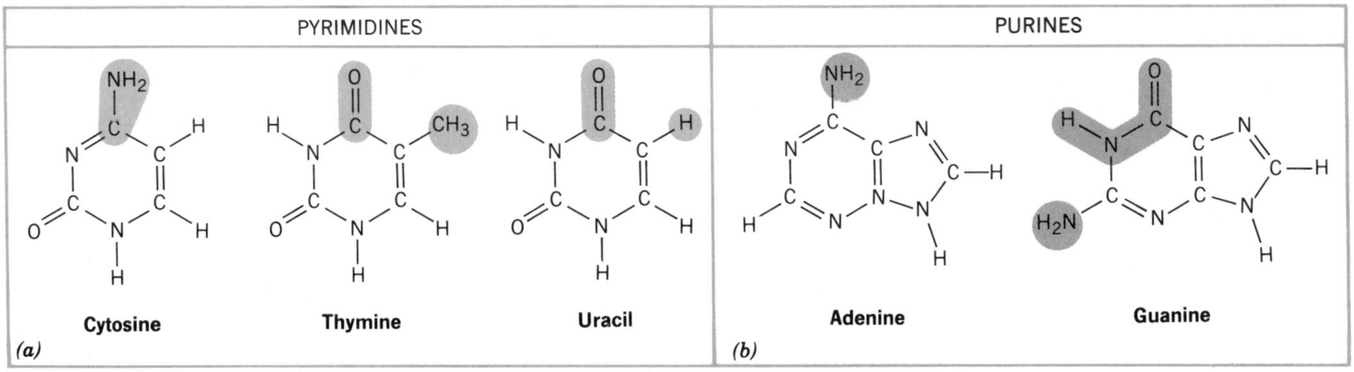

PYRIMIDINES

Cytosine

Thymine

Uracil

PURINES

Adenine

Guanine

(a)

(b)

FIG. 25.41 The amines in nucleic acids fall into two classes: (*a*) the pyrimidines: uracil, cytosine, and thymine, and (*b*) the purines: adenine and guanine.

The carbon atoms in the sugar at the center of a nucleotide are numbered from 1' to 5', as shown in Figure 25.40. The —OH group on the 3' carbon of one nucleotide can react with the phosphate attached to the 5' carbon of another to form a dinucleotide held together by phosphate ester bonds. As the chain continues to grow, it becomes a ***polynucleotide***, or nucleic acid. A short segment of a DNA chain is shown in Figure 25.42. Reading from the 5' end of this chain to the 3' end, this DNA segment contains the following sequence of amine substituents: adenine (*A*)–cytosine (*C*)–guanine (*G*)–thymine (*T*).

For many years, the role of nucleic acids in living systems was unknown. In 1944, seventy-five years after they had first been isolated, Oswald Avery presented evidence that nucleic acids were involved in the storage and transfer of the genetic information needed for the synthesis of proteins. This suggestion was actively opposed by many of his contemporaries, who believed that the structure of the nucleic acids was too regular, or too dull, to carry the information that codes for the thousands of different proteins a cell needs to survive.

At first glance, nucleic acids do appear far more regular than proteins. In retrospect, the first clue about how nucleic acids function was obtained by Erwin Chargaff, who found that DNA always contains the same relative amounts of certain pairs of bases. There is always just as much adenine as thymine and just as much guanine as cytosine.

It was not until April 1954, however, that James Watson and Francis Crick proposed a structure for DNA that clearly offered a mechanism for the storage of genetic information. Their structure consisted of two polynucleotide chains running in opposite directions and linked by hydrogen bonds between a specific purine (*A* or *G*) on one strand and a specific pyrimidine (*C* or *T*) on the other, as shown in Figure 25.43. These strands form a helix, which is not quite as tightly coiled as the α-helix Pauling and Corey proposed for proteins (shown in Figure 25.26).

This structure must be able to explain two processes — ***replication*** and ***transla-***

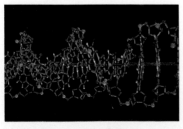

Molecular model of DNA.

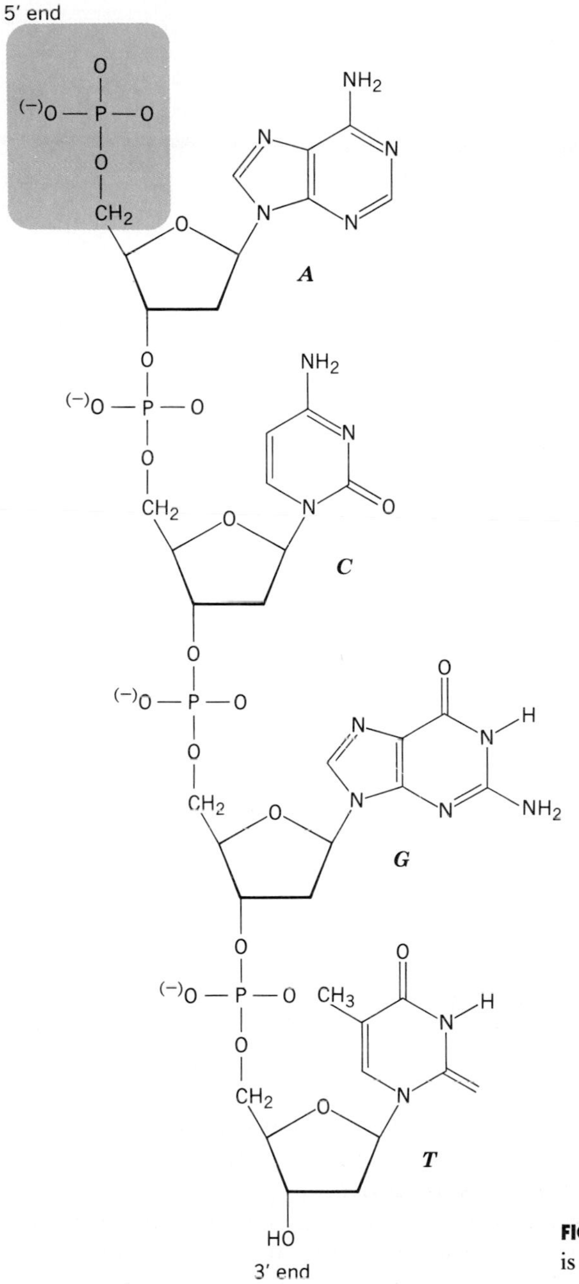

5' end

FIG. 25.42 This tetranucleotide is an example of a short segment of a single strand of DNA.

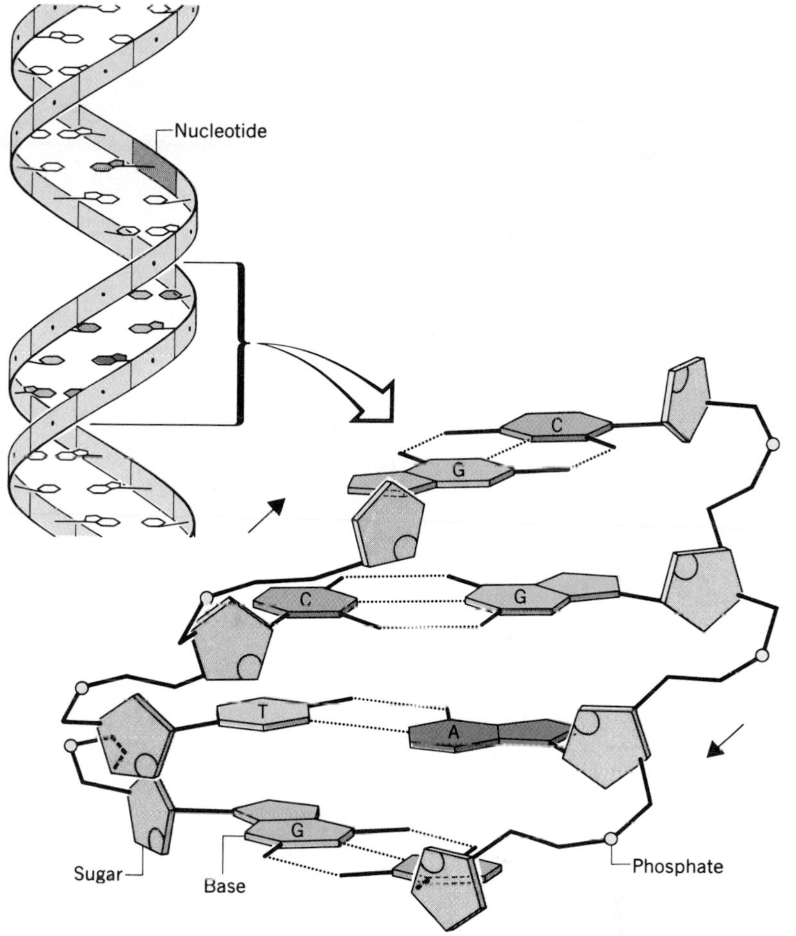

3' end

FIG. 25.43 DNA consists of two polynucleotide chains running in opposite directions and held together by hydrogen bonds between adenine and thymine or guanine and cytosine side chains.

tion. There must be some way to make perfect copies of the DNA that can be handed down to future generations (replication), and there must also be some way to translate the information coded on the DNA chain into a sequence of amino acids in a protein (translation).

Replication is fairly easy to understand. According to Watson and Crick, an adenine on one strand of DNA is always paired with a guanine on the other, and a cytosine is always paired with a thymine, as shown in Figure 25.44. The two strands of DNA therefore complement each other perfectly. The sequence of nucleotides on one strand can always be predicted from the sequence on the other. Replication

FIG. 25.44 The two strands in DNA are held together by hydrogen bonds between specific pairs of purine and pyrimidine bases. This figure shows the hydrogen bonds between *A* and *T* and *G* and *C*.

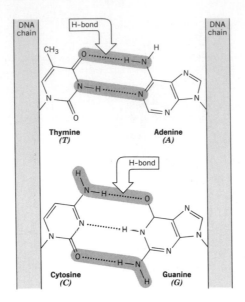

FIG. 25.45 When replication takes place, both strands of the DNA are copied simultaneously, and each of the daughter molecules obtains one strand from the parent DNA.

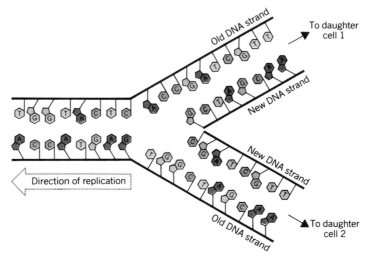

occurs, as shown in Figure 25.45, when the two strands of the parent DNA molecule separate and both strands are copied simultaneously. Thus, one strand from the parent DNA is present in each of the daughter molecules that are made when a cell divides.

25.16 PROTEIN BIOSYNTHESIS

The information that tells a cell how to build the proteins it needs to survive is coded in the structure of the DNA in the nucleus of that cell. This code can't be based on a one-to-one match between nucleotides and amino acids, because there are only four nucleotides and 20 amino acids must be coded. If the nucleotides are grouped in threes, however, there are 64 possible triplets, or *codons,* which is more than enough combinations to code for the 20 amino acids.

To understand how proteins are made, we have to divide the decoding process into two steps: *transcription* and *translation.* DNA only stores the genetic infor-

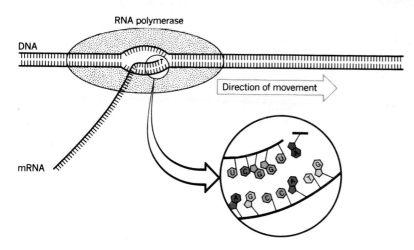

RNA polymerase

DNA

Direction of movement

mRNA

FIG. 25.46 During transcription, an RNA molecule is synthesized that is a complementary copy of one strand of the parent DNA molecule. This RNA now carries the message encoded in the DNA structure and is therefore called messenger-RNA.

mation; it is not directly involved in the process by which the information is used. The first step in protein biosynthesis therefore involves transcribing the information in the DNA structure into a useful form. In a separate step, this information is translated into a sequence of amino acids.

TRANSCRIPTION

Before the information in DNA can be decoded, a small portion of the DNA double helix must be uncoiled, as shown in Figure 25.46. A strand of RNA is then synthesized that is a complementary copy of the DNA.

Assume that the section of the DNA that is copied has the following sequence of nucleotides, starting from the 3′ end.

$$3'\quad T\text{-}A\text{-}C\text{-}A\text{-}A\text{-}G\text{-}C\text{-}A\text{-}G\text{-}T\text{-}T\text{-}G\text{-}G\text{-}T\text{-}C\text{-}G\text{-}T\text{-}G...\quad 5'\quad \text{DNA}$$

When we predict the sequence of nucleotides in the RNA complement, we have to remember that RNA uses *U* in place of the *T* in DNA. We also have to remember that base pairing occurs between two chains that run in opposite direction. The RNA complement of this DNA should therefore be as follows.

$$3'\quad T\text{-}A\text{-}C\text{-}A\text{-}A\text{-}G\text{-}C\text{-}A\text{-}G\text{-}T\text{-}T\text{-}G\text{-}G\text{-}T\text{-}C\text{-}G\text{-}T\text{-}G...\quad 5'\quad \text{DNA}$$
$$5'\quad A\text{-}U\text{-}G\text{-}U\text{-}U\text{-}C\text{-}G\text{-}U\text{-}C\text{-}A\text{-}A\text{-}C\text{-}C\text{-}A\text{-}G\text{-}C\text{-}A\text{-}C...\quad 3'\quad \text{RNA}$$

There is a remarkably large amount of apparently useless information in the DNA of higher organisms. In addition to the nucleotides that code for proteins (the **exons**), there are numerous stretches of nucleotides (the **introns**) that seem to have nothing to do with the daily work of the cell. Any of this nonsense that was copied during transcription is now snipped from the RNA chain. Since this RNA strand contains the message that was coded in the DNA, it is called **messenger-RNA,** or **m-RNA.**

TRANSLATION

The messenger-RNA now travels to a site in the cell known as the ribosome, where the message is translated into a sequence of amino acids. The 5′ end of the messenger-RNA molecule binds to the smaller subunit of the ribosome, and then the larger subunit packs on top, as shown in Figure 25.47. The amino acids that are incorporated into the protein being synthesized are carried by rela-

FIG. 25.47 The first step in translation involves the binding of the messenger-RNA to the two subunits of the ribosome. A t-RNA carrying the amino acid MET then binds to the codon on the messenger-RNA that signals the start of the protein. After that, a t-RNA carrying the next amino acid binds to the next codon on the messenger-RNA. The MET is transferred from the first t-RNA to the amino acid on the second t-RNA to form a dipeptide. This sequence continues until one of the termination codons is encountered.

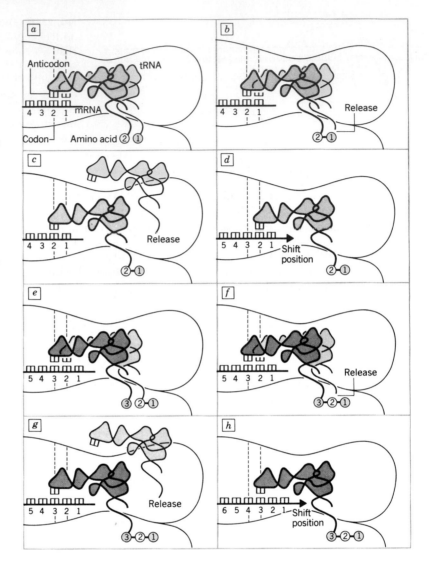

tively small RNA molecules known as **transfer-RNA,** or **t-RNA.** There are at least 60 t-RNAs in each cell, and they differ slightly in their structure. At one end of each t-RNA is a specific sequence of three nucleotides that can bind to the messenger RNA; at the other end there is a specific amino acid. Thus, each three-nucleotide segment of the messenger-RNA molecule codes for the incorporation of a particular amino acid. The relationship between these triplets, or codons, and the amino acids is shown in Table 25.5.

Exercise 25.14

Assume that the DNA chain that codes for the synthesis of a particular protein contains the triplet *A-G-T* (reading from the 3′ to the 5′ end). Predict the sequence of nucleotides in the triplet, or codon, that would be built in the messenger-RNA constructed on this DNA template. Then predict the amino acid that would be incorporated at this point in the protein.

Solution

The messenger-RNA synthesized when this DNA chain is read is a complement of the DNA. Every time a *G* appears, a *C* is placed on the complementary chain, and vice versa. In DNA, *A* and *T* are paired, but RNA uses *U* instead of *T*.

DNA:	3′	...A-G-T...	5′
RNA:	5′	...U-C-A...	3′

We now look up the *U-C-A* codon in Table 25.5. According to this table, this triplet on m-RNA codes for the amino acid serine.

The signal to start making a polypeptide chain is the triplet AUG, which codes for the amino acid methionine (MET). The synthesis of every protein therefore starts with a MET residue at the *N*-terminal end of the polypeptide chain. After the t-RNA carrying MET binds to the start signal on the messenger-RNA, a t-RNA carrying the second amino acid binds to the next codon. A dipeptide is synthesized when the MET is transferred from the first t-RNA to the amino acid on the second t-RNA. In the example shown in Figure 25.47, the dipeptide is MET-PHE (reading from the *N*-terminal to the *C*-terminal amino acid).

The messenger-RNA now moves through the ribosome, and a t-RNA carrying the third amino acid (VAL) binds to the next codon. The dipeptide is transferred to the amino acid on this third t-RNA, and a tripeptide is formed. This sequence of steps continues until one of three codons is encountered—*UAA*, *UGA*, or *UAG*. These codons give the signal for terminating the synthesis of the polypeptide chain, and the chain is cleaved from the last t-RNA residue.

TABLE 25.5

The Genetic Code

First Position	Second Position				Third Position
(5′ End)	U	C	A	G	(3′ End)
	PHE	SER	TYR	CYS	U
U	PHE	SER	TYR	CYS	C
	LEU	SER	a	a	A
	LEU	SER	a	TRP	G
	LEU	PRO	HIS	ARG	U
C	LEU	PRO	HIS	ARG	C
	LEU	PRO	GLN	ARG	A
	LEU	PRO	GLN	ARG	G
	ILE	THR	ASN	SER	U
A	ILE	THR	ASN	SER	C
	ILE	THR	LYS	ARG	A
	MET	THR	LYS	ARG	G
	VAL	ALA	ASP	GLY	U
G	VAL	ALA	ASP	GLY	C
	VAL	ALA	GLU	GLY	A
	VAL	ALA	GLU	GLY	G

a Three triplets code for termination of the polypeptide chain: *UAA*, *UGA*, and *UAG*.

The DNA we used in our example would give the following sequence of amino acids.

MET-PHE-VAL-ASN-GLN-HIS-...

This polypeptide is not necessarily an active protein. All proteins have MET first when synthesized, but not all proteins have MET first in their active form. It is often necessary to clip off this MET after the polypeptide has been synthesized to give a protein with a different *N*-terminal amino acid. If we cleave MET from this polypeptide, we are left with the first five amino acids in the *B*-chain of insulin (see Figure 25.25).

Removing MET may not be the only modification that must be made to the polypeptide before an active protein is formed. Insulin, for example, consists of two polypeptide chains connected by disulfide linkages. It should be possible to make these chains one at a time and then try to assemble them to make the final protein. Nature, however, has been far more subtle. The polypeptide chain that is synthesized contains a total of 81 amino acids. All of the disulfide bonds that will be present in insulin are present in this chain. The protein is made when a sequence of 30 amino acids is clipped out of the middle of this polypeptide chain.

SUMMARY

The term *polymer* has come to mean compounds of large molecular weight that contain long chains of relatively simple repeating units, or monomers. Polymers that are flexible, or elastic, and return to their original shapes when stretched are called elastomers. Those that flow when heated are often called plastics, or thermoplastic polymers. Polymers that take on rigid shapes that can't be changed without destruction of the material are said to be thermoset plastics. Linear (or straight-chain) and branched polymers are often plastics. As the amount of cross-linking between polymer chains increases, the polymer initially becomes more elastic. When the cross-linking becomes too extensive, however, the polymer may become rigid enough to be classified as a thermoset plastic.

Polymers that contain side chains are classified as either atatic, syndiotatic, or isotactic, depending on whether the side chains are arranged in random, alternating, or identical directions. Polymers formed simply by adding monomers to a polymer chain are addition polymers. Addition polymers are formed by free-radical, anionic, cationic, and coordination polymerization reactions. Condensation polymers are formed by reactions that condense, or eliminate, a small molecule each time another monomer is added to the polymer chain.

Naturally occurring polymers can be divided into three classes: proteins, carbohydrates, and nucleic acids. A fourth class of biologically important compounds — lipids — does not form polymers in the strictest sense of the term.

The amino acids are the building blocks from which proteins are formed. There are 20 common amino acids in protein, all but one of which has the generic formula $H_2N—CHR—CO_2H$. In aqueous solutions at neutral pH, amino acids are present as zwitterions with the generic formula $H_3N^+—CHR—CO_2^-$. The amino acids are grouped into four categories, depending on whether the *R*-groups are: (1) nonpolar, (2) polar, (3) negatively charged at neutral pH, or (4) positively charged at neutral pH. Nonpolar amino acids have *R*-groups that are hydrophobic; the other three classes have hydrophilic *R*-groups.

The sequence of amino acids in a protein defines its primary structure. As the polymer chain folds upon itself, hydrogen bonds form between peptide bonds on different portions of the protein, producing the protein's secondary structure. Interactions between the *R*-groups on different portions of the protein give rise to its tertiary structure. Proteins that consist of more than one polymer chain held together by ionic or hydrophobic interactions are said to have a quaternary structure. Anything that disrupts the secondary, tertiary, or quaternary structure of the protein is said to denature the protein.

Carbohydrates were originally defined as compounds

with the empirical formula CH_2O. Today, they are defined as polyhydroxy aldehydes or ketones. Monosaccharides, such as glucose (blood sugar) and fructose (fruit sugar), contain a single aldehyde or ketone group. Disaccharides, such as lactose (milk sugar) and sucrose (cane sugar), and polysaccharides, such as starch, glycogen, and cellulose, are formed by condensation reactions between monosaccharides.

Any biologically important compound that is soluble in a nonpolar solvent is classified as a lipid. Lipids are soluble in nonpolar solvents because they have structures that typically include long hydrocarbon chains.

They therefore contain carbon in its most highly reduced state and liberate more energy when burned to form CO_2 and H_2O than any other class of biological molecules.

Nucleic acids are divided into two categories—ribonucleic acids (RNA) and deoxyribonucleic acids (DNA). The individual nucleotides from which nucleic acids are built play an important role in the metabolism of proteins, carbohydrates, and lipids. But polynucleotides are primarily involved in the storage and expression of the genetic information in the cell.

PROBLEMS

Polymers, Plastics, and Elastomers

25-1 Define the terms *polymer, plastic,* and *elastomer.*

25-2 Use Berzelius's definition of a polymer to sort the following compounds into groups of polymers.

(a) formaldehyde, H_2CO (b) ethylene, C_2H_4 (c) glyceraldehyde, $C_3H_6O_3$ (d) 2-butene, C_4H_8 (e) cyclohexane, C_6H_{12} (f) glucose, $C_6H_{12}O_6$

25-3 Which of the following substances are polymers?

(a) Teflon (b) proteins (c) starch (d) sugars, such as cane sugar (e) poly(vinyl chloride) (f) polyethylene (g) cellulose (h) lipids

25-4 What are the characteristic properties of a plastic?

25-5 Which of the following are plastics?

(a) The rubber in an eraser (b) The celluloid used for ping-pong balls (c) The polyethylene in sandwich bags and garbage bags (d) The polystyrene in the cups used for beer and soft drinks (e) The polystyrene in disposable coffee cups

Definitions of Terms

25-6 Describe how linear, branched, and cross-linked polymers differ.

25-7 Explain the following observations.
Linear polymers, such as the polyethylene in garbage bags, tear when stretched.
Lightly cross-linked polymers, such as rubber, return to their original shape when stretched.
Highly cross-linked polymers, such as bakelite, break when "stretched."

25-8 Describe the difference between a homopolymer (such as polystyrene) and a copolymer (such as styrene-butadiene rubber).

25-9 Describe the difference between random, regular, block, and graft copolymers. Give an example of the sequence of

monomer units that would be found in a short length of each of these polymers.

25-10 Describe the differences in the structures of atactic, isotactic, and syndiotactic forms of polyvinyl chloride.

25-11 Calculate the range of molecular weights of polyethylene molecules in a sample in which individual chains contain between 500 and 50,000 $—CH_2CH_2—$ units.

25-12 Individual chains in polystyrene polymers typically weigh between 200,000 and 300,000 amu. If the formula for the monomer is $C_6H_5CH{=}CH_2$, how many monomers does the typical polymer chain contain?

25-13 Calculate the average molecular weight of the polymer chains in the cellulose nitrate in ping-pong balls if the formula for the monomer is $C_6H_9NO_7$ and the average polymer chain contains 1500 monomers.

25-14 Explain why it is possible to measure the molecular weight of linear polymers such as polyethylene but not that of cross-linked polymers such as those found in rubber.

25-15 Thermoplastic polymers flow when heated and can be molded into shapes they retain on cooling. Explain how increasing the amount of cross-linking between polymer chains can transform a thermoplastic polymer into a rigid thermoset polymer.

25-16 Describe the difference between addition and condensation polymers. Give an example of each.

25-17 Classify the following as either addition or condensation polymers.

(a) Polyethylene, $(—CH_2CH_2—)_n$
(b) Poly (vinyl chloride), $(—CH_2CHCl—)_n$
(c) Nylon, $[—CO(CH_2)_4CONH(CH_2)_6NH—]_n$
(d) Polyesters, $(—OCH_2CH_2OCOC_6H_4CO—)_n$
(e) Silicones, $[—OSi(CH_3)_2—]_n$
(f) Saran, $(—CH_2CCl_2—)_n(—CH_2CHCl)_m$
(g) Plexiglass, $[—CH_2C(CH_3)(CO_2CH_3)—]_n$

Elastomers

25-18 What are the characteristic properties of an elastomer?

25-19 Explain why elastomers must contain long, flexible molecules that are coiled in the natural state. Explain why polymers have to have some cross-links to be elastomers but can't have too many.

25-20 Natural rubber becomes too soft when heated and too hard when cooled to be of much use. Explain what happens when natural rubber is treated with sulfur that turns this material into a commercially useful product.

Free-Radical Polymerization Reactions

25-21 Describe what happens during the chain-initiation, chain-propagation, and chain-termination steps when ethylene ($H_2C=CH_2$) is polymerized by a free-radical mechanism.

25-22 Write the Lewis structures of the intermediates in the free-radical polymerization of vinyl chloride, $H_2C=CHCl$.

Ionic and Coordination Polymerization Reactions

25-23 Describe what happens during the chain-initiation, chain-propagation, and chain-termination steps in the anionic polymerization of vinyl chloride ($CH_2=CHCl$) when methyllithium (CH_3Li) is used as the chain initiator. Write the Lewis structures of all intermediates.

25-24 Describe what happens during the chain-initiation, chain-propagation, and chain-termination steps in the cationic polymerization of vinylidene chloride ($CH_2=CCl_2$) when hydrobromic acid (HBr) is used to initiate the reaction. Write the Lewis structures of all intermediates.

25-25 A typical Ziegler-Natta catalyst consists of a mixture of $TiCl_3$ and $Al(C_2H_5)_3$. Explain how the formation of a complex between a molecule of ethylene and the titanium atom in this catalyst can be thought of as an example of a Lewis acid-base reaction.

Addition Polymers

25-26 Polypropylene can be made in both atactic and isotactic forms. The atactic form is soft and rubbery, with no commercial value. The isotactic form is much more crystalline—it is hard enough, for example, to be used for furniture. Explain the difference between the physical properties of these two forms of the polymer.

25-27 Polyethylene can be made in both high-density and low-density forms. One of these polymers has a linear structure; the other is branched, with short side chains of up to five carbon atoms attached to the polymer backbone. Which structure would you expect to give the denser polymer? Which structure would give the more crystalline polymer?

25-28 Teflon, $(-CF_2CF_2-)_n$, is a waxy polymer to which practically nothing sticks. It is also the most chemically inert of all polymers. Describe at least 10 ways in which a polymer with these properties can be used.

25-29 Use the concept of van der Waals forces to explain why plastic wrap made from Saran clings to itself, whereas plastic wrap made from polyethylene does not.

Condensation Polymers

25-30 Predict the formula of the repeating unit in the condensation polymers formed by the following reactions.

(a) Dacron, $HOCH_2CH_2OH + HO_2CC_6H_4CO_2H \rightarrow$
$H_2O + ...$
(b) Nylon 6,6, $H_2N(CH_2)_6NH_2 + HO_2C(CH_2)_4CO_2H \rightarrow$
$H_2O + ...$
(c) Polycarbonate, $HOCH_2CH_2OH + (CH_3O)C=O \rightarrow$
$CH_3OH + ...$
(d) Polyaramide, $H_2NC_6H_4CO_2H \rightarrow H_2O + ...$
(e) Silicone rubber, $(CH_3)_2Si(OH)_2 \rightarrow H_2O + ...$

25-31 Explain the difference between the reactions used to prepare nylon 6,6 and nylon 6.

25-32 Explain why proteins, polysaccharides, and nucleic acids can all be classified as condensation polymers.

The Amino Acids

25-33 Explain why amino acids such as glycine, $H_2NCH_2CO_2H$, exist in aqueous solution as zwitterions ($H_3N^+CH_2CO_2^-$).

25-34 Which of the following forms of the amino acid glycine is present in strongly acidic solutions? In strongly basic solutions?

(a) $H_3N^+CH_2CO_2H$ (b) $H_3N^+CH_2CO_2^-$
(c) $H_2NCH_2CO_2^-$

25-35 The common amino acids in proteins have the generic formula $H_2NCHRCO_2H$. Classify the following amino acids as either hydrophilic or hydrophobic.

(a) Alanine: $R = -CH_3$
(b) Aspartic acid: $R = -CH_2CO_2H$
(c) Lysine: $R = -CH_2CH_2CH_2CH_2NH_2$
(d) Methionine: $R = -CH_2CH_2SCH_3$
(e) Serine: $R = -CH_2OH$

25-36 Use examples to explain why amino acids with an R group that is either positively charge or negatively charged in aqueous solution at neutral pH are hydrophilic.

Peptides and Proteins

25-37 How do amino acids, peptides, and proteins differ?

25-38 How many dipeptides can be formed when alanine, $R = -CH_3$, is condensed with cysteine, $R = -CH_2SH$? What is the difference between these dipeptides?

25-39 Describe the tripeptides that can form when alanine, $R = -CH_3$, is condensed with cysteine, $R = -CH_2SH$.

25-40 We can describe peptide chains by listing the amino acids from the N-terminal to the C-terminal residue. Write the amino acid sequences for all the tetrapeptides that can be formed from the amino acids ALA, LYS, SER, and TYR.

The Structure of Proteins

25-41 Describe how the primary, secondary, tertiary, and quaternary structures of a protein differ.

25-42 Describe the difference between the structures of the α-keratin found in hair and the β-keratin found in silk.

25-43 Describe the four factors that give rise to the tertiary structure of a protein. Give an example of each.

25-44 List the amino acids that can form hydrogen bonds in the tertiary structure of a protein.

25-45 List the amino acids that can form ionic bonds in the tertiary structure of a protein.

25-46 List the amino acids that can form the hydrophobic pockets in the tertiary structure of a protein.

25-47 Explain why each of the following can denature a protein.

(a) Increasing the temperature
(b) Changing the pH
(c) Adding a detergent
(d) Adding an oxidizing or reducing agent
(e) Adding compounds, such as urea, $(H_2N)_2C=O$, that form strong hydrogen bonds

Carbohydrates

25-48 Originally, a carbohydrate was defined as a compound with the empirical formula $C(H_2O)$. According to this definition, which of the following compounds are carbohydrates?

(a) glucose, $C_6H_{12}O_6$
(b) sucrose, $C_{12}H_{22}O_{11}$
(c) cellulose, $(C_6H_{10}O_5)_n$
(d) glyceraldehyde, $HOCH_2CHOHCHO$
(e) dihydroxyacetone, $(HOCH_2)_2C=O$
(f) ribose, $C_5H_{10}O_5$

25-49 Which of the compounds in the preceding question fit the definition of a carbohydrate as a polyhydroxy aldehyde or ketone?

25-50 Give examples of compounds that are classified as trioses, tetroses, pentoses, and hexoses.

25-51 Describe the difference between α-D-glucopyranose and β-D-glucopyranose. Describe the difference between α-D-glucopyranose and α-D-glucofuranose.

25-52 How do maltose, lactose, and sucrose differ?

25-53 How do glycogen, amylose, amylopectin, and cellulose differ?

25-54 Explain why humans are able to digest glycogen, amylose, and amylopectin but not cellulose. What would enable us to digest cellulose?

Lipids

25-55 Explain why lipids, such as fats and oils, are not soluble in water but are soluble in solvents such as chloroform, $CHCl_3$, benzene, C_6H_6, and diethyl ether, $(CH_3CH_2)_2O$.

25-56 In what ways are corn oil and olive oil similar to petroleum oil? What is the difference in these oils that allows us to digest the first two but not the other?

25-57 Draw the structure of the triglyceride formed when stearic acid combines with glycerol.

25-58 What is the difference between the structures of the fatty acids in fats and oils? How does this difference cause one of these types of compounds to be solid at room temperature and the other to be liquid?

25-59 Calculate the average oxidation states of the carbon atoms in a typical carbohydrate, such as glucose, $C_6H_{12}O_6$, and a typical fatty acid, such as palmitic acid, $C_{16}H_{32}O_2$. Use these oxidation states to explain why fatty acids give off much more energy than carbohydrates when burned.

25-60 Glucose, $C_6H_{12}O_6$, gives off 2870 kJ/mol when burned, and palmitic acid, $C_{16}H_{32}O_2$, gives off 9790 kJ/mol when burned. Use these data to estimate the energy released per gram when carbohydrates and lipids are burned.

Nucleic Acids

25-61 Draw the structure of one of the nucleotides found in nucleic acids. Show how the structures of the monomers that form DNA differ from those of the monomers that form RNA.

25-62 Classify the five common nucleic acids—adenine, cytosine, guanine, thymine, and uracil—as either purines or pyrimidines.

25-63 Explain why studies of nucleic acids invariably find roughly equivalent amounts of purine and pyrimidine nucleotides. Explain why nucleic acids that contain more adenine than guanine also contain more thymine than cytosine.

25-64 Draw the structures of the side chains on nucleic acids and show how hydrogen bonds form between adenine and thymine and between cytosine and guanine.

25-65 Describe how replication, transcription, and translation differ.

Protein Biosynthesis

25-66 Describe the difference between the roles played by m-RNA and t-RNA in protein biosynthesis.

25-67 Describe the protein, starting from the N-terminal end, that would be coded by the following sequence of DNA:

3′ -T-A-C-A-A-G-C-A-G-T-T-G-G-T-C-G-T-G- 5′

TABLES

TABLE A-1

The SI System

Base Units in the SI System

Physical Quantity	Name of Unit	Symbol
length	meter	m
mass	kilogram	kg
time	second	s
temperature	Kelvin	K
electric current	ampere	amp
luminous intensity	candela	cd
amount of substance	mole	mol

Common Derived Units in SI

Physical Quantity	Name of Unit	Symbol
energy	joule	J (kg-m^2/s^2)
frequency	hertz	Hz (cycles/s)
force	newton	N (kg-m/s^2)
pressure	pascal	P (N/m^2)
power	watt	W (J/s)
electric charge	coulomb	C (amp-s)
electric potential	volt	V (J/C)
electric resistance	ohm	Ω (V/amp)
electric conductance	siemens	S (amp/V)
electric capacitance	farad	F (C/V)

Non-SI Units in Common Use

Physical Quantity	Name of Unit	Symbol
volume	liter	L (10^{-3} m^3)
length	angstrom	Å (0.1 nm)
pressure	atmosphere	atm (101.325 kPa)
	torr	mmHg (133.32 Pa)
energy	calorie	cal (4.184 J)
	electron volt	ev (1.6022×10^{-19} J)
temperature	degree Celsius	°C (K − 273.15)
concentration	molarity	M (mol/L or mol/dm^3)

Fractions and Multipliers for Use in SI

exa, E	10^{18}	deci, d	10^{-1}
peta, P	10^{15}	centi, c	10^{-2}
tera, T	10^{12}	milli, m	10^{-3}
giga, G	10^{9}	micro, μ	10^{-6}
mega, M	10^{6}	nano, n	10^{-9}
kilo, k	10^{3}	pico, p	10^{-12}
hecto, h	10^{2}	femto, f	10^{-15}
deka, da	10^{1}	atto, a	10^{-18}

TABLE A-2

Values of Selected Fundamental Constants

Speed of light in a vacuum (c)	$c = 2.99792458 \times 10^8$ m/s
Charge on an electron (e)	$e = 1.6021892 \times 10^{-19}$ C
Rest mass of an electron (m_e)	$m_e = 9.109534 \times 10^{-28}$ g
	$m_e = 5.4858026 \times 10^{-4}$ amu
Rest mass of a proton (m_p)	$m_p = 1.6726485 \times 10^{-24}$ g
	$m_p = 1.00727647$ amu
Rest mass of a neutron (m_n)	$m_n = 1.6749543 \times 10^{-24}$ g
	$m_n = 1.008665012$ amu
Faraday's constant (F)	$F = 96{,}484.56$ C/mol
Planck's constant (h)	$h = 6.626176 \times 10^{-34}$ J-s
Ideal gas constant (R)	$R = 0.0820568$ L-atm/mol-K
	$R = 8.31441$ J/mol-K
Atomic mass unit (amu)	1 amu $= 1.6605655 \times 10^{-24}$ g
Boltzmann's constant (k)	$k = 1.380662 \times 10^{-23}$ J/K
Avogadro's constant (N)	$N = 6.022045 \times 10^{23}$ mol^{-1}
Rydberg constant (R_H)	$R_H = 1.09737318 \times 10^7$ m^{-1}
	$= 1.09737318 \times 10^{-2}$ nm^{-1}
Heat capacity of water	$C = 75.376$ J/mol-K

TABLE A-3

Selected Conversion Factors

Energy	1 J $= 0.2390$ cal $= 10^7$ erg
	1 cal $= 4.184$ J (by definition)
	1 ev/atom $= 1.6021892 \times 10^{-19}$ J/atom $= 96.484$ kJ/mol
Temperature	K $= °C + 273.15$
	$°C = 5/9(°F - 32)$
	$°F = 9/5(°C) + 32$
Pressure	1 atm $= 760$ mmHg $= 760$ torr $= 101.325$ kPa
Mass	1 kg $= 2.2046$ lb
	1 lb $= 453.59$ g $= 0.45359$ kg
	1 oz $= 0.06250$ lb $= 28.350$ g
	1 ton $= 2000$ lb $= 907.185$ kg
	1 tonne (metric) $= 1000$ kg $= 2204.62$ lb
Volume	1 mL $= 0.01$ L $= 1$ cm^3 (by definition)
	1 oz (fluid) $= 0.031250$ qt $= 0.029573$ L
	1 qt $= 0.946326$ L
	1 L $= 1.05672$ qt
Length	1 m $= 39.370$ in.
	1 mi $= 1.60934$ km
	1 in. $= 2.540$ cm (by definition)

TABLE A-4

The Vapor Pressure of Water

Temperature (°C)	Pressure (mmHg)	Temperature (°C)	Pressure (mmHg)	Temperature (°C)	Pressure (mmHg)	Temperature (°C)	Pressure (mmHg)
0	4.6	26	25.2	13	11.2	39	52.4
1	4.9	27	26.7	14	12.0	40	55.3
2	5.3	28	28.3	15	12.8	41	58.3
3	5.7	29	30.0	16	13.6	42	61.5
4	6.1	30	31.8	17	14.5	43	64.8
5	6.5	31	33.7	18	15.5	44	68.3
6	7.0	32	35.7	19	16.5	45	71.9
7	7.5	33	37.7	20	17.5	46	75.7
8	8.0	34	39.9	21	18.7	47	79.6
9	8.6	35	42.2	22	19.8	48	83.7
10	9.2	36	44.6	23	21.1	49	88.0
11	9.8	37	47.1	24	22.4	50	92.5
12	10.5	38	49.7	25	23.8		

TABLE A-5

Ionization Energies (kJ/mol)

Element		I	II	III	IV	V	VI	VII
1	H	1,312.0						
2	He	2,372.3	5,250.3					
3	Li	520.2	7,297.9	11,814.6				
4	Be	899.4	1,757.1	14,848.3	21,005.9			
5	B	800.6	2,427.0	3,659.6	25,025.0	32,825.7		
6	C	1,086.4	2,352.6	4,620.4	6,222.5	37,829.4	47,275.6	
7	N	1,402.3	2,856.0	4,578.0	7,474.9	9,444.7	50,370.4	64,358.0
8	O	1,313.9	3,388.2	5,300.3	7,469.1	10,989.2	13,326.1	71,332
9	F	1,681.0	3,374.1	6,050.3	8,407.5	11,022.4	15,163.6	17,867.2
10	Ne	2,080.6	3,952.2	6,122	9,370	12,177	15,238	19,998
11	Na	495.8	4,562.4	6,912	9,543	13,352	16,610	20,114
12	Mg	737.7	1,450.6	7,732.6	10,540	13,629	17,994	21,703
13	Al	577.6	1,816.6	2,744.7	11,577	14,831	18,377	23,294
14	Si	786.4	1,577.0	3,231.5	4,355.4	16,091	19,784	23,785
15	P	1,011.7	1,903.2	2,912	4,956	6,273.7	21,268	25,397
16	S	999.58	2,251	3,361	4,564	7,012	8,495.4	27,105
17	Cl	1,251.1	2,297	3,822	5,158	6,540	9,362	11,017.9
18	Ar	1,520.5	2,665.8	3,931	5,771	7,238	8,780.8	11,994.9
19	K	418.8	3,051.3	4,411	5,877	7,975	9,648.5	11,343
20	Ca	589.8	1,145.4	4,911.8	6,474	8,144	10,496	12,320
21	Sc	631	1,235	2,389	7,099	8,844	10,720	13,310
22	Ti	658	1,310	2,652.5	4,174.5	9,573	11,516	13,590
23	V	650	1,413	2,828.0	4,506.5	6,294	12,362	14,489
24	Cr	652.8	1,592	2,987	4,740	6,690	8,738	15,540
25	Mn	717.4	1,509.0	3,248.3	4,940	6,990	9,200	11,508

(continued)

TABLE A-5 continued

Ionization Energies (kJ/mol)

Element		I	II	III	IV	V	VI	VII
26	Fe	759.3	1,561	2,957.3	5,290	7,240	9,600	12,100
27	Co	758	1,646	3,232	4,950	7,670	9,840	12,400
28	Ni	736.7	1,752.9	3,393	5,300	7,280	10,400	12,800
29	Cu	745.4	1,957.9	3,553	5,330	7,710	9,940	13,400
30	Zn	906.4	1,733.2	3,832.6	5,730	7,970	10,400	12,900
31	Ga	578.8	1,979	2,963	6,200			
32	Ge	762.1	1,537.4	3,302	4,410	9,020		
33	As	947	1,797.8	2,735.4	4,837	6,043	12,300	
34	Se	940.9	2,045	2,973.7	4,143.4	6,590	7,883	14,990
35	Br	1,139.9	2,100	3,500	1,560	5,760	8,550	9,938
36	Kr	1,350.7	2,350.3	3,565	5,070	6,240	7,570	10,710
37	Rb	403.0	2,632	3,900	5,070	6,850	8,140	9,570
38	Sr	549.5	1,064.2	4,210	5,500	6,910	8,760	10,200
39	Y	616	1,181	1,980	5,960	7,430	8,970	11,200
40	Zr	660	1,267	2,218	3,313	7,870		
41	Nb	664	1,382	2,416	3,690	4,877	9,899	12,100
42	Mo	684.9	1,558	2,621	4,480	5,910	6,600	12,230
43	Tc	702	1,472	2,850				
44	Ru	711	1,617	2,747				
45	Rh	720	1,744	2,997				
46	Pd	805	1,874	3,177				
47	Ag	731.0	2,073	3,361				
48	Cd	867.7	1,631.4	3,616				
49	In	558.3	1,820.6	2,704	5,200			
50	Sn	708.6	1,411.8	2,943.0	3,930.2	6,974		
51	Sb	833.7	1,595	2,440	4,260	5,400	10,400	
52	Te	869.2	1,790	2,698	3,609	5,668	6,820	13,200
53	I	1,008.4	1,845.8	3,200				
54	Xe	1,170.4	2,046	3,100				
55	Cs	375.7	2,440					
56	Ba	502.9	965.23					
57	La	538.1	1,067	1,850.3				
58	Ce	527.8	1,047	1,949	3,547			
59	Pr	523	1,018	2,086	3,761	5,543		
60	Nd	530	1,035	2,130	3,900			
61	Pm	535	1,052	2,150	3,970			
62	Sm	543	1,068	2,260	3,990			
63	Eu	547	1,084	2,400	4,110			
64	Gd	592	1,167	1,990	4,250			
65	Tb	564	1,112	2,110	3,840			
66	Dy	572	1,126	2,200	4,000			
67	Ho	581	1,139	2,203	4,100			
68	Er	589	1,151	2,194	4,120			
69	Tm	596	1,163	2,285	4,120			
70	Yb	603	1,175	2,415	4,216			
71	Lu	524	1,340	2,022	4,360			

TABLE A-5 continued

Ionization Energies (kJ/mol)

Element		I	II	III	IV	V	VI	VII
72	Hf	642	1,440	2,250	3,210			
73	Ta	761						
74	W	770						
75	Re	760	1,260	2,510	3,640			
76	Os	840						
77	Ir	880						
78	Pt	870	1,791.0					
79	Au	890.1	1,980					
80	Hg	1,007.0	1,809.7	3,300				
81	Tl	589.3	1,971.0	2,878				
82	Pb	715.5	1,450.4	3,081.4	4,083	6,640		
83	Bi	703.3	1,610	2,466	4,370	5,400	8,520	
84	Po	812						
85	At							
86	Rn	1,037.0						
87	Fr							
88	Ra	509.3	979.0					
89	Ac	498.8	1,170					
90	Th	587	1,110	1,930	2,780			
91	Pa	568						
92	U	584						
93	Np	597						
94	Pu	585						
95	Am	578.2						
96	Cm	581						
97	Bk	601						
98	Cf	608						
99	Es	619						
100	Fm	627						
101	Md	635						
102	No	642						

TABLE A-6

Electron Affinities

Atomic Number	Symbol	Electron Affinity (kJ/mol)	Atomic Number	Symbol	Electron Affinity (kJ/mol)
1	H	72.8	7	N	−7
2	He	*	8	O	141.1
3	Li	59.8	9	F	328.0
4	Be	*	10	Ne	*
5	B	27	11	Na	52.7
6	C	122.3	12	Mg	*

* These elements have negative electron affinities.

(continued)

TABLE A-6   continued

Electron Affinities

Atomic Number	Symbol	Electron Affinity (kJ/mol)	Atomic Number	Symbol	Electron Affinity (kJ/mol)
13	Al	45	44	Ru	110
14	Si	133.6	45	Rh	120
15	P	71.7	46	Pd	60
16	S	200.42	47	Ag	125.7
17	Cl	348.8	48	Cd	*
18	Ar	*	49	In	29
19	K	48.36	50	Sn	121
20	Ca	*	51	Sb	101
21	Sc	*	52	Te	190.15
22	Ti	20	53	I	295.3
23	V	50	54	Xe	*
24	Cr	64	55	Cs	45.49
25	Mn	*	56	Ba	*
26	Fe	24	57–71	La–Lu	50
27	Co	70	72	Hf	*
28	Ni	111	73	Ta	60
29	Cu	118.3	74	W	60
30	Zn	0	75	Re	14
31	Ga	29	76	Os	110
32	Ge	120	77	Ir	150
33	As	77	78	Pt	205.3
34	Se	194.96	79	Au	222.74
35	Br	324.6	80	Hg	*
36	Kr	*	81	Tl	30
37	Rb	46.89	82	Pb	110
38	Sr	*	83	Bi	110
39	Y	0	84	Po	180
40	Zr	50	85	At	270
41	Nb	96	86	Rn	*
42	Mo	96	87	Fr	44.0
43	Tc	70			

TABLE A-7

Electronegativities

Atomic Number	Element	Electronegativity	Atomic Number	Element	Electronegativity
1	H	2.20	8	O	3.44
2	He	*	9	F	3.98
3	Li	0.98	10	Ne	*
4	Be	1.57	11	Na	0.93
5	B	2.04	12	Mg	1.31
6	C	2.55	13	Al	1.61
7	N	3.04	14	Si	1.90

TABLE A-7 continued

Electronegativities

Atomic Number	Element	Electronegativity	Atomic Number	Element	Electronegativity
15	P	2.19	48	Cd	1.69
16	S	2.58	49	In	1.78
17	Cl	3.16	50	Sn	1.96
18	Ar	*	51	Sb	2.05
19	K	0.82	52	Te	2.1
20	Ca	1.00	53	I	2.66
21	Sc	1.36	54	Xe	*
22	Ti	1.54	55	Cs	0.79
23	V	1.63	56	Ba	0.89
24	Cr	1.66	57–71	La–Lu	1.1–1.3
25	Mn	1.55	72	Hf	1.3
26	Fe	1.83	73	Ta	1.3
27	Co	1.88	74	W	2.36
28	Ni	1.91	75	Re	1.9
29	Cu	1.90	76	Os	2.2
30	Zn	1.65	77	Ir	2.20
31	Ga	1.81	78	Pt	2.28
32	Ge	2.01	79	Au	2.54
33	As	2.18	80	Hg	2.00
34	Se	2.55	81	Tl	2.04
35	Br	2.96	82	Pb	2.33
36	Kr	*	83	Bi	2.02
37	Rb	0.82	84	Po	2.0
38	Sr	0.95	85	At	2.2
39	Y	1.22	86	Rn	*
40	Zr	1.33	87	Fr	0.7
41	Nb	1.6	88	Ra	0.9
42	Mo	2.16	89	Ac	1.1
43	Tc	1.9	90	Th	1.3
44	Ru	2.2	91	Pa	1.5
45	Rh	2.28	92	U	1.38
46	Pd	2.20	93–103	Np–Lw	1.3
47	Ag	1.93			

Pauling electronegativities taken from A. L. Allred, *J. Inorg. Nucl. Chem.*, *17*, 215 (1961).

* An asterisk indicates a noble gas.

TABLE A-8

Radii of Atoms and Ions

Element	Ionic Radius (nm) (Charge)	Covalent Radius (nm)	Metallic Radius (nm)	Element	Ionic Radius (nm) (Charge)	Covalent Radius (nm)	Metallic Radius (nm)
Aluminum	0.050 (+3)	0.125	0.1431	Arsenic	0.222 (−3)	0.121	0.1248
Antimony	0.245 (−3)	0.141			0.058 (+3)		
	0.09 (+3)				0.047 (+5)		
	0.062 (+5)			Astatine	0.227 (−1)		
					0.051 (+7)		

(continued)

TABLE A-8   continued

Radii of Atoms and Ions

Element	Ionic Radius (nm) (Charge)	Covalent Radius (nm)	Metallic Radius (nm)	Element	Ionic Radius (nm) (Charge)	Covalent Radius (nm)	Metallic Radius (nm)
Barium	0.135 (+2)	0.198	0.2173	Molybdenum	0.062 (+6)	0.129	0.1362
Beryllium	0.031 (+2)	0.089	0.1113	Nickel	0.072 (+2)	0.115	0.1246
Bismuth	0.213 (−3)	0.152	0.1547	Nitrogen	0.171 (−3)	0.070	
	0.096 (+3)				0.013 (+3)		
	0.074 (+5)				0.011 (+5)		
Boron	0.020 (+3)	0.088	0.083	Oxygen	0.140 (−2)	0.066	
Bromine	0.196 (−1)	0.1142			0.176 (−1)		
	0.039 (+7)			Phosphorus	0.212 (−3)	0.110	0.108
Cadmium	0.097 (+2)	0.141	0.1489		0.042 (+3)		
Calcium	0.099 (+2)	0.174	0.1973		0.034 (+5)		
Carbon	0.260 (−4)	0.077		Polonium	0.230 (−2)	0.153	0.167
	0.015 (+4)				0.056 (+6)		
Cesium	0.169 (+1)	0.235	0.2654	Potassium	0.133 (+1)	0.2025	0.2272
Chlorine	0.181 (−1)	0.099		Radium	0.140 (+2)		0.220
	0.026 (+7)			Rubidium	0.148 (+1)	0.216	0.2475
Chromium	0.064 (+3)	0.117	0.1249	Scandium	0.081 (+3)	0.144	0.1606
	0.052 (+6)			Selenium	0.198 (−2)	0.117	
Cobalt	0.074 (+2)	0.116	0.1253		0.069 (+4)		
	0.063 (+3)				0.042 (+6)		
Copper	0.096 (+1)	0.117	0.1278	Silicon	0.271 (−4)	0.117	
	0.072 (+2)				0.041 (+4)		
Fluorine	0.136 (−1)	0.064	0.0717	Silver	0.126 (+1)	0.134	0.1444
	0.007 (+7)			Sodium	0.095 (+1)	0.157	0.1537
Francium	0.176 (+1)		0.27	Strontium	0.113 (+2)	0.192	0.2151
Gallium	0.062 (+3)	0.125	0.1221	Sulfur	0.184 (−2)	0.104	
Germanium	0.272 (−4)	0.122	0.1225		0.037 (+4)		
	0.053 (+4)				0.029 (+6)		
Gold	0.137 (+1)	0.134	0.1442	Tellurium	0.221 (−2)	0.137	0.1432
	0.091 (+3)				0.081 (+4)		
Hydrogen	0.208 (−1)	0.0371			0.056 (+6)		
	10^{-6} (+1)			Thallium	0.095 (+3)	0.155	0.1704
Indium	0.081 (+3)	0.150	0.1626	Tin	0.294 (−4)	0.140	0.1405
Iodine	0.216 (−1)	0.1333			0.102 (+2)		
	0.050 (+7)				0.071 (+4)		
Iron	0.076 (+2)	0.1165	0.1241	Titanium	0.090 (+2)	0.132	0.1448
	0.064 (+3)				0.068 (+4)		
Lead	0.215 (−4)	0.154	0.1750	Tungsten	0.065 (+6)	0.130	0.1370
	0.120 (+2)			Uranium	0.083 (+6)		0.1385
	0.084 (+4)			Vanadium	0.059 (+5)	0.122	0.1321
Lithium	0.068 (+1)	0.123	0.152	Zinc	0.074 (+2)	0.125	0.1332
Magnesium	0.065 (+2)	0.136	0.160	Zirconium	0.079 (+4)	0.145	0.167
Manganese	0.080 (+2)	0.117	0.124	Xenon		0.209	
	0.046 (+7)						
Mercury	0.127 (+1)	0.144	0.160				
	0.110 (+2)						

TABLE A-9

Acid-Dissociation Equilibrium Constants

Compound	Dissociation Reaction	K_a	pK_a
Acetic acid	$CH_3CO_2H \rightleftharpoons CH_3CO_2^- + H^+$	1.75×10^{-5}	4.757
Ammonium ion	$NH_4^+ \rightleftharpoons NH_3 + H^+$	5.8×10^{-10}	9.24
Arsenic acid	$H_3AsO_4 \rightleftharpoons H_2AsO_4^- + H^+$	6.0×10^{-3}	2.22
	$H_2AsO_4^- \rightleftharpoons HAsO_4^{2-} + H^+$	1.0×10^{-7}	7.00
	$HAsO_4^{2-} \rightleftharpoons AsO_4^{3-} + H^+$	3.0×10^{-12}	11.52
Arsenous acid	$H_3AsO_3 \rightleftharpoons H_2AsO_3^- + H^+$	6.0×10^{-10}	9.22
	$H_2AsO_3^- \rightleftharpoons HAsO_3^{2-} + H^+$	3.0×10^{-14}	13.52
Benzoic acid	$C_6H_5CO_2H \rightleftharpoons C_6H_5CO_2^- + H^+$	6.3×10^{-5}	4.20
Boric acid	$H_3BO_3 \rightleftharpoons H_2BO_3^- + H^+$	7.3×10^{-10}	9.14
Carbonic acid	$H_2CO_3 \rightleftharpoons HCO_3^- + H^+$	4.5×10^{-7}	6.35
	$HCO_3^- \rightleftharpoons CO_3^{2-} + H^+$	4.7×10^{-11}	10.33
Chloric acid	$HClO_3 \rightleftharpoons ClO_3^- + H^+$	5.0×10^{2}	-2.70
Chloroacetic acid	$ClCH_2CO_2H \rightleftharpoons ClCH_2CO_2^- + H^+$	1.4×10^{-3}	2.85
Chlorous acid	$HClO_2 \rightleftharpoons ClO_2^- + H^+$	1.1×10^{-2}	1.96
Chromic acid	$H_2CrO_4 \rightleftharpoons HCrO_4^- + H^+$	9.6	-0.98
	$HCrO_4^- \rightleftharpoons CrO_4^{2-} + H^+$	3.2×10^{-7}	6.50
Citric acid	$H_3Cit \rightleftharpoons H_2Cit^- + H^+$	7.5×10^{-4}	3.13
	$H_2Cit^- \rightleftharpoons HCit^{2-} + H^+$	1.7×10^{-5}	4.77
	$HCit^{2-} \rightleftharpoons Cit^{3-} + H^+$	4.0×10^{-7}	6.40
Dichloroacetic acid	$Cl_2CHCO_2H \rightleftharpoons Cl_2CHCO_2^- + H^+$	5.1×10^{-2}	1.29
Formic acid	$HCO_2H \rightleftharpoons HCO_2^- + H^+$	1.8×10^{-4}	3.75
Glycine	$H_3N^+CH_2CO_2H \rightleftharpoons H_3N^+CH_2CO_2^- + H^+$	4.5×10^{-3}	2.35
	$H_3N^+CH_2CO_2^- \rightleftharpoons H_2NCH_2CO_2^- + H^+$	2.5×10^{-10}	9.60
Hydrazoic acid	$HN_3 \rightleftharpoons N_3^- + H^+$	1.9×10^{-5}	4.72
Hydrobromic acid	$HBr \rightleftharpoons Br^- + H^+$	1×10^{9}	-9
Hydrochloric acid	$HCl \rightleftharpoons Cl^- + H^+$	1×10^{6}	-6
Hydrocyanic acid	$HCN \rightleftharpoons CN^- + H^+$	6×10^{-10}	9.22
Hydrofluoric acid	$HF \rightleftharpoons F^- + H^+$	7.2×10^{-4}	3.14
Hydroiodic acid	$HI \rightleftharpoons I^- + H^+$	3×10^{9}	-9.5
Hydrogen peroxide	$H_2O_2 \rightleftharpoons HO_2^- + H^+$	2.2×10^{-12}	11.66
Hydrogen selenide	$H_2Se \rightleftharpoons HSe^- + H^+$	1.0×10^{-4}	4.00
Hydrogen sulfide	$H_2S \rightleftharpoons HS^- + H^+$	1.0×10^{-7}	7.00
	$HS^- \rightleftharpoons S^{2-} + H^+$	1.3×10^{-13}	12.89
Hypobromous acid	$HOBr \rightleftharpoons OBr^- + H^+$	2.4×10^{-9}	8.62
Hypochlorous acid	$HOCl \rightleftharpoons OCl^- + H^+$	2.9×10^{-8}	7.54
Hypoiodous acid	$HOI \rightleftharpoons OI^- + H^+$	2.3×10^{-11}	10.64
Iodic acid	$HIO_3 \rightleftharpoons IO_3^- + H^+$	0.16	0.80
Nitric acid	$HNO_3 \rightleftharpoons NO_3^- + H^+$	28	-1.45
Nitrous acid	$HNO_2 \rightleftharpoons NO_2^- + H^+$	5.1×10^{-4}	3.29
Oxalic acid	$H_2C_2O_4 \rightleftharpoons HC_2O_4^- + H^+$	5.4×10^{-2}	1.27
	$HC_2O_4^- \rightleftharpoons C_2O_4^{2-} + H^+$	5.4×10^{-5}	4.27
Perchloric acid	$HOClO_3 \rightleftharpoons ClO_4^- + H^+$	1×10^{8}	-8
Periodic acid	$H_5IO_6 \rightleftharpoons H_4IO_6^- + H^+$	2.3×10^{-2}	1.64
Phenol	$C_6H_5OH \rightleftharpoons C_6H_5O^- + H^+$	1.0×10^{-10}	10.00
Phosphoric acid	$H_3PO_4 \rightleftharpoons H_2PO_4^- + H^+$	7.1×10^{-3}	2.15
	$H_2PO_4^- \rightleftharpoons HPO_4^{2-} + H^+$	6.3×10^{-8}	7.20
	$HPO_4^{2-} \rightleftharpoons PO_4^{3-} + H^+$	4.2×10^{-13}	12.38

(continued)

TABLE A-9 continued

Acid-Dissociation Equilibrium Constants

Compound	Dissociation Reaction	K_a	pK_a
Phosphorus acid	$H_3PO_3 \rightleftharpoons H_2PO_3^- + H^+$	1.00×10^{-2}	2.00
	$H_2PO_3^- \rightleftharpoons HPO_3^{2-} + H^+$	2.6×10^{-7}	6.59
Sulfamic acid	$H_2NSO_3H \rightleftharpoons H_2NSO_3^- + H^+$	1.03×10^{-1}	0.987
Sulfuric acid	$H_2SO_4 \rightleftharpoons HSO_4^- + H^+$	10^3	-3
	$HSO_4^- \rightleftharpoons SO_4^{2-} + H^+$	1.2×10^{-2}	1.92
Sulfurous acid	$H_2SO_3 \rightleftharpoons HSO_3^- + H^+$	1.7×10^{-2}	1.77
	$HSO_3^- \rightleftharpoons SO_3^{2-} + H^+$	6.4×10^{-8}	7.19
Thiocyanic acid	$HSCN \rightleftharpoons SCN^- + H^+$	71	-1.85
Trichloroacetic acid	$Cl_3CCO_2H \rightleftharpoons Cl_3CCO_2^- + H^+$	0.22	0.66
Water	$H_2O \rightleftharpoons OH^- + H^+$	1.8×10^{-16}	15.75

TABLE A-10

Base-Ionization Equilibrium Constants

Compound	Ionization Reaction	K_b	pK_b
Ammonia	$NH_3 + H_2O \rightleftharpoons NH_4^+ + OH^-$	1.8×10^{-5}	4.76
Aniline	$C_6H_5NH_2 + H_2O \rightleftharpoons C_6H_5NH_3^+ + OH^-$	4.0×10^{-10}	9.40
Butylamine	$CH_3(CH_2)_3NH_2 + H_2O \rightleftharpoons CH_3(CH_2)_3NH_3^+ + OH^-$	4.0×10^{-4}	3.40
Dimethylamine	$(CH_3)_2NH + H_2O \rightleftharpoons (CH_3)_2NH_2^+ + OH^-$	5.9×10^{-4}	3.23
Ethanolamine	$HOCH_2CH_2NH_2 + H_2O \rightleftharpoons HOCH_2CH_2NH_3^+ + OH^-$	3.3×10^{-5}	4.50
Ethylamine	$CH_3CH_2NH_2 + H_2O \rightleftharpoons CH_3CH_2NH_3^+ + H_2O$	4.4×10^{-4}	3.37
Hydrazine	$H_2NNH_2 + H_2O \rightleftharpoons H_2NNH_3^+ + OH^-$	1.2×10^{-6}	5.89
Hydroxylamine	$HONH_2 + H_2O \rightleftharpoons HONH_3^+ + OH^-$	1.1×10^{-8}	7.97
Methylamine	$CH_3NH_2 + H_2O \rightleftharpoons CH_3NH_3^+ + OH^-$	4.8×10^{-4}	3.32
Pyridine	$C_5H_5NH + H_2O \rightleftharpoons C_5H_5NH_2^+ + OH^-$	1.7×10^{-9}	8.77
Trimethylamine	$(CH_3)_3N + H_2O \rightleftharpoons (CH_3)_3NH^+ + OH^-$	6.3×10^{-5}	4.20
Urea	$H_2NCONH_2 + H_2O \rightleftharpoons H_2NCONH_3^+ + OH^-$	1.5×10^{-14}	13.82

TABLE A-11

Solubility Product Constants

Substance	K_{sp}	Substance	K_{sp}	Substance	K_{sp}
AgBr	5.0×10^{-13}	AuCl	2.0×10^{-13}	CaCrO_4	7.1×10^{-4}
AgCN	1.2×10^{-16}	AuCl_3	3.2×10^{-25}	CaF_2	4.0×10^{-11}
Ag_2CO_3	8.1×10^{-12}	AuI	1.6×10^{-23}	Ca(OH)_2	5.5×10^{-6}
AgOH	2.0×10^{-8}	AuI_3	5.5×10^{-46}	CdCO_3	5.2×10^{-12}
AgOAc	4.4×10^{-3}	BaCO_3	5.1×10^{-9}	Cd(CN)_2	1.0×10^{-8}
Ag_2C_2O_4	3.4×10^{-11}	BaC_2O_4	2.3×10^{-8}	Cd(OH)_2	2.5×10^{-14}
AgCl	1.8×10^{-10}	BaCrO_4	1.2×10^{-10}	CdS	8×10^{-27}
Ag_2CrO_4	1.1×10^{-12}	BaF_2	1.0×10^{-6}	CoCO_3	1.4×10^{-13}
AgI	8.3×10^{-17}	Ba(OH)_2	5×10^{-3}	Co(OH)_3	1.6×10^{-44}
Ag_2S	6.3×10^{-50}	BaSO_4	1.1×10^{-10}	α-CoS	4.0×10^{-21}
AgSCN	1.0×10^{-12}	Bi_2S_3	1×10^{-97}	β-CoS	2.0×10^{-25}
Ag_2SO_4	1.4×10^{-5}	CaCO_3	2.8×10^{-9}	Cr(OH)_3	6.3×10^{-31}
Al(OH)_3	1.3×10^{-33}	CaC_2O_4	4×10^{-9}	CuBr	5.3×10^{-9}

TABLE A-11 continued

Solubility Product Constants

Substance	K_{sp}	Substance	K_{sp}	Substance	K_{sp}
CuCl	1.2×10^{-6}	Hg_2I_2	4.5×10^{-29}	PbF_2	2.7×10^{-8}
CuCN	3.2×10^{-20}	Hg_2S	1.0×10^{-47}	$Pb(OH)_2$	1.2×10^{-15}
$CuCrO_4$	3.6×10^{-6}	HgS	4×10^{-53}	PbI_2	7.1×10^{-9}
$CuCO_3$	1.4×10^{-10}	$K_2NaCo(NO_2)_6$	2.2×10^{-11}	PbS	8.0×10^{-28}
$Cu(OH)_2$	2.2×10^{-20}	$MgCO_3$	3.5×10^{-8}	$PbSO_4$	1.6×10^{-8}
CuI	1.1×10^{-13}	MgC_2O_4	1×10^{-8}	SnS	1.0×10^{-25}
Cu_2S	2.5×10^{-48}	MgF_2	6.5×10^{-9}	$Sn(OH)_2$	1.4×10^{-28}
CuS	6.3×10^{-36}	$Mg(OH)_2$	1.8×10^{-11}	$Sn(OH)_4$	1×10^{-56}
CuSCN	4.8×10^{-15}	$MgNH_4PO_4$	2.5×10^{-13}	$SrCO_3$	1.1×10^{-10}
$FeCO_3$	3.2×10^{-11}	$MnCO_3$	1.8×10^{-11}	SrC_2O_4	1.6×10^{-7}
$Fe_2C_2O_4$	3.2×10^{-7}	$Mn(OH)_2$	2×10^{-13}	$SrCrO_4$	2.2×10^{-5}
$Fe(OH)_2$	8.0×10^{-16}	MnS	3×10^{-13}	SrF_2	2.5×10^{-9}
$Fe(OH)_3$	4×10^{-38}	$NiCO_3$	6.6×10^{-9}	$SrSO_4$	3.2×10^{-7}
FeS	6.3×10^{-18}	NiC_2O_4	4×10^{-10}	TlBr	3.4×10^{-6}
Hg_2Br_2	5.6×10^{-23}	α-NiS	3.2×10^{-19}	TlCl	1.7×10^{-4}
$Hg_2(CN)_2$	5×10^{-40}	β-NiS	1.0×10^{-24}	TlI	6.5×10^{-8}
Hg_2CO_3	8.9×10^{-17}	γ-NiS	2.0×10^{-26}	$Zn(CN)_2$	2.6×10^{-13}
$Hg_2(OAc)_2$	3×10^{-11}	$PbBr_2$	4.0×10^{-5}	$ZnCO_3$	1.4×10^{-11}
$Hg_2C_2O_4$	2.0×10^{-13}	$PbCO_3$	7.4×10^{-14}	ZnC_2O_4	2.7×10^{-8}
HgC_2O_4	1×10^{-7}	PbC_2O_4	4.8×10^{-10}	$Zn(OH)_2$	1.2×10^{-17}
Hg_2Cl_2	1.3×10^{-18}	$PbCl_2$	1.6×10^{-5}	α-ZnS	1.6×10^{-24}
Hg_2CrO_4	2.0×10^{-9}	$PbCrO_4$	2.8×10^{-13}	β-ZnS	2.5×10^{-22}

TABLE A-12

Complex Formation Equilibrium Constants

Equilibrium	K_f	Equilibrium	K_f
$Ag^+ + 2\,Br^- \rightleftharpoons AgBr_2^-$	2.1×10^7	$Fe^{2+} + 6\,CN^- \rightleftharpoons Fe(CN)_6^{4-}$	1×10^{35}
$Ag^+ + 2\,Cl^- \rightleftharpoons AgCl_2^-$	1.1×10^5	$Fe^{3+} + 6\,CN^- \rightleftharpoons Fe(CN)_6^{3-}$	1×10^{42}
$Ag^+ + 2\,CN^- \rightleftharpoons Ag(CN)_2^-$	1.3×10^{21}	$Fe^{3+} + SCN^- \rightleftharpoons Fe(SCN)^{2+}$	8.9×10^2
$Ag^+ + 2\,I^- \rightleftharpoons AgI_2^-$	5.5×10^{11}	$Fe^{3+} + 2\,SCN^- \rightleftharpoons Fe(SCN)_2^+$	2.3×10^3
$Ag^+ + 2\,NH_3 \rightleftharpoons Ag(NH_3)_2^+$	1.1×10^7	$Hg^{2+} + 4\,Br^- \rightleftharpoons HgBr_4^{2-}$	1×10^{21}
$Ag^+ + 2\,SCN^- \rightleftharpoons Ag(SCN)_2^-$	3.7×10^7	$Hg^{2+} + 4\,Cl^- \rightleftharpoons HgCl_4^{2-}$	1.2×10^{15}
$Ag^+ + 2\,S_2O_3^{2-} \rightleftharpoons Ag(S_2O_3)_2^{3-}$	2.9×10^{13}	$Hg^{2+} + 4\,CN^- \rightleftharpoons Hg(CN)_4^{2-}$	3×10^{41}
$Al^{3+} + 6\,F^- \rightleftharpoons AlF_6^{3-}$	6.9×10^{19}	$Hg^{2+} + 4\,I^- \rightleftharpoons HgI_4^{2-}$	6.8×10^{29}
$Al^{3+} + 4\,OH^- \rightleftharpoons Al(OH)_4^-$	1.1×10^{33}	$I_2 + I^- \rightleftharpoons I_3^-$	7.8×10^2
$Cd^{2+} + 4\,Cl^- \rightleftharpoons CdCl_4^{2-}$	6.3×10^2	$Ni^{2+} + 4\,CN^- \rightleftharpoons Ni(CN)_4^{2-}$	2×10^{31}
$Cd^{2+} + 4\,CN^- \rightleftharpoons Cd(CN)_4^{2-}$	6.0×10^{18}	$Ni^{2+} + 6\,NH_3 \rightleftharpoons Ni(NH_3)_6^{2+}$	5.5×10^8
$Cd^{2+} + 4\,I^- \rightleftharpoons CdI_4^{2-}$	2.6×10^5	$Pb^{2+} + 4\,Cl^- \rightleftharpoons PbCl_4^{2-}$	4×10^1
$Cd^{2+} + 4\,OH^- \rightleftharpoons Cd(OH)_4^{2-}$	4.2×10^8	$Pb^{2+} + 4\,I^- \rightleftharpoons PbI_4^{2-}$	3.0×10^4
$Cd^{2+} + 4\,NH_3 \rightleftharpoons Cd(NH_3)_6^{2+}$	1.3×10^7	$Sb^{3+} + 4\,Cl^- \rightleftharpoons SbCl_4^-$	5.2×10^4
$Co^{2+} + 6\,NH_3 \rightleftharpoons Co(NH_3)_6^{2+}$	1.3×10^5	$Sb^{3+} + 4\,OH^- \rightleftharpoons Sb(OH)_4^-$	2×10^{38}
$Co^{3+} + 6\,NH_3 \rightleftharpoons Co(NH_3)_6^{3+}$	2×10^{35}	$Sn^{2+} + 4\,Cl^- \rightleftharpoons SnCl_4^{2-}$	3.0×10^1
$Co^{2+} + 4\,SCN^- \rightleftharpoons Co(SCN)_4^{2-}$	1×10^3	$Zn^{2+} + 4\,CN^- \rightleftharpoons Zn(CN)_4^{2-}$	5×10^{16}
$Cr^{3+} + 4\,OH^- \rightleftharpoons Cr(OH)_4^-$	8×10^{29}	$Zn^{2+} + 4\,OH^- \rightleftharpoons Zn(OH)_4^{2-}$	4.6×10^{17}
$Cu^{2+} + 4\,OH^- \rightleftharpoons Cu(OH)_4^{2-}$	3×10^{18}	$Zn^{2+} + 4\,NH_3 \rightleftharpoons Zn(NH_3)_4^{2+}$	2.9×10^9
$Cu^{2+} + 4\,NH_3 \rightleftharpoons Cu(NH_3)_4^{2+}$	2.1×10^{13}		

TABLE A-13

Standard Reduction Potentials

Half-Reaction	$E°$ (V)
$3\ N_2 + 2\ H^+ + 2\ e^- \rightleftharpoons 2\ HN_3$	-3.1
$Li^+ + e^- \rightleftharpoons Li$	-3.045
$Rb^+ + e^- \rightleftharpoons Rb$	-2.925
$K^+ + e^- \rightleftharpoons K$	-2.924
$Cs^+ + e^- \rightleftharpoons Cs$	-2.923
$Ba^{2+} + 2\ e^- \rightleftharpoons Ba$	-2.90
$Sr^{2+} + 2\ e^- \rightleftharpoons Sr$	-2.89
$Ca^{2+} + 2\ e^- \rightleftharpoons Ca$	-2.76
$Na^+ + e^- \rightleftharpoons Na$	-2.7109
$Mg(OH)_2 + 2\ e^- \rightleftharpoons Mg + 2\ OH^-$	-2.69
$Mg^{2+} + 2\ e^- \rightleftharpoons Mg$	-2.375
$H_2 + 2\ e^- \rightleftharpoons 2\ H^-$	-2.23
$Al^{3+} + 3\ e^- \rightleftharpoons Al$ (0.1 M NaOH)	-1.706
$Be^{2+} + 2\ e^- \rightleftharpoons Be$	-1.70
$Ti^{2+} + 2\ e^- \rightleftharpoons Ti$	-1.63
$Zn(CN)_4{}^{2-} + 2\ e^- \rightleftharpoons Zn + 4\ CN^-$	-1.26
$Zn(NH_3)_4{}^{2+} + 2\ e^- \rightleftharpoons Zn + 4\ NH_3$	-1.04
$Mn^{2+} + 2\ e^- \rightleftharpoons Mn$	-1.029
$SO_4{}^{2-} + H_2O + 2\ e^- \rightleftharpoons SO_3{}^{2-} + 2\ OH^-$	-0.92
$Cr^{2+} + 2\ e^- \rightleftharpoons Cr$	-0.91
$TiO_2 + 4\ H^+ + 4\ e^- \rightleftharpoons Ti + 2\ H_2O$	-0.87
$2\ H_2O + 2\ e^- \rightleftharpoons H_2 + 2\ OH^-$	-0.8277
$Zn^{2+} + 2\ e^- \rightleftharpoons Zn$	-0.7628
$Cr^{3+} + 3\ e^- \rightleftharpoons Cr$	-0.74
$2\ SO_3{}^{2-} + 3\ H_2O + 4\ e^- \rightleftharpoons S_2O_3{}^{2-} + 6\ OH^-$	-0.58
$PbO + H_2O + 2\ e^- \rightleftharpoons Pb + 2\ OH^-$	-0.576
$Ga^{3+} + 3\ e^- \rightleftharpoons Ga$	-0.560
$S + 2\ e^- \rightleftharpoons S^{2-}$	-0.508
$2\ CO_2 + 2\ H^+ + 2\ e^- \rightleftharpoons H_2C_2O_4$	-0.49
$Ni(NH_3)_6{}^{2+} + 2\ e^- \rightleftharpoons Ni + 6\ NH_3$	-0.48
$Co(NH_3)_6{}^{2+} + 2\ e^- \rightleftharpoons Co + 6\ NH_3$	-0.422
$Cr^{3+} + e^- \rightleftharpoons Cr^{2+}$	-0.41
$Fe^{2+} + 2\ e^- \rightleftharpoons Fe$	-0.409
$Cd^{2+} + 2\ e^- \rightleftharpoons Cd$	-0.4026
$PbSO_4 + 2\ e^- \rightleftharpoons Pb + SO_4{}^{2-}$	-0.356
$In^{3+} + 3\ e^- \rightleftharpoons In$	-0.338
$Tl^+ + e^- \rightleftharpoons Tl$	-0.3363
$Ag(CN)_2{}^- + e^- \rightleftharpoons Ag + 2\ CN^-$	-0.31
$Co^{2+} + 2\ e^- \rightleftharpoons Co$	-0.28
$H_3PO_4 + 2\ H^+ + 2\ e^- \rightleftharpoons H_3PO_3 + H_2O$	-0.276
$Ni^{2+} + 2\ e^- \rightleftharpoons Ni$	-0.23
$2\ SO_4{}^{2-} + 4\ H^+ + 2\ e^- \rightleftharpoons S_2O_6{}^{2-} + 2\ H_2O$	-0.224
$CO_2 + 2\ H^+ + 2\ e^- \rightleftharpoons HCO_2H$	-0.20
$O_2 + 2\ H_2O + 2\ e^- \rightleftharpoons H_2O_2 + 2\ OH^-$	-0.146
$Sn^{2+} + 2\ e^- \rightleftharpoons Sn$	-0.1364
$Pb^{2+} + 2\ e^- \rightleftharpoons Pb$	-0.1263
$CrO_4{}^{2-} + 4\ H_2O + 3\ e^- \rightleftharpoons Cr(OH)_3 + 5\ OH^-$	-0.12

Standard Reduction Potentials

Half-Reaction	$E°$ (V)
$WO_3 + 6\ H^+ + 6\ e^- \rightleftharpoons W + 3\ H_2O$	-0.09
$Ru^{3+} + e^- \rightleftharpoons Ru^{2+}$	-0.08
$O_2 + H_2O + 2\ e^- \rightleftharpoons HO_2^- + OH^-$	-0.076
$Fe^{3+} + 3\ e^- \rightleftharpoons Fe$	-0.036
$2\ H^+ + 2\ e^- \rightleftharpoons H_2$	$0.0000000...$
$NO_3^- + H_2O + 2\ e^- \rightleftharpoons NO_2^- + 2\ OH^-$	0.01
$AgBr + e^- \rightleftharpoons Ag + Br^-$	0.0713
$S_4O_6^{2-} + 2\ e^- \rightleftharpoons 2\ S_2O_3^{2-}$	0.0895
$Sn^{4+} + 2\ e^- \rightleftharpoons Sn^{2+}$	0.15
$Cu^{2+} + e^- \rightleftharpoons Cu^+$	0.158
$ClO_4^- + H_2O + 2\ e^- \rightleftharpoons ClO_3^- + 2\ OH^-$	0.17
$SO_4^{2-} + 4\ H^+ + 2\ e^- \rightleftharpoons H_2SO_3 + H_2O$	0.20
$AgCl + e^- \rightleftharpoons Ag + Cl^-$	0.2223
$IO_3^- + 3\ H_2O + 6\ e^- \rightleftharpoons I^- + 6\ OH^-$	0.26
$Hg_2Cl_2 + 2\ e^- \rightleftharpoons 2\ Hg + 2\ Cl^-$	0.2682
$Cu^{2+} + 2\ e^- \rightleftharpoons Cu$	0.3402
$ClO_3^- + H_2O + 2\ e^- \rightleftharpoons ClO_2^- + 2\ OH^-$	0.35
$Ag(NH_3)_2^+ + e^- \rightleftharpoons Ag + 2\ NH_3$	0.373
$O_2 + 2\ H_2O + 4\ e^- \rightleftharpoons 4\ OH^-$	0.401
$H_2SO_3 + 4\ H^+ + 4\ e^- \rightleftharpoons S + 3\ H_2O$	0.45
$HgCl_4^{2-} + 2\ e^- \rightleftharpoons Hg + 4\ Cl^-$	0.48
$Cu^+ + e^- \rightleftharpoons Cu$	0.522
$I_3^- + 2\ e^- \rightleftharpoons 3\ I^-$	0.5338
$I_2 + 2\ e^- \rightleftharpoons 2\ I^-$	0.535
$MnO_4^- + 2\ H_2O + 3\ e^- \rightleftharpoons MnO_2 + 4\ OH^-$	0.588
$ClO_2^- + H_2O + 2\ e^- \rightleftharpoons ClO^- + 2\ OH^-$	0.59
$O_2 + 2\ H^+ + 2\ e^- \rightleftharpoons H_2O_2$	0.682
$Fe^{3+} + e^- \rightleftharpoons Fe^{2+}$	0.770
$Hg_2^{2+} + 2\ e^- \rightleftharpoons Hg$	0.7961
$Ag^+ + e^- \rightleftharpoons Ag$	0.7996
$Hg^{2+} + 2\ e^- \rightleftharpoons Hg$	0.851
$H_2O_2 + 2\ e^- \rightleftharpoons 2\ OH^-$	0.88
$ClO^- + H_2O + 2\ e^- \rightleftharpoons Cl^- + 2\ OH^-$	0.89
$2\ Hg^{2+} + 2\ e^- \rightleftharpoons Hg_2^{2+}$	0.905
$NO_3^- + 3\ H^+ + 2\ e^- \rightleftharpoons HNO_2 + H_2O$	0.94
$ClO_2 + e^- \rightleftharpoons ClO_2^-$	0.95
$NO_3^- + 4\ H^+ + 3\ e^- \rightleftharpoons NO + 2\ H_2O$	0.96
$Pd^{2+} + 2\ e^- \rightleftharpoons Pd$	0.987
$HNO_2 + H^+ + e^- \rightleftharpoons NO + H_2O$	0.99
$IO_3^- + 6\ H^+ + 6\ e^- \rightleftharpoons I^- + 3\ H_2O$	1.085
$Br_2(aq) + 2\ e^- \rightleftharpoons 2\ Br^-$	1.087
$Cr^{6+} + 3\ e^- \rightleftharpoons Cr^{3+}$	1.10
$ClO_3^- + 2\ H^+ + 2\ e^- \rightleftharpoons ClO_2^- + H_2O$	1.15
$ClO_4^- + 2\ H^+ + 2\ e^- \rightleftharpoons ClO_3^- + H_2O$	1.19
$2\ IO_3^- + 12\ H^+ + 10\ e^- \rightleftharpoons I_2 + 6\ H_2O$	1.19
$HCrO_4^- + 7\ H^+ + 3\ e^- \rightleftharpoons Cr^{3+} + 4\ H_2O$	1.195
$Pt^{2+} + 2\ e^- \rightleftharpoons Pt$	1.2

(continued)

TABLE A-13   continued

Standard Reduction Potentials

Half-Reaction	$E°$ (V)
$MnO_2 + 4\ H^+ + 2\ e^- \rightleftharpoons Mn^{2+} + 2\ H_2O$	1.208
$O_2 + 4\ H^+ + 4\ e^- \rightleftharpoons 2\ H_2O$	1.229
$O_3 + H_2O + 2\ e^- \rightleftharpoons O_2 + 2\ OH^-$	1.24
$Tl^{3+} + 2\ e^- \rightleftharpoons Tl^+$	1.247
$ClO_2 + H^+ + e^- \rightleftharpoons HClO_2$	1.27
$2\ HNO_2 + 4\ H^+ + 4\ e^- \rightleftharpoons N_2O + 3\ H_2O$	1.27
$Au^{3+} + 2\ e^- \rightleftharpoons Au^+$	1.29
$Cr_2O_7{}^{2-} + 14\ H^+ + 6\ e^- \rightleftharpoons 2\ Cr^{3+} + 7\ H_2O$	1.33
$ClO_4{}^- + 8\ H^+ + 7\ e^- \rightleftharpoons \frac{1}{2}\ Cl_2 + 4\ H_2O$	1.34
$Cl_2(g) + 2\ e^- \rightleftharpoons 2\ Cl^-$	1.3583
$ClO_4{}^- + 8\ H^+ + 8\ e^- \rightleftharpoons Cl^- + 4\ H_2O$	1.37
$Au^{3+} + 3\ e^- \rightleftharpoons Au$	1.42
$ClO_3{}^- + 6\ H^+ + 6\ e^- \rightleftharpoons Cl^- + 3\ H_2O$	1.45
$PbO_2 + 4\ H^+ + 2\ e^- \rightleftharpoons Pb^{2+} + 2\ H_2O$	1.467
$2\ ClO_3{}^- + 12\ H^+ + 10\ e^- \rightleftharpoons Cl_2 + 6\ H_2O$	1.47
$HClO + H^+ + 2\ e^- \rightleftharpoons Cl^- + H_2O$	1.49
$MnO_4{}^- + 8\ H^+ + 5\ e^- \rightleftharpoons Mn^{2+} + 4\ H_2O$	1.491
$HClO_2 + 3\ H^+ + 4\ e^- \rightleftharpoons Cl^- + 2\ H_2O$	1.56
$2\ NO + 2\ H^+ + 2\ e^- \rightleftharpoons N_2O + H_2O$	1.59
$2\ HClO_2 + 6\ H^+ + 6\ e^- \rightleftharpoons Cl_2 + 4\ H_2O$	1.63
$2\ HClO + 2\ H^+ + 2\ e^- \rightleftharpoons Cl_2 + 2\ H_2O$	1.63
$HClO_2 + 2\ H^+ + 2\ e^- \rightleftharpoons HClO + H_2O$	1.64
$MnO_4{}^- + 4\ H^+ + 3\ e^- \rightleftharpoons MnO_2 + 2\ H_2O$	1.679
$Au^+ + e^- \rightleftharpoons Au$	1.68
$PbO_2 + SO_4{}^{2-} + 4\ H^+ + 2\ e^- \rightleftharpoons PbSO_4 + 2\ H_2O$	1.685
$N_2O + 2\ H^+ + 2\ e^- \rightleftharpoons N_2 + H_2O$	1.77
$H_2O_2 + 2\ H^+ + 2\ e^- \rightleftharpoons 2\ H_2O$	1.776
$Co^{3+} + e^- \rightleftharpoons Co^{2+}$	1.842
$S_2O_8{}^{2-} + 2\ e^- \rightleftharpoons 2\ SO_4{}^{2-}$	2.05
$O_3(g) + 2\ H^+ + 2\ e^- \rightleftharpoons O_2(g) + H_2O$	2.07
$F_2(g) + 2\ H^+ + 2\ e^- \rightleftharpoons 2\ HF(aq)$	3.03

TABLE A-14

Bond Dissociation Enthalpies

Single Bond Dissociation Enthalpies (kJ/mol)

	As	B	Br	C	Cl	F	H	I	N	O	P	S	Si
As	180		255	200	310	485	300	180		330			
B		300	370		445	645		270		525			
Br			195	270	220	240	370	180	250		270	215	330
C				350	330	490	415	210	305	360	265	270	305
Cl					240	250	431	210	190	205	330	270	400
F						160	569		280	215	500	325	600
H							435	300	390	464	325	370	320
I								150		200	180		230
N									160	165			330
O										140	370	423	464
P											210		
S												260	
Si													225

Double and Triple Bond Dissociation
Enthalpies (kJ/mol)

C=C	611		C=S	477
C≡C	837		N=N	418
C=O	745		N≡N	946
C≡O	1075		N=O	594
C=N	615		O=O	498
C≡N	891		S=O	523

TABLE A-15

Standard-State Enthalpies of Formation,
Free Energies of Formation, and Absolute Entropies

Substance	ΔH_f° (kJ/mol)	ΔG_f° (kJ/mol)	S° (J/mol-K)	Substance	ΔH_f° (kJ/mol)	ΔG_f° (kJ/mol)	S° (J/mol-K)
	Aluminum				Antimony		
Al(s)	0	0	28.33	Sb(s)	0	0	45.69
Al(g)	326.4	285.7	164.54	Sb$_4$(g)	205.0	154.8	352
Al^{3+}(aq)	−531	−485	−321.7	Sb$_2$O$_5$(s)	−971.9	−829.2	125.1
Al$_2$O$_3$(s)	−1675.7	−1582.3	50.92	Sb$_4$O$_6$(s)	−1440.6	−1268.1	220.9
AlCl$_3$(s)	−704.2	−628.8	110.67	SbH$_3$(g)	145.105	147.75	232.78
AlF$_3$(s)	−1504.1	−1425.0	66.44	SbCl$_3$(s)	−382.17	−323.67	184.1
Al$_2$(SO$_4$)$_3$(s)	−3440.84	−3099.94	239.3	SbCl$_3$(g)	−313.8	−301.2	337.80
				SbCl$_5$(l)	−440.2	−350.1	301
				SbCl$_5$(g)	−394.3	−334.29	401.94
				Sb$_2$S$_3$(s)	−174.9	−173.6	182.0

(continued)

TABLE A-15 continued

Standard-State Enthalpies of Formation, Free Energies of Formation, and Absolute Entropies

Substance	ΔH_f° (kJ/mol)	ΔG_f° (kJ/mol)	S° (J/mol-K)	Substance	ΔH_f° (kJ/mol)	ΔG_f° (kJ/mol)	S° (J/mol-K)
Arsenic				*Boron*			
As(s)	0	0	35.1	$B_{10}H_{14}(s)$	-45.2	192.3	176.56
$As_4(g)$	143.9	92.4	314	$H_3BO_3(s)$	-1094.33	-968.92	88.83
$As_2O_5(s)$	-924.87	-782.3	105.4	$BF_3(g)$	-1137.00	-1120.33	254.12
$As_4O_6(s)$	-1313.94	-1152.43	214.2	$BCl_3(l)$	-427.2	-387.4	206.3
$AsH_3(g)$	66.44	68.93	222.78	$B_3N_3H_6(l)$	-541.00	-392.65	199.6
$AsF_3(g)$	-785.76	-770.76	289.10	$B_3N_3H_6(g)$	-511.75	-390.00	288.68
$AsCl_3(g)$	-261.5	-248.9	327.17				
$AsBr_3(g)$	$-130.$	-159	363.87	*Bromine*			
$As_2S_3(s)$	-169	-168.6	163.6	$Br_2(l)$	0	0	152.231
				$Br_2(g)$	30.907	3.110	245.463
Barium				$Br(g)$	111.884	82.396	175.022
Ba(s)	0	0	62.8	$HBr(g)$	-36.40	-53.45	198.695
Ba(g)	180.	146	170.243	$HBr(aq)$	-121.55	-103.96	82.4
$Ba^{2+}(aq)$	-537.64	-560.77	9.6	$BrF(g)$	-93.85	-109.18	228.97
BaO(s)	-553.5	-525.1	70.42	$BrF_3(g)$	-255.60	-229.43	292.53
Ba(OH)$_2$ ·8 H$_2$O(s)	-3342.2	-2792.8	427	$BrF_5(g)$	-428.9	-350.6	320.19
$BaCl_2(s)$	-858.6	-810.4	123.68				
$BaCl_2(aq)$	-871.95	-823.21	122.6	*Calcium*			
$BaSO_4(s)$	-1473.2	-1362.2	132.2	Ca(s)	0	0	41.42
$Ba(NO_3)_2(s)$	-992.07	-796.59	213.8	Ca(g)	178.2	144.3	154.884
$Ba(NO_3)_2(aq)$	-952.36	-783.28	302.5	$Ca^{2+}(aq)$	-542.83	-553.58	-53.1
				CaO(s)	-635.09	-604.03	39.75
Beryllium				$Ca(OH)_2(s)$	-986.09	-898.49	83.39
Be(s)	0	0	9.50	$CaCl_2(s)$	-795.8	-748.1	104.6
Be(g)	324.3	286.6	136.269	$CaSO_4(s)$	-1434.11	-1321.79	106.7
$Be^{2+}(aq)$	-382.8	-379.73	-129.7	$CaSO_4 \cdot 2H_2O(s)$	-2022.63	-1797.28	194.1
BeO(s)	-609.6	-580.3	14.14	$Ca(NO_3)_2(s)$	-938.39	-743.07	193.3
$BeCl_2(s)$	-490.4	-445.6	82.68	$CaCO_3(s)$	-1206.92	-1128.79	92.9
				$Ca_3(PO_4)_2$	-4120.8	-3884.7	236.0
Bismuth							
Bi(s)	0	0	56.74	*Carbon*			
Bi(g)	207.1	168.2	187.05	C(graphite)	0	0	5.74
$Bi_2O_3(s)$	-573.88	-493.7	151.5	C(diamond)	1.895	2.900	2.377
$BiCl_3(s)$	-379.1	-315.0	177.0	C(g)	716.682	671.257	158.096
$BiCl_3(g)$	-265.7	-256.0	358.85	CO(g)	-110.525	-137.168	197.674
$Bi_2S_3(s)$	-143.1	-140.6	200.4	$CO_2(g)$	-393.509	-394.359	213.74
				$COCl_2(g)$	-218.8	-204.6	283.53
Boron				$CH_4(g)$	-74.81	-50.752	186.264
B(s)	0	0	5.86	HCHO(g)	-108.57	-102.53	218.77
B(g)	562.7	518.8	153.45	$H_2CO_3(aq)$	-699.65	-623.08	187.4
$B_2O_3(s)$	-1272.77	-1193.65	53.97	$HCO_3^-(aq)$	-691.99	-586.77	91.2
$B_2H_6(g)$	35.6	86.7	232.11	$CO_3^{2-}(aq)$	-677.14	-527.81	-56.9
$B_5H_9(l)$	42.68	171.82	184.22	$CH_3OH(l)$	-238.66	-166.27	126.8
$B_5H_9(g)$	73.2	175.0	275.92	$CH_3OH(g)$	-200.66	-161.96	239.81

TABLE A-15   continued

Standard-State Enthalpies of Formation, Free Energies of Formation, and Absolute Entropies

Substance	ΔH_f° (kJ/mol)	ΔG_f° (kJ/mol)	S° (J/mol-K)	Substance	ΔH_f° (kJ/mol)	ΔG_f° (kJ/mol)	S° (J/mol-K)
Carbon				*Cobalt*			
$CCl_4(l)$	−135.44	−65.21	216.40	$Co(s)$	0	0	30.04
$CCl_4(g)$	−102.9	−60.59	309.85	$Co(g)$	424.7	380.3	179.515
$CHCl_3(l)$	−134.47	−73.66	201.1	$Co^{2+}(aq)$	−58.2	−54.4	−113
$CHCl_3(g)$	−103.14	−70.34	295.71	$Co^{3+}(aq)$	92	134	−305
$CH_2Cl_2(l)$	−121.46	−67.26	177.8	$CoO(s)$	−237.94	−214.20	52.97
$CH_2Cl_2(g)$	−92.47	−65.87	270.23	$Co_3O_4(s)$	−891	−774	102.5
$CH_3Cl(g)$	−80.84	−57.40	234.5	$Co(NH_3)_6^{3+}(aq)$	−584.9	−157.0	146
$CS_2(l)$	89.70	65.27	151.34				
$CS_2(g)$	117.36	67.12	237.84				
$HCN(g)$	135.1	124.7	201.78	*Copper*			
$CH_3NO_2(l)$	−113.09	−14.42	171.75	$Cu(s)$	0	0	33.150
$C_2H_2(g)$	226.73	209.20	200.94	$Cu(g)$	338.32	298.58	166.38
$C_2H_4(g)$	52.26	68.15	219.56	$Cu^+(aq)$	71.67	49.98	40.6
$C_2H_6(g)$	−84.68	−32.82	229.60	$Cu^{2+}(aq)$	64.77	65.49	−99.6
$CH_3CHO(l)$	−192.30	−128.12	160.2	$CuO(s)$	−157.3	−129.7	42.63
$CH_3CO_2H(l)$	−184.5	−389.9	159.8	$Cu_2O(s)$	−168.6	−146.0	93.14
$CH_3CO_2H(g)$	−432.25	−374.0	282.5	$CuCl_2(s)$	−220.1	−175.7	108.07
$CH_3CO_2H(aq)$	−485.76	−396.46	178.7	$CuS(s)$	−53.1	−53.6	66.5
$CH_3CO_2^-(aq)$	−486.01	−369.31	86.6	$Cu_2S(s)$	−79.5	−86.2	120.9
$CH_3CH_2OH(l)$	−277.69	−174.78	160.7	$CuSO_4(s)$	−771.36	−66.69	109
$CH_3CH_2OH(g)$	−235.10	−168.49	282.70	$Cu(NH_3)_4^{2+}(aq)$	−348.5	−111.07	273.6
$CH_3CH_2OH(aq)$	−288.3	−181.64	148.5				
$C_6H_6(l)$	49.028	124.50	172.8	*Fluorine*			
$C_6H_6(g)$	82.927	129.66	269.2	$F_2(g)$	0	0	202.78
				$F(g)$	78.99	61.91	158.754
Chlorine				$F^-(aq)$	−332.63	−278.79	−13.8
$Cl_2(g)$	0	0	223.066	$HF(g)$	−271.1	−273.2	173.779
$Cl(g)$	121.679	105.680	165.198	$HF(aq)$	−320.08	−296.82	88.7
$Cl^-(aq)$	−167.159	−131.228	56.5				
$ClO_2(g)$	102.5	120.5	256.84	*Hydrogen*			
$Cl_2O(g)$	80.3	97.9	266.21	$H_2(g)$	0	0	130.684
$Cl_2O_7(l)$	238			$H(g)$	217.65	203.247	114.713
$HCl(g)$	−92.307	−95.299	186.908	$H^+(aq)$	0	0	0
$HCl(aq)$	−167.159	−131.228	56.5	$OH^-(aq)$	−229.994	−157.244	−10.75
$ClF(g)$	−54.48	−55.94	217.89	$H_2O(l)$	−285.830	−237.129	69.91
				$H_2O(g)$	−241.818	−228.572	188.25
Chromium				$H_2O_2(l)$	−187.78	−120.35	109.6
$Cr(s)$	0	0	23.77	$H_2O_2(aq)$	−191.17	−134.03	143.9
$Cr(g)$	396.6	351.8	174.50				
$CrO_3(s)$	−589.5			*Iodine*			
$CrO_4^{2-}(aq)$	−881.15	−727.75	50.21	$I_2(s)$	0	0	116.135
$Cr_2O_3(s)$	−1139.7	−1058.1	81.2	$I_2(g)$	62.438	19.327	260.69
$Cr_2O_7^{2-}(aq)$	−1490.3	−1301.1	261.9	$I(g)$	106.838	70.50	180.791
$(NH_4)_2Cr_2O_7(s)$	−1806.7			$HI(g)$	26.48	1.70	206.594
$PbCrO_4(s)$	−930.9			$IF(g)$	−95.65	−118.51	236.17

(continued)

TABLE A-15 continued

Standard-State Enthalpies of Formation, Free Energies of Formation, and Absolute Entropies

Substance	ΔH_f° (kJ/mol)	ΔG_f° (kJ/mol)	S° (J/mol-K)	Substance	ΔH_f° (kJ/mol)	ΔG_f° (kJ/mol)	S° (J/mol-K)
	Iodine				*Magnesium*		
$IF_5(g)$	−822.49	−751.73	327.7	$Mg(s)$	0	0	32.68
$IF_7(g)$	−943.9	−818.3	346.5	$Mg(g)$	147.70	113.10	148.650
$ICl(g)$	17.78	−5.46	247.551	$Mg^{2+}(aq)$	−466.85	−454.8	−138.1
$IBr(g)$	40.84	3.69	258.773	$MgO(s)$	−601.70	−569.43	26.94
				$MgH_2(s)$	−75.3	−35.9	31.09
	Iron			$Mg(OH)_2(s)$	−924.54	−833.58	63.18
$Fe(s)$	0	0	27.28	$MgCl_2(s)$	−641.32	−591.79	89.62
$Fe(g)$	416.3	370.7	180.49	$MgCO_3(s)$	−1095.8	−1012.1	65.7
$Fe^{2+}(aq)$	−89.1	−78.90	−137.7	$MgSO_4(s)$	−1284.9	−1170.6	91.6
$Fe^{3+}(aq)$	−48.5	−4.7	−315.9				
$Fe_2O_3(s)$	−824.2	−742.2	87.40				
$Fe_3O_4(s)$	−1118.4	−1015.4	146.4		*Manganese*		
$Fe(OH)_2(s)$	−569.0	−486.5	88	$Mn(s)$	0	0	32.01
$Fe(OH)_3(s)$	−823.0	−696.5	106.7	$Mn(g)$	280.7	238.5	173.70
$FeCl_3(s)$	−399.49	−334.00	142.3	$Mn^{2+}(aq)$	−220.75	−228.1	−73.6
$FeS_2(s)$	−178.2	−166.9	52.93	$MnO(s)$	−385.22	−362.90	59.71
$Fe(CO)_5(l)$	−774.0	−705.3	338.1	$MnO_2(s)$	−520.03	−465.14	53.05
$Fe(CO)_5(g)$	−733.9	−697.21	445.3	$Mn_2O_3(s)$	−959.0	−881.1	110.5
				$Mn_3O_4(s)$	−1387.8	−1283.2	155.6
	Lead			$KMnO_4(s)$	−837.2	−737.6	171.76
$Pb(s)$	0	0	64.81	MnS	−214.2	−218.4	78.2
$Pb(g)$	195.0	161.9	175.373				
$Pb^{2+}(aq)$	−1.7	−24.43	10.5				
$PbO(s)$	−217.32	−187.89	68.7		*Mercury*		
$PbO_2(s)$	−277.4	−217.33	68.6	$Hg(l)$	0	0	76.02
$PbCl_2(s)$	−359.41	−314.10	136.0	$Hg(g)$	61.317	31.820	174.96
$PbCl_4(l)$	−329.3			$Hg^{2+}(aq)$	171.1	164.40	−32.2
$PbS(s)$	−100.4	−98.7	91.2	$HgO(s)$	−90.83	−58.539	70.29
$PbSO_4(s)$	−919.94	−813.14	148.57	$HgCl_2(s)$	−224.3	−178.6	146.0
$Pb(NO_3)_2(s)$	−451.9			$Hg_2Cl_2(s)$	−265.22	−210.745	192.5
$PbCO_3(s)$	−699.1	−625.5	131.0	$HgS(s)$	−58.2	−50.6	82.4
	Lithium				*Nitrogen*		
$Li(s)$	0	0	29.12	$N_2(g)$	0	0	191.61
$Li(g)$	159.37	126.66	138.77	$N(g)$	472.704	455.63	153.298
$Li^+(aq)$	−278.49	−293.31	13.4	$NO(g)$	90.25	86.55	210.761
$LiH(s)$	−90.54	−68.35	20.08	$NO_2(g)$	33.18	51.31	240.06
$LiOH(s)$	−484.93	−438.95	42.80	$N_2O(g)$	82.05	104.20	219.85
$LiF(s)$	−615.97	−587.71	35.65	$N_2O_3(g)$	83.72	139.46	312.28
$LiCl(s)$	−408.61	−384.37	59.33	$N_2O_4(g)$	9.16	97.89	304.29
$LiBr(s)$	−351.23	−342.00	74.27	$N_2O_5(g)$	11.35	115.1	355.7
$LiI(s)$	−270.41	−270.29	86.78	$NO_3^-(aq)$	−205.0	−108.74	146.4
$LiAlH_4(s)$	−116.3	−44.7	78.74	$NOCl(g)$	51.71	66.08	261.69
$LiBH_4(s)$	190.8	125.0	75.86	$NO_2Cl(g)$	12.6	54.4	272.15
				$HNO_2(aq)$	−119.2	−50.6	135.6

TABLE A-15 continued

Standard-State Enthalpies of Formation, Free Energies of Formation, and Absolute Entropies

Substance	ΔH_f° (kJ/mol)	ΔG_f° (kJ/mol)	S° (J/mol-K)	Substance	ΔH_f° (kJ/mol)	ΔG_f° (kJ/mol)	S° (J/mol-K)
Nitrogen				*Silicon*			
$HNO_3(g)$	−135.06	−74.72	266.38	$Si(s)$	0	0	18.83
$HNO_3(aq)$	−207.36	−111.25	146.4	$Si(g)$	455.6	411.3	167.97
$NH_3(g)$	−46.11	−16.45	192.45	$SiO_2(s)$	−910.94	−856.64	41.84
$NH_3(aq)$	−80.29	−26.50	111.3	$SiH_4(g)$	34.3	56.9	204.62
$NH_4^+(aq)$	−132.51	−79.31	113.4	$SiF_4(g)$	−1614.94	−1572.65	282.49
$NH_4NO_3(s)$	−365.56	−183.87	151.08	$SiCl_4(l)$	−687.0	−619.9	240.
$NH_4NO_3(aq)$	−339.87	−190.56	259.8	$SiCl_4(g)$	−657.01	−616.98	330.73
$NH_4Cl(s)$	−314.43	−203.87	94.6				
$N_2H_4(l)$	50.63	149.34	121.21	*Silver*			
$N_2H_4(g)$	95.40	159.35	238.47	$Ag(s)$	0	0	42.55
$HN_3(g)$	294.1	328.1	238.97	$Ag(g)$	284.55	245.65	172.97
				$Ag^+(aq)$	105.579	77.107	72.68
Oxygen				$Ag(NH_3)_2^+(aq)$	−111.29	−17.12	245.2
$O_2(g)$	0	0	205.138	$Ag_2O(s)$	−31.05	−11.20	−121.3
$O(g)$	249.170	231.731	161.055	$AgCl(s)$	−127.068	−109.789	96.2
$O_3(g)$	142.7	163.2	238.93	$AgBr(s)$	−100.37	−96.90	107.1
				$AgI(s)$	−61.84	−66.19	−115.5
Phosphorus							
$P(white)$	0	0	41.09	*Sodium*			
$P_4(g)$	58.91	24.4	279.98	$Na(s)$	0	0	51.21
$P_2(g)$	144.3	103.7	218.129	$Na(g)$	107.32	76.761	153.712
$P(g)$	314.64	278.25	163.193	$Na^+(aq)$	−240.13	−261.905	59.0
$PH_3(g)$	5.4	13.4	210.23	$NaH(s)$	−56.275	−33.46	−40.016
$P_4O_6(s)$	−1640.1			$NaOH(s)$	−425.609	−379.494	64.455
$P_4O_{10}(s)$	−2984.0	−2697.7	228.86	$NaOH(aq)$	−470.114	−419.150	48.1
$PO_4^{3-}(aq)$	−1277.4	−1018.7	−222	$NaCl(s)$	−411.153	−384.138	72.13
$PF_3(g)$	−918.8	−897.5	273.24	$NaCl(g)$	−176.65	−196.66	229.81
$PF_5(g)$	−1595.8			$NaCl(aq)$	−407.27	−393.133	115.5
$PCl_3(l)$	−319.7	−272.3	217.1	$NaNO_3(s)$	−467.85	−367.00	116.52
$PCl_3(g)$	−287.0	−267.8	311.78	$Na_3PO_4(s)$	−1917.40	−1788.80	173.80
$PCl_5(g)$	−374.9	−305.0	364.58	$Na_2SO_3(s)$	−1123.0	−1028.0	155
$H_3PO_4(s)$	−1279.0	−1119.1	110.50	$Na_2SO_4(s)$	−1387.08	−1270.16	149.58
$H_3PO_4(aq)$	−1277.4	−1018.7	−222	$Na_2CO_3(s)$	−1130.68	−1044.44	134.98
				$NaHCO_3(s)$	−950.81	−851.0	101.7
Potassium				$NaOAc(s)$	−708.81	−607.18	123.0
$K(s)$	0	0	64.18	$Na_2CrO_4(s)$	−1342.2	−1234.93	176.61
$K(g)$	89.24	60.59	160.336	$Na_2Cr_2O_7(s)$	−1978.6		
$K^+(aq)$	−252.38	−283.27	102.5				
$KOH(s)$	−424.764	−379.08	78.9	*Sulfur*			
$KCl(s)$	−436.747	−409.14	82.59	$S_8(s)$	0	0	31.80
$KNO_3(s)$	−494.63	−394.86	133.05	$S_8(g)$	102.30	49.63	430.98
$K_2Cr_2O_7(s)$	−2061.5	−1881.8	291.2	$S(g)$	278.805	238.250	167.821
$KMnO_4(s)$	−837.2	−737.6	171.76	$S^{2-}(aq)$	33.1	85.8	−14.6
				$SO_2(g)$	−296.830	−300.194	248.22

(continued)

TABLE A-15 continued

Standard-State Enthalpies of Formation,
Free Energies of Formation, and Absolute Entropies

Substance	ΔH_f° (kJ/mol)	ΔG_f° (kJ/mol)	S° (J/mol-K)	Substance	ΔH_f° (kJ/mol)	ΔG_f° (kJ/mol)	S° (J/mol-K)
	Sulfur				*Titanium*		
$SO_3(s)$	-454.51	-374.21	70.7	$Ti(s)$	0	0	30.63
$SO_3(l)$	-441.04	-373.75	113.8	$Ti(g)$	469.9	425.1	180.2
$SO_3(g)$	-395.72	-371.06	256.76	$TiO(s)$	-519.7	-495.0	34.8
$SO_4^{2-}(aq)$	-909.27	-744.53	20.1	$TiO_2(s)$ rutile	-944.8	-889.5	50.33
$SOCl_2(g)$	-212.5	-198.3	309.77	$TiCl_4(l)$	-804.2	-737.2	252.3
$SO_2Cl_2(g)$	-364.0	-320.0	311.94	$TiCl_4(g)$	-763.2	-726.8	354.8
$H_2S(g)$	-20.63	-33.56	205.79				
$H_2SO_3(aq)$	-608.81	-537.81	232.2				
$H_2SO_4(aq)$	-909.27	-744.53	20.1		*Tungsten*		
$SF_4(g)$	-774.9	-731.3	292.03	$W(s)$	0	0	32.64
$SF_6(g)$	-1209	-1105.3	291.82	$W(g)$	849.4	807.1	173.950
$SCN^-(aq)$	76.44	92.71	144.3	$WO_3(s)$	-842.87	-764.083	75.90
	Tin				*Zinc*		
$Sn(s)$	0	0	44.14	$Zn(s)$	0	0	41.63
$Sn(g)$	302.1	267.3	168.486	$Zn(g)$	130.729	95.145	160.984
$SnO(s)$	-285.8	-256.9	56.5	$Zn^{2+}(aq)$	-153.89	-147.06	-112.1
$SnO_2(s)$	-580.7	-519.76	52.3	$ZnO(s)$	-348.28	-318.30	43.64
$SnCl_2(s)$	-325.1			$ZnCl_2(s)$	-415.05	-369.39	111.46
$SnCl_4(l)$	-511.3	-440.21	258.6	$ZnS(s)$	-205.98	-201.29	57.7
$SnCl_4(g)$	-471.5	-432.2	365.8	$ZnSO_4(s)$	-982.8	-871.5	110.5

MANIPULATION OF NUMBERS EXPRESSED IN SCIENTIFIC NOTATION

B.1 MULTIPLICATION AND DIVISION

The multiplication of numbers expressed in scientific notation is based on the fact that the product of two exponential numbers with the same base is equal to the base raised to the sum of the two exponents. The product of 10^2 times 10^3, for example, is 10^5.

$$10^2 \times 10^3 = 10^{(2+3)} = 10^5$$

Similarly, the product of 10^5 times 10^{-2} is 10^3.

$$10^5 \times 10^{-2} = 10^{(5-2)} = 10^3$$

The first step in multiplying two numbers expressed in scientific notation is to group similar terms. For example, 5.0×10^3 multiplied by 1.6×10^2 is grouped as follows.

$$5.0 \times 10^3 \times 1.6 \times 10^2 = [5.0 \times 1.6] \times [10^3 \times 10^2]$$

Each term is then computed separately.

$$[5.0 \times 1.6] \times [10^{(3+2)}] = 8.0 \times 10^5$$

We can use this approach to show that the number of copper atoms per penny times the number of grams per atom is in fact equal to the weight of a penny.

$$2.4 \times 10^{22} \, \frac{\text{atoms}}{\text{penny}} \times 1.055 \times 10^{-22} \, \frac{\text{g}}{\text{atom}} = [2.4 \times 1.055] \times [10^{22} \times 10^{-22}] \, \frac{\text{g}}{\text{penny}}$$

$$= [2.5] \times [10^0] = 2.5 \, \frac{\text{g}}{\text{penny}}$$

Division with scientific notation is based on the fact that the ratio of two exponential numbers with the same base can be obtained by subtracting one exponent from the other. For example, 10^6 divided by 10^2 is 10^4.

$$\frac{10^6}{10^2} = 10^{(6-2)} = 10^4$$

Again, the first step involves grouping similar terms.

$$\frac{6.2 \times 10^5}{2.0 \times 10^3} = \frac{6.2}{2.0} \times \frac{10^5}{10^3}$$

Each term is then computed separately.

$$\frac{6.2}{2.0} \times \frac{10^5}{10^3} = 3.1 \times 10^2$$

B.2 ADDITION AND SUBTRACTION

Numbers in scientific notation can be added or subtracted only when the exponents are the same. For example, it is possible to add 1.32×10^3 and 3.46×10^3 by grouping similar terms.

$$[1.32 \times 10^3] + [3.46 \times 10^3] = [1.32 + 3.46] \times 10^3 = 4.78 \times 10^3$$

To subtract 1.02×10^2 from 2.36×10^3, however, requires converting one of these numbers to a different exponent. We can convert 1.02×10^2 into 0.102×10^3 and then subtract the numbers.

$$[2.36 \times 10^3] - [0.102 \times 10^3] = [2.36 - 0.102] \times 10^3 = 2.26 \times 10^3$$

Or we can convert 2.36×10^3 into 23.6×10^2 and subtract the numbers.

$$[23.6 \times 10^2] - [1.02 \times 10^2] = [23.6 - 1.02] \times 10^2 = 22.6 \times 10^2$$

B.3 AVOIDING ERRORS

The most common source of error in manipulating numbers in scientific notation occurs when the product of a calculation is a number such as 2357.8×10^8 or 0.005623×10^{-3}. Is the first of these numbers equal to 2.3578×10^{11} or 2.3578×10^5? Is the second number equal to 5.623 or 5.623×10^{-6}?

To answer these questions, we apply the rules for writing numbers in scientific notation. Consider 2357.8×10^8, for example. Start by writing 2357.8 in scientific notation, and then multiply by 10^8.

$$2357.8 \times 10^8 = [2.3578 \times 10^3] \times 10^8 = 2.3578 \times 10^{11}$$

The same approach can be used to evaluate 0.005623×10^{-3}. Start by writing 0.005623 in scientific notation, and then multiply by 10^{-3}.

$$0.005623 \times 10^{-3} = [5.623 \times 10^{-3}] \times 10^{-3} = 5.623 \times 10^{-6}$$

ANSWERS TO SELECTED PROBLEMS

Chapter 1
THE FUNDAMENTALS
OF MEASUREMENT

1-6   5   1-7 3.17×10^7 s/yr
1-8 3.2×10^4 oz/ton
1-9   7000 gr/lb
1-10 1.04 oz by weight per fluid oz
1-11 (a) 0.3432 (b) 0.2222
      (c) 0.500 (d) 320 (e) 0.250
1-12 2150 in^3/bushel
1-13 3.222×10^{-17} light years/ft
1-16 (a) 43 g (b) 2450 mL
      (c) 8.14 mL (d) 34.68 cm
1-17 1.9×10^{-8} cm, 1.9×10^{-10} m,
      0.19 nm, 190 pm
1-18 0.4–0.7 microns, 400–700
      millimicrons, 400–700 nm,
      $4-7 \times 10^3$ angstroms
1-19 0.4535 lb/kg
1-20 3.934 L/gal
1-21 (a) 8.13 gal (b) 3970 g
      (c) 0.0839 mL (d) 176.1 ft
      (e) 4730 mL (f) 6.5 L
1-22 324 mg   1-23 18 mL
1-24 100-m dash
1-25 117 L liquor, 159 L oil
1-26 a fifth   1-27 0.0914 cm
1-28 62.4 lb   1-29 3.99 ft^3
1-30 Honda
1-31 61.0 in^3/L, 0.0283 m^3/ft^3
1-32 720 mph
1-33 29.095 m/s, 104.6 km/hr,
      95.3 ft/s
1-34 2.9979×10^{10} cm/s,
      6.7063×10^8 mi/hr,
      5.8787×10^{12} mi/yr
1-35 1.7×10^8 mi/hr
1-36 2.8×10^9 beats, 3.7×10^7
      beats/yr
1-37 0.047 m^3/s
1-38 1.06 ton/ft^2
1-39 3.9×10^4 m/s
1-40 744 g   1-42 1000 kg/m^3
1-49 52 nickels
1-50 (a) 3 (b) 4 (c) 5 (d) 2
1-51 (a) 2 (b) 3 (c) 4 (d) 1 (e) 5

1-52 (a) 2 (b) 6 (c) 4 (d) 3 (e) 3
1-53 (a) 475 (b) 0.0680
      (c) 9.46×10^{10} (d) 30.1
1-54 0.4 g/ft^2 1-55 one-half
1-56 4187.35 m
1-57 1.8×10^{12} ton
1-58 300 tablets
1-59 12,00 cans   1-60 1 g/dose
1-61 $2.07   1-62 2.4×10^{-7} cm
1-63 1.1–1.5 ton/yd^3
1-64 226 lb, 6′ 5″
1-65 1.5×10^5 Btu/$1 for gas,
      1.4×10^5 Btu/$1 for oil
1-66 $2.56 electricity/$1.00 gas
1-67 4.8×10^{-10} J absorbed by earth
      per J given off by the sun
1-68 376 atoms
1-69 (a) 1.198×10^1
      (b) 4.6940×10^{-3}
      (c) 4.679×10^6
      (d) 7.967×10^{-6} (e) 1.98×10^0
1-70 (a) 2.126×10^2
      (b) 1.89×10^{-1}
      (c) 1.6221×10^4
      (d) 5.807×10^{-1}
      (e) 1.3276×10^2
1-71 (a) 0.00560 (b) 702500
      (c) 0.08216 (d) 900.
      (e) 0.000002
1-72 (a) 3.92 (b) 33 (c) 3.19×10^4
      (d) 4.31×10^{-2} (e) 8.05×10^5
      (f) 6.48×10^{-10}
1-73 (a) 13 (b) 43 (c) 3.90×10^{12}
      (d) 3.12×10^6 (e) 6.92×10^{-10}
      (f) 1.26×10^{10}
1-74 $P = 0.0750 (1/V)$ or $V = 0.0750$
      $(1/P)$
1-75 $\log A = -5.26 \times 10^{-5} t + 300$
      or $t = -19,000 \log A +$
      5.70×10^4
1-76 $M = 0.831 \log (I_o/I)$ or
      $\log (I_o/I) = 1.204 M$
1-81 water   1-82 6.4×10^5 g/yd^3
1-83 3907 g/gal
1-84 2500 cm^3/flask
1-85 1030 g Hg
1-86 1.75×10^8 cm

1-87 6.0×10^{27} g

1-88 6×10^{13} g/cm³   1-89 yes

1-90 (a), (d), and (e) will float on water.

1-91 Iron expands when it rusts.

1-92 153 g   1-93 yes

1-94 gold   1-97 37.0°C, 310 K

1-98 same at $-40°$, opposite signs at
11.4°F and -11.4°C

1-99 -5.98°F

Chapter 2
ELEMENTS AND COMPOUNDS

2-5 (a) nonmetal element (b) mixture of metals (c) mixture of compounds (d) mixture of compounds (e) mixture of compounds (f) mixture of compounds (g) compound (h) metal element (i) mixture of metals

2-6 mixture

2-7 (a) Sb (b) Au (c) Fe (d) Hg (e) K (f) Ag (g) Sn (h) W

2-8 (a) hydrogen (b) lithium (c) beryllium (d) boron (e) fluorine (f) neon

2-9 (a) sodium (b) magnesium (c) aluminum (d) silicon (e) phosphorus (f) chlorine (g) argon

2-10 (a) titanium (b) vanadium (c) chromium (d) manganese (e) iron (f) cobalt (g) nickel (h) copper (i) zinc

2-11 (a) molybdenum (b) tungsten (c) rhodium (d) iridium (e) palladium (f) platinum (g) silver (h) gold (i) mercury

2-24 macroscopic scale: (a), (b), (c), (f), (g), (h), (i), (j); atomic scale: (d), (e), (f)

2-25 metals: (a), (b), (c); semimetal: (d); nonmetals: (e), (f), (g), (h)

2-26 metals: (c), (f), (g), (h); semimetals: (e), (b); nonmetals: (a), (d)

2-28 O, S, Se, Te, Po

2-29 Na, Mg, Al, Si, P, S, Cl, Ar

2-30 metal: Bi; semimetals: As, Sb; nonmetals: N, P

2-31 metals: Na, Mg, Al; semimetal: Si; nonmetals: P, S, Cl, Ar

2-36 (a), (b), (c)

2-37 (a), (b), (e)

2-38 (a), (c), (e)   2-39 (a), (b)

2-40 (a), (d), (e)

2-44 (a), (b), (d), (e)

2-45 (a) 10 (b) 22 (c) 29 (d) 50 (e) 92

2-46 (a) 20 p, 18 e (b) 24 p, 21 e (c) 27 p, 25 e (d) 29 p, 27 e (e) 48 p, 46 e

2-47 (a) 7 p, 10 e (b) 16 p, 18 e (c) 35 p, 36 e (d) 52 p, 54 e

2-48 (a) 7 p, 10 e (b) 8 p, 10 e (c) 9 p, 10 e (d) 11 p, 10 e (e) 12 p, 10 e

2-51 (a) $+2$ (b) $+3$ (c) $+4$ (d) $+1$ (e) $+2$

2-52 (a) -4 (b) -3 (c) -2 (d) -1

2-53 -1 2-54 $+2$

2-55 (a) $Mg(NO_3)_2$ (b) $Fe_2(SO_4)_3$ (c) Na_2CO_3

2-56 (a) Na_2O_2 (b) $Zn_3(PO_4)_2$ (c) K_2PtCl_6

2-57 K_3N, AlN   2-58 K_2O_2

2-59 6

2-60 Fe_3O_4 is a mixture of Fe_2O_3 and FeO.

2-61 $+3$, $+3$, and $+2$

2-62 (a) -3 (b) $+5$ (c) $+3$ (d) $+2$

2-63 $+3$ in all of these compounds

2-64 $+3$ 2-65 (e), (g)

2-66 (a) 0 (b) -1 (c) $+1$ (d) $+3$ (e) $+5$ (f) $+7$

2-67 (a) -1 (b) $+1$ (c) 0 (d) $+1$ (e) $+5$ (f) $+5$ (g) $+7$ (h) $+7$

2-68 $+3$ in all compounds

2-69 (a) $+2$ (b) $+4$ (c) $+4$ (d) $+6$ (e) $+6$ (f) $+6$ (g) $+6$ (h) $+6$ (i) $+6$ (j) $+8$ (k) $+8$

2-70 $[Ba^{2+}][O_2^{2-}]$

2-71 (a) $+4$ (b) $+4$ (c) $+2$ (d) $+4$ (e) $+4$ (f) $+4$ (g) $+4$ (h) -4 (i) -4 (j) 0 (k) $+4$ (l) $+4$ (m) $+2$

2-72 (a) 0 (b) -2 (c) -2 (d) -1 (e) $+4$ (f) $+4$ (g) $+6$ (h) $+4$ (i) $+4$ (j) $+6$ (k) $+4$ (l) $+6$ (m) $+2$ (n) $+5$ (o) $+7$ (p) $+4$ (q) $+6$ (r) $+2$

2-73 (a) $+2$ (b) $+3$ (c) $+4$ (d) $+6$ (e) $+7$ (f) $+2$ (g) $+3$ (h) $+2$ (i) $+6$ (j) $+7$ (k) $+4$ (l) $+2$

2-74 (a) 0 (b) -3 (c) $+3$ (d) $+1$ (e) $+2$ (f) $+4$ (g) $+3$ (h) $+4$ (i) $+5$ (j) $+3$ (k) $+5$ (l) $+3$ (m) $+5$ (n) -3 (o) -3 (p) -1

2-75 Prussian blue, Fe^{+3}; Turnbull's blue, Fe^{+2}

2-76 (a) diphosphorus pentoxide (b) iron(III) oxide (c) sodium hydrogen carbonate (d) dichlorine monoxide or dichlorine oxide   (e) copper(II) bromide

2-77 Calcium has only one oxidation state; iron has two.

2-78 (a) P_4S_3 (b) SiO_2 (c) CS_2 (d) CCl_4 (e) PF_5

2-79 (a) SiF_4 (b) SF_6 (c) OF_2 (d) Cl_2O_7 (e) ClF_3

2-80 (a) $SnCl_2$ (b) $Hg(NO_3)_2$ (c) SnS_2 (d) Cr_2O_3 (e) Fe_3P_2

2-81 (a) BeF_2 (b) Mg_3N_2 (c) BaO_2 (d) K_2CO_3

2-82 (a) $Co(NO_3)_3$ (b) $Fe_2(SO_4)_3$ (c) $AuCl_3$ (d) MnO_2 (e) WCl_6

2-83 (a) potassium nitrate   (b) lithium carbonate   (c) barium sulfate   (d) sodium sulfate   (e) lead(II) iodide

2-84 (a) aluminum chloride   (b) sodium nitride   (c) calcium phosphide   (d) lithium sulfide   (e) magnesium oxide

2-85 (a) ammonium hydroxide   (b) hydrogen peroxide   (c) magnesium hydroxide   (d) calcium hypochlorite   (e) sodium cyanide

2-86 (a) antimony sulfide   (b) tin(II) chloride   (c) sulfur tetrafluoride   (d) strontium bromide   (e) silicon tetrachloride

2-87 (a) CH_3CO_2H (b) HCl   (c) H_2SO_4 (d) H_3PO_4 (e) HNO_3

2-88 (a) H_2CO_3 (b) HCN   (c) H_3BO_3 (d) H_3PO_4 (e) HNO_2

2-89 NaHS and $NaHSO_3$

2-90 $S_2O_3{}^{2-}$

2-91 (a) calcium fluoride   (b) lead(II) sulfide   (c) iron(II) polysulfide   (d) titanium dioxide   (e) iron(III) oxide

2-92 (a) iron oxide   (b) calcium carbonate   (c) barium sulfate   (d) silicon dioxide

Chapter 3
STOICHIOMETRY: COUNTING ATOMS AND MOLECULES

3-1   20.183 amu   3-2 Cr

3-3   Al   3-4 79.90 amu

3-5   65.40 amu   3-6 51.996 g

3-7   2:1

3-8   (a) 12.011 g   (b) 30.974 g   (c) 58.69 g   (d) 200.59 g   (e) 238.03

3-9   \$12.00, \$3.60, 30. grams

3-10 (d)

3-11 46.025 g/mol HCO_2H, 30.026 g/mol H_2CO

3-12 (a) 16.033 g/mol   (b) 180.156 g/mol   (c) 74.123 g/mol   (d) 75.135 g/mol

3-13 (a) 444.6 g/mol   (b) 46.005 g/mol   (c) 97.45 g/mol   (d) 158.032 g/mol

3-14 (a) 220.056 g/mol   (b) 241.86 g/mol   (c) 294.181 g/mol   (d) 310.17 g/mol

3-15 162.19 g/mol

3-16 169.120 g/mol

3-17 (a) 375.94 g/mol   (b) 284.75 g/mol   (c) 296.55 g/mol   (d) 444.44 g/mol

3-18 1.00 mol of each

3-19 (a) 0.925 mol   (b) 1.46 mol   (c) 0.03285 mol   (d) 7.02 mol

3-20 7.61×10^{-3} mol

3-21 0.2247 mol

3-22 195.1 g/mol

3-23 377.9 g/mol

3-24 8.95 g/mol

3-25 1.587 g/cm^3

3-26 9.99 cm^3/mol

3-27 SnF_2 3-28 Fe_3O_4

3-29 (b)   3-30 N_2O

3-31 $CuFeS_2$ 3-32 24.3 g/mol

3-33 +6   3-34 $C_{11}H_{15}NO_2$

3-35 $2\,KClO_3 \rightarrow 2\,KCl + 3\,O_2$

3-36 (a) 76.47%   (b) 68.42%   (c) 52.00%

3-37 (a) 21.20%   (b) 13.85%   (c) 16.48%   (d) 46.66%

3-38 47.44% C, 2.56% H, 50.01% Cl

3-39 31.35%   3-40 $CaCO_3$

3-41 4   3-42 $C_{10}N_{14}N_2$

3-43 two significant figures

3-44 $C_{40}H_{56}$ 3-45 $C_{20}H_{14}O_4$

3-46 $C_8H_{10}N_4O_2$

3-47 $C_{14}H_{18}N_2O_5$

3-48 $C_2HBrClF_3$

3-49 6.0×10^{23} atoms of each

3-50 9×10^{20} atoms

3-51 5.62×10^{21} atoms

3-52 1.77×10^{23} molecules

3-53 (a) 2.41×10^{23} (b) 7.53×10^{23} (c) 1.36×10^{24}

3-54 106.11 g/mol

3-61 1500 molecules, 15.0 moles

3-62 2.25 mol   3-63 1.25 mol
3-64 9.00 mol   3-65 2.0 g
3-66 27.4 g CO_2, 39.9 g O_2
3-67 2220 g   3-68 9.79 g
3-69 5   3-70 CrO_3
3-71 1.85 g O_2 3-72 0.511 lb
3-73 782 lb   3-74 MnO_2
3-75 3.73 g   3-76 11.9 g
3-77 32.1 g   3-78 315 g
3-79 250 molecules; nothing; yield
      would double
3-80 0.400 mol; nothing; P_4
      becomes limiting reagent
3-81 NO is limiting reagent;
      doubling NO would make O_2
      the limiting reagent; nothing
3-82 10.3 g; add more Cl_2
3-83 67.7 g   3-84 5.21 g
3-85 174 g
3-86 1.3×10^{-5} mol/L
3-87 14.8 mol/L
3-88 8.7 mol/L
3-89 4.14×10^{-3} M
3-90 12.5 M 3-91 0.0814 M
3-92 10.7 g   3-93 1.71 g
3-94 0.120 M 3-95 36 mL
3-96 0.172 L   3-97 2.25 M
3-100 0.0747 M 3-101 0.9803
3-102 0.575 mL   3-103 0.375 M
3-104 0.150 M 3-105 80.1%

Chapter 4
GASES

4-1 (a), (b), (c), (e)   4-5 (a), (b), (d)
4-9 1.79 lb, 14.7 lb/in²
4-10 0.9813 atm, 99,432 Pa
4-11 1.36 atm, 1030 mmHg, 1.38×10^5 Pa
4-12 760 mmHg   4-13 29.9 inHg
4-14 10.3 m   4-15 29 lb/in²
4-16 3×10^{-6} atm
4-17 324 mmHg   4-18 0.349 L
4-19 13 L   4-20 500 L
4-21 5.29 atm   4-22 186°C
4-23 four fifths   4-25 18%
4-26 0.332 L   4-27 2:1
4-28 0.690 L N_2, 2.07 L H_2
4-29 15.0 L   4-30 12.0 L NO
4-31 CO   4-32 (b)
4-33 N_2O 4-35 (e)
4-36 (b)   4-37 1206 mL-psi/mol-K
4-38 0.014¢/g   4-39 1.5 atm
4-40 7.6 L   4-41 253 K

4-42 8.0 kg   4-43 3.13 atm
4-44 yes
4-45 5.11×10^6 L, 0.164 g/L,
      5.19×10^6 g
4-46 1.3 L   4-47 342 atm
4-48 0.994 atm, 1.99 atm
4-49 24.6 L   4-50 6.85 L
4-51 233 K   4-52 9150 mmHg
4-53 3.307 g/L   4-54 0.8252 g/L
4-55 0.1787 g He/L, 1.29 g air/L
4-57 Kr   4-58 29.06 g/mol
4-59 CH_2N_2 4-60 PH_3, P_2H_4
4-61 424 mmHg   4-62 2.5 atm
4-63 18.9 atm   4-64 0.452 atm
4-65 23.8 mmHg   4-66 121 mL
4-67 284 mL   4-68 4.38 L
4-69 (e), (c), (b), (d), (a)
4-70 $O_2/Br_2 = 2.33/1$
4-71 It decreases.
4-73 nitrous oxide, N_2O; nitric
      oxide, NO
4-74 13 s   4-75 16 g/mol
4-76 row 22   4-77 234.8 g/mol
4-82 (c)   4-83 smaller
4-84 larger   4-85 an^2/V^2
4-86 high pressure, low temperature
4-87 99.87%   4-89 (f)
4-90 58.1 g/mol   4-91 B_5H_9
4-92 C_3H_6 4-93 0.543 g
4-94 CrO_3 4-95 1.66×10^4 L
4-96 997 L   4-97 2.45 L
4-98 26.7 L   4-99 Zn

Chapter 5
THE STRUCTURE OF THE ATOM

5-10 (d)   5-15 $^{52}Cr^{3+}$
5-16 19 p⁺, 20 n⁰, 18 e⁻
5-17 53 p⁺, 74 n⁰, 54 e⁻
5-18 F   5-19 $^{79}Se^{2-}$ 5-20 54
5-25 Wavelength decreases by a
      factor of two; speed of sound
      remains constant.
5-26 0.01702 m
5-27 6×10^{-7} m
5-28 4.3×10^{14} s⁻¹   5-29 red
5-30 x-rays
5-31 11.93 m, radio wave
5-32 4.95×10^{14} s⁻¹, visible
5-33 5×10^{16} s⁻¹
5-34 yellow-orange
5-35 5.089896×10^{14} s⁻¹,
      5.088262×10^{14} s⁻¹
5-36 ultraviolet   5-37 green

5-38   655 nm, 485 nm, 434 nm, 410 nm
5-39 3.03×10^{-19} J
5-40 2×10^{-26} J
5-41 2.41×10^{-7} m, UV
5-45   656.5 nm
5-46   n = 6 to n = 2   5-47 (c)
5-48   (c)   5-50 1.56×10^{-17} J
5-54   0, 1, 2, 3; shape
5-55   −2, −1, 0, 1, 2; orientation
5-56 $+\frac{1}{2}, -\frac{1}{2}$ 5-57 5   5-58 2
5-59   (d)   5-60 (b)   5-61 14
5-62   16   5-66 p 5-67 0, 7, 7
5-68   (a)   5-69 (c)
5-70   2, 8, 18, 32, 50   5-71 10
5-72   5   5-77 (a)   5-78 (c)
5-79   (e)   5-80 (b)   5-81 (d)
5-87   (b)   5-88 (c)   5-89 (b)
5-90   (b)   5-91 (d)   5-92 11
5-93   6   5-94 (d)   5-95 (f)
5-96   (d)   5-97 (b)
5-98   IVA or 14   5-99 IIA or 2
5-100 (d)   5-101 $4f$
5-102 (d)   5-103 (d)

Chapter 6
THE PERIODIC TABLE

6-6   Co−Ni, Ar−K, Te−I
6-9   (a) K_2O (b) ZnO (c) In_2O_3 (d) SiO_2 (e) V_2O_5
6-13 5, 3   6-14 5, 5
6-15 VA or 15   6-16 IA or 1
6-17 (a)   6-18 (c)
6-21 1.442×10^{-8} cm, 0.1442 nm, 144.2 pm
6-22 (c)   6-24 (d)
6-28 (b), (c), (a), (d)
6-31 Sn^{2+} is larger.
6-32 (a)−(h) and (i)−(n)
6-33 (c)   6-34 Al^{3+} 6-35 (e)
6-36 (d)   6-37 (c)   6-38 (b)
6-45 3, 0, 0, $\frac{1}{2}$
6-46 (d), (a), (e), (b), (c)
6-47 (c)   6-48 (b)   6-49 (e)
6-51 (d)   6-52 (f)   6-53 (a)
6-54 (d)   6-55 (d), (b), (e), (c), (a)
6-56 (a) +4 (b) +5 (c) +6 (d) +4 (e) +6 (f) +2
6-57 (a) +3 (b) +4 (c) +5 (d) +6 (e) +7 (f) +4 (g) +6
6-58 +2, +4   6-63 IE
6-65 (d)   6-66 (d)   6-67 (d)
6-68 IE of Na is larger than EA of Cl.
6-69 No   6-72 (d)   6-73 (a), (d)

6-76 (a) Ca   (b) Na   (c) Ca   (d) Ca (e) Al

Chapter 7
THE MAIN-GROUP METALS AND THEIR SALTS

7-2   RbX, RbH, Rb_2S, Rb_3N, Rb_3P, Rb_2O, Rb_2O_2
7-6   It would react with the water.
7-10 Pb is obviously less active.
7-11 Sn protects the Fe from reaction with acid in foods.
7-12 Al is more active.   7-14 (b)
7-15 IVA or 14   7-16 IIIA or 13
7-17 Al_2S_3 7-18 Sr_2P_3
7-19 GaAs
7-22 (a) ZnF_2 (b) AlF_3 (c) SnF_4 (d) MgF_2 (e) BiF_5
7-23 (a)   7-24 (d), (e)
7-25 (a) $MgCl_2(aq) + H_2(g)$
      (b) $CuCl_2(aq) + H_2(g)$
      (c) $MgO(s) + H_2(g)$
      (d) $CsOH(aq) + H_2(g)$
7-26 (d)
7-27 (a) MgI_2
      (b) $Al_2(SO_4)_3(aq) + H_2(g)$
      (c) $Fe + H_2O$ (d) $MgCl_2 + Ti$
7-28 (a) $KOH(aq) + H_2$
      (b) $NaOH(aq) + H_2(g)$
      (c) $Ag + Cu^{2+}$
      (d) $ZnSO_4(aq) + Cu$
7-31 reaction with Al
7-32 (a) no (b) Fe_2O_3 is reduced, CO is oxidized (c) SiO_2 is reduced, C is oxidized (d) CO_2 is reduced, H_2 is oxidized (e) CO is reduced; H_2 is oxidized
7-33 (a) Mg is oxidized, HCl is reduced (b) I_2 is oxidized, Cl_2 is reduced (c) no (d) Na is oxidized; H_2O is reduced
7-34 (a) no (b) PH_3 is oxidized, O_2 is reduced (c) no (d) P_4 is oxidized, O_2 is reduced
7-35 (a) Mg is oxidized, H_2SO_4 is reduced (b) KBr is oxidized, Cl_2 is reduced (c) no (d) H_2 is oxidized, Cl_2 is reduced (e) no
7-38 oxidizing agent (OA) = $KClO_3$, reducing agent (RA) = S_8
7-39 OA = H_2O_2, RA = HI
7-40 OA = Ag^+, RA = Cu
7-41 (a) OA = CO, RA = Al
      (b) OA = Cr_2O_3, RA = Al

(c) OA = H$^+$, RA = Al
(d) OA = I$_2$, RA = Al
7-42 (a) OA = CO$_2$, RA = Mg
(b) OA = HCl, RA = Mg
(c) OA = H$_2$O, RA = Mg
(d) OA = N$_2$, RA = Mg
(e) OA = NH$_3$, RA = Mg
7-44 (a) Na$^+$ (b) Zn^{2+} (c) H$^+$
(d) Sn^{4+} (e) H$_2$
7-45 (a) Al (b) Hg (c) H$_2$ (d) H$^-$
(e) Sn
7-51 (a), (b), (c)   7-52 (f)
7-53 (a), (b), (c), (d)
7-54 Na is stronger.
7-55 Hg, Cu, Al
7-56 Al is stronger.   7-57 yes
7-58 Mg is stronger
7-59 (b), (c), (d)
7-60 between Sn and Co
7-63 (d), (e)
7-64 S^{2-} is oxidized; O$_2$ is reduced.
7-66 reduction   7-68 (e)
7-69 (a)   7-70 (d)

Chapter 8
THE COVALENT BOND

8-2   equal
8-3   (a) 1 (b) 4 (c) 8 (d) 2 (e) 4
(f) 8
8-4   (a) 8 (b) 11 (c) 4 (d) 2 (e) 7
8-5   They all have 8 valence electrons.
8-6   All have no valence electrons.
8-9   VIA or 16   8-10 +5
8-11 (a) 24 (b) 8 (c) 8 (d) 8
(e) 32 (f) 32
8-12 (a) 14 (b) 22 (c) 26 (d) 34
(e) 40 (f) 48 (g) 54 (h) 54
8-13 (a)–(d) and (e)–(g)
8-18 Rules: (a) and (b), covalent;
(c)–(e), ionic. EN: only (c) is ionic.
8-19 Rules: (a)–(c), covalent;
(d)–(e), ionic. EN: only (c) is ionic.
8-21 F, Fr, FrF   8-22 (e)
8-23 (b)   8-24 (d)   8-25 (e)
8-26 (b), (e)   8-27 (b), (d)
8-28 none   8-29 (d)
8-30 (a)   8-46 (d)   8-47 (d)
8-50 (a), (c), (d)   8-51 (b), (c)
8-52 (a), (c), (e)   8-53 (a), (e)
8-54 1, 2   8-55 (d)   8-56 (c)
8-57 (a) 1 (b) 0 (c) 1 (d) 0
8-58 (a) 3 (b) 3 (c) 2 (d) 2
8-59 VIIA or 17   8-60 all
8-61 (b), (c)   8-63 (b), (c), (e)
8-64 (a), (e)   8-65 (a)
8-66 1.33   8-67 1.5, 1.5, 1.33
8-68 (a) +1 (b) +2 (c) +1 (d) +2
8-69 (a) 0 (b) 0 (c) −1 (d) 0 (e) 0
8-70 (a) 0 (b) +1 (c) 0 (d) −1
(e) −3
8-71 (a) 0, +1 (b) 0 (c) +1 (d) 0
(e) 0, +1 (f) +1 (g) +1
8-72 0, 0, −1/+1   8-73 −1, +2
8-74 (a) trigonal pyramidal
(b) trigonal planar (c) T-shaped (d) T-shaped
8-75 all tetrahedral
8-76 (a) bent (b) trigonal pyramidal (c) tetrahedral (d) octahedral
8-77 (a) trigonal pyramidal
(b) see-saw (c) square pyramidal (d) octahedral
8-78 (a) linear (b) T-shaped
(c) see-saw (d) square planar
(e) distorted octahedral
8-79 (a) bent (b) trigonal planar
(c) trigonal pyramidal (d) tetrahedral
8-80 (a) linear (b) linear (c) bent
(d) bent (e) trigonal planar
8-81 (a) linear (b) tetrahedral
(c) tetrahedral
8-82 (a)   8-83 (a)   8-84 (c)
8-85 (a), (c), (d), (e)
8-86 (a), (c), (e)
8-87 (a), (b), (c), (d)
8-88 (a), (b), (d)   8-89 VIIA or 17
8-90 VIIA or 17   8-91 (c)
8-92 (a) sp (b) sp^2 (c) sp^3 (d) sp^3d
(e) sp^3d^2
8-93 (a) sp^3 (b) sp^3 (c) sp^2 (d) sp^2
(e) sp
8-94 (a) sp^3 (b) sp^2 (c) sp^2 (d) sp^2
8-95 (a) sp^3d (b) sp^3 (c) sp^3d (d) sp^2
8-101 (a) sigma (b) pi (c) pi (d) pi
(e) sigma
8-102 (a) 1 (b) 2 (c) 3 (d) 2 (e) 1
8-103 less   8-105 weaker
8-106 no
8-107 (a) 1 (b) 3 (c) 3 (d) 2.5 (e) 3
8-108 (d)

Chapter 9
THERMOCHEMISTRY

9-8 3416 J/Btu
9-11 24.47 J/mol   9-21 (b), (d)

9-23 (a) heat (b) work (c) work
(d) heat and work (e) work
9-26 Energy of universe is conserved.
9-27 decreases   9-31 E, H, P, V, T
9-32 (g), (h), (i), (j)
9-33 (f), (g)   9-35 5730 J
9-36 10.6 kJ
9-37 852 kJ/mol Fe_2O_3,
426 kJ/mol Fe
9-38 bomb, q_V; cup, q_P
9-39 2657 kJ/mol
9-40 434 kJ/mol
9-41 No gas is consumed or liberated.
9-42 (b), (c)   9-44 54.2 kJ
9-45 (a)   9-46 (c)   9-50 (d)
9-51 66.4 kJ   9-52 −1123 kJ/mol
9-53 −186 kJ/mol
9-55 2.43 kJ/mol   9-56 89.8 kJ
9-57 90.3 kJ or 45.1 kJ/mol NO
9-58 −597.2 kJ   9-59 27.4 kJ
9-60 1358 kJ/mol
9-61 −2044 kJ/mol
9-62 −54 kJ/mol
9-63 (a), (f), (h)   9-64 absorbed
9-65 −822.2 kJ   9-66 49.57 kJ
9-67 −1284 kJ   9-68 −1075 kJ
9-69 −5314 kJ   9-70 −905.2 kJ
9-71 −409.2 kJ   9-72 −91.2 kJ
9-73 Fe_2O_3 + Al   9-74 −235.4 kJ
9-75 −384.9 kJ   9-76 93,000 kJ
9-77 120 kJ   9-78 −134 kJ
9-79 974 kJ/mol   9-80 −9451 kJ
9-81 −92 kJ/mol
9-82 460 kJ/mol
9-83 −1098 kJ/mol
9-84 −1384 kJ/mol
9-85 1320 kJ/mol

Chapter 10
THE CHEMISTRY
OF THE NONMETALS

10-4 (a) As (b) Se (c) S (d) As
(e) P
10-5 (d)   10-6 (e)   10-7 (c)
10-8 (c)   10-10 (a), (b), (c)
10-11 all but (e)
10-12 (a) Fe_2O_3 ox, CO red (b) H_2
ox, CO_2 red (c) CH_4 ox, O_2
red (d) H_2S ox, O_2 red
10-13 (a) PH_3 ox, Cl_2 red (b) NO ox,
F_2 red (c) Na ox, NH_3 red
(d) NO_2 ox, NO_2 red
10-14 NH_3 ox, NaOCl red
10-15 S_8 ox, SO_3^{2-} red
10-16 HOClO ox, HOClO red

10-17 (a) acid (b) OA (c) acid
(d) OA
10-18 (e)   10-19 (c)
10-20 (a) As (b) As (c) P_4 (d) S_8
(e) C
10-21 (a) Mg_3N_2 (b) Li_2O (c) Br^-
and I_2
10-22 (a) H_2SO_3 (b) HOCl + Cl^-
(c) H_2CO_3
10-23 (a) H^+ + Cl^-
(b) HNO_3 + NO (c) H_3PO_4
10-24 (a) SO_2 (b) AlI_3 (c) PF_3
10-26 HCl, H_2, NaH
10-27 (a), (b)   10-29 Zn
10-34 $KClO_3$ 10-35 (c)
10-47 (d)   10-50(b)
10-53 (b), (c)
10-55 Air is only one fifth O_2.
10-57 a glowing splint
10-59 4.5 ton HNO_3
10-60 All are involved.
10-67 reducing agent
10-68 P_4O_{10}, H_3PO_4
10-69 As = [Ar] $4s^2 3d^{10} 4p^3$
10-70 about one half   10-71 (c)
10-80 (a) linear (b) square planar
(c) square pyramidal
(d) T-shaped (e) square
pyramidal
10-86 No

Chapter 11
ACIDS, BASES AND SALTS

11-4 (c), (d)   11-5 (b), (d), (e)
11-6 all
11-10 Na_2O_2 is a base; H_2O_2 is an acid.
11-11 metal hydrides: (c), (d);
nonmetal hydrides: (a), (b), (d)
11-12 metal oxide: (d); nonmetal
oxides: (a)–(c)
11-13 (c)   11-14 (c)
11-16 (a) H_2CO_3 (b) H_3PO_4
(c) $Ca(OH)_2$ (d) H_2SO_4
(e) NaOH
11-18 (e)   11-19 (c)
11-20 metal hydroxides: (c), (d);
nonmetal hydroxides: (a), (b), (e)
11-24 Bases neutralize formic acid.
11-25 BaO is basic in water; P_4O_{10} is
an acid.
11-26 (c)   11-28 (a), (e)
11-29 (c)   11-30 (b), (e)
11-31 (a) B = H_2O, A = HSO_4^-
(b) B = OH^-, A = CH_3CO_2H
(c) B = CaF_2, A = H_2SO_4

(d) $B = NH_3$, $A = HNO_3$
(e) $B = LiCH_3$, $A = NH_3$

11-32 (a) HCl/Cl^-; H_2O/H_3O^+
(b) HCO_3^-/H_2CO_3;
H_2O/OH^- (c) NH_3/NH_4^+;
H_2O/OH^- (d) CO_3^{2-}/H_2CO_3;
HCl/Cl^- (e) O^{2-}/OH^-;
H_2O/OH^-

11-33 (b)

11-37 (a) H_2O (b) OH^- (c) O^{2-}
(d) NH_3

11-38 (a) PO_4^{3-}
(b) $Al(H_2O)_5(OH)^{2+}$
(c) CO_3^{2-} (d) S^{2-}

11-39 (a) OH^- (b) H_2O (c) H_3O^+
(d) NH_3

11-40 (a) $H_2PO_4^-$ (b) H_2CO_3
(c) HSO_4^- (d) HNO_2

11-41 (b)

11-42 $HOCH_3 + CO_3^{2-}$

11-43 $NH_3 + H_2S$

11-47 HBr is a stronger acid than
H_3O^+.

11-48 HCl is a stronger acid than
NH_4^+.

11-49 HCO_2H is a stronger acid than
H_2O.

11-51 (a) W (b) W (c) S (d) W (e) S

11-52 (d)   11-54 (c), (d), (a), (b)

11-55 (d), (c), (b), (a)   11-56 (a)

11-57 (a)   11-58 (d)   11-59 (c)

11-60 (e)   11-61 (e)   11-64 (a)

11-65 (a)   11-67 (e)

11-69 0.004 M

11-71 (a), (d), (e), (f)

11-72 all but (e)

Chapter 12
THE STRUCTURE OF SOLIDS

12-2   intramolecular

12-3   the a term

12-7   polypropylene

12-8   (a) ionic (b) ionic (c) molecular (d) molecular (e) metallic
(f) metallic (g) molecular
(h) molecular

12-9   (b), (e)   12-10 (d)

12-11 (d)   12-12 (c)

12-15 (a) 12 (b) 12 (c) 8 (d) 6

12-16 (c) or (d)   12-18 (b)

12-25 (c) and (d)

12-26 (a) 4 (b) 6 (c) 8

12-27 (a)   12-28 tetrahedral, cubic

12-31 K   12-32 TiO_2

12-33 TiC   12-34 LiH

12-35 $[Fe^{2+}][S_2^{2-}]$ 12-36 $+2$

12-37 Ti_2O_3

12-38 tetrahedral, $\frac{1}{4}$, 4

12-39 ZnS   12-40 (a)

12-41 (c)   12-42 (c)

12-43 Co_3O_4 12-44 Ti^{4+}

12-45 $+1$

12-48 cubic closest-packed

12-49 6, 6

12-50 48% empty space

12-51 32%, 26%

12-52 (a) middle of an edge
(b) center of a face (c) center
of the body (d) a tetrahedral
hole

12-55 face-centered cubic

12-56 body-centered cubic

12-57 1 NH_4^+ and 1 Cl^-

12-58 face-centered cubic, tetrahedral holes, $\frac{1}{2}$, GaAs

12-59 simple cubic, $CaTiO_3$

12-60 0.125 nm

12-61 4.25 g/cm^3

12-62 0.193 nm

12-63 10.502 g/cm^3

12-64 body-centered cubic

12-65 face-centered cubic

12-66 body-centered cubic

12-67 0.356 nm   12-68 (d)

12-69 contracts   12-70 0.435 nm

12-71 0.345 nm   12-72 0.148 nm

12-73 0.176 nm

12-74 4 Cd^{2+} and 4 O^{2-}

12-75 4 Li^+ and 4 F^-

Chapter 13
LIQUIDS AND SOLUTIONS

13-2   (a), (b), (c), (d)

13-7   remain the same

13-8   remain the same

13-10 88.4%   13-11 15.7°C

13-13 39°C   13-14 (c)

13-15 (e)   13-16 N_2

13-17 (c)   13-19 It increases.

13-24 Gasoline is less viscous.

13-25 air, gasoline, water

13-37 all nonpolar

13-38 Solubility decreases in water,
increases in nonpolar solvents.

13-39 (e)   13-40 (b)

13-43 Temperature has no effect on
molality or mole fraction, but it
decreases the molarity.

13-45 0.641 M 13-46 2.5 ppm

13-47 0.25 M

13-48 $N_2 = 0.81$, $O_2 = 0.19$

13-49 0.991   13-50 1.43 ppm

13-51 $Ni = 0.578$, $Fe = 0.243$,
       $Cr = 0.174$, $C = 0.0047$

13-52 $Zn = 0.80$, $Sn = 0.12$,
       $Cu = 0.05$, $Sb = 0.02$,
       $Pb = 0.02$

13-53 solute = 0.00180,
       solvent = 0.998

13-54 0.167

13-55 1.029 g/cm^3, 1.26 M, 1.32 m,
       0.024

13-56 99.5%   13-57 12 M

13-58 17.3 M 13-59 1.90 m

13-60 0.7 m 13-63 decreases

13-69 0.103 m 13-70 163 g/mol

13-71 (b)   13-72 (a)

13-73 $C_6H_{12}O_6$ 13-74 37.7 °C/m

13-75 -1.78°C   13-76 52.2°C

13-77 86%   13-78 -0.188°C

13-79 351.44 K   13-80 10%

13-82 73,530 g/mol

Chapter 14
GAS-PHASE REACTIONS: AN INTRODUCTION TO KINETICS AND EQUILIBRIUM

14-3   moles of NO_2 per liter of
       solution when the reaction is at
       equilibrium

14-11 (c)   14-12 (d)

14-16 5.6×10^{-6}

14-17 within experimental error,
       $K_c = 59$ in each trial

14-18 1.3×10^{-6} 14-19 8.8×10^2

14-20 1.4×10^4, 120

14-21 2.1×10^{-14} 14-22 1.0

14-24 (b)   14-25 (b)   14-26 (c)

14-27 no; to the left

14-31 Both increase by 0.250 M.

14-32 N_2 increases by 0.117 M; H_2
       increases by 0.351 M.

14-33 (b)   14-34 (e)   14-35 (b)

14-36 $[H_2] = 0.766 M$, $[NH_3] =$
       0.156 M

14-37 0.0587   14-45 (c)

14-46 K_c is neither large nor small.

14-48 $[PCl_5] = 0.96 M$; $[PCl_3]$ and
       $[Cl_2] = 0.036 M$

14-49 3.6%

14-50 $[PCl_5] = 1.0 M$, $[PCl_3] = 6.5 \times$
       $10^{-4} M$, $[Cl_2] = 0.20 M$

14-51 $[NO_2] = 0.094 M$, $[NO] =$
       $6.0 \times 10^{-3} M$, $[O_2] = 3.0 \times$
       $10^{-3} M$

14-52 $[NO_2] = 1.0 M$, $[NO] = 1.5 \times$
       $10^{-3} M$, $[O_2] = 0.05 M$

14-53 $[SO_3] = 0.40 M$, $[SO_2] = 4.8 \times$
       $10^{-5} M$, $[O_2] = 2.4 \times 10^{-5} M$

14-54 $[SO_3] = 0.40 M$, $[SO_2] = 4.8 \times$
       $10^{-5} M$, $[O_2] = 2.4 \times 10^{-5} M$

14-55 $[NO_2] = 0.0011 M$, $[N_2O_4] =$
       0.099 M

14-56 $[NO_2] = 2.4 \times 10^{-3} M$,
       $[N_2O_4] = 0.50 M$

14-57 $[N_2] = 0.15 M$, $[H_2] = 0.24 M$,
       $[NH_3] = 0.090 M$

14-58 [CO] and $[H_2O] = 0.56 M$;
       $[CO_2]$ and $[H_2] = 0.44 M$

14-59 2.25 M

14-60 $[N_2] = 0.145 M$, $[O_2] =$
       0.045 M, $[NO] = 1.4 \times 10^{-6} M$

14-62 (e)   14-63 (c)

14-64 0.0192

14-65 $SO_2Cl_2 = 1.20$ atm, SO_2 and
       $Cl_2 = 0.020$ atm

14-66 $COCl_2 = 0.110$ atm, CO and
       $Cl_2 = 0.014$ atm

14-67 $SO_3 = 0.47$ atm, $SO_2 =$
       0.018 atm, $O_2 = 0.0088$ atm

14-68 $SO_3 = 0.485$ atm, $SO_2 =$
       0.0052 atm, $O_2 = 0.0026$ atm

14-69 $SO_3 = 0.152$ atm, $SO_2 =$
       0.148 atm, $O_2 = 0.044$ atm

14-70 $NH_3 = 0.0092$ atm, $N_2 =$
       0.60 atm, $H_2 = 0.89$ atm

14-71 $NH_3 = 0.21$ atm, $O_2 = 2.6 \times$
       10^{-4} atm, $NO_2 = 0.79$ atm,
       $H_2O = 0.93$ atm

14-72 NO = 0.16 atm, $Cl_2 =$
       0.082 atm, NOCl = 0.34 atm

14-73 $NH_3 = 0.011$ atm, $N_2 =$
       1.99 atm, $H_2 = 5.97$ atm

14-74 $N_2 = 0.40 M$, $O_2 = 0.60 M$,
       $NO = 3.2 \times 10^{-5} M$

14-75 99%

14-77 (a) shift right (b) shift right
       (c) shift right (d) no effect

14-78 (a) shift right (b) shift right
       (c) no effect (d) shift left

14-79 (a) shift right (b) shift right
       (c) shift left (d) shift right

14-80 (a) shift right (b) shift right
       (c) shift left (d) shift right

14-81 Solubility increases.

14-84 Increase in volume decreases
       pressure, which causes a shift to
       the right, decreasing the
       concentration of NH_3 and O_2.

Chapter 15
ACID-BASE EQUILIBRIA

15-4   It increases.

15-5   LeChatelier's principle

15-7 6.02×10^{13}

15-8   pH = 1.46, pOH = 12.54

15-9   pH = 1.81, pOH = 12.19

15-10   pH = 6.82

15-12 $[H_3O^+] = 1.91 \times 10^{-4} M$, $[OH^-] = 5.25 \times 10^{-11} M$

15-14   (a)   15-15   (d)   15-16   (a)

15-17   HOAc, HNO$_2$, HF, HOClO

15-19   1.25   15-20   1   15-24   (a)

15-25   2.87, 1.3%; 3.37, 4.2%; 3.90, 12%

15-26 $[HCO_2H] = 0.009 M$, $[HCO_2^-]$ and $[H_3O^+] = 1.3 \times 10^{-3} M$

15-27 $[HCN] = 0.174 M$, $[CN^-]$ and $[H_3O^+] = 1.0 \times 10^{-5} M$

15-28 $6 \times 10^{-6} M$

15-29 1.3×10^{-6}

15-30   0.22 M

15-31 8×10^{-5}

15-32 5.4×10^{-4}

15-33   It increases.

15-34 $[Cl_3CCO_2H] = 0.10 M$, $[Cl_3CCO_2^-]$ and $[H_3O^+] = 0.15 M$

15-36   Adding an acid suppresses the dissociation of water.

15-37   (b)   15-38 $2.6 \times 10^{-7} M$

15-40 $[H_3O^+] = 1.2 \times 10^{-6} M$

15-41   (c)

15-43 $[HCO_2H]$ and $[OH^-] = 2.1 \times 10^{-6} M$, $[HCO_2^-] = 0.080 M$

15-44 $[HOBr]$ and $[OH^-] = 6.5 \times 10^{-4} M$, $[OBr^-] = 0.10 M$

15-45   9.31   15-47   stronger base

15-49   0.057 M 15-50   7.03

15-51 1.2×10^{-8} 15-52   8.61

15-55   (e)   15-56   (c)   15-57   (d)

15-58   (c)   15-61   5.79

15-62   3.59   15-63   5.38

15-64 $[H_3O^+]$ increases from $9 \times 10^{-6} M$ to $9.3 \times 10^{-6} M$.

15-65   0.12 mol NaHCO$_2$

15-66   0.87   15-67   1.8 g

15-68   0.045 mol/L

15-69 $[HOAc]/[OAc^-] = 1.8$

15-70 $[HPO_4^{2-}]/[H_2PO_4^-] = 0.63$

15-71   1.0 g

15-72 $[H_2CO_3] = 0.10 M$, $[H_3O^+]$ and $[HCO_3^-] = 2.1 \times 10^{-4} M$, $[CO_3^{2-}] = 4.7 \times 10^{-11} M$

15-73 $[H_2M] = 0.25 M$, $[H_3O^+]$ and

$[HM^-] = 0.0019 M$, $[M^{2-}] = 2.1 \times 10^{-6} M$

15-74   Yes!

15-75 $[H_2Gly^+] = 1.91 M$, $[H_3O^+] = [HGly] = 0.095 M$, $[Gly^-] = 2.5 \times 10^{-10} M$

15-76   (d)   15-77   (b)

15-78 $[H_2C_2O_4] = 1.0 M$, $[H_3O^+]$ and $[HC_2O_4^-] = 0.23 M$, $[C_2O_4^{2-}] = 5.4 \times 10^{-5} M$

15-79   (e)   15-80   8.36

15-81 $[CO_3^{2-}] = 0.15 M$, $[OH^-]$ and $[HCO_3^-] = 0.0056 M$, $[H_2CO_3] = 2.2 \times 10^{-8} M$

15-82   9.67   15-83   (d)

15-84   1.57; 4.10, 10.1, 12.6

15-85   acidic

15-86   just slightly acidic

Chapter 16
SOLUBILITY PRODUCT EQUILIBRIA

16-9   (c)   16-10   (e)   16-11   (e)

16-14   (d)   16-15   (c)

16-19 $[F^-] = 2[Mg^{2+}]$

16-20 $[CrO_4^{2-}] = 2[Ag^+]$

16-21 $[Bi^{3+}] = 2/3[S^{2-}]$

16-22   (e)   16-23   (b)

16-24   Ag$_2$S   16-25   PbI$_2$

16-26   Hg$_2$S

16-27 $[CrO_4^{2-}] = [Ba^{2+}]$

16-28 $6.5 \times 10^{-5} M$

16-29 $8.1 \times 10^{-5} M$

16-30 5.0×10^{-15}

16-31 2.5×10^{-9}

16-32 5.1×10^{-3}

16-33 5.2×10^{-2}

16-34 1.8×10^{-11}

16-35 1.1×10^{-10}

16-36 5.6×10^{-6}

16-37 5.5×10^{-6}

16-38 2.1×10^{-6} g

16-39 6.2×10^{-16} g

16-40   (a) 3.4×10^{-3} (b) 5.3×10^{-4}
(c) 0.046   (d) 1.4×10^{-22}
(e) 1.5×10^{-25}

16-41   (a) 2.3×10^{-3} (b) 8×10^{-19}
(c) 2.2×10^{-5} (d) 1.2×10^{-7}
(e) 2.0×10^{-3}

16-42   (d), (b), (c), (a)

16-46   (b)   16-47   (b)

16-48 $1.5 \times 10^{-19} M$

16-49 $1.5 \times 10^{-15} M$

16-50 3.5×10^{-14} g

16-51 7×10^{-23} g

16-52 (a) $1.6 \times 10^{-20} M$
     (b) $1.4 \times 10^{-38} M$
     (c) $1.3 \times 10^{-19} M$
16-53 (a), because $PbCl_2$ would precipitate from (c)
16-54 1.6×10^{-23}, 1.6×10^{-29}
16-55 5.54   16-56 $5.8 \times 10^{-7} M$
16-57 $1.5 \times 10^{-8} M$
16-59 more soluble
16-60 PbS, $5.0 \times 10^{-8} M$
16-61 AgI, $4.6 \times 10^{-12} M$
16-62 (d)   16-63 (b)   16-64 (a)
16-65 Precipitate Pb^{2+} as PbS.
16-66 strong acid
16-67 $[Zn^{2+}] = 1.6 \times 10^{-4} M$, $[Hg^{2+}] = 4 \times 10^{-33} M$, $[Bi^{3+}] = 3.2 \times 10^{-39} M$
16-68 Na_2CO_3 is a base, which generates enough OH^- ion in water to precipitate Fe^{3+} as $Fe(OH)_3$.
16-69 CdS and Fe_2S_3 but not $Fe(OH)_3$
16-70 (a), (b), (c)

Chapter 17
COMPLEX ION AND COMBINED EQUILIBRIA

17-5 (c)   17-6 (a)   17-7 (c)
17-8 (a)   17-9 (e)   17-10 (b)
17-11 (a), (b)   17-12 (b), (d), (e)
17-14 $+2, +3, +2, +2, +1$
17-19 2.9×10^{13} 17-20 5.9×10^{18}
17-21 (d)   17-22 (a)   17-23 (e)
17-25 $2.7 \times 10^{-6} M$
17-26 $2.8 \times 10^{-17} M$
17-27 $2.6 \times 10^{-11} M$
17-28 $5 \times 10^{-16} M$
17-29 $2 \times 10^{-9} M$
17-30 $7.7 \times 10^{-4} M$
17-31 $7 \times 10^{-12} M$
17-32 0.17 mol   17-33 0.66 g
17-34 yes   17-35 yes
17-36 $4 M$ 17-37 yes
17-38 23 mL 6 M NH_3 and 7.6 mL 6 M HCl
17-40 28%   17-41 72%
17-42 $8 \times 10^{-11}\%$ 17-43 99.9%
17-44 5×10^{-12}, 1×10^{-7}, 5×10^{-4}, 0.47%
17-45 (c)   17-46 (a)   17-47 (a)
17-69 0.55 M Cl^-, $3.3 \times 10^{-10} M$ Ag^+
17-70 1.3×10^{-17} 17-71 yes
17-72 HgS, CuS, PbS, CoS
17-73 yes

Chapter 18
OXIDATION-REDUCTION REACTIONS

18-10 $+3, +3, +2$ 18-11 $+3$
18-12 (e), (g)
18-13 (f), (e), (a) and (d), (b), (c)
18-14 (a) $+4$ (b) $+4$ (c) $+2$
    (d) $+4$ (e) $+4/3$ (f) $+4$
    (g) -4 (h) -2 (i) -4
    (j) -4 (k) 0 (l) $+4$ (m) $+4$
    (n) $+2$
18-15 (a) 0 (b) -2 (c) -2 (d) -1
    (e) $+4$ (f) $+4$ (g) $+6$
    (h) $+4$ (i) $+4$ (j) $+6$ (k) $+4$
    (l) $+6$ (m) $+2$ (n) $+2.5$
    (o) $+7$ (p) $+4$ (q) $+6$ (r) $+2$
18-16 (a) $+2$ (b) $+3$ (c) $+4$
    (d) $+6$ (e) $+7$ (f) $+2$
    (g) $+3$ (h) $+4$ (i) $+6$ (j) $+7$
    (k) $+4$ (l) $+2$ (m) -1
18-17 $+3, +2$ 18-19 $Ag_2S + Al$
18-20 (b) CO ox, Fe_2O_3 red (c) C ox, SiO_2 red (d) H_2 ox, CO_2 red (e) H_2 ox, CO red
18-21 (a) Mg ox, HCl red (b) I_2 ox, Cl_2 red (d) Na ox, H_2O red (e) H_2 ox, O_2 red
18-22 (b) PH_3 ox, O_2 red (d) P_4 ox, O_2 red
18-23 $3 CuO + 2 NH_3 \rightarrow 3 Cu + N_2 + 3 H_2O$
18-24 (a) $CS_2 + 3 O_2 \rightarrow CO_2 + 2 SO_2$
    (b) $CH_4 + 2 O_2 \rightarrow CO_2 + 2 H_2O$
18-25 (a) $8 SO_2 + 16 H_2S \rightarrow 3 S_8 + 16 H_2O$
    (b) $8 H_2S + 4 O_2 \rightarrow 8 H_2O + S_8$
18-26 (a) $Fe_2O_3 + 3 CO \rightarrow 2 Fe + 3 CO_2$
    (b) $2 Cu_2S + 3 O_2 \rightarrow 2 Cu_2O + 2 SO_2$
18-27 (a) $2 PbS + 3 O_2 \rightarrow 2 PbO + 2 SO_2$
    (b) $4 FeS_2 + 11 O_2 \rightarrow 2 Fe_2O_3 + 8 SO_2$
18-28 (a) $4 NH_3 + 5 O_2 \rightarrow 4 NO + 6 H_2O$
    (b) $3 NO_2 + H_2O \rightarrow 2 HNO_3 + NO$
    (c) $2 NH_4NO_3 \rightarrow 2 N_2 + 4 H_2O + O_2$
    (d) $2 HNO_3 + SO_2 \rightarrow 2 NO_2 + H_2SO_4$
18-29 (a) $4 HCl + MnO_2 \rightarrow Mn^{2+} +$

$2 Cl_2 + 2 H_2O$

(b) $2 HI + 2 Fe^{3+} \rightarrow 2 Fe^{2+} + I_2 + 2 H^+$

(c) $Cl_2 + 2 OH^- \rightarrow Cl^- + OCl^- + H_2O$

(d) $Br_2 + SO_2 + 2 H_2O \rightarrow H_2SO_4 + 2 HBr$

18-30 (a) $HIO_3 + 5 HI \rightarrow 3 I_2 + 3 H_2O$

(b) $2 HNO_2 + 2 HI \rightarrow I_2 + 2 NO + 2 H_2O$

(c) $8 HI + H_2SO_4 \rightarrow H_2S + 4 I_2 + 4 H_2O$

(d) $IO_3^- + 5 I^- + 6 H^+ \rightarrow 2 I_2 + 3 H_2O$

18-31 (a) $(NH_4)_2Cr_2O_7 \rightarrow N_2 + Cr_2O_3 + 4 H_2O$

(b) $3 PbO_2 + 4 H_2S \rightarrow 3 PbS + SO_2 + 4 H_2O$

(c) $8 NH_3 + 3 Cl_2 \rightarrow N_2 + 6 NH_4Cl$

(d) $3 H_2S + 8 HNO_3 \rightarrow 3 H_2SO_4 + 8 NO + 4 H_2O$

18-32 $3 P_4 + 10 KClO_3 \rightarrow 3 P_4O_{10} + 10 KCl$

18-33 (a) $2 HNO_3 + 6 HCl \rightarrow 2 NO + 3 Cl_2 + 4 H_2O$

(b) $2 HBr + H_2SO_4 \rightarrow SO_2 + Br_2 + 2 H_2O$

(c) $2 MnO_4^- + 10 HCl + 6 H^+ \rightarrow 2 Mn^{2+} + 5 Cl_2 + 8 H_2O$

(d) $2 S_2O_3^{2-} + I_3^- \rightarrow S_4O_6^{2-} + 3 I^-$

18-34 (a) $3 Cu + 2 HNO_3 + 6 H^+ \rightarrow 3 Cu^{2+} + 2 NO + 4 H_2O$

(b) $Cu + 2 HNO_3 + 2 H^+ \rightarrow Cu^{2+} + 2 NO_2 + 2 H_2O$

(c) $3 CuO + 2 NH_3 \rightarrow N_2 + 3 Cu + 3 H_2O$

18-35 (a) $MnO_2 + 2 HCl + 2 H^+ \rightarrow Mn^{2+} + Cl_2 + 2 H_2O$

(b) $2 HI + 2 Fe^{3+} \rightarrow I_2 + 2 Fe^{2+} + 2 H^+$

(c) $Cl_2 + 2 OH^- \rightarrow Cl^- + OCl^- + OCl^- + H_2O$

(d) $Br_2 + SO_2 + 2 H_2O \rightarrow H_2SO_4 + 2 HBr$

18-36 (a) $HIO_3 + 5 HI \rightarrow I_2 + 3 H_2O$

(b) $2 HNO_3 + 6 HI \rightarrow 3 I_2 + 2 NO + 4 H_2O$

(c) $H_2SO_4 + 8 HI \rightarrow 4 I_2 + H_2S + 4 H_2O$

(d) $IO_3^- + 5 I^- + 6 H^+ \rightarrow 3 I_2 + 3 H_2O$

18-37 (a) $8 H_2O_2 + 8 H_2S \rightarrow S_8 + 16 H_2O$

(b) $16 Fe^{3+} + 8 HS^- \rightarrow S_8 + 16 Fe^{2+} + 8 H^+$

(c) $24 HgS + 16 HNO_3 + 48 H^+ \rightarrow 24 Hg^{2+} + 3 S_8 + 16 NO + 32 H_2O$

(d) $2 S_2O_3^{2-} + I_2 \rightarrow S_4O_6^{2-} + 2 I^-$

18-38 (a) $2 CrO_4^{2-} + 6 I^- + 16 H^+ \rightarrow 2 Cr^{3+} + 3 I_2 + 8 H_2O$

(b) $Cr_2O_7^{2-} + 6 Fe^{2+} + 14 H^+ \rightarrow 6 Fe^{3+} + 2 Cr^{3+} + 7 H_2O$

(c) $16 CrO_4^{2-} + 24 H_2S + 80 H^+ \rightarrow 3 S_8 + 16 Cr^{3+} + 64 H_2O$

18-39 (a) $5 S_2O_3^{2-} + 8 MnO_4^- + 14 H^+ \rightarrow 10 SO_4^{2-} + 8 Mn^{2+} + 7 H_2O$

(b) $2 MnO_4^- + 5 H_2C_2O_4 + 6 H^+ \rightarrow 2 Mn^{2+} + 10 CO_2 + 8 H_2O$

(c) $5 PbO_2 + 2 Mn^{2+} + 4 H^+ \rightarrow 5 Pb^{2+} + 2 MnO_4^- + 2 H_2O$

18-40 (a) $MnO_4^- + 5 Ag + 8 H^+ \rightarrow Mn^{2+} + 5 Ag^+ + 4 H_2O$

(b) $MnO_4^- + 5 Fe^{2+} + 8 H^+ \rightarrow Mn^{2+} + 5 Fe^{3+} + 4 H_2O$

(c) $2 MnO_4^- + 5 SO_2 + 6 H^+ + 2 H_2O \rightarrow 2 Mn^{2+} + 5 H_2SO_4$

(d) $2 MnO_4^- + 5 H_2O_2 + 6 H^+ \rightarrow 2 Mn^{2+} + 5 O_2 + 8 H_2O$

18-41 (a) $4 Cr + 3 O_2 + 12 H^+ \rightarrow 4 Cr^{3+} + 6 H_2O$

(b) $Cr_2O_7^{2-} + 6 Fe^{2+} + 14 H^+ \rightarrow 2 Cr^{3+} + 6 Fe^{3+} + 7 H_2O$

(c) $6 BrO_3^- + 10 Cr^{3+} + 22 H_2O \rightarrow 10 HCrO_4^- + 3 Br_2 + 34 H^+$

18-42 (a) $2 MnO_4^- + I^- + H_2O \rightarrow 2 MnO_2 + IO_3^- + 2 OH^-$

(b) $MnO_4^- + NO \rightarrow MnO_2 + NO_3^-$

(c) $2 MnO_4^- + 2 NH_3 \rightarrow N_2 + 2 MnO_2 + 2 H_2O + 2 OH^-$

(d) $2 MnO_4^- + 3 CH_3OH \rightarrow 2 MnO_2 + 3 H_2CO + 2 H_2O + 2 OH^-$

18-43 (a) $2 Cr + 2 OH^- + 6 H_2O \rightarrow 2 Cr(OH)_4^- + 3 H_2$

(b) $3 H_2O_2 + 2Cr(OH)_3 +$

Chapter 19
ELECTROCHEMISTRY

Chapter 20
CHEMICAL THERMODYNAMICS

20-47 731.2 kJ, 698.6 J/K, entropy
20-48 -34.3 kJ, 75.6 J/K, makes reaction more favorable
20-49 (c)   20-50 (b)
20-51 The entropy term is unfavorable.
20-52 the second diagram
20-53 (d)   20-54 -1060.0 kJ
20-58 (a)   20-59 2.2×10^4
20-60 9.2×10^{-5}
20-61 5.8×10^5, 1.9×10^{-4}
20-62 9.8×10^{33}
20-63 1.8×10^{-5}
20-64 1.8×10^{-5}
20-65 4.4×10^{-5}, 4.7×10^{-11}
20-66 2×10^{-53}
20-67 1.8×10^{-10} 20-68 PbS
20-69 0.12 g LiF/100 g H_2O
20-70 -40.8 kJ/mol
20-71 2.9×10^6
20-72 71.18 kJ/mol   20-73 977 K
20-74 It increases.
20-75 It increases.
20-76 516 K   20-77 (c)
20-78 1.5×10^{-25}; 9.7×10^{-13}, 3×10^{-6}, 9.8×10^{-3}
20-79 64.77 kJ, -2.07 J/K, it decreases
20-80 (c)   20-81 -212.8 kJ
20-82 -88.6 kJ   20-83 65.6 kJ
20-84 3.2×10^{-12}

Chapter 21
KINETICS

21-7   the fourth graph   21-10 s^{-1}
21-11 L/M-s   21-12 L^2/M^2-s
21-13 0.0014 mol/L-s
21-18 the fourth graph
21-19 rate = $k(NO)^2(Cl_2)$
21-20 $k = 117$ L^2/M^2-s; 14.6 mol/L-s
21-21 rate = $k(NO)^2(O_2)$
21-22 3.55×10^{-7} mmHg/s
21-23 rate = $k(CH_3I)(OH^-)$
21-24 $k = 6.50 \times 10^{-4}$ L/mol-s, 3.25×10^{-7} mol/L-s
21-25 H^+ 21-26 (c)   21-27 (f)
21-28 0.0296 mol/L-s
21-29 0.020 mol/L-s
21-31 0.027 mol/L-s for both
21-32 0.0114 mol/L-s
21-34 54.5°C
21-35 yes, 26.4 s
21-36 400 L/mol-s

21-37 12.4 s   21-38 24.6 s^{-1}
21-39 It would decrease by a factor of 2.
21-40 1.1×10^9 s
21-41 1.1×10^{10} yr
21-42 85.6%   21-43 1897 yr
21-44 15.3%   21-45 23,700 hr
21-46 3900 yr   21-47 840 yr, no
21-48 41,600 yr
21-49 rate = $k(N_2O_5)$
21-50 $k = 0.0014$ s^{-1}, 0.027 M
21-51 rate = $k(NH_4^-)$
21-52 120 hr   21-53 1st-order
21-54 yes   21-55 1st-order
21-56 rate = $k(Cr(NH_3)_5Cl^{2+})(OH^-)$
21-58 1st-order   21-62 (e)
21-64 rate = $k(NO)^2(H_2)$ 21-65 (c)
21-66 (c)
21-67 rate = $k(I^-)(OCl^-)/(OH^-)$
21-70 $K = k_f/k_r$
21-71 3.5×10^4
21-72 155 kJ/mol
21-73 218 kJ/mol
21-74 51.6 kJ/mol
21-75 113 kJ/mol
21-76 0.0644 L/mol-s
21-77 0.0142 L/mol-s

Chapter 22
TRANSITION METAL COMPLEXES

22-4 $[Co(NH_3)_5Cl]Cl_2$, $[Co(NH_3)_4Cl_2]Cl$, $[Co(NH_3)_5H_2O]Cl_3$
22-9   charge
22-10 (a) 4, $+2$ (b) 6, 0 (c) 6, $+2$ (d) 4, $+2$
22-11 (a) 4, $+2$ (b) 6, $+3$ (c) 6, $+2$ (d) 6, $+3$
22-14 (b), (c)   22-17 all
22-18 all   22-19 all
22-22 (a) tetraammine copper(II) (b) hexaaquo manganese(II) (c) hexacyano ferrate(II) (d) tris(ethylenediamine) nickel(II) (e) tris(acetylacetonato) chromium
22-23 (a) dichloro diammino platinium(II) (b) tetracarbonyl nickel(O) (c) tris(ethylenediamine) cobalt(III) (d) sodium pentacarbonyl manganate($-I$)
22-24 (a) sodium hexanitrito cobaltate(III) (b) sodium tetracyano zincate(II) (c) dichlorotetra-

amminecobalt(III) chloride
(d) diammine silver(I) chloride

22-25 (a) $[Cr(NH_3)_6]Cl_3$
(b) $[Cr(NH_3)_5Cl]Cl_2$
(c) $[Co(en)_3]Cl_3$
(d) $K_2[Co(NO_2)_4(NH_3)_2]$

22-26 (b), (d)   22-27 (d), (e)

22-28 (b), (c), (d)   22-29 (d)

22-30 (c)   22-31 (a), (b), (c)

22-32 (a), (b), (c), (d)   22-33 (e)

22-34 (a) $+3$ (b) $+4$ (c) $+2, +5$
(d) $+3, +6$ (e) $+2, +4, +7$

22-35 (a) $+2, +3$ (b) $+3$ (c) $+2$
(d) $+1$ (e) $+2$

22-41 -1 22-42 -2 22-43 6

22-44 4   22-45 4

22-48 t_{2g}: xy, xz, yz; e_g: $x^2 - y^2, z^2$

22-50 no   22-56 (a)   22-59 (a)

22-63 $Fe(H_2O)_6^{3+}$ is high spin

22-66 (a) blue (b) pale yellow
(c) green (d) yellow-orange
(e) purple

22-68 wavelengths around 610 nm,
or frequencies of 4.9×10^{14} s^{-1}

22-69 indigo

22-70 longer wavelength, higher
frequency

22-71 red

Chapter 23
NUCLEAR CHEMISTRY

23-5 (a) 6 e, 6 p, 8 n (b) 33 e, 33 p,
42 n (c) 38 e, 38 p, 52 n
(d) 74 e, 74 p, 110 n (e) 94 e,
94 p, 145 n

23-6 (a) 18 e, 19 p, 21 n (b) 36 e,
34 p, 45 n (c) 42 e, 45 p,
58 n (d) 54 e, 53 p, 74 n
(e) 80 e, 83 p, 126 n

23-9 isotopes: (e), (f), (g); isobars:
(a), (b), (c); isotones: none

23-10 (a) ^{125}Te (b) ^{90}Sr (c) ^{90}Y
(d) ^{40}Ar (e) ^{224}Ra

23-11 products: (a) ^{14}N (b) ^{35}Cl
(c) ^{52}Cr (d) ^{99}Tc (e) ^{241}Am

23-12 products: (a) ^{184}Os (b) ^{200}Bi
(c) ^{220}Rn (d) ^{233}Pa (e) ^{244}Cm

23-13 products: (a) ^{7}Li (b) ^{22}Ne
(c) ^{37}Ar (d) ^{62}Ni (e) ^{128}Cs

23-14 products: (a) ^{18}O (b) ^{34}S
(c) ^{45}Sc (d) ^{73}As (e) ^{90}Nb

23-15 products: (a) ^{32}S (b) ^{11}B
(c) ^{208}Po (d) ^{125}I

23-16 (a) ^{209}Bi (b) ^{257}Unq (c) ^{96}Mo
(d) ^{220}Pu (e) ^{14}C (f) ^{62}Zn

23-17 (a) ^{4}He (b) ^{239}Pu (c) 4n
(d) ^{12}C

23-20 (a), (b)   23-21 (b), (e)

23-23 (a)   23-24 (d)

23-26 3.202 MeV, 0.537 MeV

23-27 524.8460 MeV, 8.7 MeV

23-28 238.0519 amu

23-29 ^{238}U, ^{60}Ni

23-30 925.5598 MeV

23-31 2.7131 MeV

23-32 (a) 14.3 days (b) 28.6 days
(c) 42.9 days (d) 75.2 days

23-33 55 min

23-34 (a) 0.00105 s^{-1}
(b) 2.6×10^{-8} s^{-1}
(c) 1.8×10^{-9} s^{-1}
(d) 3.8×10^{-12} s^{-1}
(e) 1.4×10^{-15} s^{-1}

23-35 2.6 yr, 0.27 yr^{-1}, 9.6 yr

23-36 36.8 days

23-37 6.25 days, 1.839 g

23-38 5.0×10^{-3}%, 72%

23-39 5.0×10^{-4} min^{-1}, 1380 min

23-40 1.76×10^4 23-41 716

23-42 (a) 1.72×10^{19}
(b) 1.55×10^{15} (c) 3.7×10^{12}
(d) 1.26×10^6

23-43 (a) 8.87 Ci (b) 27.3 Ci
(c) 0.0214 Ci
(d) 6.34×10^{-8} Ci

23-44 6.69×10^9 s

23-45 2.24×10^{-4} g

23-46 11,100 yr   23-47 23,700 yr

23-48 1897 yr   23-49 15.3%

23-50 3900 yr   23-51 840 yr, no

23-52 41,600 yr

23-53 259.5 kJ/mol, much smaller

23-54 120,000 kJ/mol, much larger

23-61 ^{212}Po, ^{212}Pb and ^{212}Bi

23-62 ^{256}Md, ^{252}Md

23-63 5, ^{252}Md and ^{256}No

23-64 ^{206}Pb

23-65 6 alpha, 4 beta

Chapter 24
THE ORGANIC CHEMISTRY OF CARBON

24-6 structural   24-7 8

24-9 2,3-dimethylpentane

24-10 3-ethyl-3,5-dimethyloctane

24-11 1-chloro-1-methylcyclopentane
24-19 (e)   24-20 (d)
24-21 (b), (d)
24-22 2,3-dimethylpent-2-ene
24-23 3-ethylhex-2-ene
24-24 2,5-dimethylhept-3-yne
24-27 2,3-dibromopentane
24-28 2-butanol   24-29 pentane
24-30 (c)   24-31 (a)
24-38 (a)   24-39 (d)
24-44 (a) aldehyde (b) alcohol
(c) ether   (d) ketone
24-45 (a) 1° (b) 2° (c) 3° (d) 2°
24-46 aromatic ring, alkene, amine,
ether, and amide
24-47 aromatic ring, amine, alkene,
ester, and ether
24-48 aromatic ring, alkene, amine,
and amide
24-53 diethyl ether
24-60 (f), (g), (a) and (b), (c) and (d), (e)
24-61 metathesis: (d); redox: (a), (b),
(c), (e)
24-62 ox: (a), (b); red: (c), (d), (e)
24-63 (a)   24-64 (c)
24-65 (b)   24-66 (d)
24-67 2-methyl-3-pentanone
24-69 C end: (b), (c), (d); O end: (a), (e)
24-77 nucleophile, C end
24-78 ethane
24-80 $(CH_3CH_2)_2CHC(OH)(CH_3)$
CH_2CH_3
24-81 (d)   24-82 (a), (b)
24-84 (e)   24-85 (d)

Chapter 25
POLYMERS: SYNTHETIC
AND NATURAL

25-2 (a), (c), (f); (b), (d), (e)
25-3 all except (d) and (h)
25-5 all except (a)
25-11 14,000 – 1,400,000 g/mol
25-12 2000 – 3000 units
25-13 310,000 g/mol
25-17 addition: (a), (b), (f), (g);
condensation: (c), (d), (e)
25-27 linear, linear   25-34 (a), (c)
25-35 hydrophilic: (b), (c), (e);
hydrophobic: (a), (d)
25-38 4, ALA—ALA, ALA—CYS,
CYS—ALA, CYS—CYS
25-39 (ALA)$_3$, (ALA)$_2$CYS,
ALA(CYS)$_2$,
ALA—CYS—ALA,
CYS(ALA)$_2$, (CYS)$_2$ALA,
CYS—ALA—CYS, (CYS)$_3$
25-40 24 quadripeptides
25-44 SER, THR, TYR, CYS, ASP,
GLU, ASN, GLN, LYS, ARG,
HIS
25-45 ASP, GLU, LYS, ARG
25-46 GLY, ALA, VAL, LEU, ILE,
PRO, MET, PHE, TRP
25-48 (a), (d), (e), (f)   25-49 all
25-59 0, − 1.75
25-60 15.9 kJ/g, 38.2 kJ/g
25-67 SER—PHE—VAL—ASN—
GLN—HIS

Absolute zero The temperature at which the volume and pressure of an ideal gas extrapolate to zero; $-273.15°C$ or 0 K.

Absorption spectrum The spectrum of dark lines against a light background that results from the absorption of selected frequencies of electromagnetic radiation by an atom or molecule.

Accuracy An estimate of the agreement between a measurement and the true value of the quantity being measured.

Acid A source of the H^+ or H_3O^+ ion; a hydrogen-ion or proton donor.

Acid-base complex The product of the reaction between a Lewis acid and a Lewis base; it contains a covalent bond formed by the donation of a pair of nonbonding electrons from the Lewis base to the Lewis acid.

Acid-base indicator A weak acid or weak base, such as litmus or phenolphthalein, that changes color when it gains or loses an H^+ ion.

Acid-dissociation equilibrium constant (K_a) A measure of the relative strength of an acid; $K_a = [H_3O^+][A^-]/[HA]$.

Activation energy The energy reactants must acquire to form an intermediate, or activated state, that can be transformed into the products of the reaction.

Active metal A metal, such as sodium or potassium, that is unusually reactive.

Activity A measure of the number of disintegrations of a radioactive nuclide per unit of time; reported in units of curies or becquerels.

Addition polymer Polymer formed by linking monomers without the loss of any atoms, such as polyethylene.

Addition reaction A reaction in which a molecule containing an $X—X$ or $X—Y$ bond is added across a $C=C$ or $C=O$ double bond; for example, the reaction between Br_2 and $H_2C=CH_2$ to form $BrCH_2CH_2Br$.

Adhesion The force of attraction between different substances, such as glass and water. See also *cohesion*.

Alcohol Compounds with an $—OH$ functional group attached to a carbon atom, such as CH_3CH_2OH.

Aldehyde Compounds with at least one hydrogen atom attached to a $C=O$ (carbonyl) group, such as formaldehyde (H_2CO).

Algorithm A set of rules for calculating something that can be taught to a reasonably intelligent system, such as a computer.

Alkali Historically, a compound that neutralizes acids.

Alkali metal A metal in Group IA (Li, Na, K, and so on).

Alkaline earth metal A metal in Group IIA (Be, Mg, Ca, and so on).

Alkaloid One of a class of organic compounds isolated from plants that contain nitrogen.

Alkane Hydrocarbon with the generic formula C_nH_{2n+2} that contains only $C—C$ and $C—H$ bonds.

Alkene An unsaturated hydrocarbon that contains one or more $C=C$ double bonds.

Alkoxide The conjugate base of an alcohol; for example, the CH_3O^- ion.

Alkyl halide Derivative of an alkane in which a hydrogen atom has been replaced by a halogen; for example, CH_3Cl.

Alkyne A hydrocarbon that contains one or more $C≡C$ triple bonds.

Allotropes Different forms of an element with different structures and therefore different chemical and physical properties, such as O_2 and O_3.

Alloy A mixture of two or more elements that acts like a metal; bronze, for example, is an alloy of copper and tin.

Alpha particle A positively charged particle consisting of two protons and two neutrons emitted by one of the radioactive elements; an alpha particle is equivalent to an He^{2+} ion.

Amide The product of the reaction between a carboxylic acid (RCO_2H) and an amine; for example, CH_3CONH_2. Also used to describe the NH_2^- ion.

Amine Compound with an $—NH_2$, $—NHR$, or $—NR_2$ substituent attached to a carbon atom.

Amino acid One of the essential building blocks out of which proteins are made. Amino acids have the generic formula $H_2NCHRCO_2H$.

Amontons's law A statement of the relationship between the temperature and pressure of a constant amount of gas at constant volume: $P \propto T$.

Amorphous Literally, without form or shape; used to describe substances that are solids but not crystals — solids that have little or no long-range order.

Amphoteric compound Compound, such as H_2O or $Al(OH)_3$, that can act as either an acid or a base.

Amplitude The height of a wave; the difference between the center of gravity of the wave and the highest (or lowest) point on the wave.

Amu Atomic mass unit; the unit in which the relative masses of atoms are expressed. The mass of a ^{12}C atom is defined as exactly 12.000... amu, and 1 amu is roughly equal to the mass of a hydrogen atom.

Angular momentum For a particle in a spherical orbit, the product of the mass of the particle times its velocity times the radius of the orbit.

Angular quantum number (ℓ) The quantum number used to describe the shape of an atomic orbital.

Anhydrous Literally, without water; used to differentiate between liquid (anhydrous) ammonia and solutions of ammonia dissolved in water, for example.

Anion A negatively charged ion, such as the Cl^- ion.

Anode The positive end of an electric field; used to describe the electrode in an electrochemical cell toward which anions flow and the electrode at which oxidation occurs.

Antibonding molecular orbital A molecular orbital in which electrons are held in a region of space that does not lie between the atoms.

Aqueous Literally, watery; used to describe solutions of substances dissolved in water.

Aromatic compound A compound, such as benzene (C_6H_6), with $C{=}C$ double bonds that does not react as an alkene does.

Arrhenius acid Any substance that dissociates when it dissolves in water to give the H^+ ion.

Arrhenius base Any substance that dissociates when it dissolves in water to give the OH^- ion.

Atactic A polymer in which the side chains are randomly distributed on either side of the polymer backbone.

Atom The smallest particle of an element that retains the properties of the element.

Atomic mass unit See *amu*.

Atomic number (Z) The number of protons in the nucleus of an atom.

Atomic orbital A region in space where the electrons on an atom can be found.

Atomic weight The weighted average of the atomic masses of the different isotopes of an element; for example, a single ^{12}C atom has a mass of 12.000 amu, but naturally occurring carbon also contains a 1.1% ^{13}C; so the atomic weight of carbon is 12.011 amu.

Aufbau principle The principle that states that atomic orbitals are filled one at a time, starting with the orbital that has the lowest energy.

Avogadro's constant The number of atoms, ions, or molecules in a mole— 6.0220×10^{23}.

Avogadro's hypothesis Hypothesis stating that equal volumes of different gases contain the same number of particles, which means that the volume of a gas is directly proportional to the number of moles of gas particles.

Axial Describes the two positions in a trigonal bipyramid that lie above and below the trigonal plane. See also *equatorial*.

Balanced equation A symbolic representation of a chemical reaction in which both sides of the equation contain equivalent numbers of atoms of each element; both charge and mass are conserved in a balanced equation.

Band of stability A narrow band of neutron-to-proton ratios that correspond to stable nuclides.

Base A source of the OH^- ion; an H^+-ion or proton acceptor.

Base-ionization equilibrium constant, K_b A measure of the relative strength of a base; $K_b = [BH^+][OH^-]/[B]$.

Beta decay A nuclear reaction in which beta particles (β^-), or electrons, are ejected from the nucleus of an atom.

Bidentate See *chelating ligand*.

Bimolecular Describes a step in a chemical reaction in which two molecules are consumed.

Binding energy The energy released when an atom is formed by combining protons, neutrons, and electrons.

Body-centered cubic structure A crystal in which the simplest repeating unit consists of nine equivalent lattice points, eight of which are at the corners of a cube and the ninth of which is in the center of the body of the cube.

Bohr model A model of the distribution of electrons in an atom used to account for the absorption and emission spectra of the hydrogen atom; based on the assumption that the electron in a hydrogen atom is in one of a limited number of circular orbits and that energy is absorbed or emitted when the electron moves from one of these orbits to another.

Boiling point The temperature at which the vapor pressure of a liquid reaches atmospheric pressure; the temperature at which a liquid boils.

Bond dissociation enthalpy The energy needed to break an $X{-}Y$ bond to give X and Y atoms in the gas phase.

Bonding electron Electron used to form a covalent bond between a pair of atoms.

Bonding molecular orbital A molecular orbital in which the electrons reside in the region in space between the atoms.

Bond order The number of bonds between a pair of atoms; for example, the bond order in N_2 is 3.

Boundary In thermodynamics, the boundary separates the system from its surroundings; heat and work are transferred across the boundary from the system to its surroundings or vice versa.

Boyle's law A statement of the relationship between the pressure and volume of a constant amount of gas at constant temperature: $P \propto 1/V$.

Bragg equation A statement of the relationship between the wavelength of the incident x-ray (λ), the distance between adjacent planes of atoms in a crystal (d), and the angle (θ) between the plane of the incident radiation and the uppermost plane of atoms in a crystal.

Breeder reactor A nuclear reactor that produces more fuel than it consumes.

British thermal unit See *Btu*.

Bromide Any compound that contains the Br^- ion or bromine with an oxidation number of -1; for example, sodium bromide ($NaBr$).

Brønsted acid Any substance that can donate an H^+ ion to a base; Brønsted acids are H^+-ion or proton donors.

Brønsted base Any substance that can accept an H^+ ion from an acid; Brønsted bases are H^+-ion or proton acceptors.

Btu British thermal unit; the heat needed to raise the temperature of 1 pound of water by $1°F$.

Buffer A mixture of a weak acid (HA) and its conjugate base (A^-) or a weak base (B) and its conjugate acid (BH^+); buffers resist changes in the pH of a solution when small amounts of acid or base are added.

Buffer capacity The amount of acid or base a buffer solution can absorb without significant changes in pH.

Calcining A process in which an ore loses a gas while being heated.

Caloric theory An obsolete theory that assumed that heat (or caloric) is a fluid that is conserved and that flows from a hot object to a cold object.

Calorie The amount of heat needed to raise the temperature of 1 gram of water by $1°C$.

Calorimeter An apparatus used to measure the heat given off or absorbed in a chemical reaction.

Calx The ashy powder formed when metals react with air; now known to be a metal oxide.

Canal rays The positively charged particles formed when electrons are removed from the gas particles in a cathode-ray tube.

Carbanion Any compound that contains a negatively charged carbon atom; for example, the CH_3^- ion.

Carbohydrate Literally, a hydrate of carbon; originally defined as any compound with an empirical formula of CH_2O; now defined as polyhydroxy aldehydes or ketones; includes the classes of compounds known as starches and sugars.

Carbonium ion Any compound that contains a positively charged carbon atom; for example, the CH_3^+ ion.

Carbonyl In organic chemistry, the $C{=}O$ functional group; in inorganic chemistry, a complex of carbon monoxide (CO).

Carboxylate ion The conjugate base of a carboxylic acid; for example, the $CH_3CO_2^-$ ion formed when acetic acid loses an H^+ ion.

Carboxylic acid Compound that contains the $-CO_2H$ functional group.

Carboxylic acid ester The product of the reaction between a carboxylic acid (RCO_2H) and an alcohol ($R'OH$); any compound that contains the RCO_2R' functional group.

Catalyst A substance that increases the rate of a chemical reaction without being consumed in the reaction; a substance that lowers the activation energy for a chemical reaction by providing an alternate pathway for the reaction.

Catenation The tendency of an element to form bonds to itself.

Cathode The negative end of an electric field; the electrode in an electrochemical cell toward which cations flow; the electrode at which reduction occurs.

Cathode rays The negatively charged particles (now recognized as electrons) that travel from the cathode toward the anode in a cathode-ray tube.

Cation A positively charged ion, such as the Mg^{2+} ion.

Chain reaction Any reaction in which one of the starting materials is regenerated in the last step of the reaction.

Chain-reaction mechanism A mechanism for a chain reaction that consists of chain-initiating, chain-propagating, and chain-terminating steps.

Charles's law A statement of the relationship between the temperature and volume of a constant amount of gas at constant pressure: $V \propto T$.

Chelating ligand A ligand that contains more than one pair of nonbonding electrons and can therefore coordinate to a metal atom more than once. These ligands are described as

bidentate, tridentate, tetradentate, or hexadentate, depending upon whether they coordinate to the metal two, three, four, or six times.

Chemical equation A symbolic representation of the relationship between the reactants and products of a chemical reaction, such as the following. $2\,H_2(g) + O_2(g) \rightarrow 2\,H_2O(g)$.

Chemical kinetics The study of the rates of chemical reactions.

Chemiluminescence A chemical reaction that gives off energy primarily in the form of light instead of heat.

Chiral Describes an object whose mirror image is not the same as itself. Compounds with four different substituents on a carbon atom are chiral.

Chloride Any compound that contains the Cl^- ion or chlorine with an oxidation number of -1; for example, hydrogen chloride (HCl).

Chromatography A technique for separating the components of a mixture on the basis of differences in their affinity for a stationary and a mobile phase.

Cis Literally, on the same side; describes isomers in which similar substituents are on the same side of a $C{=}C$ double bond or in adjacent coordination sites on a transition metal. See also *trans*.

Coal gas A gas produced when coal is heated in the absence of air; usually rich in methane (CH_4).

Codon A sequence of three nucleotides that codes for an amino acid.

Cohesion The force of attraction between molecules of the same substance. See also *adhesion*.

Colligative property Any property that depends on the number of solute particles in a solution and not their identity.

Collision theory model A model used to explain the rates of chemical reactions; assumes that molecules must collide in order to react.

Column chromatography Chromatography that uses a solid support in a vertical column or tube.

Combined equilibria Two or more equilibria that occur simultaneously and involve the same ion or molecule.

Common ion effect The decrease in the solubility of a salt that occurs when the salt is dissolved in a solution that contains another source of one of its ions.

Complex dissociation equilibrium constant (K_d) The equilibrium constant for the reaction in which a complex dissociates, such as the following. $Cu(NH_3)_4^{2+}(aq) \rightarrow$ $Cu^{2+}(aq) + 4\ NH_3(aq)$.

Complex formation equilibrium constant (K_f) The equilibrium constant for the reaction in which a complex is formed, such as the following. $Cu^{2+}(aq) + 4\ NH_3(aq) \rightarrow$ $Cu(NH_3)_4^{2+}(aq)$.

Complex ion Ion in which a ligand is covalently bound to a metal; ion formed when a Lewis acid—such as the Cu^{2+} ion—reacts with a Lewis base—such as NH_3—to form an acid-base complex—such as the $Cu(NH_3)_4^{2+}$ ion.

Compound A substance with a constant composition that contains two or more elements.

Compressibility The ability to be compressed; a characteristic of substances, such as gases, that can be compressed to fit into smaller containers.

Concentration A measure of the ratio of the amount of solute in a solution to the amount of either solvent or solution; most often expressed in units of moles of solute per liter of solution.

Condensation polymer A polymer formed when a small molecule is condensed out or lost during polymerization, such as nylon.

Conductivity apparatus An instrument used to determine whether a substance or solution can conduct an electric current.

Conjugate acid-base pair Two substances related by the gain or loss of a proton. Every Brønsted acid has a conjugate Brønsted base; an acid (such as HCl) and its conjugate base (the Cl^- ion) or a base (the OH^- ion) and its conjugate acid (H_2O) represent a conjugate acid-base pair.

Coordination complex Complex in which one or more ligands are coordinated to a metal atom.

Coordination number The number of neighboring atoms, ions, or molecules to which bonds can be formed.

Copolymer A polymer formed from two or more different monomers.

Corrosion A process in which a metal is destroyed by chemical reactions.

Covalent bond A bond between two atoms formed by the sharing of a pair of electrons.

Covalent compound A compound, such as water (H_2O) or cane sugar ($C_{12}H_{22}O_{11}$), composed of neutral molecules held together by covalent bonds.

Covalent radius The radius of an atom in a covalent bond.

Covalent solid A solid, such as diamond, in which every atom is covalently bound to its nearest neighbors to form an extended array of atoms rather than individual molecules.

Critical pressure The vapor pressure of a liquid at its critical temperature.

Critical temperature The highest temperature at which a gas can be condensed to a liquid by an increase in the pressure on the gas.

Crystal A three-dimensional solid formed by regular repetition of the packing of atoms, ions, or molecules.

Crystal field theory An extension of the valence bond theory used to explain transition metal compounds.

Cubic closest-packed structure A crystal structure formed by the stacking of closest-packed planes of atoms in an *ABCABC...* repeating pattern.

Cubic hole A hole in a simple cubic structure; atoms or ions that pack in a cubic hole are surrounded by eight other atoms or ions arranged toward the corners of a cube.

Curie A unit for measuring the activity of a radioactive nuclide; 1 Ci = 3.700×10^{10} disintegrations per second.

Cyclic hydrocarbon See *cycloalkane*.

Cycloalkane Alkane that contains a ring of carbon atoms.

Dalton's law of partial pressures A statement of the relationship between the total pressure of a mixture of gases and the partial pressures of the individual components: $P_T = P_1 + P_2 + P_3 \ldots$

Degenerate orbitals Orbitals that have the same energy; for example, the three $2p$ atomic orbitals on an isolated atom.

Dehydrating agent A reagent, such as P_4O_{10}, used to remove water.

Denature Change the three-dimensional structure of a protein through any process.

Density An intensive property of a substance equal to the mass of a sample divided by the volume of the sample.

Detergent A synthetic analog of soap that contains a long, hydrophobic tail attached to a hydrophilic $-SO_3^-$ or $-OSO_3^-$ head.

Dextrorotatory Describes compounds that rotate plane-polarized light to the right (clockwise) when viewed in the direction of the light source. See also *chiral, levorotatory,* and *optically active.*

Diamagnetic substance A substance repelled by both poles of a magnet; a substance in which all the electrons are paired.

Diatomic molecule A molecule, such as H_2 or HCl that contains two atoms.

Diffusion The movement of atoms, ions, or molecules through a gas, liquid, or solid.

Dilution The process by which a solvent is added to a solution to decrease the concentration of the solution.

Dimensional analysis An approach to solving problems that focuses on the way the units of the problem are used to set up the problem.

Dimer Literally, two parts; a compound, such as N_2O_4, produced by the combining of two identical smaller molecules, such as NO_2.

Dipole Anything with two equal but opposite electrical charges, such as the positive and negative ends of a polar bond or molecule.

Diprotic acid An acid, such as H_2SO_4, that has the potential to lose two H^+ ions.

Diprotic base A base, such as the S^{2-} ion, that can pick up two H^+ ions.

Disorder A measure of the extent to which a system differs from a perfect crystal at 0 K, where there is no disorder.

Disproportionation A reaction in which an element or compound simultaneously undergoes both oxidation and reduction; for example, the decomposition of H_2O_2 to form H_2O and O_2.

Dissociation For aqueous solutions, the process by which salts dissolve in water to give solutions that contain the corresponding ions.

Distillation A technique used to separate liquids with different boiling points.

Disulfide linkage The —S—S— linkage that can form between the —SH side chains on adjacent cysteine residues in a protein.

DNA Deoxyribonucleic acid, the nucleic acid used to store the genetic information that codes for the synthesis of proteins.

Ductile Capable of being drawn into thin sheets or wires without breaking. Metals are ductile.

Effective atomic number The number of valence electrons on a metal compared with the number of electrons on a rare gas; used to explain the bonding in transition metal complexes.

Effusion The process by which a gas escapes through a pinhole into a vacuum.

Elastic collision A collision in which no kinetic energy is lost.

Elastomer A polymer that snaps back to its original shape after being

stretched to at least twice its original length.

Electrolysis A process in which an electric current is used to decompose a compound into its elements.

Electrolytic cell An electrochemical cell in which electrolysis is done.

Electromagnetic radiation Radiation (such as radio waves, microwaves, infrared rays, light, ultraviolet rays, x-rays, or γ-rays) that contains both electric and magnetic components and travels at the speed of light.

Electron A subatomic particle with a charge of -1 and a mass of roughly 0.0005 amu.

Electron affinity (EA) The energy given off when a neutral atom in the gas phase picks up an electron to form a negatively charged ion.

Electron capture A nuclear reaction in which the nucleus captures a $1s$ electron.

Electron configuration The arrangement of electrons in atomic orbitals; for example, $1s^2\,2s^2\,2p^3$.

Electronegativity The tendency of an atom to draw the electrons in a bond toward itself.

Electron-pair acceptor See *Lewis acid.*

Electron-pair donor See *Lewis base.*

Electrophile Literally, a lover of electrons; a Lewis acid that attacks a site rich in electron density.

Element A substance that cannot be decomposed into a simpler substance by a chemical reaction; a substance composed of only one kind of atom.

Elemental analysis The process by which the percents by weight of the elements in a compound are determined.

Emission spectrum The spectrum of bright lines against a dark background obtained when an atom or molecule emits radiation when excited by heat or an electric discharge.

Empirical formula The simplest formula for a compound; the ratio of the numbers of atoms of each element in the compound.

Endothermic A process in which the system absorbs heat from the surroundings. See also *exothermic.*

End point The point at which the indicator of an acid-base titration changes color. See also *equivalence point.*

Enthalpy (H) The sum of the internal energy plus the product of the pressure times the volume of the system: $H = E + PV$.

Enthalpy of formation (ΔH_f) The enthalpy associated with the reaction that leads to the formation of a compound from its elements in their most stable states at 25°C and 1 atm.

Enthalpy of reaction (ΔH) The enthalpy associated with any chemical reaction; the difference between the sum of the enthalpies of the reactants and the products of the reaction.

Entropy (S) A measure of the disorder in a system.

Enzyme Protein that catalyzes biochemical reactions.

Equality In mathematics, a symbolic representation of two quantities that are in effect equal; for example, 12 inches = 1 foot.

Equation In mathematics, a symbolic statement that can be used to do a calculation; for example, $x = 12\,y$, where x is the number of inches in y feet.

Equation of state An equation, such as $PV = nRT$, that relates two or more of the quantities that describe the state of a system.

Equatorial Describes the three positions in a trigonal bipyramid that lie in the trigonal plane. See also *axial.*

Equilibrium The point at which there is no change in the concentrations of the reactants and the products of a chemical reaction; the point at which the rates of the forward and reverse reactions are equal.

Equilibrium constant (K_c or K_p) The quotient of the product of the concentrations of the products of a chemical reaction divided by the product of the concentrations of the reactants.

Equilibrium constant expression The expression used to calculate the equilibrium constant for a reaction.

Equilibrium region The part of a plot of the concentration of a substance versus time in which the concentration does not change; the part of this plot in which the reaction is at equilibrium.

Equivalence point The point in an acid-base titration at which equivalent amounts of acid and base have been added to the solution. See also *end point*.

Error The difference between a measurement and the true value of the quantity measured.

Ester See *carboxylic acid ester*.

Ether Compound in which an oxygen atom is attached to two carbon atoms; for example, $CH_3CH_2OCH_2CH_3$.

Excess reagent The reactant present in excess; the reaction will stop before all of the excess reagent is consumed.

Excluded volume The fraction of the volume of a gas that is not empty space; the volume of the gas actually occupied by gas particles.

Exothermic A process in which the system gives off heat to the surroundings. See also *endothermic*.

Expandibility The ability to expand; a characteristic of substances, such as gases, that can expand to fill their containers.

Extensive property A quantity that depends on the size of the sample, such as mass, weight, length, height, and width. See also *intensive property*.

Extraction A method of separating mixtures based on differences in the solubility of their components in polar versus nonpolar solvents.

Face-centered cubic structure A structure in which the simplest repeating unit consists of fourteen equivalent lattice points, eight at the corners of a cube and another six in the centers of the faces of the cube; found in cubic-closest packed structures.

Family A vertical column of elements in the periodic table; for example, the elements in Group IA — H, Li, Na, K, and so on.

Faraday The charge on a mole of electrons — 96,484.56 C.

Faraday's law A statement of the relationship between the amount of product that can be formed during electrolysis and the amount of electric current that passes through the electrolytic cell.

Fat A solid triester of glycerol and fatty acids. See also *oil*.

Fatty acid A carboxylic acid that contains a long, hydrophobic hydrocarbon chain.

Fertile nuclide A nuclide that can be converted into one that undergoes spontaneous fission; for example, ^{238}U, which can be converted into ^{239}Pu.

Filled-shell configuration An electron configuration in which a shell of atomic orbitals is filled; for example, $1s^2 2s^2 2p^6$.

First ionization energy The energy needed to remove the outermost, or highest-energy, electron from a neutral atom in the gas phase.

First law of thermodynamics A statement of the relationship between the internal energy of a system and the heat and work transferred from the system to its surroundings, or vice versa: $E = q + w$.

First-order reaction A reaction whose rate is proportional to the concentration of a single reactant raised to the first power.

Fission A nuclear reaction in which a nuclide splits into two smaller nuclides.

Fluoresce Emit light at one wavelength while being excited by radiation with a shorter wavelength.

Fluoride Any compound that contains the F^- ion or fluorine with an oxidation number of -1; for example, calcium fluoride (CaF_2).

Force The product of the mass of an object times its acceleration.

Formal charge The charge on an atom in its Lewis structure. We calculate formal change by dividing the electrons in each covalent bond between the atoms in the bond and then comparing the number of electrons that can be formally assigned to that atom with the number of electrons on a neutral atom of the element.

Free energy (*G*) The energy associated with a chemical reaction that is free to do work; the free energy of a system is the sum of its enthalpy plus the product of the temperature times the entropy of the system: $G = H - TS$.

Free radical A neutral atom or molecule with an unpaired electron.

Freezing point See *melting point*.

Frequency The number of wave crests or troughs that pass a fixed point per unit of time.

Functional group An atom or group of atoms in an organic compound that gives the compound some of its characteristic properties; for example, a $C=O$ (carbonyl) functional group.

Fusion (1) The formation of heavier nuclides by the fusing of two light nuclides. (2) The melting of a solid to form a liquid.

Galvanic cell See *voltaic cell*.

Galvanic corrosion Corrosion that occurs when two metals are in contact with each other and with water.

Gamma ray A high-energy, short-wavelength form of electromagnetic radiation emitted by the nucleus of an atom, which carries off some of the energy released in a nuclear reaction.

Gas A substance that flows freely, expands to fill its container, and can be compressed to fit into a smaller container.

Gas-phase chromatography Chromatography that involves passing a gas over a solid support.

Gas-phase reaction A chemical reaction in which all of the reactants and products are gases.

Gay-Lussac's law A statement of the fact that the ratio of the volumes of gases consumed or produced in a gas-phase chemical reaction is equal to the ratio of two simple whole numbers.

Geometric isomers Compounds that differ in the way substitutents are oriented around a $C{=}C$ bond or on a transition metal. See also *cis* and *trans*.

Graham's law The relationship between the rate at which a gas diffuses or effuses and its molecular weight: $r \propto 1/\sqrt{MW}$.

Gram The basic unit of mass in the metric system. A penny weighs roughly 2.5 grams.

Grignard reagent An alkylmagnesium halide, such as CH_3MgBr; a source of a carbanion (such as the CH_3^- ion) for use in organic synthesis.

Group A vertical column, or family, of elements in the periodic table, such as Group IA.

Group number A number that identifies a group of elements in the periodic table. Until recently, groups were labeled IA, IIA, IIIA, and so on. A new system has been proposed in which the columns or groups of the periodic table are numbered sequentially from left to right.

Haber process The industrial process used to make NH_3 from N_2 and H_2.

Half-cell One half of a voltaic cell.

Half-life The time required for the amount of a reactant to decrease to half its initial value.

Half-reaction The reaction that takes place in a half-cell.

Halide A F^-, Cl^-, Br^-, or I^- ion.

Halogen F_2, Cl_2, Br_2, or I_2.

Hard water Water with a high concentration of Ca^{2+}, Mg^{2+}, and/or Fe^{3+} ions.

Heat (q) A form of energy associated with the random motion of the elementary particles in matter.

Heat of fusion The heat that must be absorbed to melt a mole of a solid.

Heat of vaporization The heat that must be absorbed to boil a mole of a liquid.

Hess's law A law stating that the heat given off or absorbed in a chemical reaction is the same regardless of whether the reaction occurs in a single step or in many steps.

Hexadentate See *chelating ligand.*

Hexagonal closest-packed structure A structure formed by the stacking of closest-packed planes of atoms in an *ABABAB* . . . repeating pattern.

High-spin complex A transition metal complex in which the difference between the energies of the t_{2g} and e_g sets of d orbitals is smaller than the energy it takes to pair two electrons; as a result, the valence-shell d electrons on the metal atom are placed in both the t_{2g} and e_g sets of orbitals. See also *low-spin complex.*

Homonuclear diatomic molecule A diatomic molecule, such as O_2 or F_2, that contains two atoms of the same element.

Hund's rules Rules that describe how electrons are added to degenerate orbitals. The electrons are added to these orbitals with parallel spins insofar as possible; thus one electron is added to each orbital (with their spins aligned) before a second electron is placed in any orbital.

Hybrid atomic orbital Orbital formed by the mixing of two or more atomic orbitals.

Hybridization A process in which things are mixed; a resonance hybrid is a mixture, or average, of two or more Lewis structures; hybrid orbitals are formed by the mixing of two or more atomic orbitals.

Hydride Literally, a salt containing the H^- ion, such as NaH; also used to describe compounds that contain hydrogen.

Hydrocarbon Literally, a compound that contains only carbon and hydrogen, such as CH_4.

Hydrogen bond The van der Waals bond formed when the positive end of one polar molecule, such as water, is attracted to the negative end of another polar molecule.

Hydrophilic Literally, water-loving; describes polar groups that attract water molecules.

Hydrophobic Literally, water-hating; describes nonpolar groups that repel water molecules.

Hydroxide Compound that contains an OH group.

Ideal gas equation The relationship between the pressure, volume, temperature, and amount of an ideal gas: $PV = nRT$.

Immiscible Applied to liquids that are not soluble in each other; gasoline and water are immiscible.

Indicator A compound, such as phenolphthalein, that changes color when equivalent amounts of two reactants are present.

Induced dipole A short-lived separation of charge, or dipole, created by polarization of a nonpolar atom or molecule.

Induced fission Fission of a nuclide that occurs after the nuclide has absorbed another particle. See also *spontaneous fission.*

Inelastic collision A collision in which at least a portion of the kinetic energy of the colliding particles is lost.

Inert Unreactive; not able to undergo chemical reaction.

Initial rate of reaction The rate of a chemical reaction extrapolated back to the instant the reactants were mixed.

Inorganic Not containing $C{-}H$ bonds.

Insoluble Not able to dissolve in a solvent to give reasonable concentrations.

Instantaneous rate of reaction The rate of a chemical reaction at an instant in time; the infinitesimally small change in the concentration of one of the reactants (or products)—$d(X)$—that occurs over an infinitesimally small length of time—dt.

Integrated form of the rate law An alternative form of the rate law for a chemical reaction that can be used to predict the concentrations of the reactants (or products) at some moment in time.

Intensive property A quantity, such as temperature, density, or pressure, that does not depend on the size of the sample. See also *extensive property*.

Intermetallic compound A compound, such as $Cu Al_2$, with a fixed composition that results from the combination of two or more metals.

Intermolecular bonds The weak bonds between molecules in a liquid or solid, such as the hydrogen bonds between water molecules. See also *intramolecular bonds*.

Internal energy (E) The sum of the kinetic and potential energies of the particles in a system. The internal energy of the system is proportional to its temperature; for an ideal gas, $E = \frac{3}{2} RT$.

Interstitial solution A solid solution formed when solute atoms are packed in the holes, or interstices, between solvent atoms.

Intramolecular bonds The bonds that hold a molecule together, such as the covalent O—H bonds in water. See also *intermolecular bonds*.

Iodide Any compound that contains the I^- ion or iodine with an oxidation number of -1; for example, potassium iodide (KI).

Ion An atom or molecule that carries a positive or negative electrical charge, such as the Na^+, Cl^-, or SO_4^{2-} ion.

Ionic bond The bond between two ions that results from the electrostatic (or coulombic) force of attraction between particles of opposite charge.

Ionic compound A compound that contains ions.

Ionic radius The radius of the ions in an ionic compound.

Ionic solid A solid composed of ions; for example, NaCl; also called a salt.

Ionization A process in which an ion is created by addition or subtraction of electrons from a neutral atom or molecule.

Ionizing radiation Radiation with enough energy to remove an electron from a neutral atom or molecule to produce free radicals. See also *nonionizing radiation*.

Ion product (Q_{sp}) The product of the concentrations of the ions in a solution at any moment in time. When Q_{sp} is larger than the solubility product (K_{sp}), the salt precipitates.

Isobars Nuclides with the same mass number; for example, ^{40}K and ^{40}Ca.

Isoelectronic Having the same number of electrons and therefore the same electron configuration; describes atoms or ions such as C^{4-}, N^{3-}, O^{2-}, F^-, and Ne.

Isomers Compounds with the same chemical formulas but different structures and therefore different chemical or physical properties.

Isotactic polymer A polymer in which all the substituents are oriented on the same side of the polymer backbone.

Isotones Nuclides with the same number of neutrons, such as ^{13}C, ^{14}N, and ^{15}O.

Isotopes Nuclides with the same number of protons, such as ^{12}C, ^{13}C, and ^{14}C.

Isotropic Literally, the same in all directions; describes the pressure of a gas.

Joule A unit of measurement for both heat and work in the SI system; 1 J = 4.184 cal.

Ketone A compound in which a C=O (carbonyl) functional group is bound to two carbon atoms; for example, CH_3COCH_3.

Kinetic control Determination of the products of a chemical reaction by the rate of one or more steps in the reaction, not by thermodynamics. See also *thermodynamic control*.

Kinetic energy The energy associated with a moving object; the kinetic energy of the object is equal to one-half the product of its mass times its velocity squared: $KE = \frac{1}{2} mv^2$.

Kinetic molecular theory Theory stating that heat is associated with the thermal motion of particles. In the kinetic theory, heat isn't conserved —an inexhaustible amount of heat can be created when work is done on a system.

Kinetic region The portion of a plot of the concentration of a compound versus time in which the concentration either increases or decreases; the portion of this plot between the initial conditions and the point at which the system reaches equilibrium.

Latent heat Heat that cannot be detected. The heat that enters a system when ice melts to form water or when water boils to form steam is latent heat, because there is no change in the temperature of the system. See also *sensible heat*.

Lattice energy The energy given off when oppositely charged ions in the gas phase come together to form a solid; for example, the energy given off in the following reaction. $Na^+(g) + Cl^-(g) \rightarrow NaCl(s)$.

Lattice point A point about which an atom, ion, or molecule is free to vibrate in a crystal.

LeChatelier's principle A principle that describes the effect of changes in the temperature, pressure, or concentration of one of the reactants or products of a reaction at equilibrium. It states that when a system at equilibrium is subjected to a stress, it will shift in the direction that minimizes the effect of this stress.

Leveling effect The tendency of water to limit the strength of the strongest acids and bases to the strength of the H_3O^+ and OH^- ions.

Levorotatory Describes compounds that rotate plane-polarized light to the left (counterclockwise) when viewed in the direction of the light source. See also *chiral, dextrorotatory,* and *optically active*.

Lewis acid An electron-pair acceptor; a substance that acts in the same way as the H^+ ion to accept a pair of electrons.

Lewis base An electron-pair donor; a substance that acts in the same way as the OH^- ion to donate a pair of electrons.

Lewis structure A symbolic description of the distribution of valence electrons in a molecule; it uses dots to represent individual electrons and lines to represent covalent bonds.

Ligand A Lewis base that can coordinate to a metal atom.

Ligand field theory A molecular orbital description of the bonding in transition metal complexes.

Limiting reagent The reactant in a chemical reaction that limits the amount of product that can be formed — the reaction will stop when all of the limiting reagent is consumed.

Linear (straight-chain) molecule Molecule that consists of long chains of C—C bonds.

Linkage isomers Complexes with the same formula that differ in the way ligands are bound to the metal atom.

Lipid One of a diverse class of biologically important molecules that are soluble in nonpolar solvents.

Liquid A substance that flows freely and therefore conforms to the shape of the walls of its container but that cannot expand to fill the container.

Liquid crystal A substance that has some of the long-range order of a solid but the freedom of motion of a liquid.

Liter The fundamental unit of volume in the metric system; the volume of 1000 grams or 1000 cubic centimeters of water at 4°C.

Low-spin complex A transition metal complex in which the difference between the energies of the t_{2g} and e_g sets of d orbitals is larger than the energy it takes to pair two electrons; as a result, the valence-shell d electrons on the metal atom are all placed in the lower-energy set of orbitals — the t_{2g} set for octahedral complexes and the e_g set for tetrahedral complexes. See also *high-spin complex*.

Macroscopic Capable of being seen with the naked eye.

Magnetic quantum number (m) The quantum number used to describe the orientation of an atomic orbital in space.

Main-group element An element in one of the groups of the periodic table in which s and p orbitals are filled. Main-group elements are also known as representative elements.

Malleable Capable of being hammered, pounded, or pressed into different shapes without breaking. Metals are malleable.

Mass A measure of the constant amount of matter in an object.

Mass defect The difference between the mass of an atom and the sum of the masses of the protons, neutrons, and electrons that form the atom.

Mass number (M) An integer equal to the number of protons and neutrons in the nucleus of an atom.

Measurement The amount or quantity of something being measured.

Mechanism The steps by which a chemical reaction converts the reactants into the products.

Melting point The temperature at which the solid and liquid phases of a substance are in equilibrium at atmospheric pressure.

Meniscus The curved surface of a liquid, such as water or mercury, in a narrow-diameter glass tube.

Metal In general, an element, such as aluminum or gold, that is solid, has a metallic luster, is malleable and ductile, and conducts both heat and electricity.

Metallic radius Half the distance between the nuclei of adjacent atoms in a metal.

Metalloid A semimetal; an element (such as Si) with properties that fall between the properties of metals and nonmetals.

Metastable system A system that should undergo a spontaneous change but does not.

Metathesis reaction Reaction in which atoms or groups of atoms are interchanged, but none of the atoms undergo a change in oxidation number. See also *oxidation-reduction reaction*.

Meter The basic unit of length in both the metric system and SI; currently defined as the distance light travels in 1/299,792,458 second.

Metric system A system of units introduced in 1790 based on three fundamental quantities: the gram (mass), liter (volume), and meter (length).

Mixture A substance that contains two or more elements or compounds that retain their chemical identity and that can be separated into the individual components by a physical process; for example, the mixture of O_2 and N_2 in the atmosphere.

Molality (m) The number of moles of solute in a solution divided by the number of kilograms of solvent.

Molar heat capacity (C) The number of joules of heat required to raise the temperature of a mole of a substance by 1 K. See also *specific heat*.

Molarity (M) The number of moles of solute in a solution divided by the volume of the solution in liters.

Mole The amount of any substance that contains the same number of atoms, ions, or molecules as there are ^{12}C atoms in 12.000 . . . grams of ^{12}C; Avogadro's number of particles.

Molecular formula The formula of a molecule of a compound.

Molecular geometry The shape, or geometry, of a molecule as defined by the positions of its nuclei.

Molecularity Number of molecules consumed in a step of a chemical reaction. See *bimolecular* and *unimolecular*.

Molecular orbital A region in space within a molecule where electrons can be found; molecular orbitals are formed by the overlap, or mixing, of atomic orbitals.

Molecular solid A substance that consists of individual molecules held together in the solid by relatively weak intermolecular bonds.

Molecular weight (mass) The relative weight (or mass) of a molecule in amu.

Molecule The smallest particle that has any of the chemical or physical properties of an element or compound.

Mole fraction Literally, the fraction of the total number of moles due to one component of a mixture; the mole fraction of a solute, for example, is the number of moles of solute divided by the number of moles of solute plus solvent.

Mole ratio The ratio of the moles of one reactant or product compared with the moles of another reactant or product in a chemical reaction; can be determined from the balanced equation for the reaction.

Momentum The product of the mass times the velocity of an object in motion.

Monodentate See *chelating ligand*.

Monomer One of the relatively small molecules from which polymers are formed.

Monoprotic acid (H*A*) An acid, such as HCl or HCN, that can lose only one H+ ion or proton.

m-RNA Messenger RNA; the polynucleotide that codes for the synthesis of a protein. It is assembled during transcription of a chain of DNA.

Neutrino A particle with no charge and little or no mass that is ejected from the nucleus at the same time as an electron or positron.

Neutron A subatomic particle with a mass of about 1 amu and no charge.

Neutron-poor nuclide A nuclide with fewer neutrons than the stable isotope(s) of the element; a nuclide that decays by either electron capture, positron emission, or the emission of an alpha particle.

Neutron-rich nuclide A nuclide with more neutrons than the stable isotope(s) of the element; a nuclide that decays by the emission of an electron.

Nitride Any compound that contains the N^{3-} ion or nitrogen with an oxidation number of -3; for example, lithium nitride (Li_3N).

Nomenclature System of names, as for chemical compounds.

Nonbonding electron Electron in the valence shell of an atom that is not used to form covalent bonds.

Nonbonding molecular orbital A molecular orbital whose energy is more or less equivalent to the energy of the atomic orbitals from which it is formed.

Nonionizing radiation Radiation that carries enough energy to excite an electron but not enough energy to remove an electron from an atom or molecule. See also *ionizing radiation*.

Nonmetal An element, such as chlorine or oxygen, that lacks the properties generally associated with metals.

Nucleic acid A molecule of very high molecular weight used to store and process the genetic information in cells.

Nucleon A proton or neutron.

Nucleophile Literally, a lover of nuclei; describes Lewis bases that attack sites low in electron density.

Nucleophilic substitution reaction A reaction, such as the following, in which one nucleophile is substituted for another in a molecule.
$$OH^-(aq) + CH_3Br(aq) \rightarrow$$
$$CH_3OH(aq) + Br^-(aq)$$

Nucleotide The smallest repeating unit out of which nucleic acids are built.

Nucleus Literally, "little nut"; it consists of neutrons and protons, occupies an infinitesimally small fraction of the total volume of an atom, and contains almost all of its mass.

Nuclide An atom with a particular combination of protons and neutrons; for example, the ^{14}C nuclide.

Octahedral complex A complex in which six ligands are bound to the metal atom and arranged toward the corners of an octahedron.

Octahedral geometry A geometry in which six atoms, ions, or molecules

are arranged around a central atom in opposite directions along the X, Y, and Z axes of a coordinate system.

Octahedral hole A hole in a closest-packed structure surrounded by six atoms or ions arranged toward the corners of an octahedron.

Octane number A measure of the antiknock quality of a gasoline.

Octet Literally, eight; used to describe the tendency of main-group elements to have an octet of valence-shell electrons in their compounds.

Oil One of three kinds of substances: (1) mineral oils, such as crude oil from petroleum, which are mixtures of hydrocarbons; (2) animal and vegetable oils, such as corn oil, which are mixtures of triglycerides; and (3) essential oils or perfumes from plants, which are terpenes.

Olefin A common name for the class of compounds known as alkenes; compounds that contain C=C double bonds.

Optically active Describes the ability of some molecules to rotate plane-polarized light.

Organic Describes compounds that contain C—H bonds.

Osmosis The process by which one component of a solution passes through a semipermeable membrane to dilute the solution.

Osmotic pressure The pressure exerted by water or other solvents flowing into a solution through a semipermeable membrane.

Ostwald process The industrial process in which ammonia is converted into nitric acid.

Overlap The interaction between a pair of atomic orbitals that occurs when the orbitals share space.

Overvoltage The voltage over and above the voltage predicted from standard tables of reduction potentials that must be applied before an electrochemical reaction occurs.

Oxidation Any process in which the oxidation number of an atom becomes larger.

Oxidation number The charge that would be present on an atom if the element or compound in which the atom is found were ionic.

Oxidation-reduction reaction Reaction in which at least one atom undergoes a change in oxidation number. See also *metathesis reaction*.

Oxide Any compound that contains the O^{2-} ion or oxygen with an oxidation number of -2; for example, lithium oxide (Li_2O).

Oxidizing agent An atom, ion, or molecule that gains electrons in a chemical reaction, thereby oxidizing the substance with which it reacts.

Oxyacid An acid, such as H_2SO_4 or H_3PO_4, in which the acidic hydrogen atoms are attached to an oxygen atom.

Oxyanion Negatively charged ion, such as the SO_4^{2-} or PO_4^{3-} ion, formed when an oxyacid loses one or more H^+ ions.

Paramagnetic Having the capacity to be magnetized; containing one or more unpaired electrons.

Partial pressure The fraction of the total pressure of a mixture of gases that is due to one of the components of the mixture.

Particle Any object that has mass and therefore occupies space.

Peptide A relatively small polymer of amino acids; peptides are usually too small to have extensive secondary or tertiary structures.

Peptide bond The bond between amino acids in a peptide or protein.

Period A horizontal row in the periodic table; for example, the second period contains the elements Li, Be, B, C, N, O, F, and Ne.

Periodic table A matrix in which the elements are arranged across rows in order of increasing atomic number so that elements with similar chemical properties fall in the same vertical column.

Peroxide Literally, a compound that is rich in oxygen, such as a compound that contains the O_2^{2-} ion.

pH A measure of acidity; the negative of the logarithm of the H_3O^+ ion concentration: $pH = -\log [H_3O^+]$

Phase diagram A two-dimensional graph that shows the state, or phase, of a substance at any combination of temperature and pressure.

Phlogiston An imaginary element once thought to be given off when objects burned.

Phosphide Any compound that contains the P^{3-} ion or phosphorus with an oxidation number of -3; for example, calcium phosphide (Ca_3P_2).

Phospholipid A triester of glycerol with two fatty acids and one phosphate ion; polar lipids found in biological membranes.

Photon A tiny unit of electromagnetic energy.

Pi bond A bond formed by the edge-on overlap of p atomic orbitals.

Plastic Literally, a material that can flow; describes polymers that can be shaped, molded, or milled.

pOH The negative of the logarithm of the OH^- ion concentration: $pOH = -\log [OH^-]$

Poise The unit of measurement for viscosity.

Polar covalent bond A bond in a molecule, such as H_2O, that is neither strictly covalent nor strictly ionic.

Polarity The tendency of a molecule to have positive and negative poles; the unequal sharing of a pair of electrons.

Polyatomic ion An ion that contains more than one atom.

Polycrystalline solid A solid composed of many individual crystals or grains arranged in a more or less random order.

Polymer A molecule with a large molecular weight formed by the linking of 30 to 100,000 (or more) repeating units.

Polynucleotide A condensation polymer formed by the linking of nucleotides.

Polyprotic acid An acid, such as H_2SO_4 or H_3PO_4, that can lose more than one H^+ ion, or proton.

Polyprotic base A base, such as the PO_4^{3-} ion, that can accept more than one H^+ ion, or proton.

Positron The antimatter analog of an electron; positrons have the same mass as electrons but the opposite charge.

Positron emission A nuclear reaction in which a positron is emitted; positron emission leads to a decrease of 1 in the atomic number and no change in the mass number of the nuclide.

Potential A measure of the driving force behind electrochemical reactions; reported in units of volts.

Precipitation A process in which positive and negative ions combine to form a salt that precipitates out of the solution as a solid.

Precipitation hardening A process by which an alloy is hardened when an intermetallic compound, such as $CuAl_2$, is allowed to precipitate from a supersaturated solution of two metals.

Precision The agreement between individual measurements of the same quantity.

Pressure The force exerted on a surface divided by the area of the surface.

Primary structure The sequence of amino acids in a protein.

Primary valence The number of negative ions needed to satisfy the charge on a metal ion; in the $[Co(NH_3)_6]Cl_3$ complex, for example, it is three.

Principal quantum number (*n*) The quantum number that describes the size of an orbital.

Product Substance formed in a chemical reaction.

Protein A relatively large polymer of amino acids; a polypeptide that is large enough to have an extensive secondary and tertiary structure.

Proton A subatomic particle that has a charge of $+1$ and a mass of about 1 amu.

Proton acceptor An ion or molecule that can gain an H^+ ion, or proton; a Brønsted base.

Proton donor An ion or molecule that can lose an H^+ ion, or proton; a Brønsted acid.

Pseudo–first-order reaction A reaction that is first-order in two reactants but that can be made to appear first-order in only one of these reactants by use of a large excess of the other.

Pyrophoric Compounds that burst into flame in the presence of air.

Quadratic formula A formula for solving quadratic equations.

Qualitative analysis A technique used to identify the elements of a complex mixture.

Quantized Countable; oranges are quantized, orange juice is not.

Quantum number An integer that describes the size, shape, and orientation in space of an atomic orbital.

Quaternary structure The ionic or hydrophobic interactions between individual chains of amino acids in some proteins.

Rad A unit of radiation absorbed dose equal to 10^{-5} joules per gram.

Radioactivity The spontaneous disintegration of an unstable nuclide by a first-order rate law.

Radius ratio The radius of the positive ion divided by the radius of the negative ion in a salt.

Random error A source of error that limits the precision of a measurement; an error that is equally likely to give results that are too large or too small.

Raoult's law Law that describes the relationship between the vapor pressure of a solution, the mole fraction of the solute, and the vapor pressure of the solute: $P = \chi P°$.

Rate constant The proportionality constant between the rate of a step in a chemical reaction and the product of the concentrations of the reactants consumed in that step.

Rate law An equation that describes how the rate of a chemical reaction depends on the concentrations of the reactants consumed in that reaction.

Rate-limiting step The slowest step in a chemical reaction.

Rate of reaction The change in the concentration of a compound divided by the amount of time necessary for this change to occur.

RBE Radiation biological effectiveness; used to correct for differences in the danger of absorbing equivalent doses of different forms of radiation. See also *rad* and *rem*.

Reactants One of the starting materials in a chemical reaction.

Reaction coordinate The sequence of infinitesimally small steps that must be taken to convert the reactants into the products of a reaction.

Reaction quotient (Q_c or Q_p) The quotient obtained when the concentrations of the products of a reaction are multiplied and the result is divided by the product of the concentrations of the reactants. The reaction quotient can be calculated for any moment in time; when the reaction is at equilibrium, the reaction quotient is equal to the equilibrium constant for the reaction.

Redox Oxidation-reduction.

Reducing agent An atom, ion, or molecule that loses electrons in a chemical reaction, thereby reducing the substance with which it reacts.

Reduction Any process that leads to a decrease in the oxidation number of an atom.

Rem A unit of radiation absorbed dose equal to the product of the rads of absorbed radiation times the RBE.

Replication The process by which copies of DNA are made to be passed down to future generations of cells.

Resonance An averaging or mixing process that occurs when more than one satisfactory Lewis structure can be written for a molecule.

Resonance hybrid A mixture, or average, of two or more Lewis structures.

RNA Ribonucleic acid, the nucleic acid involved in transcribing the genetic information for the synthesis of proteins stored in DNA and then translating this information into the sequence of amino acids in the protein.

Roasting A process in which a metal ore (such as PbS or ZnS) is transformed into the corresponding oxide (PbO or ZnO) and SO_2.

R-process The sequence of reactions in which a number of neutrons are captured in rapid succession; it produces heavy nuclides such as ^{238}U. The R-process requires a large neutron flux, which occurs when a star becomes a supernova. See also *S-process*.

Rusting The corrosion of iron or iron-based alloys such as steel; other metals may corrode, but only iron and steel rust.

Salt See *ionic compound*.

Salt bridge A tube containing a saturated solution of potassium chloride; used to complete the electrical circuit in a voltaic cell.

Saponification The reaction between fats and oils with KOH or NaOH to produce soaps.

Saturated fatty acid A long-chain carboxylic acid that contains no C=C double bonds. See also *unsaturated fatty acid*.

Saturated hydrocarbon Hydrocarbon that contains as much hydrogen as possible; compound with the generic formula C_NH_{2N+2}.

Saturated solution A solution that contains as much solute as possible; a solution in which additional solute dissolves at the same rate at which the solute precipitates from solution.

Scientific notation A number between 1 and 10 times 10 raised to some exponent.

Secondary structure The hydrogen bonds between adjacent peptide linkages in a protein that stabilize the structure of the protein.

Secondary valence The number of ions or molecules that are covalently bound to the transition metal ion; in the $[Co(NH_3)_6]Cl_3$ complex, it is six.

Second ionization energy The energy needed to remove an electron from an M^+ ion in the gas phase.

Second-order reaction A reaction whose rate is proportional to the concentration of a single reactant raised to the second power.

Selective precipitation A technique in which one ion is selectively removed from a mixture of ions by precipitation.

Semiconductor A material whose ability to carry an electric current falls between those of metals and nonmetals.

Semimetal An element, such as silicon or arsenic, with chemical and physical properties that fall between the properties normally associated with metals and nonmetals; also called a metalloid.

Semipermeable membrane A thin, flexible solid that can pass small molecules such as water but not larger molecules such as sugar or alcohol.

Sensible heat Heat that can be sensed, or detected, by a change in the temperature of the system. See also *latent heat.*

Shell All orbitals that have the same principal quantum number.

SI The International System of Units; based on seven fundamental quantities: the meter (length), kilogram (mass), second (time), kelvin (temperature), ampere (current), mole (amount of substance), and candela (luminous intensity).

Sigma bond A cylindrically symmetrical bond formed by the overlap of *s* orbitals or the head-on overlap of *p* orbitals.

Significant figures The digits in a measurement that are reliable.

Simple cubic structure Structure in which the simplest repeating unit consists of eight equivalent atoms, ions, or molecules at the eight corners of a cube.

Sintering A process in which a finely divided ore is heated until it collects to form larger particles.

Skeleton structure A representation of the basic structure of a molecule in which lines are used to connect atoms connected by covalent bonds.

Smelting A process in which a metal oxide (such as PbO, ZnO, or Fe_2O_3) is reduced to the corresponding metal.

Soap The sodium or potassium salt of a fatty acid.

Solid A substance that does not flow and therefore does not conform to the shape of its container.

Solubility The ratio of the maximum amount of solute to the volume of solvent in which this solute can dissolve; often expressed in units of grams of solute per 100 grams of water or in moles of solid per liter of solution.

Solubility equilibria Equilibria that exist in a saturated solution, in which additional solid dissolves at the same rate that particles in solution come together to precipitate more solid.

Solubility product (K_{sp}) The number obtained when the equilibrium concentrations of the ions in an aqueous solution of a salt are multiplied.

Solute The substance that dissolves to form a solution.

Solution A mixture of one or more solutes dissolved in a solvent.

Solvent The substance in which a solute dissolves.

Specific heat The number of joules of heat required to raise the temperature of 1 gram of a substance by 1 K. See also *molar heat capacity.*

Spectrochemical series A sequence of ligands arranged in order of decreasing magnitude of the splitting between the energies of e_g and t_{2g} orbitals in transition metal complexes.

Spin quantum number (s) The quantum number used to specify the electrons that occupy an orbital.

Spontaneous fission Fission of a nuclide that occurs spontaneously. See also *induced fission.*

Spontaneous reaction A reaction in which the products are favored.

S-process The process by which heavier nuclides are built up by the slow absorption of one neutron at a time. See also *R-process.*

Square planar geometry A geometry in which four atoms, ions, or molecules lying in the same plane are bound to a central atom and arranged toward the corners of a square.

Square pyramidal geometry A geometry in which a fifth atom, ion, or molecule is added to a square planar geometry along an axis perpendicular to the plane of the other four.

Standard state State in which all concentrations are 1 M, all partial pressures are 1 atm, and the temperature is 25°C.

State (1) One of the three states of matter: gas, liquid, or solid. (2) A set of physical properties that describe a system.

State function A quantity whose value depends only on the state of the system and not its history. X is a state function if ΔX is the same regardless of the path used to go from the initial to the final state.

Stoichiometry The relationship between the weights of the reactants and the products of a chemical reaction.

Sublimation A process in which a solid goes directly to the gaseous state without passing through an intermediate liquid state.

Subshell Atomic orbitals for which the values of the n and l quantum numbers are the same; for example, the three $2p$ or five $3d$ atomic orbitals.

Successive approximations A technique in which the answer obtained from an approximate calculation is repeatedly substituted into the original equation to yield a more accurate solution.

Sulfide Any compound that contains the S^{2-} ion or sulfur with an oxidation number of -2; for example, carbon disulfide (CS_2).

Supercooled liquid Liquid cooled to a temperature below its freezing point.

Superoxide Compound that contains the O_2^- ion, such as potassium superoxide (KO_2).

Surface tension The force that controls the shape of a liquid; surface tension results from the force of cohesion between liquid molecules.

Surroundings The part of the universe not included in the system.

Syndiotactic polymer A polymer in which the substituents alternate regularly between the two sides of the polymer backbone.

Synthesis gas A mixture of CO and H_2 produced during the gasification of coal; it can be used to synthesize a wide range of materials.

System That small portion of the universe in which the investigator is interested.

Systematic error A source of error that limits the accuracy of a measurement; systematic errors tend to give results that are always too small or always too large.

Temperature An intensive property that measures the extent to which an object can be labeled hot or cold.

Tertiary structure The interactions between the side chains on amino acids in a protein that help determine the structure of the protein.

Tetradentate See *chelating ligand*.

Tetrahedral complex A complex in which four ligands are bound to the metal atom and arranged toward the corners of a tetrahedron.

Tetrahedral hole A hole in a closest-packed structure surrounded by four atoms arranged toward the corners of a tetrahedron.

Tetrahedron A geometry that resembles a trigonal-based pyramid.

Tetravalent Literally, able to form four bonds. Carbon is tetravalent because it forms four bonds in virtually all of its compounds.

Thermal conductor An object that carries, or conducts, heat easily from one side to the other or from one end to the other.

Thermal insulator An object, such as a blanket or a fur coat, that tends to slow down the rate at which heat is transferred from one object to another.

Thermochemistry The study of the heat given off or absorbed in chemical reactions.

Thermodynamic control Determination of the products of a chemical reaction by thermodynamics, not by the rate of one or more steps in the reaction. See also *kinetic control*.

Thermodynamics The study of the relationships between heat, work, and other forms of energy.

Thermonuclear reaction A fusion reaction that only occurs at high temperatures.

Thin-layer chromatography Chromatography in which a solvent is passed over a solid support that has been applied as a thin layer to a glass, metal, or plastic plate.

Thio- A prefix that describes compounds in which a sulfur atom has replaced an oxygen atom; the sulfate ion, for example, has the formula SO_4^{2-}, whereas the thiosulfate ion has the formula $S_2O_3^{2-}$.

Titration A technique used to determine the concentration of a solution.

Torr A unit of pressure equal to the pressure of a column of mercury 1 millimeter tall; 1 torr = 1 mmHg.

Trans Literally, across; describes isomers of C=C compounds in which similar substituents lie on opposite sides of the double bond or on opposite sides of a transition metal. See also *cis*.

Transcription The process by which the message encoded in DNA is transcribed to form a sequence of m-RNA that can be used as a template to make proteins. See also *translation*.

Transition metal Metals in the block of elements that serve as a transition between the two columns on the left side of the table, where *s* orbitals are filled, and the six columns on the right, where *p* orbitals are filled.

Translation The process by which the information coded in a sequence of m-RNA is translated into a sequence of amino acids in a protein. See also *transcription*.

Translational motion The net movement of an atom, ion, or molecule in one direction or another through a gas, liquid, or solid.

Transuranium elements The elements with atomic numbers larger than that of uranium.

Triad A group of three elements (such as F, Cl, and Br, or Li, Na, and K) with similar chemical properties.

Tridentate See *chelating ligand*.

Triglyceride Triester of glycerol and fatty acids found in fats and oils; also known as triacylglycerols.

Trigonal bipyramidal geometry A geometry in which two atoms, ions, or molecules are added to a trigonal planar system along an axis perpendicular to the triangular plane.

Trigonal planar geometry A geometry in which three atoms, ions, or molecules are coordinated to a central atom and arranged toward the corners of an equilateral triangle.

Trigonal pyramidal geometry A geometry in which a central atom at the apex of a pyramid is coordinated to three atoms, ions, or molecules to form a pyramid.

Triple point The combination of temperature and pressure in a phase diagram at which the substance can simultaneously exist as a gas, a liquid, and a solid.

t-RNA Transfer RNA; the relatively small polynucleotide that carries amino acids and recognizes the codon on an m-RNA chain that specifies the amino acid to be incorporated at a particular point on a protein chain.

T-shaped geometry A molecular geometry whose shape resembles the letter T.

Uncertainty The limit on the precision of a measurement; for example, analytical balances can weigh an object with an uncertainty of ± 0.001 or perhaps ± 0.0001 grams.

Unimolecular Describes a step in a chemical reaction in which one molecule is consumed.

Unit cell The simplest repeating unit in a crystal.

Unit A basis for comparing a measurement with a standard reference.

Unsaturated fatty acid A long-chain carboxylic acid that contains one or more $C{=}C$ double bonds. See also *saturated fatty acid*.

Unsaturated hydrocarbon Hydrocarbons that contains $C{=}C$ double bonds and/or $C{\equiv}C$ triple bonds and therefore contains less hydrogen than a saturated hydrocarbon.

Valence bond theory A model of the bonding in covalent compounds based on the assumption that atoms in these compounds are held together by the sharing of pairs of electrons between pairs of atoms.

Valence electron Electron in the outermost or highest-energy orbital of an atom; electron that is gained or lost in a chemical reaction.

Valence-shell electron-pair repulsion theory (VSEPR) A model in which the minimization of the repulsion between pairs of valence electrons is used to predict the shape or geometry of a molecule.

van der Waals equation An equation that corrects for the two erroneous assumptions in the ideal gas equation —that the volume occupied by the gas particles is negligibly small and that there is no force of attraction between gas particles.

van der Waals forces The weak forces of attraction between atoms or molecules that explain why gases condense to form liquids and solids when cooled.

Vapor pressure The pressure of the gas that collects above a liquid in a closed container.

Vibrational motion The stretching or bending of bonds.

Viscosity A measure of the resistance of a liquid or gas to flow.

Vital force A mythical force that was once assumed to explain the difference between animate and inanimate objects.

Vitamin Literally, vital amines; a wide range of compounds that are either water-soluble or fat-soluble and are necessary components of the diet of higher organisms, such as mammals.

Voltaic cell An electrochemical cell that uses a spontaneous chemical reaction to do work; also known as a galvanic cell.

Volume percent The percent of the volume of a mixture that is due to a particular component.

Water dissociation equilibrium constant (K_w) The product of the equilibrium concentration of the H_3O^+ and OH^- ions in an aqueous solution: $K_w = 1.0 \times 10^{-14}$ at 25°C.

Water gas A mixture of CO, CO_2, and H_2 produced from the reaction of red-hot coal with steam.

Wave A phenomenon that does not have mass and therefore does not occupy space; waves travel through space.

Wave function A mathematical equation that describes orbitals in which electrons reside.

Wavelength The distance between repeating points on a wave.

Wave-particle duality The principle that all objects are both particles and waves. Most objects have wavelengths that are too small to be detected; however, very light objects, such as photons and electrons, can have wavelengths large enough to give these objects many of the properties of waves.

Weight A measure of the gravitational force of attraction of the earth acting on an object.

Weight percent Literally, the percent by weight of an element in a compound or of a component of a mixture.

Work (w) Mechanical energy equal to the product of the force applied to an object times the distance the object is moved.

Work of expansion The work done when a system that consists of a gas expands against its surroundings.

X-ray A high-energy, short-wavelength form of electromagnetic radiation.

Ylem The primordial substance out of which the elements were synthesized in the first moments after the creation of the universe.

Zero-order reaction A reaction whose rate is not proportional to the concentration of any of the reactants.

Zeroth law of thermodynamics Law stating that two objects in thermal equilibrium with a third object are in thermal equilibrium with each other.

Zwitterion Literally, a mongrel or hybrid ion; a molecule that contains both negative and positive charges, such as the $H_3N^+CHRCO_2^-$ form of an amino acid.

CHAPTER 1 *Opener:* Pat and Tom Leeson/Photo Researchers. —*Page 2:* Daniel Zirinsky/Photo Researchers. —*Page 3:* R. Fairbanks/Photo Researchers. —*Pages 6 and 8:* Ken Karp/Omni-Photo Communications, Inc. —*Figure 1.6a:* Gary Gladstone. —*Figure 1.6b:* Ken Karp/Omni-Photo Communications, Inc. —*Figure 1.6c:* Courtesy Sartorius Company. —*Page 26:* (top) George Holton/Photo Researchers, (bottom) Ken Karp/Omni-Photo Communications, Inc.

CHAPTER 2 *Opener:* C. Raymond/Science Source/Photo Researchers. —*Pages 38 and 39:* Ken Karp/Omni-Photo Communications, Inc. —*Page 43:* (top, center and bottom left) Ken Karp/Omni-Photo Communications, Inc., (bottom center) Tom McHugh/Photo Researchers, (bottom right) Renéc Purse/Photo Researchers. —*Pages 46, 47, 49, 50 and 58:* Ken Karp/Omni-Photo Communications, Inc.

CHAPTER 3 *Opener:* E. R. Degginger. —*Page 69:* Courtesy Finnigan Corp. —*Pages 72 and 73:* Ken Karp/Omni-Photo Communications. —*Page 74:* E. R. Degginger. —*Page 86:* Ken Karp/Omni-Photo Communications, Inc. —*Page 89:* E. R. Degginger. —*Page 91:* Joseph Nettis/Photo Researchers. —*Pages 100 and 106:* Ken Karp/Omni-Photo Communications, Inc.

CHAPTER 4 *Opener:* Geiger Photo Agence Vandystadt/Photo Researchers. —*Pages 117 and 122:* Ken Karp/Omni-Photo Communications, Inc. —*Page 132:* E. R. Degginger. —*Page 133:* The Bettmann Archive.

CHAPTER 5 *Opener:* Richard Megna/Fundamental Photographs. —*Page 156:* Ken Karp/Omni-Photo Communications, Inc. —*Page 158:* Tom Hollyman/Photo Researchers. —*Figure 5.2:* The Bettmann Archive. —*Page 162:* The Bettmann Archive. —*Page 164:* From "Philosophical Magazine," 27, 703, 1914. —*Page 166:* Dohrn/Photo Researchers. —*Page 168:* Richard Megna/Fundamental Photographs. —*Page 169:* (top) E. R. Degginger, (bottom) Ken Karp/Omni Photo Communications, Inc. —*Page 171:* Ken Karp/Omni-Photo Communications, Inc. —*Figure 5.15:* E. R. Degginger.

CHAPTER 6 *Opener:* Paolo Koch/Science Source/Photo Researchers. —*Page 199:* (top) The Bettmann Archive. —*Page 210:* E. R. Degginger. —*Page 220:* Ken Karp/Omni-Photo Communications, Inc. —*Page 222:* (top) George Whiteley/Photo Researchers, (center and bottom) E. R. Degginger. —*Page 223:* Ken Karp/Omni-Photo Communications, Inc.

CHAPTER 7 *Opener:* Richard Hutchings/Photo Researchers. —*Page 229:* Ken Karp/Omni-Photo Communications, Inc. —*Page 230:* F. Gohier/Photo Researchers. —*Pages 234 and 239:* Ken Karp/Omni-Photo Communications, Inc. —*Page 241:* (top left) M. Claye Jacana/Photo Researchers, (top center) Carl Frank/Photo Researchers, (top right) George Whiteley/Photo Researchers, (bottom left and right) Charles R. Belinky/Photo Researchers, (bottom center) E. R. Degginger. —*Page 242:* Don F. Smith/Photo Researchers. —*Page 243:* Ken Karp/Omni-Photo Communications, Inc. —*Page 244:* Courtesy SPO/AC-Delco, General Motors Corp. —*Pages 247 and 257:* Ken Karp/Omni-Photo Communications, Inc. —*Page 258:* Ray Ellis/Photo Researchers. —*Page 260:* Bruce Roberts/Photo Researchers. —*Page 261:* Courtesy ALCOA.

CHAPTER 8 *Opener:* Chemical Design/Science Source/Photo Researchers. —*Figure 8.2:* from *Valence and the Structure of Atoms and Molecules,* by G. N. Lewis, published by The American Chemical Society, 1923. —*Page 282:* Ken Karp/Omni-Photo Communications, Inc. —*Figure 8.8:* Adapted from *Hemoglobin* by Dickerson and Geiss, Benjamin/Cummings, 1983. Illustration copyright Irving Geiss. —*Pages 299, 300 and 301:* George Bodner. —*Figure 8.11:* From *Acta Cryst,* 2, 238 by S. C. Abrahams, J. Monteath and J. G. White, 1949. —*Pages 306, 307 and 308:* George Bodner. —*Page 320:* Ken Karp/Omni-Photo Communications, Inc.

CHAPTER 9 *Opener:* Kees Van Den Berg/Photo Researchers. —*Page 336:* David Taylor/Science Photo Library/Photo Researchers. —*Page 346:* Dan Bernstein/Photo Researchers. —*Page 349:* E. R. Degginger. —*Page 350:* Ken Karp/Omni-Photo Communications, Inc.

CHAPTER 10 *Opener:* NASA/Science Photo Library/Photo Researchers. —*Page 367:* Ken Karp/Omni-Photo Communications, Inc. —*Page 368:* Joseph Nettis/Photo Researchers. —*Page 370:* Ken Karp/Omni-Photo Communications, Inc. —*Page 371:* NASA/Science Photo Library/Photo Researchers. —*Pages 373, 375 and 381:* Ken Karp/Omni-Photo Communications, Inc. —*Page 382:* G. Tomsich/Photo Researchers. —*Page 383:* Ken Karp/Omni-Photo Communications, Inc. —*Page 384:* Paul Silverman/Fundamental Photographs. —*Page 389:* Tom Bledsoe/Photo Researchers. —*Page 390:* Grant Heilman. —*Pages 391 and 396:* Ken Karp/Omni-Photo Communications, Inc. —*Page 398:* E. R. Degginger. —*Pages 399, 400, 403 and 406:* Ken Karp/Omni-Photo Communications, Inc. —*Page 411:* Courtesy Argonne National Laboratory. —*Page 415:*

Paul Silverman/Fundamental Photographs. —*Page 418:* John Colwell/Grant Heilman. —*Page 420:* E. R. Degginger.

CHAPTER 11 *Opener:* E. R. Degginger. —*Page 427: (top)* Andrew McClenaghan/Science Photo Library/Photo Researchers, *(center and bottom)* Ken Karp/Omni-Photo Communications, Inc. —*Page 434: (top)* E. R. Degginger, *(bottom)* Ken Karp/Omni-Photo Communications, Inc. —*Page 441:* Ken Karp/Omni-Photo Communications, Inc. —*Page 455: (top and center)* Richard Megna/Fundamental Photographs, *(bottom)* Ken Karp/Omni-Photo Communications, Inc. —*Page 456: (top)* Ken Karp/Omni-Photo Communications, Inc., *(bottom)* E. R. Degginger. —*Figure 11.12:* Courtesy The Preservation Office, The Library of Congress.

CHAPTER 12 *Opener:* Herman Eisenbeiss/Photo Researchers. —*Figure 12.2:* © copyright 1976 W. G. Davies and J. W. Moore. —*Page 472:* G. Muller Struers GMBH/Science Photo Library/Photo Researchers. —*Pages 473 and 474:* Ken Karp/Omni-Photo Communications, Inc. —*Figures 12.6, 12.7, 12.8, 12.9, 12.10, 12.11, 12.12 and 12.15:* George Bodner. —*Page 479:* Todd Gipstein/Photo Researchers. —*Figures 12.19 and 12.21:* George Bodner. —*Page 501:* David Parker/IMI/University of Birmingham High TC Consortium/Science Photo Library/Photo Researchers.

CHAPTER 13 *Opener:* Loel Martin. —*Figure 13.7:* Ken Karp/Omni-Photo Communications, Inc. —*Page 513:* Peter Aprahamian/Science Photo Library/Photo Researchers. —*Figures 13.9 and 13.10:* Ken Karp/Omni-Photo Communications, Inc. —*Figure 13.14:* © copyright 1976 W. G. Davies and J. W. Moore. —*Figures 13.18, 13.19 and 13.20:* Ken Karp/Omni-Photo Communications, Inc. —*Pages 540 and 543:* Ken Karp/Omni-Photo Communications, Inc.

CHAPTER 14 *Opener:* Tim Holt/Photo Researchers. —*Figures 14.1, 14.2, 14.3 and 14.4:* Ken Karp/Omni-Photo Communications, Inc. —*Figure 14.5:* Courtesy Milton Roy Company. —*Figure 14.9:* Ken Karp/Omni-Photo Communications, Inc.

CHAPTER 15 *Opener:* Dr. Jerrican/Science Source/Photo Researchers. —*Figure 15.1:* From *Experiments and Observations on Different Kinds of Air*, Vol. I, by Joseph Priestley, 1774. —*Figure 15.2:* Paul Silverman/Fundamental Photographs. —*Figures 15.3 and 15.5:* Ken Karp/Omni-Photo Communications, Inc.

CHAPTER 16 *Opener:* Lowell J. Georgia/Photo Researchers.

CHAPTER 17 *Opener:* Bill Longcore/Science Source/Photo Researchers. —*Figures 17.1, 17.2, 17.3, 17.6 and 17.10:* Ken Karp/Omni Photo-Communications, Inc.

CHAPTER 18 *Opener:* James L. Amos/Photo Researchers. —*Figures 18.1 and 18.2:* Ken Karp/Omni-Photo Communications, Inc. —*Page 724: (top)* E. R. Degginger, *(bottom)* Richard Megna/Fundamental Photographs. —*Page 725:* Bruno J. Zehnder/Peter Arnold. —*Pages 731, 736, 737 and 741:* Ken Karp/Omni-Photo Communications, Inc.

CHAPTER 19 *Opener:* Michael George/Bruce Coleman. —*Figure 19.1:* Walker/Peter Arnold. —*Figure 19.2:* Ken Karp/Omni-Photo Communications, Inc. —*Figure 19.8:* John Underwood. —*Figure 19.11:* Ken Karp/Omni-Photo Communications, Inc.

CHAPTER 20 *Opener:* Dr. A. Winfree/Photo Researchers. —*Figure 20.5:* Ken Karp/Omni-Photo Communications, Inc. —*Figure 20.6:* E. R. Degginger. —*Figure 20.7:* Ken Karp/Omni-Photo Communications, Inc.

CHAPTER 21 *Opener:* E. R. Degginger. —*Page 825:* Ken Karp/Omni-Photo Communications, Inc. —*Figure 21.14:* Ken Karp/Omni-Photo Communications, Inc.

CHAPTER 22 *Opener:* Paolo Koch/Photo Researchers. —*Figure 22.3:* E. R. Degginger. —*Figure 22.4:* Ken Karp/Omni-Photo Communications, Inc. —*Page 871:* Paul Silverman/Fundamental Photographs. —*Figure 22.9:* Ken Karp/Omni-Photo Communications, Inc. —*Page 880:* Chemical Design/Science Photo Library/Photo Researchers. —*Page 882:* Ken Karp/Omni-Photo Communications, Inc. —*Figure 22.24:* Ken Karp/Omni-Photo Communications, Inc.

CHAPTER 23 *Opener:* NASA/Peter Arnold. —*Figure 23.1:* The Bettmann Archive. —*Page 919:* Tom McHugh/Photo Researchers. —*Page 921:* Ken Karp/Omni-Photo Communications, Inc. —*Page 931:* Paolo Koch/Photo Researchers. —*Figure 23.15:* U.S. Department of Energy/Science Photo Library/Photo Researchers. —*Page 934: (top)* LLNL/Science Source/Photo Researchers, *(bottom)* Science Photo Library/Photo Researchers.

CHAPTER 24 *Opener:* Will and Deni McIntyre/Photo Researchers. —*Figure 24.1:* Courtesy The Edgar Fahs Smith Collection, taken from *The Development of Modern Chemistry* by Aaron Ihde, Harper & Row, 1964. —*Page 956: (top)* Rafael Macia/Photo Researchers, *(bottom)* Allen Green/Photo Researchers. —*Page 958: (top)* E. R. Degginger, *(bottom)* Richard Megna/Fundamental Photographs. —*Page 959:* Ray Ellis/Photo Researchers.

CHAPTER 25 *Opener:* Dr. Arthur Lesk, Laboratory for Molecular Biology/Science Photo Library/Photo Researchers. —*Figure 25.1:* E. R. Degginger. —*Figure 25.15:* Ken Karp/Omni-Photo Communications, Inc. —*Page 1005:* E. R. Degginger. —*Figure 25.19:* Ken Karp/Omni-Photo Communications, Inc. —*Page 1030:* Phil Degginger.